SECOND EDITION

ENGINEERING SYSTEM DYNAMICS

A Unified Graph-Centered Approach

SECOND EDITION

ENGINEERING SYSTEM DYNAMICS

A Unified Graph-Centered Approach

Forbes T. Brown

Taylor & Francis
Taylor & Francis Group
Boca Raton   London   New York

CRC is an imprint of the Taylor & Francis Group,
an informa business

MATLAB® is a trademark of The MathWorks, Inc. and is used with permission. The MathWorks does not warrant the accuracy of the text or exercises in this book. This book's use or discussion of MATLAB® software or related products does not constitute endorsement or sponsorship by The MathWorks of a particular pedagogical approach or particular use of the MATLAB® software.

CRC Press
Taylor & Francis Group
6000 Broken Sound Parkway NW, Suite 300
Boca Raton, FL 33487-2742

© 2007 by Taylor & Francis Group, LLC
CRC Press is an imprint of Taylor & Francis Group, an Informa business

No claim to original U.S. Government works
Printed in the United States of America on acid-free paper
10 9 8 7 6 5 4 3 2 1

International Standard Book Number-10: 0-8493-9648-4 (Hardcover)
International Standard Book Number-13: 978-0-8493-9648-9 (Hardcover)

Library of Congress Cataloging-in-Publication Data

Brown, Forbes T.
    Engineering system dynamics : a unified graph-centered approach / by Forbes T. Brown. -- 2nd ed.
        p. cm.
    ISBN 0-8493-9648-4 (alk. paper) -- ISBN 1-4200-0958-3 (e-book)
    1. Systems engineering. 2. Dynamics. 3. Engineering models. 4. Bond graphs. I. Title.

    TA168.B675 2006
    620.001'171--dc22   2006045235

Visit the Taylor & Francis Web site at
http://www.taylorandfrancis.com

and the CRC Press Web site at
http://www.crcpress.com

To my wife, Marjorie H. Brown
and in memory of Henry M. Paynter

Contents

Preface

Engineering system dynamics comprises the abstract modeling of actual or proposed engineering systems that involve motion, energy or power, leading to analyses of their time-dependent behaviors. It is essential to the design of complex dynamic interdisciplinary systems and is necessary, for example, when automatic control is to be employed. It encompasses most of the study of vibrations. Therefore, undergraduate programs in fields like mechanical engineering, for which this text is especially oriented, normally require at least one introductory course in system dynamics, and those who wish to focus on the design of such systems typically elect further courses, either as undergraduate or graduate students. Chapters 1–7 of this text, less certain optional sections, are suitable for a three- or four-credit introductory course, nominally at the junior level. Inclusion of the introduction to automatic control, given in Chapter 8, would require the fourth credit hour. Chapters 9–12, which contain about half of the information in the book, take advantage of the unusually powerful methodology to allow extensive advanced subsequent study that goes well beyond what is normally available, and to enhance the reference value of the work.

The structure of the second edition is different from the first, particularly with regard to the presentation of mathematical methods for linear models. Rather than the original logical development, the basic mathematics is concentrated in Chapter 4, after the introduction to modeling but before the key chapter on basic modeling. Also, better bridges are given to traditional methods, such as block diagrams, expanding the range of applicable software.

Considerable advanced material is added to Chapters 9–12, such as simulation with magnetic hysteresis, hybrid lumped/distributed simulation, simulation of multiphase thermodynamic systems including kinetic energy, and Lagrangian methods for nonholonomic systems. The entire book, which has one third more material than the first edition, has been taught as three sequenced courses. The later chapters also are particularly directed toward the practicing engineer. The author has been motivated to include the relatively comprehensive sweep of topics within a single pair of hardcovers in order to emphasize the interrelatedness of the methodology and to encourage engineering graduates to expand their expertise.

Extensive use is made of MATLAB®. A list of code that is downloadable from the Internet is given on page 1018; this includes simulations for specific problems in Chapters 9–11 that can be adapted to related situations, and general code for selected thermodynamic substances from Appendix D. Plans call for adding new items in the future, including a list of errata should the need arise.

Modeling is a nondeterministic art that is inseparable from the design process. It is the pivotal and most difficult step of the process of dynamic analysis. Accordingly, it is addressed with great respect and thoroughness, accompanied with information and advice about a wide variety of actual engineering devices and a rich collection of phenomena, examples and problems. Many of the problems are relatively concrete, including specific simulations. The development is gradual, step-by-step, which the author believes is necessary to preclude superficial learning. A variety of graphical representations is emphasized, since modeling is greatly enhanced by visual modes of thinking. Linearity is put into proper perspective, unlike in so many textbooks. This approach may appear to slow the pace relative to some alternatives, but the rate of real learning is fast and accelerates in the later chapters because the approach is unified, common to mechanical, electromechanical and thermodynamic systems. Instruction on mathematical methods, on the other hand, is relatively traditional.

Modeling and analysis is integrated and unified through the use of bond graphs. Hundreds of courses world-wide are reported on the Internet to utilize these simple graphs. Those who have been exposed to them tend to be able to analyze a broader range of systems than those restricted to more traditional methods, and with greater insight into the meaning and structure of the models they devise. They also make fewer modeling mistakes, since the graphs enforce the conservation of energy except when explicitly directed to ignore it. Further, the causal feature of bond graphs efficiently directs the assembly of complex sets of state-variable differential equations. The approach is made fully accessible to undergraduates, for whom the critical language-independent modeling concepts are new. The earlier chapters also are suitable, at a faster pace, for graduate students who have not been exposed to bond graphs. The more advanced chapters reveal the generality of the method.

A practicing engineer or engineering instructor can take advantage of commercial software that automatically produces and solves differential equations from bond graphs. The book merely cites such software, however, in the belief that learning the basic mathematics still is essential for students. Also, this software does not treat some of the more advanced bond graphs in the book.

Sections that are logically optional are indicated by asterisks. More suggestions for use of the book in courses are given in the subsequent section, To the Instructor.

The book never would have been written had I not had the privilege of contact with the late Henry Paynter, the creator of bond graphs. It also has benefitted enormously from the contributions of my colleague, N. Duke Perreira. Others who have contributed include my son, Gordon, Professors Kenneth Sawyers, Ram Chandran and Timothy Cameron, and many of my students.

Forbes T. Brown
March 1, 2006

. . . well-connected representations let you turn ideas around in your mind, envision things from many perspectives. . . that is what we mean by thinking.

Marvin Minsky

. . . there is something essential in human understanding that is not possible to simulate by any computational means.

Roger Penrose

To the Instructor

A balanced three-credit first course in system dynamics can employ the first seven chapters of the book, less most of the sections indicated with an asterisk. This may seem to comprise too many pages, but only half of them are text, case studies and example problems not designated with asterisks. This includes an introduction to linear differential equations that may not be needed, and presentation of Bode plots that could be deferred to a course in control. Most of the rest is guided problems and assignable problems, relatively few of which will be addressed by a given student. Further, the style of the first half of the book includes much that can be absorbed quickly. Instructors who wish to downplay nonlinear models in favor of linear analysis could also skip Section 3.5 and other discussions of nonlinear problems, but perhaps include Section 7.3 on matrix methods. Instructors wishing to emphasize classical methods for mechanical vibrations could include Appendix B. Chapter 8 provides an optional introduction to automatic control.

A second, probably elective course at the senior level can be more relaxed in its style, since its objectives are less constrained. A review of the material in the first course is suggested, but now could include most of the sections with asterisks. The course then could proceed to the very substantive expansion of the modeling domain offered in Chapter 9, and possibly topics from the first part of Chapter 10 on distributed-parameter models. The mathematical style of most of the remainder of the book also would admit most seniors in a third course that again greatly extends the modeling domain, although most likely the opportunity would have to be delayed until graduate school. Graduate students who have not been exposed to bond graphs also have found the first part of the book quite useable, at a relatively rapid pace. Mixing these graduate students with those who have experienced bond graphs, however, can be difficult to do effectively. An introductory graduate course could approximate the first two undergraduate courses, with the final course being common.

Modeling ultimately is a nondeterministic process, and consequently the relevant part of the text is not written in the minimalistic style of elementary textbooks in engineering science. Rather, some engineering color or practical asides are inserted from time to time. Conveying interesting practical information also develops the students' critical ability to distinguish core essentials from peripheral details. The learning objectives should depend on the maturity and level of the students, and there is considerable latitude in the sophistication of the problems available as homework assignments and for class discussion.

The book includes more concrete detail than other texts, including listings of MATLAB code for specific simulations. Most of the code in Chapters 9-11 is downloadable from the Internet (see page 1018) and can be adapted by the practicing engineer to related problems. The author hopes particularly to expand the list of thermodynamic substances treatable by the essentially non-iterative method developed in Chapter 11. (Code for refrigerant R134a is new in this edition.) You might check from time to time to see if any significant errata have been identified anywhere in the book, despite vigorous efforts to preclude them.

Most topics are introduced through case studies. Example problems help get the student started, and sometimes introduce important derivative concepts. The "guided problems" are designed to encourage active learning; you may direct students' attention to them selectively, according to your purpose. A list of suggested steps or hints are given, and a solution that follows these steps appears at the end of the list of assignable problems. The assignable problems vary considerably in their level and time required; courses can be tailored to the relative sophistication of the students. The Instructor's Manual contains solutions and certain other teaching suggestions. Over 25 physical models are addressed in identified sequences of problems over different sections, as the needed tools are developed. Assigning selected whole sequences is suggested.

Most of Chapter 1 is intended to be motivational; you may prefer to substitute your own introduction, or just suggest reading. The material on dimensions and units is hopefully remedial, but American students in particular are challenged by this critical subject. About half the problems in the book give American standard units; SI units could be provided as an alternative. Chapters 2 and 3 comprise a first pass through modeling and simulation at a fundamental level. In Chapter 4 attention then passes to analytical methods for linear models, earlier than in the first edition. A return to modeling occurs in Chapter 5, followed by mathematical analysis in Chapters 6 and 7. Modeling and analysis are intertwined in the remaining chapters.

Chapter 2 develops the seminal idea of source-load matching, which strangely is missing in traditional curricula. This emphasizes the characteristics of elements, expressed graphically rather than mathematically, part of what is meant by a "graph-centered approach." The concepts of transformer and gyrational coupling, also seminal, also are introduced without dynamics, and the idea of stability is introduced. The first three sections of Chapter 3 complete the list of primitive elements needed for most dynamic models: the compliance, the inertance and the two junctions. The bridge from model to differential equations then follows in Section 3.4. The use of graphical characteristics makes the introduction to nonlinear elements in Section 3.5 seem very natural; straight lines are simply replaced by non-straight lines. (This section nevertheless can be skipped, if associated deletions also are made later on.) Finally, Runge-Kutta simulation, particularly using MATLAB, is introduced in Section 3.6. Thus, students can easily find and plot the behavior of simple nonlinear as well as linear models.

Students with some background in linear differential equations may skip part or all of the first three sections of Chapter 4. Section 4.4 on convolution helps complete students' understanding of linear methods, but probably it will be skipped in most mechanical engineering curricula (it is considered more important in electrical engineering). The key treatment of the Laplace transform in Section 4.5 avoids the mathematical sophistication that pertains to the subject in favor of the engineer's bread-and-butter application that is so common in engineering texts. Its logical development via the Fourier transform and via convolution is added at the end of the section as an optional extra for those who have the background. The classical responses of first- and second-order models are developed in Section 4.6 with the help of the Laplace transform. Finally, the critical subject of linearization is presented in Section 4.7 from two perspectives: the linearization of individual elements, and the linearization of differential equations.

The heart of the instruction in basic modeling is given in Sections 5.1 and 5.2. Some of the material on model equivalences in Section 5.3 and all of the material on equilibrium in Section 5.4 can be skipped if little emphasis is intended on linear models. The asterisks in the table of contents and the text apply even when nonlinear models are a priority.

One of the major uses of the bond graph is its nearly automatic assembly of state differential equations from an interconnection of elements. This subject is presented in Sections 6.1 and 6.2, with emphasis on the relatively difficult under-causal cases. Students usually develop mastery here, since the procedures are deterministic, unlike in modeling. Section 6.4 presents the optional use of the loop rule to expedite the determination of transfer functions and state differential equations for linear models represented by bond graphs. The material is also the basis for some commercial bond-graph software.

The presentation of frequency response in Section 7.1 is similar to that given in textbooks on automatic control, but it applies to much more than control, including the discussion of mechanical vibrations in Section 7.2. This latter section emphasizes two and more degree-of-freedom systems; single degree-of-freedom vibration models are scattered throughout the book. Most first courses in system dynamics likely do not have time for the matrix methods given in Section 7.3, but this is an important subject in linear methods and could be included in either a first course that makes room by downplaying nonlinear systems, or in a second course. Fourier analysis, given in Section 7.4, may be considered more essential, but also could be relegated to a second course.

Certain material in Chapters 4 and 5 plus all of Chapter 8 are equivalent to about half a conventional introductory course in automatic control. A briefer but nevertheless balanced introduction can end after the first or second of the three sections in Chapter 8.

The topics of Sections 9.1 and 9.2, modulated transformers and activated bonds, are rather basic extensions of Chapter 5 that greatly expand the domain of engineering systems that can be modeled with little additional knowledge. Sections 9.3 and 9.4 allow compliances and inertances to have more than two ports, which is especially useful in modeling various kinds of power and signal transducers, and in representing certain holonomic dynamic systems. The material in Sections 9.5 and 9.6 on magnetic circuits and electric motors is new to the second edition. The method for simulating magnetic hysteresis is simple and unique. Thermal systems with heat conduction are introduced in Section 9.7.

Chapter 10 presents the seminal idea of distributed-parameter models and analysis, with restriction to one-dimensional linear media but sometimes nonlinear boundary conditions. Applications include mechanical vibrations, fluid and electrical lines, and a variety of complex systems. Half of this long chapter appears for the first time in this edition. It includes a method for hybrid lumped and pure-delay simulation that avoids the pitfalls of purely lumped simulation, the use of modal decomposition, and an elaborate case study of systems of hydraulic tubing with viscous fluid, multidimensional wall motion, curved sections, attached masses and partial constraints.

Chapter 11 presents the modeling and analyses of thermodynamic systems with flowing fluid that can be multi-phase. The first three sections deal with steady flow. The primary interest, however, starts with Section 11.4 on the thermodynamic compliance and the thermodynamic inertance (which is more complete than in the first edition). The evaluation of thermodynamic properties in Section 11.5 is the key to practical simulation, which is carried out efficiently without essential iteration. General related MATLAB programs are listed in Appendix D, and particular programs for typical problems are listed in the section. All these programs are downloadable from the Internet (see page 1018.) At present they encompass ideal gasses and real gaseous air and its principal components, refrigerants R12 and R134a, and water. Many other substances may be included in the list of downloadable substances in the future, and methods for doing this yourself are explained. The final section in the chapter deals with chemical reaction. The key ideas of chemical kinetics are expressed in simple bond-graph terms, but application to real problems is left largely undone.

Chapter 12 starts with a discussion of some practical ways to carry out the conventionally assumed lumping of distributed fields, a key subject that deserves attention. Section 12.2 presents ways to treat complex systems such as hydraulic turbines. In the remaining sections, the book integrates the Lagrange and Hamiltonian methods that are usually associated with analytical dynamics. The resulting Lagrangian bond graphs can be quite useful for a variety of otherwise difficult problems that are either holonomic or nonholonomic, including interdisciplinary systems,. A Hamiltonian bond graph also is developed for completeness, but is found to have a limited utility.

Chapter 1

Introduction

This chapter is directed primarily at undergraduates taking a first course in system dynamics and related areas. It addresses three questions: Why are you taking this course? What style of study works best? How can you reliably treat dimensions and units? Advanced students may proceed directly to Chapter 2.

Your job title is Project Engineeer, the quintessential assignment for a mechanical engineer or an engineer in an allied field such as aeronautics or agriculture. Your responsibility is to shepherd a new product idea from the conception stage to the production stage. It may be a component such as a pump or an automotive strut, it may be a machine tool, or it may be an entire production machine. You might have a small team under your direction, or you might work alone, but in any case you have access to technicians, machinists and production specialists. You are directed in turn by a management that has a limited understanding of the technical possibilities, so you also play a role in choosing the product ideas to pursue and the marketing strategy. Part of the success of the company rests on your shoulders. How do you do your job?

You discover a need, or one is presented to you. Next, you interpret this need in terms of specifications, that is a list of the specific functions the product should exhibit, including quantitative performance objectives. Then, as a good designer, you conceive possible solutions, preferably several of them. The design process you are engaged in is summarized, albeit in a very cryptic fashion, in Fig. 1.1. At this point it is unclear which of the solutions are feasible, if any, and what values of their various parameters would be necessary or desirable. If you follow the experimental option shown in the diagram, you construct and test physical models of the proposed solutions. If instead you follow the analytical option, you model the designs abstractly, and likely use a computer to help

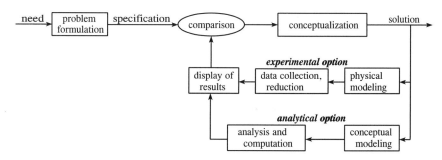

Figure 1.1: The engineering design process

determine the behavior. In either case, you display the results of the experiment or the analysis in an intelligible manner, so that you can compare them with the desired result, namely the original specifications. As designer, you must then decide what to do next; surely the discrepancies between the desired and the actual behavior require further iterative passes around the loop. At some point you decide either that the design has succeeded or failed, and you go on to another design concept or another project.

A successful conclusion is rewarding to you and your company, even if it is the abandonment of a poor idea that has been intriguing your boss. To be successful it must not take too long to accomplish, however, even if it is a revolutionary advance in its field; time costs money, and your competitors are not standing still. You will spend more time constructing and executing analytical or computer models of your designs than you will conceiving them. Efficient design implies quick analysis. How can you carry out the analysis process rapidly as well as effectively?

The answer is that you learn how to construct an abstract model of nearly any system you may design, and how to predict performance with this model far more quickly and cheaply than you could get a working physical model built and tested. Further, you should learn how to make the abstract model appropriately simple. An unnecessarily complex model, even if constructed readily, produces unnessesarily complex results and can be expensive to execute computationally. You should keep in mind the decision toward which the model is created. When you first go around the design loop, the intended decision likely is simple: whether or not to pursue the particular concept further. A crude model would be appropriate. Later, as a final design is approached, details of the system become of interest, perhaps requiring a rather fancy model.

The purpose of this text is to empower you to analyze real engineering systems, whether they exist already or are in the conceptual stage. Emphasis is placed on the process of *modeling*, that is, the construction of conceptual abstractions with well-defined meanings that behave in an appropriately approximate way to the real thing. Attention also is needed on the analytical and computational methods available to reveal this behavior. It is the modeling that is inherently difficult, however, for it is not a deterministic process. It involves intent, recognition and judgement. It is an art based on science, physical insight and experience. Together with the genius required to conceive new products, it is the "existential pleasure," the spice, that makes engineering an intensely human activity.

This text deals largely with systems that can be characterized by how they employ energy or power, either in large or small quantities. A wide variety of mechanical, fluid, electrical, thermodynamic and hybrid systems are included. Some of the systems are in a time-invariant state, particularly in Chapter 2. Most of them are time-varying, justifying the title *system dynamics*. Systems that do not operate because of energy balances, such as the informational considerations of computer systems, economic or political systems, traffic systems, etc. are not addressed, although these may use common analytical and computational methods.

1.1   Example

As an example of the process, imagine that you are a project engineer developing the basic concept of a vehicle powered by batteries and a DC electric motor. You quickly discover that a motor with a maximum power of 30kW can drive the vehicle fast enough, at a steady 45 m/s (about 100 mph). It cannot accelerate the vehicle from rest to 27 m/s (about 60 mph) in under 18 seconds, however, even with a perfectly variable transmission. To reduce this time, you consider the electrical-hydraulic drive represented diagramatically as:

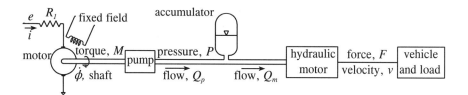

The motor has a fixed field and a 12 volt armature voltage, e, (supplied by batteries). Its shaft rotates at a variable angular velocity $\dot{\phi}$, driving a hydraulic pump, which has inertia and resistance. Fluid is stored in a hydraulic accumulator (a tank containing a compressed gas and hydraulic oil separated by a bladder or piston) at a high pressure, P. The fluid drives the vehicle at a variable speed v through the use of a hydraulic motor with a volumetric displacement (volume of fluid passed per revolution of the shaft) that is varied by the driver. The driver then can release the energy stored in the accumulator to achieve rapid acceleration. Further, he can also make the displacement negative, braking the vehicle by having the hydraulic motor pump fluid back into the accumulator. This feature, called regenerative braking, recovers the energy of motion rather than dissipating it, unlike conventional brakes. The vehicle has mass, frictional drag and an additional hill-climbing load.

You would like to have answers to such questions as: How large (and heavy) a hydraulic accumulator is necessary to accelerate the vehicle to 27 m/s in less than 6.5 seconds, without requiring a larger motor? What pressures should be used? What maximum volumetric displacements (sizes) should the pump and motor have to accomplish this? How should the volumetric displacement of the pump be automatically controlled? What would the energy efficiency of various schedules of speed be? Ultimately, is this concept worth pursuing?

As a first step in answering these questions, you make a simplified model of the system in the form of a **bond graph**:

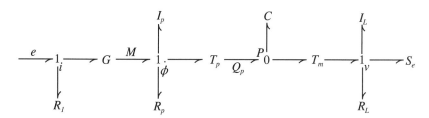

A bond graph is a minimalist representation of the implied model. Its structure also directs in routine fashion the assembly of the corresponding differential equations. Some of the components can be described by parameters (constants): R_i, G, I_p, R_p, and I_L. Others require more complex constitutive relations: C and R_L. The values of T_M and T_p are varied to control the flow of energy to or from the accumulator.

The model leads directly to a computer simulation for the dynamic performance of the system for whatever situation is of interest. Since each run takes only a couple seconds, you can simulate hundreds of different situations, if you like, adjusting the various parameters until you are satisfied with the result. Here are typical results after some of these changes have been made:

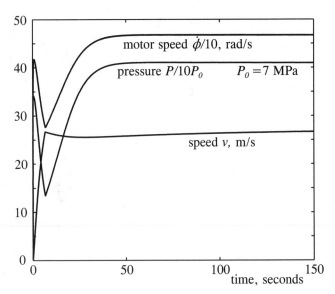

A speed of 27 m/s is indeed achieved in 6.5 seconds, by setting the volumetric displacement of the motor at a particular value. The displacement is then changed to a certain larger value, resulting in the same final steady speed. The pressure and energy in the accumulator falls during the acceleration but recovers afterward, due partly to the automatic control of the pump displacement, which can be designed partly through analysis of the bond graph. Thus a subsequent burst of acceleration also would be robust. This control also affects the speed of the electric motor. The motor is 50% efficient when its power is maximized (at 250 rad/s) and 100% efficient when its power approaches zero (at 500 rad/s). The control successfully prevents inefficient speeds (below 250 rad/s) and achieves the maximum possible 94% efficiency at the steady speed.

In another simulation, the displacement of the motor after 6.5 seconds is set to the value which is found to produce the maximum possible steady speed, which is the desired 45 m/s:

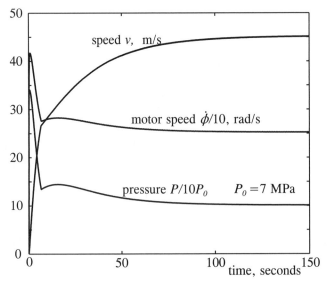

The motor speed becomes the 250 rad/s for which it produces the necessary maximum power, at 50% efficiency. The pressure drops to its minimum, which is the charging pressure of the accumulator (the pressure when it is virtually empty of hydraulic oil).

As project engineer, you are now in a position to render some judgments, which of course also involve factors not discussed here, such as cost, weight and reliability. As a student, you should be able to address some simpler problems in this manner after studying Chapters 2 and 3, and address some comparable problem as a project after completing Section 6.1. (This particular project is addressed in the problem sequence 3.60, 6.24, 6.25 and 6.26.) Most of the problems you will address in the course will be considerably shorter, however.

1.2 Modeling and Engineering Science

Real engineering systems, like the example above, do not closely resemble the pristine systems presented in introductory textbooks in mechanics, thermodynamics, fluid mechanics, etc. These textbook systems are in fact highly abstracted *models*, presented in the form of some *language*, usually largely pictoral, that convey a unique meaning. You hopefully absorbed some of the modeling idea, but what you really practiced was the application of physical principles to *translate* the given model as expressed in one language into another language, which usually was mathematical. The new expression was intended to reveal certain features or behaviors of the model which at first were hidden. This approach helped you develop a working understanding of physical principles that must be acquired before you can advance to considering real systems. But the gap between real engineering systems and approximate abstract models must be bridged.

Most non-structural engineering in the nineteenth century evaluated design concepts largely through experiment. Gradually it became apparent that the analytical option can substitute, saving time and money and sometimes reducing danger. The science of thermodynamics, for example, was born of the need to understand steam engines better. Structural science, so critical to safety, became particularly well developed. The analysis of dynamic systems lagged, because of the difficulty in solving the describing differential equations. Nevertheless, the goad of increasingly complex military hardware in World War II spawned a formal system science, based on differential equation models. Linear equations were emphasized, because only they could be solved routinely, albeit often with considerable tedium. Alternatively, components and systems were designed explicitly to act nearly linearly, sometimes just so that their behavior could be predicted.

The advent of the computer removed the computational roadblock. Today an engineer can construct a computer model of nearly any proposed system, and predict its performance far more quickly and cheaply than a prototype could be built and tested. The engineer then can adjust its parameters until the result is satisfactory, or until it appears that some different approach to the problem should be sought. Not all engineers today are competent to do this with much generality, but those who can are in a different league from those who can't. Experiment need be employed only sparingly, largely either to understand phenomena that defy analysis (often using relatively simple apparatus) or to verify a final design. The Boeing company, for example, now designs and builds its most complex aircraft without a single prototype being constructed and tested (although key components are tested separately).

1.3 Modeling Languages

A limited number of generic building blocks will be identified from which you will construct models. The language chosen to represent these elements should be general enough to encompass virtually all systems that centrally involve the management of power and energy, and narrow enough to impose the conservation of energy as the default option. Such a language allows you to take full advantage of the analogies that exist between seemingly different engineering domains. The language also should be chosen for the ease with which its objects can be reduced to mathematical form.

Perhaps the most serious hazard in modeling is the threat of making an outright mistake. Everyone is an occasional victim; the expert keeps a common-sense vigil, routinely comparing the structure of the model or its consequences with expectations. The modeling language or languages you use to represent models also greatly affects your propensity to make and discover mistakes. Mathematics and derivative computer code are the final language for most models, but these languages tend to obscure mistakes. Mathematics jokingly has been called "the universal solvent," since everything dissolves in it, becoming invisible.

This book employs a graph-based approach to modeling. Graphical languages comunicate better to your senses than do mathematics. Considerable use is made of sketches, diagrams and plots. Where practicable, the system behavior implicit in a graphically expressed model is also determined graphically. The models for most dynamic systems are converted to differential equations, which are then solved. This conversion is made as automatic as possible.

Since the subject is systems that control power and energy, a principal modeling language is desired that automatically conserves energy, unless it is directed explicitly to do otherwise. It is helpful also if the language identifies the critical geometrical and analagous constraints of such systems. Finally, it is desired that the language be both graphical and precise in its meaning and allow routine conversion to equations or equivalent computer code. Such a language should be optimally efficient and minimize mistakes.

Several languages apply to specific classes of systems. The circuit diagram is prototypical. Similar languages apply to classes of mechanical systems and to hydraulic systems. They fail to address interdisciplinary systems, however, which today means most systems. Actually, the electric circuit diagram often is applied by analogy to represent non-electrical phenomena, such as heat transfer or acoustics. It lacks adequate generality to represent all systems, however, and is awkward even when it is usable. A more generic language is sought.

Such a language exists, fortunately. **Bond graphs** were devised in their basic form in 1959 by the late Henry M. Paynter at MIT. Bond graphs are used by a significant fraction of mechanical and electrical engineers, and even by some life scientists. The language has been extended to model systems with heat fluxes, compressible fluid flow and chemical reaction. It has been promulgated by thousands of professional papers and, since 1991, by the biennial International Conference on Bond Graph Modeling (ICBGM).

Although the advantages of a knowledge of the bond graph language are well known, it has not been taught widely in undergraduate core courses and is not to be dabbled in lightly. By making its mastery a manageable and natural task, this book aims to increase its popularity for the benefit of engineering.

1.4   Modeling for Control

A model or its direct consequence is itself *built into* a control system, in the form of the controller with its control algorithms or alternatively its hardware configuration. Regardless of whether you study the introduction to control in Chapter 8, you should know that the control engineer typically succeeds or fails according to the workability of the model or models that he devises to represent the behavior of the system to be controlled.

Control systems combine a muscle part, or "plant," with a brain part. In a simple case the brain part is directed to act by some command signal, which is a measure of the desired performance of some plant variable such as a speed, an angle or a temperature. The brain usually measures this "output" variable and compares it to the command signal. The difference between these signals is an "error" which is used as the input to the controller part of the brain. The output of the controller then is sent to the plant for action. This feedback scheme also potentially helps correct for unwanted effects due to external disturbances on the system.

The controller must be tailored to the system it controls. Its design depends on a model of the system created by the engineer. Its performance hinges on the accuracy of the model. Its cost depends on the complexity of the model. Typically, the most practicable design results from a compromise between accuracy and complexity.

1.5   A Word to the Wise About Learning

Does the prospect of learning the material in a textbook like this, with its seemingly endless drawings and equations, seem daunting? The substance of this book, like many others, represents just a few patterns that theoretically ought to be expressible in a small fraction of the space. The problem is that this author, like all the others, just doesn't know how to make this work. It is up to the reader to compress the exposition down to its essential meaning. Then, and only then, is the subject truly manageable and useful. A few students manage this naturally. The following remarks are directed largely to the remaining majority.

Some subjects cannot be so distilled, such as uncorrelated lists of facts, or texts that demand memorization. There is little of this sort in this book, although some of the mathematics in Chapters 4 and 7 comes close. A good literal memory is not needed, paricularly for the core modeling concepts, and may in fact get in the way. If you can't remember much of anything verbatim, you're forced to create a mental pattern to represent the information, which is the desired result. Some patterns are logical, some relational, and some visual. Different people have different profiles of intelligences, and therefore develop different patterns, and compensate for weaknesses in different ways.

This textbook aims to arm you to address a broad range of engineering situations by placing them within a simple unified structure. Engineering has long had a tightly integrated core, which when understood allowed the engineer to extend his or her competence to new areas. Nevertheless, there used to be many specialists in various sub-disciplines, such as solid mechanics, fluid mechanics and heat transfer. Today, the roles of most of these specialists, except at the highest levels, have been replaced by commercially available software. This means that the successful engineer must be even more of a generalist, for he or she cannot take advantage of the software without a basic understanding of its meaning. He or she must be more like an orchestra conductor or even a composer, rather than the flutist or the cellist.

Your generation has made important advances in social and communication skills. At the same time old professors are keenly aware that certain basic technical skills have eroded. They wish to inspire the young with what they know is possible. The change has been caused partly by the explosion of information technology. Also, modern parents tend to fill their children's time with organized activities, believing that this will help their general education, advance their credentials, keep them out of trouble, and permit mother to work. This enhances socialization but fails to provide the free time and even boredom that is necessary for the development of independent thought and creativity. A particular decline has occurred in spatial visualization skills, due largely to schools discontinuing courses in drawing and descriptive geometry in the vain hope that the computer graphics would fill the need. Symptoms of the erosion include less attention to textbooks which are even sold at the end of the semester, greater reliance on the instructor, asking where to find the solution to a problem rather than how to solve it, conceiving of engineering as a group of loosely related procedures, struggling with drawing or spatial relationships, and expecting quick gratification.

The problem-specific mode of learning, which emphasizes detailed prescriptions for solving specific problem types through the examination of the solutions of numerous sample problems, has unfortunately ascended. The student tends to view these prescriptions as the information need for success, rather than the relatively simple underlying concepts that are gained only through a more circumspect approach. Real-world engineering covers a thousand times more situations than possibly could be addressed by sample problems. The breadth of the problems in this text may help convince you of the futility of reliance on superficial procedures, despite your earlier experience.

The alternative problem-general learning mode stores the information needed for success in an integrated network of basic concepts and procedures. Case studies, solved problems, guided problems and assigned problems are not definitive patterns to be imitated, but are suggestive illustrations that help you develop your integrated network. A small set of concepts casts myriad shadows in printed solutions to relevant problems. It is inefficient to infer reality solely from its shadows. You must, rather, *recognize* what basic concepts apply to your real-life problems. In this information age there are many ways you can find the details. The variety of situations addressable through the learning mode grows exponentially with the number of the core concepts and procedures understood. Further, the structure becomes an increasingly redundant network of ideas. If some element is forgotten, a detour around the flaw usually can be found, and the flaw itself often can be repaired without recourse to outside information. It may seem that the amount of information you need to recall actually decreases with experience. Engineering becomes heady stuff. Is your pleasure growing with each successive course?

This book complements its direct presentation of topics with four different types of problems: *case studies*, numbered and highlighted *examples*, *guided problems* and *assigned problems*. Case studies typically introduce and illustrate new ideas and should be viewed as integral parts of the text. The numbered examples, on the other hand, are mostly sample problems that demonstrate application of the basic concepts and procedures. They logically may be skipped, although you should find most of them useful. The "guided problems" in this text are a kind of compromise; they start with a problem statement and proceed with a list of "suggested steps" that is less than a unique prescription for a solution but is intended to relate the solution to the core concepts and procedures. You may then compare your solution with one given at the end of the section. Each guided problem exists for a purpose, which is briefly suggested at the outset. (Some guided problems are advanced or optional.) You defeat their primary purpose by looking at the given solution before attempting to create your own. Even when you fail to produce a solution, you will usually benefit from the attempt.

Your challenge is real, but the opportunities for excelling are great. Focus on underlying concepts rather than the nitty gritty. This takes courage at first, because you may fear that your colleagues who memorize stuff will get the upper hand during exams. Later, however, it will give you bracing self-confidence. The concepts are in fact few, while the nitty gritty is endless. Regard this and other technical textbooks as your primary professional tools; never sell one, for while you will forget the details of this formula or that diagram, you will likely remember what book it's in and probably whether it's on a right or a left facing page. And I mean decades later. Familiar textbooks become an extension of your intellect in ways the Internet cannot replace. Whether the instructor is good or not becomes a minor factor; he or she hopefully imparts some inspiration and suggests what is central but cannot convey the big picture nearly as completely, efficiently or permanently as a good textbook.

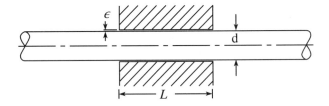

Figure 1.2: Journal bearing for case study

1.6   Treatment of Dimensions

Engineers and mathematicians write algebraic equations, in which numbers are represented
by symbols, usually letters of the Roman or Greek alphabets. These equations are solved
for *unknowns*, or *dependent variables*, which are written on the left sides of the equations,
as functions of the *knowns*, or *independent variables*, which are grouped on the right sides
of the equations. Substitution of the numbers for the symbols then produce a second type
of equation which gives numerical answers. For only the engineers, however, do the symbols
carry added meaning: physical dimensions. These dimensions, which themselves can be
represented by words or symbols, comprise the terms of a third type of equation. The
experienced engineer always keeps the dimensional equations in mind, for they give him or
her a tremendous advantage over the mathematician in the use of equations.

A case study illustrates this advantage. Consider that you are asked to find the torque
required to rotate a lightly-loaded shaft in a well-lubricated journal bearing at the angular
velocity ω. The shaft has diameter d, rotates concentrically in a sleeve or cylinder of length
L and diameter $d + 2e$, so that $e << d$ is the radial clearance between the shaft and the
sleeve, as pictured in Fig. 1.2. This space is filled with oil of absolute viscosity μ. (This is
a classical problem, which you may address in a first course in fluid mechanics, if you have
not done so already.) Perhaps you deduce the following relation to approximate the torque
or moment, M_0:

$$M_0 = \frac{\pi \mu d^2 L \omega}{2e} \tag{1.1}$$

No simple model can be *exact*, but is this one *correct*, within the several assumptions
upon which it is based? You know that most potential mistakes violate dimensional consistency, so you proceed, before substituting numbers, to substitute the dimensions of each of
the terms. You know the following dimensions:

π has no dimensions
L, d, e have dimension of *length*
ω has dimensions of $1/time$ (giving units such as rad/second)
μ has dimensions of *force× time/length squared*

Therefore, you substitute to get

$$[M_0] = \frac{force \times time}{length^2} \frac{length^2 \times length}{length} \frac{1}{time}, \tag{1.2}$$

where the square brackets around M_0 designate "dimension of." Cancellations give

$$[M_0] = \frac{force \times time \times length^3}{length^3 \times time} = force. \tag{1.3}$$

You know, however, that the torque or moment M_0 has the dimensions of *force* × *length*, not *force*. Something is wrong! The equation for M_0 must be changed to increase its dimension by one *length*.

You therefore revisit your derivation. Or, perhaps, you reason as follows: the shear stress on the surface of the shaft is proportional to the velocity of its surface, which in turn is proportional to the diameter d. The shear force also is proportional to the wetted surface area of the shaft, which also is proportional to d. Finally, the torque equals the shear force times the radius, which again is proportional to d. Therefore, the term d^2 in the numerator of equation (1.1) should be d^3. With this correction, equation (1.3) gives *force* × *length*, which you know is the correct dimensions for torque. You then return to your derivation, and find that you forgot to multiply the shear force by the radius, which is $d/2$. The proper torque therefore is

$$M_0 = \frac{\pi \mu d^3 L \omega}{4e}. \tag{1.4}$$

You are finally ready to substitute the numerical values of the parameters.

This scenario is typical of a *good* engineer. Good engineers are human and make mistakes but have gotten into the habit of catching them by checking dimensions.

In order to deal with dimensions effectively, it is necessary to work with a *minimal consistent* or *primary* set thereof. The most common and recommended primary set of dimensions in use by engineers today for treating mechanical, electric circuit and thermal systems is *force* (F), *length* (L), *time* (t), *electric charge* (Q) and *temperature* (θ). Other dimensions are called *derived*, since they can be expressed in terms of the consistent set. The names of some of these are listed in the first column of Table 1.1, and symbols therefore are listed in the second column. (The other columns are discussed later.)

A second primary set of dimensions replaces the role of *force* with *mass*, M. Specifically, whenever force F appears in Table 1.1, this set of dimensions substitues $M \cdot L/t^2$, which follows from Newton's second law: $F = ma$. This so-called "absolute" consistent set of dimensions sometimes is featured in elementary books on dynamics. Nevertheless, it is relatively awkward for most engineering systems, and in practice is rarely used. For example, engineers do not often think of pressure as having the dimensions $M/t^2 L$, or energy as having the dimensions $M \cdot L^2/t^2$.

Beginners often try to use a mixed system of dimensions, with both mass and force. This attempt prevents critical cancellations from being made. It abandons much of the benefit derived from using dimensions.

1.7   Treatment of Units

Most implementation of dimensions is carried out through the use of a specific system of units. Three systems are most common and are addressed here: SI units, American Standard or British Gravitational units, and the old English Engineering units. The first two can be treated as either force-based or mass-based. Old English Engineering units are by tradition mixed force-and-mass-based, and as a result are not listed in Table 1 but are treated separately below. Fluid mechanics texts tend to use slugs, which if not converted represent a mass-based American Standard system; dynamics texts tend to use the recommended force-based American Standard system; thermodynamics texts tend to use pounds mass as part of a mixed old English Engineering system. These texts often fail to adequately delineate force-based and mass-based use of the SI system, also. A generation of engineers have been confused, particularly when interdisciplinary systems are involved.

Table 1.1 Dimensions and Units

name	symbol	dimensions	SI units	Am. Std. units
primary set of consistent dimensions and units:				
force	F	F	newton, N	pound force, lb
length	L	L	meter, m	foot, ft
time	t	t	second, s	second, s
temperature	θ	θ	degrees Kelvin, K	deg. Rankine, R
electric charge	q	Q	coulombs, C	
derived dimensions and units:				
mass	m	$F{\cdot}t^2/L$	N·s²/m≡ kg	lb·s²/ft≡ slugs
angle	ϕ	1 (none)	(radian)	(radian)
velocity	$\dot{x}$	L/t	m/s	ft/s
angular velocity	$\dot{\phi}$	$1/t$	1/s (rad/s)	1/s (rad/s)
acceleration	$\ddot{x}$	L/t^2	m/s²	ft/s²
angular accel.	$\ddot{\phi}$	$1/t^2$	1/s²	$1/t^2$
energy	$\mathcal{V}$ or $\mathcal{T}$	$F{\cdot}L$	N·m ≡ J (joules)	ft·lb
power	$\mathcal{P}$	$F{\cdot}L/t$	N·m/s≡ W (watts)	ft·lb/s
pressure	P	F/L^2	N/m² ≡ Pa (pascals)	lb/ft²
entropy	S	$F{\cdot}L/\theta$	N·m/K≡ J/K	ft·lb/R
electric current	$i = \dot{q}$	Q/t	C/s≡ A (amperes)	
electric potential	e	$F{\cdot}L/Q$	N·m/C≡ V (volts)	

The confusion regards the treatment of mass. As advocated above, the role of mass should be replaced by force through the use of Newton's second law. Force units do not necessarily equal mass units times acceleration units, however, unlike the fundamental dimensions. Rather, force units are, in general, merely *proportional* to the units of mass times acceleration. In general,

$$F = \frac{m}{g_c} a, \tag{1.5}$$

where g_c is a constant of proportionality that may or may not have the numerical value 1.0000. The standard weight W is therefore related to the mass m by

$$W = \frac{m}{g_c} g, \qquad \text{or} \qquad \frac{m}{g_c} = \frac{W}{g}. \tag{1.6}$$

Whenever you face a mass m in an equation, you may substitute either m/g_c or W/g (and ρ/g_c or γ/g for ρ, where γ is the weight density), except in the cases of the more careful thermodynamics texts in which g_c appears already. The result is twofold: first, the units of mass will be converted automatically to equivalent units using force, as desired and as listed in Table 1.1; second, the numerical values will be properly represented, regardless of the system of units being employed. Values and units are given in the Table 1.2.

The old English Engineering system of units by tradition is a mixed system, containing both units of mass and force, which can complicate their treatment. Use of the method above automatically replaces its pounds mass (lbm) with equivalent units using pounds force (lbf, or just lb). The result is virtual conversion to the force-based American Standard system. Note that the value of g_c in the old English Engineering system is 32.174, *not* 1.0000, so use of the recommended method produces a symbol for every number (other than true conversion factors, which are dimensionless though not unitless), which is a good general policy in analysis. The system also uses British Thermal Units (Btu) for energy, rather than ft·lb, but this poses no real problem since the ratio of ft·lb to Btu is a dimensionless conversion factor (with value 778.16).

Table 1.2 Units and Values for Conversion of Mass to Force Equivalent

units of			values and units of		units of
m	L	F	g	g_c	$\dfrac{m}{g_c} = \dfrac{W}{g}$
kg	m	N	$9.81\ \dfrac{\text{m}}{\text{s}^2}$	$1.0000\ \dfrac{\text{kg·m}}{\text{N·s}^2}$	$\dfrac{\text{N·s}^2}{\text{m}}$
slug	ft	lb	$32.174\ \dfrac{\text{ft}}{\text{s}^2}$	$1.0000\ \dfrac{\text{slug·ft}}{\text{lb·s}^2}$	$\dfrac{\text{lb·s}^2}{\text{ft}}$
"sluglet"	in.	lb	$386.09\ \dfrac{\text{in.}}{\text{s}^2}$	$1.0000\ \dfrac{\text{sluglet·in.}}{\text{lb·s}^2}$	$\dfrac{\text{lb·s}^2}{\text{in}}$
lbm	ft	lb	$32.174\ \dfrac{\text{ft}}{\text{s}^2}$	$32.174\ \dfrac{\text{lbm·ft}}{\text{lb·s}^2}$	$\dfrac{\text{lb·s}^2}{\text{ft}}$

Engineers probably use inches more often than feet, such as in "psi" for pressure. The units of mass then are lb·s^2/in. The rather whimsical name "sluglet" is given in the table for this unit; it equals the weight W in pounds force divided by g in inches per second squared, or one-twelfth of a slug. This unit of mass has no official or recognized name. This namelessness underscores the preference of engineers for force rather than mass units. Many more physical phenomena directly involve force than involve mass.

The SI system of units has a particular advantage for electromechanical systems: a volt-coulomb equals a Newton-meter (equals a Joule). This book employs SI units for all such systems, as well as all electrical circuits. Therefore, Table 1.1 gives electrical units only in SI.

EXAMPLE 1.1

Evaluate the classical formula for the natural frequency, $\omega_n = \sqrt{k/m}$, for the following particular values of the mass m that is attached to ground by a linear spring with rate k: (a) $m = 2$ kg, $k = 8$ N/m; (b) $m = 2$ slugs, $k = 8$ lb/ft; (c) $m = 2$ lbm, $k = 12$ lb/ft; (d) $W = 2$ lb, $k = 1.0$ lb/in.

Solution:

(a) symbol equation : $\omega_n = \sqrt{\dfrac{k g_c}{m}}$

units equation : $[\omega_n] = \sqrt{\dfrac{\text{N}}{\text{m}} \cdot \dfrac{1}{\text{kg}} \cdot \dfrac{\text{kg·m}}{\text{N·s}^2}} = \dfrac{1}{\text{s}}$

number equation : $\omega_n = \sqrt{8 \cdot \dfrac{1}{2} \cdot 1} = 2\ \text{rad/s}$

(b) symbol equation : $\omega_n = \sqrt{\dfrac{k g_c}{m}}$

units equation : $[\omega_n] = \sqrt{\dfrac{\text{lb}}{\text{ft}} \cdot \dfrac{1}{\text{slug}} \cdot \dfrac{\text{slug·ft}}{\text{lb·s}^2}} = \dfrac{1}{\text{s}}$

number equation : $\omega_n = \sqrt{8 \cdot \dfrac{1}{2} \cdot 1} = 2\ \text{rad/s}$

(c)   symbol equation : $\omega_n = \sqrt{\dfrac{kg_c}{m}}$

units equation : $[\omega_n] = \sqrt{\dfrac{\text{lb}}{\text{ft}} \cdot \dfrac{1}{\text{lbm}} \cdot \dfrac{\text{lbm}\cdot\text{ft}}{\text{lb}\cdot\text{s}^2}} = \dfrac{1}{\text{s}}$

number equation : $\omega_n = \sqrt{12 \cdot \dfrac{1}{2} \cdot 32.2} = 13.9 \text{ rad/s}$

(d)   symbol equation : $\omega_n = \sqrt{\dfrac{kg}{W}}$

units equation : $[\omega_n] = \sqrt{\dfrac{\text{lb}}{\text{in}} \cdot \dfrac{1}{\text{lb}} \cdot \dfrac{\text{ft}}{\text{s}^2} \cdot \dfrac{\text{in}}{\text{ft}}} = \dfrac{1}{\text{s}}$

number equation : $\omega_n = \sqrt{1.0 \cdot \dfrac{1}{2} \cdot 32.2 \cdot 12} = 13.9 \text{ rad/s}$

The last case above illustrates the most common style of American engineers, except they likely would directly substitute the value and units 386 in/s^2 for g. In all the cases the result has the units s^{-1}. This ω_n is a circular frequency; a dimensionless angle is implied, which is the radian. To convert from radians per second to cycles per second (Hz), division by 2π is necessary.

Guided Problem 1.1

This first guided problem illustrates the role of g_c even when the mass does not appear explicitly in a formula being evaluated. It also shows the role of true conversion factors. Unless you already are quite competent with treating dimensions and units, you are strongly advised to at least attempt the problem before looking at the solution on the next page.

Find the velocity v of a perfect gas in the throat of a nozzle, given the formula $v = \sqrt{2\Delta h}$, and (a) $\Delta h = 10$ kJ/kg and (b) $\Delta h = 5$ Btu/lbm.

Suggested Steps:

1. Note that the Δh given is the enthalpy *per unit mass*, or the specific enthalpy. There- fore, replace this implied mass by m/g_c to modify the formula.

2. Write the units equation for the SI example. The kJ should be expressed in terms of the primary consistent units. Are the dimensions correct? Is there a need for a conversion factor? (If so, write the units for this factor in your equation.)

3. Write the numbers equation for the SI example, evaluate the velocity and note the units.

4. Write the units equation for the Btu example. The conversion factor between Btu and foot-pounds should be included. Are the dimensions and units of the answer correct?

5. Write the numbers equation for the Btu example, evaluate the velocity and note the units. There are 778.16 ft·lb per Btu.

References

Textbooks that can serve as references for this text have been authored by D.C. Karnopp, D. L. Margolis and R.C. Rosenberg[1], P. Gawthrop and L. Smith[2], J.U. Thoma[3], F.E. Cellier[4] and A. Mukherjee and R. Karmakar[5]. Notable collections and reviews of papers are given in a special issue of the *Journal of the Franklin Institute*[6] and a book edited by P.C. Breedveld and G. Dauphin-Tanguy.[7] For conference proceedings and more see www.BondGraph.com.

PROBLEMS

1.1 State the standard weight of a mass of 1 lbm.

1.2 The energy of a compressed spring accelerates a mass. Find the maximum velocity of the mass assuming a spring rate of 20 lbs/in., an initial spring compression of 5 inches from its free length and a mass weighing 10 lb. Neglect the mass of the spring.

1.3 A DC motor operates at 70% efficiency with a power supply of 24 volts and 2 amps. Find the velocity at which it can raise a 10 kg object.

1.4 Find the velocity of the flow of water (weight density 62.4 lb/ft^3) in the throat of a nozzle, given the pressure drop from the rest upstream state, $\Delta P = 10$ lb/in.^{2}, and the formula $v = \sqrt{2\Delta P/\rho}$.

1.5 Do the problem above assuming a mass density of 1000 kg/m^3 and a pressure drop of 70 kPa.

1.6 A hydraulic fluid suffers a pressure drop of 2000 psi in flowing through a valve. Neglecting any effect of compressibility or heat transfer, the dissipated energy simply raises the temperature of the fluid. Find the temperature rise assuming a specific heat of 0.5 Btu/lbm·°F and a density of 1.6 slugs/ft^3.

1.7 Do the problem above for a pressure drop of 14 MPa, density 830 kg/m^3 and specific heat 2.1 kJ/kg°C.

Solution to Guided Problem 1.1

1. $v = \sqrt{2\,\Delta h\, g_c}$

2. $[v] = \sqrt{\dfrac{\text{kN·m}}{\text{kg}} \cdot \dfrac{\text{kg·m}}{\text{N·s}^2} \cdot \dfrac{\text{N}}{\text{kN}}} = \dfrac{\text{m}}{\text{s}}$

3. $v = \sqrt{2 \cdot 10 \cdot 1 \cdot 1000} = 141.4$

4. $[v] = \sqrt{\dfrac{\text{Btu}}{\text{lbm}} \cdot \dfrac{\text{lbm·ft}}{\text{lb·s}^2} \cdot \dfrac{\text{ft·lb}}{\text{Btu}}} = \dfrac{\text{ft}}{\text{s}}$

5. $v = \sqrt{2 \cdot 5 \cdot 32.174 \cdot 778.16} = 500$ ft/s

---

[1] *System Dynamics: Modeling and Simulation of Mechatronic Systems*, 4th ed., Wiley, New York, 2006.

[2] *MetaModelling: Bond Graphs and Dynamic Systems*, Prentice-Hall, 1996

[3] *Simulation by Bondgraphs: Introduction to a Graphical Method*, Springer Verlag, Berlin, New York, 1990.

[4] *Continuous System Modeling*, Springer-Verlag, New York, 1991.

[5] *Modeling and Simulation of Engineering Systems Through Bondgraphs*, CRC Press, Boca Raton, 1999.

[6] *J. of the Franklin Institute*, guest editor P.C. Breedveld, v 328 n5/6, 1991.

[7] *Bond Graphs for Engineers*, North-Holland, Amsterdam, 1992.

Chapter 2

Source-Load Synthesis

An engineer usually starts to model an engineering system by subdividing it, conceptually, into components with explicit boundaries. In the simplest case only two components are defined: a **source** and a **load**. In somewhat more complex instances the source and the load may be interconnected by a third component. This chapter considers sources and loads that either are joined directly or have one or more components between to form a chain. Each component is characterized in isolation, and these characterizations are combined to reveal the behavior of the assembled system. The engineer also wants to know how to configure a particular system to achieve the best possible behavior.

Variables, such as force, velocity, voltage and current, generally change over time, justifying the name. This chapter, however, deals with components and systems that can be characterized in terms of variables that remain constant over time. These components or systems are said to reside in **steady state**, or in **equilibrium**. Sometimes more than one equilibrium exists for a given synthesis of components. Sometimes, further, a small disturbance on the system produces a radical change in the state of a system; such an equilibrium is said to be **unstable**.

The challenge of modeling a system in equilibrium might seem vastly simpler than the modeling of a dynamic system, in which the variables change over time. This is usually not the case, however. Most of the mistakes that modelers of dynamic systems make can be traced to the geometric and quasi-geometric constraints of the system, which apply to the equilibrium states as well. Thus you are urged to pay serious attention, even if your primary interest is dynamics.

2.1 System Reticulation

The science of thermodynamics customarily divides the **universe** into a **system** and its **surroundings**, as pictured in Fig. 2.1. The system might be **closed**, representing a fixed quantity of matter or a **control mass**. On the other hand it might be **open**, representing a region in space or **control volume** into which and/or from which matter can flow. In either case, the boundaries of the system, often called the **control surface**, might be fixed or moveable, giving a total of four different possible combinations.

Conceptual partitions of space or bodies of matter also are basic to the study of mechanics. In the study of *statics* and *dynamics* one draws a **free body diagram** of the system in question. In *fluid mechanics* one chooses a control volume. It should be no surprise, therefore, that in the general modeling of physical systems one starts by carefully defining a system, or perhaps five or twenty different sub-systems.

15

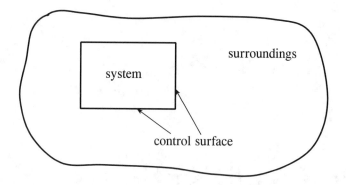

Figure 2.1: Partitioning of the universe into a system and its surroundings

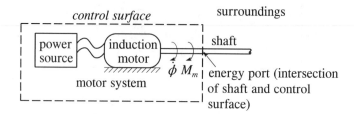

Figure 2.2: Representation of an induction motor as a one-port system

Endowing a system with structure often is called **reticulation**. The Latin stem **reti** means (fish) net; **to reticulate** means to make into or like a net. This image helps you focus not only on the sub-systems, but also on the meshpoints or bonds of the net. Reticulation is not as straightforward as sometimes it may seem in practice. It is easy to overlook the simplifying assumptions that are implicit in this act, with resultant trouble. The discussion here starts simply, as you did in the study of statics and thermodynamics, by identifying a single system and its surroundings.

2.1.1   Case Study: Induction Motor as a Source

Consider an induction motor as pictured in Fig. 2.2. This is a complicated device. The goal is merely to characterize its behavior as it applies to users, not to understand the details of its construction or operation. The entire machine and its source of power is defined as the system, letting only the shaft penetrate the system boundary or control surface. *The system is characterized in terms of appropriate variables that can be observed at the point of interconnection between it and the environment*, namely the shaft. The shaft, or more precisely its intersection with the control surface, is said to be the **energy port** or just the **port** of the system. The description "one-port model" distinguishes this representation from potentially more complex models having two or more defined ports.

The first of two variables that most engineers would identify is **angular velocity**, $\dot{\phi}$ (ϕ is the rotational angle of the shaft), usually measured in radians per second. The second variable is the torque or **moment**, M_m, in the shaft. The product of the two variables, $M_m\dot{\phi}$, is the **power**, $\mathcal{P}$, generated by the motor and propagated through the port (shaft) to the environment, assuming $\dot{\phi}$ and M_m are both defined as positive in the same sense as seen by an observer stationed near the motor:

$$\mathcal{P} = M_m\dot{\phi}. \tag{2.1}$$

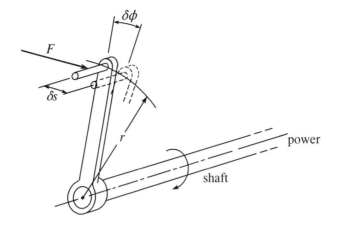

Figure 2.3: Shaft and crank

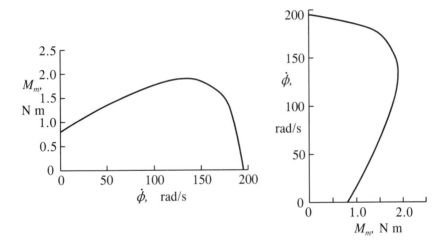

Figure 2.4: Torque-speed characteristic of the induction motor system

To establish this result, consider that the torque and the power arise from a normal force F acting on an imaginary attached crank with radius arm r, as pictured in Fig. 2.3. If the crank is displaced by an infinitesimal distance δs, the work done is $F\delta s$. But δs equals $r\,\delta\phi$, where $\delta\phi$ is the corresponding angular rotation of the crank and shaft, so the work becomes $Fr\,\delta\phi$. The power is the work per unit time, or $\mathcal{P} = Fr\,d\phi/dt = Fr\dot\phi$. Finally, the product Fr is the applied moment, M_m, establishing equation (2.1).

Imagine an experiment in which the environment is adjusted until $\dot\phi$ reaches some desired value, and then M_m is measured. This experiment is then repeated for many other values of $\dot\phi$ until a complete **characteristic**

$$M_m = M_m(\dot\phi) \tag{2.2}$$

can be plotted. A typical characteristic for an induction motor is shown in part (a) of Fig. 2.4. Alternatively, the environment is adjusted until M_m reaches some desired value, at which point $\dot\phi$ is measured. The consequence of repeating this experiment for many different values of M_m can be written with the inversion

$$\dot\phi = \dot\phi(M_m). \tag{2.3}$$

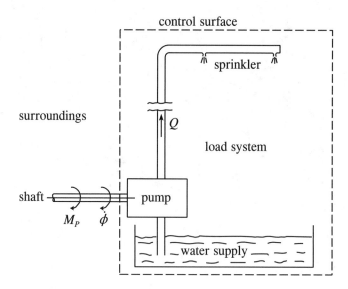

Figure 2.5: Representation of the load as a one-port system

The corresponding characteristic for the induction motor is shown in part (b) of the figure. Care must be taken because there are two values of $\dot{\phi}$ for some values of M_m.

These two plots follow the convention that the horizontal axis (abscissas) represents the independent variable on the right side of the equation, and the vertical axis (ordinate) represents the dependent variable on the left side of the equation. This convention actually is quite arbitrary, however; the two plots represent the same characteristic, with the axes merely reversed. The general idea of a characteristic overlooks any considerations of independence and dependence, that is of cause and effect. This perspective could be called existential, as opposed to causal. Thus, the two plots are completely equivalent. (Later, you will employ a causal perspective of a characteristic plot, however, for specific computational purposes.)

Either plot completely describes the motor as long as the motion is steady (no acceleration or inertial moments), the machine does not run backwards, you do not care about heat and noise, etc., and you are not worried about failure of the power or the motor. This model of the motor is described as having only one port, with the conjugate variables $\dot{\phi}$ and M_m. It is represented by the **word bond graph**

$$\text{INDUCTION} \underset{\dot{\phi}}{\overset{M_m}{\rule{4em}{0.4pt}}}$$
$$\text{MOTOR}$$

The port is penetrated by a **power bond** or, for short, **bond**, indicated by the horizontal line. The outward-directed half-arrow defines the direction of the flow of power when $\mathcal{P} = M_m \dot{\phi}$ is positive.

2.1.2   Case Study: Water Sprinkler System as a Load

Consider now that the motor drives a pump which pumps water from a basement tank through a pipe to some upper floor of a building where it passes through sprinkler heads and extinguishes a fire. This new equipment, collectively, now can be considered as the system; the motor becomes the new environment. The role of the system and the environment thus are inverted. The new system can be represented summarily by a second word bond graph:

$$\underset{\dot{\phi}}{\overset{M_p}{\rule{4em}{0.4pt}}} \underset{\text{SYSTEM}}{\text{LOAD}}$$

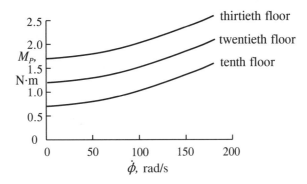

Figure 2.6: Torque-speed characteristics for the load system

Like the motor, this model of the system has only one port, which is a shaft with an angular velocity $\dot{\phi}$ and moment M_p. Note one difference: the power product $\dot{\phi}M_p$ is positive when power flows *into* the load, not *out*. This is the only difference, in fact, between a "source" and a "load."

It is now possible to run an experiment in which the environment of this load (*i.e.*, the motor) is adjusted to set any desired speed, at which the torque M_p is measured. Repeating this experiment for the whole range $\dot{\phi} > 0$ of interest produces a **load characteristic**. Three such characteristics are plotted in Fig. 2.6, for the sprinklers on the tenth, twentieth or thirtieth floors, respectively. These characteristics also could be found by inverted experiments in which M_p is set and $\dot{\phi}$ is measured, as with the motor.

This description alone does not tell you how much water flow is associated with any particular combination $\dot{\phi}$, M_p; it doesn't even recognize the existence of water. For now, just recognize that the faster the shaft turns, the greater the volume flow rate of water will be.

2.1.3 The Source-Load Synthesis: Case Study

Finally, the particular motor is joined to the particular load, as suggested by the following word bond graph:

$$\begin{array}{cc} \text{INDUCTION} & \dfrac{M}{\dot{\phi}} \quad \text{LOAD} \\ \text{MOTOR} & \quad \text{SYSTEM} \end{array}$$

These two sub-systems now share a common speed ($\dot{\phi}$) and a common moment, so that M_m equals M_p. Therefore, if their respective characteristics are plotted on common coordinates, as in Fig. 2.7, the resulting equilibrium must be represented by their intersection. For the tenth-floor case, there is only one intersection of the characteristics, and therefore only one possible steady operating speed, flow and moment. Such an equilibrium is called a **steady state**, implying that the state variables are unchanging over time. For the thirtieth floor case, there are no intersections at all. Moreover, the load torque exceeds the motor torque for all speeds $\dot{\phi} > 0$. The motor simply will not run. It might even burn itself out trying.

The case of the twentieth floor sprinkler is especially interesting. There are *two* intersections of the source and load characteristics, at the speeds labeled $\dot{\phi}_1$ and $\dot{\phi}_2$ in Fig. 2.8. Both of these intersections represent equilibrium states, which means that the source speed equals the load speed and the source torque equals the load torque. Equilibrium, however, is not a sufficient condition to insure actual operation. This is because some equilibria are unstable, that is fail to persist.

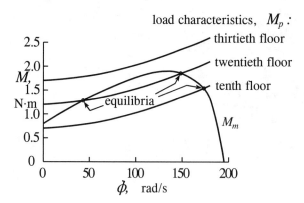

Figure 2.7: Synthesis of the source (motor) and load characteristics

Prediction of whether or not a particular equilibrium or steady state is stable requires consideration of a nearby non-equilibrium or unsteady state. For the motor-pump-sprinkler system this means a departure from the equilibrium angular velocity, so that the torques of the source and the load are different. One such pair of source and load states near the equilibrium state $\dot\phi_2$ is indicated in the figure as points (a). Since the shaft is assumed to be rigid, the source (motor) and the load have the same angular velocity, $\dot\phi$. A straight line drawn between the two points therefore must be vertical. The source and the load do not have the same torque, M. The electromagnetically induced torque of the motor can be different from the fluid-induced torque of the pump. A difference between the two torques causes the shaft, with its attached rotors in the motor and the pump, to accelerate or decelerate. Acceleration would drive the speed further above the equilibrium, indicating instability. On the other hand, deceleration would cause the speed to approach the equilibrium.

You need to know, therefore, whether the shaft accelerates or decelerates. For states (a), you can see that the motor torque, M_m, is less than the pump torque, M_p, producing a net negative torque and a decrease in the angular velocity $\dot\phi$. As the speed decreases, the torque slowing it down also decreases, so the deceleration decreases. When the speed reaches the equilibrium, $\dot\phi_2$, the net torque and the deceleration vanish. The equilibrium appears to be stable, but to make sure you ought to check what happens if the starting speed is somewhat below $\dot\phi_2$.

This check can be done by considering the non-equilibrium states (b). Here, the source

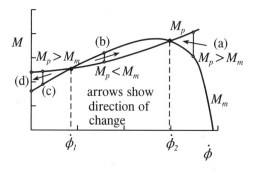

Figure 2.8: Stable and unstable equilibria for the twentieth floor sprinkler

(motor) torque exceeds the load torque, so the speed of the common shaft accelerates, as indicated by the arrow. Again, the acceleration vanishes as the speed aproaches $\dot{\phi}_2$. This observation confirms the stability of the equilibium speed $\dot{\phi}_2$. At the same time it indicates that the equilibrium speed $\dot{\phi}_1$ is unstable; no matter how close states (b) are to this equilibrium, an acceleration will drive the speed away from it. This conclusion is reinforced by consideration of states (c), which produce a deceleration, as indicated by the associated arrow.

Finally, consider the rest case (d). As with case (c), $M_m < M_p$. This means that the motor will not start, and therefore the system is unacceptable, despite the stability of the speed $\dot{\phi}_2$. (It could be made practicable, however, in the event that some auxiliary means were available to start the shaft rotating faster than $\dot{\phi}_1$. One such means is suggested in Guided Problem 2.2.)

As a more familiar example of an **unstable equilibrium**, consider a perfectly symmetrical sharpened pencil placed perfectly vertically on its tip in a perfectly still room. It is in equilibrium because there is no force to make it fall over. The slightest deviation from the vertical orientation, however, or the slightest puff of air, produces a small force in a direction away from the vertical equilibrium. As the departure from equilibrium grows, the destabilizing force grows also, and the pencil soon falls over. In practice, tiny errors in the initial state and tiny disturbances are impossible to avoid. Real pencils in real rooms never stand on sharpened tips for more than a brief moment.

2.1.4 Summary

A system is identified by its control surface. A shaft penetrating this surface defines an energy or power port of the system. The behavior of this power port can be described by the conjugate variables torque and angular velocity, the product of which is the power passing into or out from it. If the system is not influenced by any other external variables, it is said to have a single port, or be a one-port system. Its environment, similarly, could be described as another one-port system. Which is the system and which is the environment is quite arbitrary; you could simply call them system A and system B. Further, the steady-state behaviors of each such system, for any and all environments, can be represented by the relation it imposes between its conjugate variables, as determinable by experiment and as representable by an equation or a plot. Such a relation is called the system characteristic.

Two one-port systems connected at a common port experience an equilibrium state or states where they have common values of both of the conjugate variables. If the two characteristics are superimposed on a common plot, these equilibria are represented by their intersections.

An equilibrium can be stable or unstable, depending on whether an imposed departure from equilibrium spontaneously grows or shrinks. This determination requires information not contained in the characteristics themselves. For the case of the motor-pump-sprinkler this information is two-fold: the shaft is rigid, so both source and load have a common angular velocity, but the torques can be different. If the source torque exceeds the load torque, the combined system accelerates. This situation can be represented by a vertical line segment connecting the two characteristics at the particular speed, and an arrow pointing to higher speeds. The pattern of such arrows for different speeds shows which equilibria are stable and which are not.

The purpose of presenting this case study far transcends systems with shafts. Similar considerations apply to systems with interconnecting rods, fluid pipes and electrical conductors, as you shall see.

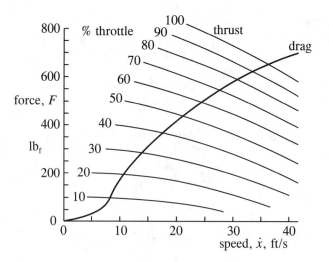

Figure 2.9: Thrust and drag for boat of Guided Problem 2.1

Guided Problem 2.1

This first guided problem involves a basic source-load synthesis. You are strongly advised to carry it out, at least roughly, before examining the solution at the end of the section.

An internal combustion (IC) engine drives a propeller that powers a boat which weighs effectively 4000 lbs. The thrust produced by the propeller depends on the throttle position and the speed of the boat, as plotted in Fig. 2.9. The drag force on the hull also depends on the speed of the boat, as shown.

(a) Plot the steady-state speed of the boat as a function of the throttle position.

(b) Assume that the throttle has been at 30% long enough for the boat to reach equilibrium speed. The throttle then is suddenly advanced to 100%. Find the instantaneous acceleration. Also, determine the acceleration when the speed of the boat has increased half way to its final equilibrium.

Suggested Steps:

1. The thrust and drag are forces that are analogous (similar) to torque, and the linear speed is analogous to angular velocity. To answer part (a), determine the intersections of the thrust curves with the drag curves, cross-plot the speeds of these points against the throttle positions, and connect the resulting points with a smooth curve.

2. Get the net force which accelerates the craft by subtracting the drag of the hull at the equilibrium speed for 30% throttle from the thrust at the same speed corresponding to 100% throttle.

3. Use $F = ma$ to determine the acceleration.

4. To answer the final question, find the average of the initial and final speeds, and repeat steps 2 and 3.

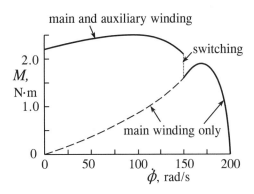

Figure 2.10: The motor of Guided Problem 2.2

Guided Problem 2.2

This problem illustrates stable and unstable equilibria as well as basic source-load synthesis. Its value depends on your effort.

Clearly the water-sprinkler system you have examined is unsatisfactory for the twentieth floor and above. One solution is to replace the motor, which is of the standard shaded-pole design common for small induction motors. The proposed motor, characterized in Fig. 2.10, is of the capacitor-start induction type. It has two windings; the one with a series capacitance is switched out of the circuit by a centrifugal switch at 75% of synchronous speed. Ignore potential problems with overheating.

(a) Estimate the maximum possible powers below and above the switching speed for the proposed motor, and compare to the original motor.

(b) Find the equilibrium speeds for the sprinklers on the tenth, twentieth and thirtieth floors, respectively, using the proposed motor.

(c) Discuss the stability and any peculiar operation of these equilibria.

(d) As project engineer for the system, recommend a course of action.

Suggested Steps:

1. Power is the product $M\dot\phi$. The maximum powers can be estimated by trial-and-error guesses of points on the characteristics. Later, you will see some graphical means that will help you zoom in on the maxima quickly.

2. All you really need to find the equilibrium points is the motor characteristic, represented by the solid line in Fig. 2.10, and the three load characteristics given in Fig. 2.7. Superimpose the load characteristics on the motor characteristics, and find the intersections.

3. Carry out a stability analysis as in Fig. 2.8. Note that the vertical segment of the load charateristic merely connects its two end points; operation cannot occur at an intermediate point. Instead, switching (rapid?) may occur.

4. Consider whether the nominal solutions seem practical; use your common sense to make a suggestion beyond the formal scope of the analysis as presented.

PROBLEMS

2.1 When the boat of Guided Problem 2.1 and Fig. 2.9 is traveling at a steady 20 ft/s the throttle is abruptly reset at 80%.

   (a) Find the thrust and drag forces before the velocity has time to change.

   (b) Find the instantaneous acceleration of the boat.

2.2 Estimate the maximum power $\mathcal{P}_{max}$ the induction motor characterized in Fig. 2.4 can deliver, and the associated torque M and angular velocity $\dot{\phi}$. Also, find the ratio of this power to the power $\mathcal{P}_{max\,M}$ delivered when the *torque* is maximized.

2.3 Consider that the motor of Guided Problem 2.2 and Fig. 2.10 drives the twentieth-floor sprinkler system. The load has an effective inertia of 0.015 kg·m^2.

   (a) Estimate the average net torque available for acceleration over the speed range $0 < \dot{\phi} < 100$ rad/s.

   (b) Repeat (a) for the speed ranges $100 < \dot{\phi} < 130$ rad/s and $130 < \dot{\phi} < 150$ rad/s.

   (c) Estimate the time required for the system to accelerate from rest to the switching speed of 150 rad/s.

2.4 An induction motor and load have the characteristics shown below. The shaft and the load are rigidly connected, but have some inertia. Determine the stabilities of the three equilibrium points.

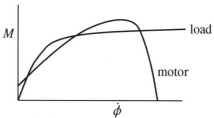

SOLUTIONS TO GUIDED PROBLEMS

Guided Problem 2.1

1.

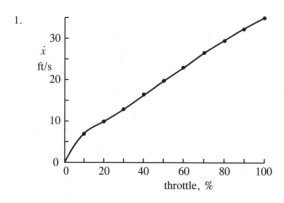

2. The 100% thrust curve must be extrapolated to estimate the thrust at 13 ft/s: about 870 lb. Subtracting 270 lb for the drag force at the same speed gives a net force of 600 lbs.

3. $a = \dfrac{Fg}{mg} = \dfrac{600 \times 32.2}{4000} = 4.83 \text{ft/s}^2$.

4. The average of the initial and final speeds is $\dfrac{13 + 35}{2} = 24$ ft/s. At this speed, the difference between the maximum thrust and the drag is $770 - 490 = 280$ lbs, which gives an acceleration of $\dfrac{280 \times 32.2}{4000} = 2.25$ ft/s^2.

Guided Problem 2.2

1.

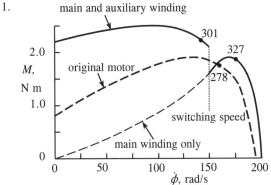

The three maximum powers are at the points indicated by dots and have the powers, $M\dot{\phi}$, in Watts, indicated by the numbers. They can be found by trial-and-error. Note that the points must lie on the descending portions of the respective characteristics.

2, 3.

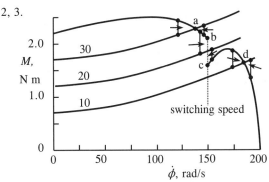

Points a and d are seen to be stable equilibria. The intersection of the 20th-floor characteristic with the dotted switching line would be a stable equilibrium if the dotted line represented a real characteristic. In reality, only states b and c exist, so the motor switches back and forth rapidly between these two states.

4. Something must be done to eliminate the rapid switching for the 20th-floor case, which would burn out the switch quickly. The simplest solution employs a more powerful motor and likely would not be any more expensive than fancier solutions that retain one of the existing motors. Better matching between one of these motors and the existing pump nevertheless presents an alternative class of solution, as discussed in Section 2.4.

2.2 Generalized Forces and Velocities

The power that is transferred through the bond between the two sub-systems in the case study of Section 2.1 equals the product of the torque M on the shaft and the angular velocity $\dot{\phi}$ of the shaft. This idea of factoring power into a product of two conjugate variables also applies to mechanical translation, mechanical shear, electrical power, fluid power, heat transfer and other energy domains. Observation of the scalar or vector nature of the resulting conjugate variables establishes *analogies* (special similarities) between variable types across these different domains. As a result, analogies can be drawn also between the

Figure 2.11: Convention distinguishing generalized force (effort) from generalized velocity (flow)

various types of *components* that the bonds interconnect. The variables and components in an electric circuit, for example, behave functionally in the same way as the corresponding variables in an analogous mechanical "circuit." These analogies enable you to transfer your knowledge of one type of physical or engineering system to another. They also reduce the variety of functionally different components in a model of an interdisciplinary system.

2.2.1   Efforts and Flows

The power factor variables for a generic (or general abstract) bond are called its **effort** or **generalized force**, labeled as "e," and the **flow** or **generalized velocity**, labeled as "$\dot{q}$".[1] Thus the power becomes

$$\boxed{\mathcal{P} = e\dot{q}.} \tag{2.4}$$

Equation (2.4) represents the first of the two parts of the definition of efforts and flows. Thus, for example, since $\mathcal{P} = M\dot{\phi}$, M and $\dot{\phi}$ qualify as effort and flow. Which is the effort and which is the flow are determined by the second part of the definition.

   Power flows along a bond in one direction or the other. Flows are normally defined as directed, also, while efforts are normally either true scalars or are treated as scalars. (There are infrequent instances in which some advanced bond graphers may consciously invert the roles of effort and flow, but you should overlook this possibility for now.) Scalars such as voltage, pressure and temperature are efforts. Directed quantities such as electric current and volume fluid flow are flows.

   Conventionally, the symbol chosen to represent an effort variable is written above a horizontal bond or to the left of a vertical bond, and the symbol chosen to represent the power conjugate flow variable is written below the horizontal bond or to the right of the vertical bond:

$$\frac{e}{\dot{q}} \qquad\qquad e \,\bigg|\, \dot{q}$$

The half-arrows on the bonds indicate the direction that power flows when $\mathcal{P} > 0$. If e is a true scalar, this is also the direction of positive flow, $\dot{q}$. One detail: according to a standard set by a committee during the 11th IMACS World Congress (1985), the half-arrows should be placed on the flow side of the bond. (Some authors do not adhere to this convention.) Consider the examples of Fig. 2.11. In case (a) the positioning of u above and to the left of the bond, and v below and to the right, signals that u is the effort (generalized force) while v is the flow (generalized velocity). In case (b), however, only the location of the half-arrow resolves an ambiguity to show the same association.

---

[1]The traditional bond graph symbol for flow is f. This book substitutes $\dot{q}$ to emphasize its relation to the displacement, q, and to reflect the practice in analytical dynamics.

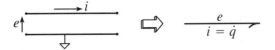

Figure 2.12: Power propagated by a pair of electric conductors

2.2.2 Electric Conductors

Each power flow in an *electric circuit* is represented by the product of **voltage**, e, and **current**, i:

$$\boxed{\mathcal{P} = e\, i.} \tag{2.5}$$

Voltage is a symmetric or scalar variable with respect to the conductor; it is identified therefore as a generalized force or effort, represented in this book also by the symbol e (the symbol v also could be used). Current, on the other hand, is anti-symmetric with respect to the conductor; it flows one way or the other. It is identified therefore as a generalized velocity or flow, represented by either the symbol $\dot{q}$ or the symbol i.

A pair of electric conductors, one grounded, is shown in Fig. 2.12. Note that the voltage e is the *difference* in electric potentials of the two conductors, while the current $\dot{q} = i$ is equal in magnitude and opposite in direction. In practice one usually does not bother to draw the grounded conductor in a circuit diagram, but nevertheless tacitly assumes its existence. The corresponding bond graph symbol also is shown, including the power convention half-arrow.

The time integral of the electric current,

$$\int_1^2 \dot{q}\, dt = q_2 - q_1, \tag{2.6}$$

equals the net **electric charge** or **electric displacement** over the interval, which happily is traditionally represented by the same symbol, q, as mechanical displacement.

2.2.3 Longitudinal Mechanical Motion

Consider an inextensible cable or a push-rod that connects two sub-systems, as illustrated in Fig. 2.13. The cable has a tension, F_t, with dimensions of force. The push-rod can sustain either tension or compression; the choice of variable shown is compression, F_c. If the velocity of the cable or rod, $\dot{x}$, is defined as positive in the rightward direction, the power conveyed from the left to the right becomes

$$\boxed{\mathcal{P} = -F_t\dot{x} \qquad \text{or} \qquad \mathcal{P} = F_c\dot{x}.} \tag{2.7}$$

Note that if the cable or the rod is in tension ($F_t > 0$ or $F_c < 0$) and $\dot{x} > 0$, the power is *negative, that is being propagated to the left*. The "power convention" ($\mathcal{P} > 0$ for power propagated to the right) is indicated by the half-arrow on the word bond graph.

The variable $\dot{x}$ is a real velocity aligned with the bond. The generalized velocity or flow is directed similarly. Therefore, F_t or F_c must be the generalized force or effort. *Force* is defined classically as an action *on a system* describable by a *vector*. The tension F_t or compression F_c, on the other hand, refers to both the action on the system and the reaction on its environment. In this symmetric view the tension or the compression acts between two systems like a *scalar*. A "generalized force" or effort is a generalized concept of such a variable; voltage is another example.

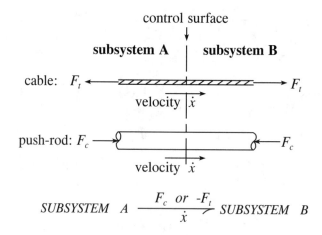

Figure 2.13: Sub-systems interconnected by a cable or push-rod

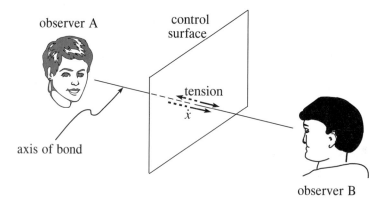

Figure 2.14: The perspectives of two observers at opposite sides of a bond

A powerful way to determine whether a variable has the directedness of a flow or generalized velocity or the symmetry of an effort or generalized force is to consider two observers facing each other head-to-head and toe-to-toe on either side of the control surface or either end of the bond, as suggested in Fig. 2.14. They both perceive the variable in question at the control surface through which the bond penetrates. If their perceptions are identical, the variable can be said to be symmetric with respect to the observers, and the variable is an effort or generalized force. In our immediate example, if one observer senses tension, so does the other; if instead he senses compression, so does the other. If the variable is $\dot{x}$, however, one sees a motion directed *away* from himself, while at the same time the other sees motion directed *toward* himself. Anti-symmetry implies a flow or generalized velocity.

2.2.4   Incompressible Fluid Flow

The flow of a fluid through a pipe or tube can be represented at any cross-section or control surface as the sum of the flows through a bundle of infinitesimal stream tubes, each with cross-sectional area dA, as suggested in Fig. 2.15. Each stream tube acts much like a push-rod with compression force $dF_c = P\,dA$, where P is the static pressure, and velocity $\dot{x} = v$.

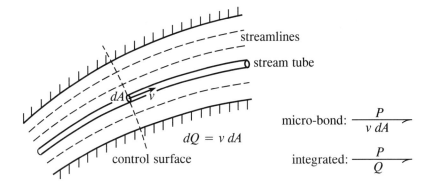

Figure 2.15: Channel with elemental stream tubes of area dA

Thus its energy flow rate or power is

$$dP = \dot{x}\, dF_c = Pv\, dA. \tag{2.8}$$

Assuming P is uniform over the section, integration of this power over the section gives the total power[2]

$$\boxed{\mathcal{P} = PQ; \qquad Q = \int v\, dA,} \tag{2.9}$$

The symbol Q stands for **volume flow rate**. This representation is particularly useful when the density of the fluid can be assumed to be constant, for then Q is uniform along the tube or pipe and across any throttling restrictions. (The analysis of compressible flow in Chapter 11 employs mass flow rate instead.)

The pressure P is the same whether the observer is facing upstream or downstream; this implies symmetry and identifies P as an effort or generalized force. It is representable by the generic symbol e, like simple force, moment or voltage. Conversely, Q has a direction and thus is a flow or generalized velocity, representable by the generic symbol $\dot{q}$, like simple velocity, angular velocity and electric current. Note that the time integral of Q (representable by the generic symbol q) is the net volume of fluid which has penetrated the control surface. Sometimes, as a result, the symbol V is preferred for this volume or generalized displacement, and the symbol $\dot{V}$ is preferred for the volume flow rate.

2.2.5   Rotational Motion

The power transmitted along a rotating shaft was seen in Section 2.1 to be the product of the moment, M, and angular velocity, $\dot{\phi}$:

$$\boxed{\mathcal{P} = M\dot{\phi}.} \tag{2.10}$$

Which of the two power factor variables, M or $\dot{\phi}$, is the effort and which is the flow? To answer this question, imagine the two observers facing each other at the two ends of the shaft, as shown in Fig. 2.16. If one sees clockwise rotation, the other sees counterclockwise

---

[2]The approximation of equations (2.8) and (2.9) omits the transport of both internal and kinetic energies, which can be important particularly when P is small. Nevertheless, they apply with considerable accuracy to most hydraulic "fluid power" systems. More complete representations for compressible flow systems are presented in Chapter 11, where it is shown that it is more accurate to substitute the stagnation pressure for the static pressure.

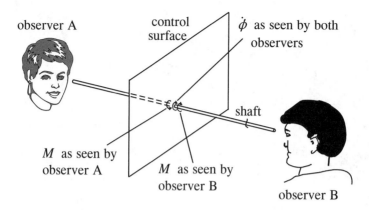

Figure 2.16: Observers viewing rotating shaft

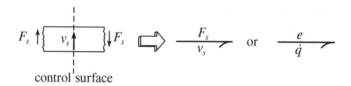

Figure 2.17: Power propagated by lateral motion

rotation. (Check this out by rotating your pencil in one direction and observing it from both ends.) Thus, $\dot\phi$ is a *flow* variable. Now, let the two observers sense the torque or moment. Both will feel the same direction, either clockwise or counterclockwise. (Check this out, also, by placing a clockwise moment on one end of your pencil with your left hand, as seen by your left wrist, and balancing that moment with your right hand. Your right hand then also exerts a clockwise moment, as seen from its wrist.) Thus, moment M is an *effort* or *generalized force*.

2.2.6   Lateral Mechanical Motion

Transverse or lateral motion of a member transmits power through a control surface if a corresponding shear force is present. As suggested in Fig. 2.17, *which represents a top view*, the power is the product of the shear force, F_s, and the lateral velocity, v_s:

$$\boxed{\mathcal{P} = F_s v_s.}\tag{2.11}$$

Observers facing each other at the two ends of the bond see the shear force in the same sense, to their respective right sides or respective left sides. If one observer sees the motion as rightward, on the other hand, the other observer sees it as leftward, and vice-versa. Thus, the force is the effort and the velocity is the flow.

   Should the shear motion be vertical rather than horizontal, however, the roles of the two variables become reversed: the velocity is seen as the same by both observers, while the shear force has opposite sense. How the observers see transverse vectors is pictured in Fig. 2.18. But do you want the distinction between effort and flow to hinge on whether the motion is lined up with the axes of the observers? In exceptional cases like this, when neither variable is a true scalar, you are permitted to *choose* which is the effort and which is the flow. (Most modelers choose the force as the effort, and the velocity as the flow.) The

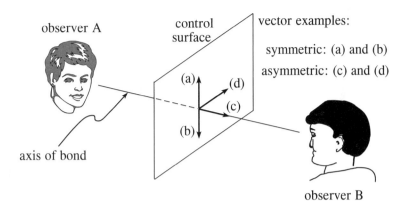

Figure 2.18: Transverse vectors as seen by two observers at opposites sides of a bond

convention of the power half-arrow is inviolate, on the other hand: the arrow points in the direction that the power flows when the product of the effort and flow variables happens to be positive.

2.2.7 Microbonds

The bonds addressed so far can be called **macrobonds**, in contrast to **microbonds** which describe the power propagated through a microport of infinitesimal cross-sectional area, dA. A bond or macrobond is thus the sum or integral of an infinite number of microbonds; its existence implies one or more simplifying assumptions. For the incompressible fluid flow above (Fig. 2.15), the power of a microbond is given directly by equation (2.8). Thus, P is the effort and $v\,dA$ is the flow. This equation can be integrated to form the macrobond only if either P or v is uniform over the channel. If P is uniform or is assumed to be uniform (the more common assumption), the equation integrates to give the product PQ, where $Q = \int v\,dA$, as noted above. To the extent that neither P nor v is uniform, however, the macrobond does not properly model the situation.

Microbonds for longitudinal motion share the flow variable $\dot{x}$ with the microbond. Therefore, the effort on the microbond is the compressive force $-\sigma\,dA$, where $-\sigma$ is the normal compressive stress. If $\dot{x}$ is uniform throughout the cross section, the effort of the macrobond becomes the integral of the stress, namely the force. If $\dot{x}$ is not uniform, the macrobond cannot model the situation precisely.

The rotating shaft also can be considered to be the sum or integration of an infinite set of microports. At the microport level, the rotation is a lateral motion, with the power density being the product of the shear stress τ and the shear velocity v_s. Neither of these quantities is common to all area elements dA. The moment M is τ times the moment arm integrated over the area,

$$M = \int_0^a \tau r[2\pi r\,dr],\qquad\qquad (2.12)$$

where r is the radius variable and a is the radius of the shaft. M doesn't exist at a point. The power propagated is the integral of the shear stress times the shear velocity, v_s, over the area, or

$$\mathcal{P} = \int_0^a \tau v_s[2\pi r\,dr]; \qquad\qquad v_s = r\dot{\phi}. \qquad\qquad (2.13)$$

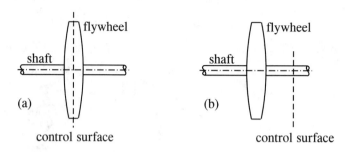

Figure 2.19: The effect of power factoring on accuracy, due to the choice of control surface or port location

A comparison of equations (2.12) and (2.13) shows that $\mathcal{P} = M\dot{\phi}$ is correct only if you assume that $\dot{\phi}$ is uniform and thus can be taken forward of the integral. This happens when the cross section of the shaft rotates as a rigid unit. To the extent that $\dot{\phi}$ is not uniform, the very concept of the macroport is in error.

Only in extremely rare cases would the assumption of uniform $\dot{\phi}$ not be entirely appropriate. The point is, however, that whenever you employ a macroscopic variable (such as M or $\dot{\phi}$) you are inevitably introducing some measure of approximation, or inexactness. Sometimes the error is not small.

As an example, consider a "shaft" that is a thin but large-diameter flywheel, as shown in Fig. 2.19 part (a). The boundary between the system and the environment is assumed to lie right in the middle of the flywheel. The center and the periphery of this flywheel are much more likely to experience different angles of rotation than would the center and the periphery of a small shaft. If the periphery were suddenly braked, for example, the core might continue to rotate through an additional small angle. Or, if the core is oscillated rotationally at a high frequency, the periphery might not follow in lock-step; in fact, a "natural frequency" might well exist at which resonance occurs. Should this phenomenon possibly exist, you would be well advised to place the system boundary to the left or to the right of the entire flywheel, as shown in Fig. 2.19 part (b).

It is a good idea, then, to partition a system from its environment at a location where the interaction can be described *simply* (in the present case by the variables $\dot{\phi}$ and M). There are no inviolate rules for doing this; consequently, modeling ultimately becomes an *art*. (You should not be discouraged by this fact, for any subject that can be reduced to rules can become very dull, and its practitioner is subject to being replaced by software.) Partitioning a motor through its middle, for example, likely would cause you grief, even though it would not be incorrect.

2.2.8   Analogies

The effort and flow symbols for the various media that have been examined are summarized in Table 2.1. The various effort variables are said to be analogous to one another, as are the various flow variables, due to their common definitions with respect to power and directivity. The implications will gradually become clearer, as you identify common types of elements that represent various classes of interactions between the efforts and flows or their time integrals or time derivatives.

Table 2.1 Effort and Flow Analogies

	generalized force or effort, e	generalized velocity or flow, $\dot{q}$
Fluid, incompressible:		
approximate[†]	P	$Q = \dot{V}$
micro-bond[†]	P	$v\,dA$
Mechanical, longitudinal:		
approximate[†]	F_c	$v = \dot{x}$
micro-bond[†]	$-\sigma dA$	$v = \dot{x}$
Mechanical, transverse:		
rotation	M	$\dot{\phi}$
translation (shear)	F_s	v_s
micro-bond	τdA	v_s
Electric conductor:	e	i

[†] The powers associated with the transport of kinetic and potential energies are added in Chapter 11. For most mechanical and incompressible fluid cases they are small.

The **direct analogy** between mechanical and electric circuit variables associates mechanical force and torque with electric voltage, and linear or rotary mechanical velocity with electric current.[3] This analogy can help the student well grounded in mechanical systems transfer his knowledge to electric circuits. Similarly, it can help the student well grounded in electric circuits to extend his understanding into mechanical "circuits." Similar conceptual and computational advantages in the transference of knowledge accrue from the other analogies.

A few physical interactions are not best represented by simple bonds or product conjugate variables. The example of compressible fluid flow is treated in Chapter 11.

2.2.9 Summary

The power flowing through a simple port or simple bond customarily is factored into a product of two variables, $e\dot{q}$. The "effort" or "generalized force" e normally is symmetric, and the "flow" or "generalized velocity" $\dot{q}$ normally is anti-symmetric with respect to two observers facing each other from opposite ends of the bond. This simple factoring applies to longitudinal, lateral and rotational motions of a rod, cable or shaft, to electric conductors and to the flow of an incompressible fluid, establishing important analogies between variables used in these different domains. The direction of positive power flow is indicated on a bond by a half-arrow.

A standard port or macro-port and its bond can be considered to be the sum or integration of a bundle of micro-ports and their micro-bonds, each representing an infinitesimal cross-sectional area. This summing or integration inevitably introduces some degree of approximation. All use of (macro) bonds, and indeed the very use of the word "modeling," implies approximation. Only pure mathematics is exact. The practical question is not the presence of error, but rather its extent and significance.

---

[3]Electrical engineers sometimes draw the "Firestone analogy" between force and current on one hand and velocity and voltage on the other. This gives a greater apparent similarity between circuit diagrams and some corresponding mechanical "circuits." For basic physical reasons including its extensibility to non-circuit situations, however, the author strongly prefers the direct analogy.

PROBLEMS

2.5 A cable attached to the top of the machine shown on the left at the top of the next page is pulled upward, thereby producing some energetic effect. Represent the interaction with a bond, annotated with a power half-arrow and effort and flow variables, which should be defined.

2.6 A lever attached to the machine shown above right is rotated upward. Represent the interaction by an energy bond, annotated by symbols for effort and flow which you should define and by a power convention half-arrow.

2.7 Heat conduction is a special kind of power that can be treated as the product of an effort and a flow, like the other power types above. The effort variable is absolute temperature, which you can label as T. The power itself is the rate of heat transfer, which you can label as $\dot{Q}$, where Q is a quantity of thermal energy called heat. Determine the flow variable, $\dot{q}$, in terms of these variables, and express its integral as the displacement variable, q. Relate this displacement variable to a common variable used in thermodynamics. (This question presupposes that you have been introduced to thermodynamics. Further consideration of heat transfer and more general thermodynamics is postponed until Chapters 9 and 11.)

2.3   Generalized Sources, Sinks and Resistances

The case study of Section 2.1 was represented by the word bond graph

$$\begin{array}{cc} INDUCTION & \underline{\quad M \quad} \quad LOAD \\ MOTOR & \dot{\phi} \quad\; SYSTEM \end{array}$$

Both the induction motor and the load system were modeled by particular static relations between the common shaft moment, M (or M_m or M_p), and the shaft angular velocity, $\dot{\phi}$. In Section 2.2 such static relationships were generalized, so that the form of this system becomes

$$S \xrightarrow[\dot{q}]{e} R$$

in which S stands for "source" and R stands for "resistance." Sources normally emanate power, whereas resistances normally absorb or dissipate power.

The effort e and flow $\dot{q}$ in this graph could represent any of the energy types discussed in the last section. The graph could represent an electrical system, a system with incompressible fluid flow, or a mechanical system with either longitudinal, lateral or rotational motion. The source and resistance elements in each case might be characterized in one of several special ways, which now will be discussed.

2.3.1   Independent-Effort and Independent-Flow Sources and Sinks

An **independent-effort source**, usually called simply an **effort source**, and an **independent-effort sink**, usually called simply an **effort sink**, are defined to have efforts that are independent of their flows. This means either that e is a *constant*, or in general a function

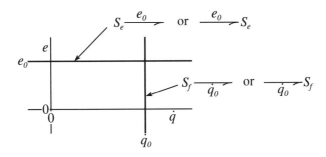

Figure 2.20: Characteristics of effort and flow sources and sinks

of time, $e = e(t)$. The elements provide or absorb *any* flow that the attached system may require in response to the imposed effort. This behavior is represented in the plot of Fig. 2.20 by a horizontal line in the e–$\dot{q}$ plane. Both elements are designated by the symbol S_e. The only difference between them is that the source is designated with an outward-directed power arrow, and the sink with an inward-directed power arrow:

$$\text{effort source}: \quad S_e \longrightarrow$$
$$\text{effort sink}: \quad \longrightarrow S_e$$

They are really the same element, since there is no absolute requirement that the power flows in the direction of the power arrow, that is $e\dot{q} > 0$; it is just nice to have different words to express the different functions normally served by the element.

Physical examples approximated by the S_e element include a force source as from a weight in a gravity field, a torque source, a voltage source as from an ideal battery, and a pressure source or sink from a large body of water at some elevation. These examples are suggested in Fig. 2.21. Coulomb or "dry" friction also can be modeled as a force source, as long as the motion never reverses direction, as discussed below.

The **independent-flow source** and the **independent-flow sink**, more simply called the **flow source** and the **flow sink**, are similar to the effort source and the effort sink except that the flow is independent of the effort rather than the effort being independent of the flow. As this reversal of roles suggests, the flow may be either constant or an independent function of time. The elements provide whatever effort that the system requires in response to the independent flow. Flow sources and sinks are designated by the symbol S_f:

$$\text{flow source}: \quad S_f \longrightarrow$$
$$\text{flow sink}: \quad \longrightarrow S_f$$

The generalized characteristic is plotted in Fig. 2.20 as a vertical line.

A **synchronous motor** turns with a constant speed, determined by the frequency of the driving AC voltage (usually 50 or 60 Hz), as long as the load torque does not exceed some limit. A source S_f element models this behavior but does not recognize the torque limits. Any other constant velocity source, constant current source or constant fluid flow source also can be represented by the element.

Notice again that a current sink of 5 amps is the same as a current source of -5 amps, as a pressure source of 5 psi is the same as a pressure sink of -5 psi. Thus, you are permitted to avoid effort and flow sinks in favor of effort and flow sources, or vice-versa, although most modelers prefer to use both types.

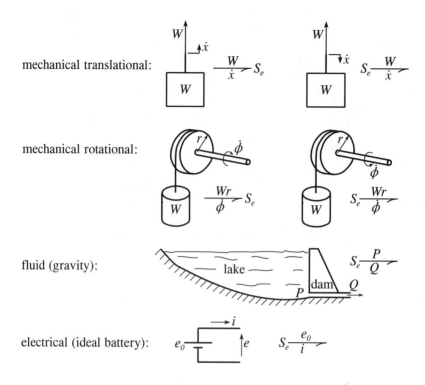

Figure 2.21: Example effort sources and sinks

2.3.2   General Sources and Sinks

A general source (S——————) can represent any prescribed static (or algebraic) relationship
between its effort and its flow. Thus, as noted above, the induction motor of Section 2.1
can be represented as a general source. Other examples include a model of a battery for
which the voltage depends on the current, a model of a velocity-dependent force applied to
a mechanical system and a model of a pressure source that depends on the fluid flow. In
the most general case, the relationship itself could be an explicit function of time.

The general sink element (——————S) is not employed *per se* in this book, in favor of the
equivalent term **resistance** and its symbol R, as discussed below. You will see in Section
5.3 that any general source can be represented well by an effort source or a flow source in
combination with a simple resistance. Ultimately, the general source is employed, if at all,
only as a convenient shorthand.

2.3.3   Linear Resistances

The generalized resistance, indicated as

$$\frac{e}{\dot{q}}\ R$$

can have any functional dependency between its effort and its flow. It is helpful to distinguish different classes of these elements.

The simplest class is the **linear resistance** which, in a plot of effort vs. flow as illustrated in part (a) of Fig. 2.22, is represented by a straight line drawn through the origin.
The slope of this characteristic is defined as the **modulus** of the resistance, or for short

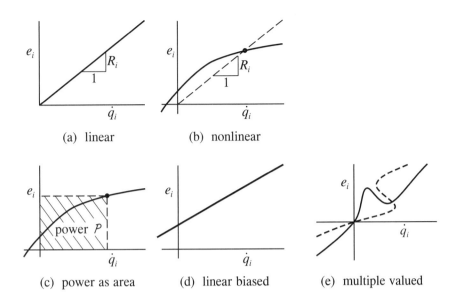

Figure 2.22: Types of resistance characteristics

its "resistance," which in this book is given the same symbol, R. Thus, a symbol R_i in a bond graph indicates that a resistance relation applies between the associated effort e_i and flow $\dot{q}_i$. It does not specify the details of the relation, however. If R_i is a constant, that is if the element is linear, the algebraic expression $e_i = R_i \dot{q}_i$ applies. Otherwise, the relationship must be expressed separately. *Bond graph elements never include algebra.* The two meanings of "resistance" are distinct.

The electrical resistor may be the most familiar approximation to a linear resistance. The effort is the voltage drop e across the resistor, the flow is the current i which passes through it, and $e = Ri$. This and other linear resistances are pictured in Fig. 2.23.

A common linear *fluid* resistance results from the assumption of viscous laminar fully-developed steady flow through a circular tube. In your fluid mechanics course you will derive the relation that the pressure drop P across the tube is related to the volume flow rate Q through it by

$$\boxed{P = RQ; \qquad R = \frac{8\mu L}{\pi a^4},} \tag{2.14}$$

where a is the inner radius of the tube, L is its length, and μ is the absolute viscosity of the assumed incompressible fluid. More generally, whenever the viscous forces overwhelm the inertial forces in an incompressible flow through a passage, that is when the Reynold's number is low, the resistance can be approximated as linear, and its modulus as a constant. Examples include flow through a thin slit such as a leakage path in a machine, flow through a porous plug and "creeping" flow through an orifice.

Mechanical devices that can be approximated by a linear resistance are hard to make, but are very commonly assumed in analysis. They are called **dashpots**, and are given the special symbol shown in part (c) of Fig. 2.23. The ratio of the force to the velocity traditionally is designated by the symbol b, so the modulus of the bond graph element is $R = b$. One way to construct a linear translational dashpot is to convert the relative velocity of the two translating members to the flow of a liquid, which can be done by employing a piston and cylinder. Then, as shown in the example, this liquid is forced through a passage that behaves as a linear fluid resistance according to equation (2.14).

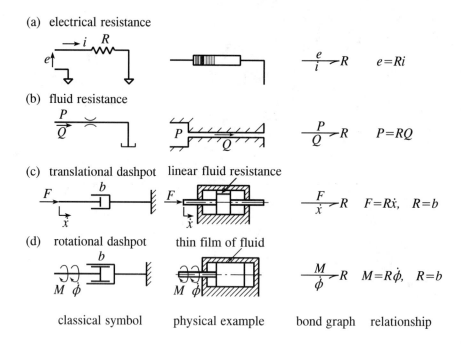

Figure 2.23: Devices approximated by linear resistances

A rotational dashpot is designated by a similar symbol, as shown in part (d) of the figure. Translational and rotational dashpots often are described as "viscous dampers" because it is hard to approximate linearity mechanically without using a viscous fluid. The shear flow of a viscous liquid in a very thin annulus between a rotating cylinder and a cylindrical shell, for example, can produce a torque proportional to angular velocity, that is a rotational linear resistance. As a practical matter, however, the small amount of fluid trapped in the annulus is apt to become hot, dropping its viscosity, and possibly causing the fluid to oxidize and turn black, like the motor oil in your car.

2.3.4   Nonlinear Resistances

Any component with a single effort or generalized force that is a function of an associated flow or generalized velocity, but is not proportional to it, can be described as a **nonlinear resistance**. The general resistance is the same as the general sink, and is used in its stead. Flow through an orifice at anything higher than creeping Reynold's numbers is an example. If Bernoulli's equation is used to model the flow between an upstream state with zero velocity and the point in the throat where it achieves maximum velocity, and complete dissipation of the kinetic energy is assumed to occur downstream, the result is

$$P = \frac{1}{2}\rho v^2 = \frac{1}{2}\rho \left(\frac{Q}{c_d A_0} \right)^2 = \frac{\rho}{2c_d^2 A_0^2} Q^2. \qquad (2.15)$$

Here, ρ is the fluid density and c_d is a **flow coefficient** which represents the fact that the flow separates from the walls of the orifice to form a narrower *vena contracta* or effective orifice area $c_d A_0$. (For a sharp-edged orifice, $c_d = 0.611$; in most other cases it is larger but less than 1.0.) The equation often is used even when the assumptions above are recognized

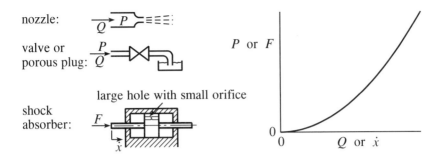

Figure 2.24: Nonlinear fluid resistance and application in a mechanical damper

Figure 2.25: A mechanical brake and its ideal resistance characteristic

as inaccurate; in this case, c_d is adjusted to accommodate the error and likely is determined experimentally. The characteristic given by equation (2.15) is plotted in Fig. 2.24.

Replacing the linear fluid resistance in the translational viscous damper in Fig. 2.23 with an orifice flow produces a nonlinear mechanical damper, as suggested in Fig. 2.24. The shock absorbers in your car are essentially of this construction. Not only is the nonlinear behavior easier to obtain, it is desirable; for small bumps you want the shock absorber to act gently, whereas for large bumps you prefer that it act more vigorously so as not to bottom out (reach its mechanical limit).

A mechanical brake is another kind of nonlinear resistance. As shown in Fig. 2.25, the classical model for such a resistance comprises a constant or **coulomb friction** torque for forward motion, associated with the kinetic coefficent of friction, and an equal but opposite torque for reverse motion. This assumes application of a fixed force to the brake calipers. When there is no motion, the resistance torque balances whatever torque is applied, as long as it does not exceed the "breakaway" torque which is associated with the static coefficient of friction. If the disk always rotates in one direction, however, this model gives nothing but a constant torque, and can be modeled more simply as an effort sink. You also will see more exotic friction models in Chapter 5, such as the one that makes a violin sing or a dry door hinge squeal.

The resistance characteristic shown in part (b) of Fig. 2.22 is nonlinear in two respects: it is curved, and it does not pass through the origin, or is **biased**. In general, a resistance can be described mathematically by either of the forms

$$\boxed{e = e(\dot{q}) \quad \text{or} \quad \dot{q} = \dot{q}(e).} \tag{2.16}$$

Equation (2.15)is an example of the former; solved for Q rather than P, it becomes an example of the latter. There is no need to define a modulus for the resistance, but this can be done if one wishes[4]. The power dissipated is the area under the curve, as indicated in

---

[4]The modulus or "resistance" of a nonlinear resistance, R, can be defined as the *slope of the chord* from the origin to the state point of interest, as shown in part (b) of Fig. 2.22. This **chordal resistance** therefore

part (c) of Fig. 2.22. This is the same as for a linear resistance; for both,

$$\boxed{\mathcal{P} = e\dot{q}.}$$ (2.17)

A **linear biased** characteristic is shown in part (d) of Fig. 2.22. Biased characteristics are represented in Section 5.3 as combinations of an effort or flow source and an unbiased linear or nonlinear (curved) characteristic. Finally, multiple-valued characteristics are shown in part (e). The characteristics illustrated, nevertheless, are **globally passive**, unlike the biased curves (b) and (d), since they reside only in the first and third quadrants where the power product $e\dot{q}$ is positive everywhere. An example of a multiple-valued characteristic is given in Guided Problem 2.3. Note that any bias destroys passivity, but a lack of bias does not assure passivity.

2.3.5   Source-Load Synthesis

The interconnection of a rotary source to a rotary load, by a shaft, was treated in Section 2.1.3 by finding the intersection of the two torque-speed characteristics. The same type of synthesis applies to any simple interconnection in which the power is represented by the product of a pair of conjugate generalized forces and velocities.

> **EXAMPLE 2.1**
>
> The output pressure of an impeller pump reduces the net flow below that of the geometrically slip-free flow. The result is a nonlinear source characteristic, as plotted below. The particular load characteristic plotted in the figure also is nonlinear, typical of the flows through sharp restrictions as discussed above and as employed in the water-sprinkler nozzles of Section 2.1. Determine the equilibrium state for the combination of these elements, and its stability.

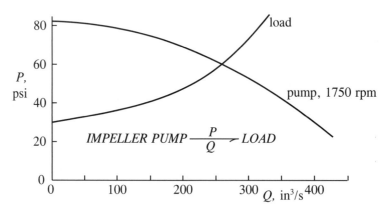

is not constant, but rather is a function of $\dot{q}$ or e:

$$e = R\dot{q}; \qquad R = R(\dot{q}) \quad \text{or} \quad R = R(e).$$

These are the same equations that apply to the linear case, which is the primary reason for defining the chordal resistance, except that R becomes constant. Further, the power dissipated in the resistance, for either the linear or the nonlinear case, becomes

$$\mathcal{P} = e\dot{q} = R(\dot{q})\dot{q}^2 = (1/R(e))e^2.$$

Solution: The equilibrium state is given by the intersection of the characterisitics. This equilibrium is stable, since if operation is attempted at a larger flow, the pressure of the load would exceed the pressure of the pump, forcing the flow back toward equilibrium. If operation is attemped at a smaller flow, the pressure of the pump would exceed that of the load, again forcing the flow toward equilibrium.

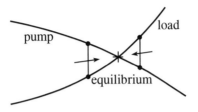

2.3.6 Power Considerations

The power that is transmitted from a source to a load often is of interest. Loci of constant power can be constructed on an $e - \dot{q}$ plot. One such locus is shown in part (a) of Fig. 2.26. The power equals the cross-hatched area of the rectangle of height e and width $\dot{q}$. All points on the locus must have the same area, as suggested by the two other rectangles shown. The curve is a rectilinear hyperbola.

The slope of a chord drawn from the origin to any point on a locus of constant power equals the magnitude of the slope of the tangent to the locus drawn at the same point. This fact, illustrated in part (b) of the figure, aids the sketching of constant power loci. Note also that the chord and the tangent form isosceles triangles with the $\dot{q}$ axis.

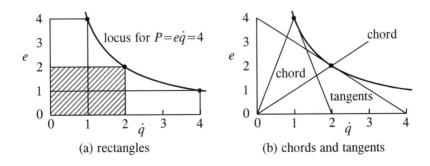

Figure 2.26: Loci of constant power as rectilinear hyperbolas.

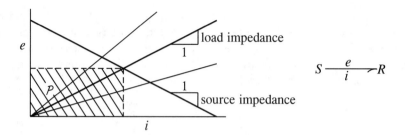

Figure 2.27: Synthesis of linear source and load to maximize power transfer

EXAMPLE 2.2

Locate the point of maximum power on the torque-speed characteristic of the induction motor as plotted below.

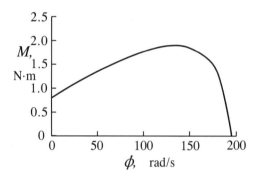

Solution: A family of loci for various constant powers is superimposed on the plot below. The locus representing the maximum power, shown by a dashed rectilinear hyperbola, is tangent to the characteristic at the operating point for that power. All other points on the characteristic have less power, including the point of maximum moment. A simpler way to estimate the location of the maximum power equates the magnitude of the slope of the chord from the origin to the point, and the slope of the tangent to the characteristic at that point. These straight lines are drawn dashed.

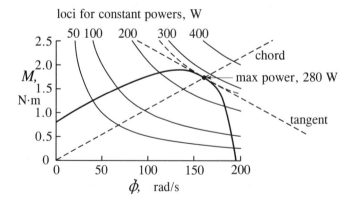

The source-load synthesis of a linear electrical source and linear electrical load is given in Fig. 2.27. The magnitude of the slope of the source characteristic is known as the **source**

impedance. The slope of the load characteristic, which is also the resistance of the load, is similarly known as the **load impedance**. Maximum power is transferred from source to load at the midpoint of the source characteristic. Not only does this point maximize the area of the rectangle between it and the origin, but the chord from the origin to this point has the same magnitude of slope as the tangent to the load characteristic, which is the load impedance itself. Thus, maximum power is transmitted when the load impedance equals the source impedance. Such "impedance matching" is standard practice for connecting audio amplifiers to audio speakers, for example.

2.3.7 Summary

One-port elements with constant effort independent of flow, but possibly dependent explicitly on time, are designated by the bond-graph symbol S_e. They are called constant-effort sources if their power arrow is drawn outward, and constant-effort sinks if it is drawn inward. One-port elements with constant flow independent of effort, but possibly dependent explicitly on time, are designated by the bond-graph symbol S_f. They are called constant-flow sources if their power arrow is drawn outward, and constant-flow sinks if it is drawn inward. General sources have other relationships between the effort and flow, and are designated by the bond-graph symbol S.

General sinks normally are called resistances, and are designated by the bond graph symbol R. Resistances occur whenever electrical, mechanical or fluid energy is dissipated into heat. The slope of a chord drawn from the origin to a point on the characteristic curve of a resistance is called the chordal resistance or just the resistance, and is designated by the same symbol, R. Resistance characteristics that pass through the origin of the effort-flow coordinates are called unbiased; others are called biased. An unbiased resistance is called passive if its characteristic remains in the first and third quadrants of the coordinates, since then the power $\mathcal{P} = e\dot{q}$ is always positive into the element, regardless of the operating point. It will be shown in Section 5.3 that both general sources and resistances can be represented by a combination of an effort or flow source or sink and an unbiased resistance. The simplest resistance is linear and unbiased; its value of R is constant.

Linear fluid resistances depend of the viscosity of the fluid. Nonlinear fluid resistances depend rather on the density of the fluid. Linear mechanical resistances or dashpots often are called "viscous dampers," and often depend on the viscosity of a fluid. Mechanical friction, as in a brake, is an unbiased nonlinear resistance, but may be approximated as an effort source or sink if its motion is always in one direction.

The source-load synthesis illustrated in Section 2.1 and again in Example 2.1 can be generalized for any of the power types considered. The synthesis maximizes the power transmitted if the chordal resistance of the load equals the slope of the source characteristic at the equilibrium point.

Guided Problem 2.3

This problem offers you needed experience in source-load synthesis, and reveals an important phenomenon called **hysteresis**, or history-dependent behavior including sudden jumps in state.

The IC engine and propeller drive from Guided Problem 2.1 is placed in a small boat with a planing-type hull. Unlike the displacement-type hull considered before, the drag characteristic of this hull, plotted in Fig. 2.28, has a pronounced local maximum (at about 24 feet per second). This peak is due to the wave generated by the motion. At the speed of its maximum, the bow is considerably elevated over the stern, so in effect the boat is continually trying to climb a hill of water. At higher speeds the boat "planes," riding high and nearly level, with the generated wave crest being amidships.

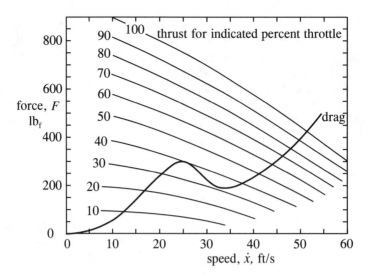

Figure 2.28: Thrust and drag characteristics for the boat in Guided Problem 2.3

Find and plot the steady-state relationship between the throttle position and the speed of the boat. Clearly indicate any hysteretic behavior.

Suggested Steps:

1. Draw a word bond graph for the system, naming a source, a load, a generalized velocity and a generalized force.

2. Were it not already done, you would superimpose plots of the source characteristics and the load characteristic, making sure the dimensions, units and scales are compatible.

3. Cross-plot all intersections of the source and load characteristics onto coordinates with speed on the ordinate and throttle position on the abcissas.

4. If and where it is not clear how to connect the points on your cross-plot with a continuous curve, generate more points by interpolating more source characteristics.

5. Identify any jumps in behavior with arrows on the cross-plot. Note explicitly the range of throttle positions for which there are multiple speeds. In this range the actual speed depends on the history of the throttle position as well as its current state.

Guided Problem 2.4

This type of problem is familiar to electrical engineers. It requires determination of a source characteristic. Your instructor may advise you on its relevance to your situation.

An n-channel enhancement MOSFET is driven by a 12 v battery and has a load resistance of 1.5 kΩ. Given the characteristics plotted in Fig. 2.29, plot the drain current i_D and voltage e_D as a function of the gate voltage e_G.

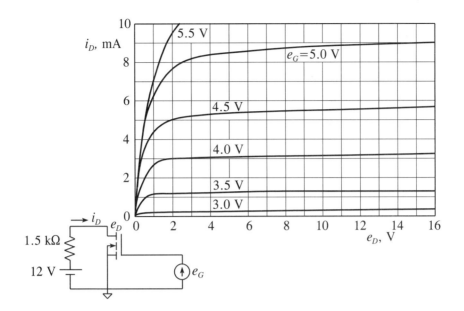

Figure 2.29: Characteristics of a MOSFET for Guided Problem 2.4

Suggested Steps:

1. Consider the battery and resistor as the power "source," and the given characteristics as the "load." Superimpose the source characteristic on the given plot. Hint: It is a straight line with negative slope; its intersections with the axes are readily found.

2. Cross-plot the intersections of the source line and the given load characteristics onto plots of i_D vs. e_G and e_D vs. e_G.

PROBLEMS

2.8 A shaft drives a load that resists with a constant torque regardless of speed, assuming that the shaft never reverses direction. Indicate what type of bond-graph element most simply represents this frictional behavior. Label the effort and flow of its bond with symbols that indicate their meanings, and show the power-convention half-arrow.

2.9 As noted in the text and pictured in Fig. 2.23, a viscous fluid damper can be made by placing a cylindrical hole through the piston of a fixed double-rod-end cylinder. Fluid leakage around the piston and friction between the piston and the cylinder walls may be neglected. (These seemingly contradictory assumptions could be reasonable if low-friction polymer seals are used between the piston and the cylinder.) You will need to define relevant physical parameters (constants) in order to carry out the following:

(a) Find the relationships between the force and the pressure difference across the piston, and between the velocity and the flow.

(b) Sketch-plot the force vs. the velocity, and express the resistance R relative to the velocity as a function of fixed parameters to the extent possible. Is this approximate resistance a constant?

2.10 Answer the above problem when the cylindrical hole is replaced by a short orifice of area A_o, to create a standard type of shock absorber.

2.11 A shaft of radius a, shown below, rotates concentrically in a fluid-filled journal with radial clearance $\epsilon << a$ and length L.

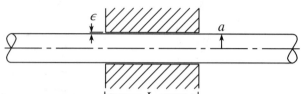

(a) Find the steady-state resistance relative to the angular velocity of the shaft.

(b) Discuss any major practical deviation from the assumed behavior.

2.12 Write the chordal resistance of the flow through an orifice, as modeled by equation (2.15) (p. 38), in the form $R = R(Q)$. (Chordal resistance is defined in the footnote on pp. 39-40.)

2.13 The impeller pump of Example 2.1 (p. 40) drives a load for which $P = 20 + 0.10\,Q$ psi, where Q has units of in^3/s, in place of the plotted load. Estimate the equilibrium pressure and flow.

2.14 Estimate the maximum power that the impeller pump of Example 2.1 (p. 40) could deliver if the load could be changed arbitrarily. Give the corresponding pressure and flow rate, also.

2.15 The hydraulic load plotted in Example 2.1 (p. 40) is driven by a hydraulic source described by $P = 80 - 0.2Q$ psi, where Q has units of in^3/s, in place of the impeller pump. Estimate the equilibrium pressure and flow.

2.16 Estimate the maximum power that the internal combustion engine characterized in Fig. 2.39 (p. 63) can produce. (This figure appears in Section 2.5, but you should disregard everything in the figure except the engine characteristic.)

2.17 An npn transistor in the common emitter configuration has the characteristics plotted below for various values of the base current, i_b.

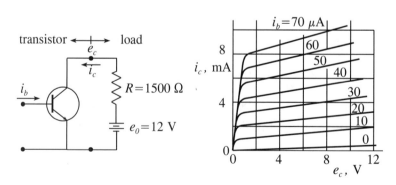

(a) Superimpose on the plot the characteristic for the load which comprises the voltage source e_0 and the resistor R.

(b) Plot the load (emitter) current, i_c, as a function of the base current, and approximate this relation algebraically.

2.18 The boat of Guided Problem 2.3 and Fig. 2.28 pulls one or more water skiers. The force in the tow line for one skier is shown plotted below as a function of speed; note that a planing phenomenon exists which is similar to that of the boat hull, but more pronounced.

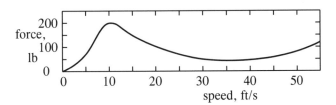

(a) Carefully estimate the maximum steady speed for towing a single skier.

(b) Estimate the maximum number of identical skiers that can be accelerated from rest simultaneously, and estimate their maximum speed.

(c) In a fancy maneuver, skiers are transferred to the boat from another boat, at full speed. Estimate the maximum number of skiers that can be towed successfully after the transfer, and estimate their maximum speed.

SOLUTIONS TO GUIDED PROBLEMS

Guided Problem 2.3

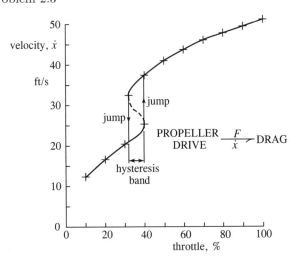

Guided Problem 2.4

1.
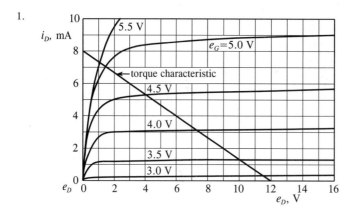

2.
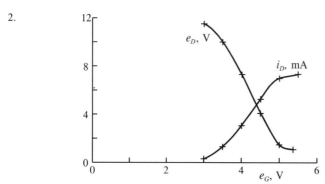

2.4   Ideal Machines: Transformers and Gyrators

An ideal machine transmits work into one of its two "ports" to work out from the other "port." Although the effort and the flow usually are changed, energy is not stored, generated or dissipated. Entropy also is not generated. Levers, gears, electric motors and piston pumps are examples of engineering components that may be approximated as ideal machines.

In modeling physical or engineering systems one often neglects the dissipation of energy, or more properly the conversion of energy to heat. This simplification is done despite the fact that leakage or slippage in the presence of friction or analogous generalized forces inevitably has at least a small effect. For example, a gear train might be assumed to be frictionless, an electric transformer might be assumed to have perfect coupling and zero resistance, an electric motor might be assumed to be 100% efficient in converting electrical energy to mechanical form, and a fluid pump might similarly be approximated as converting mechanical energy to fluid energy completely. Even if energy dissipation is not neglected, often it is represented separately. For example, a real electric motor can be represented as a frictionless or ideal motor plus an external resistance to represent electrical losses plus another external resistance to represent mechanical friction.

2.4.1   Ideal Machines

All of these devices can be described as "machines," in a generalized sense, and models of them that neglect energy loss are called **ideal machines**. There are two particular generic (or generalized) types of ideal machine that command special attention. One is called a **transformer**, and the other a **gyrator**. (The familiar electrical transformer is an electrical approximation of this transformer, but only if it is idealized so as to work perfectly at all frequencies, including DC, unlike a real electrical transformer.)

An ideal machine can be represented by the word bond graph

$$\xrightarrow[\dot{q}_1]{e_1} \quad \begin{array}{c} IDEAL \\ MACHINE \end{array} \quad \xrightarrow[\dot{q}_2]{e_2}$$

It is defined to conserve energy, that is to consume or produce zero net power at all times. Thus the input power equals the output power. With the power convention of this graph,

$$e_1\dot{q}_1 = e_2\dot{q}_2. \tag{2.19}$$

The ideal machine can be run in either direction. Although a positive power flows from left to right according to the power convention arrows chosen arbitrarily above, the power could be negative, and therefore flow from right to left. As well as being able to run forwards or backwards, the ideal machine is thermodynamically *reversible*, which means that it generates no entropy. As a result, neither power can represent the flow of heat.

2.4.2   Transformers

Equation (2.19) allows for many different types of ideal machines. The most common type, called the **transformer**, satisfies the additional relation

$$\boxed{\dot{q}_2 = T\dot{q}_1.} \tag{2.20}$$

Thus T is defined as the ratio of the generalized velocity or flow on the bond with the outward power arrow to the generalized velocity or flow on the bond with the inward power convention arrow. This ratio, called the **modulus** of the transformer, may be constant, or in general may be a function of the displacement q_1 or some other displacement. Until Chapter 9, attention is restricted to tranformers with constant moduli. Constant modulus or not, the symbol T *within a bond graph* designates this *type* of relationship. Other authors usually elaborate with the designation TF, or, if the modulus is not a constant, with MTF (for *modulated transformer*):

$$\xrightarrow[\dot{q}_1]{e_1} T \xrightarrow[\dot{q}_2]{e_2} \qquad \xrightarrow[\dot{q}_1]{e_1} TF \xrightarrow[\dot{q}_2]{e_2} \qquad \xrightarrow[\dot{q}_1]{e_1} MTF \xrightarrow[\dot{q}_2]{e_2}$$

Subscripts can be added to the symbol T to distinguish different transformers that might appear in the same model. (Most other authors do not designate a common symbol for the modulus of transformers.) You must not confuse the bond-graph symbol T with the modulus T; only the modulus can participate in an equation, such as $T = 5$.

Substitution of equation (2.20) into the conservation-of-energy equation (2.19) gives a simple but profound result:

$$\boxed{e_1 = Te_2.} \tag{2.21}$$

In words, *the ratio of the generalized forces of an ideal transformer equals the inverse of the ratio of the respective generalized velocities.*

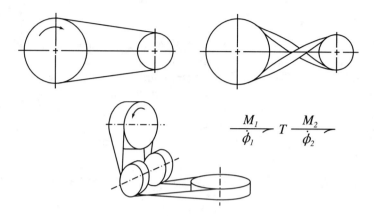

Figure 2.30: Examples of pulley drives

2.4.3 Gyrators

The second major possibility for the ideal machine is the **gyrator**, which by definition satisfies

$$e_2 = G\dot{q}_1. \tag{2.22}$$

Thus, *the modulus of the gyrator, G, is defined as the ratio of the effort on one of the bonds — either one — to the flow on the other bond.* Substitution of equation (2.22) into (2.19) gives

$$e_1 = G\dot{q}_2, \tag{2.23}$$

The gyrator in a bond graph is designated by the symbol G; other authors more commonly employ GY or, if the modulus is not a constant, MGY:

$$\frac{e_1}{\dot{q}_1} \mathrel{G} \frac{e_2}{\dot{q}_2} \qquad \frac{e_1}{\dot{q}_1} \mathrel{GY} \frac{e_2}{\dot{q}_2} \qquad \frac{e_1}{\dot{q}_1} \mathrel{MGY} \frac{e_2}{\dot{q}_2}$$

Like T, the modulus G need not be a constant, although consideration of non-constant moduli is deferred until Chapter 9.

2.4.4 Mechanical Devices Modeled as Transformers

Most mechanical drives can be represented as transformers with constant moduli, as long as frictional losses are neglected. Such drives can be based either on non-sliding friction (shear forces) or positive action (normal forces). Friction drives include belt drives, as suggested in Fig. 2.30, and rolling drives, as suggested in Fig. 2.31. Friction drives are inexpensive, and can damp vibrations. On the other hand, the transmitted forces are limited by the frictional properties of the materials.

Teeth added to a belt gives a **timing-belt drive**, and a chain of links and pins substituted for the belt gives a **chain-and-sprocket drive**. Similarly, teeth added to the wheels of a rolling drive produce a **gear drive**. These positive-action drives can carry heavy loads and high power efficiently. Examples are shown in Fig. 2.32. All of them can be designed to operate very smoothly, for example by using gear teeth with involute shape. Nevertheless, slight manufacturing errors or deflections due to heavy loads occasionally produce significant vibrations not included in a constant-modulus transformer model.

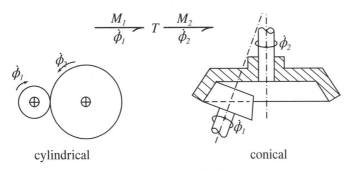

cylindrical   conical

Figure 2.31: Examples of rolling contact drives

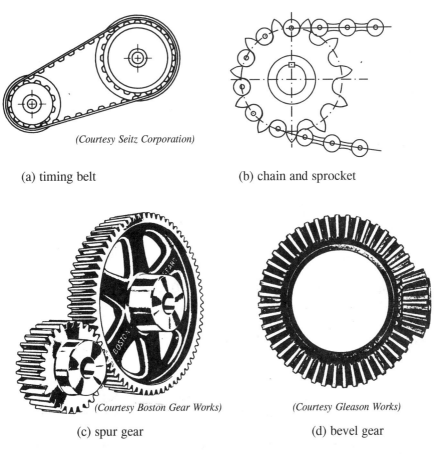

(Courtesy Seitz Corporation)

(a) timing belt   (b) chain and sprocket

(Courtesy Boston Gear Works)   (Courtesy Gleason Works)

(c) spur gear   (d) bevel gear

Figure 2.32: Examples of toothed drives

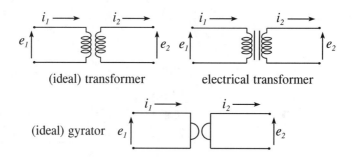

Figure 2.33: Standard electric circuit symbols for transformers and gyrators

The nominal kinematics of most toothed drives can be analyzed as though the teeth or sprockets weren't there. The virtual surface of contact for a rotating member is known as the **pitch surface**. A cylindrical pitch surface possesses a **pitch radius**, r, and a **pitch diameter**. For a pulley, wheel or gear, the common velocity of the two coupled members 1 and 2 at the point of contact is $v = r_1\dot{\phi}_1 = r_2\dot{\phi}_2$, where $\dot{\phi}_1$ and $\dot{\phi}_2$ are the respective angular velocities. The ratio of the angular velocities of the two pulleys or two gears etc. therefore is the inverse of the ratio of their pitch radii or diameters, *i.e.*, $T = \dot{\phi}_2/\dot{\phi}_1 = r_1/r_2$.

2.4.5   Electrical Transformers

An electrical transformer acts like an ideal transformer only for frequencies of excitation that are neither too small nor too large, and only if resistance losses are neglected. When unqualified, the term *transformer* refers to the ideal element as discussed above and therefore does not include the real electrical transformer. Circuit diagrams, however, sometimes refer to idealized behavior that is equivalent to the (ideal) transformer. Electric circuit symbols for both the real electric transformer and the (ideal) transformer are given in Fig. 2.33. The non-ideal behavior of an electrical transformer is modeled in Section 9.3 by combining the (ideal) transformer with other elements.

Purely fluid devices (fluid-in and fluid-out) that act like transformers without any moving mechanical parts are almost nonexistent. Nevertheless, devices with internal moving parts that behave like fluid transformers can be synthesized from two component devices that act like transducing transfomers or gyrators. An example is given in Section 2.5, after the idea of transducing transformers and gyrators is introduced.

2.4.6   Transducers Modeled as Transformers

A two-port device that converts (or *transduces*) energy directly from one energy domain to another, without employing a thermal engine, is called a **transducer**. Electrical and fluid motors, actuators, generators, pumps, compressors and turbines are **power transducers** that convert substantial power from one domain to another. Some people reserve the word "transducer" for **instrument transducers**, such as microphones, tachometers, strain gages and instruments that measure temperature, pressure, acceleration, etc. These convert relatively little power into electrical form, since their function is to provide information. Some transducers are transformational, while others are gyrational.

EXAMPLE 2.3

Show that the positive-displacement mechanical-to-fluid power transducer in the form of a **piston-and-cylinder** or **ram**, as shown below, can be modeled as a transformer if leakage and friction are neglected. Find the modulus of the transformer in terms of physical constant(s).

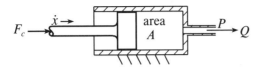

Solution: The volume flow rate of the fluid, Q, equals the velocity of its shaft, $\dot{x}$, times the area of the piston, A:

$$Q = \dot{x}A.$$

Further, the input power on one side equals the output power on the other side,

$$F_c \dot{x} = PQ.$$

Substitution gives

$$F_c = (Q/\dot{x})P = AP.$$

If the device is modeled as a transformer, its bond graph must be

$$\frac{F_c}{\dot{x}} \; T \; \frac{P}{Q}$$

The transformer modulus T is defined as the ratio $Q/\dot{x}$, and therefore equals the area A. The transformer modulus also must equal the ratio F_c/P, which has been verified. Therefore, the transformer model is proper.

In the example above, the relation $F_c = AP$ is derived from a velocity constraint combined with the conservation of energy. A simple force balance on the piston would also give this result. Students who have taken a course in statics often tend to recognize this result more readily, in fact, than they recognize the velocity constraint. It is important to get in the habit of identifying such **geometric constraints** directly, however, for in more difficult situations, particularly when dynamics is involved, the force balances can be relatively awkward. In any case, use of the principle of the conservation of energy such as in this example precludes the need to perform *both* a force balance and a geometric constraint analysis. You can choose, or do both as a check.

The power in a transformer can flow in either direction, regardless of the orientation of the power convention half-arrow. The piston-cylinder device can be operated either as a pump or as an **actuator** (its most common role). Rotary limited-angle actuators, such as drawn in Fig. 2.34, also are common. They, for example, articulate the revolute joints of hydraulic robots.

The term **pump** usually refers to machines that permit continuous rotation, such as the **gear pump** of Fig. 2.35 part (a), the **vane pump** of part (b), or the **piston pump** of part (c). This type of machine is characterized geometrically by the volume of fluid which passes through it per radian of shaft rotation, known as the **radian displacement** of the machine, D. Its definition assumes no leakage flow between the inlet and outlet chambers. (The effects of modest leakage and friction on nominal positive displamencent machines is considered in Chapter 5.) It also ignores the unsteadiness of the flow that exists in most

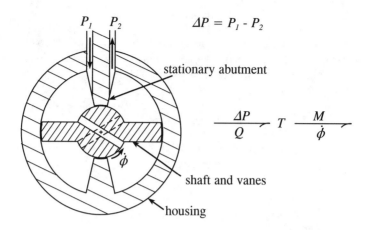

$$\Delta P = P_1 - P_2$$

Figure 2.34: Rotary actuator

pumps. With these idealizations,

$$Q = D\dot{\phi} \tag{2.24}$$

A pump approximated as an ideal machine also has no friction, so power out equals the power in. Thus, if ΔP is the pressure *rise* from the fluid inlet to the fluid outlet, and M is the torque on the shaft, then

$$M\dot{\phi} = Q\,\Delta P. \tag{2.25}$$

EXAMPLE 2.4

Show that the idealized positive-displacement pump is a transformer. Determine the modulus of this transformer, and find the relation between the moment on the shaft and the pressure rise.

Solution: Substituting equation (2.24) into equation (2.25) gives the desired relation between the moment on the shaft and the pressure rise:

$$M = D\,\Delta P.$$

This result together with equation (2.24) defines the transformer

$$\frac{M}{\dot{\phi}} \quad T \quad \frac{\Delta P}{Q}$$

The modulus of the transformer is the ratio $Q/\dot{\phi}$, which equals the ratio $M/\Delta P$, so that $T = D$, the volumetric displacement per radian.

What happens to the pump if ΔP is *negative*, corresponding to a pressure *drop* instead of a rise? The equations above and the bond graph still apply, so a *negative* value of the torque M results. This implies that the fluid power is flowing *into* the machine, whereas the mechanical power is flowing *out*. A device operating this way is not usually called a pump; rather, it is called a **hydraulic motor**. Thus, barring any practical problems such as rubber seals being over-stressed and failing, a positive displacement pump run in reverse becomes a motor, and vice-versa, including the machines shown in Fig. 2.35. The transformer representation nicely represents this fact, for it is neutral on the question of which way the power flows. In practice, however, one might prefer to reverse the power

(a) gear pump or motor

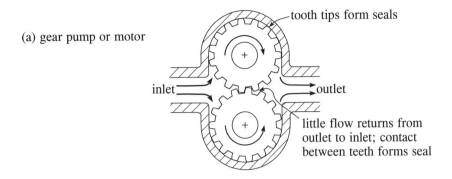

(b) vane pump or motor

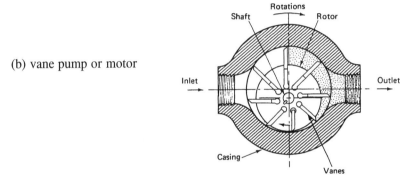

drawings (b) and (c): Reprinted with permission and courtesy of Eaton Corporation

(c) axial piston pump or motor

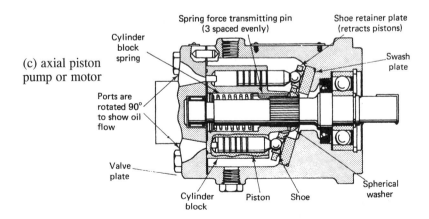

Figure 2.35: Common types of positive-displacment pumps and motors

arrows when use as a motor is intended, so that both ΔP and M become positive and $T = 1/D$.

The pump/motors being considered are called **positive displacement machines**, since packets of fluid with fixed volume are conveyed mechanically from one port to the other. In the gear machine the packets are the fluid trapped between adjacent teeth. In the vane pump they are the fluid trapped between successive vanes; the vanes slide in and out of slots to prevent leakage. In the piston pump the packets are the fluid volumes that enter and subsequently are expelled from each cylinder. The axial design pictured has several cylinders bored around a common barrel, shown cross-hatched, which rotate with the drive shaft like the barrel of a six-shooter. Pivotable piston shoes rotate with the pistons, and are also given a sinusoidal axial motion by sliding on a stationary angled cam plate. When the volume of a cylinder is expanding, it is connected to the inlet port through a kidney-shaped slot in a fixed valving plate. When the volume is contracting, it is connected to the outlet port through another kidney-shaped slot. (These inlet and output ports or lines actually are centered at angles rotated 90° about the machine axis from those pictured.)

Positive-displacement hydraulic machines are widely used in industry and transportation because they handle vastly larger forces and powers than their electromechanical counter-parts (e.g., motors and solenoids) for a given size and weight. Their use comprises the major part of the area of engineering known as **fluid power**. The remainder of fluid power encompases **pneumatics**, which uses compressed air or other gases in similar linear and rotary machines.

Dynamic hydraulic machines do not have fixed displacements, and are not represented as two-port elements until Section 12.2, since their behavior is more complex. Dynamic machines typically are used in relatively low pressure applications, varying from tiny im-peller pumps and fans to enormous hydraulic turbines. In many cases, however, it suffices to represent a dynamic machine as a one-port fluid device, presuming a fixed speed or some other fixed condition for the rotor. Thus, an impeller pump was modeled in the last section as a one-port general source.

2.4.7   Mechanical Devices Modeled as Gyrators

The word "gyrator" may conjure up in your mind the word "gyroscope." This association is appropriate, for when stripped of its complexities, which are beyond our present interest, a gyroscope indeed exhibits gyrational coupling between two axes of rotation that are both normal to each other and normal to the principal axis of rotation. As Fig. 2.36 attempts to indicate, a *torque* $F_1 L$ on one of these axes, where L is the length of the shaft, produces a proportional *angular velocity* $\dot{x}_2/L$ on the other axis, and vice versa. As an example, a gyroscope with a horizontal shaft supported only at one end will not fall down if it is allowed to *precess*, that is rotate slowly about a vertical axis through the support point (see Problem 2.25, p. 60).

The strength of the gyrational coupling, that is the magnitude of the modulus G, equals the angular momentum of the rotation about the principal axis. As a result, any mechanical system with a significant angular momentum has a potential for gyrational coupling. Such coupling can be enormous if the angular momentum is large. Nevertheless, most mechanical systems do not combine large angular momentum with a rotation of its axis, justifying the neglect of gyrational coupling.

The electric circuit symbol for a gyrator is included in Fig. 2.33 (p. 52). No simple passive electrical element for low-frequency circuit use acts like a gyrator, however; the symbol is used mostly to represent idealized electric circuit analogies for mechanical systems.

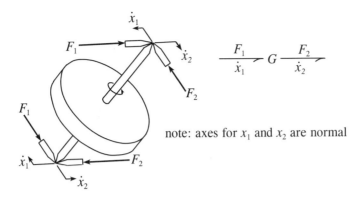

Figure 2.36: Gyroscope with idealized model

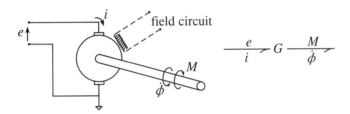

Figure 2.37: Idealized DC motor and gyrator model

2.4.8 Transducers Modeled as Gyrators

The force on an electrically charged particle moving in a magnetic field is proportional to the charge, to the field and to the particle velocity. The force on an electrical conductor therefore is proportional to the electrical current and the field. In a simple **DC electric motor**, the armature wire is wound and the commutation so arranged that the force produces a torque on a shaft. This torque therefore is proportional both to the current in the armature circuit and to the magnetic field. If, further, the magnetic field is constant, as in a permanent magnet motor or a motor with a field circuit driven by a fixed voltage or current, the torque is simply proportional to the armature current. Thus, as suggested in Fig. 2.37,

$$M = Gi, \tag{2.26a}$$
$$e = G\dot{\phi}, \tag{2.26b}$$

where G is a constant dependent on the geometry of the motor. Equation (2.26b) follows from the assumed conservation of energy. It directly suggests the idea of the **tachometer** or speedometer, an instrument in which the angular velocity of a shaft produces a proportional voltage that can be displayed, with an appropriate conversion factor, by a voltmeter. A large tachometer qualifies as an electric generator. Thus, the electric motor and the electric generator are the same machine simply operated backwards, much like the hydraulic motor and the hydraulic pump.

A **translating coil device**, such as that examined in Guided Problem 2.6, is an electrical motor with translational rather than rotational motion. It also can be modeled as a gyrator. There are many similar electromechanical devices based on magnetic fields, such as solenoids, although most are complicated by varying energy storage and so are addressed later (in Chapter 9). On the other hand, the class of electromechanical devices based on electric fields rather than magnetic fields can be shown to give *transformational* coupling. Examples include capacitor microphones and piezoelectric transducers; detailed consideration is again deferred to Chapter 9 because of complications due to energy storage.

Transducing gyrators also are useful in modeling dynamic fluid machines. Their modeling in this manner goes beyond the scope of this text. Simpler representations are given in Section 12.2.

2.4.9   Summary

Ideal two-port machines are by definition conservative, reversible and without stored energy. The two simplest forms are the transformer and the gyrator. Many actual machines can be approximately represented by these forms, and more complicated models of these and other devices often have transformers or gyrators imbedded within them. You will examine how these elements can be combined with one another and with other elements, such as the sources and loads already introduced, beginning in the following section.

The output flow of a transformer equals the product of a modulus, T, and the input flow. The input effort equals the product of the same T and the output effort. The effort on each of the two ports of a gyrator equals the product of a common modulus G and the flow of the opposite port. Engineering approximations to the transformer include pulley and gear pairs, electrical transformers, hydraulic rams and hydraulic pump/motors. Engineering approximations to the gyrator include gyroscopes, DC motors and other moving coil devices.

Guided Problem 2.5

This problem and the next should enhance your understanding of the nature and application of transformers and gyrators by having you evaluate the moduli of particular devices modeled as ideal machines, and examine their bilateral behavior.

Evaluate the transformer modulus for the rotary actuator shown in Fig. 2.34 (p. 54). The depth of the working chamber may be taken as w, and other parameters may be defined as appropriate.

Suggested Steps:

1. Define with symbols the radius of the shaft and the radius of the vanes/inner housing, and the depth of the vanes and chamber.

2. Choose whether to relate $\dot{\phi}$ to Q (easier) or M to ΔP, and evaluate T as their ratio in terms of (fixed) parameters.

3. Check to see that the relation not chosen agrees with the modulus of step 2.

Guided Problem 2.6

A translating coil device comprises a coil wrapped on a tube and positioned in a radial magnetic field. The example illustrated in Fig. 2.38 employs a permanent magnet, as in a loudspeaker. The axial force on the coil is

$$F = 2\pi r N B i$$

where $2\pi r$ is the circumference of the coil, N is the number of turns, B is the magnetic field strength and i is the electric current.

Model this device as a form of an ideal transducer, neglecting losses. Give its modulus and the relation between the voltage e and the velocity $\dot{x}$.

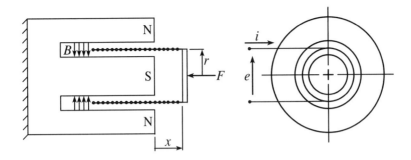

Figure 2.38: The translating coil device of Guided Problem 2.6

Suggested Steps:

1. Identify the power conjugate variables on the mechanical and electrical bonds which are related by this device.

2. Identify the type of ideal machine and its modulus from the given equation; connect the two bonds with the associated standard bond-graph symbol.

3. Find the relation between e and $\dot{x}$ from this bond graph. Note that it follows from the conservation of energy.

PROBLEMS

2.19 Identify a set of variables for a transformer model of the pulley drive systems of Fig. 2.30 (p. 50), define the transformer modulus, and express this modulus in terms of physical constants that also should be defined.

2.20 Answer the question of the preceding problem for the cylindrical roller drive shown on the left side of Fig. 2.31 (p. 51).

2.21 Answer the question of the preceding two problems for the bevel or conical roller drive system shown on the right side of Fig. 2.31 (p. 51).

2.22 You are given a hydraulic rotary actuator such as pictured in Fig. 2.34 (p. 54), but are not given its specifications. Propose and describe an experiment that could determine the modulus of the transformer that models the actuator.

2.23 Find the transformer modulus for a steady-flow model of the piston pump/motor shown in Fig. 2.35 part (c) (p. 55). You may define the radius of each piston as r_p, the distance between the machine centerline and a piston centerline as r_b, the number of pistons as n and the angle of the cam plate from the normal to the axis as θ.

2.24 A tachometer produces 1 volt for each 10 revolutions per second. If the machine is used instead as a motor, find the current needed to produce a torque of 0.02 N·m. Assume the behavior is consistent with that of an ideal machine, and note whether it is a transformer or a gyrator.

2.25 The shaft of a rapidly spinning rotor (or gyroscope) is horizontal and is simply supported at a distance L from its center of mass. The axis of the shaft is observed to precess about a vertical axis at a steady rate $\dot{\phi}$. The axis remains horizontal. The rotor has mass m and spins at ω rad/s. Knowing that this device can be represented by a gyrator with modulus $I\omega$, where $I = mr_g^2$ is the mass moment of inertia and r_g is the radius of gyration, determine the rate of precession.

2.26 A bicyclist negotiates a turn with a 20-foot radius at 24 feet per second. Each of the wheels weighs five pounds, has a radius of 1.1 feet and has a radius of gyration of 1.0 feet.

  (a) The gyrator modulus connecting the vertical axis of the turn with the axis of tilt equals the angular momentum of the two wheels. Determine this modulus. Neglect the angle of tilt.

  (b) Determine the moment produced by the rotation of the wheels that partly counteracts the centrifugally-induced tilting moment.

SOLUTIONS TO GUIDED PROBLEMS

Guided Problem 2.5

1. Radius of shaft: r_s; radius of vanes or inner housing: r_v; depth of the vanes and chamber: w.

2. The volume of one chamber is $\pi(r_v^2 - r_s^2)(\phi/2\pi)w$. Therefore, the rate of change of the volume of one chamber is $w(r_v^2 - r_s^2)\dot{\phi}/2$. The flow Q enters *two* of the four chambers (and leaves the other two), so $Q = (1/T)\dot{\phi}$ with $1/T = w(r_v^2 - r_s^2) = w(r_v - r_s)(r_v + r_s)$.

3. The relation $M = (1/T)\Delta P$ can be checked most easily in the limit when $r_v - r_s = \Delta r$ is much smaller than either r_v or r_s. The result is $M = 2rF$ where $r = (r_s + r_v)/2$ and $F = \Delta P \Delta r$, which can be seen to be correct.

Guided Problem 2.6

1. Mechanical force and velocity: F and $\dot{x}$, respectively. Electrical generalized force and velocity: e and i, respectively.

2. $\dfrac{e}{i} \rightharpoonup G \dfrac{F}{\dot{x}}$.   From the given equation, $G = 2\pi rNB$. Therefore, $e = 2\pi rNB\,\dot{x}$.

2.5   Systems with Transformers and Gyrators

Sections 2.1 and 2.3 describe how to predict the equilibria of sources attached to loads. Section 2.4 introduces the transformer and the gyrator. How transformers and gyrators act when cascaded head-to-tail, and how they can be placed between a source and load to improve the behavior, is now explored.

2.5.1 Cascaded Transformers

A pair of transformers can be cascaded:

$$\frac{e_1}{\dot{q}_1} \;\rightharpoonup\; T_1 \;\frac{e_2}{\dot{q}_2}\;\rightharpoonup\; T_2 \;\frac{e_3}{\dot{q}_3}\;\rightharpoonup$$

The definition of a transformer requires

$$\dot{q}_3 = T_2\dot{q}_2 = T_2T_1\dot{q}_1, \qquad (2.27a)$$

$$e_1 = T_1e_2 = T_1T_2e_3, \qquad (2.27b)$$

which says simply that the cascade reduces to a single effective transformer with modulus equal to the product of the component moduli:

$$\frac{e_1}{\dot{q}_1} \;\rightharpoonup\; T \;\frac{e_3}{\dot{q}_3}\;\rightharpoonup \qquad\qquad T = T_1T_2$$

Thus, if n transformers are cascaded, the result is a single effective transformer with modulus equal to the product of all the component moduli:

$$\dot{q}_{n+1} = T\dot{q}_1, \qquad (2.28a)$$

$$e_1 = Te_{n+1}, \qquad (2.28b)$$

$$T = T_1T_2 \ldots T_n. \qquad (2.28c)$$

EXAMPLE 2.5

Show that gear trains, such as those pictured below, can be represented as a cascade of transformers if friction is neglected, and that they act as if there were only a single gear pair.

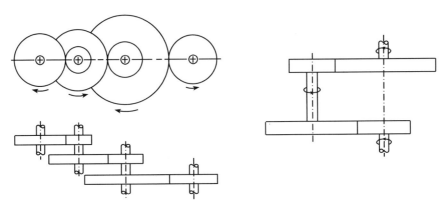

Solution: The transformer modulus for each pair of meshing gears equals the ratio of their numbers of teeth or of their diameters, as noted in Section 2.4.4. The overall train acts like a single gear pair, or single transformer, with a transformer modulus or gear ratio equal to the product of all the individual transformer moduli, that is all the component gear ratios. The inverse torque ratio has the same value only if friction is neglected; the model is invalid otherwise.

2.5.2   Cascaded Gyrators

Cascading two gyrators produces a *transformer* with a modulus that equals the ratio of the two gyrator moduli. This result can be deduced step-by-step as follows:

$$\xrightarrow{\;\dfrac{e_1}{\dot{q}_1}\;} G_1 \xrightarrow{\;\dfrac{e_2}{\dot{q}_2}\;} G_2 \xrightarrow{\;\dfrac{e_3}{\dot{q}_3}\;}$$

$$e_1 = G_1 \dot{q}_2 = \frac{G_1}{G_2} e_3 = T e_3; \qquad T = \frac{G_1}{G_2}, \qquad (2.29a)$$

$$\dot{q}_3 = \frac{1}{G_2} e_2 = \frac{1}{G_2} G_1 \dot{q}_1 = T \dot{q}_1. \qquad (2.29b)$$

Cascading a gyrator with a transformer, however, gives a gyrator. (You should take a few moments to demonstrate this fact for yourself.) As a result, the cascading of any *even* number of gyrators produces a transformer, while the cascading of any *odd* number of gyrators produces a gyrator.

EXAMPLE 2.6

Show that if the two shafts of ideal DC electric motors with constant fields are connected together (below left), the voltages and currents of the two pairs of electrical terminals are related like those of an ideal electrical transformer. In addition, show that if the two pairs of electrical leads are interconnected instead (below right), the two shafts have torques and speeds related like those of a transformer.

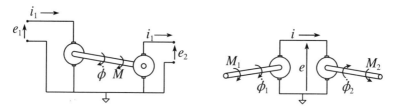

Solution: An ideal DC electric motor with constant magnetic field has been represented as a gyrator. Equation (2.29) shows that connecting together the shafts of two such motors produces a transformer:

$$\xrightarrow{\;\dfrac{e_1}{i_1}\;} G_1 \xrightarrow{\;\dfrac{M}{\dot{\phi}}\;} G_2 \xrightarrow{\;\dfrac{e_2}{i_2}\;} \quad \Rightarrow \quad \xrightarrow{\;\dfrac{e_1}{i_1}\;} T \xrightarrow{\;\dfrac{e_2}{i_2}\;} \quad T = \frac{G_1}{G_2}$$

This device has an important advantage over the usual electrical transformer: it works at DC. For the second case, the result is a mechanical transmission known as a Ward-Leonard system:

$$\xrightarrow{\;\dfrac{M_1}{\dot{\phi}_1}\;} G_1 \xrightarrow{\;\dfrac{e}{i}\;} G_2 \xrightarrow{\;\dfrac{M_2}{\dot{\phi}_2}\;} \quad \Rightarrow \quad \xrightarrow{\;\dfrac{M_1}{\dot{\phi}_1}\;} T \xrightarrow{\;\dfrac{M_2}{\dot{\phi}_2}\;} \quad T = \frac{G_1}{G_2}$$

Energy losses are addressed in Section 5.2.4.

2.5.3   Case Study of a Transformer Connecting a Source to a Load

Transformers routinely interconnect sources and loads. Transducing transformers do this intrinsincally. Non-transducing transformers often are chosen explicitly to improve the

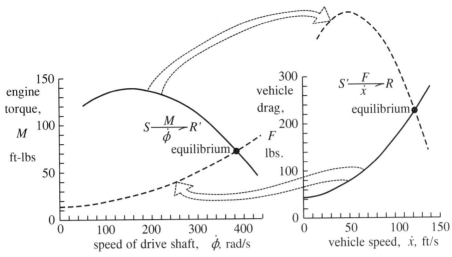

Figure 2.39: Automobile source and load characteristics with synthesis

matching of the source to the load. The key ideas perhaps are best learned by a familiar example.

Consider an automobile IC engine with a fixed throttle position which drives a vehicle through a fixed transmission along a level road. The engine can be modeled as a source; the torque-speed characteristics are plotted on the left side of Fig. 2.39. The force-speed characteristics of the load are plotted on the right side of the figure. (This force is the sum of the air drag and the rolling drag.) The drive train between the engine and drive axle comprises a transmission, drive shaft and gear differential. This drive train converts engine torque and speed to vehicle thrust force and speed. Losses in this system are neglected here, so these components can be modeled, collectively, as a transformer, shown as T_d below:

$$\text{ENGINE} \xrightarrow{\frac{M}{\dot\phi}} \text{DRIVETRAIN} \longrightarrow \text{WHEELS} \xrightarrow{\frac{F}{\dot x}} \text{VEHICLE}$$

$$S \xrightarrow{\frac{M}{\dot\phi}} T_d \longrightarrow T_w \xrightarrow{\frac{F}{\dot x}} R$$

$$S \xrightarrow{\frac{M}{\dot\phi}} T \xrightarrow{\frac{F}{\dot x}} R$$

The drive axles in turn rotate the wheels. This action is modeled by the transformer T_w. The modulus of this transformer represents the ratio of the linear motion of the vehicle to the rotary motion of the wheels, which is the radius of the wheels.

The two cascaded transformers can be telescoped into one, labeled as T. The modulus of this combined transformer equals the ratio of the velocity of the vehicle to the angular velocity of the engine shaft, or the distance moved per radian of rotation of the shaft. Assume the vehicle is known to move two feet for every rotation of the shaft; therefore, $T = 2/2\pi = 1/\pi$ ft/rad.

To determine the equilibrium speeds of the vehicle and the engine, you need either to transform the load curve into the torque-speed coordinates of the engine plot, or to transform the engine characteristics into the force-velocity characteristics of the vehicle. Both of these transformations are shown in the figure, using dashed lines. In the first case, the transformer and the load are effectively combined into an equivalent load, labeled as R'. In the second case, the engine and the transformer are effectively combined into an

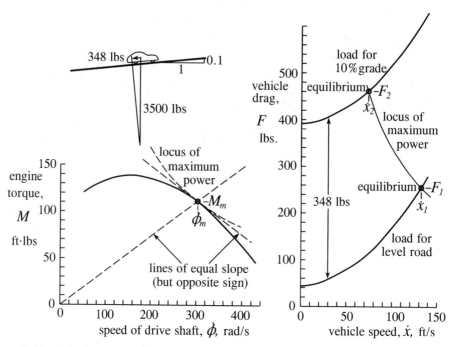

Figure 2.40: Calculation of the transformer modulus for maximum power transmission

equivalent source, labeled as S'. The transformations are carried out by choosing individual points along the characteristic to be transformed. The relations $\dot{\phi} = (1/T)\dot{x}$ and $M = TF$ (first case) or $\dot{x} = T\dot{\phi}$ and $F = (1/T)M$ (second case) are used.

The equilibrium state is given by the intersections of the characteristics. The two approaches give the same answer, as they should: $\dot{x} = 123$ ft/s, $F = 226$ lbs, $\dot{\phi} = 386$ rad/s, $M = 72$ ft·lb.

The transmission ratio of a real car changes when you shift gears to address a changed loading condition. Presume you wish to find the drive train ratio T_d that maximizes the steady-state speed of the vehicle for a given load characteristic and a given radius of the wheels. You can reason as follows: maximum load speed means maximum load power. Since the transmission is assumed to be lossless, this means maximum engine power. Thus, you start by finding the point on the engine characteristic that gives the greatest power. The power is the product of the torque and the speed; you can simply use trial-and-error. Better yet, you can use one of the methods described in Section 2.3.6, which are implemented on the left-hand side of Fig. 2.40 by dashed lines. Note specifically that the rectangular hyperbola representing maximum power is tangent to the characteristic at the maximum power point, and that the slope of the chord from the origin to this point has the same slope as the tangent to the characteristic at this point. The maximum power is approximately $\mathcal{P} = M_m\dot{\phi}_m = 111.5$ ft·lb $\times 304$ rad/s $= 33,900$ ft·lb/s.

The locus for maximum power can be transformed into the right-hand plot, where its intersection with the load characteristic gives the corresponding maximum speed and force: approximately $\dot{x}_1 = 132.4$ ft/s and $F_1 = 256$ lbs. The transformer ratio which permits this to occur is therefore $T = \dot{x}_1/\dot{\phi}_1 = M_m/F_1 = 0.4355$ ft/rad. (Clearly, you've made a mistake if both calculations don't give you the same answer.) Finally, the required transmission ratio T_d is the ratio of T/T_w, where T_w is the radius of the wheel. For example, if this radius is

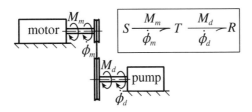

Figure 2.41: Addition of pulley drive to the induction motor system

1.1 feet, there follows $T_d = 0.396$. (This drive train ratio in turn is factored into the ratio for the transmission and the ratio for the gear differential.)

Achieving maximum speed on an incline requires a lower transmission ratio (downshifting). Assume that the vehicle and its payload weigh 3500 lbs. and are climbing a long 10% grade, as shown in the figure. The weight times the sine of the angle gives an additional drag force of 348 lb. This is added to the air drag and wheel drag to give a new load characteristic and a new solution[5]: $T = \dot{x}_2/\dot{\phi}_m = 73.5/304 = 0.242$ ft/rad or $T = M_m/F_1 = 111.5/461 = 0.242$ ft/rad.

2.5.4 Second Case Study of a Transformer Connecting a Source to a Load

The matching of a source and a load often is improved through the purposeful introduction of a coupling that acts like a transformer, as the gears in the transmission of the case-study vehicle above illustrate. As a second case study, consider inserting an ideal pulley or gear drive between the induction motor and the pump of the water sprinkler system discussed earlier in Section 2.1. The new system, shown in Fig. 2.41, can be represented by the bond graph

$$S \xrightarrow[\dot{\phi}_m]{M_m} T \xrightarrow[\dot{\phi}_d]{M_d} R$$

The objective is acceptable operation for the twentieth and thirtieth-floor sprinklers, for which the direct drive failed, without using a larger motor.

The angular velocity and the torque of the ideal pulley drive are

$$\dot{\phi}_d = T\dot{\phi}_m, \tag{2.30a}$$

$$M_d = \frac{1}{T}M_m, \tag{2.30b}$$

where T is the ratio of the diameter of the motor pulley to the diameter of the pump pulley. The motor and the pulley system can be considered as a single equivalent source, S':

$$S' \xrightarrow[\dot{\phi}_d]{M_d} R$$

(An acceptable alternative combines the pulley system and the pump system into an equivalent one-port load.) The torque-speed characteristic of this new source depends on the value of T. Characteristics for four different values are plotted in Fig. 2.42. This is much like the choice you have in shifting gears of an automobile with a manual transmission. A

---

[5]The solution as presented overlooks a small correction for the case of the 10% slope that results from the fact that the normal load on the pavement is reduced by about one-half percent below the weight. Since about 40 pounds of the load force is rolling friction which would be approximately proportional to this force, the drag should be reduced by about 0.2 lbs.

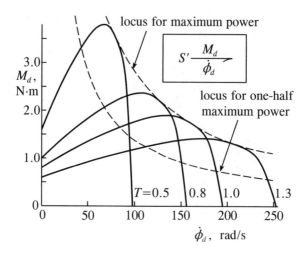

Figure 2.42: Characteristics of the source combining the induction motor and the pulley drive

large value of T (as with "fourth gear") gives a high speed at low torque, whereas a small value of T (as with "first gear") gives low speed at high torque, all for the same motor speed and torque.

The characteristic for $T = 1$ is identical to that of the motor itself; the ideal pulley acts like a direct drive. The other characteristics are transformations of this basic characteristic. Each point on the basic characteristic is mapped onto a corresponding point on the transformed characteristic according to equation (2.30). The power associated with the point is unchanged by the transformation; *i.e.* $M_d\dot{\phi}_d = M_m\dot{\phi}_m$. The transformed point lies on a locus of constant power that passes through the original point. This locus is a hyperbola with vertical and horizontal asymptotes, the same rectilinear hyperbola discussed in the prior example. Two such loci are drawn in Fig. 2.42 as dashed lines. Rapid sketching is aided by the fact, as noted above, that the magnitude of the negative slope at each point on the hyperbola equals the positive slope of the chord from the origin to that point.

The locus for maximum power just graces each characteristic. All of these points of tangency represent the same maximum power the motor can deliver. The various speed ratios and torque ratios equal the transformer moduli or their reciprocals. The other locus drawn in the figure represents one-half the maximum power. This locus intersects each source characteristic twice. The respective lower-speed intersections have the same speed and torque ratios as both the upper-speed intersections and the maximum power points. It is strongly suggested that you choose some arbitrary point or points on one of the characteristics and, using only knowledge of the values of T, find and validate the corresponding point or points on another characteristic.

The largest possible load flow corresponds to the largest possible load power, which in turn equals the largest possible source power. The maximum power that can be delivered by the source is indicated in Fig. 2.42 by the locus of maximum power. In Fig. 2.43 this locus is seen to intersect the load characteristic for the thirtieth floor when $T = 0.8$. For the twentieth floor, the intersection is close to $T = 1$, and for the tenth floor it is about $T = 1.1$ (interpolating). Are these solutions acceptable?

The candidate design solution with $T = 1$ for the twentieth floor already has been eliminated, since unfortunately the pump won't start. The solutions $T = 0.8$ and 0.5 for the thirtieth floor suffer the same problem. For $T = 1.1$, the starting torques for the motor

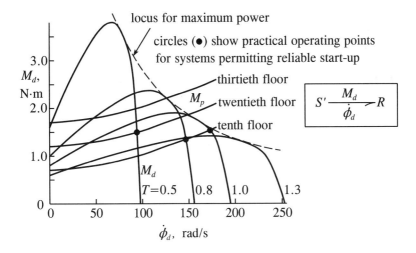

Figure 2.43: Matching of motor/pulley system to pump/sprinkler load

and pump are about equal; starting is too problematical to justify the choice. Thus, unless you were to replace the motor, for example using the "capacitor-start" feature highlighted in Guided Problem 2.2, you must settle for a lower output flow and power in all these cases. Allowing about a 10% margin to cover errors inherent in the modeling, including friction which is being neglected, the solution for $T = 0.5$ is seen to be acceptable for the twentieth floor. For the tenth floor, both $T = 0.8$ and 1.0 are acceptable; the latter gives virtually maximum possible flow. The acceptable solutions are marked by heavy dots in the figure.

EXAMPLE 2.7

A DC motor drives a frictionless and leak-free positive displacement pump which forces a liquid through a flow restriction to a load:

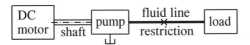

The motor has the torque-speed characteristics plotted below left, and the pressure drop across the restriction is plotted below right as a function of the flow through it.

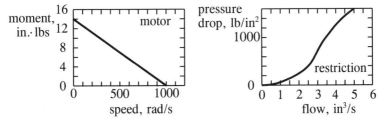

Assuming the pressure of the load is zero, find the displacement of the pump that (approximately) maximizes the flow to the load, and the associated flow. Then, repeat the solution when the load pressure is known to be 1000 psi.

Solution: Maximizing the flow to the load implies maximizing the power transfer. Since the pump is 100% efficient, this means maximizing the power from the motor. Maximum power from the motor occurs at the mid-point of the straight-line characteristic, as discussed in Section 2.3.6 and shown specifically in Fig. 2.27 (p. 42). Thus the motor should be run at 500 rad/s with a torque of 7 in.·lb, producing the power $500 \times 7 = 3500$ in.·lb/s. A locus with this power is superimposed on the pressure drop plot below left, revealing an intersection at a pressure of 1000 psi and flow of 3.5 in^3/s. This is the solution when the load pressure is zero, so that the pressure seen by the pump equals the pressure drop across the fluid restriction.

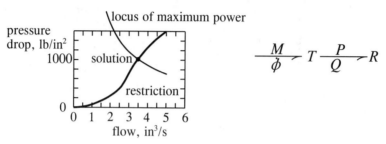

The pump acts as a transformer, as shown above right. Its volumetric displacement is the modulus of the transformer, which can be deduced from the ratio $3.5/500 = 0.007$ in^3/rad or the ratio $7/1000 = 0.007$ in^3/rad.

The pressure generally seen by the load is the sum of the pressure drop across the restriction and the load pressure. This sum is plotted below for the second case of a 1000 psi load pressure.

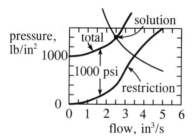

The intersection with the same power locus as before is now at 1400 psi and 2.5 in^3/s. The modulus of the transformer that produces this result is the volumetric displacement $2.5/500 = 0.005$ in^3/rad or $7/1400 = 0.005$ in^3/rad.

2.5.5   Case Study of a Gyrator Connecting a Source to a Load

Consider a DC motor with constant field and moment $M = ai$, $a = 0.15$ N m/amp, which drives a load. The electric power source for the motor has the voltage-current characteristic shown on the left side of Fig. 2.44, and the mechanical load has the torque-speed characterisitic shown on the right side. You wish to find the equilibrium torque and speed, neglecting losses.

The system can be modeled by the bond graph

$$S \xrightarrow[\;i\;]{\;e\;} G \xrightarrow[\;\dot{\phi}\;]{\;M\;} R$$

The given information $M = ai$ establishes the gyrator modulus as a, so you also know that $e = a\dot{\phi}$.

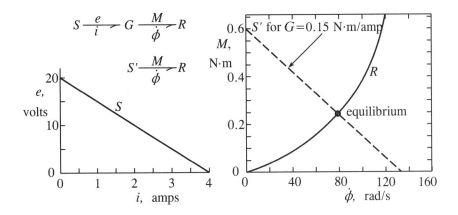

Figure 2.44: Source and load characteristics for DC motor system

You could combine the source and the gyrator into an equivalent source, so that the source characteristic is mapped into the load plot:

$$S' \xrightarrow{\quad \frac{M}{\dot\phi} \quad} R$$

Alternatively, you could combine the gyrator and the load into an equivalent load, so that the load characteristic is mapped into the source plot:

$$S \xrightarrow{\quad \frac{e}{i} \quad} R'$$

The source characteristic is a straight line in the present case, so the first approach is easier. The mapping is shown in the figure by the dashed line labeled S'. Note that the left end of the source characteristic becomes the right end of its transformation, and vice-versa. You now simply identify the intersection of the two characteristics as the equilibrium point.

2.5.6 Transmission Matrices*

The use of matrix notation expedites the telescoping of a chain of linear two-port elements into a single equivalent two-port element. This idea is introduced here for the special cases involving only transformers and gyrators. Further use of the method is given in Chapter 5 and beyond.

The state of a bond is treated as a two-element vector, with its effort first and its flow second. The equations describing a transformer of modulus T and a gyrator of modulus G become

$$\xrightarrow[\dot q_1]{e_1} T \xrightarrow[\dot q_2]{e_2} \qquad \begin{bmatrix} e_1 \\ \dot q_1 \end{bmatrix} = \begin{bmatrix} T_1 & 0 \\ 0 & 1/T_1 \end{bmatrix} \begin{bmatrix} e_2 \\ \dot q_2 \end{bmatrix} \qquad (2.31a)$$

$$\xrightarrow[\dot q_1]{e_1} G \xrightarrow[\dot q_2]{e_2} \qquad \begin{bmatrix} e_1 \\ \dot q_1 \end{bmatrix} = \begin{bmatrix} 0 & G \\ 1/G & 0 \end{bmatrix} \begin{bmatrix} e_2 \\ \dot q_2 \end{bmatrix} \qquad (2.31b)$$

The cascading of two transformers, as in Section 2.5.1,

$$\xrightarrow[\dot q_1]{e_1} T_1 \xrightarrow[\dot q_2]{e_2} T_2 \xrightarrow[\dot q_3]{e_3}$$

gives

$$\begin{bmatrix} e_1 \\ \dot{q}_1 \end{bmatrix} = \mathbf{T}_1 \begin{bmatrix} e_2 \\ \dot{q}_2 \end{bmatrix}; \qquad \begin{bmatrix} e_2 \\ \dot{q}_2 \end{bmatrix} = \mathbf{T}_2 \begin{bmatrix} e_3 \\ \dot{q}_3 \end{bmatrix}, \tag{2.32a}$$

$$\begin{bmatrix} e_1 \\ \dot{q}_1 \end{bmatrix} = \mathbf{T}_1 \mathbf{T}_2 \begin{bmatrix} e_3 \\ \dot{q}_3 \end{bmatrix} = \begin{bmatrix} T_1 T_2 & 0 \\ 0 & 1/T_1 T_2 \end{bmatrix} \begin{bmatrix} e_3 \\ \dot{q}_3 \end{bmatrix}. \tag{2.32b}$$

Denoting the transmission matrix of the ith element in the cascade of n transformers as $\mathbf{T}_i$, there results

$$\begin{bmatrix} e_1 \\ \dot{q}_1 \end{bmatrix} = \mathbf{T}_1 \mathbf{T}_2 \cdots \mathbf{T}_n \begin{bmatrix} e_n \\ \dot{q}_n \end{bmatrix} = \begin{bmatrix} T & 0 \\ 0 & 1/T \end{bmatrix} \begin{bmatrix} e_{n+1} \\ \dot{q}_{n+1} \end{bmatrix}; \quad T = T_1 T_2 \cdots T_n, \tag{2.33}$$

which agrees with equation (2.28) (p. 61).

The cascade of two gyrators,

$$\frac{e_1}{\dot{q}_1} \longrightarrow G_1 \frac{e_2}{\dot{q}_2} \longrightarrow G_2 \frac{e_3}{\dot{q}_3} \longrightarrow$$

gives

$$\begin{bmatrix} e_1 \\ \dot{q}_1 \end{bmatrix} = \begin{bmatrix} 0 & G_1 \\ 1/G_1 & 0 \end{bmatrix} \begin{bmatrix} e_2 \\ \dot{q}_2 \end{bmatrix}; \qquad \begin{bmatrix} e_2 \\ \dot{q}_2 \end{bmatrix} = \begin{bmatrix} 0 & G_2 \\ 1/G_2 & 0 \end{bmatrix} \begin{bmatrix} e_3 \\ \dot{q}_3 \end{bmatrix}, \tag{2.34a}$$

$$\begin{bmatrix} e_1 \\ \dot{q}_1 \end{bmatrix} = \begin{bmatrix} 0 & G_1 \\ 1/G_1 & 0 \end{bmatrix} \begin{bmatrix} 0 & G_2 \\ 1/G_2 & 0 \end{bmatrix} \begin{bmatrix} e_3 \\ \dot{q}_3 \end{bmatrix} = \begin{bmatrix} G_1/G_2 & 0 \\ 0 & G_2/G_1 \end{bmatrix} \begin{bmatrix} e_3 \\ \dot{q}_3 \end{bmatrix}, \tag{2.34b}$$

which agrees with equation (2.29) (p. 62).

2.5.7   Summary

Transformers cascaded head-to-tail can be represented by a single transformer with modulus equal to the product of the individual moduli. A pair of gyrators in series, on the other hand, is equivalent to a transformer with a modulus equal to the ratio of the individual moduli. A gyrator connected to a transformer gives a gyrator. This means that an even number of gyrators in cascade gives an equivalent transformer, while an odd number of gyrators in cascade gives an equivalent gyrator.

Models often interconnect a source to a load by a transformer or a gyrator. The source then can be combined with the transformer or gyrator to give an equivalent source, which can be directly synthesized with the load. Alternatively, the transformer or gyrator can be combined with the load to give an equivalent load, which can be directly synthesized with the source. Graphical representations of the characteristics enables you to visualize the solutions of a variety of problems, such as determination of the moduli that give maximum power transfer from the source to the load.

Matrix relations can be used to represent the characteristics of linear two-port elements such as transformers and gyrators. This approach is helpful but not necessary.

Guided Problem 2.7

Solving this problem and the next should help you understand how to choose the modulus of a coupler so as to maximize the performance of an electrical, mechanical or fluid system.

Choose the modulus of an electric transformer that connects an amplifier with a 24-ohm resistive output impedance to a speaker with a 6-ohm resistive impedance so as to maximize the power transfer.

Suggested Steps:

1. Represent the source, transformer and load with a bond graph. Find a relation to characterize the source.

2. Combine the transformer and the load and draw a new bond graph. Evaluate the modulus of the new load.

3. Express the power delivered in terms of the bond current and the parameters.

4. Find the current that maximizes this power.

5. Find the transformer modulus that gives this maximized power.

Guided Problem 2.8

A hydraulic power supply drives a hydraulic ram (piston/cylinder) against a mechanical load. The characteristics of the power supply and the load are plotted in Fig. 2.45. Find the area of the piston that maximizes the power delivered to the load. Neglect friction.

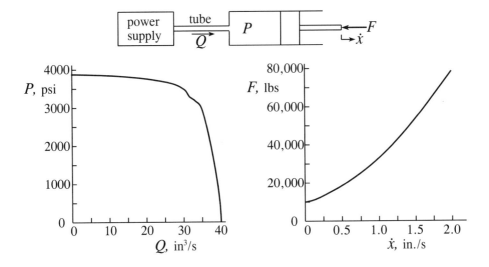

Figure 2.45: The system of Guided Problem 2.8

Suggested Steps:

1. Draw a bond graph of the system.

2. Find the maximum power point on the source characteristic.

3. Find the corresponding point on the load characteristic.

4. The transformer modulus sought is the ratio $\dot{x}/Q$ of these two points. Relate this ratio to the area of the piston.

5. Check that the transformer modulus is also the ratio P/F for these two points.

PROBLEMS

2.27 Model the gear train below as a cascade of transformers. Find the moduli of the component transformers and the overall equivalent single transformer.

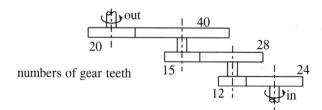

numbers of gear teeth

2.28 Find the relationships between the boundary variables for the following cascade. Can this system be represented by an equivalent single transformer or gyrator, and, if so, what would its modulus be?

$$\frac{e_1}{\dot{q}_1} \;\; T \;\; \frac{e_2}{\dot{q}_2} \;\; G \;\; \frac{e_3}{\dot{q}_3}$$

2.29 Carry out the previous problem using the method of transmission matrices.

2.30 The vehicle considered in Section 2.5.3 (which has $T = 1/\pi$ ft/rad and $mg = 3500$ lbs) is traveling at 100 ft/s on a level road. Using only the data given by the solid lines in Fig. 2.39 (p. 63),

    (a) find the drag force.

    (b) find the engine speed, moment and power.

    (c) find the thrust force (from the engine power and vehicle speed).

    (d) find the acceleration of the vehicle.

2.31 The capacitor-start motor described in Guided Problem 2.2 (p. 23) is substituted for the basic induction motor used in the system with the motor, pulley drive and pump load analyzed in Section 2.5.4. Determine which of the candidate solutions $T = 0.5, 0.8, 1.0$ and 1.3 are acceptable for the tenth, twentieth and thirtieth floor sprinklers.

2.32 The analysis of the system with the motor, pulley drive and pump load given in Section 2.5.4 employs an equivalent source bonded to the actual load:

$$S' \;\; \frac{M_d}{\dot{\phi}_d} \;\; R$$

Draw an annotated sketch of the replacement for Fig. 2.43 (p. 67) that corresponds to the bonding of the actual motor to an equivalent load:

$$S \;\; \frac{M_m}{\dot{\phi}_m} \;\; R'$$

2.33 An ideal hydraulic power supply (comprising a motor, pump and relief valve) has the pressure-flow characteristic as plotted below left. It drives a piston-cylinder with the characteristic shown below right. Find the maximum possible speed of the piston, and the area of the piston that results in this speed. Neglect friction.

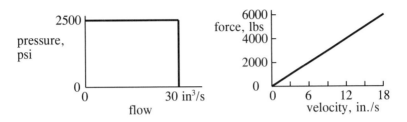

2.34 A hydraulic power supply (comprising a motor-driven pump and a relief valve) has the pressure-flow characteristic plotted below left. It drives a positive-displacement hydraulic motor which in turn rotates a shaft that drives some mechanical equipment. Frictional and leakage losses in the hydraulic motor may be neglected. The characteristics of the mechanical load are plotted below right.

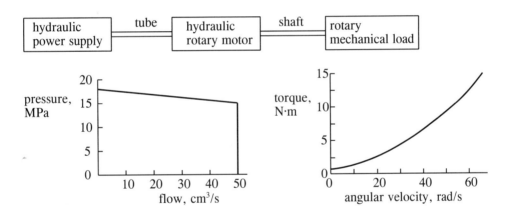

(a) Define key variables, and place them on both the drawing and a bond-graph model of the system.

(b) Find the maximum possible speed of the load.

(c) Determine the volumetric displacement per revolution of the hydraulic motor that achieves the maximum speed.

2.35 An IC engine with the torque-speed characteristic given in Fig. 2.39 (p. 63) drives a frictionless positive-displacement pump with a volumetric displacement of 3 in^3/rev. The hydraulic oil then raises a weight of 30,000 lb using a frictionless piston/cylinder of unspecified area A, as shown at the top of the following page.

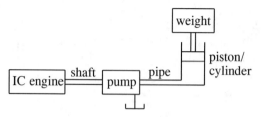

(a) Define key variables, and place them on both the drawing and a bond-graph model of the system.

(b) Find the value of A that maximizes the upward velocity of the weight.

2.36 A motor drives a fan through a belt drive. The torque-speed characteristics of the motor and the fan are plotted below. The pulley ratio is not specified, but the motor should not be run steadily with a torque higher than its continuous rating of 4.0 in.·lb (to prevent overheating). *Continued in Problem 5.19 (p. 297).*

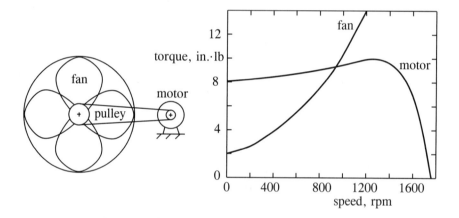

(a) Draw a bond graph model of the system, neglecting energy storage elements.

(b) Determine the speed of the motor that maximizes its allowable power, and find this power.

(c) Determine the corresponding speed and torque of the fan, and the pulley ratio that achieves them. Neglect belt losses, so power is conserved.

2.37 An electrical power supply drives a DC motor that in turn drives a rotational load. The power supply is regulated to give constant voltage up to a limiting current, as plotted below. The motor has a constant magnetic field. The load has the torque-speed characteristic, also plotted below.

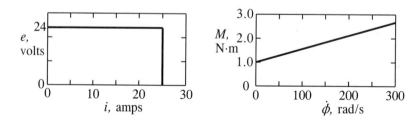

(a) Draw a bond graph model for the system. Neglect losses in the motor.

(b) Determine the maximum speed that the load can achieve.

(c) Evaluate the modulus that describes the basic property of the motor.

2.38 A DC motor is connected to an electrical source described by $e = 20 + 5.5i - i^2$, using units of amps and volts. Its output shaft rotates a wisp in a thick batter such that $M = 2\dot{\phi}^2$, where $\dot{\phi}$ has units rad/s and M has units N·m. Neglecting electrical resistance and friction, model the system with a bond graph, and determine the key parameter of the motor that maximizes the speed.

2.39 A pump with the characteristics given in Fig. 2.45 (p. 71) drives a rotary hydraulic motor with volumetric displacement D per radian, in place of the piston load shown in the figure. The load torque on the shaft of the motor is $M = a\dot{\phi}$ where $a = 100$ in.·lb s. Draw a bond graph model, and find the value of D that maximizes the speed of the load.

2.40 Consider the system studied in Section 2.5.5, which comprises an electrical source, an ideal DC motor with a constant magnetic field and a particular mechanical load with characteristics given in Fig. 2.44 (p. 69). In the present case, the modulus a is not specified.

(a) Determine the maximum power that can be transmitted to the load.

(b) Determine the value of a that accomplishes this transmission.

2.41 The armature circuit of an ideal DC machine with constant magnetic field is completed with an electrical resistance R_e, as shown to the right. The machine is characterized by the relation $M = ai$, and is otherwise ideal. Show that the system acts like like a linear rotary mechanical dashpot, and determine its mechanical resistance, R_m.

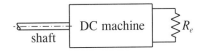

2.42 Repeat Problem 2.9 (p. 45) using the reticulated model $\dfrac{F}{\dot{x}} \!-\! T \!-\! \dfrac{P}{Q} \!-\! R.$

2.43 Repeat Problem 2.10 (p. 46) using the reticulated model of the problem above.

SOLUTIONS TO GUIDED PROBLEMS

Guided Problem 2.7

1. $S \dfrac{e = R_L T^2 i}{i} \!-\! T \dfrac{R_L T i}{T i} \!-\! R_L \qquad e = e_0 - R_a i; \qquad R_a = 24\ \Omega$

2. $S \dfrac{e}{i} \!-\! R_L'; \qquad R_L' = R_L T^2$

3. $\mathcal{P} = ei = (e_0 - R_a i)i$

4. $\dfrac{d\mathcal{P}}{di} = 0 = e_0 - 2R_a i.$ Therefore, $i = \dfrac{e_0}{2R_a}, \quad e = e_0/2.$

5. $e = R_L' i = R_L T^2 i$. Solving for T and using the above,

$$T = \sqrt{\frac{e}{iR_L}} = \sqrt{\frac{e_0/2}{(e_0/2R_a)R_L}} = \sqrt{\frac{R_a}{R_L}} = \sqrt{\frac{24}{6}} = 2.0.$$

Guided Problem 2.8

1. $\quad S \xrightarrow{\ \frac{P}{Q}\ } T \xrightarrow{\ \frac{F}{\dot{x}}\ } R$

2,3.

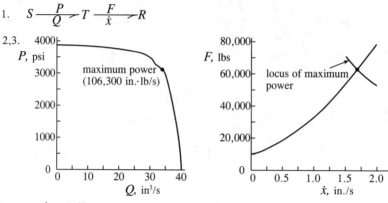

4. $\quad T = \dfrac{\dot{x}}{Q} = \dfrac{1.69}{34.0} = 0.050 \text{ in}^2.$ The piston area is $1/T = 20.1 \text{ in}^2.$

5. $\quad 1/(\text{piston area}) = P/F,$ also.

Chapter 3

Simple Dynamic Models

The systems considered thus far respond instantaneously with a new state whenever their environment or boundary conditions change. When a real physical system is subjected to a sudden change in boundary conditions, however, it might in fact respond noticeably slowly. The variables that define its state would approach their new equilibrium either monatonically (non-oscillatorily) or oscillatorily, assuming stability. Such systems, or the models that represent them, are called **dynamic**.

Dynamic physical systems contain mechanisms that store *energy* temporarily, for later release. The dynamics can be thought of as a sloshing of energy between different energy storage mechanisms, and/or a gradual dissipation of energy in resistances of the type you have examined already. Section 3.1 introduces a major class of energy storage mechanisms called **compliances**. Section 3.2 continues by introducing the other major category of energy storage mechanisms, called **inertances**.

This chapter makes a first pass through the major ideas in the modeling and simulation of dynamic systems. Attention is restricted to simple models comprising, at most, a source, a resistance, a compliance and an inertance joined together by a **junction**. Junctions are the only simple elements that can interconnect three or more bonds, and therefore become the very heart of the structure of a bond graph. They are introduced in Section 3.3. Dynamic models are represented mathematically by differential equations, as contrasted to static or steady-state models that are represented mathematically by algebraic equations. Conversion of the bond graph models to differential equations in presented in Section 3.4. Nonlinear compliances and inertances, previously omitted, are introduced in Section 3.5. This leads, finally, to numerical simulation in Section 3.6. The analytical solution of linear differential equation models is the subject of Chapter 4, with continuation in Chapter 7.

3.1 Compliance Energy Storage

Elements that store energy by virtue of a generalized displacement are called **compliances**. Physical approximations include mechanical compliances (*e.g.,* springs), fluid compliances (*e.g.,* gravity tanks) and electrical compliances (*e.g.,* capacitors), among others. Compliances are designated in bond graphs by the capital letter C.

3.1.1 Linear Springs and Energy

The **mechanical spring** is a familiar example of a one-port element that is totally different from the sources, sinks and resistances considered in Chapter 2. As illustrated in Fig. 3.1,

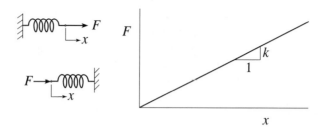

Figure 3.1: Linear spring and spring characteristic

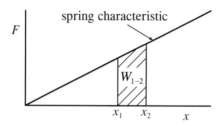

Figure 3.2: Work as area under characteristic

the spring is characterized by a relationship between its generalized force and its generalized *displacement*, rather than its *velocity*:

$$\frac{F}{\dot{x}} \!\!-\!\!-SPRING$$

$$F = kx. \tag{3.1}$$

The coefficient k is known variously as the **spring rate**, the **spring constant** or the **spring stiffness**. Note that in equation (3.1), x is defined so as to equal zero when the spring force is zero.

Springs do not dissipate energy, as resistances do; they *store* energy. This can be shown explicitly by integrating the power flowing into a spring over time, so as to compute the input work, $W_{1 \to 2}$, required to compress or stretch the spring from $x = x_1$ to $x = x_2$:

$$W_{1 \to 2} = \int_1^2 F\dot{x}\, dt = \int_1^2 F\frac{dx}{dt}\, dt = \int_{x_1}^{x_2} F\, dx = \int_{x_1}^{x_2} kx\, dx = \frac{1}{2}k(x_2{}^2 - x_1{}^2). \tag{3.2}$$

As suggested in Fig. 3.2, this work is the area under the spring characteristic between the two values of x. If the process is subsequently reversed and the spring is returned to its original position, the additional work is exactly the negative of the original work:

$$W_{2 \to 1} = \frac{1}{2}k(x_1{}^2 - x_2{}^2) = -W_{1 \to 2}. \tag{3.3}$$

It is customary to say that *energy is conserved* by the spring; the energy it stores is

$$\mathcal{V} = \frac{1}{2}kx^2 = \frac{1}{2}\frac{1}{k}F^2, \tag{3.4}$$

so that, in general,

$$W_{1 \to 2} = \mathcal{V}_2 - \mathcal{V}_1. \tag{3.5}$$

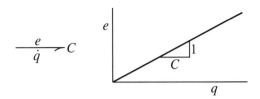

Figure 3.3: The generalized linear compliance

A **rotary spring** acts like a compression or tension spring, except its effort is a moment, M, and its displacment is an angle, ϕ, which is the time integral of the angular velocity, $\dot\phi$:

$$M = k\phi. \tag{3.6}$$

3.1.2   The Generalized Linear Compliance

A "generalized linear spring" results from replacing the force F of the compression or tension spring in equation (3.1) by the generalized force or effort e, and replacing the displacement x by the generalized displacement q. The **constitutive relation** is

$$e = \frac{1}{C}q. \tag{3.7}$$

The same result applies when the M and ϕ of the rotary spring characterized by equation (3.6) are replaced by e and q, respectively. In bond graph form this is written

$$\frac{e}{\dot q} \!\!\!\longrightarrow C$$

and is called a **compliance element** or just a **compliance**. "Compliance" means the inverse of "stiffness," so $C = 1/k$, a constant, for both translational and rotational mechanical springs. The result is summarized graphically in Fig. 3.3.

Note that the modulus of a compliance is defined in a manner similar to the definition of the modulus of a resistance. Further, just as the symbol R is used for both the resistance and its modulus, the symbol C is used for both the compliance and its modulus.

The energy stored in the linear generalized compliance follows from substitution of the generalized variables into equation (3.4):

$$\mathcal{V} = \frac{1}{2}\frac{1}{C}q^2 = \frac{1}{2}Ce^2. \tag{3.8}$$

This is known as a **potential energy**. The traditional symbol is capital V, which in this text is given as script $\mathcal{V}$ (to distinguish it from volume).

3.1.3   Electric Circuit Compliance

An electric **capacitor** is indicated by its traditional symbol in Fig. 3.4. Its applied voltage e and **charge** or electrical displacement q are modeled as

$$q = Ce. \tag{3.9}$$

This equation agrees precisely with equation (3.7), revealing that the capacitor is modeled as an electric circuit compliance, and that the **capacitance** C is analogous to the compliance

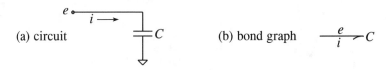

Figure 3.4: Capacitor as a compliance

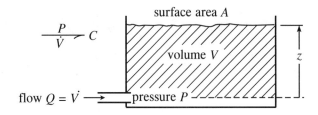

Figure 3.5: Liquid gravity tank as a compliance

C. The coincidence that "capacitance" and "compliance" both begin with the letter "c" contributed to the selection of that letter to designate the generic ratio of generalized displacement to generalized effort. A capacitance usually is considered to be constant; the model is *linear*.

3.1.4   Linear Fluid Compliance Due to Gravity

The mechanical spring and the electrical capacitor have a generalized force which is proportional to a generalized displacement. A fluid compliance should have the same thing: a pressure proportional to a volumetric displacement. Think for a moment what this could be.

The most common answer is a fluid tank of constant area, independent of the depth of liquid in it, as shown in Fig. 3.5. A fluid port enters at or near the bottom of the tank. The depth of the liquid equals the fluid displacement V, which is the volume of liquid in the tank above the port, divided by the cross-sectional area of the tank:

$$z = V/A. \tag{3.10}$$

The pressure in the tank at the level of the port equals the weight density of the liquid times the depth:

$$P = \rho g z. \tag{3.11}$$

The compliance of the tank, C, is the ratio of the volume to the pressure:

$$\boxed{C \equiv \frac{q}{e} = \frac{V}{P} = \frac{V}{\rho g V / A} = \frac{A}{\rho g}.} \tag{3.12}$$

Note that the time derivative of the volume of liquid in the tank is the volume flow rate into the tank, which can be designated either as $\dot{V}$ or Q.

3.1.5   Fluid Compliance Due to Compressibility

A second type of fluid compliance is associated with the compressibility of the fluid. Assuming small changes in density only, such as found in considerations of air acoustics and nearly all problems involving the compressiblity of liquids, it is conventional to assume

$$dP = \beta \, \frac{d\rho}{\rho}, \tag{3.13}$$

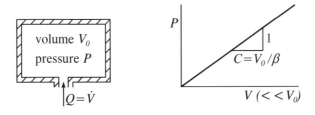

Figure 3.6: Fluid-filled rigid chamber as a compliance

in which the **bulk modulus**, β, is treated as a constant. This equation says, for example, that a one percent change in density is associated with a pressure change of virtually one percent of β; for water with no gas bubbles this gives about one percent of 300,000 psi, or 3000 psi.

The equation of state for a perfect gas is $P = \rho R \theta$, where θ is absolute temperature. Differentiation of this equation and comparison with equation (3.13) reveals that $\beta = P_0$ for small isothermal changes; P_0 is the mean absolute pressure. In the absense of heat transfer, the pressure and density of a perfect gas with constant specific heats are related by P/ρ^k =constant. Upon differentiation, this gives $\beta = kP_0$, where k is the ratio of specific heats. A one percent change in volume of air at atmospheric conditions therefore, in the absence of heat transfer, would be associated with a pressure change of about $.01*1.4*14.7 = 0.206$ psi.

A rigid chamber of volume V_0 filled with a slightly compressible fluid is shown in Fig. 3.6. The fluid compliance of this system is

$$C \equiv \frac{q}{e} = \frac{dq}{de} = \frac{V_0 \, d\rho/\rho}{dP} = \frac{V_0}{\beta}. \tag{3.14}$$

Should the volume of the chamber expand with increasing pressure, the effective compliance would be larger. Note that this compliance is approximated as a constant.

EXAMPLE 3.1

A uniform rectangular block has half the mass density as the water in which it floats. Find the compliance associated with small rotations induced by an applied moment about an axis normal to the page.

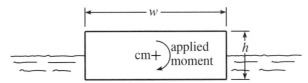

Solution: The center of mass is on the waterline, so it remains fixed when the block is rotated. For a clockwise rotation through a small angle ϕ, a triangular area shown shaded below emerges above the waterline on the left side, and an equal triangular area is submerged on the right side:

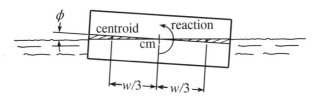

The bouyancy forces on the two sides change by the area of this triangle, $(w\phi/2)(w/4) = w^2\phi/8$, times the weight density of the water, ρg. These forces act through the centroids of the triangles to give a pure moment or couple. The couple equals the force times the separation between the centroids, or

$$M = \frac{\rho g w^3}{12}\phi.$$

This is the moment per unit length of the block in the direction normal to the page. The corresponding compliance, by definition, is the ratio of the angular deflection to the applied moment, or

$$C \equiv \frac{\phi}{M} = \frac{12}{\rho g w^3}.$$

3.1.6 Summary

The reversible storage of potential energy can be represented by one-port elements for which the effort is a function of the generalized displacement. The generalized displacement is the time integral of the generalized velocity or flow. Such elements are designated by the bond graph symbol C, which stands for *compliance*. This symbol doubles as the modulus of the compliance, that is $C = q/e$, which is the inverse slope of the characteristic plotted on e, q coordinates. Examples modeled by compliances include the mechanical spring with its force or moment and linear or angular displacement, the electrical capacitor with its voltage and charge, the open fluid tank with its pressure and liquid volume, and a container confining a compressible fluid with its pressure and volume displacement.

Nonlinear compliances are discussed in Section 3.5.

Guided Problem 3.1

It is essential that you grasp the idea that the definition and value of a compliance depends on the displacement variables to which it is referenced. This first problem is basic in this regard, while the second problem is more sophisticated.

The rotation of a lever is resisted by a spring with rate k, as shown in Fig. 3.7. Find the compliance of the system with respect to the displacement x and force F.

Suggested Steps:

1. Find the displacement of the spring, which you could call x_s, as a function of the parameters a and b and the displacement variable x.

2. Write the energy stored in the spring as a function of x_s.

3. Combine the results of steps 1 and 2 to get the energy of the spring as a function of x.

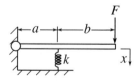

Figure 3.7: Guided Problem 3.1

4. Equate the result of step 3 to the known form, $\mathcal{V} = (1/2C)x^2$, and solve for C.

5. You might also wish to solve for C using a more familiar force-based analysis, although presently it is more important that you learn the method above. Find the force on the spring, which you could call F_s, as a function of the parameters a and b and the effort variable F. Then, solve this equation for F. Solve the equation from step 1 for x. Finally, compute the ratio $C = x/F$. The answer, of course, should contain only parameters, and no variables.

Guided Problem 3.2

A long slender-walled cylinder expands under the influence of an internal pressure, creating an effective compliance relative to an incompressible fluid within. The internal radius is r, and wall thickness is $t \ll r$ and the length is $L \gg r$. The wall material has Young's modulus E and Poisson's ratio μ.

(a) Estimate the compliance of the chamber, neglecting effects of its ends.

(b) Estimate the effective overall compliance relative to a fluid which is itself compressible, having a bulk modulus β.

Suggested Steps:

1. Draw a cross-sectional end view of one-half of the cylinder. Place the pressure forces across the diameter and the resisting peripheral forces $t\sigma$ on the shell. Balance the forces, and solve for the peripheral normal stress.

2. Draw a side view, and find the normal axial stress.

3. Use the results of steps 1 and 2 and the elastic parameters of the material to find the peripheral and axial strains.

4. Find the ratio of the change in volume to the original volume, using the results of step 3 and the assumption of small strains. Neglect the effects of the ends of the cylinder.

5. Evaluate the compliance, which is the ratio of the change in volume to the applied pressure.

6. Find the compliance for the different case in which the shell is rigid but the fluid has a finite bulk modulus, β.

7. If both the shell and the fluid are elastic, then there are two compliances. The total or effective compliance seen at the port is the sum of the two individual compliances (as will be seen more clearly later on). Find this combined compliance.

PROBLEMS

3.1 Derive the compliance of a solid body floating on the surface of a liquid with respect to small vertical displacements and forces. Define parameters as needed.

3.2 Find the compliance of a shaft in torsion with respect to its overall angle of twist. Define parameters as needed. (You may refer to your textbook on the mechanics of materials.)

3.3 A pendulum comprises a point mass attached to the end of a massless rigid rod which is pivoted about its upper end. Interest is restricted to small angles of deflection from the vertical equilibrium.

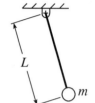

  (a) Find the gravity-induced compliance by considering the correcting moment for an angular deflection from the vertical.

  (b) Find the compliance by considering the gravity energy associated with the deflection. Hint: Use the approximation $\cos\theta \simeq 1 - \frac{1}{2}\theta^2$.

3.4 Solve the preceding problem substituting a uniform slender rod of mass m and length L for the massless rod and point mass.

3.5 For the pictured rolling disk of mass m and radius r,

  (a) Define a variable to describe the displacement of the disk from its equilibrium position. Find the potential energy of the disk in terms of this displacement.

  (b) Find the compliance of the disk with respect to the displacement, assuming it has only small values. *Continued in Problems 3.13, 3.53 and 3.58.*

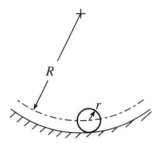

3.6 Find the compliance of the piston/cylinder shown opposite with respect to F, x. Assume $x << L$ and that the air acts as an ideal gas, neglect friction and leakage, and presume (a) isothermal behavior (b) adiabatic behavior.

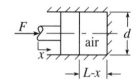

3.7 Verify that the bulk modulus for acoustic (small) and adiabatic compression of a perfect gas is $\beta = kP_0$, where k is the ratio of specific heats and P_0 is the mean absolute pressure.

SOLUTIONS TO GUIDED PROBLEMS

Guided Problem 3.1

1. $x_s = \frac{a}{a+b}x$.

2-4. $\mathcal{V} = \frac{1}{2}kx_s^2 = \frac{1}{2}k\left(\frac{a}{a+b}\right)^2 x^2 = \frac{1}{2C}x^2$; therefore, $C = \left(\frac{a+b}{a}\right)^2 \frac{1}{k}$.

5. $kx_s = F_s = \frac{a+b}{a}F$, so that $F = \left(\frac{a}{a+b}\right)F_s = \left(\frac{a}{a+b}\right)kx_s$.

   From step 1, $x = \frac{a+b}{a}x_s$.

   Finally, $C \equiv \frac{x}{F} = \frac{1}{k}\left(\frac{a+b}{a}\right)^2$, which is the same as above.

Guided Problem 3.2

1.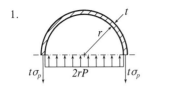

   $2t\sigma_p = 2rP$, therefore $\sigma_p = Pr/t$

2.

   $2\pi rt\sigma_a = \pi r^2 P$, therefore
   $\sigma_a = Pr/2t = \sigma_p/2$

3. peripheral strain: $\epsilon_p = \dfrac{1}{E}(\sigma_p - \mu\sigma_a) = \left(1 - \dfrac{\mu}{2}\right)\dfrac{Pr}{Et}$

   axial strain: $\epsilon_a = \dfrac{1}{E}(\sigma_a - \mu\sigma_p) = \left(\dfrac{1}{2} - \mu\right)\dfrac{Pr}{Et}$

4. volume: $V = \pi[r(1 + \epsilon_p)]^2 L(1 + \epsilon_a) \simeq \pi r^2 L(1 + 2\epsilon_p + \epsilon_a)$

   change in volume: $\Delta V \simeq \pi r^2 L(2\epsilon_p + \epsilon_a) = (2.5 - 2\mu)\dfrac{\pi r^3 LP}{Et}$

5. For the chamber alone, neglecting end effects, $C_c = \dfrac{\Delta V}{P} = (2.5 - 2\mu)\dfrac{\pi r^3 L}{Et}$

   $\Longrightarrow \dfrac{P}{Q = \dot{V}}\ C_c$

6. For the fluid alone, $C_f = \dfrac{V}{\beta} = \dfrac{\pi r^2 L}{\beta}.$ $\Longrightarrow \dfrac{P}{Q = \dot{V}}\ C_f$

7. For combined compliance when both the walls and the fluid are elastic,

   $C = C_c + C_f = \pi r^2 L\left[(2.5 - 2\mu)\dfrac{r}{Et} + \dfrac{1}{\beta}\right]$ $\Longrightarrow \dfrac{P}{Q = \dot{V}}\ C$

3.2 Inertance Energy Storage

Inertance energy storage, also known as generalized kinetic energy, is the second of the two categories of energy storage. Most models of this storage are representable as one-port elements, designated by the symbol I. (More complicated models are considered in Chapter 10.)

3.2.1 Mass, Momentum and Kinetic Energy

Kinetic energy, as its name suggests, is associated with motion, unlike potential energy, which is associated with displacement. Like potential energy, however, kinetic energy is potentially recoverable. Consider the mass m of Fig. 3.8, which is accelerated by a force $F(t)$ from rest in an inertial reference frame to a velocity, $\dot{x}$, and a corresponding linear momentum, $p = m\dot{x}$. Newton's law gives

$$\frac{d\dot{x}}{dt} = \frac{F}{m}, \tag{3.15}$$

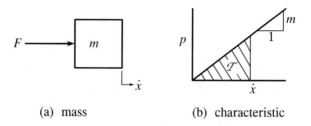

(a) mass   (b) characteristic

Figure 3.8: Accelerating mass and characteristic

which can be integrated to give

$$\dot{x} = \frac{\int F\, dt}{m} = \frac{p}{m}. \tag{3.16}$$

The kinetic energy is designated traditionally by a capital T, and is given in this text by script $\mathcal{T}$ (to distinguish it from transformer moduli). It equals the work done by the force:

$$\mathcal{T} = \int F\, dx = \int m \frac{d\dot{x}}{dt} \dot{x}\, dt = m \int \dot{x}\, d\dot{x} = \frac{1}{2} m \dot{x}^2 = \frac{1}{2m} p^2. \tag{3.17}$$

This energy is represented by the area under the characteristic, as shown in part (b) of the figure.

The same type of development that gives equation (3.5) for the work associated with a change of displacement of a compliance gives the work required to change the velocity or momentum from state 1 to state 2:

$$\mathcal{W}_{1\to 2} = \mathcal{T}_2 - \mathcal{T}_1. \tag{3.18}$$

3.2.2   The Generalized Linear Inertance

The generalized compliance has been described by

$$e = q/C, \tag{3.19}$$

which is a linear relation between the generalized displacement and the effort. The corresponding "constitutive relation" for the generalized inertance is

$$\boxed{\dot{q} = p/I,} \tag{3.20}$$

which is a linear relation between the generalized momentum and the flow. This is precisely equation (3.16) if $\dot{q}$ is replaced by $\dot{x}$ and the **generalized inertance**, I, is replaced by m. The **generalized momentum** is defined by

$$\boxed{p \equiv \int e\, dt \quad \text{or} \quad e \equiv \dot{p}.} \tag{3.21}$$

The relationship of equation (3.20) is represented in part (b) of Fig. 3.8 by changing the labels $\dot{x}$ to $\dot{q}$ and m to I. The bond graph symbol for inertance uses the letter I in the same way that the compliance uses the letter C:

$$\frac{e = \dot{p}}{\dot{q}} \longrightarrow I$$

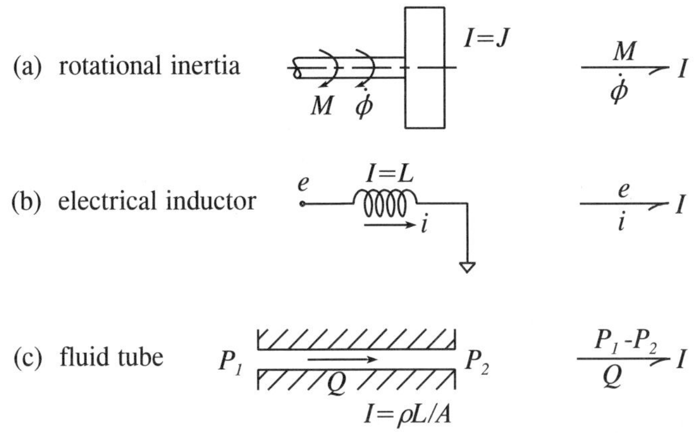

Figure 3.9: Examples of constant inertances

The momentum of the mass in Fig. 3.8 is proportional to the velocity, so that $I = m$ is a constant; the inertance is said to be linear. The generalized kinetic energy of a constant inertance is

$$\boxed{\mathcal{T} = \frac{1}{2}I\dot{q}^2 = \frac{1}{2I}p^2,}$$

(3.22)

which can be compared with equation (3.17).

3.2.3 Common Engineering Elements Modeled by Constant Inertances

A rigid body that rotates about a fixed axis, such as a flywheel, resists angular acceleration because of its mass moment of inertia, I, as suggested in Fig. 3.9 part (a). (Some authors use the symbol J to represent the mass moment of inertia, in which case you can write $I = J$ for the modulus. You should not write J in the bond graph, however, since it is not a defined bond graph symbol.) In this application, p is the *angular* momentum. The kinetic energy is $\frac{1}{2}I\dot{\phi}^2$ or $\frac{1}{2}p^2/I$, consistent with equation (3.22).

EXAMPLE 3.2

Deduce a formula for the value of I for a rigid mass rotating about a fixed axis. Use a momentum method and then an energy method; the results should agree.

Solution: The angular momentum is

$$p = \int r\,v\,dm = \int r(r\dot{\phi})dm = \left[\int r^2 dm\right]\dot{\phi},$$

and the kinetic energy is

$$\mathcal{T} = \int \frac{1}{2}v^2 dm = \frac{1}{2}\int (r\dot{\phi})^2 dm = \frac{1}{2}\left[\int r^2 dm\right]\dot{\phi}^2,$$

where r is the radial distance of the element dm from the axis, of rotation, and the integrations are carried out over the entire body. Either comparing the first equation to equation(3.20) or the second equation to equation (3.22), with $\dot{q} = \dot{\phi}$, the result is

$$I = \int r^2 dm.$$

This is the mass moment of inertia, familiar from introductory dynamics courses. It gives, for example, mR^2 for a slender hoop of radius R, $\frac{1}{2}mR^2$ for a uniform disk of radius R and $(1/12)m(2R)^2$ for a rod of length $2R$, all about their respective normal centroidal axes.

An inductor in an electric circuit impedes the rate of change of current. It also can be modeled in bond graph language by an inertance, as shown in part (b) of Fig. 3.9. In this application, p is still $\int e\, dt$, the modulus I equals the inductance, L, and the generalized kinetic energy is $\frac{1}{2}Li^2$ or $\frac{1}{2}p^2/L$. Many engineers use the symbol I for inductance anyway.

The mass of a fluid in a tube impedes the rate of change of the volume flow rate. Its use in a fluid circuit also can be represented in a bond graph by an inertance, as shown in part (c) of Fig. 3.9. The standard variables $e = P$ and $\dot{q} = Q$ can be used[1].

EXAMPLE 3.3

Deduce the inertance of an incompressible fluid of density ρ in a uniform channel of area A and length L. Use both a momentum approach and a separate energy approach.

Solution: The momentum approach can be represented by the successive steps in the multiple equation below:

$$I = \frac{p}{Q} = \frac{dp/dt}{dQ/dt} = \frac{P_1 - P_2}{A\, dv/dt} = \frac{\rho L}{A},$$

The final step above directly uses Newton's law; $P_1 - P_2$ equals the net force divided by A times the mass ρAL and the acceleration dv/dt. The resulting formula is approximate, and in fact somewhat underestimates the inertance, since it assumes that the velocity is uniformly equal to Q/A whereas the velocity is more apt to be nonuniform over the cross section.

Approaching this result from the perspective of kinetic energy,

$$\mathcal{T} = \frac{1}{2}mv^2 = \frac{1}{2}(\rho AL)\left(\frac{Q}{A}\right)^2 = \frac{1}{2}\left(\frac{\rho L}{A}\right)Q^2.$$

Here, m is the total mass of fluid in the tube and v is its common velocity. The term on the right within the parentheses is the desired inertance, and agrees with the right-end term of the previous equation.

The inertances of two cascaded constant-area channels sum, as suggested in Fig. 3.10.

---

[1] Use of $e = P$ neglects the dynamic pressure, as noted before. Therefore, the inertance relation presented here will not fully represent the behavior when the dynamic pressure is significant relative to the static pressure; in particular, entrance and exit losses may need to be added.

Figure 3.10: Inertance of two cascaded fluid channels

EXAMPLE 3.4

Find an appproximate expression for the inertance of a fluid channel of varying area $A(s)$, where s is the distance along the channel.

Solution: A channel of continuously varying area can be considered as a cascade of infinitesimally long segments, each of which has the inertance $\rho\,ds/A$, as suggested below:

The overall inertance is the sum of these, or

$$I = \rho \int_0^L \frac{1}{A}\,ds.$$

The result of Example 3.3 is the special case for $A = $ constant.

3.2.4 Tetrahedron of State*

The four essential variables of state for physical systems are represented by the vertices of the **tetrahedron of state**, pictured in Fig. 3.11. Many new bond graph modelers use this diagram to help them remember these variables and the R, C and I elements that are defined by relationships between them. It is a conception of Henry M. Paynter.

The edge of the tetrahedron between the vertices for the generalized displacement q and the flow or generalized velocity $\dot{q}$ represents a simple time differentiation/integration relationship. The edge between the vertices for generalized momentum p and the effort or

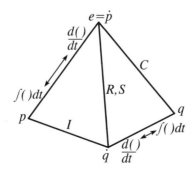

Figure 3.11: Tetrahedron of state

generalized force $e = \dot{p}$ also represents a simple time differentialtion/integration relationship. The edge between the vertices for e and $\dot{q}$ suggests the relationship that characterizes a resistance, ——— R, or a general source or sink, S ——— or ——— S. The edge between the vertices for e and q suggests the relationship that characterizes a compliance, ——— C. Finally, the edge between the vertices for p and $\dot{q}$ suggests the relationship that characterizes an inertance, ——— I. (The hidden remaining edge has no significance.)

3.2.5   Summary

A linear one-port inertance or generalized kinetic energy can be characterized by the proportionality between its flow or generalized velocity and its generalized momentum, where the generalized momentum is the time integral of the effort. Thus, the generalized momentum, p, is to the effort, e, as the generalized displacement, q, is to the generalized velocity (or flow) $\dot{q}$. The inertance relationship is represented in a bond graph by the symbol I. The proportionality constant is the modulus of the inertance, and also is called the inertance and is designated by the symbol I. The generalized potential and kinetic energies equal the areas under their respective characteristics, which for the linear cases considered are, respectively, $\mathcal{V} = \frac{1}{2}(1/C)q^2 = \frac{1}{2}Ce^2$ and $\mathcal{T} = \frac{1}{2}I\dot{q}^2 = \frac{1}{2}(1/I)p^2$. Nonlinear inertances are introduced in Section 3.5.

Mass, rotational inertia, fluid inertia and electrical inductance are examples of a generalized inertance or generalized kinetic energy. General formulas have been developed for the rotational and fluid cases. Table 3.1 summarizes some of the more primitive physical elements commonly modeled as compliances and inertances.

Table 3.1 Primitive Physical Compliances and Inertances

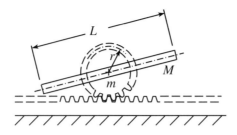

Figure 3.12: Guided problem 3.3

Guided Problem 3.3

It is esssential that you grasp the idea that the definition and value of an inertance depends on the generalized velocity to which it is referred. This problem offers a basic example.

A disk-shaped pinion of radius r and mass m is engaged to a fixed straight rack, as shown in Fig. 3.12. A slender rod of length L and mass M is welded to the pinion, as shown. Find the inertance of the unit relative to the horizontal velocity of its center, $\dot{x}$.

Suggested Steps:

1. Write the kinetic energy of the unit due to the *horizontal velocity* of its center of mass as a funtion of $\dot{x}$.

2. Find the angular velocity of the unit as a function of $\dot{x}$.

3. Find the kinetic energy of the unit due to its *angular velocity* as a function of this angular velocity, and then use the result of step 2 to convert this to a function of $\dot{x}$.

4. Add the energies of steps 1 and 3, and set the sum equal to $\frac{1}{2}I\dot{x}^2$. Solve for I, which should be the desired answer.

PROBLEMS

3.8 Define the symbols L, E, I, G, r, A, ρ, β and V_0 in Table 3.1 as they pertain to the (a) cantilever beam, (b) floating block, (c) torsion rod, (d) pendulum, (e) compliance volume, (f) gas accumulator, (g) gravity tank and (h) inertance channel.

3.9 A disk of mass m and radius r rolls without slipping down an incline with angle θ relative to the horizontal. Represent the situation with a bond graph, using the velocity of the center of the disk as the flow. Determine the inertance, and find the acceleration of the center of the disk. Suggestion: Use the concept of kinetic energy to find the inertance, as in Guided Problem 3.3.

3.10 A tube of length L and internal area A is filled with, and its entrance surrounded by, an inviscid liquid of density ρ. The pressure at its entrance is a constant $P = \rho g h$ and is zero at the other end. Represent the situation with a bond graph, using the volume flow rate Q as the flow variable but neglecting entrance and exit losses. Find the rate of change of the flow.

3.11 A pendulum comprises a point mass m attached to the lower end of a massless rod of length L pivoted at its upper end, as in Problem 3.3 (p. 84).

(a) Find the moment required to produce an angular acceleration, neglecting the gravity compliance property, and thereby deduce the inertance with respect to angular velocity.

(b) Check to make sure that $\frac{1}{2}I\dot{\phi}^2$, with your I from part (a), is the correct kinetic energy.

3.12 Solve the preceding problem substituting a uniform slender rod of mass m and length L for the massless rod and point mass (as in Problem 3.4, p. 84).

3.13 For the rolling disk of Problem 3.5 (p. 84):

(a) Find the kinetic energy of the disk, first in terms of any velocity but finally in terms of $\dot{\phi}$, where ϕ is the angle between vertical and a line drawn from the center of curvature to the center of the cylinder.

(b) Use the result of step (a) to find the inertance of the disk relative to $\dot{\phi}$. *Continued in Problems 3.53 and 3.58.*

SOLUTION TO GUIDED PROBLEM

Guided Problem 3.3

1. $T_x = \frac{1}{2}(m + M)\dot{x}^2$.

2. $\dot{\phi} = \dot{x}/r$.

3. $T_\phi = \frac{1}{2}\left(\frac{1}{2}mr^2 + \frac{1}{12}ML^2\right)\dot{\phi}^2 = \frac{1}{2}\left(\frac{1}{2}m + \frac{1}{12}\frac{ML^2}{r^2}\right)\dot{x}^2$.

4. $T_x + T_\phi = \frac{1}{2}I\dot{x}^2 = \frac{1}{2}\left[\frac{3}{2}m + \left(1 + \frac{L^2}{12r^2}\right)M\right]\dot{x}^2$.

   Therefore, $I = \frac{3}{2}m + \left(1 + \frac{L^2}{12r^2}\right)M$.

3.3   Junctions

The models presented thus far exhibit no **branching**. Most models of engineering systems require branching, which usually is accomplished through the use of simple **junctions**. The conventional bond-graph abstractions for junctions are given below first, showing that there are precisely two types. Mechanical, electrical and fluid constraints are then modeled by these junctions, and a single junction is used at the heart of models for very simple systems. Application to a variety of more complex systems is made in Chapter 5 and subsequent chapters.

3.3.1   Junction Types

An arbitrary number of bonds can be joined by a junction, identified here by the temporary symbol J:

$$e_2 \Big| \dot{q}_2$$

$$\frac{e_1}{\dot{q}_1} \longrightarrow J \xrightarrow{} \frac{e_3}{\dot{q}_3}$$

$$e_n \Big| \dot{q}_n$$

The power convention half-arrows in this example are chosen arbitrarily, with the first two being inward and all others being outward.

The following rules are imposed to define two types of junctions:

Rule (i): For both types, the junction is ideal, neither storing, creating nor dissipating energy.

Rule (ii): For the 0-junction type, the efforts on all the bonds are equal. For the 1-junction type, the flows on all bonds are equal.

The first rule implies that the input power equals the output power, requiring

$$e_1\dot{q}_1 + e_2\dot{q}_2 = e_3\dot{q}_3 + \ldots + e_n\dot{q}_n. \tag{3.23}$$

For the **0-junction** (spoken *zero*-junction), represented as

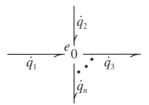

the second rule requires

$$e_1 = e_2 = e_3 = \ldots = e_n = e, \tag{3.24}$$

which when introduced into equation (3.23) gives

$$\dot{q}_1 + \dot{q}_2 = \dot{q}_3 + \ldots + \dot{q}_n. \tag{3.25}$$

The labels $e_1 \ldots e_n$ have been removed from the respective bonds, not of necessity (they could be left on) but because the effort e, written above and to the left of the junction symbol, is recognized as being common to all the bonds.

The 0-junction is the **common-effort junction**. It is often more useful to recognize its property described by equation (3.25), so it is also the **flow-summing junction**.

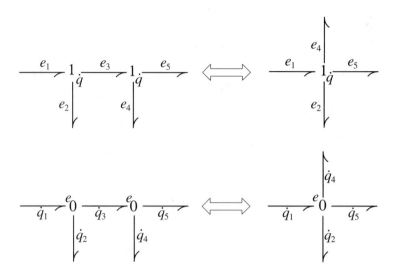

Figure 3.13: Bond-graph junction equivalences

For the **1-junction**, represented as

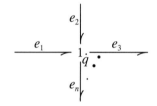

the second rule requires

$$\dot{q}_1 = \dot{q}_2 = \dot{q}_3 = \ldots = \dot{q}_n = \dot{q}, \tag{3.26}$$

which through equation (3.23) gives

$$e_1 + e_2 = e_3 + \ldots + e_n. \tag{3.27}$$

The labels $\dot{q}_1 \ldots \dot{q}_n$ have been removed from the respective bonds, again because they would be redundant. The common flow or generalized velocity is indicated by the $\dot{q}$ (or a specialized symbol, if one prefers) written below and to the right of the 1.

The 1-junction is called the **common flow junction**. It also is called the **effort-summing junction**. It is the *dual* of the 0-junction, since the constraints imposed on the efforts in one case are the same as the constraints imposed on the flows in the other, and vice-versa. The concept and notation for these junctions were introduced by H.M. Paynter [2] and were the most crucial step in his creation of bond graphs.

Two junctions of the same type connected by a bond can be collapsed into a single junction, with the bond vanishing, as shown in Fig. 3.13. The only difference between the expanded and contracted versions of the graphs shown is the presence or absence of the variable e_3 or $\dot{q}_3$, respectively, which may or may not be of interest.

3.3.2   Mechanical Constraints Modeled by 1-Junctions

Common-flow junctions abound in models of engineering and natural systems. A mechanical example for translation is shown in Fig. 3.14. A push-rod with compression force F_t drives

---
[2]H.M. Paynter, *Analysis and Design of Engineering Systems*, The MIT Press, Cambridge, Mass., 1961.

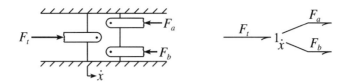

Figure 3.14: Junction for mechanical translation

two push-rods with compression forces F_A and F_B. All three push rods share the common velocity $\dot{x}$, as implemented in the bond graph by the 1-junction. The conservation of power requires

$$F_t \dot{x} = F_A \dot{x} + F_B \dot{x}, \tag{3.28}$$

so that

$$F_t = F_A + F_B. \tag{3.29}$$

Equation (3.29) can be found alternatively from a *free-body analysis* which is traditionally taught in courses entitled "statics" and "dynamics." Generalized force balances represent the "conservation of momentum" which is one of three great principles used to analyze combinations of mechanical elements. The other two are the geometric constraints between the displacements or velocities (sometimes called "continuity"), and the conservation of energy. Any two of the three are sufficient; the third is redundant. It is extremely advantageous for you to be able to employ any one of the three consequent approaches. When the modeling becomes complex or confusing, the use of the energy balance and the geometric constraints tends to be clearer and simpler than either combination using the force balances. It also is more generalizable to non-mechanical phenomena.

EXAMPLE 3.5

Model the parallel combination of two masses, a spring and a dashpot shown below with a bond graph. Reduce the graph, if possible, so only one element of each type remains.

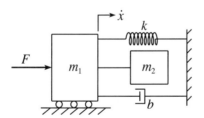

Solution: The 1-junction of the bond graph model below represents the fact that all four elements share a common velocity. The two inertances (masses) sum to give an equivalent inertance (mass). Constant inertances arrayed around a 1-junction always can be summed in this manner. Elements of different types *cannot* be merged.

EXAMPLE 3.6

Model the parallel combination of three dashpots below by a bond graph. Also, show how the individual characteristics, as plotted by solid lines, can be combined into a single resistance characteristic.

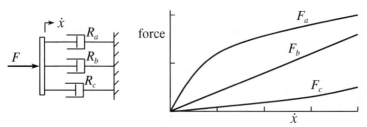

Solution: All three dashpots experience a common velocity ($\dot{x}$), whereas the force F is the *sum* of the three individual forces. Therefore, as shown by the dashed line, the three forces sum for each value of $\dot{x}$ to give the overall characteristic.

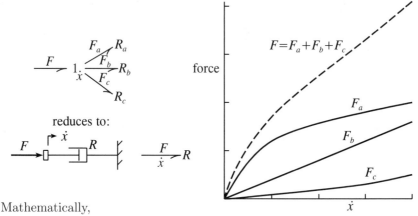

reduces to:

Mathematically,

$$F = F_a + F_b + F_c = R_a\dot{x} + R_b\dot{x} + R_c\dot{x} = R\dot{x},$$

It follows that the equivalent overall resistance R equals the *sum* $R_a + R_b + R_c$. This equation is most easily understood if the resistances are constants and the characteristics are straight lines through the origin. It is quite correct, nevertheless, if the resistances are functions of $\dot{x}$ and the characteristics are curved lines.

EXAMPLE 3.7

Each of the three springs below has a constant spring rate. Determine the modulus of a single compliance element, C, that can represent the combination.

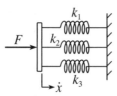

Solution: The three individual forces sum to give the total force, as with the dashpot example above. These forces are proportional to the three spring-rate constants, which are inversely proportional to the three compliances. The combination rule therefore is

$$\frac{1}{C} = \frac{1}{C_1} + \frac{1}{C_2} + \frac{1}{C_3}.$$

In general, the 1-junction represents the **geometric constraint** in which three or more bonds share a common generalized velocity or flow. The presence of a 1-junction may not be always obvious, as the following example suggests.

EXAMPLE 3.8

Model the fire-extinguishing system of Chapter 2 with an additional load driven from the same shaft, as shown. The "pump" includes its fluid load.

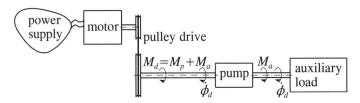

This auxiliary load might be a siren, for example, that is operated at the same time as the pump, or it might be a separate pump. Also, show how the individual characteristics of the pump and the auxiliary load, as plotted below, combine to give an equivalent overall resistive load.

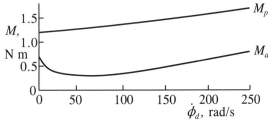

Solution: To represent the combination of the drive and the two one-port devices, first answer this: do these three elements share one of their respective conjugate variables? The answer is *yes*; they have the same angular velocity, $\dot{\phi}_d$, because they have a common shaft. This fact is recognized by a 1-junction:

$$\text{MOTOR} \underset{\dot{\phi}_m}{\overset{M_m}{\longrightarrow}} \begin{array}{c}\text{BELT}\\\text{DRIVE}\end{array} \underset{\dot{\phi}_d}{\overset{M_d}{\longrightarrow}} 1. \begin{array}{c} \overset{M_p}{\nearrow} \text{PUMP}\\ \underset{M_A}{\searrow}\\ \text{AUXILIARY LOAD}\end{array}$$

For emphasis, all three bonds are annotated with "$\dot{\phi}_d$." This is redundant; "$\dot{\phi}_d$" need be noted only below and to the right of the "1" symbol, also as shown.

The powers delivered to the load A, load B and their combination, respectively, are

$$\mathcal{P}_p = M_p \dot{\phi}_d; \qquad \mathcal{P}_a = M_a \dot{\phi}_d,$$
$$\mathcal{P}_{\text{total}} = M_p \dot{\phi}_d + M_a \dot{\phi}_d = (M_p + M_a)\dot{\phi}_d = M_d \dot{\phi}_d.$$

Thus, the total torque provided by the belt drive is

$$M_d = M_p + M_a.$$

The summation of the last equation above may be implemented graphically by "vertical" addition of the two component characteristics, as suggested below, to give the characteristic drawn with a dashed line. This curve is the effective one-port characteristic of the equivalent combined load. It can be used as before to carry out a source-load analysis or optimization.

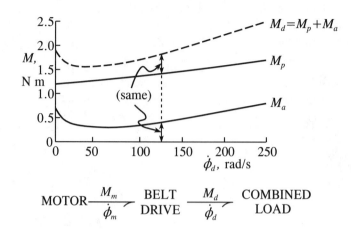

3.3.3 Electrical Circuit Constraints Modeled by 1-Junctions

Common-flow implies common-current for electric circuits, and common-current implies *series* interconnection of components. Suppose that three components arbitrarily labelled Z_1, Z_2 and Z_3 are connected in series, as shown in Fig. 3.15. The bond graph follows directly. The total voltage of the source equals the *sum* of the voltages (or voltage drops) of the three components; the sum of the voltage drops around a loop equals zero. Thus if the elements happen to be resistances, the total resistance of the circuit is the sum $R_1 + R_2 + R_3$. Electrical resistances in series combine like mechanical dashpots in parallel. If the elements are inductors, the total inductance of the circuit is the sum $I_1 + I_2 + I_3$. The same is analogous to any other type of inertances bonded to a common 1-junction, such as the masses discussed earlier. If the elements are capacitors, the summation of the voltages implies summing the *reciprocals* of the capacitances, or $1/C_1 + 1/C_2 + 1/C_3$. This is like the springs bonded to a common 1-junction as discussed earlier.

3.3.4 Fluid Circuit Constraints Modeled by 1-Junctions

Fluid flow is analogous to electric current. A *series* interconnection of components such as pumps, valves, hydraulic motors and reservoirs can be represented by a 1-junction with a bond radiating outward to the appropriate 1-port element for each individual component. The order of these bonds is immaterial.

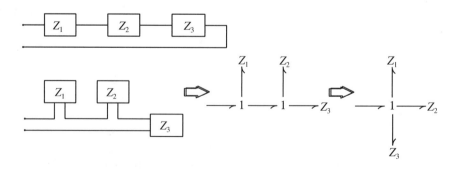

Figure 3.15: Electrical example of a 1-junction: series circuit

Sometimes the effect of an element is small enough to justify its removal from a model. This action can leave a 1-junction with only two bonds attached. In this case, if the two remaining power arrows are directed unilaterally, it is permissible to remove the junction altogether and connect the two dangling bonds.

Example 3.9

Model the hydraulic system pictured below. You may represent the four elements in the bond graph by the words PUMP, VALVE, MOTOR and RESERVOIR. Also, describe how the model can be simplified if the pressure drop across the valve is negligible or the reservoir pressure is virtually atmospheric.

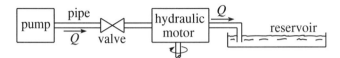

Solution: The first step is to recognize that all four elements have the same flow, Q. Thus, their bonds are joined by a 1-junction:

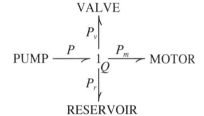

Note that the pressure on the valve bond, P_v, is the pressure *drop* across the valve. The pressure P_m is similarly the pressure *drop* across the motor. The sum of these two pressure drops plus the reservoir pressure equals the pump pressure. This means that the resistances of the components also sum, despite the nonlinearities.

If the valve is wide open, $P_v \simeq 0$, and the power dissipated in the valve is virtually zero. In this case the valve and its bond simply can be removed from the bond graph model. Similarly, the reservoir pressure P_r may be considered to be zero gage, so if atmospheric pressure is employed as the reference (to give gage pressure), the reservoir bond carries no power. In this case, the bond and the reservoir element may be removed similarly. Finally, if only two bonds are left on the junction, the junction itself can be removed and the two bonds simply connected together, assuming the two bonds have power convention half-arrows in the same direction.

3.3.5 Mechanical Constraints Modeled by 0-Junctions

A dashpot is the idealization of a damper with zero mass; it therefore always has the same force at its two ends. Thus, a dashpot can be represented by a three-ported 0-junction and a resistance, as shown in Fig. 3.16. The force is a function of the velocity *difference* $\dot{x}_1 - \dot{x}_2$, which therefore is the flow on the resistance bond.

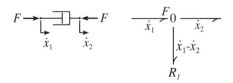

Figure 3.16: Mechanical example of a 0-junction; two-ended dashpot

EXAMPLE 3.10

Show how a cascade of dashpots can be modeled by a bond graph, and how that graph can be simplified by combining the characteristics of the individual dashpots.

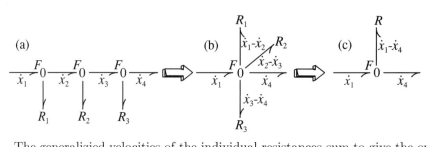

Solution: The bonds of the successive elements can be joined as shown in (a) below, because of the common velocities at the connections. Then, the three 0-junctions can be coalesced as shown in (b), since their bonds all have the same effort (force). Finally, since each dashpot has the same force, the bonds of their respective resistances can be joined together with a 0-junction, as in (c).

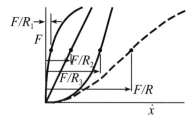

The generalizied velocities of the individual resistances sum to give the overall generalized velocity. Therefore, the dashed line representing the overall resistance characteristic is found by a *horizontal* summation of the individual characteristics:

Mathematically,

$$\dot{x}_1 - \dot{x}_4 = (\dot{x}_1 - \dot{x}_2) + (\dot{x}_2 - \dot{x}_3) + (\dot{x}_3 - \dot{x}_4)$$
$$= \frac{F}{R_1} + \frac{F}{R_2} + \frac{F}{R_3}$$
$$= \frac{F}{R}$$

Thus, the overall *reciprocal* of the resistance, $1/R$, called the overall **conductance**, equals the sum $1/R_1 + 1/R_2 + 1/R_3$ of the individual conductances. The same statement applies if the resistances are nonlinear, as long as the individual resistances are expressed as functions of the common effort e rather than the non-common velocities.

An idealized spring has zero mass, so like a dashpot it has the same force at each end. It therefore can be represented similarly, by a three-ported 0-junction and a compliance, as shown in Fig. 3.17. The deflection which causes the force (or torque, in the case of a rotary spring) is proportional to the *difference* between the displacements at the two ends, $q_1 - q_2$.

Should one end of the spring be motionless, there is no power on the associated bond, which therefore may be erased. Part (d) of Fig. 3.17 shows such a case. The 0-junction

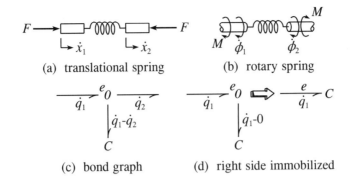

Figure 3.17: Double-ended springs

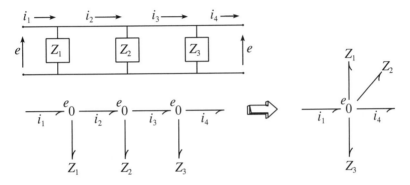

Figure 3.18: Electrical example of a 0-junction: parallel connection

is left with only two bonds, and therefore serves no purpose. It can be removed, and its two bonds joined (assuming the power conventions of the two bonds are compatible). The result is a single bond and a compliance, identical to the case which introduced springs in Section 3.1.

EXAMPLE 3.11

Determine the combination rule for combining a cascade of springs with individual compliances C_1, C_2, C_3 etc. into a single compliance, C.

Solution: Since the forces are common and the relative displacements sum, and a relative displacement is proportional to a compliance for a given force, the compliances sum:

$$C = C_1 + C_2 + C_3 + \cdots.$$

3.3.6 Electric and Fluid Circuit Constraints Modeled by 0-Junctions

A common-voltage current-summing junction is typified by a soldered joint or a parallel combination of elements, as suggested in Fig. 3.18. *Conductances* sum in the same way as they do for mechanical 0-junctions. *Capacitances* sum similarly, satisfying the same equation as the springs of Example 3.10, which applies to any type of compliance. A 0-junction for a fluid circuit is typified by the pipe tee, or interconnection of fluid lines at a point of common pressure. An example is shown in Fig. 3.19. Finally, you need to consider how inertances connected to a common 0-junction can be combined.

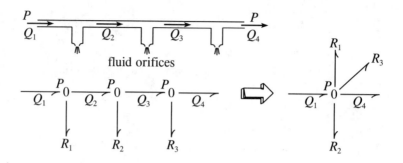

Figure 3.19: Fluid examples of a 0-junction: parallel connection

EXAMPLE 3.12

Determine the combination rule for multiple inertances bonded to a common 0-junction.

Solution: The effort is common to all the inertances, while the flow (and its rate of change) sums. The rate of change for the flow of each inertance is inversely proportional to that inertance. Therefore, the combination rule is

$$\frac{1}{I} = \frac{1}{I_1} + \frac{1}{I_2} + \frac{1}{I_3} + \cdots.$$

This means that the overall inertance is *less* than any individual inertance. Should any of the inertances be zero, for example, the overall inertance is zero also, causing the effort on the 0-junction to be zero.

The "geometric constraints" of mechanical circuits are analogous to the "continuity" or "conservation of mass" constraints of fluid circuits. (The reader aware of the difference between the stagnation and the static pressures should recall that this difference is being neglected for the time being.) Since flows sum, conductances, compliances and reciprocal inertances (called **susceptances**) sum.

3.3.7   Simple IRC Models

Mechanical, electrical and fluid examples of simple IRC systems and bond graph models thereof are given in Fig. 3.20. All of these models include a source, a resistance, a compliance and an inertance arrayed around a common junction. Consideration of more complex models is deferred to Chapter 5, although the R, C and I elements could be the result of combining two or more elements of the same type. Special RC and IR models result from deletion of the I and C elements, respectively.

These examples are the major vehicle used in this chapter to introduce the modeling and analysis of dynamic systems. *You are urged to cover up the bond graphs in the figure and practice modeling these systems yourself.*

3.3.8   Summary

0 and 1-junctions normally are employed as the only means of constructing a branched model, and therefore play a major role in bond-graph modeling. The **junction structure** of a model comprises its bonds, junctions, and (as will be illustrated later) transformers and gyrators. The complete model merely adds one-port elements to this junction structure: sources, resistances and energy storage elements.

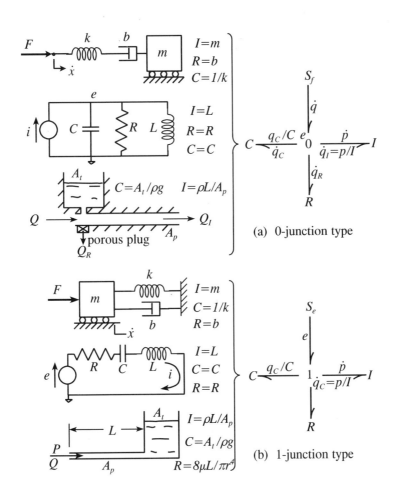

Figure 3.20: Simple IRC models

New modelers often are uncertain which junction to use in a particular situation. This dilemma is reliably resolved only by careful identification of the key variables in the physical system, which is a required step no matter what type of modeling language is used. The bonds on a 0-junction have a common effort and flows that sum. The bonds on a 1-junction have a common flow and efforts that sum. Trying instead to remember which type of junction applies to series connections and which applies to parallel connections usually leads to mistakes, since the answer depends in too complicated a way on whether the system is mechanical, electrical or fluid.

Multiple resistances bonded to a common 1-junction can be combined to give an overall resistance greater than any of its components. Multiple resistances bonded to a common 0-junction combine to give an overall resistance smaller than any of its components. The same statements apply to inertances. On the other hand, two or more compliances bonded to a common 0-junction produce an equivalent overall compliance larger than any of its components. Two or more compliances bonded to a common 1-junction, conversely, produce an equivalent compliance smaller than any of its components. The difference results from the fact that resistance and inertance are proportional to impedance, whereas compliance is proportional to admittance, the reciprocal of impedance.

You are expected at this point to be able to find appropriate bond graph models for simple systems such as those pictured in Fig. 3.20. Extracting the behavior of these models comes next.

Guided Problem 3.4

Needed practice dealing with compliances as well as junctions is provided by this problem. Part (b) should become routine.

Three constant compliances are combined two different ways:

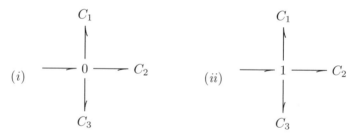

(a) illustrate mechanical, electrical and fluid implementations, and (b) find the compliance of the equivalent integrated compliance, ——▸C, in terms of the moduli C_1, C_2 and C_3.

Suggested Steps:

1. For part (a) focus on the type of variable (e or q) which *sums* and the type of variable which is *common* to all three compliances.

2. For part (b) sum the variables of the appropriate type to get a total value, and compare the result with the desired form.

Guided Problem 3.5

This problem requires the consolidation of distinct inertances into a single inertance.

Water pumped from a river to cool the steam in a power plant passes through the circular channels shown in Fig. 3.21. Estimate the overall inertance relative to the total volume flow.

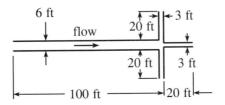

Figure 3.21: Guided Problem 3.5

Suggested Steps:

1. Find the inertances of the four separate channels relative to their individual flows.

2. Combine the three inertances which share a common pressure.

3. The inertance resulting from step 2 shares a common flow with the fourth channel. Combine the inertances to get the overall inertance.

Guided Problem 3.6

This is the first of three short, basic modeling problems which you should be able to do at this point.

A drum viscometer comprises a cylinder with a rotating drum submerged in the fluid being tested, as pictured in Fig. 3.22. The rate at which the drum coasts to a stop is observed. Model the system with a bond graph, and evaluate the moduli of its elements. The mass moment of inertia of the drum is known to be 0.12 in lb s^2, and the viscous drag on it produces the moment

$$M = \frac{2\pi L r^3 \mu}{w}\dot{\phi}$$

where $L = 6$ in, $r = 3$ in, $w = 0.02$ in and $\mu = 1.4 \times 10^{-7}$ lb s/in^2.

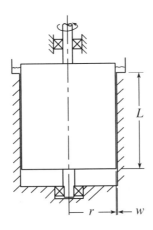

Figure 3.22: Drum viscometer for Guided Problem 3.6

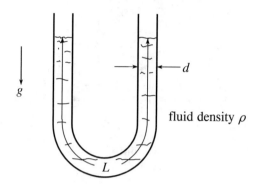

Figure 3.23: U-tube for Guided Problem 3.7

Suggested Steps:

1. Identify the key flow variable, and label this on a 1-junction.

2. Identify the important physical mechanisms that perpetuate and those that dissipate energy. Represent these as bond graph elements appended to the 1-junction.

3. State or find the moduli of the elements of part 2, using the given information.

Guided Problem 3.7

A U-tube is filled with water, as shown in Fig. 3.23. The motion of the water following a non-equilibrium initial condition is of ultimate interest. Model the system with a bond graph, and evaluate the parameters of your model in preparation for an analysis. Neglect the effects of viscosity.

Suggested Steps:

1. Compliance and inertance elements are present. Identify the variable which is common to them, and deduce which kind of junction is proper. Bond the C and I elements to the junction.

2. Express the moduli of the I and C elements in terms of the parameters given in the drawing.

3. You may have more than one energy storage element of the same type in your model. If so, combine these elements into a single element of the same type, determine its modulus, and draw the reduced bond graph.

Guided Problem 3.8

A current source drives an essentially inductive load. Small perturbations in the current source are inevitable, and produce undesirable variations in the voltage. A shunt resistor and capacitor are added to the circuit, as shown in Fig. 3.24, to moderate these pulsations. You are asked to model the system, in preparation for an analysis directed at determining appropriate values for the resistance and capacitance.

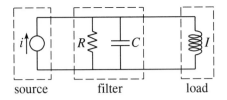

Figure 3.24: Guided Problem 3.8

Suggested Steps:

1. Determine the effort or flow variable that is common to the four elements, and the effort or flow variables that sum. Draw and label a junction accordingly.

2. Bond the four elements to this junction.

PROBLEMS

3.14 An engine drives both a virtually ideal DC generator with constant magnetic field and a mechanical load on the same shaft. Model each of the components by an appropriate bond graph element and assemble a corresponding complete bond graph model.

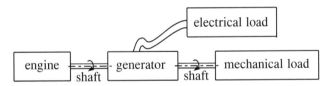

3.15 The bond graphs below can be reduced to a single resistance, ——▶ R'. Find the modulus of R' assuming constant moduli as given.

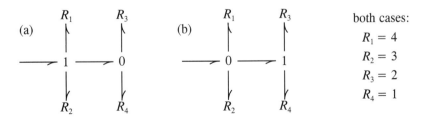

3.16 Draw a bond graph to represent the electric circuit below. Reduce the load seen by the voltage source to a single resistance, R, and find its modulus.

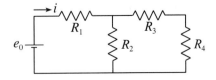

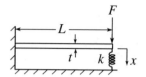

3.17 An applied force F is resisted by the bending of a slender cantilevered beam of length L, thickness t and width w as well as a spring with rate k, as shown.

    (a) Find the moduli of the two component compliances in terms of fixed parameters. (You may use references in your personal library.)

    (b) Represent with a bond graph how the two compliances combine, labeling F and $\dot{x}$ appropriately. (Ask yourself: which of these variables is common to the two compliances, and which is the result of a summation?)

    (c) Find the combined compliance.

3.18 Two linear springs with rates k_1 and k_2 are connected in (a) series (b) parallel. Find the compliance of the combinations.

3.19 Find the capacitance of two capacitors, C_1 and C_2, connected in (a) series (b) parallel.

3.20 A physical damper has significant mass, which can be assumed to be divided equally between its two ends. Give a bond graph model for this damper, and relate the moduli of its elements to physical parameters.

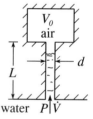

3.21 A vertical tube of diameter d and length L is capped with a closed chamber of volume V_0, as shown. Water enters from the bottom, and air is trapped above. Estimate the effective compliance of the system with respect to the entering water; the following steps are suggested.

    (a) Evaluate separately the compliances due to gravity and the compression of the air. Assume that the water never enters the upper chamber, and that the compression of the air is (i) isothermal (*i.e.* slow) (ii) adiabatic (*i.e.* fast).

    (b) Represent how the compliances combine with a bond graph, labeling P and $\dot{V}$ appropriately. (Ask yourself: which of these variables is common to the two compliances, and which is the result of a summation?)

    (c) Find the overall compliance.

3.22 A member with negligible mass rotates within a viscous bearing that offers a resistance to motion of 10 N per rad/s. The member also is attached to one end of a rotary spring with stiffness 15 N·m/rad; the other end of the spring is attached to ground. The member is excited by an applied torque that can be either clockwise (positive) or counter-clockwise. Model the system with a bond graph in preparation for a dynamic analysis. *Continued in Problems 4.56, 7.8 and 7.45.*

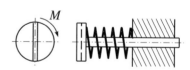

3.23 A disk with radius $r = 0.05$ m and mass moment of inertia $I = 0.10$ kg $\cdot$ m^2 rotates at angular velocity $\dot{\phi}$ about a journal bearing with resistive torque equal to $b\dot{\phi}$, where $b = 0.02$ N $\cdot$ m $\cdot$ s. A cable wrapped around the the cylinder has tension force F. Model the system with a bond graph, and give the values of the parameters. *Continued in Problem 4.21.*

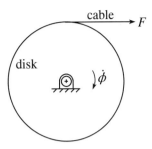

3.24 Define variables and model the electric circuits below with bond graphs. *Continued in Problems 3.40 and 4.17.*

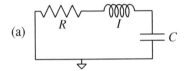 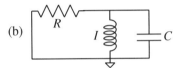

3.25 Define variables and model the mechanical circuits below with bond graphs. *Continued in Problems 3.41 and 4.18.*

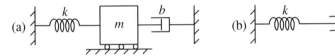 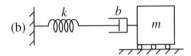

3.26 Model the tank and tube shown below left, neglecting the inertance of the fluid. Specify the moduli of your bond graph elements. *Continued in Problems 3.42, 3.59 and 4.51.*

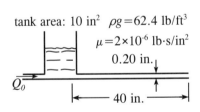 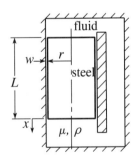

3.27 A steel plunger ($\rho_s g = 0.283$ lb/in^3) is released from rest in a fluid-filled cylinder, as shown above right. The fluid can flow unimpeded through an auxiliary channel. Model the system in preparation for an analysis to determine its motion. Neglect fluid inertia, and assume the plunger remains centered. Specify the values of the moduli of your bond graph elements. *Continued in Problems 3.43 and 4.52.*

3.28 The lower end of an open-ended vertical circular tube with a known inner radius a is located a known distance L below the surface of liquid of known density ρ and viscosity μ in a large tank. The liquid level within the tube is nominally the same as in the tank (the surface tension force is negligible) but is given an initial condition slightly below this level. The level subsequently is seen to oscillate with a decaying amplitude. Model the system with a bond graph, and relate the moduli of the elements to know parameters. The resistance of a tube (ratio of pressure drop to volume flow in steady flow) due to viscosity is approximately $R = 8\mu L/\pi a^4$ (book equation 2.14). *Continued in Problems 3.44 and 4.57.*

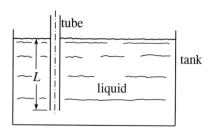

3.29 A water tank of area 2.0 m^2 drains through a porous plug in its bottom with a fluid resistance of 98.7 kN s/m^5. It also drains through a tube of length 10 m and diameter 2.0 cm. Model the system with a bond graph, and evaluate its parameters. Neglect the effect of viscosity. *Continued in Problems 3.45 and 4.53.*

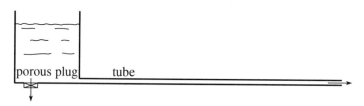

3.30 Often it is impractical to reduce the vibrational forces of a machine to an acceptable level, but it is practical to isolate the machine from its foundation, and thereby prevent large oscillations from shaking the floor, creating noise, etc. A standard scheme, valid for y-motion only, is shown below. Model this system in preparation for an analysis which will be used to choose the springs and the dashpot. If your model has separate elements to represent the separate springs, combine these elements and give the modulus of the new element in terms of k. *Continued in Problems 3.46 and 4.54.*

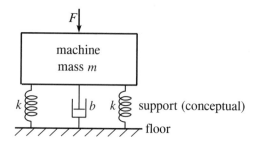

3.31 Presume that the *LRC* filter shown below drives an instrument with a virtually infinite input impedance. The relation between the voltages e_i and e_0 is of interest, since the latter is intended to be a smoothed but not smothered version of the former. Model the system with a bond graph in preparation for an analysis. *Continued in Problems 3.47 and 4.55.*

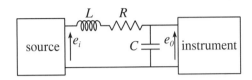

3.32 A slender rod of length $L = 2$ ft and weight $W = 4$ lb is held upright by two springs, each with rate $k = 1$ lb/ft.

(a) Model the system with a bond graph, labeling any key velocity or displacement variable.

(b) Evaluate the moduli or the relevant bond graph elements, valid for small motions about the equilibrium only. (Make sure the units are included.)

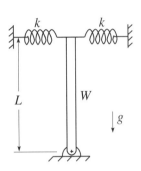

Continued in Problem 4.58.

SOLUTIONS TO GUIDED PROBLEMS

Guided Problem 3.4

(a) (i) common effort   mechanical:

         electrical:

         fluid:

(ii) common displacement   mechanical:

         electrical:

         fluid:

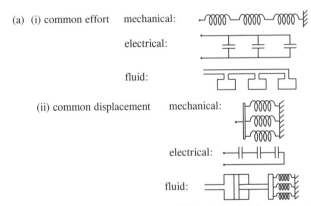

(b) The moduli need not be constants if chordal definitions are used.

(i) $q = \Sigma q_i = \Sigma C_i e \quad \Longrightarrow \quad C = C_1 + C_2 + C_3$

(ii) $e = \Sigma e_i = \Sigma (1/C_i) q \quad \Longrightarrow \quad 1/C = 1/C_1 + 1/C_2 + 1/C_3$

Guided Problem 3.5

1. Large channel: $I_L = \dfrac{\rho L}{A} = \dfrac{4\rho L}{\pi d^2} = \dfrac{4 \times 1.94 \times 100}{\pi (6)^2} = 6.86$ lb s^2/ft^5.

     Each small channel: $I_s = \dfrac{4 \times 1.94 \times 20}{\pi (3)^2} = 5.49$ lb s^2/ft^5.

2. Combination of three small channels: $I = \dfrac{1}{3} 5.49 = 1.83$ lb s^2/ft^5.

3. Combination of all four channels: $I = 6.86 + 1.83 = 8.69$ lb s^2/ft^5.

Guided Problem 3.6

1,2. The key flow variable is the angular velocity of the drum, $\dot{\phi}$. There is kinetic energy $\frac{1}{2}I\dot{\phi}^2$ associated with this velocity, where I is the mass moment of inertia of the drum. Energy dissipation takes place because of the viscous drag in the narrow annulus. The shear force is a function of angular velocity, so this gives a resistance, R. The bond graph that represents these observations is

$$I \longleftarrow 1 \underset{\phi}{\overset{M}{\longrightarrow}} R$$

3. The mass moment of inertia is the integral $I = \int r^2\, dm = 0.12$ in lb s^2. The resistance is the ratio of the torque M due to viscous drag to the angular velocity $\dot{\phi}$, or $R = 2\pi L r^3 \mu / w = (2\pi \times 6 \times 3^3 \times 1.4 \times 10^{-7}/0.02) = 0.00713$ in·lb·s.

Guided Problem 3.7

1. The potential or compliance energy is a function of the displacement of the water, and the kinetic or inertance energy is a function of the velocity of the water. One could choose as the generalized velocity either the common vertical velocity of the surfaces of the water or the volume flow of the water. The associated generalized displacements are the common vertical displacements of the surfaces of the water and the volume displacement of the water, respectively. The latter choice is made here, with symbols $Q = \dot{V}$ and V, since equations (3.12) (p. 80) give the compliances and Example 3.3 (p. 88) gives the inertance with reference to these variables. Note there are *two* compliances, one for each free surface, as shown in the bond graph.

2. From equation (3.12), each compliance is $C = A/\rho g = \pi d^2/4\rho g$. From Example 3.3, the inertance is $I = \rho L/A = 4\rho L/\pi d^2$.

3. The two compliances share a common flow or displacement, so their efforts or pressures sum. This means that the reciprocal of the combined compliance equals the sum of the reciprocals of the individual compliances, so that the combined compliance is $C' = C/2 = \pi d^2/8\rho g$. Note that half the compliance means twice the stiffness; the two stiffnesses sum.

Guided Problem 3.8

1,2. All four elements share the same voltage difference. Therefore, all four associated bond graph elements share a common 0-junction. The common voltage at the bottom ends of the elements is taken as ground, or zero.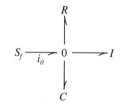

3.4   Causality and Differential Equations

The dynamic behavior of a well-posed model with energy storage elements is found by expressing it as a differential equation or set of equations, and solving the equations numerically — called simulation — or analytically. The procedures are largely deterministic, unlike in modeling, requiring care but no special insight. They can be expedited by software. This section concerns the determination of the differential equations; later sections deal with simulation and analytical solutions.

It is possible to define all the efforts and flows in a bond graph with symbols, and write one equation for each one-port element, two equations for each transformer and gyrator, and

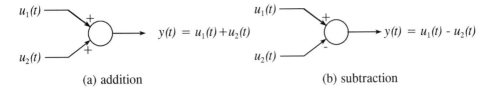

Figure 3.25: Block diagrams with summation

one equation for the variables which sum about each junction. This usually large number of equations then can be combined, eliminating the variables of minor interest, to give a solvable set of differential equations. This procedure is excessively complex in practice, however, unless an effective over-arching plan is applied. The same complication occurs if the modeling is done by other than bond-graph means.

One first-order differential equation is wanted for each independent energy storage element. This **state-space formulation** is suited ideally for both numerical simulation techniques and, in the case of linear models, analytical solution. The ideal plan for deducing these equations from the primitive model also ought to be so automatic that it can be programmed into software. Sophisticated bond graph processors are indeed available[3], and can be very efficient for practicing engineers. Users of such software do not have to write the differential equations themselves or even find their solutions, which are directly plotted. This rapidly evolving software is not discussed further in this book, however, in the belief that students need first to develop a basic understanding of the analytic procedures and the use of more general-purpose and widely used software, and to avoid restricting the choices of practicing engineers. Further, not all bond graphs given in this book are treated by this software.

3.4.1 Operational Block Diagrams

Operational block diagrams, usually just called **block diagrams**, are used universally to represent automatic control systems, and also are used widely to represent any dynamic system or system of differential equations. They expedite the assembly of a set of primitive equations into a solvable set of differential equations. They are treated automatically by widely available software introduced in Section 3.6.3. They are presented here largely as a bridge to the application of what is called **causality** to bonds graphs. They also are used in their own right in Chapter 8, which introduces automatic control. They are further related intimately with the signal flow diagrams of Section 6.3.

Block diagrams contain small circles, triangles, rectangular blocks, and interconnecting straight lines with arrowheads. Two or more **input signals** or **variables** are directed into a circle by arrows, as depicted in Fig. 3.25. The single *output variable* in each case is the sum or difference of the input variables. Plus and/or minus signs are shown alongside the circles, or sometimes inside them. These signs correspond to the signs in the corresponding equations, as shown. Note that the input variables are the terms on the right sides of the equations, known mathematically as the independent variables, while the output variables

---

[3]Controllab Products B.V. (Netherlands) has a program called 20-sim; information can be found at http://www.rt.el.utwente.nl/20sim/clp.htm. Lorenz Simulation SA (Belgium) has a package called MS1; information can be found at http://www.lorsim.be. Cadsim Engineering (California) has a product called CAMP-G; information is available at http://www.bondgraph.com. A. Mukherjee initiated a major family of software based on bond graphs called SYMBOLS 2000; information is available at www.symbols2000.com/home/htm. IMAGINE SA has a related type of software based on bond graph submodels depicted as icons, called AMESim; information is available at www.amesim.com.

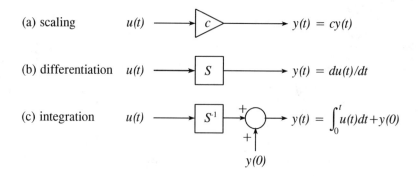

Figure 3.26: Use of triangles and blocks

are the terms on the left sides of the equations, known as the dependent variables. Some block diagrams also employ circles to represent multiplication or division of the input variables, as indicated by a multiplication or division symbol written inside them, but these nonlinear operations are not used here.

A single input variable can be directed into a triangle or block, producing a single output variable, as illustrated in Fig. 3.26. Case (a) represents the input variable being multiplied by a constant to produce the output variable. The constant is written inside a triangle inspired by the traditional symbol for an amplifier. Case (b) has an S written inside of a block. This special symbol is reserved not for a constant, but to designate *time differentiation*; the output is the time derivative of the input. It is called the **differential operator**, and in general the contents of a block are said to *operate* on the input to produce the output. These contents can be rather elaborate, but only the primitive cases shown here are needed for now.

In case (c) the input variable is time-integrated to produce the ouput variable. The symbol S^{-1}, or $1/S$, the inverse of the differential operator, is known as the **integration operator**. The integration starts at time 0, so note how a non-zero value of the output variable at that time (the constant of integration) can be introduced through the use of a circle or summation.

Individual equations for linear models of systems can be represented by block diagrams using the elements above. Then, the separate diagrams can be assembled together by recognizing that some of the inputs to each diagram are the outputs to other diagrams. The assembled block diagram can be used to find an overall relation between those inputs that are not the output of any component diagram and any other variable. This includes the writing of individual first-order differential equations or equations of higher order for the whole system or a larger part of it. Also, the block diagram can be inserted into a block diagram software package which can do the work.

3.4.2   Causal Bond Graphs

Bond graphs can be annotated to serve like block diagrams. This gives them double-duty as both models closely related to the physics and computational diagrams that give state differential equations, which is a great advantage. Each bond has of course two variables, and effort and a flow. Consider first an effort source. The diagram on the left below comprises the bond graph symbol for an effort source plus superimposed full arrows, annotated with e and $\dot{q}$. These arrows represent the corresponding block diagram. The arrow for the effort is pointed away from the source, because the source uniquely specifies or produces the effort; it is its *cause*. The arrow for the flow is pointed toward the source, because it is a consequence of whatever the source is bonded to (unshown) on the right-hand end of the bond. The opposite directedness of the two arrows is called *bilateral causality*.

$$S_e \xrightarrow{\quad e \rightarrow \quad}_{\leftarrow \dot{q}} \qquad \text{or} \qquad S_e \xrightarrow{\quad e \quad}|\; \dot{q}$$

The diagram on the right has the same combined meaning of bond graph plus block diagram. The vertical **causal stroke** is shorthand for effort directed toward the stroke and flow directed away from it.

Similarly, the flow source S_f specifies the flow $\dot{q}$; the (unshown) system to the right. This situation can be depicted by either the combined bond graph plus block diagram on the left below, or the bond graph with an added causal stroke on the right below:

$$S_f \xrightarrow[\dot{q} \rightarrow]{\leftarrow e} \qquad \text{or} \qquad S_f \;|\!\xrightarrow{\quad e \quad}_{\dot{q}}$$

Note that causal stroke is on the opposite end of the bond than with the effort source.

The key idea now is that *every bond has bilateral causality*, which can be represented by a causal stroke. These strokes are recognized internationally and are always drawn perpendicular to the bond at one of its ends:

$$\xrightarrow[\dot{q} \rightarrow]{\leftarrow e} \quad \Longrightarrow \quad |\!\!\text{———} \qquad\qquad \xrightarrow[\leftarrow \dot{q}]{e \rightarrow} \quad \Longrightarrow \quad \text{———}\!|$$

The causal orientation is completely independent of the power convention orientation, which is indicated by the half-arrow on the bond, which could be directed either way. The power convention arrows are directly associated with the physics and the definitions of the variables. Causality, on the other hand, does not always represent physical necessity, apart from the effort and flow sources, but generally indicates a computational scheme. Sometimes there are choices, and not all permitted choices have the same utility.

The use of block diagrams to represent physical systems has the same arbitrariness. One can readily construct a block diagram from a causal bond graph, but one cannot generally construct a bond graph from a block diagram. Block diagrams, in fact, do not necessarily represent physical systems with energy, etc. at all, whereas bond graphs are constrained to do so (unless one or more bonds are *activated*, as described in Section 9.2).

A resistance element with constant modulus R can be represented by the impedance relation $e = R\dot{q}$ or the admittance relation $\dot{q} = (1/R)e$. Bond graphs and equivalent block diagrams for these two cases are given in part (a) of Fig. 3.27. While the structures of the two block diagrams are different, the only difference for the bond graphs is the location of the causal stroke. A compliance with constant C is shown in part (b) of the figure. The first case shows integral causality, which happens also to be impedance causality (flow input). The second case shows differential causality, which happens also to be admittance causality (effort input). The difference regards practical computation, not the physical meaning. (Note that the constant of integration is missing or taken as zero; it could be added, as in Fig. 3.26.) An inertance with constant I is shown in part (c) of the figure. Note that integral causality also is admittance causality, the opposite of the compliance.

Cases with transformers are shown in part (d) of the figure. Note that each case has two primitive block diagrams, because they represent two equations. Both causal strokes must be placed on either the right or left ends of their respective bonds, because of their definitions; thus if you know one, you can directly place the other. Cases with gyrators are shown in part (e). Here, the two causal strokes must be placed symmetrically with respect to the element, giving either impedance or admittance causality. Like the transformer, therefore, if you know the location of one of the causal strokes you can place the other.

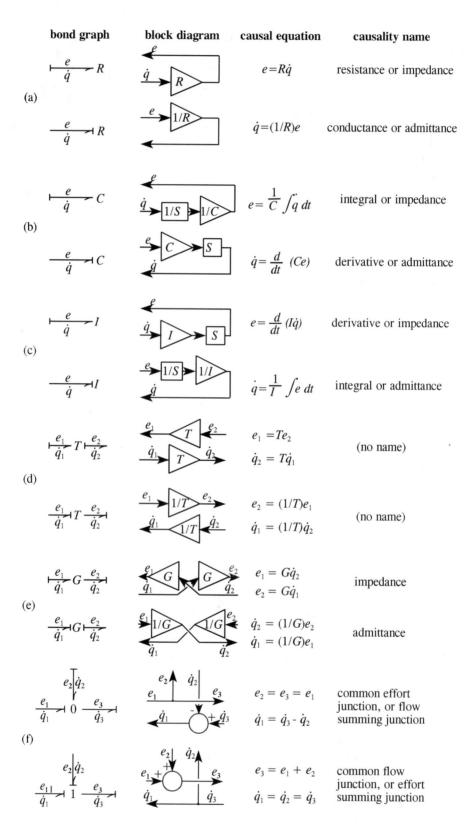

Figure 3.27: Linear bond graph and block diagram elements

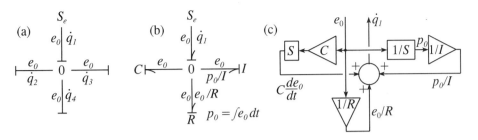

Figure 3.28: Effort and flow sources bonded to a 0-junction to give uncoupled behavior

Note in general that the orientation of any bond in a bond graph or any line in a block diagram is arbitrary; only the interconnectedness matters.

Examples with junctions are shown in part (f) of Fig. 3.27. All bonds on the 0-junction have the same effort, by its definition. This effort is specified by only one outside cause, or bond. In the case shown, this is chosen to be e_1. Also, the flows sum, with appropriate signs that depend on the power convention arrows. Carefully study the ways the two types of diagrams represent these facts. There could be more bonds and inputs. Regardless of the number, precisely one causal stroke must appear adjacent to the junction. All bonds on the 1-junction, on the other hand, have the same flow. Therefore, all its causal stokes except one must appear adjacent to the junction. Next, you will see how to take advantage of these rules.

3.4.3 Junctions with Elements Having Uncoupled Behavior

An effort source is bonded to a 0-junction in part (a) of Fig. 3.28. Since the effort, e_0, is specified by the effort source, and all bonds on the 0-junction have the same effort, placement of the causal strokes on the other three bonds are mandated as shown. The causal stroke on the effort source bond is said to *propagate* automatically to the other bonds. The flows $\dot{q}_2$, $\dot{q}_3$ and $\dot{q}_4$, on the other hand, must be caused separately by whatever (unshown) elements are attached to the outer ends of these bonds. The flow $\dot{q}_1$, then, must be the sum of $\dot{q}_2$, $\dot{q}_3$ and $\dot{q}_4$, with signs determined by whatever power convention half-arrows are part of the model.

The three bonds are now attached, in part (b) of the figure, to a C, an I and an R element, respectively. The flow on the R bond then becomes e_0/R, the flow on the C bond becomes $C\, de_0/dt$, and the flow on the I bond becomes $\int e_0\, dt/I = p_0/I$. There is no choice; all three elements have imposed *admittance causality*. This means that their flows respond to an induced effort and are expressed as functions of that effort or its time integral, p_0. The behavior of the three elements is said to be *uncoupled*, which means that the flow on each of them is independent of the size or even the presence of the others. The flow on the source bond completes the picture; it is the sum $\dot{q}_1 = e_0/R + C\, de_0/dt + p_0/I$. Note that the C element has differential causality, since its causal output is a function of the time derivative of its causal input, and the I element has integral causality, since its causal output is a function of the time integral of its causal input.

The block diagram perspective of the situation, given in part (c) of the figure, may help you understand the bond graph and its causal strokes. It was constructed with help from the bond graph, but with a little more effort it could have been drawn directly from the constitutive equations of the elements.

EXAMPLE 3.13

By analogy to the example above, explain why the causal strokes in the bond graph below left are drawn where they are. Then explain the annotations for the efforts and flows given on the bond graph below right, including giving appropriate names to the causalities of the R, C and I elements. Are the behaviors of these elements coupled or uncoupled?

$$S_f$$
$$e_1 | \dot{q}_0$$

$$\frac{e_2}{\dot{q}_0} \dashv 1 \vdash \frac{e_3}{\dot{q}_0}$$

$$e_4 | \dot{q}_0$$

$$S_f$$
$$e_1 | \dot{q}_0 \qquad I \frac{d\dot{q}_0}{dt}$$

$$C \xleftarrow{q_0/C} \dashv 1 \vdash \xrightarrow{\dot{q}_0} I$$

$$R\dot{q}_0 | \dot{q}_0$$
$$R$$

Solution: A flow source is bonded to a 1-junction. All the bonds and the elements are forced to have, in common, the flow $\dot{q}_0$ imposed by the source. This fact is underscored by the placements of the causal strokes on the bonds. In general, all the bonds on *any* 1-junction must have their causal strokes at their junction ends, except for precisely one which has its causal stroke at the remote end. Similarly, all the bonds on any 0-junction must have their causal strokes at the remote end, except for precisely one which has its causal stroke adjacent to the junction. You will find that these facts make causal strokes very useful.

Three of the bonds are attached to C, I and R elements with the respective efforts expressed as $R\dot{q}_0$, $(1/C)\int \dot{q}_0\,dt = q_0/C$, $I\,d\dot{q}_0/dt$ and $R\dot{q}_0 + q_0/C + I\,d\dot{q}_0/dt$. These elements are said to have *impedance causality*, since their efforts are functions of the flow or its integral or its derivative. They are said to be uncoupled, as in the 0-junction example, because each effort is independent of the presence of the other two elements. The C element also is said to have integral causality, and the I element is said to have differential causality.

3.4.4   Junctions with Elements Having Coupled Behavior

The system represented in Fig. 3.29 differs from that in Fig. 3.28 in that the source does not directly determine either the efforts or the flows on any bonds other than the source bond itself. One of the bonds for the R, C and I elements must have its causal stroke placed adjacent to the junction; since there is only one effort, the other two causal strokes must be placed at the outer ends of their bonds. But which of the three possible patterns do you use?

The short answer to this question is that, given a choice, you should use integral causality, regardless of whether you are going to work by hand or use either bond-graph oriented or block-diagram-oriented software. The reasons should become clear later. For the present model, integral causality is applied to both the C and I elements in part (b) of Fig. 3.28. This forces impedance causality on the R element.

The next thing you do in finding the appropriate differential equations for the system is to annotate the effort and flow sides of the C and I bonds as follows, depending in general on the directions of the power convention half-arrows:

$$\vdash \frac{q_i/C_i}{(\dot{q}_i)} \rightarrow C_i \qquad \text{or} \qquad \vdash \frac{-q_i/C_i}{(\dot{q}_i)} \leftarrow C_i$$

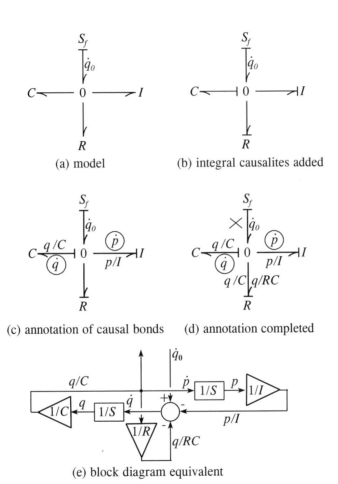

(a) model

(b) integral causalites added

(c) annotation of causal bonds

(d) annotation completed

(e) block diagram equivalent

Figure 3.29: Flow source bonded to a 0-junction to give coupled behavior

$$\frac{\overline{\dot{p}_i}}{p_i/I_i} \dashv I_i \quad \text{or} \quad \frac{\overline{\dot{p}_i}}{-p_i/I_i} \dashv I_i$$

The author has found that it helps students if they draw circles around the $\dot{q}_i$ and $\dot{p}_i$. Although this is not standard bond graph procedure, it is done throughout this book for reasons explained later. Note that $\dot{p}_i$ is a way of writing e_i. Note also how the sign changes compensate for the reversal in the power convention half-arrow. These annotations are made in part (c) of Fig. 3.28.

The next step is to annotate the efforts and flows for the remaining bonds, as shown in part (d) of the figure. The causal stroke on the C bond, combined with its effort, dictates that the effort on the R bond be q/C. The R-element responds, with its admittance causality, by giving the flow q/CR.

The only remaining unannotated effort or flow is the effort on the source bond. This is strictly a causally output variable, and need not be labeled unless it is of interest. Here, it is simply labeled with a large cross, signifying disinterest.

The equivalent block diagram is given in part (e) of the figure. Since integral causality is used, blocks with the integration operator $1/S$ are used. The input signal when this operator refers to the inertance is labeled as the effort $\dot{p}$, and when this operator refers to the compliance is labeled as the flow $\dot{q}$.

EXAMPLE 3.14

Apply causal strokes to the bond graph below, which substitues an effort source for the flow source in Fig. 3.28, and substitues a 1-junction for the 0-junction. If possible, apply integral causality to both energy-storage elements. Then, annotate all the efforts and flows except for the flow on the source bond, which may be labeled with a large X to signify disinterest.

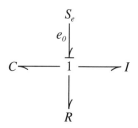

Solution: All but one of the four bonds can and must have its causal stroke placed adjacent to the 1-junction, since all the flows are the same and its specification should come only from one place. Since this reasoning allows a choice, you can choose integral causality for the C bond, and then integral causality for the I bond. The R bond is forced to have impedance causality.

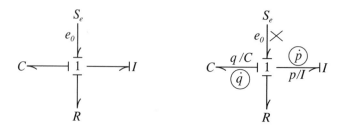

The causal strokes are shown above left. Next, you place the designations for the effort and flow on the bonds that have integral causality, as shown above right. Finally, to complete the annotation of all the efforts and flows in the graph, you observe that the causal input to the R element is its flow, and this flow comes, causally, from the I element, and equals p/I. The effort on the R element is the causal output of the element, and is written as R times the flow. The final result is shown at the right.

3.4.5 Writing Differential Equations

Once the bonds in a bond graph have all their efforts and flows annotated according to the the rules given above, or the equivalent block diagram has been drawn and annotated, the writing of the associated differential equations is a simple matter. There is one differential equation for each C or I element with integral causality. The number of such elements or equations is called the **order** of the model. The left side of each respective equation comprises the term which is circled in the annotation of its bond, or is the input to an integration block in a block diagram. The right side comprises what this term is equal to, as dictated by either the bond graph or the block diagram with their annotations.

For the 0-junction case of Fig. 3.29, the flow $\dot{q} = dq/dt$ is a causal output of the junction, and equals the sum of the causal inputs $\dot{q}_0$ (with plus sign due to its power convention half-arrow), p/I (with minus sign due to its power convention half-arrow), and q/CR (with minus sign). The rate of change of momentum $\dot{p} = dp/dt$ is another causal output of the junction, which because there is only one effort on the junction equals the causal input, q/C. The same relations can be seen directly from the block diagram. Thus, the equations become

$$0 - \text{junction case}: \qquad \frac{dq}{dt} = \dot{q}_0 - \frac{1}{I}p - \frac{1}{RC}q, \qquad (3.32a)$$

$$\frac{dp}{dt} = \frac{1}{C}q. \qquad (3.32b)$$

EXAMPLE 3.15

Write the state differential equations for the bond graph of Example 3.14, using the method described and illustrated above.

Solution: The effort $\dot{p} = dp/dt$ is a causal output of the junction, and equals the sum of the causal inputs e_0 (with plus sign), q/C (with minus sign) and Rp/I (with minus sign). The rate of change of flow $\dot{q} = dq/dt$ is another causal output of the junction, which because there is only one flow on the junction equals the causal input, q/C. The equations therefore are

$$1 - \text{junction case}: \qquad \frac{dp}{dt} = e_0 - \frac{1}{C}q - \frac{R}{I}p,$$

$$\frac{dq}{dt} = \frac{1}{I}p.$$

The differential equations above represent the mathematical models of their systems. The variables p and q are called the **state variables**, since their derivatives are the only

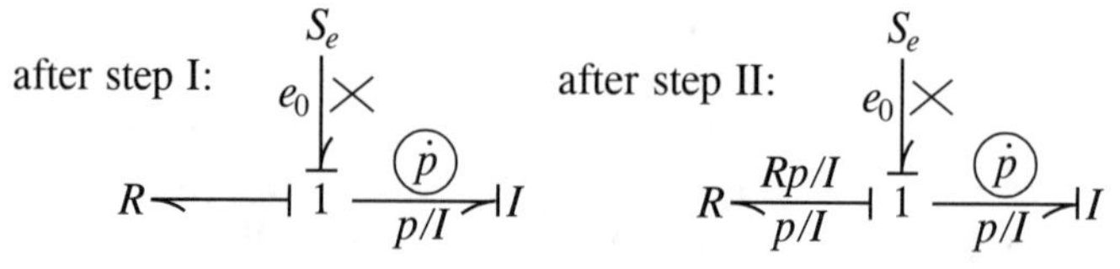

(a) IR model

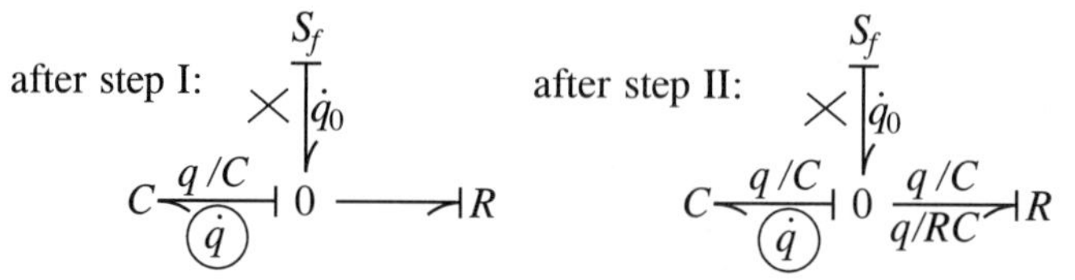

(b) RC model

Figure 3.30: First-order models

dependent variables in the equations. The only other variables are the independent or causal input variables $\dot{q}_0$ and e_0, which could be either constants or specified functions of time. The only other letters in the equations represent parameters, or constants: R, C and I. Each case has therefore two equations and two unknowns, and the pairs of equations are solvable. Numerical solutions are considered in Sections 3.6 and 3.7, and analytical solutions in Chapter 4.

Should you be interested in some variable other than the state variables, you first solve for the state variables, and then express the variable of interest in terms of them. For example, if you are interested in the flow on the 1-junction, you use $\dot{q} = p/I$; should you want the effort on the zero junction, you use $e = q/C$. The annotated bond graph provides adequate information.

Do you get differential equations for the uncoupled cases of Fig. 3.25 and Example 3.13? The answer is *yes*, you get one for the single instance of integral causality in each case. For the 0-junction case you get

$$\frac{dq}{dt} = \dot{q}_0, \tag{3.33}$$

and for the 1-junction case you get

$$\frac{dp}{dt} = e_0. \tag{3.34}$$

The fact of uncoupled behavior vastly simplifies these cases; they can be considered to be degenerate. Most problems of interest do not involve such uncoupled behavior.

Should one of the C or I elements in Fig. 3.26 or Example 3.14-3.15 be zero, or missing, the number of differential equations reduces to one, and the model is said to be of first-order. First-order models without coupling between the C or I element and the R element are presented in Fig. 3.30. The methodology is the same, and so are the results, except for the missing terms. Models often are called by the elements they contain: RC models, IR models and ICR models. This description is not unique, however, unless you specify whether or not their elements are coupled, or which junction type joins them.

The RC model represented in Fig. 3.29 gives, on inspection, the single differential equation

$$\frac{dq}{dt} = \dot{q}_0 - \frac{1}{RC}q, \tag{3.35}$$

which represents the only variable that is encircled.

EXAMPLE 3.16

Find the single differential equation for the IR model given in Fig. 3.30.

Solution: The only variable that is encircled is $\dot{p}$. This is placed on the left side of the equation, and terms on the right side are dictated by the efforts annotated on the other bonds, which according to the causal strokes are causal inputs to the junction. Their signs are dictated by the power convention arrows. The result is

$$\frac{dp}{dt} = e_0 - \frac{R}{I}p.$$

The concept of causality is particularly crucial in the writing of differential equations when more complex models are considered, including the use of transformers, gyrators and multiple junctions. At that point (Chapter 6) the logical underpinning of the procedures introduced here should become more apparent.

3.4.6 Summary

Each bond in a bond graph may be assigned a bilateral causality, as indicated by a transverse mark called a causal stroke. The effort on the bond is causal in one direction, and the flow in the other. In the equivalent block diagram the two signals are represented by two lines with oppositely directed arrows. This causality refers only to computation, that is the determination of independent and dependent variables, and may or may not be related to physical causes and consequences.

The use of integral causality for the energy storage elements — the C's and I's — expedites the writing of the state differential equations for a system. The causal inputs to these elements equal the time derivatives of the state variables, that is the generalized velocity dq_i/dt for each C element and the generalized force dp_i/dt for each I element. Each causal output becomes a function of its state variable, specifically q_i/C_i or p_i/I_i. The causal strokes on the other bonds in a system usually allows these state functions and any input variables to be propagated to the other bonds in the model; each effort and flow thus can be indicated as a function of the state variables and/or the input variables. The same thing happens in block diagrams through directed paths. (Exceptions will be noted in Chapter 6.) Usually it is desirable to annotate all the bonds or all the block diagram signals before attempting to write the differential equations. You should *never* annotate an effort or flow in violation of the causal strokes, even if what you would write is functionally correct; such an action frustrates the purpose of the causal method.

One differential equation of first order results from equating the derivative dp_i/dt or dq_i/dt, for each energy storage element with integral causality, to the causal input of its bond or the signal input to the integration block. The signs are determined by the power convention half-arrows or the signs for the summation circles in the block diagram. The number of first-order differential equations that model an entire system therefore equals the number of energy storage elements with integral causality and is called the order of the model.

Differential causality sometimes is forced. Its treatment for uncoupled models has been addressed. Its treatment for other cases is deferred.

The use and meaning of the causal method is expanded upon in Chapter 6. The present applications involve models with a single junction, which is a very special case.

Figure 3.31: Guided Problem 3.9

Guided Problem 3.9

Solving this problem and the next can be considered simple and essential first steps in your writing of differential equations from bond graph models.

A simple RC model is excited by an indepedent effort source, as shown in Fig. 3.31. Define the state variables, and write a corresponding first-order differential equation(s).

Suggested Steps:

1. Apply the mandatory causal stoke to the effort source bond, and then (since it is allowed) apply integral causality to the energy-storage element.

2. Annotate the efforts and flows on the source bond and the bond with integral causality.

3. Annotate the effort and flow on the remaining bond according to the mandates of the causal strokes.

4. (optional) Draw the corresponding block diagram, labeling the derivative of the state variable and the signs around the summation circle.

5. Write the differential equation by equating the circled term on the bond graph to the sum of terms dictated by the causal strokes, or noting from the block diagram what the derivative of the state variable equals.

Guided Problem 3.10

Write differential equation(s) for the U-tube modeled in Guided Problem 3.7 (pp. 106, 112).

Suggested Steps:

1. Apply the causal strokes, using integral causality if possible. Identify the order of the model.

2. Annotate the energy storage bonds in the standard way, *or* draw the block diagram, including annotating the derivatives of the state variables.

3. Complete the annotation of any other bonds or block diagram lines, if necessary.

4. Write the differential equations by equating the circled terms to whatever the bond graph dictates, *or* employing the block diagram.

PROBLEMS

3.33 Complete the causal strokes on each of the bond graphs below, in so far as they are mandated or represent integral causality. Leave blank otherwise. Whenever integral causality occurs, write and circle the appropriate variable $\dot{p}$ or $\dot{q}$.

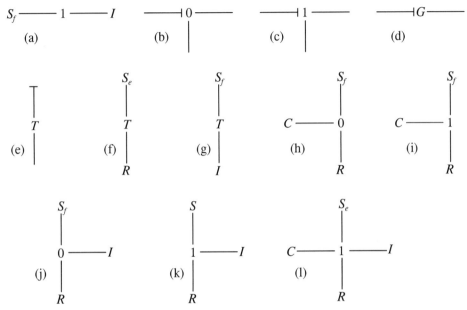

3.34 Determine whether or not the behaviors of the R and I elements shown below left are coupled. Determine their behaviors for $\dot{q}_0$ = constant.

3.35 Determine whether or not the behaviors of the R and C elements shown above right are coupled. Determine their behaviors for e_0 = constant.

3.36 Draw a block diagram for the system of Example 3.13 (p. 118), and state why this diagram reveals that the system is coupled or uncoupled.

3.37 Repeat the above problem for the system of Example 3.14 (p. 120).

3.38 I and C elements are bonded to a common junction which is excited by an effort or flow source. The behaviors or the two energy storage elements are uncoupled. Identify the possible combinations of junction and source types, and describe mathematically the relations between the behaviors of the sources and the I and C elements.

3.39 Define state variable(s) and write the corresponding differential equations for the model represented by the bond graph opposite.

3.40 Write appropriate differential equation(s) for the electric circuits of Problem 3.24 (p. 109). Use both causal bond graph and block diagram methods. *Continued in Problem 4.17.*

3.41 Write appropriate differential equation(s) for the mechanical circuits of Problem 3.25 (p. 109). Use both causal bond graph and block diagram methods. *Continued in Problem 4.18.*

3.42 Write appropriate differential equation(s) for the fluid tank and pipe of Problem 3.26 (p. 109). *Continued in Problems 3.59 and 4.51.*

3.43 Write appropriate differential equation(s) for the plunger of Problem 3.27 (p. 109). Choose your method. *Continued in Problem 4.52.*

3.44 Write appropriate differential equation(s) for the tank-and-tube of Problem 3.28 (p. 110). *Continued in Problem 4.57.*

3.45 Write appropriate differential equation(s) for the fluid system of Problem 3.29 (p. 110). Choose your method. *Continued in Problem 4.53.*

3.46 Write appropriate differential equation(s) for the vibration isolator of Problem 3.30 (p. 110). Use the bond graph only. *Continued in Problem 4.54.*

3.47 Write appropriate differential equation(s) for the filter circuit of Problem 3.31 (p. 111). Use the bond graph only. *Continued in Problem 4.55.*

SOLUTIONS TO GUIDED PROBLEMS

Guided Problem 3.9

1.

The model has one energy storage (the compliance); this can be given integral causality, and therefore must be given this causality. As a result, the order of the model is one.

2.   3.   4.

5. $\dfrac{dq}{dt} = \dfrac{1}{R}e_0 - \dfrac{1}{RC}q$

Guided Problem 3.10

1.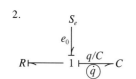

This bond graph assumes the two original compliances have been combined into one, with one-half the compliance of the individual compliances.

2. 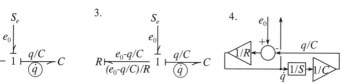   or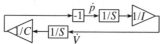

3. This step is not necessary.

4. $\dfrac{dV}{dt} = \dfrac{1}{I}p$

   $\dfrac{dp}{dt} = -\dfrac{1}{C}V$ (Make sure you have the minus sign.)

3.5 Nonlinear Resistances, Compliances and Inertances

Nonlinear resistances have been introduced in Chapter 2 but have yet to be considered in the writing of differential equations. The compliances and inertances considered thus far are constant and therefore give linear relations between effort and displacement or momentum and velocity.

Modeling systems with nonlinear one-port resistances, compliances and/or inertances is hardly more difficult than modeling linear systems. Further, finding the associated nonlinear differential equations also is hardly more difficult. In addition, simulating a system numerically and plotting the results usually is no harder if that system is nonlinear, although some nonlinearities complicate the results. Despite all this, nonlinear systems have the reputation of being relatively difficult. The historical reason is that *analytical* solutions to nonlinear differential equations are notoriously elusive. They are rarely sought today, however; computer-generated simulations suffice.

Nonlinear transformers and gyrators are considered in Chapter 9, along with R, C and I elements that have more than one port.

3.5.1 Nonlinear Resistances

Nonlinear resistances with a single port or bond are characterized in Section 2.3 by either of the relations

$$e = e(\dot{q}) \quad \text{or} \quad \dot{q} = \dot{q}(e). \tag{3.51}$$

The procedures described in Section 3.4 for annotating the bond graph or drawing a block diagram and writing differential equations need only small adjustments. If a resistance R has impedance causality (the output variable is the effort e), you write $e(\dot{q})$ for the effort rather than $R\dot{q}$. For admittance causality, you write $\dot{q}(e)$ for the flow in place of e/R. These functions may be left with this general notation until you actually carry out a simulation, or could be given the actual assumed algebraic form. You then write the differential equations from the bond graph the same way as before, substituting the new notations. If you use a block diagram, replace the triangles that otherwise would have R or $1/R$ as a scaling factor with blocks that generate the proper function $e(\dot{q})$ or $\dot{q}(e)$ before writing the differential equations. You may designate the nonlinear box with the letters NL, to stand for nonlinear algebraic relation, or write the actual nonlinear function.

Example 3.22

The linear system from Fig. 30 of the previous section is reproduced below. Make modifications to both the bond graph and the block diagram to account for the resistance becoming some nonlinear function $\dot{q}_R = \dot{q}_R(e)$, where e and $\dot{q}_R$ are defined as the effort and flow on the resistance bond, respectively. Then, write the resulting differential equations in terms of the given information.

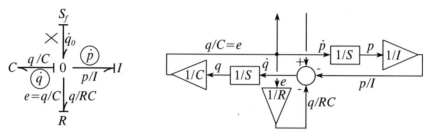

Solution: The annotations on the bond graph and the block diagram become

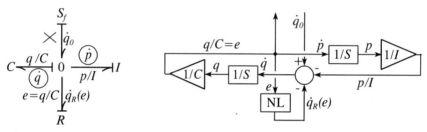

which give equations

$$\frac{dq}{dt} = \dot{q}_0 - \frac{1}{I}p - \dot{q}_R(e); \quad e = q/C,$$
$$\frac{dp}{dt} = \frac{1}{C}q.$$

Note that the functional relation $\dot{q}_R(e)$ must be spelled out before the equations become solvable.

3.5.2   Nonlinear Compliances

The spring shown in Fig. 3.32 comprises a uniform strip or beam of elastic material, cantilevered at one end, that wraps around a fixed curved member as it deflects. Without the

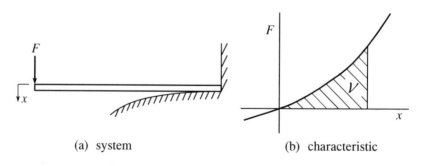

(a) system   (b) characteristic

Figure 3.32: Example of a nonlinear spring

curved member it would have a linear characteristic for small displacements. When present, this member foreshortens the free length of the beam as x increases. Therefore the spring becomes progressively stiffer as it deflects. This type of characteristic is represented by a curved line as in part (b) of the figure. The energy stored in the spring equals the area under the curve, or

$$\mathcal{V}(x) = \int F\,dx. \tag{3.51}$$

which also is true of the linear special case.

A generalized nonlinear compliance can be described by a nonlinear characteristic, which can be expressed algebraically either in a form appropriate for integral causality or in a form appropriate for differential causality. For the preferred integral causality,

$$\boxed{\vdash \frac{e_i(q_i)}{(\dot{q}_i)} \; C_i \qquad e_i = e_i(q_i); \quad q_i = \int \dot{q}_i\,dt} \tag{3.52}$$

Thus, the causal input $\dot{q}$ is integrated to give the state variable q, and the causal output is expressed as the appropriate (nonlinear) function of q.

The relation $e_i = q_i/C_i$, which is used in the linear case, can be retained for nonlinear cases, if you prefer to define an actual compliance[4]. Computing C_i involves about the same effort as the direct computation of e_i using equation (3.52). As a practical matter, therefore, there is usually no advantage to it.

Placing differential causality on a nonlinear compliance should be avoided, if possible. Nevertheless, it can be treated directly; applications of differential causality for both linear and nonlinear compliances and inertances are given in Chapter 6. The differential relation implied for the general compliance is

$$\frac{e_i}{\dot{q}_i(e_i)} \dashv C_i \qquad \dot{q}_i = \frac{d}{dt} q_i; \quad q_i = q_i(e_i). \tag{3.54}$$

The constant-effort source, $S_e \,\underline{\quad\quad}$, could be thought of as a very special type of nonlinear compliance, since it serves to store and dispense potential energy and can be represented by an effort-displacement characteristic. From a practical viewpoint, however, it is useful to consider the constant-effort source as a unique element.

3.5.3 Nonlinear Fluid Compliance Due to Gravity

The gravity compliance of a tank of liquid is nonlinear if the pressure at the port near the bottom of the tank is not proportional to the volume of the liquid stored above this level.

---

[4] This C_i is the **chordal compliance**, defined as shown below as the inverse slope of the chord drawn from the origin to the point of interest on the characteristic.

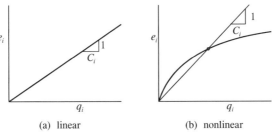

(a) linear

(b) nonlinear

This definition, analogous to that of the chordal resistance, allows the writing of differential equations as though the element were linear. The computational requirement is that C_i be a function of the state variable q_i, that is $C_i = C_i(q_i)$.

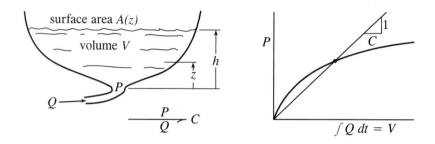

Figure 3.33: Nonlinear fluid compliance due to gravity

This happens, as illustrated in Fig. 3.33, if the surface area of the tank varies with the depth of the liquid; the pressure is proportional to the depth, but the volume is not. The general relation is $P = P(V)$. For an arbitrary nonlinear compliance with generalized displacement q and effort e, the relation is $e = e(q)$, as given by equation (3.52). The corresponding particular relation for the tank is found by using the relations $P = \rho g h$, where h is the depth of the water, and $V = \int_{z=0}^{z=h} A(z)\,dz$, where z is a running variable for the depth.

EXAMPLE 3.23

A conical-shaped tank ends in a small orifice of area A_0. Determine its compliance relation $P = P(V)$, where V is the volume of liquid in the tank. Then, give a fully annotated bond-graph model, and write a solvable differential equation in terms of V, with Q_{in} treated as an input.

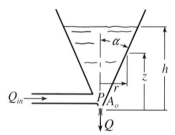

Solution: The compliance relation can be found as follows:

$$V = \int_0^h \pi r^2\,dz = \int_0^h \pi z^2 \tan^2 \alpha\,dz = \frac{\pi \tan^2 \alpha\, h^3}{3} = \frac{\pi \tan^2 \alpha}{3}\left(\frac{P}{\rho g}\right)^3 .$$

This result is inverted to get the desired form $P = P(V)$:

$$P = \rho g \left(\frac{3V}{\pi \tan^2 \alpha}\right)^{1/3} .$$

Should you wish to carry out the unnecessary step of finding a chordal compliance, as described in the footnote on the preceding page,

$$C(V) = \frac{V}{P(V)} = \frac{1}{\rho g}\left(\frac{\pi \tan^2 \alpha\, V^2}{3}\right)^{1/3} .$$

The fully annotated bond graph below includes the effort $P(V)$ above, and the orifice flow resistance relation $Q(P)$.

Using Bernoulli's equation for the nonlinear resistance that gives $Q(P)$,

$$\frac{dV}{dt} = Q_{in} - Q(P) = Q_{in} - A_0\sqrt{\frac{2}{\rho}P} = Q_{in} - A_0\sqrt{2g\left(\frac{3V}{\pi\tan^2\alpha}\right)^{1/3}}.$$

This equation is solvable, since the only unknowns on its right side are the given excitation Q_{in} and the state variable V.

3.5.4 Nonlinear Compressibility Compliance

A **hydraulic accumulator**, shown in Fig. 3.34 part (a), is a rigid chamber into which a liquid such as water or oil enters or leaves, and which also contains a fixed amount of a gas such as air or nitrogen (the latter to reduce the fire hazard with mineral oils). The gas may be separated from the liquid by gravity or, more effectively, by a piston or a flexible barrier such as a diaphragm or a bladder. In any case the pressures of the two fluids are virtually equal. Accumulators serve the purposes of storing significant amounts of liquid and energy, and reducing pressure and flow surges in hydraulic systems.

An accumulator can be modeled at least crudely over a wide variation in pressure by a *nonlinear* compliance. The flow on the compliance bond refers to the flow of the liquid, which is relatively incompressible. It is positive for flow directed inward. The integral of this flow equals minus the change of volume of the gas. The resulting pressure can be estimated from an appropriate equation of state for the gas. Adiabatic conditions for an ideal gas give $P/\rho^k =$ constant, from which the characteristic in the desired form is

$$P = \frac{P_0}{(1 - V/V_0)^k}. \tag{3.55}$$

Here, P_0 is the "charging pressure" that exists when the accumulator is empty of liquid but contains the full charge of gas, and V_0 is the total volume of the chamber. Isothermal conditions give the same equation except that the ratio of specific heats, $k = c_p/c_v$, is replaced by 1. The general shape of the characteristic is shown in part (b) of the figure.

Compressing or expanding a gas by a large proportion changes its temperature significantly, potentially causing heat transfer with the surroundings. Heat transfer over a significant temperature difference is irreversible. Adiabatic conditions are present only to the extent that there is too little *time* for significant heat transfer to occur. Isothermal conditions, for which heat transfer is reversible, are present only when the charging and discharging is so *slow* that significant temperature differences do not develop. Actual charging and discharging rates usually lie somewhere between these extremes. The system no longer conserves mechanical energy, and cannot be modeled precisely as a pure compliance. For the present this complication is neglected. The correction is addressed in Section 9.7.8.

3.5.5 Junctions With Multiple Bonded Compliances

The characteristics of three arbitrary compliances are given in part (a) of Fig. 3.35. When these compliances are bonded to a common 1-junction, the three efforts sum for a given

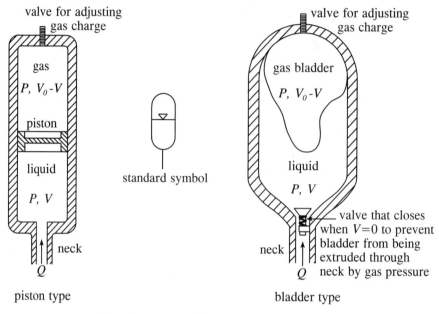

(a) major construction types

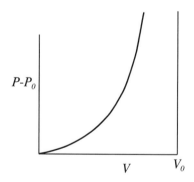

(b) compliance characteristic

Figure 3.34: Hydraulic accumulators

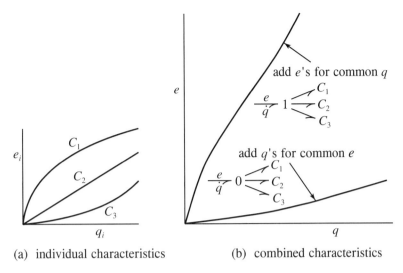

(a) individual characteristics   (b) combined characteristics

Figure 3.35: Graphical reduction of multiple compliances bonded to a junction

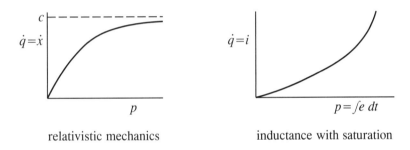

relativistic mechanics   inductance with saturation

Figure 3.36: Nonlinear inertances

common displacement; the combined characteristic is the vertical sum of the three characteristics, as shown in part (b). When the compliances are bonded to a common 0-junction, on the other hand, the three displacements sum for a given common effort; the combined characteristic is the horizontal sum of the three characteristics[5], as also shown.

3.5.6 Nonlinear Inertances

An inertance is defined as nonlinear if p is not proportional to $\dot{q}$. This is analogous to a nonlinear compliance in which q is not proportional to e. A happy fact: in classical mechanics, including fluid mechanics, masses are not functions of their velocities, and therefore inertances are linear. You will find nonlinear inertances only in relativistic mechanics, and (more importantly to engineers) in the magnetic domain. The shapes of the characteristics for these cases are shown in Fig. 3.36. (The axes have been reversed to make them compatible with the next figure.) The electrical inductor has a limiting generalized momentum because the magnetic flux of real materials reaches a limit; the phenomenon is called **magnetic saturation**. (The weight of many electronic packages such as stereos often resides mostly in their transformers, which contain considerable iron in order to minimize magnetic

---

[5]The chordal definitions of compliance allow the combination rules for linear compliances to be retained for nonlinear compliances, assuming dependencies are expressed as functions of the common variable. Therefore, for the 0-junction, $C = \sum_i C_i(e)$, and for the 1-junction, $1/C = \sum_i 1/C_i(q)$.

saturation.)

The preferred integral causality and the associated computation are similar to those for the compliance:

$$\frac{(\dot{p}_i)}{\dot{q}_i(p)} \dashv I_i \qquad \dot{q}_i = \dot{q}_i(p_i); \quad p_i = \int \dot{p}_i \, dt. \tag{3.56}$$

For differential causality they are

$$\frac{\frac{d}{dt}p(\dot{q}_i)}{\dot{q}_i} I_i \qquad e_i = \frac{dp_i}{dt}; \quad p_i = p_i(\dot{q}_i). \tag{3.57}$$

3.5.7   Kinetic and Potential Energies and Co-Energies

The power flowing into a simple compliance or inertance, linear or nonlinear, is the product of the effort and the flow, or

$$\mathcal{P} = e\dot{q} = \dot{p}q. \tag{3.58}$$

Since this power is flowing into a reversible energy-storing one-port element, the energy stored is its time integral:

$$\int \dot{p}\dot{q} \, dt = \int e \, dq = \int \dot{q} \, dp. \tag{3.59}$$

The integral $\int e \, dq$ above is integrable, as you have seen, if e is a function of q; this is called *potential energy* and is given the traditional symbol $\mathcal{V}$. Using the dummy variable of integration q',

$$\boxed{\mathcal{V}(q) = \int_0^q e(q') \, dq'.} \tag{3.60}$$

Sometimes $\mathcal{V} = \mathcal{V}(q)$ can be determined directly, and the result $e = e(q)$, which corresponds to integral causality, found therefrom:

$$e = e(q) = \frac{d\mathcal{V}(q)}{dq}; \qquad \text{used for} \quad \vdash \frac{e}{\dot{q}} \dashrightarrow C \tag{3.61}$$

In similar fashion, the integral $\int \dot{q} \, dp$ above also is integrable if $\dot{q}$ is a function of p; this is called **generalized kinetic energy** and is given the traditional symbol[6] $\mathcal{T}$:

$$\boxed{\mathcal{T}(p) = \int_0^p \dot{q}(p') \, dp'.} \tag{3.62}$$

When $\mathcal{T} = \mathcal{T}(p)$ is found directly, the result $\dot{q} = \dot{q}(p)$ can be calculated therefrom, consistent with integral causality:

$$\dot{q} = \dot{q}(p) = \frac{d\mathcal{T}(p)}{dp}; \qquad \text{used for} \quad \frac{e}{\dot{q}} \dashv I \tag{3.63}$$

The dual nature of the functions $e = e(q)$ and $\dot{q} = \dot{q}(p)$ and the energies $\mathcal{V}$ and $\mathcal{T}$ are suggested in Fig. 3.37. Note that linearity is just a special case. (The axes of the inertances are inverted from those given before in order to emphasize that momentum p plays the same role for kinetic energy that displacement q plays for potential energy. Both are state variables, assuming integral causality, as you know already.)

---

[6] As noted before, roman (non-script) V and T traditionally also are used for potential and kinetic energy, respectively.

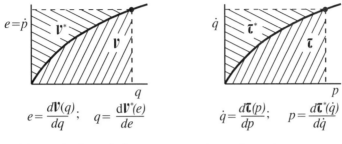

(a) constitutive relations, energy and complementary energy

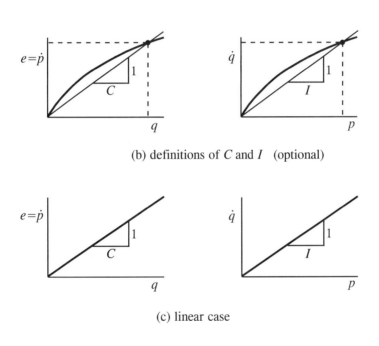

(b) definitions of C and I (optional)

(c) linear case

Figure 3.37: Generalized compliances and inertances

The chordal definitions for compliance and inertance[7] are given in part (b) of the figure for completeness, but may be ignored.

On the infrequent occasions when you wish to use differential causality, results similar to those above can be found by invoking the **complementary energy** (or **co-energy**) functions $\mathcal{V}^*$ and $\mathcal{T}^*$. These functions also represent areas shown in part (a) of Fig. 3.37. They can be related to $\mathcal{V}$ and $\mathcal{T}$ by what are known as **Legendre transformations**:

$$\mathcal{V}^*(e) = eq - \mathcal{V} = \int_0^e q(e')\, de', \tag{3.64a}$$

$$\mathcal{T}^*(\dot{q}) = p\dot{q} - \mathcal{T} = \int_0^{\dot{q}} p(\dot{q}')\, d\dot{q}'. \tag{3.64b}$$

If you know or can find $\mathcal{V}^*$ or $\mathcal{T}^*$, the results are

$$\dot{q} = \frac{dq}{dt}; \quad q = q(e) = \frac{d\mathcal{V}^*(e)}{de}; \qquad \text{used for} \quad \frac{e}{\dot{q}} \dashv C \tag{3.65a}$$

$$e = \frac{dp}{dt}; \quad p = p(\dot{q}) = \frac{d\mathcal{T}^*(\dot{q})}{d\dot{q}}; \qquad \text{used for} \quad \frac{e}{\dot{q}} \dashv I \tag{3.65b}$$

Constant inertances are linear. It is not necessary for an inertance to be constant for its relation between $\dot{q}$ and p to be linear, however; a change in something else, not infrequently a displacement, might also cause an inertance to change. For example, the fluid inertance of a channel that is empty toward the downstream end depends on the *displacement* of the fluid interface. In another example, the inductance (electrical inertance) of the coil in a solenoid depends on the *displacement* of the moving mechanical member. Consideration of displacement-dependent inertances is deferred to Section 9.4.

Constant compliances are linear. Similarly to inertances, however, some linear compliances vary because of changes in a variable other than the two used to define the linearity. For example, the capacitance (electrical compliance) of a capacitance microphone varies with the mechanical displacement of its diaphragm. Potential energies dependent on *two* displacments are considered starting in Section 9.3, where they are treated as compliances with two ports or bonds.

3.5.8   Summary

When an effort and its associated displacement are proportional to one another, the associated C is constant, and the compliance is said to be linear. When proportionality does not exist, $C = C(q)$ and the compliance is said to be nonlinear. In nonlinear cases it is usually more convenient mathematically to use the particular relation in the form $e = e(q)$ rather than $e = q/C(q)$, which refers to a chordal compliance. When the generalized force is independent of the displacement, you should substitute an effort source, S_e, for the compliance.

Two or more compliances bonded to a common 0-junction produce an equivalent overall compliance larger than any of its components. Two or more compliances bonded to a common 1-junction, conversely, produce an equivalent compliance smaller than any of its components

---

[7]The **chordal inertance**, $I = I(p)$, is defined as the reciprocal of the slope of the chord, so the two chordal definitions are

$$C \equiv \frac{q}{p} = \frac{q}{e}, \qquad I \equiv \frac{p}{\dot{q}}.$$

The slopes of the chords, that is the *reciprocals* of the compliance and the inertance, are called **generalized stiffness** and **generalized susceptance**, respectively.

In classical mechanics, including fluid mechanics, inertances are linear. In engineering practice, most nonlinear inertances occur because of magnetic saturation, such as in electrical inductors. Analogous to the compliances, these cases usually are most readily treated computationally using $\dot{q} = \dot{q}(p)$ rather than $\dot{q} = p/I(p)$ which uses a chordal inertance, although either of these forms imply the favored integral causality. The use of differential causality is treated in Chapter 6.

Sometimes a generalized kinetic energy or potential energy depends on more than one independent variable. These cases are deferred to Chapter 9.

Guided Problem 3.11

This problem gives needed practice in determining a mechanical compliance relation from a force balance and also from the use of potential energy.

A simple pendulum comprises a slender uniform rod of mass m and length L pivoted about its upper end. Find the compliance relation between the moment and the angle without restricting the size of the angle. Also, find the state differential equations that describe the oscillation. These equations are solved numerically in Example 3.24.

Suggested Steps:

1. Draw a bond graph model for the system. It is suggested that you start with a 1-junction labeled with the angular velocity.

2. Perform a force-moment balance to get the relation between the gravity-induced moment and the angle.

3. Get the relation between the moment and the angle a different way: express the gravity energy as a function of the angle, and take its derivative with respect to the angle to get the moment. Make sure the two answers agree.

4. Place causal strokes on the bond graph, and annotate the efforts and flows using the results above.

5. Write the nonlinear state differential equations. Note that the methods introduced thus far cannot solve these equations.

PROBLEMS

3.48 A well-advertised mattress gets abruptly stiffer above a certain deflection. Suggest how this can be accomplished with special shaping and mounting of coil springs, and briefly discuss its desirability.

3.49 Evaluate the nonlinear compliance relation for the two-dimensional parabolic tank shown below, assuming integral causality. The shape of the tank is described by $y = ax^2$ and the width, w.

3.50 Solve the problem above for the tank being axisymmetric about the y axis.

3.51 The fluid tank of Example 3.23 (p. 130) is now formed in the shape of a two-dimensional "V" rather than an axysymmetrical cone. Repeat the analysis for this case.

3.52 The conical tank of Example 3.23 (p. 130), but with no supply tube, drains to atmosphere through a vertical tube of area A and length L. Write a set of state differential equations to model the system. Viscous effects may be neglected, and the resistance assumed to correspond to a single head loss in the tube (giving a pressure drop of $\frac{1}{2}\rho v^2$). Include the inertance effects of the tube.

3.53 The rolling disk of Problem 3.5 (p. 84) and Problem 3.13 (p. 92) is given a *large* initial displacement, so the compliance is not constant.

    (a) Determine the compliance relation.

    (b) Write a complete set of first-order state differential equations.

Continued in Problem 3.58.

SOLUTION TO GUIDED PROBLEM

Guided Problem 3.11

1.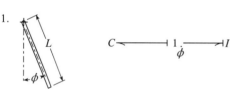

2. $M = (mgL/2)\sin\phi$

3. $\mathcal{V} = mg\dfrac{L}{2}(1 - \cos\phi); \quad M = \dfrac{d\mathcal{V}}{d\phi} = mg\dfrac{L}{2}\sin\phi$

4. $C \xleftarrow{\frac{(mgL/2)\,\sin\phi}{(\dot\phi)}} 1 \xrightarrow{\frac{\mathcal{p}}{p/I}} I \qquad\qquad I = mL^2/3$

5. $\dfrac{dp}{dt} = -(mgL/2)\sin\phi$

   $\dfrac{d\phi}{dt} = \dfrac{1}{I}p$

3.6 Numerical Simulation

Only linear differential equations and rather simple nonlinear differential equations allow analytical solution. Until the 1940s other differential equations could be tackled only through limited and laborious graphical or numerical procedures, often using grindingly slow mechanical calculators. This impacted the modeling of dynamic engineering systems and, indeed, the very design of those systems. In particular, great emphasis was placed on linear or linearized models, differential equations and the systems they represent. Linear systems indeed possess some intrinsic virtue that commends them for engineering design. Nevertheless, slavish focusing on them is no longer warranted or necessary.

The first major tool developed to solve time-based differential equations automatically was the **analog computer**. In the 1950s these "simulators" evolved away from their mechanical origins to become largely electronic circuits based on the operational amplifier in which various voltages behaved in time as did the associated variables in the equations. A maze of patch cords gave structure to a particular model, and banks of adjustable resistors allowed parameters to be set. Analog computers were used into the 1970s because of their potentially high speed, which was realized most fully when they were hybridized with digital computers.

The convenience, flexibility, accuracy and increasing speed of the digital computer finally pushed the analog computer into obsolescence. Even if you have little facility for solving differential equations analytically, you can utilize them effectively in the analysis of dynamic systems. You don't even need to understand the numerical methods that the software employs, including the MATLAB integrators described below and elsewhere. On the other hand, such integrators can behave erratically; some knowledge of their algorithms is worth the modest effort required.

3.6.1 State-Variable Differential Equations

A model first should be reduced to a set of n independent first-order differential equations. This form results naturally from the procedure described in Section 3.5 and used throughout this book. The use of vector notation is helpful, partly for its economy when a model is of high order, and partly because the MATLAB simulator that is used below is based on it. This "state-space" notation is written as

$$\boxed{\frac{d\mathbf{x}}{dt} = \mathbf{f}(\mathbf{x}, t).}$$

(3.66)

Boldface is used to indicate vectors; when written by hand, the typesetters' symbol for boldface is recommended: a wavy underline. The **state vector**, $\mathbf{x}$, is a vector of the state variables, or $[x_1 \ x_2 \ \ldots \ x_n]^T$. The number of state variables, n, is called the order of the system.

3.6.2 Simulation With ODE Routines of MATLAB®[8]

MATLAB is a commercial software package that competes with and can be used in conjunction with C, C++ and Fortran. It is especially easy to use and has features that commend it particularly for the analysis of dynamic systems. It is employed frequently in this book. Before you proceed further it is desirable that you study Appendix A.

The MATLAB simulators presented below "integrate" any set of linear or nonlinear differential equations of the form given by equation (3.66). MATLAB also has a simulator for linear models, presented in Section 4.1.7, that is much more efficient computationally because it employs analytic solutions.

The first step in carrying out a simulation is to code the coupled first-order state differential equations (equations (3.66)) themselves. This is done in a special subprogram called a **function M-file**. The first active line of this file is as follows:

```
function f = <name>(t,x)
```

Unless you make special provision, the only arguments that are communicated to this subprogram from the main program are the time **t** and the state vector **x**, which comprises the

---

[8]MATLAB is a registered trademark of The MathWorks Inc.

state variables $x(1)$, $x(2)$,...,$x(n)$. The derivatives dx_1/dt, dx_2/dt, $\cdots$, dx_n/dt are represented by the vector f, which has components $f(1)$, $f(2)$,..., $f(n)$. The subprogram normally ends with the differential equations in the form

```
f(1) = <first function of t, x₁, x₂,···,xₙ >
f(2) = <second function of t, x₁, x₂,···,xₙ >
.

.

f(n) = <nth function of t, x₁,x₂,···,xₙ >
f=f'
```

The last statement above converts a row vector to a column vector. Unless you make special provision, only the elements of this vector, $f(1)$, $f(2)$,..., $f(n)$, are communicated back to the main program.

With this basic scheme, any parameters that are needed to evaluate the derivatives must be defined within the function file. This file may be executed thousands of times during a single simulation, so the procedure can be quite inefficient. To avoid the repetition, values can be communicated back and forth between the main program and the function M-file through the use of a "global" declaration. You also can write your main program as a special "script M-file," again using the global declarations, rather than merely entering it into the command window. These optional elements of good programming practice are indicated in Appendix A under the heading Communication Between Files.

The function M-file is called with its file name, not the name given on its first line, but you should make them the same. All M-files must be given the extension .m and be properly opened, in order to be recognized. They may reside on your floppy disk, or better yet on the hard drive. Their directory location should be entered into the path that MATLAB uses. Procedural details are given in Appendix A under the heading Script Files.

Before the simulation can be executed, the initial conditions must be specified as a vector. For example, the initial state vector $x_1(0) = 1$, $x_2(0) = 0$, $x_3(0) = 2$ can be entered as

```
  x0 = [1 0 2];
```

Next, the command to carry out a simulation using a particular function file for the differential equations can be issued directly:

```
[t,x]=ode45('<name>',[0 10],x0)
```

The command "ode" stands for ordinary differential equation; the "45" is a descriptor of the algorithm used (several alternative available integrator algorithims are described under "help: Function functions and ODE solvers"). The second argument on the right is a two-element vector comprising the initial time and the time the simulation should stop. The third argument, x0, is the initial condition noted above.

Further optional arguments can be added to control the accuracy of the simulation. Accuracy is increased, at the expense of execution time, by decreasing the size of the time steps. These, in turn, are set automatically with respect to error indices that you can specify. Default values are set, otherwise. Details are found by clicking on "ODESET" within the help window.

Mistakes in the differential equations often result in instabilites in the solution which take almost forever to execute. Should you wish to abort a simulation that seems endless, type control-C. Often it is wise to try a very small value for the final time, until you are sure the behavior is reasonable.

Securing a plot of the results is probably your final step. The command

```
plot(t,x)
```

will plot all the variables, with a common scale, as a function of time. The color code (which depends on the version of MATLAB being used) gives you the state vectors in the order they have been defined. Likely different variables have different magnitudes, and some may be of little interest. Should you wish to plot x(1) and 100 times x(3), for example, enter

```
plot(t,x(:,1),t,x(:,3)*100)
```

The state vector x is stored as a three-column matrix; the colon (:) is a wild card. The time is stored as a vector of the same length. The plot command will execute only if the number of rows in t equals the number of rows in x. For information on how to control the colors of individual plots, or to designate individual data points by various symbols, use the "help" window.

EXAMPLE 3.24

The pendulum of Guided Problem 3.11 (pp. 137, 138) exhibits interesting behavior because of the nonlinearity of the gravity-induced compliance. Simulate the system, starting from rest at various angles. Plot the angle as a function of time for a period of three seconds. The differential equations are

$$\frac{dp}{dt} = -mg\frac{L}{2}\sin\phi,$$
$$\frac{d\phi}{dt} = \frac{1}{I}p.$$

Solution: Letting the state vector be $\mathbf{x} = [p\ q]'$ and the parameters be $L = 1$ ft, $mg = 1$ lb and $g = 32.2$ ft/s², the function m-file becomes

```
function f = pendulum(t,x);
L=1; W=1; g=32.2; I=L^2*W/(3*g);
f(1) = -W*(L/2)*sin(x(2));
f(2) = x(1)/I;
f=f';
```

The final statement above converts a row vector into the necessary column vector. The main program, entered into the command window with its prompt >>, could comprise only two lines:

```
>> [t,x] = ode23('pendulum',[0 3],[0 pi/18]);
>> plot(t,x(:,2)*180/pi)
```

This produces a simulation and plot of the angle, in degrees, from 0 to 3 seconds, with initial conditions comprising zero velocity and an angle of $\pi/18 = 10°$. Should you wish to superimpose plots for simulations with different initial conditions, enter the command

```
>>hold on
```

Some results are plotted at the top of the next page. The frequency of the oscillation for the 20-degree swings can be seen to be almost the same as the frequency for the 10-degree swings. This shows that the assumption of linearity, with its natural frequency of $\omega_n = 1/\sqrt{IC} = \sqrt{3*32.2/2} = 6.9498$ rad/s, or one cycle in 0.904 seconds, is approximately valid for angles as large as 20°. The frequency for an initial angle of 90°, on the other hand, is noticeably smaller. As the angle approaches 180°, the frequency approaches zero. Notice the hesitation when the angle is near its maximum. Does this agree with your intuitive sense?

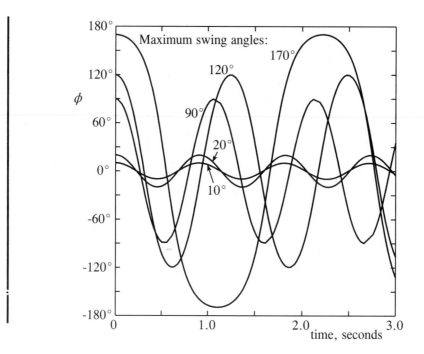

3.6.3   Simulation with Simulink®*[9]

Simulink is an appendage to MATLAB, which it requires. It comprises a graphical interface oriented around block diagrams, serving as an alternative to working directly from the differential equations with the ODE and other MATLAB simulation routines. It is presented here partly because it is widely used, especially by those who do not take advantage of bond graphs. Its use is never necessary, however; your instructor may give you guidance. The introduction below is limited to simple block diagrams, but the copious help functions that accompany Simulink should enable you to expand to relatively fancy ones, that for example can include complex variables.

You type *simulink* in the MATLAB command window to get the **Simulink Library Browser** window. A new model can be created by selecting **New** from the **File** menue, or clicking on an icon that resembles a sheet of paper. An **Untitled** window will open. Later, you will use the file menu for open, close, save and print operations on model files, which have the extension *mdl*.

The Library Browser window has a column of libraries on the left; clicking one of them displays its sub-library in a column on the right. The selected libraries and sub-libraries referred to below are listed in Fig. 3.38. If you click and drag on Integrator in the Continuous library, for example, you can place an integrator block in your new model window. Similarly, scaling triangles (Gain blocks) can be dragged from the Math Operations library, where summer circles (Sum blocks) also can be found. The default summers have three inputs: from the top, left and bottom. You select the signs by double-clicking on the circle, and in the List of Signs window type the signs you want in this order. To eliminate an input, type the spacer |.

Horizontal default arrows proceed from left to right. To reverse them, right-click on the desired block, then select **Format** and **Flip Block**. To interconnect the various blocks, click on the arrow and drag it to the desired location. Alternatively, you can click on

---

[9]Simulink is a registered trademark of The MathWorks, Inc.

menu item	selected contents
Continuous	Integrator block
Math Operations	Gain block
	Sum block
	Sine block
	Saturation block
	Coulomb Friction block
Signal Routing	Mux block
Sinks	Scope block
Sources	Sine Wave block
	Square Wave block
	Constant block
User Defined Functions	User Defined block

Figure 3.38: Selected libraries and sub-libraries in the Simulink Library Browser window

the first block, hold down the Control key, and click on the second block. The result will comprise vertical and/or horizontal line segments. To create a branching with a small dot, called a **takeoff point**, you can click on the new block to be connected and drag to the desired location for the dot on the existing path.

To secure a time plot of a variable, you place a **Scope** with its single input arrow. This is found on the **Sinks** menu. Should you wish to plot more than one variable, they can be fed into the left side of a **Mux** box; the right side has a single arrow that you direct to the Scope. The Mux box is within the **Signal Routing** menu. "Mux" is an abbreviation for "multiplexer."

The value of the multiplication factor in a gain block can be set by double-clicking on the block, which opens the **Block Parameters** window and allows you to enter a numeric value. The initial value for the output of an integrator is set using the same procedure, as can the parameters of the various sources to be found in the **Sources** menu, such as the **Sine Wave** block, the **Square Wave** block, and the **Constant** block. The start and stop times can be set from the **Configuration Parameters** item in the **Simulation** window accessible from the toolbar. This window also has a **Start** item, which starts the simulation after you press OK; the **Start** icon on the toolbar (a black triangle) will do the same.

A plot of the results appears, which can be self-scaled by clicking on the binocular icon, and directly printed to paper or a file.

Simulink has many nonlinear functions available, such as certain trigonometric functions, a saturation function and a Coulomb friction function. These can be found in the **Math Operations** window. If you do not find what you want there, you can construct your own by clicking on the User Defined category. Click and place this block, then double-click, and type whatever function you want, using (u) for the input variable.

EXAMPLE 3.25

Repeat the simulation of Example 3.24 (p. 141-142) for the pendulum of Guided Problem 3.13 using Simulink. Then, add Coulomb friction at the joint with the magnitude 0.05 ft·lb · s.

Solution: The following diagram results from the the procedure outlined above:

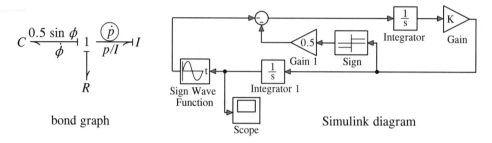

bond graph   Simulink diagram

The plots without the friction resemble those given in the solution to Example 3.23. The plot with the friction and an initial angle of 90 degrees is given below:

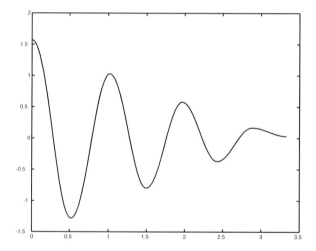

3.6.4   Integration Algorithms

The numerical integration of differential equations is a major subject in itself. There are many different methods available, even for the ordinary differential equations of concern here. Rather than survey these methods, the presentation below is limited to the simple class of Runge-Kutta single-step algorithms. This choice is based on their simplicity and robustness, plus their reasonable if not optimum efficiency. Should you at some time do a great deal of numerical simulation, you ought to avail yourself of current literature and software packages.

The method presented below is a basic **Runge-Kutta** integration scheme. The fourth-order version with automatic setting of the time increment is a satisfactory workhorse for most problems the engineer is likely to encounter. Runge-Kutta algorithms estimate the state $\mathbf{x}(t_0 + h)$ based solely on knowledge of $\mathbf{f}(\mathbf{x}, t)$ and $\mathbf{x}(t_0)$, where h is the increment between successive times at which the state is computed. Once $\mathbf{x}(t_0 + h)$ is found, t_0 is reset to equal the previous $t_0 + h$, and the "single-step" process is repeated. Throwing away the known history in this manner reduces the potential efficiency of the process but makes the method self-starting, robust and easy to program. The ease with which the step size, h,

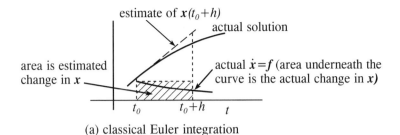

(a) classical Euler integration

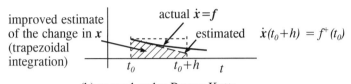

(b) second-order Runge-Kutta

Figure 3.39: First and second-order Runge-Kutta integration

can be varied at any stage in the solution is a special advantage. This allows the time step to be made as large as possible, reducing computer time, without the error exceeding set limits.

The fourth-order Runge-Kutta method will be approached inductively, starting with the lowest order, which is

$$\mathbf{x}(t_0 + h) \simeq \mathbf{x}_0 + h\mathbf{f}(\mathbf{x}_0, t_0), \tag{3.67}$$

where $\mathbf{x}_0 \equiv \mathbf{x}(t_0)$. This is the classical Euler formula; the estimated state is a direct extrapolation of the slope $\dot{\mathbf{x}}(t_0)$, as shown in part (a) of Fig. 3.39.

3.6.5 Second-Order Runge-Kutta

A far more accurate estimate of $\mathbf{x}(t_0 + h)$ results if the term $\mathbf{f}(\mathbf{x}_0, t_0)$ in equation (3.67) is replaced by the average of $\mathbf{f}$ evaluated at t_0 and $\mathbf{f}$ evaluated at $t_0 + h$. The latter is not known in advance, but can be estimated using a linear extrapolation of $\dot{\mathbf{x}} = \mathbf{f}$, as shown in part (b) of Fig. 3.39. Hence, the estimate of $\mathbf{f}$ at time $t_0 + h$ is

$$\mathbf{f}^+(t_0) = \mathbf{f}[\mathbf{x}^+(t_0), t_0 + h], \tag{3.68}$$

in which

$$\mathbf{x}^+(t_0) = \mathbf{x}_0 + h\mathbf{f}(\mathbf{x}_0, t_0) \tag{3.69}$$

is the same extrapolation as in the lowest-order method (equation (3.67)). Thus the second-order symmetrical Runge-Kutta algorithm becomes

$$\mathbf{x}(t_0 + h) \simeq \mathbf{x}(t_0) + \frac{h}{2}\left[\mathbf{f}(\mathbf{x}_0, t_0) + \mathbf{f}^+(t_0)\right], \tag{3.70}$$

where $\mathbf{f}^+(t_0)$ is found from equations (3.68) and (3.69).

A more formal method of derivation, which can be extended to higher orders, employs a Taylor's series expansion for $\mathbf{x}(t_0 + h)$:

$$\mathbf{x}(t_0 + h) = \mathbf{x}(t_0) + \mathbf{f}(t_0)h + \left[\left(\frac{\partial \mathbf{f}}{\partial \mathbf{x}}\right)_{t_0}\mathbf{f}(t_0) + \left(\frac{\partial \mathbf{f}}{\partial t}\right)_{t_0}\right]\frac{h^2}{2} + \dots. \tag{3.71}$$

Here, $(\partial \mathbf{f}/\partial \mathbf{x})_{t_0}$ is an $n \times n$ Jacobian matrix of derivatives. Equation (3.71) can be approximated by the general second-order Runge-Kutta formula

$$\mathbf{x}(t_0 + h) \simeq \mathbf{x}(t_0) + \lambda_1 \mathbf{f}(t_0)h + \lambda_2 \mathbf{f}[(\mathbf{x}(t_0) + \mu_1 \mathbf{f}(t_0)h, t_0 + \mu_2 h]h, \qquad (3.72)$$

in which λ_1, λ_2, μ_1 and μ_2 are constants to be determined. A second Taylor's series expansion, this time for the rightmost term of equation (3.70), permits the equation to become

$$\mathbf{x}(t_0 + h) \simeq \mathbf{x}(t_0) + (\lambda_1 + \lambda_2)\mathbf{f}(t_0)h + \lambda_2 \left[\mu_1 \left(\frac{\partial \mathbf{f}}{\partial \mathbf{x}} \right)_{t_0} \mathbf{f}(t_0) + \mu_2 \left(\frac{\partial \mathbf{f}}{\partial t} \right)_{t_0} \right] h^2, \qquad (3.73)$$

which can be compared directly with equation (3.71). Equating the terms through second order gives

$$\lambda_2 = 1 - \lambda_1, \qquad (3.74)$$

$$\mu_1 = \mu_2 = \frac{1}{2(1 - \lambda_1)}, \qquad (3.74b)$$

where $\lambda_1 \neq 1$ but otherwise is arbitrary. Infinitely many choices exist, but a highly stable and common choice is the symmetrical form

$$\lambda_1 = \lambda_2 = \tfrac{1}{2} \qquad (3.75a)$$

$$\mu_1 = \mu_2 = 1 \qquad (3.75b)$$

which gives precisely the result of equations (3.68) to (3.70).

3.6.6 Fourth-Order Runge-Kutta

The symmetrical fourth-order Runge-Kutta formulae probably are in most common use. Two equivalent forms of this are

$$\mathbf{x}(t_0 + h) \simeq \mathbf{x}(t_0) + \frac{1}{6}(\mathbf{z}_1 + 2\mathbf{z}_2 + 2\mathbf{z}_3 + \mathbf{z}_4), \qquad (3.76)$$

$$\mathbf{x}(t_0 + h) \simeq \mathbf{x}(t_0) + \mathbf{z}_2 - \mathbf{e}, \qquad (3.77a)$$

$$\mathbf{e} = \frac{1}{6}(-\mathbf{z}_1 + 4\mathbf{z}_2 - 2\mathbf{z}_3 - \mathbf{z}_4), \qquad (3.77b)$$

in which

$$\mathbf{z}_1 = h\mathbf{f}(\mathbf{x}_0, t_0), \qquad (3.78a)$$

$$\mathbf{z}_2 = h\mathbf{f}[\mathbf{x}(t_0) + \tfrac{1}{2}\mathbf{z}_1, t_0 + \tfrac{1}{2}h], \qquad (3.78b)$$

$$\mathbf{z}_3 = h\mathbf{f}[\mathbf{x}(t_0) + \tfrac{1}{2}\mathbf{z}_2, t_0 + \tfrac{1}{2}h], \qquad (3.78c)$$

$$\mathbf{z}_4 = h\mathbf{f}[\mathbf{x}(t_0) + \mathbf{z}_2, t_0 + h]. \qquad (3.78d)$$

The error $\mathbf{e}$ in equations (3.77) is proportional to the fifth power of the time interval, h, and the cumulative error is proportional to the fourth power of h. Thus, for equivalent accuracy to the second-order formulae, much longer time intervals can be used and the process usually is more efficient.

The magnitude of the vector $\mathbf{e}$ or its square $\mathbf{e}^T\mathbf{e}$, where the superscript T means transpose, is a very useful gage for the expected error. Thus if $\mathbf{e}^T\mathbf{e}$ is greater than some preset maximum $\mathbf{e}_0^2$, the entire calculation for the time step can be thrown away, the time interval halved, and the calculation repeated. Similarly, if $\mathbf{e}^T\mathbf{e}$ is less than some small preset fraction of $\mathbf{e}_0^2$, the time interval for the succeeding time step can be doubled. There is no point in this latter case for throwing away the previous calculation, which is merely more accurate than necessary.

A good integrator such as the fourth-order Runge-Kutta often uses very small time steps near the beginning of a simulation, which subsequently increase by a large factor. A fixed-step integrator would either be relatively inaccurate at first, or would eat up excessive run time. The results should be insensitive to large changes in the error gage, although an excessively large error gage will produce excessive time steps and resulting error. The problem is more acute for systems demonstrating sudden changes such as occur when a variable is abruptly limited, as in a mechanical impact.

The MATLAB integration algorithms are largely hidden from the user. The algorithm ode23 has some of the features of a second-order Runge-Kutta, and some of the features of a third-order. The algorithm ode45 is a hybrid fourth-to-fifth order Runge Kutta. MATLAB also offers other integrators with special virtues. MATLAB also offers several other integrators with significant particular virtues, which are described in the "help" window. Some further discussion of the use of these programs is given in Section 6.2 and Appendix A (p. 1017).

3.6.7 Summary

Computer simulation is an alternative to analytical solution of the differential equations that describe the behavior of dynamic models. It is especially useful for nonlinear models, where analytic solution is typically extremely difficult or impossible. Sometimes it is even used merely to find the *equilibrium* of a complex system in order to avoid having to solve difficult algebraic equations, which itself might require iterative numerical methods with uncertain success.

Basic fourth-order Runge-Kutta is the most practical algorithm presented. Although it is not the most efficient scheme available, it is simple and robust, and possesses reasonable efficiency for a very broad range of systems and conditions. The simulators ode23 and ode45 are Runge-Kutta algorithms of between second and third-order and between fourth and fifth-order respectively; they are amongst several MATLAB integrators that are simple to use but not fully documented.

Guided Problem 3.12

This first simulation problem for you to do is about as simple as a nonlinear system can get, and allows comparison of simulated and analytical solutions.

A water tank has an area of 6 ft^2 and an initial depth of 2 ft. A drain orifice has an effective area of 1.0 in^2. You are asked to simulate the emptying of the tank, using various error indices, and to compare the result to an analytical solution.

Suggested Steps:

1. Model the system with a bond graph.

2. The relation between the orifice flow and the tank pressure is nonlinear. Model it assuming Bernoulli's equation. The "effective" area is meant to include the effect of the vena contracta, or flow coefficient.

3. Annotate the graph in the usual way, noting the nonlinear resistance. Find a differential equation relating the volume of water left in the tank during draining to time.

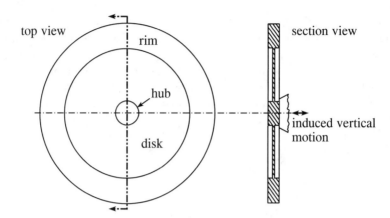

Figure 3.40: Disk-and-rim vibration absorber for Guided Problem 3.13

4. Write the function m-file, which can be done in as few as two lines of code.

5. Simulate the system from the given initial volume until it is empty. Since you do not know *a priori* how long this takes, choose a short time to start with, and extend it with subsequent runs. What does your model say if the volume were to become *negative*?

6. Plot your result.

7. Solve the differential equation analytically. This is a rare example of a nonlinear equation that is simple enough to be solved: the method of separation of variables applies. Compare the result to your simulation.

8. Investigate errors that result from choosing a large error index, RelTol, as used in the example on page 1017. You could change the index by factors of ten. Its default value for ode23 is .001, and for ode45 is 1e-6 (see MATLAB *help*, "ODESET," for further information).

Guided Problem 3.13

This problem gives you practice simulating a nonlinear second-order model, without having to do much coding. The resulting behavior is surprisingly exotic.

A vibration absorber comprises a flexible disk with a fused rigid hub and fused outer ring, as pictured in Fig. 3.40. The hub is vibrated vertically with the applied sinusoidal force, as shown. The disk acts like a spring, and the rim acts like a mass at the end of the spring. You may neglect the masses of the hub and the disk in your analysis. The rim weighs 3.0 pounds. The force-deflection characteristic of the disk is nonlinear:

$$F = kx + cx^3, \qquad k = 300\text{lb/in}, \qquad c = 30,000\text{lb/in}^3.$$

The linear coefficient k results from the bending stiffness of the disk. The nonlinear coefficient c results from radial or membrane stresses set up in the disk because the rather rigid rim resists being pulled inward as the disk bends.

The effect of non-zero initial conditions does not decay to reveal the forced part of the behavior unless some damping is introduced. Add a damping term to the force-balance equation with a damping coefficient $b = 0.02$ lb s/in.

To explore a little of the behavior of the disk, you are asked first to simulate its behavior, starting from rest and proceeding for 0.5 seconds, for a small hub displacement equal to $h_0 \sin \omega_n t$ with the small amplitude $h_0 = 0.0001$ inch at the calculated natural frequency. Then, repeat the simulation for an amplitude ten times higher and a frequency 15% higher, for two seconds or longer. Next, carry out a third simulation that is the same as the second, except for introduction of the initial values $h = x = 0.058$ inch (compression, with rim position neutral). Finally, repeat the last case, increasing the 0.058 inches to 0.059 inches. Plot the position of the rim as a function of time in each case. Comment on your observations.

Suggested Steps:

1. Model the system with a bond graph, omitting the damping. Note particularly that while the force on the disk spring and the rim is common, the absolute deflection of the rim is not equal to the relative deflection of the disk spring.

2. Apply causal strokes to the bond graph in the standard manner, and annotate the bonds according to the standard procedure.

3. Write the differential equations. Add the damping term. It is necessary to introduce the rim position as a third state variable simply to have it available for plotting.

4. For very small deflections, the cubic term with the coefficient c in the force characteristic can be neglected. Calculate the resulting resonant frequency, which is virtually unaffected by the damping: $\omega_n = 1/\sqrt{IC}$.

5. Code the differential equations in a MATLAB function file. Set the frequency at the frequency computed in step 4, and the displacement associated with the excitation velocity at 0.0001 inch.

6. Run the simulation with zero initial conditions for the specified 0.5 seconds. Use integrator `ode45` with the default error indices. Plot the resulting deflection as a function of time. The amplitude of the oscillation of a linear oscillator excited with a sinusoidal signal at its natural frequency grows linearly in time, except for the effect of damping. Does your result seem to agree with this description?

7. Increase the amplitude and the frequency as indicated, and carry out the second simulation for two, three or four seconds. Plot the result. The amplitude of motion for a sinusoidal signal at any frequency other than the natural or resonant frequency should ultimately reach an equilibrium value. Does it?

8. Repeat the second simulation, but specify the rather large initial deflections indicated. Plot the result. The response of a stable linear model approaches the same steady-state oscillation regardless of the initial condition. Does it in this case?

9. Repeat step 8 with the almost infinitesimally larger indicated initial conditions. The results should be drastically different, however. Can you explain this?

PROBLEMS

3.54 Consider the vehicle of Section 2.5.3 (pp. 62-65) accelerating on level ground from an initial velocity of 20 ft/s until equilibrium is reached. It weighs 3500 lbs.

(a) Approximate the characteristics given in Fig. 2.39 (p. 63) by simple polynomial expressions.

(b) Write the differential equation for the speed.

(c) Carry out a numerical simulation.

3.55 Carry out the same steps as in the previous problem for the acceleration from rest of the motor of Guided Problem 2.2 (p. 23). The load comprises an inertia of 2.0 kg·m^2 and a linear resistance of 0.005 N·m·s.

3.56 Carry out the same steps as in the previous two problems for the acceleration of the boat of Guided Problem 2.3 (pp. 43-44) from an initial velocity of 15 ft/s. Assume 50% throttle and a total effective weight of 3000 lbs.

3.57 The differential equation for the conical tank of Example 3.23 (p. 130-131), with zero input flow, can be transformed to the form $dV'/dt' = -(V')^{1/6}$ by defining a nondimensional "volume" $V' = V/V_0$, where V_0 is the initial fluid volume in the tank, and an appropriate nondimensional "time" t'. Simulate the emptying of the tank numerically, and compare the result with the analytic solution (which is readily found despite the nonlinearity).

3.58 Simulate the motion of the rolling disk of Problems 3.5 (p. 84), 3.13 (p. 92) and 3.53 (p. 138) for release angles of 10°, 30° and 60°. Assume $r = 1$ cm, $R = 10$ cm and $m = 0.1$ kg. Plot the angle for two seconds of time.

3.59 Simulate the response of the fluid tank system of Problems 3.26 (p. 109) and 3.42 (p. 126) for an initially empty tank with the flow $Q_0 = 0.5$ in^3/s starting abruptly. (An analytic solution is sought in Problem 4.51 on p. 214.)

3.60 *Project on the electro-hydraulic vehicle drive system in Chapter 1:* This project is to investigate the system described on pages 2-5, aiming to answer the questions asked there. The parameters below are close to those used for the example given. The work is divided into four phases; phases two and three are assigned as Problems 6.24 and 6.25, and phase four (the design phase) as Problem 6.26. Your instructor may ask you to work in small groups.

The vehicle plus payload weighs 3000 lbs (or 14 kN in an SI version of the project). Assume there is a single hydraulic motor with a volumetric displacement of 2.2 in^3/rev (36 cm^3) (or two motors each with half this, or four each with one-quarter). Neglect friction and leakage in this motor. The diameter of the wheels is 15 in (38 cm). Assume (temporarily) that the pressure of the hydraulic fluid entering the hydraulic motor is 2000 psig (14 MPa)(so that you can disregard the entire upstream system, including the electric motor, pump and the accumulator). The air drag on the vehicle corresponds to a frontal area of 20 ft^2 (1.8 m^2) with a drag coefficient of 0.32; the density of air is 0.0024 lb·s^2/ft^4 (1.25 kg/m^3). Let the ratio of the rolling friction force to the weight be $0.005 + 0.000034v$, where v is the velocity in ft/s (1 ft equals 0.305 m).

Simulate the acceleration of the vehicle from rest, and plot the resulting velocity for perhaps 500 seconds. (A word to the wise: considerable care is required to keep the units straight, particularly with the American standard units. Check their consistency for the various terms in your coding very carefully. A single error usually makes the result look like garbage.)

SOLUTIONS TO GUIDED PROBLEMS

Guided Problem 3.12

1.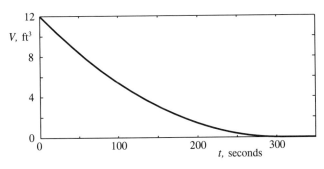

2. $Q = A_0 \sqrt{\dfrac{2P}{\rho}}$ where $A_0 = 1$ in^2.

3.

$$C = \frac{A}{\rho g}; \quad \frac{dV}{dt} = -A_0 \sqrt{\frac{2V}{\rho C}} = -A_0 \sqrt{\frac{2Vg}{A}}$$

   (Note that the density ρ does not affect the result.)

4.
```
function f = tank(t,V)
A0=1/144; A=6; g=32.2;
f=-A0*sqrt(2*V*g/A);
```

5,6. Enter into the command window:

```
[t,V]=ode23('tank',[0 100],12);
plot(t,V)
```

   The result shows that the final time needs to be extended. The plot for 350 seconds (rather than the 100 seconds above) is as follows:

   If the volume is started negative, or becomes slightly negative due to numerical errors, the square root of the negative value is treated as zero. Normally, on the other hand, MATLAB treats the square root of a negative number as an imaginary number.

7.

$$\frac{dV}{\sqrt{V}} = -A_0 \sqrt{\frac{2g}{A}}\, dt$$

$$2\sqrt{V}\Big|_{V_0}^{V} = -A_0 \sqrt{\frac{2g}{A}}\, t\Big|_0^t$$

$$2(\sqrt{V} - \sqrt{V_0}) = -A_0 \sqrt{\frac{2g}{A}}\, t$$

$$V = V_0 \left(1 - A_0 \sqrt{\frac{g}{2V_0 A}}\, t\right)^2 = 12 \left(1 - \frac{1}{144}\sqrt{\frac{32.2}{2 \times 12 \times 6}}\, t\right)^2$$

$$= 12(1 - 0.003284\, t)^2$$

The tank empties at $t = 1/.003284$ seconds. The plot of this equation is indistinguishable from that above found using ode23. The equation can be plotted with the MATLAB instructions

```
t=[0:0.5:304.5];
V=12*(1-0.003284*t).^2; % Note the array power (the dot before the ^)
plot(t,V)
```

8. Increasing the error index RelTol by a factor of 1000 produces a negligible difference. This result applies to first-order models only; the behavior of high-order models is apt to be more sensitive to the eror index. The most common cause of a significant error is an abrupt change in a forcing function, such as at the instant of a mechanical impact.

Guided Problem 3.13

1.

2.

3. $\dfrac{dp}{dt} = kq + cq^3 + \text{damping force} = kq + cq^3 + d\dot{q}$

$$\frac{dq}{dt} = \dot{x} - \frac{1}{I}p; \quad \dot{x} \equiv \frac{dx}{dt} = \frac{d}{dt}(x_0 \sin \omega t) = x_0 \omega \cos \omega t$$

$$\frac{dy}{dt} = \frac{1}{I}p$$

$$\frac{dx}{dt} = \omega_0 \omega \cos \omega t$$

4. $\omega_n = 1/\sqrt{IC} = \sqrt{k/m} = \sqrt{kg/W} = \sqrt{300 \times 386/3} = 196.47$ rad/s

5. Let the state vector be $x = [p, q, y, x]'$. Then,

```
f=function disk(t,x)
k = 300;  c = 30000;  b = 0.02;  I = 3/386;  x0 = 0.0001;  om = 196.47;
f(4) = x0*om*cos(om*t);
f(3) = x(1)/I;
f(2) = f(4)-f(3);
f(1) = k*x(2)+c*x(2)^3+b*f(2);
f=f'
```

6. Enter into the command window:

```
[t,x]=ode45('disk',[0 0.5],[0 0 0 0]);
plot(t,x(:,4),t,x(:,2))
```

The resulting plot (with some notational refinements) indeed shows a growing response:

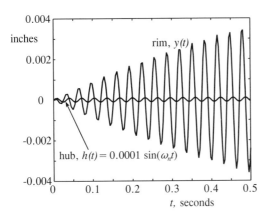

7. Changing x_0 to 0.001 in, increasing ω by the factor 1.15 and increasing the final time to 4 seconds gives the following result. The ratio of the amplitudes of the rim to the hub approaches a constant value of 4.0.

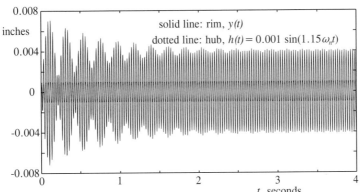

8. Introducing the initial condition [0 0.058 0 0.058] gives the following result. Although the amplitude starts large, it ultimately decays to the same ±0.004 inches as when zero initial conditions are specified.

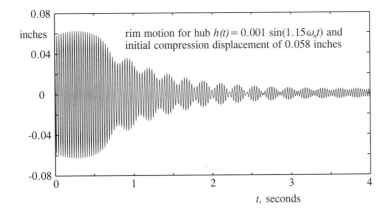

9. Increasing the number 0.058 inches slightly, to 0.059 inches, causes a major change in behavior: the rim amplitude does not decay, but approaches a value about 17 times larger than before.

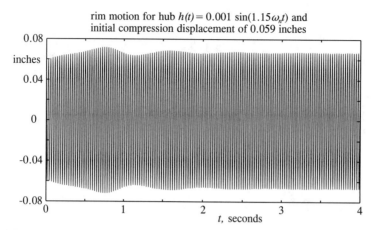

rim motion for hub $h(t) = 0.001 \sin(1.15\omega_n t)$ and initial compression displacement of 0.059 inches

There are in fact two grosssly different resulting rim amplitudes for the same hub amplitude, depending on the initial conditions. Should the rim motion be momentarily damped, for example, the large amplitude would vanish, leaving only the smaller motion, a phenomenon known as the **jump effect**. The short explanation is that increasing amplitude means increasing stiffness, due to the nonlinear force term with the coefficient c. Increasing stiffness in turn means increasing natural frequency.

Therefore, if the amplitude is large enough at first, the natural frequency will rise to match the applied frequency, 15% larger than the small-amplitude natural frequency, and the disk resonates with a rim-to-hub amplitude ratio of about 67. The behavior depends on the history of the state, or demonstrates **hysteresis**, as the resonance diagram opposite suggests.

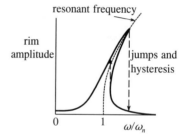

Chapter 4

Analysis of Linear Models, Part 1

Linear models always have enjoyed a special status in engineering, despite nature itself being mostly nonlinear. The use of linear models was paramount before the days of digital computers, because usually only they possess analytical solution. Much more important today is the fact that the behavior of linear models can be characterized in vastly simpler ways than almost all nonlinear systems, regardless of whether mathematics is employed. The significance of this fact has not waned; well engineered systems *ought* to behave in simple, predictable ways, so linear behavior remains a frequent *object* of design. Third, strategies for the control of dynamic systems predominantly depend on the use of at least approximate linear models. Therefore, even nonlinear models are often **linearized**, despite the approximateness of the result. Finally, systems in which the variables range over only a small fraction of that which is possible, such as with most acoustics and other vibrations, are very nearly linear anyway.

This chapter develops the ideas above, with emphasis on scalar as opposed to vector differential equations. Such a formulation can be deduced readily from the state-vector differential equations that follow naturally from bond graphs and primitive block diagrams. Emphasis is placed also on the **time domain**, which simply means that behavior is expressed as functions of time. Responses to disturbances that occur during a brief segment of time are called **transient reponses**, since their dynamic effects generally decay over time. An alternative perspective, behavior expressed as functions of frequency, is developed in Chapter 7, along with vector and matrix methods.

An introduction to the idea of superposition and linearity is given first, followed by operator notation and related MATLAB® simulation techniques. A discussion of common analytic functions leads into a review of classical means for solving linear differential equations and an optional presentation of convolution, which directly implements the idea of superposition. The Laplace transform method, which is the primary analytical workhorse in this book, is then presented, followed by a discussion of the most common linear models and their behavior. Finally, methods for linearization are presented in some detail.

4.1   Linear Models and Simulation

Bond graphs with elements having constant moduli are linear models. Dynamic linear models can be represented by linear differential equations. Operational notation allows these equations to be manipulated like algebraic equations. Since linear models can be solved

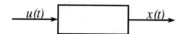

(a) "black box" system with excitation and response

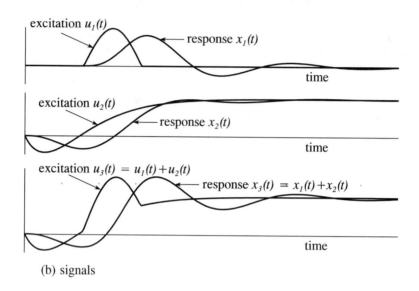

(b) signals

Figure 4.1: Property of superposition

analytically, computer solutions based on them are more accurate and computationally efficient than simulations such as the Runge Kutta class based on numerical methods. These ideas are amongst those developed in this section.

4.1.1   Superposition and Linearity

Linear systems can be identified merely by their response to disturbances; it is not necessary to have any analytic representation. Consider a "black box" system, as represented by the block diagram in Fig. 4.1 part (a), with a single disturbance or input, $u(t)$, and a single response or output, $x(t)$. Suppose that you can choose $u(t)$ and observe $x(t)$ for a variety of separate "runs." First, you choose two arbitrary functions, $u_1(t)$ and $u_2(t)$, and observe the respective responses $x_1(t)$ and $x_2(t)$, as illustrated in part (b) of the figure. Then you choose a function $u_3(t)$ which equals the *sum* of $u_1(t)$ and $u_2(t)$:

$$u_3(t) = u_1(t) + u_2(t). \tag{4.1}$$

The **property of superposition** is said to pertain to the system if the observed response $x_3(t)$ equals the sum of $x_1(t)$ and $x_2(t)$, that is if

$$x_3(t) = x_1(t) + x_2(t). \tag{4.2}$$

More precisely, the property applies if this relation is true for *arbitrary* $u_1(t)$ and $u_2(t)$. Systems often have a domain of u, x within which superposition is satisfied and the system acts linearly. Outside of this domain, usually for sufficiently large signals, the system is nonlinear.

A convenient special test employs

$$u_2(t) = u_1(t), \tag{4.3}$$

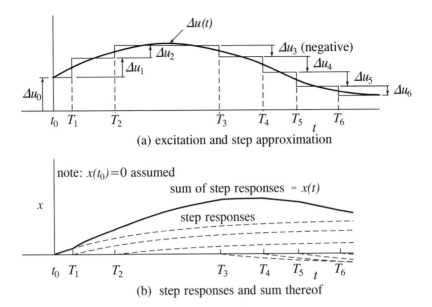

(a) excitation and step approximation

(b) step responses and sum thereof

Figure 4.2: Approximation of a response of a system using steps and superposition

which implies

$$u_3(t) \equiv u_1(t) + u_2(t) = 2u_1(t), \tag{4.4a}$$

$$x_2(t) = x_1(t). \tag{4.4b}$$

The property of superposition, if it pertains, requires

$$x_3(t) = x_1(t) + x_2(t) = 2x_1(t). \tag{4.5}$$

In words, superposition requires that the doubling of an excitation merely doubles the response at each and every time, t. Further, this happens regardless of the size of $u_1(t)$, leading to the conclusion that the response to the excitation $cu_1(t)$, where c is any constant, is $cx_1(t)$. This is what is meant by linearity.

The simplest case of all eliminates time as an independent variable, leaving an algebraic relation between the excitation and the response. Now, if u_1 produces x_1, $2u_1$ must produce $2x_1$, so in general $x = ku$ where $k = x_1/u_1$. A plot of x versus u is a straight line through the origin; any other relation violates superposition and mandates the name *nonlinear*.

An example of the direct use of the property of superpostion is illustrated in Fig. 4.2. An arbitrary excitation $u(t)$ is approximated roughly by a sum of seven steps, each having the amplitude Δu_n, $n = 0, \cdots, 6$. The responses to these individual steps, plotted in part (b) as dashed lines, are all the same, except for different amplitude factors and different initiation times. (The assumed system model is first order, which has a response that is discussed in detail in Section 4.6.1.) The individual responses are added up to give the solid line. This sum represents a fairly good approximation of the response to the actual excitation, since the step response results in marked smoothing. If the calculated response is thought to be too approximate, the error can be reduced indefinitely by increasing the number of steps.

4.1.2 Linearity and Differential Equations

You may be most familiar with linearity in the context of algebraic and differential equations, such as the nth order differential equation

$$
\begin{aligned}
a_n \frac{d^n x}{dt^n} + a_{n-1}\frac{d^{n-1}x}{dt^{n-1}} + a_{n-2}\frac{d^{n-2}x}{dt^{n-2}} + \cdots + a_1 \frac{dx}{dt} + a_0 x \\
= f(t) = b_m \frac{d^m u}{dt^m} + b_{m-1}\frac{d^{m-1}u}{dt^{m-1}} + \cdots + b_1 \frac{du}{dt} + b_0 u,
\end{aligned}
\tag{4.6}
$$

in which the coefficients a_i and b_j are constants. This represents a linear relation between the independent function $u(t)$ and the dependent function $x(t)$ because each term satisfies superposition. For example, doubling $x(t)$ doubles each term, and therefore corresponds to doubling $u(t)$. Were any term raised to any power other than unity, for example, the linearity and its property of superposition would be lost.

More explicitly, equation (4.6) expresses a **stationary linear model**. The model becomes **nonstationary** if one or more of the coefficients a_i become explicit functions of the independent variable of time, t. Superposition still applies, however; the model is still linear.

The fact that the shape or form of the response of a linear system depends only on the shape or form of the disturbance and not on its magnitude allows the system to be characterized simply in terms of its operational behavior. This property of linear systems is developed in this and later chapters to give powerful analytic results. By comparison, characterization of most nonlinear systems in terms of their operational behavior is very difficult and usually impracticable; the response to a given disturbance may for example vary wildly depending on the initial condition. (The modern theory which goes by the name "chaos" actually is based on deterministic nonlinear dynamics.) Despite this complexity, nonlinear models may be deduced *from direct considerations of the physics* almost as easily as linear models. Identifying a dynamic nonlinear model from observation of *dynamic behavior* rather than knowledge of the internal structure of the system, on the other hand, is apt to be much more difficult than for a linear model of comparable complexity.

The use of causality on a particular bond graph results directly in a set of first-order differential equations. When such a set of equations is linear and stationary, it can be placed in the canonical or standard matrix form

$$
\frac{d\mathbf{x}}{dt} = \mathbf{A}\mathbf{x} + \mathbf{B}\mathbf{u},
\tag{4.7}
$$

in which $\mathbf{A}$ and $\mathbf{B}$ are matrices of constants. The general excitation is the vector $\mathbf{u} = \mathbf{u}(t)$ and the response is the vector $\mathbf{x} = \mathbf{x}(t)$, which is known as the **state vector**. Note again that the term "stationary" does not imply "static," but rather means that the dynamic characteristics are invariant in time. Thus, characteristics such as natural frequencies and time constants remain unchanged. In matrix notation, nonstationary linear models are represented by the same equation with $\mathbf{A} = \mathbf{A}(t)$ and $\mathbf{B} = \mathbf{B}(t)$. This book considers analytical solutions for stationary models only. As you probably have noticed, bond graphs in which the moduli of all the elements are constants directly produce linear algebraic and/or linear stationary differential equations, with all their conceptual and computational advantages.

A model described by equation (4.7) can be called *complete*, since the behavior of all the state variables is represented. Equation (4.6), by contrast, describes only the relation between a single input variable and a single output variable of the system. This incomplete model may nevertheless be all the modeler needs, and is the focus of the solutions developed in the rest of this chapter. Matrix methods are deferred largely, though not exclusively, to

Figure 4.3: Transfer function in a block diagram

Chapter 7. The relation between input and output variables, for example, establishes the **transfer function** between these variables, which is sufficient for vibration analysis or classical control calculations.

When the state variable representation is used, it is customary to represent variables of special interest, called **output variables**, by weighted sums of the state variables and the input vector. Calling these output variables $\mathbf{y}(t)$,

$$\mathbf{y}(t) = \mathbf{Cx} + \mathbf{Du}. \tag{4.8}$$

For the special case $\mathbf{y} = \mathbf{x}$, $\mathbf{C}$ equals the unit diagonal matrix, $\mathbf{I}$, and $\mathbf{D} = \mathbf{0}$. If there is only a single output variable of interest, $\mathbf{C}$ and $\mathbf{D}$ become row matrices. Most commonly, further, $\mathbf{D} = \mathbf{0}$.

4.1.3 Operator Notation

The manipulation and solution of linear differential equations is simplified by the use of **operator notation**. The **derivative operator**, S, is defined by[1]

$$S \equiv \frac{d}{dt}, \tag{4.9}$$

consistent with its use in block diagrams. With this notation, equation (4.6) becomes

$$(a_n S^n + a_{n-1} S^{n-1} + \cdots + a_1 S + a_0)x$$
$$= (b_m S^m + b_{m-1} S^{m-1} + \cdots + b_1 S + b_0)u. \tag{4.10}$$

This equation also is written in terms of the **transfer function operator**, $G(S)$:

$$x = G(S)u; \quad G(S) = \frac{b_m S^m + b_{m-1} S^{m-1} + \cdots + b_1 S + b_0}{a_n S^n + a_{n-1} S^{n-1} + \cdots + a_1 S + a_0}, \tag{4.11}$$

Transfer functions often are written inside the boxes of block diagrams. This is shown in Fig. 4.3 using a generic transfer function $G(s)$. Simulink makes powerful use of this notation, as illustrated in Section 4.1.8 below.

4.1.4 Transformation from State-Space to Scalar Form

One use of the operator notation facilitates the reduction of the state-space equation (4.7) to the single-variable form of equation (4.6) and its operator representations given by equations (4.10) and (4.11).

---

[1]Some authors prefer to use the symbol D for the time derivative operator, which historically is known as the Heaviside operator. Others use lower-case s, which in this book is reserved for the Laplace operator and also a constant complex number.

EXAMPLE 4.1

Find a transfer function relating the input $u(t)$ to the output $q(t)$ for the following pair of state differential equations, and give the corresponding second-order differential equation:

$$\frac{dp}{dt} = -2p + 2q + 24u(t),$$

$$\frac{dq}{dt} = 4p - 9q + 4\frac{du(t)}{dt} + 2u(t).$$

Solution: In operator notation, these equations become

$$(S+2)p = 2q + 24u,$$
$$(S+9)q = 4p + (4S+2)u.$$

The variable p can be solved for in terms of q, *algebraically*:

$$p = \frac{2q + 24u}{S+2}.$$

This result then can be substituted into the second operator equation, giving the following equation from which p has been eliminated:

$$(S+9)q = \frac{4(2q + 24u)}{S+2} + (4S+2)u.$$

This equation can be multiplied by the denominator factor $(S+2)$ to give

$$[(S+9)(S+2) - 8]q(t) = [96 + (4S+2)(S+2)]u(t),$$

from which

$$G(S) = \frac{96 + 4S^2 + 10S + 4}{S^2 + 11S + 18 - 8} = \frac{4S^2 + 10S + 100}{S^2 + 11S + 10}.$$

This transfer function can be interpreted to give the differential equation

$$\frac{d^2q}{dt^2} + 11\frac{dq}{dt} + 10q = 4\frac{d^2u}{dt^2} + 10\frac{du}{dt} + 100u.$$

Notice that algebraic operations have replaced derivative operations, which is a great convenience.

4.1.5 Transformation from Scalar to State-Space Form*

It is possible also to transform a scalar differential equation into the state-space form of equations (4.7) and (4.8). The result is not unique; it depends upon what definitions of $x_1, x_2, \cdots, x_n$ are chosen. The simplest scheme assumes $y_1 = x$ and $a_n = 1$, and employs

$$\mathbf{A} = \begin{bmatrix} -a_{n-1} & -a_{n-2} & \cdots & -a_1 & -a_0 \\ 1 & 0 & \cdots & 0 & 0 \\ 0 & 1 & \cdots & 0 & 0 \\ \cdot & \cdot & & \cdot & \cdot \\ \cdot & \cdot & & \cdot & \cdot \\ 0 & 0 & \cdots & 1 & 0 \end{bmatrix} \qquad (4.12a)$$

$$\mathbf{B} = \begin{bmatrix} 1 \\ 0 \\ \cdot \\ \cdot \\ 0 \end{bmatrix} \tag{4.12b}$$

$$\mathbf{C} = [b_{n-1} \;\; b_{n-2} \;\; \cdots \;\; b_0] \tag{4.12c}$$

$$\mathbf{D} = 0 \tag{4.12d}$$

which applies as long as $m \le n - 1$.

EXAMPLE 4.2

Find a state-space differential equation consistent with the scalar differential equation

$$\frac{d^3x}{dt^3} + 3\frac{d^2x}{dt^2} + 4\frac{dx}{dt} + 2x = 2\frac{du}{dt} + u.$$

Equation (4.12) gives

$$\frac{d}{dt}\begin{bmatrix} x_1 \\ x_2 \\ x_3 \end{bmatrix} = \begin{bmatrix} -3 & -4 & -2 \\ 1 & 0 & 0 \\ 0 & 1 & 0 \end{bmatrix}\begin{bmatrix} x_1 \\ x_2 \\ x_3 \end{bmatrix} + \begin{bmatrix} 1 \\ 0 \\ 0 \end{bmatrix} u(t); \quad y = [0\;2\;1]\begin{bmatrix} x_1 \\ x_2 \\ x_3 \end{bmatrix}.$$

4.1.6 Transformations Using MATLAB®

MATLAB provides translation from the state-space format to the scalar format. This is accomplished with the command

```
[num,den]=ss2tf(A,B,C,D,i)
```

The `ss2tf` can be read "state-space to transfer function" (which really means *scalar* transfer function). The state-space matrices `A`, `B`, `C`, `D` must be defined numerically before this command is issued. The argument `i` is an index designating which of possibly multiple inputs is to be considered; if you have only one input, as in Example 4.1 above, insert 1. The output argument `num` (which could be designated with any symbol) is a row vector of the coefficients of the numerator polynomial of $G(S)$. The output argument `den` is a row vector of the denominator polynomial. If the matrix `C` has more than one row, more than one transfer function is being requested. Each has the same denominator, which therefore is reported only once. The respective numerator polynomials are reported as the rows in `num`.

EXAMPLE 4.3

Use MATLAB to find the transfer function from the input $u(t)$ to the output $x_2(t)$ for the third-order model

$$\frac{dx_1}{dt} = -4x_1 + \frac{1}{2}x_2 + u(t),$$

$$\frac{dx_2}{dt} = 2x_1 - 2x_2 + 2x_3,$$

$$\frac{dx_3}{dt} = x_2 - 4x_3,$$

Solution: The MATLAB entry

```
A=[-4 1/2 0;2 -2 2;0 1 -4];
B=[1;0;0];
C=[0 1 0];
D=[0];
[num,den] = ss2tf(A,B,C,D,1)
```

gives the response

```
num =
            0           0        2.0000    8.0000
den =
        1.0000    10.0000    29.0000    20.0000
```

which indicates the transfer function

$$G(s) = \frac{2S + 8}{S^3 + 10S^2 + 29S + 20}.$$

An inverse transformation is produced by the command

```
[A,B,C,D]=tf2ss(num,den)
```

in which `tf2ss` can be read "transfer function to state-space."

4.1.7   Simulation of Linear Models Using MATLAB*

Numerical solutions to linear scalar differential equations are plotted, with little effort on your part, by the MATLAB command

```
lsim(num,den,u,t)
```

Related options given below include the treatment of matrix differential equations. `lsim` stands for "linear simulation." The argument `t` is a row vector of the times at which the response is to be computed. The argument `u` is a row vector of the input values at precisely those same times; it must have the same length as `t`. Normally, the program assumes that the actual input for intermediate times is given by a linear interpolation of the immediately surrounding given values. (In some cases it concludes that a step-like zero-order-hold function is intended, and excercises that assumption.) Exact analytical solutions of the differential equations are evaluated numerically, a process which is potentially both more accurate and more efficient than the numerical simulation scheme presented in Section 3.7.

EXAMPLE 4.4

Secure a plot of the solution of the differential equation used in Example 4.2, repeated here:

$$\frac{d^3x}{dt^2} + 3\frac{d^2x}{dt^2} + 4\frac{dx}{dt} + 2x = 2\frac{du}{dt} + u,$$

with the particular excitation

$$u = \begin{cases} 5\sin(2\pi t/10), & 0 \le t \le 5, \\ 0, & t \ge 5. \end{cases}$$

Consider the range $0 \le t \le 10$ seconds. Superimpose a plot of $u(t)$.

Solution: You start by defining a vector of discrete times at which the solution will be found. The commands

```
t1 = [0:.1:5]; t2 = [5.1:.1:10]; t=[t1 t2]
```

establish vectors for the two segments of time and the combined time duration of 0 to 10 seconds. Next, you specify a corresponding vector for the excitation signal, $u(t)$:

```
u1 = 5*sin(2*pi*t1/10); u2 = zeros(size(t2)); u=[u1 u2];
```

The coefficients on the right side of the differential equation, which are given in the numerator of the transfer function format for the model, are recognized by the statement

```
num = [2 1];
```

Similarly, the coefficients on the left side of the differential equation or the denominator of the transfer function are recognized by the statement

```
den = [1 3 4 2];
```

A plot of the response results from the added commands

```
x=lsim(num,den,u,t); plot(t,x)
```

The plot below results. The excitation $u(t)/2$ is added by the commands

```
hold on
plot(t,u/2,'- -')
```

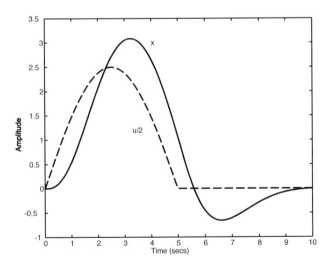

The command `lsim` handles matrix differential equations through use of the standard `A,B,C,D` notation. The options below include creating an accessible file of the results by adding a left side to the statement:

```
[y,x] = lsim(A,B,C,D,u,t);
[y,x] = lsim(A,B,C,D,u,t,x0);
y = lsim(num,den,u,t);
```

The second case allows you to introduce non-zero initial conditions by first defining the vector x0. Leaving off the left side of any of the statements results directly in plots of both the input(s) and the output(s).

CHAPTER 4. ANALYSIS OF LINEAR MODELS, PART 1

EXAMPLE 4.5

Repeat Example 4.4 using the state-space formulation (as given in Example 4.2).

Solution: The matrices A, B, C, D are defined with the commands

```
A = [-3 -4 -2;1 0 0;0 1 0];
B = [1;0;0];
C = [0 2 1];
D = 0
```

The plot results from the command

```
lsim(A,B,C,D,u,t)
```

or you could use the command with the left side [y,x] and then type

```
plot(t,y)
```

lsim converts the differential equation into a difference equation which for initial-value problems is accurate regardless of the size of the time interval. (It does this using a matrix exponential, which is described in Sections 7.3.1 and 7.3.2.) Because of the assumed linear interpolation of $u(t)$, however, the time interval might have to be rather small to give acceptable accuracy.

4.1.8   Simulation of Linear Models Using Simulink® *

Simulink offers the use of scalar transfer functions and state-space representations. Both of these can be found in the Continuous menu. Double-click on a **Transfer Fcn** block to set the coefficients of the numerator and denominator polynomials; for example, the polynomial $S^3 + 3S^2 + 4S + 1$ would be indicated by typing $[1, 3, 4, 1]$. To set the values for the matrices A, B, C and D of a **State-Space** block, double-click on it and type in the values using brackets, commas and semicolons the same way you would for matrices in the basic MATLAB. If there is more than one input, the associated arrows must be placed in the proper order. If there is more than one output, all are carried on the same arrow to a Scope.

Transfer function and state-space blocks can be used in conjunction with nonlinear elements.

EXAMPLE 4.6

Repeat Example 4.4 using Simulink.

Solution: The input signal for this example can be represented as the sum of a sine wave and the same sine wave delayed by one-half a cycle; this produces a zero sum for all time greater than the on-half cycle. The sine wave is in the Source menu. The delay function is given as **Transport Delay** in the Continuous menu; the delay time of 5 seconds is set by double-clicking on it. The block diagram below also shows the use of a Mux block to plot the input signal as well as the output. The result is the same as given in

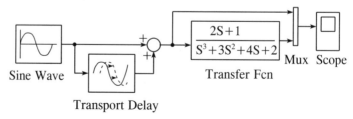

4.1.9 Summary

Systems that exhibit a domain of input and response variables satisfying the property of superposition are, by definition, linear within that domain. They may be represented by linear algebraic or differential equations. The domain cannot be infinite for a real physical system, of course; if the variables get large enough there must be an irreversible failure, which is a nonlinearity of the most severe type.

Operator notation allows a linear differential equation to be treated like an algebraic equation. This expedites the transformation of differential equations from one format to another, including the elimination of unwanted variables. Transfer function and state-space formats are of particular interest. MATLAB carries out these transformations with little effort on the part of the analyst. Operator notation also leads to analytic solutions using transform methods, as is developed later. MATLAB capitalizes on some of this with the linear simulation command `lsim`, which you can use even before you understand its mathematical basis. This simulation is more efficient than a corresponding one using a nonlinear integrator such as a Runge Kutta algorithm, but it applies only to linear models.

Guided Problem 4.1

Reduce the model described by the two state-space differential equations

$$\frac{dp}{dt} = -2p - q$$

$$\frac{dq}{dt} = 7p - q$$

into a single differential equation with the variable p.

Suggested Steps:

1. Write the two equations with operator notation.

2. Solve one of the equations for q as a funciton of S and p.

3. Substitute the result of step 2 into the other equation so as to eliminate q.

4. Cast the result of part 3 into the form $f(S)p = 0$. The function $f(S)$ should be a polynomial.

5. Interpret the result of step 4 as a differential equation with dependent variable p.

Guided Problem 4.2

This guided problem gives needed practice in the reduction of state-space differential equations into differential equations with a single dependent variable, using both direct and operator analytical methods, and using MATLAB. It also offers practice in linear simulation with `lsim`.

The model of an electromechanical system considered in Section 6.1.5 gives the differential equations

$$\frac{dq}{dt} = \frac{T}{G}\left(e - \frac{R_1 T}{GC}q\right) - \frac{1}{I}p,$$

$$\frac{dp}{dt} = \frac{1}{C}q - \frac{R_2}{I}p.$$

Combine these equations directly to get two single second-order differential equations with dependent variables p and q, respectively. Then use operator methods (but not MATLAB) to accomplish the same objective. Next, find the matrices A, B, C, D, assuming that the input variable is the voltage e, and that the output variable of interest is the angular velocity of the shaft. Use MATLAB to find the two second-order differential equations a third way. Finally, use lsim to plot the shaft speed as a function of time for 1.0 second, assuming $e = e_0 \sin \omega t$ with $e_0 = 5$ volts and $\omega = 8\pi$ rad/sec. The parameters are $T = 0.2$, $G = 1.0$ volt sec, $C = 0.05$ N^{-1}m^{-1}, $I = 0.2$ kg m^2, $R_1 = 4$ ohms and $R_2 = 1.0$ N m s.

Suggested Steps:

1. Place the given equations into the forms $p = p(q,\ dq/dt)$ and $q = q(p,\ dp/dt)$, respectively.

2. Take the devivatives of the equations found in step 1. Then, substitute the results of step 1 into these equations to get two resulting differential equations of second order, each having only one dependent variable, p or q, respectively.

3. Rewrite the original differential equations using operator notation. Combine these algebraic equations to get one equation with dependent variable q only, and one equation with dependent variable p only.

4. Interpret the results of step 3 in terms of differential equations, and check to see that they agree with the results of step 2.

5. Place the original equations in the state-space matrix form, extract the coefficients A and B and evaluate these matrices numerically.

6. Relate the desired output variables to the state variables p and q, so as to define the coefficients C and D. Evaluate these matrices numerically.

7. Use the MATLAB function ss2tf with the results of steps 5 and 6; check to see that the differential equation that emerges agrees with those of steps 2 and 4.

8. Define a new vector of times from 0 to 1.000 seconds with an interval of perhaps 0.005 seconds, using MATLAB. Then, express the excitation voltages at these times as a row vector.

9. Employ the lsim command to secure the resulting plot. You may annotate this plot by hand if you do not care to learn MATLAB labeling procedures. Are the results reasonable?

PROBLEMS

4.1 A system responds as shown below to the given excitation. Examine the data carefully and report whether there are any signs of nonlinearity.

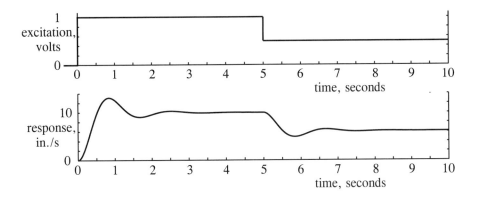

4.2 A linear model gives the following representation:

$$x(t) = G(S)u(t) \qquad G(S) = \frac{S+3}{(S+2)(S^2+2S+5)}$$

Give the corresponding differential equation relating $x(t)$ to $u(t)$.

4.3 The excitation $u(t) = 2t,\ 0 \le t \le 1;\ u(t) = 2,\ t > 1$ is applied to the system of the previous problem. Plot the response $x(t),\ 0 \le t \le 5$, using `lsim`.

4.4 Consider the *IRC* model of Fig. 3.29 (p. 119), as represented by equations (3.32) (p. 121).

(a) Combine these equations to get single second-order differential equations with dependent variables p and q, respectively.

(b) Repeat part (a), using operator methods (but not MATLAB).

(c) Find the state-space matrices A, B, C and D, assuming that $\dot{q} = p/I$ and q are the output variables of interest.

(d) Use MATLAB to convert your answer to part (c) to single differential equations in $\dot{q}_I$ and q_C, using $R = 1$, $C = 2$ and $I = 3$. Verify whether your answers are consistent with those of part (a).

4.5 Answer the questions of the preceding problem for the *IRC* model of Examples 3.14 and 3.15 (pp. 120-121).

4.6 A model is described by the following state differential equations:

$$\frac{dp}{dt} = -3p + q + \frac{du(t)}{dt}$$
$$\frac{dq}{dt} = 2p - 2q + 5u(t)$$

(a) Find the transfer function $G(S)$ that relates the excitation $u = u(t)$ to the response $q(t)$.

(b) For the input $u = t$, $0 \leq t \leq 2$; $u(t) = 4 - t$, $2 \leq t \leq 4$; $u(t) = 0$, $t > 4$, use `lsim` to plot the response $q(t)$, $0 \leq t \leq 8$.

4.1.10   SOLUTIONS TO GUIDED PROBLEMS

Guided Problem 4.1

1. $(S+2)p = -q$,
   $(S+1)q = 7p$.

2. From the first equation, $q = -(S+2)p$.

3. Substituting the result above into the second equation, $-(S+1)(S+2) = 7p$.

4. Rearranging, $(S^2 + 3S + 9)p = 0$.

5. $\dfrac{d^2p}{dt^2} + 3\dfrac{dp}{dt} + 9p = 0$.

Guided Problem 4.2

1. $p = -\dfrac{R_1 T^2 I}{G^2 C} q - I\dfrac{dq}{dt} + \dfrac{IT}{G}e$ (a)

   $q = \dfrac{CR_2}{I}p + C\dfrac{dp}{dt}$ (b)

2. $\dfrac{dp}{dt} = -\dfrac{R_1 T^2 I}{G^2 C}\dfrac{dq}{dt} - I\dfrac{d^2q}{dt^2} + \dfrac{IT}{G}\dfrac{de}{dt}$ (c)

   $\dfrac{dq}{dt} = \dfrac{CR_2}{I}\dfrac{dp}{dt} + C\dfrac{d^2p}{dt^2}$ (d)

   Substitution of equations (b) and (d) into (a) gives

   $p = -\dfrac{R_1 T^2 I}{G^2 C}\left(\dfrac{CR_2}{I}p + C\dfrac{dp}{dt}\right) - I\left(\dfrac{CR_2}{I}\dfrac{dp}{dt} + C\dfrac{d^2p}{dt^2}\right) + \dfrac{IT}{G}e$

   or, collecting terms,

   $IC\dfrac{d^2p}{dt^2} + \left(CR_2 + \dfrac{R_1 T^2 I}{G^2}\right)\dfrac{dp}{dt} + \left(1 + \dfrac{R_1 R_2 T^2}{G^2}\right)p = \dfrac{IT}{G}e$

   Substitution of equations (a) and (c) into (b) gives

   $q = \dfrac{CR_2}{I}\left(-\dfrac{R_1 T^2 I}{G^2 C}q - I\dfrac{dq}{dt} + \dfrac{IT}{G}e\right) + C\left(-\dfrac{R_1 T^2 I}{G^2 C}\dfrac{dq}{dt} - I\dfrac{d^2q}{dt^2} + \dfrac{IT}{G}\dfrac{de}{dt}\right)$

   or, collecting terms,

   $IC\dfrac{d^2q}{dt^2} + \left(\dfrac{R_1 T^2 I}{G^2} + CR_2\right)\dfrac{dq}{dt} + \left(1 + \dfrac{R_1 R_2 T^2}{G^2}\right)q = \dfrac{CIT}{G}\dfrac{de}{dt} + \dfrac{CR_2 T}{G}e$

3. $\left(S + \dfrac{R_1 T^2}{G^2 C}\right)q + \dfrac{1}{I}p = \dfrac{T}{G}e$ and $\left(s + \dfrac{R_2}{I}\right)p - \dfrac{1}{C}q = 0$

   Solving the second equation for p and substututing into the first,

   $\left(S + \dfrac{R_1 T^2}{G^2 C}\right)q + \dfrac{1}{I}\dfrac{q/C}{S + R_2/I} = \dfrac{T}{G}e$

   Multiplying both sides by $(S + R_2/I)$ gives

   $\left[S^2 + \left(\dfrac{R_1 T^2}{G^2 C} + \dfrac{R_2}{I}\right)S + \left(\dfrac{1}{IC} + \dfrac{R_1 R_2 T^2}{G^2 CI}\right)\right]q = \left(\dfrac{T}{G}s + \dfrac{R_2 T}{IG}\right)e$

   Solving the second equation for q, on the other hand, and substituting into the first,

   $\left(S + \dfrac{R_1 T^2}{G^2 C}\right)C\left(S + \dfrac{R_2}{I}\right)p + \dfrac{1}{I}p = \dfrac{T}{G}e$

or $\quad \left[S^2 + \left(\dfrac{R_1 T^2}{G^2 C} + \dfrac{C R_2}{I} \right) S + \left(\dfrac{R_1 R_2 T^2}{G^2 I C} + \dfrac{1}{C I} \right) \right] p = \dfrac{T}{GC} e$

4. The differential equations resulting from step 3 are the same as those resulting from step 2 (apart from the multiplicative factor IC in part 2).

5. $\dfrac{d}{dt} \begin{bmatrix} q \\ p \end{bmatrix} = \begin{bmatrix} -R_1 T^2 / G^2 C & -1/I \\ 1/C & -R_2/I \end{bmatrix} \begin{bmatrix} q \\ p \end{bmatrix} + \begin{bmatrix} T/G \\ 0 \end{bmatrix} e$

Substituting values into the two matrices, $\mathbf{A} = \begin{bmatrix} -3.2 & -5 \\ 20 & -5 \end{bmatrix}$ and $\mathbf{B} = \begin{bmatrix} 0.2 \\ 0 \end{bmatrix}$.

The output variable is $\dot{\phi} = p/I$, so $\mathbf{C} = [0 \ 5]$ and $D = 0$.

7. The MATLAB programming and its response is as follows:

```
A = [-3.2 -5;20 -5]; B = [.2;0]; C = [0 5]; D=0;
[z,p] = ss2tf(A,B,C,D)

z =

        0         0    20.0000

p =

   1.0000    8.2000  116.0000
```

This means that $\dot{\phi} = \dfrac{20}{S^2 + 8.2S + 116} e$

or $\ (S^2 + 8.2S + 116)p = 20Ie(t) = 4e(t)$, which agrees with the result above.

8.-9. The MATLAB programming and response is as follows:

```
t=[0:.005:1];
e=5*sin(8*pi*t);
lsim(A,B,C,D,e,t)
```

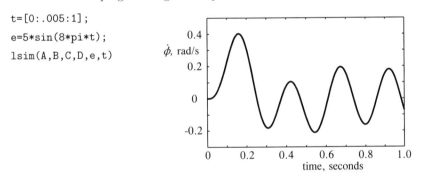

$\dot{\phi}$, rad/s

time, seconds

4.2 Common Functions in Excitations and Responses

This section reviews common functions that are employed as excitations of systems or describe the responses of linear models to these excitations. It is important that you be able to sketch-plot these functions, and recognize them from plots.

4.2.1 Exponential Functions

The exponential function

$$y(t) = e^{st}, \tag{4.13}$$

where t is time and s has the dimensions of inverse time, occurs naturally in the response of linear systems to simple excitations. The constant s can be real, imaginary or complex.

Positive real values of s produce functions that grow exponentially in time, whereas negative real values produce functions that decay exponentially, as indicated in Fig. 4.4. The function te^{-t}, which occurs frequently in the response of dynamic systems, is plotted

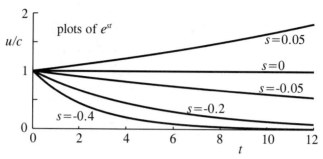

Figure 4.4: Real exponential waveforms

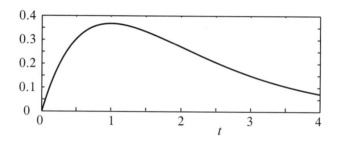

Figure 4.5: The function te^{-t}

in Fig. 4.5. It starts at zero and peaks at $t = 0$. This t is dimensionless; you can replace it with t/τ to make it dimensional.

An imaginary value of $s = j\omega$, where ω is the constant magnitude and $j = \sqrt{-1}$ gives the complex function[2]

$$y(t) = e^{\pm j\omega t} \equiv \cos(\omega t) \pm j \sin \omega t. \tag{4.14}$$

No physical variable is actually complex, but sums of complex functions are often quite useful representations of real physical variables. The sum and difference

$$y_1(t) = \frac{1}{2}\left[e^{j\omega t} + e^{-j\omega t}\right] \equiv \cos(\omega t); \tag{4.15a}$$

$$y_2(t) = \frac{1}{2j}\left[e^{j\omega t} - e^{if\omega t}\right] \equiv \sin(\omega t); \tag{4.15b}$$

in which the imaginary terms cancel, are extremely useful ways to represent the cosine and sine functions, respectively. The frequency of these functions is ω, given as radians per unit time. Recall that a unitless angle always is given in radians.

The weighted sum of a cosine wave and a sine wave,

$$y(t) = c_r \cos \omega t - c_i \sin \omega t, \tag{4.16}$$

is still sinusoidal but is shifted in time, as illustrated in Fig. 4.6.

Often the exponential decay function is multiplied by the sine or cosine function to produce a decaying sine or cosine wave. The example

$$y(t) = \frac{1}{2}\left[(c_r + jc_i)e^{(-a+j\omega)t} + (c_r - jc_i)e^{(-a-j\omega)t}\right]$$
$$= e^{-at}(c_r \cos \omega t - c_i \sin \omega t), \tag{4.17}$$

---

[2]Mathematicians usually use the symbol i for the unit imaginary number, but engineers have reserved this symbol for electric current and prefer j. MATLAB recognizes either letter.

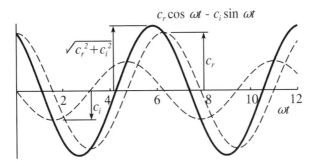

Figure 4.6: General sinusoidal waveform

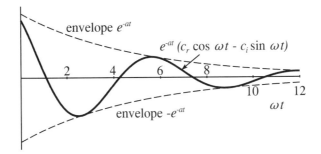

Figure 4.7: Waveform for complex conjugate pair

in which the real part of s is $a = 1/\tau = \omega/2\pi$, is shown in Fig. 4.7. The decaying exponential is called the **envelope** of the function, which has the magnitude $\sqrt{c_r^2 + c_i^2}\, e^{-at}$.

The excitations or inputs to engineering systems frequently are assumed to be sinusoidal, particularly when the concern is vibrations. The responses to many excitations include steady and/or decaying sinusoids.

4.2.2 Singularity Functions

A family of singularity functions that is zero for all time $t < 0$ is particularly useful.

The **unit step function** $u_s(t)$, pictured in Fig. 4.8, is defined as

$$u_s(t) = \begin{cases} 0 & \text{for } t \leq 0 \\ 1 & \text{for } t > 0 \end{cases} \tag{4.18}$$

Step changes of other amplitudes or dimensions can be represented by multiplying the unit step by a constant.

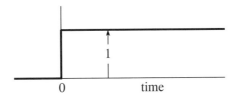

Figure 4.8: Unit step waveform

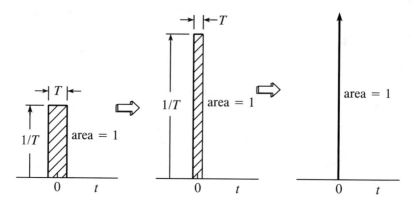

Figure 4.9: Unit impulse waveform

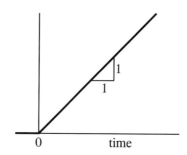

Figure 4.10: Unit ramp waveform

The **unit impulse function** $\delta(t)$ approximates a sharp pulse disturbance. As pictured in Fig. 4.9, it is defined as the limit of the square pulse of width T and integral (area) 1 as $T \to 0$:

$$\delta(t) = \lim_{T \to 0} \delta_T(t); \qquad \delta_T(t) = \begin{cases} 0 & \text{for } t \le 0 \\ 1/T & 0 < t \le T \\ 0 & \text{for } t > 0. \end{cases} \tag{4.19}$$

The time integral of the unit impulse is the unit step:

$$\int_{-\infty}^{t} \delta(t)\, dt = u_s(t). \tag{4.20}$$

Conversely, the time derivative of the unit step can be considered to be the unit impulse, although discontinuous functions are not differentiable in the formal sense:

$$\delta(t) = \frac{du_s}{dt}. \tag{4.21}$$

The **unit ramp function** $u_r(t)$, pictured in Fig. 4.10, is defined as increasing linearly with derivative (slope) 1 from the origin:

$$u_r = \begin{cases} 0 & \text{for } t \le 0 \\ t & \text{for } t > 0 \end{cases} \tag{4.22}$$

The ramp is the integral of the unit step, and the unit step is the derivative of the ramp:

$$u_r = \int_{-\infty}^{t} u_s(t)\, dt; \qquad u_s(t) = \frac{du_r}{dt}. \tag{4.23}$$

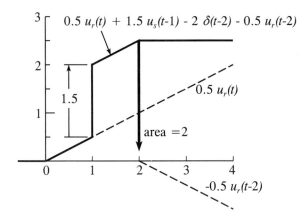

Figure 4.11: Time-shifted singularites and waveform synthesized therefrom

A singularity can be located at a time $t = t^*$ different from $t = 0$ by writing $t - t^*$ in place of t (or $t - 0$) as the argument of the singularity function. An example is shown in Fig. 4.11. This allows functions composed of a series of steps, ramps and impulses to be represented as a sum of singularity functions. The property of superposition then can be invoked to represent the response of a particular system to such a function as the sum of the responses of the individual singularity functions. This possibility considerably enhances the usefulness of singularity functions.

4.2.3 Summary

A short list of functions occurs over and over again in mathematical descriptions of commonly assumed excitations for linear models and the responses to these excitations. Exponential functions can be made to include sinusoidal and decaying sinusoidal oscillations by employing imaginary and complex numbers. Impulses, steps and ramps represent the other most important class of excitation; the step is the time integal of the impulse, and the ramp is the time integral of the step.

PROBLEMS

4.7 Sketch-plot the following functions, without using any calculator or computer:

(a) $x(t) = 5\,e^{-2t}$

(b) $x(t) = 2\cos(3t + 45°)$

(c) $x(t) = 10\,e^{-2t}\cos(3t + 45°)$

(d) $x(t) = t - 1 + e^{-t}$

4.8 Write approximate analytic expressions for the functions plotted below:

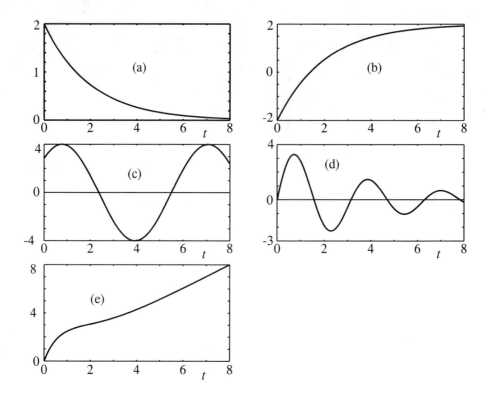

4.9 For the function plotted below,

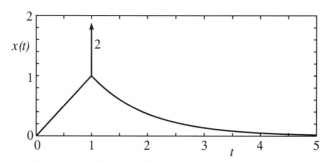

(a) Estimate the function $x(t)$ in analytical form.

(b) Determine analytically and plot $y(t) = \int x(t)dt$.

4.3   Direct Solutions of Linear Differential Equations

This section reviews classical methods for the solution to the general nth-order scalar linear ordinary equation with constant coefficients, which is given as equation (4.6) and is repeated here:

$$
\boxed{
\begin{aligned}
&a_n \frac{d^n x}{dt^n} + a_{n-1} \frac{d^{n-1} x}{dt^{n-1}} + a_{n-2} \frac{d^{n-2} x}{dt^{n-2}} + \cdots + a_1 \frac{dx}{dt} + a_0 x \\
&= f(t) = b_m \frac{d^m u}{dt^m} + b_{m-1} \frac{d^{m-1} u}{dt^{m-1}} + \cdots + b_1 \frac{du}{dt} + b_0 u.
\end{aligned}
}
\tag{4.24}
$$

These methods tend to be more ponderous than the use of the more sophisticated LaPlace transform; your instructor may direct skipping part or all of this section, depending on your background and the objectives of your course. The consideration of impulses as disturbances or input excitations are most likely to be new to the student, but are also especially amenable to the Laplace transform approach.

Some results are given without derivation; you are referred to almost any mathematical textbook on linear differential equations for a rigorous development.

The **homogenous equation** is the differential equation above with $f(t) = 0$. Its solution, designated $x_h(t)$, plus any **particular solution** of the full equation, designated as $x_p(t)$, equals the general solution to equation (4.24):

$$
\boxed{x(t) = x_h(t) + x_p(t).}
\tag{4.25}
$$

The **existence and uniqueness theorem** states that this solution is unique, provided that precisely n independent conditions are specified. These conditions most commonly comprise values at the initial time: $x(0), \dot{x}(0), \cdots, d^{n-1}x/dt^{n-1}|_{t=0}$.

4.3.1 The Homogeneous Solution

The homogeneous solution is the complete solution when there is no excitation $f(t)$. This solution contains undetermined coefficients equal in number to the order of the system. Their values are determined by an equal number of conditions that must be specified to make the solution unique. Most commonly, these conditions are specified at time $t = 0$ and are called initial conditions. The most common initial conditions specified are the values of x and its first $n - 1$ derivatives. If all these conditions are zero, all the coefficients and the solution are zero.

When the excitation $f(t)$ is non-zero, the homogeneous solution forms a part of the complete solution, as indicated by Equation (4.25). In this case the undetermined coefficients depend partly on the excitation, and usually are non-zero even if all the initial conditions are zero.

The homogeneous differential equation can be written in an algebraic form through the use of the operator notation introduced in Section 4.1.3 (p. 159):

$$
(a_n S^n + a_{n-1} S^{n-1} + a_{n-2} S^{n-2} + \cdots + a_1 S + a_0) x_h = 0.
\tag{4.26}
$$

Since the general $x_h \neq 0$, the following **characteristic equation** results:

$$
\boxed{a_n S^n + a_{n-1} S^{n-1} + a_{n-2} S^{n-2} + \cdots + a_1 S + a_0 = 0.}
\tag{4.27}
$$

This is an nth order polynomial in S which possesses n solutions, or roots, that are labeled here as $r_1, r_2, \ldots, r_n$. It is necessary to find these roots, regardless of whether you use the direct solution technique or a fancier transform technique.

The homogeneous solution comprises a sum of terms associated with the individual roots. A *real unrepeated* root r_i contributes the term $c_i e^{r_i t}$ to this solution, where c_i is an undetermined coefficient. The simplest example is the first-order differential equation, for which the only root is $r = -a_0/a_1$, giving the exponential decay

$$
\boxed{x_h = c e^{-rt}.}
\tag{4.28}
$$

Real roots r_1 and r_2 of a second-order equation give a solution with two time constants and exponential decays:

$$x_h = c_1 e^{-r_1 t} + c_2 e^{-r_2 t}. \tag{4.29}$$

Second and higher-order charateristic equations often have complex roots, which come in complex-conjugate pairs:

$$r_{1,2} = -a \pm j\omega. \tag{4.30}$$

The homogeneous solution becomes

$$x_h = c_1 e^{(-a+j\omega)t} + c_2 e^{(-a-j\omega)t}. \tag{4.31}$$

To be meaningful physically, this solution must be real. This happens if and only if the coefficients c_1 and c_2 are also complex conjugate:

$$c_{1,2} = \frac{1}{2}(c_r \pm j c_i). \tag{4.32}$$

With the use of the identities

$$e^{j\alpha} \equiv \cos\alpha + j\sin\alpha, \tag{4.33a}$$
$$e^{-j\alpha} \equiv \cos\alpha - j\sin\alpha, \tag{4.33b}$$

there results

$$\boxed{x_h = e^{-at}(c_r \cos\omega t - c_i \sin\omega t),} \tag{4.34}$$

which is the decaying oscillation of equation (4.17) and Fig. 4.7.

A *multiple real root* of the characteristic equation, r_i, repeated m times, contributes to the homogeneous solution the terms

$$(c_{i0} + c_{i1}t + c_{i2}t^2 + \cdots + c_{im}t^m)e^{r_i t}.$$

The example of a second-order equation with two equal roots $S_{1,2} = -\omega$ gives

$$\boxed{x_h = (c_1 + c_2 t)e^{-\omega t}.} \tag{4.35}$$

A repeated pair of complex roots (which doesn't happen very often) produces a combined contribution of the form

$$[c_{i0}\sin(\omega_i t + \beta_{i0}) + c_{i1}t\sin(\omega_i t + \beta_{i1}) + \cdots]e^{-at},$$

similar to the case with repeated real roots.

EXAMPLE 4.7

Find the homogeneous solution of the fifth-order differential equation

$$\frac{d^5 x}{dt^5} + 9\frac{d^4 x}{dt^4} + 43\frac{d^3 x}{dt^3} + 71\frac{d^2 x}{dt^2} - 24\frac{dx}{dt} - 100x = 2 + 3t^2.$$

Solution: The characteristic equation is

$$S^5 + 9S^4 + 43S^3 + 71S^2 - 24S - 100 = 0.$$

No general analytic solution exists for a fifth-order polynomial; resorting to a numerical solution is necessary. To solve by MATLAB, you enter and receive

```
>> roots([1 9 43 71 -24 -100])
ans=
   -3.0000   +4.0000i
   -3.0000   -4.0000i
   -2.0000
   -2.0000
    1.0000
```

The root $S = 1.0000$ contributes the term $c_1 e^t$ to the homogeneous solution. Since this root is positive, $x_h \to \infty$ as $t \to \infty$, and the model is *unstable*. The pair of roots $S = -2.0000$ contributes the terms $(c_2 + c_3 t)e^{-2t}$, and the pair of roots $S = -3.0000 \pm 4.0000i$ contributes $(c_4 \sin 4t + c_5 \cos 4t)e^{-3t}$ which also can be written as $c_4 \sin(4t + \beta)e^{-3t}$. Collecting these terms, the homogeneous solution can be written in either of the equivalent forms

$$x_h = c_1 e^t + (c_2 + c_3 t)e^{-2t} + c_4 e^{-3t}\sin(4t + \beta),$$
$$x_h = c_1 e^t + (c_2 + c_3 t)e^{-2t} + e^{-3t}(c_4 \sin 4t + c_5 \cos 4t),$$

which have five undetermined coefficients.

A method for determining the constant coefficients is illustrated in Section 4.3.3. Should there be no excitation or input disturbance, the procedure described there can be applied directly. If there is an input, however, the determination of the coefficients must be deferred until the particular solution is added to the homogeneous solution.

4.3.2   The Method of Undetermined Coefficients

This is the classical method for finding the particular solutions to linear ordinary differential equations with the various types of inputs $f(t)$ detailed in Section 4.2. The particular solution can be *any* solution to the complete differential equation; it might as well be the simplest possible solution. The forms of the terms in the particular solution that correspond to the various input terms are given in Table 4.1. All that needs to be done is to affix initially undetermined coefficients to each of the applicable terms, substutite them for x_p in the differential equation, and evaluate the coefficients by satisfying the resulting equation for all time t.

Table 4.1   **Method of Undetermined Coefficients**

term in $f(t)$	corresponding terms in x_p
1	c
e^{st}	ce^{st}
t^n	$c_n t^n + c_{n-1} t^{n-1} + \cdots + c_1 t + c_0$
$\sin \omega t$ or $\cos \omega t$ or $\cos(\omega t + \beta)$	$c_1 \sin \omega t + c_2 \cos \omega t$ or $c \cos(\omega t + \beta + \phi)$

The particular solution for the step input can be seen to be $x_p = $ constant for $t > 0$.

The exponential input $u(t) = u_0 e^{st}$ gives the form $f(t) = f_0 e^{st}$. This corresponds to the second entry in Table 4.1; the particular solution is of the form $x_p(t) = cu_0 e^{st}$. Carrying out the method of undetermined coefficients by substituting these into the differential equation (equation (4.24)) and performing the indicated differentiations,

$$(a_n s^n + a_{n-1} s^{n-1} + \cdots + a_0) c u_0 e^{st} = (b_m s^m + b_{m-1} s^{m-1} + \cdots + b_0) u_0 e^{st}. \qquad (4.36)$$

Solving for c and thus $x_p(t)$,

$$c = \frac{b_m s^m + b_{m-1} s^{m-1} + \cdots + b_0}{a_n s^n + a_{n-1} s^{n-1} + \cdots + a_0} = G(s), \qquad (4.37a)$$

$$\boxed{x_p(t) = G(s) u_0 e^{st}.} \qquad (4.37b)$$

Thus c equals the same transfer function $G(s)$ as given by the operator notation of equation (4.11) in Section 4.1.3 (p. 159) to represent the general linear differential equation with constant coefficients, except that the original argument and derivative operator S is replaced by the exponential coefficient s. *This fact is very useful*; it is employed extensively, particularly for sinusoidal inputs.

EXAMPLE 4.8

Find the particular solution to the differential equation given in Example 4.7.

Solution: The terms on the right side of the example, namely $f(t) = 2 + 3t^2$, are of the power-law form t^n, with $n = 0$ and $n = 2$. Table 4.1 gives the form of the particular solution as $x_p = c_2 t^2 + c_1 t + c_0$. Substitution of this solution into the differential equation gives

$$71(2c_2) - 24(2c_2 t + c_1) - 100(c_2 t^2 + c_1 t + c_0) = 2 + 3t^2,$$

from which

$$-100c_2 = 3; \quad -48c_2 - 100c_1 = 0; \quad 142c_2 - 24c_1 - 100c_0 = 2.$$

The result is

$$x_p = -0.03t^2 + 0.0144t - 0.046056.$$

4.3.3   Application of Initial Conditions

The undetermined coefficients in the homogeneous solution must be found after the complete solution is formed by adding the particular solution, if it is non-zero. The particular solution itself should at this point have no undetermined coefficients. The undetermined coefficients in the homogeneous part of the solution are resolved through the introduction of the same number of independent conditions. In the most common case these conditions are given at $t = 0$ and therefore are called initial conditions. This case is illustrated below. When one or more conditions are specified at a different time, such as a final time, the procedure is essentially the same.

EXAMPLE 4.9

Solve the following second-order differential equation for the given initial conditions:

$$\frac{d^2x}{dt^2} + 2\frac{dx}{dt} + 5x = 3(1 - e^{-2t}); \qquad x(0) = 0; \quad \dot{x}(0) = 0.$$

Solution: The characteristic equation and its roots are

$$S^2 + 2S + 5 = (S+1)^2 + (2)^2 = 0; \qquad S_{1,2} = -1 \pm j2,$$

so that the homogeneous solution is

$$x_h = e^{-t}(c_1 \cos 2t + c_2 \sin 2t).$$

Treating the input as $u(t) = u_1(t) + u_2(t)$ with $u_1(t) = 1$ and $u_2(t) = -e^{-2t}$, the particular solution can be found from equation (4.37b) as follows:

$$x_p = 3G(0) - 3G(-2)e^{-2t} = \frac{3}{5} - \frac{3}{s^2 + 2s + 5}e^{-2t}\bigg|_{s=-2} = 0.6(1 - e^{-2t}).$$

Note that u_1 is in the proper exponential form with $s = 0$, giving simply $G(0) = 0.6$ for its response. Thus the complete solution and its time derivative are

$$x = x_h + x_p = e^{-t}(c_1 \cos 2t + c_2 \sin 2t) + 0.6(1 - e^{-2t}),$$

$$\frac{dx}{dt} = e^{-t}[(2c_2 - c_1)\cos 2t - (2c_1 - c_2)\sin 2t] + 1.2e^{-2t}.$$

Substitution of the initial conditions into these equations specifies c_1 and c_2:

$$0 = c_1 + 0 \qquad \text{or} \quad c_1 = 0$$
$$0 = 2c_2 - c_1 + 1.2 \qquad \text{or} \quad c_2 = -0.6.$$

Substitution of these values gives the unique result

$$x = 0.6(1 - e^{-2t} - e^{-t}\sin 2t).$$

This is plotted below, along with its homogeneous and particular components. Note that the given initial conditions are satisfied, and that virtually any other initial conditions also could be satisfied by different weightings of the two components.

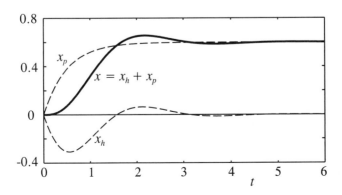

4.3.4   Solutions to Impulse Inputs

The impulse function $f(t) = \delta(t)$ is not included in Table 4.1, since it acts in a special way. The entire excitation lasts but an instant; it is useful to say that its duration is between time $t = 0^-$ and time $t = 0^+$. The excitation is zero for $t > 0^+$, which means that the total solution has identically the same form as the unforced or homogeneous solution. The effect of the impulse is to impart an initial condition for this homogeneous solution at time $t = 0^+$ that is different from the condition at time $t = 0^-$.

The jump $x(0^+) - x(0^-)$ can be determined by integrating the full differential equation between the times $t = 0^-$ and $t = 0^+$. For the first-order case, $dx/dt + (1/\tau)x = c\delta(t)$, this integration is

$$\int_{0^-}^{0^+} \frac{dx}{dt} dt + \frac{1}{\tau} \int_{0^-}^{0^+} x\, dt = \int_{0^-}^{0^+} c\,\delta(t)\, dt, \tag{4.38}$$

which gives

$$x\big|_{0^-}^{0^+} + \frac{1}{\tau} \int_{0^-}^{0^+} x\, dt = c. \tag{4.39}$$

Note the use of the fact that the integral across a unit impulse is 1. The integral term in equation (4.39) must be zero unless x were infinite over the vanishingly small interval between the limits of integration, which it is not.

Since $x(0^-)$ is by definition zero, the result is $x(0^+) = c$, which gives the complete solution

$$x = ce^{-t/\tau}, \qquad t \geq 0^+. \tag{4.40}$$

For the nth-order differential equation,

$$\int_{0^-}^{0^+} \frac{d^n x}{dt^n} dt + \int_{0^-}^{0^+} a_{n-1} \frac{d^{n-1} x}{dt^{n-1}} dt + \cdots + a_0 \int_{0^-}^{0^+} x\, dt = \int_{0^-}^{0^+} c\,\delta(t)\, dt, \tag{4.41}$$

Only the highest order derivative contributes to the left side of this equation, with the result that the impulse imparts a non-zero value to only this term:

$$\frac{d^{n-1} x}{dt^{n-1}}\bigg|_{t=0^-}^{0^+} = c. \tag{4.42}$$

EXAMPLE 4.10

Find the response of the general underdamped second-order model to an impulse of amplitude c, for both general initial conditions and for zero initial conditions.

Solution: The complete solution (for $t \geq 0^+$) is the same as the homogeneous solution, as given by equation (4.34) and repeated here:

$$x = e^{-at}(c_r \cos \omega t - c_i \sin \omega t), \qquad t \geq 0^+.$$

This gives

$$x(0^+) = c_r$$
$$\dot{x}(0^+) = -ac_r - \omega_d c_i$$

Since only the highest-order derivative changes across the impulse, $x(0^+) = x(0^-)$. From equation (4.42), $\dot{x}(0^+) - \dot{x}(0^-) = c$. Therefore,

$$c_r = x(0^-),$$
$$c_i = -\frac{1}{\omega}[ax(0^-) + \dot{x}(0^-) + c].$$

Should the initial conditions $x(0^-)$ and $\dot{x}(0^-)$ be zero, the solution simplifies to $x = (1/\omega_d)e^{-at} \sin \omega_d t$.

4.3.5 Differentiation and Integration Properties

If an excitation $u_2(t)$ of a system is the time derivative of an excitation $u_1(t)$ to the same system, then the response $x_2(t)$ is the time derivative of the response $x_1(t)$. That is,

$$\text{if} \quad u_2 = \frac{du_1}{dt} \quad \text{then} \quad x_2 = \frac{dx_1}{dt}. \qquad (4.43)$$

Thus, for example, the response of a given model to a unit impulse input, traditionally written $g(t)$, equals the time derivative of the reponse to a unit step, traditionally written $h(t)$, since the impulse is the time derivative of the step; $g(t) = d\,h(t)/dt$. Inversely, if the excitation $u_2(t)$ of a system is the time integral of the excitation $u_1(t)$ of the same system, the response $x_2(t)$ is the time integral of the response $x_1(t)$, or

$$\text{if} \quad u_2 = \int u_1(t)\,dt \quad \text{then} \quad x_2(t) = \int x_1(t)\,dt. \qquad (4.44)$$

Thus, similarly, the response of a given model to a unit step input equals the time integral of the response to a unit impulse; $h(t) = \int g(t)\,dt$.

These properties allow you to extend a known solution to certain other cases. Rather than go through the analysis above to find the response to an impulse, for example, you have the option of finding the response to a step, and then computing its derivative, taking care not to overlook any discontinuities.

4.3.6 Summary

The classical representation for the solution of a linear differential equation with constant coefficients comprises the general solution to the homogeneous equation (with zero right side) plus any particular solution to the complete equation. The homogeneous solution is found using the roots of the characteristic polynomial formed from the coefficients of the various terms. The particular solution can be found, for the cases of primary interest, using the method of undetermined coefficients. The final step in finding the unique solution to a particular input is the determination of the coefficients of the homogeneous part of the solution through the use of the initial conditions.

Singularity waveforms including impulses, steps and ramps, and exponential and power-law waveforms have been featured as simple system inputs with readily calculable particular solutions. These waveforms are particularly useful, because they can be combined to represent much more complicated waveforms.

Guided Problem 4.3

This straightforward problem reviews the classical method for solving linear differential equations. It also illustrates an important phenomenon.

Find and plot the solution $x(t)$ to the following differential equation, subject to the initial conditions $x(0) = 0$, $\dot{x}(0) = 0$:

$$\frac{1}{\omega_n^2} \frac{d^2 x}{dt^2} + x = \begin{cases} 0 & t < 0 \\ \sin(0.95\omega_n t) & t > 0. \end{cases}$$

Suggested Steps:

1. Note that the equation can be simplified by employing $t' = \omega_n t$ rather than t as the independent variable.

2. Write the characteristic equation and find its roots. Use these to write the homogeneous solution with its two unknown constants.

3. Use the method of undetermined coefficients as summarized in Table 4.1 (p. 178) to find the particular solution.

4. Substitute the initial conditions into the total solution to evaluate the unknown constants.

5. Plot the results using MATLAB or some other software package. The result demonstrates a "beating" phenomenon between motion at the natural frequency and motion at the slightly different forcing frequency, which is perpetuated because of the absence of damping.

PROBLEMS

4.10 Solve the following differential equation with the initial condition $x(0) = 0$ and the forcing function $u(t) = 10$. Sketch-plot your answer.

$$\frac{dx}{dt} + 5x = u(t)$$

4.11 Answer the preceding question substituting $u(t) = 10 \sin 2t$.

4.12 Answer the preceding question substituting $u(t) = 10(1 - e^{-t})$.

4.13 A model and its excitation is described by the following linear differential equation. Solve the equation for $x(t)$ given the initial condition $(dx/dt)_{t=0} = 9$, and sketch-plot the answer roughly to scale in both t and x.

$$\frac{dx}{dt} + 3x = 6.$$

4.14 Solve the following differential equation with the initial conditions $x(0) = 0$ and $\dot{x}(0) = 0$. Sketch-plot your answer.

$$\frac{d^2x}{dt^2} + 4\frac{dx}{dt} + 8x = 10$$

4.15 Solve the preceding problem with the coefficient 4 replaced by 6.

4.16 Solve the preceding problem retaining the coefficient 6 but changing the coefficent 8 to 9.

4.17 The capacitors of the electric circuits of Problems 3.24 (p. 109) and Problem 3.40 (p. 126) have an initial charge q_0. Find the currents through the inductors as a function of time.

4.18 The masses of Problems 3.25 (p. 109) and 3.41 (p. 126) are given an initial velocity $\dot{x}_0$, but the springs are initially unstressed. Find the position of the mass as a function of time for (i) system (a), (ii) system (b). Assume $\zeta < 1$.

4.19 A linear model of a system results in the following representation:

$$x(t) = G(S)u(t); \qquad G(S) = \frac{S}{S^2 + 2S + 5}.$$

    (a) Give the corresnponding differential equation relating $u(t)$ to $x(t)$.

    (b) Give the resulting homogeneous solution for arbitrary initial conditions.

    (c) Find the particular solution when $u(t) = \sin 2t$.

    (d) Find $x(t)$ when $u(t) = 3u_s(t)$ and all initial conditions are zero.

4.20 A particular system responds to a step excitation with a ramp response. Characterize the dynamics of the system generally.

4.21 *Continuation of Problem 3.23 (p. 109):* (a) Apply causal strokes to the bond graph of the disk, and annotate the variables in the conventional fashion. (b) Write the state differential equation(s). (c) Determine and sketch-plot the resulting angular velocity of the disk as a function of time, assuming it starts at rest and is excited subsequently by the force $F = 6$ N.

4.22 The excitation $u(t)$ and impulse response $g(t)$ of a linear system are plotted below.

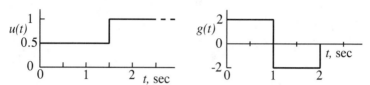

(a) Carefully sketch and label the step response, $h(t)$, of the system.

(b) Represent the input $u(t)$ as a sum of steps.

(c) Sketch the response of the system as the superposition of the responses to the individual steps in part (b).

4.23 An input to a system is as plotted below:

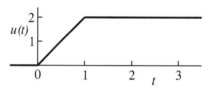

(a) Represent this waveform as a sum of singularity functions.

(b) Find the response to the first-order system $dx/dt + x = u(t)$.

4.24 Carry out the above problem for the input plotted below:

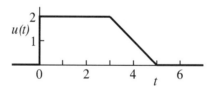

4.25 A spring pushes a piston which forces a viscous liquid through a tube, as shown.

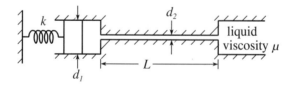

(a) Find the effective resistance of the system with respect to the motion of the piston, in terms of the symbols given. Assume fully developed flow. The parameters are $k = 50$ lb/in., $d_1 = 1$ in., $d_2 = 0.05$ in., $L = 40$ in. and $\mu = 2.9 \times 10^{-6}$ lb·s/in^2 (water at 70°F).

(b) Neglecting inertia, the characteristic equation has only a single root. Find its reciprocal, which later will be described as a "time constant" that characterizes how fast a system approaches a new steady state. Assume x moves a certain distance from an initial state to a new equilibrium state. Determine what fraction of this distance it moves during one time constant.

4.26 Find $F(t)$ for the system shown with $x = x_0 u_s(t)$. $\qquad F(t) \longrightarrow$

4.27 One way to measure viscosity is to observe the velocity v at which small spheres settle in still fluid. The terminal velocity sometimes is not reached quickly, however, due to the inertia of the sphere and some of the surrounding fluid.

Consider a sphere of radius $r = 0.13$ cm and density $\rho_s = 7840$ kg/m^3 (steel) settling in a fluid of density $\rho = 1200$ kg/m^3 and viscosity $\mu = 0.20$ N·s/m^2. Assume Stokes law, which gives a force equal to $6\pi\mu r v$ and applies with considerable accuracy for Reynold's number $N_R = 2\rho v r/\mu < 2$, and can be adequate for $2 < N_R < 10$.

(a) Find the resistance with respect to the velocity v.

(b) Find the terminal velocity of the sphere. Check the Reynold's number to make sure Stokes relation applies.

(c) Find the effective inertance, which includes the mass of the sphere plus the "virtual mass" of the fluid that also is accelerated. The latter approximately equals the mass of fluid that occupies a volume equal to one-half that of the sphere.

(d) Find the time at which v reaches 95% of the terminal velocity.

(e) Find the corresponding distance.

SOLUTION OF GUIDED PROBLEM

Guided Problem 4.3

1. With $t' = \omega_n t$, $\qquad \dfrac{d^2 x}{dt'^2} + x = 0 \qquad t' < 0$

$$= \sin(0.95t') \qquad t' > 0$$

2. The characteristic equation is $s^2 + 1 = 0$, which has roots $s_{1,2} = \pm j$.

   The resulting homogeneous solution is $x_h = c\sin(t' + \phi)$.

3. Since $f(t) = \sin(0.95t')$, $x_p = x_{p1}\sin(0.95t') + x_{p2}\cos(0.95t')$,

$$\frac{d^2 x_p}{dt'^2} = -(0.95)^2 [x_{p1}\sin(0.95t') + x_{p2}\cos(0.95t')].$$

   Substituting into the differential equation,

   $[-(0.95)^2 + 1][\sin(0.95t') + x_{p2}\cos(0.95t')] = \sin(0.95t')$.

   Therefore, $x_{p2} = 0$, $x_{p1} = \dfrac{1}{1 - (0.95)^2} = 10.256$,

   giving $x = c\sin(t' + \phi) + 10.256\sin(0.95t')$.

4. $\dfrac{d^2 x}{dt'^2} = -c\sin(t' + \phi) - 9.256\sin(0.95t')$.

   At $t' = 0$, $x = 0 = c\sin\phi$; therefore, $\phi = 0$.

   At $t' = 0^+$, $\dfrac{dx}{dt} = 0 = c\cos t' + 9.744\cos(0.95t') = c + 9.744$,

   therefore, $c = -9.744$, giving $x = -9.744\sin t' + 10.256\sin(0.95t')$.

5.

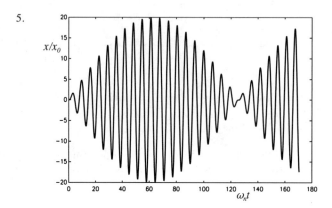

4.4   Convolution*

This section is helpful in gaining a full understanding of superposition, and in accessing certain exact analytical and approximate numerical applications thereof, including MATLAB programs. It is optional, however, in the sense that the Laplace transform of the following section is sufficient to solve most of the relevant problems. Your instructor may wish to save time by skipping ahead.

Convolution is a basic process of applying superposition, in which the input signal to a linear system is decomposed into a sum of steps or a sum of square pulses or impulses equally spaced over time. If a limited number of steps or pulses are used, the decomposition may be approximate; if an infinite number of steps or pulses is used, the decomposition becomes exact. The response to a unit step or pulse is then determined, from which the responses to each individual input step or pulse is found. Finally, these responses are summed to give the response to the original input.

4.4.1   Decomposing Signals into a Sum of Steps

An arbitrary function of time $f(t)$ can be approximated by either a sum of pulses or a sum of steps. Steps may be preferred, since fewer are required to form a decent approximation, as suggested in Fig. 4.12. If the successive steps are distinguished by a subscript, k, their various amplitudes are designated by Δf_k and the times at which the jumps occur are designated by T_k, the entire step-function approximation for $t > t_0$ can be written

$$f(t) \simeq f(t_0) + \sum_k \Delta f_k \, u_s(t - T_k). \tag{4.45}$$

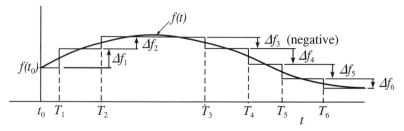

Figure 4.12: Approximation of a continuous function by a series of step functions

The steps in the figure are placed at varying intervals of time, as appropriate considering the varying slope of $f(t)$.

Alternatively, the steps could be uniformly spaced with a constant interval ΔT. In this case, multiplying and dividing the summation term by ΔT (so as not to change the equation) gives

$$f(t) \simeq f(t_0) + \sum_k \frac{\Delta f_k}{\Delta T} u_s(t - T_k)\Delta T. \tag{4.46}$$

Now as $\Delta T \to dT$ (*i.e.*, becomes infinitesimally small), the ratio $\Delta f/\Delta T$ becomes the *slope* df/dT and the summation becomes an exact integration:

$$f(t) = f(t_0) + \int_{t_0}^{\infty \text{ or } t} \frac{df}{dT} u_s(t - T)\, dT. \tag{4.47}$$

The interval $t < T < \infty$ contributes nothing to the integral (since $\mu_1(t - T) = 0$ for $T > t$), so the upper limit of integration can be reduced from ∞ to t if desired.

4.4.2 Discrete Convolution

An arbitrary input or disturbance $u(t)$ is to be imposed onto a general stationary linear system as represented by its operator $G(S)$. Either a crude step-wise approximation or an exact respresentation of $u(t)$ can be used. In both cases, it is assumed that $u(t)$ is zero for $t < 0$. In the case of the approximation,

$$u(t) \simeq \sum_k \Delta u(T_k)\, u_s(t - T_k). \tag{4.48}$$

The key idea is that the linear operator $G(S)$ satisfies superposition. The response to the sum of steps is simply the sum of the responses to the individual steps:

$$\boxed{x(t) \simeq \sum_k \Delta u(T_k)\, h(t - T_k).} \tag{4.49}$$

EXAMPLE 4.12

Consider the system with the response to a unit step, $h(t)$, as measured experimentally (perhaps following the throwing of a switch) and plotted in part (a) of Fig. 4.13. The system is excited by the signal $u(t)$ as also plotted. Use step decompostion of the input with $\Delta t = 0.04$ seconds and discrete convolution to approximate the response.

Solution: The first step is to approximate $u(t)$ by a sum of steps. This is done graphically in part (a) of the figure. The times T_k and amplitudes of the steps $\Delta u * T_k$ also are given in the first numerical row of Table 4.2. The use of only five steps may seem rather crude, but in view of the sluggishness of the step response its approximateness is not bad for most purposes.

The next step is to find the responses to the five component steps. These are shown by dashed lines in part (b) of the figure and are given in the five associated columns in Table 4.2. The sum of these responses is the approximate system response, $x(t)$, plotted as a solid line and tabulated in the right-most column of the table.

An alternative discrete convolution is based on decomposing the excitation signal into a sum or sequence of square pulses of amplitude $u(T_k)\Delta T$, which is the product of the height and the width of the kth pulse. Therefore,

$$u(t) \simeq \sum_k u(T_k)\Delta T\, \delta_T(t - T_k). \tag{4.50}$$

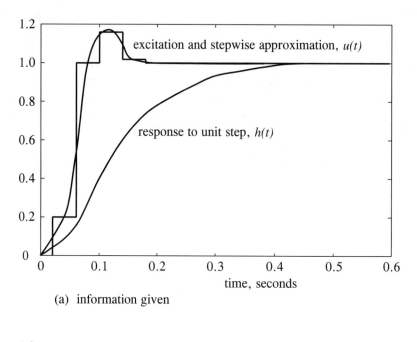

(a) information given

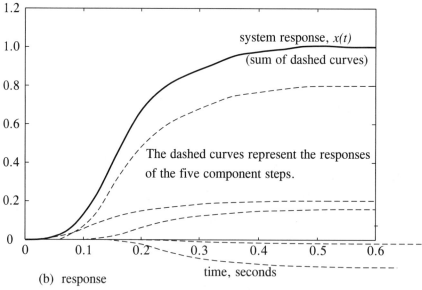

(b) response

Figure 4.13: Example 4.12

Table 4.2   Equation (4.49) Applied to Example 4.12

unit step at origin		time	$\Delta u(T_k)h(t - T_k)$ for $T_k/\Delta u(T_k)$					sum
t, sec	$h(t)$	t, sec	.02/.20	.06/.80	.10/.16	.14/-.14	.18/-.02	$x(t)$
0.02	0.04	0.04	0.008					0.008
0.06	0.16	0.08	0.032	0.032				0.064
0.10	0.40	0.12	0.080	0.128	0.006			0.214
0.14	0.60	0.16	0.120	0.320	0.026	-0.006		0.460
0.18	0.74	0.20	0.148	0.480	0.064	-0.022	-0.001	0.669
0.22	0.82	0.24	0.164	0.592	0.098	-0.056	-0.003	0.793
0.26	0.88	0.28	0.176	0.656	0.118	-0.084	-0.008	0.858
0.30	0.94	0.32	0.118	0.704	0.131	-0.104	-0.012	0.907
0.34	0.96	0.36	0.192	0.752	0.141	-0.115	-0.015	0.955
0.38	0.98	0.40	0.196	0.768	0.150	-0.123	-0.016	0.975
0.42	1.00	0.44	0.200	0.784	0.154	-0.132	-0.018	0.988
0.46	1.00	0.48	0.200	0.800	0.157	-0.134	-0.019	1.004
0.50	1.00	0.52	0.200	0.800	0.160	-0.137	-0.019	1.004
0.54	1.00	0.56	0.200	0.800	0.160	-0.140	-0.020	1.000
0.58	1.00	0.60	0.200	0.800	0.160	-0.140	-0.020	1.000

The response to a single pulse of unit amplitude, centered at time T_k, is defined as $g_T(t - T_k)$. Therefore, the overall response is

$$x(t) \simeq \sum_k g_T(t - T_k)u(T_k)\Delta T. \qquad (4.88)$$

EXAMPLE 4.13

Apply pulse convolution to the signals of Example 4.12, using the unit impulse response $g(t)$ plotted in part (a) of Fig. 4.14 and the same excitation $u(t)$ as repeated in the figure. Use the same time intervals.

Solution: Fifteen square pulses as represented in the figure serve to approximate $u(t)$. Their times and areas are listed in the first two numerical rows of the main table given in Table 4.3. Values of $g(t)$ are listed just under the figure title. The responses to the individual pulses are plotted with dashed lines in part (b) of the figure, and listed in columns in the table. The sums of the responses at a given time t represent the complete response $x(t)$ at that time; these are shown in the plot by dots and in the table by the right-most column.

4.4.3   Discrete Convolution by MATLAB

The MATLAB command `conv(p,q)` carries out discrete convolution of the signals p and q, which must be vectors of equal length. For time convolution based on steps, p represents the vector of values of Δu, and q represents the vector of values of h. For time convolution based on pulses, p represents the vector of values of the pulse areas $u\Delta t$, and q represents the vector of values of g. The order of p and q can be reversed.

The computation produces a vector for the output that is twice the length of the proscribed vectors. For purposes here, the second half of this vector is spurious, and should be discarded.

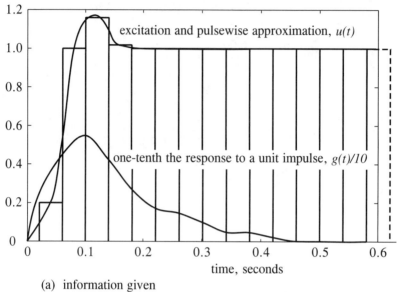

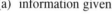

(a) information given

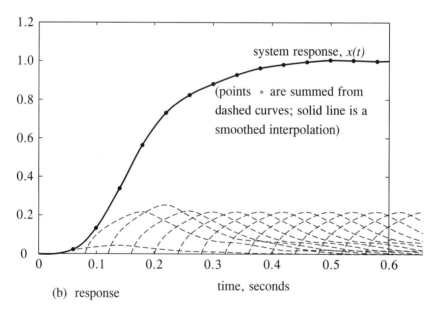

(b) response

Figure 4.14: Example 4.13

Table 4.3 Equation (4.51) Applied to Example 4.12
for unit pulse centered at $t = 0$

t,sec	.02	.06	.10	.14	.18	.22	.26	.30	.34	.38	.42	.46	.50	.54	.58
$g(t)$	2.5	4.5	5.5	4.25	2.75	1.75	1.5	1.0	0.5	0.5	.25	0	0	0	0

center of pulse T_k, sec (1st line); pulse area (2nd line)

T_k, s	.04	.08	.12	.16	.20	.24	.28	.32	.36	.40	.44	.48	.52	.56	.60
area	.008	.04	.0464	.0408	.04	.04	.04	.04	.04	.04	.04	.04	.04	.04	.04

t, s	over $\Delta u(T_k)g(t - T_k)$															$x(t)$
.06	.020															0.0200
.10	.036	.10														0.1360
.14	.044	.18	.1160													0.3400
.18	.034	.22	.2088	.1020	values sum (horizontally) to give $x(t)$											0.5648
.22	.022	.17	.2552	.1836	.10											0.7308
.26	.014	.11	.1972	.2244	.18	.10										0.8256
.30	.012	.07	.1276	.1734	.22	.18	.10									0.8830
.34	.008	.06	.0812	.1122	.17	.22	.18	.10								0.9314
.38	.004	.04	.0696	.0714	.11	.17	.22	.18	.10							0.9650
.42	.004	.02	.0464	.0612	.07	.11	.17	.22	.18	.10						0.9816
.46	.002	.02	.0232	.0408	.06	.07	.11	.17	.22	.18	.10					0.9960
.50	0	.01	.0232	.0204	.04	.06	.07	.11	.17	.22	.18	.10				1.0036
.54	0	0	.0116	.0204	.02	.04	.06	.07	.11	.17	.22	.18	.10			1.0020
.58	0	0	0	.0102	.02	.02	.04	.06	.07	.11	.17	.22	.18	.10		1.0002
.62	0	0	0	0	.01	.02	.02	.04	.06	.07	.11	.17	.22	.18	.10	1.0000

EXAMPLE 4.14

Carry out the step-wise convolution of Example 4.12 using MATLAB.

Solution: The MATLAB coding

```
du = [.2 .8 .16 -.14 -.02 0 0 0 0 0 0 0 0 0 0];
h = [.04 .16 .4 .6 .74 .82 .88 .94 .96 .98 1 1 1 1 1];
x = conv(du,h)
```

in which du is really Δu, produces the response

```
x =
Columns 1 through 7
0.0080   0.0640   0.2144   0.4600   0.6688   0.7928   0.8584
Columns 8 through 14
0.9076   0.9552   0.9748   0.9884   1.0036   1.0036   1.0004
Columns 15 through 21
1.0000   0.8000   0        -0.1600  -0.0200  0         0
Columns 22 through 28
0        0        0        0        0        0         0
Columnn 29
0
```

The first 15 numbers are correct; all the remaining numbers should equal 1.0000, but do not. This is because the vector h is truncated at 15 numbers, whereas the real function continues at unity. The vector can be truncated to exclude the spurious tail by entering x = x(1:15), and plotted by entering plot(x).

EXAMPLE 4.15

Carry out the pulse-wise convolution of Example 4.13 using MATLAB.

Solution: The coding

```
udt=[.2 1 1.16 1.02 1 1 1 1 1 1 1 1 1 1 1]*.04;
g=[2.5 4.5 5.5 4.25 2.75 1.75 1.5 1.0 0.5 0.5 0.25 0 0 0 0];
x=conv(udt,g)
```

gives virtually the same response as the previous step-wise case.

4.4.4   Convolution Integrals

For an exact analysis you can let $\Delta u \to du$ and let T become continuous, so equation (4.49) (p. 187) gives the pair of equivalent equations

$$x(t) = \int_{t_0}^{t} h(t-T)\,\dot{u}(T)\,dT = \int_{0}^{t-t_0} h(\tau)\,\dot{u}(t-\tau)\,d\tau. \tag{4.52}$$

The dots over the u's in these expressions imply differentiation with respect to T and $\tau \equiv t-T$, respectively. These key equations are known as the superposition or **convolution integrals** of the time functions $h(t)$ and $u(t)$, often written $h(t) * \dot{u}(t)$, which means that *the output equals the convolution of the step response with the time rate of change of the input.*

Equations (4.52) can be integrated by parts to give an alternative pair of convolution integrals that are more widely used. Recall the formula

$$\int u\,dv = uv - \int v\,du, \tag{4.53}$$

and let

$$du = -\dot{h}(t-T)\,dT = -g(t-T)\,dT, \tag{4.54a}$$
$$dv = u(T)\,dT. \tag{4.54b}$$

The first term on the right side of equation (4.53) becomes

$$h(t-T)\,u(T)\big|_{t_0}^{t} = h(0)\,u(t) - h(t-t_0)\,u(t_0) = 0, \tag{4.55}$$

which vanishes because $u(t_0)$ was given as zero, and, for a causal system, $g_s(0) = 0$. Thus equation (4.53) gives

$$x(t) = \int_{t_0}^{t} g(t-T)\,u(T)\,dT = \int_{0}^{t-t_0} g(\tau)\,u(t-\tau)\,d\tau. \tag{4.56}$$

These express the convolution $g(t) * u(t)$, which represents that *the output equals the time convolution of the impulse response with the excitation signal.*

The discrete convolution based on square pulses, as given by equation (4.56), corresponds to the approximation of equation (4.51) for uniformly separated discrete times.

EXAMPLE 4.16

A simple system is described by its impulse response:

$$g(t) = e^{-at}, \qquad t > 0.$$

Use a convolution integral to compute its response to the input signal

$$u(t) = 1 - e^{-bt}, \qquad t > 0.$$

These signals are plotted in parts (a) and (b) of Fig. 4.15.

Solution: From equation (4.56), the response is

$$x(t) = \int e^{-a(t-T)} \left(1 - e^{-bT}\right) dT = e^{-at} \int \left[e^{aT} - e^{(a-b)T}\right] dT$$

$$= e^{-at} \left(\frac{e^{aT}}{a} - \frac{e^{(a-b)T}}{a-b}\right)\Bigg|_0^t = \frac{1}{a} + \frac{be^{-at}}{a(a-b)} - \frac{1}{a-b}e^{-bt}$$

The meaning of convolution is presented visually in part (c) of the figure. The functions $g(t - T)$ and $u(T)$ are plotted vs. the common dummy time T; their product also is shown. The shaded area of the product is the value of $x(t)$. The first function $g(t - T)$ can be moved right or left to give any desired time t. It can be perceived as a function which weights the contributions to the present state of the various parts of the excitation signal as a function of the age of those parts, that is how long before the present time they occurred. In the example, the older the segment of excitation $u(T)$ considered, the less it contributes to the response $x(t)$ (regardless of the function $u(T)$), since $g(t - T)$ decays monatonically as T is reduced backward in time from the current time t. (The other three convolution integrals can be interpreted similarly.)

4.4.5 Summary

The superposition property dictates that the response to a signal comprising a sum of component signals equals the sum of the responses to the individual component signals. Approximate implementation of this property can be carried out by decomposing a disturbance into a sum of discrete steps, and summing the responses to each of these steps, in a process known as discrete convolution. An alternative discrete convolution employs the impulse response instead of the step response, and the input signal directly in place of the changes in the input signal. In both cases the result approaches the exact solution as the steps are made smaller and smaller; the summations become integrations known as convolution integrals.

Guided Problem 4.4

This is a mandatory problem involving superposition which can be done graphically, without mathematics.

A system $G(s)$ responds to a unit impulse with a square pulse of amplitude 2.0 and duration 1.0 seconds. The system is excited with a double-step signal as shown in Fig. 4.16. Find and plot the response of the system.

Suggested Steps:

1. Since the excitation comprises steps, it is most convenient to work with the response

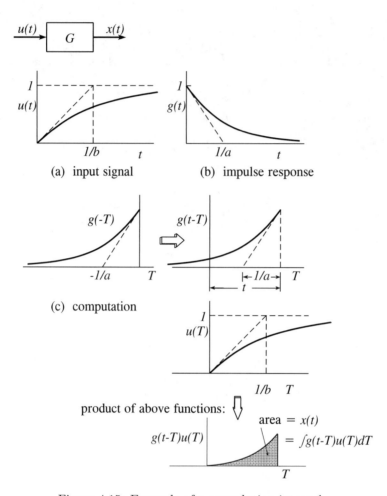

Figure 4.15: Example of a convolution integral

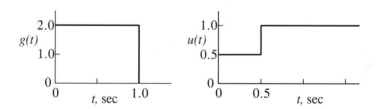

Figure 4.16: Excitation signal for Guided Problem 4.4

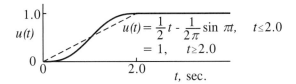

Figure 4.17: Excitation signal for Guided Problem 4.9

of the system to steps. Integrate the unit impulse response to give the unit step response.

2. Find the response of the system to the individual steps of the input signal.

3. Superimpose the responses of step 2 to give the complete response.

Guided Problem 4.5

The system in the previous guided problem is excited by the signal shown in Fig. 4.17. Find its response. Analytical convolution is intended.

Suggested Steps:

1. Since the excitation is not a finite sum of steps, the superposition must be carried out either by representing the excitation approximately by a sum of small steps or square pulses, or by carrying out a convolution integration. Choose the integration so as to get an exact answer to the problem as posed. You have the choice of computing the convolution $g(t) * u(t)$ or $h(t) * \dot{u}(t)$. If you opt for the latter, find $\dot{u}(t)$ and $h(t)$.

2. Set up one of the four convolution integrals as given in equations (4.52) and (4.56) (p. 192). Sketch the factors in the integrand versus the variable of integration. Take special care of the limits of integration.

3. Complete the integration and sketch the result. Does it make sense?

PROBLEMS

4.28 The response of a particular linear system to a *unit impulse* is $2\sin 3t$.

(a) Find the response of the system to a *unit step*, assuming zero initial conditions.

(b) The system is excited by the square pulse shown below. Find the smallest value of T for which the response of the system becomes zero for $t > T$.

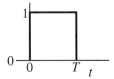

4.29 Carry out suggested step 9 of Guided Problem 4.2 (pp. 165-166) using the MATLAB command `conv` rather than `lsim`.

4.30 Find and plot the response of the system $G(S)$ in Problem 4.2 (p. 167) to the excitation $u(t) = te^{-3t}$, using the MATLAB command `conv`. Let $0 < t < 6$.

4.31 Do part (b) of Problem 4.6 (p. 168) using the MATLAB command `conv` rather than `lsim`.

4.32 Use MATLAB to plot an approximate response $x(t)$, $t \le 8$ seconds, of the model which has the step response $h(t)$ and the time derivative $\dot{u}(t)$ of the input disturbance $u(t)$, as given below.

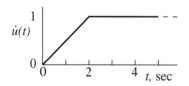 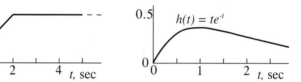

4.33 Find the exact analytic solution to the preceding problem.

SOLUTIONS TO GUIDED PROBLEMS

Guided Problem 4.8

1. $h(t) = \int g(t)\,dt = \begin{cases} 0, & t \le 0 \\ 2t, & 0 \le t \le 1 \\ 2, & t \ge 1 \end{cases}$

2. Response to step at $t = 0$: $x_1 = 0.5\,h(t) = \begin{cases} 0, & t \le 0 \\ t, & 0 \le t \le 1 \\ 1, & t \ge 1 \end{cases}$

   Response to step at $t = 0.5$ s:$x_2 = 0.5\,h(t - 0.5) = \begin{cases} 0, & t \le 0.5 \\ t - 0.5, & 0.5 \le t \le 1.5 \\ 1, & t \ge 1.5 \end{cases}$

3. $x(t) = x_1(t) + x_2(t) = \begin{cases} 0, & t \le 0 \\ t, & 0 \le t \le 0.5 \\ 2t - 0.5, & 0.5 \le t \le 1 \\ t + 0.5, & 1 \le t \le 1.5 \\ 2, & t \ge 1.5 \end{cases}$

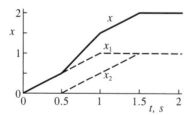

Guided Problem 4.9

1. As in the previous problem, $g(t) = 2[u_s(t) - u_r(t - 1)]$ and $h(t) = 2[u_r(t) - u_r(t-1)]$ where $u_r(t)$ is a ramp with slope 1 which starts at $t = 0$. The function
$$u(t) = \begin{cases} \frac{1}{2}\left(t - \frac{1}{\pi}\sin \pi t\right), & t \le 2 \\ 1 & t \ge 2 \end{cases} \text{ is given, from which}$$
$$\dot{u}(t) = \begin{cases} \frac{1}{2}(1 - \cos \pi t) & t \le 2 \\ 0 & t \ge 2 \end{cases}$$

2. For the choice $x(t) = \int_0^t g(t-T)\,u(T)\,dT,$

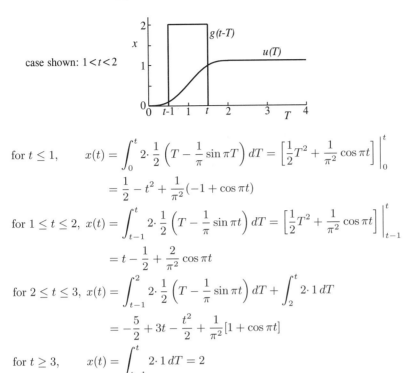

case shown: $1 < t < 2$

for $t \leq 1$, $\quad x(t) = \int_0^t 2 \cdot \frac{1}{2}\left(T - \frac{1}{\pi}\sin \pi T\right) dT = \left[\frac{1}{2}T^2 + \frac{1}{\pi^2}\cos \pi t\right]\Big|_0^t$

$$= \frac{1}{2} - t^2 + \frac{1}{\pi^2}(-1 + \cos \pi t)$$

for $1 \leq t \leq 2$, $x(t) = \int_{t-1}^t 2 \cdot \frac{1}{2}\left(T - \frac{1}{\pi}\sin \pi t\right) dT = \left[\frac{1}{2}T^2 + \frac{1}{\pi^2}\cos \pi t\right]\Big|_{t-1}^t$

$$= t - \frac{1}{2} + \frac{2}{\pi^2}\cos \pi t$$

for $2 \leq t \leq 3$, $x(t) = \int_{t-1}^2 2 \cdot \frac{1}{2}\left(T - \frac{1}{\pi}\sin \pi t\right) dT + \int_2^t 2 \cdot 1 \, dT$

$$= -\frac{5}{2} + 3t - \frac{t^2}{2} + \frac{1}{\pi^2}[1 + \cos \pi t]$$

for $t \geq 3$, $\quad x(t) = \int_{t-1}^t 2 \cdot 1 \, dT = 2$

Plot of the result:

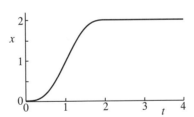

4.5 The Laplace Transform

The Laplace transform is a formal extension of the operator methods introduced starting in Section 4.1.3. It aids the solution of ordinary linear differential equations, which it converts to algebraic equations. The Laplace transform also is used in Chapter 10 to aid the solution of partial differential equations, which it converts to ordinary differential equations.

The Laplace transform is most logically developed from the Fourier transform; it also can be developed logically from the convolution integral. Since the former is not presented until Chapter 7 and the latter is treated as optional, however, these developments are not given until the end of the present section, and themselves are treated as optional. They are not necessary for the level of understanding needed to address the problems given in this book, although they are quite helpful in gaining a fuller understanding of the Laplace transform and its relation to other methods.

4.5.1 Definition and Inverse

The Laplace transform $F(s)$ of a function of time $f(t)$ usually is defined as

$$F(s) \equiv \mathcal{L}[f(t)] = \int_0^\infty f(t)e^{-st}dt, \qquad s = \sigma + j\omega. \tag{4.57}$$

This transform exists (does not become infinite) for all physically possible (finite) signals if σ has a real value larger than some minimum number. Note that the transform is unaffected by the history of $f(t)$ for $t < 0$. Either you restrict its use to functions that are zero for all $t < 0$, or you represent that history solely by its effect on $f(0)$ and the time derivatives $f'(0)$, $f''(0)$, etc. The **inverse Laplace transform** is written as

$$f(t) = \frac{1}{2\pi j} \int_{\sigma-j\infty}^{\sigma+j\infty} F(s)e^{+st}ds. \tag{4.58}$$

The general evaluation of this integral is a major topic in the study of the functions of a complex variable, which is not addressed here since you need to know only a restricted set of easily computed transforms. A set of transform pairs sufficient to handle linear differential equations with constant coefficients is given in Table 4.4, assuming they are used with what is called partial fraction expansions. Several of these pairs are derived below. A more extensive set of Laplace transform pairs, given in Appendix C, reduces the need for partial fraction expansions.[3]

4.5.2 The Derivative Relations

The Laplace transform of the time derivative of a function is related to the Laplace transform of the function itself in a special way that can be found by an integration by parts:

$$\begin{aligned}
\mathcal{L}\left[\frac{df(t)}{dt}\right] &= \int_0^\infty e^{-st}\frac{df(t)}{dt}dt \\
&= s\int_0^\infty e^{-st}f(t)dt + e^{-st}f(t)\Big|_0^\infty \\
&= sF(s) - f(0). \tag{4.59}
\end{aligned}$$

As a result, s often is called the **time derivative operator**, a description which is most apt when $f(0) = 0$. It can be viewed as a formalization of the operator S introduced in Section 4.1.3. Applying this property again to find the Laplace transform of the second derivative,

$$\mathcal{L}\left[\frac{d^2 f(t)}{dt^2}\right] = s\mathcal{L}\left[\frac{df(t)}{dt}\right] - f(0) = s^2 F(s) - sf(0) - \dot{f}(0). \tag{4.60}$$

Thus, for the nth derivative,

$$\mathcal{L}\left[\frac{d^n f(t)}{dt^n}\right] = s^n F(s) - s^{n-1}f(0) - s^{n-2}\dot{f}(0) - \ldots - (d^{n-1}f/dt^{n-1})_{t=0}. \tag{4.61}$$

The inverse of s can be considered to be the **time integral operator**:

$$\frac{1}{s} \longleftrightarrow \int_0^t (\ )dt. \tag{4.62}$$

---

[3]Laplace transform pairs 34 to 44 given on the last page of Appendix C apply to distributed-parameter models such as in Chapter 10.

Table 4.4   A Short Table of Laplace Transform Pairs

time function, $t \geq 0$	Laplace transform
$a\,f(t)$	$a\,F(s)$
$f(t) + g(t)$	$F(s) + G(s)$
$\dfrac{d^n}{dt^n} f(t)$	$s^n F(s) - s^{n-1} f(0) - s^{n-2} \dot{f}(0) - \cdots - (d^{n-1} f/dt^{n-1})_{t=0}$
$\displaystyle\int_{-\infty}^{t} f(t)dt$	$\dfrac{1}{s} F(s) + \dfrac{1}{s} \displaystyle\int_{-\infty}^{0} f(t)$
$t^n f(t)$	$(-1)^n \dfrac{d^n}{ds^n} F(s)$
$e^{-at} f(t)$	$F(s + a)$
$f(t - T)$	$e^{-Ts} F(s)$

time function, $t \geq 0$	Laplace transform
unit impulse $\delta(t)$	1
unit step $u_s(t)$	$1/s$
$\dfrac{1}{n!} t^n \quad (n = 1, 2, 3, \ldots)$	$1/s^{n+1}$
e^{-at}	$1/(s + a)$
$\dfrac{1}{n!} t^n e^{-at} \quad (n = 1, 2, 3, \ldots)$	$1/(s + a)^{n+1}$
$2e^{p_r t}[r_r \cos(p_i t) - r_i \sin(p_i t)]$	$\dfrac{r_r + r_i j}{s - p_r - p_i j} + \dfrac{r_r - r_i j}{s - p_r + p_i j}$ $= \dfrac{2r_r(s - p_r) - 2r_i p_i}{(s - p_r)^2 + p_i^2}$
$Ae^{-at} \cos(\omega t - \phi)$ $= Ae^{-at} \sin(\omega t + \psi)$	$(bs + c)/[(s + a)^2 + \omega^2]$

$$A = \sqrt{b^2 + [(c - ab)/\omega]^2} \qquad \phi = \tan^{-1}[(c - ab)/b\omega] \qquad \psi = \tan^{-1}[b\omega/(c - ab)]$$

The lower limit of integration in equation (4.57) and the transform pair (4.62) is given as $t = 0$; there also is $f(0), f'(0), \ldots, f^{n-1}(0)$ in the general pairs for first and higher-order derivatives as given in Table 4.4. This time is ambiguous in cases in which the function jumps from one value to another at precisely $t = 0$. The two values can be distinguished by referring to the one approached from negative time as occurring at time 0^-, and the one approached from positive time as occurring at time 0^+. In practice one may choose to use *either* 0^- or 0^+ in a given equation, but it is essential that all terms in the equation be interpreted consistently. When one deals with functions $f(t)$ that equal zero for $t < 0$, it is simplest to employ the 0^- option, since then the values $f(0^-), f'(0^-), \ldots, f^{n-1}(0^-)$ are all zero.

4.5.3   Singularity Functions and Discontinuities

It is also important to have at your disposal the Laplace transforms of a handful of key functions. The transform of the *unit step* $u_s(t)$ is

$$\mathcal{L}[u_s(t)] = \int_0^\infty e^{-st} dt = -\frac{1}{s} e^{-st} \Big|_0^\infty = \frac{1}{s}. \tag{4.63}$$

The unit impulse $\delta(t)$ is the time derivative of the unit step, as you have seen. Thus, equation (4.57) gives

$$\mathcal{L}[\delta(t)] = \mathcal{L}\left[\frac{du_s(t)}{dt}\right] = 1 - u_s(0). \tag{4.64}$$

But what is $u_s(0)$? At time $t = 0$ the function jumps from 0 to 1; $u_s(0^-) = 0$ and $u_s(0^+) = 1$. Which do you use? The answer, again, is that *either* is correct, as long as you choose the *same* 0^- or 0^+ in the Laplace transform of all terms. In a particular case, one may be more useful than the other, however. In the present case the use of 0^+ gives $\mathcal{L}[\delta(t)] = 0$, which isn't very useful. Examination of the defining integral reveals that the impulse then resides entirely *outside* of the integration interval. To include the impulse in that interval you use the 0^- option, which gives $\mathcal{L}[\delta(t)] = 1$. This result may be checked by direct calculation:

$$\mathcal{L}[\delta(t)] = \int_{0^-}^{\infty} \delta(t)e^{-st}dt = \int_{0^-}^{0^+} \delta(t)dt = 1. \tag{4.65}$$

The use of the 0^+ option in effect transfers the role of an excitation to that of an initial condition, which in some cases may be helpful.

4.5.4   Other Key Relations

The Laplace transforms of other key functions given in Table 4.4 are computed here. For the exponential decay e^{-at},

$$\mathcal{L}\left[e^{-at}\right] = \int_0^{\infty} e^{-at}e^{-st}dt = \int_0^{\infty} e^{-(a+s)t}dt$$

$$= \left.\frac{-e^{-(a+s)t}}{a+s}\right|_0^{\infty} = \frac{1}{s+a}. \tag{4.66}$$

For a product of the exponential decay and a function $f(t)$,

$$\mathcal{L}\left[e^{-at}f(t)\right] = \int_0^{\infty} e^{-(a+s)t}f(t)dt = F(s+a). \tag{4.67}$$

For the product $t^n f(t)$,

$$\mathcal{L}[t^n f(t)] = \int_0^{\infty} t^n f(t)e^{-st}dt = (-1)^n \frac{d^n}{ds^n}\int_0^{\infty} f(t)e^{-st}dt$$

$$= (-1)^n \frac{d^n}{ds^n} F(s); \tag{4.68}$$

Finally, for a function shifted in time by the delay T,

$$\mathcal{L}[f(t-T)] = \int_0^{\infty} e^{-st}f(t-T)dt$$

$$= e^{-Ts}\int_{-T}^{\infty} e^{-s(t-T)}f(t-T)d(t-T)$$

$$= e^{-Ts}\int_0^{\infty} e^{-s\tau}f(\tau)d\tau$$

$$= e^{-Ts}F(s), \qquad \text{where } f(t) = 0 \text{ if } t < 0. \tag{4.69}$$

The function e^{-Ts} is known as the **time delay operator**, since when multiplied by the Laplace transform of a function of time it gives the Laplace transform of the same function delayed in time by T. Note that the switch in the lower limit of integration from $-T$ to 0 requires that the function be zero-valued for $t < 0$.

4.5.5   Finding Laplace Transforms of Output Variables

The primary interest now is using the Laplace transform to solve the general linear scalar differential equation with constant coefficients, which is repeated from equation (4.6) or (4.24):

$$
\boxed{
\begin{aligned}
a_n \frac{d^n x}{dt^n} &+ a_{n-1}\frac{d^{n-1}x}{dt^{n-1}} + a_{n-2}\frac{d^{n-2}x}{dt^{n-2}} + \cdots + a_1\frac{dx}{dt} + a_0 x \\
&= f(t) = b_m\frac{d^m u}{dt^m} + b_{m-1}\frac{d^{m-1}u}{dt^{m-1}} + \cdots + b_1\frac{du}{dt} + b_0 u,
\end{aligned}
}
\tag{4.70}
$$

The first of two steps in solving this equation for $x(t)$ is finding the Laplace transform $X(s)$. The second step is evaluating $x(t)$ from this transform. The first step is addressed in this subsection. The second step is simple if a directly applicable Laplace tansform pair is available in a table.

The first case of interest is the initially quiescent system subject to a transient excitation. This is especially simple to treat, because all of the initial-value terms $x(0)$, $\dot{x}(0)$, $\cdots$, $(d^{n-1}x/dt^{n-1})_{t=0}$ and $u(0)$, $\dot{u}(0)$, $\cdots$, $(d^{m-1}u/dt^{m-1})_{t=0}$ are zero. Therefore, the Laplace transform of the differential equation becomes

$$
(a_n s^n + a_{n-1}s^{n-1} + \cdots + a_1 s + a_0)X(s) = (b_m s^m + b_{m-1}s^{m-1} + \cdots + b_0)U(s), \tag{4.71}
$$

which can be written in the shorthand notation

$$
\boxed{
\begin{aligned}
X(s) &= G(s)U(s), \\
G(s) &= \frac{b_m s^m + b_{m-1}s^{m-1} + \cdots + b_1 s + b_0}{a_n s^n + a_{n-1}s^{n-1} + \cdots + a_1 s + a_0}.
\end{aligned}
}
\tag{4.72}
$$

The **Laplace transfer function** $G(s)$ therefore is the same as the transfer function operator of equation (4.11) (p. 159), except a lower-case s replaces the upper-case S. $G(s)$ also is noted in Section 4.3.2 (equation (4.37), p. 178) to represent the ratio of the response $x(t)$ to the excitation e^{st}. Finally, $G(j\omega)$ is employed in Section 7.1.1 to compute the responses to sinusoidal disturbances.

Another major case is the initial-value problem in which $f(t) = 0$ for $t \geq 0$. In this case, the Laplace transform of equation (4.70) gives

$$
\begin{aligned}
X(s) = \frac{1}{p(s)}\Big[&(a_n s^{n-1} + a_{n-1}s^{s-2} + \cdots + a_1)x(0) + (a_n s^{n-2} + a_{n-2}s^{n-3} \\
&+ \cdots + a_2)\dot{x}(0) + (a_n s^{n-3} + \cdots + a_3)\ddot{x}(0) + \cdots \\
&+ (a_n s + a_n - 1)\left(\frac{d^{n-2}x}{dt^{n-2}}\right)_{t=0} + a_n\left(\frac{d^{n-1}x}{dt^{n-1}}\right)_{t=0}\Big].
\end{aligned}
\tag{4.73}
$$

In almost all examples of interest, very few of the terms within the square brackets above are non-zero; usually the terms are best found by direct application of the third entry in Table 4.4 to each term in the differential equation.

More general cases that involve a combination of non-zero initial conditions and a non-zero disturbance for $t > 0$ also can be treated by applying the third entry in Table 4.4 to each term in the differential equation.

To address the linear state-variable formulation, recall the matrix notation:

$$
\boxed{\frac{d\mathbf{x}(t)}{dt} = \mathbf{A}\mathbf{x}(t) + \mathbf{B}u(t).}
\tag{4.74}
$$

Taking the Laplace transform of both sides gives

$$s\mathbf{X}(s) - \mathbf{x}(0) = \mathbf{A}\mathbf{X}(s) + \mathbf{B}\mathbf{U}(s). \tag{4.75}$$

Solving for $\mathbf{X}(s)$,

$$\boxed{\begin{aligned} \mathbf{X}(s) &= \mathbf{G}(s)\mathbf{U}(s) + (s\mathbf{I} - \mathbf{A})^{-1}\mathbf{x}(0) \\ \mathbf{G}(s) &= (s\mathbf{I} - \mathbf{A})^{-1}\mathbf{B}. \end{aligned}} \tag{4.76}$$

Setting $\mathbf{B} = \mathbf{0}$ or $\mathbf{u}(t) = \mathbf{0}$ leaves an *initial value problem*, and setting $\mathbf{x}(0) = \mathbf{0}$ leaves a *forced response problem*.

A more general case traditionally is cited in the literature. Repeating from equation (4.8), the output variable is defined as

$$\boxed{\mathbf{y}(t) = \mathbf{C}\mathbf{x}(t) + \mathbf{D}\mathbf{u}(t).} \tag{4.77}$$

In this case,

$$\boxed{\begin{aligned} \mathbf{Y}(s) &= \mathbf{H}(s)\mathbf{U}(s) + \mathbf{C}(s\mathbf{I} - \mathbf{A})^{-1}\mathbf{x}(0), \\ \mathbf{H}(s) &\equiv \mathbf{C}(s\mathbf{I} - \mathbf{A})^{-1}\mathbf{B} + \mathbf{D}. \end{aligned}} \tag{4.78}$$

The matrix $\mathbf{H}(s)$ is simply a more generalized transfer function which allows an output, $\mathbf{y}(t)$, to be different from the state vector, $\mathbf{x}(t)$.

The *elements* of the matrices $\mathbf{G}(s)$ and $\mathbf{H}(s)$ are *scalar* transfer functions or, in the language of most elementary textbooks in automatic control which do not employ matrices, just "transfer functions." The subsequent development extended this result to frequency response, using $s = j\omega$. These facts are closely related.

EXAMPLE 4.17

Find the response of a system with the scalar transfer function

$$G(s) = \frac{2}{(s+1)(s+5)}$$

to an input unit step, using the Laplace transform table in Appendix C.

Solution: The Laplace transform of the unit step is

$$U(s) = \frac{1}{s}.$$

The Laplace transform of the response $x(t)$ is the product of $G(s)$ and $U(s)$, or

$$X(s) = \frac{2}{s(s+1)(s+5)}.$$

This is proportional to transform #15 in the table of Laplace transforms in Appendix C, with the coefficient values $a = 1$ and $b = 5$. The corresponding time function therefore is

$$x(t) = 0.4 - 0.5e^{-t} + 0.1e^{-5t}.$$

EXAMPLE 4.18

Use Laplace transforms to find the response $x(t)$ of the general underdamped second-order model to an impulse $c\,\delta(t)$, in the presence of general initial conditions $x(0)$ and $\dot{x}(0)$.

Solution: The Laplace transform of the differential equation gives

$$[(s+a)^2 + \omega_d^2]X(s) - s\,x(0) - \dot{x}(0) - 2a\,x(0) = c; \qquad a \equiv \zeta\omega_n.$$

Solving for $X(s)$,

$$X(s) = \frac{1}{(s+a)^2 + \omega_d^2}\,[(s+a)x(0) + c + \dot{x}(0) + a\,x(0)]\,,$$

which from items #18 and #19 in Appendix C (p. 608) gives

$$x(t) = \left[x(0)\cos\omega_d t + \frac{c + \dot{x}(0) + a\,x(0)}{\omega_d}\sin\omega_d t\right]e^{-at}.$$

EXAMPLE 4.19

Find the behavior of the state vector of a third-order autonomous linear system described by its matrix $\mathbf{A}$ and initial conditions $\mathbf{x}(0)$:

$$\mathbf{A} = \begin{bmatrix} -4 & 1/2 & 0 \\ 2 & -2 & 2 \\ 0 & 1 & -4 \end{bmatrix}; \qquad \mathbf{x}(0) = \begin{bmatrix} 1 \\ 0 \\ -1 \end{bmatrix}.$$

Solution: From equation (4.113),

$$\mathbf{X}(s) = (s\mathbf{I} - \mathbf{A})^{-1}\mathbf{x}(0).$$

Carrying out the indicated operations,

$$\mathbf{X}(s) = \frac{1}{p(s)}\begin{bmatrix} s^2 + 6s + 6 & s/2 + 2 & 1 \\ 2s + 8 & (s+4)^2 & 2s + 8 \\ 2 & s + 4 & s^2 + 6s + 7 \end{bmatrix}\begin{bmatrix} 1 \\ 0 \\ -1 \end{bmatrix};$$

$$p(s) \equiv |s\mathbf{I} - \mathbf{A}| = s^3 + 10s^2 + 29s + 20 = (s+1)(s+4)(s+5).$$

Carrying out the indicated matrix product,

$$\mathbf{X}(s) = \frac{1}{p(s)}\begin{bmatrix} s^2 + 6s + 5 \\ 0 \\ -(s^2 + 6s + 5) \end{bmatrix}x(0)$$

$$= \frac{1}{(s+1)(s+4)(s+5)}\begin{bmatrix} (s+1)(s+5) \\ 0 \\ -(s+1)(s+5) \end{bmatrix}x(0)$$

$$= \frac{1}{s+4}\begin{bmatrix} 1 \\ 0 \\ -1 \end{bmatrix}x(0).$$

The inverse transform, from either the tenth item in Table 4.4 or item #5 in the table of Appendix C, is

$$\mathbf{x}(t)_{=}x(0)\begin{bmatrix} e^{-4t} \\ 0 \\ -e^{-4t} \end{bmatrix}.$$

From the present perspective the cancellation of so many numerator and denominator factors to give so simple an answer seems rather an incredible coincidence. Later, in Section 7.3.4, this problem will be shown to correspond to an excitation of a "mode of motion" of a particular system. One rarely chooses initial conditions corresponding to a mode of motion accidentally; they are very special.

The MATLAB command [num,den]=ss2tf(A,B,C,D,i) gives the numerator and denominator polynomials of the matrix $\mathbf{H}(s) = \mathbf{C}(x\mathbf{I} - \mathbf{A})^{-1}\mathbf{B} + \mathbf{D}$ given in equation (4.78). Thus, the coefficients of the nine polynomials in $p(s)$ above are given (as well as the denominator coefficients) by setting $\mathbf{C}$ equal to the unit diagonal matrix and $\mathbf{D}$ equal to the null matrix. The coefficients of $X(s)$ above result from setting $\mathbf{B}$ equal to the vector $\mathbf{x}(0)$ and $\mathbf{D}$ equal to the null vector.

4.5.6   Finding Inverse Transforms: Partial Fraction Expansions

When the Laplace transform of the solution to a linear differential equation of interest is not available in a table of transforms, it is necessary to break up the transform into a sum of terms that are indeed listed. The solution is then the sum of the individual inverse transforms. The process of the break-down is called partial fraction expansion. The Laplace transforms to be expanded comprise ratios of polynomials in s. The usual expansion comprises terms of first and second order only, which are indeed available in the tables.

The partial fraction expansion of a transfer function $G(s)$ with distinct poles is

$$G(s) = \frac{r_1}{s - p_1} + \frac{r_2}{s - p_2} + \cdots + \frac{r_n}{s - p_n} + k(s). \tag{4.79}$$

A pole p_j of multiplicity m expands as follows:

$$\frac{r_j}{s - p_j} + \frac{r_{j+1}}{(s - p_j)^2} + \cdots + \frac{r_{j+m-1}}{(s - p_j)^m} \tag{4.80}$$

The MATLAB command residue can be used to carry out this expansion. This requires that you have previously defined the numerator and denominator polynomials of $G(s)$, num and den, in the standard way. The following example illustrates the procedure:

EXAMPLE 4.20

Use MATLAB to find the partial-fraction expansion of the fifth-order transfer function

$$G(s) = \frac{200,000}{s^5 + 26s^4 + 2225s^3 + 4900s^2 + 240,500s + 200,000}$$

Solution: The coding

```
num = 200000;
den = [1 26 2225 42700 240500 200000];
[r,p,k] = residue(num,den)
```

produces the response

```
r =                          p =                              k=
    0.0234 + 0.0072i             -2.5000 + 44.6514i           [ ]
    0.0234 - 0.0072i             -2.5000 - 44.6514i
   -1.2838                      -10.0000
  -10.8401                      -10.0000
    1.2370                       -1.0000
```

The term k is blank; there is no remainder. Thus,

$$G(s) = \frac{0.0234 + 0.0072j}{s + 2.5 - 44.6514j} + \frac{0.0234 - 0.0072j}{s + 2.5 + 44.6514j} - \frac{1.2838}{s + 10} - \frac{10.8401}{(s + 10)^2} + \frac{1.2370}{s + 1}.$$

The respective terms can be found in Table 4.4, leading to the result

$$\begin{aligned} g(t) =& 2e^{-2.5t} \left[0.0234 \cos(44.6514\,t) - 0.0072 \sin(44.6514\,t) \right] \\ &- 1.2838 e^{-10t} - 10.8401\, t e^{-10t} + 1.2370 e^{-t}. \end{aligned}$$

Partial fraction expansions also can be found analytically. A transfer function $G(s)$ that is a ratio of polynomials in s, i.e., $G(s) = N(s)/D(s)$, with the order of $N(s)$ being no higher than that of $D(s)$, can be expanded with no remainder. In the unusual cases where the order of $N(s)$ exceeds that of $D(s)$, $D(s)$ can be divided into $N(s)$ to give a new $N(s)$ with the same order as $D(s)$ plus a remainder polynomial in s. This first step, if necessary, is presumed below to have been executed already.

In the simplest cases the polynomial $D(s)$ has distinct real roots, as in equations (4.79), and the remainder $k(S)$ is zero. The coefficients r_i are called **residues**. They can be determined by multiplying both sides of equation (4.79) by $s - r_i$ and then setting s equal to r_i.

EXAMPLE 4.21

Evaluate the residues r_1, r_2 and r_3 of the transfer function

$$G(s) = \frac{2}{s(s + 1)(s + 5)} = \frac{r_1}{s} + \frac{r_2}{s + 1} + \frac{r_3}{s + 5},$$

which (from the tables) has the impulse response

$$g(t) = r_1 + r_2 e^{-t} + r_3 e^{-5t}.$$

Solution: To find r_1, for which $s_1 = 0$, multiply both sides of the equation by s and then set $s = 0$:

$$\lim_{s \to 0} sG(s) = \lim_{s \to 0} \frac{2}{(s + 1)(s + 5)} = \lim_{s \to 0} \left(r_1 + \frac{r_2 s}{s + 1} + \frac{r_3 s}{s + 5} \right),$$

This gives $r_1 = 2/(0 + 1)(0 + 5) = 0.4$. To find r_2, for which $s_2 = -1$, multiply $G(s)$ by $s + 1$ and set $s = -1$:

$$\lim_{s \to -1} (s + 1)G(s) = \lim_{s \to -1} \frac{2}{s(s + 5)} = \lim_{s \to -1} \left[\frac{r_1(s + 1)}{s} + \frac{r_2}{s} + \frac{r_3(s + 1)}{s + 5} \right].$$

This gives $r_2 = 2/(-1)(-1 + 5) = -0.5$. The same procedure gives $r_3 = 0.1$.

When a pair of distinct roots are complex conjugate, the same procedure can be applied to get the real and imaginary terms r_r and r_i of the next-to-last entry in Table 4.4 (p.

199). It is simpler to get these coefficients, however, by leaving the sum of the complex conjugate terms of the transfer function in the alternative real form, which also is given. A compromise strategy starts by evaluating the residues for any distinct real poles using the method above. Then, multiply both sides of the equation by the denominator of the transfer function (so the left side of the equation becomes the original numerator) and proceed to evaluate the coefficients. This procedure will provide one redundant check on the results for each real root previously evaluated.

EXAMPLE 4.22

Find the partial fraction expansion and the unit impulse response for the model with the transfer function

$$G(s) = \frac{5s^2 + 22s + 39}{(s^2 + 2s + 5)(s + 2)}$$

Compare the results using both the next-to-last entry in Table 4.4 and the last entry; which of these choices do you prefer?

Solution: Expand $G(s)$ as follows:

$$\frac{5s^2 + 22s + 39}{(s^2 + 2s + 5)(s + 2)} = \frac{as + b}{(s + 1)^2 + 4} + \frac{c}{s + 2}.$$

Next, find c by multiplying both sides by $s + 2$ and then setting $s = -2$:

$$c = \frac{20 - 44 + 39}{4 - 4 + 5} = 3.$$

Then, multiply both sides of $G(s)$ by the denominator:

$$5s^2 + 22s + 39 = (as + b)(s + 2) + 3(s^2 + 2s + 5)$$
$$= (3 + a)s^2 + (2a + b + 6)s + (2b + 15).$$

A comparison of the coefficients of the polynomials in s gives $a = 2$ and $b = 12$, plus a check. (Had c not been found first, this comparison would have given three equations with three unknowns, permitting c to be found anyway, without the check.)

The part of the impulse response corresponding to the second-order pole is found using the next-to-last entry in Table 4.4 (p. 199) gives

$$2r_r = a = 2; \qquad 2r_i = \frac{-b - 2r_r p_r}{p_i} = -5.$$

The complete impulse response becomes

$$g(t) = 3e^{-2t} + (2\cos 2t + 5\sin 2t)e^{-t}.$$

The final entry in Table 4.3 gives the two alternative results

$$f(t) = 3e^{-2t} + \sqrt{29}e^{-t}\cos(2t - 68.2°);$$
$$f(t) = 3e^{-2t} + \sqrt{29}e^{-t}\sin(2t + 21.8°).$$

All three answers are in fact equal to one another. The first version is somewhat easier to find, but either of the others reveal the behavior more clearly, since they contain only a single sinusoidal term.

The same method can be used in the case of multiple real roots, except for the use of different entries in the table.

EXAMPLE 4.23

Find the partial fraction expansion and the unit impulse response for a model with the transfer function

$$G(s) = \frac{s^2 + 2s + 3}{(s+1)^3}.$$

Solution: First, $G(s)$ is expanded according to equation (4.80) (p. 204):

$$\frac{s^2 + 2s + 3}{(s+1)^3} = \frac{a_3}{(s+1)^3} + \frac{a_2}{(s+1)^2} + \frac{a_1}{(s+1)};$$

Multiplying by the denominator,

$$(s+1)^3 G(s) = s^2 + 2s + 3 = a_3 + (s+1)a_2 + (s+1)^2 a_1$$
$$= a_1 s^2 + (2a_1 + a_2)s + (a_1 + a_2 + a_3).$$

Therefore,

$$a_1 = 1; \qquad 2a_1 + a_2 = 2; \qquad a_1 + a_2 + a_3 = 3,$$

from which $a_1 = 1$, $a_2 = 0$, $a_3 = 2$. The impulse response is, from Table 4.4,

$$g(t) = t^2 e^{-t} + 0 + e^{-t} = (1 + t^2)e^{-t}, \qquad t \geq 0.$$

4.5.7 Initial and Final Value Theorems

Sometimes you may be satisfied with a quick calculation for an initial value, $x(0)$, and a final value, $x(\infty)$, rather than the more difficult determination of the entire function, $x(t)$. These values also can be used as a partial check on the complete response. The **initial value theorem** is

$$\lim_{t \to 0} x(t) = \lim_{s \to \infty} [sX(s)], \qquad (4.81)$$

in which $t = 0^-$ or 0^+ according to the choice employed in the original Laplace transform. The **final value theorem**, which assumes stable behavior (no poles outside the left-half plane), is

$$\lim_{t \to \infty} x(t) = \lim_{s \to 0} [sX(s)]. \qquad (4.82)$$

EXAMPLE 4.24

Apply the initial and final value theorems to the $X(s)$ of Example 4.17 (p. 202):

$$X(s) = \frac{2}{s(s+1)(s+5)}.$$

Solution:

$$x(0) = \lim_{s \to \infty} \left[\frac{2}{(s+1)(s+5)} \right] = 0,$$

$$x(\infty) = \lim_{s \to 0} \left[\frac{2}{(s+1)(s+5)} \right] = 0.4,$$

which agree with the $x(t)$ found in the example.

4.5.8   Development of the Laplace Transform from the Fourier Transform*

The Fourier transform of $u(t)$, presented in Section 7.4.3, fails when the integral $\int_{-\infty}^{\infty}|u(t)|dt$ doesn't exist. Thus, signals that do not decay as $t \to$ have no Fourier transform, but many of these signals, such as a simple step function, are of keen interest. The Laplace transform can be viewed as a Fourier transform tweaked so as to make it exist for such functions. The tweaking is accomplished by multiplying $u(t)$ by $e^{-\sigma t}$, geatly increasing the likelihood that the Fourier transform of the *product* does exist, at least for some range of σ; that is, $\int_{-\infty}^{\infty}|u(t)e^{-\sigma t}|dt < \infty$. This is virtually always true for signals that are zero for $t < 0$, which is the traditional transient signal; it is merely necessary that σ be greater than some finite positive number. Therefore, the function

$$U(\sigma + j\omega) = \mathcal{F}\left[u(t)e^{-\sigma t}\right] = \int_{-\infty}^{\infty} u(t)e^{-\sigma t}e^{-j\omega t}dt, \qquad (4.83)$$

or

$$U(s) = \int_{-\infty}^{\infty} u(t)e^{-st}dt \equiv \mathcal{L}[u(t)]; \qquad s = \sigma + j\omega, \qquad (4.84)$$

is defined. This is known as the **two-sided Laplace transform** of $u(t)$, since it is the same as equation (4.57) (p. 198) except that the lower limit of integration is $-\infty$ rather than 0. The inverse formula is

$$u(t)e^{-\sigma t} = \mathcal{F}^{-1}[\mathcal{L}(\sigma + j\omega)], \qquad (4.85)$$

or, from equation (7.138a) (p. 506),

$$u(t) = \frac{e^{\sigma t}}{2\pi} \int_{-\infty}^{\infty} U(\sigma + j\omega)e^{j\omega t}d\omega = \frac{1}{2\pi j}\int_{-j\infty}^{j\infty} U(\sigma + j\omega)e^{(\sigma+j\omega)t}d(j\omega), \qquad (4.86)$$

or

$$u(t) = \frac{1}{2\pi j}\int_{\sigma-j\infty}^{\sigma+j\infty} U(s)e^{st}ds \equiv \mathcal{L}^{-1}[U(s)]. \qquad (4.87)$$

This is identical to equation (4.58), except that $f(t)$ applies to $t < 0$ as well as $t > 0$.

Two-sided Laplace transforms are used infrequently because a convergence problem often arises when $u(t) \neq 0$ for $t < 0$. For the cases of usual interest in which $u(t) = 0$ for $t < 0$, however, they reduce to the common **one-sided Laplace transform**, which is the only Laplace transform referred to in most textbooks and given by equation (4.57), repeated here:

$$\boxed{U(s) = \int_{0}^{\infty} u(t)e^{-st}dt.} \qquad (4.88)$$

The inverse formula, equation (4.58), unfortunately is not simplified, although a formal derivation of it is now given that was absent before.

The response of the linear process $G(s)$ to the input $u(t)$ now can be generalized from equation (7.141) (p. 506):

$$\boxed{\begin{aligned} x(t) &= \mathcal{L}^{-1}[X(s)]; \\ X(s) &= G(s)U(s). \end{aligned}} \qquad (4.89)$$

This is the key result that you have already used to compute the response of linear systems to explicit excitations. Note that the Laplace transfer function $G(s)$, the Heaviside operator $G(S)$, the special $G(s)$ used for exponential functions e^s and the frequency transfer function $G(j\omega)$ are all intimately related transfer functions, and have identical functional forms.

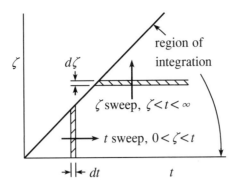

Figure 4.18: Changing the order of integration in equation (4.93)

4.5.9 Development of the Laplace Transform from the Convolution Integral*

Equation (4.89) (and its counterpart for Fourier transforms) also can be derived from a convolution integral. Presuming a *causal* process, that is excluding outputs that occur before their causal inputs, the convolution integral for the response of a system with impulse response $g(t)$ to the excitation $u(t)$ can be written as (Section 4.4.4)

$$x(t) = \int_0^t u(t - \zeta)\, g(\zeta)\, d\zeta. \tag{4.90}$$

Note that ζ is a dummy variable; any symbol could be used. The one-sided Laplace transforms of the excitation $u(t)$ and the response to a unit impulse $g(t)$ are, respectively,

$$U(s) = \int_0^\infty e^{-st} u(t)\, dt; \qquad G(s) = \int_0^\infty e^{-st} g(t)\, dt. \tag{4.91}$$

The objective is to derive the second of equations (4.89), with the added observation that $G(s)$ *is the Laplace transform of the impulse response.* Therefore, the product

$$G(s)U(s) = \int_0^\infty e^{-\zeta s} g(\zeta) d\zeta \int_0^\infty e^{-\xi s} u(\xi) d\xi$$
$$= \int_0^\infty \int_0^\infty e^{-s(\xi+\zeta)} u(\xi) g(\zeta) d\zeta \tag{4.92}$$

is computed, taking care to use *dummy* times, ζ and ξ. The definition $t \equiv \xi + \zeta$ gives

$$G(s)U(s) = \int_0^\infty \left(\int_0^\infty e^{-st} u(t - \zeta) g(\zeta) dt \right) d\zeta. \tag{4.93}$$

The key idea is changing the order of integration, as suggested in Fig. 4.18. This gives

$$G(s)U(s) = \int_0^\infty \left(\int_0^t e^{-st} u(t - \zeta) g(\zeta) d\zeta \right) dt$$
$$= \int_0^\infty e^{-st} \left\{ \int_0^t u(t - \zeta) g(\zeta) d\zeta \right\} dt. \tag{4.131}$$

The integral in the brackets { } above equals $x(t)$, as indicated by equation (4.90). Therefore,

$$G(s)U(s) = \int_0^\infty e^{-st} x(t) dt = X(s), \tag{4.95}$$

establishing the objective.

4.5.10   Summary

Laplace transforms reduce ordinary differential equations to algebraic equations. In particular, the Laplace transform of the response of a linear system with zero initial conditions equals the product of the transfer function, with its argument interpreted as the Laplace variable s, and the Laplace transform of the excitation. The inverse Laplace transform of the transfer function itself is the impulse response of the model. Both problems with excitations and with non-zero initial values are treated by finding the Laplace transform of the desired output and then computing its inverse.

The Laplace transfer functions for lumped models with constant moduli are ratios of polynomials in s. The same is true for the Laplace transforms of most of the simple excitations signals considered herein, such as impulses, steps, ramps and sinusoidal signals that commence at time $t = 0$. (Signals involving pure delays are exceptions, as noted in Section 7.1.7. Another class of exceptions, illustrated in Guided Problem 4.7 below, is developed for distributed-parameter models in Chapter 10.) In these cases the defining integrals for the transform and its inverse can be avoided by the expedient of table look-up, using a partial fraction expansion if necessary.

Multiplying a Laplace transform by s is equivalent to taking the time derivative of the corresponding time function. This fact permits simple determination of the Laplace transform of a differential equation. A transfer function $G(s)$ is algebraically equal to the function $G(s)$ found earlier for the ratio of the response of the system to the exponential excitation e^{st}. Initial and final value theorems permit rapid evaluation of a time function at $t = 0$ and $t = \infty$, respectively, from examination of its Laplace transform.

Software packages such as MATLAB dispatch most analytical drudgery.

The Laplace transform can be viewed as a Fourier transform modified so that it can treat nearly any transient signal. The fact that $G(s)$ is the Laplace transform of the *impulse response* of the system also follows directly from the superposition property.

Guided Problem 4.6

Find the Laplace transform of the function $a(1 - e^{-bt})$, which is one of the most commonly encountered in engineering and science.

Suggested Step:

1. Substitute this function for $u(t)$ in equation (4.57) (p. 198) and integrate.

Guided Problem 4.7

The purpose for this optional problem is to suggest that Laplace transforms apply to a much broader class of signals than is considered in this chapter.

Find the Laplace transform of the function $a\sqrt{t}$.

Suggested Steps:

1. Look up the integral that defines the transform in a table of definite integrals.

2. Note that the result is *not* a ratio of polynomials, unlike the other Laplace transforms considered in this volume of the text (except for the pure-delay operator).

Guided Problem 4.8

This problem illustrates the application of the transform of a pure delay.
    Find the Laplace transform of the truncated ramp:

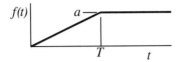

Suggested Steps:

1. Represent the function as the sum of an upward sloping ramp and a delayed downward sloping ramp.

2. Compute the Laplace transform of a single ramp.

3. Use the delay operator e^{-Ts} to assemble the Laplace transform for the sum of the two ramps.

Guided Problem 4.9

You are urged to start your experience solving differential equations by the Laplace transform method with a straight-forward scalar initial-value problem.
    Consider the example

$$\frac{d^3x}{dt^3} + 5\frac{d^2x}{dt^2} + 8\frac{dx}{dt} + 4x = F(t).$$

Find the Laplace transform of $x(t)$ given the values $x(0) = 1$, $x(0) = 2$ and $x(0) = -3$. Then, for $F(t)$ being a unit impulse, find $x(t)$. Do not refer to the Laplace transform table in Appendix C.

Suggested Steps:

1. Take the Laplace transform of the differential equation using the formula for $d^n x(t)/dt^n$ given in Table 4.4.

2. Substitute the function for $\mathcal{L}[F(t)]$ and the initial values as given above.

3. Solve for $\mathcal{L}[x(t)] \equiv X(s)$.

4. Use a partial fraction expansion to find terms in the forms given in Table 4.4. Note the existence of a repeated root.

5. Use the table to evaluate $x(t)$.

Guided Problem 4.10

This problem uses the matrix formulation, and compares the procedure for a forced response with that for an equivalent initial-value problem.
    Solve the third-order system example of equations (7.92) and (7.93) (in Chapter 7, p. 478) for the responses of the three state variables to an impulse of force $F(t)$ of magnitude

$\mathcal{I}$. It is not necessary for you to have studied Chapter 7. Set this up in two ways: as a forced response, and as an initial value problem.

Suggested Steps:

1. Define the matrices **B**, **C** and **D**, and note **A** from equation (7.93).

2. For the forced response, find $U(s)$ and compute $\mathbf{Y}(s)$ from equations (4.78) (p. 202).

3. For the initial value problem, find $\mathbf{x}(0)$ by integrating the differential equation over the infinitesimal time interval 0^- to 0^+. Show that the first of equations (4.78) gives the same result as it does for the forced response.

4. Find the inverse transform of $\mathbf{Y}(s)$ using one of the tables. A partial fraction expansion allows the use of simpler tables.

Guided Problem 4.11

This problem includes both a sinusoidal steady-state component and a transient component in the response to a disturbance. The system is identified by its response to a step excitation.

   A linear system gives the response $3(1 - e^{-2t})$ when excited by a step signal at $t = 0$ of amplitude 2.0. Find the response of the system to the sine wave $4\sin(3t)$ which starts abruptly at $t = 0$. Distinguish the steady-state component from the transient component.

Suggested Steps:

1. Find the Laplace transforms of the step and its response using the table in Appendix C. Find the transfer function $G(s)$ which is their ratio.

2. Find the Laplace transform of the sine wave. Multiply this by $G(s)$ to get the Laplace transform of the desired response.

3. Find the desired response from its transform. A partial fraction expansion may help.

4. The response cannot suffer a discontinuity at $t = 0$, and therefore must be zero there. Verify this with the initial value theorem. Note that the final value theorem fails, however, since the response has no asymptotic value as $t \to \infty$.

5. The transient response decays indefinitely, leaving only the steady-state or particular solution when $t \to \infty$.

PROBLEMS

4.34 Find the response of the system $\dot{x} + 4x = u(t)$, $x(0) = 4$, for the input shown to the right.

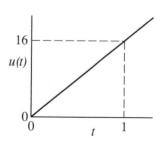

4.35 A system is modeled by the differential equation and initial conditions

$$\ddot{x} + 5\dot{x} + 6x = u(t), \quad \dot{x}(0) = 0, \quad x(0) = 0.$$

Find $x(t)$ for

    (a) $u(t) = \delta(t)$.

    (b) $u(t) = 2u_s(t)$.

    (c) $u(t) = 2\sin 3t$.

4.36 Find $x(t)$ given $\ddot{x} + 4\dot{x} + 3x = \dot{u} + 2u$ and $u(t) = \delta(t)$, $x(0^-) = 0$, $\dot{x}(0^-) = 0$.

4.37 Repeat the above problem given $u(t) = u_s(t)$.

4.38 Find the response of the model
$\ddot{x} + 3\dot{x} = u(t)$, $x(0) = 0$, $\dot{x}(0) = 3$,
for $t \geq 0$ and $u(t)$ as plotted to the
right.

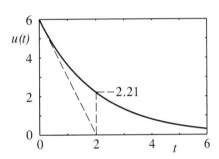

4.39 The system $\ddot{x} + 2\dot{x} + 5x = 0$ has initial conditions $x(0) = 2$, $\dot{x}(0) = 4$. Find $x(t)$, $t \geq 0$.

4.40 Carry out a partial fraction expansion for

$$G(s) = \frac{4s^2 + 10s + 100}{(s+1)(s+10)}.$$

4.41 A model is given by (same as in Problem 4.2)

$$x(t) = G(S)u(t); \quad G(S) = \frac{S+3}{(S+2)(S^2+2S+5)}.$$

    (a) Find $x(t)$ for zero excitation and $x(0) = x_0$, $\dot{x}(0) = \dot{x}_0$.

    (b) Find $x(t)$ for $u(t) = \delta(t)$, $x(0) = 0$, $\dot{x}(0) = 0$.

4.42 A model characterized by the transfer function $G(s) = s/(s^2 + 2s + 2)$ has zero initial conditions, and is excited by the following transfer functions. Find the respective responses.
(a) $u(t) = 3$, (b) $u(t) = 2e^{-3t}$, (c) $u(t) = 2\cos 3t$.

4.43 Find the inverse transform of $X(s) = \frac{2s^2 + 8s + 10}{(s+2)^3}.$

4.44 Find the Laplace transform of $F(t)$ defined by

$$
\begin{aligned}
f(t) &= \quad 0, \quad\; t < 0 \\
&= \; t\sin\omega t, \quad t \geq 0
\end{aligned}
$$

4.45 Find the Laplace transform of $F(t)$ for a single sinusoidal pulse defined by

$$
\begin{aligned}
f(t) &= \quad 0, \quad\quad t < 0 \\
&= \; \sin\omega t, \quad 0 < \omega t < \pi \\
&= \quad 0 \quad\quad \omega t, > \pi
\end{aligned}
$$

4.46 Find the solution of the differential equation

$$
\dot{x} + ax = b\sin\omega t, \quad x(0) = 0
$$

4.47 A second-order system with damped natural frequency ω and envelope decay rate e^{-at} is excited by a transient signal $f(t) = e^{-bt}$ with $b \neq a$. Find the response.

4.48 A model defined by the differential equation

$$
\frac{d^2 x}{dt^2} + 2\frac{dx}{dt} + 8x = u(t)
$$

is excited by the disturbance
$$
u(t) = 3(1 - e^{-t/2}).
$$
Find an analytic expression for $x(t)$.

4.49 Find the Laplace transform of the square pulse

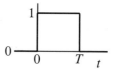

4.50 The square pulse of the above problem excites a system with the transfer function $G(S) = b/(S + a)$. Find and plot the response.

4.51 The fluid tank of Problems 3.26 (p. 109), 3.42 (p. 126) and 3.59 (p. 150) has the flow $Q_0 = 0.5$ in.^{3}/s begin abruptly. Find the subsequent depth of the water in the tank.

4.52 The plunger of Problems 3.27 (p. 109) and 3.43 (p. 126) is released from rest. Find its subsequent velocity.

4.53 The water tank of Problems 3.29 (p. 110) and 3.45 (p. 126) has an initial depth of 1.0 m. The tube is full of water that has a zero initial velocity. Find the subsequent depth as a function of time.

4.54 The vibration isolator of Problems 3.30 (p. 110) and 3.46 (p. 126) is released from rest 0.02 m from its equilibrium position, but suffers no other excitation. Find and sketch-plot the position of the mass as a function of time. Assume $k = 900$ N/m, $m = 2$ kg and $b = 12$ N s/m.

4.55 The filter circuit of Problems 3.31 (p. 111) and 3.47 (p. 126) experiences a step change in the voltage e_i, from 0 to 1 volt. Find the output voltage e_0, assuming $L = 1.0$ mh, $C = 10$ μf and $R = 2.0$ ohms.

SOLUTIONS TO GUIDED PROBLEMS

Guided Problem 4.6

$$U(s) = \int_0^\infty a(1 - e^{-bt})e^{-st}dt = a\left(\frac{-1}{s} + \frac{1}{s+b}e^{-bt}\right)e^{-st}\Bigg|_0^\infty = a\left(\frac{1}{s} - \frac{1}{s+b}\right)$$

Guided Problem 4.7

$$U(s) = \int_0^\infty a\sqrt{t}e^{-st}dt = \frac{a}{2s}\sqrt{\frac{\pi}{s}}$$

(This result is given in Appendix C as transform pair #36.)

Guided Problem 4.8

1. $f(t) = (a/T)[u_r(t) - u_r(t-T)]$

2. For the first ramp, $F_1(s) = \dfrac{a/T}{s^2}$

3. $F(s) = \dfrac{a/T}{s^2}(1 - e^{-Ts})$

Guided Problem 4.9

1. $\dfrac{d^3x}{dt^3} + 5\dfrac{d^2x}{dt^2} + 8\dfrac{dx}{dt} = \delta(t)$

    $\left[s^3X(s) - s^2x(0) - s\dot{x}(0) - \ddot{x}(0)\right] + 5\left[s^2X(s) - sx(0) - \dot{x}(0)\right]$
    $+ 8\left[sX(s) - x(0)\right] + 4\left[X(s)\right] = \mathcal{L}\left[\delta(t)\right]$

2-3. $(x^3 + 5s^2 + 8s + 4)X(s) = (s^2 + 5s + 8)x(0) + (s+5)\dot{x}(0) + \ddot{x}(0) + 1$

$$X(s) = \frac{(s^2 + 5s + 8)1 + (s+5)2 + (-3) + 1}{x^3 + 5s^2 + 8s + 4} = \frac{s^2 + 7s + 16}{(s+1)(s+2)^2}$$

$$= \frac{a}{s+1} + \frac{b}{(s+2)^2} + \frac{c}{s+2}$$

4. Can use the MATLAB command `[c,p,r]=residue(num,den)`, or solve by hand as follows:

$$\underset{s \to -1}{L}\left[\frac{s^2 + 7s + 16}{(s+2)^2}\right] = a = 10$$

$$s^2 + 7s + 16 = 10(s+2)^2 + b(s+1) + c(s+1)(s+2)$$
$$= (10 + c)s^2 + (40 + b + 3c)s + (40 + b + 2c)$$

Therefore, $c = -9$, $b = -6$ and check: $40 - 6 - 18 = 16$

$$X(s) = \frac{10}{s+1} - \frac{6}{(s+2)^2} - \frac{9}{s+2}$$

5. $x(t) = 10e^{-t} - (6t + 9)e^{-2t}$

Guided Problem 4.10

1. Equation (7.93 gives $\mathbf{A} = \begin{bmatrix} 0 & 0 & 1 \\ 0 & -2 & 1 \\ -8 & -6 & -3 \end{bmatrix}$. From equation (7.92),

$$\mathbf{B} = \begin{bmatrix} 0 \\ 1 \\ 0 \end{bmatrix}$$

Presuming all three state variables are of interest,

$$\mathbf{C} = \begin{bmatrix} 1 & 0 & 0 \\ 0 & 1 & 0 \\ 0 & 0 & 1 \end{bmatrix}, \ \mathbf{D} = \mathbf{0}$$

2. $U(s) \equiv \mathcal{L}[u(t)] = \mathcal{I}$

   $\mathbf{Y}(s) = \mathbf{H}(s)U(s) = \mathcal{I}\mathbf{H}(s)$

$$\mathbf{H}(s) = \begin{bmatrix} s & 0 & -1 \\ 0 & s+2 & -1 \\ 8 & 6 & s+3 \end{bmatrix}^{-1} \begin{bmatrix} 0 \\ 1 \\ 0 \end{bmatrix} = \frac{1}{p(s)} \begin{bmatrix} 6 \\ s^2 + 3s + 8 \\ 8 \end{bmatrix};$$

   $p(s) = s^3 + 5s^2 + 20s + 16$

3. $x(0) = \begin{bmatrix} 0 \\ \mathcal{I} \\ 0 \end{bmatrix}; \qquad \mathbf{Y}(s) = \mathbf{C}(s\mathbf{I} - \mathbf{A})^{-1}\mathbf{x}(0) = $ same as above

4. The denominator of the Laplace transforms for p_a and p_b can be factored into the product of a first-order term and a second-order term. Transform pair #32 in the table of Appendix C then gives $p_a(t)$ directly. Finding $p_b(t)$ the same way is more awkward, because of the numerator term s^2; a summation of terms from pair #32, pair #33 and the time derivative of pair #33 are necessary. Alternatively, the use of the MATLAB command [c,p,r]=residue(num,den) carries out a partial fraction expansion, after which the associated time function can be found more readily. The development below carries out the partial fraction expansions by hand (which is relatively uneconomical in engineering practice):

$$p_a = \mathcal{L}^{-1}\left[\frac{6}{s^3 + 5s^2 + 20s + 16}\right] = \mathcal{L}^{-1}\left[\frac{k}{s+1} + \frac{bs+c}{s^2+4s+16}\right]$$

$$k = \underset{s \to -1}{L} \left(\frac{6}{s^2+4s+16}\right) = \frac{6}{13}$$

$$6 = \frac{6}{13}(s^2 + 4s + 16) + (s+1)(bs+c)$$

$$= \left(\frac{6}{13} + b\right)s^2 + \left(\frac{24}{13} + b + c\right)s + \left(\frac{96}{13} + c\right)$$

Therefore, $b = -\frac{6}{13}; \quad c = -\frac{24}{13} + \frac{6}{13} = -\frac{18}{13}; \quad$ check : $\frac{96}{13} - \frac{18}{13} = \frac{78}{13} = 6$

Thus: $p_a(t) = \frac{1}{13}\mathcal{L}^{-1}\left[\frac{6}{s+1} - \frac{6s+18}{(s+2)^2 + 12}\right]$

$$= \frac{6}{13}e^{-t} - \sqrt{\frac{3}{13}}e^{-2t}\sin(2\sqrt{3}t + \tan^{-1} 2\sqrt{3})$$

Similarly, $p_b(t) = \mathcal{L}^{-1}\left[\frac{s^2+3s+8}{s^3+5s^2+20s+16}\right] = \mathcal{L}^{-1}\left[\frac{k}{s+1} + \frac{bs+c}{(s+2)^2+12}\right]$

$$k = \underset{s \to -1}{L} \left[\frac{s^3+3s+18}{s^2+4s+16}\right] = \frac{16}{13}$$

$$s^2 + 3s + 18 = \frac{16}{13}(s^2 + 4s + 160) = (s+1)(bs+c)$$

$$= \left(\frac{16}{3} + b\right)s^2 + \left(\frac{64}{13} + b + c\right)s + \left(\frac{256}{13} + c\right)$$

Therefore, $b = -\dfrac{3}{13}$; $\quad c = -\dfrac{22}{13}$

$$p_b(t) = \frac{16}{13}e^{-t} - \sqrt{\frac{91}{39}}e^{-2t}\sin\left[2\sqrt{3}t - \tan^{-1}(3\sqrt{3}/16)\right]$$

Finally, $q(t) = \dfrac{8}{6}p_a(t) = \dfrac{8}{13}e^{-t} - \dfrac{4}{\sqrt{39}}e^{-2t}\sin(2\sqrt{3}t + \tan^{-1}2\sqrt{3})$

Guided Problem 4.11

1. $G(s) = \dfrac{\mathcal{L}[3(1 - e^{-2t})u_s(t)]}{\mathcal{L}[2u_s(t)]} = \dfrac{3}{2/s}\left(\dfrac{1}{s} - \dfrac{1}{s+2}\right) = \dfrac{3}{2}\left(1 - \dfrac{s}{s+2}\right) = \dfrac{3}{s+2}$

2. $U(s) = \dfrac{4 \times 3}{s^2 + 3^2}$; $\qquad X(s) = G(s)U(s) = \dfrac{36}{(s^2 + 3^2)(s + 2)}$

3. Transform pair #32 can be used directly, with $a = 0$, $c = 2$ and $\omega = 3$. Alternatively, the following partial fraction expansion (which could be found using MATLAB) can be used:

$$X(s) = \frac{k}{s+2} + \frac{bs + c}{s^2 + 3^2}$$

$$\underset{s \to -2}{L}\left[\frac{36}{s^2 + 3^2}\right] = k = \frac{36}{13}$$

$$36 = \frac{36}{13}(s^2 + 9) + (s + 2)(bs + c) = \left(\frac{36}{13} + b\right)s^2 + (2b + c)s + \left(\frac{324}{13} + 2c\right)$$

Therefore, $b = -\dfrac{36}{13}$; $c = \dfrac{72}{13}$; $36 = \dfrac{324}{13} + \dfrac{2 \times 72}{13} = \dfrac{468}{13}$ (check)

From Table 4.4 (p. 199),

$$x(t) = \frac{36}{13}e^{-2t} + \frac{12}{\sqrt{13}}\sin(3t + \psi), \quad \psi = \tan^{-1}\left(\frac{-3}{2}\right) = -\tan^{-1}\left(\frac{3}{2}\right)$$

4. $x(0) = L_{s\to\infty}\left[sX(s)\right] = 0$ $\quad$ Note that the time function above agrees.

$L_{s\to 0}\left[sX(s)\right] = 0$ $\quad$ This is *not* $x(\infty)$, however, since there are two poles on the origin.

5. As $t \to \infty$, $x(t) \to \dfrac{12}{\sqrt{13}}\sin\left[3t - \tan^{-1}\left(\dfrac{3}{2}\right)\right]$

which is the steady-state or particular solution.

4.6 Responses of Primitive Linear Models

This section emphasizes key definitions and features of the responses of first and second-order linear models with constant coefficients to initial conditions and to impulse, step and ramp excitations. These models occur by themselves in engineering more often than any others. More importantly, higher-order models usually decompose into sums of first and second-order models, particularly using the partial-fraction expansions, and excitations can be decomposed into sums of steps or impulses through the property of superposition. The responses to be discussed therefore characterize the behavior of most linear models. They are described in terms of time constants, natural frequencies, damping ratios and damped natural frequencies.

Behavior for sinusoidal excitations and decompositions based thereon are deferred to Chapter 7.

MATLAB commands for impulse and step responses also are given.

The Laplace transform of a scalar response $x(t)$ to a scalar excitation $u(t)$ for a model of any order can be put in the form

$$X(s) = \frac{P(s)}{p(s)} + G(s)U(s); \qquad G(s) = \frac{N(s)}{p(s)} \tag{4.96}$$

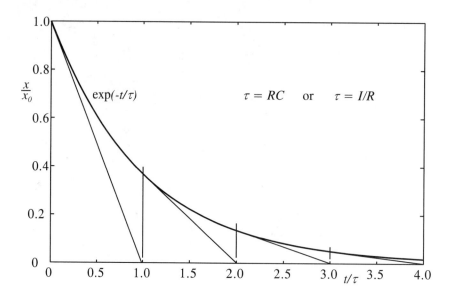

Figure 4.19: Response of first-order model to initial condition or impulse

in which $p(s)$ is the characteristic polynomial, $G(s)$ is the transfer function of the model which has $N(s)$ as its numerator, and $P(s)$ is a polynomial in s that represents any non-zero initial conditions.

4.6.1   Responses of First-Order Models

The first-order model has the characteristic polynomial

$$p(s) = a_1 s + a_0. \tag{4.97}$$

When the model has an initial value of $x(0)$ but no excitation, the solution can be written as

$$x(t) = x(0)e^{-a_0 t/a_1} = x(0)e^{-t/\tau}; \quad \tau = -a_0/a_1, \tag{4.98}$$

which is plotted in Fig. 4.19. The symbol τ is known as a **time constant**; its meaning is indicated in the figure. A tangent drawn to the curve at any time t intersects the axis for $x = 0$ at the time $t + \tau$. Three of these are shown in the figure. During any time interval equal to the time constant the response decays by the factor $e = 2.718$, which makes its value $0.368x_0$ after one time constant, $0.135x_0$ after two time constants, $0.050x_0$ after three time constants, and $0.018x_0$ after four time constants. If you know a time constant, therefore, you can sketch the response fairly readily. Conversely, if you have a plot of experimental data, you can recognize that the behavior is like a first-order model, and estimate its time constant fairly accurately.

If the model is excited by an impulse of amplitude u_0 but has a zero initial condition, the Laplace transform of the response is the same as above, with u_0/a_1 subsituting for $x(0)$. Thus, the response has the same shape. What happens, in fact, is that the impulse kicks the model abruptly to a new state that acts like an intial condition of a subsequently unexcited system, since in fact there is no excitation for $t > 0$.

EXAMPLE 4.24

The combination of two springs each with $k = 1$ lb/ft and a dashpot with $b = 5$ lb·s/ft, shown to the right, are hit by a hammer with the impulse $\int F \, dt = 2$ lb·s. Find and plot the resulting position $x(t)$.

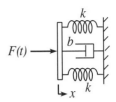

Solution: The annotated bond graph below leads to the first-order differential equation

$$\frac{dx}{dt} + \frac{1}{RC}x = \frac{1}{R}F,$$

which has the Laplace transform

$$\left(s + \frac{1}{RC}\right)X(s) = \frac{1}{R}\int F \, dt,$$

that gives

$$X(s) = \frac{\int F \, dt/R}{s + 1/RC}.$$

$$S_e \xrightarrow{F = \mathscr{I}\delta(t)} 1 \xrightarrow{x/C} C$$

$$F\text{-}x/C \Big| (F\text{-}x/C)/R$$

$$R$$

$$C = 1/2k; \quad R = b$$

Since the impulse is $\int F \, dt = 2$ lb·s, $R = b = 5$ lb·s/ft, and $\tau = RC = b/2k = 5/(2 \times 1) = 2.5$ s, the resulting displacement is

$$x(t) = 0.4e^{-t/2.5} \text{ ft},$$

where t is in seconds. The plot of Fig. 4.19 applies.

A step change of amplitude c has a Laplace transform of c/s. Therefore, the response of a first-order model with $G(s) = 1/(s + 1/\tau)$ to a suddenly applied constant excitation of amplitude c has the Laplace transform and time response

$$X(s) = \frac{c}{s(s + 1/\tau)} = \frac{c_1}{s} + \frac{c_2}{s + 1/\tau} = c\tau\left(\frac{1}{s} - \frac{1}{s + 1/\tau}\right), \qquad (4.99)$$

$$x(t) = c\tau(1 - e^{-t/\tau}), \qquad (4.100)$$

This classic step response is plotted in part (a) of Fig. 4.20. Note the decomposition, via the partial fraction expansion, into the sum of a step and a decaying exponential with time constant τ. Note also that since the step is the time integral of an impulse, the step response is the time integral of the impulse response. Note finally that the magnitude of the step can be found simply from the final value theorem, without carrying out the partial fraction expansion, etc: $x(\infty) = \lim_{s \to 0}[sX(s)] = c\tau$.

EXAMPLE 4.25

The spring-dashpot system of Example 4.24 above is excited by the sudden application of a force $F_0 = 4$ lb. Find and plot the resulting displacement.

Solution: The impulse $\int F \, dt$ of Example 4.24 is replaced by the step F_0/s, to give

$$X(s) = \frac{F_0/R}{s(s + 1/\tau)}.$$

This is the same as equations (4.99) and (4.100) and the plot of Fig. 4.20 with $c = F_0/R = 4/5 = 0.8$ ft. The time constant remains at 2.5 seconds.

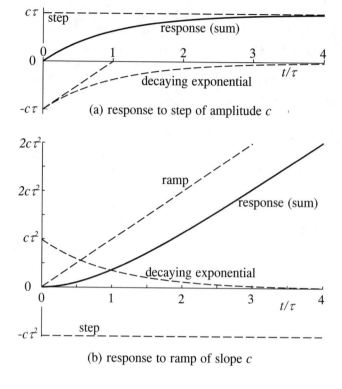

(a) response to step of amplitude c

(b) response to ramp of slope c

Figure 4.20: Step and ramp responses of a first-order model

The excitation for a ramp excitation $u(t) = ct$ is the time integral of the step excitation $u(t) = c$. Therefore, the response of the first-order model is

$$X(s) = \frac{c}{s^2(s+1/\tau)} = \frac{c_1}{s^2} + \frac{c_2}{s} + \frac{c_3}{s+1/\tau} \qquad (4.101)$$

The coefficients of the partial fraction expansion can be found as follows. Multiply both sides of equation (4.101) by s^2 and set $s = 0$ to get $c_1 = c\tau$. Then, multiply both sides by $s+1/\tau$ and set $s = -1/\tau$ to get $c_3 = c\tau^2$. Finally, multiply both sides by $s^2(s+1/\tau)$ to get $c = (c\tau + sc_2)(s+1/\tau) + s^2 c\tau^2$. Equating the coefficients of the s^2 term gives $c_2 = -c\tau^2$; the other terms serve as a check. The result is

$$x(t) = c\tau^2 \left(t/\tau - 1 + e^{-t/\tau} \right), \qquad (4.102)$$

which alternatively can be found by directly integrating the step response. The result is plotted in part (b) of Fig. 4.20.

4.6.2   Responses of Second-Order Models to Initial Conditions

Most linear scalar second-order models with constant coefficients satisfy the equation

$$a_2 \frac{d^2 x}{dt^2} + a_1 \frac{dx}{dt} + a_0 x = b_1 \frac{du}{dt} + b_0 u. \qquad (4.103)$$

The Laplace transform of this equation can be placed into the form

$$(s^2 + 2\zeta\omega_n s + \omega_n^2)X(s) = (s + 2\zeta\omega_n)x(0) + \dot{x}(0) + [(b_1/a_2)s + b_0/a_2]U(s) - (b_1/a_2)u(0), \qquad (4.104)$$

by defining the **natural frequency**, ω_n, and the **damping ratio**, ζ, as

$$\omega_n \equiv \sqrt{\frac{a_0}{a_2}}; \qquad \zeta \equiv \frac{a_1}{2a_2\omega_n}. \qquad (4.105)$$

Consider as a first special case the initial value problem with $x(0) = x_0$, $\dot{x}(0) = 0$, $\zeta = 0$ and no excitation ($u(t) = 0$). This gives

$$X(s) = \frac{s x_0}{s^2 + \omega_n^2}, \qquad (4.106a)$$

$$x(t) = x_0 \cos \omega_n t, \qquad (4.106b)$$

which is the curve in part (a) of Fig. 4.21 designated as $\zeta = 0$. The natural frequency, therefore, is the actual frequency of oscillation of the model in radians per unit time.

EXAMPLE 4.26

The floating, pitching block of Example 3.1 (p. 81-82) has a mass moment of inertia about its center of mass $I = m(w^2 + h^2)/12$, where m is its mass. Characterize the behavior of the block after it is released from rest at an angle of ϕ_0. Neglect damping and the effect of any virtual inertia due to the water.

Solution: The model comprises a compliance with the value found in Example 3.1, and an inertance with the value above, bonded to a common 1-junction with the angular velocity $\dot{\phi}$:

$$C \xrightarrow[\;\;\widehat{\phi_i}\;\;]{\phi/C} 1 \xrightarrow[\;\;p_i/I_i\;\;]{\widehat{p_i}} I$$

Since the mass equals the mass of the displaced water, or $m = \frac{1}{2}wh\rho$, the inertance is $I = hw(w^2 + h^2)\rho/24$. The differential equations are

$$\frac{dp}{dt} = -\frac{1}{C}\phi,$$

$$\frac{d\phi}{dt} = \frac{1}{I}p.$$

To combine these equations and eliminate the variable p, you can take the derivative of the second equation and substitute the first:

$$\frac{d^2\phi}{dt^2} = \frac{1}{I}\frac{dp}{dt} = -\frac{1}{IC}\phi.$$

This is in the form of equation (4.103) with $u(t) = 0$, the damping ratio $\zeta = 0$ and the natural frequency, from equation (4.105),

$$\omega_n = \frac{1}{\sqrt{IC}} = \sqrt{\frac{2g}{h[1 + (h/w)^2]}}.$$

The solution for zero initial velocity, from equation (4.106), is

$$\phi = \phi_0 \cos \omega_n t.$$

The motion is a sinusoidal oscillation at the natural frequency, ω_n.

When the coefficient a_1 and therefore the damping ratio is non-zero, the response is

$$X(s) = \frac{(s + 2\zeta\omega_n)x_0}{s^2 + 2\zeta\omega_n s + \omega_n^2}. \qquad (4.107)$$

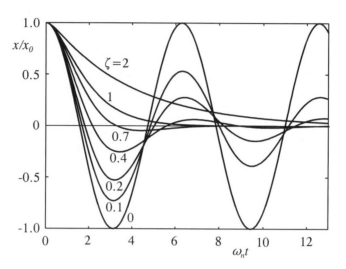

(a) homogeneous response for $x(0)=1$, $\dot{x}(0)=0$

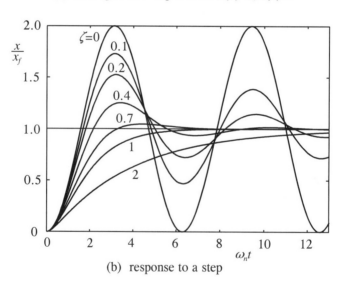

(b) response to a step

Figure 4.21: Response of the second-order model to an initial displacement and to a step excitation

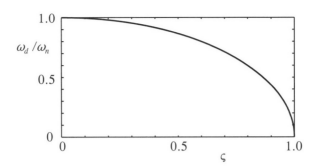

Figure 4.22: Comparison of Damped and Undamped Natural Frequencies

When $a_1 \neq 0$ it is convenient to define the **damped natural frequency**, ω_d:

$$\boxed{\omega_d \equiv \sqrt{1 - \zeta^2}\,\omega_n, \qquad \zeta < 1.}$$ (4.108)

This is done only when the resulting ω_d is a real number, that is when the damping ratio, ζ, is less than 1. With this notation,

$$X(s) = \frac{(s + 2\zeta\omega_n)x_0}{(s + \zeta\omega_n)^2 + \omega_d^2},$$ (4.109a)

$$x(t) = Ae^{-\zeta\omega_n t}\cos(\omega_d t - \phi); \quad A = \frac{x_0}{\sqrt{1 - \zeta^2}} \quad \phi = \tan^{-1}\left(\frac{\zeta^2}{1 - \zeta^2}\right).$$ (4.109b)

Equation (4.109b) comes from the last entry in Table 4.4. It is also possible to represent $x(t)$ by the weighted sum of a sine function and a cosine function, times the exponential decay envelope, by using either the next-to-last entry in Table 4.4 or items # 18 and #19 in the Laplace transform table of Appendix C.

Responses for a few values of ζ are plotted in part (a) of Fig. 4.21. A time-shifted cosine wave is multiplied by a decaying exponential. Note that the time-shift is set so as to give $\dot{x}(0) = 0$, which was given above as part of the initial condition. The frequency of oscillation, ω_d, is less than the undamped natural frequency. The difference may be so small as to be virtually imperceptible, however, which happens when the damping ratio ζ is small enough that many cycles appear before substantial decay sets in. The ratio ω_d/ω_n, taken from equation (4.105), is plotted in Fig. 4.22. The time constant of the decay envelope is $\tau = 1/\zeta\omega_n$, which is long when ζ is small.

One often wishes to examine the ratio of the amplitudes of succeeding oscillations, which is a constant dependent on the damping ratio. Noting that a period of the oscillation is $T = 2\pi/\omega_d$, this ratio is

$$\frac{x_n}{x_{n+1}} = \frac{e^{-\zeta\omega_n t_n}}{e^{-\zeta\omega_n(t_n+T)}} = \frac{1}{e^{-\zeta\omega_n T}} = e^{2\pi\zeta/\sqrt{(1-\zeta^2)}}.$$ (4.110)

The natural logarithm of this ratio is known as the **logarithmic decrement**:

$$\text{logarithmic decrement} = 2\pi\zeta/\sqrt{1 - \zeta^2}.$$ (4.111)

The damping ratio ζ can be estimated from an experimental plot of the decay through the use of the logarithmic decrement. It may be most accurate to observe the decay ratio over several cycles. The log of the ratio for m cycles is simply m times the log of the ratio for one cycle. This gives

$$\zeta = \frac{\ln(x_n/x_{n+m})/m}{\sqrt{4\pi^2 + [\ln(x_n/x_{n+m})/m]^2}}. \tag{4.112}$$

Note that this applies regardless of the values of the initial conditions $x(0)$ and $\dot{x}(0)$.

EXAMPLE 4.27

The mass-spring-dashpot system below exhibits the plotted response:

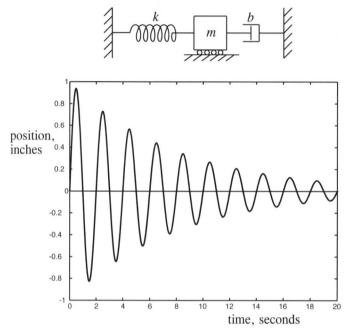

Estimate the natural frequency and the damping ratio. Then estimate the spring constant and the damping coefficient, given that the mass is 4 kg.

Solution: Without even looking at the picture, the response resembles that of a second-order system. With this assumption, the decay ratio over nine cycles is approximately 1.8/0.019=9.5 (using peak-to-peak values, which is most accurate). Equation (4.112) gives

$$\zeta = \frac{\ln(9.5)/9}{\sqrt{4\pi^2 + [\ln(9.5)/9]^2}} = 0.040$$

There are 10 cycles in 20 seconds, so the damped natural frequency is $2\pi \times 10/20 = \pi$ rad/s. The ratio of the damped natural frequency to the undamped natural frequency is 0.999, according to equation (4.108), which is indistinguishable from 1.000, so the natural frequency also is essentially π rad/s.

Another way to estimate the damping ratio starts with an estimate of the time constant of the envelope. One can sketch the envelope and estimate the time at which it has dropped to $e^{-1} = 0.368$ of its initial value, or sketch a tangent to the envelope at $t = 0$ and observe the time at which it crosses the time axis. Either way, the answer is essentially 8.0 s. Knowing that the natural frequency is essentially π rad/s, this time constant gives

$$\zeta = 1/\tau\omega_n \simeq 1/8.0\pi = 0.040.$$

The system is the same as in Examples 3.14 and 3.15 (pp. 120-121) with q being the displacement of the mass, p being its momentum, and the excitation e_0 being eliminated. The mass is $m = I$, the spring constant is $k = 1/C$, and the dashpot coefficient is $b = R$. The state-space equations therefore are

$$\frac{dp}{dt} = -kq - \frac{b}{m}p,$$
$$\frac{dq}{dt} = \frac{1}{m}p.$$

These equations combine to give

$$\frac{d^2q}{dt^2} + \frac{b}{m}\frac{dq}{dt} + \frac{k}{m}q = 0.$$

Comparing this to equation (4.103) gives $a_2 = 1$, $a_1 = b/m$ and $a_0 = k/m$. From equation (4.105), $\omega_n = \sqrt{k/m}$ (a famous result you might remember) and $\zeta = b/2\sqrt{km}$. Thus,

$$k = m\omega_n^2 \simeq 4\pi^2 = 39.5\,\text{N/m}; \quad b = 2\zeta\sqrt{km} \simeq 2 \times 0.040\sqrt{39.5 \times 4} = 1.0\text{N}\cdot\text{s/m}$$

The very special case $\zeta = 1$ is known as **critical damping**. The characteristic polynomial factors into two equal roots; equation (4.107) becomes

$$X(s) = \frac{(s + 2\omega_n)x_0}{(s + \omega_n)^2}, \tag{4.113}$$

which gives the response, again for $\dot{x}(0) =$, of

$$x(t) = (1 + \omega_n t)e^{-\omega_n t}x_0. \tag{4.114}$$

For supercritical damping, $\zeta > 1$, the characteristic polynomial factors into two real roots, so the response comprises the sum of two exponential decay terms. The time constants (reciprocals) of these roots are

$$\boxed{\tau_{1,2} = \frac{1}{\omega_n}\left(\zeta \pm \sqrt{\zeta^2 - 1}\right); \quad \zeta > 1.} \tag{4.115}$$

The unexcited solution for $\dot{x} = 0$ becomes

$$X(s) = \frac{(s + 2\omega_n)x(0)}{(s + 1/\tau_1)(s + 1/\tau_2)} = \left[\frac{2\omega_n\tau_1 - 1}{(\tau_1/\tau_2 - 1)(\tau_1 s + 1)} + \frac{2\omega_n\tau_2 - 1}{(\tau_2/\tau_1 - 1)(\tau_2 s + 1)}\right]x_0, \tag{4.116a}$$

$$x(t) = \left[\left(\frac{2 - \zeta + \sqrt{\zeta^2 - 1}}{2\sqrt{\zeta^2 - 1}}\right)e^{-t/\tau_1} + \left(\frac{\zeta - 2 + \sqrt{\zeta^2 - 1}}{2\sqrt{\zeta^2 - 1}}\right)e^{-t/\tau_2}\right]x_0. \tag{4.116b}$$

These results are plotted in Fig. 4.21 part (a) for the cases $\zeta = 1$ and $\zeta = 2$. The larger ζ becomes, the slower the response. In the limit of a virtually infinite value, the response deteriorates to essentially that of a first-order model.

4.6.3 Responses of Second-Order Models to Step and Impulse Excitations

Consider next the case with zero initial conditions and $\zeta < 1$, but with an abrupt imposition of a steady excitation that produces a final steady-state value $x(\infty) = x_f$:

$$\frac{d^2x}{dt^2} + 2\zeta\omega_n\frac{dx}{dt} + \omega_n^2 x = \omega_n^2 x_f, \tag{4.117a}$$

$$X(s) = \frac{\omega_n^2 x_f}{s[(s + \zeta\omega_n)^2 + \omega_d^2]}, \tag{4.117b}$$

$$x(t) = 1 - \frac{1}{\sqrt{1 - \zeta^2}} e^{-\zeta\omega_n t} \sin(\omega_d t - \phi); \quad \phi = \tan^{-1}\frac{\sqrt{1 - \zeta^2}}{\zeta}. \tag{4.117c}$$

A few cases with various values of ζ (including critically damped and overdamped) are plotted in part (b) of Fig. 4.21. Note that these curves have the same shape as those in part (a) for the initial value problem, except that they are inverted. If you look at the Laplace transform equation (4.104), which covers both cases, you can see why. After an excitation ceases or becomes constant, the model demonstrates its common natural behavior.

EXAMPLE 4.28

A system is described by the second-order differential equation

$$\frac{d^2x}{dt^2} + 6\frac{dx}{dt} + 9x = 18,$$

in which the units of time are seconds. Find the natural frequency and damping ratio of the system, and find $x(t)$ assuming $x(0) = 1$, $\dot{x}(0) = 0$.

Solution: The natural frequency and damping ratio are, from equation (4.105),

$$\omega_n = \sqrt{a_0/a_2} = \sqrt{9/1} = 3 \text{ rad/s}; \quad \zeta = a_1/2a_2\omega_n = 6/(2 \times 1 \times 3) = 1.$$

Since $\zeta = 1$, which is not addressed specifically above for the forced case, the response is

$$X(s) = \frac{18/s + (s + 2\omega_n)x(0)}{(s + 3)^2} = \frac{s^2 + 6s + 18)}{s(s + 3)^2} = \frac{c_1}{s} + \frac{c_2}{s + 3)} + \frac{c_3}{(s + 3)^2}$$

$$c_1 = (0 + 0 + 18)/(0 + 3)^2 = 2; \quad c_3 = \frac{(-3)^2 + 6(-3) + 18}{(-3)} = -3;$$

$$s^2 + 6s + 18 = c_1(s + 3)^2 + c_2 s(s + 3) + c_3 s = (c_1 + c_2)s^2 + (6c_1 + 3c_2 + c_3)s + 9c_1$$

$$= (2 + c_2)s^2 + (12 + 3c_2 - 3)s + 18, \text{ which gives } c_2 = -1,$$

$$x(t) = 2 - (1 + 3t)e^{-3t}.$$

Note that the first term is the readily determined steady-state or particular solution $x_p = 18/9 = 2$.

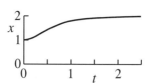

The simplest response to find is that for an impulse excitation, $u(t) = \delta(t)$. Again treating the underdamped case,

$$\frac{d^2x}{dt^2} + 2\zeta\omega_n\frac{dx}{dt} + \omega_n^2 x = \frac{b_0}{a_2}\delta(t), \tag{4.118a}$$

$$X(s) = \frac{b_0/a_2}{(s + \zeta\omega_n)^2 + \omega_d^2}, \tag{4.118b}$$

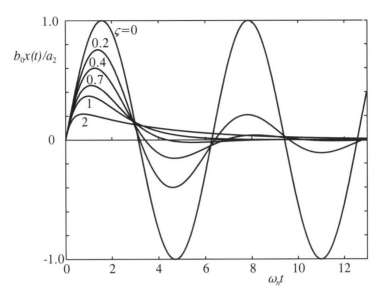

Figure 4.23: Response of a Second-Order Model to an Impulse Excitation

$$x(t) = [e^{-\zeta\omega_n t}\sin(\omega_d t](b_0/a_2\omega_d). \tag{4.121c}$$

This result is plotted in Fig. 4.23 for a few values of ζ. Note that the response both starts and ends at $x = 0$, as it must. The response for an initial velocity and zero excitation can be made identical, for the effect of the impulse is to produce such a condition, with no subsequent input.

4.6.4 Step and Impulse Responses Using MATLAB

MATLAB automatically computes the step and impulse responses of linear models with constant coefficients. You enter the coefficients $b_m, \cdots, b_0$ of the numerator polynomial and the coefficients $a_n, \cdots, a_0$ of the denominator polynomial, and then simply call the routine of interest. The program chooses its own (default) scales for the axes (which you can override).

EXAMPLE 4.29

Using MATLAB, plot the unit impulse and step responses of the model

$$\frac{d^3x}{dt^3} + 3\frac{d^2x}{dt^2} + 102\frac{dx}{dt} + 100x = 50\frac{du}{dt} + 200.$$

Solution: The transfer function is

$$G(S) = \frac{50S + 200}{S^3 + 3S^2 + 102S + 100}.$$

To get the plots below, you enter

```
>>num = [50 200];
>>den = [1 3 102 100];
>>step(num, den)
>>impulse(num,den)
```

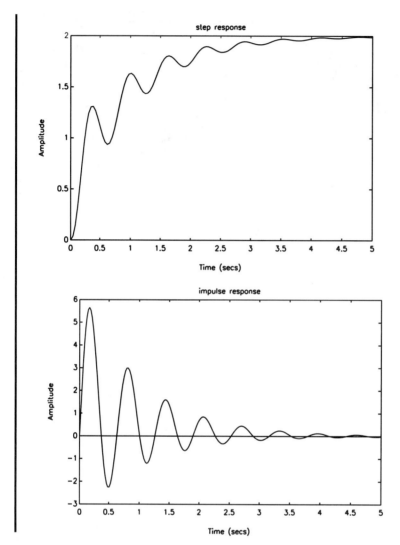

Pressing the enter key will erase the present plot, unless you enter hold on, which will superimpose subsequent plots. Further options to control the details of the plots and secure data files for further use are available. If you substitute [y,x,t]=step(num, den) or [y,x,t]=impulse(num,den), the times used are stored in a vector t, the output values are stored in a vector y and the state variables chosen by the routine are stored in a matrix x. Values of t specified in advance will be used if t is used as the third argument of the command, *i.e.*, step(num,den,t). To specify 201 values of time evenly distributed over ten seconds, for example, you enter either t=linspace(0,10,201) or t=[0:.05:10].

4.6.5   Summary

Responses to initial conditions and to impulse and step excitations share the same dynamics, since the impulse serves only to impart abruptly a new initial condition, and the step serves only to change the final steady-state value. The response to a unit step is the time integral of the response to a unit impulse, and the response to a unit ramp is the time integral of the response to a unit step.

The responses of a first-order model are characterized by its time constant. An underdamped second-order model exhibits an exponentially decaying envelope that also can be described by its time constant. The response of an over-damped second-order model exhibits two exponential decays and time constants. A second-order model also is characterized by its natural frequency, damping ratio and, if under-damped, its damped natural frequency. The response of a critically damped model includes a special term of the form $te^{-t/\tau}$.

MATLAB has commands that will plot the impulse and step responses of linear models of any order with constant coefficients.

Guided Problem 4.12

The current source of the electric circuit of Guided Problem 3.8 (pp. 106-107, 112) undergoes a step change in time, from 0 amps to 1 amp. The load inductance is 0.01 Henry. The voltage should not exceed 15 V. Choose values for the resistance, R, and capacitance, C which satisfy the constraint and provide a good response.

Suggested Steps:

1. Apply the causal strokes to the bond graph and annotate all the efforts and flows according to the procedure developed in Section 3.4.

2. Write the two coupled first-order differential equations.

3. You are interested in the load current. The limiting voltage is proportional to the rate of change of this current. Therefore, choose the state variable proportional to current as the variable for your combined differential equations.

4. Combine the equations so the variable chosen in step 3 becomes the dependent variable.

5. Compare your differential equation to equation (4.103) (p. 220). Compute the values of the natural frequency and the damping ratio from equation (4.105) (p. 221).

6. Find the Laplace transform of the current.

7. Find the current as a function of time using a partial fraction expansion, if needed, and using a Laplace transform table.

8. The limiting voltage effectively sets the maximum rate of change of the current. Examine the nature of the solution with this constraint in mind to choose an effective damping ratio, perhaps using plots given in the book for various damping ratios. Only one undetermined parameter should remain at this point.

9. Use the limitation of the maximum allowed voltage to find an equation that determines the remaining parameter. You should now be able to give values for both C and R, and plot the current as a function of time.

PROBLEMS

4.56 *Continuation of Problem 3.22 (p. 108).* Evaluate any critical time constant or natural frequency for the rotational system. *Continued in Problems 7.8 and 7.41.*

4.57 *Continuation of Problems 3.28 (p. 110) and 3.44 (p. 126).* Determine the natural frequency of the model of the tank-and-tube model in terms of bond graph moduli and then in terms of physical parameters.

4.58 *Continuation of Problem 3.32 (p. 111).* Find the natural frequency of the spring-mass model for small displacements about the equilibrium.

4.59 The U-tube of Guided Problems 3.7 and 3.10 (pp. 106, 112, 124, 126-127) has a diameter $d = 0.25$ inches and a length $L = 10$ inches; the fluid is water ($\rho g = 62.4$ lb/ft^3). Determine the natural frequency of the motion following a non-equilibrium initial condition.

4.60 The bond graph below describes a mechanical system with linear displacement x and constant parameters R and C. The displacement responds to the sudden application of a constant force, $F_0 = 1$ lb, as plotted below.

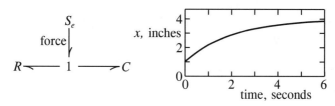

(a) Annotate the bond graph in standard fashion, and write the state differential equation(s) for the model.

(b) Estimate the values of R and C (don't forget the units).

4.61 The drum viscometer of Guided Problem 3.6 (p. 105) coasts to a halt from an initial velocity of 1.0 radian per second. Find and sketch-plot its speed and position as functions of time.

4.62 Find the natural frequency and damping ratio for the model described in Fig. 3.30 (p. 122) and equations (3.32) (p. 121).

4.63 A model has the transfer function $G(S) = 12/(S^2 + 10S + 169)$. Determine its damping ratio and damped and undamped natural frequencies.

4.64 A model is described by the given differential equations. Determine the values of whichever of the following parameters that apply: τ, ω_n, ω_d, ζ.
$$\frac{dp}{dt} = 3q - 6,$$
$$\frac{dq}{dt} = -3q - 3p.$$

4.65 Answer the questions of the preceding problem for
$$\frac{dp}{dt} = 4q,$$
$$\frac{dq}{dt} = -\frac{25}{4}p - 6q.$$

4.66 Answer the questions of the preceding two problems for
$$\frac{dp}{dt} = 2q - 4,$$
$$\frac{dq}{dt} = -5q - 3p.$$

4.67 Find the response of a critically damped second order system to an impulse excitation. Compare with the plot for $\zeta = 1$ in Fig. 4.23 (p. 227).

4.68 Find the response of a second order system with $\zeta = 2$ to an impulse. Compare with the corresponding plot in Fig. 4.23 (p. 227).

4.69 The cylinder shown opposite has a radius and height $r = h = 10$ cm and density equal to 750 kg/m^3. It bobs vertically in a tub of water of density 1000 kg/m^3 with an observed frequency of 1.35 Hz. Damping is to be neglected.

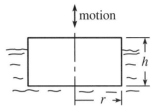

(a) Find the change in the bouyancy force resulting from a vertical displacement to the disk from its equilibrium position. Interpret this result to give a compliance.

(b) Solve for the effective inertance, using the observed natural frequency and the value of the compliance determined in part (a).

(c) By comparing the result of part (b) with the mass of the cylinder, determine the "virtual mass" of the water that effectively moves with the cylinder.

(d) Find the length of a cylinder of water of the same radius r that has a mass equal to the virtual mass. (Note: In estimating the inertance of the fluid inside a circular tube with thin-walled ends surrounded by fluid, this length also should be added to each end. Such a correction is widely used in acoustics, and is especially important if the tube is short. If the end of the tube is surrounded by a flange, the end correction should be about 39% greater. These cases are discussed further in Section 10.1.8.)

4.70 A long tunnel or *penstock* of area $A_p = 100$ ft^2 and length $L = 3000$ ft carries water from a large lake to a turbine. In an emergency shutdown the flow through the turbine, initially 1000 ft^3/s, ceases abruptly. A large surge chamber of area $A_s = 1200$ ft^2 relieves the pressure surge that occurs.

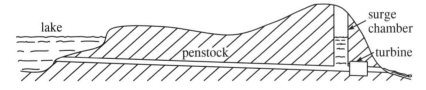

(a) Define variables and parameters and draw a bond graph model for the system. Neglect all damping and the inertia of the water in the surge chamber.

(b) Write differential equations of motion.

(c) Solve these equations analytically and sketch-plot the results. How high does the water in the surge tank rise, and when does it peak?

4.71 A model with the transfer function below has three roots. Find these roots, and state any applicable damping ratio, natural frequency and damped natural frequency.

$$G(S) = \frac{1000}{10S^3 + S^2 + 1000S}$$

SOLUTION TO GUIDED PROBLEM
Guided Problem 4.12

1.

2. $\dfrac{dq}{dt} = i_0 - \dfrac{1}{RC}q - \dfrac{1}{I}p$

   $\dfrac{dp}{dt} = \dfrac{1}{C}q$

3. The load current i equals p/I. Therefore, q will be eliminated from the equations above in favor of p.

4. $\dfrac{d^2p}{dt^2} = \dfrac{1}{C}\dfrac{dq}{dt} = \dfrac{1}{C}i_0 - \dfrac{1}{RC}\dfrac{q}{C} - \dfrac{1}{CI}p = \dfrac{1}{C}i_0 - \dfrac{1}{RC}\dfrac{dp}{dt} - \dfrac{1}{IC}p$

5. Multiplying this equation by C and substituting $p = Ii$ gives

   $$IC\dfrac{d^2i}{dt^2} + \dfrac{I}{R}\dfrac{di}{dt} + i = i_0$$

   Comparing this with equation (4.103) (p. 220) and using equation (4.105) gives

   $$\omega_n = \dfrac{1}{\sqrt{IC}}; \quad \zeta = \dfrac{\omega_n}{2}\dfrac{I}{R} = \dfrac{1}{2R}\sqrt{\dfrac{I}{C}}.$$

6. Assuming $\zeta < 1$,

   $$I(s) = \dfrac{i_0 \omega_n^2}{s[(s + \zeta\omega_n)^2 + \omega_d^2]} \qquad \omega_d = \sqrt{1 - \zeta^2}\,\omega_n$$

7. From Appendix C entry # 22,

   $$i(t) = [1 - (\omega_n/\omega_d)e^{-\zeta\omega_n t}\sin(\omega_d t - \phi)]i_0; \quad \phi = \tan^{-1}(\omega_d/\omega_n\zeta)$$

8. The voltage that must not exceed 15 V equals $q/C = dp/dt = I\,di/dt$. The plots in part (b) of Fig. 4.21 (p. 222) reveal that the larger ζ is chosen, the larger ω_n can be without the slope of the curve exceeding any particular maximum; a large ω_n is desired. When ζ is greater than 1, however, the curves approach their final value with an unnecessarily long tail. Therefore, $\zeta = 1$ is a good solution.

9. Unfortunately, the equations above fail for this critical damping. The solution becomes

   $$\mathcal{L}[i(t)] = \dfrac{i_0\omega_n^2}{s(s + \omega_n)^2} = \dfrac{c_1}{s} + \dfrac{c_2}{s + \omega_n} + \dfrac{c_3}{(s + \omega_n)^2}$$

   $$c_1 = \dfrac{\omega_n^2}{\omega_n^2} = 1, \quad \omega_n^2 = (s + \omega_n)^2 + c_2 s(s + \omega_n) + c_3 s = (1 + c_2)s^2 + (2\omega_n + c_2\omega_n + c_3)s + \omega_n^2$$

   Therefore, $c_2 = -1$ and $c_3 = -\omega_n$, giving $i(t) = [1 - (1 + \omega_n t)e^{-\omega_n t}]i_0$

   $$\dfrac{di}{dt} = \omega_n^2 t\, e^{-\omega_n t} i_0$$

   $$\dfrac{d^2i}{dt^2} = i_0(\omega_n - \omega_n^2 t)e^{-\omega_n t}$$

   The maximum voltage occurs when d^2i/dt^2 is zero, or when $\omega_n t = 1$. At this time,

   $$\left(I\dfrac{di}{dt}\right)_{max} = I\omega_n e^{-1} = 15\text{V} = \sqrt{\dfrac{I}{C}}e^{-1}$$

   from which $\omega_n = 15e/I = 15 \times 2.718/.01 = 4077$ rad/s $= 649$ Hz, and

   $$R = \dfrac{1}{2}\sqrt{\dfrac{I}{C}} = \dfrac{1}{2}\sqrt{\dfrac{0.1}{6.015 \times 10^{-6}}} = 20.4 \text{ ohms.}$$

$$C = \frac{1}{\omega_n^2 I} = \frac{1}{(4077)^2 \times .01} = 6.015 \ \mu\text{f};$$

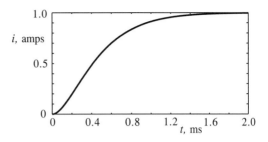

4.7 Linearization

Nature is nonlinear. Nevertheless, for small perturbations about some nominal state, as is typically the case for vibrations, most nonlinear systems act as though they were linear. The simplicity of the characterization and analysis of linear models further motivates the modeler to overlook nonlinearities. Methods for approximating a nonlinear system by a linear model are considered in this section.

A model may be linearized before or after it is converted to differential equations. In most cases the latter procedure is preferred. The meaning of linearization usually is clearer when it is carried out directly on the individual elements of the model, however, so this approach is presented first. It also allows a graphical alternative approach, and therefore sometimes is preferred.

4.7.1 Case Study With Linearization of a Resistance

Consider a fluid tank with uniform area A_t and a Bernoulli square-law orifice flow, as shown in Fig. 4.24. The pressure at the bottom of the tank is related to the orifice flow by

$$P = \frac{1}{2}\rho \left(\frac{Q}{A_0}\right)^2, \tag{4.119}$$

where Q is the volume flow through the orifice and A_0 is its effective area (including the *vena contracta*). This characteristic is plotted in part (b) of the figure. A bond graph model of the system is given in part (c). It shows that the flow is computed with the causality $Q = Q(P)$, inverting equation (4.119). An analytic solution to the nonlinear differential equation

$$\frac{dV}{dt} = Q_{in} - A_0 \sqrt{\frac{2Vg}{A_t}} \tag{4.120}$$

happens to exist if the input flow Q_{in} is a constant, or changes in discrete steps. This is not usually the case, however; simulation would be the natural recourse.

The solution of the corresponding problem with a *linear* or constant resistance is much simpler. Further, the property of superposition would apply if the model were linear, enabling the response to a complicated excitation $Q_{in}(t)$ to be assembled from a sum of the responses to simple components of that excitation. Can an approximate linear model be substituted for the nonlinear model in order to gain these advantages?

The answer to this critical question depends on how greatly the depth of the water varies, and how accurate the solution needs to be. A **linearization** now will be carried out, and the behavior of the linearized model compared with that of the nonlinear model.

A linearized model describes the behavior of **perturbations**, or changes, in the state variables relative to some nominal condition. The water tank has a single state variable,

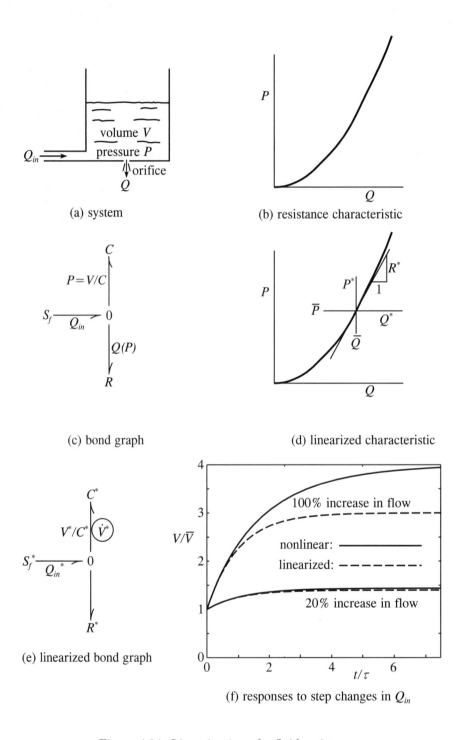

Figure 4.24: Linearization of a fluid resistance

the volume of water in the tank, V. This volume is represented as the sum of a nominal volume, $\overline{V}$, and a perturbation volume, V^*:

$$V = \overline{V} + V^*. \tag{4.121}$$

The nominal volume is the equilibrium state for a particular flow rate, which also is designated by an overbar: $\overline{Q}$. Note that at equilibrium, the input flow equals the output flow. Further, the pressure at the bottom of the tank then has its nominal value, designated as $\overline{P}$. The non-equilibrium values of these variables are designated as

$$Q_{in} = \overline{Q} + Q_{in}^*, \tag{4.122a}$$
$$Q = \overline{Q} + Q^*, \tag{4.122b}$$
$$P = \overline{P} + P^*. \tag{4.122c}$$

where Q_{in}^*, Q^* and P^* are the perturbations. The three pressures and three output flows are indicated in part (d) of Fig. 4.24.

The linearization corresponds to replacing the curved characteristic for the resistance by a straight line, as shown in the figure. The usual choice is a tangent to the nonlinear characteristic passing through the equilibrium point. The linearized characteristic is described by

$$P^* \simeq R^* Q^*; \qquad R^* = \left(\frac{dP}{dQ}\right)_{Q=\overline{Q}}. \tag{4.123}$$

The slope R^* is called the **linearized or perturbation or tangential resistance**[4]. The value of R^* can be estimated by drawing the tangent with a pencil, and measuring its slope. It also can be computed mathematically by evaluating the derivative, as given in equation (4.123), from equation (4.119).

The linearized or perturbation variables and parameters can be placed on a special linearized bond graph, as shown in part (e) of Fig. 4.24. This graph gives a linear differential equation in terms of the perturbation state variable, V^*:

$$\frac{dV^*}{dt} = Q_{in}^* - \frac{1}{R^* C} V^* \quad \text{or} \quad \tau \frac{dV^*}{dt} + V^* = \tau Q_{in}^*, \qquad \tau = R^* C. \tag{4.124}$$

Solutions to the cases in which the input flow is increased by 20% and by 100% from initial equilibrium values are plotted in part (f) of Fig. 4.24. The nonlinear solutions are found by simulation. The linear solutions are plots of the analytic solution of equation (4.124) for $Q_{in}^* = $ constant, namely

$$V^* = (1 - e^{-t/\tau}) \tau Q_{in}^*. \tag{4.125}$$

The solution of the linearized model for the 20% increase in flow is seen to be close to that of the nonlinear model; agreement would be better for smaller perturbations. The solutions for the 100% increase, on the other hand, are significantly different; the results of the linear model are only of qualitative usefulness.

Nonlinear models with small disturbances about some equilibrium act virtually the same as their simpler linearized models predict, in most cases. Thus, for example, linearized models are used almost exclusively in the field of acoustics, which deals largely with very small pressure and velocity perturbations. The same situation commonly applies regarding mechanical vibrations. The engineering analyst must excercise vigilance, however, for nonlinear effects that are large enough to require abandonment of the simplicities of linear analysis.

---

[4]The tangential resistance is very different from the *chordal* resistance described in the footnote on p. 39-40 and shown in Fig. 2.22 part (b) (p. 37).

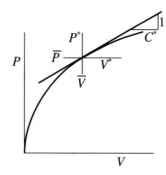

Figure 4.25: Linearization of a compliance

4.7.2   Linearization of a Function of One Variable

The variables x and y in the function

$$y = f(x), \tag{4.126}$$

can be expressed as

$$x = \bar{x} + x^*, \tag{4.127a}$$

$$y = \bar{y} + y^*, \tag{4.127b}$$

where $\bar{x}$ and $\bar{y}$ are *nominal* values of x and y that satisfy

$$\bar{y} = f(\bar{x}), \tag{4.128}$$

and x^* and y^* are the differences between the actual and the nominal values. Assuming $f(x)$ is continuous in x, y can be represented by the Taylor's series expansion

$$y = f(\bar{x}) + \left.\frac{df}{dx}\right|_{x=\bar{x}} (x - \bar{x}) + \left.\frac{d^2 f}{dx^2}\right|_{x=\bar{x}} \frac{(x - \bar{x})^2}{2} + \text{higher– order terms.} \tag{4.129}$$

With the notation of equation (4.127), this becomes

$$y^* = \left.\frac{df}{dx}\right|_{x=\bar{x}} x^* + \left.\frac{d^2 f}{dx^2}\right|_{x=\bar{x}} \frac{x^{*2}}{2} + \text{higher– order terms.} \tag{4.130}$$

This relation can be approximated, at least for small values of x^* and y^* described as perturbations, by truncating the series after the first, or linear, term. The resulting linearization can be written

$$y^* \simeq ax^*; \qquad a = \left.\frac{df}{dx}\right|_{x=\bar{x}}. \tag{4.131}$$

The linearization of the resistance relation applies this more general result; in that case, $x = Q$ and $y = P$. Another application is to linearization of a compliance relation between a generalized displacement $x = q$ and a generalized force $y = e$. In this case, the derivative $(df/dq)_{q=\bar{q}}$ equals the reciprocal of the linearized compliance, or $1/C^*$. This linearization is pictured in Fig. 4.25 for the special case of a gravity fluid compliance for a tank of nonuniform area, in which x equals the volume of liquid in the tank, V, and y equals the pressure at the bottom, P. The derivative can be found either analytically or graphically (by drawing the tangent to the curve at the nominal point, and measuring its slope). The latter procedure does not require an analytical expression for the nonlinear function, but can proceed based on a plot created from experimental data.

EXAMPLE 4.29

Linearize the compliance and the resistance of the conical tank of Example 3.23 (p. 130) about some nominal volume, $\overline{V}$. The relation between the volume of the liquid in the tank and the pressure at its bottom is reproduced here for convenience:

$$V = \frac{\pi \tan^2 \alpha}{3} \left(\frac{P}{\rho g} \right)^3.$$

Solution: The slope of the characteristic at a nominal volume $\overline{V}$ is

$$\frac{1}{C^*} = \left. \frac{dP}{dV} \right|_{V=\overline{V}} = \frac{\rho g}{3} \left(\frac{3}{\overline{V}^2 \pi \tan^2 \alpha} \right)^{1/3}.$$

The corresponding linearized resistance becomes, from equations (4.119) and (4.123),

$$R^* = \left. \frac{dP}{dQ} \right|_{Q=\overline{Q}} = \frac{\rho \overline{Q}}{(c_d A)^2} = \frac{\sqrt{2\rho \overline{P}}}{c_d A} = \frac{2\rho^{3/2}}{c_d A} \left(\frac{3\overline{V}}{\pi \tan^2 \alpha} \right)^{1/6}.$$

EXAMPLE 4.30

Write the linearized differential equation corresponding to the results of Example 4.29 above, and identify the time constant that characterizes the response to perturbations of the input flow.

Solution:

The compliance and the resistance of the combined system are at equilibrium if the flow into the tank, Q_{in}, equals the flow out at its bottom, Q. A linearized bond graph for perturbations about this equilibrium is as follows:

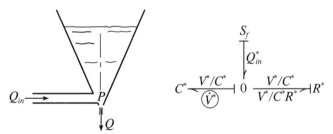

Although this has the same structure as the bond graph for the actual variables, the products of its conjugate efforts and flows are not the true powers and are not even proportional to the true powers. Such a graph sometimes is called a **pseudo bond graph** to underscore the distinction. The graph gives

$$\frac{dV^*}{dt} = Q_{in}^* - \frac{1}{C^* R^*} V^*.$$

This is a linear first-order differential equation with a characteristic equation in the form of equation (4.97) (p. 218) with a time constant, given by equation (4.98), of

$$\tau = R^* C^* = \frac{6}{A_0 \sqrt{2g}} \left(\frac{\pi \overline{V}^5 \tan^2 \alpha}{3} \right)^{1/6}.$$

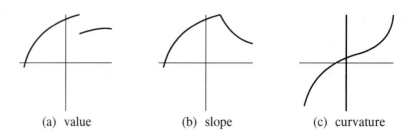

(a) value   (b) slope   (c) curvature

Figure 4.26: Types of discontinuities

4.7.3 Essential Nonlinearities

One case in which the above model for orifice flow may not be considered acceptably accurate for *any* departure of the arguments from their nominal values occurs when linearization is performed for perturbations about zero pressure drop and zero flow, that is about the origin of the characteristic plot. The calculation in Example 4.29 gives $R^* = 0$ or $1/R^* = \infty$, which implies no resistance whatever. This is correct; for *precisely* zero pressure drop and flow, the linearized pressure-vs.-flow characteristic is horizontal. Thus, if one truly is interested only in extremely small flows one could neglect the orifice resistance altogether. (This assumption is indeed commonly made in acoustics, although the inertance is not neglected.) On the other hand, the resistance departs from zero very rapidly as the flow increases. You might or might not decide that this characteristic comprises an **essential nonlinearity**, that is a condition for which linearization is not justified.

A characteristic that suffers a significant *discontinuity in value* at a singular point clearly gives an essential nonlinearity at that point. This case is illustrated in part (a) of Fig. 4.26. If, instead, it suffers a *discontinuity in a first derivative* at the point, as shown in part (b) of the figure, the linearization still cannot be carried out formally by taking the derivative. If a discontinuity of value or of slope is small enough, however, one might choose to overlook it in favor of some linear compromise. The issue faced at zero pressure drop for the orifice and in part (c) of Fig. 4.26 is even more subtle: a *discontinuity of a second derivative*, which is a kind of borderline essential nonlinearity. Discontinuities in the third or higher order derivatives are not considered essential nonlinearities.

A possible resolution of the question for the orifice results from use of a more accurate model of the orifice flow. The discontinuity in the second derivative disappears, and linearization about the origin becomes fully justified for small disturbances. This happens because when the flow is very small the effect of viscosity, which is neglected in equation (4.119), is no longer negligible compared with the effect of inertia. The actual characteristic for a fixed orifice would resemble the plot shown in Fig. 4.27. The resistance to flow (slope of the characteristic) never goes below some value associated with "creeping viscous flow." Details can be found in texts on fluid mechanics or hydraulics.

Generalizing, some apparent essential nonlinearities are consequences of our modeling abstractions rather than the physics, although the more common danger is the overlooking of significant phenomena through excessively crude abstractions. Modeling, as noted before, is an art.

4.7.4 Linearization of a Function of Two Variables

The function

$$y = f(x, u) \tag{4.132}$$

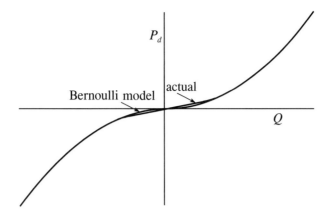

Figure 4.27: Correction of Bernoulli model for viscous effects

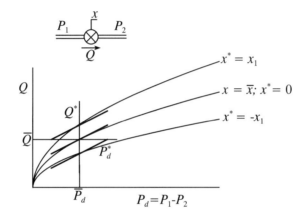

Figure 4.28: Characteristics of variable orifice (valve) and linearization

can be linearized about the nominal values $x = \overline{x}$, $u = \overline{u}$ and $y = \overline{y} = f(\overline{x}, \overline{u})$ by retaining only the first or linear terms in a Taylor's series expansion:

$$y \simeq \overline{y} + \left.\frac{\partial f}{\partial x}\right|_{\substack{x=\overline{x} \\ u=\overline{u}}} (x - \overline{x}) + \left.\frac{\partial f}{\partial u}\right|_{\substack{x=\overline{x} \\ u=\overline{u}}} (u - \overline{u}). \tag{4.133}$$

Defining perturbation variables as before, this result can be written

$$y^* \simeq c_x x^* + c_u u^*; \qquad c_x = \left.\frac{\partial f}{\partial x}\right|_{\substack{x=\overline{x} \\ u=\overline{u}}}; \qquad c_u = \left.\frac{\partial f}{\partial u}\right|_{\substack{x=\overline{x} \\ u=\overline{u}}}. \tag{4.134}$$

It is applied below both to single resistances and entire differential equations.

Resistances not infrequently vary in response to changes in some displacement. Such cases are perhaps best understood in graphical as opposed to algebraic terms. The **adjustable hydraulic valve**, as represented in Fig. 4.28, is an example. Note that flow is plotted on the vertical axis and pressure on the horizontal axis, contrary to the normal usage in this book but consistent with the valve industry.

Such a valve is essentially an adjustable orifice. The area of the *vena contracta* is considered a function of some valve displacement, labeled here as x:

$$A = A(x). \tag{4.135}$$

The pressure-flow characteristics, assuming the model $A = A_0(x/x_0)^{1.5}$, are plotted in the figure for three equally spaced values of x. This plot could be refined by adding any number of added characteristics for interpolated or extrapolated values of x. Mathematically, it is of the form

$$Q = Q(P_d, x), \tag{4.136}$$

which is a function of two variables.

The family of characteristics may be approximated in a region about some nominal pressure difference $\overline{P}_d$, valve displacement $\overline{x}$ and associated flow $\overline{Q}$ by a set of parallel, equally spaced straight-line characteristics, as shown in the figure. These represent a linearized model. Mathematically, equation (4.27) can be applied to give

$$Q^* = \frac{P_d^*}{R_*} + kx^*, \tag{4.137a}$$

$$x^* \equiv x - \overline{x}, \tag{4.137b}$$

$$\frac{1}{R_*} = \left.\frac{\partial Q}{\partial P_d}\right|_{\substack{P_d=\overline{P}_d \\ x=\overline{x}}}, \tag{4.137c}$$

$$k = \left.\frac{\partial Q}{\partial x}\right|_{\substack{P_d=\overline{P}_d \\ x=\overline{x}}} = \left.\frac{\partial Q}{\partial A}\right|_{\substack{P_d=\overline{P}_d \\ x=\overline{x}}} \left.\frac{dA}{dx}\right|_{x=\overline{x}}. \tag{4.137d}$$

For most cases this linearization is accurate within any desired bound for sufficiently small departures of x, P_d and Q from the nominal $\overline{x}$, $\overline{P}_d$ and $\overline{Q}$, respectively. How much inaccuracy to accept when the departures are not small, in order to reap the benefits of linearity, is a matter of engineering judgment. An engineer may elect to start an analysis with a quick, rough and simple linearized model, and, when satisfied with the general nature of the result, refine the analysis with a relatively expensive-to-use nonlinear model.

EXAMPLE 4.31

Linearize the drag and thrust characteristics of the boat described in Guided Problem 2.3 (pp. 43-44) about the equilibrium condition at 45 feet per second. Incorporate the results into a linear state-space model of the boat, assuming a virtual inertia 50% higher than the mass of the boat and its contents to approximate the effect of the water which is accelerated along with the boat. The boat and its contents weigh 2500 pounds. Finally, for how large a change in thrust would you judge that the linearization gives a reasonable representation?

Solution: The first step is to draw tangents to the drag and thrust curves at the given equilibrium point directly on the given plot. Then, a set of equally spaced parallel lines can be drawn to approximate the thrust for other values:

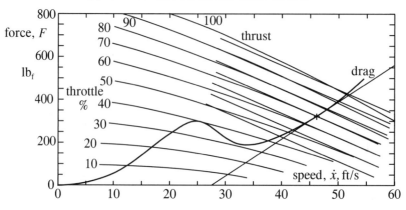

This plot helps you represent the linearized characteristics with linear algebraic equations. Let the thrust be F_1 and the drag be F_d, both in pounds force, and the percent throttle be P:

$$F_1 = 324 + 5.59(P - 70) - 11.6(\dot{x} - 46.2) = 5.59\,P - 11.6\,\dot{x} + 4.69$$
$$F_d = 17.3(\dot{x} - 27.5) = 17.3\,\dot{x} - 476$$

The differential equation for the speed of the boat simply represents that the acceleration equals the ratio of the net thrust $(F_1 - F_d)$ to the effective inertia:

$$\frac{d\dot{x}}{dt} = \frac{F_1 - F_d}{m} = -0.248\,\dot{x} + 0.0480\,P + 8.11 \text{ ft/s}^2.$$

The linearization applies exclusively in the region where the boat planes, working fairly well for speeds between 42 and 52 ft/s. It totally fails to represent the decrease in drag associated with the onset of planing.

4.7.5 Linearization of a First-Order Differential Equation

The differential equation for a first-order dynamic model with state variable x and excitation $u(t)$ can be written

$$\frac{dx}{dt} = f(x, u(t)). \tag{4.138}$$

The term on the left side equals a function of two variables, so this equation is in the form of equation (4.132) and can be linearized in the same way. Thus, the variables x and u are decomposed into two parts:

$$x(t) = \overline{x}(t) + x^*(t), \tag{4.139a}$$
$$u(t) = \overline{u}(t) + u^*(t). \tag{4.139b}$$

The novel feature is that x and u are functions of *time*, but this does not upset the linearization. The terms $x^*(t)$ and $u^*(t)$ are generally small variations or perturbations of $x(t)$ and $u(t)$ about their nominal values $\overline{x}(t)$ and $\overline{u}(t)$. These nominal values, as always, must satisfy the original nonlinear equation. In general, they represent some nominal "trajectory" in time about which the linearization is taken. Usually, the engineer is interested in perturbations about an *equilibrium* or steady state, in which $\overline{u}(t) =$ constant and the derivative $d\overline{x}/dt = 0$. This is the only case considered in detail below, although the more general case is hardly more involved. All time-dependent phenomena are therefore represented by $x^*(t)$ and $u^*(t)$. The first order of business is to solve for the $\overline{x}$ which corresponds to the given $\overline{u}$. For the case of a constant equilibrium, this implies solving the algebraic equation

$$f(\overline{x}, \overline{u}) = 0. \tag{4.140}$$

The second step applies equation (4.134) to give the linearization

$$\boxed{\frac{dx^*}{dt} \simeq Ax^* + Bu^*, \qquad A = \left.\frac{\partial f}{\partial x}\right|_{\substack{x=\overline{x}\\u=\overline{u}}}; \quad B = \left.\frac{\partial f}{\partial u}\right|_{\substack{x=\overline{x}\\u=\overline{u}}}.} \tag{4.141}$$

One approach to linearization linearizes the individual nonlinear elements before constructing the differential equation. This has been done above for the example of the conical tank and orifice, with its nonlinear resistance and nonlinear compliance. A simpler and more reliable approach, in most cases, is to find the nonlinear differential equation and then linearize. This approach now is illustrated.

EXAMPLE 4.32

Return to the conical tank and orifice of Examples 3.23 (p. 130), 4.29 (p. 237) and 4.30 (p. 237). Linearize directly the nonlinear differential equation, which is

$$\frac{dV}{dt} = Q_{in} - \frac{A_0\sqrt{2g}}{(\pi \tan^2 \alpha)^{1/6}} (3V)^{1/6},$$

and compare the result to that found in Example 4.30 by linearizing the resistance and compliance elements individually.

Solution: The equilibrium solution corresponding to $Q_{in} = \overline{Q}$ is

$$0 = \overline{Q} - A_0\sqrt{2g} \left(\frac{3\overline{V}}{\pi \tan^2 \alpha} \right)^{1/6}.$$

Using equation (4.141), the linearized differential equation becomes

$$\frac{dV^*}{dt} \simeq Q_{in}^* - \frac{A_0\sqrt{2g}}{6} \left(\frac{3}{\pi \overline{V}^5 \tan^2 \alpha} \right)^{1/6} V^*.$$

This is a first-order linear differential equation. The time constant, from equation (4.98) (p. 218), is

$$\tau = \frac{6}{A_0\sqrt{2g}} \left(\frac{\pi \overline{V}^5 \tan^2 \alpha}{3} \right)^{1/6}.$$

This agrees with the results of Example 4.30, and is reasonably valid for perturbations in the volume, $V^*(t)$, that are considerably smaller than the mean volume, $\overline{V}$.

4.7.6   Linearization of State-Variable Differential Equations

Linearization of a higher-order model usually is best carried out on the individual first-order differential equations in state-variable form, or

$$\frac{d\mathbf{x}}{dt} = \mathbf{f}\left(\mathbf{x}, \mathbf{u}(t), t\right). \tag{4.142}$$

The discussion that follows drops the final argument t, restricting consideration to stationary models. In this case the equilibrium and perturbation variables are

$$\mathbf{x} = \overline{\mathbf{x}} + \mathbf{x}^*, \tag{4.143a}$$

$$\mathbf{u} = \overline{\mathbf{u}} + \mathbf{u}^*, \tag{4.143b}$$

The equilibrium solution to equation (4.143) is

$$0 = \mathbf{f}\left(\overline{\mathbf{x}}, \overline{\mathbf{u}}\right). \tag{4.144}$$

The Taylor's series expansion of equation (4.143) is

$$\frac{d\mathbf{x}}{dt} = \mathbf{0} + \frac{d\mathbf{x}^*}{dt} = \mathbf{f}\left(\overline{\mathbf{x}}, \overline{\mathbf{u}}\right) + \left.\frac{\partial \mathbf{f}}{\partial \mathbf{x}}\right|_{\substack{\mathbf{x}=\overline{\mathbf{x}}\\\mathbf{u}=\overline{\mathbf{u}}}} \mathbf{x}^* + \left.\frac{\partial \mathbf{f}}{\partial \mathbf{u}}\right|_{\substack{\mathbf{x}=\overline{\mathbf{x}}\\\mathbf{u}=\overline{\mathbf{u}}}} \mathbf{u}^* + \text{higher order terms}. \tag{4.145}$$

This result can be cast in traditional form as follows:

$$\frac{d\mathbf{x}^*}{dt} \simeq \mathbf{A}(t)\,\mathbf{x}^* + \mathbf{B}(t)\,\mathbf{u}^*(t),$$

$$\mathbf{A} = \left.\frac{\partial \mathbf{f}}{\partial \mathbf{x}}\right|_{\substack{x=\bar{x}\\u=\bar{u}}},$$

$$\mathbf{B} = \left.\frac{\partial \mathbf{f}}{\partial \mathbf{u}}\right|_{\substack{x=\bar{x}\\u=\bar{u}}}. \tag{4.146}$$

The matrices of derivatives that comprise $\mathbf{A}$ and $\mathbf{B}$ are called **Jacobian matrices**.

EXAMPLE 4.33

A mass rests on a spring which in turn rests on a rigid foundation. The spring satisfies the nonlinear stiffening relation

$$F_s = k_1 x + k_3 x^3; \qquad k_1 = 2000\ rmN/m; \qquad k_3 = 40,000\ \text{N/m}^3$$

(in order to support a range of masses and applied forces without bottoming out). The particular mass of interest compresses the spring 0.100 meters when the applied force F is zero.

Model the system, determine the mass, find a single linearized differential equation relating $F = F^*$ to x^*, and determine the linearized natural frequency.

Solution: The bond graph

gives the state-space model

$$\frac{d}{dt}\begin{bmatrix} x \\ p \end{bmatrix} = \begin{bmatrix} -p/m \\ F + mg - k_1 x - k_3 x^3 \end{bmatrix}.$$

The equilibrium condition for $F = F^* = 0$ is

$$0 = 0 + mg - k_1\bar{x} - k_3\bar{x}^3,$$

from which

$$m = \frac{1}{g}(k_1\bar{x} + k_3\bar{x}^3) = \frac{2000(.1) + 40,000(.1)^3}{9.81} = 24.5\ \text{kg}$$

The linearized differential equation is

$$\frac{d}{dt}\begin{bmatrix} x^* \\ p^* \end{bmatrix} = \begin{bmatrix} 0 & -1/m \\ -k_1 - 3k_3\bar{x}^2 & 0 \end{bmatrix}\begin{bmatrix} x^* \\ p^* \end{bmatrix} + \begin{bmatrix} 0 \\ 1 \end{bmatrix}F^*,$$

so that

$$\frac{d^2 x^*}{dt^2} = \frac{1}{m}\frac{dp^*}{dt} = -\frac{k_1 + 3k_3\bar{x}^2}{m}x^* + \frac{1}{m}F^*.$$

The linearized natural frequency therefore is

$$\omega_n = \sqrt{\frac{k_1 + 3k_3\bar{x}^2}{m}} = \sqrt{\frac{2000 + 3 \times 40000(.1)^2}{24.5}} = 11.44\ \text{rad/s} = 1.820\ \text{Hz}.$$

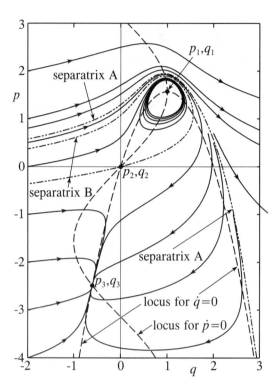

Figure 4.29: Phase-plane plot of system with three different types of equilibria

4.7.7   Case Study With Three Different Types of Equilibria

The second-order system

$$\frac{dp}{dt} = -q + \sin p$$
$$\frac{dq}{dt} = p - \pi q \left(1 - \frac{q}{2}\right) \tag{4.147}$$

has three equilibria, each of a different type. The behavior can be studied globally, using simulation, to achieve a picture of the overall behavior. With much less effort, the behaviors for states close to the equilibria can be determined using linearization. The two approaches are developed and compared below.

Equations (4.147) represent an autonomous system, that is with no excitation. The entire behavior depends on whatever initial state is imposed. A **phase-plane plot** of the system is shown in Fig. 4.29. **Trajectories** in state-space are given by solid lines with arrows that designate direction with increasing time. These trajectories were found by simulation of the equations using the standard MATLAB routine ode45, using various initial states on the boundaries of the plot. All trajectories starting above the line designated separatrix A sweep across the state space, remaining above the separatrix, and continue toward infinite values. This region is unstable.

A second region in state space is virtually surrounded by the line designated as separatrix B. Any initial state in this region produces a trajectory that approaches and then encircles point p_1, q_1 in the clockwise sense. This point and this region are called **meta-stable**, since no trajectory ever reaches the point itself, but rather encircles it indefinitely. Looked at as functions of time, both p and q oscillate virtually sinusoidally. All trajectories in the third and remaining region, which lies between separatrix A and separatrix B, end at point p_3, q_3. This point and region are stable.

Equilibrium points satisfy the conditions $dp/dt = 0$ and $dq/dt = 0$. Labeling such points as $\bar{p}$ and $\bar{q}$, equations (4.147) give

$$\bar{q} = \sin \bar{p}, \tag{4.148a}$$

$$\bar{p} = \pi \bar{q}(1 - \bar{q}/2). \tag{4.148b}$$

These two equations are plotted in Fig. 4.29 by dashed lines. Their intersections represent the equilibrium points. In addition to points $\bar{p}_1, \bar{q}_1 = \pi/2, 1$ and $\bar{p}_3, \bar{q}_3 = -2.4886, -0.60757$, noted above, there is point $\bar{p}_2, \bar{q}_2 = 0, 0$. This is an unstable equilibrium point, because any initial condition represented by a point near but not at it produces a trajectory that heads away from it.

The behaviors in the local neighborhoods of the three equilibrium points can be determined by linearization, without any need for extensive numerical computations such as those that produced the trajectories and separatrixes shown in the figure. Equations (4.147) linearize about any nominal state or state trajectory $\bar{p}, \bar{q}$ to become

$$\frac{dp^*}{dt} = -q^* + (\cos \bar{p})p^*, \tag{4.149a}$$

$$\frac{dq^*}{dt} = p^* + \pi(\bar{q} - 1)q^*. \tag{4.149b}$$

(You are urged to derive these equations yourself.) This result now will be particularized to give linearizations about the three equilibrium points.

Substituting the values $\bar{p} = \pi/2, \bar{q} = 1$ for the first equilibrium into equations (4.149), there results

$$\frac{dp^*}{dt} = -q^*$$
$$\frac{dq^*}{dt} = p^*. \tag{4.150}$$

Combining these two equations to form a second-order differential equation in p^*,

$$\frac{d^2 p^*}{dt^2} + p^* = 0. \tag{4.151}$$

This equation describes an oscillation with natural frequency $\omega_n = 1$ and damping ratio $\zeta = 0$. The absense of damping renders this equilbirum meta-stable.

Substituting the values $\bar{p} = 0$, $\bar{q} = 0$ for the second equilibrium into equations (4.149), there results

$$\frac{dp^*}{dt} = p^* - q^*,$$
$$\frac{dq^*}{dt} = p^* - \pi q^*. \tag{4.152}$$

In operator notation these become

$$(S - 1)p^* = -q^*,$$
$$(S + \pi)q^* = p^*. \tag{4.153}$$

Combining these two equations to give an equation in $p^*(t)$,

$$[S^2 + (\pi - 1)S + (1 - \pi)]p^* = 0 \qquad \text{or} \qquad \frac{d^2p^*}{dt^2} + (\pi - 1)\frac{dp^*}{dt} + (1 - \pi)p^* = 0. \quad (4.154)$$

The solution is

$$p^* = p_1^* e^{-2.8841t} + p_2^* e^{0.7425t}, \qquad\qquad (4.155)$$

where p_1^* and p_2^* are arbitrary constants dependent on the initial conditions. The second term, with its positive exponent, reveals the instability.

Substituting the values $\bar{p} = -2.4886, \bar{q} = -0.60757$ for the third equilibrium into equation (4.149), there results

$$\frac{dp^*}{dt} = -0.7943p^* - q^*,$$

$$\frac{dq^*}{dt} = p^* - 5.0503q^*. \qquad\qquad (4.156)$$

Combining these equations (the use of operators methods as above is suggested) gives

$$\frac{d^2p^*}{dt^2} + 5.1297\frac{dp^*}{dt} + 1.4011p^* = 0. \qquad\qquad (4.157)$$

This has a natural frequency of 1.184 rad/s, but no oscillations appear because it is over-damped with $\zeta = 2.167$. It is rock stable.

4.7.8   Summary

Most nonlinear systems behave as though they were linear for sufficiently small disturbances about some equilibrium, which together with the advantages of superposition motivates their linearization. Linearization usually is best applied to the set of nonlinear state differential equations. Alternatively, nonlinear resistances and compliances can be linearized directly, before differential equations are formulated, by replacing their nonlinear effort-flow or effort-displacement characteristics, respectively, with linear or straight-line characteristics. The linearized characteristic passes through the origin defined by these incremental or perturbation variables, which is located at the point of tangency of the two characteristics and may be different from the origin of the full variables. Graphs that act much like bond graphs may be constructed using the linearized variables, but such pseudo bond graphs do not represent the actual power flows and energy storages, or even their incremental variations.

Analytical linearization of a term in an algebraic equation or of a differential equation is carried out formally by truncating a Taylor's series expansion after the linear or first-order terms. Graphical linearization is carried out by replacing curved characteristics by straight lines, and families of curved characteristics by familes of equally spaced straight lines. The graphical option is especially useful when you have experimental data rather than an analytical model. Even if you perform a linearization analytically, a comparison of the graphical representations of the nonlinear and approximate linear characteristics conveys a sense of how the error imposed by the linearization varies with the span of the changes in the state variables.

A third way to linearize a system is to expand the energy storage functions $\mathcal{V}$ and $\mathcal{T}$ and the power dissipation function $\mathcal{P}$ in Taylor's series, and drop all terms of *third* and higher order. This method is illustrated in Guided Problem 4.15.

Guided Problem 4.13

This first linearization problem employs the straightforward analytical procedure on a set of differential equations.

Linearize the following state differential equations about the equilibrium state. The variables are p and q; the coefficients $a, \dots, f$ are constants.

$$[1 + (a + bq)^2]\frac{dp}{dt} = e_1 - cq + d(a + bq)p^2$$

$$\frac{dq}{dt} = ep - f\sqrt{q}$$

Suggested Steps:

1. Find the equilibrium state or states by setting the two time derivatives equal to zero. Label these as $\bar{p}$ and $\bar{q}$.

2. Expand each term in a Taylor's series expansion. Note that the equilibrium value of dp/dt is zero. (Derivatives do not always have a zero equilibirum value, however.)

3. Truncate the expansions to the linear terms in p^* and q^*.

4. Subtract the terms of the original differential equation, with $\bar{p}$ and $\bar{q}$ substituted for p and q, from the truncated expansion, to leave only terms proportional to the incremental variables p^* and q^*.

Guided Problem 4.14

This second linearization problem employs graphical linearization with a subsequent mathematical analysis, a very common procedure.

An induction motor drives a load that has the resistive characteristic plotted in Fig. 4.30 plus a rotational inertia of 0.05 kg·m². The drive may be considered to be inflexible. Perturbations are imposed on the load torque equal to $M = 2.0 \sin(40\ t)$ N·m. You are asked to make a close estimate of the resulting perturbations of the angular velocity.

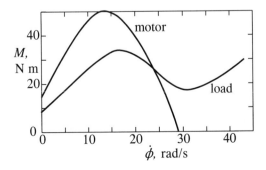

Figure 4.30: Guided Problem 4.15

Suggested Steps:

1. Linearize the steady-velocity characteristics of the motor and the load, determining numerical values for the resistances.

2. Draw a linearized bond graph for the system, including the inertance and the disturbance. Apply causality, and define the state variable.

3. Write the differential equation.

4. This text has not yet addressed the solution to differential equations with sinusoidal excitations. The differential equation is only of first order, however, and the part of the solution you want is so simple that the most primitive method taught in your course on differential equations can be applied. Substitute an assumed solution of interest $p_1^* \sin 40t + p_2^* \cos 40t$ into the differential equation, and determine the coefficients p_1^* and p_2^*. The answer you want is the magnitude of the sinusoidal motion $\dot{\phi}^*$, which equals $(1/I)\sqrt{p_1^* + p_2^*}$.

5. Check that the resulting perturbations of the angular velocity are not large enough to impart significant error to the linearization.

Guided Problem 4.15

This third linearization problem is intended to show how linearization can be carried out in terms of energy relations.

A simple pendulum has a moment of inertia I about its center of mass and $I_0 = I + md^2$ about its center of rotation, where m is its mass and d is the distance between the centers of rotation and mass.

(a) Neglecting dissipation, find a nonlinear characteristic for the gravity-based compliance, and sketch a plot of this characteristic.

(b) Linearize this characteristic, and plot on the same axes as the nonlinear characteristic for comparison.

(c) Write the nonlinear and linear state-based differential equations.

Suggested Steps:

1. Sketch the pendulum, and find its gravity energy $\mathcal{V}$ relative to its minimum for the vertical equilibrium, using the angle of departure from vertical, ϕ, as the displacement.

2. Compute the torque $M = d\mathcal{V}/d\phi$ and plot M versus ϕ.

3. Carry out the linearization *two* different ways. First, directly linearize the relation between M and ϕ, and plot.

4. For the second approach, do a Taylor's series expansion of the function $\mathcal{V} = \mathcal{V}(\phi)$. Drop terms higher than the quadratic. The result gives the linear relation between $M = d\mathcal{V}/d\phi$ and ϕ.

5. Write the state differential equations, using as state variables either the standard pair $(p, q = \phi)$ or an alternative $(\phi, \dot{\phi})$.

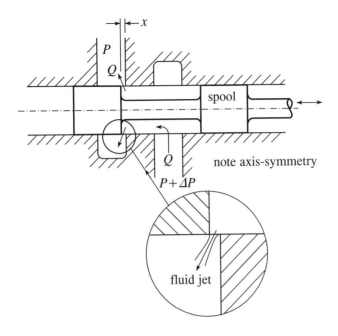

Figure 4.31: Guided Problem 4.17

Guided Problem 4.16

The last guided problem involves linearization in two variables.

The resistance to the flow of oil through a hydraulic **spool valve** is associated with an orifice in the shape of a very short segment of a cylinder of diameter d, as shown in Fig. 4.31. The fluid can be considered to be incompressible with density ρ, and viscous effects can be neglected.

(a) Find the volume flow rate Q through the valve as a function of the pressure drop P and the spool position, x. Sketch the characteristics using appropriate dimensionless coordinates.

(b) Linearize the characteristics about some non-zero operating pressure drop and flow. Add the resulting linearized characteristics to your plot.

Suggested Steps:

1. Assume the Bernoulli equation for the flow, representing the area of the orifice as the area of the segment of the cylinder of length x. A flow coefficient c_d can be applied, which has a theoretical value of 0.611.

2. Group the variables and parameters such that a dimensionless group proportional to the flow Q is a function of a dimensionless group proportional to the pressure drop P and a dimensionless group proportional to x. An arbitrary maximum or reference pressure P_s can be defined to make this possible.

3. Sketch-plot the results with the first group above as the ordinate, the second group as the abcissas and the third group as the parameter which distinguishes three of four different curves.

4. Carry out the linearization about some operating point $\overline{Q}$, $\overline{P}$, $\overline{x}$. Equation (4.137) (p. 240) can be adapted to the dimensionless groups.

5. Draw a set of equally spaced parallel lines that correspond to the linearized characteristics. Note the limited area of applicability.

PROBLEMS

4.72 A dashpot obeys the relation $F = a\dot{x}^{3/2}$. Find its linearized resistance R^* for perturbations about a nominal velocity $\overline{\dot{x}}$.

4.73 A "stiffening" spring satisfies the force-displacement relation

$$F = kx + ax^3.$$

Find its linearized compliance C^* for perturbations about the displacment $x = \overline{x}$.

4.74 A vehicle weighing 3000 lbs. is driven at a constant speed of 50 ft/s. The thrust force then is increased suddenly by 10% and maintained at that level. You are asked to find how the speed of the vehicle increases with time. The drag characteristic is plotted below.

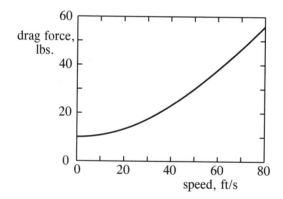

(a) Estimate the initial and terminal thrust forces.

(b) Linearize the drag characteristic, and estimate the terminal speed.

(c) Draw a linearized pseudo-bond graph and write a differential equation describing the transient behavior.

(d) Solve for the velocity as a function of time; sketch-plot the result.

4.75 A series-wound DC motor has the torque-speed characteristic plotted near the top of the next page. It drives a load with a total mass moment of inertia of 0.25 in.·lb·s^2 and steady-state torque of 5.0 in.·lb at its equilibrium speed.

(a) Determine the equilibrium speed, $\dot{\phi}_0$.

(b) At time $t = 0$ a brake applies an additional torque of 1.0 in.·lb. Determine its final speed.

(c) Draw a linearized bond graph and write a linear differential equation that when solved approximates the transient change in speed, $\dot{\phi}^*(t)$, for time $t > 0$. State the values of all constants.

(d) Sketch the speed vs. time, noting the values of any time constants or natural frequencies that exist.

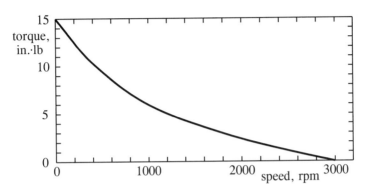

4.76 A mixer (essentially a motor with a stirring rod or whisk) mixes a viscous substance ("goop"). A simple model is given below (the compliance of the whisk is included but the effect of density is not).

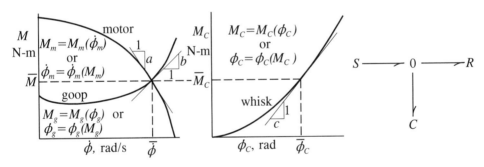

(a) Identify the equilibrium state, and determine whether it is stable or unstable.

(b) Apply causal strokes to the bond graph.

(c) Identify which three of the six functional forms given for the various nonlinear characteristics are to be used in writing the state differential equations, and write these in the appropriate places on the bond graph.

(d) Complete the annotation of the graph, and write the nonlinear state differential equations.

(e) Write appropriate linearized relations for the three nonlinear characteristics, using definitions given above.

(f) Determine linearized state differential equations.

(g) Relate the time constant or natural frequency and damping ratio (whichever applies) to the given parameters.

4.77 Briefly describe under what circumstances the following situations give rise to essential nonlinearities:

   (a) coulomb (dry) friction

   (b) fluid check valve

   (c) electrical diode

   (d) finite-sized fluid gravity tank of uniform area

   (e) a simply mounted compression spring

4.78 Linearize the following differential equation about the equilibrium corresponding to $\bar{u} = 4$.

$$\frac{dx}{dt} = -6\sqrt{x} + 3u$$

4.79 Linearize the following differential equation about the equilibrium corresponding to $u(t) = 2$:

$$\frac{dx}{dt} + x^{3/2} = 4u.$$

4.80 For the differential equation

$$\frac{dx}{dt} + \sin x = \sqrt{u},$$

   (a) find the smallest positive equilibrium value of x corresponding to $u = 0.25$.

   (b) find the linearized differential equation corresponding to perturbations about the equilibrium of part (a).

4.81 Linearize the differential equation below about the equilibrium $\bar{u} = 4$.

$$\frac{d^2x}{dt^2} + 2\left(\frac{dx}{dt}\right)^{3/2} + e^{2x} = u(t)$$

4.82 Linearize the following set of state differential equations about the smallest positive equilibrium corresponding to $\bar{u} = 2$.

$$\frac{dq}{dt} = 2\sqrt{25 - p^2} - \frac{4pq}{\pi} - u$$

$$\frac{dp}{dt} = 10\cos q$$

4.83 A system has been modeled using a bond graph and characteristics as shown below.

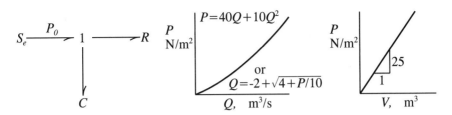

(a) Annotate the bond graph as needed and find the state differential equation(s).

(b) Determine the equilibrium state(s) of the system if $P_0 = 50$ N/m^2.

(c) Linearize the resulting differential equation(s) for disturbances about the equilibrium state.

4.84 The model described by

$$\frac{d^2x}{dt^2} + 2\left(\frac{dx}{dt}\right)^{3/2} + 2x = u(t)$$

has a nominal input $u(t) = \bar{u}(t) = 2 + 2t$.

(a) Find the nominal response, $\bar{x}(t)$. Hint: try $\bar{x} = a + bt$.

(b) Linearize the differential equation for the responses $x^*(t) = x(t) - \bar{x}(t)$ to the disturbances $u^*(t) = u(t) - \bar{u}(t)$.

SOLUTIONS TO GUIDED PROBLEMS

Guided Problem 4.14

1. If $\dfrac{dp}{dt} = 0$ and $\dfrac{dq}{dt} = 0$, then $\bar{p} = \dfrac{f}{e}\sqrt{\bar{q}}$ and

$$0 = e_1 - cq_0 + d(a + b\bar{q})\frac{f^2}{e^2}\bar{q},$$

or $\quad \dfrac{dbf^2}{e^2}\bar{q}^2 + \left(\dfrac{daf^2}{e^2} - c\right)\bar{q} + e_1 = 0,$

or $\quad \bar{q}^2 + \left(\dfrac{a}{b} - \dfrac{ce^2}{dbf^2}\right)\bar{q} + \dfrac{e^2e_1}{dbf^2} = 0.$

Therefore, $\quad \bar{q} = -\dfrac{1}{2}\left(\dfrac{a}{b} - \dfrac{ce^2}{dbf^2}\right) \pm \sqrt{\dfrac{1}{4}\left(\dfrac{a}{b} - \dfrac{ce^2}{dbf^2}\right)^2 - \dfrac{ee_1}{dbf^2}}.$

2,3. Since $\dfrac{d\bar{p}}{dt} = 0,\quad \left[1 + (a + b\bar{q})^2\right]\dfrac{dp^*}{dt} \simeq e_1 - c(\bar{q} + q^*) + d(a + b\bar{q})\bar{p}^2$
$$+ d(a + b\bar{q})2\bar{p}p^* + db\bar{p}^2q^*.$$

Since $\dfrac{d\bar{q}}{dt} = 0,\quad \dfrac{dq^*a}{dt} \simeq (\bar{p} + p^*) - f\sqrt{\bar{q}} - \dfrac{f}{2\sqrt{\bar{q}}}q^*.$

4. After the equilibrium terms are substracted, this gives

$$\left[1 + (a + b\bar{q})^2\right] \frac{dp^*}{dt} \simeq -cq^* + 2d(a + b\bar{q})\bar{p}p^* + db\bar{p}^2 q^*,$$

$$\frac{dq^*}{dt} \simeq ep^* - \frac{f}{2\sqrt{\bar{q}}} q^*.$$

Guided Problem 4.15

1.

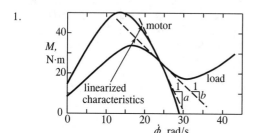

From the linearized characteristics superimposed on the plots above,

motor: $M_m^* = -a\dot{\phi}^*$, $a = 3.94$ N·m·s

load: $M^* = -b\dot{\phi}^*$, $b = 1.86$ N·m·s

2.

MOTOR $\xrightarrow{-a\dot{\phi}^*}$ $\left.\right|1.\left.\right|$ $\xrightarrow{-b\dot{\phi}^*}$ LOAD

$\qquad\qquad\qquad\qquad$ $I = 0.05$ kg·m^2

3. $\dfrac{dp^*}{dt} = M^* - a\dot{\phi}^* - (-b\dot{\phi}^*) = M^* - \dfrac{a - b}{I} p^*$

Therefore, $\dfrac{dp^*}{dt} + 41.6\, p^* = 2\sin 40t$.

4. The suggested substitution gives

$40(p_1^* \cos 40t - p_2^* \sin 40t) + 41.6(p_1^* \sin 40t + p_2^* \cos 40t) = 2\sin 40t$, from which

$$40p_1^* + 41.6p_2^* = 0,$$
$$41.6p_1^* - 40p_2^* p2 = 2,$$

giving $p_1^* = 0.025$ and $p_2^* = 0.024$ so that the magnitude of the sinusoidal response of $\dot{\phi}^*$ equals $\sqrt{(0.025)^2 + (0.024)^2}/0.05 = 0.693$ rad/s.

5. From the plot it can be seen that the magnitude of the perturbations $\dot{\phi}^*$ are so small that the linearization introduces negligible error.

Guided Problem 4.16

1.

$$\overline{T} = mgy = mgd(1 - \cos\theta)$$

2. $M = \dfrac{d\overline{T}}{d\theta} = mg\sin\theta$

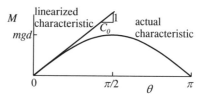

3. $\dfrac{1}{C_0} = \left(\dfrac{dM}{d\theta}\right)_{\theta=0} = mgd\cos\theta = mgd$; therefore, $C_0 = \dfrac{1}{mgd}$

4. $T = mgd \left(1 - 1 + \dfrac{\theta^2}{2!} - \dfrac{\theta^4}{4!} + \dfrac{\theta^6}{6!} - \cdots \right) = \dfrac{1}{2} mgd\,\theta^2$ for small θ.

$M = \dfrac{dT}{d\theta} = mgd\,\theta$

5. $I_0 \begin{array}{c} -\dot{p} = \theta/C_0 \\ \hline p/I_0 = \dot\theta \end{array} C_0$ linearized: $\left| \begin{array}{l} \dfrac{dp}{dt} = \dfrac{-1}{C_0}\theta = -mgd\,\theta \\[2mm] \dfrac{d\theta}{dt} = \dfrac{1}{I_0} p \end{array} \right.$ or $\left| \begin{array}{l} \dfrac{d\dot\theta}{dt} = -\dfrac{mgd}{I_0}\theta \\[2mm] \dfrac{d\theta}{dt} = \dot\theta \end{array} \right.$

Guided Problem 4.17

1. The area of the orifice is $A = x\pi d$. The volume flow rate, according to Bernoulli's equation, is

$$Q = c_d A \sqrt{2\Delta P / \rho}$$

where ΔP is the pressure drop between the inlet and outlet channels, and $c_d = 0.611$.

2. The dimensionless flow rate is

$$Q_d = \frac{\sqrt{\rho/2P_s}\,Q}{c_d x_m \pi d} = \frac{x}{x_m} \sqrt{\frac{\Delta P}{P_0}}$$

where x_m and P_0 are the nominal maximum valve opening and the maximum applied pressure, respectively.

3.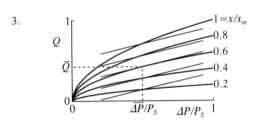

4. $Q = Q_0 + \left(\dfrac{\partial Q}{\partial(\Delta P/P_0)} \right)_{P_0, x_0} \left(\dfrac{\Delta P - P_0}{P_s} \right) + \left(\dfrac{\partial Q}{\partial(x/x_m)} \right)_{P_0, x_0} \left(\dfrac{x}{x_m} - \dfrac{x_0}{x_m} \right)$

For the special case $\dfrac{x}{x_m} = 0.6$ and $\dfrac{\Delta P}{P_0} = 0.5,$

there results $Q_0 = 0.6\sqrt{0.5} = 0.424,$

$\left(\dfrac{\partial Q}{\partial(\Delta P/P_s)} \right)_{P_0, x_0} = \dfrac{1}{2}\dfrac{x}{x_m}\sqrt{\dfrac{P_s}{P_0}} = \dfrac{1}{2}(0.6)\dfrac{1}{\sqrt{0.5}} = 0.424;$

$\left(\dfrac{\partial Q}{\partial(x/x_m)} \right)_{P_0, x_0} = \sqrt{\dfrac{P_0}{P_s}} = 0.707.$

5. See the set of parallel lines superimposed in the plot above (step 3).

Chapter 5

Basic Modeling

This chapter represents a second pass at the modeling of dynamic systems. The physical systems considered require bond graph models with more than the single junction assumed in Chapter 3. Models for the simple circuits considered in Section 5.1 employ junction structures comprising any number of bonds and junctions. Transformers and gyrators are added in Section 5.2, with special attention on hydraulic-mechanical, electromechanical and mechanism-based systems.

Models often can be simplified or transformed to make them more amenable to analysis, without losing meaning. The most common model equivalences that accomplish this objective are presented in Section 5.3. This leads to a consideration of the equilibrium state of dynamic models in Section 5.4. Stability, instability and limit cycle behavior are critical properties that emerge.

The translation of the bond graphs to differential equations is deferred until Chapter 6.

5.1 Simple Circuits

Circuits comprise components, each of which has two sides, interconnected at simple junctions. Either the generalized forces or the generalized velocities are common at each junction. The structure of a curcuit is represented by its pattern of interconnections, that is its topology. The geometric orientations of the components, on the other hand, either are immaterial or are constrained in a simple universal way. Electric, fluid and mechanical circuits are addressed below. Some of the systems considered in Section 5.2 are circuits, also.

5.1.1 Simple Electric Circuits

Simple electric circuits comprise interconnections of resistors, capacitors, inductors, voltage and current sources. Three-port and higher ported elements such as transistors are excluded, and transformers are deferred to the following section. The junction structures of the bond graphs are independent of the nature of its one-port elements. As a result, one-port elements sometimes are designated in this book with the generic symbol Z, in both the circuit diagrams and the bond graphs.

The following routine procedure can be employed for simple circuits without transformers:

1. Represent each node by a zero-junction, since the voltage is common to all joined conductors. It is helpful to label the junction with a symbol for its voltage.

257

2. Represent each branch i by the combination

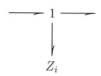

where Z_i represents the ith element (source, resistor, capacitor or inductor). This says that the current entering the element on one side emerges unchanged on the other side. It is helpful to label the junction with a symbol for this current.

3. Discard all bonds for which either $e = 0$ (ground or reference voltages) or $i = 0$, since they represent zero power. (An exception is the rare case in which capacitors prevent *all* of the conductors from having any current, and you still are interested in the distribution of voltages.)

4. Eliminate all junctions from which only two bonds emanate, and join the bonds. Coalesce all directly bonded 0-junctions and directly bonded 1-junctions into single junctions of the respective types.

Steps 3 and 4 do not change the meaning of the bond graph but serve to simplify its appearance.

Recall that the electrical and bond-graph symbols for resistance and for capacitance are R and C, respectively, whereas the electrical symbol for inductance is commonly L and the bond-graph symbol is I. *Do not write L in the bond graph; it is undefined.* Instead, write $I = L$ off to one side.

EXAMPLE 5.1

Apply the four steps above to get a bond graph for the electric circuit below:

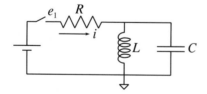

Solution: The bond graph after step 2:

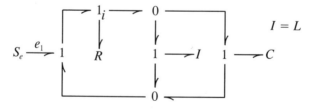

After step 3:

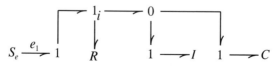

and finally after step 4:

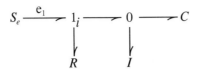

After some practice it is possible to draw the final bond graph directly, without separating the four steps.

EXAMPLE 5.2

Draw a bond graph for the Wheatstone bridge with arbitrary elements Z_i. The bond graph elments also may be labeled Z_i (as word bond graph elements). This circuit could comprise any combination of types of elements, although in usual practice the four legs are of the same type.

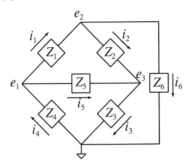

Solution: The bond graph after step 2 is

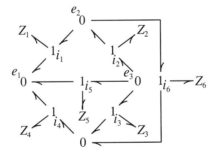

and after step 3 it is

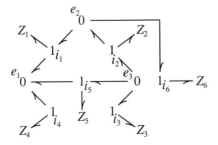

Finally after step 4 it is

You are urged to try this problem using a different node as the reference or ground; although the benzene-ring-like form of the resulting bond graph will remain, the order of the elements will change. This change corresponds to the different choices of variables and does not imply different physical behavior.

An electric battery used for signal-level purposes often is assumed to be a simple effort source. A more accurate model comprises a capacitance in series with a resistance, as shown in part (a) of Fig. 5.1. The capacitace describes the equilibrium potential, as represented for lead-acid batteries by the Nernst equation, and the resistance represents irreversibility described as the overpotential and represented by the Tafel's equation. A yet more accurate model of a single electrode is given by Esperilla et al.[1], which also gives references to the literature on electrochemical behavior. This model, shown in part (b) of the figure, recognizes a secondary reaction that causes gas evolution; i_m and i_s are the main and secondary currents. Since there is both a cathode and an annode in a real battery, the complete model has two of these combinations, as shown in part (c) of the figure. The compliances and resistances are nonlinear. All of these models assume isothermal conditions. Thermal effects, which can be quite important when heavy charging or discharging currents are involved, are considered in Section 9.7.9.

5.1.2   Fluid Circuits

Fluid circuits include gravity tanks, compression chambers and fluid restrictions such as orifices or valves interconnected by pipes or channels that may exhibit significant resistance and/or inertance. A very large gravity tank might be approximated as a constant-pressure source, called a **reservoir** or **sump**.

The modeling steps for fluid circuits are analogous to those for electric circuits. A fluid restriction is modeled like an electrical resistor, with an R element bonded to a 1-junction, since the input and output flows are the same. A junction of three pipes (a pipe "tee") or more is represented by a 0-junction, like an electric circuit junction, since the pressures are common and the flows sum.

---

[1]J.J. Esperilla, C. Vera, J. Felez, J.M. Mera, "Electrochemical cells modelling by means of the bond graph technique. Application to lead-acid batteries," *Proceedings of the ICBGM 2003 Simulation Series* 35 (2003) 293-299.

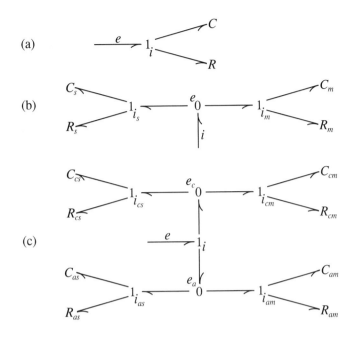

Figure 5.1: Isothermal models of electric batteries

EXAMPLE 5.3

Draw a bond-graph model for the hybrid mechanical/fluid sytem pictured below:

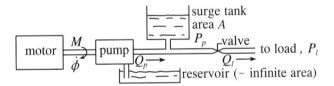

Solution: The gravity tank is analogous to a *shunt* capacitor in an electrical circuit. (*Series* fluid compliances cannot be made without energy transduction to mechanical form, which you will see later.) The pressure in the reservoir is neglected compared to other pressures.

$$S \xrightarrow{\frac{M}{\dot{\phi}}} T \longrightarrow 1 \underset{Q_p}{\overset{P_p}{\longrightarrow}} 0 \longrightarrow 1 \underset{Q_l}{\overset{P_l}{\longrightarrow}} \text{LOAD}$$

C, R above the 0 and 1 junctions

T = displacement per radian

$C = A/\rho g$

since $S_e = 0$, its bond can be dropped

5.1.3   Mechanical Circuits

An assemblage of masses or inertances, springs and dashpots that move either in simple translation or simple rotation is called a mechanical circuit. A translational example is given in Example 5.4 below. Two or more components join at a "mechanical node," where they have a common linear or angular velocity. This is different from an electrical node, at which the efforts (voltages) rather than the flows or generalized velocities (currents) are common. The two sides of a spring or dashpot have the same force but different velocities, which also is unlike electrical capacitors or resistors that have common flows rather than efforts. On the other hand, the two sides of a mass have the same velocities but different forces, like the two sides of an inductor that have the same currents but different voltages.[2] Steps similar to those for electric circuits are as follows:

1. Represent each mechanical junction with a 1-junction; if inertia is associated with the generalized velocity of a junction, add its inertance:

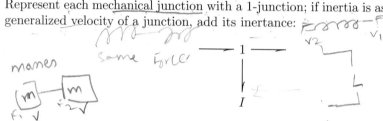

   It is usually desirable to label the linear or angular velocity with a customized symbol.

2. Represent each spring and each dashpot, respectively, with a 0-junction bonded to the C or R element:

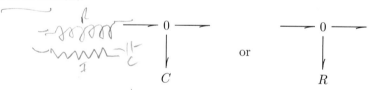

   Again, it is usually desirable to label the common force or torque with a particularized symbol.

3. Coalesce bonded junctions of the same type into a common junction.

4. Represent each input by S, S_f or S_e as appropriate.

5. Delete any bonds for which either the generalized force or velocity is zero.

6. Eliminate all junctions from which only two bonds emanate, and join the bonds.

---

[2]Some modelers use the Firestone analogy between mechanical and electrical variables in order to make the modeling diagrams of similar assembledges of electrical and mechanical components look alike. In this analogy, velocity and voltage are known as "potential" or "across" variables, and force and current are known as "flow" or "through" variables. Inertial elements are treated as having zero velocity on one side. Fluid flow is treated like a current or a force, and pressure is treated like a voltage or a velocity. This analogy turns a hydraulic cylinder into a gyrator, and a DC motor into a transformer. "Line graphs" (originally known as "linear graphs") represent the models. These graphs are more complicated than bond graphs, particularly when transformers or gyrators are employed. This author considers the identification of force as a flow and velocity as a potential to be unnatural, and prefers to define variables by their nature at a single cut in space rather than with respect to a circuit element with its two sides. Furthermore, the effort-flow concept which emerges from the single-cut idea extends better to thermodynamic systems and distributed-parameter models.

The example produces a loop of bonds, called a **mesh**. This and similar four-bond meshes can be reduced to tree-like bond graphs, as is shown in Section 4.3.

EXAMPLE 5.4

Translate the system drawn below in the language of mechanical circuits into the language of bond graphs:

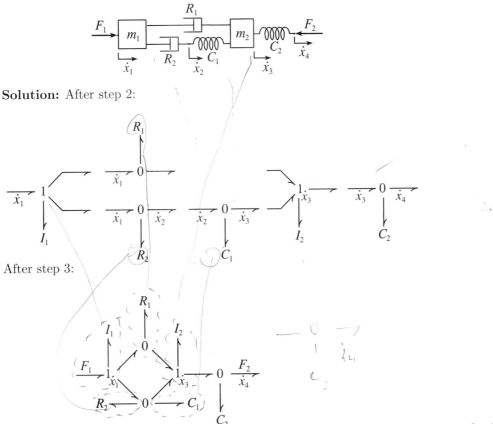

Solution: After step 2:

After step 3:

Steps 4, 5 and 6 do not apply to the model as pictured. If the right end were fixed to ground, however, the bond with $\dot{x}_4$ would be deleted (step 5), leaving a two-port 0-junction that could be eliminated by bonding C_2 directly to the 1-junction for $\dot{x}_3$ (step 6).

5.1.4   Use of Energy Integrals

Concepts of energy often enable an otherwise tough problem to be addressed with ease. The determination of a lumped parameter to approximate a distributed entity is an important case that is elaborated on in Section 12.1. The following example develops a classical result for the effective mass of a spring that is fixed at one end. The critical limitation is to low frequency behavior; at very high frequencies, a spring can exhibit travelling waves. The method employs integration over the element, eliminating the waves by approximating how the strains are distributed. The high-frequency waves are addressed in Chapter 11.

EXAMPLE 5.5

A spring with stiffness k connects a mass M to ground, as shown below. Investigate how the behavior is affected by the mass of the spring, m_s, assuming that this mass, though not insignificant, is small enough not to upset significantly the uniformity of the strain from one end of the spring to the other under dynamic conditions. This implies that the frequencies of interest are not very high; the primary interest is the (lowest) natural frequency of the mass-spring combination.

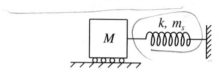

Solution: This problem is ideally suited to the use of the concept of the kinetic energy in the spring. Start by defining a position variable y such that the left end of the spring represents $y = 0$, and for the right end, $y = L$:

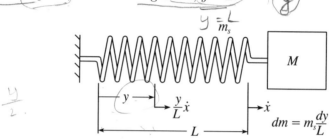

Under the asumption of uniform strain, each infinitesimal segment dy has mass $m_s(dy/L)$. Similarly, the velocity of the segment is $(y/L)\dot{x}$, where $\dot{x}$ is the velocity of the large rigid mass. Evaluate, by integration, the kinetic energy of the entire spring relative to the velocity $\dot{x}$:

$$T = \int_0^L \frac{1}{2}\left(\frac{y}{L}\dot{x}\right)^2 \left(m_s\frac{dy}{L}\right) = \frac{1}{2}\frac{m_s\dot{x}^2}{L^3}\int_0^L y^2\, dy = \frac{1}{2}\left(\frac{m_s}{3}\right)\dot{x}^2 \equiv \frac{1}{2}I_s\dot{x}^2.$$

This energy is expressed in terms of the virtual mass of the spring, I_s, which can be seen to be $I_s = m/3$, that is, one-third of the actual mass of the spring. This is a famous result, since it occurs frequently in mechanical systems.

The *total* inertance relative to $\dot{x}$, defined here as I, includes the mass M of the block:

$$I = M + I_s = M + m_s/3.$$

As a result, the (lowest) natural frequency of the system is

$$\omega_n = \sqrt{\frac{1}{IC}} = \sqrt{\frac{k}{M + m_s/3}}.$$

Higher natural frequencies also exist, but these frequencies are well beyond the range of most interest unless the mass of the spring is large compared to the mass of the block.

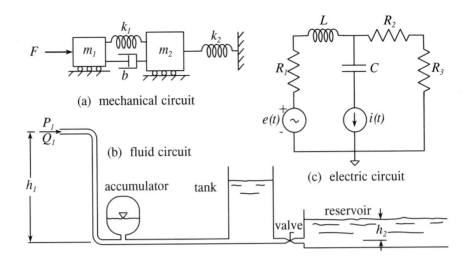

(a) mechanical circuit

(b) fluid circuit

accumulator   tank

h_1

P_1
Q_1

valve

reservoir

h_2

(c) electric circuit

Figure 5.2: Guided Problem 5.1

5.1.5 Summary

Simple electric, fluid and mechanical circuits comprise assemblies of simple components each having two ports with a common effort or a common flow. The components thus can be modeled with a bond graph having two boundary bonds, a 0-junction or a 1-junction and an R, C or I element. Interconnections also occur at 0 or 1-junctions. It is strongly recommended that you define symbols for all common variables, and place these symbols on the system sketch, first, and on the corresponding bond graph junctions, second. Bonds with either zero effort or zero flow, and therefore zero power, may be erased from the graph.

Guided Problem 5.1

The three circuit diagrams in Fig. 5.2 imply unique models. Translate the language of these diagrams into the language of bond graphs. It is critical that you attempt these simple cases before you examine the solutions.

Suggested Steps:

1. Define symbols for whichever variable is common for the two ports of each element, and place them on the diagrams.

2. Carry out the steps for electric circuits (which apply by analogy to fluid circuits also) or mechanical circuits, as given in this section. Label each junction with the proper symbols from step 1. Do not attempt more than a crude modeling of the accumulator.

Guided Problem 5.2

Part (a) and step 1 of this problem are a simple example of the use of junctions. The balance of the problem gives insight into some of the virtues and liabilities of major conventional hydraulic circuits.

Hydraulic motors often are driven from simple fixed-displacement pumps run at constant speed so they produce virtually constant flow. See Fig. 5.3, which employs standard fluid

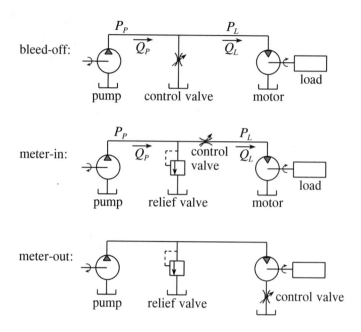

Figure 5.3: Hydraulic system of Guided Problem 5.2

power symbols. The speed of the motor usually is controlled by an adjustable control valve in one of three ways. In **bleed-off control**, excess pump flow is diverted through the valve. In **meter-in control**, the valve admits only the desired load flow, while the excess, if any, is diverted through a **relief valve**. The relief valve is constructed so as to limit the pressure of the pump, P_p. **Meter-out control** is similar, but the control valve is placed downstream of the hydraulic motor. The small symbol at the bottom of each vertical line represents the vented reservoir or tank, which is common. (This is analoguous to the ground symbol in electrical circuits.)

(a) Draw bond graphs to model the systems, neglecting leakage and friction and assuming constant pump flow. The terms "RELIEF VALVE," "CONTROL VALVE" and "MOTOR/LOAD" may be used for the relevent components.

(b) Find the overall efficiency of these systems when operation is at the load-to-relief pressure ratio (P_L/P_R) of 0.1 and 0.9, and the load-to-pump flow ratio (Q_L/Q_P) is 0.2 and 0.8 (for a total of four combinations), neglecting leakage and friction.

Suggested Steps:

1. Draw bond graphs for the systems, using 0-junctions for pipe tees and 1-junctions for series interconnections.

2. Plot the pump characteristics, as modified by the relief valve (if any), on a map of flow vs. pressure. Also, plot the four operating load points.

3. Find and plot on the map the corresponding points at which the pump operates.

4. Compute the efficiencies η as the ratios of the motor power to the pump power, power being the area of the rectangles defined by the operating points and the origin.

PROBLEMS

5.1 Find the bond graph for the circuit of Example 5.1, moving the ground from the lower end of the inductor to the uppper end. Is the structure of the graph changed? Are the meanings of the effort and flow variables changed? (Pay particular attention to signs.)

5.2 Model the systems below with bond graphs. Identify variables and parameters where appropriate on copies of the diagrams and on the bond graphs.

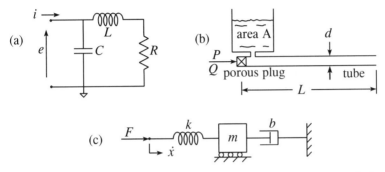

5.3 A hydraulic circuit includes a positive-displacement pump, rigid fluid lines, a rotary hydraulic motor and a load with a rotational inertia. Give the structure for a model of the system in bond-graph form. Describe in words what kinetic energies are represented by the inertance(s) in your model.

5.4 Model the translational mechanical system below with a bond graph.

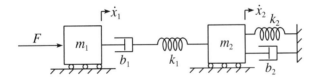

5.5 Model the rotational mechanical system below with a bond graph.

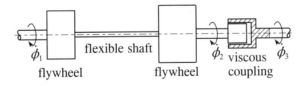

5.6 Develop a bond graph model for the mechanical system opposite. Identify all variables on both the drawing of the physical system and on the graph, and relate the moduli of bond graph elements to the physical parameters.

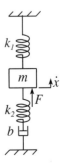

5.7 Draw a bond graph for the circuit shown below.

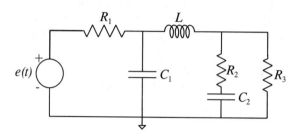

5.8 Infections of the middle ear frequently are stubborn and potentially serious, particularly for small children. It is common to insert a small tube surgically through the eardrum to drain the infectious material. (The tube is rejected by the body a few months later.) The direct effect of the tube on the hearing, presuming the tube and middle ear are free of liquid, is the present concern. The mechanical loading of the eardrum by the tube appears to be negligible; the question is the effect on the vibration of the eardrum by the acoustic bypass hole through the tube.

Assume that a pressure drop of 1 lb/ft^2 across the eardrum produces a volumetric displacement of 0.5×10^{-6} ft^3. Also, assume that the effect of the small bones which transmit the vibration is simply to increase the effective mass of the eardrum by 50%. Other parameters are listed below.

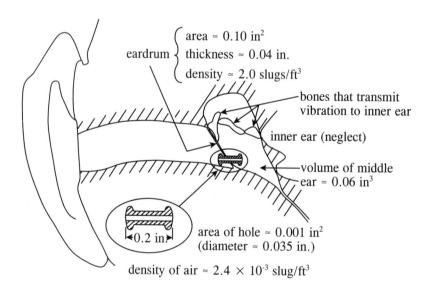

Describe the physical significance of each of the elements in the simple bond-graph model shown below. Relate the moduli of its elements to the parameters of the physical system, and evaluate those parameters. The motion of the upper 1-junction is transmitted to the inner ear by an "activated" transformer which you need not consider. *Coninued in Problems 6.43 and 7.6; the latter requests the response of the system to sinusoidal excitations, both with and without the drain tube.*

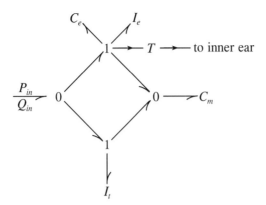

5.9 *This problem and the next might be deferred until the end of Section 6.1.* A positive-displacement hydraulic pump produces a flow, Q_p, with a steady component, $Q_p = 2 \text{ in}^3/\text{s}$, and a small periodic perturbation, Q_p^*, due to the discrete pistons, vanes or gear teeth. A small accumulator is used to reduce the transmission of pressure pulses to the load. Its pressure-volume relation is given by equation (3.55), reproduced below, where P_0 is the charging pressure (pressure when empty of oil) and V_0 is the value of the total volume:

$$P = \frac{P_0}{(1 - V/V_0)^k}, \quad P_0 = 1000 \text{ psi}, \quad V_0 = 10 \text{ in}^3, \quad k = 1.4.$$

The flow passes through a valve with the pressure-flow characteristic as plotted on the right below, and finally to an assumed frictionless and leakless piston-cylinder and inertial load. A bond graph model of the system also is given below.

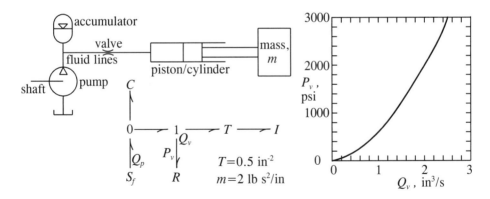

(a) Write the nonlinear differential equations for the model. The nonlinearity of the valve may be represented by $P_v = P_v(Q_v)$ or $Q_v = Q_v(P_v)$ if the argument (in parentheses) is identified in terms of parameters and state variables.

(b) Determine the equilibrium value of the flow through the valve, $\overline{Q}_v$, the pressure in the accumulator, $\overline{P}$, and the oil volume in it, $\overline{V}$.

(c) Linearize the differential equations for the perturbations in the variables that respond to the perturbations in the pump flow. Estimate numerical values for all coefficients.

5.10 A member which moves at a controlled speed $\dot{x}$ pushes one end of a linear spring, the other end of which pushes a weight at velocity $\dot{y}$. The weight slides over a horizontal surface with the friction force as plotted below.

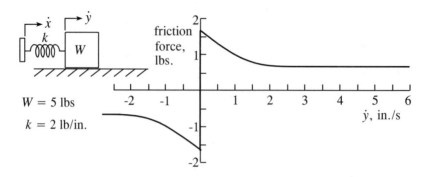

(a) Draw a bond graph for the system.

(b) For the case $\dot{x} = 1$ in/sec, determine whether the corresponding equilibrium value $\dot{y} = 1$ in/sec is stable, *neglecting the effect of the inertia*.

(c) Linearize the friction characteristic for the equilibrium state of part (b), and estimate the corresponding incremental or linearized resistance, R^*.

(d) Determine whether the inertance affects the stability of the equilibrium state, and describe the character of the behavior.

SOLUTIONS TO GUIDED PROBLEMS

Guided Problem 5.1

(a) **mechanical circuit**

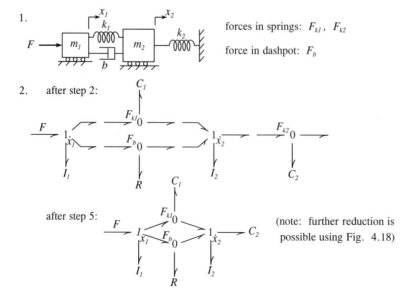

(b) fluid circuit

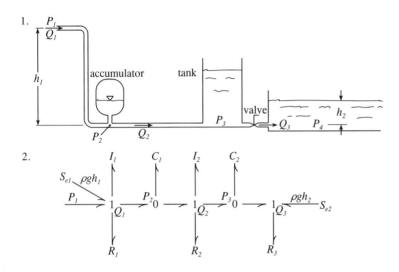

(c) electric circuit

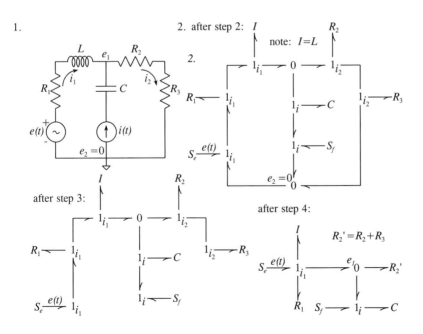

Guided Problem 5.2

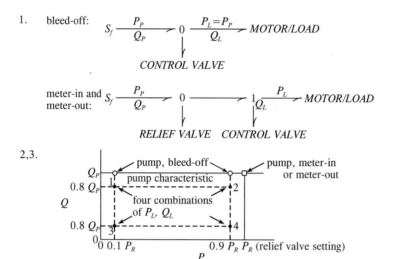

1. bleed-off:

    $$S_f \xrightarrow{\quad \dfrac{P_P}{Q_P} \quad} 0 \xrightarrow{\quad \dfrac{P_L = P_P}{Q_L} \quad} MOTOR/LOAD$$

    CONTROL VALVE

    meter-in and
    meter-out:

    $$S_f \xrightarrow{\quad \dfrac{P_P}{Q_P} \quad} 0 \xrightarrow{\quad\quad\quad} 1 \dfrac{P_L}{Q_L} \xrightarrow{\quad} MOTOR/LOAD$$

    RELIEF VALVE   CONTROL VALVE

2,3.

4. Bleed-off: $\eta_1 = \eta_2 = 0.8$; $\eta_3 = \eta_4 = 0.2$. Meter-in or out: $\eta_1 = 0.1 \times 0.8 = 0.08$; $\eta_2 = 0.9 \times 0.8 = 0.72$; $\eta_3 = 0.1 \times 0.2 = 0.02$; $\eta_4 = 0.9 \times 0.2 = 0.18$. (Bleed-off control gives higher efficiency, but meter-in or meter-out control is more direct and less susceptible to errors in the pump flow.)

5.2   System Models with Ideal Machines

The models of the previous section comprise one-port elements (S, S_e, S_f, R, C, I) interconnected by **junction structures** of bonds, 0-junctions and 1-junctions. Constant-parameter transformers and gyrators (T, G) are now added to the junction structures to permit modeling a much broader set of systems. Circuit-like systems are considered first. Certain commonly occuring two- and three-dimensional geometric constraints are then recognized.

5.2.1   Electric Circuits

An idealized electric transformer is shown in Fig. 5.4. The modulus of the transformer, T, equals the turns ratio of the two coils. The presence of four wires but only two different

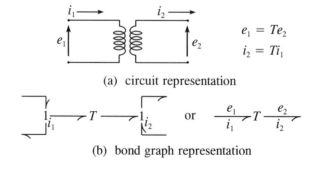

$$e_1 = T e_2$$
$$i_2 = T i_1$$

(a) circuit representation

(b) bond graph representation

Figure 5.4: Ideal electrical transformer

currents is represented by the use of two 1-junctions. The four-port model reduces to the two-port model if the voltage *differences* across the terminal pairs are adopted as the effort variables. This simpler model also applies directly if the voltages on the lower terminals are considered as ground. The bonds with zero voltage carry no power and may be excised from the graph.

This model implies that the transformer behaves the same at all frequencies, including DC. In fact, a *real* transformer fails completely at DC and exhibits deteriorated performance for very rapid changes or high frequencies. These limitations are corrected in Section 9.3.1; for the present they are neglected.

Step 2 of the procedure for electric circuits given in the preceding section is expanded to include placement of the four-ported combination of the transformers and 1-junctions.

EXAMPLE 5.6

Model the following circuit with a bond graph:

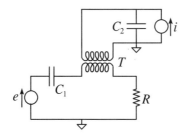

Solution: The bond graph after step 2 is

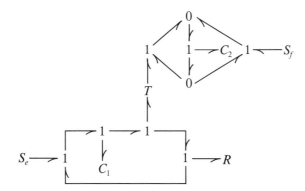

The final graph with the grounds accounted for reduces to

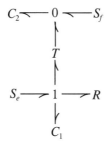

5.2.2   Fluid/Mechanical Circuits

The first example below models a fluid/mechanical device that was built for laboratory experiments. Standard positive-displacement fluid-to-mechanical machines are considered in subsequent examples.

EXAMPLE 5.7

The device shown below inserts high-frequency sinusoidal waves into a liquid within a very long pipe that is shown in cross section. The far end of the pipe is made reflectionless, and as a result the impedance of the pipe as seen locally is virtually a pure resistance R_s known as the *surge* or *characteristic impedance*. (This phenomenon is addressed in Chapter 11.)

The bellows (made from a metal) must be fairly thin in order to be sufficiently flexible, but it nevertheless must withstand a pressure of 800 psi. This pressure requires a large steady force to be placed on the bellows, in addition to the much smaller oscillating force from the electromechanical shaker. This force conceptually could be provided by a spring, but in practice such a spring would be too massive. The solution was to employ a second bellows, identical to the first, but instead filled with air, which is very compliant. The pressure of the air is maintained at the mean pressure in the pipe by allowing the liquid to pass through a small hole into the bottom of an accumulator, with the air on top.

The present assignment is the modeling of the system sufficient for a subsequent dynamic analysis. Define key variables and parameters, and place them on the drawing. Identify any critical constraints between your variables, and then model the system with a bond graph. Relate the moduli of your bond graph elements to the physical parameters.

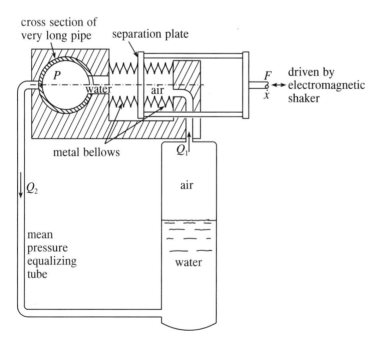

Solution:

Key variables already shown in the drawing include the velocity of the separation plate, $\dot{x}$, force from the shaker, F, volume flow into the pipe from the bellows, Q_1, volume flow through a small hole in the pipe into the pressure equalizing tube, Q_2, pressure in the pipe, P, and pressure of the air, $P_a ir$. A key constraint is $Q_1 = T\dot{x}$, where T is the area of the bellows. Parameters in the bond graph model below are the mechanical compliance of the bellows, C_m, the compliance of the air in the tank, $C_{air} \simeq V_0/kP_0$, the (surge) resistance of the fluid in the pipe to axial flow, R_s, the resistance of the small hole, R_0, the mass of the separation plate assembly and some of the liquid, I_m, and the inertance of the small hole and pressure equalizing tube, I_0.

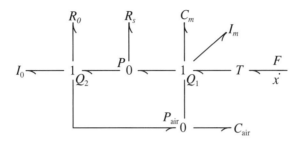

This bond graph is reduced to tree-like form in Example 5.14 (p. 315). Transfer functions are found in Probolem 6.51.

The piston-and-cylinder idealized in Example 2.3 (p. 53) has a potentially serious practical problem: if the piston fits too tightly, friction is apt to be severe; if it fits too loosely, leakage is bound to result. Further, high pressure expands the cylinder walls, opening or widening the leakage path. Some leakage across the piston must be tolerated. Rather than allow puddles on the floor, common practice creates chambers on both sides of the piston, rendering any leakage as "internal" and therefore acceptable, within limits. The symmetric **double-rod** cylinder, shown in Fig. 5.5, is one example of such a design. Neglecting the leakage, it can be modeled by a combination of two 1-junctions (to equate the inlet and outlet flows and the motions of the two rod ends) and an ideal transformer, much like the electical transformer. (Leakage is modeled in the subsequent sub-section.) The modulus of this transformer is the annulus area of the cylinder. If the power conventions arrows across T are reversed, the modulus becomes the reciprocal of the annulus area.

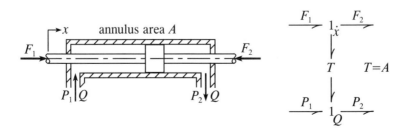

Figure 5.5: Double-rod-end cylinder as an ideal machine

EXAMPLE 5.8

Draw a bond graph to model three symmetric double-rod cylinders placed in series:

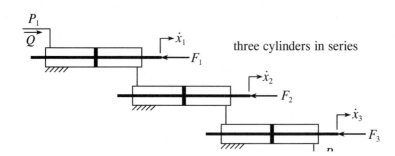

three cylinders in series

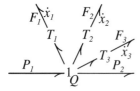

Solution: Since all three cylinders have the same flow rate Q, their fluid bonds should emanate from a common-flow 1-junction. Also, no mechanical forces are placed on the left sides of any of the piston rods, eliminating the need for any 1-junctions for the mechanical bonds.

EXAMPLE 5.9

The **single-rod** cylinder is cheaper and more common than the double-rod cylinder. The two ports of this cylinder sometimes are interconnected, as shown, to give a **regenerative circuit** with a single external fluid port. Model this system with a bond graph, neglecting friction and leakage.

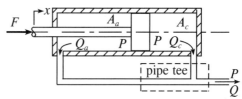

Solution: The two faces of the cylinder have the same velocity, which is recognized in the bond graph below by a 1-junction. The flows Q_a and Q_c are different in magnitude (as well as direction), however, because the areas of the two faces are different. Two transformers represent these geometric relations, one having its modulus equal to the total area of the piston, while the other modulus equals minus the area of the annulus. Finally, the two flows are joined at a pipe tee; the flows sum and the pressure is common, as represented by the 0-juction.

The regenerative circuit also can be viewed as an integrated unit with a single fluid port and a single mechanical port. The flows Q_c and Q_d sum to give Q, which means that the single modulus T on the reduced bond graph, drawn on the right side above, equals the sum $T_c + T_a$, which is the area of the shaft. This result can be deduced directly. The sum of the volumes of the fluid and the rod within a fixed control volume drawn around the unit is a constant. Therefore, every net cubic inch of fluid pushed into the cylinder pushes out one cubic inch of rod, and vice-versa.

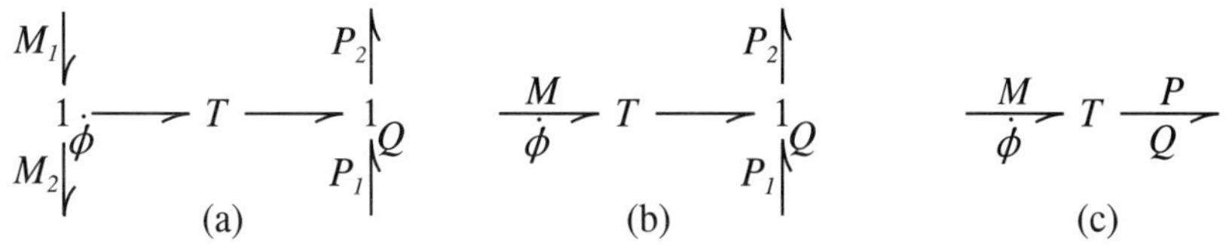

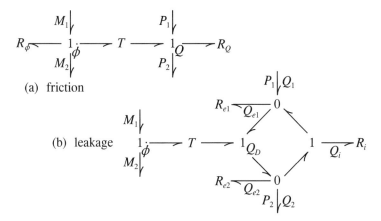

Figure 5.6: Ports of idealized model of a hydraulic motor/pump

Figure 5.7: Models for friction and leakage in positive displacement fluid/mechanical machines

A rotary hydraulic motor/pump with a shaft that emerges at both ends can be modeled as a four-port machine, like the electric transformer and the double-rod-end cylinder. The now familiar structure has two 1-junctions and an ideal transformer, as shown in Fig. 5.6 part (a). If the shaft does not penetrate through the machine, the three-port version of part (b) of the figure applies. If one of the pressures and one of the torques are zero, the model reduces to a two-port ideal transformer, as shown in part (c).

5.2.3   Losses in Positive Displacement Machines*

The models of positive displacement machines presented thus far neglect the effects of friction and leakage. Mechanical friction and fluid friction (pressure drops due to orifice flow, etc.) can be modeled by bonding the resistances R_ϕ and R_Q to the respective terminal 1-junctions, as shown in Fig. 5.7 part (a). Two types of leakage flows are represented in part (b) of the figure. Q_{e1} and Q_{e2} represent **external leakage**; the 0-junctions indicate the literal diversion of fluid, which either drips on the ground or floor or preferably is carried back to the reservoir by a special drain line at essentially zero (gage) pressure. The leakage resistances are R_{e1} and R_{e2}. The **internal leakage** Q_i occurs as back-flow across the piston face or the vanes or the gear teeth, etc, from the high pressure side of the machine to the low pressure side. The resistance to this leakage is R_i. The displacement flow of the machine, labeled Q_D, is the theoretical flow which would exist were there no leakage.

Any of the five resistances R_ϕ, R_Q, R_{e1}, R_{e2} and R_i can be included or excluded from the complete model, depending on the magnitudes of the phenomena and how accurate you wish to be. Note that the frictional resistances R_ϕ and R_Q can be removed if their moduli are sufficiently *small*, whereas the leakage resistances R_{e1}, R_{e2} and R_i can be removed if their moduli are sufficiently *large*.

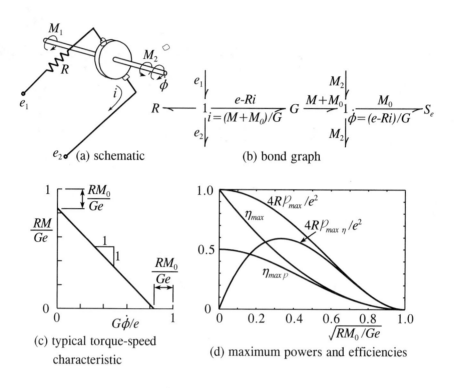

Figure 5.8: Model of electrical/mechanical machine with losses

5.2.4   Losses With DC Motor/Generators*

The resistance of the armature circuit of a DC motor/generator, given as R in the model of Fig. 5.8, almost always is important. The brushes of a conventional machine impose a nearly constant frictional torque on the shaft; the bearings contribute some additional torque. The bond graph of part (b) of the figure therefore shows an effort-source element S_e with its constant torque labeled as M_0. Viscous resistance usually is not significant enough to be considered and is omitted in the analysis below, although it can be appended readily to the bond graph as an R element.

Note the familiar four-port structure of the bond graph. As with the other machines, one or two of these ports may be removed if its effort is zero. The net applied voltage and torque are taken below as e and M, respectively.

The bond graph reveals that $i = (M + M_0)/G$ and $\dot{\phi} = (e - Ri)/G$. These two relations can be combined to eliminate the current i, giving the nondimensionalized torque-speed relationship

$$\frac{RM}{Ge} = 1 - Q - \frac{G\dot{\phi}}{e}; \quad Q \equiv RM_0/Ge, \tag{5.1}$$

in which the parameter Q characterizes the quality of the machine. This equation plots as a straight line in part (c) of the figure. As with other such straight-line source characteristics, maximum output power occurs at its midpoint, for which $RM/Ge = \dot{\phi}G/e = (1 - Q)/2$, so that

$$\mathcal{P}_{max} = (1 - Q)^2 \frac{e^2}{4R}. \tag{5.2}$$

The energy efficiency of this operating point, defined as $M\dot{\phi}/ei$, is

$$\eta_{max\,\mathcal{P}} = \frac{1}{2}\left(1 - Q\right)^2 \left(1 + Q\right)^{-1}. \tag{5.3}$$

The maximum energy efficiency is higher and occurs at a higher speed and lower power:

$$\eta_{max} = \left(1 - \sqrt{Q}\right)^2, \tag{5.4a}$$

$$\frac{G\dot{\phi}_{max\,\eta}}{e} = 1 - \sqrt{Q}, \tag{5.4b}$$

$$\mathcal{P}_{max\,\eta} = \sqrt{Q}\left(1 - \sqrt{Q}\right)^2 \frac{e^2}{R}. \tag{5.4c}$$

(These results follow from setting the derivative of an expression for the efficiency equal to zero.) The maximum powers and efficiencies are plotted in part (c) of the figure as functions of the quality index Q.

Most electric motors will overheat if run continuously at high torque and low speed. Complete specifications include a coefficient of proportionality between the power being dissipated and temperature. Forced air convection increases the range of allowable operation.

Specifications for permanent magnet DC motors ideally include values for e, R, M_0 and G. The modulus G is typically described as a "torque constant" or "torque sensitivity" with the units oz-in./amp, or as a "back EMF constant" with units of volts/rpm. You should convert these to SI units to take advantage of the fact that one Newton-meter/second equals one Watt and one volt-amp; they should then give the same number. Often, the torque, speed and current are specified instead at two (or more) conditions, perhaps including stall, the no-load condition, or a "rated" condition, all at some voltage, e. This information is sufficient for deducing the values of R, M_0 and G.

5.2.5   Case Study with Source and Load*

The DC motor of Fig. 5.9 is driven by an electrical source with characteristic plotted as S in part (c). It drives a mechanical load, with characteristic plotted as R_3 in part (d), through a viscous coupler. The motor is characterized by a torque constant of $G = 0.15$ N m/amp, armature resistance of $R_1 = 1.67$ ohms as plotted in part (c), and friction torque $M_0 = 0.08$ N m. The viscous coupler has a slip velocity proportional to the torque, with coefficient $R_2 = 0.025$ N m s/rad as plotted in part (d).

A bond graph model is shown in part (b) of the figure. The viscous coupler has three bonds, one for its input power, one for its output power and one for the the difference or dissipated power. The torque (effort) is common to the three bonds, while the angular velocities (flows) sum. Therefore, the bonds are joined by a 0-junction.

To find the equilibrium, the source S and the armature resistance R_1 are combined in part (c) of the figure to give an equivalent source, S'. It is found by subtracting the effort for R_1 from the effort for S, since $e_3 = e_1 - e_2$ and $e_2 = R_1 i$. Next, the load characteristic R_3 and the viscous coupler characteristic R_2 are combined to give the lower dashed line plotted in part (d) of the figure. This characteristic is found by a *horizontal* addition of the characterisitics R_2 and R_3, since $\dot{\phi}_1 = \dot{\phi}_2 + \dot{\phi}_3$ and $\dot{\phi}_2 = M/R_2$. The frictional torque M_0 then is added to this characteristic, *vertically* because of the 1-junction, to give the higher dashed characteristic labelled R. This represents the overall load and is transcribed into part (e).

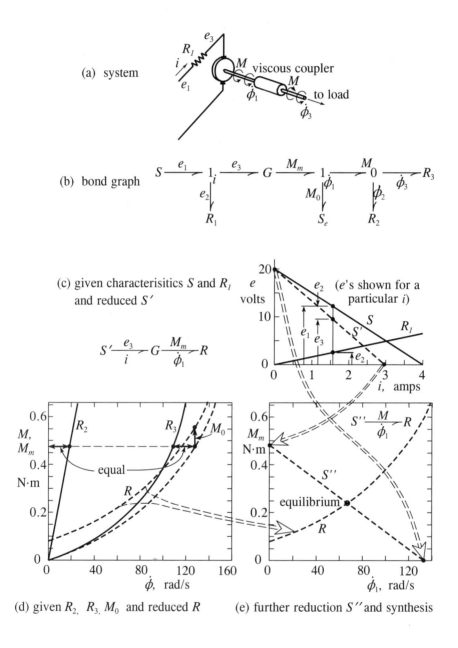

Figure 5.9: Example of DC motor with a particular source and load

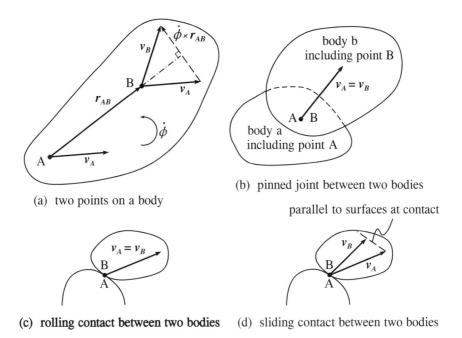

(a) two points on a body

(b) pinned joint between two bodies

(c) rolling contact between two bodies

(d) sliding contact between two bodies

parallel to surfaces at contact

Figure 5.10: Kinematical constraints for rigid bodies

The result at this point is a source characteristic S' bonded through a gyrator to a load characteristic R. Next, the combination $S' \longrightarrow G \longrightarrow$ is reduced to $S'' \longrightarrow$, as indicated and implemented in part (e) of the figure. The definition of the gyrator ($e_2 = G\dot{\phi}_1$ and $M = Gi$) is used to establish the end points (or any other points) of the S'' characteristic, as shown by the dashed arrows from part (c) to part (e).

The equilibrium operating state is given, finally, by the intersection of the charactertistics R and S'': $M_m = 0.24$ N m (which gives $M = 0.16$ N m) and $\dot{\phi}_1 = 67$ rad/s (which gives $\dot{\phi}_3 = 60$ rad/s).

5.2.6 Two- and-Three-Dimensional Geometric Constraints

The complexity of two- and-three-dimensional geometric constraints is not shared by electrical or other circuit-like systems, and is one of the fascinations of mechanical engineering. It is not the purpose of this book to present a course in kinematics, but certain key ideas are reviewed.

One approach develops direct relations between the *displacements* of the inputs and the outputs of the mechanisms. The relations between velocities are found by differentiating the resulting equations with respect to time. A different approach deals directly with velocities rather than the displacements. Sometimes this is easier, particularly when there are multiple independent velocities or displacements. The approach is based largely on the following key kinematical relations, as illustrated in Fig. 5.10:

1. The vector velocities for arbitrary points A and B on a rigid body are related by

$$\mathbf{v}_B = \mathbf{v}_A + \dot{\phi} \times \mathbf{r}_{AB}, \qquad (5.5)$$

where $\mathbf{r}_{AB}$ is a geometric vector from point A to point B, and $\dot{\phi}$ is the angular velocity vector for the body.

2. Point A on one member and point B on another member have the same velocities if both points are located coextensively at a pinned or swivel joint between the members; i.e.,

$$\mathbf{v}_A = \mathbf{v}_B. \tag{5.6}$$

3. Two instantaneously contacting points A and B which belong to separate members in rolling contact also satisfy equation (4.6). (The accelerations of these two points are different, however, unlike the corresponding points for pinned members.)

4. Two instantaneously contacting points A and B which belong to separate members in sliding contact have zero relative velocity in the direction normal to the surfaces in contact. That is, if $\mathbf{n}$ is a vector normal to the surfaces of contact,

$$(\mathbf{v}_A - \mathbf{v}_B) \cdot \mathbf{n} = 0. \tag{5.7}$$

The development of a bond graph model normally starts with identification of the critical velocity variables, including their vector orientations. These should first be noted directly on a drawing of the physical system. In the bond graph, then, a 1-junction is introduced to represent the magnitude and sense of each such velocity, and the corresponding label is attached to the junction. If the orientation of a velocity is not constrained by the geometry of the system, separate 1-junctions may represent its orthogonal components. Next, the constraints interrelating the various 1-junction velocities are sought and represented by proper junction structures. Finally, any inputs or outputs are represented by bonds or sources, and any dashpots, springs and inertias, etc. are modeled by attached R, C or I elements.

The procedure may be understood best by examples. This chapter focuses only on geometric constraints that can be modeled with junctions and constant-moduli transformers. More complex mechanisms are addressed in Chapter 9.

5.2.7   Case Study: Pulley System

Consider the pulley and parallel cables shown in Fig. 5.11. Points A and B are common to both the cable and the pulley. The velocities of points A, B and C are assumed to be horizontal.

The velocities of points along a vertical diameter of the pulley are parallel. They form a trapezoidal vector diagram as shown in part (b) of the figure, where $\dot{x}_A$ and $\dot{x}_B$ are chosen arbitrarily. The velocities $\dot{x}_A$ and $\dot{x}_B$ are related to the velocity of the center of the pulley, $\dot{x}_C$, and the angular velocity, $\dot{\phi}$, by

$$\dot{x}_A = \dot{x}_C - r\dot{\phi}, \tag{5.8a}$$
$$\dot{x}_B = \dot{x}_C + r\dot{\phi}. \tag{5.8b}$$

The constraints represented by equations (5.8) are incorporated into the bond graph shown in part (c) of the figure. Specifically, equation (5.8a) is represented by the upper 0-junction, and equation (5.8b) is represented by the lower 0-junction. Study this graph carefully; its construction is the critical step. The applied forces F_A, F_B and F_C and the moment M are added in part (d) of the figure.

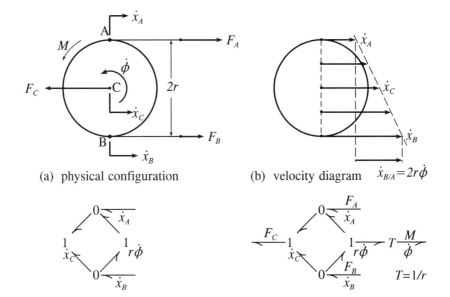

(a) physical configuration   (b) velocity diagram $\dot{x}_{B/A} = 2r\dot{\phi}$

(c) bond graph with kinematical constraints   (d) bond graph with forces added

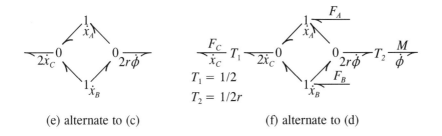

(e) alternate to (c)   (f) alternate to (d)

Figure 5.11: Pulley system

An alternative perspective results from adding equations (5.8a) and (5.8b) and solving for $\dot{x}_C$, and subtracting one from the other and solving for $r\dot{\phi}$:

$$\dot{x}_C = \frac{1}{2}(\dot{x}_A + \dot{x}_B), \tag{5.9a}$$

$$r\dot{\phi} = \frac{1}{2}(\dot{x}_B - \dot{x}_A). \tag{5.9b}$$

These equations are represented directly by the bond graph in part (e) of the figure. Specifically, the first equation is represented by the left-hand 0-junction, and the second by the right-hand 0-junction. The forces and moments are added in part (f) of the figure. Both bond graph models must represent correct relationships between the forces and the moments, since the kinematical relationships and the boundary powers are proper. Which ought to be preferred in practice depends on additional circumstances. For example, if the moment is zero, the resulting erasures from the bond graphs of parts (e) and (f) lead to a simpler interpretation.

EXAMPLE 5.9

The floating lever below represents a minor modification of the pulley system, as long as the lever remains nearly vertical. The only substantive difference is the addition of a force applied at point D. Model the system with a bond graph.

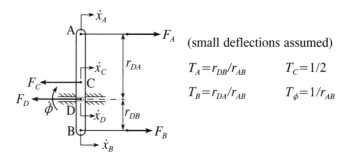

(small deflections assumed)

$T_A = r_{DB}/r_{AB}$ $\qquad$ $T_C = 1/2$

$T_B = r_{DA}/r_{AB}$ $\qquad$ $T_\phi = 1/r_{AB}$

Solution: The bond graph of part (f) of Fig. 5.10 is adapted below to represent the entire system except for the presence of a force at point D. The key geometric constraint regarding point D is

$$\dot{x}_D = T_A\dot{x}_A + T_B\dot{x}_B; \qquad T_A = \frac{r_{DB}}{r_{AB}}, \qquad T_B = \frac{r_{DA}}{r_{AB}}.$$

This constraint is implemented by a 0-junction in the bond graph:

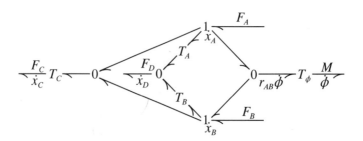

EXAMPLE 5.10

An epicyclic gear train comprises a sun gear, three kinematically redundant planet gears, a ring gear and a spider which contains the bearings for the planet gears. All these members can rotate about their centers, and the centers of the planet gears also move with the spider. Model the system with a bond graph.

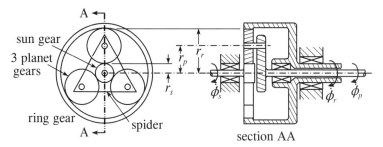

Solution: The key velocities are defined in the diagram below left. The planet gears share a common velocity with the sun gear at their point of mesh, $\dot{x}_s$, and a common velocity with the ring gear at their point of mesh, $\dot{x}_r$. The kinematical relations are

$$\dot{x}_s = r_s \dot{\phi}_s,$$
$$\dot{x}_p = r_p \dot{\phi}_p,$$
$$\dot{x}_r = r_r \dot{\phi}_r,$$
$$\dot{x}_p = \tfrac{1}{2}(\dot{x}_r + \dot{x}_s).$$

These four equations are represented in the bond graph by four transformers. The last equation also requires a 0-junction to represent its summation of velocities.

Note: The 1-junctions are written and labeled first to aid in drawing the graph. They may be removed later since they have only two ports.

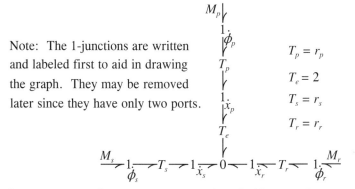

In usual practice, one of the three input shafts is held motionless at any moment in time, by some external means, reducing the system to a net two-port transmission.

5.2.8 Model Structure from Energy Expressions

When the structure of a bond graph representing geometric constraints is not obvious, the initial writing of expressions for stored energy, or sometimes the dissipated power, can be a powerful guide. The following example illustrates this idea.

EXAMPLE 5.11

A system comprises two identical pendulums with point masses connected by a weak spring and weak dashpot. Model the system with a bond graph. Consider only small deflections from normally vertical equilibrium positions of the pendulums.

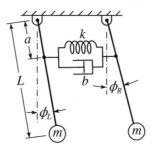

Solution: An energy approach could start with exact expressions for the kinetic energy and potential energy:

$$\mathcal{T} = \frac{1}{2}m(L\dot{\phi}_L)^2 + \frac{1}{2}m(L\dot{\phi}_R)^2.$$

$$\mathcal{V} = mgL(1 - \cos\phi_L) + mgL(1 - \cos\phi_R) + \frac{1}{2}kx^2.$$

The first equation is already in the form of a linear model. The second is not, and use of the spring deflection x is not related explicitly to the deflections ϕ_L and ϕ_R because the exact expression is very complicated and not required. Approximating x and using $\cos\phi \simeq 1 - \phi^2/2$ for small deflections gives

$$\mathcal{V} \simeq \frac{1}{2}mgL\phi_L^2 + \frac{1}{2}mgL\phi_R^2 + \frac{1}{2}k(a\phi_L - a\phi_R)^2.$$

These expressions suggest using $\dot{\phi}_L$, $\dot{\phi}_R$, and $a(\dot{\phi}_L - \dot{\phi}_R)$ as the velocity variables. This choice in turn gives the following bond graph and its parameter values:

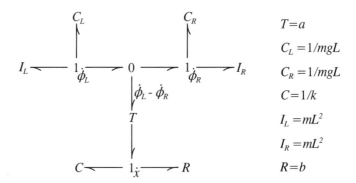

State differential equations can be deduced readily from this graph. Its interesting behavior, without the dashpot, is explored in Guided Problem 7.7 (p. 491). Another example, with vehicle dynamics, is given in Section 6.2.7.

5.2.9   Modeling Guidelines

There is no universal modeling procedure. Nevertheless, the steps below serve as a useful guideline for most cases.

1. Identify any input and output variables, define symbols for these variables, and place these symbols appropriately on a sketch of the system being modeled. If any of these variables are efforts, also define and label the corresponding flow variables. Use arrows to show directionality.

2. Define and label on the sketch all other variables that appear to be important in relating the input to the output variables.

3. Identify whatever generalized geometric constraints exist between the various flow variables you have defined. This can be done in the form of equations and often involves defining parameters such as the lengths of levers or the areas of pistons. This step may utilize knowledge gained in your engineeering science courses or in some of the examples given in this book.

4. Start to form a bond graph by drawing any boundary bonds for the input variables. Then draw a 1-junction without bonds for each other flow variable identified in step 2, and annotate it with your symbol for that variable. These may be placed on the sheet, in isolation, wherever you imagine will leave enough room for the missing bonds and elements.

5. Implement the constraint equations of step 3 by placing appropriate junctions, bonds and transformers. Identify the meaning or values of any transformer moduli. At this point you may have a completed *junction structure*, that is an interconnection between all the flows that you have defined and are needed to relate the input variables to the output variables.

6. If one or more of the key variables are not yet interconnected by bonds, the system may possess one or more gyrational couplings, that is proportionalities between efforts and flows. Identify these, and represent them by gyrators. Note their moduli.

7. Should there still be a key dangling variable or two, examine each of the steps above to identify what is missing.

8. Identify the one-port effort and flow sources, resistances, compliances and inertances that you believe are important. Represent each by a bond graph element S_e, S_f, R, C or I, using distinguishing subscripts if necessary, and place these where they belong on the bond graph. This step and the next may draw heavily from your knowledge of engineeering science. Whether to include a possible element depends on the relative significance of its power or energy compared with the power or energy of the other bonds or elements.

9. Identify the *constitutive relations* for each 1-port element introduced in step 8; that is, give the values of constant moduli and equations or plots for nonlinear characteritstics. Consideration of power or energy may help here, also.

10. If the display of your bond graph is messy or awkward, redraw it to suit your taste. Eliminate any unnecessary 2-port junctions. You may wish also to simplify the graph in preparation for analysis. (Some simplifications are presented in Section 5.3.)

It is important to complete steps 1–3 *before* starting to draw the bond graph. Most of the critical modeling decisions are made here and have nothing to do with bond graphs or whatever other modeling language you plan to use. The essential virtue of the bond graph is that its junction structure imposes the conservation of energy, saving you from having to find the relationships between the various efforts independently. You can substitute such relationships in the modeling process for the generalized geometric constraints between the associated flows, but this approach often is harder and is not recommended as a general procedure. Examination of force relationships is most helpful as a check; this becomes an optional step 11.

5.2.10 Tutorial Case Study

The steps above are applied to the hydraulic/mechanical system pictured in part (a) of Fig. 5.12. Hydraulic fluid drives a single-rod cylinder which is connected to a weight through a cable and a pulley. The flow emerging from the rod-end of the cylinder drives a rotary hydraulic actuator which in turn drives a flywheel. Friction and leakage are to be neglected in the model you are asked to create.

Step 1 asks for the key variables to be defined. The input volumetric flow rate Q and pressure P are given already. The velocity of the piston and cable is labeled in part (b) of the figure as $\dot{x}$. The angular velocity of the output shaft of the rotary actuator is labeled as $\dot{\phi}$. These are the key input and output variables. The flow rate between the cylinder and the rotary actuator also appears to be a useful internal generalized velocity, so it also is labeled (Q_r) as part of Step 2.

Step 3 asks for the key geometric constraints. The velocity $\dot{x}$ is related to the flow Q by $A_p \dot{x} = Q$, where A_p is the area of the piston. The flow Q_r is in turn related to $\dot{x}$ by $Q_r = A_a \dot{x}$, where A_a is the area of the annulus (the area A_p minus the area of the shaft). This flow also is related to the angular velocity $\dot{\phi}$ by $Q_r = D\dot{\phi}$, where D is the volumetric displacement of the rotary actuator per radian of rotation. If desired, this displacement can be related more specifically to the diameters and width of the actuator chamber, or found from experiment (Problem 2.22, p. 59).

Only now that the key variables and constraints are identified ought one start to draw a bond graph. The input boundary bond and 1-junctions to allow the key generalized velocities to be registered are shown in part (c) of the figure, following the instructions of Step 4. Their placement on the sheet is arbitrary, but enough space for likely further development should be allowed. Step 5 now directs that the constraints found under Step 3 be implemented, as shown in part (d), in the form of three transformers. This structure implies relations between the pressures, forces and torques that have not been directly considered. In place of these relations, however, the conservation of energy has been assumed implicitly. As a result, these relations must be correct. This is an automatic benefit of using the bond graph.

Steps 6 and 7 do not apply to the present situation. Step 8 starts by asking about effort sources. The mass at the right is pulled downward by a constant gravity force, which is represented in part (e) of the figure by an effort source. Acceleration of the mass and the piston requires or imparts an "acceleration force," which is represented by an inertance element bonded to the 1-junction for $\dot{x}$. The flywheel itself represents a substantive inertia, which is added to the 1-junction for $\dot{\phi}$ by a second inertance. The constitutive relations for these added 1-port elements, requested by Step 9, are $S_e = mg$, $I_m = m$ (ignoring the mass of the piston and piston rod, etc.) and $I_\phi = J$, where J is the mass moment of inertia of the flywheel.

The bond graph is now complete, barring any refinements to represent friction, leakage, inertive or compliance phenomena. Step 10 suggests you consider streamlining or simplifying the graph. One possibility, shown in part (f) of the figure, telescopes the transformers T_2 and T_3 into a single transformer with modulus $T_2 T_3$, removes the unnecessary 1-junction for $\dot{\phi}$, and straightens out the layout. Subsequent analysis is considerably simplified by combining the two inertances into a single inertance, as you will see later.

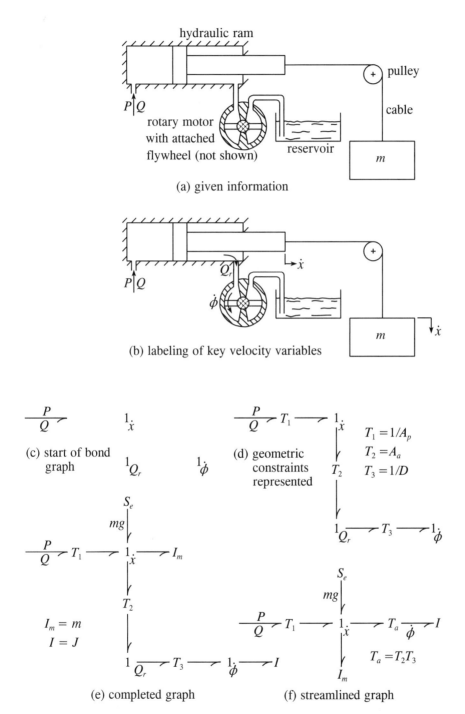

(a) given information

(b) labeling of key velocity variables

(c) start of bond graph

(d) geometric constraints represented

$T_1 = 1/A_p$
$T_2 = A_a$
$T_3 = 1/D$

$I_m = m$
$I = J$

(e) completed graph

$T_a = T_2 T_3$

(f) streamlined graph

Figure 5.12: Tutorial example

5.2.11   Common Misconceptions

Should you experience difficulty in constructing bond-graph models, the following list of commmon misconceptions and comments may help.

Misconception 1:   Each physical component is represented by a bond graph element.

Comment: Some physical components require several bond graph elements for their representation. On the other hand, one bond-graph element sometimes can represent more than one physical component. The bond graph is an integration of a set of relationships rather than a set of components. *Never* invent ad hoc bond graph elements such as three-ported T or G elements or (before reading Chapter 10, at least) two-ported R, I or C elements; they have not been defined. And, never use *ad hoc* symbols in a bond graph, such as L for an inductance or b for a dashpot.

Misconception 2:   A "LOAD" always can be represented by a resistance element.

Comment: Elements indicated by a word in a word bond graph do not have their type of behavior uniquely specified, unlike the standard bond graph elements. A "LOAD," for example, could include inertance or compliance effects or a source.

Misconception 3:   One can usually draw a bond graph before defining the key variables.

Comment: Defining the key variables on a drawing of the physical system is the key step in the modeling process, regardless of the choice of the modeling language. The definitions implicitly represent many of the assumptions being made. You also should determine the relations between these variables before drawing the bond graph.

Misconception 4:   Gravity implies potential energy, and therefore is always represented by a compliance element.

Comment: Oftentimes an S_e element, which also can store energy, is much more appropriate, particularly to represent the (constant) weight of a solid object. In general, first determine the relation between the generalized force and the generalized displacement. Also, remember that the *inertia* of a mass is treated separately from its weight.

Misconception 5: Each unique kinetic energy, T, has a unique inertance, I.

Comment: The unique kinetic energy equals $\frac{1}{2}I\dot{q}^2$, presuming linearity applies, but there may be several different possible definitions of $\dot{q}$. Thus, I is *referred* to $\dot{q}$. The same is true of resistances.

Misconception 6: Each unique potential energy, V, has a unique compliance, C.

Comment: C depends on the q it is referred to, just as I depends on its $\dot{q}$. Sometimes the choice is important; for example, the formula $\omega_n = 1/\sqrt{IC}$ requires that C and the I are based on the *same* q and its time derivative, $\dot{q}$, respectively.

Misconception 7:   The bonds around a junction need not have either a common effort or a common flow.

Comment: You know better than this when you think about it, but are you *always* consistent? Following the 10 steps of the modeling guidelines will minimize this type of accident.

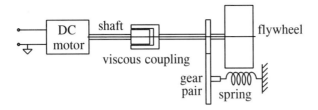

Figure 5.13: Guided Problem 5.3

5.2.12   Summary

Couplers such as electric transformers, fixed-field DC motors and positive displacement fluid/mechanical transducers often are modeled with two efforts but with a single flow at one or both of their ends. The corresponding bond graphs have a 1-junction and two boundary bonds at these ends, joined by a transformer or a gyrator. Frictional losses can be represented by resistances bonded to 1-junctions, while slip or leakage losses can be represented by conductances (or resistances) bonded to 0-junctions.

A hard mechanical constraint between two velocities within a machine can be represented by a transformer. Such kinematical constraints often are between the sums or differences of the velocities used as the variables of the model. Such sums and differences can be represented by a structure of 0 and 1-junctions. If the structure is based on known velocity constraints, the conservation of energy implicit in the structure automatically ensures that the forces are properly related, apart from frictional losses, which can be added in the form of resistances, and kinetic energy storage, which can be added in the form of inertances. If the structure is based on known force relationships in the absence of friction and inertia, the proper velocity constraints also must result. Most students who have studied elementary mechanics tend to favor force-based approaches at first but ultimately recognize that velocity-based approaches are more powerful. You are encouraged to develop skill in dealing with constraints in terms of velocities, and in practice to determine the relationships before attempting to draw a bond graph.

Guided Problem 5.3

This is a relatively simple problem with geometric constraints. It is important that you attempt it before viewing the author's solution.

For the system pictured in Fig. 5.13, define variables for a dynamic model, draw the bond graph for this model and annotate the graph with the variables.

Suggested Steps:

1. Identify distinct components and the variables used in the definition of their characteristics.

2. Represent the components by bond graph models. If you have not done so already, identify which variable is common in the key junction of the model, and represent this by a 1-junction.

3. Assemble the elements of your model into a complete bond graph.

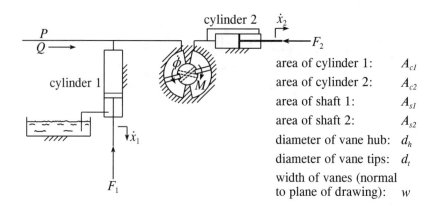

Figure 5.14: The hydraulic system of Guided Problem 5.4

Guided Problem 5.4

This straightforward problem involves finding a steady-state model for a circuit with hydraulic actuators.

The hydraulic system shown in Fig. 5.14 drives two hydraulic cylinders and one rotary actuator. One of the cylinders is connected regeneratively, while the other has one side connected directly to a reservoir at a negligible pressure. Draw a bond graph for the system, identify its elements in terms of the geometric parameters listed in the figure, and relate the input pressure P and flow Q to the forces F_1 and F_2, moment M, and velocities $\dot{x}_1$, $\dot{x}_2$ and $\dot{\phi}$. Neglect friction and leakage.

Suggested Steps:

1. Draw a bond graph for the system, representing each of the actuators as a simple transformer. (The details of the regenerative circuit are not addressed at this point.) Pay special attention to the need for any 0 or 1-junctions. Label the variables P, Q, F_1, $\dot{x}_1$, F_2, $\dot{x}_2$, M and $\dot{\phi}$.

2. Find the modulus of the transformer for cylinder 1, which by convention equals the ratio of the *output* generalized velocity to the *input* generalized velocity *as indicated by the power convention half-arrows* that you have chosen. (Both half-arrows must be oriented the same way.)

3. Find the modulus of the transformer for cylinder 2, which is regenerative. This can be done either of two ways:

   (a) The brute-intelligence way employs a bond graph for the subsystem, as in Example 5.7 (pp. 274-275), with a 0-junction and a 1-junction. Note the *negative* transformer modulus (which could be made positive by redirecting the two relevant power-convention arrows).

   (b) The insightful genius way notes that for every net unit volume of oil that enters the cylinder, an equal volume of steel shaft must emerge.

4. Find the modulus of the transformer for the rotary actuator. This is directly related to the volumetric displacement, D, defined as the volume of fluid displaced per radian of rotation.

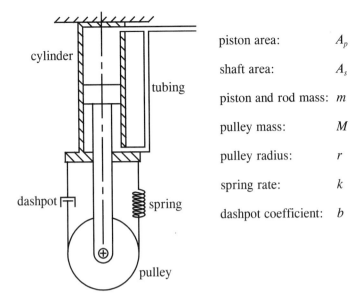

Figure 5.15: Guided Problem 5.5

5. The three loads are attached to the core of the model by 0- or 1-junctions. Write expressions for the efforts and flows on the load bonds connected to these junctions in terms of the load variables and the transformer moduli.

6. Perform the appropriate summations for these junctions in order to find expressions for the source pressure and flow in terms of the load efforts and velocities.

Guided Problem 5.5

This problem includes a regenerative hydraulic cylinder as in the problem above but emphasizes mechanical constraints with rotation and dynamics.

Incompressible fluid is ported to both sides of a piston, as shown in Fig. 5.15. A belt with a spring at one end and a dashpot at the other end is wrapped without sliding around a pulley comprising a disk of uniform thickness. Parameters are listed in the figure; phenomena not represented by these parameters, other than gravity, may be neglected. Model the system, considering the pressure of the fluid, P, to be an independent excitation and the motions of the solid parts to be the responses of interest.

Suggested Steps:

1. The piston/cylinder is connected in a regenerative circuit; model it accordingly.

2. The velocity of the piston is an obvious state variable. Something about the rotation of the pulley also is needed; choose between its angular velocity and the velocity at its periphery relative to the velocity of the piston. Place corresponding 1-junctions on your paper labeled with symbols for the chosen velocities.

3. Write an equation to represent the kinematical constraint between the velocities you have chosen, and represent it by a structure of bonds, junctions and transformers. Take care with the power convention arrows.

4. At this point you have completed the junction structure of the model, which is the difficult part of the modeling process. Complete your model by adding R, I and C elements, as suggested by the drawing and the list of parameters. Also, add something to represent gravity.

Guided Problem 5.6

This is a typical problem regarding the interpretation of sales literature for a constant-field DC motor.

A permanent-magnet DC motor is advertised as behaving as follows for a rated armature voltage of 40 VDC: the no-load speed is 1153 rpm, for which the armature current is 0.27 amps; at a "rated" speed of 900 rpm, the torque is 120 oz-in and the current is 3.00 amps. Find the values of G, R and M_0. Also, estimate the stall torque, the maximum power the motor can deliver and the associated speed and efficiency. Finally, estimate the maximum efficiency the motor can deliver and the associated speed and power. Assume that the model given in Fig. 5.8 (p. 278) applies.

Suggested Steps:

1. Extrapolate the current to the stall condition on a plot of current vs. speed. Compute the armature resistance, R.

2. Compute G as the ratio of the change in torque to the change in current.

3. Compute M_0 from the no-load current and the value of G.

4. Compute the stall torque by extrapolating the characteristic as defined by the two given conditions to zero speed, or use the plot in part (c) of the figure or equation (5.1) (p. 278).

5. The speed for maximum power lies at the mid-point of the characteristic. The associated power is the product of the speed and the torque. It can be found from equation (5.2), also. The associated efficiency can be found from equation (5.3) or picked off from the plot in part (d) of Fig. 5.8.

6. The maximum efficiency and the associated speed and power can be found from equations (5.4). The efficiency and power also can be picked off from the plots.

PROBLEMS

5.11 Model the hydraulic system below with a bond graph.

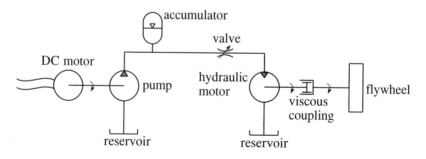

5.12 Represent the electric circuit below with a bond graph.

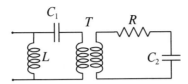

5.13 A motor drives a pump to supply a hydraulic cylinder that lifts a 3000-lb weight. The torque-speed characteristic of the motor is plotted on the next page. The displacement of the pump is 0.50 in^3/rad, and the area of the cylinder is 5.0 in^2. The flow passes through a valve with the characteristic also plotted.

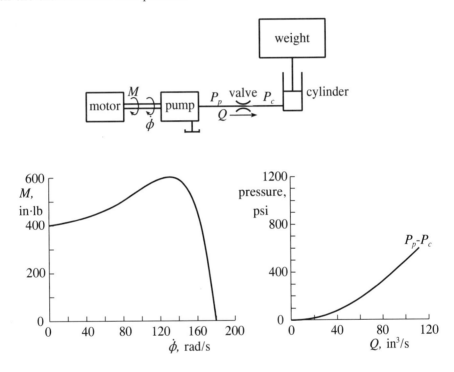

(a) Model the system with a bond graph, neglecting inertia, friction and leakage.

(b) Determine the pressure P_c in the cylinder, and add it to the plotted pressure $P_p - P_c$ to get an effective load characteristic as seen by the pump.

(c) Transform the given torque-speed motor characteristic to plot an equivalent pressure-flow source characteristic at the outlet of the pump on the same axes as the plot of part (b).

(d) Determine the equilibrium pressure P_p and flow Q, the speed at which the weight rises and the torque and angular velocity of the motor.

5.14 A motor, gear pump, mechanical load and two hydraulic loads are interconnected as shown below.

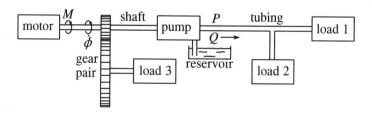

(a) Model the system with a bond graph, identifying the loads as resistances.

(b) Relate the moduli of any transformers in your bond graph to physical parameters of the system.

(c) Express the torque on the motor as a function of the shaft speed and the various parameters.

5.15 Modify the solution of Guided Problem 5.4 (pp. 292, 305) to include friction in the three actuators and internal leakage in the rotary actuator. Write the corresponding equations, leaving the friction and leakage relations as unspecified functions.

5.16 A double-rod-end cylinder exhibits internal leakage across its piston proportional to the pressure difference across it. Model the device with a bond graph.

5.17 The three-port hydraulic system below can be used to increase or decrease pressure or flow. (It is used as the heart of a device known as a hydraulic "intensifier.") Model the system for steady velocity with a bond graph, neglecting leakage, inertia and friction. Start by labeling the flows for all three ports. Relate all transformer moduli to the areas A_{LP}, A_{SP} and A_{SH}.

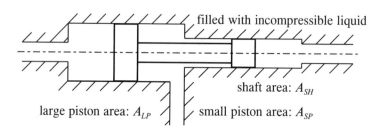

5.18 Water is pumped from a well to an elevated tank without the use of electric power. The level of water in a well is significantly above the level of a nearby pond. Some of the well water (flow Q_m) drives a hydraulic motor on the way to the pond. The motor and pump have displacements of D_m and D_p, respectively. Assume steady-state conditions; the heights z_w and z_t may be assumed to be given. The water has density ρ.

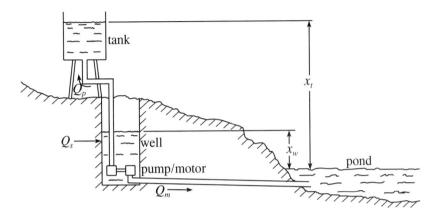

Find a steady-state bond-graph model. Identify any parameters in terms of the information given. Neglect friction and leakage. The seepage into the well (flow Q_s) may be represented by a general source.

5.19 *Repeat and continuation of Problem 2.36 (p. 74).* A motor drives a fan through a belt drive. The torque-speed characteristics of the motor and the fan are plotted below. The pulley ratio is not specified, but the motor should not be run for long with a torque higher than its continuous-duty rating of 4.0 in.·lb (to prevent overheating).

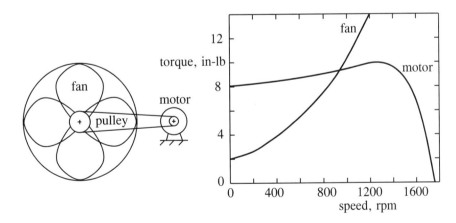

(a) Draw a bond graph model of the system, neglecting energy storage elements.

(b) Determine the speed of the motor that maximizes its allowable power.

(c) Determine the corresponding speed and torque of the fan, and the pulley ratio that achieves them. Neglect belt losses.

(d) Redraw the bond graph, including the effects of kinetic energy associated with the motor shaft and with the fan shaft.

(e) Estimate the torque available to accelerate the system, reflected to the motor shaft, for speeds less than one-half the equilibrium.

(f) The motor and fan have rotational inertias of 0.2 and 1.2 lb·in.·s², respectively. Estimate ($\pm 10\%$) how long it takes the system to go from rest to one-half its steady speed.

5.20 A battery powers a DC motor with armature resistance R_E and modulus G. The motor shaft has negligible friction and rotates a large mass with inertia J. Develop a bond graph model. Identify all key and parameters and variables in both the drawing and the bond graph. Place causal strokes consistent with the preferred computational approach.

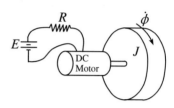

5.21 Derive equations (5.4) (p. 279) that describe the operating point with maximum efficiency of a DC motor with constant field.

5.22 You are asked to purchase a permanent-magnet DC motor which will be driven by a power supply that has a maximum voltage of 24VDC. The maximum load is to be 10 oz·in. at 2000 rpm. A wide assortment of fair-quality motors with $Q \equiv RM_0/Ge = 0.04$ are available.

(a) Determine the values of G, R and M_0 and the associated stall torque and no-load speed that would represent the smallest motor possible (which would therefore have to operate at maximum power).

(b) Determine the values of G, R and M_0 and the associated stall torque and no-load speed that would require the smallest power supply possible (which would therefore have to operate at maximum efficiency).

5.23 A commercial permanent-magnet DC motor is advertised as having a speed of 2200 rpm at 24 VDC and no load, a 4 oz·in. friction torque, a "torque sensitivity" of 14.8 oz·in./amp, a "back EMF" of 11.0 V/Krpm and an armature resistance of 1.9 ohms.

(a) Find the value of G in volt-seconds from the "back EMF," and compare to the value of G in N·m/amp found from the "torque sensitivity." Do they agree?

(b) Find the quality parameter $Q \equiv RM_0/Ge$, assuming the model of Fig. 5.8 (p. 278).

(c) Use the information from parts (a) and (b) to compute the no-load speed. Does this approximately agree with the advertised value?

(d) Estimate the maximum power this motor can deliver, and the associated speed and efficiency.

(e) Estimate the maximum efficiency this motor can provide, and the associated speed and power.

5.24 A shunt-controlled DC motor is shown below. The gyrator modulus for the ideal machine part of this system is proportional to the magnetic field and thus to the current i_f passing through the field circuit; you may assume $G = a i_f$ where a is a constant.

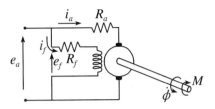

(a) Assuming constant supply voltage e_a, find the torque M and sketch-plot as a function of the shaft speed $\dot{\phi}$.

(b) Represent this system (with constant e_a) by a bond graph with fixed-parameter elements. (Hint: Two versions are possible, one using a 1-junction and the other using a 0-junction; you might try to find both. No gyrator or transformer is needed.)

5.25 A Ward-Leonard drive comprises two DC motors with their electrical leads interconnected. As pointed out in Example 2.6 (p. 62), the combination acts like a rotary mechanical transformer. The transformer modulus can be varied through adjustments in one of the circuits for the magnetic fields. Investigate the effects of the total of the resistance in the armature curcuits, R, and the brush frictions M_{01} and M_{02}. The gyrator moduli can be written as G_1 and G_2, and assumed to be known. The following steps are suggested:

(a) Draw a bond graph of the system, neglecting energy storage. Label the input and output speeds and torques, the friction torques and the electrical current.

(b) Use the diagram to deduce expressions for the two moments as functions of the speeds and the parameters. (The use of causal strokes can provide helpful guidance in assembling the equations for the elements.)

(c) Plot the output moment as a function of the output speed, assuming constant input speed. The use of nondimensional ratios is preferred.

(d) For the particular "quality" values $Q_1 \equiv R M_{01}/G_1^2 \dot{\phi}_1 = 0.01$ and $Q_2 \equiv R M_{02}/G_1 G_2 \dot{\phi}_1 = 0.01$, find and plot the efficiency of the drive as a function of the dimensionless speed $r \equiv G_2 \dot{\phi}_2/G_1 \dot{\phi}_1$. Also, find the peak efficiency, and compare the power delivered at this efficiency to the peak power that can be transmitted.

5.26 The pulley system shown in Fig. 5.11 (p. 283) has negligible moment, M (implying negligible mass moment of inertia, since M includes inertial torque). Give a reduced bond graph model. (Does it make sense?)

5.27 The pulley system shown in Fig. 5.11 (p. 283) has negligible force, F_C (implying negligible mass, since F_C includes inertial force.) Give a reduced bond graph model. (Does it make sense?)

5.28 A cylinder rolls without sliding. One cable is wrapped around the periphery; another is attached to a small central shaft.

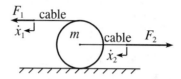

(a) Model the system with a bond graph, using the appropriate geometric constraint(s) directly. Omit all inertance phenomena. Give the modulus of any T or G element.

(b) Find a bond graph by a different route: specialize the result of Fig. 5.11 (p. 283) by noting that $\dot{x}_b = 0$. Is the result compatible with that of part (a)?

(c) Introduce translational and rotational kinetic energy into your model.

(d) Reduce the model of part (c) so it contains a *single* inertance, and give its modulus. (Reference: Guided Problem 3.3, p. 91.)

5.29 A cord with a spring is wrapped around a cylinder. No horizontal excitation is imposed, so the cord and the spring remain vertical. Define key variables by placing them on the drawing, and model the system with a bond graph, placing the variables there also. Relate the elements in your graph to the physical parameter indicated on the drawing.

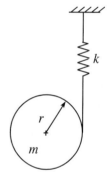

5.30 Model the mechanical system below with a bond graph, and relate the elements in the graph to the given parameters. Assume that the link rotates no more than a small angle.

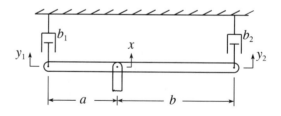

5.31 Determine the speed ratio of the rotating shafts of the epicyclic gear train of Example 5.10 (p. 285) when the ring gear is clamped. Also, find the result when the spider is clamped, instead.

5.32 The lever pictured can be assumed to be massless, and to rotate only through small angles from that shown. The physical dimensions of the mass m are small relative to the lengths r_1 and r_2. Define variables, and model the system with a bond graph. Identify the moduli of the model elements in terms of the given parameters.

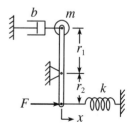

5.33 Repeat the above problem with the lever assumed to be slender and uniform with mass M. Let $m = 0$.

5.34 Oil passes through a valve to drive a hydraulic cylinder which in turn raises a cable that lifts a heavy steel ingot, as pictured. You are asked to define appropriate variables, using symbols placed directly on the drawing and explained by words. Also, relate the moduli of all elements to physical parameters of the system which you should define. Include the effect of the principal inertia.

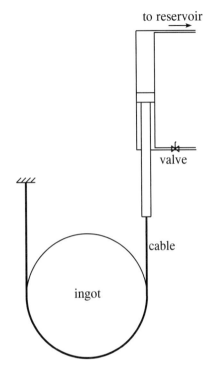

5.35 A DC motor with constant field, armature resistance R and brush friction M_0 rotates a winch of diameter d that drives a vertical cable. The cable raises one end of a uniform heavy beam of mass m and length L that is pivoted about its other end, as shown below. Nevertheless, the beam remains not far from the horizontal position. Define variables and model the system with a bond graph, including the principal inertial effect. Relate the moduli of bond graph elements to the given parameters.

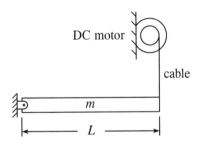

5.36 A rotary spring unwinds to drive a DC generator through a linear viscous coupling. The electrical load comprises a resistor and capacitor, as shown. Model the system with a bond graph, and relate all of its moduli to physical parameters in so far as possible.

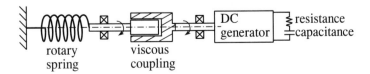

5.37 Hydraulic fluid at a known pressure passes
to a cylinder through a valve. The piston does
not fit the cylinder perfectly, allowing some leakage. The cylinder rod pulls a cable that rotates a
pulley. A spring and a weight are attached to a
second cable, as shown.

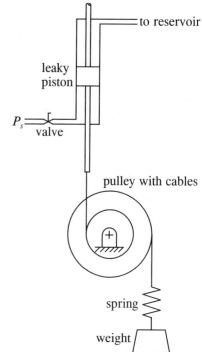

(a) Define key variables, placing them on the
drawing.

(b) Represent a model of the system with a
bond graph, annotated with the same variables. Included appropriate inertance(s).

(c) Define key parameters of the system,
keying them to the drawing where appropriate.

(d) Relate the moduli of the bond graph to
the parameters that you have defined.

5.38 A testing machine produces an extremely high pressure
pulse in a liquid-filled chamber of volume V by dropping a
weight with mass M on a piston of small mass m and small
area A. Bouncing of the weight is inhibited, and the plulse
length extended, by inserting a spring with rate k and a linear
dashpot with coefficient b between the mass and the piston, as
pictured to the right. The liquid has bulk modulus β.

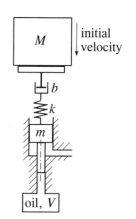

(a) Define key variables, and place them on the drawing.

(b) Model the system with a bond graph, assuming contact
of the mass M.

(c) Relate the bond graph parameters to the physical parameters mentioned above.

5.39 Define variables for an approximate model of the system pictured below, and place
them on both the drawing and a bond-graph model of the system. Relate the moduli of
the elements to physical parameters, insofar as practicable.

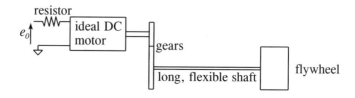

5.40 A manometer has a mechanical dial output, with rotational inertia I_d about its pinned axis. The gravity-balanced dial rotates when the level of the fluid changes, due to a small float of area A_f and mass m and a link of negligible mass. The manometer tube has internal area A_t and contains a fluid with density ρ. Note that the level of the float relative to the water level is variable. Model the system with a bond graph, neglecting the effect of fluid inertia. Label your generalized velocities on the drawing and on the graph, and relate its moduli to physical parameters. Suggestion: Use the depth of the fluid when the float is in equilibrium as a generalized diplacement.

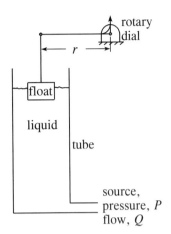

5.41 A DC motor with armature resistance rotates a pinion which drives a rack supported by a linear ball bearing. The rack in turn drives a second gear that is attached to both a flywheel and a spring. Note that the second gear can translate as well as rotate. Define appropriate generalized variables and label the sketch accordingly, state the geometric constraints between these variables, draw a bond graph model, and relate the moduli of your model to the given primitive physical parameters and any others which you define. Inertial effects apart from the mass m may be neglected. (Hint: Make sure you identify *four* distinct mechanical velocities.)

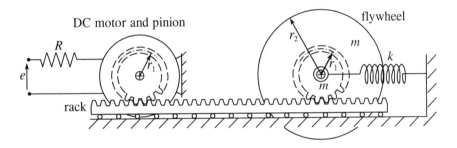

5.42 A motor drives a cable with a winch of radius r_w, as depicted below. The cable is wrapped around a massless pulley of radius r_p, and is terminated in a dashpot of modulus b. The pulley pulls a weight W_1 which is supported by two rollers that each weigh W_2.

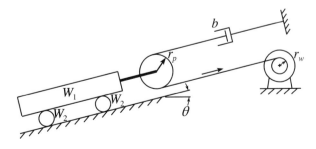

(a) Define key velocities and forces on a copy of the drawing. Note particularly that the centers of the rollers W_2 do not move at the same velocity as the slab W_1.

(b) Draw a bond graph of the system, including the effects of both the weights and the inertias of the slab and the rollers.

(c) Evaluate the moduli of the elements in your bond graph in terms of the given parameters. Suggestion: The kinetic energy method may be particularly helpful for the rollers.

5.43 A proposed sound film movie camera must have a steady flow of film past the sound head despite the discontinuous loads and motion at 24 Hz (plus multiples) of the gate, with its claw drive, which acts on the sprocket holes. One path for vibrations to reach the sound head from the claw drive is through the pulley system. To minimize this transmission, flexible belts are proposed on the pulleys, as shown.

Treat the claw mechanism simply as an independent unsteady force on the top of the claw drive system. Find a bond graph model and identify each component with a couple words (or more). Neglect variations in the magnetic torque in the motor.

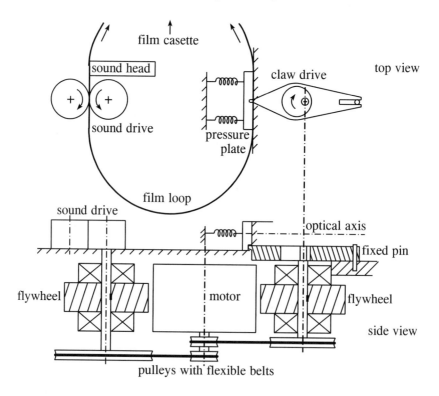

SOLUTIONS TO GUIDED PROBLEMS

Guided Problem 5.3

1. Motor shaft: effort is torque, M_m; flow is angular velocity, $\dot{\phi}_m$.

   Flywheel: effort is torque, M_f; flow is angular velocity, $\dot{\phi}_f$.

   Larger gear and spring: effort is torque M_s; flow is angular velocity $\dot{\phi}_s$.

2.

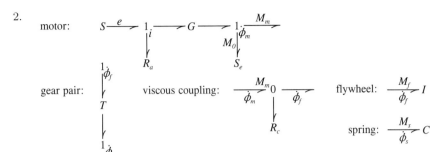

motor:

gear pair: viscous coupling: flywheel:

spring:

3.

assembled bond graph:

Guided Problem 5.4

1.

Notes: Q_1 is the flow to actuator 1, and Q_2 is the *common* flow through actuator 3 to actuator 2. The power to the reservoir shown is neglected, because its pressure is virtually zero.

2. $T_1 = \dfrac{\dot{x}_1}{Q_1} = \dfrac{1}{A_{c1}}$

3. (a)

$$T_a \equiv \frac{\dot{x}_2}{Q_a} = \frac{1}{A_{c2}}; \qquad T_b \equiv \frac{\dot{x}_2}{Q_b} = -\frac{1}{A_{c2} - A_{s2}};$$

$$T_2 \equiv \frac{\dot{x}_2}{Q_2} = \frac{\dot{x}_2}{Q_a + Q_b} = \frac{1}{Q_a/\dot{x}_2 + Q_b/\dot{x}_2} = \frac{1}{A_{c2} - (A_{c2} - A_{s2})} = \frac{1}{A_{s2}}$$

(b) $Q_2 = A_{s2}\dot{x}_2$; therefore, $T \equiv \dfrac{\dot{x}_2}{Q_2} = \dfrac{1}{A_{s2}}$

4. $Q_2 = D\dot{\phi}$; therefore, $T_3 = \dfrac{\dot{\phi}}{Q_2} = \dfrac{1}{D} = \dfrac{1}{w(r_v^2 - r_s^2)}$, where r_v is the radius to the tip of the vanes, r_s is the radius of the shaft and w is the width (not the thickness) of the vanes. See Guided Problem 2.5 (p. 69) and its solution for further explanation.

5. - 6.

$P = T_1 F_1 = T_2 F_2 + T_3 M$

$Q = \dfrac{\dot{x}_1}{T_1} + \dfrac{\dot{\phi}}{T_3} = \dfrac{\dot{x}_1}{T_1} + \dfrac{\dot{x}_2}{T_2}$ (choice)

$T_2 F_2 + T_3 M$

$Q_2 = \dot{\phi}/T_3 = \dot{x}_2/T_2$

$T_2 \dfrac{F_2}{\dot{x}_2}$

$T_1 F_1 \big| Q_1 = \dot{x}_1/T_1$

$T_3 M$

T_1

$F_1 \big| \dot{x}_1$

T_3

$M \big| \dot{\phi}$

Guided Problem 5.5

1. detailed model:

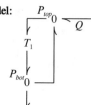

alternative detailed model:

$$T_a = 1/A_p; \quad T_b = A_p\text{-}A_s$$

$$T_1 = \frac{A_p\text{-}A_s}{A_s}$$

$$T_2 = \frac{A_s}{A_p\text{-}A_s}$$

A_p = area of the piston
A_s = area of the shaft

desirable simple model:

$$T = \frac{1}{A_s} \quad \left(= \frac{1}{\frac{1}{T_a} - T_b} \right)$$

2 - 3.

$$\dot{y}_d = \dot{y} + r\dot{\phi}$$

$$\dot{y}_s = \dot{y} - r\dot{\phi}$$

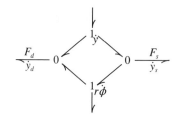

4.

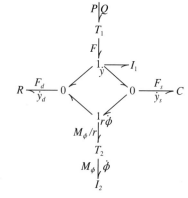

$$T_1 = 1/A_s$$

$$T_2 = 1/r$$

$$I_1 = m + M$$

$$I_2 = \tfrac{1}{2} M r^2$$

$$C = 1/k$$

$$R = b$$

Guided Problem 5.6

1.

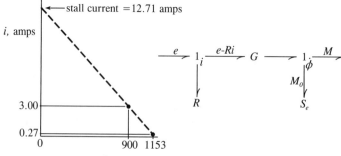

$$R = \frac{e}{i_{stall}} = \frac{40}{12.71} = 3.147 \text{ ohms}$$

2. $G = \dfrac{\text{change of torque}}{\text{change of current}} = \dfrac{120}{(3 - 0.27) \times 16 \times 39.37 \times 0.22481}$

$= 0.3104$ N·m/amp

The citing of the 120 oz·in. of torque is redundant. The value of G can be found without this datum as follows:

$e - Ri = G\dot{\phi}$; therefore,

$$G = \frac{e - Ri}{\dot{\phi}} = \left(\frac{e - Ri}{\dot{\phi}} \right)_{no-load} = \frac{40 - 3.147 \times 0.27}{1153 \times 2\pi/60} = 0.3242 \text{ volt·s/rad}$$

These units are the same as N·m/amp. The value is slightly different from that found using the given torque. The difference could be due to round-off by the advertiser, or errors in the assumed model.

3. When $M = 0$, $M_0 = Gi = 0.3104 \times 0.27 = 0.084$ N·m.

4. $M_{stall} = Gi_{stall} - M_0 = 0.3104 \times 12.71 - M_0 = 3.86$ N·m. The linear extrapolation gives the same value.

5. $\mathcal{P}_{max} = \dfrac{3.86}{2} \times \dfrac{1153}{2} \times \dfrac{2\pi}{60} = 117$ watts

6. $\dfrac{RM_0}{Ge} = \dfrac{3.145 \times 0.084}{0.3104 \times 40} = 0.0213$

$$\eta_{max} = \left(1 - \sqrt{\frac{RM_0}{Ge}} \right)^2 = 0.73$$

$$\mathcal{P}_{max\ \eta} = \sqrt{\frac{RM_o}{Ge}} \left(1 - \sqrt{\frac{RM_0}{Ge}} \right)^2 \frac{e^2}{R} = \sqrt{0.0213}(1 - \sqrt{0.0213})^2 \frac{40^2}{3.145} = 54.2 \text{ watts}$$

5.3 Model Equivalences

The bond graph models developed thus far can often be simplified readily, without any loss in accuracy despite some loss in detail. Such simplifications usually are highly desirable for two reasons: the essential character of the simpler model is easier to grasp, aiding design, and any subsequent analysis is likely simplified.

The most important reduction step has been treated already: the combining of multiple R, C and I elements bonded to a single 1-junction or 0-junction. It is essential to remember that each element can be combined only with other elements of the same type that happen to be arrayed on the same junction. Remember also that two junctions of the same type connected by a bond can themselves be reduced to a single junction. Further, a junction with only two bonds can be removed and the bonds connected, as long as the two power half-arrows are directed similarly. If they are not so directed, you may re-direct the half-arrow on one of the bonds and compensate by changing the sign of one of its variables and the sign of the modulus of any associated one-port element.

Further key reductions are given below. First, however, an important *elaboration* is considered which allows you to dispense with all one-port elements with characteristics that do not pass through the origin, except for the special constant-effort and constant-flow sources.

5.3.1 Thevenin and Norton Equivalent Sources and Loads

This book has repeatedly represented sources with characteristics that are biased, that is do not pass through the origin. An electric motor, for example, has a torque at zero speed, and a speed at zero torque; an impeller pump unit has pressure at zero flow, and a flow at zero pressure. Resistive loads with biased characteristics also have been employed. The pressure of the water sprinkler system of Section 2.1, for example, is not zero at zero flow, and the drag of a vehicle does not vanish when the speed approaches zero. Rather than directly to employ a source with these characteristics, it is usually better, for purposes of analysis or simulation, to construct the source by combining three more primitive elements: a *constant-effort* or a *constant-flow* source, an *unbiased* resistance and a junction. The same type of structure also can represent biased loads. The same approach also can replace compliances having characteristics that do not pass through the origin with those that do.

These replacements are particularly useful for straight-line characteristics. The resulting resistance components have constant moduli, instead of functions. The sources also are constants, or independent functions of time. The conversion from a bond graph to differential equations is expedited, as you will experience in the next chapter.

The creation of a biased sink from an unbiased resistance is shown in part (a) of Fig. 5.16. Linear examples are shown by solid lines, and nonlinear examples by dashed lines. In the first example of a sink, the effort e_0 for $\dot{q} = 0$ is marked by a heavy dot. The Thevenin equivalent sink is based on the resistance characteristic that results when the whole curve, including this point, is shifted vertically until the point is at the origin. The biased characteristic is reconstituted by adding the effort e_0 to this resistance curve with its effort e_R. This summation, suggested by vertical arrows in the figure, is achieved in the bond graph by bonding the unbiased resistance element R to an effort source, S_e, using a common-flow or 1-junction.

The Norton equivalent is an alternative that employs flow summation rather than effort summation. The second case shown in the figure has the same characteristic as the first. A heavy dot is drawn when the curve intersects the $\dot{q}$ axis, defining the negative flow $-\dot{q}_0$. A passive resistance R with flow $\dot{q}_R$ results when this curve is shifted *horizontally* until the dot lies at the origin. The biased characteristic is reconstituted by summing $\dot{q}_R$ with $-\dot{q}_0$, as suggested by the horizontal arrows. In the bond graph, this is accomplished by bonding the element R to the flow source S_f with a constant-effort or 0-junction. Note that the resistance elements for the Thevenin and Norton equivalents are the same for linear cases, but are distinctly different for nonlinear cases.

Most sinks have characteristics that pass through the origin and therefore need no Thevenin or Norton equivalents. Sources, on the other hand, virtually always benefit from these equivalents. Examples are shown in the bottom half of Fig. 5.16. In the Thevenin version, a resistive effort e_R is *subtracted* from a source effort e_0 to get the total effort. Effort summations are vertical, again, and accomplished by a 1-junction. In the Norton version, a resistive flow $\dot{q}_R$ is subtracted from a source flow $\dot{q}_0$ to get the total flow. Flow summations are horizontal, as always, and are accomplished by a 0-junction. Note again that the Thevenin and Norton resistances are the same if and only if they are linear.

Compliance characteristics sometimes do not pass through the origin. A Thevenin decomposition can be applied here, also, as shown in Fig. 5.17. A Norton equivalent does not exist, since *displacement* sources are not defined. Nevertheless, it is possible to redefine the displacement variable of a compliance such that the characteristic passes through the origin, which is tantamount to horizontal shifting of the characteristic.

5.3.2   Passivity With Respect to a Point on a Characteristic*

The Thevenin and Norton structures can be viewed as translating the origin of the characteristics in the effort-flow plane. Translating the origin of source and load characteristics to the equilibrium point at which they intersect can aid dynamic analysis. Information regarding stability can result, for example, as will be suggested in Section 5.4.7. Consequences regarding general dynamic behavior appear later. For the present, consider only how such a translation can be accomplished. If the origin is to be displaced along a characteristic,

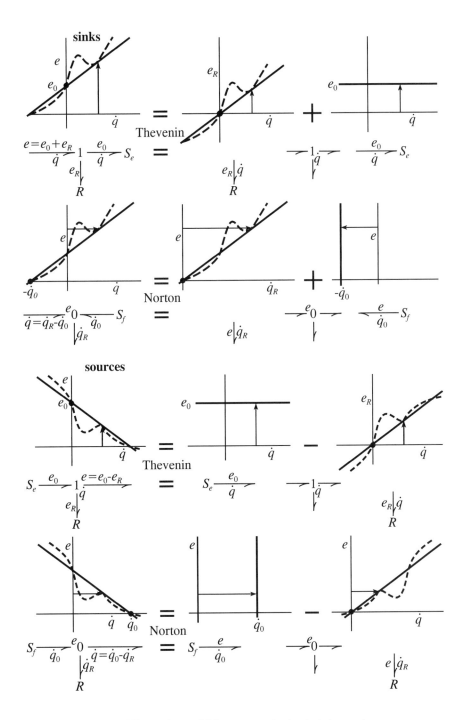

Figure 5.16: Thevenin and Norton equivalent sinks and sources

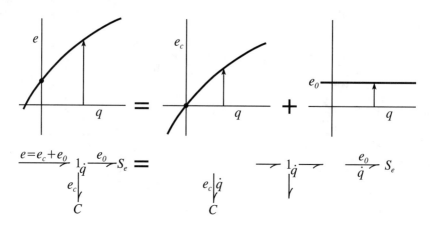

Figure 5.17: Replacing a biased compliance with an unbiased compliance

translation of both the effort and flow axes are necessary. This suggests combining the Thevenin and Norton structures.

One way to combine them is illustrated in Fig. 5.18. The original characteristic, shown in part (a), employs the variables e_R and $\dot{q}_R$. The shifted characteristic employs the variables e and $\dot{q}$. The new axes for the point labelled "b" are shown by dashed lines. Note that the only difference between the sink or load version, shown in part (b), and the source version, shown in part (c), lies in the signs of the flows.

A load or sink has been defined as *passive* if it lies entirely in the first and third quadrants, that is if it never generates power because the product $e\dot{q}$ is nowhere negative. This property can change when the origin is moved along the characteristic; passivity becomes a property of the point chosen as the origin as well as being a function of the shape of the characteristic. Consider the given characteristic with point b as the origin, as shown in part (b) of Fig. 5.18. The characteristic resides partly in quadrants two and four, and therefore is said to be active or non-passive *with respect to this origin*. The same conclusion applies to any point between points a and c. If, on the other hand, a point to the left or below point a or to the right or above point c is chosen as the origin, the characteristic is said to be passive with respect to that origin.

The characteristic in part (c) of the figure is identical to that of part (b), except that the sign of the flow is inverted by employing a source rather than a sink power convention to the bond for e and $\dot{q}$. The resulting plot is a left-to-right flip of that given in part (b). The property of passivity does not change as a result of this change in perspective. The test for passivity above, however, assumes the power convention of a sink; the power convention of a source must be inverted before the test is applied.

5.3.3   Truncation of Transformers and Gyrators Bonded to R, C or I Elements

Frequently a one-port element is attached to a system model by a transformer or a gyrator. The combination of two elements can be replaced by a single equivalent element. This option simplifies subsequent analysis, and can enhance your grasp of the fundamental structure of a system.

Table 5.1 gives the equivalences. A transformer does not alter the type of the combination; a resistance remains a resistance, a compliance remains a compliance and an inertance remains an inertance. On the other hand, a gyrator bonded to a compliance yields an equivalent inertance, and a gyrator bonded to an inertance yields an equivalent compli-

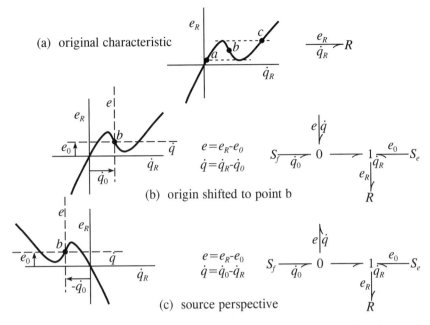

Figure 5.18: Shifting the origin of a source or sink characteristic along its path

ance. Although a gyrator bonded to a resistance remains a resistance, its modulus becomes inverted.[3]

You need not memorize these equivalences, since you can reconstruct them readily at least two different ways. In the first, or direct way, you first label either the effort or the flow on the left-hand bond with a symbol. Then, you employ the definitions of the transformer or gyrator and the one-port element to find the corresponding conjugate variable. Finally, you compare this result to the corresponding conjugate variable for the reduced case, and deduce the equivalent modulus.

Table 5.1 Equivalences for One-Port Elements Bonded to Transformers and Gyrators

—T—R	$\Longleftrightarrow$	—R'	$R' = T^2 R$
—G—R	$\Longleftrightarrow$	—R'	$R' = G^2/R$
—T—C	$\Longleftrightarrow$	—C'	$C' = C/T^2$
—G—C	$\Longleftrightarrow$	—I'	$I' = G^2 C$
—T—I	$\Longleftrightarrow$	—I'	$I' = T^2 I$
—G—I	$\Longleftrightarrow$	—C'	$C' = I/G^2$

[3]The gyrator is said to be a *dualizing* coupler. Thus, for example, if two inertances are bonded together by a gyrator (I_1—G—I_2), each inertance "sees" the other inertance as a compliance. As a result, this system would exhibit oscillatory behavior, since an inertance bonded to a compliance — for example a mass attached to a spring — is an oscillator. This is why two fluid vortices, neither of which has any "springiness," interact in an oscillatory fashion.

Consider the following two examples, which are the first and the fourth entries in the table, assuming the parameters are constants:

$$\underset{\dot{q}}{\overset{e = T^2 R\dot{q}}{\longrightarrow}} T \; \underset{T\dot{q}}{\overset{RT\dot{q}}{\longrightarrow}} R \qquad\qquad \underset{\dot{q}=p/CG^2}{\overset{e}{\longrightarrow}} G \; \underset{e/G}{\overset{p/CG}{\longrightarrow}} C$$

In the left-hand example, you start by labeling the flow variable on the left as $\dot{q}$. As a consequence, the flow variable on the right must be $T\dot{q}$. The resistance element then requires the effort variable on the right to be $RT\dot{q}$, and finally the transformer requires the effort variable on the left to be $TRT\dot{q}$. Since by the definition of R' this effort variable also equals $R'\dot{q}$, it follows that $R' = T^2 R$. Alternatively you could start by labeling the effort variable on the left as e, and proceed around the bonds in inverse order. In the right-hand example, you start the procedure by labeling the effort on the left as e or the flow as $\dot{q}$. If you start with the effort, the gyrator requires the flow on the right to be e/G, the compliance then requires the effort on the right to be p/CG (recall that $p \equiv \int e\, dt$), and finally the gyrator requires the flow on the left to be p/GCG. Since the inertance I' is defined by the relation $\dot{q} = p/I'$, its value is $I' = G^2 C$.

The second method for deducing the equivalences is based on power or energy, a superior discipline. For the first example, the power dissipated in the resistances equals both $\frac{1}{2}R'\dot{q}^2$ and $\frac{1}{2}R(T\dot{q})^2$, so that $R' = T^2 R$. For the second example, the energy stored in I', namely $\frac{1}{2}I'\dot{q}^2$, equals the energy stored in C, namely $\frac{1}{2}C(G\dot{q})^2$, so that $I' = G^2 C$.

5.3.4 Reduction of Two-Pair Meshes

Cascaded bonds interconnecting alternate 0 and 1-junctions to form a closed path are called a simple bond-graph **mesh**. Meshes occur often. Meshes of the types shown in Fig. 5.19 can be reduced to equivalent tree-like bond graphs, simplifing subsequent analysis.

A mesh is called **even** if the number of mesh bonds which have the same power direction convention in the clockwise sense is even, and **odd** if the number is odd. The same statement applies to the bonds with counterclockwise power direction convention, since the total number of mesh bonds is even. All the meshes in Fig. 5.19 are even. It can be shown that such evenness or oddness is a property of the mesh.[4] This means that reassigning the power convention arrows and compensating by changing the signs of the efforts or flows, in a compatible fashion, always leaves the evenness or oddness unchanged. Two meshes that are identical, except for the property of evenness vs. oddness, therefore model markedly different systems.

Four-bond even meshes, as shown in the left side of Fig. 5.19 part (a), can be reduced to the tree-like forms shown in the right side of the figure. In the first case, the efforts on the left-most and right-most bonds are the same, because they both equal the sum of the efforts on the two 0-junctions. In the second case, the flows on the left-most and right-most bonds are the same, because they both equal the sum of the flows on the two 1-junctions. These reductions are extremely useful, because tree-like graphs are easier to analyze than graphs with meshes, as you will see later.

Nested even meshes each having four mesh bonds, as shown in part (b) of the figure, can be reduced similarly, as shown. These equivalences apply to the junction structure regardless of whether the elements bonded to it are resistances, sources or something else. The figure employs the symbol Z to represent unspecified one-port elements, which could be resistances, compliances, inertances, boundary bonds or something more complex.

Odd meshes are not reducible in this manner. As a result, their analysis is more involved. They tend to occur less often, fortunately.

---

[4]F.T. Brown, "Direct Application of the Loop Rule to Bond Graphs," *J. Basic Engineering, ASME Trans.*, v 94 n 3 p 253-261, Sept. 1972.

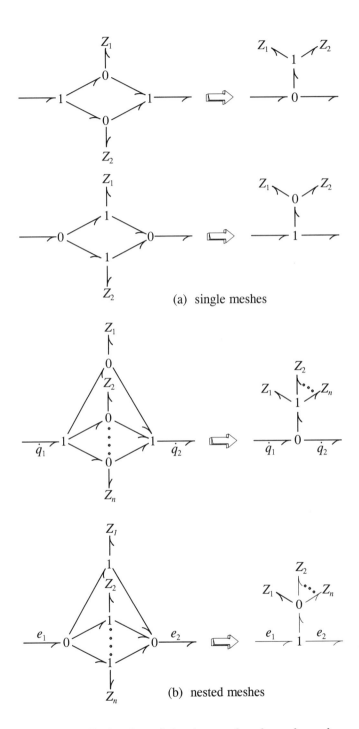

Figure 5.19: Reduction of simple even bond-graph meshes

EXAMPLE 5.12

Apply the standard steps to modeling the circuit below with a bond graph, and show that an even mesh results. Then, reduce the graph to an equivalent tree-like structure.

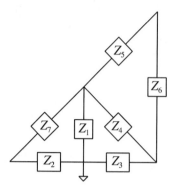

Solution: The standard procedure gives the following results after steps 2 and 4 (p. 258), respectively:

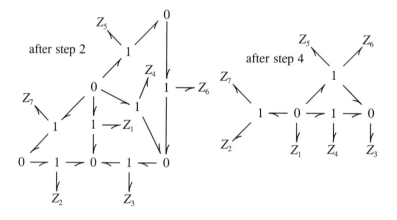

The mesh within this graph is even, since two of its mesh bonds have clockwise-directed power arrows and the other two are counterclockwise. The 0-junction is first split into two 0-junctions in order to remove element Z_1 from the mesh. The mesh then is in the form of the second mesh given in Fig. 5.19, and so the equivalence given there is used to get the final graph.

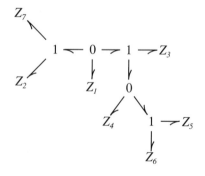

EXAMPLE 5.14

Find an equivalent tree-like bond graph for the mesh bond graph found in Example 5.7 (p. 274-275) (which can expedite subsequent analysis). *Continued in Problem 6.51.*

Solution: The mesh bond graph, reproduced on the left below, is even, and therefore reducible. The mesh bonds all have counterclockwise power orientations, however, unlike those in Figure 5.19 (p. 313). The first transformation shown below changes the sign of the effort (pressure) P_a on the lower-right 0-junction, with a compensating change in the power convention arrows on the three bonds on that junction. The convention on the bond for the compliance C_a is then returned to its preferred outward direction by changing the sign of the flow on that bond. The elements C_m and I_m also are moved away from the 1-junction for Q_1 and changed in value to account for the transformer. The 1-junction for the pair I_0 and R_0 can be expanded into two bonded 1-junctions, similarly, to avoid confusion, although this step is not shown explicitly. Now the mesh bonds have the same power-convention arrows as in Figure 5.19, giving the final tree-like equivalence.

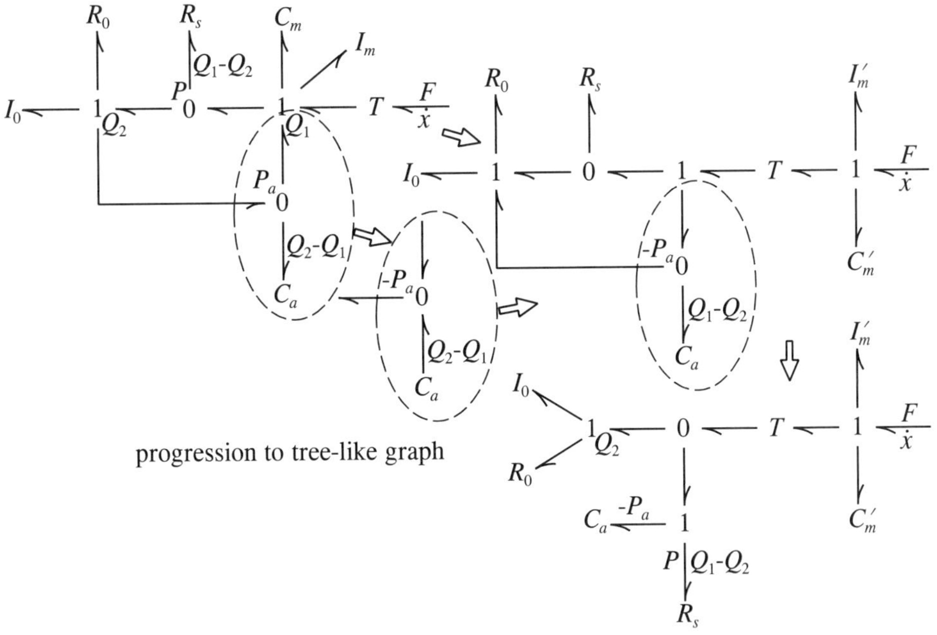

progression to tree-like graph

5.3.5 Transmission Matrix Reduction of Steady-State Models*

Often a two-ported model or sub-model comprises a chain-like structure with no energy-storage elements. A common case is shown in Fig. 5.20. Such cases can be represented as

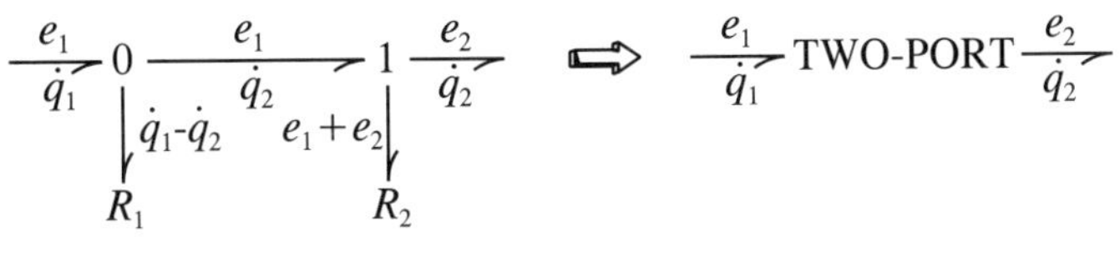

Figure 5.20: Reduction of a chain of junctions with stub elements

a two-port element described by a single transmission matrix, which for some purposes is a significant simplification. The transmission matrix concept is introduced in Section 2.5.6 (pp. 69-70).

The transmission matrices for the shunt and series resistances are readily deduced as

$$\begin{array}{c}\xrightarrow[\dot{q}_1]{e_1} \quad 0 \quad \xrightarrow[\dot{q}_2]{e_2} \\ \downarrow \\ R\end{array} \qquad \Longrightarrow \qquad \begin{bmatrix} e_1 \\ \dot{q}_1 \end{bmatrix} = \begin{bmatrix} 1 & 0 \\ 1/R & 1 \end{bmatrix} \begin{bmatrix} e_2 \\ \dot{q}_2 \end{bmatrix};$$

$$\begin{array}{c}\xrightarrow[\dot{q}_1]{e_1} \quad 1 \quad \xrightarrow[\dot{q}_2]{e_2} \\ \downarrow \\ R\end{array} \qquad \Longrightarrow \qquad \begin{bmatrix} e_1 \\ \dot{q}_1 \end{bmatrix} = \begin{bmatrix} 1 & R \\ 0 & 1 \end{bmatrix} \begin{bmatrix} e_2 \\ \dot{q}_2 \end{bmatrix}.$$

The cascade of the two such structures in Fig. 5.20 can be reduced by multiplying their respective transmission matrices:

$$\begin{bmatrix} e_1 \\ \dot{q}_1 \end{bmatrix} = \mathbf{T} \begin{bmatrix} e_2 \\ \dot{q}_2 \end{bmatrix};$$

$$\mathbf{T} = \mathbf{T}_1 \mathbf{T}_2 = \begin{bmatrix} 1 & 0 \\ 1/R_1 & 1 \end{bmatrix} \begin{bmatrix} 1 & R_2 \\ 0 & 1 \end{bmatrix} = \begin{bmatrix} 1 & R_2 \\ 1/R_1 & 1+R_2/R_1 \end{bmatrix}.$$

A cascade of such elements of any length can be telescoped in this manner. More general steady-state couplers are discussed in Section 12.2.

EXAMPLE 5.15

Find an equivalent transmission matrix for the static two-port model:

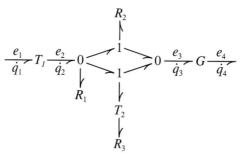

Solution: The combination of T_2 and R_3 is first reduced to an equivalent resistance R_3'. Also the left-hand 0-junction is expanded into two bonded 0-junctions, in order to isolate the even mesh clearly:

The next intermediate graph implements the standard reduction of this mesh. In the third intermediate graph, the resistances R_2 and R'_3 are combined to give R_4:

The central portion of the bond graph, with its 0- and 1-junctions and attached resistances, is now precisely in the form of Fig. 5.19:

The transmission matrix approach allows the transfomer and gyrator parts of the graph, as well as this central portion, to be telescoped in one step:

$$\begin{bmatrix} e_1 \\ \dot{q}_1 \end{bmatrix} = \begin{bmatrix} T_1 & 0 \\ 0 & 1/T_1 \end{bmatrix} \begin{bmatrix} 1 & 0 \\ 1/R_1 & 1 \end{bmatrix} \begin{bmatrix} 1 & R_4 \\ 0 & 1 \end{bmatrix} \begin{bmatrix} 0 & G \\ 1/G & 0 \end{bmatrix} \begin{bmatrix} e_4 \\ \dot{q}_4 \end{bmatrix}$$
$$= \begin{bmatrix} T_1 R_4/G & T_1 G \\ (1+R_4/R_1)/GT_1 & G/T_1 R_1 \end{bmatrix} \begin{bmatrix} e_4 \\ \dot{q}_4 \end{bmatrix}$$

5.3.6   Summary

Subsequent analysis is simplified if general sources and resistive loads are replaced by Thevenin or Norton equivalents, if compliance characteristics similarly are made to pass through the origin, if transformers and gyrators bonded to one-port elements are replaced by equivalent one-port elements, if simple even meshes are replaced by three-like structures, and if steady-state portions of models comprising junction structures and resistances are reduced to simpler one- or two-port models. Bond-graph equivalences and analytical reduction techniques have been presented toward this end.

These equivalences are first used in the next section to help find the equilibium states of models. The use of graphical methods for nonlinear structures is developed, also.

Guided Problem 5.7

This problem typifies simple analytical reduction of a model comprised of steady-state elements.

For the system of dashpots and a lever shown in Fig. 5.21, (a) draw a bond graph model, (b) draw an analogous electric circuit, and (c) find an equivalent overall resistance. The lever may be assumed to rotate only through a very small angle.

Suggested Steps:

1. Follow the procedure suggested in Sections 5.2.6 and 5.2.7 for finding the junction structure, and note that the lever acts as a transformer.

2. Reduce any mesh, should your bond graph have one, to a tree-like graph.

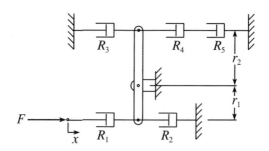

Figure 5.21: Guided Problem 5.7

3. Invert the procedure of Section 5.1.1 (p. 258) to find the equivalent electric circuit.

4. Reduce the graph one junction-structure element at a time, starting presumably at the right end. At each step find the new terminal equivalent resistance. The final result should be the desired answer.

5. As a check you might reduce the electric circuit by some other means.

PROBLEMS

5.44 A constant voltage source $e_0 = 12$ V drives a DC motor with armature resistance $R_a = 6$ ohms and negligible friction that drives a nonlinear load R_L. As indicated in the bond graphs below, the source and the armature resistance comprise a Thevenin source, which can be collapsed into a source S'. The load characteristic is plotted.

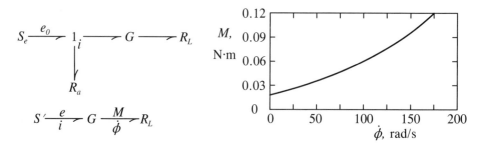

(a) Plot the characteristic of the equivalent source S'.

(b) Give the Norton equivalent, including parameter values.

(c) Estimate the value of G that gives the maximum speed $\dot{\phi}$.

5.45 Represent the hydraulic power supply of Fig. 2.45 (p. 71) by (i) Thevenin and (ii) Norton equivalences. Sketch the characteristics of the resistances.

5.46 Continue Problem 5.18 (p. 297) with the well and pump. The flow Q_s into the well decreases linearly as the level of the well is raised; represent this source with a Thevenin or Norton equivalent. Also, add to your model the compliances of the well and the tank, the resistance due to friction in the pump/motor subsystem, and an output flow Q_t drawn from the tank.

5.47 Determine the criterion for a linear (constant) resistance to be active with respect to any point on the characteristic.

5.48 The flywheel with rotational inertia I_f depicted below is driven through a pair of gears with ratio $r > 1$. The gears and bearings may be presumed to be frictionless and massless. Draw a bond graph for the system. Also, reduce the model to a simple equivalent inertia, with no transformer, and find its modulus in terms of I_f and r.

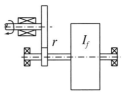

5.49 Show that the translating coil device of Guided Problem 2.6 (p. 58-59) becomes a dashpot when the electric circuit is completed by an added resistor to give a total electric resistance R. Also, determine the dashpot coefficient, b, in terms of the given parameters.

5.50 A permanent magnet DC motor can be made into a rotary dashpot by completing the armature circuit with a resistor.

(a) If the motor satisfies the relation $M = ai$ and the electrical resistance of the armature curcuit is R, determine the dashpot coefficient, b.

(b) Resistors come in different *physical* sizes depending on the heat they need to reject. What power rating is needed if $b = 0.1$ N·m·s and the dashpot must tolerate a steady speed of 5 rev/sec?

(c) State a *functional* advantage this device has over a viscous rotary damper of the type pictured in Fig. 2.23 part (d) (p. 38).

5.51 Reduce the bond-graph mesh in Fig. 5.7 (p. 277) to a tree-like structure. Also, combine the two bond graphs to get a single model representing a positive-displacement pump or motor with both friction and leakage.

5.52 Determine whether any of the meshes in the bond graphs of Fig. 5.11 (p. 283) and Example 5.9 (p. 284) are reducible to tree-like structures.

5.53 Reduce the mesh for the bond graph of Problem 5.8 (pp. 268-269).

5.54 Derive the second and third entries in the right-hand column of Table 5.1 (p. 311) using (i) direct interpretation of the elements, and (ii) considerations of power or energy.

5.55 A rotary spring with rate k, a massless and frictionless pair of gears with diameters d_1 and d_2, and a flywheel with moment of inertia I are connected as shown to the right.

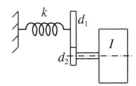

(a) Model the system with a bond graph. Relate all moduli to physical parameters.

(b) Find an equivalent simpler bond graph with only two elements. Relate the new modulus to physical parameters.

(c) Find the natural frequency at which this system oscillates following a non-zero initial condition.

5.56 Show that the bond graph below represents an oscillator, and find its natural frequency, ω_n.

$$C_1 \longleftarrow 1 \longrightarrow G \longrightarrow C_2$$

5.57 A large charged capacitor drives an ideal DC motor (neglect armature resistance and friction) after a switch is closed. The motor is connected either to a (i) flywheel, or a (ii) spring. Certain parameters are defined on the drawings.

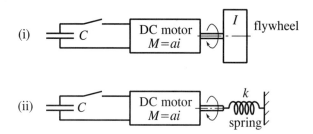

(a) Model each case with a bond graph, and state whether or not it is oscillatory in nature.

(b) If either (or both) of the systems are oscillatory, find its natural frequency in terms of the parameters defined.

5.58 The electrical terminals of a DC motor/generator for which $e = a\dot{\phi}$ are connected to a large inductor with inductance L. In version (i), the shaft is connected to a flywheel with rotational moment of inertia I. In version (ii), the shaft is connected to a rotational spring with constant k, as shown below. Electrical resistance and mechanical friction can be neglected for purposes of the following questions:

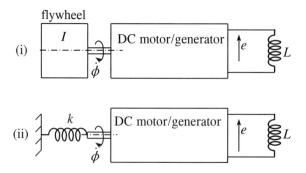

(a) Model each case with a bond graph, and state whether or not it is oscillatory in nature.

(b) If either (or both) of the systems is oscillatory, determine its natural frequency in terms of the defined parameters.

5.59 A motor with a known torque drives a frictionless and leak-free positive-displacement pump that in turn drives a system that includes a frictionless and leak-free single-rod-end cylinder and two valves (that might be adjustable). Assume that all parameters are known, and that the load force F is known. Gravity may be neglected.

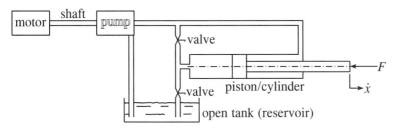

(a) Define parameters and variables and draw an appropriate bond graph model.

(b) Write whatever relations for the elements that are not explicit in your annotated bond graph.

(c) Solve for $\dot{x}$ in terms of the known information.

5.60 The energy of an aircraft landing on an aircraft carrier is dissipated in an "arresting engine" located below decks. A hook on the aircraft grabs a cable that is connected to this machine by pulleys, which are not shown below. The cable wraps nine times around the two sets of sheaves (pulleys) on the machine itself. The nine sheaves at one end rotate about a fixed shaft; the nine pullleys at the other end rotate about a common shaft which is attached to a single large piston via a moveable crosshead. As a result, hydraulic fluid is forced out of the cylinder and flows through a flow restriction to an accumulator. The restriction is adjusted continuously throughout the process in order to maintain the desired pressure in the cylinder and therefore the desired tension in the cable. The accumulator contains a free piston to prevent mixing of the oil and the air (or nitrogen); the pressures of the oil and air are virtually identical. One end of the cable is anchored.

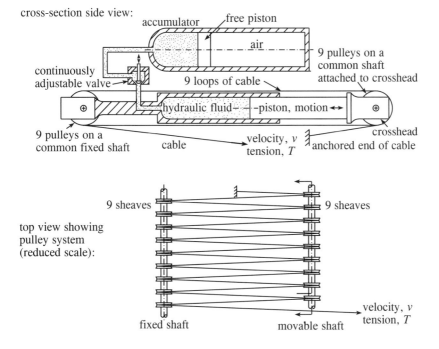

(a) Find a model for the system, considering the tension T as the input variable and the velocity v as the output. Define parameters and variables as needed. The inertias of the crosshead assembly and the pulleys are important, but losses due to mechanical friction can be neglected.

(b) Reduce your model to give a single equivalent resistance, compliance and inertance. Relate these parameters to your primitive parameters.

(c) Evaluate the compliance relationship assuming (i) adiabatic or (ii) isothermal conditions. The initial pressure of the gas, p_0, can be taken as given. The resistance can be presumed to be a known function of the position of the cable.

SOLUTION TO GUIDED PROBLEM
Guided Problem 5.7

1.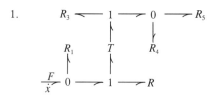

2. The bond graph above is tree-like. However, should you draw the electric circuit on the right below, the formal procedure gives a mesh that is best replaced by an equivalent tree-like graph.

3.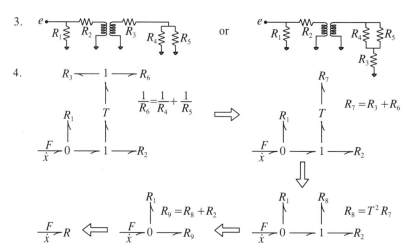

4.

5. This is left to the student, but the net result should be the same:

$$R = \frac{R_1 R_9}{R_1 + R_9} = \frac{R_1\{R_2 + T^2[R_3 + R_4 R_5 /(R_4 + R_5)]\}}{R_1 + R_2 + T^2[R_3 + R_4 R_5 /(R_4 + R_5)]}$$

5.4 Equilibrium

An equilibrium state of a model is called stable if it is reached as a steady-state after time-varying excitation ceases. A model with one or more unstable equilibria might never settle down to a stable equilibrium, even if one exists, but oscillate in what is known as a **limit-cycle oscillation**. The *location* of an equilibrium can be affected by compliances, but not by inertances. The *stabilities* of the equilibrium states, on the other hand, are apt to be strongly affected by both inertances and compliances.

This section first addresses the equilibria of models comprising only non-dynamic elements. It then proceeds to consider the effect of compliances. How inertances affect the stability of equilibria is deferred until the next chapter.

5.4.1 Reduction of Steady-State Models with a Single Source; Case Study

A steady-state model (having no energy-storage elements) with a single unspecified source or boundary bond can be reduced to a Thevenin or Norton equivalent load regardless of its original complexity. Such a reduction allows the location and stability of its equilbrium state(s) to be determined for any source characteristic. Analytical and graphical options for carrying out the reduction are illustrated below. The analytical approach implies the simultaneous solution of algebraic equations for the various elements of the model. It is the method of choice for linear models; the constancy of the moduli of the various sources, resistances, transformers and gyrators lead to efficient matrix methods. Analytical methods for dealing with complicated nonlinear models may need to be quite fancy, or may create confusion, particularly when more than one equilibrium exists. Graphical methods, on the other hand, may be simpler in practice, provide greater insight, and allow direct dealing with experimental data. Thus, even if you plan to employ an analytical method ultimately, a preliminary rough graphical analysis can be well worth the effort. Either procedure employs step-by-step coalescence of cascades, meshes and parallel combinations of elements, until a single source-load synthesis remains. Section 5.3 above describes most of the individual steps in a reduction.

As a case study, consider the system of dashpots given in Fig. 5.22. As a first step, you may choose to combine the resistances R_2 and R_3 to give R_5, as shown in part (c) of the figure. Alternatively, you may defer this coalescence, and proceed as shown in part (d) by adding a new 0-junction so as to isolate the mesh in preparation for its reduction. The mesh reductions are carried out with the aid of Fig. 5.19 (p 313). The resulting tree-like graphs are shown in parts (e) and (f); the latter is reduced to the former by the delayed combining of R_2 and R_3. The two resulting 0-junctions are combined in part (g). Next, R_1 and R_5 are combined to give R_6, as shown in part (h). Finally, R_4 and R_6 are combined to give the overall resistance, R, as shown in part (i). Note that the force is common to both boundary bonds. R is the resistance of an effective overall dashpot.

Carrying out the indicated steps is especially simple if the resistances are constants. Each of the pairs R_2, R_3 and R_4, R_6 have a common effort, so

$$\frac{1}{R_5} = \frac{1}{R_2} + \frac{1}{R_3}; \qquad \frac{1}{R} = \frac{1}{R_4} + \frac{1}{R_6}.$$

The pair R_1, R_5 has a common flow, so

$$R_6 = R_1 + R_5.$$

Nonlinear characteristics can complicate the algebraic approach considerably. You may prefer to substitute graphical methods. Example characteristics for the five original dashpots are plotted in parts (a), (b) and (c) of Fig. 5.23. Horizontal addition is employed in

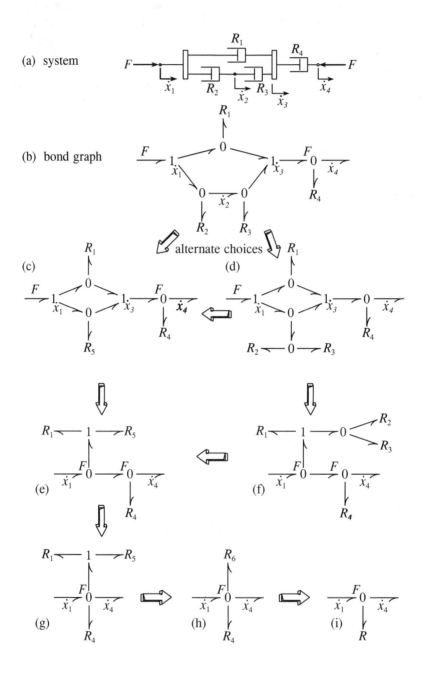

Figure 5.22: Determination of the equilibrium for a steady-state model

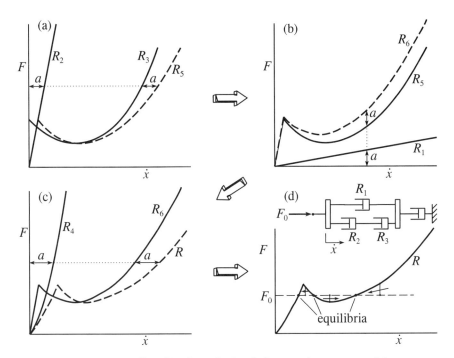

Figure 5.23: Graphical analysis of the steady-state model

part (a) to form R_5 from R_2 and R_3. Vertical addition is employed in part (b) to form R_6 from R_1 and R_5. Finally, horizontal addition is employed in part (c) to form R from R_4 and R_6.

The characteristic for the particular R_3 shown in the figure allows no motion until a threshold of force is exceeded; this is typical of dry friction. Then, the force *decreases* for a range of small velocities, before again increasing. This pattern is typical of some bearings and rubbing surfaces that become better lubricated as the sliding velocity increases. The resulting characteristic for the overall system also has a dip-then-rise. In part (d) of the figure, a constant force F_0 is applied to the left end of the system, and the right end is grounded. The source-load synthesis, shown graphically, reveals *three* equilibria. All are valid, but are they *stable*?

To answer this question it is necessary, as always, to consider non-equilibrium states. If the resistance force of the overall equivalent dashpot exceeds F_0, rapid leftward acceleration ensues, impeded only by the inertia of the parts. This case is illustrated by the states associated with the arrows on the left and on the right. If, on the other hand, the resistance force is less than F_0, rapid rightward acceleration ensues, illustrated by the states associated with the center arrow. This analysis reveals that the center equilibrium is unstable and the outer two equilibria are stable. Which of the two stable states comes into being depends on the history of the force and the initial velocity of the dashpot.

5.4.2 Alternative Approaches to Reducing Steady-State Models

Not all very complex systems can be reduced completely through use of the above procedures. Recourse can be made to iterative numerical procedures applied to the set of corresponding equations. In the special case of linear systems, non-iterative matrix methods can be used. Such approaches are the subject of textbooks and monographs on algebra

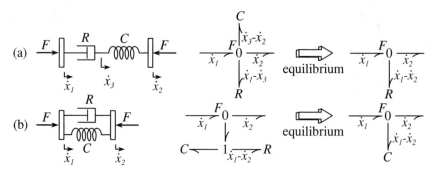

Figure 5.24: Removal of compliances and resistances for the equilibrium state

and numerical methods, and are not addressed here, in favor of a simpler alternative.

This alternative converts the static model to a dynamic or time-varying model. This is done by adding either real or fake compliances to 0-junctions, or adding inertances to 1-junctions. The revised model leads to differential equations, with time as the independent variable, in place of the original algebraic equations. Routine numerical simulation then applies. The solution usually settles down to a stable equilibrium state that is independent of the presence of the added compliances and inertances. If there are multiple equilibria, you may need to start with an initial condition somewhat close to the desired equilibrium. Occasionally, the solution may continue to oscillate rather than settle down to an equilibrium state. You should then change the assumed compliances or inertances until stability is achieved.

The availability of this option sharply reduces your need to become proficient in the more exotic methods available for reducing complex algebraic or steady-state systems. It is a by-product of the modeling and analysis of dynamic systems, which is the major subject of this book.

5.4.3   Removal of Elements for Equilibrium

Equilibrium implies that each bond has a time-invariant effort and flow. An inertance with a constant flow has zero effort. Whenever the objective is merely the determination of equilibria, therefore, all inertances may be removed from a bond graph. Further, if an inertance is bonded to a 0-junction, the effort on the junction becomes zero, so all other bonds attached to it also may be excised.

The situation is more complicated for compliances. A compliance at equilibrium has a time-invariant effort and therefore a constant displacement and zero flow. When both resistance and compliance elements are bonded to a common 0-junction, and therefore have a common effort, the flow of each compliance element is zero, whereas the flows on the resistance elements are not. The compliance elements therefore can be removed from the graph; this action leaves the flow state of the system unchanged. This situation is illustrated in Fig. 5.24 part (a). The force compressing the spring equals the total applied force. If this F is constant, a constant displacement $x_3 - x_2$ results, so $\dot{x}_3 = \dot{x}_2$. As a consequence, the power on the compliance bond is zero, and this bond may be removed from the graph.

In the situation illustrated in part (b) of the figure, on the other hand, both resistance and compliance elements are bonded to a common 1-junction. Since both elements have a common flow, this flow must be zero if the appled force is constant. Otherwise, the displacement of each compliance would increase without limit, requiring infinite force. Since the equilibrium flow is zero, the effort on each resistance is zero. Thus, removal of the resistances from the graph leaves the equilibrium state unchanged. Retention of the compliances

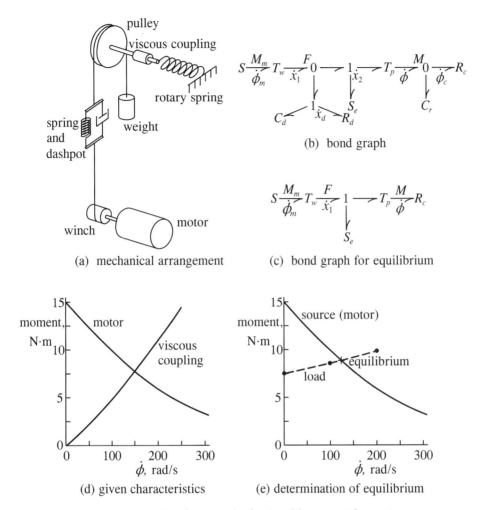

Figure 5.25: Case study for equilibrium with motion

then gives the proper displacements.

The above erasures, when combined with the other steps for bond-graph reduction, eliminate either all resistance or all compliance from all or major portions of the bond graph. If all compliance is gone, the resulting equilibrium involves a steady flow. If instead all resistance is gone, the resulting equilibrium is static. Problems regarding the deflection of static structures, for example, are of this latter type.

You usually can tell whether to delete compliances or resistances by appealing to physical reasoning. The examples of Fig. 5.24 are no exception.

5.4.4 Case Study with a Steady-Velocity Equilibrium

The objective is to find the equilibrium state of the motor-driven mechanical system shown in Fig. 5.25, disregarding the dynamics and assuming the weight does not interfere with the pulley before the equilibrium becomes meaningful. The model comprises some components with constant moduli and others with nonlinear characteristics. The winch has a radius of 0.05 m and the pulley has a radius of 0.10 m. The weight weighs 150 N. The motor and the viscous coupling have nonlinear characteristics, as plotted.

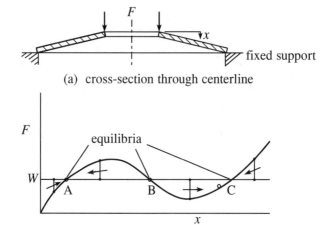

(a) cross-section through centerline

(b) force-deflection characteristic and stability
analysis for applied weight, W

Figure 5.26: Case study of Belleville springs with stable and unstable equilibria

The bond graph given in part (b) of the figure includes a compliance C_r and a resistance R_c bonded to a common 0-junction. When $M = $ constant, the compliance C_r has a fixed deflection (zero relative velocity), and therefore is removed from the graph. The compliance C_d and the resistance R_d are bonded to a common 1-junction, so when $F = $ constant the velocity $\dot{x}_d = 0$; as a result, R_d is removed. The compliance C_d now becomes bonded directly to the 0-junction, since there is no significance to a two-ported 1-junction. The velocity $\dot{x}_2$ equals $\dot{x}_1$, so C_d also can be removed. Part (c) of the figure shows the resulting reduced bond graph for equilibrium.

Given $W = 150$ N, $T_w = 0.05$ m and $T_p = 1/0.10 = 10$ m, the bond graph gives

$$\dot{\phi}_m = \frac{1}{T_w T_p}\dot{\phi} = 2.0\dot{\phi} \qquad (5.10a)$$

$$M_m = T_w F = T_w(W + T_p M) = 0.05(150 + 10M) = 7.5 + 0.5M \qquad (5.10b)$$

Values of the viscous coupling moment $M = 0$, 2.5 and 5.0 N m are now chosen, and the corresponding values of $\dot{\phi}_c$ read from the plot for the coupling. Equation (5.10a) gives the corresponding values of $\dot{\phi}_m$, and equation (5.10b) gives the corresponding values of M_m. The three resulting pairs of values of M_m, $\dot{\phi}_m$ are plotted as heavy dots in part (e) of the figure. Fitting a smooth curve through these points gives a "load" characteristic that intersects with the "source" (motor) characteristic to give a close estimate of the equilibrium state.

5.4.5   Case Study with Stable and Unstable Static Equilibria

Belleville springs are washer-like disks with a conical shape when they are unloaded, as drawn in Fig. 5.26. The force–deflection characteristic exhibits a rise-fall-rise shape as shown. Should a particular given weight W be placed on the spring, there are three equilibria. The stability of these equilibria can be determined, as usual, by considering off-equilibrium states. Examples of these, given in part (b) of the figure, indicate acceleration resisted only by inertia. The analysis reveals that equilibria A and C are stable, whereas equilibrium B is unstable. This is similar to the stability analyses carried out in Chapter 2 for constant-velocity equilibria.

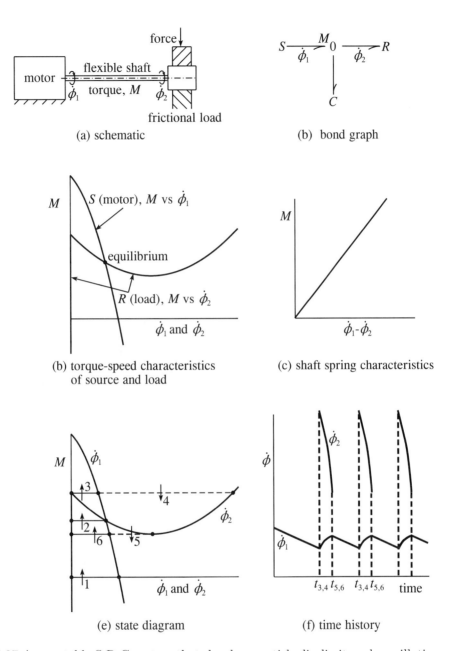

Figure 5.27 An unstable S-R-C system that develops a stick-slip limit-cycle oscillation

5.4.6 Case Study with Limit-Cycle Behavior

The gear-head electric motor in Fig. 5.27 is presumed to have the plotted torque-speed characteristic. The motor drives a load, represented by another torque-speed charactertistic, that is typical of a fluid-filled bearing heavily loaded in a transverse direction. The bearing friction drops as the speed increases from zero and a lubricant begins to act. For higher speeds the torque again increases as viscous drag takes over. The shaft connecting the motor to the bearing is presumed to have significant torsional flexibility, modeled as a rotational spring. The model of the system comprises a source for the motor, a resistance for the load, a compliance for the spring and a 0-junction to recognize that all three components experience the same torque.

The single equilibrium state is defined by the intersection of the source and load characteristics. This equilibrium is unstable! The spring allows the motor to run at a different speed from the load, but requires that the torques at its two ends be equal. The state of the system at a moment in time therefore can be described by a horizontal line segment, with the state of the motor at one end and the state of the load at the other. If the system is started from rest, it will initially jump to the pair of states labeled as 1 in part (e) of the figure. (All inertia of the rotor and gears is being neglected, so $\dot{\phi}_1$ jumps to the given value virtually instantly.) The velocity difference $\dot{\phi}_1 - \dot{\phi}_2$ now causes the twist angle $\phi_1 - \phi_2$ to increase in time. Thus the spring winds up, and the torque M increases. When $\dot{\phi}_1$ reaches its equilibrium value, at state 2, the load still hasn't budged; $\dot{\phi}_2$ is far from equilibrium. Thus the wind-up of the spring continues, at only a slightly smaller rate. When the moment attempts to exceed that of state 3, however, the load finally is forced to break away. It accelerates virtually instantaneously (no inertia, again) to state 4. At this point $\dot{\phi}_2 \gg \dot{\phi}_1$, so the spring rapidly unwinds and the torque decreases. It is still decreasing when it reaches state 5, at which point the load speed has no choice but to suddenly become zero, giving state 6. State 6 soon becomes state 2, and then state 3, and so on; the cycle is repeated.

The endless cycling from states 6 to 3 to 5 to 6 is known as a **limit cycle** oscillation. This particular type of limit cycle often is called a **stick-slip instability**. It is also known as a common type of "chatter." Should the system be started somehow at its equilibrium condition, any small disturbance will trigger the instability, and the same limit cycle results. In fact, any initial condition results almost immediately in the same limit cycle, as you can verify.

This problem is continued in Section 6.1.4, with the inevitable load inertia added. Computer simulation and analysis will show how this inertia modifies and, if large enough, eliminates the limit cycle.

EXAMPLE 5.16

An induction motor and load have the characteristics plotted below (which are the same as in Problem 2.4, p. 28.) The shaft and the load are connected by a *spring coupling* (rather than the rigid shaft of Problem 2.4). Determine the stability of the three equilibria, neglecting the effect of all inertias. Compare the result to the case in which the springiness is neglected and inertia assumed.

Solution: Model the system with a bond graph, taking care to get the right junction type:

The *moment* is common; moment is plotted on the vertical axis, so the vertical coordinates of the two states are the same, which means that a line segment drawn between the two states is horizontal. The arrows, which indicate the direction of change, are pointed upward if $C\,dM/dt = \dot{\phi}_M - \dot{\phi}_L > 0$; otherwise, they are downward. Equilibrium 1 is seen to be unstable, while both equilibria 2 and 3 are stable.

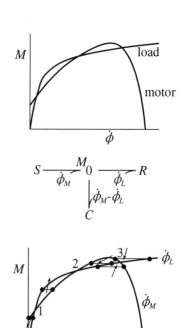

If the initial value of $\dot{\phi}_M$ is between that of equilibrium 1 and the peak of the curve, equilibrium 2 is approached. If it is to the right of the peak, equilibrium 3 is approached. If it is less than that of equilibrium 1, the system stalls altogether.

For inertia but no spring, the 0-junction is replaced by a 1-junction and the C element by an I element. The horizontal line segments are replaced by *vertical* line segments. These indicate that equilibria 1 and 3 are stable, whereas 2 is unstable.

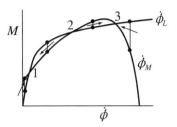

5.4.7 Necessary Condition for Instability or Limit-Cycle Oscillation*

The stability of an equilibrium point between a static source characteristic and a static load characteristic and the absence of a limit-cycle oscillation are assured if both characteristics are *passive with respect to the equilibrium point*. This passivity is defined in Section 5.3.2 (pp. 308, 310). Whether an equilibrium point that fails this test is actually unstable depends on the magnitudes of any compliances and inertances that may be present. Further, an equilibrium point could be stable in the presence of small impulsive disturbances but respond to a large impulsive disturbance with a limit-cycle oscillation. The tools necessary to make this determination analytically, without simulation, are developed in later chapters.

All points on the source characteristic in Fig. 5.27 are passive. On the other hand, the non-passivity of any point on the negatively sloped portion of the load characteristic can be observed readily. The graphical analysis presented with that example is sufficient to predict the limit cycle instability associated with any equilibrium point in this range. Points in the positively sloped portion of the load characteristic with moments below the break-away moment are non-passive, also, but the analysis reveals neither an instability nor a limit-cycle oscillation for an equilibrium in that range.

5.4.8 Summary

The equilibrium states for bond graphs that model constant velocity systems can be found analytically or graphically. When these methods become excessively awkward, the simultaneous solution of a set of nonlinear algebraic equations by iterative numerical means is a major alternative but is beyond the scope of this text. A more straightforward approach adds real or fake compliances and/or inertias to the model, leading to a solvable set of differential equations with time as the independent variable. The equilibrium state is approached asymptotically in time, assuming the model is stable, using numerical simulation (as introduced in Section 3.6).

The equilibria of models with both compliances and resistances can be found by eliminating resistance elements that have zero generalized force, and eliminating compliance elements with a non-zero generalized velocity. In some cases equilibria with steady velocity result. In others, static equilibria result.

An equilibrium state between source and load characteristics is known to be stable if both characteristics are passive at that state. Otherwise, determination of the stability requires a detailed examination of compatible non-equilibrium states. The presence of an unstable equilibrium may simply favor one or more stable equilibria. In other cases a limit cycle results.

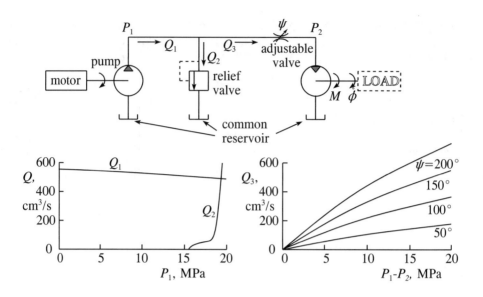

Figure 5.28 The system of Guided Problem 5.8

Guided Problem 5.8

This problem typifies graphical reduction of a model comprising steady-state elements.

Hydraulic oil is pumped to high pressure by a motor-driven pump. The fluid in turn drives a hydraulic motor with an unspecified attached load, as shown in Fig. 5.28. A relief valve limits the pressure, partly for safety, and an adjustable "meter-in" valve can be set to control the torque-speed characteristic seen by the load. (The drawing uses standard fluid-power symbols.) Characteristics of the motor/pump, relief valve and adjustable valve, including the attached fluid line, are plotted.

The motor has a displacement $D = 60$ cm/rad and suffers an internal leakage of 0.2 cm/s for each MPa of pressure drop across it. It also has a static or break-away friction of 60 Nm and an added viscous friction of 5.0 Nm/(rad/s).

You are asked to find and plot the torque-speed characteristics of the system as seen by the load, for various angular positions, ψ, of the adjustable valve.

Suggested Steps:

1. Construct a bond graph model, treating the motor/pump as a one-port source. Presumably you will include several junctions, several resistances and possibly a fixed-effort source. This is probably the most critical step in your solution. Use this graph to guide the steps below.

2. Combine the given characteristics of the motor/pump and the relief valve to plot a source characteristic pertaining to a location between the relief valve and the adjustable valve.

3. Combine the results of step 2 with the characteristics of the adjustable valve to plot a family of source characteristics, each for a different angle ψ, pertaining to a location between the adjustable valve and the hydraulic motor.

4. Modify the results of step 3 to refer to the flow which actually displaces the motor but does not leak past it.

5. Convert the characteristics plots of step 4 to get torque vs. angular velocity. Due to the simple transformer type of coupling, this can be done by merely replacing the scales of the ordinate and abcissas.

6. Correct the characteristics of step 5 to account for the static and viscous friction on the shaft. The result should be the desired family of source characteristics as seen by the load. Subsequent specification of a load characteristic and an angle ψ would give a specific operating condition.

PROBLEMS

5.61 Slightly compressible fluid enters a cylinder, shown below, pushing a piston against a linear dashpot with modulus R. The bond graphs ignore inertia.

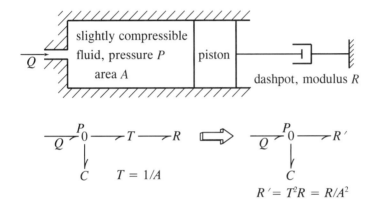

(a) Add the piston inertia to the bond graph. Determine the moduli of the elements in your graph in terms of expressed and implied parameters.

(b) Simplify the model as much as possible; find the moduli of any new elements.

(c) Give the simplest bond graph valid for equilibrium, and the associated algebraic relationship.

5.62 The bond graphs below have constant parameters except for the sources, which can be characterized by either of the known functions $e_s = f(\dot{q}_s)$ or $\dot{q}_s = g(e_s)$. In each case, determine the values of e_s and $\dot{q}_s$ at equilibrium in terms of the given parameters. *Continued in Problems 6.39 and 6.40.*

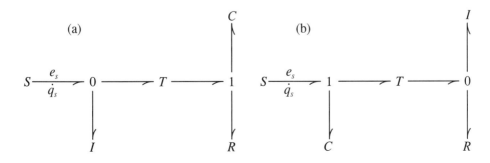

5.63 A motor drives a winch of radius 1.0 ft which both lifts a 50 lb weight and pulls a massless pulley of radius 3.0 ft with spring and damper attached, as shown below. Characteristics of the components are plotted.

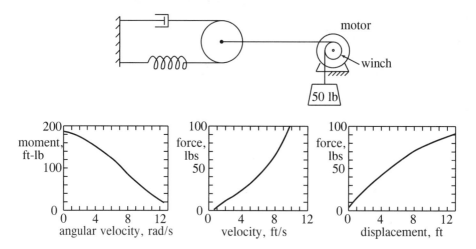

(a) Define appropriate variables, and identify and label the plots correspondingly.

(b) Draw a bond graph of the system, and label its variables consistent with part (a).

(c) Estimate the moduli of any elements not described by the plots.

(d) Find the equilibrium state of the system.

5.64 A gear motor (relatively slow speed) with torque-speed characteristic below rotates a winch which rolls a cylinder of weight $W = 10$ lbs. up a 30° incline at steady speed, without slipping, against a restraining dashpot with resistance coefficient $b = 0.25$ lb·s/in. The indicated radii are $r_w = 6$ in., $r_1 = 10$ in. and $r_2 = 5$ in. *Continued in Problem 6.36.*

(a) Define variables, draw a bond graph for the system, and find the values of the parameters (moduli) therein.

(b) Estimate the equilibrium angular velocity of the winch and linear velocity of the center of the roller.

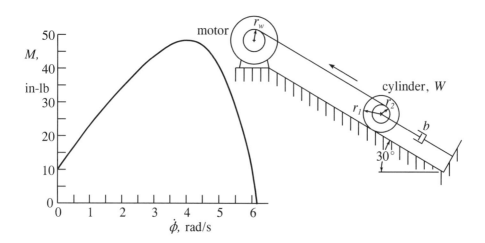

5.65 *Design problem.* A four-passenger vehicle is to be given an engine scaled to an engine with a known displacement, plus a manual transmission. Most parameters can be taken as given below. You are asked to size the engine and propose the values of between three and five transmission ratios. Factors such as vehicle loading, hill-climbing capability, acceleration and fuel economy are left to your discretion. The work can be divided into two phases:

(a) Model the system, presumably with a bond graph. Get explicit constants or functions that detail the meanings of the various one-port elements in the model. Some interpolation is necessary. Include hill-climbing and the possibility of acceleration. The result should give the vehicle speed and engine torque which results from any specified gear ratio, inclination of the road and vehicle acceleration.

(b) The design phase of the problem includes your compromises regarding acceleration, hill-climbing, fuel consumption and relative cost. In addition to stating the engine displacement and the various gear ratios, describe the performance of the vehicle in a form that an informed consumer would want and understand.

Vehicle weight: 2200 lb., exclusive of the engine and accessories.
Air drag coefficient: 0.35. Density of the air: 0.0024 lb·s^2/ft^4.
Frontal area: 15.4 ft^2.
Tires: 1032 rev/mile; constant rolling resistance force equals 0.65% of the total road weight.
Overall power train losses: 25% at 10 ft·lb engine torque; 15% at 20 ft·lb; 6% at 90 ft·lb.
Axle gear ratio: reduction of 2.6:1.
Engine: The weight, available torque, fuel consumption and needed power for the accessories are proportional to the engine displacement. An engine with a displacement of 2.0 liters weighs 240 lbs including accessories, and has the characteristics plotted below. The solid lines represent constant values of the brake-specific fuel consumption (BSFC), defined as pounds of fuel per brake horsepower hour. ("Brake" refers to the means of measurement.) The power to run the accessories (alternator, cooling fan, power steering and brakes, etc.) varies linearly from 2 HP at zero speed to 6 HP at 80 mph.

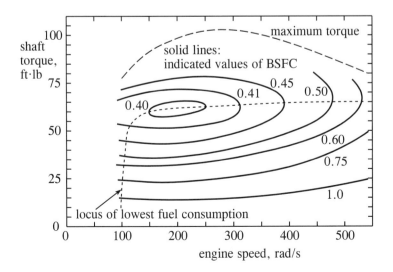

5.66 For the Belleville spring shown in Fig. 5.26 (p. 328), sketch the *stored energy* as a function of the displacement. Describe how the force is related to this plot.

5.67 Identify the portion of the load curve for the case study in Fig. 5.27 (p. 329) that is non-passive. Also, identify which portion of the load curve would be unstable or produce a limit-cycle oscillation should the source (motor) characteristic be modified to give an equilibrium there. Assume the motor characteristic is more negatively sloped than the load characteristic.

5.68 A submerged turbulent jet of water discharges into a region of still water, passing below a stand-pipe with a sharp edge or lip as shown below. The jet entrains a constant rate of fluid, Q_e, from the bottom of the stand pipe. Also, if the constant pressure of the still water, P_s, is higher than the pressure, P, at the bottom of the stand pipe (just above the jet), with a correction for the weight of the water in-between, the jet bends upward; if $P_s < P$, it bends downward. Considerable upward bending causes a return flow Q_r to be peeled off by the edge. If on the other hand the jet is not bent upward very much, a pressure difference $P_s - P$ causes a flow Q_p to enter or leave the stand-pipe region through the gap between the jet and the edge. These flows are plotted.

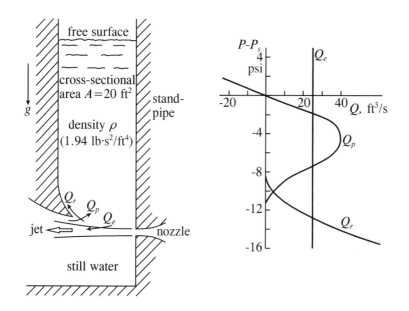

(a) Draw a bond graph for the system. The stand-pipe acts like a compliance, the sum of the flows $Q_r + Q_p$ acts like like a multivalued source of fluid to this compliance, and the flow Q_e acts like a (negative) flow source.

(b) Replace the individual characteristics for Q_r and Q_p with their sum, according to the bond graph model.

(c) Determine the equilibrium states. (Hint: there are three of them.)

(d) Determine the stability of each equilibrium state. (Postulate compatible non-equilibrium state pairs in the usual fashion, and identify which direction the pressure is induced to change as a result.)

5.69 For the problem above with its standpipe and fluid jet,

(a) linearize the characteristic for the upward flow for the region $-2.5 < P - P_s < 0$ psi.

(b) represent the dynamics of the system by a bond graph, allowing the linearized characteristic to substitute for the actual characteristic. Inertial effects in the standpipe may be neglected.

(c) the jet is suddenly turned on, following a quiescent state. Compute the transient and steady-state response of the level of the free surface, using the model of part (b).

5.70 A type of reverse check valve, constructed as shown below left, has the pressure-flow characteristic as plotted. The fluid is incompressible. This valve can be used as the heart of a "hydraulic ram" (see Problem 5.77, p. 340).

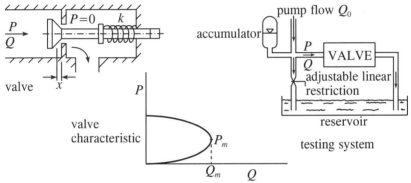

(a) It is desired to test this valve over as much of its characteristic as possible, without inducing an instability. A pump with constant flow Q_0 and the traditional accumulator is used with an adjustable linear restriction placed as shown above right. Choose a value of $Q_0 > Q_m$ and plot this point on the Q axis of the plot above. Then, construct two or three useful source characteristics, corresponding to different restrictions, on the plot. Deduce the region of the (load) valve characteristic that can be tested stably with your flow source.

(b) Remove the accumulator and repeat (a). (A tiny fluid inertance remains.)

(c) The orifice in the valve has area $2\pi r x/\sqrt{2}$, where r is a constant effective radius. The valve opening x equals x_0 when the spring (stiffness k) has no force. The fluid has density ρ. Find the characteristic of the valve in terms of the parameters given, and compute P_m and Q_m.

5.71 Consider the two systems (i) and (ii) below, in which the charactistics of the source and the resistance are plotted. In each case, and for equilibrium points 1 and 2, indicate whether the equilibrium is stable or unstable. Show your reasoning.

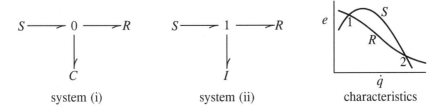

5.72 Do the preceding problem, substituting the characteristics opposite.

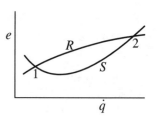

5.73 A motor drives a load through a *spring* coupling. Assuming the characteristics given below and neglecting all inertia, sketch-plot the angular velocities of the motor and the load as funtions of time. Hint: limit-cycle behavior exists for a wide range of starting conditions.

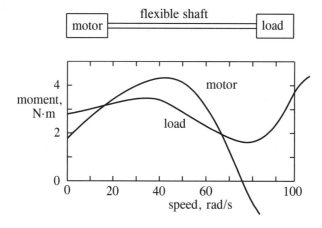

5.74 Axial-flow fans and compressors typically demonstrate an S-shaped characteristic. A fan with the characteristic plotted below supplies air to a chamber of volume $V = 4$ ft^3. An orifice in the chamber exhausts the air to the environment; two different possible characteristics are plotted. Assume atmospheric conditions. Determine the stability of the respective equilibria, and detail any limit-cycle operation that results. *Continued in Problem 6.17.*

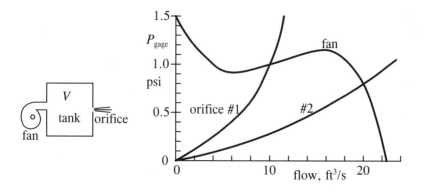

5.75 A motor with a given characteristic and negligible inertia drives a load with a given characteristic and very small inertia through a gear pair with negligible friction and inertia.

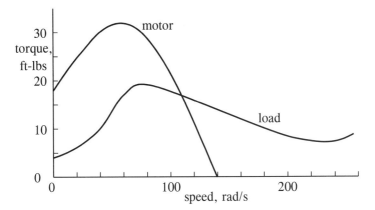

(a) Model the system with a bond graph, assuming that the coupling is quite rigid.

(b) Determine whether the equilibrium is stable when the gear ratio is 1.

(c) Estimate the gear ratio that gives the largest possible stable equilitrium speed, starting from rest.

(d) Model the system with a bond graph again, now assuming that the coupling is quite flexible.

(e) Answer (b) for the new model.

(f) Answer (c) for the new model.

5.76 A cylindrical workpiece is machined on a lathe by a cutting tool fed axially by a screw drive at a constant velocity $\dot{x}$, as shown on the next page. The tool is somewhat flexible, however, so its feed rate $\dot{y}$ at the cutting edge can differ from $\dot{x}$. The reaction force on the tool in the same direction, plotted as a function of $\dot{y}$, has a region of negative slope (due to heating and softening of the material). The concern is that strong vibration might result (tool "chatter").

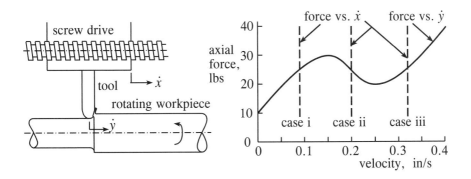

(a) Model the process with a bond graph, including only phenomena mentioned above. (For example, exclude inertias and any variation in the rotation rate of the workpiece.)

Hint: Your model should include but not be limited to a compliance, a resistance and a junction.

(b) Find the equilibrium drive forces for the three drive characteristics plotted by the dashed lines above: case (i), $\dot{x} = 0.09$ in/s; case(ii), $\dot{x} = 0.20$ in/s; case (iii), $\dot{x} = 0.32$ in/s.

(c) Determine whether each of the three equilibria is stable. Show your reasoning. Also, should any limit cycle exist, show how the force and speed of the workpiece change.

(d) If you were to add one more phenomenon or parameter to the model, what would it be? (There are several good answers.)

5.77 A penstock such as in Problem 4.70 (p. 232), but smaller, is terminated in a reverse check valve such as given in Problem 5.70 (p. 337) to form what is known as a *hydraulic ram*. A drawing is shown below. Water accelerates through the long tube until its velocity is so great that the valve slams shut, whereupon the pressure rises and opens a simple check valve to a load tank with a constant high pressure P_L, pumping water to that tank. After the flow has stopped, the check valve closes to prevent downward flow, the reverse check valve opens automatically, and the cycle repeats. Water passing through the reverse check valve is dumped into a stream at that level. The device is attractive when no electrical power is available and a cheap, simple system is preferred.

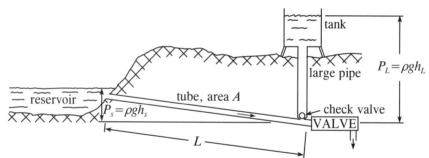

Find the quantity of water pumped to the load tank in one cycle, neglecting losses to make the problem relatively simple, in terms of the parameters given above and in Problem 4.70. Also, find the cycle time.

5.78 Review question: Make a table of the dimensions of the moduli of bond graph elements. Label five rows as follows: R, C, I, T, G. Label four columns as follows: mechanical translation, mechanical rotation, incompressible fluid flow, electric circuits. Assume the transformers and gyrators involve no transduction. Employ the symbols L for length, F for force, t for time and Q for electric charge, as given in Chapter 1.

5.79 Answer the previous question using M for mass and eliminating F for force.

5.80 Answer the previous question substituting SI units for the dimensions. Use Newtons, meters, seconds and coulombs only.

5.81 Answer the previous question using kilograms but not Newtons.

5.82 Answer the above question in terms of inches, pounds force and seconds. Omit the electric circuit column.

SOLUTION TO GUIDED PROBLEM

Guided Problem 5.8

1.

$$MOTOR/PUMP \xrightarrow[Q_1]{P_1} 0 \longrightarrow 1 \xrightarrow[Q_3=Q_1-Q_2]{P_2} 0 \xrightarrow[Q_m]{} T \xrightarrow{M_i} 1 \xrightarrow{M} LOAD$$

R_r: resistance to flow of the relief valve.

R_s: resistance to flow of the adjustable valve.

R_L: resistance of the hydraulic motor to leakage flow, 0.2 cm^3/(s MPa).

R_f: viscous frictional torque resistance on the shaft, 5.0 N m/(rad/s).

S_e: break-away frictional torque, 60 N m.

2.

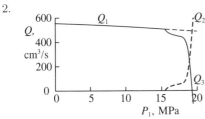

3. $P_2 = P - 1 - (P_1 - P_2)$. Therefore, subtract the (horizontal) values of $P_1 - P_2$ given in the plot for R_a from those of P_1 above, to get

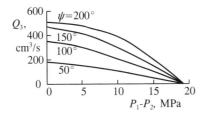

4. Subtract $Q_L = P/R_L = P/0.2$ from Q_3 to get Q_m:

5. $\dfrac{\dot{\phi}}{Q_m} = T = \dfrac{1}{D} = \dfrac{1}{60}\dfrac{\text{rad/s}}{\text{cm}^3/\text{s}}$; therefore $\dot{\phi} = Q_m/60$

and ideal torque $M_i = 60P_2\dfrac{\text{cm}^3 \cdot \text{MN}}{\text{m}^2} = 60 \times 10^{-6}P_2 \times 10^6 \dfrac{\text{m}^3}{\text{m}^2}\dfrac{\text{MN}}{\text{MN}}\dfrac{\text{N}}{\text{}} = 60P_2 \text{ N·m}$

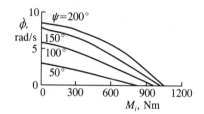

6. $M = M_i - S_e - R_f \dot{\phi} = M_i - 60 - 5.0\dot{\phi}$ (Subtract torque loss from plot above.)

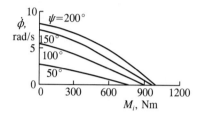

Chapter 6

Mathematical Formulation From Bond Graphs

The writing of differential equations from bond graphs is extended in this chapter from the simple cases given in Chapter 3 to the much broader class of models considered in Chapter 5. Models are categorized according to their causal status, and whether they are linear or nonlinear. Section 6.1 focuses on the writing of differential equations for the simplest causal classification, and Section 6.2 focuses on more difficult cases.

Section 6.3 introduces the optional use of the loop rule, developed from the theory of signal flow graphs, which allows transfer functions for linear models with constant coefficients to be found from bond graphs with great efficiency. The block diagrams seen earlier are a form of signal flow graph.

6.1 Causality and Differential Equations

The application of causal strokes to bond graphs leads to a categorization of models according to the types of equations that they produce. After these categories are identified, the discussion focuses on the basic category of "causal models," for which a set of state-variable differential equations suitable for analysis or simulation is produced directly.

6.1.1 Applying Causal Strokes

Certain steps must be followed in applying causal strokes to every bond of a bond graph to assure that a subsequent procedure gives a proper set of state differential equations. In most cases the following steps are sufficient:

1. Apply the mandatory causal strokes for each effort or flow source (but not general source) that is present.

2. Apply any mandatory causal strokes that follow from the presence of 0- and 1-junctions, transformers and gyrators.

3. Apply integral causality to the bond of one of the remaining unspecified compliance or inertance elements.

4. Repeat steps 2 and 3 as many times as possible.

343

This procedure produces one of the following outcomes:

1. Every bond has assigned causality, and each compliance and inertance element is assigned integral causality. These models are called **causal**. They are the easiest to treat.

2. Every bond has assigned causality, but one or more compliance or inertance elements is assigned differential causality. These models are called **over-causal**. They can be treated directly, or reformulated to become causal models.

3. Some bonds are not assigned any causality. These models are called **under-causal**. They are more difficult to treat than causal models.

6.1.2   Differential Equations for Causal Models

Once causal strokes are placed on every bond, a four-step procedure produces a complete set of state-variable differential equations. The procedure is a modest generalization of that introduced in Section 3.4. Before you start you should remove any variable designations that may already be written on the bond graph or, since these designations may still be valuable to you, start with a fresh unannotated rendering of the graph.

(I) Annotate the input effort on each effort source and the input flow on each flow source with some appropriate symbol. Also, if the conjugate variable is of no interest, you may place a large cross ($\times$) in the appropriate place to register your disinterest.

(II) Annotate the flow side of the bond of each element C_i with $\dot{q}_i$, and place a circle around this designation. Annotate the effort side of the bond with q_i/C_i or, if nonlinear, a functional designation $e_i(q_i)$. Similarly, annotate the effort side of the bond of each element I_i with $\dot{p}_i$, and place a circle around this designation. Annotate the flow side of the bond with p_i/I_i or, if nonlinear, a functional designation $\dot{q}_i(e_i)$. It is better if you use more specific symbols than the generic q_i and p_i to represent the respective generalized displacements and momenta, which become the state variables.

(III) That part of the bond graph that remains without effort and flow annotations has, as causal inputs, only the input terms from step I and the terms q_i/C_i and p_i/I_i (or their substitutes). Propagate these variable designations throughout the bond graph, *following the dictates of the causal strokes* and using the power-convention half-arrows to determine signs. This key procedure is highlighted in the examples below. Note that the circled terms are "bottled up" and do not migrate into the graph.

(IV) Write one first-order differential equation for each circled $\dot{q}_i$ and $\dot{p}_i$. The circled term appears alone as the left side of the differential equation. The right side of the equation comprises functions of the input and state variables as dictated by the annotations and power-convention half-arrows on the bonds in the immediate vicinity of the circled term. For example, if the circled term is bonded to a junction, the right side comprises the sum of the properly signed causal inputs of the proper type (effort or flow) of the other bonds on that junction.

6.1.3   Case Study: A Linear Circuit

Consider the electric circuit shown in Fig. 6.1. The bond graph has two junctions. It is driven by an effort (voltage) source, which demands the causal stroke as shown in part (b)

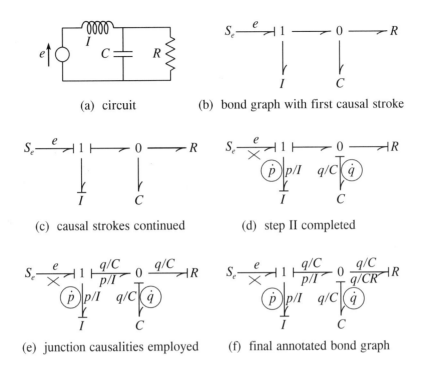

(a) circuit   (b) bond graph with first causal stroke

(c) causal strokes continued   (d) step II completed

(e) junction causalities employed   (f) final annotated bond graph

Figure 6.1: Electric circuit example

of the figure. Next, integral causality is placed on the inertance (inductance) bond. This in turn forces the causality on the bond to the right side of the 1-junction, as shown in part (c). Integral causality now can and must be placed on the compliance bond, forcing admittance causality on the resistance bond, as shown in part (d). This completes the assignment of causal strokes, and the model is seen to be causal, since both energy-storage elements have integral causality. Part (d) of the figure also shows the application of steps I and II for the writing of differential equations; the input effort is designated as e, the associated flow is marked with a $\times$, and the efforts and flows on the bonds for the C and I elements are designated as directed. Step III starts by observing that the causal stokes around the 1-junction dictate the labeling of the flows on the left and center horizontal bonds as p/I, and the causal strokes around the 0-junction dictate that the efforts on the center and right bonds be labelled with q/C, as shown in part (e) of the figure. This leaves only the flow on the R bond unannotated. The presence of impedance causality for the R element forces this effort to be written as q/RC, as shown in part (f).

The first-order state differential equations now can be written (step IV). The circled term dp/dt is written on the left side of one of the differential equations, and the causal inputs of the attached 1-junction provide the terms on the right side:

$$\frac{dp}{dt} = e - \frac{1}{C}q. \tag{6.1}$$

The circled term dq/dt is written on the left side of the other differential equation, and the causal inputs of the attached 0-junction provide the terms on its right side:

$$\frac{dq}{dt} = \frac{1}{I}p - \frac{1}{RC}q, \tag{6.2}$$

The first term on the right side of this equation is positive because the power convention half-arrow on the center horizontal bond is directed toward the junction. The second term is negative because the power convention half-arrow on the R bond is oppositely directed.

There are two dependent variables (p and q) and the same number of first-order differential equations. As a result, the equations are solvable as a set. This is always the outcome when the procedures are applied to a causal model.

Taking the derivative of equation (6.2), substituting the right side of equation (6.1) for the resulting term dp/dt and rearranging, gives

$$IC\frac{d^2q}{dt^2} + \frac{I}{R}\frac{dq}{dt} + q = Ce. \tag{6.3}$$

This is a classical second-order linear system of the form of equation (4.103) (p. 220). Comparing the coefficients, the natural frequency is seen to be $\omega_n = 1/\sqrt{IC}$ and the damping ratio is seen to be $(R/2)\sqrt{C/I}$. The steady-state solution, which emerges when the two derivative terms in the differential equation are set equal to zero, is $q = Ce$.

EXAMPLE 6.1

A machine of mass m is isolated from its foundation by a spring and a dashpot, as shown below (and in Problems 3.30 (p. 110), 3.46 (p. 126) and 4.54 (p. 215)). In the present case, the foundation itself has a known displacement $z_i = z_i(t)$, and the resulting displacement of the machine, $z_o(t)$, is to be found. The bond graph below has been drawn as the first step in the analysis. As the second step, define state variables, find a set of state differential equations, and relate z_o to the state variables.

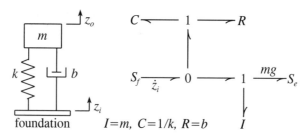

Solution: The causal strokes are drawn following the four steps given in Section 6.1.1, starting with the mandatory strokes on the bonds for S_f and S_e. Integral causalities on the bonds for C and I produce no causal conflicts at the junctions, and therefore also are required. Since there is no choice in any of the causal stokes, the model is *causal*. Steps I and II of the procedure for writing the differential equations, as given in Section 6.1.2, also are shown below as annotations on the bonds for S_f, S_e, C and I:

The output variable is related to the state variables p and q by $z_i = \int \dot{z}_i \, dt = (1/I) \int p \, dt$. As shown at the top of the next page, the lower 1-junction and its causal strokes force the writing of p/I for the flow on the bond to the left of the junction. The causal stokes on the bonds for the 0-junction mandate that the flow on the vertical bond be written as $\dot{z}_i - p/I$, which also becomes the flow on the bond for R. The effort on the bond for R becomes $R(\dot{z}_i - p/I)$, and finally the effort on the vertical bond and the bond to the right of the 0-junction becomes $q/C + R(\dot{z}_i - p/I)$.

$$C \xrightarrow{q/C} 1 \xrightarrow[\dot{z}_i - p/I]{R(\dot{z}_i - p/I)} R$$

$$q/C + R(\dot{z}_i - p/I) \mid \dot{z}_i - p/I$$

$$S_f \xrightarrow[\dot{z}_i]{\times} 0 \xrightarrow[p/I]{q/C + R(\dot{z}_i - p/I)} 1 \xrightarrow{mg} S_e$$

$$\overset{\cdot}{(p)} \Big\downarrow p/I$$

$$I$$

The two differential equations result from the respective causal strokes and annotations on the bonds for the respective 1-junctions:

$$\frac{dq}{dt} = \dot{z}_i(t) - \frac{1}{I}p$$

$$\frac{dp}{dt} = \frac{1}{C}q - \frac{R}{I}p + R\dot{z}_i(t) - mg.$$

6.1.4 Case Study: Nonlinear Stick-Slip

The stick-slip system analyzed in Section 5.4.6 (p. 329) is now revisited, introducing the effect of the rotational inertance of the load that was neglected in Fig. 5.27. The system is re-drawn in part (a) of Fig. 6.2, and the given nonlinear torque-speed characteristics of the motor and the resistance (frictional) load are reproduced in part(b). The bond-graph model shown in part (c) of the figure is the same as given before, except for the added inertia, which also requires the use of a 1-junction to register the fact that the load resistance and load inertance refer to the same angular velocity.

The model is now too complex for its behavior to be deduced directly; differential equations are needed. The source S is neither an effort source nor a flow source, so step 1 of the procedure for drawing causal strokes (p. 343) does not apply. In step 2, integral causality is applied to the compliance, as shown in part (d) of the figure. Step 3 recognizes the causal mandate of the zero junction, as shown in part (e). No such mandate exists for the 1-junction, however, so one proceeds to step 4 and applies integral causality to the inertance, as shown in part (f). Finally, the causal mandate of the 1-junction dictates the causal stroke on the final bond, leaving the bond graph of part (g).

Step I of the procedure for writing differential equations (p. 344) does not apply. Step II annotates both the efforts and the flows of the energy-storage elements; it is included in part (g) of the figure. In step III, the causal stroke on the compliance bond now dictates the annotation of the efforts of the left-hand and center bonds, and the causal stroke on the inertance bond dictates the annotation of the flows of the center and right-hand bonds. For convenience, the effort on the source is also labelled as M_S, and the flow on the resistance is also labeled as $\dot{\phi}_R$, to give the situation shown in part (h) of the figure. Finally, the admittance causality of the nonlinear source gives a function in the form $\dot{\phi}_S(M_S)$ for its flow, and the impedance causality of the resistance gives a function in the form $M_R(\dot{\phi}_R)$ for its effort. All the efforts and bonds are now annotated, as shown in part (i) of the figure.

The writing of the differential equations (step IV) now can proceed. The circled term $\dot{\phi}_C$ is set equal to the sum of the efforts on the 0-junction:

$$\frac{d\phi_C}{dt} = \dot{\phi}_S(M_S) - \frac{1}{I}p; \quad \dot{\phi}_S = \dot{\phi}_S(M_S); \quad M_S = \phi_C/C. \tag{6.4}$$

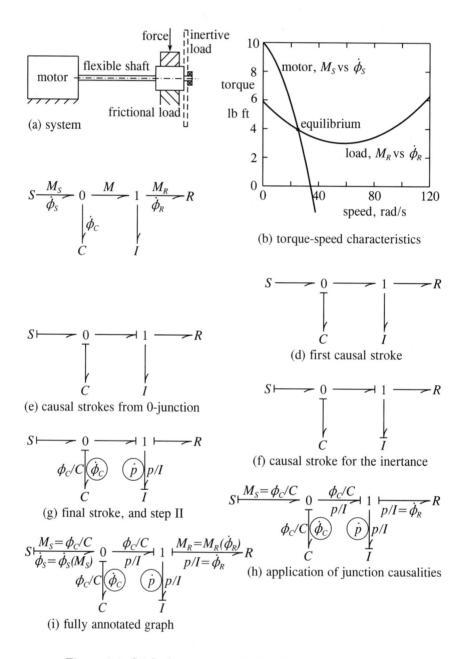

Figure 6.2: Stick-slip system with load inertia included

Similarly, the circled term $\dot{p}$ is set equal to the sum of the flows on the 1-junction:

$$\frac{dp}{dt} = \phi_C/C - M_R; \quad M_R = M_R(\dot{\phi}_R); \quad \dot{\phi}_R = \frac{p}{I}. \tag{6.5}$$

These equations are solvable, presuming the functions $\dot{\phi}_S(M_S)$ and $M_R(\dot{\phi}_R)$ are known, since the only variables are the state variables ϕ_C and p. They are not readily solvable analytically, however, due to the nonlinearities of these functions.

The source and load torque-speed characteristics, as plotted, are given by

$$M_S = M_{S0} - M_{S1}\dot{\phi}_s - M_{S2}\dot{\phi}_s^2, \tag{6.6a}$$

$$M_R = \text{sign}(p)M_{R0} - M_{R1}|p/I| + M_{R2}(p/I)^2; \quad p/I = \dot{\phi}_R, \tag{6.6b}$$

$$M_{S0} = 10.0 \text{ ft·lb} \quad M_{R0} = 5.9 \text{ ft·lb} \quad M_{S1} = M_{R1} = 0.10 \text{ ft·lb·s}$$

$$M_{S2} = 0.0056 \text{ ft·lb·s}^2; \quad M_{R2} = 0.00086 \text{ ft·lb·s}^2; \quad C = 0.1 \text{ rad}/(\text{ft·lb}). \tag{6.6c}$$

Noting that $M_S = M = \phi_C/C$, equation (6.6a) can be solved for $\dot{\phi}_S$:

$$\dot{\phi}_S = -\frac{M_{S1}}{2M_{S2}} + \sqrt{\left(\frac{M_{S1}}{2M_{S2}}\right)^2 + \frac{M_{S0}}{M_{S2}} - \frac{\phi_C}{CM_{S2}}}. \tag{6.7}$$

Substitution of equation (6.7) into equation (6.4) and equation (6.6b) into equation (6.5) gives an integrable state-space formulation. A corresponding MATLAB function file follows. The particular value of I given ($I = 0.00001$ lb-ft s^2) corresponds to virtually negligible inertia.

```
function f=stkslp(t,x)
MS0=10; MR0=5.9; MS1=0.1; MR1=0.1; MS2=0.0056; MR2=0.00086;
C=0.1; I=0.00001;
f(1)=-MS1/(2*MS2)+sqrt(MS1/(2*MS2))^2+MS0/MS2-x(1)/(C*MS2))
-x(2)/I;
f(2)=x(1)/C-sign(x(2))*MR0+MR1*abs(x(2)/I)-MR2*(x(2)/I)^2;
```

The model can be simulated and the output shaft speed $\dot{\phi}_C = p/I$ plotted for 0.1 second, from a resting start, with the following commands:

```
[t,x]=ode23('stkslp',0,0.1,[0 0]);
plot(t,x(:,2/.00001))
```

The result, the first plot of Fig. 6.3, reveals a limit-cycle behavior almost the same as deduced graphically in Fig. 5.27, but with a very small effect of the inertia. The corresponding result for $I = .0010$ ft-lb s^2, shown next, reveals a lower limit-cycle frequency and a reduced fraction of time spent sticking. When the inertia is increased to $I = 0.0021$ ft-lb s^2, the sticking portion is seen to have virtually vanished. When the inertia is increased marginally to 0.0022 ft-lb s^2, however, the simulation reveals that the shaft speed converges toward the equilibrium speed, which is

$$\dot{\phi}_{equib} = \sqrt{\frac{M_{S0} - M_{R0}}{M_{R2} + M_{S2}}} = 25.19 \text{ rad/s}. \tag{6.8}$$

Thus the instability has abruptly given away to stability. A further increase in inertia increases the decay rate for oscillations without any effect on $\dot{\phi}_{equib}$. The example of $I = 0.01$ ft-lb s^2 shows little oscillation.

It is instructive to plot one of the state variables of a second-order model versus the other. This is called a **phase-plane plot**. In Fig. 6.4, phase-plane plots of the torque

Plots of load speed vs time:

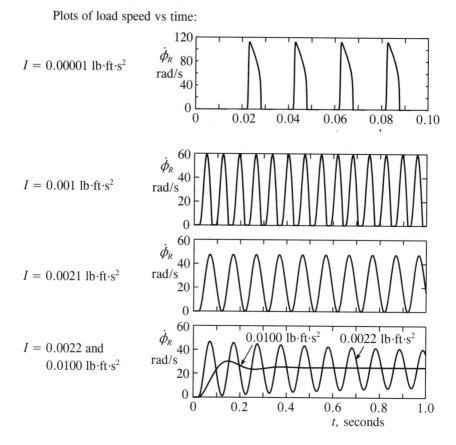

Figure 6.3: Simulation results for stick-slip model

$M_S = \phi_C/C$ versus the angular velocity p/I are given for the various cited cases. Time is supressed in these plots, but the directions of increasing time of the **trajectories** is indicated by arrows. The underlying source and load torque-speed curves are shown by dashed lines.

The criterion for stability can be found without the trial-and-error of simulations by linearizing the differential equations. The linearization of the differential equation (6.4) with the algebraic equation (6.7) substituted, and differential equation (6.5) with the algebraic equation (6.6b) substituted, is

$$\frac{d\phi_C^*}{dt} = \frac{1}{2}\frac{-(1/CM_{S2})\phi_C^*}{\sqrt{(M_{S1}/2M_{S2})^2 + M_{S0}/M_{S2} - \overline{\phi_C}/CM_{S2}}} - \frac{1}{I}p*, \tag{6.9a}$$

$$\frac{dp^*}{dt} = \frac{1}{C}\phi_C^* + \frac{M_{R1}}{I}p^* - \frac{2M_{R2}\dot{\phi}_{equib}}{I}p^*. \tag{6.9b}$$

Substituting all the given parameter values except for the intertance I gives

$$\frac{d\phi_C^*}{dt} = -26.17\phi_C^* - (1/I)p^*, \tag{6.10a}$$

$$\frac{dp^*}{dt} = 10\phi_C^* + (0.0567/I)p^*. \tag{6.10b}$$

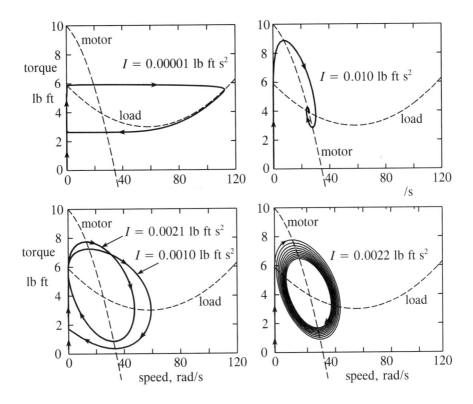

Figure 6.4: Phase-plane trajectories for the stick-slip system

or, in operator notation,

$$(S + 26.17)\phi_c^* + (1/I)p^* = 0, \tag{6.11a}$$
$$(S - 0.0517/I)p^* - 10\phi_C^* = 0. \tag{6.11b}$$

Solving one of these for either ϕ_C^* or p^* and substituting into the other gives the characteristic equation

$$S^2 + (26.17 - 1/I)S + (10 - 26.7 \times 0.0567/I) = 0. \tag{6.12}$$

The coefficient of the first term in parentheses equals $2\zeta\omega_n$, which is positive for positive damping and negative when the system is unstable. The criterion for stability therefore is that this coefficient be non-negative, or

$$I \geq 0.0567/26.17 = 0.00217 \text{ in} \cdot \text{lb} \cdot \text{s}^2, \tag{6.13}$$

which is indeed between the values of 0.0021 and 0.0022 as determined by the simulations. Finally, the second term of equation (6.12) in parentheses equals ω_n^2, so the frequency of the oscillation with zero damping is $\sqrt{10 - 26.17 \times 0.0517/0.00217} = 62.6$ rad/s $= 9.97$ Hz, which also agrees with the simulation plots.

EXAMPLE 6.2

Repeat Example 6.1 (p. 346-347), substituting the nonlinear constitutive equations $e_C = a\sqrt{q_C}$ for the compliance and $e_R = b\dot{q}_R^2$ for the resistance.

Solution: The only differences from the solution given for Example 6.1 are in the notations for the efforts on the bonds for C and R:

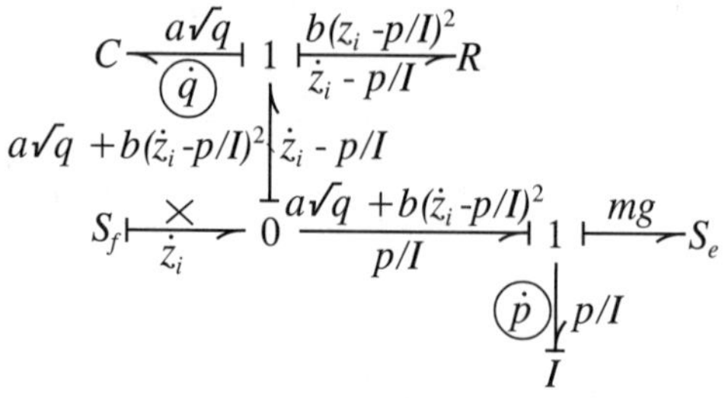

$$C \xleftarrow[\textcircled{q}]{a\sqrt{q}} 1 \xrightarrow[\dot{z}_i - p/I]{\frac{b(\dot{z}_i - p/I)^2}{\dot{z}_i - p/I}} R$$

$$a\sqrt{q} + b(\dot{z}_i - p/I)^2 \Big| \dot{z}_i - p/I$$

$$S_f \xrightarrow[\dot{z}_i]{\times} 0 \xrightarrow[p/I]{\frac{a\sqrt{q} + b(\dot{z}_i - p/I)^2}{p/I}} 1 \xrightarrow{mg} S_e$$

$$\textcircled{$\dot{p}$} \Big\downarrow p/I$$
$$I$$

The differential equations therefore become

$$\frac{dq}{dt} = -\frac{1}{I}p - \dot{z}_i \quad (equation \ unchanged)$$

$$\frac{dp}{dt} = a\sqrt{q} - b\left(\dot{z}_i(t) - \frac{p}{I}\right)^2.$$

6.1.5   Case Study with Transformers and Gyrators

The causal strokes on the two bonds of a transformer must be aligned the same, that is be asymmetrical with respect to the element, as shown in Fig. 6.5. This follows directly from the definition of the transformer; specifying the flow on one side also specifies the flow on the other side, and similarly for the efforts. The causal strokes on the two bonds of a gyrator must be aligned oppositely, that is be symmetrical with respect to the element, also as shown. Either both flows must be specified and the two efforts follow directly, or vice-versa. These causal constraints direct the propagation of causal strokes through transformers and gyrators in bond graphs.

The electromechanical system shown in Fig. 6.6 illustrates the use of causal strokes with transformers and gyrators. A DC motor with a resistance in the armature circuit and a fixed field drives a flexible shaft through a gear reduction. The shaft in turn drives a flywheel. The non-idealites of the motor are ignored, along with the inertia of all parts except the relatively massive flywheel. The springiness of the shaft is assumed to be important, however, and is represented by a compliance. The resulting bond graph model is shown in part (b) of the figure.

Placing integral causality on the compliance determines the causalities of all except the two right-most bonds. You should carefully see that this happens, step-by-step, starting with the two horizontal bonds off the 0-junction. Placing integral causality on the inertance bond determines the final two causalities. Step I of the procedure for writing differential

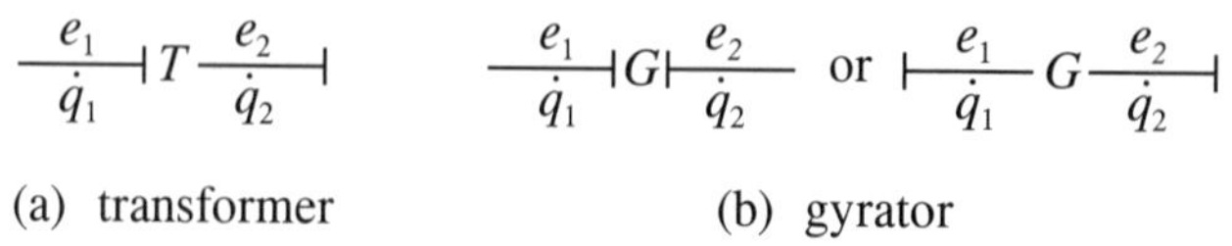

$$\frac{e_1}{\dot{q}_1} \dashv T \frac{e_2}{\dot{q}_2} \dashv \qquad \frac{e_1}{\dot{q}_1} \dashv G \vdash \frac{e_2}{\dot{q}_2} \ \text{ or } \ \vdash \frac{e_1}{\dot{q}_1} G \frac{e_2}{\dot{q}_2} \dashv$$

(a) transformer   (b) gyrator

Figure 6.5: Causal constraints for bonds on transformers and gyrators

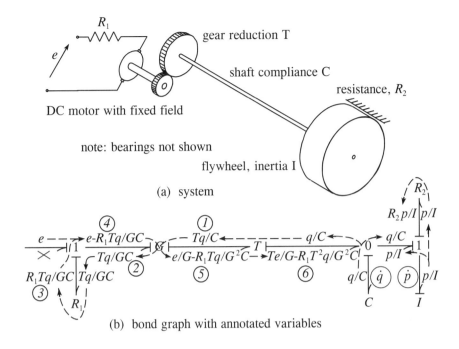

Figure 6.6: DC motor driving a flywheel through a flexible shaft

equations (p. 344) places the effort e on the left-hand bond, and is completed by placing an $\times$ on the effort side of the bond for the flow source. Step II places the standard notations $\dot{q}$ and q/C on the compliance bond, and $\dot{p}$ and p/I on the inertance bond.

Step III, annotating the efforts and flows on all the bonds according to the dictates of the causal strokes, demands particular attention. First, the effort on all the bonds of the 0-junction are labeled as q/C, since this is the causal input to that junction. Similarly, the flow on all the bonds of the rightmost 1-junction are labeled as p/I, since this is the causal input to that junction. The resistance R_2 has one of these flows as its causal input; its causal output must be labeled as $R_2 p/I$, consistent with its impedance causality. Dashed lines with arrows have been placed on the bond graph to underscore the meanings of the causal strokes and the consequent sequence of the determinations of the variables written next to each bond. You may find this practice helps you handle confusing situations.

You now proceed to label the remaining efforts and flows, following a sequence dictated by the causal strokes. The effort ① is a causal output from the transformer; it equals T times the causal input effort on the right side of the transformer. The flow ② then follows from the causal strokes on the gyrator. This flow is the causal input to the left-hand 1-junction, and therefore is written also on the bond for R_1. The resistance R_1 has impedance causality, like R_2; its causal output is the effort ③. The effort ④ is the causal output effort of the 1-junction; it equals the difference between the effort e and the effort ③. Note that the power convention half-arrows determine the *signs* of the terms, while the causal strokes direct their content. Next, the causal strokes on the gyrator bonds dictate the flow ⑤. Finally, the transformer converts this to the flow ⑥.

The differential equations for the system now can be written upon inspection of the annotated bond graph (step IV). The circled $\dot{q}$ is seen, from the causal strokes and power-convention half-arrows around the 0-junction, to equal the flow entering the 0-junction from the left minus the flow exiting to the right, or

$$\frac{dq}{dt} = \frac{T}{G}\left(e - \frac{R_1 T}{GC}q\right) - \frac{1}{I}p. \tag{6.14}$$

The circled $\dot{p}$ is similarly seen, from the causal strokes and power-convention half-arrows around the right-most 1-junction, to equal the effort of the zero junction minus the effort of the resistance R_2, or

$$\frac{dp}{dt} = \frac{1}{C}q - \frac{R_2}{I}p. \tag{6.15}$$

These equations are solved for particular parameter values in Guided Problem 4.2 given in Section 4.1 (pp. 165-166, 168-169).

EXAMPLE 6.3

The bond graph model below comprises elements with constant moduli. Apply causal strokes, define state variables and write a set of state differential equations.

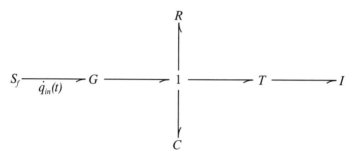

Solution: The causal stroke on the left-most bond is mandated by the flow source, and the gyrator forces the causality of the bond to its right, as given at the top the next page. The 1-junction allows integral causalities for the compliance and the inertance, which therefore must be chosen. Note the constraint on causalites imposed by the transformer. The state variables therefore are the displacement q on the compliance and the momentum p on the inertance; these are placed on the respective bonds, with dots to indicate differentiation and surrounding circles as flags for their significance. The conjugate variables for the I and C bonds also are shown.

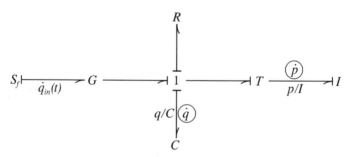

The causal strokes now dictate the writing of the remaining efforts and flows. As always, all uncircled annotations must include no variables other than the state variables and input variables. (Parameters such as I, R, G, etc. are constants in this problem, not variables.) The result is

$$S_f \xrightarrow[\dot{q}_{in}(t)]{Gp/TI} G \xrightarrow[p/TI]{I\dot{q}_{in}(t)} 1$$

R

$$Rp/TI \Big| p/TI$$

$$\underset{p/TI}{\overset{G\dot{q}_{in}(t)\text{-}Rp/TI\text{-}q/C}{\longrightarrow}} T \; \underset{p/I}{\overset{\textcircled{$\dot{p}$}}{\longrightarrow}} I$$

$$q/C\,\textcircled{$\dot{q}$}$$

C

from which the differential equations can be written directly:

$$\frac{dq}{dt} = \frac{1}{TI}p,$$
$$\frac{dp}{dt} = -\frac{1}{TC}q - \frac{R}{T^2 I}p + \frac{G}{T}q_0(t).$$

6.1.6   Models Reducible to Causal Form; Order of a Model

The causal constraints for effort and flow sources, junctions, transformers and gyrators sometimes prevent one or more compliances or inertances from being assigned integral causality. These over-causal models can be reduced to causal form, simplifying the subsequent determination of their differential equations, although sometimes this reduction is not practicable.

The energy stored in any element found to have differential causality is not independent of the energy stored in the elements given integral causality. Specifically, the displacements on the compliances with integral causality and the momenta on the inertances with integral causality not only determine the energy stored in their respective members. They also are sufficient, collectively, to determine the energy stored in the inertances and/or compliances with differential causality. The number of independent first-order differential equations, that is the order of the system, equals the number of energy storage elements with integral causality only. This assumes that the standard procedure which favors integral causality has been followed.

Consider two compliances bonded to a commmon 0-junction, as shown in Fig. 6.7. One of them must be assigned differential causality because of the causal constraint of the junction. The two displacements are not independent, because they both are proportional to the same effort. As a result, the two energy storages are dependent. This situation is easily simplified by combining the two compliances into a single compliance, as shown in part (b) of the figure: $C = C_1 + C_2$.

In contrast, consider two compliances bonded to a common 1-junction, as shown in part (c) of the figure. Both can be assigned integral causality; two state variable displacements can be defined, and two differential equations can be written. The contribution to the order of the overall system by these compliances therefore is two. Nevertheless, it is possible to combine the two compliances into a single compliance: $1/C = 1/C_1 + 1/C_2$. If this is done, the order of the system is legitimately reduced by one. You can say that the **minimum order** of the system (with the combined compliance) is less than the nominal or actual order of the system when the compliances are treated as distinct.

Parts (d) and (e) of the figure show pairs of compliances that are less obviously combinable because of the presence of transformers. The causal strokes reveal the same pattern as before, however: with a 0-junction, only one independent differential equation can be

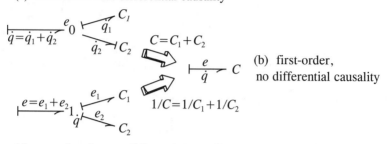

(a) first-order with differential causality

$C=C_1+C_2$

(b) first-order, no differential causality

$1/C=1/C_1+1/C_2$

(c) second-order, no differential causality

$C=C_1+C_2/T^2$

$1/C=1/C_1+T^2/C_2$

(d) first-order with transformer   (e) second-order with transformer

Figure 6.7: Combining dependent compliances

written, whereas with a 1-junction, two differential equations may be written, although the number may be reduced to one by combining the two compliances and the transformer into a single equivalent compliance. In either case the combined compliance can be found either of two ways. In the first method, the compliance C_2 is combined with the transformer to give an equivalent compliance with modulus C_2/T^2 (as noted in Table 5.1 on page 311), which then is combined with C_1. In the second method, the efforts or flows of the two compliances are related to a common effort or flow. Their energies, expressed as a function of this commmon effort or flow, are summed to give the energy of the combined compliance. The second method tends to be more powerful and general, and therefore is preferred.

Similar considerations apply to two inertances bonded to a common junction, as shown in Fig. 6.8. As you have seen before, the combination rules for inertances are the dual of those for compliances; those for a 0-junction in one case act like those for a 1-junction in the other case. It is not necessary or even desirable to memorize these rules. You should work out each case by focusing on its variables, noting particularly which variable is common and which variable sums, and utilizing the constitutive relations of the elements.

In some cases the reduction of an over-causal model to a causal model is relatively difficult to accomplish. You have the option of foregoing this reduction, and proceeding directly to the writing of differential equations. This option is discussed in Section 6.2.

(a) first-order with differential causality

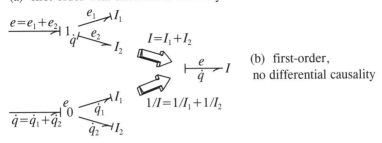

(b) first-order,
    no differential causality

(c) second-order, no differential causality

(d) first-order with transformer   (e) second-order with transformer

Figure 6.8: Combining dependent inertances

EXAMPLE 6.4

Determine the order of the following model that comprises elements with constant moduli:

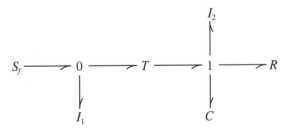

Solution: Causal strokes are placed, starting with the mandated flow-source bond. Any two of the three energy storage elements can have integral causality assigned, but the third must have differential causality. One of the three possibilities is

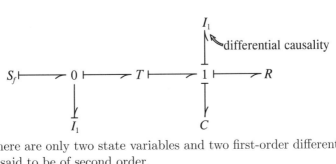

As a result, there are only two state variables and two first-order differential equations. The model is said to be of second order.

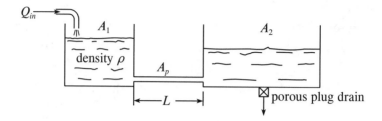

Figure 6.9: Two-tank system for Guided Problem 6.1

6.1.7 Summary

A four-step procedure for assigning causal strokes has been given which is sufficient for causal and over-causal models. Every compliance and inertance element in a causal model is given integral causality by this procedure; the over-causal models are those for which one or more differential causalities result. The model is called under-causal whenever the procedure fails to assign causalities to all the bonds.

Another four-step procedure is given for subsequently writing a set of state-variable differential equations for a causal bond graph. This procedure annotates the efforts and flows on all the bonds, before the equations themselves are written, following the special causal rules. Previously written annotations may violate these rules, and should be removed or set aside whenever this happens. Even if an annotated effort or flow is functionally correct, if it violates the causal strokes it will frustrate the purpose of the causal method. Definitions of variables using different symbols can be reintroduced *after* the differential equations are written.

One differential equation of first order results from equating the derivative dp_i/dt or dq_i/dt, for each energy storage element, to the causal input of its bond. The signs are determined by the power convention half-arrows.

The energy storages with differential causality in an over-causal model can be expressed as functions of the same state variables in terms of which the energy storages with integral causality are described. They are therefore not independent, and the number of differential equations and the order of the model equals the number of energy storage elements with integral causality, assuming the standard procedure that maximizes this number has beeen followed. It is advisable to reduce over-causal models to causal form whenever the reduction is not very difficult to carry out; simple examples have been given. Otherwise it is still possible to proceed, as described in the following section, which also addresses under-causal models.

Guided Problem 6.1

Needed practice in finding differential equations for linear models is provided by this problem. Fig. 6.9 shows two fluid tanks with an independent supply of liquid to the left-hand tank, an interconnecting tube with both fluid resistance and inertance, and a resistive drain from the right-hand tank. Draw a bond graph for this system. The resistances depend on viscosity, and may be identified by unspecified R's. Other moduli should be expressed in terms of physical parameters such as areas, lengths and fluid density. Identify the order of the system, and write a corresponding number of state differential equations.

Suggested Steps:

1. Identify five flows: the input flow Q_{in} at the left, the net flow that fills the left-hand

tank, the flow between the two tanks, the net flow that fills the right-hand tank and the flow which leaves the drain. Use these flows to establish a junction structure for the bond graph, and complete the graph.

2. Apply causal strokes to the graph, using the proper priorities. Is integral causality possible for all the energy storage elements? Identify the order of the system.

3. Label each compliance bond with $\dot{q}_i$ and q_i/C_i, and each inertance with $\dot{p}_j$ and p_j/I_j, where the subscripts are chosen to distinguish different elements of the same type. Circle the terms $\dot{q}_i$ and $\dot{p}_j$. Place $\times$ for the effort variable associated with the input flow Q_{in}.

4. Propagate the causal inputs Q_{in}, q_i/C_i and p_i/I_i through the bond graph until all efforts and flows are properly annotated.

5. Write the differential equations, using the causal stokes and power convention arrows to direct your work. You are not asked to combine these equations into a single equation of higher order, but note that this could be done.

Guided Problem 6.2

Dynamics are added to a steady-state model in this problem, and the corresponding state differential equations are found. An important way to eliminate differential causality is demonstrated, which allows the differential equations to be found.

   Augment the model of the hydraulic system of Guided Problem 5.4 (pp. 292, 305) by adding masses or rotational inertia and resistances to the rotary actuator and the two pistons. Define state variables and write the associated set of state differential equations.

Suggested Steps:

1. Add 1-junctions, inertances and resistances to the bond graph to represent the three inertances and three resistances.

2. Attempt to add integral causality to these inertances, and observe the violation of causal constraints.

3. Compute the kinetic energy of the offending inertance, and express it as a function of the generalized velocity of one or both of the remaining inertances. Use this calculation to transfer the removed inertance to one or both of these inertances, augmenting the value(s) of the inertance(s) appropriately.

4. Compute the energy dissipation of the corresponding resistance, and similarly express it as a function of the generalized velocity of one or both of the same inertances. Use this calculation to transfer the removed resistance to one or both of these locations, augmenting the resistance there appropriately.

5. Reapply integral causality (it works now), and prepare the energy storage bonds with the usual notations.

6. Write the differential equations, employing the causal strokes in the usual manner.

7. Write auxiliary equations for the outputs Q and $\dot{\phi}$ as functions of the state variables, in case these are of interest.

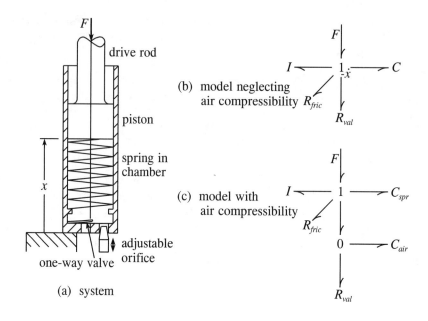

Figure 6.10: Guided problem 6.3

Guided Problem 6.3

Even if your instructor suggests that you forgo carrying out this more sophisticated problem, you should at least study it and its given solution. It involves the determination of nonlinear differential equation models for the mechanical snubber shown in part (a) of Fig. 6.10, which is available commercially. The piston is made with a low-friction coating, and fits closely with the glass cylinder to give little leakage. An air valve acts like a small orifice for out-flow (which in practice can be adjusted), but a one-way valve (check valve) presents a much larger effective orifice for in-flow. A soft spring fits inside the cylinder to resist compression. You are asked to find a differential equation model which can give the position $x(t)$ in response to an applied force $F(t)$. The mechanical rubbing can be characterized by a dry friction force F_0, the fluid leakage across the piston by a constant resistance R_p (based on laminar flow), the spring by its rate k and free length x_0, the mass of the piston assembly by m, the nominal density of the air by ρ_0, the effective area of the orifice for out-flow by A_o, the effective area of the orifice for in-flow by A_i and the area of the piston by A_p.

(a) The bond graph in part (b) of the figure neglects the compressibility of the air. Annotate the graph with symbols, and find an approximate differential equation model. Note: the symbol $\text{sign}(y)$ may be useful; it equals $+1$ when $y > 0$ and -1 when $y < 0$.

(b) The bond graph in part (c) of the figure recognizes compression of the air. Find an approximate differential equation model. The air compliance is more difficult to implement than most nonlinear compliances, so significant details are proposed below. It is suggested that you retain the approximation $Q \sim \sqrt{(2/\rho)|P - P_a|}$ for valve flow, since the interest is in pressures P which do not differ radically from the atmospheric pressure P_a. (A more accurate modeling requires considerations of compressible fluid flow, which are addressed in Chapter 11.) Further, the compression may be considered to be isentropic.

Suggested Steps:

1. Apply causal strokes to the bond graph of part (b), define state variables in the conventional way and annotate the conjugate forces or velocities on the I and C bonds with the standard computationally realizable functions.

2. The resistance R_{fric} acts something like a force source, but its sign reverses with the direction of the motion. Write a computationally realizable function for this force adjacent to the bond.

3. Use an orifice equation to approximate the flow through the valve, add the flow for the leakage across the piston, and invert to find the pressure as a function of the total flow. Then relate this flow to a state variable. Also, relate the force F_{air} to the pressure and thereby to a state variable.

4. Write the two first-order differential equations, with guidance from the causal strokes.

5. To start the solution of the model with compression of the air, apply causal strokes to part (c) of the figure and adapt the prior model, to the extent that it applies. Note that the causality on the leakage resistance element is the inverse of what you used for the incompressible model, so that you do not need to invert the equation for leakage flow.

6. Note that the assumption of an isentropic process means that the pressure in the chamber can be written as $P = P_a(q/x)^k$, where q is the portion of the piston displacement associated with compression of the air and $k = 1.4$ (not to be confused with the spring constant). Use this relation to express the effort on the 0-junction as a function of state variables. Then, get expressions for the leakage flow in terms of this effort and fixed parameters, depending on the sign of the velocity.

7. Write all three state differential equations. Make sure the only variables that appear on the right sides of the equations are, directly or by substitution, the excitation variable F and the state variables x, q and p.

PROBLEMS

6.1 Systems are modeled with the constant-parameter bond graphs shown below. Apply causal strokes, using integral causality. Define state variables, and write the corresponding set of state differential equations.

6.2 Answer the question above for the bond graph shown below left.

6.3 Answer the question above for the bond graph shown above right.

6.4 Find electric circuits that correspond to the bond graphs of (a) Problem 6.2 and (b) Problem 6.3.

6.5 Find mass-spring-dashpot systems with x-motion only that correspond to the bond graphs of (a) Problem 6.1 and (b) Problem 6.2.

6.6 Derive a differential equation for a system modeled by the bond graph below left.

6.7 Derive a differential equation for a system modeled by the bond graph above right.

6.8 Write the state differential equations for the model developed in Guided Problem 5.3 (pp. 291, 304-305). The electric current is specified as a function of time.

6.9 Answer the questions of Problem 6.1 for the bond graph given below:

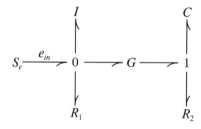

6.10 For the bond graph below, in which the causal input is e_0,

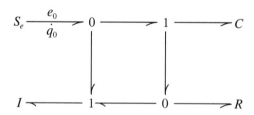

(a) Apply causal strokes, choose state variables and label all the efforts and flows accordingly.

(b) Write the corresponding set of state differential equations.

(c) The output variable of interest is $\dot{q}_0$. Relate this to the state variables and the input variable, e_0.

(d) Express, in terms of the given parameters, any time constants, natural frequencies or damping ratios that apply.

6.11 Define state variables and write a set of state differential equations for the adjacent model:

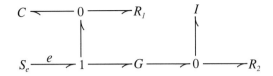

6.12 Answer the previous question for the adjacent model:

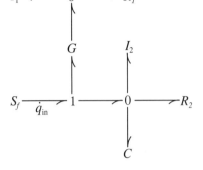

6.13 Define state variables and write differential equations for the adjacent model. The moduli of the elements are constants except for the resistance, which can be expressed as either $e_R = \sqrt{a_R}$ or $\dot{q}_R = ae_R^2$.

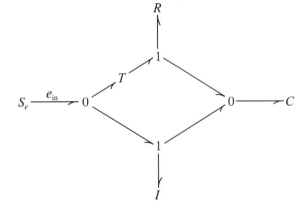

6.14 Identify the order of the following bond graph by applying causal strokes. All moduli are constants.

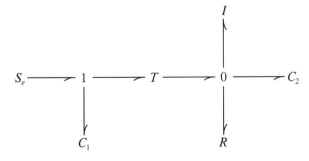

6.15 Find systems with (i) springs and dashpots, and (ii) electric capacitors and resistors, that are analogous to the two-tank system of Fig. 6.9 (p. 358). Assume constant parameters.

6.16 Identify the order of the following bond graph by applying causal strokes. All moduli are constants.

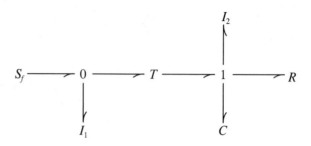

6.17 The system with an axial-flow fan and tank given in Problem 5.74 (p. 338) is modified by placing a duct of length $L =$ and cross-sectional area 3×3 inches, as pictured below. Consider the case of orifice #1.

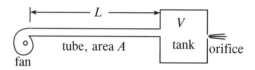

(a) Model the system with a bond graph.

(b) Write a set of differential equations for the model. At this point you may leave the source and load charateristics as unspecified functions of specific variables.

(c) Linearize the model of part (b). Take particular care with the signs.

(d) Determine the natural frequency and damping ratio of the linearized model.

(e) Show that a small value of L produces instability (and by inference a limit-cyle oscillation), and a large value of L produces stability.

(f) Determine the borderline value of L.

6.18 A simple $I - C$ (or $L - C$) filter connects a voltage source to a resistive load, as shown below. Find a differential equation relating the load voltage e_L to the source voltage e_S.

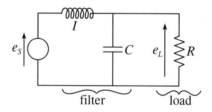

6.19 Continue Guided Problem 5.5 (pp. 293, 306) by defining a minimum set of state variables, considering the fluid pressure, P, to be an independent excitation and the motions of the solid parts to be responses. Find the corresponding state differential equations in a proper form for integration. You may employ parameters of your own devising if they are defined explicitly.

6.20 Consider the mechanical system described in Fig. 5.25 (p. 327) and the accompanying text. Define state variables and write a set of first-order differential equations in a form suitable for solution. Continue to ignore the inertias of the moving parts. The nonlinear functions may be defined as unspecified functions, *e.g.* $M_m = M_m(\dot{\phi}_m)$ or $\dot{\phi}_m = \dot{\phi}_m(M_m)$.

6.21 A hydraulic shock absorber of the type described in Fig. 2.24 (p. 39) is employed as the dashpot in the system of Problem 5.2 part (c) (p. 267). Write the state differential equations for the resulting nonlinear system.

6.22 Write a set of state differential equations for the aircraft carrier "arresting engine" of Problem 5.60 (p. 321).

6.23 *Design Problem.* Carpenters often use pneumatically driven nailers for framing, roofing and finish nails, powered by a large air compressor through a rather large hose. Battery-powered nailers have been considered, but the power a reasonably-sized battery with solenoid or electric motor can deliver is vastly less than the instantaneous power needed to drive a large nail in one motion. The nail must be driven home in a few milliseconds, for otherwise the recoil momentum is too great. Some means is needed to store energy for quick release. Springs weigh too much. Kinetic energy is one possibility. The energy storage of a compressed gas is another, which is the subject of this problem.

It is proposed to use a small motor-driven hyralic pump to force oil into a small hydraulic accumulator. The oil also pushes on a piston which, when a trigger release is actuated, drives the nail. Some of the details are suggested in the schematic drawing below. The plunger is returned to its initial position by a small spring, not shown.

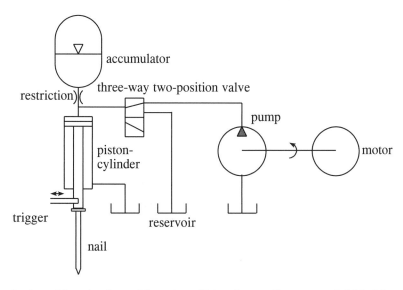

The principal problem is that although sufficiently small commercial bladder accumulators exist, they have a valve mechanism to prevent the gas pressure from extruding the bladder through the neck of the accumulator. (See Fig. 3.34, p. 132.) This presents an excessive resistance to the flow of oil that surges out of the accumulator to drive the nail. You are asked to investigate the problem and make design recommendations. The questions regard the size of the accumulator, the gas pre-charge pressure, the minimum allowable effective area of the flow restriction in question, and the diameter of the hydraulic actuator.

It is suggested that a computer simulation would be an effective tool in addressing these questions. Details about the electric motor, pump, valving and mechanism details are not to be addressed.

The nails are the large 16-d size, 3.25 inches in length and 0.022 lbs in weight. They are resisted by a force that grows linearly with the penetration of the nail into the wood, for a total work of 100 ft-lbs. It has been decided that the accumulator can hold a pressure of up to 3000 psi, the charging pressure of the nitrogen (the pressure when no oil is in the accumulator) should be at least one-third of the maximum pressure to be used, to prevent excessive flexure of the bladder. It is suggested that the inertia of the nail and the other moving parts is important enough to be included in any dynamic model. Energy lost because of spent momentum and orifice flow must be kept small so that the battery and electric motor not become too heavy.

6.24 *Phase 2 of project started in Problem 3.60 (p. 150) on the electro-hydraulic vehicle drive system in Chapter 1.* This second phase of the project involves adding the accumulator and achieving a working simulation under an assumed supply of hydraulic fluid.

Start by adding the accumulator, which for present purposes you may assume has a total volume of 2.0 ft^3 (0.057 m^3), acts adiabatically and has a charging pressure (gas pressure when empty of oil) of 1014.7 psia (1000 psig) (7.00 MPa absolute). Write an expression for the oil pressure as a function of the quantity of oil in the accumulator. Then, assume that the pump flow Q_p is a constant 53.0 in^3/s (0.87 × 10^{-3} m^3/s), and keep the motor displacement at 2.2 in^3/rev (36 × 10^{-6} m^3/rev. Write a pair of first-order differential equations, representing in solvable form the volume of oil in the accumulator and the speed of the vehicle. Finally, simulate the acceleration of the vehicle, assuming that the accumulator starts empty of oil. Plot the speed of the vehicle, the quantity of oil in the accumulator, and the pressure there.

Safety requires that the pressure in the accumulator be prevented from exceeding about 4000 psi (28 MPa). You may simply seek to avoid this condition in the later phases of the project. Better, you could insert a relief valve that dumps excess fluid from the accululator into the sump. This can be accomplished with a conditional statement that zeros the net flow into the accumulator whenever the pressure there exceeds the pressure limit *and* the flow otherwise would be positive.

6.25 *Phase 3 of the electro-hydraulic vehicle drive system.* This phase completes the modeling by adding the electic motor and pump to your model, and achieving a working simulation.

Add the DC electric motor, assuming a non-load speed of 500 rad/s (4775 rpm) and a stall torque of 288 Nm (212 ft·lb)(which gives the maximum power of 36 kW (48.2 hp)). Neglecting brush friction, infer from this information the values of the gyrator modulus and the electrical resistance in the bond-graph model. The effective rotational inertia of the rotor is 0.044 ft·lb·s^2 (0.060 kg·m^2).

Now add the pump, temporarily assuming that its volumetric displacement is 0.90 in^3/rev (14.7 cm^3/rev), and that the linear frictional resistance of shaft rotation is 2.3×10^{-4} ft·lb/rpm (3.16 × 10^{-4} N·m/rpm). Infer from this information the value of the resistance R_p. Then, write the state differential equations, assuming that the input voltage, e, is 12 V. Be *extremely* careful with the units; the value of G or T can be different for the two directions if you are switching from inches to feet or N·m to ft·lb.

Carry out a simulation, again for about 500 seconds, using the same zero initial volume of oil in the accumulator as in phase 2, zero initial motor speed, and zero initial vehicle speed. Plot the resulting motor speed, accumulator pressure and vehicle speed.

6.26 *Final phase of the electro-hydraulic vehicle drive system.* This is the design phase of the project. You should choose a total volume for the accumulator as a compromise between performance and weight (which could be considerable). The maximum pressure in the accumulator (limited in practice by the safety or relief valve that you don't want to open, dumping energy, unless you are experiencing a major braking) should not be greater than four times the charging pressure (to protect the bladder within); about 4000 and 1000 psi (28 and 7 MPa), respectively, are acceptable values. You may select three values of pump displacement, one for the initial rapid acceleration phase, the second for the subsequent phase in which the speed approches an equilibrium speed in the range of 60 to 70 mph (97 to 113 km/hr), and a third for which the speed approaches the maximum possible. (Note that negative values give regenerative braking whenever the accumulator pressure is lower than the relief vale setting.)

The major design issue is the control algorithm for the displacement of the hydraulic pump. This is to be implemented automatically; the driver directly controls only the displacement of the hydraulic motors. The automatic control affects the speed and power of the electric motor, and how much oil flows toward the accumulator. The greater the power, the lower the efficiency. Note that the electric motor achieves maximum power when it runs at one-half its no-load speed, and you never want it to operate more slowly because you can always get the same power with a higher effieiency at a higher speed. (These matters are discussed in Section 5.2.4.) When the vehicle is running at steady speed on the level, its most common condition, it is best that the motor power and pump flow match the needed load, rather than cycle up and down, charging and discharging the accumulator. Such a control can be achieved by making the pump displacement depend in an appropriate way on the pressure observed in the accumulator; the displacement can be a fairly simple function of this pressure. You may try a few different functions, but a deductive approach based on the desired dependency and anticipated loads appears to be almost necessary for a real success.

Your final report should be succinct and should not include unsatisfactory trials, although it should indicate the process you went through as well as the final result. Give simulations for your design that correspond to the two in the book, and include your MATLAB® coding. Finally, it should be recognized that the model you have developed is still relatively crude. You should attempt to identify the most significant sources of error in it, and what qualitative bias they introduce into your results.

SOLUTIONS TO GUIDED PROBLEMS

Guided Problem 6.1

1-3.

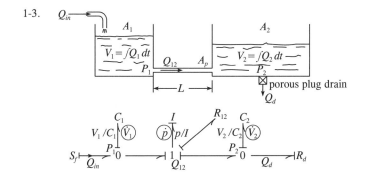

Integral causality can be applied to the inertance and both compliances without producing a causal conflict. The order of the system, therefore, is three. The state variables are V_1, V_2 and p.

4.

5. $\dfrac{dV_1}{dt} = Q_{in} - Q_{12} = Q_{in} - \dfrac{1}{I}p$

$\dfrac{dV_2}{dt} = Q_{12} - Q_d = \dfrac{1}{I}p - \dfrac{1}{R_d}P_2 = \dfrac{1}{I}p - \dfrac{1}{R_dC_2}V_2$

$\dfrac{dp}{dt} = P_1 - P_2 - R_{12}Q_{12} = \dfrac{1}{C_1}V_1 - \dfrac{1}{C_2}V_2 - \dfrac{R_{12}}{I}p$

Guided Problem 6.2

1-2.

3. $\mathcal{T} = \dfrac{1}{2}I_3\dot{\phi}_3^2 = \dfrac{1}{2}I_3(T_3Q_2)^2 = \dfrac{1}{2}I_3(T_3\dot{x}_2/T_2)^2 = \dfrac{1}{2}I_3'\dot{x}_2^2$

Therefore, $I_3' = (T_3/T_2)^2 I_3$.

Combine this inertance with I_2 to give the overall inertance

$I_2' = I_2 + (T_3/T_2)^2 I_3$.

4.-5. $\mathcal{P}_3 = R_3\dot{\phi}_3^2 = R_3(T_3/T_2)^2\dot{x}_2^2 = R_3'\dot{x}_2^2$; therefore, $R_3' = (T_3/T_2)^2 R_3$.

Combine R_3' with R_2 to get $R_2' = R_2 + (T_3/T_2)^2 R_3$.

6. $\dfrac{dp_1}{dt} = \dfrac{1}{T_1}P - \dfrac{R_1}{I_1}p_1$

$\dfrac{dp_2}{dt} = \dfrac{1}{T_2}P - \dfrac{R_2'}{I_2'}p_2$

7. Compare the two bond graphs above to get

$Q = \dfrac{1}{T_2I_2'}p_2 + \dfrac{1}{T_1I_1}p_1$

$$\dot{\phi} = \frac{T_3}{T_2 I_2'} p_2$$

Guided Problem 6.3

1.

2. Let $\dot{x} > 0$ for upward motion.
$$F_{fric} = -\text{sign}(\dot{x})F_0 = \text{sign}(p)F_0$$
$$C = 1/k; \quad I = m$$

3. Volume flow:

$$Q = A_0 \sqrt{\frac{2}{\rho}(P - P_a)} + \frac{1}{R_p}(P - P_a), \quad p > 0$$

$$Q = -A_i \sqrt{\frac{2}{\rho}(P_a - P)} - \frac{1}{R_p}(P_a - P), \quad p < 0$$

Therefore, $F_{air} = A_p(P - P_a) = A_p \left(-\frac{R_p A_0}{\sqrt{2\rho}} + \sqrt{\frac{R_p^2 A_0^2}{2\rho} + \frac{R_p A_p}{I} p} \right)^2, \quad p > 0$

$$F_{air} = -A_p \left(\frac{-R_p A_i}{\sqrt{2\rho}} + \sqrt{\frac{R_p^2 A_i^2}{2\rho} + \frac{R_p A_p}{I} p} \right)^2, \quad p < 0$$

4. Differential equations (note that p is *downward* momentum):

$$\frac{dx}{dt} = -\frac{1}{m} p$$

$$\frac{dp}{dt} = F - \frac{1}{C}(x - x_0) - F_{fric} - F_{air}$$

$$= F + k(x - x_0) - F_0 - A_p \left(\frac{-R_p A_0}{\sqrt{2\rho}} + \sqrt{\frac{R_p^2 A_0^2}{2\rho} - \frac{R_p p}{A_p I}} \right)^2, \quad p > 0$$

$$= F + k(x - x_0) + F_0 + A_p \left(\frac{-R_p A_i}{\sqrt{2\rho}} + \sqrt{\frac{R_p^2 A_i^2}{\sqrt{2\rho}} + \frac{R_p p}{A_p I}} \right)^2, \quad p < 0$$

5.

6. $P = P_a(q/x)^k$

Therefore, $F_{air} = A_p P_a[(q/x)^k - 1]$

and $Q = A_0 \sqrt{\frac{2F_{air}}{\rho A_p}} + \frac{F_{air}}{R_p A_p}, \quad p > 0$

$$Q = -A_i \sqrt{\frac{2F_{air}}{\rho A_p}} + \frac{F_{air}}{R_p A_p}, \quad p < 0$$

7. The differential equations are as given below, with the expressions given in the above step substituted for F_{air} and Q.

$$\frac{dx}{dt} = -\frac{1}{m} p$$

$$\frac{dp}{dt} = F - \frac{1}{C_{spr}}(x - x_0) - F_{fric} - F_{air}$$

$$= F + k(x - x_0) - \text{sgp}(p)F_0 - F_{air}$$

$$\frac{dq}{dt} = \frac{1}{I} p - \frac{Q}{A_p}$$

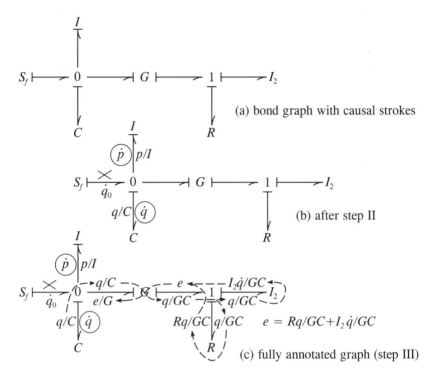

Figure 6.11: Example treatment of an over-causal model

6.2   Over-Causal and Under-Causal Models

Over-causal models are described by fewer differential equations than there are energy storage elements, as noted above. As a result, direct use of the bond graph produces equations in which derivatives appear on both sides of one or more of the first-order differential equations. Under-causal models produce a typically more difficult problem: some of the equations produced are algebraic rather than differential. In many cases of both types, the equations can be reduced to a simple set of differential equations. The problem of a mixed set of generally nonlinear differential-algebraic equations (commonly known as DAEs) sometimes must be faced, however. Special software is available. Also, a method for treating these situations is described which approximates an under-causal model by a causal model, permitting use of standard methods for ordinary differential equations.

6.2.1   Treatment of Over-Causal Models; Case Study

Differential equations can be found directly from most over-causal bond graphs. The same four steps given for causal models in Section 6.1.2 (p. 344) are used, with careful interpretation, and a fifth step is added. The care is in properly implementing the differential causalities in the third and fourth steps. The fifth step is the rearrangement of the resulting differential equations to insure that the derivatives of the state variables lie only on the left sides.

   An example is shown in Fig. 6.11. After the mandatory causal stroke is placed on the bond for the flow source, integral causality is placed on the compliance. This placement forces the causalities of all the other bonds, as shown in part (a) of the figure. The inertance I_1 receives integral causality, but the inertance I_2 is forced to have differential causality. As a result, the model is only second order; the energy in I_2 is expressible in terms of the state variables for the other energy storage elements. The modeler must decide at this point whether to find an equivalent bond graph with only two energy storage elements, or whether to press ahead with the existing graph. The following presses ahead.

Step I starts with placing the notations $\dot{q}_0$ and $\times$ on the bond for the flow source. Then, step II proceeds with the placement of notations adjacent to the bonds describing integral causality: $\dot{q}$ and q/C on the compliance bond and $\dot{p}$ and p/I_1 on the inertance bond. This completes the step; the inertance I_2 does not participate, since it does not have the required integral causality. The situation at this stage is shown in part (b) of the figure.

The propagation of the terms $\dot{q}_0$, q/C and p/I onto the remainder of the efforts and flows, step III, is shown in part (c) of the figure. Dashed lines with arrows describe the sequence. First, the remaining effort off the 0-junction is annotated as q/C. Second, the gyrator causalities are implemented by writing the flow on the 1-junction as q/GC. Third, this flow is replicated as the causal inputs to both the resistance R and the inertance I_2. Fourth, the causal output of the resistance is written as Rq/GC, as mandated by its impedance causality. The causal output of the inertance I_2 is written as I_2 times the *time derivative* of the flow, or $I_2\dot{q}/GC$, as shown. This appearance of the time derivative of a state variable is peculiar to the differential causality, although it does not depart from the procedure. Fifth, the effort noted as e (to save space) now is written as shown to the right of the graph. This term is dictated by the causal strokes; the signs of the sub-terms are determined by the power-convention half-arrows. Finally, the flow to the left of the gyrator is written as e/G, completing step III.

The two state differential equations for the system now can be written (step IV). The equation for $\dot{p}$ is

$$\frac{dp}{dt} = \frac{1}{C}q. \tag{6.16}$$

The equation for q differs in an important respect:

$$\frac{dq}{dt} = \dot{q}_0 - \frac{1}{I}p - \frac{1}{G}e = \dot{q}_0 - \frac{1}{I}p - \frac{I_2}{G^2C}\frac{dq}{dt} - \frac{R}{G^2C}q. \tag{6.17}$$

The difference is that *the derivative term appears on both sides of the equation.* This problem is rectified by collecting the two terms together on the left side:

$$\left(1 + \frac{I_2}{G^2C}\right)\frac{dq}{dt} = \dot{q}_0 - \frac{1}{I}p - \frac{R}{G^2C}q. \tag{6.18}$$

This final step can be viewed as a usually simple step V that is necessary whenever derivative causality is implemented. Equations (6.16) and (6.18) comprise the state differential equation model of the system.

This model is somewhat more complex than the corresponding model that results from subsuming the dependent inertance element into the independent compliance element. This is because different state variables are employed.

The propagation of the state variables can be carried out even when the elements are nonlinear. You need to be especially careful in these cases to follow the dictates of the causal strokes meticuously. There is one potentially serious complication, however: if the causal output of an element with differential causality is a nonlinear function of the derivative of a state variable, it may be difficult or impossible to carry out step V. This rare situation sometimes can be avoided by proper choice of the energy storage element given differential causality. In the case considered above, for example, the element I_2 could have been given integral causality, rather than the element C.

EXAMPLE 6.5

Find a differential equation relating the displacement of the compliance in the model below to the excitation $e_i(t)$. All elements have constant moduli.

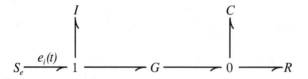

Solution: Either the inertance or the compliance element may be given integral causality, but not both; the model is over-causal. Since the question asks for the displacement on the compliance, written here as q, it is simplest to make this the sole state variable by assigning integral causality to the compliance. The result at this point is

Propagating the variables according to the dictates of the causal strokes gives the effort on the right side of the gyrator, the flow on the left side of the gyrator, and the flow on the inertance, as shown below. The differential causality on the inertance tells you that its effort is I times the *time derivative* of the flow, or $I\dot{q}/GC$:

The completed annotations then become

The sole differential equation becomes, at first,

$$\frac{dq}{dt} = \frac{1}{G}e_i(t) - \frac{I}{G^2C}\frac{dq}{dt} - \frac{1}{RC}q,$$

in which $\dot{q}$ has been written with the more formal notation dq/dt in order to emphasize that this term appears on both sides of the equation.

Collecting these terms,

$$\left(1 + \frac{I}{G^2C}\right)\frac{dq}{dt} = \frac{1}{g}e_i(t) - \frac{1}{RC}q.$$

The literal answer to the question is this equation with both sides divided by the constant within the parentheses.

6.2.2 Equations for Under-Causal Models

Under-causal models result when the four-step procedure for placing causal strokes (Section 6.1.1, p. 343) leaves some bonds without strokes. The author advocates a special procedure to handle these cases. As Step 5, a **virtual inertance** is bonded to a 1-junction that has some strokeless bonds, or a **virtual compliance** is bonded to a 0-junction that has some strokeless bonds. These added elements normally will be considered to have zero moduli, so as not to change the meaning of the model. Integral causality is applied to the added virtual element, and the causal strokes are propagated as far as possible, consistent with the usual rules. Step 5 is repeated, if necessary, using other junctions with strokeless bonds, until all bonds have causal strokes.

Step 5 is tentative, contingent on its not producing a causal conflict. In some cases involving a bond-graph mesh such a conflict appears at a junction; what to do if this happens is described and illustrated in Section 6.2.7.

The number of virtual energy storages can be called the degree of under-causality. For each degree, one *algebraic* equation results. Taken together with the differential equations in the model, they form a set of differential-algebraic equations (DAEs) that are said to be in semi-explicit form:

$$\frac{d\mathbf{x}}{dt} = \mathbf{f}(\mathbf{x}, \mathbf{y}, t), \tag{6.19a}$$

$$0 = \mathbf{g}(\mathbf{x}, \mathbf{y}, t). \tag{6.19b}$$

6.2.3 Algebraic Reduction Method; Case Study

The algebraic equations (6.19b) potentially can be used to eliminate the variables $\mathbf{y}$ from equations (6.19a), that is to reduce the DAE to an ODE (ordinary differential equation). DAE models for which this reduction is possible are called **index zero** models, which is the simplest type of DAE to solve.

An example is shown in Fig. 6.12. Steps 1-4 of the procedure for placing causal strokes leave four bonds without causal strokes, as shown in part (a) of the figure. The new Step 5 adds either a virtual inertance to the 1-junction or a virtual compliance to the 0-junction. The first option (it makes little difference) is implemented in part (b) of the figure. All four bonds are given causal strokes as a result; the degree of under-causality is one.

The four-step procedure for writing differential equations for causal bond graphs (p. 344) now is implemented, with a critical difference that recognizes the zero value of the added virtual inertance, I_v. Specifically, in place of writing a circled $\dot{p}_v$ on the effort side of the virtual inertance bond and p_v/I_v on the flow side, a circled 0 is written on the effort side to represent its zero effort, and a symbol f_v is written on the flow side to represent its non-zero flow. The result after completion of step II is shown in part (c) of the figure, and after part III of the procedure is shown in part (d).

The equations for the two real and one virtual energy storage elements are now written (Step IV) as

$$\frac{dq}{dt} = f_v, \tag{6.20a}$$

$$\frac{dp}{dt} = R_2 \left(\dot{q}_0 - \frac{f_v}{T} - \frac{p}{I}, \right) \tag{6.20b}$$

$$0 = \frac{R_2}{T} \left(\dot{q}_0 - \frac{f_v}{T} - \frac{p}{I} \right) - \frac{q}{C} - R_1 f_v. \tag{6.20c}$$

The virtual inertia has given algebraic equation (6.20c) which, since it is linear, can be solved for the flow f_v in terms of the input variable $\dot{q}_0$ and the state variables q and p. This

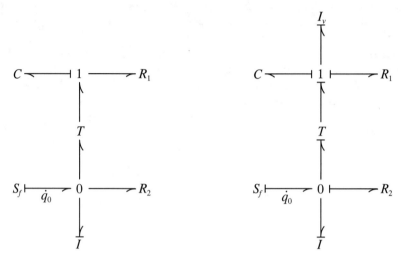

(a) steps 1-4 for placing causal strokes   (b) step 5 with virtual inertance

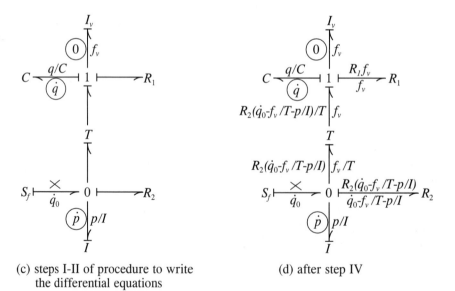

(c) steps I-II of procedure to write   (d) after step IV
    the differential equations

Figure 6.12: Linear under-causal example

f_v is then substituted where it appears in the differential equations. The final result is the ODE

$$\frac{dq}{dt} = \frac{T}{1 + R_1 T^2/R_2}\left[\dot{q}_0 - \frac{1}{I}p - \frac{T}{R_1 C}q\right], \qquad (6.21a)$$

$$\frac{dp}{dt} = \frac{T}{1 + R_1 T^2/R_2}\left[R_1 T \dot{q}_0 - \frac{R_1 T}{I}p + \frac{1}{C}q\right]. \qquad (6.21b)$$

EXAMPLE 6.6

Write a differential equation for the sole state variable in the following model. All moduli are constants.

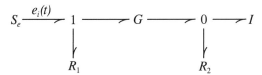

Solution: Only the left-most and the right-most bonds have mandated causal strokes, so the model is under-causal. You have the choice of adding a virtual inertance to the 1-junction or a virtual compliance to the 0-junction. Doing the latter, the flow on the virtual compliance is written as a zero and is encircled, and the effort, which is the non-zero ratio of a zero displacement to a zero compliance, is written as e_v:

$$\begin{array}{c}
C_v \\
e_v \left(0\right) \\
S_e \xrightarrow{e_i(t)} 1 \longrightarrow G \longmapsto 0 \longrightarrow I \\
R_1 \qquad\qquad\qquad R_2
\end{array}$$

The annotations on all the bonds now can be completed to give the result

$$\begin{array}{c}
C_v \\
e_v \left(0\right) \\
\overbrace{}^{p} \\
S_e \xrightarrow[e_v/G]{e_i(t)} 1 \xrightarrow[e_v/G]{e_i - R_1 e_v/G} G \xleftarrow[e_i/G - R_1 e_v/G^2]{e_v} 0 \xrightarrow[p/I]{} I \\
R_1 e_v/G \big| e_v/G \\
R_1 \qquad\qquad\qquad\qquad\qquad R_2
\end{array}$$

Setting the encircled zero flow equal to the sum of the input flows to the 0-junction gives the algebraic equation

$$0 = \frac{1}{G}e_i - \frac{R_1}{G^2}e_v - \frac{1}{R_2}e_v - \frac{1}{I}p,$$

which can be solved for e_v. Setting the encircled $\dot{p}$ equal to the effort of the 0-junction gives the answer

$$\frac{dp}{dt} = e_v = \frac{1}{R_1/G^2 - 1/R_2}\left(\frac{1}{G}e_i(t) - \frac{1}{I}p\right).$$

6.2.4   Differentiation Method; Case Study*

Equation (6.20c) is solvable for f_v because it is linear. Multiple linear equations also are solvable if they are non-singular. In other cases, however, different methods usually are needed.

The ground-effect machine or GEM (sometimes called a Hovercraft), shown in Fig. 6.13, is a nonlinear example. This vehicle hovers over land or especially over water, without direct contact, due to air pressure underneath caused by a large fan. A skirt around the periphery constricts the outflow of air from the interior plenum, or chamber. If the craft is displaced downward, the restriction for this airflow increases, increasing the pressure and causing the craft to rise. The GEM pictured, with its single plenum, does not correct for the tilting motions of roll and pitch; embellishments to the design are needed for these, as described later. For now, attention is focused on vertical translation only.

The vehicle shown measures 4×10 meters, with a plenum depth of 0.8 meters. The pressure-flow characteristic of the fan, as plotted, is represented by the equation

$$Q_f = Q_0 - a_1 P + a_2 P^2 - a_4 P^4, \tag{6.22a}$$

$$Q_0 = 180; \ \mathrm{m^3/s}; \quad a_1 = 0.08 \ \mathrm{m^5/N\,s}; \quad a_2 = 2.0 \times 10^{-5} \ \mathrm{m^7/N^2\,s};$$

$$a_4 = 1.482 \times 10^{-12} \ \mathrm{m^9/N^3\,s}, \tag{6.22b}$$

where the pressure P is in $\mathrm{N/m^2}$ (Pa) and the flow Q is in $\mathrm{m^3/s}$. The pressure-flow characteristic for the skirt flow can be approximated by applying Bernoulli's equation:

$$Q_s = Lyc_d\sqrt{\frac{2}{\rho}P}. \tag{6.23}$$

The periphery of the GEM has length $L = 28$ m, and y is the elevation of the bottom of the skirt above the surface. The coefficient $c_d = 0.65$ represents a discharge flow coefficient, and $\rho = 1.23 \ \mathrm{kg/m^3}$ is the density of the air. The fan is represented in the bond graph of part (b) by a general source, and the resistance to skirt flow by a resistance. The effect of the compressibility of the air in the plenum is slight, and is neglected, so that the excess of the fan flow over the skirt flow, $Q_f - Q_s$, produces the vertical velocity of the vehicle, $\dot{y}$. The relation is represented in the bond graph by the transformer with modulus $T = 1/A$, where $A = 40 \ \mathrm{m^2}$ is the area of the vehicle. The weight of the craft is represented by the element S_e, and its mass by the element $I = 4000$ kg.

Step 4 of the procedure for placing causal strokes (p. 343) ends without designations for the source and resistance bonds, as shown in part (c) of the figure. Therefore, step 5 (p. 313) is invoked, as shown in part (d), by adding a virtual compliance C_v to the 0-junction. Steps I–III for writing the equations (p. 344) also are included in part (d). The resulting equations become

$$\frac{dp}{dt} = P/T - mg, \tag{6.24a}$$

$$\frac{dy}{dt} = \frac{1}{I}p, \tag{6.24b}$$

$$0 = Q_f - Q_s - \frac{1}{TI}p$$

$$= Q_0 - a_1 P + a_2 P^2 - a_4 P^4 - Lyc_d\sqrt{\frac{2}{\rho}P} - \frac{1}{TI}p. \tag{6.24c}$$

The momentum of the vehicle, p, is a state variable. The fact that the resistance R is a function of y makes y a state variable and requires the inclusion of equation (6.24b). (In

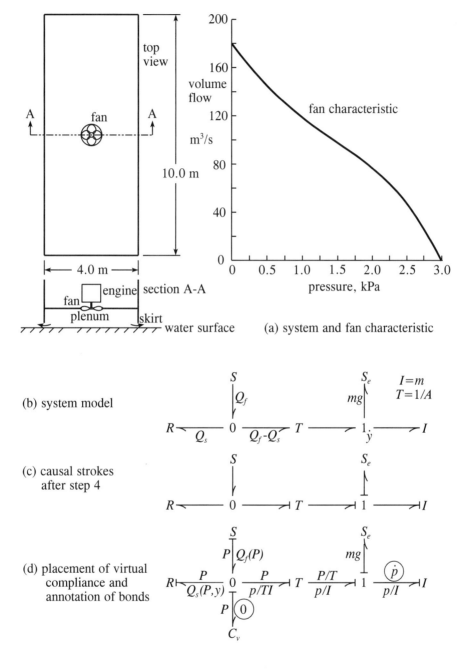

(a) system and fan characteristic

(b) system model

(c) causal strokes after step 4

(d) placement of virtual compliance and annotation of bonds

Figure 6.13: Primitive GEM for vertical stabilization

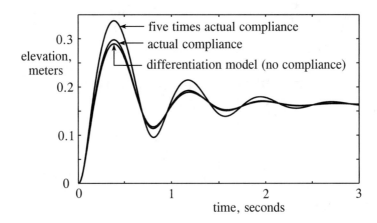

Figure 6.14: Three simulations of the lift-off of the single-plenum GEM

general, state variables are contributed both by energy-storage elements and by resistances and transformers that are functions of displacments; other cases are illustrated in Chapter 9.) Equation (6.24c) theoretically can be solved for the pressure in the plenum, P, in terms of the net flow $Q_f - Q_s = p/TI$, and the result substituted into the differential equations. As a result, the DAE model would be of index zero, and P would not be a true state variable. Unfortunately, however, an analytic solution of such a high-order polynomial equation is not known.

Algebraic equations become differential equations upon differentiation. The index of a DAE is defined as the number of differentiations needed to solve for the time derivatives of the variables represented by the algebraic equations. The higher the index, the more difficult a DAE is to solve, regardless of the method used. The derivative of equation (6.24c) can be solved for dP/dt as a function of p, x and P. This differential equation can be used as the third in a solvable set of ODEs. Therefore P is treated as though it is a third state variable, despite that the minimum order of the model is two.

Solving the derivative of equation (6.24c) for dP/dt gives

$$\frac{dP}{dt} = -\frac{Lc_d\sqrt{2P/\rho}(dy/dt) + (1/TI)(dp/dt)}{a_1 - 2a_2P + 4a_4P^3 + Lyc_d/\sqrt{2\rho P}}. \tag{6.25}$$

A solvable third-order ODE results when this equation is combined with equations (6.24a) and (6.24b). The three dependent variables are p, y and P. The elevation y is plotted in Fig. 6.14 for a simulation that starts at lift-off ("differentiation model"). The initial conditions are $p(0) = 0$, $y(0) = 0.0001$ m (to prevent division by zero) and $P(0) = 3000$ Pa. Note that the number that represents the fan flow at $P = 0$ (180 m^3/s) was lost when equation (5.19c) was differentiated; its effect is recovered through the use of the proper initial condition $P(0) = 3000$ Pa. This value is mandatory. No other value has a legitimate meaning, given the values of $p(0)$ and $y(0)$, since the model has minimum order two, permitting only two *independent* initial conditions.

6.2.5 Method of Non-Zero Virtual Energy-Storages; Case Study Continued

Regardless of the index of a DAE, it is possible to generate an *approximate* ODE substitute by imbuing the virtual compliances C_v and inertances I_v (in the step 5 above) by small but non-zero values. Each algebraic equation is then converted to an approximate differential equation, so any ODE solver can be used. This procedure usually is easier than the differentiation method.

For the current example, the annotations on the bond for the virtual compliance C_v are changed to $\dot{q}_v$ on the flow side and q_v/C_v on the effort side, like any real compliance. Therefore, $P = q_v/C_v$, and equations (6.24) are replaced by

$$\frac{dp}{dt} = \frac{1}{C_v T} q_v, \tag{6.26a}$$

$$\frac{dy}{dt} = \frac{1}{I} p, \tag{6.26b}$$

$$\frac{dq_v}{dt} = Q_0 - \frac{a_1}{C_v} q_v + \frac{a_2}{C_v^2} Q_v^2 - \frac{a_3}{C_v^3} q_v^3 - L y c_d \sqrt{\frac{2}{\rho C_v} q_v} - \frac{1}{TI} p. \tag{6.26c}$$

The principal questions now are what value should C_v be given, and how significant an error this compliance introduces into the model and its solution.

A *more* accurate model of the system, it so happens, would recognize the compressibility of the air in the plenum, resulting in an actual compliance in place of the virtual compliance. Treated as a constant, this compliance would have the approximate value $V_0/kP_0 = 0.265 \times 10^{-3}$ m^5/N, where V_0 is the nominal volume of the plenum (a little greater than its minimum value, due to the mean elevation y of the craft), P_0 is the mean absolute pressure (virtually 101 kPa) and k is the ratio of specific heats (1.4).

The result of a simulation with this model also is given in Fig. 6.14. It is barely different from the solution with $C_v = 0$, demonstrating that compressibility is of virtually negligible effect. Execution of the simulation is relatively slow, however, due to the smallness of C_v. (In absolute terms, however, the simulation is not excessively slow for this simple model.) Pressure excursions from the nominal moving equilibrium have very rapid dynamics relative to the basic motion of the vehicle. Simulations must proceed at a rate dictated by the need to have many time steps per cycle of the fastest phenomenon in the model, even if this phenomenon is not of interest. The execution can be sped up by using a larger value of C_v, but at a cost of inaccuracy. A faster simulation with five times the proper compliance is included in Fig. 6.14. The error is significant, although it might be acceptable for some purposes. A trade-off between accuracy and computing time is characteristic of the method of non-zero virtual compliance or inertance.

Models combining both relatively very rapid and very slow phenomena are called "stiff." The MATLAB "help" feature suggests Runge-Kutta alternatives to `ode45`. Alternatively, multi-step integrators typically are superior to single-step Runge-Kutta integrators for handling stiff models.

6.2.6 Commercial Software for DAEs

The solving of DAEs in general is a major enterprise. Should you deal with them often, you probably should avail yourself of a software package designed explicitly for them. Two presently popular packages are DASSL[1] and LSODI. [2] These are based on multistep backward-differentiation formulas (BDF). Extensive discussions of these formulas and the programs are available.[3] These programs work well for most index zero and index one sys-

---

[1]L.R. Petzold," A description of DASSL: A differential/algebraic system solver," *Scientific Computing*, eds. R.S. Stepleman et al., North-Holland, Amsterdam, 1983, p. 65-68.

[2]A.C. Hindmarsh, "LSODE and LSODI, two new initial value ordinary differential equation solvers," *ACM-SIGNUM Newsletters*, v 15, 1980, p. 10-11.

[3]K.E.Brenan, S.L. Campbell and L.R. Petzold, *Numerical Solution of Initial-Value Problems in Differential-Algebraic Equations*, North-Holland, Elsevier, New-York, 1989.

tems, and some index two systems. They are beyond the scope of this book, however, which focuses on modeling rather than numerical methods. The MATLAB® integrator `ode15s` is a variable-order integrator for "stiff" systems that you can readily use to handle a class of DAEs, however.[4]

6.2.7   Case Study With Meshes*

Some bond-graph meshes, or closed loops of bonds, produce special complications. Sophisticated or basic methods can be used to overcome these complications. For simple even meshes a sophisticated method substitutes a tree-like equivalent bond graph, as given in Fig. 5.19 (p. 313). (Care must attend the treatment of the power-convention half-arrows should they not be aligned initially in the manner illustrated.) More generally, an even or odd mesh can be removed by reformulating the model with different state variables, based upon examination of relations for the stored energy, as described in Chapter 9. For the present, no such cleverness is attempted for a complex bond graph with a mesh: the basic method for finding differential equations is applied directly to the given graph.

Step 5 of the procedure for designating causal strokes (the introduction of a virtual compliance or virtual inertance when a model is under-causal) sometimes produces an uncorrectable causal conflict when a bond-graph mesh is present. In these cases, the offending virtual element must be removed. Some different junction should be chosen, the proper virtual energy-storage element appended, and the resulting graph checked for causal compatability. Occasionally, an apparent causal conflict can be corrected by switching the causality of an energy-storage element from integral to differential. You might prefer, however, to retain the original state variable by selecting a different virtual element, if this is possible.

The classical model for the small-motion dynamics of an assumed rigid vehicle shown in part (a) of Fig. 6.15 illustrates the problem and its solution. Rolling and sideways motions are omitted, so there are two "degrees of freedom": that is, two generalized displacements and their derivatives are required to specify the state of the system. The two displacements could be y_1 and y_2, which are the vertical displacements of the two axles. However, the vertical displacement of the center of mass, y_{cm}, and the pitch angle, ϕ, are more convenient, since the kinetic and potential energies are more readily expressed in terms of these and their time derivatives. The ground under the front and rear axles has vertical displacements $y_{g1}(t)$ and $y_{g2}(t)$, respectively, relative to some reference level; the time derivatives of these displacements will be considered to be the excitations of the system. The suspension at each axle (springs, tires and shock absorbers) is modeled simply (and rather crudely) by a parallel spring-dashpot combination, with parameters k_1, b_1, k_2 and b_2, as shown. The vehicle has a known mass, m, and known moment of inertia, J, about a known center of mass.

The geometric constraints between the velocities of the axles and the center of mass are

$$v_{cm} = \frac{L_2}{L_1 + L_2}\dot{y}_1 + \frac{L_1}{L_1 + L_2}\dot{y}_2, \tag{6.27a}$$

$$\dot{\phi} = \frac{1}{L_1 + L_2}(\dot{y}_2 - \dot{y}_1). \tag{6.27b}$$

These lead directly to the junction structure of the bond graph shown in part (b) of the figure.

---

[4]L.F. Shampine, M.W. Reichelt and J.A. Kierzenka, "Solving Index-1 DAE's in MATLAB and Simulink," *SIAM Review*, v. 41, 1999, p. 538-552.

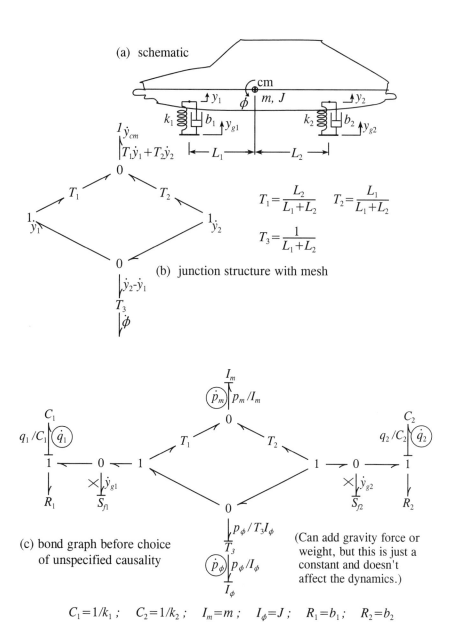

(a) schematic

$$T_1 = \frac{L_2}{L_1 + L_2} \quad T_2 = \frac{L_1}{L_1 + L_2}$$

$$T_3 = \frac{1}{L_1 + L_2}$$

(b) junction structure with mesh

(c) bond graph before choice
of unspecified causality

(Can add gravity force or
weight, but this is just a
constant and doesn't
affect the dynamics.)

$$C_1 = 1/k_1 ; \quad C_2 = 1/k_2 ; \quad I_m = m ; \quad I_\phi = J ; \quad R_1 = b_1 ; \quad R_2 = b_2$$

Figure 6.15: Model of vehicle dynamics with bond graph representation

Expressions for the kinetic and potential energies of the model in terms of the defined displacements and velocities are written as the next step in the modeling process. These are

$$T = \frac{1}{2}mv_{cm}^2 + \frac{1}{2}J\dot{\phi}^2,$$

(6.28)

$$V = \frac{1}{2}k_1 q_1^2 + \frac{1}{2}k_2 q_2^2,$$

(6.29)

where q_1 and q_2 are the extensions of the springs from their nominal length:

$$q_1 = y_1 - y_{g1},$$

(6.30a)

$$q_2 = y_2 - y_{g2}.$$

(6.30b)

These allow the bond graph to be completed as shown in part (c) of the figure.

Integral causality is applied successfully to all four energy storage elements, as also shown in part (c), revealing that the model is indeed fourth order. The causalities of many bonds, including the mesh bonds, are not mandated by this causality, however, indicating an under-causal model. Therefore, a virtual energy-storage element is to be bonded to one of the causally incompleted junctions. If you try to append a virtual inertance to either of the mesh 1-junctions, however, a causal conflict results, as illustrated in part (a) of Fig. 6.16. This conflict is correctable by switching the causality of the I_ϕ element to differential form. Although acceptable, this approach is more complicated than necessary; a simpler method exists.

As a second attempt, a virtual compliance is added to the upper mesh 0-junction in part (b) of the figure. This produces no causal conflict, but still leaves the bond graph under-causal. Thus, a second virtual compliance is added to one of the remaining 0-junctions with incomplete causality, as illustrated in part (c). Now, the causal strokes are completed without conflict, and the efforts and flows on all the bonds can be annotated in the usual manner, as shown. To reduce the clutter on the graph, it is helpful to define the flows f_1 and f_2 as functions of the state variables and the efforts e_1 e_2 of the virtual compliances:

$$f_1 = f_2 - \frac{1}{T_3 I_\phi},$$

(6.31a)

$$f_2 = \dot{y}_{g2} - \frac{e_1 + e_2}{R_2} - \frac{q_2}{R_2 C_2}.$$

(6.31b)

The algebraic equations associated with the circled zeros on the bond graph are

$$0 = f_1 - \frac{e_2}{R_1} + \frac{q_1}{R_1 C_1} - \dot{y}_{g1},$$

(6.32a)

$$0 = T_1 f_1 + T_2 f_2 - \frac{p_m}{I_m}.$$

(6.32b)

Substituting equation (5.26a) into equations (6.32), and using the fact that

$$T_1 + T_2 = 1$$

(6.33)

reduces these equations to

$$0 = f_2 - \frac{p_\phi}{T_3 I_\phi} - \frac{e_2}{R_1} + \frac{q_1}{R_1 C_1} - \dot{y}_{g1},$$

(6.34a)

$$0 = f_2 - \frac{T_1 p_\phi}{T_3 I_\phi} - \frac{p_m}{I_m}.$$

(6.34b)

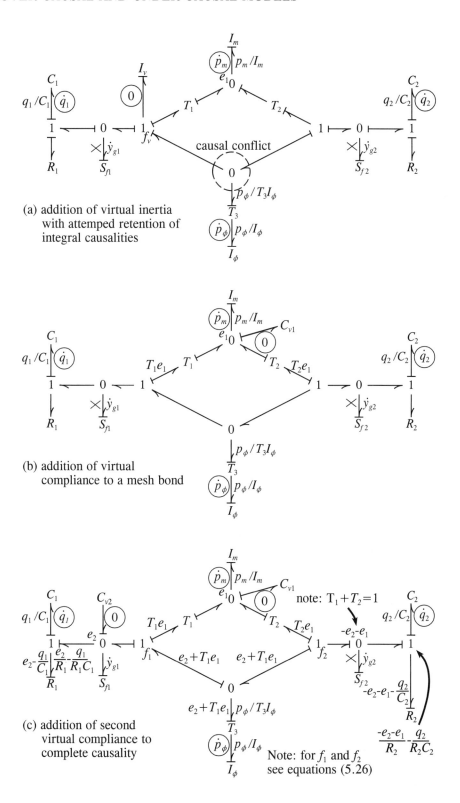

Figure 6.16: Continuation of analysis from Fig. 6.15

Solving equation (6.32a) for e_2, substituting equation (6.31b) for f_2 and noting equation (6.33) gives the effort e_2 in terms of state variables, as desired:

$$e_2 = -\frac{T_2 R_1 p_\phi}{T_3 I_\phi} + \frac{R_1 p_m}{I_m} + \frac{q_1}{C_1} - R_1 \dot{y}_{g1}. \tag{6.35}$$

To find the corresponding equation for e_1, equation (6.31b) can be substituted into equation (6.34b) to eliminate f_2, giving

$$e_1 = R_2 \dot{y}_{g2} - e_2 - \frac{q_2}{C_2} - \frac{T_1 R_2 p_\phi}{T_3 I_\phi} - \frac{R_2 p_m}{I_m}. \tag{6.36}$$

Substituting equation (6.35) therein gives the effort e_1 in terms of the state variables, as desired:

$$e_1 = R_2 \dot{y}_{g2} + R_1 \dot{y}_{g1} + \frac{(T_2 R_1 - T_1 R_2) p_\phi}{T_3 I_\phi} - \frac{(R_1 + R_2) p_m}{I_m} - \frac{q_1}{C_1} - \frac{q_2}{C_2}. \tag{6.37}$$

Finally, the four state differential equations, which employ e_1 and e_2 on their right sides, become

$$\frac{dp_m}{dt} = e_1, \tag{6.38a}$$

$$\frac{dp_\phi}{dt} = \frac{1}{T_3}(e_2 + T_1 e_1), \tag{6.38b}$$

$$\frac{dq_1}{dt} = \frac{e_2}{R_1} - \frac{q_1}{R_1 C_1} = -\dot{y}_{g1} - \frac{T_2 p_\phi}{T_3 I_\phi} + \frac{p_m}{I_m}, \tag{6.38c}$$

$$\frac{dq_2}{dt} = -\frac{e_1 + e_2}{R_2} - \frac{q_2}{R_2 C_2} = -\dot{y}_{g2} + \frac{T_1 p_\phi}{T_3 I_p hi} + \frac{p_m}{I_m}. \tag{6.38d}$$

The choice of variables above and the resulting differential equations are not the only ones possible. A recast model presented in Fig. 9.14 (p. 620) for this problem has no mesh, saving effort at the cost of increased required competence.

EXAMPLE 6.7

Determine whether the three mesh-type bond graphs below are reducible to tree-like equivalences. Then, determine whether each of the graphs is causal, over-causal or under-causal. Finally, define state variables and write the corresponding set of state differential equations for case (ii).

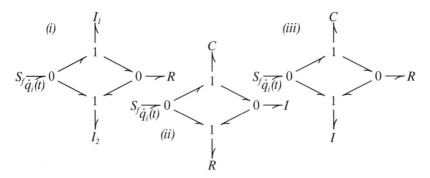

Solution: Three of the mesh bonds have clockwise power arrows and one has a counterclockwise power arrow in all three cases. Therefore, the meshes are *odd* and not reducible to tree-like structures. The mandated causal strokes are as follows:

In case (i) the causalities of the two inertances could be inverted, but regardless of the choice only one of them can have integral causality, so the model is over-causal (and of first order). In case (ii), several bonds have no mandated causalities, so the model is under-causal. In case (iii), both energy storage elements have integral causality and no bonds are left unspecified, so the model is causal.

The under-causal case (ii) is the most difficult to address. An attempt to place a virtual inertance on the lower 1-junction produces a causal conflict, as shown below left. A second attempt, this time placing a virtual compliance on the left-hand 0-junction, produces a satisfactory result, as shown below right. Therefore, the efforts and flows on this bond graph are annotated in the standard fashion, as shown.

The algebraic equation associated with the circled zero is

$$0 = q_i(t) - \frac{4e_v}{R} + \frac{2q}{RC} - \frac{p}{I},$$

which gives

$$e_v = \frac{q}{2C} + \frac{Rq_i(t)}{4} - \frac{Rp}{4I}.$$

The state differential equations become

$$\frac{dq}{dt} = \frac{2e_v}{R} - \frac{q}{RC} - \frac{p}{I} = \frac{1}{2}\dot{q}_i(t) - \frac{3}{2I}p,$$

$$\frac{dp}{dt} = e_v - \frac{q}{C} = -\frac{1}{2C}q + \frac{R}{4}\dot{q}_i(t) - \frac{R}{4I}p.$$

6.2.8 Summary

Over-causal models are those for which it is impossible to apply integral causality compatibly to all the inertance and compliance elements. The differential causality on an energy-storage element implies that its energy is not independent of the state variables for the other energy-storage elements. The basic 4-step method for writing differential equations still applies, but requires special attention to insure faithful application. Time derivative terms appear on both sides of the differential equation that involves an element with derivative causality. In most cases these terms can be collected on one side, and the equation solved for the derivative. The result is a set of differential equations equal in number to the number of energy storage elements with integral causality plus the number of resistances or inertances that depend on a displacement that is not otherwise a state variable.

Under-causal models are those for which the basic four-step procedure for placing the causal strokes (p. 343) fails to define all of them. A fifth step appends a virtual compliance to an affected 0-junction, or a virtual inertance to an affected 1-junction. When a bond-graph mesh is present, this step sometimes produces a causal conflict. Should this happen, a different junction must be chosen for appending a virtual inertance or compliance. One algebraic equation is produced for each virtual energy storage element in a bond graph, because of the zero value of its modulus.

The algebraic equations emerging from step 5 may or may not be solvable for the defined effort or flow in terms of the input and state variables. If they are so solved, the writing of differential equations can proceed normally. Otherwise, the model comprises a set of one or more algebraic equations as well as differential equations, called a DAE (differential-algebraic equation).

DAEs are more difficult to solve numerically than ODEs (ordinary differential equations). Special software is widely available, including the MATLAB integrator `ode15s`. Alternatively, it is possible to approximate a DAE by an ODE by conferring small non-zero values to whatever virtual compliances and inertances have been introduced by the procedure for placing causal strokes. The smaller the moduli chosen, the more accurate the approximation but the more difficult the solution. Simulation based on multistep BDF algorithms (backward-differentiation formulas) usually work better than single-step Runge-Kutta algorithms.

In a more accurate procedure, algebraic equations within a DAE set are differentiated to give differential equations, and the resulting expanded set of differential equations is simulated with a simple ODE integrator. Care must be taken to employ the proper initial conditions for the added differential equations.

Guided Problem 6.4

The system of Fig. 6.17 is used here to illustrate the two basic methods for handling dependent energy storages. Start by drawing a detailed bond graph for the system, and defining the moduli of all elements. Then apply causal strokes, and note that differential causality appears for one of the inertances. Proceed to find the corresponding differential equations, using the standard method. Then, as an alternative approach, find a bond graph which has only one inertance; also find its differential equations, and compare with the first set.

Suggested Steps:

1. Carry out the first part of the solution; it is suggested that the angular velocity $\dot{\phi}$ describe one of the 1-junctions, and that the two radii become transformer moduli. To get the correct sign for the force source, note whether the input work is positive or negative for your defined velocity. Note that the inertance bond with differential causality has no state variable; its effort should be labeled as the product of the inertance and the *time derivative* of the flow. Annotate both sides of all bonds, and then write the differential equations.

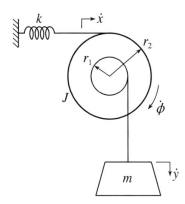

Figure 6.17: System of Guided Problem 6.4

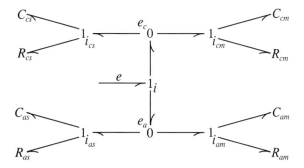

Figure 6.18: Isothermal model of a cell of a lead-acid battery

2. Drop one of the inertances from the bond graph. To let the remaining inertance represent both of the original inertances, write an expression for its stored energy in terms of its velocity, and compare to an expression for the sum of the original two stored energies expressed as a function of the same velocity. Find the differential equations.

3. Are the differential equations the same? Are they equivalent? Which set more clearly reveals the behavior?

Guided Problem 6.5

This problem shows different ways the under-causal model for a cell of a lead-acid battery, as discussed in Section 5.1 (p. 260-261) and repeated in Fig. 6.18, can be treated computationally. Write state differential equations through the use of small virtual inertias (which is the approach used by the cited authors). Also, investigate the possibility of using one or more virtual compliances. Finally, indicate your preferred approach, for which you should also develop a set of DAE equations for which the moduli of any virtual elements is identically zero.

Suggested Steps:

1. Place the mandated causal strokes on the bond graph, using integral causality. Verify that the model is undercausal.

2. Place a single virtual inertia in a central location, in hopes of completing the causality. Propagate the causality insofar as mandated. Are bonds left still without causal strokes?

3. Place more virtual inertias until the causal strokes are complete. How many have you needed?

4. Try the alternative idea of placing one or more virtual compliances. How many are needed? Which approach do you prefer?

5. Annotate the efforts and flows in the preferred bond graph. The compliances and resistances are nonlinear.

6. Write a set of DAE equations from the annotated graph. Note when this could be reduced readily to a minimal set of differential equations.

7. Note how the algabraic equation(s) could be approximated by (an) additional differential equation(s).

Guided Problem 6.6

This relatively complex problem compares the the use of virtual energy-storage elements with the differentiation method in the treatment of under-causality. It extends the modeling and analysis for the ground-effect machine addressed in Sections 6.2.4 and 6.2.5.

Roll stability is imparted to the GEM by introducing two internal skirts, as shown in Fig. 6.19. The lower edges of these skirts are in the same plane as the external skirts. When the GEM rolls to the right, the resulting increase in the resistance to flow for the right-side skirt and the decreased resistance to flow for the left-side skirt produces a higher pressure in the right-side plenum than the left-side plenum. These pressures impart a correcting moment, or roll stiffness, to the vehicle. You are asked to model the system with a bond graph that ignores the small compliances of the plenums and proceed to develop the corresponding describing equations. You will see that the algebraic equations are too complex to permit direct reduction of the model to a minimal set of state differential equations.

The simplest approach confers non-zero values to the virtual compliances, increasing the order of the model by three. The differentiation approach is more difficult and also increases the order of the model by three, but does not modify the model. The effort required for you to carry out the details may not be justified, but it is instructive for you to establish a

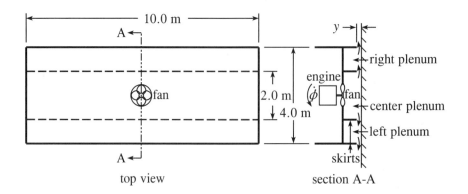

Figure 6.19: GEM with internal skirts for Guided Problem 6.6

plan. The steps below and the solution at the end of the section outline the procedure and describe the solutions, which illustrate important conclusions.

The two models are simulated first with no roll angle, partly to determine the equilibrium elevation. The physical compliances and five times these values are employed separately. Next, the behavior following an initial roll velocity of 0.05 rad/s at the equilibrium elevation is simulated. The radius of gyration of the GEM in roll is taken as 1.5 meters; other parameters are as before or as shown in the figure. The various results are compared with respect to the behavior of the physical design, the accuracies of the simulations and the computational time required.

Suggested Steps:

1. Define variables and parameters. Draw a bond graph for the system, omitting compliances. Each of the three plenums merits a 0-junction, and there are 1-junctions and inertias for both elevation and roll.

2. Apply the first four steps for placing causal strokes. Note that the model is under-causal.

3. Carry out step 5 of the procedure for causal strokes.

4. Annotate the bond graph in the recommended manner.

5. Write the state differential equations and associated algebraic equations.

6. The algebraic equations cannot be solved analytically for the non-state variables in terms of the state variables. The simplest approach employs non-zero values of the virtual energy-storage elements. Write the corresponding set of state differential equations.

7. Find the differential equations for the derivative method. This is done by taking the derivatives of the algebraic equations. A 3×3 matrix equation can be written, with the time derivatives of the three pressures as the dependent variable. This equation is readily solved by MATLAB.

8. Establish the initial conditions for lift-off (with no roll) for the three models. Also find the initial conditions when the physical compliance is increased by a factor of five. Note that the pressures in all three plenums is non-zero, and that the center pressure is higher than the others. A very small initial elevation is required to prevent the differentiation model from trying to invert a singular matrix (which would make it an "index two" problem).

9. Carry out three-second simulations for lift-off, and compare with each other and with the response in Fig. 6.14 (p. 378). Note how many seconds your computer takes to carry out the simulation.

10. Establish the initial conditions for equilibrium height and roll angle but the specified roll angular velocity.

11. Carry out the respective simulations, and again compare.

PROBLEMS

6.27 The element Z is either an I, C or R in the two configurations shown. Determine whether the six cases are causal, undercausal or overcausal.

(a) 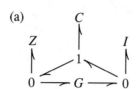   (b)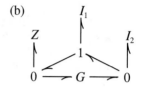

6.28 Augment the adjacent bond graph with causal strokes, define state variables, and write the corresponding set of state variable differential equations.

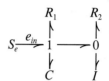

6.29 For the bond graph given below left:

(a) Apply causal strokes to find the order of the system, assuming constant parameters.

(b) Find the set of state differential equations corresponding to your causal strokes.

(c) Reduce the number of energy storage elements to a minimum and find the corresponding state variables and differential equations. Check to make sure they agree with the results of part (b).

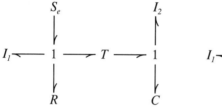   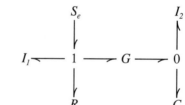

6.30 Answer the questions of the preceding problem for the bond graph above right.

6.31 The bond graph below left has constant parameters.

(a) Assign causality, identify state variables, and write state differential equation(s).

(b) Find the output flow $\dot{q}_0$ as a function of the input effort e_0.

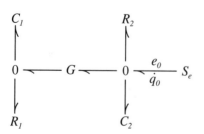   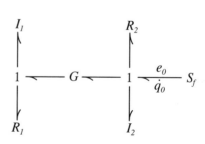

6.32 Answer the questions of the preceding problem for the bond graph above right, except note that the input is $\dot{q}_0$ and the output is e_0.

6.33 A hydraulic shock absorber of the type described in Fig. 2.24 (p. 39) is employed as the dashpot in the system of Problem 5.25 (p. 327). Write the state differential equations for the resulting nonlinear model. (This is a modification of Problem 6.20, p. 365.)

6.34 Write the state differential equation(s) for the model shown in part (d) of Fig. 6.7 (p. 356), using the differential causality given. Compare the result to that for the equivalent model with a single energy storage element.

6.35 Repeat the above problem for the model shown in part (d) of Fig. 6.8 (p. 357).

6.36 The motor of Problem 5.64 (p. 334) has a rotational inertia $J = 0.3$ in.·lb·s^2 and the cylinder has a radius of gyration $r_g = 8$ in.

(a) Repeat part (a) of that problem to include dynamic effects.

(b) Assign causal strokes to the bonds. Note the presence of differential causality. State the order of the model.

(c) Carry out the procedure illustrated in Guided Problem 6.4 to combine the energy of the element with differential causality with that of a different element, so as to eliminate the differential causality.

(d) Write state variable differential equation(s) for the model. The electromagnetic torque on the motor may be left in the form $M = M(\dot{\phi})$.

(e) Simulate the response of the system, starting from an initial motor speed of 1.5 rad/s. Plot the angular velocity for the motor and the displacement of the center of the roller as functions of time until equilibrium is virtually reached. The function $M(\dot{\phi})$ can be approximated by $M = 10 + 16\dot{\phi} - 3.5423\dot{\phi}^2 + 1.146\dot{\phi}^3 - 0.1667\dot{\phi}^4$.

6.37 Augment the adjacent bond graph with causal strokes, define state variables, and write the corresponding state variable differential equation(s).

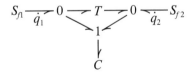

6.38 The bond graph model below comprises elements with constant moduli.

(a) Define state variables, and write a set of state differential equations.

(b) Sketch an engineering system for which the bond graph could be a reasonable model, and relate the parameters of the bond graph to those of the engineeering system. Suggestion: Let the source be electrical, and the right-hand portion of the model represent a mechanical subsystem.

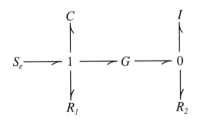

6.39 The adjacent bond graph has constant parameters I, C, R, T. The source is characterized by a known function $e_s = f(\dot{q}_s)$ or $\dot{q}_s = g(e_s)$.

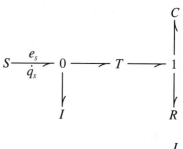

(a) Determine the values of e_s and $\dot{q}_s$ at equilibrium in terms of the given parameters and/or functions. *Repeat from Problem 5.62.*

(b) Define state variables, and find a set of state differential equations.

6.40 Answer the preceding question for the adjacent bond graph.

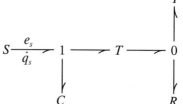

6.41 The bond graph model below comprises elements with constant moduli.

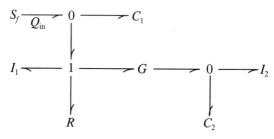

(a) Apply causal strokes and determine the order of the model. Is it causal, over-causal or under-causal?

(b) Identify state variables, and annotate all the efforts and flows on the bond graph consistent with the causal strokes.

(c) Write a solvable set of first-order differential or DAE equations. The differential equations should have the derivatives only on their left sides, in standard fashion.

6.42 The bond graph at the right comprises elements with constant moduli. Identify state variables, annotate the graph and write a solvable set of state differential or DAE equations.

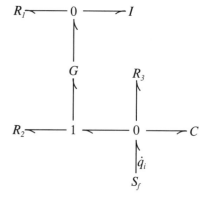

6.43 Consider the bond graph given in Problem 5.8 (p 268) for the effect of a drain tube on human hearing, except omit the transformer. *Continued in Problem 7.6.*

(a) Define state variables and write differential equations without changing the given bond graph with its mesh. (Note that integral causality specifies the causalities of all the bonds, making this a relatively simple problem.)

(b) Apply a bond graph equivalence from Fig. 5.19 (p. 313) to eliminate the mesh. Find the differential equations in terms of the same state variables as in part (a), and compare to make sure they are the same.

6.44 Repeat the problem above for the adjacent bond graph. Note that integral causality does not specify the causalities of all the bonds, making this a more difficult problem.

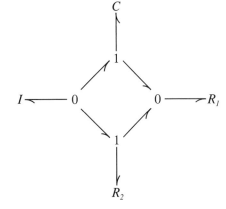

6.45 The bond-graph model below, comprising elements with constant moduli, is closely related to that for the vehicle dynamics in Fig. 6.15 (p. 381). Apply causal strokes and determine the order of the model and whether it is causal, over-causal or under-causal. Identify state variables, annotate the bonds and write a set of first-order differential or DAE equations.

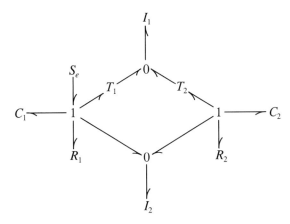

6.46 Consider the ground effect machine (GEM) described in Section 6.2.4 (pp. 376-379) and Fig. 6.13 and modeled by equations (6.22)–(6.24).

(a) Determine the equilibrium values of the state variables y and p.

(b) Linearize the differential equations for perturbations about the equilibrium.

(c) Determine the natural frequency and damping ratio of the vertical (heave) behavior. Compare these results to the corresponding nonlinear simulation in Fig. 6.14.

6.47 Consider that the ground-effect machine discussed in Sections 6.2.4 and 6.2.5 hovers over water that rises and falls uniformly according to the formula $z = z_0 \sin \omega t$. Carry out simulations for frequencies $\omega_1 = 2$ rad/s, $\omega_2 = 8$ rad/s and $\omega_3 = 32$ rad/s, all with the amplitude $z_0 = 0.05$ meters. The initial conditions can be for equilibrium conditions.

6.48 The ground-effect machine discussed in Sections 6.2.4 and 6.2.5 lacks roll stability and pitch stability, and the improvement in Guided Problem 6.6 lacks pitch stability. The design pictured below employs five plenums with internal and external skirts all having their lower edges in the same plane. Model this system with a bond graph, as a first step toward an analysis.

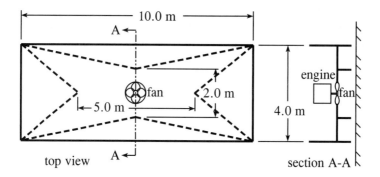

6.49 Guided Problem 6.6 (pp. 388-389, 397-400) finds nonlinear models for a ground-effect machine (GEM) with roll stiffness. Linearize the model with the actual compliances for roll motion. Compute and plot the roll angle for an initial condition that is at equilibrium except for an initial angular velocity of 0.05 rad/s, and compare with the plot on page 400. (The parameter values give in the solution to the guided problem may be used.

SOLUTIONS TO GUIDED PROBLEMS

Guided Problem 6.4

1.

$$C \xrightarrow[\dot{x}]{x/C} T_x \xleftarrow{x/T_xC} 1 \underset{\substack{\downarrow p/I_\phi \\ \dot{p} \\ I_\phi}}{\overset{T_y e}{\underset{p/I_\phi}{\phi}}} T_y \xleftarrow[\substack{I_yT_y\dot{p}/I_\phi}]{\overset{e}{T_yp/I_\phi}} 1 \xleftarrow{mg} S_e$$

$$\begin{aligned}
T_x &= r_2\\
T_y &= r_1\\
I_\phi &= J\\
I_y &= m\\
C &= 1/k
\end{aligned}$$

$$e = (I_yT_y/I_\phi)\dot{p} - mg$$

$$\frac{dx}{dt} = \frac{T_x}{I_\phi}p = \frac{r_2}{J}p \quad (a)$$

$$\frac{dp}{dt} = -\frac{T_x}{C}x - T_y\left(\frac{I_yT_y}{I_\phi}\frac{dp}{dt} - mg\right) \quad (b)$$

Collecting terms in equation (b),

$$\left(1 + \frac{I_yT_y^2}{I_\phi}\right)\frac{dp}{dt} = \frac{T_x}{C}x + T_ymg$$

or $\left(1 + \dfrac{mr_1^2}{J}\right)\dfrac{dp}{dt} = kr_2x + r_1mg$ (c)

Equations (a) and (c) are a complete set of state differential equations.

Note that $\dot{\phi} = p/J$.

2.   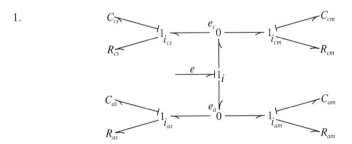

$\mathcal{V} = \dfrac{1}{2}kr_2^2\phi^2 = \dfrac{1}{2C'}\phi^2$

therefore, $C' = 1/kr_2^2$

$\mathcal{T} = \dfrac{1}{2}J\dot{\phi}^2 + \dfrac{1}{2}m\dot{y}^2 = \dfrac{1}{2}(J + mr_1^2)\dot{\phi}^2 = \dfrac{1}{2}I\dot{\phi}^2$; therefore, $I = J + mr_1^2$

$\dfrac{d\phi}{dt} = \dfrac{1}{I}p' = \dfrac{1}{J + mr_1^2}p'$

$\dfrac{dp'}{dt} = r_1mg - \dfrac{1}{C'}\phi = r_1mg - kr_2^2\phi$

3.   These are the same equations, since $\dot{\phi} = \dfrac{1}{J + mr_1^2}p' = \dfrac{1}{J}p$

or $p' = \left(1 + \dfrac{mr_1^2}{J}\right)p$ and $\phi = x/r_2$.

Although the first method shows the structure in more detail, the second method is easier and shows the behavior more clearly.

Guided Problem 6.5

1.

There is no propagation of the causal strokes, so the model is indeed under-causal.

2.

Only two additional bonds have mandated causalities, so the single virtual inertia is insufficient.

3.

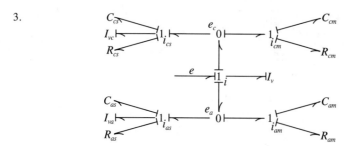

Two more virtual inertias are needed to mandate all causal strokes.

4.

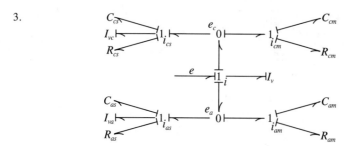

Placing a virtual compliance on just one of the zero junctions is sufficient to complete all causal assignments. This should be the preferred approach.

5.

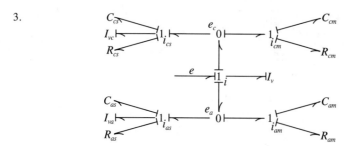

$$e_{Rcs} = e_v - e_{cs}$$

$$e_{Rcm} = e_v - e_{cm}$$

$$e_{Ras} = e - e_v - e_{as}$$

$$e_{Ram} = e - e_v - e_{am}$$

6.   The nonlinear DAE equations become, using the functional dependencies and equations above,

$$\frac{dq_{cs}}{dt} = e_v - e_{cs}$$

$$\frac{dq_{cm}}{dt} = e_v - e_{cm}$$

$$\frac{dq_{as}}{dt} = e - e_v - e_{as}$$

$$\frac{dq_{am}}{dt} = e - e_v - e_{am}$$

$$0 = i_{as} + i_{am} - i_{cs} - i_{cm}$$

The algebraic equation readily could be solved for e_v if the parameters were constants. In that case, substitution of this e_v into the differential equations would give the minimum and precise state variable model. As it is, however, analytical solution is highly unlikely.

7. The algebraic equation can be converted into an approximate fifth differential equation by replacing the left side with dq_v/dt and, in the right side, replacing e_v with q_v/C_v.

Guided Problem 6.6

1.

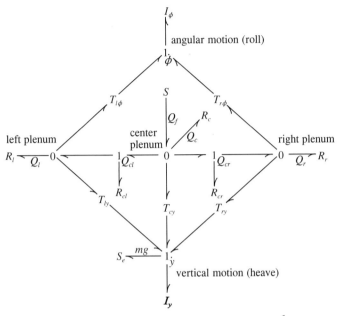

The area of the deck is $A = 40$ m^2, and $T_{ry} = T_{ly} = \dfrac{1}{A/4} = 0.100$ m^{-2}; $T_{cy} = \dfrac{1}{A/2} = 0.050$ m^{-2}.

The width of the deck is $w = 4$m, and $-T_{l\phi} = T_{r\phi} = \dfrac{1}{(A/4)(3w/8)} = \dfrac{1}{15}$ m^{-3}.

From the given information, $I_y = 4000$ kg; $I_\phi = mr_g^2 = 9000$ kg·m^2.
Repeating equation (5.17), $Q_f(P_c) = Q_0 - a_1 P_c + a_2 P_c^2 - a_4 P_c^4$;

$Q_0 = 180$ m^3/s; $a_1 = 0.08$ m^5/N·s; $a_2 = 2.0 \times 10^{-5}$ m^7/N^2·s; $a_4 = 1.482 \times 10^{-12}$ m^9/N^3·s.

The center plenum has a periphery of *outer* skirt of length $L_c = 4$m, with flow $Q_c = L_c y c_d \sqrt{\dfrac{2}{\rho} P_c}$.

Outer skirt area of the left-hand plenum: $A_l = y \left(L + \dfrac{w}{2} \right) - \phi w \left(\dfrac{L}{2} + \dfrac{3w}{16} \right)$

Outer skirt area of the right-hand plenum: $A_r = y \left(L + \dfrac{w}{2} \right) + \phi w \left(\dfrac{L}{2} + \dfrac{3w}{16} \right)$

The respective flows are $Q_l = A_l c_d \sqrt{\dfrac{2}{\rho} P_l}$ and $Q_r = A_r c_d \sqrt{\dfrac{2}{\rho} P_r}$.

The flows across the internal skirts are $Q_{cl} = L \left(y - \dfrac{w\phi}{4} \right) c_d \sqrt{\dfrac{2}{\rho}(P_c - P_l)}$

and $Q_{cr} = L \left(y + \dfrac{w\phi}{4} \right) c_d \sqrt{\dfrac{2}{\rho}(P_c - P_r)}$

2.

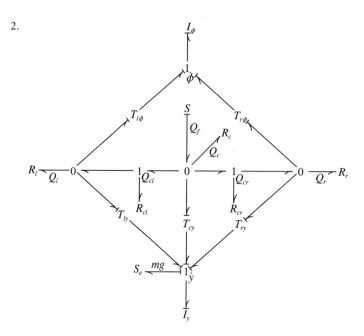

3-4.

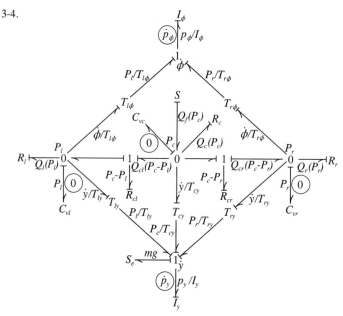

The four state variables p_y, p_ϕ, y and ϕ give four differential equations in terms of the functions defined in step 1:

$$\frac{dy}{dt} = p_y/I_y$$

$$\frac{d\phi}{dt} = p_\phi/I_\phi$$

$$\frac{dp_y}{dt} = \frac{1}{T_{ly}}P_l + \frac{1}{T_{cy}}P_c + \frac{1}{T_{ry}}P_r - mg$$

$$\frac{dp_\phi}{dt} = \frac{1}{T_{l\phi}}P_l + \frac{1}{T_{r\phi}}P_r$$

There are also three algebraic equations:

$$0 = Q_{cl}(P_c - P_l, y, \phi) - Q_l(P_l, y, \phi) - \dot{\phi}/T_{l\phi} - \dot{y}/T_{ly}$$
$$0 = Q_{cr}(P_c - P_r, y, \phi) - Q_r(P_r, y, \phi) - \dot{\phi}/T_{r\phi} - \dot{y}/T_{ry}$$
$$0 = Q_f(P_c) - Q_c(P_c, y) - Q_{cl}(P_c - P_l, y, \phi) - Q_{cr}(P_c - P_r, y, \phi) - \dot{y}/T_{cy}$$

6. The three state variables V_l, V_r and V_c are added to represent the three volumetric compressions in m^3. In the differential equations of step 5, the pressure P_l is replaced by V_l/C_l, P_r is replaced by V_r/C_r and P_c is replaced by V_c/C_c.

The following three state differential equations replace the derivatives of the pressures:

$$\frac{dV_l}{dt} = Q_{cl} - Q_l - \frac{1}{T_{ly}I_y}p_y - \frac{1}{T_{l\phi}I_\phi}p_\phi$$
$$\frac{dV_r}{dt} = Q_{cr} - Q_r - \frac{1}{T_{ry}I_y}p_y - \frac{1}{T_{r\phi}I_\phi}p_\phi$$
$$\frac{dV_c}{dt} = Q_f - Q_c - Q_{cl} - Q_{cr} - \frac{1}{T_{cy}I_y}p_y$$

Note that C_c represents half of the total compliance, and C_l and C_r each represent one-quarter. The total compliance is about 0.265×10^{-3}m^5/N, as computed in Section 5.2.5.

7.
$$\begin{bmatrix} dP_c/dt \\ dP_l/dt \\ dP_r/dt \end{bmatrix} = \mathbf{F}^{-1}\mathbf{f}, \text{ where } \mathbf{f} = \begin{bmatrix} f_3 \\ f_6 \\ f_{10} \end{bmatrix} \text{ and } \mathbf{F} = \begin{bmatrix} f_1 & -f_2 & 0 \\ f_4 & 0 & -f_5 \\ f_7 & f_8 & f_9 \end{bmatrix}$$

$$f_1 = \frac{\partial Q_{cl}}{\partial(P_c - P_l)}; \quad f_2 = \frac{\partial Q_{cl}}{\partial(P_c - P_l)} + \frac{\partial Q_l}{\partial P_l};$$

$$f_3 = \left(\frac{\partial Q_l}{\partial y} - \frac{\partial Q_{cl}}{\partial y}\right)\dot{y} + \left(\frac{\partial Q_l}{\partial \phi} - \frac{\partial Q_{cl}}{\partial \phi}\right)\dot{\phi} + \frac{1}{T_{l\phi}}\frac{d\dot{\phi}}{dt} + \frac{1}{T_{ly}}\frac{d\dot{y}}{dt}$$

$$f_4 = \frac{\partial Q_{cr}}{\partial(P_c - P_l)}; \quad f_5 = \frac{\partial Q_{cr}}{\partial(P_c - P_l)} + \frac{\partial Q_r}{\partial P_r};$$

$$f_6 = \left(\frac{\partial Q_r}{\partial y} - \frac{\partial Q_{cr}}{\partial y}\right)\dot{y} + \left(\frac{\partial Q_r}{\partial \phi} - \frac{\partial Q_{cr}}{\partial \phi}\right)\dot{\phi} + \frac{1}{T_{r\phi}}\frac{d\dot{\phi}}{dt} + \frac{1}{T_{ry}}\frac{d\dot{y}}{dt}$$

$$f_7 = \frac{\partial Q_f}{P_c} - \frac{\partial Q_c}{P_c} - \frac{\partial Q_{cl}}{\partial(P - c - P_l)} - \frac{\partial Q_{cr}}{\partial(P_c - P_r)}; \quad f_8 = \frac{\partial Q_{cl}}{\partial(P_c - P_l)};$$

$$f_9 = \frac{\partial Q_{cr}}{\partial(P_c - P_r)}; \quad f_{10} = \left(\frac{\partial Q_c}{\partial y} + \frac{\partial Q_{cl}}{\partial y} + \frac{\partial Q_{cr}}{\partial y}\right)\dot{y} + \left(\frac{\partial Q_{cl}}{\partial \phi} + \frac{\partial Q_{cr}}{\partial \phi}\right)\dot{\phi} + \frac{1}{T_{cy}}\frac{d\dot{y}}{dt}$$

During lift-off, the pressures in the side plenums briefly drop below zero because of the rapid rise of the craft caused by the high pressure in the center plenum. Consequently, it is necessary to employ sign and absolute value functions in computing the flows Q_l and Q_r and their derivatives. This problem doesn't quite arise in the simulation with the compliances, since these compliances reduce the pressure decrease enough to keep the pressures positive.

8. At equilibrium, the flow leaving one of the side plenums equals the flow entering. Therefore, for $\phi = 0$ and any y,

$$L\sqrt{P_c - P_s} = \left(L + \frac{w}{2}\right)\sqrt{P_s},$$

where P_s is the pressure in the side plenum and P_c is the pressure in the center plenum. Therefore,

$$\frac{P_c}{P_s} = 1 + \left(1 + \frac{w}{2L}\right)^2 = 2.44.$$

When $y = 0$ the fan flow is zero, so the pressure in the center is 3000 Pa. The reasoning above gives the pressures in the sides as 1042 Pa. The initial compression of the center plenum is therefore $V_c(0) = \frac{1}{2} \times 0.265 \times 10^{-3} \times 3000 = 0.399$ m^3, and in the side plenums is $V_l(0) = V_r(0) = \frac{1}{4} \times 0.265 \times 10^{-3} \times 1229.6 = 0.0818$ m^3.

9-10.

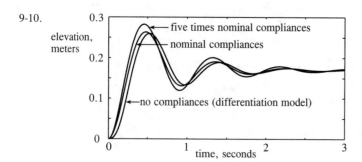

The lift-offs with and without the compliances differ little, although increasing the compliances by five times gives a significant error. The responses are similar to those found in the absence of internal skirts, although the lowered pressures in the outer plenums reduces the equilibrium height. The final elevation is 0.1703 m, which is used as an initial condition for the simulation with roll. The final pressures in the center and side plenums, respectively, are 1392 Pa and 570 Pa. The roll rate of 0.05 rad/s corresponds to an initial condition for the angular momentum of $9000 \times 0.05 = 450$ kg·m/s. For this roll, the proper initial pressure in the left plenum is 13 Pa greater than the 570 Pa, and the initial pressure in the right plenum is 13 Pa less. (These corrections can be found by trial-and-error; any other values gives an equilibrium roll angle different from zero. Note that the minimum order of the differential model is four, so that only four initial conditions can be selected independently.)

11.

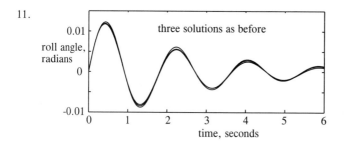

The simulations for roll reveal significantly less damping than those for lift-off, a weakness in the performance. The three simulations are nearly indistinguishable, unlike for lift-off. The reason is that the three pressures do not vary nearly as much as in the case of lift-off.

The six-second simulation for roll with the nominal compliances required 32 seconds of computer time on an old PC computer, using an error factor of 0.0001 with the MATLAB ode23 in version 4. (On a modern PC, with ode45 in version 7, it takes a small fraction of a second, but the point here is the ratios of compute times.) Increasing the compliances by a factor of five reduced the computation time to 5 seconds. The derivative model took only about 2 seconds, despite the fact that each iteration required roughly twice the computation. Had the efficiencies of the computations been increased through the use of script files and global variables, the relative advantage of the derivative model would have been even greater. Nevertheless, it is hard to justify the effort required to carry out its programming. Whenever computer time becomes excessive for the virtual energy-storage method to give adequate accuracy, the use of special software such as DASSL can be justified.

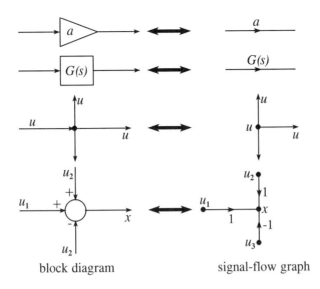

Figure 6.20: Equivalences for block diagrams and signal flow graphs

6.3 The Loop Rule*

You have employed the causal strokes on bond graphs to direct the writing of state differential equations and employed matrix methods to extract differential equations with a single dependent variable. This procedure is sufficient to address any linear lumped-parameter model. An optional graph method for deducing transfer functions of linear models is available, however, which facilitates the deduction of transfer functions without the use of matrices, and lends greater insight into the structures of particular and general system models. It also can be used to find the state differential equations rapidly.

This method is based on the **loop rule**. As originally presented by Shannon and elucidated by Lorens,[5] this method applies directly to **signal flow graphs**. Bond graphs can be converted to signal flow graphs readily, although no general inverse procedure is known. The author has extended the loop rule to apply directly to bond graphs,[6] with consequent simplifications. The description below starts with the definition of the signal flow graph and proceeds to the conversion of a bond graph to this form. The loop rule is then presented. Application of the loop rule to find transfer fuctions directly from bond graphs follows. The procedure is particularly simple to apply to the determination of the state differential equations, which is the final subject.

6.3.1 Signal Flow Graphs

The signal flow graph is a variation of the block diagram, as suggested in Figure 6.20. The block diagram uses triangles or boxes in which coefficients and operators are placed; signal flow graphs write these symbols next to a simple arrowhead instead. The block diagram uses circles to indicate summation, and places plus or minus signs next to these circles to indicate the signs of the variables being summed. The signal flow graph replaces the

---

[5]C. Lorens, *Flowgraphs for the Modeling and Analysis of Linear Systems*, McGraw Hill, 1964.

[6]F.T. Brown, "Application of the Loop Rule to Bond Graphs," *ASME Transactions, J. of Dynamic Systems, Measurement and Control*, v. 94 n. 3 (Sept. 1972) p. 253-261. This work is used as a basis for the Lorenz Simulation software package MS1 (see footnote p. 136).

circles with conspicuous dots and assumes all plus signs in the summations. A negative sign is accommodated by placing a minus sign in front of the coefficient of the particular signal, and labeling a coefficient of -1 if there is no explicit coefficient in the block diagram. Symbols written adjacent to a summation dot represent the output variable or sum itself. Finally, the block diagram lives up to its name by employing vertical and horizontal lines almost exclusively. The tradition of signal flow graphs employs curved lines, which helps one visually identify loops more easily.

A signal flow graph, like the block diagram, therefore represents a set of linear differential and algebraic equations.[7] None of the power and energy constraints assumed with bond graphs are required. One or more directed lines emerging from a node carry the value of the variable represented by that node:

The linear algebraic relation $x = u_1 + u_2 + \ldots + u_n$ is represented by a node with incident lines that represent each of the component terms:

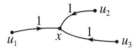

The linear algebraic relation $x = au$ is represented by the arrow itself on the associated line; the symbol a, or whatever symbol represents the coefficient, is labeled next to the arrow. Minus signs can be used where appropriate. If the line conveys a variable directly, without any modifying coefficient a, the symbol 1 is substituted:

$$\overset{a}{\underset{u}{\longrightarrow}} \quad x=au$$

Finally, the differential operation $x = a\,du/dt$ is represented by placing the notation as next to the associated arrow, and the integration operation $x = a\int u\,dt$ is represented by placing the notation a/s next to the arrow:

$$\overset{as}{\underset{u}{\longrightarrow}} \quad x=a\frac{du}{dt} \qquad \overset{a/s}{\underset{u}{\longrightarrow}} \quad x=a\int u\,dt$$

Example 6.8

Draw a signal flow graph for the following set of linear algebraic, differential and integro-differential equations, in which the coefficients $a \ldots k$ are constants:

$$x_1 = \int ax_3\,dt + b\,\frac{dx_3}{dt} + c\,u_2;$$

$$x_2 = -d\,\frac{dx_6}{dt};$$

$$x_3 = e\,x_5 - f\,\frac{dx_6}{dt};$$

$$x_4 = g\,x_5 + h\,u_1;$$

$$x_5 = i\,x_2 + j\,u_2;$$

$$x_6 = k\,x_1 + x_2.$$

---

[7]Nonlinear algebraic functions also can be incorporated, but are avoided in this text.

Solution: These equations are assembled in the signal flow diagram below. Each variable is represented by a node, and the constant coefficients appear next to the arrows, along with the operator s to represent differentiation and $1/s$ to represent integration. You ought to identify the individual equations embedded in the graph.

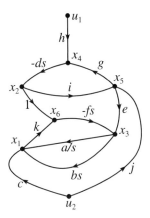

6.3.2 The Loop Rule for Signal Flow Graphs

The loop rule gives the transfer function between any input and any output in one, sometimes simple and sometimes giant, step. If the input in question is labeled as u_j, and the output of interest as x_i, the loop rule can be expressed as

$$x_i = \frac{1}{\Delta} \sum_k L_{ijk} \Delta_{ijk}. \tag{6.39}$$

The symbol L_{ijk} represents the **path gain** for the kth directed path from the input to the output. This is the product of all the coefficients and operators on the path. Paths are distinct if any one segment is distinct; separate paths may traverse some common segments. One entire path does not qualify as part of another path, however. Identifying all the paths and writing a list of their path gains is a major part of the procedure for implementing the loop rule.

The symbol Δ in equation (6.39) is called the **graph determinant**. The graph determinant is in general a complicated function of the gains of **loops** inside the signal flow graph:

$$\Delta = \sum_\ell (1 - L_\ell^1 + L_l^2 - L_\ell^3 + \ldots). \tag{6.40}$$

A loop is a closed directed path from one node back to itself; the gain of the ℓth loop is L_ℓ^1. Therefore, the term $\sum_\ell L_\ell^1$ represents the sum of the gains of all the loops in the graph, where separate loops are identified the same way as are separate paths. The superscript 2 means the *products* of loop gains for *pairs of non-touching loops*. Loops are said to touch if they share one or more common nodes. Loops so often touch one another that, even if there are several loops in a graph, there may be few pairs of non-touching loops. The superscript 3 means the *products* of loop gains for *sets-of-three non-touching loops*. Different sets-of-three loops may share two loops but have distinct third loops; nevertheless, these combinations are usually few, if any. The subscripts continue, representing set-of-four and more non-touching loops, although in practice such combinations are rare. Finding the graph determinant is apt to be the hardest and therefore most critical part of the application of the loop rule and must be done with care.

The final symbol in equation (6.39), Δ_{ijk}, is called a *path determinant*. This represents the determinant of the graph in which the kth path from the input node u_j to the output node x_i is expunged. Erasing this path usually eliminates most of the terms in the overall graph determinant, particularly those for products of loop gains. If there are no loops left at all, the unity first term in the series survives, nevertheless.

Example 6.9

Find the transfer function between the input u_2 and the output x_3 of the graph of Example 6.8 (p. 403).

Solution: First, you identify the three loops and their gains:

$$\text{Loop 1:} \qquad (-ds)ig = -dig\,s;$$
$$\text{Loop 2:} \quad k(-fs)(a/s) = -kfa;$$
$$\text{Loop 3:} \quad k(-fs)(bs) = -kfb\,s^2.$$

Loops 2 and 3 touch (share one or more common nodes); loop pairs 1-2 and 1-3 are not touching, and no set-of-three nontouching loops exists. The system determinant becomes

$$\Delta = 1 + dig\,s + kfa + kfb\,s^2 + digkfa\,s + digkfb\,s^3.$$

The first path from the input u_2 to the output x_3 has the gain $-ckf\,s$. This path touches loops 2 and 3, so when it is erased it leaves the path determinant $\Delta_1 = 1 + dig\,s$. The only other path has the gain je. It touches all of the loops, so $\Delta_2 = 1$. The resulting transfer function therefore is

$$G_{32} = \frac{je - ckf\,s(1 + dig\,s) + igdf\,s^2}{(1 + kfa) + (dig + digkfa)s + kfb\,s^2 + digkfb\,s^3}.$$

The value of x_3 responds to input u_1 also, of course. Its transfer function G_{31} shares the same graph determinant, however, and both of its paths (with gains $-hdie\,s$ and $hdf\,s^2$) touch all three loops. Thus both path determinants are unity, and

$$G_{31} = \frac{-hdie\,s + hdf\,s^2}{(1 + kfa) + (dig + digkfa)s + kfb\,s^2 + digkfb\,s^3}.$$

The overall behavior of the system becomes

$$x_3 = G_{31}u_1 + G_{32}u_2.$$

6.3.3   Converting Bond Graphs to Signal Flow Graphs

The conversion from bond graphs to block diagrams is described in Section 3.4.2, particularly Fig. 3.27 (p. 116), and the equivalences between block diagrams and signal flow graphs have been given in the past few pages. It is better to convert directly from the bond graph to the signal-flow graph, however; the following material therefore should be helpful, although it is not logically necessary.

Each non-activated bond[8] in a bond graph represents a bilateral pair of signal flows:

---

[8]Activated bonds are introduced in Section 9.2. They are included here for completeness but may be disregarded unless you wish to use them.

An activated bond cuts one of the signal flows:

A transformer T_i simply represents a coefficient T_i:

A gyrator is very similar:

The 0-junction has a common effort, which is represented in the signal flow graph by a node with a single input and two or more outputs, as dictated by the causal strokes on the bonds. There are, on the other hand, two or more input flows and one output flow; the latter is represented by a node with a single output. The result is

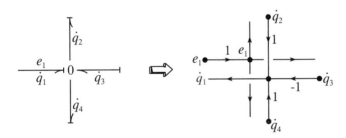

The 1-junction is the dual of the 0-junction. Its signal flow graph is the same as for a 0-junction, therefore, except the roles of effort and flow are inverted:

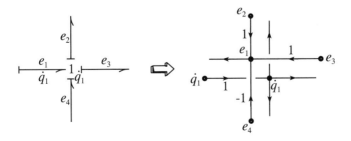

Example 6.10

Consider a two-tank system such as in Guided Problem 6.1 (p. 299), but with the inertance of the tube that interconnects the two tanks included in the model, as well as the resistance. Find the transfer function that relates the input flow, Q_i, to the volume of the second tank, V_2.

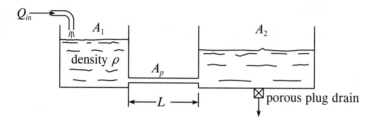

Solution: A linear bond-graph model is drawn below left, and integral causality is applied. The corresponding signal-flow graph is drawn beside it to the right.

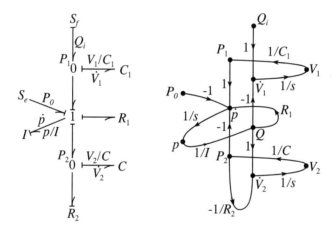

This graph contains four loops, but only two non-touching pairs and no set-of-three non-touching loops. The gains of the loops and the system determinant are

gains of loops: $\dfrac{-1}{C_1 sIs}$; $\dfrac{-1}{R_2 Cs}$; $\dfrac{-R_1}{Is}$; $\dfrac{-1}{CsIs}$

non-touching pairs

graph determinant: $\Delta = 1 + \dfrac{1}{R_2 Cs} + \dfrac{R_I}{Is} + \dfrac{1}{C_1 Is^2} + \dfrac{1}{CIs^2} + \dfrac{1}{R_2 CIs^2} + \dfrac{R_I}{R_2 C_1 CIs^3}$

There is only one path from the single input, Q_i, to the output $\dot{V}_2$; this touches all loops, so the corresponding path determinant, Δ_2, is unity. There also is only one path from Q_i to the output $\dot{V}_1$, but this path touches only one of the four loops; as a consequence, its path determinant, Δ_1, is the same as the determinant Δ except for the deletion of the two terms that involve the gain of that loop ($-1/IC_1 s^2$):

path determinants: $\Delta_I = 1 + \dfrac{1}{R_2 Cs} + \dfrac{R_I}{Is} + \dfrac{1}{CIs^2} + \dfrac{1}{R_2 CIs^2}$; $\Delta_2 = 1$

The resulting transfer functions are

$$\frac{V_1(s)}{Q_i(s)} = \frac{\dot{V}_1(s)}{sQ_i(s)} = \frac{s^2 + (\frac{R_1}{I} + \frac{1}{R_2 C})s + (\frac{1}{IC} + \frac{R_1}{R_2 IC})}{s^3 + (\frac{R_1}{I} + \frac{1}{R_2 C})s^2 + (\frac{1}{IC} + \frac{1}{IC_1} + \frac{R_1}{R_2 IC})s + \frac{1}{R_2 ICC_1}}.$$

$$\frac{V_2(s)}{Q_i(s)} = \frac{\dot{V}_2(s)}{sQ_i(s)} = s^2 \left[s^3 + (\frac{R_1}{I} + \frac{1}{R_2 C})s^2 + (\frac{1}{IC} + \frac{1}{IC_1} + \frac{R_1}{R_2 IC})s + \frac{1}{R_2 ICC_1} \right]^{-1}$$

6.3.4 Direct Application of the Loop Rule to Bond Graphs Without Meshes

The loop rule as presented here does not apply to bond graphs with multiport fields (as introduced in Chapter 9 and beyond), but it does allow activated bonds (as introduced in Section 9.2).

The first step is to apply causal strokes to the graph. Impedance or admittance causalities may be chosen in any compatible combination. If you prefer to find polynomials in s directly, rather than polynomials in $1/s$ as were first found in the example above, employ *differential* causality for the C and I elements.

You wish to forgo the drawing of the signal flow graph. The principal task is the determination of the system determinant. Five steps can be distinguished: (i) locating each loop; (ii) determining which loops touch; (iii) determining the *magnitudes* of the gains of the loops; (iv) determining the *signs* of the gains of the loops; and (v) accounting for the effect of any bond activation that may be present. In practice these steps can be carried out virtually simultaneously.

Bond graphs without meshes, which are tree-like, produce only what are called **flat loops**. A flat loop includes *both* signal-flow paths for each involved bond. Thus, these bonds form a simple cascade with one-port non-source elements at each end, as illustrated in Fig. 6.21. The signal flow paths in this figure and the next two figures are shown as dashed lines superimposed on the bond graph. This practice is not customary, but occasionally is instructive. The causality of bonds in the chain must be directed unilaterally through each 0 and 1-junction. Reversals in causality, in fact, occur only at gyrators. Each R, C and I element can be singled out as a potential termination for a cascade of bonds corresponding to a flat loop. Proper causal paths to other R, C and I elements are sought. Any bond activation in a cascade voids the loop. If carried out to completion, this process will identify each flat loop twice.

Flat loops touch if and only if they involve a common 0- or 1-junction. The loop gains include the impedance (R, $1/Cs$ or Is) or the admittance ($1/R$, Cs or $1/Is$) of the terminal R, C and I members, according to the indicated causalities. They also include either T_i^2 or $1/T_i^2$ for each transformer T_i in the cascade, and G_i^2 or $1/G_i^2$ for each gyrator G_i in the cascade, again according to their causalities. Finally, *the gains of all flat loops are negative*, presuming only that the power convention directions on each of the two one-port elements are directed into the element.

Inspection of the bond graph in Example 6.10 reveals directly the existence of flat loops between C_1 and I, I and R_1, I and C_2, and C_2 and R_2. Only two pairs of these four loops do not use a common junction and therefore do not touch. I and R_2 have admittance causality; the others have impedance causality. There are no transformers or gyrators. The system determinant follows directly from these facts.

Paths from an input to an output variable can be seen directly from the causal consistency of their cascaded bonds. In Example 6.10, the path from Q_i to $\dot{V}_1$ simply passes

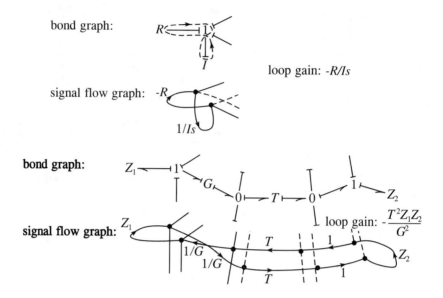

bond graph:

loop gain: -R/Is

signal flow graph:

bond graph:

signal flow graph:

loop gain: $-\dfrac{T^2 Z_1 Z_2}{G^2}$

Figure 6.21: Examples of flat loops

through the upper zero junction; the power convention arrows on the two bonds are aligned similarly, so there is no sign change and the gain is 1. The path from Q_i to $\dot{V}_2$ starts the same way, but passes through C_1 (impedance causality) and back through the upper zero junction. Since this second pass is on the effort sides of the bonds, there is again no sign change. The path then proceeds down through the 1-junction to the I element; the causal strokes around the 1-junction allow no choice. The path continues through the I element, which has admittance causality. It returns through the 1-junction on the flow side; again there is no change in sign. The path then passes through the lower 0-junction, remaining on the flow side of the bonds, and emerges as the desired $\dot{V}_2$. There is no sign change again, this time because the power convention arrows of both bonds have the same orientation. The path gain becomes $(1/C_1 s)(1/Is)$. This path and gain can also be seen in the signal flow graph.

To generalize, there is no sign change for a path traversing a 0-junction on the effort side or a 1-junction on the flow side. Similarly, there is no sign change if the two power convention arrows are aligned the same way. Since these two conditions are independent, on average one in four traverses of junctions produce a sign change.

Note that the path to $\dot{V}_1$ clearly touches only one loop, whereas the path to $\dot{V}_2$ clearly touches all four loops. Thus the path determinants Δ_1 and Δ_2 can be found and the transfer functions completed.

6.3.5 Bond Graphs with Meshes

The presence of one or more meshes in the bond graph considerably complicates the application of the loop rule. As a result, you usually are well advised to apply any conversion that would eliminate such meshes. Otherwise, the following procedure can be used.

Each mesh produces a pair of counter-directed **open loops**. Some open loops involve a **mesh stub**, such as shown in Fig. 6.22. Other open loops do not; a stub is involved if and only if the mesh bonds on either side of the associated junction have opposite causalities in the clockwise-counterclockwise sense. The magnitude of an open loop equals the product

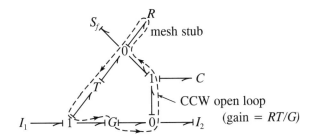

Figure 6.22: An open loop with a mesh stub

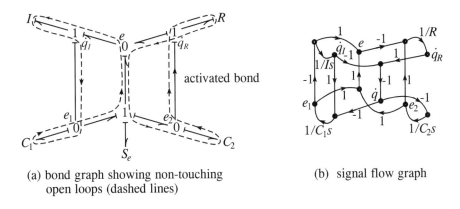

(a) bond graph showing non-touching open loops (dashed lines)

(b) signal flow graph

Figure 6.23: Meshes with two non-touching open loops

of the moduli of each mesh transformer or gyrator, or its inverse as determined by the causalities, and the impedance or admittance of each mesh stub. *Both counter-directed loops of a mesh without gyrators have the same gain, which is positive if the mesh is even and negative if it is odd.* (Recall that an even mesh has an even number of branches between successive junctions with a clockwise or counterclockwise causality; evenness and oddness are immutible properties of a mesh.) Loop gains must be dimensionless, a fact that helps identify mistakes.

A gyrator inserted into a branch of an otherwise even or odd mesh forces a mixed causality for that branch. The result is called a **neutral mesh**, for which *the two gains of the open loops have the same magnitude but opposite sign.* The loop with the positive gain is directed similarly as are the power convention arrows of those non-gyrator branches which are odd in number. A second gyrator inserted into the mesh converts it back to even or odd status, and a third gyrator reconverts it to neutral status, and so on. Neutrality is also a property of a mesh; it cannot be changed without changing the meaning of the model.

The two open loops of a mesh touch each other if and only if one or more mesh stubs are present. Two open loops of different meshes might not touch one another even if they share a common bond, however. An example is shown in Fig. 6.23.

This example also illustrates the fact that *activation* of a mesh bond cuts out one of its two open loops. (Activation of bonds is not considered until Section 9.2, but reference is made here for efficiency and because this section can be addressed before the present section.)

Example 6.11

Reconsider the heaving and pitching vehicle presented in Fig. 6.15 (p. 381), substituting an exciting force $F(t)$ on the left axle for the original positional disturbances. Determine the transfer functions from this force to the elevation of the center of mass, y_{cm}, and to the pitch angle, ϕ. To get you started, a bond graph of the situation is given below, with integral causality applied. The associated signal flow graph is drawn next to it, for reference. It may be tempting to employ this graph for the analysis, but at this point you are urged to work directly from the bond graph instead. Once you are used to identifying paths and loops directly from the bond graph augmented with causal strokes, your process becomes quicker and less prone to error.

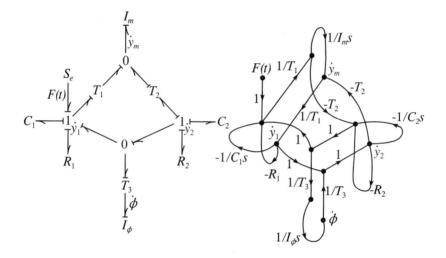

Solution: All four mesh branches have the same clockwise or counterclockwise causalities, so the mesh is even and involves no mesh stubs. The rules above tell you that the gains of both mesh loops are $-T_2/T_1$, and that these loops do not touch one another. The bond graph also has eight flat loops, with gains $-1/C_1 T_1^2 I_m s^2$, $-R_1/T_1^2 I_m s$, $-T_2^2/C_1 T_1^2 T_3^2 I_\phi s^2$, $-R_1 T_2^2/T_1^2 T_3^2 I_\phi s$, $-1/C_2 T_3^2 I_\phi s^2$, $-R_2/T_3^2 I_\phi s$, $-1/C_2 T_1^2 I_m s^2$ and $-R_2/T_1^2 I_m s$; only the first and second do not touch the fifth and sixth, and all eight touch both open loops. As a result, the system determinant is

$$\Delta = 1 + 2\frac{T_2}{T_1} + \frac{T_2^2}{T_1^2} + \frac{1/C_1 s + R_1}{T_1^2 I_m s} + \frac{(1/C_1 s + R_1)T_2^2}{T_1^2 T_3^2 I_\phi s}$$
$$+ \frac{1/C_2 s + R_2}{T_3^2 I_\phi s} + \frac{1/C_2 s + R_2}{T_1^2 I_m s} + \frac{(1/C_1 s + R_1)(1/C_2 a + R_2)}{T_1^2 T_3^2 I_\phi s^2}.$$

The path from the input force, F, to the velocity of the center of mass, $\dot{y}_m$, passes through the transformer T_1 and the inertance I_m. Its gain is therefore $1/T_1 I_m s$. This path touches the clockwise mesh loop, but not the counterclockwise loop. It also touches all of the flat loops except those that pass between C_2 and I_ϕ and between R_2 and I_ϕ. Its path determinant is therefore

$$\Delta_{ym} = 1 + \frac{T_2}{T_1} + \frac{1/C_2 s + R_2}{T_3^2 I_\phi s},$$

and the transfer function becomes

$$\frac{\dot{y}_m}{F} = \frac{1}{\Delta}\frac{\Delta_{ym}}{T_1 I_m s}.$$

The path from F to the angular velocity, $\dot{\phi}$, passes through T_1, T_2, T_3 and finally the inertance I_ϕ. The gain changes sign at the right-most 1-junction, since the path is on the effort side of a common-flow junction and the power convention arrows reverse direction relative to the path. The path gain, therefore, is $-T_2/T_1 T_3 I_\phi s$. This path also does not touch the counterclockwise mesh loop but does touch all eight of the flat loops. Thus its path determinant is

$$\Delta_\phi = 1 + \frac{T_2}{T_1},$$

and the transfer function becomes

$$\frac{\dot{\phi}}{F} = -\frac{1}{\Delta}\frac{T_2 \Delta_\phi}{T_1 T_3 I_\phi s}.$$

6.3.6 Determination of State Differential Equations

Finding state differential equations usually is considerably easier than finding transfer functions. The reason is that most if not all of the loops in the full signal flow graph are excised.

The procedure presumes that integral causality has been applied to the bond graph, unlike above where integral causality is merely one option. The application of standard state-variable notation also is presumed. Next, *all I and C elements are removed.* This can be indicated by crossing out these elements with Xs. The flow p_i on the ith inertance and the effort q_i/C_i on the ith compliance are treated as *inputs* to the system. The corresponding dp_i/dt for the inertances and dq_i/dt for the compliances are treated as *outputs.* The expression for each output in terms of all the inputs becomes the associated state differential equation.

Example 6.12

Write the state differential equations for the two-tank system of Example 6.10 (p. 406).

Solution: The causal bond graph with the I and C elements removed is

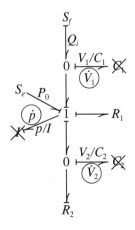

Removal of the elements C_1, C_2 and I eliminates all three loops, so the system determinant as well as all path determinants equals unity. The path gains from the five causal inputs to the three outputs are assembled in a table given below left. These directly give the differential equations given to the right.

output	Q_i	P_0	V_1/C_1	V_2/C_2	p/I
$\dot{V}_1$	1	0	0	0	-1
$\dot{V}_2$	0	0	0	$-1/R_2$	1
$\dot{p}$	0	1	1	-1	$-R_1$

$$\frac{dV_1}{dt} = Q_i - \frac{1}{I}p$$

$$\frac{dV_2}{dt} = \frac{1}{I}p - \frac{1}{R_2 C_2}V_2$$

$$\frac{dp}{dt} = \frac{1}{C_1}V_1 - \frac{1}{C_2}V_2 - \frac{1}{I}p - P_0$$

Example 6.13

Write the state differential equations for the heaving and pitching vehicle of Example 6.11 (p. 410).

Solution: The causal bond graph with the energy-storage elements crossed out is given at the top of the next page, with identification of the loops. Removal of the four energy-storage elements eliminates all eight flat loops, but leaves the pair of mesh loops intact. These loops do not touch one another, so the system determinant becomes

$$\Delta = 1 + 2\frac{T_2}{T_1} + \frac{T_2^2}{T_1^2} = \left(1 + \frac{T_2}{T_1}\right)^2$$

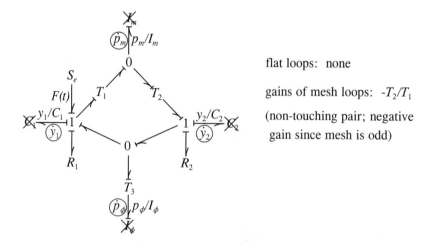

flat loops: none

gains of mesh loops: $-T_2/T_1$

(non-touching pair; negative gain since mesh is odd)

The path gains and path determinants are given in the table below. Some paths touch both loops and therefore have a path determinant equal to 1. Others touch only one path, and therefore have a path determinant equal to $1 + T_2/T_1$.

output	input				
	y_1/C_1	y_2/C_2	p_m/I_m	p_ϕ/I_ϕ	$F(t)$
$\dot{y}_1$	0	0	$1/T_1$ $\Delta=\Delta_1$	$-T_2/T_1$ $\Delta=\Delta_1$	0
$\dot{y}_2$	0	0	$1/T_1$ $\Delta=\Delta_1$	$1/T_3$ $\Delta=\Delta_1$	0
$\dot{p}_m$	$-1/T_1$ $\Delta=\Delta_1$	$-1/T_1$ $\Delta=\Delta_1$	$-R_1/T_1^2$ $\Delta=1$	$-R_2/T_3T_1$ $\Delta=1$	$-1/T_1$ $\Delta=\Delta_1$
$\dot{p}_\phi$	T_2/T_1T_3 $\Delta=\Delta_1$	$-1/T_3$ $\Delta=\Delta_1$	R_1T_2/T_1T_3 $\Delta=1$	$-R_2/T_3^2$ $\Delta=1$	$-T_2/T_1T_3$ $\Delta=\Delta_1$

$$\Delta_1 \equiv 1 + T_2/T_1$$

The resulting differential equations are

$$\frac{dy_1}{dt} = \frac{\Delta_1}{\Delta}\left[\frac{1}{T_1 I_m}p_m - \frac{T_2}{T_1 T_3 I_\phi}p_\phi\right]$$

$$\Delta = (1 + T_2/T_1)^2$$

$$\frac{dy_2}{dt} = \frac{\Delta_1}{\Delta}\left[\frac{1}{T_1 I_m}p_m + \frac{1}{T_3 I_\phi}p_\phi\right]$$

$$\frac{dp_m}{dt} = \frac{1}{\Delta}\left[-\frac{\Delta_1}{T_1 C_1}y_1 + \frac{\Delta_1}{T_1 C_2}y_2 - \frac{R_1}{T_1^2 I_m}p_m - \frac{R_2}{T_1 T_3 I_\phi}p_\phi + \frac{\Delta_1}{T_1}F(t)\right]$$

$$\frac{dp_\phi}{dt} = \frac{1}{\Delta}\left[\frac{\Delta_1 T_2}{T_1 T_3 C_1} - \frac{\Delta_1}{T_3}y_2 + \frac{R_1 T_2}{T_1 T_3}p_m - \frac{R_2}{T_3^2 I_\phi}p_\phi - \frac{\Delta_1 T_2}{T_1 T_3}\right]$$

6.3.7 Summary

The loop rule facilitates the writing of state differential equations for linear systems. It also is efficient in finding one or two transfer functions for a complex system, precluding the need to deal with the matrix equations. Should one wish to find several transfer functions for a particular system, however, the basic matrix approach may be preferred, particularly if a program such as MATLAB is available.

The rules for applying the loop rule to causal bond graphs are based on the special energy constraints implied by the graphs. They aid in finding paths and loops, determining which loops touch, and finding the loop gains without laborious examination of the corresponding signal flow paths.

Guided Problem 6.7

Consider the drain tube placed through an ear drum as described in Problem 5.8 (pp. 268-269), including the given simple model that neglects damping and other effects. Find the transfer function relating the input pressure to the displacement of the eardrum, and compare to the case with no drain tube. The following parameter values may be used: $C_e = 0.5 \times 10^{-6}$ ft^5/lb, $I_e = 9.6$ lb·s^2/ft^5, $C_m = 1.17 \times 10^{-8}$ ft^5/lb, and $I_t = 5.76$ lb·s^2/ft^5.

(The effect of the tube on hearing is addressed in Problem 7.6.)

Suggested Steps:

1. Since the mesh is even, it is appropriate to find the equivalent tree-like bond graph.

2. Apply causal strokes in any consistent fashion. *Differential* causality actually simplifies the intermediate expressions somewhat. Identify and label the input and output variables.

3. Find the loops (which are all flat), write their loop gains and identify pairs of non-touching loops, etc.

4. Find the graph and path determinants.

5. Write the transfer functions with and without the tube in place.

Guided Problem 6.8

This problem offers a fairly general and complex bond graph; it is appropriate only if you have already studied Section 9.2 on activated bonds.

  Repeat Guided Problem 9.5 (pp. 609-611) using the loop rule method as applied to bond graphs.

Suggested Steps:

1. Copy the causal bond graph from part (d) of Fig. 9.11, or spread it out as shown in the solution to Guided Problem 9.5. Annotate the appropriate bonds with the time derivatives of the state variables and the conjugate variables, as also shown in the solution. Cross out or erase the energy storage elements.

2. Identify the flat loops, if any. Note that any potential flat loop must involve two of the three resistances, since the graph has no other non-source stubs.

3. Identify the open loops, if any. Note that a mesh stub with a deleted energy storage element will preclude any open loop that would use that stub. Also, pay attention to the activated bonds, which eliminate one of an otherwise pair of open loops.

4. Find the graph determinant using the results of steps 2 and 3.

5. Make a table of the various path gains from the six graph inputs to the five graph outputs. This is the most difficult part of the solution, and this is a complex circuit. First, locate the various paths starting at each input and following those bonds which have compatible causal strokes. Hint: there are a total of twenty paths, including more than one from certain inputs to certain outputs. Second, find the *magnitudes* of the path gains, and enter these into the table. Only the resistances and the transformers cause the magnitudes to be different from unity. Third, find the *signs* of each path, and enter minus signs where appropriate in the table. In this endeavor it is helpful to note that signs change only when a path goes through a junction, and then only if an effort path passes through a 1-junction or a flow path passes through an effort junction *and* the power sign arrows of the bonds on either side of the junction are oppositely directed.

6. Find the path determinant for each path, and enter these into the table above. Path determinants are the graph determinants of the graph with the path deleted; since the determinant of the full graph is quite simple, these path determinants are simple also, a majority equaling unity.

7. Write the desired differential equations directly by assembling the information found in steps 4, 5 and 6. This should be the simplest part of the job. It is suggested that you check your answers.

PROBLEMS

6.50 A permanent-magnet DC motor drives a gear pump through a reduction gear, forcing a virtually incompressible fluid from a vented sump, past a hydraulic accumulator, through a long rigid tube, into a piston-cylinder load connected in parallel to an effective mass, spring, and dashpot, and from the other side of the piston back to the sump. The elements can be assumed to act linearly.

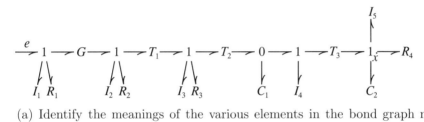

(a) Identify the meanings of the various elements in the bond graph model above. Also, identify on the graph the flow that emerges from the gear pump as Q_1, the flow through the rigid tube as Q_2, and the velocity of the load as $\dot{x}$.

(b) Simplify the bond graph by changing the definitions of some of the one-port elements so as to eliminate the need for transformers.

(c) Draw a signal-flow graph for the bond graph of part (b), considering the voltage e as the input and the velocity $\dot{x}$ as the output.

(d) Find the transfer function between the input and the output.

(e) Repeat (d) using the bond graph directly.

(f) Is oscillatory behavior still possible if all compliances are removed? Why?

6.51 The device modeled in Example 5.7 (pp. 274-275) contains an even mesh that is converted to a more convenient tree-like mesh in Example 5.14 (p. 315).

(a) Find the transfer function $P(s)/F(s)$ directly from both graphs. They should be the same.

(b) Draw the signal-flow graphs that correspond to the application of differential causality to both bond graphs. Check your answers to (a) from these new graphs.

(c) Find a set of state variable equations. (Work directly from one of the bond graphs, but use integral causality in the standard fashion.)

6.52 The sound movie camera of Problem 5.43 (p. 304) could be modeled as shown below.

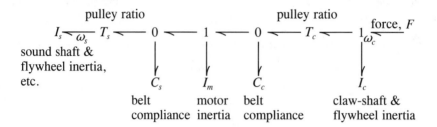

(a) Apply differential causality, insofar as possible, and find the transfer function between the input force, F, and the output fluctuations in the angular velocity of the sound shaft, ω_s.

(b) The force perturbations presumably are well above both natural frequencies. Does the analysis suggest good performance under this assumption? (This part of the problem requires some knowledge of frequency response as presented in Section 7.1.)

6.53 Determine the transfer function between the input effort e and the output effort F for the system described by the bond graph below, using the loop rule. All parameters are constants.

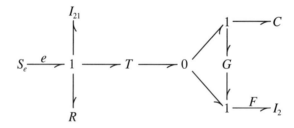

SOLUTIONS TO GUIDED PROBLEMS

Guided Problem 6.7

1-2.

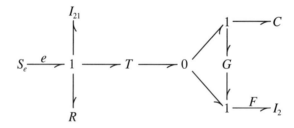

3. Loop gains: $-I_t C_e s^2$; $-I_t C_m s^2$; $-I_e C_e s^2$

The first loop above touches each of the others, but the other two do not touch one another.

4. $\Delta = 1 + I_t C_e s^2 + I_t C_m s^2 + I_e C_e s^2 + I_t C_m I_e C_e s^4$

Path from P to $\dot{x}$: $(+C_m s)(+I_t s)(+C_e s)T = C_m I_t C_e T s^3$

The path touches all three loops, so the path determinant is 1.

5. $\dfrac{\dot{X}(s)}{P(s)} = \dfrac{C_m I_t C_e T s^3}{1 + (I_t C_e + I_t C_m + I_e C_e)s^2 + (I_t C_m I_e C_e)s^4}$

$= \dfrac{3.37 \times 10^{-14} T s^3}{3.24 \times 10^{-13} s^4 + 7.702 \times 10^{-6} s^2 + 1}$

Absence of the tube gives $I_t \to \infty$, resulting in

$\dfrac{\dot{X}(s)}{P(s)} = \dfrac{C_m C_e T s}{(C_e + C_m) + C_m I_e C_e s^2} = \dfrac{5.85 \times 10^{-15} T s}{5.62 \times 10^{-14} s^2 + 0.5117 \times 10^{-6}}$

6. No tube present: double-pole at $\dfrac{\omega}{2\pi} = \dfrac{1}{2\pi}\sqrt{5.801 \times 10^7} = 1212.2$ Hz

Tube present: double poles at

$$\omega^2 = \frac{3.6568 \times 10^{-6}}{2 \times 5.22 \times 10^{-14}} \pm \sqrt{\left(\frac{3.6568 \times 10^{-6}}{2 \times 5.22 \times 10^{-14}}\right)^2 - \frac{1}{5.22 \times 10^{-14}}}$$

or $\dfrac{\omega}{2\pi} = 83.4$ Hz, 1329.5 Hz

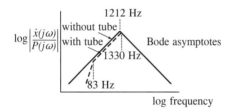

The response is greatest in the mid-range; the inner ear must compensate for this. The tube has no effect at high frequencies (> 1330 Hz), but causes a small loss at lower frequencies (a factor of 1.2 which implies 1.6 db) and a severe loss at extremely low frequencies (< 80 Hz).

Guided Problem 6.8

1.

2. There are no sequences of bonds with unilateral causality interconnecting any of the three pairs of resistances, and therefore no flat loops exist.

3. Loops for the left-hand mesh are blocked causally at both the upper 0- and 1-junctions. The right-hand mesh has a clockwise loop with gain $-1/T_1 T_2$; the counterclockwise loop is blocked by the bond activation.

4. The graph determinant is $\Delta_g = 1 + 1/T_1 T_2$.

5-6. In each entry below the path determinant is given below the path gain.

outputs	S_e	$\dfrac{q_1}{C_1}$	inputs $\dfrac{q_2}{C_2}$	$\dfrac{q_3}{C_3}$	$\dfrac{p_1}{I_1}$	$\dfrac{p_2}{I_2}$
$\dot q_1$	$\dfrac{1}{R_1}$ Δ_g	$\dfrac{-1}{R_1}$ Δ_g	— —	— —	$-1;\ \ 1$ $\Delta_g;\ \ \Delta_g$	— —
$\dot q_2$	— —	— —	— —	— —	— —	1 Δ_g
$\dot q_3$	— —	— —	— —	— —	1 Δ_g	— —
$\dot p_1$	1 1	$1;\ -1$ $\Delta_g;\ 1$	— —	-1 1	$-R_3;\ \dfrac{-R_2}{T_2^2}$ $1;\ \ 1$	$\dfrac{-R_2}{T_2^2}$ 1
$\dot p_2$	$\dfrac{-1}{T_1T_2}$ 1	$\dfrac{1}{T_1T_2}$ 1	-1 Δ_g	$\dfrac{1}{T_1T_2}$ 1	$\dfrac{R_3}{T_1T_2};\ \dfrac{-R_2}{T_2^2}$ $1;\ \ 1$	$\dfrac{-R_2}{T_2^2}$ 1

7. $\dfrac{dq_1}{dt} = \dfrac{S_e}{R} - \dfrac{1}{R_1 C_1} q_1$

$\dfrac{dq_2}{dt} = \dfrac{1}{I_2} p_2$

$\dfrac{dq_3}{dt} = \dfrac{1}{I_1} p_1$

$\dfrac{dp_1}{dt} = \dfrac{S_e}{\Delta_g} + \dfrac{1}{\Delta_g T_1 T_2 C_1} q_3 - \dfrac{1}{\Delta_g C_3} q_3 + \dfrac{1}{\Delta_g I_1}\left(-R_3 - \dfrac{R_2}{T_2^2}\right) p_1 - \dfrac{R_2}{\Delta_g T_2^2 I_2} p_2$

$\dfrac{dp_2}{dt} = -\dfrac{S_e}{\Delta_g T_1 T_2} + \dfrac{1}{\Delta_g T_1 T_2 C_1} q_1 - \dfrac{1}{C_2} q_2 + \dfrac{1}{\Delta_g T_1 T_2 C_3} q_3$

$\qquad + \dfrac{1}{\Delta_g I_1}\left(\dfrac{R_3}{T_1 T_2} - \dfrac{R_2}{T_2^2}\right) - \dfrac{R_2}{\Delta_g T_2^2 I_2} p_2$

Chapter 7

Analysis of Linear Models, Part 2

The first three sections of this chapter focus on the **frequency domain**, which means the behavior of linear models with sinusoidal excitations and excitations based on sums of sinusoidal excitations. This is in contrast to Chapter 4, which focuses largely on the time domain, or the behavior of linear models in response to excitations characterized as functions of time. Section 7.1 deals with simple sinusoidal excitations, developing the powerful phasor method which uses complex numbers. Section 7.2 focuses on special aspects of vibrating mechanical systems. The idea of decomposing total motion into the sum of simply behaving modes is introduced. Section 7.3 applies matrix methods to systems with vibration and other behaviors, extending the modal concept to systems with damping. Certain matrix-based computational methods also are developed, including the basis of the MATLAB command `lsim`.

Section 7.4 starts with the representation of periodic excitations and responses by discrete sums of sinusoids, called Fourier series. It then continues by representing pulse-like signals and responses as continuous sums of sinusoids, called Fourier transforms.

7.1 Sinusoidal Frequency Response

Periodic excitations of natural and engineered systems abound. Aerodynamic and water wave forces vibrate bodies, acoustic signals comprise music and speech, and rotational imbalance shakes machines. The simplest representation of excitations such as these is sinusoidal. It will be shown in Section 7.2 that any periodic waveform can be decomposed into a sum of discrete sinusoidal oscillations at multiples of the base frequency, and a large class of much more general waveforms can be decomposed into an infinite sum of sinusoidal oscillations. The Laplace transform, presented in Chapter 4, can be viewed as a twist on the Fourier transform.

Regardless of whether their excitation is periodic, systems often are characterized by how they respond to sinusoidal inputs. Further, physical experiments are carried out using electrical function generators and mechanical shaker tables that sweep an excitation over a range of frequencies.

This section is concerned with the response of linear models to the sinusoidal excitation

$$\boxed{u(t) = u_0 \cos(\omega t + \beta).} \tag{7.1}$$

The focus is on the particular solution, which often is called the **steady-state response** since it persits as long as the input persists. The homogeneous part of a complete solution, on the other hand, usually decays quickly. All that is needed is stability and a pinch of damping. Without stability, there is no point in studying any particular solution, which would quickly be overwhelmed by the exponential growth of the homogeneous solution. The tacit assumption is made below, therefore, that the system is stable, which means that the real parts of all the roots of the characteristic equation are non-positive.

7.1.1   The Phasor Method

The sinusoidal excitation of equation (7.1) can be recast as the sum of two complex conjugate terms:

$$
\begin{aligned}
u(t) &= \frac{1}{2}u_0 \left[e^{j(\omega t+\beta)} + e^{-j(\omega t+\beta)} \right] \\
&= \frac{1}{2}u_0 \left[e^{j\beta}e^{j\omega t} + e^{-j\beta}e^{-j\omega t} \right].
\end{aligned}
\tag{7.2}
$$

The equivalence of equations (7.1) and (7.2) follows from the trigonometric identities

$$
\begin{aligned}
e^{j\alpha} &= \cos\alpha + j\sin\alpha \\
e^{-j\alpha} &= \cos\alpha - j\sin\alpha,
\end{aligned}
\tag{7.3}
$$

which also were used in Section 4.2.1. Each of the terms in the excitation of equation (7.2) has the exponential form ce^{st}, with c being a complex number and $s = \pm j\omega$. According to the analysis in Section 4.3.2 (equation (4.37b), p. 178), therefore, the response or particular solution of the system with the transfer function $G(s)$ to this input is $ce^{\pm j\omega t}$, giving

$$
x_p(t) = \frac{1}{2}u_o \left[e^{j\beta}G(j\omega)e^{j\omega t} + e^{-j\beta}G(-j\omega)e^{-j\omega t} \right].
\tag{7.4}
$$

$G(j\omega)$ is called the **frequency transfer function**. The terms $G(j\omega)$ and $G(-j\omega)$ are complex conjugate, so that they can be written

$$
G(j\omega) = |G(j\omega)|e^{j\phi}; \qquad G(-j\omega) = |G(j\omega)|e^{-j\phi},
\tag{7.5}
$$

where $\phi \equiv \angle G(j\omega)$ is called the **phase angle** of $G(j\omega)$, which has as its tangent

$$
\tan\phi = \frac{Im[G(j\omega)]}{Re[G(j\omega)]}.
\tag{7.6}
$$

Equation (7.4) becomes

$$
x_p(t) = \frac{1}{2}|G|u_0 \left[e^{j(\omega t+\beta+\phi)} + e^{-j(\omega t+\beta+\phi)} \right],
\tag{7.7}
$$

or

$$
\boxed{
\begin{aligned}
x_p(t) &= |G(j\omega)|u_0 \cos(\omega t + \beta + \phi), \\
\phi(\omega) = \angle G(j\omega) &= \tan^{-1}\left[\frac{Im[(G(j\omega)]}{Re[(G(j\omega)]} \right].
\end{aligned}
}
\tag{7.8}
$$

   Thus, $x_p(t)$ is sinusoidal with an amplitude $|G(j\omega)|$ times the amplitude of the excitation; $|G(j\omega)|$ is called the **amplitude ratio** or **magnitude ratio** or **gain**. The phase angle of $x(t)$ is advanced relative to that of $u(t)$ by the angle $\phi(\omega)$, known as the **phase shift**. Most commmonly this angle is negative; $-\phi$ is known as the **phase lag**.

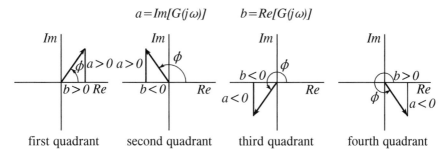

Figure 7.1: Determination of the quadrant of $G(j\omega)$

Examples of the complex number $G(j\omega)$ are drawn in Fig. 7.1 as vectors or **phasors** in the complex plane. It is essential that ϕ be recognized in its proper quadrant. For example, if both $\Im[G(j\omega)]$ and $\Re[G(j\omega)]$ are negative, ϕ lies in the third quadrant, or $180° < \phi < 270°$. You are urged to find the quadrant of phasors by first determining the signs of both terms, and then sketching the phasor. Be warned that many simple calculator and computer routines automatically assume the first or fourth quadrant.

EXAMPLE 7.1

Find the steady-state response of the model with the transfer function

$$G(S) = \frac{50S + 200}{S^3 + 3S^2 + 102S + 100}$$

to the excitation

$$u(t) = 2\sin(10t).$$

Solution: At the specified frequency,

$$|G(j10)| = \left| \frac{50(j10) + 200}{(j10)^3 + 3(j10)^2 + 102(j10) + 100} \right|$$

$$= \left| \frac{200 + 500j}{(100 - 300) + j(1020 - 1000)} \right| = \sqrt{\frac{(200)^2 + (500)^2}{(200)^2 + (20)^2}} = 2.679$$

$$\phi = \tan^{-1}\left(\frac{500}{200}\right) - \tan^{-1}\left(\frac{20}{-200}\right) = 68.20° - 174.29° = -106.09°$$

The numerator of $G(j10)$ is a phasor in the first quadrant, whereas the denominator is a phasor in the second quadrant. The phase angle of $G(10j)$ equals the angle for the numerator minus the angle for the denominator. The particular solution is

$$x_p(t) = 2 \times 2.679\sin(10t - 106.09°) = 5.358\sin(10t - 106.09°).$$

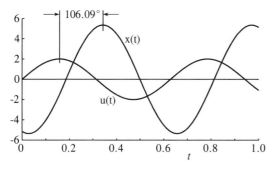

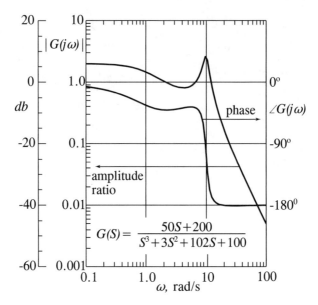

Figure 7.2: Example Bode plot

7.1.2   Bode Plots

Magnitude ratios and phase angles expressed as functions of frequency serve to characterize a linear transfer function completely. Plots of the magnitude ratio or gain and the phase angle as functions of frequency convey to the analyst not only the meaning of a frequency transfer function, but also the very character of the dynamic model, in a clear, visual way. The information can be deduced from direct experiment on the physical system, with no use of models or mathematics; the response to sinusoidal excitation at many different frequencies is observed and plotted. A **shaker table** often is used to produce sinusoidal disturbances on mechanical systems.

Both linear and logarithmic axes have been used widely for the plots. An example using logarithmic axes for the frequency and the magnitude ratio and a linear axis for the phase angle is given in Fig. 7.2. Such a representation is known as a **Bode plot**. The transfer function chosen is the same as in Example 7.1 above. You should check to see that the magnitude ratio and the phase angle as plotted at 10 rad/s agree with the calculated values.

The particular Bode coordinates have become the standard for system analysts interested in dynamics and control. This is largely because they allow a model and its transfer function to be decomposed into components and reassembled graphically, as you will see later. Further, the use of the logarithmic axes allows small but important amplitudes to be seen clearly despite the presence of vastly larger amplitudes at other frequencies. Finally, widely available software, including MATLAB, employs Bode coordinates.

The **decibel scale** has become a conventional representation of the logarithmic magnitude ratio. If the magnitude ratio is m, the decibel scale, abbreviated "db," is[1]

$$\boxed{db \equiv 20 \log_{10} m} \tag{7.10}$$

Both the m and the db scales are shown in part (a) of the figure.

---

[1]Some older references use $10 \log_{10} m$. The use of the multiplier 20 rather than the more natural 10 appears to be a carryover from the prior and continuing use of decibels to represent the power of an acoustic wave or an acoustic field. Power is proportional to the *square* of the sound pressure (or the product of pressure and velocity, as we have seen), and the logarithm of a square gives a multiplier of 2.

EXAMPLE 7.2

Use MATLAB to produce a Bode plot for the model of Example 7.1.

Solution:

```
>> num = [50 200];
>> den = [1 3 102 100];
>> bode(num,den)
```

Compare the resulting plot with that of Fig. 7.2:

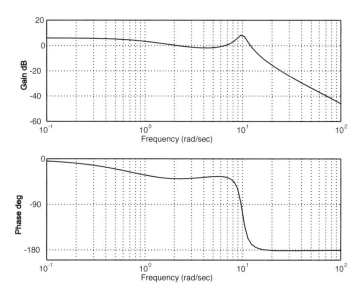

The alternate command [mag,phase,w] = bode(num,den) directs that the calculated values of frequency, magnitude and phase be stored in the vectors w, mag and phase, respectively. It also lets you specify the frequencies. For example, to specify 200 values of frequency with even logarithmic spacing over the band from 1 to 40 rad/s, you enter w=logspace(1,40,200).

7.1.3   Models Comprising a Single Pole or Zero

The transfer function $G(S) = S$ is said to have a **zero** at the origin, since $G(0) = 0$. The model $G(S) = 1/S$ is said to have a **pole** at the origin, since it becomes infinity when $S = 0$. Bode plots for these cases are given in Fig. 7.3. For the equal magnitude scales used, the slope of the magnitude curve is $+1$ in the first case and -1 in the second case. (With the decibel format, it would be ± 20 db/decade.) The phase angle is a constant $90°$ in the first case and a constant $-90°$ in the second case. You should verify that the substitution $S \rightarrow j\omega$ gives these results.

Multiplying the transfer functions by a constant factor k, to give $G(S) = kS$ and $G(S) = k/S$, would merely shift the magnitude plots up or down, leaving their slopes unchanged. The phase plots would be unchanged.

The transfer function $G(S) = k(1 \pm S/\omega_z)$ is described as a zero at $S = \mp\omega_z$, since $G(\mp\omega_z) = 0$. Bode plots for this model are given in Fig. 7.4 for the special case of $k = 1$. As before and for all transfer functions, a value of k different from zero simply shifts the magnitude curve up or down and leaves the phase curve unchanged.

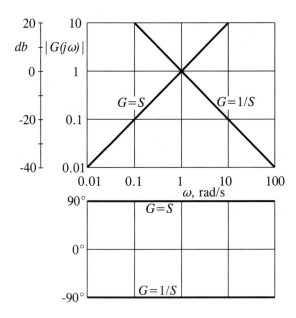

Figure 7.3: Bode plots for a pole or a zero at the origin

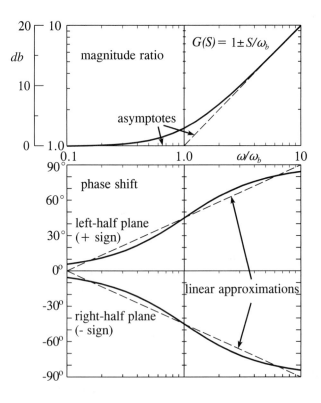

Figure 7.4: Bode plots for single real non-zero zeros

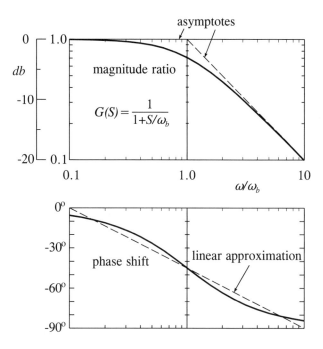

Figure 7.5: Bode plot for single real non-zero pole

These plots follow from the substutution $S = j\omega$:

$$\log_{10}|G(j\omega)| = \log_{10}|1 \pm j\omega/\omega_z| = \log_{10}\sqrt{1 + (\omega/\omega_z)^2}, \qquad (7.11a)$$

$$\angle G(j\omega) = \pm\tan^{-1}\left(\frac{\omega}{\omega_z}\right), \qquad (7.11b)$$

Note that the magnitude plot is the same regardless of whether the plus or the minus sign is used. On the other hand, the phase angle increases when the sign is plus, and decreases when it is minus. This is a critical difference.

The transfer function $G(S) = k/(1 + S/\omega_p)$ is described as a pole at $S = -\omega_p$, since $G(-\omega_p) \to \infty$. Its Bode plot, again for $k = 1$, is given in Fig. 7.5. Again, these plots follow from the substutution $S = j\omega$:

$$\log_{10}|G_{(}j\omega)| = \log_{10}\left|\frac{1}{1 + j\omega/\omega_p}\right|$$

$$= \log_{10}\frac{1}{\sqrt{1 + (\omega/\omega_p)^2}} = -\log_{10}\sqrt{1 + (\omega/\omega_p)^2}, \qquad (7.12a)$$

$$\angle G_{pi}(j\omega) = -\tan^{-1}\left(\frac{\omega}{\omega_p}\right). \qquad (7.12b)$$

Note that only a plus sign is used in the transfer function, rather than the $\pm$ sign used in the transfer function for the zero. The reason is that a minus sign would represent an unstable system. As noted before, the particular solutions of unstable systems are of no interest.

The magnitude plots for the models with the single pole and the single zero are top-to-bottom flips of one another. So are the phase angles, for the common plus-sign case. The magnitude plots for the real non-zero poles and zeros can be approximated by asymptotes, as shown in Figs. 7.4 and 7.5 by dashed lines. The low-frequency asymptote corresponds to the neglect of the term ω/ω_z or ω/ω_p relative to 1; it produces a horizontal straight line. The high-frequency asymptote corresponds to the exact opposite: neglecting the term 1 relative to the ω/ω_z or ω/ω_p. The high-frequency asymptote therefore is the same as a pole or zero at the origin, scaled so that it equals 1 at the **break frequency** $\omega = \omega_z$ or $\omega = \omega_p$. The two asymptotes intersect at the break frequency. This fact enables you to closely approximate a break frequency if you are given the plot (e.g., experimental data) rather than the equation, a common situation for an engineer. The greatest departure of the actual magnitude from the closest asymptote occurs at the break frequency, where it equals the ratio $\sqrt{2}$ or $1/\sqrt{2}$, or approximately ± 6 db.

Asymptotes also exist for the phase angles, corresponding to the neglect of the corresponding terms. The asymptote for low frequency is $0°$, and for high frequency is $\pm 90°$. The greatest departure between the actual value and the value of the nearest asymptote is $45°$, again at the break frequency. The phase asymptotes represent the actual phase more poorly than the magnitude asymptotes represent the actual magnitude, however. A much better approximation for the frequency range $0.1\omega_i < \omega < 10\omega_i$ is a straight line (in the semi-log coordinates of this plot) between the two asymptotes at the two respective frequencies a decade below and a decade above the break frequency. These linear approximations are represented in Figs. 7.4 and 7.5 by dashed lines.

EXAMPLE 7.3

A mass-and-spring system is excited by a force $F = F_0 \sin \omega t$. Determine the resulting steady-state velocity $\dot{x}_p(t)$.

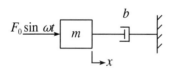

Solution: The bond graph above right leads to

$$\frac{dp}{dt} = F_0 \sin \omega t - \frac{R}{I} p,$$

which gives

$$G(S) = \frac{1}{S + R/I} = \frac{1/b}{1 + S/\omega_b}; \quad \omega_b = \frac{I}{R} = \frac{m}{b},$$

so that the phasor $S = j\omega$ yields

$$\dot{x}_p(t) = \frac{F_0/b}{\sqrt{(1 + (\omega/\omega_b)^2}} \sin(\omega t + \phi); \quad \phi = -\tan^{-1}\left(\frac{\omega}{\omega_b}\right).$$

This gives the plot of Fig. 7.5, scaled to the magnitude of F_0/b at zero frequency.

7.1.4 Models Comprising a Pair of Complex Poles or Zeros

The model with the transfer function

$$G = \frac{k}{1 + (2\zeta_n/\omega_p)S + (S/\omega_n)^2} \tag{7.13}$$

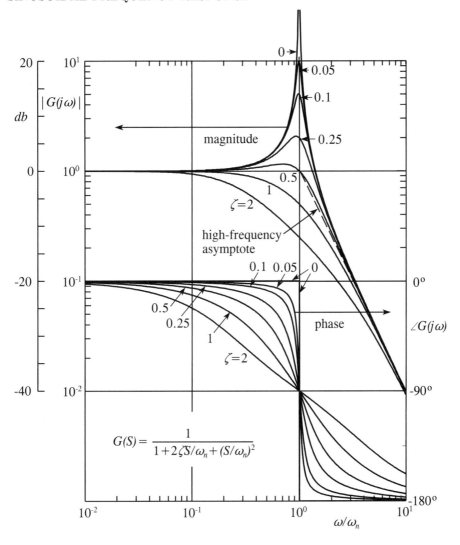

Figure 7.6: Bode plots for second-order poles

has a pair of complex conjugate poles whenever the damping ratio $\zeta < 1$. The Bode plot variables are

$$\log_{10}|G(j\omega)| = \log_{10}\left|\frac{1}{1 - (\omega/\omega_n)^2 + j2\zeta\omega/\omega_n}\right|$$

$$= -\log_{10}\sqrt{\left[1 - \left(\frac{\omega}{\omega_n}\right)^2\right]^2 + \left(\frac{2\zeta\omega}{\omega_n}\right)^2}, \qquad (7.14a)$$

$$\angle G(j\omega) = -\tan^{-1}\left[\frac{2\zeta\omega/\omega_n}{1 - (\omega/\omega_n)^2}\right]. \qquad (7.14b)$$

Plots for various values of ζ are given in Fig. 7.6. Although cases with $\zeta \geq 1$ are included, these cases are better computed by factoring the transfer function into a product of two real poles, as described later.

The low frequency asymptote for magnitude corresponds to the neglect of all terms in any sum except those proportional to the lowest power of ω. This leaves simply a horizontal line with value 1, regardless of ζ (which is not shown explicitly on the plot). The high-frequency asymptote neglects all terms in any sum except those proportional to the highest power of ω. This leaves a straight line with a slope of -2 that intersects the low-frequency asymptote at the break frequency, which equals the natural frequency ω_n regardless of ζ. This is drawn as a dashed line in the plot.

The damping ratio ζ strongly affects the behavior only for frequencies near the break or natural frequency. For low damping ratios, the magnitude peaks or resonates at approximately the break frequency. At precisely the break or natural frequency, the magnitude ratio equals $1/2\zeta$, which can be deduced simply from equation (7.14a). Use of this fact plus the asymptotes allows you to sketch the magnitude curve for a particular damping ratio with adequate accuracy for most purposes.

The phase curves are harder to sketch with accuracy (you will likely refer to this plot for this purpose), but observe that the drop in phase totals $180°$ and is relatively abrupt for small damping ratios and gradual for large damping ratios. The straight-line approximation used for real poles applies only to the special case of $\zeta = 1$, which corresponds to the product of two equal real poles. The total phase drop is double the $90°$ for a single pole. (This approximation is omitted from the plot to avoid confusion, but you might wish to pencil it in.)

A system with a pair of complex-conjugate zeros rather than poles has a transfer function that is the reciprocal of the transfer function above. The logarithm of the magnitude of this transfer function therefore equals precisely minus the logarithm of the transfer function above; this relationship is the same as you saw before with first-order zeros. Consequently, the magnitude plot for the complex pair of zeros is exactly the top-to-bottom flip of the magnitude plot for the complex pair of poles. For this reason, and because such second-order zeros occur relatively infrequently, a separate plot is not given here. The phase angle plots for the zeros are similarly top-to-bottom flips of the phase angle plots for the corresponding poles; the phase increases by a total of $180°$ rather than decreases. On the other hand, the case of the zeros permits the damping ratio to be negative without causing an instability. In this case, the phase plot is identical to that of the pole with positive damping.

EXAMPLE 7.4
Find the response of a classic mass-spring-dashpot system to an imposed sinusoidal force of amplitude 10 N and an arbitrary frequency. The mass is 2 kg, the spring has stiffness 50 N/m and the dashpot has coefficient 5 N·s/m.

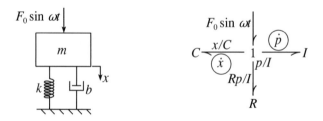

Solution: The bond graph above right gives

$$\frac{dp}{dt} = F_0 \sin \omega t - \frac{1}{C}x - \frac{R}{I}p,$$
$$\frac{dx}{dt} = \frac{1}{I}p.$$

Combining,

$$\frac{d^2x}{dt} = \frac{1}{I}\frac{dp}{dt} = \frac{1}{I}F_0 \sin\omega t - \frac{R}{I}\frac{dx}{dt} - \frac{1}{IC}x,$$

which gives

$$G(S) = \frac{1/I}{S^2 + 2\zeta\omega_n S + \omega_n^2};$$

$$\omega_n = \frac{1}{\sqrt{IC}} = \sqrt{\frac{k}{m}} = \sqrt{\frac{50}{2}} = 5\,\text{rad/s}; \quad \zeta = \frac{R}{2}\sqrt{\frac{C}{I}} = \frac{b}{m\sqrt{km}} = \frac{5}{2\sqrt{50 \times 2}} = 0.25,$$

so that

$$x_p(t) = \frac{F_0/I}{\sqrt{(\omega_n^2 - \omega^2)^2 + (2\zeta\omega_n)^2}}\sin(\omega t + \phi); \quad \phi = -\tan^{-1}\left(\frac{2\zeta\omega_n\omega}{\omega_n^2 - \omega^2}\right),$$

$$= \frac{10/2}{\sqrt{(5^2 - \omega^2)^2 + (2 \times 0.25 \times 5)^2}}\sin\left[\omega t + \tan^{-1}\left(\frac{2 \times 0.25 \times 5\omega^2}{\sqrt{5^2 - \omega^2}}\right)\right]\,\text{meters},$$

$$= \frac{5}{\sqrt{(25 - \omega^2)^2 + 6.25}}\sin\left[\omega t + \tan^{-1}\left(\frac{2.5\omega}{\sqrt{25 - \omega^2}}\right)\right]\,\text{meters}.$$

This result is plotted by the curves for $\zeta = 0.25$ in Fig. 7.6, except that the amplitude needs to be shifted by the factor $5/\sqrt{625 + 6.25} = 0.199$ (downward shift of 14.02 db).

7.1.5 Factorization of Higher-Order Models

Transfer functions of second- and higher-order models are profitably factored as follows:

$$G(S) = \frac{k(S - z_1)(S - z_2)\cdots(S - z_m)}{S^N(S - p_1)(S - p_2)\cdots(S - p_{n-N})}. \tag{7.15}$$

The denominator of $G(S)$ is the characterisitic polynomial, and the values $p_1,\cdots,p_{n-N}$, known as the **poles** of the transfer function, are precisely the roots $S_1,\cdots,S_{n-N}$ of the characteristic equation. The integer N represents the number of additional poles with zero value (or at the *origin*), if any. The numerator is factored similarly; the values $z_1,\cdots,z_m$ are the roots of the numerator polynomial, and are known as the **zeros** of the transfer function. Should there be one or more zeros at the origin, equation (7.15) still applies, but N becomes a negative integer.

Individual poles and zeros can be real, imaginary or complex numbers. They often are represented in an **S-plane plot**, in which the abcissas is the real part and the ordinate is the imaginary part. Poles are indicated by crosses ($\times$), and zeros by (0), as illustrated in Fig. 7.7. The pole labeled 1 in this example is real and negative. It contributes a term to the homogeneous solution proportional to e^{-2t}; the minus sign in the exponent is associated with the presence of the pole in the left-half of the S-plane. Right-half-plane poles produce instability, but left-half-plane poles do not. The pole 2 is at the origin. It allows the homogeneous solution to contain a constant (or $c\,e^{0t}$).

Imaginary and complex poles and zeros come in complex-conjugate pairs. The pair of poles labeled 3a and 3b in the example S-plane lie in the left-half plane; they contribute a term porportional to $e^{-t}\sin(3t + \beta)$ to the homogeneous solution. Again, their presence in the left-half plane is associated with the minus sign in the exponent and the stability of their contribution to the unforced behavior. The imaginary parts ± 3 give the damped natural frequency, ω_d, and the real part -1 gives the exponent $a = -\zeta\omega_n$. The length of the chord from the origin to either pole is the natural frequency $\omega_n = \sqrt{a^2 + \omega_d^2}$.

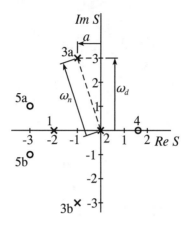

Figure 7.7: Example S-plane plot

The zero labeled 4 is real and positive; those labeled 5a and 5b are complex with conjugate with negative real parts. Zeros, unlike poles, can exist in either half plane without affecting stability.

For most purposes it is convenient to replace the factors for a complex-conjugate pair by its real product, as was done above for the underdamped second-order model. They are then described in terms of its damped natural frequency and damping ratio. Also, the non-zero poles and zeros are interpreted in terms of postive frequencies $\omega_{zi} = |z_i|$, $\omega_{pi} = |p_i|$:

$$G(S) = \frac{k \prod_{i=1}^{j}(1 \pm S/\omega_{zi}) \prod_{i=j+1}^{m} [1 + (2\zeta_{zi}/\omega_{zi})S + (S/\omega_{zi})^2]}{S^N \prod_{i=1}^{k}(1 + S/\omega_{pi}) \prod_{i=k+1}^{n-N} [1 + (2\zeta_{pi}/\omega_{pi}S + (S/\omega_{pi})^2]}. \tag{7.16}$$

The notation $\prod_{i=1}^{j}(1 \pm S/\omega_{zi})$ implies the product $(1 \pm S/\omega_{z1})(1 \pm S/\omega_{z2})$ $\cdots(1 \pm S/\omega_{zj})$, with plus signs being used for left-half plane zeros and minus signs for right-half-plane zeros. The parameters $z_1, \cdots, z_j$ and $p_1, \cdots, p_k$ are *real* zeros and poles. The damping ratios $\zeta_{z,j+1}, \cdots, \zeta_{zm}$ and $\zeta_{p,k+1}, \cdots, \zeta_{n-N}$ all represent subcritical behavior; they are less than 1.

MATLAB® will factor a transfer function with the numerator and denominator specified in terms of the polynomials num and den with the following command:

```
[z,p,k] = tf2zp(num,den)
```

The "tf2zp" can be read as the "transfer function to zero-pole" transformation. Should you start with the state-space format A, B, C, D, the command

```
[z,p,k] = ss2zp(A,B,C,D,i)
```

produces the same result. The index i is an integer that indicates which of possibly several scalar transfer functions is being requested.

EXAMPLE 7.5: Find a factored transfer function for the model described by

$$\frac{d\mathbf{x}}{dt} = \begin{bmatrix} -4 & 0.5 & 0 \\ 2 & -2 & 2 \\ 0 & 1 & -4 \end{bmatrix} \mathbf{x} + \begin{bmatrix} 1 \\ 0 \\ 0 \end{bmatrix} u(t)$$

with the output $y = x_2(t)$.

Solution: The MATLAB coding

```
A=[-4 1/2 0;2 -2 2;0 1 -4];
B=[1;0;0]; C=[0 1 0]; D=[0];
[z,p,k] = ss2zp(A,B,C,D,1)
```

gives the response

```
z =
         -4.0000
p =
         -1.0000
         -4.0000
         -5.0000
k =
      2
```

which indicates the factored result $G(S) = \dfrac{2(S+4)}{(S+1)(S+4)(S+5)}$.

Notice that the example has both a zero and a pole for which $S = -4$. These should be canceled, but MATLAB is not that smart (yet). In practice, poles and zeros rarely cancel *exactly.* Cancellation of an approximately equal left-half-plane pole-zero pair produces little error, but cancellation of an approximately equal right-half-plane pole-zero pair covers up an instability.

7.1.6   Bode Plots for Higher-Order Models*

Bode plots for models with any number of poles and zeros can be assembled readily from the Bode plots of the individual poles and zeros. The ordinate of the magnitude Bode plot equals the logarithm of the magnitude of $G(j\omega)$, or $\log_{10}|G(j\omega)|$. From equation (7.16), this becomes

$$\log_{10}|G(j\omega)| = \log_{10} k + \sum_{i=1}^{j} \log_{10}|1 \pm j\omega/\omega_{zi}|$$

$$+ \sum_{i=j+1}^{m} \log_{10}|1 - (\omega/\omega_{zi})^2 + j2\zeta_{zi}\omega/\omega_{zi}| - N\log|j\omega|$$

$$- \sum_{i=1}^{k} \log -10|1 + j\omega/\omega_{pi}| - \sum_{i=K+1}^{n-N} \log_{10}|1 - (\omega/\omega_{pi})^2 + j2\zeta_{pi}\omega/\omega_{pi}|.$$

$$(7.17)$$

The ordinate of the phase Bode plot is the angle of the phasor $G(j\omega)$, with mathematical symbol $\angle G(j\omega)$:

$$\angle G(j\omega) = \sum_{i=1}^{j} \angle(1 \pm j\omega/\omega_{zi}) + \sum_{i=1+j}^{m} \angle[/1 - (\omega/\omega_{zi})^2 + j2\zeta_{zi}\omega/\omega_{zi}] - N\frac{\pi}{2}$$

$$- \sum_{i=1}^{k} \angle(1 + j\omega/\omega_{pi}) - \sum_{i=k+1}^{n-N} \angle[1 - (\omega/\omega_{pi})^2 + j2\zeta_{pi}\omega/\omega_{pi}]. \qquad (7.18)$$

These equations expedite the determination and interpretation of Bode plots. Equation (7.17) shows that the ordinates of both the magnitude and phase plots equal the sum of the ordinates of the component poles and zeros. The magnitude plot also is shifted vertically by $\log_{10} k$. Decibels sum directly. These summations apply to the asymptotes and linear approximations as well as to the exact curves.

The two examples below regard the third-order model

$$a_3\frac{d^3x}{dt^3} + a_2\frac{d^2x}{dt^2} + a_1\frac{dx}{dt} + a_0x = f_0\cos(\omega t + \beta). \qquad (7.18)$$

EXAMPLE 7.6: Find analytic expressions for the Bode plot variables in the case of three real roots with break frequencies ω_1, ω_2 and ω_3. Then, for the special case of $\omega_1 = 10$ rad/s, $\omega_2 = 50$ rad/s and $\omega_3 = 200$ rad/s, plot the asymptotic approximation for the magnitude curve and the approximation for the phase curve, and sketch the complete Bode curves.

Solution: The transfer function becomes

$$G(S) = \frac{f_0}{a_3S^3 + a_2S^2 + a_1S + a_0} = \frac{k}{(1 + S/\omega_1)(1 + S/\omega_2)(1 + S/\omega_3)}.$$

The Bode plot variables are, in light of the equations above for individual poles,

$$\log|G(j\omega)| = \log_{10} k - \log_{10}\sqrt{1 + \left(\frac{\omega}{\omega_1}\right)^2} - \log_{10}\sqrt{1 + \left(\frac{\omega}{\omega_2}\right)^2}$$

$$- \log_{10}\sqrt{1 + \left(\frac{\omega}{\omega_3}\right)^2},$$

$$\angle G(j\omega) = -\tan^{-1}\left(\frac{\omega}{\omega_1}\right) - \tan^{-1}\left(\frac{\omega}{\omega_2}\right) - \tan^{-1}\left(\frac{\omega}{\omega_3}\right).$$

The Bode plots below are for the given values of ω_1, ω_2 and ω_3 and $k \equiv f_0/a_0 = 1$. The sum of the asymptotes for the magnitudes of the individual poles is shown by a dashed line. Below the first break frequency all the asymptotes equal 1. Between the first and the second break frequency only the asymptote for the first pole is sloped, and therefore the sum of the three has the slope of -1. For frequencies between the second and third breaks both the asymptotes for the first two breaks have slope -1; their sum has slope -2. Above the third break frequency all three poles have asymptotes with slope -1, so their sum has slope -3.

The linear approximations to the three individual phase curves also are shown. Recall that these approximations equal $0°$ for frequencies less than one-tenth their break frequencies, $-90°$ for frequencies greater than ten times their break frequencies, and a straight-line interpolation in between. The sum of these three approximations can be seen to approximate the actual phase shift rather nicely.

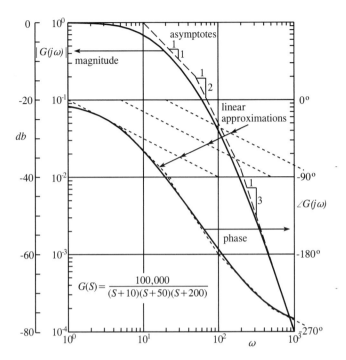

A generalization regarding the magnitude asymptotes: in the absence of zeros, the slope of the asymptotic approximation to the magnitude curve at any frequency equals -1 times the number of poles that have lower break frequencies, including poles at the origin.

EXAMPLE 7.7

Find analytic expressions for the Bode plot variables for the case of one real root at ω_1 and a pair of complex roots, with damping ratio ζ, at ω_2. Then, set $\omega_1 = 10$ rad/s, $\omega_2 = 100$ rad/s and $\zeta = 0.065$, plot the asymoptotic approximation for the magnitude curve and the approximation for the phase curve, and sketch the complete Bode curves.

Solution: The transfer function can be factored as

$$G(S) = \frac{k}{(1 + S/\omega_1)[1 + (2\zeta/\omega_2)S + (S/\omega_2)^2]}.$$

The Bode variables become

$$\log_{10}|G(j\omega)| = \log_{10} k - \log_{10}\sqrt{1 + \left(\frac{\omega}{\omega_1}\right)^2}$$

$$- \log_{10}\sqrt{\left[1 - \left(\frac{\omega}{\omega_2}\right)^2\right]^2 + \left(\frac{2\zeta\omega}{\omega_2}\right)^2},$$

$$\angle G(j\omega) = -\tan^{-1}\left(\frac{\omega}{\omega_1}\right) - \tan^{-1}\left[\frac{2\zeta\omega/\omega_2}{1 - (\omega/\omega_2)^2}\right].$$

The case with the given parameters and $k = 1$ is plotted below:

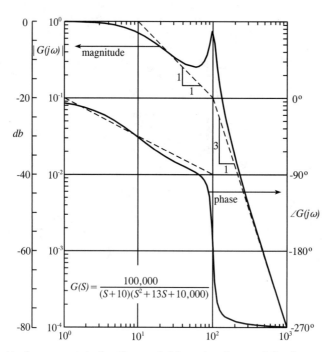

$$G(S) = \frac{100{,}000}{(S+10)(S^2+13S+10{,}000)}$$

The magnitude asymptote for the model has the slope -1 for frequencies between ω_1 and ω_2, and the slope -3 for frequencies above ω_2; the drop in slope of -2 results from the *double break*, or two poles, located at $\omega = \omega_2$. The model resonates about this natural frequency ω_2. The value of the magnitude at the break frequency is almost exactly a factor of $1/2\zeta = 7.69$ greater than the intersection of the asymptotes at that frequency; the effect of the break frequency ω_1 produces only about one-half percent error.

The linear approximation to the phase curve can be seen to apply reasonably well for $\omega < \omega_2$. The phase drops precipitously by 180° in the vicinity of $\omega = \omega_2$, because the damping is fairly small. The total phase shift at infinite frequency is the total number of poles times $-90°$, or $-270°$. Note that the same result applies to the other third-order models in Examples 7.6 and 7.8.

EXAMPLE 7.8

Repeat Example 7.7 for the case in which ω_1 and ω_2 are switched.

Solution The procedure is the same as in the previous example. The resulting plots are given below. The resonance now appears at 10 rad/s, and the first-order pole appears at 100 rad/s. In general, the lowest-order singularies tend to be the more important. In the present case, the presence of the first-order pole in this case might be neglected for practical purposes (as also noted below).

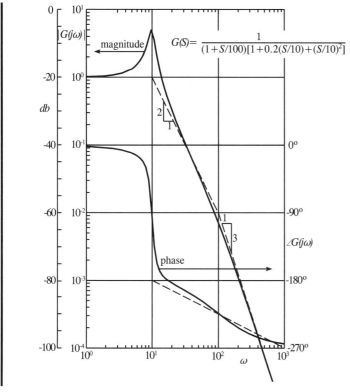

A time constant in a homogeneous solution or an impulse or step response is the reciprocal of the break frequency for a real pole, and a natural frequency is the break frequency for a complex conjugate pair of poles. Step and impulse responses for the cases of Examples 7.6 and 7.7 are given in Fig. 7.8, and for the cases of Example 7.8 are given in Fig. 7.9.

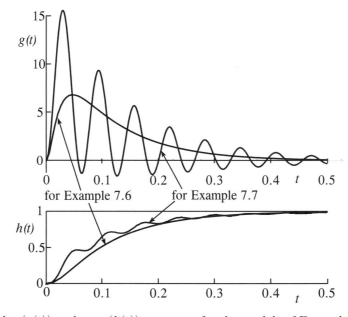

Figure 7.8: Impulse ($g(t)$) and step ($h(t)$) responses for the models of Examples 7.6 and 7.7

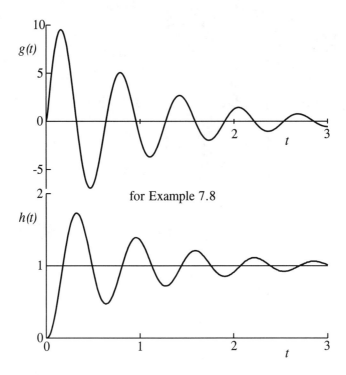

Figure 7.9: Impulse $(g(t))$ and step $(h(t))$ responses for the model of Exmple 7.8

These plots can be secured readily through the MATLAB commands step and impulse. For the case of Example 7.6, the dominant time constant is $\tau_1 = 1/\omega_1 = 0.1$ second, which can be seen in both the impulse and step responses. For the case of Example 7.7, the oscillations occur at the damped natural frequency $\omega_2\sqrt{1-\zeta^2} = 98.2$ rad/s $= 15.63$ Hz, and decay according to the damping ratio $\zeta = 0.065$. In addition, the effect of the time constant $\tau = 0.1$ second is similar to that of case (a). For the case of Example 7.8, the oscillations occur at the lower resonant frequency of virtually 10 rad/s and decay with the damping ratio $\zeta = 0.1$. The time constant in this case is so relatively short, only $\tau = 1/\omega_2 = 0.01$ seconds, that it effects the responses imperceptibly. For many purposes, therefore, this model could be simplified by omitting the real pole (as also noted above).

EXAMPLE 7.9

Sketch-plot the Bode curves for the fifth-order model

$$G(S) = \frac{200,000}{(S+1)(S+10)^2(S^2+5S+2000)}$$

Also, use MATLAB to plot the impulse response of the model.

Solution: This case includes a pair of identical first-order poles, which is the same as a second-order pole with critical damping. There is also a first-order pole at low frequency and a second-order pole at high frequency. The impulse response shows the effects of all these poles, but the pole at the lowest frequency produces the dominant relatively slow exponential delay. The double break at $\omega = 10$ controls the rise in the first one-half second. The resonance at about $\omega = 45$ produces the small rapid oscillation superimposed on an otherwise smooth response.

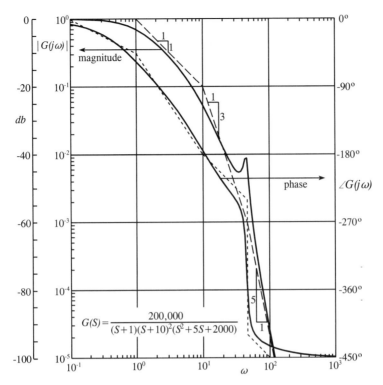

MATLAB coding that gives the complete Bode plot (although with decibels) as well as the impulse response is as follows:

```
num = [200000];
den = [1 26 2225 42700 240500 200000];
bode(num,den)
impulse(num,den)
```

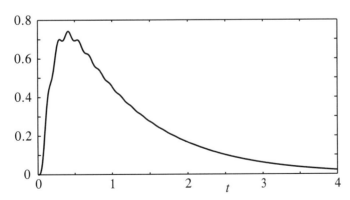

Zeros always contribute an increasing value to a magnitude plot as frequency is increased. A transfer function with a zero produces a magnitude Bode plot that is unchanged if the sign of the real part of the zero is changed, that is if the zero is moved an equal distance into the opposite half-plane. On the other hand, the right-half-plane zero contributes an increasing phase as the frequency is increased, whereas the left-half-plane zero contributes a decreasing phase (or increasing phase lag) as the frequency is increased. Therefore, you can use the phase plot to determine the signs of the real parts of zeros. Models with no zeros in the right-half plane are called **minimum phase-lag** models. It is possible to deduce the transfer function of a known minimum-phase-lag model from its magnitude plot alone. A model with one or more right-half-plane zeros is called **non-minimum phase-lag**.

EXAMPLE 7.10

The Bode plots below correspond respectively to the following transfer functions with zeros:

$$G_a(S) = \frac{1 \pm 0.1S}{S^2 + 0.2S + 1}; \qquad G_b(S) = \frac{0.01(S^2 \pm 2S + 100)}{S^2 + 0.2S + 1}$$

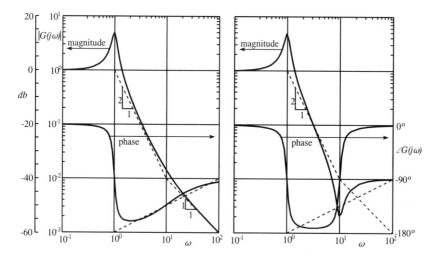

Resolve whether the ± signs should be plus or minus.

Solution: In both cases the behavior in the vicinity of $\omega = 1$ is caused by the identical poles (in the denominator). Both zeros have break frequencies of $\omega = 10$. In both cases the breaks cause the slopes of the asymptotes to increase, from -2 to -1 in the first-order case and from -2 to 0 in the second-order case. Therefore, both ± signs should be +.

The magnitude asymptotes and phase approximations also given in the figure themselves are sufficient to determine the transfer functions.

Models often are classified according to their number of poles or zeros at the origin. The generic designation is **type N**, where N is the net number of poles at the origin. The models presented previously have been *type zero*. Type-one models (with the factor S in the denominator) are also fairly common; type-two and type-minus-one models (with the factor S in the numerator) are not uncommon, but others are rare. The presence and number of poles or zeros at the origin is readily identifiable by the slope of the magnitude curve as the frequency approaches zero, which is $-N$ or $-20N$ db/decade. Note that the phase angle at zero frequency is directly affected also; it is $-N \times 90°$.

EXAMPLE 7.11

Verify that the model with the first Bode plot below has a single pole at the origin, and that the model with the second Bode plot has a single zero at the origin.

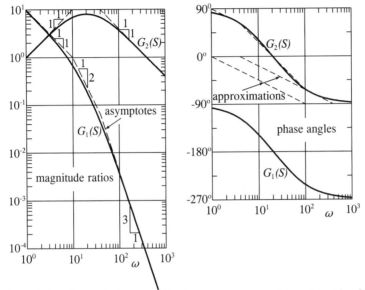

Solution: The slope of the magnitude curve as $\omega \to 0$ is -1 in the first case and $+1$ in the second, directly leading to the indicated conclusions. Further, the phase angle for $\omega \to 0$ is $-90°$ in the first case and $90°$ in the second, corroborating the conclusions.

The magnitude asymptotes and the phase approximations are also shown above. These are sufficient to give the complete transfer functions.

A summary of the Bode asymptotes and linear phase approximations for first-order poles and zeros is given in Fig. 7.10. Recall that an approximation to any higher-order model results from appropriately summing these asymptotes and approximations, and shifting the magnitude result vertically to account for the coefficient k. The phase of a second-order pole or zero that has low damping is a partial exception, however. Rather than spread its 180° phase change over two decades of frequency, it is better to concentrate the change in an abrupt jump at the break frequency. You would also wish at least to approximate the resonance or anti-resonance in the magnitude of such a case, which is not included in the asymptotes, by noting the factor of $1/2\zeta$ at the break frequency.

It is possible to estimate a transfer function from a Bode plot drawn using experimental data, as suggested in Examples 7.10 and 7.11. An understanding of the asymptotes and the discrepancies may have its greatest utility in the interpretation of experimental data, since known transfer functions can be plotted readily by software programs including MATLAB. The slope of the magnitude asymptote at infinite frequency for any transfer function equals -1 times the number of poles minus the number of zeros, called the **pole-zero excess**. You guess the type and locations of the break frequencies, draw the associated asymptotes and the phase approximations, and note the discrepancies between the approximation and the actual curves using Figs. 7.4, 7.5 and 7.6. You then make adjustments until a satisfactory fit results. This process is an example of what is called the **experimental identification** of a dynamic model.

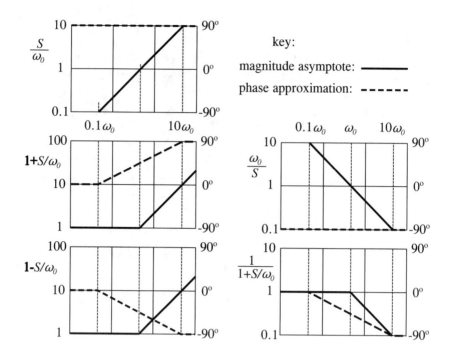

Figure 7.10: Summary of Bode asymptotes and linear phase approximations

EXAMPLE 7.12

The Bode plot below might represent experimental data. Approximate the associated transfer function, and thereby write the associated differential equation.

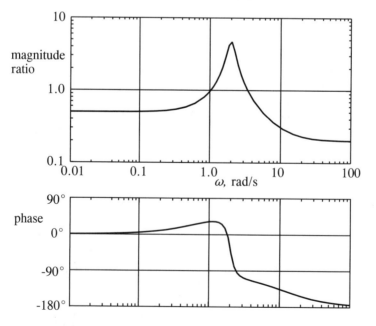

Solution: First, examine the magnitude ratio and observe that the low-frequency asymptote is horizontal. This means you have a type-zero model; $N = 0$. Next, observe that you also have a horizontal asymptote at high frequency; the pole-zero excess is zero, or the number of zeros equals the number of poles. The most prominent feature of the magnitude plot is an apparent resonance at about 2 rad/s. You confirm the presence of a double pole by examination of the phase plot, which shows the telltale rapid drop of well over 90° in the vicinity of 2 rad/s. The magnitude starts to rise noticeably for frequencies lower than one-quarter of this, however, which doesn't happen for sharp resonances, as can be verified by examination of Fig. 7.6 (p. 427). You also know that there has to be at least two zeros in the model (because of the pole-zero excess). Finally, observe that the phase rises at low frequencies, which only can happen because of a zero there. Puting this information together, you annotate the magnitude plot as follows:

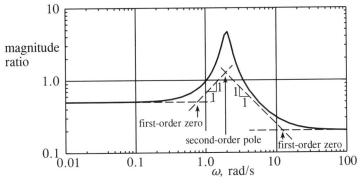

The break frequency for the zero is merely guessed; in any case it is followed by an asymptote with slope +1, as shown.

The break for the second zero must occur at some frequency above the resonance, for otherwise the magnitude curve would not bend from a sharply downward slope toward zero slope. Therefore, you try an asymptote with slope −1 before this pole, as also shown by a dashed line above. You note with pleasure that the two sloped asymptotes have a difference in slopes of −2, which is precisely what the second-order pole demands. Further, the intersection of these asymptotes is close to the peak of the resonance at about 2 rad/s.

You now make adjustments to the asymptotes to make their intersection more precisely align with the resonant frequency, and to recognize the fact that at their respective breaks the deviation of the actual curves from the intersection of the first-order zeros is $\sqrt{2}$, or about 3 db, plus something more for being near the second-order pole:

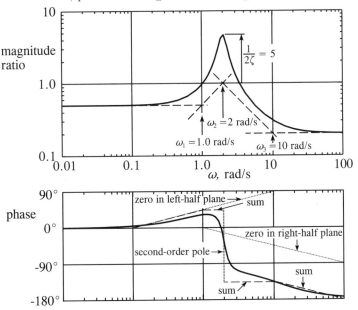

Your value of $k = 0.5$ is determined so that $G(j0) = 0.5$, as plotted. Note also that your transfer function gives $G(j\infty) = 0.2$, a confirmation.

The damping ratio can be estimated by comparing the value of the magnitude at the resonance to the intersection of the two asymptotes there. The ratio of the values is 5:1, which equals $1/2\zeta$, so that $\zeta = 0.1$.

There is one ambiguity left to be resolved: the sign of the zeros. The phase increases in the vicinity of the lower zero, so it must be in the left-half plane. On the other hand, the phase change due to the higher-frequency zero is negative, which dictates the right-half plane. The presence of one zero in the right-half plane is confirmed by the fact that the overall phase change is $-180°$. Were this a miniumum-phase-lag system, the final phase would be zero, because the pole-zero-excess is zero. The phase approximations associated with the zeros are shown above by dotted lines. The sum of these, plus the $180°$ phase drop approximating the pole, is shown by dashed lines. The differences between the actual curves and this approximation are consistent with the plots of Figs. 7.4 and 7.6 (pp. 424, 427).

The resulting transfer function can be interpreted to give the following differential equation, which in turn could be used to solve for the response of the system to any other input:

$$\frac{d^2 x}{dt^2} + 0.2\frac{dx}{dt} + 4x = -0.2\frac{d^2 u}{dt^2} + 1.8\frac{du}{dt} + 2u.$$

The procedure used in Example 7.12 is one of many that would arrive at the same answer. The information is redundant. You should not attempt to discover one procedure for all cases. If you understand the individual effects, you should be able to identify the whole picture, as with a picture puzzle having several pieces.

7.1.7   The Pure Delay Operator*

Imagine a **pure-delay model**, for which the step response is a step delayed by T seconds, and the impulse response is an impulse delayed by T seconds, etc. Transport and wave-like phenomena are among those often approximated by this model. To find its transfer function, imagine a sine wave of frequency ω which also is delayed by T seconds. If the frequency is one cycle in T seconds, or $\omega = 2\pi/T$ rad/s or $T\omega = 2\pi$ rad, the phase shift is $-360°$ or -2π radians. If for example the frequency is halved, the phase shift is halved. Thus the phase shift, in general, equals $T\omega$ radians, and the frequency transfer function is

$$G(j\omega) = e^{-jT\omega}. \tag{7.19}$$

This is representable by a phasor of unit length and angle $T\omega$, since the pure delay neither amplifies nor distorts the signal. The more general **pure-delay operator** becomes both

$$G(S) = e^{-TS} \quad \text{and} \quad G(s) = e^{-Ts}, \tag{7.20}$$

where S is the general derivative operator and s is its Laplace counterpart, so that

$$x(t) = e^{-TS}u(t) = e^{-Ts}u(t) = u(t - T). \tag{7.21}$$

Since the magnitude of the pure delay operator is 1 while the phase lag is non-zero, it must be a non-minimum-phase-lag operator. That is, it must have zeros in the right-half plane. There are an infinity of such zeros, in fact, each of which is matched by a pole of the same magnitude in the left-half plane.

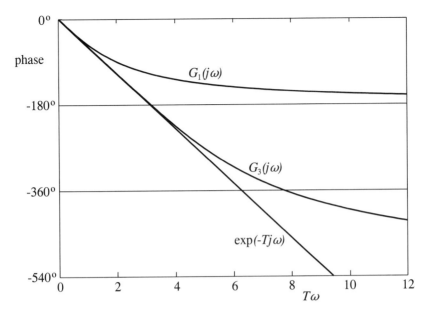

Figure 7.11: Phase lags of a pure delay and Pade approximants

A crude approximation for a pure delay phasor is offered by the combination of a single right-half-plane zero in combination with a single left-half-plane pole of equal magnitude:

$$G_1(j\omega) = \frac{j\omega - 2/T}{j\omega + 2/T}, \tag{7.22a}$$

$$|G_1(j\omega)| = 1, \tag{7.22b}$$

$$\angle G_1(j\omega) = -2\tan^{-1}\left(\frac{T\omega}{2}\right). \tag{7.22c}$$

This is the crudest of a series of approximations to the pure delay in the form of ratios of polynomials known as the **Pade approximants** (pronounced Pah-day). The Pade approximant with third-order numerator and denominator is

$$G_3(j\omega) = \frac{-(Tj\omega)^3 + 12(Tj\omega)^2 - 60Tj\omega + 120}{(Tj\omega)^3 + 12(Tj\omega)^2 + 60Tj\omega + 120}, \tag{7.23}$$

which also has unity magnitude for all frequencies, identical to the exact phasor $e^{-jT\omega}$. The phase angles are reasonably valid only for a restricted band of frequencies which increases with the order of the approximant[2], as indicated in Fig. 7.11.

7.1.8 Summary

Sinusoidal excitations abound in the physical and engineering world, so the response of engineering systems to these inputs is a major concern. The transfer function $G(S)$, when evaluated with $S = j\omega$, gives a complex number $G(j\omega)$ called a frequency transfer function. Its magnitude equals the ratio of the amplitude of the sinusoidal response $x(t)$ to the sinusoidal input $u(t)$ at the frequency ω. Its phase angle equals the phase shift between the

---

[2]The Control System Toolbox of the professional version of MATLAB includes a command **pade** which generates Pade approximants of any order.

sinusoidal response and the input. Plots of the logarithm of the magnitude ratio $|G(j\omega)|$ and of the phase shift versus the logarithm of the frequency are called Bode plots. The quantity $20\log_{10}|G(j\omega)|$, called decibels (db), often is used to represent the magnitude, including by the bode routine of MATLAB.

Transfer functions for systems described by ordinary linear differential equations are ratios of polynomials, which can be factored to reveal poles and zeros that can be real and/or complex. Complex poles and zeros come in complex conjugate pairs; it is convenient to leave their polynomials in unfactored quadratic form. Each pair is describable in terms of its natural frequency and damping ratio. Real poles occur in all first, third and other odd-ordered models, and contribute exponential behavior that is either damped or explosively unstable. Fourth-order models, then, have two natural frequencies and two damping ratios, although if a damping ratio exceeds unity it is preferable to factor its associated second-order polynomial into two first-order polynomials.

Bode plots can be sketched quickly by first drawing their straight-line asymptotes and phase approximations. Conversely, the transfer function corresponding to a given Bode plot can be estimated by approximating these asymptotes and phase approximations. As the frequency rises past a pole, the slope of the magnitude asymptote breaks downward by an added -1; for a zero it breaks upward by $+1$. This applies even to poles and zeros at the origin. Complex poles cause resonances in the vicinity of their second-order break frequencies, while complex zeros cause notches. Each left-half-plane pole reduces the phase asymptote by $90°$, as does right-half-plane zeros. The opposite changes occur for right-half-plane poles and left-half-plane zeros, although only zeros can be in the right-half-plane without causing instability. As a consequence, models with only left-half-plane singularities are known as minimum-phase; the phase angles follow uniquely from the magnitudes, and vice-versa.

Models with right-half-plane zeros are non-minimum phase. The pure delay function includes an infinity of right-half-plane zeros. Its approximations have a finite number of such zeros.

Guided Problem 7.1

This problem regards the use of the phasor method to find a frequency response. Combining the four first-order differential equations into a single fourth-order differential equation is the most difficult part, unless matrix software is used; both the use of MATLAB and detailed manual steps are indicated below. Plotting is particularly simple if MATLAB is used.

The mechanical system shown in part (a) of Fig. 5.2 (p. 265) is excited with the force $F = F_0\sin\omega t$. Find the frequency transfer function for the displacement of the mass m_1 relative to the exciting force, and plot its magnitude and phase in Bode coordinates. The parameters are as follows: $m_1 = 1$ kg, $m_2 = 2$ kg, $k_1 = 25$ N/m, $k_2 = 100$ N/m, $R = 0.25$ N s/m.

Suggested Steps:

1. Find a bond graph model for the system. (This is done in the solution to Guided Problem 5.1, p. 270.)

2. Apply causal strokes, and write the four first-order differential equations in the standard fashion.

3. If you wish to use MATLAB for finding the scalar transfer function of interest, find the elements A, B, C, D of the state-variable formulation. Then, use the MATLAB function ss2tf to find the answer.

4. If you wish to practice manual determination of the scalar transfer function, start by converting the differential equations to algebraic equations, using the operator $S \equiv d/dt$. Steps 5–8 below complete this part of the problem.

5. At this point you wish to get a single fourth-order equation in terms of the momentum of the first mass, which equals $m_1 \dot{x}_1$. Start by multiplying the two most complex algebraic equations by S, and substitute the other two equations to eliminate two of the four variables. You should be left with two equations in terms of the two momenta.

6. Solve one of the equations to get the second momentum as a function of the first (and the operator S), and substitute this into the other equation to get the first momentum as a function of the input excitation.

7. Write p_1/F as a ratio of polynomials in S. Convert this to the ratio x_1/F by noting that $S x_1 = p_1/m_1$.

8. Substitute $j\omega$ for S. The result is the frequency transfer function $G(j\omega)$.

9. To compute and plot the frequency response, you can apply equations (7.8) (p. 420), or use the **bode** command of MATLAB as illustrated above. You have now completed what, if done analytically in an introductory course in vibrations, is usually considered a very difficult problem.

Guided Problem 7.2

This is a basic problem in the identification of a linear model from its frequency response.

A Bode magnitude plot for a model is given in Fig. 7.12. Assuming the model is minimum phase, (a) sketch-plot an approximation for the corresponding Bode phase plot, and (b) estimate the corresponding frequency transfer function.

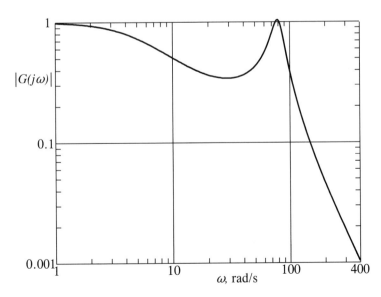

Figure 7.12: Bode plot for Guided Problem 7.2

Suggested Steps:

1. Estimate the asymptotes of the magnitude plot, noting that their slopes must be whole integers. The deviations of the curves from the asymptotes can be appreciated with the help of Figs. 7.4, 7.5 and 7.6 (pp. 424-427).

2. Identify the break frequencies from your magnitude asymptotes, and start the phase diagram by sketching its corresponding horizontal asymptotes.

3. Note the sloped straight-line approximation for the phase lag of first-order poles and zeros given in Figs. 7.4, 7.5 and 7.10 (p. 440). Incorporate this improvement into your sketch wherever it applies.

4. Complete your phase estimate with a smooth curve.

5. Use the break frequencies you have already identified to estimate the transfer function.

Guided Problem 7.3

This problem illustrates the effect of right-half-plane zeros on the frequency response of linear models.

Three stable models have the same Bode magnitude plot given in Fig. 7.13, but different phase plots as shown. Estimate the corresponding transfer functions.

Suggested Steps:

1. Estimate the asymptotes of the given magnitude plot. Note that a second-order zero has precisely the same form in Bode coordinates as the corresponding second-order pole except for an opposite sign. Identify the damping coefficient from the degree of inverse peaking. Write the transfer function or functions, assuming all poles and zeros affect the plot. Recognize any ambiguity which nevertheless results in a stable system, which the existance of the plot assumes.

2. Predict the phase angles using the results of step 1. Can a non-minimum phase lag exist? Compare with the given plots for the three cases, and draw conclusions.

3. Additional phase lag with no effect on the magnitude can result from symmetric placement of one or more pairs comprising a zero in the right-half-plane and a pole in the left-half-plane, as in the Pade approximants for a pure delay. A pure delay itself is included as a possiblility. Estimate the actual case by focusing on the differences between the phase lag of the remaining unknown system and the phase lags of the two known systems.

PROBLEMS

7.1 Estimate the impulse response in analytical form for the system characterized in Bode form in Guided Problem 7.2 (above).

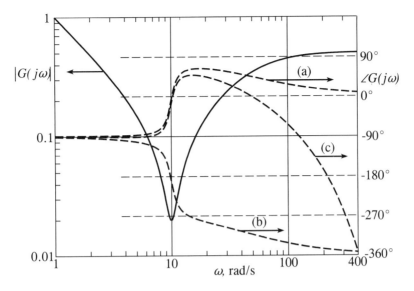

Figure 7.13: Bode plot for Guided Problem 7.3

7.2 Estimate the step responses in analytical form for the three systems characterized by Bode plots in Guided Problem 7.3 (above).

7.3 Evaluate, directly from the given transfer function, the magnitude ratio and phase angle of the steady-state response of the model given in Fig. 7.2 (p. 422) for sinusoidal excitations at 1 and 100 rad/s. Check to see that they agree with the plots.

7.4 A linear model is described by the transfer function $G(S) = \dfrac{S}{S^2 + 2S + 5}$. Find the steady-state response when the excitation is $u(t) = \sin(2t)$.

7.5 Answer the above question for $G(S) = \dfrac{S+3}{(S+2)(S^2 + 2S + 5)}$.

7.6 *Continuation of Problem 5.8 (p. 268) and Problem 6.43 (p. 393) on the effect of a drain tube placed through an eardrum.* Find the transfer function from the input P_{in} to the output p_e/I_e. (You may use the MATLAB command `ss2tf`.) Find and sketch the Bode magnitude asymptotes for the model, which ignores dissipation, and compare to the corresponding model without the tube. Briefly discuss the anticipated consequences the tube has on hearing. Also, speculate on the way the inner ear compensates for the natural response, using whatever knowledge you may have about hearing.

7.7 Find the steady-state force $F(t)$ for the system shown below with $x = x_0 \sin \omega t$.

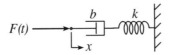

7.8 *Continuation of Problems 3.22 (p. 108) and 4.56 (p. 229).* The applied torque M is sinusoidal at frequency ω. Determine the frequency transfer function. *Continued in Problem 7.45.*

7.9 The model with the transfer function $G(S) = 1000/(10S^3 + S^2 + 1000S)$ has three roots.

(a) Determine the natural frequency, damped natural frequency and damping ratio for the two non-zero roots.

(b) Determine the amplitude of the steady-state output for an input of unit amplitude at a frequency of 10 rad/s. Also, determine the phase shift of the output relative to the input.

(c) Sketch-plot the Bode diagram for the model. Base this largely on straight-line asymptotes and approximation plus the answer to part (b).

7.10 A model has the transfer function $G(S) = \dfrac{12}{S^2 + 10S + 169}$.

(a) Determine the damped and undamped natural frequencies, and the percent of critical damping.

(b) Determine the frequency at which the steady-state response has its greatest amplitude, and find that amplitude and its phase relative to the input.

(c) Sketch-plot the Bode diagram for the model, taking advantage of straight-line asymptotes and approximations and the answers to part (b).

7.11 An instrument transducer produces a signal that contains an unwanted periodic noise signal at 60 Hz. The part of the signal of interest lies below 5 Hz. The signal is received by a recording and display instrument that requires negligible signal current; *i.e.*, its input impedance is very high. A simple potential solution employs an RC filter, as shown below. Standard electrolytic capacitors of 2, 5, 10 and 20 mf are available, as are the standard resistors of 10, 12, 15, 22, 27, 30, 33, 39, 43, 47, 51, 56, 62, 68, 75, 82, 91 kΩ and all factors of 10 lower and higher.

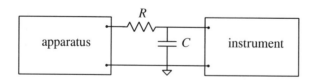

Find a design, if it exists, that reduces the unwanted signal by a factor of at least 10 without reducing any of the wanted signal by more than 40%, while keeping the current small. The instrument is inaccurate if more than 0.1 ma peak-to-peak is drawn from the transducer for a signal of 4 volts peak-to-peak at the 5 Hz.

7.12 A duct leads from an acoustic source to a resistive load; an exhaust pipe from an IC engine is an example. In order to reduce the propagation of acoustic pulsations in the flow, a **Helmholtz resonator** is attached, which comprises a chamber with fluid compliance, C, attached to the duct by a narrow entrance with fluid inertance, I, as shown.

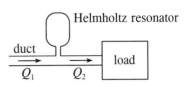

(a) Model the system with a bond graph.

(b) Write differential equations.

(c) Combine the differential equations to get a second-order differential equation with a momentum as the dependent variable.

(d) Relate this momentum to the input flow, Q_1. Also, relate the output flow Q_2 to this momentum and Q_1, in order to get a transfer function between Q_1 and Q_2.

(e) Sketch, or plot using MATLAB, the Bode plot for the particular case in which $I = R^2 C/4$. (The frequency $\omega_n = 1/\sqrt{IC}$ can be used as normalizing frequency to give a dimensionless frequency ratio.)

(f) Use the above results to suggest the design of a muffler that would be effective over a range of frequencies.

7.13 A system with the transfer function

$$G(S) = \frac{100(S+1)}{S^2 + 2S + 100}$$

is excited by the disturbance

$$u(t) = 4\sin\left(\frac{2\pi}{5}t\right) \qquad 0 \le t \le 2.5\text{sec}$$

$$= 0 \qquad t \ge 2.5\text{sec}$$

(a) Find the homogeneous solution $x_h(t)$.

(b) Find the particular solution $x_p(t)$ valid for $0 \le t \le 2.5$ seconds. (Use of the phasor method is suggested.)

(c) Sum $x_h(t)$ and $x_p(t)$ to get the complete solution valid for $0 \le t \le 2.5$ seconds. Evaluate the undetermined coefficients using the known initial conditions $x(0) = 0$, $\dot{x}(0) = 0$.

(d) Find $x(2.5)$ and $\dot{x}(2.5)$, and use these as initial conditions for the solution for $t \ge 2.5$ seconds.

7.14 Find an approximate numerical solution to the previous problem using the `lsim` command of MATLAB.

7.15 The model represented by the Bode plot in Fig. 7.2 (p. 422) is excited by the same disturbance as the two previous problems.

(a) Find the response analytically, following the steps given for the first of these problems.

(b) Find the response numerically, following the steps given for the second of these problems.

7.16 The impedance of a spring is $1/CS$ in terms of the operator S, and $1/Cj\omega$ in terms of the frequency ω. The factor $1/j = -j$ describes the fact that the force lags the velocity by $90°$, or is in phase with the displacement. Sometimes a servo system is designed to imitate a spring, but produces a force that lags the displacement somewhat, or lags the velocity by a little more than $90°$; it can be described as a "spring with a lag." Your task is to determine the stability of a system comprising a mass connected to ground through such a "spring," and what effect is produced by adding a dashpot. The following steps are intended to help, but are not mandatory.

(a) Represent the system (without the dashpot) with a bond graph. You can invent a symbol, such as Z, to represent the "spring."

(b) Characterize the "spring" by its impedance as a function of frequency; choose some reasonable assumption to represent the small added phase lag.

(c) Transform the result of part (b) into a function of S.

(d) Write the differential equation for the system in operational form, using the result of part (c).

(e) Use the result of part (d) to find the characteristic values that determine the stability.

(f) Modify your model to represent the addition of a dashpot between the mass and ground. Repeat step (d) and (e).

(g) Repeat part (f), instead placing the dashpot between the mass and the spring.

7.17 Sketch the Bode diagram for the model described by the following transfer function:

$$G(S) = \frac{S^2 + 20S}{S^3 + 4S^2 + 104S + 200}$$

7.18 Sketch Bode plots for the models described by the following minimum-phase and nonminimum-phase transfer functions. Find and plot their responses to a unit step input.

$$G_1(S) = \frac{1}{S}(1 + TS); \qquad G_2(S) = \frac{1}{S}(1 - TS)$$

7.19 Find the transfer functions for the two systems characterized by Bode plots in Example 7.11 (p. 439). Note that magnitude asymptotes and straight-line phase approximations have been drawn already.

7.20 Estimate the transfer function for the system that produces the Bode plot below.

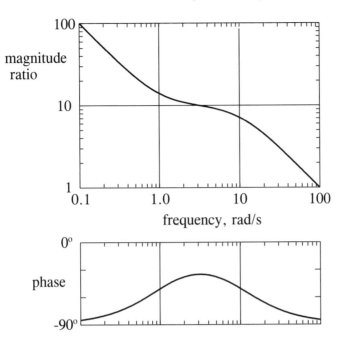

7.21 Answer the previous question for the Bode plot below.

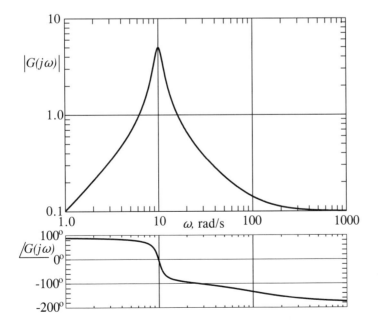

7.22 Answer the previous question for the Bode plot below.

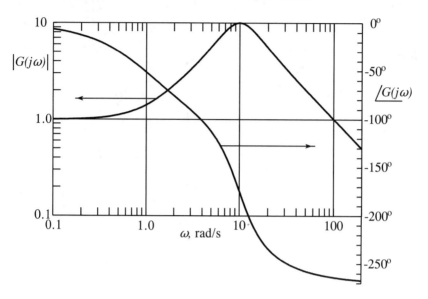

SOLUTIONS TO GUIDED PROBLEMS

Guided Problem 7.1

1. From the solution to part (a) of Guided Problem 5.1 (p. 270):

$I_1 = m_1 = 1$ kg $C_1 = 1/k_1 = 1/25$ m/N

$I_2 = m_2 = 2$ kg $C_2 = 1/k_2 = 1/100$ m/N

$R = 0.25$ N s/m

$e = R(p_1/I_1 - p_2/I_2)$

2. $\dfrac{dq_1}{dt} = \dfrac{1}{I_1}p_1 - \dfrac{1}{I_2}p_2$

$\dfrac{dx_2}{dt} = \dfrac{1}{I_2}p_2$

$\dfrac{dp_1}{dt} = F - \left(\dfrac{1}{C_1}q_1 + e\right) = F - \dfrac{1}{C_1}q_1 - \dfrac{R}{I_1}p_1 + \dfrac{R}{I_2}p_2$

$\dfrac{dp_2}{dt} = \dfrac{1}{C_1}q_1 + e - \dfrac{1}{C_2}x_2 = \dfrac{1}{C_1}q_1 - \dfrac{1}{C_2}x_2 + \dfrac{R}{I_1}p_1 - \dfrac{R}{I_2}p_2$

3. $\dfrac{d}{dt}\begin{bmatrix} q_1 \\ x_2 \\ p_1 \\ p_2 \end{bmatrix} = \begin{bmatrix} 0 & 0 & 1/I_1 & -1/I_2 \\ 0 & 0 & 0 & 1/I_2 \\ -1/C_1 & 0 & -R/I_1 & R/I_2 \\ 1/C_1 & -1/C_2 & R/I_1 & -R/I_2 \end{bmatrix}\begin{bmatrix} q_1 \\ x_2 \\ p_1 \\ p_2 \end{bmatrix} + \begin{bmatrix} 0 \\ 0 \\ 1 \\ 0 \end{bmatrix} F.$

Therefore, $\mathbf{A} = \begin{bmatrix} 0 & 0 & 1 & -0.5 \\ 0 & 0 & 0 & 0.5 \\ -25 & 0 & -.25 & 0.125 \\ 25 & -100 & .25 & -.125 \end{bmatrix}$ and $\mathbf{B} = \begin{bmatrix} 0 \\ 0 \\ 1 \\ 0 \end{bmatrix}^T$.

Let $y = p_1 = [0\ 0\ 1\ 0]\mathbf{x}$, so that $\mathbf{C} = [0\ 0\ 1\ 0]$ and $D = 0$.

With these values, the MATLAB command [z,p]=ss2tf(A,B,C,D) gives the response

```
z =
      0 1.0000 0.1250 62.5000 0

p = 1.0e+003 *
      0.0010 0.0004 0.0875 0.0125 1.2500
```

The accuracy of the second coefficient in the last line above can be increased by typing p(2), which gives the response 0.375.

4. $Sq_1 = \dfrac{1}{I_1}p_1 - \dfrac{1}{I_2}p_2$

$Sx_2 = \dfrac{1}{I_2}p_2$

$Sp_1 = F - \dfrac{1}{C_1}q_1 - \dfrac{R}{I_1}p_1 + \dfrac{R}{I_2}p_2$

$Sp_2 = \dfrac{1}{C_1}q_1 - \dfrac{1}{C_2}x_2 + \dfrac{R}{I_1}p_1 - \dfrac{R}{I_2}p_2$

5. $S^2p_1 = SF - \dfrac{1}{C_1}\left(\dfrac{1}{I_1}p_1 - \dfrac{1}{I_2}p_2\right) - \dfrac{RS}{I_1}p_1 + \dfrac{RS}{I_2}p_2$

$S^2p_2 = \dfrac{1}{C_1}\left(\dfrac{1}{I_1}p_1 - \dfrac{1}{I_2}p_2\right) - \dfrac{1}{C_2I_2}p_2 + \dfrac{RS}{I_1}p_1 - \dfrac{RS}{I_2}p_2$

or, combining terms,

$$\left(S^2 + \dfrac{R}{I_1}S + \dfrac{1}{C_1I_1}\right)p_1 = \left(\dfrac{R}{I_2}S + \dfrac{1}{C_1I_2}\right)p_2 + SF$$

$$\left(S^2 + \dfrac{R}{I_2}S + \dfrac{1}{C_1I_2} + \dfrac{1}{C_2I_2}\right)p_2 = \left(\dfrac{R}{I_1}S + \dfrac{1}{I_1C_1}\right)p_1$$

6. Since it is desired to keep p_1, solve the second equation above for $p_2 = p_2(p_1)$:

$$p_2 = \dfrac{(R/I_1)S + 1/I_1C_1}{S^2 + (R/I_2)S + 1/I_2C_1 + 1/I_2C_2}p_1 = \dfrac{0.25S + 25}{S^2 + 0.125S + 62.5}p_1$$

Substituting this into the first equation above gives

$$\left[S^2 + .25S + 25 - \dfrac{(0.125S + 12.5)(0.25S + 25)}{S^2 + 0.125S + 62.5}\right]p_1 = SF$$

7. $\dfrac{x_1}{F} = \dfrac{p_1}{SF} = \dfrac{S^2 + 0.125S + 62.5}{S^4 + 0.375S^3 + 12.5S + 1250}$

8. $G(j\omega) = \dfrac{(j\omega)^2 + 0.125j\omega + 62.5}{(j\omega)^4 + 0.375(j\omega)^3 + 87.5(j\omega)^2 + 12.5j\omega + 1250}$

9. The MATLAB commands

```
num=[1 .125 62.5];
den=[1 .375 87.46875 12.5 1250];
bode(num,den)
```

produce the screen display below, which shows two modestly damped resonant frequencies:

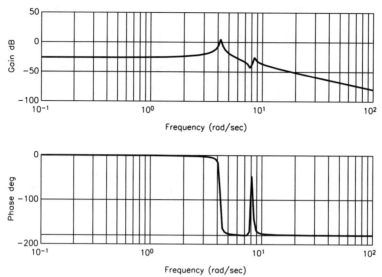

Frequency (rad/sec)

Guided Problem 7.2

1.

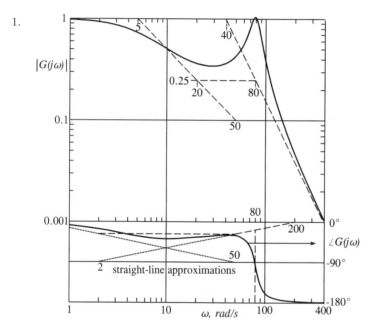

The magnitude plot clearly indicates the presence of a first-order pole at roughly 5 rad/s and an underdamped second-order pole pair at approximately 80 rad/s. Attempting this combination by itself produces an asymptote for the high frequencies which is far below the data, however. A closer look also reveals that the asymptote for the upper frequency range has a slope of −2, not the −3 associated with three poles and no zero. These facts identify the presence of a first-order zero. A good approach to locating all the poles and zeros starts with approximating the asymptotes immediately after the first pole and after the complex pole pair. Then, a horizontal asymptote is drawn to locate the resonance at about 80 rad/s. The intersection of this asymptote with the first asymptote locates the zero at about 20 rad/s. Finally, the peak of the resonace is at a factor of roughly 4 above the intersection of the corresponding asymptotes. This gives a damping ratio of $\zeta = 1/(2 \times 4) = 1/8$.

2.-4. The horizontal asymptotes for the phase drop down $90°$ at the first pole (5 rad/s.), rise back to $0°$ at the zero (20 rad/s), and drop to $-180°$ at the complex pole pair (80 rad/s). The straight-line approximations for the first-order pole and zero, shown in short-dashed lines, spreads each rise or fall over two decades. The sum of these approximations is given by long-dashed straight-line segments. The actual phase curve now can be estimated, as shown by a curved solid line. The behavior near the second-order pole pair is estimated with the help of Fig. 7.2, recognizing the value of ζ.

5. The transfer function is very close to $G(S) = \dfrac{1600(S + 20)}{(S + 5)(S^2 + 20S + 6400)}$.

Note that the number 6400 is ω_n^2, the number 1600 is chosen to give $G(j0) = 1$, and the coefficient 20 for the damping equals $2\zeta\omega_n$. The substitution $S = j\omega$ can be used to check the magnitude plot and possibly refine the transfer function, and to refine the phase plot.

Guided Problem 7.3

1. Five features of the given magnitude ratio are evident. First, the asymptote for low frequencies has a slope of -1, which means that $G(S)$ has a factor of $1/S$. Second, there is a second-order zero at 10 rad/s, which means that the numerator of $G(S)$ has a factor of $S^2 + 2\zeta 10S + (10)^2$. The corresponding asymptote in the region immediately above 10 rad/s has a slope of 1. Third, the value of $1/2\zeta$ is 5, since this is the ratio of the asymptotes at $\omega = 10$ to the actual magnitude at that frequency. Therefore, $\zeta = 0.1$. Fourth, the asymptote for high frequency is horizontal at the magnitude of 0.5. The intersection of this asymptote with the asymptote in the region immediately above the zero gives a pole with break frequency 50 rad/s. Fifth, the value of $|G(S)|$ at $\omega = 1$ is 1.0. Putting these five conclusions together gives

$$G(S) = \frac{S^2 \pm 2S + 100}{2S(S + 50)} \quad \text{so that} \quad G(j\omega) = \frac{100 - \omega^2 \pm j2\omega}{2j\omega(j\omega + 50)}.$$

The $\pm$ in the numerator recognizes the possibility that the zero could lie in the right-half plane. The pole, on the other hand, must lie in the left-half plane; otherwise, the system would be unstable and the frequency transfer function never could have been measured. This transfer function overlooks the possibility of pole-zero combinations for which the magnitudes cancel, but the phase angles do not.

2. $\angle G(j\omega) = \tan^{-1}\left(\dfrac{\pm 2\omega}{100 - \omega^2}\right) - \tan^{-1}\left(\dfrac{50}{-\omega}\right)$.

Note that a negative denominator for the argument of an inverse tangent places the angle in the second or third quadrant, whereas a negative numerator places the angle in the third or fourth quadrant. A check of a few values shows that the plus sign in the numerator gives the proper phase for case (a), and the minus sign gives the proper phase for case (b).

3. The phase for case (c) is virtually identical to that for case (a) for low frequencies, but a gradual discrepancy apprears shortly above the notch or anti-resonance which appears to accelerate for higher frequencies. Apparently some magnitude-cancelling pole-zero pairs exist. If the phase difference between the two cases is examined as a function of *linear* frequency, on the other hand, it can be seen to increase at a *steady* rate of 0.9 degrees per radian/second, reaching $90°$ at 100 rad/s and $360°$ at 400 rad/s. This reveals a pure delay of $T = 2\pi/400$ seconds, so the transfer function becomes

$$G(S) = \frac{S^2 + 2S + 100}{2S(S + 50)} e^{-(\pi/200)S}.$$

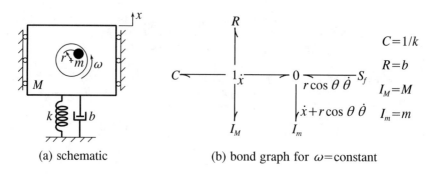

(a) schematic   (b) bond graph for $\omega=$constant

(c) fully annotated bond graph

Figure 7.14: Machine with unbalanced rotor

7.2 Mechanical Vibrations

The vibrations of simple underdamped mechanical ystems that are produced either by non-zero initial conditions or impulse or step excitations have been considered in Chapter 4. The mathematical models of the previous section with their sinusoidal forcing and responses also apply to vibrating mechanical systems, amongst others.

This section focuses on certain issues specific to vibrating mechanical systems, mostly through the use of case studies. Systems with rotating unbalanced masses are considered, and undamped and damped vibration absorbers are introduced. The "modes of motion" of undamped systems with more than one "degree of freedom" are investigated. More advanced topics regarding mechanical vibrations are presented in Section 7.4 and Chapters 9 and 10, and a different stylistic approach is introduced in Appendix B.

7.2.1 Case Study: Rotating Unbalanced Mass

Fig. 7.14 shows a system comprising an unbalanced rotor representable by a point mass m at radius r mounted with bearings in a machine of mass M and mounted on a rigid floor with intervening spring with rate k and dashpot with coefficient b. The spring and dashpot are intended to reduce the vibratory force on the floor. The machine is constrained to move vertically in order to simplify the analysis; sideways motion likely would be allowed in practice, but would not be so important and would be essentially uncoupled from the more important vertical motion and forces.

The bond graph in part (b) of the figure is found partly from an expression for the kinetic energy:

$$\mathcal{T} = \frac{1}{2}M\dot{x}^2 + \frac{1}{2}m(\dot{x} + r\cos\theta\,\dot{\theta})^2 + \frac{1}{2}m(r\sin\theta\,\dot{\theta})^2. \qquad (7.24)$$

The first term within parentheses is the vertical component of the velocity of the mass m, and the second term within parentheses is the horizontal component. The angle of the rotor is defined as θ, and its time derivative is the angular velocity $\omega \equiv \dot{\theta}$. The angular acceleration $\ddot{\theta}$ is assumed to be zero in order to simplify the analysis.[3] This makes the velocity $r \cos \theta \, \dot{\theta}$ an independent variable, representable by the velocity source S_f.

Causal strokes now may be placed on the bond graph and the annotations completed, as depicted in part (c) of the figure. Note that derivative causality is required of the inertance I_m. The state differential equations become

$$\frac{dp}{dt} = mr \sin \theta \, \dot{\theta}^2 - \frac{m}{M}\frac{dp}{dt} - \frac{b}{M}p - kx, \tag{7.25a}$$

$$\frac{dx}{dt} = \frac{1}{M}p. \tag{7.25b}$$

Solving equation (7.25a) for dp/dt, and setting $\dot{\theta} = \omega$, $\theta = \omega t$,

$$\frac{dp}{dt} = \frac{M}{M+m}\left(mr\omega^2 \sin \omega t - \frac{b}{M}p - kx\right). \tag{7.26}$$

Combining equations (7.25b) and (7.26) to eliminate p, and rearranging,

$$\frac{d^2x}{dt^2} + \frac{b}{M+m}\frac{dx}{dt} + \frac{k}{M+m}x = \frac{mr\omega^2}{M+m}\sin \omega t. \tag{7.27}$$

This result is consistent with a mass of $M+m$, spring constant k, damping coefficient b and excitation force $mr\omega^2 \sin \omega t$ (equal to the centripetal force of a rotary mass about a fixed point). In conventional operator form it becomes

$$(S^2 + 2\zeta\omega_n S + \omega_n^2)x = \frac{mr\omega^2}{M+m}\sin \omega t; \quad \omega_n = \sqrt{\frac{k}{M+m}}; \quad \zeta = \frac{b}{2\sqrt{k(M+m)}}. \tag{7.28}$$

The steady-state or particular solution is

$$x_p = \frac{mr\omega^2/(M+m)}{\sqrt{(\omega_n^2 - \omega^2)^2 + (2\zeta\omega_n\omega)^2}}\sin(\omega t + \phi); \quad \phi = -\tan^{-1}\left(\frac{2\zeta\omega_n\omega}{\omega_n^2 - \omega^2}\right) \tag{7.29}$$

Note the resonance at approximately $\omega = \omega_n$, and that when $\omega \to \infty$, $|x_p| \to |mr/(M+m)|$. Note also that the force acting on the floor is $F_p = kx_p$. This leads to the plots in part (a) of Fig. 7.15.

The **force transmissibility** T_F is defined as the magnitude ratio of the force transmitted to the floor to the excitation force, in this case $|F_p|/mr\omega^2$, which becomes

$$T_F = \frac{k|x_p|}{mr\omega^2} = \frac{\omega_n^2}{\sqrt{(\omega_n^2 - \omega^2)^2 +)2\zeta\omega_n\omega)^2}}. \tag{7.30}$$

This ratio is plotted in part (b) of Fig. 7.15. As $\omega \to \infty$, it rapidly approaches zero, unlike the absolute force $|F_p|$, which approaches a constant.

The plots suggest a strategy for keeping the force on the floor small for a given rotational speed, ω: make the spring soft enough so that $\omega_n >> \omega$, and make sure b and therefore ζ are small. On the other hand, making ζ small may produce excessive vibration at start-up if the speed of the rotor is not passed through the region $\omega \simeq \omega_n$ rapidly on the way to its equilibrium speed. There may be a compromise, consequently, between low steady-state vibration and low transient vibration.

---

[3]Setting $\ddot{\theta} = 0$ requires a pulsating torque on the shaft. A better model likely would have a constant or zero torque, which renders the system nonlinear, although the result may be a little different. The effective mass becomes $M + m\sin^2 \theta$ instead of the constant $M + m$.

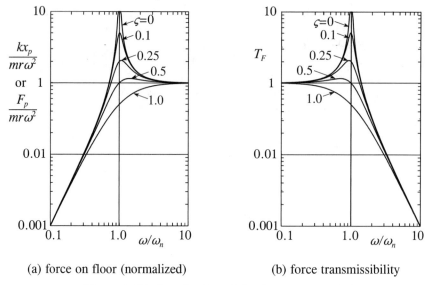

(a) force on floor (normalized)   (b) force transmissibility

Figure 7.15: Amplitudes of displacements and forces

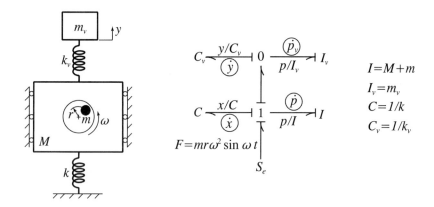

Figure 7.16: Addition of tuned vibration absorber; machine damping removed

7.2.2   Case Study: Tuned Vibration Absorber

One design strategy for reducing the motion x_p at a given rotational speed is the **tuned vibration absorber** pictured in Fig. 7.16. A separate mass, m_v, is attached to the machine through a separate spring, k_v. The dashpot to ground is removed in the interest of simplicity; it is reinstated in Section 7.4 where more powerful matrix methods are employed. The force in the added spring can cancel the force of the unbalance, even at the frequency of resonance $\omega = \omega_n$, eliminating the motion of the mass M. Unfortunately, however, this introduces resonances at other frequencies, as shown below.

The bond graph shown in the figure takes advantage of the knowledge developed above that the unbalance produces a sinusoidal force on an equivalent single mass $I = M + m$. The state differential equations corresponding to the four energy storage elements are

$$\frac{dx}{dt} = \frac{1}{I}p, \tag{7.31a}$$

$$\frac{dy}{dt} = \frac{1}{I}p - \frac{1}{I_v}p_v, \tag{7.31b}$$

$$\frac{dp}{dt} = -\frac{1}{C_v}y - \frac{1}{C}x + F, \tag{7.31c}$$

$$\frac{dp_v}{dt} = \frac{1}{C_v}y. \tag{7.31d}$$

The dependent variable of primary interest is x; all others should be eliminated, to give a single fourth-order differential equation. This can be done various ways, with or without the use of operator notation, retaining the unspecified values for the parameters. None of these are nearly as easy as using the MATLAB function ss2tf, which requires specific values of the parameters.

The values $I_1 = 1$, $C_1 = 1$ will be assigned. These give $\omega_n = 1$ rad/s, so that regardless of the *units* of the individual parameters, the results can be interpreted with time considered as the nondimensional $\omega_n t$ and frequency considered as the nondimensional ω/ω_n. Since the natural frequency of the added mass-spring is to equal that of the machine and its mount, $I_v C_v = 1$. The only substantive choice is the ratio of the two masses; the larger the added mass, the less sensitive the vibration absorption is to a discrepancy between the two natural frequencies. On the other hand, a large mass would be expensive and awkward. As a compromise, the added mass will be set at 20% of the machine mass, giving $I_v = 0.2$ and $C_v = 5$. The results can be applied to any values of the parameters that satisfy the 20% condition.

The matrices needed for ss2tf are

$$\mathbf{A} = \begin{bmatrix} 0 & 0 & 1 & 0 \\ 0 & 0 & 1 & -5 \\ -1 & -0.2 & 0 & 0 \\ 0 & 0.2 & 0 & 0 \end{bmatrix}; \quad \mathbf{B} = \begin{bmatrix} 0 \\ 0 \\ 1 \\ 0 \end{bmatrix}; \tag{7.32a}$$

$$\mathbf{C} = [1\,0\,0\,0]; \quad \mathbf{D} = 0. \tag{7.32b}$$

The MATLAB coding to get the transfer function from the force F to the force on the floor is

```
A=[0 0 1 0;0 0 1 -5;-1 -.2 0 0;0 .2 0 0];
B=[0;0;1;0]; C=[1 0 0 0]; D=0;
[num, den]=ss2tf(A,B,C,D)
```

The response is

```
num = 0 0 1.0000 0 1.0000
den = 1.0000 0 2.2000 0 1.0000
```

which means that

$$G(S) = \frac{S^2 + 1}{S^4 + 2.2S^2 + 1} \tag{7.33a}$$

$$G(j\omega) = \frac{-\omega^2 + 1}{\omega^4 - 2.2\omega^2 + 1}. \tag{7.33b}$$

The force therefore vanishes when $\omega = 1$ (or in the general interpretation, when $\omega/\omega_n = 1$).

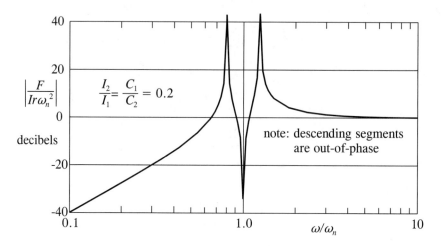

Figure 7.17: Behavior of system with tuned vibration absorber

This transfer function has as its denominator the excitation force, but this force increases with the square of the frequency. Therefore, the $G(S)$ needs to be multiplied by the ratio $(\omega/\omega_n)^2$, which can be accomplished by multiplying the numerator by S^2, changing it to $S^4 + S^2$. The MATLAB coding continues:

```
num=[1 0 1 0 0];
bode(num,den)
```

The resulting magnitude plot, with some annotations, is given in Fig. 7.17. The existence of two resonant freqencies is revealed, one at 80% of ω_n and the other at 125%. Therefore, the idea of the vibration absorber has merit only when the excitation frequency is fixed or nearly fixed. The two resonances can be spread further apart in frequency by increasing the mass $I_v = m_v$ (and decreasing the compliance C_v correspondingly). Similarly, if the mass were reduced, the separation would decrease and in the limit of zero mass the behavior (and the system) would be identical to that of the original system.

The tuned vibration absorber is analyzed in the alternative style of classical vibrations in Appendix B. This style somewhat simplifies the analysis for the restricted class of models to which it applies, making it easier to retain unspecified parameters.

7.2.3 Modes of Motion

The case study above with the tuned vibration absorber has two **degrees of freedom**, and two **modes of motion**. The number of degrees of freedom of a mechanical system is defined as the minimum number of variables needed to uniquely specify its position. In the present case, the variables could be x and y. A system behaves with a mode of motion when it is unexcited and the initial conditions are such that all parts of the system move at the same frequency and the same decay time constant. The number of modes that exist in general equal the number of degrees of freedom. The discussion here is limited to the case with zero damping; a mode of motion then comprises sinusoidal behavior of all elements at a single natural frequency. For the vibration absorption system these are 80% and 125% of the frequency defined as ω_n, which is the natural frequency of the system before the vibration absorber was added. The general unexcited motion of a system comprises a weighted sum of the individual modes of motion. It is helpful to understand these modes, therefore, for multiple reasons.

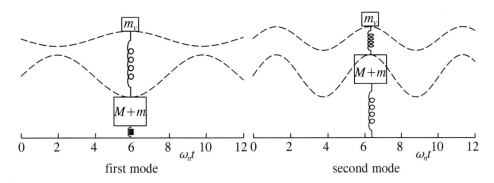

Figure 7.18: Modal behaviors for the tuned vibration absorber system

Since the model of the vibration absorber has no damping, the substitution $s = j\omega$ can be made in the homogeneous equation:

$$(\mathbf{A} - j\omega I)\mathbf{x} = 0, \qquad \mathbf{x} = \begin{bmatrix} x \\ y \\ p_1 \\ p_2 \end{bmatrix}. \qquad (7.34)$$

The values of $j\omega$ that satisfy this equation are known as the **eigenvalues** of the matrix $\mathbf{A}$. They comprise two complex conjugate pairs, which because of the absence of damping means two frequencies that could be called ω_1 and ω_2. These are the natural frequencies at which the forced response became infinite. Setting the determinant of the matrix equal to zero,

$$\begin{vmatrix} -j\omega & 0 & 1 & 0 \\ 0 & -j\omega & 1 & -5 \\ -1 & -0.2 & -j\omega & 0 \\ 0 & -.2 & 0 & -j\omega \end{vmatrix} = \omega^4 - 2.2\omega^2 + 1 = 0, \qquad (7.35)$$

gives the resonant frequencies $\omega_1 = 0.801\omega_n$ and $\omega_2 = 1.248\omega_n$. When ω is set equal to ω_1, the vector $\mathbf{x} = \mathbf{x_1}$ is called the first eigenvector, and when $\omega = \omega_2$, $\mathbf{x} = \mathbf{x_2}$ is the second eigenvector. Any one non-zero element in an eigenvector can be set arbitrarily, since the the left side of equation (7.34) equals zero, but the proportions of the elements are fixed, and represent the **modal shapes**.

The first and third equations within equation (7.35) give, respectively,

$$p_1 = j\omega x, \qquad (7.36a)$$
$$-x - y = j\omega p_1. \qquad (7.36b)$$

Substituting equation (7.36a) into (7.36b) yields

$$y = (\omega^2 - 1)x. \qquad (7.37)$$

This gives $y = -0.358x$ for the first modal shape, and $y = 0.558x$ for the second modal shape. The proportions of the momenta are the same, as can be verified by using the remaining equations within equation (7.34). This modal behavior is plotted in Fig. 7.18.

EXAMPLE 7.13

Two identical masses and three identical springs are mounted as shown below:

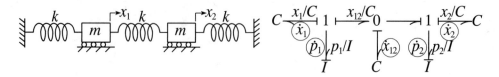

Determine the natural frequencies and the modes of motion.

Solution: The state differential equations, found from the bond graph above, are

$$\frac{dx_1}{dt} = \frac{1}{I}p_1,$$

$$\frac{dx_2}{dt} = \frac{1}{I}p_2,$$

$$\frac{dx_{12}}{dt} = \frac{1}{I}(p_1 - p_2),$$

$$\frac{dp_1}{dt} = -\frac{1}{C}(x_1 + x_{12}),$$

$$\frac{dp_2}{dt} = \frac{1}{C}(x_{12} - x_2).$$

Taking the derivative of the first equation substituting the fourth, and taking the derivative of the second equation substituting the fifth, and noting that $x_{12} = x_1 - x_2$, gives

$$\frac{d^2x_1}{dt^2} = -\frac{1}{IC}(2x_1 - x_2),$$

$$\frac{d^2x_2}{dt^2} = -\frac{1}{IC}(2x_2 - x_1).$$

With operator notation and the substitution $S = j\omega$, this becomes

$$\begin{bmatrix} 2/IC - \omega^2 & -1/IC \\ -1/IC & 2/IC - \omega^2 \end{bmatrix} \begin{bmatrix} x_1 \\ x_2 \end{bmatrix} = 0.$$

Setting the determinant of the matrix equal to zero gives

$$\omega^4 - \frac{2}{IC}\omega^2 + \frac{3}{IC} = 0,$$

which gives the solutions $\omega_1 = 1/\sqrt{IC} = \sqrt{k/m}$, $\omega_2 = \sqrt{3/IC} = \sqrt{3k/m}$.

Substituting these values into equation (7.40) establishes that the first mode has $x_2 = x_1$, which means that both masses oscillate back and forth at ω_1 maintaining a constant distance apart, and that the second mode has $x_2 = -x_1$, which means that the masses oscillate symmetrically in opposite directions at ω_2.

The particular configuration is so simple that you can see what the modes ought to be on inspection, and with this knowledge readily confirm the two resonant frequencies.

Further examples of modal behavior are given in Section 7.3.

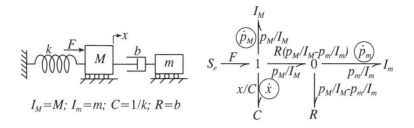

Figure 7.19: Untuned viscous damper

7.2.4 Case Study: Untuned Viscous Damper

The tuned vibration absorber works only if the frequency of excitation is known fairly accurately and is constant. The simple **untuned viscous damper**, shown in Fig. 7.19, replaces the spring of the tuned vibration absorber with a dashpot. It provides damping over a broader band of frequency, although it is not nearly as effective as the tuned vibration absorber very near the center of the band. The transmissibility of the applied force $F = F_0 \sin \omega t$ to the force on the ground $F_g = kx$ is the principal concern. The values of m and b should be chosen to achieve a desirable ratio over a band of excitation frequencies.

The bond graph gives

$$\frac{dp_M}{dt} = F - \frac{1}{C}x - \frac{R}{I_M}p_M + \frac{R}{I_m}p_m, \tag{7.38}$$

so that

$$p_m = \frac{I_m}{R}\left[\left(S + \frac{R}{I_M}\right)p_M + F + \frac{1}{C}x\right]. \tag{7.39}$$

The bond graph also gives

$$Sp_m = \frac{R}{I_M}p_M - \frac{R}{I_M}p_m, \tag{7.40}$$

so that

$$\left(S + \frac{R}{I_m}\right)\frac{I_m}{R}\left[\left(S + \frac{R}{I_M}\right)p_M - F + \frac{1}{C}x\right] = \frac{Rp_m}{I_M}, \tag{7.41}$$

or

$$\left[\frac{I_m}{R}S^2 + \left(1 + \frac{I_m}{I_M}\right)S\right]p_M = \left(\frac{I_m}{R}S + 1\right)\left(F - \frac{x}{C}\right). \tag{7.42}$$

Returning to the bond graph,

$$Sx = \frac{1}{I_M}p_M = \frac{\frac{1}{I_M}\left(\frac{I_M}{R}S + 1\right)\left(F - \frac{x}{C}\right)}{\frac{I_m}{R}S^2 + \left(1 + \frac{I_m}{I_M}\right)S}, \tag{7.43}$$

Combining to solve for x in terms of the force F,

$$x = \frac{(I_mC)^{3/2}\left(\frac{1}{I_M}S + \frac{R}{I_MI_m}\right)}{(I_mC)^{3/2}\left[S^3 + \left(\frac{R}{I_m} + \frac{R}{I_M}\right)S^2 + \frac{1}{I_MC}S + \frac{R}{I_mI_MC}\right]}F. \tag{7.44}$$

With the definitions

$$\omega_n^2 \equiv \frac{1}{I_MC}; \qquad \Omega \equiv \left(\frac{\omega}{\omega_n}\right); \qquad \zeta \equiv \frac{R}{2}\sqrt{\frac{C}{I_M}}; \qquad r \equiv \frac{I_m}{I_M}, \tag{7.45}$$

and the assumptions

$$F = F_0 \sin \omega t; \qquad x = x_0 \sin(\omega t + \phi); \qquad S = j\omega, \tag{7.46}$$

there results

$$\left(\frac{x_0}{CF_0}\right)^2 = \frac{\Omega + 4\zeta^2/r^2}{\Omega(1-\Omega)^2 + (4\zeta^2/r^2)[1-(1+r)\Omega]^2}. \tag{7.47}$$

This complicated expression is proportional to the square of the force transmissibility of interest. It varies considerably with the values of $I_m = m$ and $R = b$ chosen. What to do?

If the excitation frequency could be anything, the best design may be the one that minimizes the peak value of the force transmissibility. In this case, the derivatives of the right side of equation (7.47) with respect to Ω and with respect to ζ should vanish. Tackling the latter first, note that if the numerator and the denominator of equation (7.47) are defined as u and v, respectively, the requirement is that $v(du/d\zeta^2) = u(dv/d\zeta^2)$. Letting $\Omega = \Omega_p$ at this point, this gives

$$\left\{\Omega_p(1-\Omega_p)^2 + \frac{4\zeta^2}{r^2}[1-(1+r)\Omega_p]^2\right\}\frac{4}{r^2} = \left(\Omega_p + \frac{4\zeta^2}{r^2}\right)\frac{4}{r^2}[1-(1+r)\Omega_p]^2. \tag{7.48}$$

Many terms cancel, leaving

$$1 - \Omega_p = \pm[1-(1+r)\Omega_p]. \tag{7.49}$$

The option of the plus sign requires $r = 0$, which is not meaningful. For the minus sign, the value of the frequency at the peak being minimized is

$$\Omega_p = \frac{2}{2+r}. \tag{7.50}$$

Substituting this value into equation (7.47) gives, after a bit of shaking, that the value of this peak is

$$\left(\frac{x_0}{CF_0}\right)_p = 1 + \frac{2}{r}. \tag{7.51}$$

Thus the maximum value of the force transmissibility is considerably greater than unity, which may sound like a failure until you recognize that without the damper it is infinity. The ratio of masses r must not be very small, furthermore; the added mass m must be quite significant. For a maximum transmissibility of 5, for example, the mass m must be one-eighth the mass M.

For the derivative of the right side of equation (7.47) with respect to Ω to be zero, $v(du/d\Omega) = u(dv/d\Omega)$. Carrying out this instruction, and setting $\Omega = \Omega_p$ in the result,

$$\Omega_p(1-\Omega_p)^2 + \frac{4\zeta^2}{r^2}[1-(1+r)\Omega_p]^2$$
$$- \left(\Omega_p + \frac{4\zeta^2}{r^2}\right)\left\{(1-\Omega_p)^2 - 2\Omega_p(1-\Omega_p) - \frac{8\zeta^2}{r^2}[1-(1+r)\Omega_p](1+r)\right\} = 0. \tag{7.52}$$

Shaking this equation down, including making major cancellations, finally gives

$$\zeta = \frac{r}{\sqrt{2(2+r)(1+r)}}. \tag{7.53}$$

This leads directly to the proper choice of the damper coefficient, b.

Large vibration dampers are sometimes placed near the top of modern tall buildings. To make them more effective than either the tuned vibration absorber or the untuned viscous damper, active sensing, control and actuation is employed, which is beyond the scope of the present discussion.

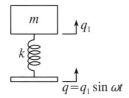

Figure 7.20: System for Guided Problem 7.4

7.2.5 Summary

Although the vibrations of mechanical systems can be treated by the same methods as any physical system, there are a few special considerations. Rotating shafts with unbalanced rotors produce sinusoidal forces that may be analyzed with frequency-response techniques. Vibrations of a mass excited at a known constant frequency may be virtually eliminated by introducing a tuned vibration absorber, which comprises a mass and a spring that resonate at the excitation frequency. The new system has one more resonant frequency than the original system, but the frequencies are changed. Adding a mass and a dashpot, instead, introduces relatively little reduction in vibration at any one frequency, but acts over a broader range of frequencies without introducing a new resonance.

The unexcited behavior of a linear undamped system comprises a weighted sum of the motions for its individual modes, the number of which equals the degrees of freedom of the system. In modal motion, all members of the system oscillate at the same frequency, which is the resonance frequency for that mode. The frequencies are the eigenvalues of the $\mathbf{A}$ matrix, and the mode shapes are represented by the associated eigenvectors.

The concept of modes is extended in the Section 7.4 to systems with damping. Matrix methods are further developed there for systems with and without damping.

Guided Problem 7.4

This is a classical elementary problem involving the undamped vibration of the single-degree-of-freedom system shown in Fig. 7.20.

The objective is to compare the amplitude of the steady-state displacement of the block, q_1, with the amplitude of the displacement of the bottom end of the springs, $q = q_0 \sin \omega t$. Their ratio is an example of a **displacement transmissibility**.

Suggested Steps:

1. Model the system with a bond graph. Note that its input is the time derivative of $q(t)$.

2. Write the state-variable differential equations for your model.

3. Combine the differential equations to give a second-order equation in terms of a momentum.

4. Express the output variable of interest, q_1, as a function of the momentum.

5. Determine a transfer function between the input variable, q, and q_1. Convert this to a frequency transfer function.

6. Either sketch the Bode plot corresponding to the transfer function, or use MATLAB.

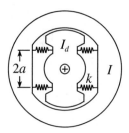

Figure 7.21: Torsional tuned vibration absorber for Guided Problem 7.5

7. Presuming that the mass and the frequency are given, observe the approximate range of spring constants, k, that would (i) keep the amplitude ratio $|q_1|/|q|$ to within ten percent of unity, and (ii) would keep the amplitude of $|q_1|/|q|$ to less than 0.1.

Guided Problem 7.5

The system of Fig. 7.21 can be used as a torsional tuned vibration absorber. A shaft with angular velocity $\dot{\phi}$ and applied moment M drives the central member of a torsional vibration absorber that has a moment of inertia $I_d = 0.1$ ft·lb·s^2. This member in turn drives, through four springs as pictured below (bearings not shown), a flywheel of moment of inertia $I = 1.0$ ft·lb·s^2. Each spring has stiffness $k = 500$ lb/ft, and the length a is 0.25 ft. Determine the behavior of the system in response to sinusoidal excitation at various frequencies.

Suggested Steps:

(a) Model the system with a bond graph.

(b) Evaluate the parameters of your model.

(c) Find a single differential equation relating $\dot{\phi}$ to M.

(d) Find the steady-state amplitude of $\dot{\phi}$ in response to an M that is oscillatory at frequency ω. Sketch-plot this amplitude as a function of ω with the help of asymptotes.

(e) Identify the frequency at which this system acts as an ideal vibration absorber.

PROBLEMS

7.23 The spring-mass-dashpot system with mass m_1 to the right is at equilibrium initially. A mass m_2 is then placed on top at time $t = 0$. Modeling this system in a conventional way gives the second-order differential equation $(m_1 + m_2)\ddot{x} + b\dot{x} + kx = m_2 g$. The values of the parameters are $m_1 = 0.796$ kg, $m_2 = 0.204$ kg, $m_2 g = 2$ N, $k = 65$ N/m and $b = 2$ N·s/m.

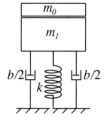

(a) Determine the natural frequency, damping ratio and damped natural frequency of the model.

(b) Determine the transfer function $G(s)$ between an applied force, F, and the downward displacement, $x(t)$

(c) Determine the Laplace transform of the excitation corresponding to the placement of the weight $m_2 g$, and of the response, $X(s)$.

(d) Determine and sketch-plot the response $x(t)$.

7.24 Consider the classical case of sinusoidal position excitation of a mass-spring-dashpot system shown to the right. The objective is to choose springs and the dashpot so that the motion of q produces little motion of q_1. (Note that this problem is similar to Guided Problem 7.4 (p. 465-466), except that a dashpot has been introduced. The use of similar steps is suggested.)

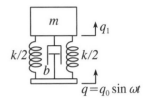

(a) Model the system with a bond graph. (Hint: Two junctions are needed.)

(b) Determine a transfer function between the input, $q(t)$, and the output, $q_1(t)$. Write the corresponding frequency transfer function.

(c) Determine whether b should be large or small if the frequency is (i) much smaller than the natural frequency of the system, ω_n, (ii) about the same as ω_n, and (iii) much higher than ω_n. Also, describe a design strategy if k as well as b can be chosen. Suggestion: Consider the nature of the Bode plots for large and small amounts of damping, b.

7.25 Show for what ratios of parameters the torsional system shown to the right corresponds to the tuned vibration absorber shown in Section 7.2.2. (pp. 458-461).

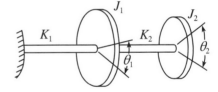

7.26 A torsional system is modeled for purposes of vibration analysis as a flexible shaft with a compliance of 0.001 rad/ft·lb driven at one end and attached to a rigid disk with rotational inertia 2.0 ft·lb·s^2 at the other end. It is excited torsionally at a frequency of 6 Hz. It is proposed to minimize the resulting vibrations of the disk by fixing the inner mass (with $I_d = 0.1$ ft · lb · s^2) of the tuned vibration absorber used in Guided Problem 7.5 to the rigid disk. (The outer mass has $I = 1.0$ ft · lb · s^2.) The stiffnesses k of the four spings must be changed, however. Determine the proper value. (You are *not* asked to find the response at other frequencies, although that would be interesting.)

7.27 Determine the transmission of sound through a wall comprising a single sheet of plaster 3/8 inch thick, which is very thin compared with the acoustic wavelengths of interest. The specific gravity of plaster is 1.8, and its material damping is virtually nil. The acoustic impedance for plane waves entering a large space such as a room is a resistance of magnitude 2.67 lb·s/ft^3, which represents a ratio of pressure change to velocity change.

(a) Model the wall so as to permit estimation of the ratio of the transmitted to incident sound pressure, P_t/P_i. Neglect the effect of motion of the plaster on the incident sound pressure, and neglect the effect of any mechanical supports for the plaster.

(b) Evaluate the ratio for (i) 100 Hz (deep bass) (ii) 2000 Hz (soprano). Express the answers in decibels (db), which are defined as $20\log_{10}(P_t/P_i)$. (Note that humans can detect loud sounds despite attenuations of 50 db and occasionally even 80 db.) What would rock music sound like on the quiet side?

(c) A double wall is constructed with the same plaster sheets and a 1 inch air space in between. Repeat parts (a) and (b) above, assuming no mechanical connection between the two layers. Do your results suggest the acoustic significance of the mechanical connections conventionally used to strengthen the wall?

(d) *Design problem.* Design a practical non-load-bearing interior wall for sound isolation, using plaster sheets and steel supports. Defend your design in terms of strength and other practical considerations as well as its acoustic properties.

7.28 A tall building sways excessively in the wind. It is proposed to place a heavy mass on the roof that can move back and forth on rollers, and which is constrained by a hydraulic damper as shown:

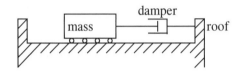

A soft spring (not shown) tends to center the mass, without materially affecting the dynamics, so the damper never gets over or under-extended. The building sways in its first mode of vibration, which has a period of ten seconds; the wind force produces a synchronous component. The stiffness of the building referenced to the lateral motion at the roof is 4.0×10^8 lb/ft.

(a) Model the system with no internal damping and no added mass and damper, and evaluate the effective mass.

(b) Add the inherent internal damping which gives a damping ratio of 0.01, and find the ratio of the maximum displacement at the roof level to the amplitude of an effective sinusoidal force applied at that level.

(c) Introduce the added mass and damper, and write the corresponding differential equations. Solve for the ratio of the displacement amplitude of the top of the building to the amplitude of the applied force.

7.29 *Design problem.* Specify the parameters of a damping system for the building of the previous problem. Tuned and untuned damper designs are possible. An equivalent damping ratio (for a building with no damping system) of 0.02 is desired (doubling the inherent damping), and 0.03 is excellent. There is concern that the added mass not be too heavy for the structure to support, and that the motion of the mass relative to the building not be too great. Use the information given and whatever judgements you can defend. Describe the behavior of your solution quantitatively, and report the method by which it was chosen.

SOLUTIONS TO GUIDED PROBLEMS

Guided Problem 7.4

1-2.

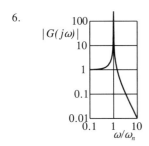

Don't confuse q with $q_c = q - q_1$.

$$\frac{dp}{dt} = \frac{1}{C}q_c$$

$$\frac{dq_c}{dt} = \dot{q}(t) - \frac{1}{I}p.$$

3. $\quad \dfrac{d^2p}{dt^2} = \dfrac{1}{C}\dfrac{dq_c}{dt} = \dfrac{1}{C}\left[\dot{q}(t) - \dfrac{1}{I}p\right]$

In operator form, $\left(S^2 + \dfrac{1}{IC}\right)p(t) = \dfrac{S}{C}q$

4. $\quad q_1 = q - q_c = q - CSp$

5. $\quad q_1 = q - CS\dfrac{S_C}{S^2 + 1/IC}q = \left(1 - \dfrac{S^2}{S^2 + 1/IC}\right)q = \left(\dfrac{1/IC}{S^2 + 1/IC}\right)q$

$$|G(j\omega)| = \left[\frac{1}{1 - (\omega/\omega_n)^2}\right]q; \qquad \omega_n = \frac{1}{\sqrt{IC}}$$

6.

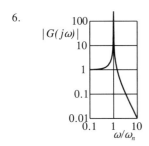

7. (i) For the amplitude ratio to be within ten percent of 1, the frequency square of the ratio, $(\omega/\omega_n)^2$, should be less than about 0.1. (The small range of alternative possible values at slightly above 1.0 is likely of no practical use.) Therefore, $k > 10m\omega^2$.

(ii) For the amplitude ratio to be less than 0.1, $(\omega/\omega_n)^2$ must be greater than about 3.3. Therefore, $k < 0.3m\omega^2$.

Guided Problem 7.5

1.

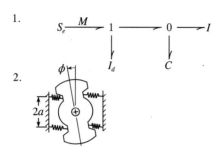

2.

Displacement of each spring: $a\phi$.
Change of force, each spring: $ka\phi$.
Total torque: $4ka^2\phi$.
Therefore, the angular compliance is $C = 1/4ka^2$
$= 1/(4 \times 500(.25)^2 = 0.0080$ rad/ft-lb.

3.

$$\frac{d\phi_C}{dt} = \frac{1}{I_d}p_d - \frac{1}{I}p$$

$$\frac{dp}{dt} = \frac{1}{C}\phi_C$$

Combining, $S^2\phi_C = \dfrac{S}{I_d}p_d - \dfrac{1}{IC}\phi_C$, so that $\phi_C = \dfrac{S/I_d}{S^2 + 1/IC}p_d$.

$$\frac{dp_d}{dt} = M - \frac{1}{C}\phi_C = M - \frac{S/CI_d}{S^2 + 1/IC}p_d$$

Therefore, $S(S^2 + \dfrac{1}{IC})p_d = \left(S^2 + \dfrac{1}{IC}\right)M - \dfrac{S}{I_dC}p_d$.

Collecting terms, $\left[S^3 + \dfrac{1}{IC}\left(\dfrac{1}{I_d} + \dfrac{1}{I}\right)S\right]p_d = \left(S^2 + \dfrac{1}{IC}\right)M,$

giving finally $\dot\phi = \dfrac{p_d}{I} = \dfrac{1}{I}\dfrac{S^2 + 1/IC}{S[S^2 + (1/C)(1/I_d + 1/I)]}M.$

4. Let $S = j\omega$, $M = M_0\cos\omega t$, $\dot\phi = \dot\phi_0\sin\omega t$.

$$\dot\phi_0 = \frac{-\omega^2 + 1/IC}{I\omega[-\omega^2 + (1/IC)(1/I_d + 1/I)]}M_0 = \frac{-\omega^2 + 1/0.008}{\omega[-\omega^2 + (1/0.008)(1 + 1)]}M_0 = \frac{(125 - \omega^2)M_0}{\omega(250 - \omega^2)}$$

As $\omega \to 0$, $\dfrac{\dot\phi_0}{M_0} \to \dfrac{125}{250\omega} = \dfrac{0.5}{\omega}$

As $\omega \to \infty$, $\dfrac{\dot\phi_0}{M_0} \to \dfrac{-1}{\omega}$

For $\omega = \sqrt{125} = 11.18$ rad/s, $\dfrac{\dot\phi_0}{M_0} = 0$

For $\omega = \sqrt{250} = 15.81$ rad/s, $\dfrac{\dot\phi}{M_0} \to \infty$

The asymptotes are shown as dashed lines. The system acts as an ideal tuned vibration absorber for $\omega = 11.18$ rad/s.

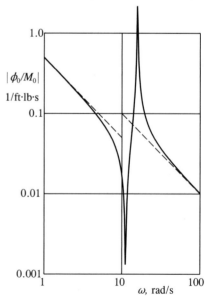

7.3   Matrix Representation of Dynamic Behavior*

The previous chapters and sections have made a restricted use of matrix concepts, despite frequent use of matrix notation and MATLAB commands. In this section, matrix exponential solutions to linear models and the modal behavior of these models will be developed. The result is an enhanced understanding of the behavior of linear models as well as the development of certain practical computational methods.

The linear state differential equations are, as before,

$$\boxed{\begin{aligned} \frac{d\mathbf{x}}{dt} &= \mathbf{A}\mathbf{x}(t) + \mathbf{B}\mathbf{u}(t), \\ \mathbf{y}(t) &= \mathbf{C}\mathbf{x}(t) + \mathbf{D}\mathbf{u}(t). \end{aligned}} \tag{7.54}$$

The special case of the stationary model, for which $\mathbf{A}$, $\mathbf{B}$, $\mathbf{C}$ and $\mathbf{D}$ are constants, is assumed. The *autonomous* case, for which $\mathbf{u}(t) = 0$, is treated first.

7.3.1   The Matrix Exponential

The general solution of equation (7.54) for the homogeneous case $\mathbf{u} = \mathbf{0}$ can be written as

$$\boxed{\mathbf{x}(t) = e^{\mathbf{A}t}\mathbf{x}(0),} \tag{7.55}$$

in which $e^{\mathbf{A}t}$ is known as a **matrix exponential**. It has the same series expansion as a scalar exponential:

$$e^{\mathbf{A}t} = \mathbf{I} + \mathbf{A}t + \frac{1}{2!}\mathbf{A}^2 t^2 + \frac{1}{3!}\mathbf{A}^3 t^3 + \ldots . \tag{7.56}$$

This equation can be used to evaluate the matrix exponential, but there are much better approaches that do not require evaluating many terms and do not suffer a truncation error. Mathematics texts give various ways for evaluating functions of matrices in general and the matrix exponential in particular. A particular method for finding the matrix exponential is developed below that also elucidates the behavior of the system.

First, however, it is important to recognize two uses of the matrix exponential. The obvious use is to give analytic functions of time for the various states in $\mathbf{x}(t)$. The second use employs a fixed time interval, Δt, which can be chosen arbitrarily:

$$\mathbf{x}(t + \Delta t) = e^{\mathbf{A}\Delta t}\mathbf{x}(t). \tag{7.57}$$

In this application, the matrix exponential $e^{\mathbf{A}\Delta t}$ is evaluated *numerically*. Then, $\mathbf{x}(\Delta t)$ is found from equation (7.57). After this, $\mathbf{x}(2\Delta t)$ is found the same way, that is by premultiplying $\mathbf{x}(\Delta t)$ by the same numerical matrix exponential. Again, $\mathbf{x}(3\Delta t)$ equals $\mathbf{x}(2\Delta t)$ premultiplied by the matrix exponential, which can be called a **state transition matrix**. This process may be repeated indefinitely, to evaluate $\mathbf{x}(n\Delta t)$ for $n = 1, 2, 3, \ldots$ It is important to note that regardless of the magnitude of Δt, this process entails no inherent error (beyond round-off error), assuming the matrix exponential is evaluated properly. By contrast, general simulation routines such as the Runge-Kutta method presented in Section 3.6 suffer an error that grows rapidly as the time increment is increased. The general simulation schemes, on the other hand, have the advantages of handling arbitrary forcing functions as well as nonlinearities.

7.3.2   Response to a Linearly Varying Excitation

Equation (7.54) with a linearly varying excitation

$$\mathbf{u} = \mathbf{u}_0 + \mathbf{u}_1 t, \tag{7.58}$$

frequently is employed to represent linear interpotation of an excitation between known data points $\mathbf{u}(t_k)$. This is the basis of the MATLAB function lsim, the use of which is described in Section 4.1.7 (pp. 162-164). Note that the assumption of a zero-order hold corresponds to $\mathbf{u_1} = \mathbf{0}$.

The Laplace transform of equation (7.58) is

$$\mathbf{U}(s) = \frac{\mathbf{u_0}}{s} + \frac{\mathbf{u_1}}{s^2}. \tag{7.59}$$

The Laplace transform of the equation (7.54a) (same as equation (4.74, p. 202) is given by equation (4.75), and its solution for $\mathbf{X}(s)$ is given by equation (4.76). Substituting equation (7.59) into this result and finding the inverse transform gives

$$\mathbf{x}(t) = \int_0^t e^{\mathbf{A}t'}\mathbf{B}\mathbf{u_0}\,dt' + \int_0^t \int_0^{t'} e^{\mathbf{A}t''}\mathbf{B}\mathbf{u_1}\,dt'' + e^{\mathbf{A}t}\mathbf{x}(0), \tag{7.60}$$

in which t' and t'' are dummy variables of integration. The desired result follows from carrying out the integrations:

$$\mathbf{x}(t) = \mathbf{A}^{-1}(e^{\mathbf{A}t} - \mathbf{I})\mathbf{B}\mathbf{u_0} + \mathbf{A}^{-2}(e^{\mathbf{A}t} - \mathbf{I})\mathbf{B}\mathbf{u_1} - \mathbf{A}^{-1}\mathbf{B}\mathbf{u_1}t + e^{\mathbf{A}t}\mathbf{x}(0). \tag{7.61}$$

The accuracy of lsim can be seen to depend on the fidelity of the linear interpolation of the excitation between the beginning and end of each time step. When the excitation is constant or varies precisely linearly, there is virtually no error. The command lsim also has a feature that is not well documented: it attempts to identify inputs that involve zero-order holds (step-wise excitations), and to represent them accordingly. This can be useful in treating systems that have sample-and-hold control signals, for example, but also is a potential source of error. For more information, see the Help instructions that accompany MATLAB.

7.3.3 Eigenvalues, Eigenvectors and Modes

Rather than giving the solution to the autonomous or homogeneous problem in the form of equation (7.55), the inverse transform of equation (4.76) (p. 202) gives in general

$$\mathbf{x}(t) = \sum_i \mathbf{x}_i(t) = \sum_i e^{s_i t}\mathbf{x}_i(0), \tag{7.62}$$

assuming that the roots s_i of the characteristic polynomial are distinct. Each term in this summation, called a **mode of motion**, satisfies the equation

$$(s_i\mathbf{I} - \mathbf{A})\mathbf{x}_i = \mathbf{0}, \tag{7.63}$$

in which s_i is an **eigenvalue** that satisifies the characteristic equation

$$p(s) \equiv \det(s\mathbf{I} - \mathbf{A}) = 0, \tag{7.64}$$

and has its corresponding **eigenvector**, $\mathbf{x}_i$. Equation (7.62) reveals that the individual modes of motion can be superimposed in any proportion, depending on the initial conditions, without interaction with one another. It can be very instructive to grasp the significance of the individual eigenvalues and eigenvectors. The eigenvalues customarily are assembled in a square diagonal **eigenvalue matrix, S**:

$$\mathbf{S} = \begin{bmatrix} s_1 & 0 & \cdots & 0 \\ 0 & s_2 & \cdots & 0 \\ \vdots & \vdots & \ddots & \vdots \\ 0 & 0 & \cdots & s_n \end{bmatrix}. \tag{7.65}$$

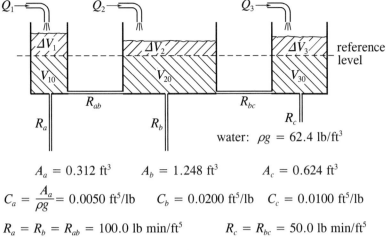

(a) system and parameters

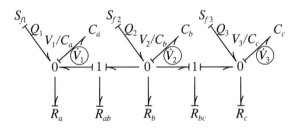

(b) bond graph with causal strokes

Figure 7.22: Three-tank system

If one or more of the roots of the polynomial are repeated, equation (7.62) produces fewer than the requisite number of terms. For a double root s_i the associated terms are

$$e^{st}[\mathbf{x}_i(0) + \mathbf{v}_i t],\qquad(7.66)$$

where $\mathbf{v}_i(0)$ is a vector of constants. For a root repeated m times, this generalizes to

$$e^{st}\left[\mathbf{x}_i(0) + \mathbf{v}_{i1}t + \mathbf{v}_{i2}t^2 + \ldots + \mathbf{v}_{im}t^m\right].\qquad(7.67)$$

The other terms in equation (7.62) are unchanged.

7.3.4   Case Study: Three Fluid Tanks

As a case study, consider the system shown in Fig. 7.22, which has three independent energy storage elements and thus is of third order.

Applying integral causality in the usual manner, as shown, gives the state equations

$$\frac{d}{dt}\begin{bmatrix}V_1\\V_2\\V_3\end{bmatrix} = \mathbf{A}\begin{bmatrix}V_1\\V_2\\V_3\end{bmatrix} + \begin{bmatrix}Q_1\\Q_2\\Q_3\end{bmatrix},\qquad(7.68a)$$

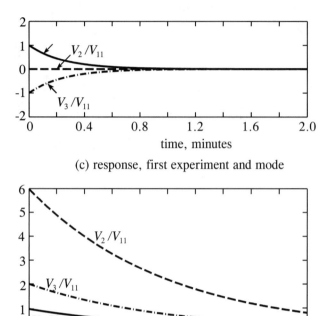

(c) response, first experiment and mode

(d) response, second experiment and mode

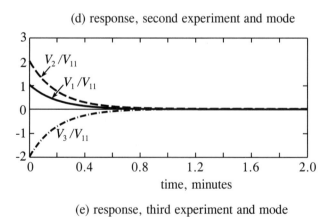

(e) response, third experiment and mode

Fig. 7.22 continued

$$A_{11} = -\frac{1}{C_a}\left(\frac{1}{R_{ab}} + \frac{1}{R_a}\right); \quad A_{12} = \frac{1}{R_{ab}C_b}; \quad A_{13} = 0,$$

$$A_{21} = \frac{1}{R_{ab}C_a} \quad A_{22} = -\frac{1}{C_c}\left(\frac{1}{R_b} + \frac{1}{R_{ab}} + \frac{1}{R_{bc}}\right); \quad A_{23} = \frac{1}{R_{bc}C_c},$$

$$A_{31} = 0; \quad A_{32} = \frac{1}{R_{bc}C_b}; \quad A_{33} = -\frac{1}{C_c}\left(\frac{1}{R_{bc}} + \frac{1}{R_c}\right). \tag{7.68b}$$

To make equation (7.68) correspond to equation (7.54) (p. 471), the vector $\mathbf{Q} = [Q_1, Q_2, Q_3]^T$ somehow must be removed. This is accomplished by defining *constant* values $\mathbf{Q}_0 = [Q_{10}, Q_{20}, Q_{30}]^T$ and $\mathbf{V}_0 = [V_{10}, V_{20}, V_{30}]^T$ such that

$$\mathbf{Q} = \mathbf{Q}_0 + \Delta\mathbf{Q}(t); \qquad \mathbf{V} = \mathbf{V}_0 + \Delta\mathbf{V}(t). \tag{7.69}$$

Thus all changes in time reside in the Δ terms. The volumes V_{10}, V_{20}, V_{30} can be chosen such that they represent equal levels of water in the three tanks, as indicated in part (a) of the figure. The corresponding depth of the water is $V_{10}/A_a = V_{20}/A_b = V_{30}/A_c$, which is proportional to $V_{10}/C_a = V_{20}/C_b = V_{30}/C_c$. Thus,

$$Q_{10}R_a = Q_{20}R_b = Q_{30}R_c, \tag{7.70}$$

or

$$Q_{10} = Q_{20} = \frac{1}{2}Q_{30}. \tag{7.71}$$

The equilibrium equation

$$0 = \mathbf{A}\mathbf{V}_0 + \mathbf{Q}_0 \tag{7.72}$$

now can be subtracted from equation (7.68) to give the desired result[4]

$$\frac{d\,\Delta\mathbf{V}}{dt} = \mathbf{A}\,\Delta\mathbf{V} + \Delta\mathbf{Q}. \tag{7.73}$$

The immediate interest is the special case of $\Delta\mathbf{Q} = \mathbf{0}$. In particular, how do the three levels approach their equilibrium with its equal depths if they start with different depths?

The numerical parameters in the figure give the matrix $\mathbf{A}$ and characteristic polynomial already used in Example 4.18 (p. 203). The roots of the characteristic polynomial therefore are

$$s_1 = -4; \qquad s_2 = -1; \qquad s_3 = -5 \quad \text{min}^{-1}. \tag{7.74}$$

The order of these roots is arbitrary.

The solution can be found from equation (7.62):

$$\begin{bmatrix} V_1(t) \\ V_2(t) \\ V_3(t) \end{bmatrix} = \begin{bmatrix} V_{11} \\ V_{21} \\ V_{31} \end{bmatrix} e^{-4t} + \begin{bmatrix} V_{12} \\ V_{22} \\ V_{32} \end{bmatrix} e^{-t} + \begin{bmatrix} V_{13} \\ V_{23} \\ V_{33} \end{bmatrix} e^{-5t}. \tag{7.75}$$

The essential character of the dynamics can be seen to be represented partly by three different exponential decay rates. Decay rates of linear models, as noted before, traditionally are expressed by $e^{-t/\tau}$, where τ has the dimensions of time and is called a time constant. In the present system the three time constants are 0.25, 1.0 and 0.20 minutes, respectively.

The solution to equation (7.75) also is characterized by three eigenvectors given in the square brackets on the right side. Each eigenvector can be factored into a product of a scalar

---

[4]The approach is consistent with linearization, as discussed in Section 4.7. Equation (7.68) is "linear biased," however, and as a result the model of equation (7.73) introduces no additional error.

and a normalized eigenvector. For simplicity it is assumed here that the first element in the normalized eigenvector is unity. (Later it will be seen that MATLAB makes a different choice to get around the possibility that the first element is zero.) The normalized second eigenvector of the three-tank problem, for example, can be found by substituting the second eigenvalue, $s_2 = -1$, into equation (7.63) (p. 472), as follows:

$$\begin{bmatrix} 3 & -1/2 & 0 \\ -2 & 1 & -2 \\ 0 & -1 & 3 \end{bmatrix} \begin{bmatrix} 1 \\ V_{22}/V_{12} \\ V_{32}/V_{12} \end{bmatrix} = 0. \tag{7.76}$$

The first of the three equations imbedded here requires $V_{22}(0)/V_{12}(0) = 6$. The third then gives $V_{32}(0)/V_{12}(0) = 2$. The second is redundant, but affords an opportunity to check the results. Repeating the same procedure for the two other normalized eigenvectors and assembling the results into a **modal matrix, P**, gives

$$\mathbf{P} = \begin{bmatrix} 1 & 1 & 1 \\ V_{21}/V_{11} & V_{22}/V_{12} & V_{23}/V_{13} \\ V_{31}/V_{11} & V_{32}/V_{12} & V_{33}/V_{13} \end{bmatrix} = \begin{bmatrix} 1 & 1 & 1 \\ 0 & 6 & -2 \\ -1 & 2 & 2 \end{bmatrix}. \tag{7.77}$$

With this result equation (7.75) can be written

$$\begin{bmatrix} V_1(t) \\ V_2(t) \\ V_3(t) \end{bmatrix} = \begin{bmatrix} 1 \\ 0 \\ -1 \end{bmatrix} V_{11}e^{-4t} + \begin{bmatrix} 1 \\ 6 \\ 2 \end{bmatrix} V_{12}e^{-t} + \begin{bmatrix} 1 \\ -2 \\ 2 \end{bmatrix} V_{13}e^{-5t}. \tag{7.78}$$

Imagine the following experiment. Initially, the three tanks are at equilibrium, all at the same level with the rate of flow out of each equal to the rate of flow in. Suddenly, you remove a bucket of water of volume V_{11} from the third tank and dump it into the first tank. This establishes the initial condition

$$\begin{bmatrix} V_1(0) \\ V_2(0) \\ V_3(0) \end{bmatrix} = \begin{bmatrix} 1 \\ 0 \\ -1 \end{bmatrix} V_{11}, \tag{7.79}$$

which, when compared with equation (7.78), shows that V_{12} and V_{13} equal zero. The subsequent behavior must be

$$\begin{bmatrix} V_1(t) \\ V_2(t) \\ V_3(t) \end{bmatrix} = \begin{bmatrix} 1 \\ 0 \\ -1 \end{bmatrix} V_{11}e^{-4t}, \tag{7.80}$$

which is plotted in Fig. 7.22 part (c). Note that the entire response involves only the first time constant, and the ratios of the three excess volumes always retain the proportions $[1, 0, -1]$, respectively. This is the behavior of the first **mode**; the *vector* $[V_1(t), V_2(t), V_3(t)]^T$ remains *parallel* to its initial direction in **state space**, and decays at the rate e^{-4t} dictated by the first eigenvalue.

This same initial condition was given in Example 4.19 (pp. 203-204). The cancellation of the numerator and denominator factors of $\mathbf{X}(s)$ is now explained, leading to the simple result for the $\mathbf{x}(t)$ of that problem.

If as a second experiment you suddenly dump one unit of water of arbitrary volume V_{12} into tank 1, six units into tank 2, and two units into tank 3, the resulting behavior would be

$$\begin{bmatrix} V_1(t) \\ V_2(t) \\ V_3(t) \end{bmatrix} = \begin{bmatrix} 1 \\ 6 \\ 2 \end{bmatrix} V_{12}e^{-t}, \tag{7.81}$$

which is plotted in part (d) of the figure. This is behavior of the *second mode*; the vector $[V_1(t), V_2(t), V_3(t)]^T$ remains parallel to its initial direction in state space, and behaves as directed by the second eigenvalue, s_2, that is the decay rate e^{-t}.

The fastest decay would result from a third experiment that established conditions for the third mode of behavior, which is plotted in part (e) of the figure.

Equation (7.78) can be generalized to

$$\boxed{\mathbf{x}(t) = \mathbf{Pz}(t);} \tag{7.82}$$

$$\boxed{\mathbf{z}(t) = \mathbf{E}_{St}\mathbf{z}(0);} \tag{7.83}$$

$$\boxed{\mathbf{E}_{St} = \begin{bmatrix} e^{s_1 t} & 0 & \cdots & 0 \\ 0 & e^{s_2 t} & \cdots & 0 \\ \vdots & \vdots & \ddots & \vdots \\ 0 & 0 & \cdots & e^{s_n t} \end{bmatrix}.} \tag{7.84}$$

Equation (7.82) represents a *transformation of variables* between $\mathbf{z}(t)$ and $\mathbf{x}(t)$. The variables $\mathbf{z}(t)$ are very special, as revealed by equations (7.83) and (7.84): each component $z_i(t)$, $i = 1, \cdots, n$, acts *independently* of the other components, employing only its unique time constant, $\tau_i = -1/s_i$. Thus $\mathbf{z}(t)$ is the set of **modal variables**. Inverting equation (7.82) and setting $t = 0$,

$$\mathbf{z}(0) = \mathbf{P}^{-1}\mathbf{x}(0). \tag{7.85}$$

Substituting this result into equation (7.80) and the resulting $\mathbf{z}(t)$ into equation (7.82), gives the final general result

$$\boxed{\mathbf{x}(t) = \mathbf{P}\mathbf{E}_{St}\mathbf{P}^{-1}\mathbf{x}(0).} \tag{7.86}$$

A comparison of this result to equation (7.55) (p. 471) reveals that the matrix exponential can be computed by evaluating

$$\boxed{e^{\mathbf{A}t} = \mathbf{P}\mathbf{E}_{St}\mathbf{P}^{-1}.} \tag{7.87}$$

For the three-tank problem, equation (7.85) gives

$$\begin{bmatrix} V_{11} \\ V_{12} \\ V_{13} \end{bmatrix} = \mathbf{y}(0) = \frac{1}{24}\begin{bmatrix} 16 & 0 & -8 \\ 2 & 3 & 2 \\ 6 & -3 & 6 \end{bmatrix}\begin{bmatrix} V_1(0) \\ V_2(0) \\ V_3(0) \end{bmatrix}$$

$$= \frac{1}{24}\begin{bmatrix} 16V_1(0) + 0 - 8V_3(0) \\ 2V_1(0) + 3V_2(0) + 2V_3(0) \\ 6V_1(0) - 3V_2(0) + 6V_3(0) \end{bmatrix}, \tag{7.88}$$

which can be substituted into equation (7.78) to give the general result for that problem.

7.3.5   Case Study With Complex Roots

The three-tank problem is rather special in that its roots are all real. The method above is applied next to a problem with complex roots.

Consider the system shown in Fig. 7.23, which has three independent energy storage elements and is thus of third order. Applying integral causality in the usual manner, as

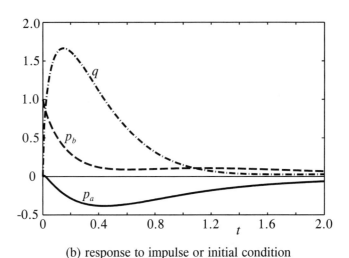

(a) bond graph

(b) response to impulse or initial condition

Figure 7.23: Third-order system example

shown, gives the state equations

$$\frac{d}{dt}\begin{bmatrix} p_a \\ p_b \\ q \end{bmatrix} = \begin{bmatrix} 0 & 0 & 1/C \\ 0 & -R_b/I_b & 1/C \\ -1/I_a & -1/I_b & -1/R_aC \end{bmatrix}\begin{bmatrix} p_a \\ p_b \\ q \end{bmatrix} + \begin{bmatrix} 0 \\ F(t) \\ 0 \end{bmatrix}. \tag{7.89}$$

The numerical parameters in the figure give

$$\mathbf{A} = \begin{bmatrix} 0 & 0 & 1 \\ 0 & -2 & 1 \\ -8 & -6 & -3 \end{bmatrix}, \tag{7.90}$$

so the characteristic equation (7.64) becomes

$$p(s) \equiv \begin{vmatrix} s & 0 & -1 \\ 0 & s+2 & -1 \\ 8 & 6 & s+3 \end{vmatrix} = 0, \tag{7.91a}$$

or

$$p(s) \equiv s^3 + 5s^2 + 20s + 16 = (s+1)(s^2 + 4s + 16) = 0. \tag{7.91b}$$

The roots of the characteristic polynomial are

$$s_1 = -1; \qquad\qquad s_{2,3} = -2 \pm 2j\sqrt{3}, \tag{7.92}$$

where j is the unit imaginary number, $\sqrt{-1}$.

Even if you don't continue the analysis further, these eigenvalues tell you the essential character of the dynamics of the system, as represented by the decay rates e^{-t} and e^{-2t} with their time constants of 1.0 and 0.5 seconds, respectively, and the damped natural frequency, $\omega_d = 2\sqrt{3}$ rad/sec.

You find the first normalized eigenvector by substituting the first eigenvalue, -1, into equation (7.64), and setting x_{11} equal to unity:

$$\begin{bmatrix} -1 & 0 & -1 \\ 0 & 1 & -1 \\ 8 & 6 & 2 \end{bmatrix} \begin{bmatrix} 1 \\ x_{12} \\ x_{13} \end{bmatrix} = 0. \tag{7.93}$$

The first of the three equations imbedded here requires $x_{13} = -1$. The second then gives $x_{12} = -1$. The third equation is redundant, but affords an opportunity to check the results. Repeating the same procedure for the two other normalized eigenvectors and assembling the results into the modal matrix gives

$$\mathbf{P} = \begin{bmatrix} 1 & 1 & 1 \\ -1 & 1+j/\sqrt{3} & 1-j/\sqrt{3} \\ -1 & -2+2j\sqrt{3} & -2-2j\sqrt{3} \end{bmatrix}. \tag{7.94}$$

The determinant $|\mathbf{P}|$ is $-2j\sqrt{3}$, and the inverse of $\mathbf{P}$ is

$$\mathbf{P}^{-1} = \frac{\text{Adj}(\mathbf{P})}{\det(\mathbf{P})} = \frac{1}{26} \begin{bmatrix} 16 & -12 & 2 \\ 5-j\sqrt{3} & 6-j\sqrt{3} & -1-2j\sqrt{3} \\ 5+j\sqrt{3} & 6+j\sqrt{3} & -1+2j\sqrt{3} \end{bmatrix}. \tag{7.95}$$

As an example, consider as the initial condition the result of an *impulse* of the force $F(t)$ which occurs at the time $t = 0$. The force is designated as $\mathcal{I}\delta(t)$, where $\mathcal{I}$ is the magnitude of the impulse or time integral $\mathcal{I} = \int F\, dt$. This time integral is a step of amplitude $\mathcal{I}$ and can be designated as $\mathcal{I}u_s(t)$.

The immediate effect of the impulse excitation appears only on the state variable p_2, as can be seen by integrating the differential equation (7.73) over the infinitesimal time interval from $t = 0$ to $t = 0^+$, which by definition includes the entire impulse. The result is

$$p_a(0^+) = 0; \qquad p_b(0^+) = \mathcal{I}; \qquad q(0^+) = 0. \tag{7.96}$$

Equation (7.85) now gives

$$\mathbf{z}(0) = \mathbf{P}^{-1} \begin{bmatrix} 0 \\ \mathcal{I} \\ 0 \end{bmatrix} = \frac{\mathcal{I}}{26} \begin{bmatrix} -12 \\ 6-j\sqrt{3} \\ 6+j\sqrt{3} \end{bmatrix}. \tag{7.97}$$

(Note that if you knew in advance that only $p_b(0^+)$ would be non-zero, you could forgo computing the first and third columns of $\mathbf{P}^{-1}$, saving significant effort if hand calculation is used. For most practical situations above the order two, however, the calling of the MATLAB inversion command, inv(P), is recommended.)

Next, you find the matrix $\mathbf{PE}_{St}$, the separate columns of which represent the separate modes at time t:

$$\mathbf{PE}_{St} = \begin{bmatrix} e^{s_1 t} & e^{s_2 t} & e^{s_3 t} \\ -e^{s_1 t} & (1+j/\sqrt{3})e^{s_2 t} & (1-j/\sqrt{3})e^{s_3 t} \\ e^{s_1 t} & (-2+2j\sqrt{3})e^{s_2 t} & (-2-2j\sqrt{3})e^{s_3 t} \end{bmatrix}. \tag{7.98}$$

Multiplying the vector of modal amplitudes, $\mathbf{y}(0)$, given by equation (7.97), by this matrix gives

$$\mathbf{x}(t) = \mathbf{P}\mathbf{E}_{St}\mathbf{P}^{-1}\mathbf{x}(0)$$

$$= \frac{\mathcal{I}}{26} \begin{bmatrix} -12e^{-t} + (6 - j\sqrt{3})e^{(-2+2j\sqrt{3})t} + (6 + j\sqrt{3})e^{(-2-2j\sqrt{3})t} \\ 12e^{-t} + (7 + j\sqrt{3})e^{(-2+2j\sqrt{3})t} + (7 - j\sqrt{3})e^{(-2-2j\sqrt{3})t} \\ 12e^{-t} - (6 - 14j\sqrt{3})e^{(-2+2j\sqrt{3})t} - (6 + 14j\sqrt{3})e^{(-2-2j\sqrt{3})t} \end{bmatrix}$$

$$= \frac{\mathcal{I}}{13} \begin{bmatrix} -6e^{-t} + e^{-2t}[6\cos(2\sqrt{3}t) + \sqrt{3}\sin(2\sqrt{3}t)] \\ 6e^{-t} + e^{-2t}[7\cos(2\sqrt{3}t) - \sqrt{3}\sin(2\sqrt{3}t)] \\ 6e^{-t} - e^{-2t}[6\cos(2\sqrt{3}t) - 14\sqrt{3}\sin(2\sqrt{3}t)] \end{bmatrix}. \tag{7.99}$$

These results are plotted in part (b) of Fig. 7.23. As a partial check, you can verify that they satisfy the initial conditions given by equations (7.96).

When numerical as opposed to analytical results are satisfactory, MATLAB can relieve the analyst of most of the drudgery above. When analytical results such as equation (7.99) are desired, it is possible to avoid having to deal with the complication of complex numbers. An optional method is presented next.

7.3.6   Modified Method for Complex Eigenvalues*

The complex conjugate nature of the terms in the modal matrix permits them to be coded with real numbers, saving effort. For example, the **modified modal matrix** for the third-order system above is

$$\mathbf{P}_m = \begin{bmatrix} 1 & 1 & 0 \\ -1 & 1 & 1/\sqrt{3} \\ -1 & -2 & 2\sqrt{3} \end{bmatrix}. \tag{7.100}$$

The first eigenvector or column is real, so it is reproduced without change from equation (7.100). The second and third eigenvalues are complex conjugates of one another; their common real parts are placed in the second column, and the imaginary parts, with the signs of the second eigenvector, are placed in the third column. In general, complex eigenvectors are paired off, the real parts being placed in one column and the imaginary parts being placed in the adjacent column to the right. This modified modal matrix is readily inverted to give

$$\mathbf{P}_m^{-1} = \frac{1}{13} \begin{bmatrix} 8 & -6 & 1 \\ 5 & 6 & -1 \\ 3\sqrt{3} & \sqrt{3} & 2\sqrt{3} \end{bmatrix}. \tag{7.101}$$

The key result of equation (7.86) (p. 477) can be modified to apply to these matrices as follows:

$$\boxed{\mathbf{x}(t) = \mathbf{P}_m\mathbf{E}_{Smt}\mathbf{P}_m^{-1}\mathbf{x}(0),} \tag{7.102}$$

where, for the present problem, the matrix $\mathbf{E}_{Smt}$ is

$$\mathbf{E}_{Smt} = \begin{bmatrix} e^{-t} & 0 & 0 \\ 0 & e^{-2t}\cos(2\sqrt{3}t) & e^{-2t}\sin(2\sqrt{3}t) \\ 0 & -e^{-2t}\sin(2\sqrt{3}t) & e^{-2t}\cos(2\sqrt{3}t) \end{bmatrix}. \tag{7.103}$$

In general, the exponential terms for the real roots appear on the diagonal, and the terms for each complex conjugate pair appear as a 2×2 array, centered on the diagonal, and given in the form

$$e^{-\alpha t}\cos\omega t \quad e^{-\alpha t}\sin\omega t$$

$$-e^{-\alpha t}\sin\omega t \quad e^{-\alpha t}\cos\omega t$$

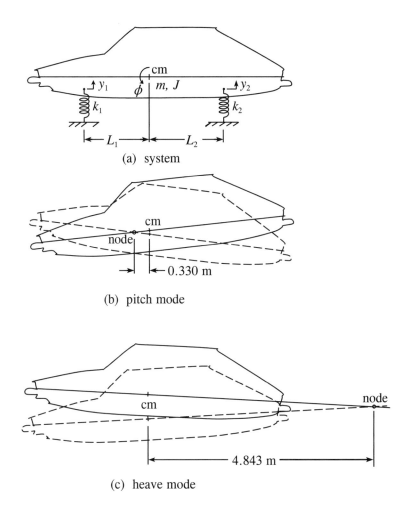

(a) system

(b) pitch mode

(c) heave mode

Figure 7.24: Undamped vehicle modes

where α is the real part and ω is the imaginary part of the associated eigenvalue. Application of equation (7.102) to the third-order system now gives the modal decomposition and the function $\mathbf{x}(t)$ relatively simply.

Some software automatically reports eigenvectors in this modified form, and computes only the inverses of real matrices, further motivating its use. MATLAB, on the other hand, handles complex eigenvectors and modal and other matrices directly. The modified form has the advantage of giving analytic solutions directly in terms of real functions.

7.3.7   Case Study: Vehicle Dynamics

As a final problem regarding autonomous motion, consider the vehicle dynamics of Figs. 6.15 and 6.16 (pp. 381, 383), which has a fourth-order model with two degrees of freedom. The excitation (the road bumps) is removed to make it autonomous, and the damping is removed to clearly reveal the modal behavior. (The damping is reinstated in Section 7.3.11

below.) The reduced model is shown in Fig. 7.24. The equation of motion becomes

$$\frac{d}{dt}\begin{bmatrix} y_1 \\ y_2 \\ p_m \\ p_\psi \end{bmatrix} = \begin{bmatrix} 0 & 0 & 1/m & -L_1/J \\ 0 & 0 & 1/m & L_2/J \\ -k_1 & -k_2 & 0 & 0 \\ L_1 k_1 & -L_2 k_2 & 0 & 0 \end{bmatrix}\begin{bmatrix} y_1 \\ y_2 \\ p_m \\ p_\psi \end{bmatrix}. \tag{7.104}$$

The parameters are given as $m = 1250\,\text{kg}$, $L_1 = 1.4\,\text{m}$, $L_2 = 1.6\,\text{m}$, $J = mr^2$ with $r^2 = 1.6\,\text{m}^2$, $k_1 = 30\,\text{kN/m}$ and $k_2 = 32\,\text{kN/m}$. Therefore,

$$s\mathbf{I} - \mathbf{A} = \begin{bmatrix} s & 0 & -0.0008 & 0.0007 \\ 0 & s & -0.0008 & -0.0008 \\ 30\,000 & 32\,000 & s & 0 \\ -42\,000 & 51\,200 & 0 & s \end{bmatrix}, \tag{7.105}$$

and

$$\begin{aligned} p(s) &= |s\mathbf{I} - \mathbf{A}| \\ &= s^4 + \frac{1}{m}\left(k_1 + k_2 + \frac{L_1^2 k_1 + L_2^2 k_2}{r^2}\right)s^2 + \frac{k_1 k_2}{m^2 r^2 (L_1 + L_2)^2} = 0, \end{aligned} \tag{7.106}$$

or

$$p(s) = s^4 + 119.96 s^2 + 3456 = (s^2 - 71.88)(s^2 - 48.08) = 0, \tag{7.107}$$

$$s_{1,2} = \pm 6.934 j; \qquad s_{3,4} = \pm 8.478 j. \tag{7.108}$$

The ith normalized eigenvector $[1, P_{2i}, P_{3i}, P_{4i}]^T$ satisfies

$$\begin{bmatrix} s_i & 0 & -0.0008 & 0.0007 \\ 0 & s_i & -0.0008 & -0.0008 \\ 30\,000 & 32\,000 & s_i & 0 \\ -42\,000 & 51\,200 & 0 & s_i \end{bmatrix}\begin{bmatrix} 1 \\ P_{2i} \\ P_{3i} \\ P_{4i} \end{bmatrix} = 0. \tag{7.109}$$

The fourth scalar equation embedded above gives

$$P_{2i} = (42\,000 - s_i m_{4i})/51\,200, \tag{7.110}$$

which, when substituted into the third equation, yields

$$P_{4i} = -(1250 s_i^2 + 56\,250)/(0.25 s_i). \tag{7.111}$$

The first equation gives

$$P_{3i} = 1250 s_i + \tfrac{7}{8} m_{4i}. \tag{7.112}$$

Equation (7.111) can be used to find the fourth element in all four eigenvalues, whereupon equation (7.110) gives the second element and (7.109) gives the third. It is simpler to compute only the first and third eigenvectors this way, however, and note that the second and fourth eigenvectors must be the complex conjugates of the first and third, respectively. Were this not true the state variables themselves would become non-real, which is impossible. The resulting modal matrix is

$$\mathbf{P} = \begin{bmatrix} 1 & 1 & 1 & 1 \\ 0.5195 & 0.5195 & -1.8047 & -1.8047 \\ 6724j & -6724j & -3274j & 3274j \\ -2221j & 2221j & -15\,853j & 15\,853j \end{bmatrix}, \tag{7.113}$$

and the corresponding modified modal matrix is

$$\mathbf{P}_m = \begin{bmatrix} 1 & 0 & 1 & 0 \\ 0.5195 & 0 & -1.8047 & 0 \\ 0 & 6724 & 0 & -3274 \\ 0 & -2221 & 0 & -15\,853 \end{bmatrix}. \tag{7.114}$$

These matrices reveal that the first complex conjugate pair of roots combine to give a mode of motion in which the amplitude of $y_2(t)$ is 1.8047 times the amplitude of $y_1(t)$, with a $180°$ phase difference. This requires that a **node** (point with zero velocity) be located between the two points, as shown in Fig. 7.24 part (b), suggesting the designation *pitch mode*. Similarly, the second complex conjugate pair of roots combine to give a mode of motion in which the amplitude of $y_2(t)$ is 0.5195 times the amplitude of $y_1(t)$, with a $0°$ phase difference. This requires that a node be located at some distance from the vehicle, as shown in part (c) of the figure, suggesting the designation *heave mode*.

The eigenvalue matrix is

$$\mathbf{S} = \begin{bmatrix} j\omega_1 & 0 & 0 & 0 \\ 0 & -j\omega_1 & 0 & 0 \\ 0 & 0 & j\omega_2 & 0 \\ 0 & 0 & 0 & -j\omega_2 \end{bmatrix}, \tag{7.115a}$$

$$\omega_1 = 6.934 \text{ rad/s}, \qquad \omega_2 = 8.478 \text{ rad/s}, \tag{7.115b}$$

from which

$$\mathbf{E}_{St} = \begin{bmatrix} e^{j\omega_1 t} & 0 & 0 & 0 \\ 0 & e^{-j\omega_1 t} & 0 & 0 \\ 0 & 0 & e^{j\omega_2 t} & 0 \\ 0 & 0 & 0 & e^{-j\omega_2 t} \end{bmatrix}; \tag{7.116a}$$

$$\mathbf{E}_{Smt} = \begin{bmatrix} \cos\omega_1 t & \sin\omega_1 t & 0 & 0 \\ -\sin\omega_1 t & \cos\omega_1 t & 0 & 0 \\ 0 & 0 & \cos\omega_2 t & \sin\omega_2 t \\ 0 & 0 & -\sin\omega_2 t & \cos\omega_2 t \end{bmatrix}. \tag{7.116b}$$

The inverses of the modal matrices are

$$\mathbf{P}^{-1} = \begin{bmatrix} 0.11176 & -0.21513 & 0.9753j \times 10^{-5} & 2.9526j \times 10^{-5} \\ 0.11176 & -0.21513 & -0.9753j \times 10^{-5} & -2.9526j \times 10^{-5} \\ 0.3882 & 0.21513 & -6.961j \times 10^{-5} & 1.4376j \times 10^{-5} \\ 0.3882 & 0.21513 & 6.961j \times 10^{-5} & -1.4376j \times 10^{-5} \end{bmatrix}; \tag{7.117a}$$

$$\mathbf{P}_m^{-1} = \begin{bmatrix} 0.22352 & -0.43026 & 0 & 0 \\ 0 & 0 & -0.1951 \times 10^{-4} & -0.5905 \times 10^{-4} \\ 0.77648 & 0.43026 & 0 & 0 \\ 0 & 0 & 1.3922 \times 10^{-6} & -0.2875 \times 10^{-4} \end{bmatrix}. \tag{7.117b}$$

The matrix exponential can be found from either equation (7.87) (p. 477) or (7.102) (p. 480):

$$e^{\mathbf{A}t} = \mathbf{P}\mathbf{E}_{St}\mathbf{P}^{-1} = \mathbf{P}_m\mathbf{E}_{Smt}\mathbf{P}_m^{-1}. \tag{7.118}$$

This square matrix can be used either to find analytic functions of time in responses to given analytic excitations, or it can be used as a state transition matrix to compute the responses at times $n\Delta t$ numerically. As an example of the latter, consider $\Delta t = 0.1$ sec. This gives

$$e^{0.1\mathbf{A}} = \begin{bmatrix} 0.74506 & 0.04623 & 0.7436 \times 10^{-4} & -0.6265 \times 10^{-4} \\ 0.04333 & 0.68565 & 0.7263 \times 10^{-4} & -0.7037 \times 10^{-4} \\ -2788.3 & -2905.4 & 0.7622 & -0.0208 \\ 1104.9 & 605.66 & -0.0332 \times 10^{-4} & 0.6685 \end{bmatrix}. \tag{7.119}$$

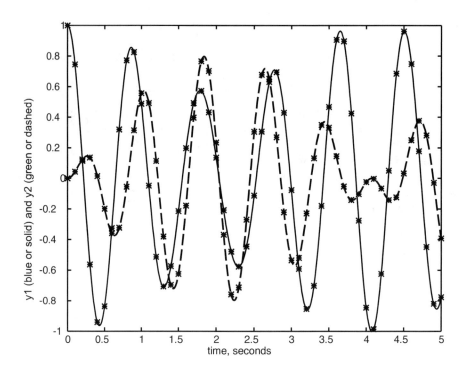

Figure 7.25: Motion of undamped vehicle

Further, consider as an initial condition $y_1(0) = 1$, $y_2(0) = 0$, $p_m(0) = 0$, $p_\psi(0) = 0$. This gives the analytic solution

$$\begin{bmatrix} y_1(t) \\ y_2(t) \\ p_m(t) \\ p_\psi(t) \end{bmatrix} = \begin{bmatrix} 0.22352\cos\omega_1 t + 0.77648\cos\omega_2 t \\ -0.40338\cos\omega_1 t + 0.40338\cos\omega_2 t \\ 731.8\sin\omega_1 t - 5221\sin\omega_2 t \\ 3543.5\sin\omega_1 t + 1724.6\sin\omega_2 t \end{bmatrix}, \tag{7.120}$$

which is plotted in Fig. 7.25. Successive premultiplication of the state vector by the constant matrix exponential above gives the highlighted points. The coarseness of the time interval itself produces no error in the calculation.

7.3.8   Application of MATLAB

Eigenvalue problems are simply treated with MATLAB. The procedures are illustrated here by application to the three problems above.

Once the matrix **A** is entered, the command `eig(A)` then gives the eigenvalues. To get the modal matrix as well as the eigenvalues, you enter `[P,S] = eig(A)`. The response is the modal matrix

```
P =
        0.7071   -0.3333   -0.1562
       -0.0000    0.6667   -0.9370
       -0.7071   -0.6667   -0.3123
```

and the diagonal eigenvalue matrix

```
S =
     -4.0000     0           0
      0        -5.0000       0
      0          0        -1.0000
```

The three eigenvectors are normalized to unity length; that is, the sum of the squares of their elements is 1. This is different from the modal matrix of equation (7.77) (p. 476), in which the elements in the top row were arbitrarily chosen as 1. It accomodates the use of a zero value in this row, among other advantages, but would have entailed extra calculation if done manually.

The diagonal matrix e^{St} can be found for the specified value of $t = 0.1$ by entering ES = expm(.1*S). The corresponding matrix exponential of equation (7.87) can be found by entering P*ES*inv(P). It is simpler, however, merely to enter expm(.1*A). Note that expm(X) returns the matrix exponential e^X, whereas exp(X) returns with e^x for each of the elements in X.

Equation (7.86) (p. 477) allows the determination of the analytic response as a function of t. Finding this result is expedited by determining the vector $\mathbf{z}(0)$ from equation (7.85) by invoking z0=inv(M)*x0. The initial state vector $\mathbf{x}(0) = x0$ must have been defined first.

MATLAB treats systems with imaginary or complex eigenvalues and eigenvectors the same way; standard notation is used. For example, the modal matrix corresponding to equation (7.94) (p. 479) but normalized to give real unit vectors is given as

```
P =
      0.5774 -0.1666 + 0.1637i    -0.1666 - 0.1637i
     -0.5774 -0.2611 + 0.0675i    -0.2611 - 0.0675i
     -0.5774 -0.2339 - 0.9044i    -0.2339 + 0.9044i
```

The normalized vector $\mathbf{z}(0)$ of equation (7.97) is found by typing z0=inv(P)*[0;1;0]. Construction of the analytic solution of equation (7.99) (p 480) would be directly in the first or complex form of the equation.

The use of the modified modal matrices with real elements is not necessary but might be preferred as a means of finding analytic results directly in the form of real functions. This is an advantage for finding the motion of the vehicle as expressed in equation (7.120), for example.

The highlighted points in the plot of Fig. 7.25 can be computed through the repeated use of the numeric matrix exponential of equation (7.119). The points and plot can be programmed in MATLAB as follows:

```
A = [0 0 .0008 -.0007;0 0 .0008 .0008;-30000 -32000 0 0;
42000 -51200 0 0];

EX = expm(0.1 * A);
x(:,1) = [1;0;0;0];
t(1) = 0;

for k = 2:51
      x(:,k) = EX * x(:,k-1);
      t(k) = 0.1 * k;
end

xlabel ('time, seconds')
ylabel ('y1 (blue) and y2 (green)')
plot (t,x(1:2,:),'*')
```

Note that only the first two of the four elements in x are chosen for plotting. Some student versions of MATLAB restrict individual matrices to 8192 elements. This permits the time interval for the five-second run above to be reduced to 0.005 seconds, more than enough to give virtually continuous curves. In this case, of course, the asterisk should be left out of the plot statement.

7.3.9   Response to Exponential and Frequency Excitations

The response of scalar differential equations to exponential and to frequency excitations was addressed in Section 4.3. For exponential excitations of the form $u_0 e^{st}$, the response, taken from equation (4.37b) (p. 178), is

$$x(t) = G(s)u_0 e^{st}. \tag{7.121}$$

For frequency excitations of the form $u_0 \cos(\omega t + \beta)$, the response, taken from equation (7.8) (p. 420), is

$$x(t) = u_0 |G(j\omega)| \cos(\omega t + \beta + \phi) \tag{7.122a}$$

$$\phi(\omega) = \tan^{-1}\left\{ \frac{\Im[G(j\omega)]}{\Re(G[j\omega]} \right\}. \tag{7.122b}$$

These results can be generalized to the A, B, C, D matrix formulation of equation (4.74) and (4.77) (p. 202), with the only changes being that the scalar transfer fuction or Laplace transform $G(s)$ is replaced by the array of transfer functions $\mathbf{H}(s)$ of equation (4.78) (p. 202), and s becomes either the coefficient of the time exponent or the frequency $j\omega$ of the excitation. MATLAB offers the command [mag,phase,w]=bode(A,B,C,D,i) and related variants, in which the number i designates which of possibly several inputs is to be used.

Example 7.14

The undamped vehicle discussed in Section 7.3.7, with parameters given by equations (7.104) and (7.105) (pp. 482), is excited by the force F_1 on the forward axle and the force F_2 on the rear axle, where

$$F_1 = F_{10} \cos(\omega t), \qquad F_2 = F_{20} \cos(\omega t + \beta).$$

Find the matrix transfer function $\mathbf{H}(s)$ that relates the state variables y_1, y_2, p_m and p_ψ to these forces. Further, plot the magnitude of the transfer functions between these forces and the two displacements.

Solution: The equation of motion is the same as equation (7.104) except for the added forcing terms:

$$\frac{d}{dt}\begin{bmatrix} y_1 \\ y_2 \\ p_m \\ p_\psi \end{bmatrix} = \mathbf{A}\begin{bmatrix} y_1 \\ y_2 \\ p_m \\ p_\psi \end{bmatrix} + \mathbf{B}\begin{bmatrix} F_1 \\ F_2 \end{bmatrix},$$

where $\mathbf{A}$ is given by equation (7.105), and

$$\mathbf{B} = \begin{bmatrix} 0 & 0 \\ 0 & 0 \\ 1 & 1 \\ -L_1 & L_2 \end{bmatrix} = \begin{bmatrix} 0 & 0 \\ 0 & 0 \\ 1 & 1 \\ -1.4 & 1.6 \end{bmatrix}.$$

(You might wish to verify the added terms by placing bonds

$$S_e \xrightarrow{F_1} \qquad \text{and} \qquad S_e \xrightarrow{F_2}$$

on the respective 1-junctions for $\dot{y}_1$ and $\dot{y}_2$, and carrying out the standard procedure.) Since all four state variables are considered to be the output variables of interest, $\mathbf{C} = \mathbf{I}$, $\mathbf{D} = \mathbf{0}$ and $\mathbf{H}(s) = \mathbf{G}(s)$.

The matrix transfer function $\mathbf{H}(s)$ becomes, using equation (4.78) (p. 202),

$$\mathbf{H}(s) = \mathbf{G}(s) = (s\mathbf{I} - \mathbf{A})^{-1}\mathbf{B} = \frac{1}{p(s)}[\text{Adj}(s\mathbf{I} - \mathbf{A})]\mathbf{B}$$

$$= \frac{1}{p(s)}\left\{\text{Adj}\begin{bmatrix} s & 0 & -0.0008 & 0.0007 \\ 0 & s & -0.0008 & -0.0008 \\ 30,000 & 32,000 & s & 0 \\ -42,000 & 51,200 & 0 & s \end{bmatrix}\right\}\begin{bmatrix} 0 & 0 \\ 0 & 0 \\ 1 & 1 \\ -1.4 & 1.6 \end{bmatrix}.$$

The first two columns of $(s\mathbf{I} - \mathbf{A})^{-1}$ or $\text{Adj}(s\mathbf{I} - \mathbf{A})$ do not affect the result, because of the zeros in the matrix $\mathbf{B}$, and need not be found. The result is the 2×4 matrix

$$\mathbf{H}(s) = \frac{1}{p(s)}\begin{bmatrix} 0.00178s^2 + 0.1152 & -0.00032s^2 \\ -0.00032s^2 & 0.00208s^2 + 0.108 \\ s^3 + 76.8s & s^3 + 63s \\ -1.4s^3 - 76.8s & 1.6s^3 + 72s \end{bmatrix},$$

$$p(s) = s^4 + 119.96s^2 + 3456.$$

The resulting displacements are

$$y_1(t) = |H_{11}|F_{10}\cos(\omega t + \theta_{11}) + |H_{12}|F_{20}\cos(\omega t + \beta + \theta_{12}),$$
$$y_2(t) = |H_{21}|F_{10}\cos(\omega t + \theta_{21}) + |H_{22}|F_{20}\cos(\omega t + \beta + \theta_{12}),$$

$$H_{ij} = H_{ij}(j\omega), \qquad \theta_{ij} = \angle H_{ij}(j\omega).$$

Only even powers of s appear in the numerator and the denominator polynomials of each H_{ij}, resulting in a real number. If this number is positive, the phase angle ϕ_{ij} is zero; if negative, ϕ_{ij} is $\pm 180°$. (There is no difference between $+180°$ and $-180°$.) The magnitudes of $|H_{11}|$ and $|H_{21}|$ are plotted in Bode coordinates below:

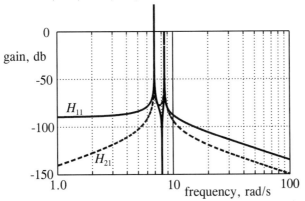

The two natural frequencies, $\omega = 8.478$ and 6.934 rad/s, which were found in equations (7.108) (p. 482) as the imaginary eigenvalues, appear as resonant frequencies with infinite amplitude responses. The eigenvalues or poles are $s = \pm 6.934j$ and $s = \pm 8.478j$. The zeros of G_{11} are at $\omega = 8.045$ rad/s ($s = \pm 8.045j$) and of G_{22} are at $\omega = 7.206$ rad/s ($s = \pm 7.206j$). A phase angle jumps by $180°$ whenever the frequency passes through either a pole or a zero, since there is no damping.

Example 7.15

Introduce the damping resistances R_1 and R_2 into the model of the vehicle above, to make it more respectable. Determine the modified matrix of transfer functions for the special case $R_1 = R_2 = 4000$ N s^2/m, and make Bode plots of the positional responses to the force on the first axle.

Solution: The matrix $\mathbf{A}$ is modified to become

$$
\mathbf{A} = \begin{bmatrix}
0 & 0 & 1/m & -L_1/mr^2 \\
0 & 0 & 1/m & L_2/mr^2 \\
-k_1 & -k_2 & -(R_1+R_2)/m & -(L_2R_2 - L_1R_1)/mr^2 \\
L_1k_1 & -L_2k_2 & -(L_2R_2 - L_1R_1)/m & -(L_2^2R_2 + L_1^2R_1)/mr^2
\end{bmatrix}.
$$

With the given values of the parameters, this becomes

$$
s\mathbf{I} - \mathbf{A} = \begin{bmatrix}
s & 0 & -0.0008 & 0.0007 \\
0 & s & -0.0008 & -0.0008 \\
30\,000 & 32\,000 & s+6.4 & 0.4 \\
-42\,000 & 51\,200 & 0.64 & s+9.04
\end{bmatrix},
$$

from which the values of H_{11}, H_{12}, H_{21} and H_{22} are

$$
\mathbf{H}(s) = \frac{1}{p(s)} \begin{bmatrix}
0.00178s^2 + 0.0144s + 0.1152 & -0.00032s^2 \\
-0.00032s^2 & 0.00208s^2 + 0.0144s + 0.1082
\end{bmatrix},
$$

$$
p(s) = s^4 + 15.44s^3 + 177.56s^2 + 892.8s + 3456.0.
$$

The Bode plots comprising the new magnitude ratios $|\mathbf{H}_{ij}|$ and phase angles ϕ_{ij} for $i = 1,2$ and $j = 1$ are as shown opposite:

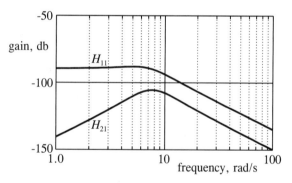

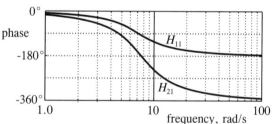

These responses are clearly preferable from the viewpoints of comfort and safety. The resonances are sufficiently damped to be almost unnoticeable. The transfer functions still have poles, which correspond to the eigenvalues

$$
s_{1,2} = -3.161 \pm 6.179j, \qquad s_{3,4} = -4.559 \pm 7.139j.
$$

The various scalar transfer functions still have zeros, also, which are either real or complex. For example, the numerator of H_{11} vanishes for $s = z_{1,2} = -4.045 \pm 6.954j$.

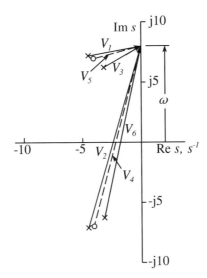

Figure 7.26: Pole zero plot for $H_{11}(s)$ for the damped vehicle

7.3.10 Representation in the s-Plane

It is customary to plot the poles and zeros of a transfer function in the s-plane, which has the real part of s as its abcissas and the imaginary part of s as its ordinate, as first illustrated in Fig. 7.7 (p. 430). Such a plot is given in Fig. 7.26 for the H_{11} of the example above. Poles are indicated by $\times$s and zeros by Os.

An s-plane plot can be used to visualize the response to an exponential disturbance $u = u_0 e^{st}$. This is illustrated in the plot for the special case of frequency reponse; *i.e.,* $s = j\omega$. Solid vectors are drawn from the poles to the point on the imaginary axis that represents the particular sinusoidal disturbance of interest. Dashed vectors are drawn, also, from the zeros to the same point. The polynomials in the numerator and in the denominator of the transfer function can be factored into products of first-order polynomials:

$$H_{11}(s) = \frac{(s-z_1)(s-z_2)}{(s-s_1)(s-s_2)(s-s_3)(s-s_4)}. \tag{7.123}$$

The solid vectors represent the denominator factors, and the dashed vectors represent the numerator factors, both with $s = j\omega$. Labeling the vectors $V_1, V_2, \ldots, V_6$ as shown, the magnitude and phase angle of $G_{11}(j\omega)$ becomes

$$|H_{11}(s)| = \frac{|V_5|\,|V_6|}{|V_1|\,|V_2|\,|V_3|\,|V_4|}, \tag{7.124a}$$

$$\angle H_{11}(j\omega) = \angle V_5 + \angle V_6 - \angle V_1 - \angle V_2 - \angle V_3 - \angle V_4. \tag{7.124b}$$

It is possible, therefore, to determine the magnitude ratio by measuring the lengths of the vectors and multiplying and dividing as indicated, and to determine the phase angle by measuring the angles of the vectors and adding and subtracting as indicated. The frequency ω can be changed and the process repeated.

It is more efficient to compute the transfer function algebraically, and MATLAB does it automatically. It can be very instructive, nevertheless, to be able to look at an s-plane display of poles and zeros and *visualize* the form of the frequency transfer function.

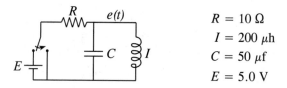

$$R = 10 \ \Omega$$
$$I = 200 \ \mu\text{h}$$
$$C = 50 \ \mu\text{f}$$
$$E = 5.0 \ \text{V}$$

Figure 7.27: Guided Problem 7.6

7.3.11 Summary

An unexcited linear system with nonequilibrium initial conditions exhibits behavior that can be decomposed into a sum of simple modal behaviors. The dynamics of each mode is represented by its eigenvalue. The shape of each mode, that is the proportions of its component variables as represented by its eigenvector, remains fixed in time. Thus the contributions of each mode to the total behavior depends on the initial conditions. Standard matrix methods for treating the details have been presented and illustrated. When complex roots are present, you can choose to use either real or complex matrices. MATLAB accommodates complex matrices, whereas certain other software packages use real matrices only.

Initial value problems can be addressed in a more routine if somewhat less insightful way using Laplace transforms, as presented in the previous section. Cases with repeated roots are best handled this way, so details of their modal behavior have been omitted here.

The simplest approach for finding the autonomous or forced behavior of a linear system employs the numeric matrix exponential. These solutions obscure modal behavior, but particularly in heavily damped systems the modal decompostions may be of minor interest. MATLAB allows the matrix exponential to be requested directly, or indirectly through the use of the command `lsim` which treats excitations that vary linearly in time over each time interval.

Exponential and frequency responses for a model represented by a matrix transfer function are found in essentially the same way as a model represented by a scalar transfer function. The matrix transfer function is treated as an array of scalar transfer functions.

Guided Problem 7.6

The first guided problem deals with a second-order system and can be treated analytically without undue difficulty so as to reinforce the key concepts.

The two-way switch in the circuit in Fig. 7.27 replaces the battery at $t = 0$ with a conductor. Find the subsequent voltage $e(t)$.

Suggested Steps:

1. Represent the after-switch circuit with a bond graph. Apply integral causality and define state variables in the usual way.

2. Write the state differential equation; find the matrix $\mathbf{A}$.

3. Solve $|s\mathbf{I} - \mathbf{A}| = 0$ to get the characteristic polynomial. Solve for its roots.

4. Write the general solution for the charge, $q(t)$, of the capacitor.

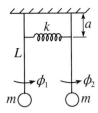

Figure 7.28: Guided Problem 7.7

5. This problem is simple enough that you can avoid having to evaluate the eigenvectors and modal matrix, etc. Instead, specialize the result of step 4 by applying the known boundary conditions $q(0) = q(\infty) = 0$.

6. Find the remaining constant by noting that the initial current through the inductor is E/R and $p(0)/I$, and $p(0)$ is related to $q(t)$.

7. Sketch-plot the resulting $e(t) = q(t)/C$ to see if it makes sense.

Guided Problem 7.7

This problem involves a fourth-order, two-degree-of-freedom vibration model. It is treated analytically, and therefore may be too ambitious for your needs. The beating phenomenon that occurs is highly instructive, however, so at least you ought to examine the given solution. As an alternative, you could apply some reasonable set of parameters, making sure the spring is weak, and find the behavior using MATLAB.

Two otherwise simple oscillators are not infrequently weakly interconnected by a compliance. One example of this is two identical pendulums connected by a weak spring, as shown in Fig. 7.28. (The modeling for this system, plus an added dashpot, is carried out in Example 5.11, p. 286.) Considering small motions only, examine the modal behavior of this system following arbitrary initial conditions.

Suggested Steps:

1. Draw a bond graph, label variables, and find the consequent differential equations. Evaluate any inertances and compliances. If you write a differential equation for the energy storage associated with the spring, note that it is not independent of the other differential equations and does not introduce a fifth state variable.

2. Convert the result of step 1 to give the matrix $\mathbf{A}$.

3. Find the characteristic polynomial, using $|s\mathbf{I} - \mathbf{A}| = 0$.

4. Solve for the four characteristic values, which can be written as $j\omega_1, -j\omega_1, j\omega_2, -j\omega_2$.

5. Find the four eigenvectors, letting $m_{1i} = 1$. Assemble into either the standard complex modal matrix or the modified real modal matrix.

6. Find the eigenvalue matrix $\mathbf{E}_{St}$ or $\mathbf{E}_{Smt}$.

7. Compute the matrix product $\mathbf{P}\mathbf{E}_{St}$ or $\mathbf{P}_m\mathbf{E}_{Smt}$. Each column of this matrix represents the behavior of its respective mode. Therefore, the general solution can be written as the sum of an unspecified coefficient times the content of each column.

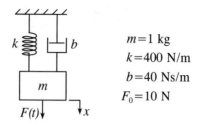

$m=1$ kg
$k=400$ N/m
$b=40$ Ns/m
$F_0=10$ N

Figure 7.29: Guided Problem 7.8

If you used the complex matrices to get the general solution, the sum of the first two columns and the sum of the other two columns must be real; this means that the unspecified coefficients must be complex conjugate pairs. Combine these pairs of columns to get the functions in real terms (sines and cosines). This awkward step is unnecessary if you use the real matrices from the start.

8. Interpret the results. Describe a motion for which only the ω_1 terms exist. Repeat for the ω_2 terms. Now, note that if the spring coupling is weak, ω_1 and ω_2 are almost the same. The sum of two sine or cosine waves of nearly equal frequency can be represented by a sine or cosine wave of frequency equal to the average of the two frequencies modulated in amplitude by an envelope that is a sine or cosine wave with a frequency equal to the difference between the two frequencies. This is called a *beating* phenomenon.

Guided Problem 7.8

This frequency response problem employs a familiar simple model, but asks you to treat it by starting with the matrix format.

The prototypical vibration problem with a single degree of freedom comprises a mass, spring and dashpot excited with a force $F(t) = F_0 \sin \omega t$, as shown in Fig. 7.29. Find the response of this system in the form $x(t) = x_0(\omega) \sin[\omega t + \phi(\omega)]$ and sketch the corresponding Bode plot for the parameters listed in the figure.

Suggested Steps:

1. Draw a bond graph, write the state differential equations and place in standard matrix form. Find the matrices $\mathbf{A}$ and $\mathbf{B}$, using symbols rather than numbers.

2. Extract the scalar transfer function $G(s)$ that relates the force of amplitude F_0 to the displacement of amplitude x_0 from the more general matrix $\mathbf{G}(s)$ given by equation (4.76) (p. 202).

3. Substitute $j\omega$ for s in $G(s)$, and find the magnitude $|G(j\omega)|$ and phase angle $\angle G(j\omega)$ as functions of the frequency ω and the parameters.

4. Substitute the values of the parameters and employ the value of F_0 to find the functions $x_0(\omega)$ and $\phi(j\omega)$.

5. Evaluate these functions for several values of ω in the range of 1.0 to 100 rad/s. Plot in Bode coordinates. You may prefer to use MATLAB.

Guided Problem 7.9

The model needed for this second freqency response problem is of higher order; as a result, the use of the state-space matrix methods is more efficient than scalar methods.

Consider again the two weakly interconnected pendulums of Guided Problem 7.7. A small horizontal excitation force $F(t) = F_0 \sin \omega t$ is applied to the left mass. Find the response of this mass (ϕ_1) and represent it by a Bode plot for the parameters $L = 0.5$ m, $a = 0.15$ m, $m = 1.0$ kg, $k = 8.0$ N/m and $F_0 = 1.0$ N.

Suggested Steps:

1. You already have the matrix $[s\mathbf{I} - \mathbf{A}]$ and its determinant, and $\mathbf{B}$ is a vector with a single non-zero element. Thus only one element in the inverse of the matrix $[s\mathbf{I} - \mathbf{A}]$ needs to be found to determine the transfer function $G(s)$. Determine which element this is.

2. One possibility is to solve this problem using MATLAB. Otherwise, follow the steps below.

3. Find $G(s)$ using the determinant $p(s) = |s\mathbf{I} - \mathbf{A}|$ and the adjoint matrix $\text{Adj}(s\mathbf{I} - \mathbf{A})$.

4. Substitute the values of the parameters and substitute $j\omega$ for s; evaluate the magnitude and the phase angle of $G(j\omega)$.

5. Plot $F_0 G(j\omega)$ in Bode coordinates, including a frequency band starting at least one decade below the natural frequencies and ending at least one decade above.

PROBLEMS

7.30 The mass-spring system shown to the right, known as a **Wilberforce spring**, exhibits both translational and rotational motion. The parameters include the radius of gyration, r_g. The static equations for force and torque include a weak coupling coefficient, $k_{x\phi}$, which is typical of coil springs:

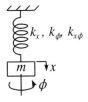

$$M = k_\phi \phi + k_{x\phi} x; \qquad F = k_{x\phi} \phi + k_x x$$

(a) Model the system in standard matrix form.

(b) Find a modal matrix and describe the individual modal motions.

(c) Find analytical expressions for the response given initial deviations from equilibrium of $x = 0.1$ m, $\phi = 0$, $\dot{x} = 0$, $\dot{\phi} = 0$, and parameter values $k_x = 400$ N/m, $k_\phi = 0.400$ N m/rad, $k_{x\phi} = 1.00$ N/rad, $m = 1.00$ kg and $r_g = 0.013$ m.

7.31 Address the problem above using MATLAB with the initial conditions and parameter values given in part (c).

(a) Do parts (a) and (b) numerically.

(b) Find the numerical state transition matrix for a time step of 0.1 second.

(c) Plot $x(t)$ and $\phi(t)$ for several cycles.

7.32 The double pendulum shown at the right can be modeled as two point- masses of weight 1 lb at the ends of two massless wires each of length 10 inches. The angles ϕ_1 and ϕ_2 should be assumed to be small, permitting a linear analysis. Answer the same questions as stated in the preceding problem, substituting ϕ_1 and ϕ_2 for the displacements x and ϕ, respectively. Let $\phi_1(0) = 0.1$ rad, and $\phi_2(0) = p_1(0) = p_2(0) = 0$.

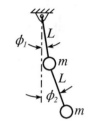

7.33 A five-story building is assumed to rest on a rigid foundation, but each floor can move laterally with a relative deflection per floor proportional to the shear force. Model the system with a bond graph, define state variables, find the matrix $\mathbf{A}$, compute the natural frequencies of vibration, and describe the positional modal shapes. Use of MATLAB is suggested. The stiffness per floor is 1×10^8 N/m for all five floors. The mass is concentrated in the floors, and equals 5×10^5 kg per floor. Damping may be neglected.

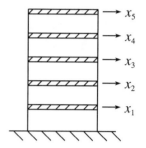

7.34 The building of the above problem is subject to a sinusoidal lateral force of amplitude 2×10^6 N at various frequencies, uniformly distributed over the height. You are to investigate the deflection of the top floor.

(a) Evaluate the matrices $\mathbf{A}, \mathbf{B}, \mathbf{C}$ and $\mathbf{D}$.

(b) Give a magnitude Bode plot for the displacement of the fifth floor.

7.35 Repeat the above problem if, instead of a wind force, an earthquake shakes the ground laterally with a sinusoidal amplitude of 0.2 meters.

7.36 The building of the above problems has a damping coefficient per floor of 1×10^6 N·s·m. Find the decay rates and damped natural frequencies of the individual modes.

7.37 Repeat Problem 7.34 assuming the damping added in Problem 7.36.

7.38 Repeat Problem 7.35 assuming the damping added in Problem 7.36.

7.39 *Project on vehicle heave and pitch, phase 1. Background:* A linear model for the heave and pitch dynamics of a vehicle is developed in Section 6.2.7 (pp. 380-384). This model contains only one rigid body, and one spring and dashpot (or shock absorber) at each of the two axles. The model therefore is of fourth order; a set of state differential equations is given in equation (6.38). Later, in Section 7.3.7 (pp. 481-485), the dashpots are removed and the modes of motion are investigated, using parameters given on p. 482. Finally, in Examples 7.14 and 7.15 (pp. 486-488), the responses of first the undamped and then the damped model to sinusoidal forces on the two axels are investigated, using MATLAB.

Problem: Your task is to introduce the effect of the masses of the axles with the wheels and tires and the effective spring compliance that the tires introduce between the road and the axels. Specifically, you may retain the assumed linearity, and find the responses to a wavy road and to an abrupt small step. You will have the opportunity also to adjust the shock absorber to see if you can improve the performance. The use of MATLAB is a practical necessity.

In this first phase, start by adding the axle/wheel mass and tire compliance to the model, using bond-graph notation. You may retain nearly all of the annotated causal bond graph at the bottom of Fig. 6.16. The ground displacements y_g1 and y_g2 that had been directly under the previously massless axles become the displacements under the two tires, and are considered to be the input disturbances. Revise or expand the state vector as needed, and write the state differential equations. Then, evaluate the matrices $\mathbf{A}$ and $\mathbf{B}$ in terms of symbols and then numbers, using the parameter values on page 482 and the third line of Example 7.15 (p. 488), except increasing the spring constants to $k_1 = 240$ kN/m and $k_2 = 200$ kN/m. The masses of each of the two axels are $m_a = 55$ kg, and the stiffness of the pair of tires relative to the ground is 900 kN/m. The output variables of interest are the vertical velocity of the center of mass of the body, and the angular velocity of the body. It will be necessary also to evaluate matrices $\mathbf{C}$ and $\mathbf{D}$. Print out the four matrices.

Finally, compute the eigenvalues and the modal matrix of the matrix $\mathbf{A}$, using the command [P,S]=eig(A). Do this with the coefficients of the dashpots set equal to zero, and then to their given values. Briefly interpret their meanings.

7.40 *Continuation of the project above.* There are two parts of this final phase of the project, which can be done in either order: the wavy road excitation and the step rise in the pavement, such as often encountered at construction sites.

Consider the wavy road to be sinusoidal with a half-amplitude of 0.025 meters, and wavelengths of (i) 1.5 meters and (ii) 2.0 meters. Convert this to frequency-dependent excitations of the vertical velocities under each wheel. Frequencies of interest range from 1 to 1000 rad/s. These can be reduced to a single excitation by noting that the ground motions under the two wheels have equal frequencies and amplitudes but are in-phase and 180° out-of-phase in the two cases, respectively; find the corresponding simplified matrices $\mathbf{B}$ which become column vectors. Compute and plot Bode plots for the two outputs for each of the two cases, using MATLAB. Interpret the frequency axes with the vehicle velocity, v; you may add a second axis by hand. Do these reponses represent good design?

Some details and suggestions: The C matrix with its two rows for the two outputs of interest, the vertical and angular velocities, can be retained. The corresponding D matrices become column vectors comprising two zeros. The difficulty now is that the input excitation is the vertical velocity under the tires, which is the time derivative of the vertical position. You therefore must multiply the transfer function by $S = j\omega$ to get velocity, but this can't be done using the bode(A, B, C, D) command. You can, however, convert to the num, den format using the MATLAB command ss2tf. The multiplication by S is accomplished by simply adding a zero to the end of the num vector, and using the Bode command bode(num,den) to get the desired plots. Should you wish more control over the frequency band, you may add an argument to the bode command, and define this argument as a vector of frequencies, best using the logspace command as described on p. 423. You will want separate Bode plots for each of the four cases of interest, to allow clear interpretation.

For the problem with the step rise in the pavement, note that the second axle is impacted by a time delay from the first that depends on the velocity; use the values $v = 8$ m/s and $v = 18.75$ m/s, which are particularly interesting, as two cases. Plot the responses of the two state variables, and comment on the virtues of the design.

Some details and suggestions: Find the responses to the two axles separately, and then sum them. This is impractical with the MATLAB functions step and impulse, since they generate unsynchronized data files. Instead, it is suggested that you convert the problem into initial-value form with no subsequent excitation, and multiply the state vector by the state transition matrix `expm(A*dt)` 2000 times, with `dt=0.001` seconds, to generate a two-second run. The initial compression of the tires under an impacted axle is 0.05 meters. Only two vectors of identical length can be summed in a MATLAB plot command, so make sure that the second response starts with the proper number of zeros to represent the time delay, and that both responses end at the same time. The command `zeros(1,k)` will give a string of k zeros, which can be added to the front of a vector "old" by writing `new=[zeros(1,k) old]`. Both the vertical velocity and the angular velocity for a given speed can be placed on the same plot, so only two plots are requested.

Compare the various results for the two types of excitations to see when large velocities occur and when they don't. Each case is chosen to emphasize one of the velocities. Finally, consider whether the damping coefficients should be increased or decreased to give better performance. You may keep the two coefficients equal to one another. State your conclusions.

SOLUTIONS TO GUIDED PROBLEMS

Guided Problem 7.6

1.

$$R \mathrel{\vdash}\!\!\!\frac{q/C}{q/RC}\; 0 \;\frac{\widehat{p}}{p/I}\mathrel{\dashv} I$$
$$q/C\,\widehat{\dot{q}}$$
$$C$$

2. $\begin{cases}\dfrac{dq}{dt} = -\dfrac{1}{RC}q - \dfrac{1}{I}p \\[2mm] \dfrac{dp}{dt} = \dfrac{1}{C}q\end{cases}$ therefore, $\dfrac{d}{dt}\begin{bmatrix} q \\ p \end{bmatrix} = \mathbf{A}\begin{bmatrix} q \\ p \end{bmatrix}$,

$$\mathbf{A} = \begin{bmatrix} -1/RC & -1/I \\ 1/C & 0 \end{bmatrix} = \begin{bmatrix} -2000 & -5000 \\ 20000 & 0 \end{bmatrix}$$

3. $\begin{vmatrix} s+2000 & 5000 \\ -20000 & s \end{vmatrix} = 0$, or $s^2 + 2000\,s + 1\times 10^8 = 0$

    Therefore, $s = -1000 \pm \sqrt{(1000)^2 - 10^8} = -1000 \pm j9950$ s^{-1}

4. $q = q_0 e^{-1000t}\sin(9950t + \alpha)$

5. $q(0) = 0 = q_0 \sin\alpha$; therefore, $\alpha = 0°$

6. $\dfrac{p}{I} = -\dfrac{1}{RC}q - \dfrac{dq}{dt} = [-(2000+1000)\sin 9950t - 9950\cos 9950t]\,q_0 e^{-1000t}$

    $\dfrac{p(0)}{I} = \dfrac{E}{R} = -9950\,q_0$; therefore, $q_0 = -50.25\times 10^{-6}$ amps

7. $e(t) = \dfrac{1}{C}q(t) = -\dfrac{50.25}{50}e^{-1000t}\sin 9950t$

Guided Problem 7.7

1.

spring: $\dfrac{1}{2}k(\text{deflection})^2 = \dfrac{1}{2}k[a(\phi_1 - \phi_2)]^2 = \dfrac{1}{2C_{12}}(\phi_1 - \phi_2)^2;$

therefore, $C_{12} = \dfrac{1}{ka^2}$

gravity: $\mathcal{V} = (1 - \cos\phi)Lmg \simeq \dfrac{1}{2}\phi^2 Lmg = \dfrac{1}{2C}\phi^2;$ therefore, $C_1 = C_2 = \dfrac{1}{mgL}$

inertia: $\dfrac{1}{2}mv^2 = \dfrac{1}{2}m(L\dot\phi)^2 = \dfrac{1}{2}I\dot\phi^2;$ therefore, $I_1 = I_2 = mL^2$

$\dfrac{d\phi_1}{dt} = \dfrac{1}{I_1}p_1$

$\dfrac{d\phi_2}{dt} = \dfrac{1}{I_2}p_2$

$\dfrac{dp_1}{dt} = -\dfrac{1}{C_1}\phi_1 - \dfrac{1}{C_{12}}(\phi_1 - \phi_2)$

$\dfrac{dp_2}{dt} = -\dfrac{1}{C_2}\phi_2 + \dfrac{1}{C_{12}}(\phi_1 - \phi_2)$

2. The state variables p_1 and p_2 are replaced below with $\dot\phi_1 = p_1/I_1$ and $\dot\phi_2 = p_2/I_2$ since this helps clarify the results.

$$\frac{d}{dt}\begin{bmatrix}\phi_1\\\phi_2\\\dot\phi_1\\\dot\phi_2\end{bmatrix} = \begin{bmatrix}0 & 0 & 1 & 0\\0 & 0 & 0 & 1\\-1/I_1C_1 - 1/I_1C_{12} & 1/I_1C_{12} & 0 & 0\\1/I_2C_{12} & -1/I_2C_2 - 1/I_2C_{12} & 0 & 0\end{bmatrix}\begin{bmatrix}\phi_1\\\phi_2\\\dot\phi_1\\\dot\phi_2\end{bmatrix}$$

Therefore, $\mathbf{A} = \begin{bmatrix}0 & 0 & 1 & 0\\0 & 0 & 0 & 1\\-a_1 & a_2 & 0 & 0\\a_2 & -a_1 & 0 & 0\end{bmatrix},$ $\begin{aligned}a_1 &= \dfrac{g}{L} + \dfrac{ka^2}{mL^2}\\a_2 &= \dfrac{ka^2}{mL^2}\end{aligned}$

3. $|s\mathbf{I} - \mathbf{A}| = \begin{vmatrix}s & 0 & -1 & 0\\0 & s & 0 & -1\\a_1 & -a_2 & s & 0\\-a_2 & a_1 & 0 & s\end{vmatrix} = s^4 + 2a_1 s^2 + (a_1^2 - a_2^2) = 0$

4. $s^2 = -\omega^2 = -a_1 \pm \sqrt{a_1^2 - (a_1^2 - a_2^2)} = -a_1 \pm a_2$

therefore, $\omega^2 = \dfrac{g}{L} + \dfrac{2ka^2}{mL^2}, \dfrac{g}{L},$ which gives $\omega_1 = \sqrt{\dfrac{g}{L} + \dfrac{2ka^2}{mL^2}};$ $\omega_2 = \sqrt{\dfrac{g}{L}}$

5. $\begin{bmatrix}j\omega_i & 0 & -1 & 0\\0 & j\omega_i & 0 & -1\\a_1 & -a_2 & j\omega_i & 0\\-a_2 & a_1 & 0 & j\omega_i\end{bmatrix}\begin{bmatrix}1\\m_{2i}\\m_{3i}\\m_{4i}\end{bmatrix} = 0$

From the first equation represented in this matrix relation, $P_{3i} = j\omega_i$. From the third equation, $P_{2i} = (a_1/a_2) - (\omega_i^2/a_2)$. The second equation then gives $P_{4i} = (j\omega_i/a_2)(a_1 + \omega_i^2)$. The fourth equation serves as a check. Substitution of $\pm\omega_1$ and $\pm\omega_2$ simplifies the result to

$$\mathbf{P} = \begin{bmatrix}1 & 1 & 1 & 1\\-1 & -1 & 1 & 1\\j\omega_1 & -j\omega_1 & j\omega_2 & -j\omega_2\\-j\omega_1 & j\omega_1 & j\omega_2 & -j\omega_2\end{bmatrix};$$ $$\mathbf{P}_m = \begin{bmatrix}1 & 0 & 1 & 0\\-1 & 0 & 1 & 0\\0 & \omega_1 & 0 & \omega_2\\0 & -\omega_1 & 0 & \omega_2\end{bmatrix}$$

6. $\mathbf{E}_{St} = \begin{bmatrix} e^{j\omega_1 t} & 0 & 0 & 0 \\ 0 & e^{-j\omega_1 t} & 0 & 0 \\ 0 & 0 & e^{j\omega_2 t} & 0 \\ 0 & 0 & 0 & e^{-j\omega_2 t} \end{bmatrix}$

$\mathbf{E}_{Smt} = \begin{bmatrix} \cos\omega_1 t & \sin\omega_1 t & 0 & 0 \\ -\sin\omega_1 t & \cos\omega_1 t & 0 & 0 \\ 0 & 0 & \cos\omega_2 t & \sin\omega_2 t \\ 0 & 0 & -\sin\omega_2 t & \cos\omega_2 t \end{bmatrix}$

7. $\mathbf{PE}_{St} = \begin{bmatrix} e^{j\omega_1 t} & e^{-j\omega_1 t} & e^{j\omega_2 t} & e^{-j\omega_2 t} \\ -e^{j\omega_1 t} & -e^{-j\omega_1 t} & e^{j\omega_2 t} & e^{-j\omega_2 t} \\ j\omega_1 e^{j\omega_1 t} & -j\omega_1 e^{-j\omega_1 t} & j\omega_2 e^{j\omega_2 t} & -j\omega_2 e^{-j\omega_2 t} \\ -j\omega_1 e^{j\omega_1 t} & j\omega_1 e^{-j\omega_1 t} & j\omega_2 e^{j\omega_2 t} & -j\omega_2 e^{-j\omega_2 t} \end{bmatrix}$

$\mathbf{P}_m \mathbf{E}_{Smt} = \begin{bmatrix} \cos\omega_1 t & \sin\omega_1 t & \cos\omega_2 t & \sin\omega_2 t \\ -\cos\omega_1 t & -\sin\omega_1 t & \cos\omega_2 t & \sin\omega_2 t \\ -\omega_1 \sin\omega_1 t & \omega_1 \cos\omega_1 t & -\omega_2 \sin\omega_2 t & \omega_2 \cos\omega_2 t \\ \omega_1 \sin\omega_1 t & -\omega_1 \cos\omega_1 t & -\omega_2 \sin\omega_2 t & \omega_2 \cos\omega_2 t \end{bmatrix}$

The result in either case can be written

$$\begin{bmatrix} \phi_1 \\ \phi_2 \\ \dot\phi_1 \\ \dot\phi_2 \end{bmatrix} = \begin{bmatrix} b_1\cos\omega_1 t + b_2\sin\omega_1 t + b_3\cos\omega_2 t + b_4\sin\omega_2 t \\ -b_1\cos\omega_1 t - b_2\sin\omega_1 t + b_3\cos\omega_2 t + b_4\sin\omega_2 t \\ -b_1\omega_1\sin\omega_1 t + b_2\omega_1\cos\omega_1 t - b_3\omega_2\sin\omega_2 t + b_4\omega_2\cos\omega_2 t \\ b_1\omega_1\sin\omega_1 t - b_2\omega_1\cos\omega_1 t - b_3\omega_2\sin\omega_2 t + b_4\omega_2\cos\omega_2 t \end{bmatrix}$$

in which $b_1,\ldots,b_4$ are coefficients determined by the initial conditions. Specifically,

$$\begin{bmatrix} b_1 \\ b_2 \\ b_3 \\ b_4 \end{bmatrix} = \mathbf{P}_m^{-1}\mathbf{x}(0) = \frac{1}{2}\begin{bmatrix} 1 & -1 & 0 & 0 \\ 0 & 0 & 1/\omega_1 & -1/\omega_1 \\ 1 & 1 & 0 & 0 \\ 0 & 0 & 1/\omega_2 & 1/\omega_2 \end{bmatrix}\begin{bmatrix} \phi_1(0) \\ \phi_2(0) \\ \dot\phi_1(0) \\ \dot\phi_2(0) \end{bmatrix}$$

8. Note that the third and fourth rows above equal the derivatives of the first two rows, respectively, which we anticipate anyway since $\dot\phi_1 = d\phi/dt$ and $\dot\phi_2 = d\phi_2/dt$. When $b_3 = b_4 = 0$, only the first mode with ω_1 exists. In this case $\phi_2 = -\phi_1$, indicating motion that is symmetric about the vertical centerline, as shown below left. This implies that the center of the spring does not move, so each pendulum sees an effective spring rate of $2k$, which explains the natural frequency $\omega_1^2 = g/L + 2ka^2/mL^2$. When $b_1 = b_2 = 0$, on the other hand, only the second mode with ω_2 exists. In this case $\phi_1 = \phi_2$, indicating

no deflection or force whatever exists in the spring, as shown below right. The natural frequency of the motion therefore is the same as a single free pendulum, or $\omega_2^2 = g/L$. The general solution is of course the sum of both modes. The beating phenomenon referred to in the suggested steps is particularly interesting.

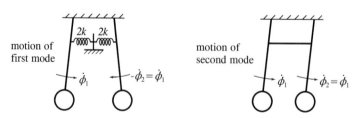

Guided Problem 7.8

1.

$$\frac{dp}{dt} = F(t) - \frac{1}{C}x - \frac{R}{I}p$$

$$\frac{dx}{dt} = \frac{1}{I}p$$

$$\frac{d}{dt}\begin{bmatrix} p \\ x \end{bmatrix} = \mathbf{A}\begin{bmatrix} p \\ x \end{bmatrix} + \mathbf{B}F(t); \text{ therefore, } \mathbf{A} = \begin{bmatrix} -R/I & -1/C \\ 1/I & 0 \end{bmatrix}, \ \mathbf{B} = \begin{bmatrix} 1 \\ 0 \end{bmatrix}$$

2. $\mathbf{G}(s) = (s\mathbf{I} - \mathbf{A})^{-1}\mathbf{B} = \begin{bmatrix} s + r/I & 1/C \\ -1/I & s \end{bmatrix}^{-1}\begin{bmatrix} 1 \\ 0 \end{bmatrix} = \frac{1}{p(s)}\begin{bmatrix} s \\ 1/I \end{bmatrix}$

$p(s) = s^2 + (R/I)s + 1/IC$

$x = GF; \qquad G(s) = \dfrac{1/I}{p(s)}$

3. $G(j\omega) = \dfrac{1/I}{1/IC - \omega^2 + (R/I)j\omega}$

$$|G(j\omega)| = \frac{1/I}{\sqrt{(1/IC - \omega^2)^2 + (R\omega/I)^2}}; \qquad \angle G(j\omega) = -\tan^{-1}\left(\frac{R\omega/I}{1/IC - \omega^2}\right)$$

4. $x = x_0 \sin[\omega t + \phi(\omega)]$

$$x_0 = F_0|G(j\omega)| = \frac{10}{\sqrt{(400 - \omega^2)^2 + (40\omega)^2}}; \qquad \phi(\omega) = -\tan^{-1}\left(\frac{40\omega}{400 - \omega^2}\right)$$

5. The following plots are given by MATLAB:

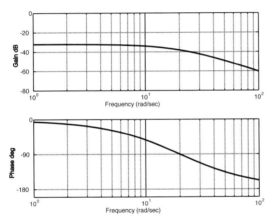

Guided Problem 7.9

1. The third differential equation has a forcing term added:

$$\frac{dp}{dt} = -\frac{1}{C_1}\phi_1 - \frac{1}{C_{12}}(\phi_1 - \phi_2) + LF(t)$$

or $\dfrac{d\dot\phi}{dt} = -\dfrac{1}{C_1 I_1}\phi_1 - \dfrac{1}{C_{12}I_1}(\phi_1 - \phi_2) + \dfrac{L}{I_1}F(t)$

Therefore, $\mathbf{B} = \begin{bmatrix} 0 \\ 0 \\ L/I_1 \\ 0 \end{bmatrix} = \begin{bmatrix} 0 \\ 0 \\ 1/mL \\ 0 \end{bmatrix}$ You want $\phi_1(t)$, which is the first state variable, so you need to find G_{13}

2. After the matrices $\mathbf{A}, \mathbf{B}, \mathbf{C}, \mathbf{D}$ are entered into MATLAB, the command `bode(A,B,C,D,1)` produces plots like those below. Alternatively, the actual plots were produced using the results of steps 3 and 4 below with the commands

```
w=[1:.05:4 4.01:.01:4.42 4.429 4.4297 4.43:.01:4.58];
w=[w 4.588 4.5889 4.59:.01:5 5.1:.1:10];
num = [2/3 0 40.68/3];
den =[1 0 40.68 0 413.2];
bode(num, den, w)
```

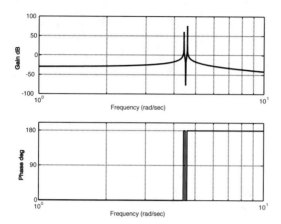

3. From step 3 of the solution to Guided Problem 8.5, $p(s) = s^4 + 2a_1 s^2 + (a_1^2 - a_2^2)$

$$G(s) = (s\mathbf{I} - \mathbf{A})^{-1}\mathbf{B} = \frac{1}{p(s)}\text{Adj}(s\mathbf{I} - \mathbf{A})\mathbf{B})$$

You need only the element of $\text{Adj}(s\mathbf{I} - \mathbf{A})$ in the first row and the third column, which is the cofactor of the element in the third row and first column. Since

$$(s\mathbf{I} - \mathbf{A}) = \begin{bmatrix} s & 0 & -1 & 0 \\ 0 & s & 0 & -1 \\ a_1 & -a_2 & s & 0 \\ -a_2 & a_1 & 0 & s \end{bmatrix} \text{ this is } \begin{vmatrix} 0 & -1 & 0 \\ s & 0 & -1 \\ a_1 & 0 & s \end{vmatrix} = (a_1 + s^2)$$

Therefore, $G_{13}(s) = \dfrac{(s^2 + a_1)/mL}{s^4 + 2a_1 s^2 + (a_1^2 - a_2^2)}$

4. $a_2 = \dfrac{ka^2}{mL^2} = \dfrac{8.0 \times (.15)^2}{1 \times (0.5)^2} = 0.720 \text{ s}^{-2}$

$a_1 = a_2 + \dfrac{g}{L} = 0.720 + \dfrac{9.81}{0.5} = 20.34 \text{ s}^{-2}$ $\qquad \dfrac{1}{mL} = \dfrac{1}{1 \times 1.5} = \dfrac{2}{3}$

$G_{13}(j\omega) = \dfrac{0.6667(-\omega^2 + 20.34)}{\omega^4 - 40.68\omega^2 + 413.2}$

This is a real number for all ω; its phase angle is 0 when the number is postive and $\pm 180°$ when it is negative.

5. See plots given above. The poles and zero were pinpointed for better plotting by the MATLAB commands r=roots(den) and r=roots(num). The zero occurs at the frequency $\omega = \sqrt{a_1} = \sqrt{\dfrac{g}{L} + \dfrac{ka^2}{mL^2}}$. It means that the forced pendulum is immobile at that frequency; the spring and second pendulum resonate and therefore appear to the forced pendulum as an infinite impedance.

7.4   Fourier Analysis

An arbitrary signal is decomposed into a sum of steps or impulses in the time-domain analysis emphasized in Chapter 4. The response of a linear model to such a signal is reconstructed by summing the individual responses to the individual steps or impulses. A different approach is now introduced, which often is called **frequency-domain** analysis. Instead of steps or impulses, the signal is decomposed into a sum of sine and cosine waves. The superposition property is invoked, as before: the response of the linear model equals the sum of the responses to the individual sine and cosine waves.

A signal which is periodic in time comprises a sum of sinusoidal waves at discrete frequencies, called a **Fourier series**. A non-periodic signal comprises a sum of sinusoidal waves at all frequencies, called a **Fourier transform**.

7.4.1 Fourier Series

Consider a *periodic* function of time $u(t)$ with period T, such as that shown in Fig. 7.30. The frequency of the periodic function is

$$\omega_0 = \frac{2\pi}{T}. \tag{7.125}$$

Any such function can be represented by a sum of sine and cosine waves, or by an equivalent sum of phase-shifted cosine waves:

$$
\begin{aligned}
u(t) &= a_0 + \sum_{n=1}^{\infty} \left[2a_n \cos(n\omega_0 t) + 2b_n \sin(n\omega_0 t) \right], \\
u(t) &= a_0 + \sum_{n=1}^{\infty} 2c_n \cos(n\omega_0 t + \theta_n), \\
c_n &= \sqrt{a_n^2 + b_n^2}; \quad \theta_n = -\tan^{-1}\left(\frac{b_n}{a_n} \right).
\end{aligned}
\tag{7.126}
$$

These sums are known as Fourier series.

The example of a particular square wave is plotted in Fig. 7.31. This is an *even* function of time, that is symmetric about time $t = 0$. As a result, only the Fourier cosine series coefficients are non-zero. (All cosine waves are *even*, and all sine waves are *odd*, or antisymmetric functions of time.) These coefficients are plotted in part (b) of the figure, and the first three harmonics are plotted in part (c). Their sum, plus the time-average term, is compared to the actual square wave in part (d). The sum of the first seven harmonics also is shown. The more harmonics that are added, the closer the sum approaches the square-wave function.

You need to evaluate the Fourier coefficients a_n and b_n in order to use a Fourier series. The general formula for computing the coefficients a_n results from changing the index n in equation (7.126a) to k, multiplying both sides by $\cos(n\omega_0 t)$, and integrating over one period. To compute the coefficients b_n, you replace the cosine with a sine. The respective integrations are

$$\int_{-T/2}^{T/2} u(t) \cos(n\omega_0 t)\, dt = \int_{-T/2}^{T/2} a_0 \cos(n\omega_0 t)\, dt$$

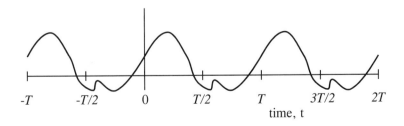

Figure 7.30: Periodic signal with period T

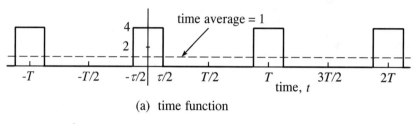

(a) time function

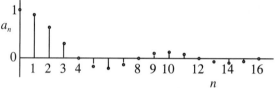

(b) Fourier coefficients (for $\tau = T/4$, time average $=1$)

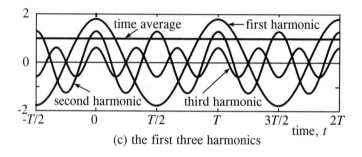

(c) the first three harmonics

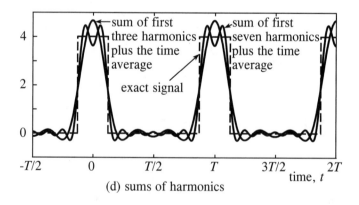

(d) sums of harmonics

Figure 7.31: A particular square-wave periodic signal

$$+ \sum_{k=1}^{\infty} \left[2a_k \int_{-T/2}^{T/2} \cos(n\omega_0 t)\cos(k\omega_0 t)\, dt + 2b_k \int_{-T/2}^{T/2} \cos(n\omega_0 t)\sin(k\omega_0 t)\, dt \right],$$

$$(7.127a)$$

$$\int_{-T/2}^{T/2} u(t)\sin(n\omega_0 t)\, dt = \int_{-T/2}^{T/2} a_0 \sin(n\omega_0 t)\, dt$$

$$+ \sum_{k=1}^{\infty} \left[2a_k \int_{-T/2}^{T/2} \sin(n\omega_0 t)\cos(k\omega_0 t)\, dt + 2b_k \int_{-T/2}^{T/2} \sin(n\omega_0 t)\sin(k\omega_0 t)\, dt \right].$$

$$(7.127b)$$

The period chosen here is the symmetric interval $-T/2 < t < T/2$. All terms on the right-hand side of these equations equal zero, except for those that integrate the *square* of a cosine or sine function, which exist only when $k = n$. The terms that vanish are all sinusoids with zero mean value integrated over an integral number of cycles. The multiplied functions are said to be **orthogonal** over the time interval. The functions

$$\cos^2(n\omega_0 t) = \frac{1}{2}[1 + \cos(2n\omega_0 t)], \qquad (7.128a)$$

$$\sin^2(n\omega_0 t) = \frac{1}{2}[1 - \cos(2n\omega_0 t)], \qquad (7.128b)$$

on the other hand, have an average value of $1/2$; their integrals over the time interval T equal $T/2$. As a result, equations (7.127) give

$$\boxed{\begin{aligned} a_n &= \frac{1}{T} \int_{-T/2}^{T/2} u(t)\cos\left(\frac{2\pi n t}{T}\right) dt; \\ b_n &= \frac{1}{T} \int_{-T/2}^{T/2} u(t)\sin\left(\frac{2\pi n t}{T}\right) dt. \end{aligned}} \qquad (7.129)$$

Equations (7.126) and (7.129) comprise the **Fourier series pair**.

EXAMPLE 7.16

Find the Fourier series expression for the square-wave signal addressed in Fig. 7.31.

Solution: Equations (7.129) give

$$a_n = \frac{1}{T} \int_{-\tau/2}^{\tau/2} 4\cos\left(\frac{2\pi n t}{T}\right) dt = \frac{4}{T}\frac{T}{2\pi n}\sin\left(\frac{2\pi n t}{T}\right)\Bigg|_{-\tau/2}^{\tau/2}$$

$$= \frac{4\tau}{T}\frac{\sin(\pi n\tau/T)}{\pi n\tau/T}; \quad n \geq 1$$

$$b_n = 0.$$

Substituting $\tau = T/4$ gives the Fourier series

$$u(t) = 1 + 2\sum_{n=1}^{\infty} \frac{\sin(\pi n/4)}{\pi n/4}\cos(2\pi n t/T).$$

An alternative expression for the Fourier series pair employs complex numbers:

$$u(t)= \sum_{n=-\infty}^{\infty} U_n e^{jn\omega_0 t},$$

$$U_n= \frac{1}{T} \int_{-T/2}^{T/2} u(t)\, e^{-jn\omega_0 t} dt. \qquad (7.130)$$

Since $u(t)$ is a real function, it is necessary that U_{-n} be the complex conjugate of U_n. In particular,

$$U_n= a_n - jb_n, \qquad n \geq 1;$$

$$U_{-n}= a_n + jb_n, \qquad n \geq 1. \qquad (7.133)$$

which gives the explicit relationship between the coefficients U_n and a_n, b_n. This more compact representation of the Fourier series pair is used below to generate the Fourier transform, and in Section 4.5.8 (p. 208) to generate the Laplace transform.

7.4.2   Response of a Linear System to a Periodic Excitation

Recall that the response of a linear system with the transfer function $G(S)$ to an input signal $u(t) = \cos(\omega t + \theta)$ is $x(t) = |G(j\omega)| \cos(\omega t + \alpha)$, where $\alpha = \angle G(j\omega)$ is the angle of the phasor $G(j\omega)$. Each term in a Fourier series is of this form, with $\omega = n\omega_0$. The property of superposition, then, requires the response to the Fourier series of equation (7.126b) to be

$$x(t) = X_0(0) + 2 \sum_{n=0}^{\infty} |X_n| \cos(n\omega_0 t + \alpha_n),$$

$$|X_n| = |G(jn\omega_0)|c_n; \qquad \alpha_n = \theta_n + \angle G(jn\omega_0). \qquad (7.132)$$

Note that the constant time term is treated as a cosine of zero frequency. For the more compact notation of equation (7.130), the equivalent result is

$$x(t) = \sum_{n=-\infty}^{\infty} X_n e^{jn\omega_0 t}; \qquad X_n = G(jn\omega_0)U_n. \qquad (7.133)$$

EXAMPLE 7.17
Find the steady-state response of a linear system with the transfer function

$$G(S) = \frac{2}{S^2 + 6S + 5}$$

to the square-wave excitation of Fig. 7.31 and Example 7.16.

Solution: The response, from equation (7.132), is

$$x(t) = G(0) \times 1 + 2 \sum_{n=1}^{\infty} \frac{\sin(\pi n/4)}{\pi n/4} |G(2\pi jn/T)| \cos(2\pi nt/T + \alpha_n);$$

$$\alpha_n = \angle G(2\pi jnt/T).$$

The magnitudes and phase angles of the transfer function contained in these equations are, from the given $G(S)$,

$$|G(2\pi jn/T)| = \frac{2}{\sqrt{[5 - (2\pi n/T)^2]^2 + [12\pi n/T]^2}};$$

$$\angle G_{21}(2\pi jn/T) = -\tan^{-1}\left[\frac{12\pi n/T}{5 - (2\pi n/T)^2}\right].$$

The response $x(t)$ is plotted below for the case $T = 5$ minutes. Its mean value is $G(0) \times 1 = 0.4$. Each of the first three harmonics is also shown, and it can be seen that their sum plus the mean value comprises the bulk of the complete solution.

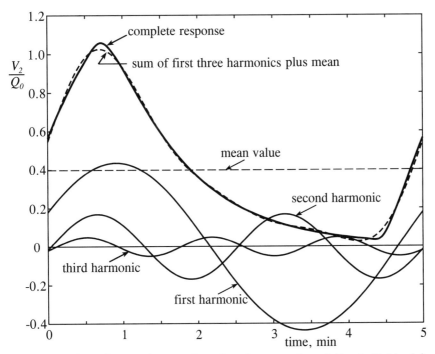

Care must be taken to insure that the phase angles of the individual harmonics are identified in their proper quadrants. Use of a standard inverse tangent routine to evaluate the phase angle might place the angle for $n = 1$ in the fourth quadrant and all the other angles in the first quadrant. The angles should be in the third quadrant for $n \geq 2$, however, so it is necessary to add (or subtract) π to the calculated angles in these cases.

7.4.3 Fourier Transform

A non-periodic signal can be viewed as part or all of a single cycle of a periodic signal with an extremely long period, T. If you let $T \to \infty$, any signal can be so represented. In this case,

$$\omega_0 = \frac{2\pi}{T} = \Delta\omega \to 0, \qquad\qquad (7.134a)$$

$$n\omega_0 \to \omega. \qquad\qquad (7.134b)$$

The discrete frequencies $n\omega_0$ become the continuous frequency ω. Equation (7.126) becomes

$$u(t) = \lim_{\Delta\omega\to 0} \sum_{n=-\infty}^{\infty} U_n e^{jn\Delta\omega t}$$

$$= \lim_{\Delta\omega\to 0} \frac{1}{2\pi} \sum_{n=-\infty}^{\infty} \left(\frac{U_n}{\Delta\omega/2\pi}\right) e^{jn\Delta\omega t}\Delta\omega, \tag{7.135}$$

or

$$u(t) = \frac{1}{2\pi}\int_{-\infty}^{\infty} U(j\omega)e^{j\omega t}d\omega; \tag{7.136a}$$

$$U(j\omega) \equiv \lim_{\Delta\omega\to 0}\left(\frac{U_n}{\Delta\omega/2\pi}\right) = \lim_{\Delta\omega\to 0}(U_n T). \tag{7.136b}$$

Here U_n is known as the **frequency** or **spectral density**[5], which is the *content per Hz* (per cycle per second). The formula for $U(j\omega)$ as a function of $u(t)$ is found from equation (7.130b):

$$U(j\omega) = \int_{-\infty}^{\infty} e^{-j\omega t}u(t)\,dt. \tag{7.137}$$

This computation of $U(j\omega)$ is known as the **Fourier transform** of $u(t)$, and the computation of $u(t)$ using equation (7.136a) is known as the **inverse Fourier transform** of $U(j\omega)$. Indicating the Fourier transform with the symbol $\mathcal{F}$,

$$U(j\omega) = \mathcal{F}[u(t)];$$
$$u(t) = \mathcal{F}^{-1}[U(j\omega)]. \tag{7.138}$$

The response $x(t)$ of the linear system G to the input $u(t)$ can be found by modifying equation (7.133) after the fashion of equations (7.136) and (7.137):[6]

$$x(t) = \mathcal{F}^{-1}[X(j\omega)];$$
$$X(j\omega) = G(j\omega)U(j\omega). \tag{7.139}$$

Although the Fourier transform is very powerful in addressing many signals and the responses of linear systems to these signals, it has a serious limitation because it is not defined (becomes infinite) if $\int_{-\infty}^{\infty}|u(t)|dt = \infty$. Thus, for example, the method fails for step functions and the responses of most systems to steps, and for oscillatory functions that do not decay. The transform applies only to "pulse-like" signals. This limitation is remedied, in Section 4.5, by employing a modification of the Fourier transform known as the Laplace transform. This text does not ask you to solve problems analytically using the Fourier transform. Should you ever wish to do so, however, many common time functions have transforms that are given in widely available tables, relieving the analyst of analytical drudgery.

---

[5]Energy or power is usually proportional to a signal squared, and thus $|U(j\omega)|^2$ is known as the **energy density spectrum**.

[6]This result also can be proven by applying the convolution integral, noting that $G(j\omega) = \mathcal{F}[g(t)]$, where $g(t)$ is the impulse response. A similar proof is given in Section 4.5.9 for the Laplace transform.

7.4.4 Digital Spectral Analysis*

The availability of an extremely powerful experimental tool known as the **dynamic analyzer** or **spectrum analyzer** is sufficient to justify learning the concepts of the Fourier transform and partly justifies the development above. In its basic form, one voltage represents an excitation signal $u(t)$, and a second voltage represents a system response $x(t)$, as measured by instrument transducers. These voltages are connected to the input terminals of an **analog-to-digital converter** ("A/D converter"), which generates N digital values to represent each of these voltages at equally spaced times over an interval of time, T. In a basic analyzer, $N = 2^{10} = 1024$ or $N = 2^{11} = 2048$. The Fourier series for each signal is then computed, using an algorithm called a **fast Fourier transform** (FFT), which is much more efficient than a direct approximation of equation (7.127). By using a dedicated computer chip these calculations are completed in a small fraction of a second. A sampling theorem limits the number of possible frequencies $f_n = \omega_n/2\pi = n/T$ for which U_n and X_n can be estimated to $N/2$, or 512 for $N = 1024$ or 1024 for $N = 2048$. In practice only perhaps the first 400 for $N = 1024$ and 800 for $N = 2048$ are found, since the remaining harmonics are increasingly susceptible to error.

The work of a dynamic analyzer can be performed on a general purpose computer coupled to an A/D converter, also, particularly when high speed is not critical. Software is widely available. MATLAB, as described below, can compute the fast Fourier transform and its inverse for data provided to it.

The computation is more generally called a discrete Fourier transform (DFT). The signal need not have any periodicity. On the other hand, the duration of the measurements is finite, so this duration also can be viewed as the period, T, of a periodic signal; the transform then becomes an approximate Fourier *series*, with the fundamental harmonic frequency $1/T$. This fact is exploited below to allow the FFT statements in MATLAB to closely estimate the results plotted in Figs. 7.31 part (b) and 7.32. In practice, most dynamic analyzers offer a choice of special automatic "windowing" modifications to minimize errors that result when a non-periodic signal overlaps the time boundaries of accepted data, appearing to give it an abrupt start and stop. The details are beyond the scope of current interest.

The analyzer can display the real and imaginary parts of $U(2\pi j f_n)$ and $X(2\pi j f_n)$, or alternatively their magnitudes and phase angles, as a function of the frequency $f_n = n/T$ Hz. With its typically 400 or 800 points or steps per signal, the displays resemble continuous functions. Numerical values also can be displayed and digitally read out.

As an example, assume that the square-wave signal of Fig. 7.31 and Examples 7.16 and 7.17 has a period of 5 minutes and is sampled over a time of $T = 200$ minutes.[7] The true Fourier transform would show peaks at the fundamental frequency of $1/5$ cpm and multiples thereof. The values between the peaks would be zero. The actual screen display for $N = 1024$ and $1/T = 0.005$ cpm, as shown in Fig. 7.32, is different from the ideal shown in part (b) of Fig. 7.31, however. The first ten harmonics appear properly, up to the maximum frequency of $400/T = 2$ cpm; twenty would appear (up to 4 cpm) if $N = 2048$. The principal difference is the appearance of much smaller components at many other frequencies. These result from the discretization of the signals. Note that in the case shown, the signal is an even function of time so that the imaginary part of the transform is zero.

On the other hand, a non-periodic signal with a single peak, such as shown in Fig. 7.33 part (a), gives a transform that is essentially continuous, such as shown in part (b) of the figure. A very important special case is the impulse (of zero width).

---

[7]Commercially available dynamic analyzers do not usually take data at such a slow rate. At the upper end, a rate of at least 256,000 data points per second and a display frequency of 100,000 Hz is common.

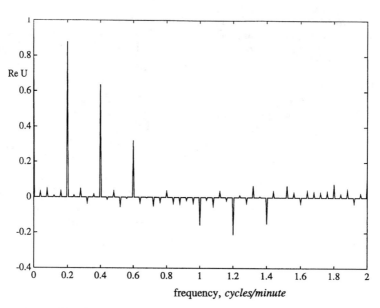

Figure 7.32: Display on the dynamic analyzer for the square-wave signal

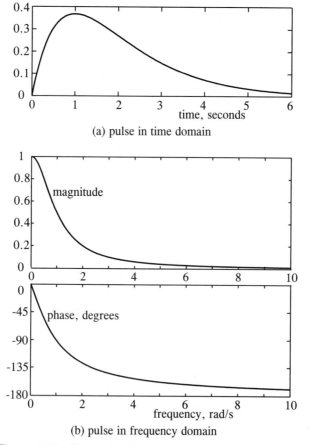

(a) pulse in time domain

(b) pulse in frequency domain

Figure 7.33: Example of a pulse and its Fourier transform

EXAMPLE 7.18

Find and plot the Fourier transform of the unit impulse $\delta(t)$:

(a) time domain

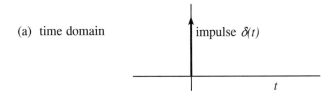

Solution: Equation (7.137) gives

$$U(j\omega) = \int_{-\infty}^{\infty} e^{-j\omega t} \delta(t)\, dt.$$

Since the integrand is non-zero only for the instant at which $t = 0$, the factor $e^{j\omega t} \to e^0 = 1$. Further, the integral of a unit impulse is identically 1, so the result is

$$U(j\omega) = 1.$$

(b) frequency domain

note: phase angle is zero
($\mathcal{F}$ is positive real)

Thus, *an impulse contains all frequencies equally,* and a hammer blow, which approximates an impulse, excites all frequencies below some limit associated with its actual non-zero duration. This makes the combination of a dynamic analyzer and a hammer, instrumented with an accelerometer to give an electrical pulse when it strikes an object, a powerful means for testing mechanical vibration.

Equation (7.139b) gives the frequency transfer function

$$G(j\omega) = \frac{X(j\omega)}{U(j\omega)}. \tag{7.140}$$

If the analyzer is used to measure $x(t)$ and $u(t)$ synchronously, it also can compute this ratio. In practice, for each of the 400 or 800 frequencies, the computations

$$|G(2\pi j f_n)| = \frac{|X(2\pi j f_n)|}{|U(2\pi j f_n)|}; \tag{7.141a}$$

$$\angle G(2\pi j f_n) = \alpha_x(2\pi j f_n) - \alpha_u(2\pi j f_n). \tag{7.141b}$$

are made and displayed. The results are meaningful for each frequency f_n that has an experimentally significant content $U(2\pi j f_n)$. For a hammer blow, for example, this means all frequencies, showing its special virtue (aside from ease of use).

Digital spectral analysis is a powerful tool for examining the frequency content of time-based signals, and determining experimentally the transfer function of a linear system or process by monitoring its input and output. It can predict the responses of systems. The discussion above serves merely as an introduction to the possibilities, techniques and limitations.

7.4.5   Fourier Analysis Using MATLAB®*

The commands U=fft(u) and U=fft(u,m) return the discrete Fourier transform (DFT) of the vector u, with a complex conjugate pair of numbers for each frequency. It is theoretically impossible to compute more than half as many frequency components as there are data points, so if the integer m exceeds the length of the vector u, U is padded up to the length m with zeros. If m is less than the length of the vector u, U is simply truncated at m elements. If m is a power of two, a radix-1 fast Fourier transform (FFT) algorithm is employed. This choice is recommended, for otherwise a much slower mixed-radix algorithm is used. The commands ifft(U) and ifft(U,m) return the inverse transform.

The discrete Fourier transform is the same as an approximate Fourier *series* for a periodic signal of period equal to the duration of the data vector u. One cycle of the square wave of Fig. 7.31 and Examples 7.16 and 7.17, for example, could be entered as 1024 data points evenly distributed over the $T = 5$ minutes as follows:

```
u1=4*ones(1,128);
u2=zeros(1,768);
u=[u1 u2 u1];
```

The command U=fft(u) returns the Fourier series, with each element multiplied by 1024. The first element in U is for zero frequency and therefore is real. The second element in U is for the frequency $1/T$ cyles per second or $2\pi/T$ radians per second, the third is for the frequency $2/T$ Hz, and so on up to the 513th which is for the frequency $512/T$ Hz. The remainder of the numbers up to the 1024th are redundant: the complex conjugates of the 2nd through the 512th numbers. They are in inverse order; the 514th is the complex conjugate of the 512th, and the 1024th is the complex conjugate of the 2nd, which means the fundamental frequency $1/T$ Hz. The instruction U(1:16)/1024 displays the first sixteen elements as follows:

```
1.0000   0.9003 + 0.0028i    0.6366 + 0.0039i    0.3001 + 0.0028i
0         -0.1800 - 0.0028i   -0.2122 - 0.0039i   -0.1286 - 0.0028i
0          0.1000 + 0.0028i    0.1273 + 0.0039i    0.0818 - 0.0028i
0         -0.0692 - 0.0028i   -0.0909 - 0.0039i   -0.0600 - 0.0028i
```

The small imaginary components correspond to a time shift of one-half a sampling interval, and are of little significance. Nevertheless, they can be eliminated with the instruction U=real(U). The result agrees with that of Example 7.16 and part (b) of Fig. 7.17, up to at least four significant digits for the first several Fourier harmonics.

The Fourier series can be multiplied by the frequency transfer function $G(j\omega)$ to get the Fourier series of the ouput signal, $x(t)$. The function $G(j\omega)$ must have 1024 elements corresponding to the same sequence of frequencies as U, however. Therefore, you can construct a frequency vector w:

```
w0=2*pi/5;
for i=1:513
        w(i)=(i-1)*w0;
end
```

The 512 values of $G(j\omega)$ for the case addressed in Example 7.17 are found as follows:

```
num=[2];
den=[1 6 5];
G1=freqs(num,den,w);
```

The complex conjugates of the second through the last values, in inverse order, are needed to complete the vector G:

```
for i=1:511
      G2(i)=conj(G1(513-i));
end
G=[G1 G2];
```

The theoretical limit of $1024 \div 2 = 512$ frequencies means that there can be 511 complex values of U (the second through the 512th, and the 514th through the 1024th) plus two real values (the first and the 513th). To be completely consistent, therefore, you should eliminate the imaginary part of the 513th value of G:

```
G(513)=real G(513);
```

Now you are ready to compute the Fourier series for the output signal:

```
X=U.*G;
```

Finally, a plot consistent with the complete response as plotted in Example 7.17 results from the inverse Fourier transform command

```
plot(real(ifft(V)))
```

The inverse Fourier routine produces very small spurious imaginary parts due to round-off errors. The plotting routine requires real numbers, explaining the use of the sub-command `real`. Should you ever find a non-trivial imaginary part of an inverse transform, look for a mistake.

7.4.6 Summary

A periodic signal can be represented by a Fourier series comprising a sum of sine and cosine waves. The frequency of each component is an integral multiple of the frequency of periodicity. The response of a linear system to a periodic signal is the sum of the responses to the individual sine and cosine waves. The different frequencies are apt to be filtered quite differently, so that the ouput signal might look quite different from the input signal. High-frequency noise often is purposely stripped from a measured signal, for example, by passing it through a "low-pass filter"; what emerges is the lower-frequency signal of interest.

A broad class of aperiodic signals, such as pulses, can be represented by a Fourier transform. The criterion for its existence is that the time integral of the absolute value of the signal over its entire duration be finite, ruling out step-like functions and oscillations that do not decay. The Fourier transform is the same as the Fourier series except it includes a continuous distribution of all frequencies rather than discrete multiples of a base frequency. The response of a linear system to such a signal is therefore the sum (or actually the integral) of the responses to the individual frequency components.

Digital spectral analysis may be carried out by a special instrument called a frequency or spectral analyzer, or on a general-purpose computer. MATLAB has a capability in this regard. A discrete Fourier transform is computed for a train of numbers that represents a signal measured over a period of time. The result approximates the Fourier transform for a non-periodic signal, and a Fourier series for a periodic signal with period equal to the duration of the data train. The analyzer can process both excitation and response signals simultaneously. It then can compute and display the ratio of the two, expressed as a function of frequency. This is the frequency transfer function, $G(j\omega)$.

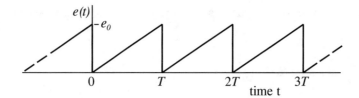

Figure 7.34: Disturbance for Guided Problem 7.10

Guided Problem 7.10

This problem provides a basic experience in computing a Fourier series for a periodic signal, and then finding the periodic response when this signal is applied as the excitation to a linear first-order system.

The saw-tooth disturbance plotted in Fig. 7.34 acts on a system modeled as

$$\tau \frac{dq}{dt} + q = Ce$$

(which corresponds to the RC system of Guided Problem 3.9, p 124). Find the response $q(t)$.

Suggested Steps:

1. Find the average value of the disturbance $e(t)$ and the consequent constant term in the response of $q_C(t)$.

2. Find an analytic expression for $e(t)$ valid for at least one complete period of the disturbance.

3. Evaluate the Fourier coefficients a_n and b_n for the disturbance using the function $e(t)$. Write the Fourier series using the cosine form with its coefficients c_n and phase angles β_n.

4. Find the response of the system to a cosine wave of arbitrary phase angle.

5. Assemble the response of the system to the complete disturbance by combining the results of steps 3 and 4.

Guided Problem 7.11

This problem both illustrates basic concepts regarding the use of Fourier series and transforms and explores a very practical application of the dynamic analyzer. Should you have access to a dynamic analyzer with an instrumented hammer, you would likely benefit from carrying out a similar experiment.

The spectrum of the response of a linear system to a perfect impulse is proportional to its transfer function $G(j\omega)$, as described in the text. Experimentally, one attempts to generate the impulse with an instrumented hammer. The acceleration or force signal from this hammer and the signal from an accelerometer or other motion transducer attached to the object under test are sent to a dynamic analyzer. The analyzer displays the Fourier transforms of these two signals, as shown in Fig. 7.35. Describe qualitatively in what way the actual excitation pulse differs from a perfect impulse. Also, sketch-plot the transfer

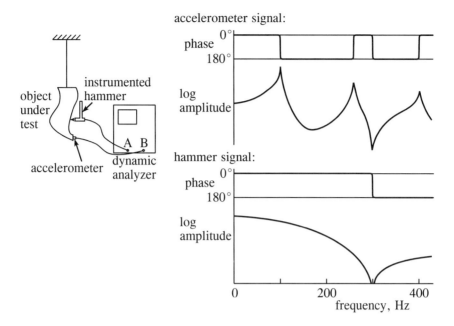

Figure 7.35: Guided Problem 7.11

function (which the analyzer would do for you if asked). Discuss briefly conditions under which the analyzer with this excitation is prone to give large errors in $G(j\omega)$.

Suggested Steps:

1. A perfect impulse would have a constant magnitude spectrum. The actual signal decays with frequency and goes negative at a particular frequency. Show that a square pulse of duration T seconds could produce a hammer signal similar to that given; evaluate T. (The actual pulse likely would be rounded, but the difference is not significant for the low frequencies shown.)

2. There are no scales given for the magnitude plots, so the best you can do is sketch-plot $G(S)$ to the same scale with no absolute amplitude specified. The interest presumably is in the dynamic character of the response, that is the natural frequencies and relative magnitudes of the various resonances. Note that the log of a ratio of two quantities equals the difference between the logs of the quantities, and sketch-plot the magnitude ratio.

3. The phase of $G(j\omega)$ equals the difference between the phases of the numerator and the denominator. Sketch.

4. Note that large errors in a computed ratio occur when both the measured numerator and denominator are small.

PROBLEMS

7.41 A linear system is excited by the signal $u(t)$ and responds with the output variable $x(t)$. The system is described by

$$\frac{d^2x}{dt^2} + \frac{dx}{dt} + 81\,x = u(t),$$

and the excitation, plotted below, is the periodic signal (valid for all t)

$$u(t) = 2\cos(3t) + \cos(9t).$$

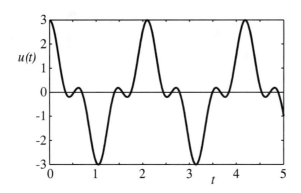

(a) Find $G(j\omega)$ for the system.

(b) Evaluate $|G(j\omega)|$ for the frequency(ies) appropriate to this problem.

(c) State which frequency component predominates in the response $x(t)$.

(d) Evaluate $\angle G(j\omega)$ for the frequencies appropriate to this problem.

(e) Assemble the response $x(t)$.

7.42 Find the steady-state response of the second-order system

$$\frac{1}{\omega_n^2}\frac{d^2x}{dt^2} + \frac{2\zeta}{\omega_n}\frac{dx}{dt} + x = u(t)$$

to the square-wave disturbance $u(t)$ shown in Fig. 7.31 (p. 502) and described with the Fourier coefficients found in Example 7.16 (p. 503). Assume $T = 4\pi/\omega_n$ and $\zeta = 0.2$.

(a) Note the time-average value of the disturbance, and find the response to this zero-frequency component.

(b) Write expressions for the magnitude and phase angle of the nth harmonic component of the response.

(c) Plot the first five harmonic components for one cycle. Which component is the largest, and why? (The use of MATLAB is suggested.)

(d) Plot the sum of a least the first five harmonics to approximate the overall response.

7.43 Find the response of the problem above using the fast Fourier transform capability of MATLAB.

7.44 Water with a density of 1000 kg/m^3 flows through a linear resistance of $R = 5 \times 10^5$ N·s/m^5 to a tank of area 0.06 m^2. The applied pressure comprises an infinite series of pulses, each of magnitude 6 kPa·s, 2 seconds apart.

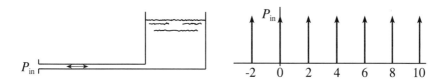

(a) Evaluate the compliance of the tank.

(b) Model the system with a bond graph.

(c) Evaluate any critical time constant or natural frequency.

(d) Determine the average applied pressure and the resulting average depth of the water in the tank.

(e) Represent the applied pressure by a Fourier series.

(f) Determine the first three Fourier terms in an expression for the depth of the tank. Plot their sum.

7.45 *Continuation of Problems 3.22 (p. 108), 4.56 (p. 229) and 7.8 (p. 448).* An impact wrench alternately strikes the member with clockwise impulses of amplitude 8 ft·lb·s and counterclockwise impacts of 4 ft·lb·s, as plotted to the right.

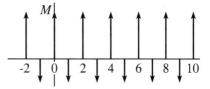

(a) Determine the time-average torque applied to the member, and the resulting angular displacement.

(b) Show the form of a Fourier series representation of the applied impacts, omitting terms known to be zero.

(c) Evaluate the first two Fourier coefficients in your series, and write a corresponding expression for the angular response of the member.

7.46 A linear system responds to a unit step excitation with the output $2(1 - e^{-t/\tau})$. Find the response of the system to the symmetric square wave shown below. Let $T = 2\tau$.

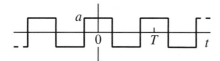

(a) Evaluate the first three non-zero Fourier series components of $u(t)$.

(b) Find the transfer function of the system, or the corresponding differential equation.

(c) Find the response of the system to a sinusoidal excitation.

(d) Find the first three Fourier series components of $x(t)$. Plot these and their sum; either sketch-plots or the use of MATLAB is acceptable.

(e) Since the signal comprises repeated upward and downward steps of equal amplitude, the exact solution to this problem can be solved by assuming an undetermined initial state, using the given step response to find the consequent state at the end of one cycle, and setting the two states equal to one another in order to evaluate the actual state at the beginning and end of each cycle. Carry out this procedure, and compare the result to that of part (d).

7.47 Find the response of the problem above using the fast Fourier transform capability of MATLAB.

7.48 A mass-dashpot system is excited by a triangular periodic force, as plotted below. Find the consequent velocity as a function of time, neglecting any transient behavior associated with the initiation of the force.

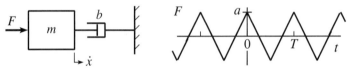

(a) Write the differential equation with $\dot{x}(t)$ as the state variable; determine **A** and **B**, which in this case are scalars.

(b) Deduce $G(S)$.

(c) Find the Fourier series representation for the applied force $F(t)$. The calcuation can be simplified by noting that $F(t)$ is an even function, so the interval of integration can be limited to positive values of time, and only the cosine terms are non-zero. You may wish to use a table of integrals; otherwise, integration by parts can be used. Note finally that the even numbered harmonics vanish.

(d) Find the response of the system to the nth harmonic. Represent this in terms of a phase-shifted cosine or sine wave rather than separate sine and cosine waves, since this reduces the number of summations if you intend to make a sketch-plot. Sum these responses to get an expression for the overall response.

(e) For $T = 1$ s, $m = 1$ kg, $a = 1$ N and $b = 10$ N s/m, evaluate the magnitude and phase of the first harmonic of $\dot{x}$ and all others that are as large or larger than one percent of the first harmonic. Sketch these terms and their sum. Note that very few terms contribute significantly.

7.49 Plot the velocity $\dot{x}$ of the problem above, for the values given in part (e), using the fast Fourier transform capability of MATLAB.

7.50 An instrumented hammer strikes two specimens that are freely suspended and instrumented with accelerometers. With one of the specimens, the phase angle between the excitation and the response is 0° or ±180° at virtually all frequencies, with abrupt jumps in between. With the other, the corresponding phase angle varies continuously with frequency.

(a) Characterize the difference between the two transfer functions, assuming both can be represented by ratios of polynomials.

(b) What causes the difference?

(c) What categorical difference would you anticipate between the two magnitude Bode plots?

7.51 An accelerometer placed on the rim of an axisymmetric disk-and-rim vibration absorber, shown below left in cross-section, produces the display shown below right on a spectrum analyzer when shaken with a purely sinusoidal force at frequency ω_0. Both vertical and horizontal axes have linear scales.

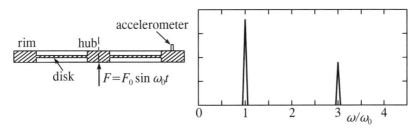

(a) State if and why the data indicates linear or nonlinear behavior.

(b) The neglect of damping allows the data to describe two different time functions. Find and sketch-plot these accelerations, and state why the ambiguity exists.

(c) (*extra credit*) Explain, qualitatively, the cause of the non–fundamental harmonic revealed in the display. Also, suggest any practical implications this might have for the absorber. [Hint: Consider the nature of the spring characteristic between the hub and the rim, and particularly the effect of radial as well as bending stresses. Does the natural frequency of the absorber vary with the amplitude of its motion?]

SOLUTIONS TO GUIDED PROBLEMS

Guided Problem 7.10

1.

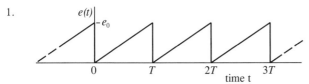

The average value of the disturbance is seen from the plot to be $e_0/2$.

2. $e(t) = \dfrac{e_0}{T}t + e_0, \ -T < t < 0; \ e(t) = \dfrac{e_0}{T}t, \ 0 < t < T$

3. From equation (7.131),

$$a_n = \frac{1}{T}\int_{-T/2}^{0}\left(\frac{e_0}{T}t + e_0\right)\cos\left(\frac{2\pi nt}{T}\right)dt + \frac{1}{T}\int_{0}^{T/2}\left(\frac{e_0}{T}t\right)\cos\left(\frac{2\pi nt}{T}\right)dt$$

$$= \frac{e_0}{T^2}\int_{-T/2}^{T/2}t\cos\left(\frac{2\pi nt}{T}\right)dt + \frac{e_0}{T}\int_{-T/2}^{0}\cos\left(\frac{2\pi nt}{T}\right)dt$$

$$= \frac{e_0}{T^2}\left[\left(\frac{T}{2\pi n}\right)^2\cos\left(\frac{2\pi nt}{T}\right) + \frac{tT}{2\pi n}\sin\left(\frac{2\pi nt}{T}\right)\right]\Bigg|_{-T/2}^{T/2} + \frac{e_0}{T}\left[\frac{T}{2\pi n}\sin\left(\frac{2\pi nt}{T}\right)\right]\Bigg|_{-T/2}^{0}$$

$$= e_0\left\{\frac{1}{(2\pi n)^2}[\cos(\pi n) - \cos(-\pi n)] + \frac{1}{4\pi n}[\sin(\pi n) - \sin(-\pi n)]\right\}$$

$$+ \frac{e_0}{2\pi n}[\sin(0) - \sin(-\pi n)] = 0.$$

This answer can be anticipated since $e(t)$ is an *odd* function of time and a_n represents only the cosine or *even* parts of the signal.

$$b_n = \frac{c_0}{T^2}\int_{-T/2}^{T/2} t\sin\left(\frac{2\pi nt}{T}\right)dt + \frac{e_0}{T}\int_{-T/2}^{0}\sin\left(\frac{2\pi nt}{T}\right)dt$$

$$= \frac{e_0}{T^2}\left[\left(\frac{T}{2\pi n}\right)^2 - \frac{tT}{2\pi n}\cos\left(\frac{2\pi nt}{T}\right)\right]\Bigg|_{-T/2}^{T/2} - \frac{e_0}{T}\left[\frac{T}{2\pi n}\cos\left(\frac{2\pi nt}{T}\right)\right]\Bigg|_{-T/2}^{0}$$

$$= e_0\left\{\frac{1}{(2\pi n)^2}[\sin(\pi n) - \sin(-\pi n)] - \frac{1}{2\pi n}[\cos(0) - \cos(-\pi n)]\right\} = \begin{cases} -e_0/\pi n & n \text{ odd} \\ 0 & n \text{ even} \end{cases}$$

Therefore,

$$e(t) = \frac{e_0}{2} - \sum_{n=1}^{\infty}\frac{2e_0}{\pi n}\sin\left(\frac{2\pi nt}{T}\right) = \frac{e_0}{2} + \sum_{n=1}^{\infty}\frac{2e_0}{\pi n}\cos\left(\frac{2\pi nt}{T} + 90°\right).$$

4. $\tau\dfrac{dq_C}{dt} + q_C = Ce.$ The steady-state is $q_C = \dfrac{Ce_0}{2}$. $\tau = RC.$

If $e = e_1\cos(\omega t + \beta - tan^{-1}(\tau\omega))$, then $q_C = \dfrac{C}{\sqrt{(\tau\omega)^2 + 1}}\cos[\omega t + \beta - \tan^{-1}(\tau\omega)]$.

5. $q_C = \dfrac{Ce_0}{2} + \sum_{n=1}^{\infty}\dfrac{2Ce_0/\pi n}{\sqrt{1 + (2\pi n\tau/T)^2}}\cos\left(\dfrac{2\pi nt}{T} + 90° + \phi_n\right);\ \phi_n = -\tan^{-1}\left(\dfrac{2\pi n\tau}{T}\right).$

Result for $\tau/T = 1$:

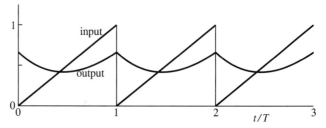

Guided Problem 7.11

1.   A crude model of the pulse is
$$U(j\omega) = \int_{-T/2}^{T/2} e^{-j\omega t}dt = 2\int_{0}^{T/2}\cos\omega t\,dt = \frac{2}{\omega}\sin\left(\frac{\omega}{\ }\right.$$
This resembles the given spectrum for the hammer if $\omega t/2 = \pi$ for $\omega = 2\pi \times 300$, or $T = 1/300$ s.

2-4.

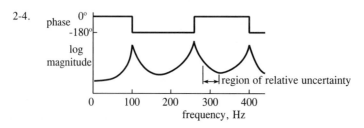

Chapter 8

Introduction to Automatic Control

The purposeful management of the state of an engineering system over time is called **system control** or, to emphasize that the details are carried out by machine rather than by a person, **automatic control**. In one scenario, the state is varied in some desired manner: a vehicle is accelerated or turned automatically, a machine tool is directed to follow a prescribed path, or a metal casting is cooled in such a manner that its properties achieve a desired standard. In the other most common scenario, the state is held constant despite buffeting from the environment: a boat is prevented from rolling in the waves, a rolling mill for sheet steel produces uniform thickness product despite thermal expansion and contraction of the rollers, or the temperature in your house is held at 72°F despite outside temperature fluctuations. These two scenarios often coexist.

Although ancient examples of automatic control exist, the modern discipline of control system science can be dated to World War II with its special impetus for gunnery and vehicle control. The science comprises many mathematical and computer techniques. Its application to engineering systems requires the development of mathematical or computer models of these systems. This is an important application of the modeling that is the primary subject of this book. Often, the control science that needs to be added to the modeling is relatively modest. This chapter, which focuses on continuous as opposed to discrete control, suffices for many situations, despite presenting only a small fraction of the discipline of system control. Should you wish to learn more, you can elect a course or series of courses devoted exclusively to it, or study any of a score of textbooks.

Simulink® is very popular amongst control engineers, but is not used here in favor of basic application of MATLAB®. Special roles that bond graphs can play in control system design are noted briefly at the end of the chapter (Section 8.3.7, pp. 574-575).

8.1 Open and Closed-Loop Control

The major components of an automatic control system are the the system to be controlled, the controller, which is the brains, the actuator, which is the muscle, and signal transducers, which are the senses. The actuator together with the system being controlled is known as the **plant**. A block digram showing the basic interconnections is given in Fig. 8.1. This is classic feedback control: the behavior of the system as observed by a signal transducer or sensor is compared with the desired or command value of that variable, and the difference or error is sent to a controller, which in turn directs the actuator to reduce the error. In

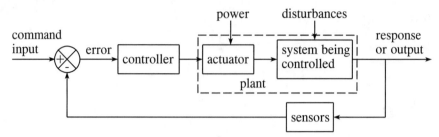

Figure 8.1: Basic feedback control system

complex systems the actuators themselves are imbedded control subsystems, with their own brains, sensors and muscle.

8.1.1   Example Plant

An example plant is used repeatedly in this chapter to help develop the basic concepts of control systems. In this example, shown in Fig. 8.2, the position of a rotational load with inertia I is to be controlled. The muscle or actuator of the system is a DC motor with gyrator modulus $G = a$ and armature resistance R coupled to a gear drive with reduction ratio $T = r$. The mechanical friction in the motor and the gear drive will be neglected in order to give a linear model. The plant is the combination of the actuator and the inertial load. The transfer function from the input to this plant, which is the armature voltage v, to the angular velocity $\dot{\phi}$, is shown to be

$$G_1(S) = \frac{r/a}{(1 + \tau S)} \qquad \tau = \frac{r^2 R I}{a^2}. \tag{8.1}$$

This transfer function with its time-derivative operator[1] S is represented in the block diagram in part (c) of the figure. The angular velocity $\dot{\phi}(t)$ is then integrated in time, as indicated by the block with transfer function $1/S$, to give the angle $\phi(t)$. The overall transfer function of the plant relates the input $v(t)$ to the output $\phi(t)$:

$$\phi(t) = \frac{G_1}{S} v(t) \equiv G_2(S) v(t), \qquad G_2(S) = \frac{r/a}{S(1 + \tau S)}. \tag{8.2}$$

8.1.2   Open-Loop and Optimal Control

To move the load from rest in one position to rest in another position in minimum time, maximum allowable voltage of the appropriate sign is applied for a certain period of time, followed by maximum allowable voltage of the opposite sign for a second period of time. The result for a particular maximum voltage and angle shift is plotted in part (d) of the figure. The minimization of some performance index, in this case the total time for the move from the first state to the second, is called **optimal control**. It is relatively difficult to compute the periods of maximum effort, even in the present simple case; **suboptimal control** strategies are more commonly adopted because they typically require much less on-line computation and are otherwise easier to implement. Nevertheless, the optimal control serves as a benchmark against which simpler control schemes can be judged.

Most practical control schemes involve one or more feedback loops. The more primitive scheme without any feedback loop is called **open loop control**. This scheme is addressed first.

---

[1]The reader may substitute the Laplace s for S throughout this chapter.

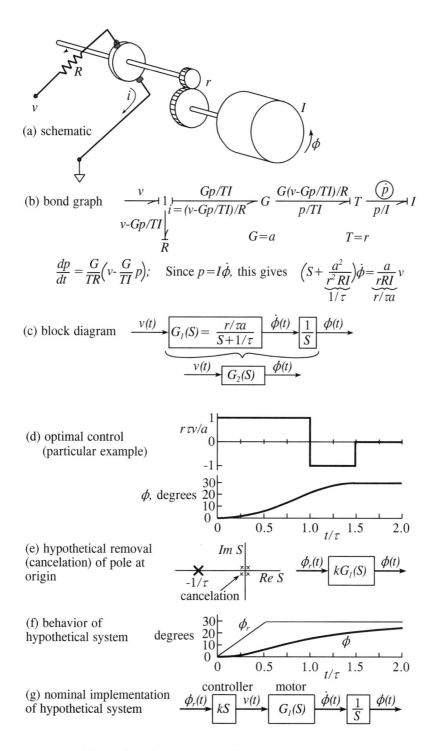

(a) schematic

(b) bond graph

$$\frac{dp}{dt} = \frac{G}{TR}\left(v - \frac{G}{TI}p\right); \quad \text{Since } p = I\dot{\phi}, \text{ this gives} \quad \left(S + \underbrace{\frac{a^2}{r^2RI}}_{1/\tau}\right)\dot{\phi} = \underbrace{\frac{a}{rRI}}_{r/\tau a}v$$

(c) block diagram

(d) optimal control
(particular example)

(e) hypothetical removal
(cancelation) of pole at
origin

(f) behavior of
hypothetical system

(g) nominal implementation
of hypothetical system

Figure 8.2: Open-loop position control system

The steady-state response of the motor-load combination to a constant voltage input is an output with constant angular *velocity*. This fact is associated directly with the pole at the origin in the transfer function $G_2(S)$. If this pole somehow were eliminated, to give the transfer function $kG_1(S)$ (with k being a constant), as shown in part (e) of Fig. 8.2, a single real pole with $S = -1/\tau$ would be left, and the desired constant angular *position* would result instead. The remaining pole would impose an exponential approach to the final position, with time constant τ, as shown in part (f) of the figure. (The rate of change of the command signal $\phi_c(t)$ is limited in the plot, producing the initial ramp, to prevent the armature voltage from exceeding the limit imposed in the earlier optimal control. This allows a meaningful comparison of the results. Without the limit, the initial voltage would be infinite.) Could this desirable situation be achieved?

The open-loop approach achieves the goal, at least theoretically, by superimposing a zero directly on top of the undesired pole, cancelling it. The controller, therefore, comprises a differentiator; its transfer function can be written kS. This controller operates on the command signal to produce the armature voltage, as suggested in the block diagram of part (g) of the figure. The resulting overall transfer function between the command signal and the output signal therefore is the desired $kG_1(S)$. If the command signal is taken to be the desired dimensionless angular position of the load (ϕ_c), the ratio kr/a equals 1, so that

$$\phi(t) = \frac{1}{1 + \tau S}\phi_r(t). \tag{8.3}$$

As a practical matter, however, this control scheme is unworkable. A real controller would not be a perfect differentiator, and the pole at the origin would be imperfectly cancelled. The load either would not come to a perfect stop following an extended zero value of the command signal (that is would "drift" unacceptably), or friction would cause it to stick prematurely at the wrong angle. Errors would accumulate. The idea of open-loop control itself allows no correction for the inevitable errors that exist because no controller is matched perfectly to its system.

Most systems also are susceptible to disturbances from the environment, against which an open-loop control system is impotent. For example, the rotary output member in the example might interact with its environment such that a small torque is placed on it. Some means to compare the actual output of the system (the angular position of the load) with the command input is needed to address both this problem of disturbances and the problem of imperfect controllers. This means is called feedback control.

8.1.3   Feedback Control

Measurement of the actual position of the load, by some means such as a rotary potentiometer or a digital resolver, makes known the difference $e(t) \equiv \phi_r(t) - \phi(t)$ between input or desired position and the actual position. Since minimization of this *error* is the objective, it makes sense to use $e(t)$ as the input signal to the controller. This scheme is represented by the block diagram given in part (a) of Fig. 8.3. The choice of the controller is left to the control engineer. The class of linear controllers, the simplest and most common, is assumed here.

The simplest linear controller is a constant coefficient, here given as k. With this **proportional controller**, shown in part (b) of the figure, the transfer function between the error $e = \phi_r - \phi$ and the output angle ϕ is $kG_2(S)$.

This block diagram is generalized in part (c) of the figure as the universal single-loop system with an input or reference variable $r(t) = \phi_r(t)$, output or controlled variable $c(t) = \phi$, an error signal $e(t) = r(t) - c(t)$ and **unity feedback**. Since

$$c(t) = G(S)e(t) = G(S)[r(t) - c(t)], \tag{8.4}$$

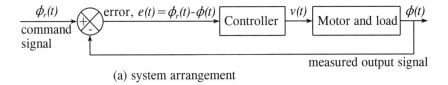

(a) system arrangement

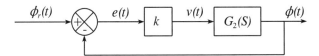

(b) example with proportional control

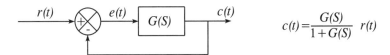

$$c(t) = \frac{G(S)}{1+G(S)} r(t)$$

(c) generalized unity-feedback loop

Figure 8.3: Position control system with feedback

the universal transfer function relation between the input and output signals is

$$c(t) = \frac{G(S)}{1+G(S)} r(t). \tag{8.5}$$

For the particular case at hand, substitution $G(S) = kG_2(S)$ gives

$$\phi(t) = \frac{1}{1 + (a/rk)S + (\tau a/rk)S^2} \phi_r(t). \tag{8.6}$$

You should recognize this as a classical linear second-order model with two poles and no zeros. Most important, there is no **steady-state error**; when ϕ_r is constant, the output angle ϕ is equal to it. The transient behavior is represented by the two poles. For small values of k these poles are real, so the response to a step change contains only exponential time-constant terms, and is smooth with no **overshoot** of the final value. For large values of k these poles are complex conjugate, and can be represented by a natural frequency and a damping ratio. The response to a step oscillates and exhibits overshoot. In general, one can say that too small a value of k renders the response too sluggish, and too high a value of k renders it too oscillatory. What is the best value of k?

Creating a **root locus plot** helps answer this question in general for scalar unity-feedback loops. The locations of the poles of the closed-loop system $G/(1 + G)$ in the complex S-plane are plotted as functions of the gain k. The fixed poles of the open-loop system G also are plotted. The root-locus plot for the present case is shown in Fig. 8.4. The two closed-loop poles emerge from the open-loop locations as k grows from zero, at first approaching one another along the real axis. After they meet, further increases in k drive these poles apart into complex territory, symmetrically, along a vertical line that will be shown is the perpendicular bisector of the line segment connecting the two open-loop poles. Thus the envelope decay coefficient $\zeta\omega_n$ remains constant, while the damping ratio decreases and the natural frequency increases. As $k \to \infty$, $\zeta \to 0$ and $\omega_n \to \infty$.

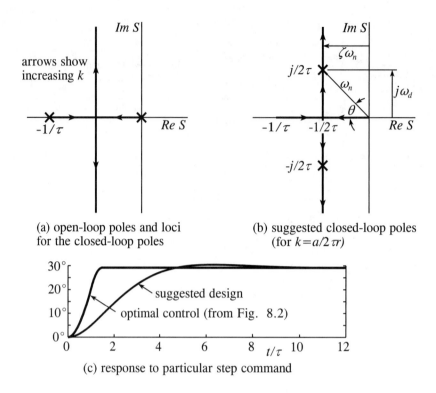

(a) open-loop poles and loci
for the closed-loop poles

(b) suggested closed-loop poles
(for $k = a/2\,\pi r$)

(c) response to particular step command

Figure 8.4: Root loci and behavior of position control system with feedback

Responses generally are dominated by the pole nearest the origin, which contributes the slowest transient term. Usually this means that you seek to keep the closed-loop poles as far from the origin as possible, particularly with regard to their real parts. For the example, therefore, the value of k ought to be chosen such that the real parts of the two closed-loop poles lie on the vertical part of the locus, which has the real part $-1/2\tau$. It is less clear how large the imaginary parts of the poles should be; different practitioners could disagree. The high natural frequency that follows from large imaginary parts is attractive, but the envelope function $e^{-\zeta\omega_n t}$ that bounds the decaying oscillation and largely establishes the **settling time** is a function only of the fixed real parts. Many engineers would choose a damping ratio of about 0.7 as a compromise between fast initial response, modest overshoot and rapid settling time. The settling time is defined as the last moment at which the error of the step response is larger than some percentage, perhaps 2% or 5%, of the step size.

The cosine of the angle between the real axis and a radial line from the origin of the S-plane equals the damping ratio of the pole pair. Therefore, the value $\zeta = 1/\sqrt{2} = 0.707$ makes the angle θ, shown in part (b) of Fig. 8.4, equal 45°. With this value, $\omega_d = 1/2\tau$ and $k = a/2\tau r$. The associated step response of this suboptimal strategy is plotted in part (c) of the figure.

The response for the sample step is inferior to that of the optimal response. The most important contributing reason for the inferiority is that never more than one-quarter of the armature voltage allowed in the optimal control is in fact used. Since the system is linear, the control effort (and the armature voltage) is proportional to the size of the command signal. For example, were the command signal four times larger (116.9° instead of 29.2°), the initial armature voltage would equal the maximum allowed for the optimal control, and the response curve would simply be four times larger but no slower. The settling time for the optimal control, on the other hand, would increase markedly relative to its given plot, although the response would remain somewhat superior to that for the proportional control. Since the optimal control exerts maximum effort regardless of the size of the command step, it completes small excursions much faster than large excursions. The linear controller will exceed the allowable voltage for yet larger command signals; some kind of nonlinear limit must be imposed either naturally or by design to prevent damage.

The summation employed in a feedback control system sometimes is carried out mechanically or with a pneumatic/mechanical system, but more often it is carried out electrically. The same is true of the proportional controller action discussed above, and the other types of controller actions discussed below. The various actions can be carried out with continuous electrical signals using circuits based on operational amplifiers; see for example Ogata.[2] Some can be carried out with passive electrical circuits. Alternatively, the use of analog-to-digital and digital-to-analog converters allows the use of digital controllers.

EXAMPLE 8.1

A plant with the transfer function $G(S) = 1/(S^2 + 2S - 3)$ is unstable by itself, since its homogeneous solution or unforced response to non-zero initial conditions includes a term ae^t that grows exponentially. Place this plant in a unity-feedback loop with an adjustable proportional gain k, and find the transfer function of the overall system. Also, find the value of k above which this sytem is stable. Then, plot the root locus for the system as k varies from 0 to ∞, using the example given in Fig. 8.4 as a guide. Finally, choose a value of k that gives responses that are relatively fast yet not excessively oscillatory, and sketch-plot the response of the system with this value to a unit step excitation.

Solution: The closed-loop transfer function $G_c(S)$ is, from equation (8.6),

$$G_c(S) = \frac{G(S)}{1 + G(S)} = \frac{k}{S^2 + 2S + (k - 3)}.$$

One root is positive for $k < 3$, causing instability, but both roots have negative real parts for all $k > 3$. The open-loop poles are at the zeros of $S^2 + 2S - 3 = (S+3)(S-1)$, namely $S = \pm 1, -3$:

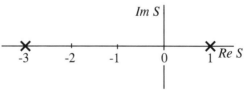

This is the same pattern as the open-loop poles in Fig. 8.4, except they are shifted to the right, moving one of them into the right-half plane. The root locus therefore has the same pattern:

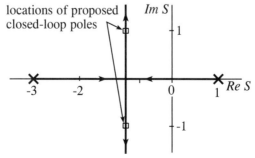

---

[2] K. Ogata, *Modern Control Engineering*, Third Edition, Prentice Hall, 1997, especially p. 270.

When $k = 4$, both roots equal -1, which is close to the best response but slower than what can be achieved with a larger value. As k becomes very large, its behavior becomes increasingly oscillatory. A good compromise would be $k = 5$, for which the roots are $S + -1 \pm j$. This gives

$$G_c(S) = \frac{5}{S^2 + 2S + 2},$$

which dictates the same response to a unit step as the example in Fig. 8.4 if $-1/2\tau$ is interpreted as -1 s^{-1}, that is $\tau = 0.5$ seconds, and if the steady-state is replaced by $5/2$:

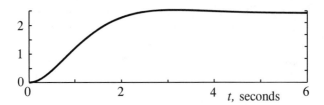

8.1.4 Response to Disturbances

Unpredictable disturbances in the form of torques on the load may produce undesired errors. These disturbances are labeled as M_d in the bond graph and the block diagram of Fig. 8.5. If they are considered to be the input to the system, rather than ϕ_r, the system can be viewed as having a forward path transfer function or gain (from the summation point to the output ϕ) of $G(S) = G_2(S)$, and a feedback gain (from the output to the summation with the input) of $H(S) = k$. Such non-unity feedback loops occur frequently and are generalized in part (c) of the figure. Equation (8.5) for the simpler unity-feedback system

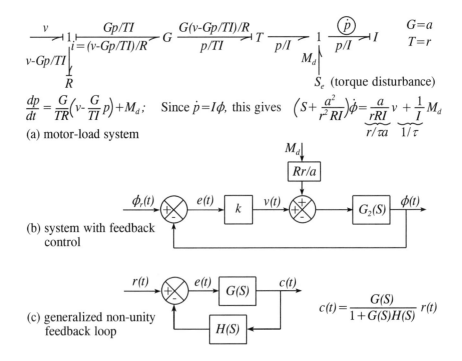

Figure 8.5: Position control system with disturbances

is modified to become

$$c(t) = G(S)[r(t) - H(S)c(t)], \tag{8.7}$$

which when solved for $r(t)$ gives the general result

$$\boxed{c(t) = \frac{\text{forward path gain}}{1 + \text{loop gain}} = \frac{G(S)}{1 + G(S)H(S)}r(t).} \tag{8.8}$$

For the particular system under consideration, this gives

$$\phi(t) = \frac{rR/ak}{1 + (a/rk)S + (\tau a/rk)S^2}M_D(t). \tag{8.9}$$

The poles of this transfer function, and therefore the dynamics of the behavior, are identical to those for the response to ϕ_r. On the other hand, the output $\phi(t)$ of equation (8.9) can be considered to be an error, and increasing k decreases this error. If the error is critical to the success of the system, you could justify using a larger value of k than that chosen previously, despite the reduction in damping.

8.1.5 Root Locus Basics

Control engineers tend to become experts at sketching root locus plots, following several detailed rules that are given in virtually any first text on automatic control. The discussion below, in contrast, gives only the basics. Instead, emphasis is placed on the use of the MATLAB command `rlocus`, which does the detailed work for you.

The root locus is a plot of the values of S which satisfy

$$1 + \frac{k\,\text{num}(S)}{\text{den}(S)} = 0, \tag{8.10}$$

for values of k ranging from zero to infinity. For the general single-input single-output feedback system with a proportional controller having gain k, comparison to the denominator of equation (8.8) gives

$$\frac{k\,\text{num}(S)}{\text{den}(S)} = G(S)H(S), \tag{8.11a}$$

$$\text{num}(S) = (S - z_1)(S - z_2)\cdots(S - z_m), \tag{8.11b}$$

$$\text{den}(S) = (S - p_1)(S - p_2)\cdots(S - z_n), \tag{8.11c}$$

Other controllers are considered later.

The locus as a whole therefore is defined by the phase condition

$$\angle[\text{num}(S)/\text{den}(S)] = (1 \pm 2n)180°. \qquad n = 1, 2, \cdots \tag{8.12}$$

Points on the locus corresponding to a particular value of k have the additional magnitude restriction

$$\left|\frac{\text{num}(S)}{\text{den}(S)}\right| = \frac{1}{k}. \tag{8.13}$$

EXAMPLE 8.2

Show that both the horizontal and vertical segments of the root locus of the postion control example of Figs. 8.3 and 8.4 satisfy the defining relation of equation (8.12). Then, use equation (8.13) to find the values of k corresponding to the points $S_1 - 1/4\tau$ and $S_2 = (-1+j)/2\tau$ on the locus.

Solution The segment of the real axis between the two open-loop poles is part of the root locus, since num$(S) = 1$ and, as shown on the left below for any point thereon, den(S_1) equals the product of the vector $S_1 - p_2$ (a positive real number) and the vector $S_1 - p_1$ (a negative real number) and therefore satisfies equation (8.13). The perpendicular bisector of this line segment also is on the locus, since for any of those values of S_2 the sum of the angle of the vector $S_2 - p_1$ and the angle of the vector $S_2 - p_2$ equals either $180°$ or $-180°$.

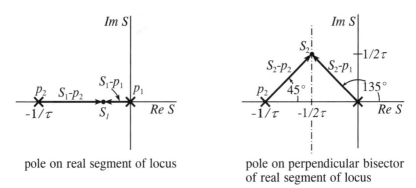

pole on real segment of locus   pole on perpendicular bisector of real segment of locus

The value of k in equations (8.10)-(8.13) equals the product of the magnitudes of the two vectors. For the particular point $S_1 = -1/4\tau$, the resulting value of k is $3/16\tau^2$; for $S_2 = (-1+j)/2\tau$ it is $1/2\tau^2$. This k incorporates the factor $r/\tau a$ included in $G_1(S)$ and $G_2(S)$; the respective values of the k of equations (7.10) and before are $3a/16r\tau$ and $a/2r\tau$.

Certain general rules are especially helpful in anticipating the migrations of the closed-loop roots as k progresses from zero to infinity. The locus for a closed-loop pole emerges from each open-loop pole as k grows from zero. One of the loci is terminated, as $k \to \infty$, at each finite open-loop zero. Let the number m represent the number of open-loop poles minus the number of open-loop zeros; m normally is assumed to be zero or positive. The m loci which do not terminate at a finite zero terminate at infinity. Specifically, they become asymptotic to radial lines focused at a particular point on the real axis, and they are arrayed uniformly about the full $360°$ of arc and symmetrically with respect to the real axis. If m is odd, one asymptote lies on the negative real axis, and, if it is even, none lie on the real axis. For cases with $m \geq 2$, the centroid of all closed-loop poles remains fixed, independent of k. The focus of the asymptotes is at the centroid of all open-loop poles and zeros, with zeros treated as negative poles.

The position-control system has two poles and no finite zeros, so $m = 2$. Examples of root loci with $m = 0, 1, 3$ and 4 are offered in Fig. 8.6.

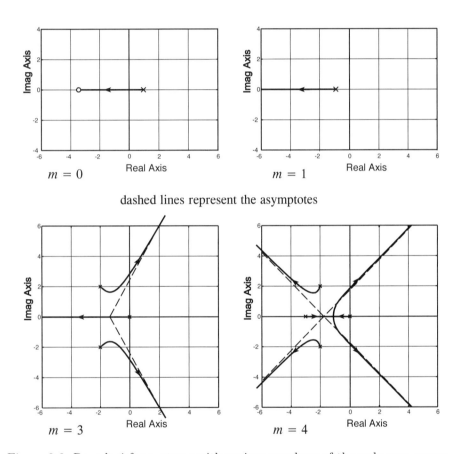

Figure 8.6: Root loci for systems with various numbers of the pole-zero excess

8.1.6   Use of MATLAB®

The command `rlocus` is included in the student version of MATLAB and resides in the Controls Toolbox of the professional version. It has the following forms:

```
rlocus(num,den)
rlocus(num,den,k)
rlocus(A,B,C,D)
rlocus(A,B,C,D,k)
[r,k]=rlocus(num,den)
[r,k]=rlocus(num,den,k)
[r,k]=rlocus(A,B,C,D)
[r,k]=rlocus(A,B,C,D,k)
```

When invoked, the vector k is user defined; otherwise, a default vector is used. When the left-hand arguments are not specified, the root loci are plotted directly; otherwise, the values of the complex closed-loop poles and the associated values of gain are stored in the vectors r and k, and are displayed in response to typing r or k. The command

```
plot(r,' ')
```

plots the root loci. The alternative forms

```
plot(r,'o')      or      plot(r,'x')
```

marks each pole with a small circle or cross. In regions where the pole locations change slowly with increasing k, these markings will be densely spaced; where the pole locations change rapidly, they will be sparsely spaced.

> **EXAMPLE 8.3**
>
> Plot the root loci for the transfer function $G(S) = (S + 2)/(S^2 + 4)$, using MATLAB. Make sure that the entire region of interest is included. Then, mark the locations of the closed loop poles for several values of the gain, k.
>
> **Solution:** The MATLAB commands
>
> ```
> >> num=[1 2];
> >> den=[1 0 4];
> >> rlocus(num,den)
> ```
>
> produce the plot

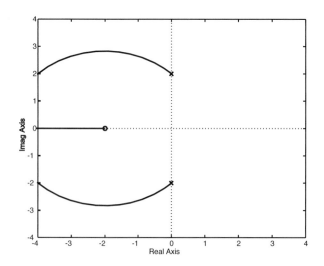

The axes happen not to include the entire region of interest. This problem can be remedied by the `axis` command, which allows the extent of the axes to be specified. Then, specific closed-loop poles for specific values of the gain k can be added to the plot through the use of the command `r=rlocus(num,den,k)`. The coding below

```
>> v=[-8 2 -4 4];
>> axis(v)
>> hold
   Current plot held
>> grid
>> rlocus(num,den)
>> k=[2:2:12];
>> rlocus(num,den,k);
>> plot(r,'o')
>> title('Root-locus plot of k(S+2)/(S\^{}2+4)')p
```

produces the result

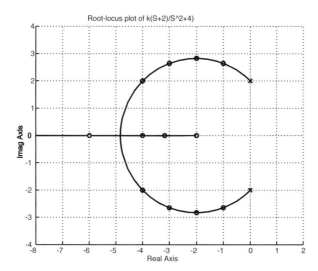

Perhaps the best behavior is that associated with the poles for $k = 8$, namely $r = -4 \pm 2j$.

EXAMPLE 8.4

Repeat the above example for the case $G(S) = S/(S^3 + 3S^2 + S + 3)$.

Solution: The codings on the left produce the plots on the right below.

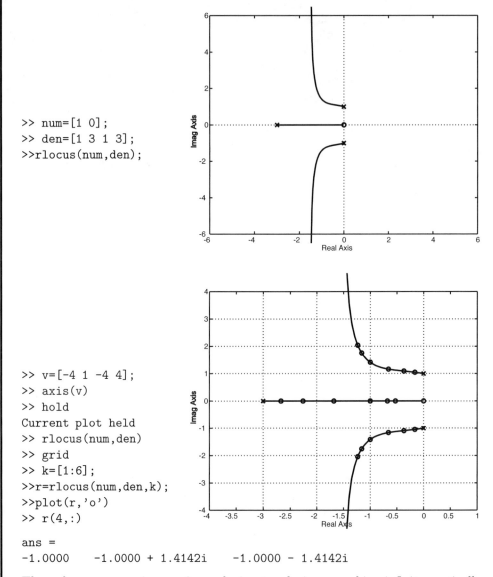

```
>> num=[1 0];
>> den=[1 3 1 3];
>>rlocus(num,den);
```

```
>> v=[-4 1 -4 4];
>> axis(v)
>> hold
Current plot held
>> rlocus(num,den)
>> grid
>> k=[1:6];
>>r=rlocus(num,den,k);
>>plot(r,'o')
>> r(4,:)
```

```
ans =
-1.0000    -1.0000 + 1.4142i    -1.0000 - 1.4142i
```

The pole-zero excess is $m = 2$, producing two loci approaching infinity vertically with the asymptote passing through the centroidal location $S = -1.5$. (The zero is treated as a negative entity in the calculation of this centroid.) The explicit closed-loop locations pinpointed in the second plot suggest that $k = 4$, $r = -1$, $-1 \pm 1.414j$ would be a good choice.

EXAMPLE 8.5

Repeat the above example for the case $G(S) = (S + 0.4)/(S^4 + 6S^3 + 8S^2)$.

Solution: The codings on the left below produce the plots on the right.

```
>> num=[1 .4];
>> den=[1 6 8 0 0];
>>rlocus(num,den)
```

```
>> v=[-5 1 -3 3];
>> axis(v)
>> hold
Current plot held
>> grid
>> rlocus(num, den)
>> k=[2:2:12];
>> r=rlocus(num,den,k);
>> plot(r,'o')
>> r=rlocus(num,den,7)
r = -4.5502      -0.3746 + 0.8590i
                 -0.3746 - 0.8590i      -0.7006
```

This example has a double pole at the origin, from which two root loci emerge vertically. The pole-zero excess is $m = 3$, so three loci extend to infinity at uniformly separated angles. The zero near the origin acts as an attractor to the complex-conjugate loci. The larger-scale drawing with explicit pole locations shown for $k = 2$, 4, 6, 8, 10 and 12 suggests that the optimal value is about $k = 7$, giving $r = -5.55$, -0.701, $-.375 \pm 0.859i$.

8.1.7 Criteria for Stability

Feedback control can make an unstable plant stable, and can make a stable plant unstable. A root locus clearly reveals whether or not all the poles of the closed-loop transfer function are in the left-half plane, as they must be for stability. Alternatively, with less effort, you can determine qualitative stability by examining the characteristic equation directly. Use of the `root` command of MATLAB will determine the poles readily, as noted before, if you have numerical values for the coefficients of the characteristic equation. If on the other hand one or more of the coefficients is to be chosen, or if you don't have a computer handy, use of the **Routh-Hurwitz algorithm** will determine the number of poles in each half-plane. This algorithm is given in virtually any textbook that goes beyond an introduction to automatic control. Rather than reproduce the full algorithm here, certain special results are given.

The characteristic equation for linear lumped models can be written in the form

$$a_n S^n + a_{n-1} S^{n-1} + \cdots + a_1 S + a_0 = 0. \qquad (8.14)$$

A necessary condition for stability is that all coefficients $a_n, a_{n-1}, \cdots, a_0$ *have the same sign.* This condition suffices for the first order system, in which it is tantamout to requiring the time constant to be positive, and for the second order system, in which it requires either the two time constants be positive or the natural frequency be real and the damping ratio be positive. For higher-order systems, however, the condition is not *sufficient* for stability.

Stability is assured for a third-order system if and only if an additional condition is satisfied:

$$\boxed{a_2 a_1 - a_3 a_0 \geq 0.} \qquad (8.15)$$

Stability is assured for a fourth-order system if and only if two additional conditions are satisfied:

$$\boxed{a_3 a_2 - a_4 a_1 \geq 0; \qquad a_3 a_2 a_1 - a_4 a_1^2 - a_3^2 a_0 \geq 0.} \qquad (8.16)$$

EXAMPLE 8.6

A fourth-order system satisfies the characteristic equation

$$S^4 + 5S^3 + 10S^2 + (k+5)S + k = 0,$$

in which k is to be chosen. Find the range of k for which the system is stable.

Solution: The requirement that the fifth coefficient have the same sign as the others demands $k \geq 0$. This automatically satisfies the requirement that the fourth coefficient also be non-negative. The first of equations (8.16) gives the additional requirement $5 \times 10 - 1 \times (k+5) \geq 0$; combining these requirements, $0 \leq k \leq 45$. The second of equations (8.16) gives the further requirement $5 \times 10(k+5) - 1(k+5)^2 - 5^2 k \geq 0$, which gives $k \leq 24.27$. The complete result therefore is $0 \leq k \leq 24.27$.

8.1.8   Summary

Feedback is the key concept of automatic control. In a simple system, it is desired that some output variable match a reference signal, which either is a fixed set point or a dynamically varied input. The difference, or error, between the desired and the measured and "fed back" output variable is used as the basis for actuation. This error becomes the input to a controller. The output of the controller commands the actuator to move, generally in the direction of reduced error.

Outside disturbances can buffet a system, causing undesired changes in its output. The presence of a feedback loop acts to minimize these changes. Control systems without feedback can be designed but cannot effect these corrections (unless they directly sense the disturbances and act on this information). The controller of such an open-loop control

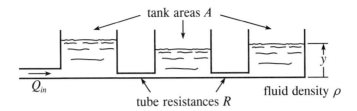

Figure 8.7: Guided Problem 8.1

system also would be ineffective unless it is based on a precise model of the system being controlled. By contrast, a feedback system is comparatively forgiving, since the error itself is independent of the model upon which the controller is designed.

The location of the poles of the transfer function of a complete or closed-loop linear system is the most critical indicator of the quality of its behavior. The zeros influence the relative contributions of the various poles, but not their essential time constants, natural frequencies and damping ratios which depend on the poles. The pole locations depend on the constants chosen for the controller, as well as the parameters of the system being controlled. In deciding how large to make a particular controller constant, it is helpful to plot the loci of these poles as the constant is varied from zero to infinity. A few of the more basic rules for constructing these root loci have been given, but principal reliance has been placed on use of the MATLAB command `rlocus`. A few constraints on the coefficients of the characteristic equation also have been given.

The assumption of linearity leads to many relatively simple analytical methods. including those discussed above. The control engineer must keep his eye out for gross errors that this assumption can produce. A common error occurs when the controller commands the actuator to act faster than its capability.

The example of a DC motor acting through a pair of gears on an inertive load (such as a flywheel) has been used to illustrate the ideas above.

Guided Problem 8.1

This first guided problem gives basic practice with proportional feedback control, using the MATLAB command `rlocus` as the major tool.

Three fluid gravity tanks, each of area $A = 0.1$ m^2, are interconnected by two flow restrictions, each with linear resistance $R = 9.81 \times 10^6$ N s/m^5, as pictured in Fig. 8.7. The object is to control the level in the third tank by controlling the flow Q_{in} into or out from the first tank. The relation between this flow and the level of the third tank is readily deduced to be

$$y = \frac{a^2/A}{S(S+a)(S+3a)} Q_{in}; \qquad a = \frac{\rho g}{RA} = 0.01 \text{ s.}$$

You are asked to investigate the possibility of simple proportional feedback control, based on measurement of the level in the third tank. The value of a reasonable gain, k, is desired.

Suggested Steps:

1. Draw a block diagram of the system, including the feedback. Label all variables, and pay particular attention to their signs. Determine the units of the gain k (which is not a pure number).

2. Sketch-plot the three poles of the open-loop system on S-plane coordinates. Since there are three poles and no finite zeros, the pole-zero excess is $m = 3$. Note what this says about the asymptotes of the closed-loop root loci as $k \to \infty$.

3. Use MATLAB to plot the loci of the closed-loop poles. Note that too large a value of k produces instability, whereas too small a value produces an unnecessarily sluggish response.

4. The response, y, to changes in the desired level, y_r, is faster when two of the poles are complex-conjugate, but the damping ratio should not be so small that excessive oscillations result. A good compromise is $\zeta = 0.5$, which gives an angle between the real axis and a radial line drawn from the origin to one of the poles equal to $\cos^{-1}(0.5) = 60°$, assuming the vertical and horizontal axes have the same scale. Enter an appropriate vector of gains k into the MATLAB command `rlocus`, and estimate the desired value of k and the corresponding closed-loop transfer function.

5. Use the MATLAB command `step` to demonstate the step response of the feedback system.

PROBLEMS

8.1 A closed-loop system has the characteristic equation $S^3 + 3S^2 + kS + 6 = 0$. Determine by the simplest method available the range of k for which the system is stable.

8.2 The angular velocity of the motor-load system considered in this section is to be controlled, rather than the angular position. Thus, the role previously taken by $G_2(S)$ is replaced by $G_1(S)$. The parameters are $a = 0.006$ V s, $\tau = 1.0$ s and $r = 0.2$, and the maximum motor voltage is $v_m = 3.0$ V.

(a) Determine the optimal open-loop control signal which produces an angular velocity of 2000 degrees per second, starting from rest, in as short a time as possible.

(b) Draw a block diagram for linear proportional feedback control. Label the variables.

(c) Draw the root locus for the closed-loop pole(s). What does this plot say about the choice of the gain k?

(d) Find the closed-loop transfer function for the motor voltage, $v(t)$. Determine the maximum value of k for which this voltage does not exceed the stated limit for the command signal of part (a).

(e) Find the closed-loop transfer function for the angular velocity. Plot the step response for the value of k determined in part (e).

(f) Find the closed-loop error between the command and response angular velocities for the case considered in parts (d) and (e). Do you expect this error is acceptable, or should design changes be sought? If the largest command inputs were smaller, could the value of k be acceptably changed to give a reduced error?

8.3 A proportional controller and unity feedback is applied to the system with transfer function $G(S) = S/(S^2 + 1)$.

(a) Determine whether the open-loop system is stable.

(b) Plot the root loci for the controller gain k increasing from zero to infinity.

(c) Determine the range of k for which the closed-loop system is stable.

(d) Choose a value of k which gives a good combination of speed of response and relative stability. Determine the corresponding closed-loop transfer function.

(e) Determine from the transfer function whether the response of the system to a step input has a non-zero final value. If it does, report the steady-state error between that value and the unity input value, if any.

(f) Plot the step response of the closed-loop system, using your value of k.

8.4 Answer the questions of Problem 8.2 for $G(S) = 1/S(S+1)^2$.

8.5 Answer the questions of Problem 8.2 for $G(S) = (S+1)/S^2(S+4)$.

8.6 Answer the questions of Problem 8.2 for $G(S) = (S^2+1)/(S+4)(S-1)$.

8.7 Answer the questions of Problem 8.2 for $G(S) = (S+0.2)/S(S-1)(S+4)^2$.

SOLUTION TO GUIDED PROBLEM

Guided Problem 8.1

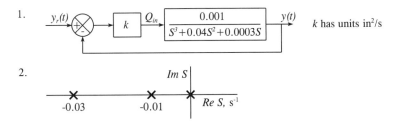

1. $y_r(t)$ — k — Q_{in} — $\dfrac{0.001}{S^3+0.04S^2+0.0003S}$ — $y(t)$ k has units in^2/s

2. *Im S* / *Re S*, s^{-1} ; poles at -0.03, -0.01

Since $m = 3$, one asymptote is the negative real axis, and the other two have angles of $\pm 60°$ relative to the positive real axis. (When the scales of the real and imaginary axes are not the same, as below, the slopes of the asymptotes remain at $\sqrt{3}$.) The intersection of these three asymptotes is at the centroid of the three open loop poles, or $-4a/3 = -0.0133$ rad/s.

3. The MATLAB coding is

```
num = 0.001;
den = [1 0.04 0.0003 0];
rlocus(num,den)
```

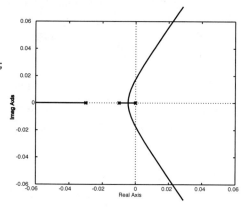

4. The following MATLAB commands give results for 13 values of k:

```
k = [0:.25:3]*1e-3;
[r,k] = rlocus(num,den,k);
plot(r,'x')
r
```

Lines drawn at the slope $\pm\sqrt{3}$ through the origin show that the value $k = 1.5 \times 10^{-3}$ m^2/s gives virtually the desired $\zeta = 0.5$. The associated values of the closed-loop roots are $r_1 = -0.0325$ s^{-1}, $r_{2,3} = -0.00377 \pm 0.00642j$ s^{-1}. Therefore, the closed-loop transfer function is

$$\frac{G(S)}{1 + G(S)} = \frac{1.8 \times 10^{-6}}{S^3 + 0.04S^2 + 0.003S + 1.8 \times 10^{-6}}.$$

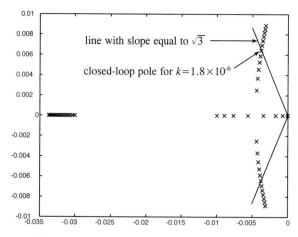

5. The MATLAB commands

```
n = 1.8*1e-6;
d = [1 .04 .003 1.8*1e-6];
step(n,d)
```

give the step response shown here:

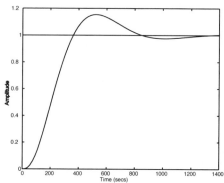

8.2 Dynamic Compensation

The proportional controller employed above is but the simplest in a pantheon of possibilities. The most common family of controllers in use today directly sums integral or derivative actions (or both) to the proportional control. Phase lead and phase lag compensators also are simple and popular. These examples of compensators are introduced in this section and applied to the motor-driven position control system of the previous section, along with other examples, to illustrate the richness of the possibilities for improving over proportional control.

8.2.1 Proportional-Plus-Integral Control

The steady-state error in the output angle of the motor-driven position control system caused by a constant disturbance torque (such as produced by friction) has been investigated in Section 8.1.4. This error can be eliminated by integrating it over time, which can be accomplished using an operational amplifier. Should the error persist indefinitely, its integral would grow toward infinity. This integral is then multiplied by a constant and summed with the proportional action of the controller to produce the control signal which drives the system toward zero error. The transfer function relation of the PI controller becomes

$$u(t) = k(1 + 1/T_iS)e(t), \qquad (8.17)$$

in which T_i is known as the **integral time**. A block diagram for the system is given in part (a) of Fig. 8.8. The overall transfer function from the moment disturbance to the angle of the output becomes

$$\phi(t) = \frac{(rR/a)G_2(S)}{1 + k(1 + 1/T_iS)G_2(S)}M_d(t)$$
$$= \frac{(rRT_i/ka)S}{1 + T_iS + (aT_i/rk)S^2 + (\tau aT_i/rk)S^3}M_d(t). \qquad (8.18)$$

The derivative operator S in the numerator operates directly on the disturbance moment $M_d(t)$. If the moment is constant, it produces a zero value of the steady-state response of $\phi(t)$. Otherwise, the response is proportional to the numerator coefficient rRT_i/ka; increasing k and decreasing T_i reduces its magnitude. Unfortunately, increasing k and decreasing T_i also reduces the relative stability of the system and can even produce instability. A compromise is necessary.

Let the particular value of k chosen earlier ($k = a/2r\tau$) be assumed. It is helpful to plot the locus of the poles of the closed loop system as the inverse time $1/T_i$ is increased from zero toward infinity. Such a root locus can be plotted if the characteristic equation (the denominator of the closed-loop transfer function) is placed in the form of equation (8.10), as always; the k of that equation is now τ/T_i. The numerator coefficient τ makes this ratio dimensionless; the product τS also becomes a dimensionless derivative operator. The characteristic equation can indeed be placed in this form by dividing the denominator of the transfer function in equation (8.18) by the operator $S[1 + (a/rk)S + (\tau a/rk)S^2]$:

$$1 + \frac{1/T_i}{S[1 + (a/rk)S + (\tau a/rk)S^2]} = 1 + \frac{\tau/T_i}{2(\tau S)^3 + 2(\tau S)^2 + \tau S} = 0. \qquad (8.19)$$

The "open-loop" poles of this equation include one pole at the origin plus precisely the same two complex-conjugate poles of the closed-loop system with proportional control only. They are plotted in part (b) of Fig. 8.8. The resulting root locus shows the pole at the origin migrating to the left, which is desirable. It also shows the complex pair of poles

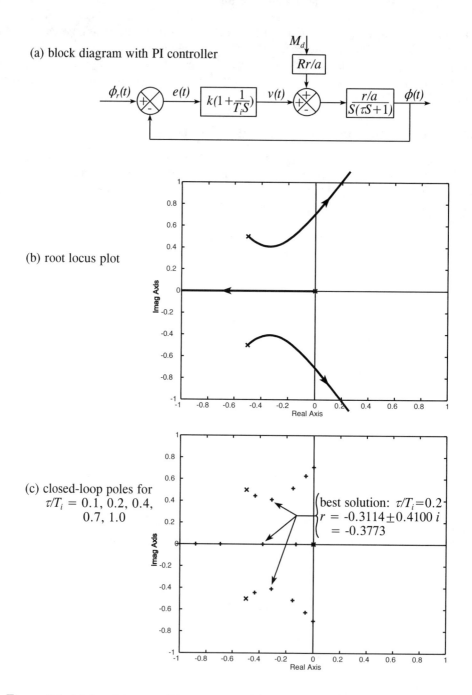

Figure 8.8: Motor-driven position system with proportional-plus-integral control

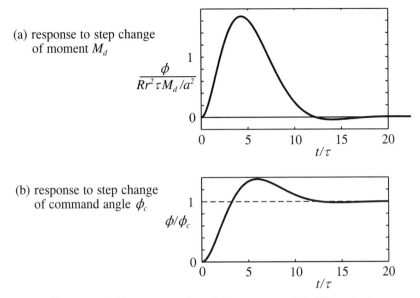

(a) response to step change of moment M_d

$$\frac{\phi}{Rr^2\tau M_d/a^2}$$

(b) response to step change of command angle ϕ_c

ϕ/ϕ_c

Figure 8.9: Responses of position system with PI control

moving toward the right, which reduces the damping. The sum of the three closed-loop poles remains constant (remember the rule for cases in which the pole-zero excess is 2 or more), so the complex poles cross the imaginary axis, and the system becomes unstable when the real pole has moved to the location $\tau S = -1$. The closed-loop poles are replotted in part (c) of the figure for five specific values of the gain τ/T_i: 0.1, 0.2, 0.4, 0.7 and 1.0. The MATLAB coding that produces this plot and reports the values of the respective roots is as follows:

```
num = 1;
den = [2 2 1 0];
k = [.1 .2 .4 .7 1];
[r,k] = rlocus(n,d,k)
rlocus (n,d,k)
```

Thus, $\tau/T_i = 1.0$ (or $T_i = \tau$) is the case of incipient instability. The best overall behavior results with $\tau/T_i \simeq 0.2$ (or $T_i \simeq 5\tau$). The associated responses of the system to a step disturbance of moment and to a step command are plotted in Fig. 8.9.

These responses are probably quite satisfactory, although they are distinctly inferior to optimal responses, largely because they do not utilize much of the available torque of the motor. Choosing a larger value of k would increase the speed of response, but at a probably unacceptable cost of decreased relative stability and potential for insensitivity to moment disturbances. A much smaller value of k would make the response more sluggish. If some means of increasing the damping of the system were introduced, on the other hand, the values of k and $1/T_i$ could be increased, giving significantly faster responses without sacrificing stability.

EXAMPLE 8.7

A first-order thermal plant satisfies the differential equation

$$\tau\frac{d\theta}{dt} + \theta = \frac{1}{H}\dot{Q}_{in} + \theta_e; \qquad \tau = \frac{mc}{H},$$

where θ is the temperature of the plant, θ_e is the temperature of the environment, $\dot{Q}_{in}$ is the heat input used to control θ, H is the heat-transfer coefficient between the plant and the environment, m is the mass of the plant and c is its specific heat. The plant is placed in a feedback loop that includes a measurement system with a simple lag and a proportional-plus-integral controller, as shown below:

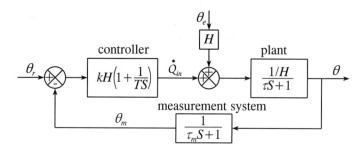

Consider the case in which the measurement lag time τ_m equals τ. Find the transfer functions between the reference input θ_r and θ, and between θ_e and θ. For the case of simple proportional control, *i.e.*, $T \to \infty$, find the smallest possible steady-state errors without the damping ratio of the system being less than 0.5, and the associated value of k. Then reintroduce the integral control, show that it eliminates the steady-state errors, and find acceptable values of k and T.

Solution: The transfer functions can be found using the basic equations (8.5) and (8.8):

$$\frac{\theta}{\theta_r} = \frac{k(TS+1)/(\tau S+1)TS}{(1 + k(TS+1)/(\tau S+1)TS(\tau_m S+1) + k(TS+1)}$$

$$= \frac{k(TS+1)(\tau S+1)}{(\tau S+1)TS(\tau_m S+1) + k(TS+1)},$$

$$\frac{\theta}{\theta_e} = \frac{1/(\tau S+1)}{1 + k(TS+1)/(\tau S+1)TS(\tau_m S+1)}$$

$$= \frac{TS(\tau S+1)}{(\tau S+1)TS(\tau_m S+1) + k(TS+1)}.$$

When $T \to \infty$ and $S = 0$, the steady-state responses for simple proportional control result:

$$\left(\frac{\theta}{\theta_r}\right)_{steady-state} = \frac{k}{1+k} = 1 - \frac{1}{1+k}; \qquad \left(\frac{\theta}{\theta_e}\right)_{steady-state} = \frac{1}{1+k}.$$

(These result by dividing the numerators and denominators by T before substituting $T = 0$.) Both error ratios are the same and are reduced by making k as large as possible. With $\tau_m = \tau$, the characteristic equation becomes $\tau^2 S^2 + 2\tau S + (1+k) = 0$, which describes a second-order system with a natural frequency $\omega_n = \sqrt{1+k}/\tau$ and damping ratio of $\zeta = 1/\sqrt{1+k}$. For $\zeta \geq 0.5$ as required, $k \leq 3$. For $k = 3$, the steady-state error ratios are 0.25, which likely aren't very satisfactory, and $\omega_n = 2/\tau$.

With the integral control reintroduced, both steady-state errors become zero regardless of the values of k and T. The characteristic equation is third order; it becomes $(\tau S)^3 + 2(\tau S)^2 + (1 + k)(\tau S) + k\tau/T = 0$. This can be factored as $(\tau S + \tau p)[\tau S)^2 + 2\zeta\omega_n(\tau S) + (\tau\omega_n)^2] = 0$. A large value of ω_n requires a small value of p; to maximize the distance of all poles from the origin, set $p = \omega_n$. Keeping $\zeta = 0.5$, this gives $(\tau S)^3 + 2\tau\omega_n(\tau S)^2 + 2(\tau\omega_n)^2(\tau S) + (\tau\omega_n)^3 = 0$. Comparing term by term, $\tau\omega_n = 1$, $k = 1$ and $T = \tau$. Thus, the elimination of the steady-state error without reducing the effective damping comes at a cost of reducing the natural frequency by a factor of two.

8.2.2 Proportional-Plus-Derivative Control

The addition of derivative action to the controller is one of the oldest, simplest and most effective means of increasing the damping of a feedback control system. This is considered first in the absence of integral action. The transfer function relation of the PD controller is

$$u(t) = k(1 + T_d S)\, e(t), \tag{8.20}$$

in which k is the coefficient for the proportional action of the controller, as before, and T_d is the **derivative time**. The term $kT_dS\, e(t) \equiv kT_d\dot{e}(t)$ is proportional to the instantaneous rate of change of $e(t)$. (When implemented by an analog integrator network, the transfer function of the derivative term is actually $T_dS/(\tau_d S + 1)$, where τ_d is set so small that its effect can be neglected.) The transfer function equals zero for $S = -1/T_d$; this zero is represented in the S-plane of the overall forward-path transfer function $G(S)$ by a symbol $\circ$ drawn at this location on the negative real axis.

Application of the PD controller to the motor-driven position control system gives the forward-path transfer function

$$G(S) = \frac{kr(1 + T_dS)/a}{S(1 + \tau S)}, \tag{8.21}$$

and the closed-loop transfer function relation

$$\phi(t) = \frac{G(S)}{1 + G(S)} = \frac{1 + T_dS}{1 + (T_d + a/rk)S + (a\tau/rk)S^2}\phi_r(t). \tag{8.22}$$

The introduction of the derivative control action has left the order of the system unchanged at two. The natural frequency remains at $\sqrt{k/a\tau}$. A positive value of T_d is seen to increase the damping ratio by increasing the middle coefficient of the denominator quadratic. This equation shows that since the designer can freely choose the values of both k and T_d, he has *carte blanche* to achieve any natural frequency and damping ratio he or she wishes. **Design by pole placement** is frequently practiced by control engineers. So why not achieve an almost instantaneous response by setting k at an enormous value?

Think for a moment. Why should this behavior be better than that of the so-called optimal control presented in Section 8.1.2?

The answer is that a very large value of k would produce an impractically large value of the armature voltage for any but the smallest command step. The optimal system recognized a practical limit to this voltage, presumably imposed by the regulated voltage of the electrical power supply.[3] The optimal system in fact responds exceedingly fast for very small command steps.

---

[3]Damage to the motor alternatively could be prevented by use of a power supply and amplifier that limits the armature *current* rather than its voltage. The result would be somewhat better performance, at a higher cost.

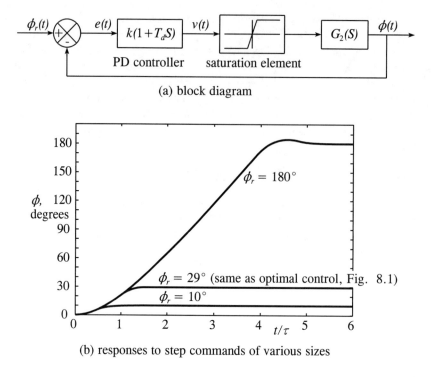

(a) block diagram

(b) responses to step commands of various sizes

Figure 8.10: Position control system with a PD controller and saturation

What if the output voltage of the linear PD controller were similarly limited? Could an extremely large value of k then indeed be chosen, along with an appropriate time T_d? This scheme, shown in Fig. 8.10, features a nonlinear **saturation element.** Responses are plotted for the same step as before (29.2°), a smaller step (10°) and a much larger step (180°). These were found by nonlinear simulation. The value of k used (1000) is so large that either the positive or the negative saturation limit is imposed almost all the time, giving almost the so-called **bang-bang** or switching control characteristic of the optimal scheme. The derivative time T_d is chosen at $T_d = 0.226\tau$, which gives barely zero overshoot for the 29.2° step. The response to this step is virtually indistinguishable from that of the optimal control as plotted in part (d) of Fig. 8.2. Larger steps produce a modest transient overshoot of the final position, whereas smaller steps produce a tail in which the response creeps up to its final value more slowly than an optimal control.

Raising T_d to $\tau \log(2) = 0.693\tau$ (or higher) prevents overshoot for all step responses, but also increases the overdamped tails for small steps. The responses are not critically dependent on the value of T_d, however. Note that the time required to approach the final angle increases markedly as the step size increases. This nonlinear behavior has the distinct advantage over linear behavior of taking advantage of the full available motor torque, no matter how small the command step.

EXAMPLE 8.8

A controllable force, F, acts on a mass $m = 1$ kg of a simple mass-spring system with spring rate $K = 4$ N/m so as to control its position, x:

$$(mS^2 + K)x = F.$$

The position x is monitored by a transducer, the output signal of which is compared to a command input, x_r, to produce an error signal. This signal acts on a controller that produces the force. Proportional control gives imaginary roots, a virtually uncontrollable condition. A PD controller is proposed:

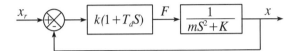

Select values of T and k, and discuss the resulting behavior.

Solution: The transfer function of the forward path is

$$\frac{x}{F} = \frac{k(TS + 1)}{S^2 + 4} = \frac{kT(S + 1/T)}{S^2 + 4},$$

and the closed-loop transfer function is

$$\frac{x}{x_r} = \frac{kT(S + 1/T)}{S^2 + kTS + (k + 4)},$$

which has a steady-state value of $k/(k + 4)$. Thus, a large value of k is desired.

A root locus for the case of $T = 0.5$ seconds is given in Example 8.3 (pp 530-531). The value of kT suggested there is 8, or $k = 16$, giving a steady-state error of 20%. The dynamic behavior is attractive: the transfer function is $8(S + 2)/(S^2 + 8S + 20)$, which has the fast and well-damped roots $-4 \pm 2j$ s^{-1}. This solution would be satisfactory if compensation can be provided for the steady-state error. Otherwise, the steady-state error can be reduced and the response made faster by increasing k. The value of T is reduced to keep the response from becoming over-damped. For example, $k = 96$ and $T = 1/8$ second produces a 4% steady-state error with a natural frequency of 10 rad/s. There is no limit to the apparent improvement available in the response. There must be a catch, and it is this: the larger k, the more powerful and expensive the controller must be. Further, as with the motor-driven position control system, a rapid change in x_r will saturate any real controller; the linear model then gives a false prediction of the response. The solution with $T = 0.5$ and $k = 16$ may be acceptable, however.

8.2.3   Proportional-Plus-Integral-Plus-Derivative Control

The elimination of the steady-state error produced by integral action is unaffected by the presence of derivative action. The combination of proportional, integral and derivative control actions is called PID control. Adjustable PID controllers are available commercially, and represent by far the single largest type of dynamic compensation in use today. They are particularly prevalent in the field of process control, which largely refers to chemical processes, where they are called **process controllers** or **three-mode controllers**.

The presence of the derivative action directly imparts an impulse to the output of the controller whenever the command signal is given a step change. This impulse inevitably saturates the plant being controlled, resulting in nonlinear action. This saturation was utilized in the previous section on the PD controller to give nearly optimal responses to input steps. Nevertheless, most users of PID control do not consider nonlinearities. Applications usually are restricted to systems for which the simple superposable behavior of linearity is favored over the quickness of response of the near-optimal nonlinearity. In practice, steps of enough amplitude to cause marked saturation are rarely encountered anyway; the consideration of step inputs becomes more of a conceptual tool than a physical reality. Certain rules-of-thumb are given in textbooks on control for achieving satisfactory PID control of processess, based on approximate observations of the behavior of the plant being controlled. Part of the popularity of these controllers stems from the simplicity of their application; it is not even necessary to have a detailed model of the plant.

Designers of most control systems prefer linear behavior to the fastest behavior possible, which typically is nonlinear. One reason is that consistent maximum effort, regardless of the size of the commands or disturbances, would reduce the life of the hardware. Saturation is normally desired only for the relatively infrequent occurrences of the largest disturbances anticipated; the power of the system is sized to accommodate these disturbances. Other reasons for favoring linearity include the simplicity of analysis, and the consistency of the behavior in terms of the classical measures of natural frequencies and damping ratios.

Non-PID controllers are often designed to avoid saturation and thereby validate linear behavior. Thus, simple derivative actions are avoided. The motor control system under consideration is in fact not a good candidate for PID control, unless saturation and the resulting nonlinearity is directly recognized and utilized. A different kind of controller is introduced below, instead.

EXAMPLE 8.9

The steady-state error of the system addressed in Example 8.8 is eliminated by adding an integral action to the controller, which already has a proportional gain $k = 16$ and derivative time $T_d = 0.5$ second. Find the integral time T_i that produces marginal stability, and show that a good response results from a value three times larger.

Solution: The controller has the transfer junction $k(1 + T_d S + 1/T_i S)$, and the plant has a transfer function $1/(S^2 + 1)$, giving for the values indicated above,

$$\frac{x}{x_c} = \frac{8T_i S^2 + T_i S + 16}{S^3 + 8T_i S^2 + (4 + 16T_i)S + 16}.$$

Equation (8.15) gives the condition for marginal stability: $8T_i(4 + 16T_i) = 16$, or $T_i = 0.25$ second. For three times that, or $T_i = 0.75$ second, the roots of the characteristic equation are $S = -2, -2 \pm 2j$, which is reasonably fast without significant oscillation ($\zeta = 0.707$). Increasing T_i to 1.0 second gives $S = -4, -2, -2$ (double pole), which is slower and somewhat inferior. Increasing T_i further continues to slow the response. The proposed solution may be quite acceptable despite the saturation caused by the derivative action when the commanded position suffers an abrupt change.

8.2.4   Phase Lead Controllers

A **phase lead controller** possesses a single zero and a single pole, plus a gain, for a total of three parameters. Its transfer function is

$$G_c = k\frac{S + z}{S + \alpha z}; \qquad \alpha > 1. \tag{8.23}$$

The pole and zero are plotted in part (a) of Fig. 8.11. The requirement $\alpha > 1$ means that the phase angle of the transfer function of the controller for sinusoidal excitation is positive,

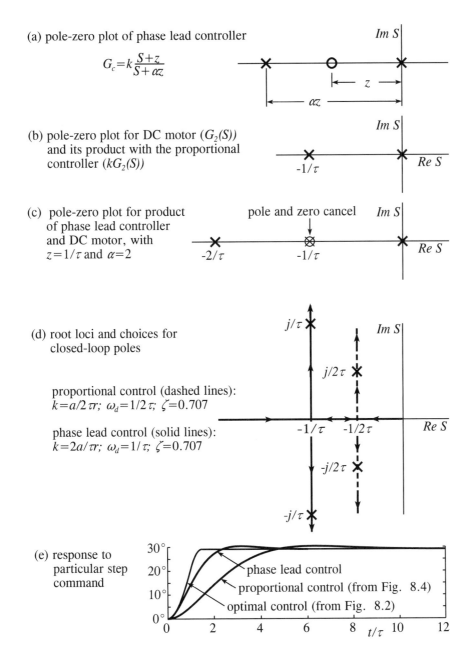

(a) pole-zero plot of phase lead controller

$$G_c = k\frac{S+z}{S+\alpha z}$$

(b) pole-zero plot for DC motor $(G_2(S))$
and its product with the proportional
controller $(kG_2(S))$

(c) pole-zero plot for product
of phase lead controller
and DC motor, with
$z = 1/\tau$ and $\alpha = 2$

(d) root loci and choices for
closed-loop poles

proportional control (dashed lines):
$k = a/2\tau r;\ \omega_d = 1/2\tau;\ \zeta = 0.707$

phase lead control (solid lines):
$k = 2a/\tau r;\ \omega_d = 1/\tau;\ \zeta = 0.707$

(e) response to
particular step
command

Figure 8.11: Phase lead compensation applied to the DC motor system

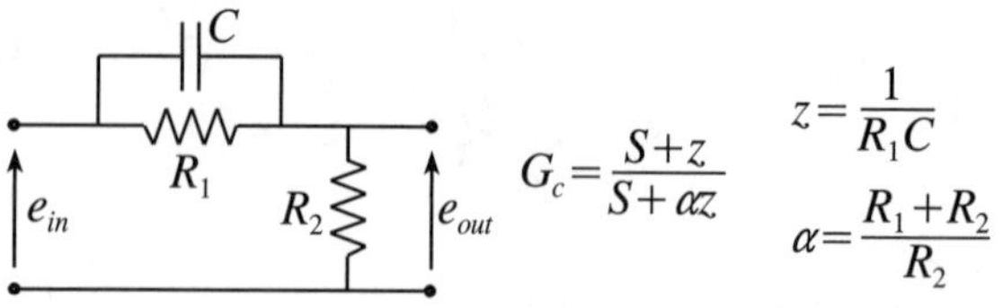

Figure 8.12: RC circuit for phase lead compensator

that is a phase lead, giving the controller its name. The response of a phase lead controller to a step input is rather like that of a highly muted PD controller. Rather than an impulse followed by a constant value, a constant value is preceded by a noninfinite but larger value of non-zero duration, blended by an exponential decay. It is no surprise, then, that the consequences of the two controllers are similar.

The phase lead control can be achieved by the simple passive RC circuit shown in Fig. 8.12, explaining part of its historical popularity. The gain k must be added through use of an amplifier.

A phase lead controller can improve the performance of the motor-driven position control system. The transfer function for the DC motor plant, taken from equation (8.1), is

$$G_2(S) = \frac{r/a\tau}{S(S+1/\tau)}. \tag{8.24}$$

Its pair of poles is plotted in part (b) of Fig. 8.11. When a phase lead controller acts directly on this plant, the overall open-loop transfer function becomes

$$G(S) = \frac{(kr/a\tau)(S+z)}{(S+\alpha z)(S+1/\tau)}. \tag{8.25}$$

The simplest and most desirable application sets $a = 1/\tau$, so that the zero of the compensator cancels the non-zero pole of the motor. The result is a pair of poles, one at the origin and the other at $S = -\alpha z$. The example with $\alpha = 2$ is shown in part (c) of the figure. Note that it is acceptable to ignore any slight error in this cancellation, since the pole and zero lie in the left-half plane; should they have been in the right-half plane such an error would directly represent an instability.

The root loci for the closed-loop poles are shown in part (d) of the figure. The loci for the original system with proportional control only are shown by dashed lines, and the loci for the system with the lead compensator is shown by solid lines, again for the special case of $z = 1/\tau$ and $\alpha = 2$. In both cases the closed-loop poles are chosen to give damping ratios of $1/\sqrt{2}$. The net effect of the phase lead compensator can be seen to double the speed of the system, as represented by the distance of the poles from the origin (the natural frequency).

The resulting responses for the 29.2° step considered earlier are compared in part (e) of the figure. The fact that the response is somewhat more sluggish than the optimal indicates that saturation is not a problem. Were the value of αa made much larger, however, or were the step made much bigger, saturation would in fact occur, and the linear prediction would be in error.

The larger the value of α, the more closely the phase lead controller $k(S+z)/(S+\alpha z)$ can resemble the PD controller $k_d(1+T_dS)$. The resemblance otherwise could be said to be closest when the responses of the two controllers to a unit ramp disturbance approach each other as $t \to \infty$. This happens when $k = k_d\alpha$ and $z = (\alpha-1)/\alpha T_d$. One result is equal steady-state errors.

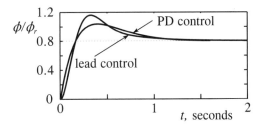

$$G_c = \frac{1}{\beta}\frac{S+\beta p}{S+p} \qquad p = \frac{1}{\beta R_1 C}$$
$$\beta = \frac{R_1 + R_2}{R_2}$$

Figure 8.13: RC circuit for lag compensator

EXAMPLE 8.10

Replace the PD controller employed in Example 8.8 ($k_d = 16$, $T_d = 0.5$ s) with the phase lead controller giving the "closest" equivalence, as described above, and compare the two step responses. Let $\alpha = 10$.

Solution: The "closest" equivalence gives $k = k_d\alpha = 160$ and $z = (\alpha - 1)/\alpha T_d = 1.8$ s^{-1}. The two transfer functions become

$$\text{PD controller}: \quad \frac{x}{x_r} = \frac{8(S+2)}{S^2 + 8S + 20},$$

$$\text{phase lead controller}: \quad \frac{x}{x_r} = \frac{k(S+z)}{mS^3 + \alpha zmS^2 + (K+k)S + z(\alpha K + k)}$$

$$= \frac{160(S+1.8)}{S^3 + 18S^2 + 164S + 360}.$$

The two step responses are

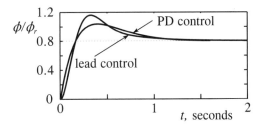

Unless you already have designed a PD controller, however, it is better to design a phase lead controller directly, for example by using a root locus. The rough equivalence above deteriorates when α is small, mandating a separate analysis.

8.2.5 Phase Lag Controllers

A **phase lag controller** resembles a phase lead controller, but the pole is closer to the origin rather than farther away:

$$G_c = k\frac{S + \beta p}{S + p}; \qquad \beta > 1. \tag{8.26}$$

Phase lag controllers are used to reduce steady-state errors; they are to integral controllers what phase lead controllers are to PD controllers. They can also be fabricated with a simple *RC* circuit, apart from the amplifier needed to achieve the gain k, as shown in Fig. 8.13.

A phase lag controller can be used with the motor-driven position control system, substituting for the PI controller used in Section 8.2.1 (Fig. 8.8, p. 540). This is carried out in Fig. 8.14. The overall transfer function from the moment disturbance M_d to the angle of

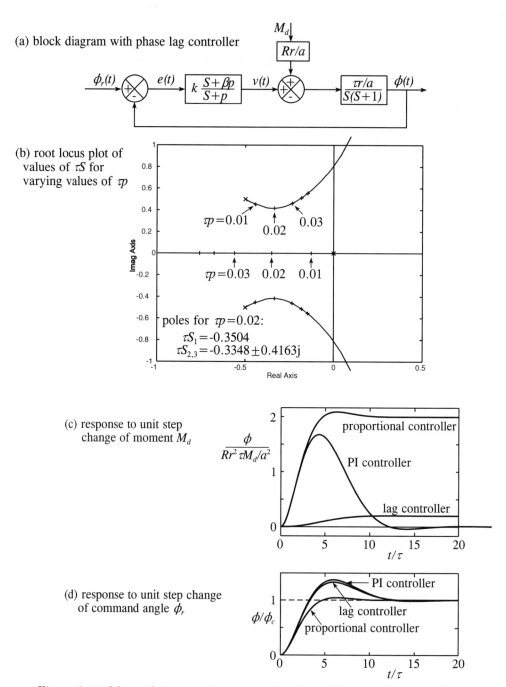

(a) block diagram with phase lag controller

(b) root locus plot of values of τS for varying values of τp

$\tau p = 0.01$ 0.02 0.03

$\tau p = 0.03$ 0.02 0.01

poles for $\tau p = 0.02$:
$$\tau S_1 = -0.3504$$
$$\tau S_{2,3} = -0.3348 \pm 0.4163j$$

(c) response to unit step change of moment M_d

$$\frac{\phi}{Rr^2\,\tau M_d/a^2}$$

proportional controller

PI controller

lag controller

(d) response to unit step change of command angle ϕ_r

$$\phi/\phi_c$$

PI controller

lag controller

proportional controller

Figure 8.14: Motor-driven position control system with phase lag controller

output ϕ becomes

$$\frac{\phi}{M_d} = \frac{Rr^2 z/\tau a^2}{S^3 + (p+1/\tau)S^2 + (p/\tau + rk/\tau a)S + rkp\beta/\tau a}. \tag{8.27}$$

Substitution of $S \to 0$ gives the steady-state response, or error due to the disturbance, of $Rr/ak\beta$. This is the same as with proportional control alone, as originally determined, reduced by the factor of β. Phase lag controllers generally do not eliminate steady-state error, unlike PI controllers.

To be consistent with the analysis for the PI control, the value $k = a/2r\tau$ is assumed. To make the lag control effective without an unreasonably large value of β, the value $\beta = 10$ will also be assumed, reducing the error to one-tenth of its original value. The value of p remains to be determined. In order to plot a root-locus for increasing values of the dimensionless ratio τp, the characteristic equation is solved for τp and then placed in the usual form of equation (8.10): $1 + \tau p\,\text{num}(S)/\text{den}(S)) = 0$. This gives

$$\frac{\text{num}(S)}{\text{den}(S)} = \frac{(\tau S)^2 + (\tau S) + \beta/2}{(\tau S)^3 + (\tau S)^2 + (1/2)(\tau S)} \tag{8.28}$$

The critical part of the resulting root locus plot near the origin is shown in part (b) of Fig. 8.14. One pole migrates to the left along the real axis for increasing values of τp, while two other complex poles approach and eventually enter the right-half plane. The value $\tau p = 0.02$ is close to the best compromise.

Using this value gives the response of ϕ to a step change of the disturbance M_d plotted in part (c) of the figure. The corresponding responses of the system with the proportional-plus-integral controller (from Fig. 8.9) and of the system with the original proportional controller are plotted for comparison. The PI controller has the advantage of reducing the steady-state response to zero, while the lag controller has the advantage of a smaller maximum excursion. The system without dynamic compensation behaves relatively poorly.

The final set of plots in part (d) of the figure shows the comparative responses of the three systems to a step change in the command angle, ϕ_r. Neither system with dynamic compensation behaves quite as well as the original system without dynamic compensation. This is because the dynamic compensators were designed explicitly to address the problem of sensitivity to disturbances in M_d. This is unlike most design situations, where dynamic compensation is employed to improve the response to the command signal.

8.2.6 Phase Lead-Lag Controllers

Phase lead and lag actions often are used together, under the name of lead-lag compensation; the combination is akin to the PID control:

$$G_c = k\frac{1 + \alpha\tau_1 S}{1 + \tau_1 S}\frac{1 + \tau_2 S}{1 + \beta\tau_2 S}. \tag{8.29}$$

The simple RC circuit shown in Fig. 8.15, plus an amplifier, serves as a simple implementation for the case when $\beta = \alpha$. Note that the impedance of any one element can be set independently, and the values of the other elements are determined by the desired values of the time constants τ_1 and τ_2 and α. The impedances cover an excessive range if α gets too large; the value 10 sometimes is cited as a practical limit if passive analog components are used. Digital and operational-amplifier implementations avoid this limitation. In practice, the phase lag portion of the controller is used to reduce steady-state error; the value of τ_2 normally is large, so the frequency range at which it acts is lower than the principal frequencies associated with the plant dynamics. The value of τ_1, on the other hand, is chosen to be considerably larger, so the phase lead action provides damping at frequencies where it is needed, permitting a larger gain and increased bandwidth.

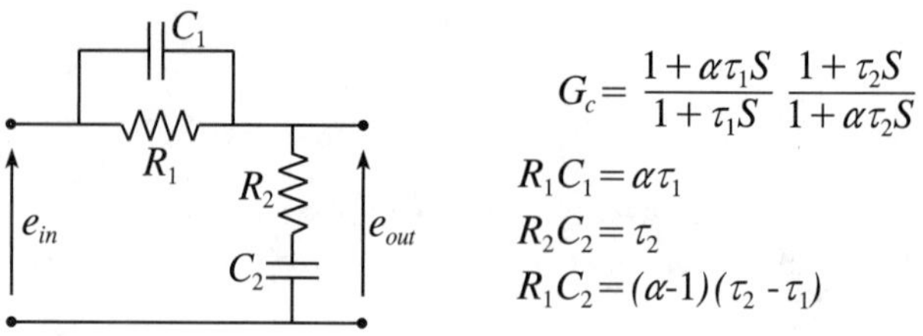

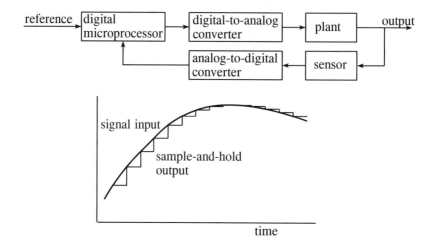

$$G_c = \frac{1+\alpha\tau_1 S}{1+\tau_1 S}\,\frac{1+\tau_2 S}{1+\alpha\tau_2 S}$$

$$R_1 C_1 = \alpha\tau_1$$

$$R_2 C_2 = \tau_2$$

$$R_1 C_2 = (\alpha-1)(\tau_2 - \tau_1)$$

Figure 8.15: RC circuit for lag-lead controller

Figure 8.16: Digital control system with sampled data signals

8.2.7 Digital Control Systems

A majority of control systems today employ digital processors that are coupled to the actuators and systems controlled by analog-to-digital and digital-to-analog converters, as suggested in the block diagram of Fig. 8.16. This allows greater flexibility in the choice of control algorithms and avoids the need for analog circuits with their impedance limitations, amongst other advantages.

An analog-to-digital converter produces a train of numbers at discrete intervals of time. These **sampled data** signals normally are treated as though the signal were step-wise in time, as illustrated by the plot given in the figure. The signal sent by the digital-to-analog converter to the plant is also in the **sample and hold** format. If the sample time is considerably shorter than any of the significant time constants of the plant, the effective time delays produced in the sampling process are insignificant, and the analysis results of this chapter can be applied directly. This is often the case today, since processor speeds have been increasing for many years.

When the sampling interval is not greatly shorter than the smallest significant time constant of the plant, the time delays impose phase lags (nonminimum phase-lag processes) that normally reduce the relative stability of a system or cause outright instability. Thus, at the least you should allow an extra margin of stability if you are applying the analog methods of this chapter. Better yet, you can roughly predict the effects of sample-and-hold using the frequency response methods of Section 8.3 below. Best of all, you can employ an analysis based on the z-plane, where $z = e^{st}$ with s being the Laplace operator. The formal study of this methodology including the design of digital compensators involves the **z-transform**, a modified form of the Laplace transform. This subject appears in many textbooks on control.

8.2.8 Summary

Many types of dynamic compensators or controllers are used in feedback systems. Introducing an additional integral action to a proportional controller usually eliminates any steady-state error that results from an unwanted disturbance. Large values of the gain and the coefficient of the integral action desirably reduce the steady-state error. They also generally decrease the relative stability of the system, with the consequence of undesirable resonances and overshoot and, for large enough values, outright instability. The introduction of an additional derivative action to the controller is one way to increase the damping of the system.

Whether a control system is linear or nonlinear is the most fundamental distinction. Since the power of a system is limited, the responses to small commands or disturbances potentially can be faster than the responses to larger commands or disturbances. A system that implements this potential is nonlinear. The use of derivative control also invokes nonlinearities in the response to a step command. Nevertheless, linearity is usually preferred by control engineers, to save on wear and to give simple, consistent behavior.

A phase lead controller acts rather like the proportional-plus-derivative (PD) controller, but has less tendency to saturate. The phase lead controller has a gain, a zero, and a pole of larger value than the zero. The controller typically is used to replace a real pole with a real pole of a larger value, increasing the speed of response of the system without decreasing relative stability. A phase lag controller serves to reduce steady-state error, acting partly like a proportional-plus-integral (PI) controller, which usually eliminates steady-state error. A phase lead-lag controller combines both actions, somewhat like a PID controller.

All these controllers can be implemented with digital, operational amplifier or mechanical/pneumatic circuits. The phase lead, lag and lead-lag functions also can be implemented by passive RC circuits.

Guided Problem 8.2

The first part of the final guided problem relates to the control system that keeps a rocket vertical at launch despite the applied force being applied at the lower end. Without control the rocket would fall over. The second part of the problem relates to the position control of the rocket. In both cases proportional-plus-derivative control is central to success. It is not necessary to carry out the second part to benefit from the first. Both parts give practice with root loci as implemented by MATLAB.

The inverted pendulum system shown in Fig. 8.17 comprises two equal masses separated by a rigid massless rod, a constraint such that the lower mass moves horizontally only, and a means of applying a control force to that mass. The parameters are $mg = 1$ lb, $g = 32.2$ ft/s^2 and $L = 0.644$ ft.

(a) The relation between the force F and the angle ϕ is

$$\left(S^2 - \frac{2g}{L}\right)\phi = \frac{1}{Lm}(-F),$$

assuming linearity (small angles). Proportional feedback control cannot stabilize this system, but PD control can. You are asked to find appropriate values of the gain k and derivative time T_d, and the associated poles of the closed-loop system.

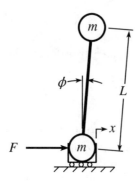

Figure 8.17: Inverted pendulum for Guided Problem 8.2

(b) The relation between ϕ and x, again assuming small angles, is

$$S^2 x = -L \left(S^2 - \frac{g}{L} \right) \phi.$$

Unless an appropriate additional control is introduced, the position x can wander widely in response to disturbances. Place the system resulting from part (a) inside a second feedback loop. Again, apply PD compensation (proportional control again won't work) so that $x(t)$ stably follows a command input, $x_r(t)$.

Part (b) of this problem is revisited in Guided Problem 8.3 at the end of the chapter.

Suggested Steps:

1. Find the poles of the transfer function between $-F$ and ϕ, and sketch-plot them in the S-plane. Sketch the simple root locus that would result from the use of proportional feedback, showing that it fails to produce the desired stability for any value of the gain k.

2. Draw a block diagram for a system with unity feedback. Include a reference signal ϕ_r, even though your interest may be only for a zero value. Also, include a block for the compensator, and label all variables. Does the compensator relate the error to F or to $-F$?

3. A PD controller has the transfer function $k(1 + T_d S)$, which gives a single zero in the left-half plane. It is tempting to superimpose this zero on one of the poles of the system, canceling it to leave a very simple system. This is acceptable, but somewhat misleading since perfect cancelation never really happens; further, better results follow from the choice of a zero somewhat to the left of this pole. Place the zero 50% further out on the real axis, and plot the resulting root loci as the gain increases from zero to infinity. Use MATLAB.

4. Choose points on the root loci which are close to being the best. Use MATLAB also to specify these points and the associated gain precisely. You now have completed part (a).

5. Extend your block diagram to give the position x, and place the result inside a second loop with unity feedback, a reference signal $x_r(t)$ and a second controller.

6. Use MATLAB to give the root loci for this complete system, using proportional feedback. You should discover that no value of the second feedback gain gives stable behavior.

7. Replace the proportional controller with a PD controller. Its zero will have to be rather close to the origin (a large derivative time constant on the order of one-quarter to one-half second) to be effective. Repeat the root locus procedure, and note that a range of satisfactory gain values exists. Choose points on the loci and find the associated gain. Determine the roots of the characteristic equation for the overall system.

8. Write the resulting transfer function between $x_r(t)$ and $x(t)$, and use the **step** command of MATLAB to plot the response to a unit step change. Note that x starts by moving *away* from its final value, rather than toward it, unlike most of the systems you have investigated. Does this make physical sense?

PROBLEMS

8.8 Show that PI or PID control acting on any plant $G(S)$ with unity feedback produces zero steady-state error in response to an input and to a disturbance acting directly on the plant, as long as $G(0) \neq 0$.

8.9 A motor-load system of the type featured in this chapter is reconfigured to control angular velocity, rather than angular position. A feedback loop with proportional-plus-integral control is proposed in order to reduce or eliminate the steady-state error that proportional control alone gives (as found in Problem 8.2, p. 536). Let $a = 0.006$ V·s, $\tau = 1.0$ s and $r = 0.2$, and restrict the motor voltage to 3.0 V for commanded step changes from zero to 2000 degrees/s (as in Problem 8.2). The proportional gain k and the integral time T_i are to be chosen to give good performance.

(a) Draw a block diagram of the system. Include the motor voltage $v(t)$ as an internal variable.

(b) Find transfer functions from the command input $\dot{\phi}_r(t)$ to the output velocity $\dot{\phi}(t)$ and from the command input to the voltage $v(t)$.

(c) Is there a steady-state error in the output velocity for a step change in the command velocity? Hint: There is no such error if the transfer function approaches 1.00 when $S \to 0$, which corresponds to the frequency response for zero frequency.

(d) Determine the natural frequency and the damping ratio in terms of k and T_i.

(e) Set the damping ratio equal to a reasonable value, such as 0.4. Determine the resulting relation between k and k/T_i.

(f) Note that the largest allowable value of the ratio k/T_i gives the fastest response. Program in MATLAB the numerator and denominator polynomials in the transfer function with $v(t)$ as its output, substituting the relation from part(e) for every k that appears alone, so the ratio is the only unknown. Multiply the numerator by the command input corresponding to 2000 degrees per second, and use the **step** command to secure a trace of the voltage as a function of time for various values of the ratio. Determine the value that gives the largest allowed voltage. Then determine the values of k and T_i individually.

(g) Use MATLAB to plot the output $\dot{\phi}(t)$. If you have done part (a) of Problem 8.2, compare this response with the optimal.

8.10 Determine which of the closed-loop systems in Problems 8.3–8.7 have steady-state errors in response to a step input. For each of these, specify the gain and the integral time of an integral controller that eliminates the error while giving acceptable dynamics.

8.11 Repeat part (a) of Guided Problem 8.2 (p. 553), substituting a lead controller for the PD controller.

8.12 The system of Problem 8.7 (p. 537) exhibits either small damping or negative damping (instability) for all values of the gain k. Determine the gain and derivative time of a substitute PD controller that gives better responses.

8.13 Solve the previous problem using a phase lead controller in place of the PD controller. The ratio of the magnitudes of the pole to the zero of the controller should not exceed 5:1.

8.14 The steady-state error in the thermal system of Example 8.7 (pp. 542-543) was eliminated by the introduction of integral action, but at a cost of reducing the natural frequency. Expand the controller by adding derivative action (making a PID controller) so as to regain the original $\omega_n = 2/\tau$ while retaining $\zeta = 0.5$. Find the "best" values of the derivative time T_d, integral time T_i and the gain k.

8.15 A correspondence between a phase lag controller and a PI controller can be defined in the sense that their responses to a unit step agree for small values of time. Find the correspondence assuming a value of β is chosen already. Apply the result to the PI controller of the motor-driven position control system, as carried out in Section 8.2.1 (pp. 539-541), finding the associated value of τp for the corresponding lag controller. Compare the result to the lag controller with $\beta = 10$ for the same plant but chosen by other means in Section 8.2.5 (pp. 541-551), namely $\tau p = 0.020$.

8.16 Derive the relations for the values of the following RC circuits as given in the respective figures. (Note: The simplest method starts with a bond graph, and then finds the transfer function using the loop rule as developed in Section 7.4.2–7.4.4.)

(a) The circuit of Fig. 8.12 (p. 548) for the phase lead compensator.

(b) The circuit of Fig. 8.13 (p. 549) for the phase lag compensator.

(c) The circuit of Fig. 8.15 (p. 552 for the phase lead-lag compensator.

SOLUTION TO GUIDED PROBLEM

Guided Problem 8.2

1. The open-loop poles are at $S = \pm\sqrt{2g/L} = \pm 10$ rad/s. The gain of the plant is $50/(S^2 - 100)$. For a small feedback gain, the closed-loop poles lie on the real axis, with one in the right-half plane indicating instability. For a large gain, both closed-loop poles lie on the imaginary axis, indicating oscillation without necessary damping or decay.

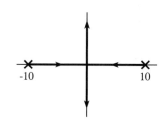

2.

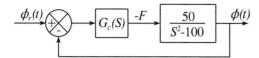

The output of the compensator is defined above as $-F$.

3. $G_c(S) = k(1 + T_d S)$

$T_d = 1/1.5\sqrt{2g/L} = 1/15$ s

The open-loop gain becomes

$\dfrac{k50(S/15 + 1)}{S^2 - 100}$

The coding for the root locus plot is

```
num = 50*[1/15 1];
den = [1 0 -100]
rlocus(r,k)
```

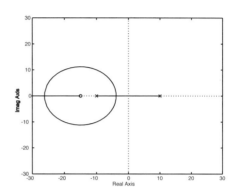

4. The following additional coding gives the plot shown opposite, showing the closed-loop poles for the values $k = 3, 6, 9, 12, 15, 18,$ and 21. The value $k = 15$ is chosen, for which $r = -25 \pm 5j$ rad/s.

```
k = [3:3:21]:
[k,r] = rlocus(n,d,k);
plot(r,'x')
```

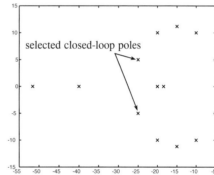

5. The block diagram below gives the open-loop gain

$\dfrac{-32.2k(S^3 + 15S^2 - 50S - 750)}{S^4 + 50S^3 + 650S^2}$

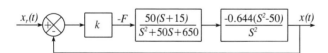

6. The root locus shows two closed-loop poles in the right-half plane, producing an instability, for all values of the gain, k.

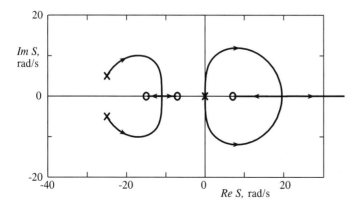

7. The zero of the PD controller is placed at $S = -a$. This multiplies the numerator of the open-loop transfer function by the factor $(S + a)$ to give $-32.2k[S^4 + (15 + a)S^3 + (15a - 50)S^2 - (50a + 750)S - 750a]$. The root-locus plot below is for $a = 3$ rad/s. One root approaches $-\infty$ on the real axis before the critical pole pairs near the origin emerge very far along their paths in complex territory. The root reappears at $+\infty$ on the real axis, and ends at the zero in the right-half plane. This pole must be confined to the left-half plane to give stability. A reasonable solution is $k = 0.02$, for which $r = -87$ rad/s, -20.1 rad/s, and $-0.39 \pm 1.50j$ rad/s.

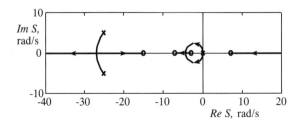

8. $\dfrac{x}{x_r} = \dfrac{0.644(-S^4 - 18S^3 + 5S^2 + 900S + 2250)}{0.356S^4 + 38.41S^3 + 653.2S^2 + 579.6S + 1449}$.   The step response plot is on p. 656.

8.3   Frequency Response Methods

The design of feedback controllers has been based thus far on the placement of the zeros and especially the poles of the open and closed-loop transfer functions, with emphasis on the root locus plot. The use of frequency response methods is a valuable complement or alternative. You have been exposed already, through the extensive material in Section 6.1, to the concept of frequency response and its primary graphical tool, the Bode plot. In this section the Bode plot, and its siblings the polar (Nyquist) and magnitude-vs.-phase (Nichols) plots, will be used to predict closed-loop performance from open-loop performance, and to aid in the design of controllers. Should you wish a deeper exposition or illustration of the methods, many standard textbooks are available.

8.3.1   Polar or Nyquist Frequency Response Plots

The Bode plots comprise separate representations for the magnitude and phase angle as functions of frequency. Often it is desirable to combine magnitude and phase angle into a single curve. One way to do this is the polar plot, also called the Nyquist diagram. The x-axis of a Nyquist plot of $G(j\omega)$ equals $Re[G(j\omega)]$, and the y-axis equals $Im[G(j\omega)]$. Therefore, the magnitude $|G(j\omega)|$ equals the distance or radius from the origin, and the phase angle is the angle from the real axis. Since frequency is not used as an axis, it is helpful to place tick marks along the curve, labeled with the corresponding values of ω.

EXAMPLE 8.10

Draw polar or Nyquist plots for the functions $G_1(S) = S + 1$ and $G_2(S) = 1/(S+1)$.

Solution: The function $G_1(j\omega)$ can be plotted directly. The function $G_2(j\omega) = 1/(j\omega + 1)$ is its reciprocal, that is has reciprocal magnitudes and oppositely signed phase angles for corresponding values of ω. The latter curve can be shown to be a semi-circle.

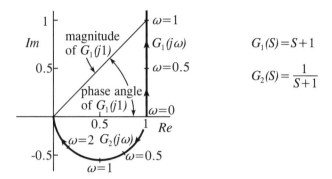

MATLAB draws polar plots in response to the command Nyquist(num,den) or Nyquist(A,B,C,D), where num, den or A,B,C,D are defined as before. MATLAB includes the plot for negative values of frequency, which has the same real part and minus the imaginary part of the corresponding values of frequency; it is the same curve flipped about the real axis. The reason for including negative frequencies will appear shortly.

EXAMPLE 8.11

Use MATLAB to plot the Nyquist diagram for the unstable open-loop transfer function $G(S)H(S) = k(S + 2)/(S^2 + S + 1)(S - 0.5)$ with $k = 0.2$, 0.4 and 0.6.

Solution: The following code produces the plots as given (except for the annotations). See the "help" window for optional arguments of Nyquist, including a frequency vector.

```
den=[1 .5 .5 -.5];
for i=1:3
    k=.2*i;
    num=k*[1 2];
    Nyquist(num,den)
    hold on
end
```

8.3.2 The Nyquist Stability Criterion

The objective is to determine the stability of a closed-loop system from knowledge of its open-loop frequency response, without having to carry out computations. The open-loop frequency reponse could be from experimental data, with no analytic model. In particular,

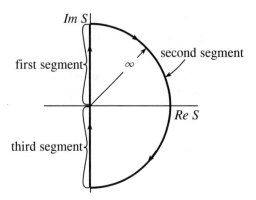

Figure 8.18: Contour in the S-plane that encircles all potential unstable roots (zeros) of the characteristic equation

it is desired to know whether any of the roots or zeros of the closed-loop characteristic equation

$$F(S) \equiv 1 + P(S) \equiv 1 + G(S)H(S) = 0 \qquad (8.30)$$

lie in the right-half plane, and therefore represent an instability. A famous theorem by Cauchy refers to a closed path or contour in the S-plane for a function such as $F(S)$, and the plot of the complex function $F(S)$ for the same path values of S. It states:

> *If a clockwise contour in the S-plane encircles Z zeros and P poles of the complex function $F(S)$ without directly passing through any of them, the mapping of the contour in a plot of $F(S)$ encircles its origin $Z - P$ times in the clockwise direction.*

An encirclement of the origin in a plot of $F(S)$ becomes an encirclement of the point $S = -1$ in a plot of the open-loop transfer function $P(S) = F(S) - 1$.

Since the stability test regards the entire right-half of the S-plane, the contour of choice encircles the entire right-half of that plane, as shown in Fig. 8.18. Part of the contour traverses the imaginary axis in two segments, as labeled. The first segment is $S = j\omega$, with ω ascending from zero to infinity. The third segment is its complex conjugate, $S = -j\omega$, with ω proceeding from infinity to zero. The mapping of this third segment in the $P(S)$ plane is the complex conjugate of the mapping of the first segment, which means it is the mirror image about the real axis. This is why MATLAB includes both segments in its Nyquist plots. A further simplification results whenever the second segment of the contour gives zero values of $P(S)$ for all such infinite values of S, which happens whenever the denominator polynomial of $P(S)$ is of higher order in S than is its numerator polynomial.

EXAMPLE 8.12

Interpret the Nyquist plots for the three cases in Example 8.11 to give the stability of the closed-loop system. Also, plot a root locus that shows the locations of the closed-loop poles for the three cases. Finally, use the Routh-Hurwitz criteria for stability (pp. 533-534) to determine the range of k for which the system is stable.

Solution: There is one open-loop pole in the right-half plane of the open-loop transfer function, so $P = 1$. For the case with $k = 0.2$, the -1 point is not encirled by the Nyquist contour, so $Z - P = 0$. Thus, the number of zeros in the right-half plane for the characteristic equation is $Z = 0 + 1 = 1$, which means that the closed-loop system has one pole in the right-half plane and therefore is unstable. For the case with $k = 0.4$, the -1 point is encircled once *counterclockwise*, so $Z - P = -1$ and $Z = -1 + 1 = 0$. There are therefore no closed-loop ploes in the right-half plane, and the system is stable. For $k = 0.6$, on the other hand, the -1 point is encircled once *clockwise*, so $Z - P = 1$ and $Z = 1 + 1 = 2$. There are two poles in the right-half plane of the closed-loop system, therefore, and the system is again unstable.

The root locus plot below shows how the poles migrate in the S-plane as k is increased. It shows that all poles are in the left-half plane only for a single band of values of k.

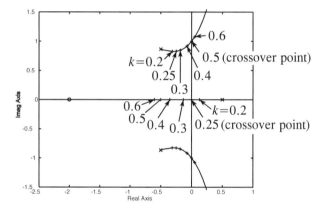

The closed-loop transfer function is $G(S)/(1 + G(S)H(S))$, which gives the characteristic equation

$$(S^2 + S + 1)(S - 0.5) + k(S + 2) = 0.$$

Comparing this to the standard form

$$a_3 S^3 + a_2 S^2 + a_1 S + a_0 = 0$$

gives

$$a_3 = 1; \qquad a_2 = 0.5; \qquad a_1 = 0.5 + k; \qquad a_0 = 2k - 0.5.$$

The necessary condition for stability that all coefficients have the same sign requires $a_0 > 0$ or $k > 0.25$. Equation (8.15) (p. 534) additionally requires

$$a_2 a_1 - a_3 a_0 = 0.25 + 0.5k - (2k - 0.5) > 0,$$

which is satisfied only for $k < 0.75/1.5 = 0.50$. Therefore, the system is stable only between the **crossover points** at which $k = 0.25$ and $k = 0.5$.

The Cauchy theorem requires that the contour in the S-plane not pass through any of the poles of the open-loop transfer function. The contour must be modified, therefore, if one or more poles lie on the imaginary axis. The usual strategy employs a 180-degree

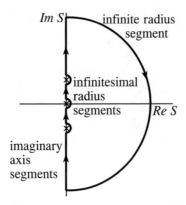

Figure 8.19: Modified contour in the S-plane to avoid poles on the imaginary axis

arc of vanishingly small radius as a detour around such poles, excluding them from the encompassed region and the number P, as illustrated in Fig. 8.19. The mappings of these counterclockwise arcs in the S-plane become clockwise arcs of infinite radius in the $P(S)$ plane, as the next three examples demonstrate. MATLAB does not display any segments having infinite radius, however, so it is up to you to determine their effect on the number of encirclements of the -1 point.

EXAMPLE 8.13

Plot the Nyquist diagram for the type 1 system having $G(S) = k/S(S^2 + S + 1)$ and $H(S) = 1$ when $k = 1$. Determine from the plot what values of k give stability.

Solution: The Nyquist plot at the right assembles a directly-drawn pseudo-infinite arc with a MATLAB-created plot. Since there is one pole at the origin in the S-plane of a type-1 system, the infinite arc traverses $180°$. It swings clockwise, as can be verified by checking the mid-point on the real axis $S = \epsilon \to 0$ and $G(\epsilon) \simeq 1/\epsilon \to \infty$.

The contour passes directly through the -1 point; for $k > 1$ it would encircle the point clockwise once, indicating one unstable pole. The criterion of stability, therefore, is $k < 1$.

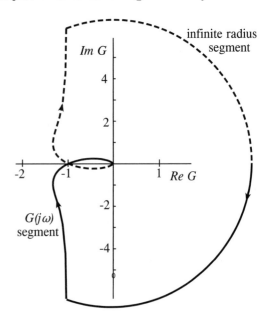

EXAMPLE 8.14

Repeat the above example for the type 2 system $G(S) = k/S^2(S+1)$ and $H(S) = 1$ with $k = 1$.

Solution: The presence of two poles at the origin produces an infinite arc of 360° in the Nyquist plot, as shown. It is critical that you attach it correctly to the finite portion of the contour; placing clockwise arrows on the infinite segment helps. There are two encirclements of the -1 point, regardless of the value of k. This system is unstable regardless of the value of k; it has two right-half-plane poles.

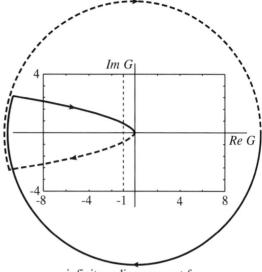

infinite radius segment from infinitesimal arc on S-plane

EXAMPLE 8.15

Plot the Nyquist diagram for the system with

$$G(S) = \frac{kS}{(S^2+1)(S+1)(S+3)}; \qquad H(S) = 1; \qquad k = 1.$$

Determine the range of k for stability of the closed-loop system, and determine the answer independently using the results of Section 8.1.7 (p. 534)

Solution: The MATLAB coding is

```
num=[1 0]; den=[1 4 4 4 3];
nyquist(num,den)
axis([-.3 .3 -.3 .3])
```

The resulting contour approaches infinity at each of the two open-loop poles on the the imaginary axis. The arcs of infinitesimal radius around these poles in the S-plane become 180° clockwise arcs of infinite radius in the $G(S)$ plane, and are added below to the finite segments of the locus produced by MATLAB to complete the locus. The -1 point is not encircled for the value $k = 1$ of the locus, but would be encircled, twice, if the plot were expanded by the factor $k = 8$ or more. The criterion for stability, therefore, is $k < 8$.

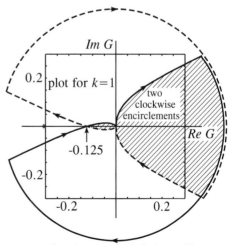

circular arcs at infinite radii

The characteristic equation $1 + G(S)H(S) = 0$ can be written

$$S^4 + 4S^3 + 4S^2 + (4+k)S + 3 = 0$$

so that $a_4 = 1$, $a_3 = 4$, $a_2 = 4$, $a_1 = 4 + k$ and $a_0 = 3$. The first of equations (8.16) gives $16 - 4 - k > 0$ or $k < 12$. The second of equations (8.16) gives $64 + 16k - (16 + 8k + k^2) - 48 > 0$, from which $k < 8$. The latter is controlling, and agrees with the Nyquist conclusion.

8.3.3   Measures of Relative Stability

A control system must not only be stable, it must also be reasonably **robust**, which means insensitive to at least modest unplanned variations in its parameters. Further, the dynamics should not be excessively oscillatory. These requirements can be addressed effectively from the perspective of the frequency response.

Most type 0 and type 1 open-loop systems produce stable closed-loop systems for small values of the proportional gain k, and become unstable for large values. The **gain margin** for a particular design is defined as the factor by which k could be increased before the system becomes unstable. Most designers would like to see a gain margin of 2 or larger, although they are willing to shave this in special circumstances. The gain margin is readily apparent from the Nyquist plot; it is the reciprocal of the magnitude of the contour at the point it crosses the negative real axis. The phase lag of the open-loop system traverses $-180°$ at this point, so the gain margin also is readily determined from a Bode plot. Its value often is given in decibels. An example is given in Fig. 8.20.

The **phase margin** is defined as the additional phase lag of $G(j\omega)$, at the frequency ω where $|G(j\omega)| = 1$, that would barely produce instability. This also is shown in Fig. 8.20. A phase margin of $45°$ or more is desirable, although sometimes it is reduced to as little as $30°$. The phase margin also can be read from the Bode plot, as shown.

A third and particularly meaningful measure of relative stability is the maximum value, M_m, of the magnitude ratio, M, of the closed-loop frequency response. Assuming unity feedback ($H(S) = 1$) as before, the ideal value at all frequencies is $M = 1$. The deviation frequency is the steady-state error. A value $M_m = 2$ may be tolerated, depending on the application. The value of M_m can be determined upon inspection of the Nyquist plot, although this is not as obvious as the gain and phase margins.

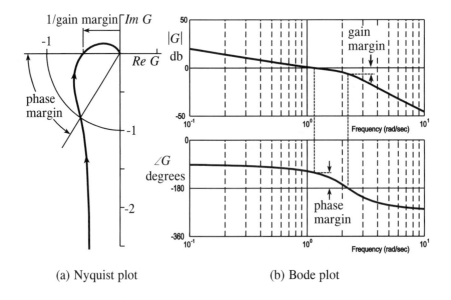

(a) Nyquist plot   (b) Bode plot

Figure 8.20: Example system showing gain and phase margins

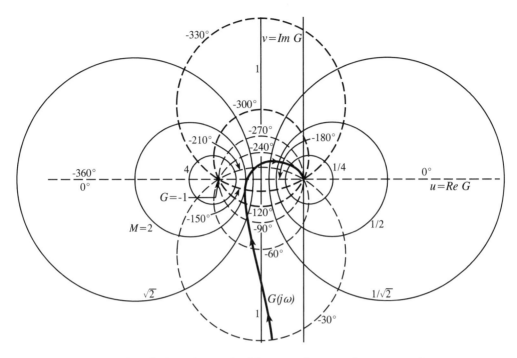

Figure 8.21: M and N circles in the Nyquist plane, with an example contour

Let $P(j\omega) = u + jv$. Then, the square of the magnitude ratio of the closed-loop system is

$$M^2 = \left| \frac{G(j\omega)}{1 + G(j\omega)} \right|^2 = \frac{u^2 + v^2}{(1 + u)^2 + v^2}, \tag{8.31}$$

which can be rearranged as

$$\left(u - \frac{M^2}{1 - M^2} \right)^2 + v^2 = \left(\frac{M}{1 - M^2} \right)^2. \tag{8.32}$$

For a fixed value of M, this is the equation for a circle in the Nyquist or u, v plane with radius $M/(1 - M^2)$ that is centered at $v = 0$ and $u = M^2/(1 - M^2)$. A resulting family of **M-circles** is plotted in Fig. 8.21. A sample Nyquist contour also is plotted, which has the value $M_m = 2$.

A second family of circles is plotted by dashed lines in Fig. 8.21. These are lines of constant closed-loop phase angle, ϕ, called **N-circles** where $N \equiv \tan\phi$. The relation

$$\phi = \tan^{-1}\left(\frac{v}{u} \right) - \tan^{-1}\left(\frac{v}{1 + u} \right) \tag{8.33}$$

can be rearranged into the form

$$(u + .05)^2 + \left(v - \frac{1}{2N} \right)^2 = \frac{1}{4}\left(1 + \frac{1}{N^2} \right), \tag{8.34}$$

which shows that the loci for constant values of N are indeed circles, centered at $u = 0.5$, $v = 1/2N$ and having radius $0.5\sqrt{1 + 1/N^2}$.

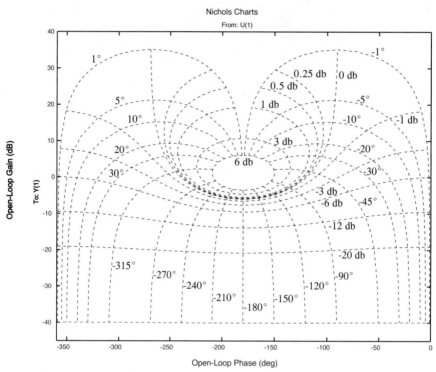

Figure 8.22: Nichols chart (MATLAB version 5.3, with annotations for the closed-loop gain and phase added)

8.3.4   Nichols Charts

A particular set of coordinates is more helpful for design than either the Bode or Nyquist coordinates. The MATLAB version of the **Nichols chart** is shown in Fig. 8.22. A more detailed paper version also is available. The log of the magnitude of $G(S)$ is plotted vs. the linear phase of $G(S)$. The M and N "circles" for the closed-loop transfer function $G(S)/(1+G(S)$ are mapped into this plane, where they no longer are circular. The principal advantage of the Nichols coordinates is that the value of k can be changed merely by translating the contour vertically. If done with a paper chart, the contour can be placed on an overlay that is slid vertically. In MATLAB, you start by requesting the Nichols contour for $G(j\omega)$, which appears without the M and N contours. You then right-click on the figure and select "grid" to get these contours. You likely will wish to change the boundaries of the plot using the `axis` command.

Thus, to achieve a particular value of M_m for a particular system, you plot the Nichols contour for $k = 1$ and translate it vertically until it is tangent to the M-contour with the value of M_m. The distance translated, in db, is the gain sought. You then can observe values of M and N (or ϕ) at several frequencies, thereby rapidly gaining an appreciation of the entire closed-loop frequency response.

EXAMPLE 8.16

Use the Nichols chart to find the closed-loop gain k for the unity-feedback system with open-loop transfer function $G(S) = k/S(S^2 + S + 1)$. Let the criterion be a maximum peaking ratio $M_m = 1.5$ (3 db).

Solution: The MATLAB coding below produces the Nichols plot of the open-loop system, but does not directly show the M and N contours. Right-click on the figure, and select " grid." Then, to expand the plot vertically, issue the command `axis([-360 0 -45 40])`, which gives the result below. (This plot was made with version 7.1 of MATLAB, which labels the db values of the M contours, but not the angles of the N contours. Subsequent plots were made with version 5.3, which labels neither. See Fig. 8.22 whenever the values are uncertain.)

```
>>num=1;
>>den=[1 1 1 0];
>>nichols(num,den)
```

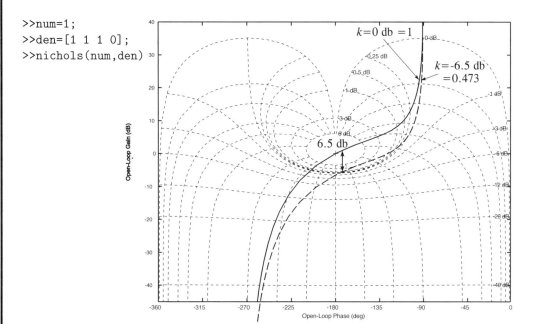

The vertical translation of $G(j\omega)$ needed to make the contour tangent to the locus for a maximum closed-loop magnitude of 3 db can be seen to be about -6.5 db, giving $k = 0.47$. The plot of $G(j\omega)$ corresponding to this value is plotted as a dashed line; you can superimpose this plot by issuing the command `hold on`, followed by a repeat of the `nichols` command with `num=.47`. You have now maximized the bandwidth of the closed-loop system, consistent with the specifications.

8.3.5 Dynamic Compensation Using Nichols Charts

The Nichols chart allows you to design a dynamic compensator visually. Information is given below to help you select phase lead, phase lag, phase lead-lag, PD, PI and PID controllers. The phase lead controller has the transfer function (from equation (8.23))

$$G_{lead} = k\frac{S + z}{S + \alpha z}; \qquad \alpha > 1, \tag{8.35}$$

and the phase lag controller has the transfer function (from equation (8.23))

$$G_{lag} = k\frac{S + \beta p}{S + p}; \qquad \beta > 1. \tag{8.36}$$

Recall that values of α or β much larger than 10 likely would be impractical, due to impedance considerations, and values less than two would have little effect. Nichols plots for three sizes of each of the controllers are plotted in Fig. 8.23. Varying k merely shifts these plots vertically, so the value $k = 1$ is chosen in all cases. Use of the dimensionless frequencies $\Omega = \omega/z$ and $\Omega = \omega/p$ allows these curves to represent all values of z and p. The scale of this plot is the same as the scale of Fig. 8.22

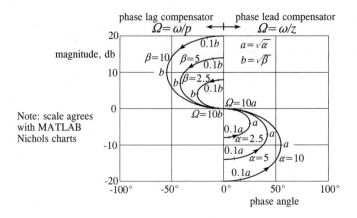

Figure 8.23: Nichols plots for phase lead and phase lag controllers

The transfer function $G(S)$ is the product of the transfer function of the controller and that of the plant. The magnitude at a particular frequency is the product of the two magnitudes, and the phase angle is the sum of the two phase angles. With the Nichols axes, therefore, both the vertical and horizontal components of $G(j\omega)$ are the sums of the vertical and horizontal components of the controller and the plant.

This fact is illustrated in Fig. 8.24 for the system of Example 8.10 (p. 549) in which a lead compensator with $\alpha = 10$ is applied to the mass-spring system of Example 8.8 (p. 545). The gain $k = 160$ in this example is taken as part of the original system transfer

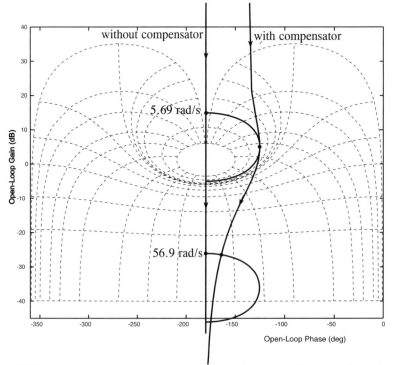

Figure 8.24: Nichols plots for a mass-spring system without and with a phase lead controller

function. The branch of the frequency transfer function shown represents frequencies above the 2 rad/s resonant frequency of the system, for which the phase angle is a constant $-180°$ and its Nichols plot is a vertical line through the -1 point. The compensated open-loop frequency function is pushed to the right and can be seen to nicely wrap around the -1 point, keeping the closed-loop gain below 2 db for all frequencies. The frequency transfer function of the controller is shown in two postions: the upper and lower positions represent its effect at the "nose" frequency of 5.79 rad/s and at ten times that frequency, respectively.

The phase lead controller exerts its maximum desired rightward push for the frequency at the nose of its Nichols contour. The contoller also changes the magnitude of the contour, in a direction that usually is not helpful. As a result, the controller is less useful when the slope of the plant contour is small. When successful, the lead controller allows the gain k to be increased, making the system faster.

EXAMPLE 8.17

A plant has the transfer function

$$G_p = \frac{1}{S^3 + 2S^2 + 2S + 1}.$$

When unity feedback is employed with proportional feedback and gain $k = 1$, the resulting steady-state error is 50%. It is not possible to reduce this error much by increasing k, without causing relative or absolute instability. Investigate what can be done using a phase lag controller, with $\alpha \leq 10$.

Solution: The commands below produce the Nichols chart for the plant below, presuming you right-click on the figure and select "grid" before issuing the **axis** command.

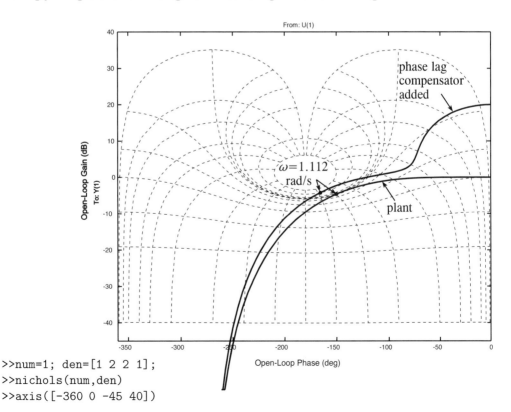

```
>>num=1; den=[1 2 2 1];
>>nichols(num,den)
>>axis([-360 0 -45 40])
```

The largest allowed lag compensator should be chosen, which has $\beta = 10$. The "nose" frequency should be large from the perspective of the speed of response of the system, and small from the perspective of stability. As a compromise, the point having the open-loop phase angle of $-150°$ is chosen to have ten times the nose frequency. This point is labeled on the chart as "$\omega = 1.112$ rad/s." This frequency value is found by entering

```
[m,p,w]=nichols(num,den);
[log10(w),p]
```

and looking down the resulting table to find the frequency "w" listed opposite $p = -150°$. (The use of the common logarithm precludes scaling problems in the table; you raise 10 to the value of the log to get the frequency.) At this frequency, $\omega/p = 0.1b = 0.1\sqrt{\beta}$, from which $p = 0.035$ rad/s. The phase lag controller therefore has the transfer function (equation (8.36))

$$G_c = \frac{S + 0.35}{S + 0.035},$$

and the overall open-loop transfer function is

$$G(S) = kG_cG_p = \frac{S + .35}{S^4 + 2.035S^3 + 2.07S^2 + 1.07S + 0.035}.$$

Entering "**hold on**" and then setting these new polynomials gives the second plot above labeled "phase lag compensator added." Leaving the present value $k = 1$ seems a reasonable compromise between reducing steady-state error and preserving relative stability. The closed-loop transfer function becomes

$$\frac{G(S)}{1 + G(S)} = \frac{S + .35}{S^4 + 2.035S^3 + 2.07S^2 + 2.07S + .385}.$$

The step response, found using **step** in MATLAB, is plotted below. The steady-state error is 9.1%, which is rather large but may be acceptable. If a smaller error is needed, some other compensator is needed; a PID controller would do admirably.

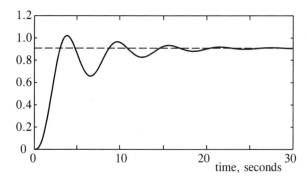

The transfer function for the PD, PI and PID compensators are

$$\text{PD}: \quad G_c(S) = k(1 + T_dS), \tag{8.37a}$$

$$\text{PI}: \quad G_c(S) = k\left(1 + \frac{1}{T_iS}\right) = k\left(\frac{T_iS + 1}{T_iS}\right), \tag{8.37b}$$

$$\text{PID}: \quad G_c(S) = k\left(1 + T_dS + \frac{1}{T_iS}\right) = k\left[1 + r\frac{(\tau S)^2 + 1}{\tau S}\right]; \tag{8.37c}$$

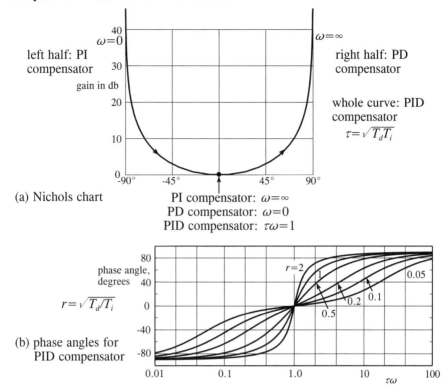

Figure 8.25: Frequency characteristics of PD, PI and PID compensators

$$\tau = \sqrt{T_d T_i}; \quad r = \sqrt{\frac{T_d}{T_i}}. \tag{8.37d}$$

Nichols plots for these compensators, with $k = 1$, are plotted in part (a) of Fig.8.25. Note that the contours for the PI and PD controllers are mirror images of each other, and the PID contour traverses precisely their combined contour. The mid-points in the phase angles of the PI and PD compensators, at $-45°$ and $45°$, respectively, occur when the dimensionless frequencies $T_d\omega$ and $T_i\omega$ equal 1. The gain and phase shifts for the PID controller are zero at its central frequency $\tau\omega = 1$, with $\tau \equiv \sqrt{T_d T_i}$. How rapidly this dimensionless frequency changes with changes in the phase angle depends on the "strength" of the controller, $r = \sqrt{T_d/T_i}$, as plotted in part (b) of the figure.

EXAMPLE 8.18

The open-loop plant

$$G_p = \frac{50(-S + 100)}{S^3 + 101S^2 + 150S + 5000}$$

is particularly difficult to control because of its right-half-plane zero and the small damping of its second-order pole pair. It is desired nevertheless to eliminate steady-state error. Design a controller.

Solution: The procedure of the above examples for plotting the Nichols contour of the plant gives the solid line below:

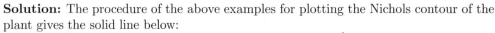

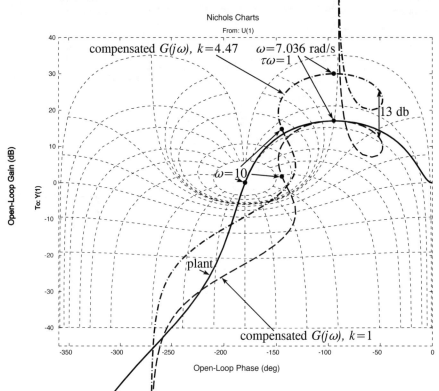

Integral action is needed to eliminate the steady-state error. Taken alone, however, this would require a very small value of k, on the order of 0.1, to give a reasonable relative stability. The system would become very sluggish. To improve the response significantly, it is proposed to include derivative action as well. The information needed to describe the resulting PID controller is given in Fig. 8.25. The point at which $\tau\omega = 1$ is chosen to lie at the maximum amplitude point, as labeled. The controller leaves the contour unchanged at this point, assuming $k = 1$; to the right of this point, the phase is shifted leftward, and to the left of this point it is shifted rightward. The magnitude is shifted upward in both cases, slightly for modest phase shifts and greatly for phase shifts in excess of 45°, as the plot in part (a) of the figure shows.

The choice of the controller shown was made in one pass, without iteration, by visual examination of the plant contour and the characteristics given in Fig. 8.25. The point at which the contour passes through the -1 point (labelled as $\omega = 10$) was chosen for a 35°-40° rightward phase shift. This decision is implemented by proper selection of the "strength" parameter, r. It was first necessary to determine the frequencies of the -1 point and the maximum amplitude point. This was done with the commands

```
[m,p,w]=nichols(num,den);
[log10(w),20*log10(m),p]
```

The resulting table shows that the peak occurs when $\log_{10}(\omega) = 0.8477$, or $\omega = 10^{0.8477} = 7.036$ rad/s, and the -1 point occurs where the phase angle is $-180°$, which happens to be 10.00 rad/s. Since $\tau\omega = 1$ at the maximum point, $\tau\omega = 10.00/7.036 = 1.421$ at the -1 point. The plot in part (b) of the Fig. 8.25 indicates that $r \equiv 1$ to achieve the 35°-40° phase shift. Using this value then gives $T_d = T_i = 0.1421$ seconds.

The controller becomes, from equation (8.34c),

$$G_c = k\frac{0.02019S^2 + 0.1421S + 1}{0.1421S},$$

and its product with the plant transfer function becomes

$$G(S) = k\frac{(0.02019S^2 + 0.1421S + 1)50(-S + 100)}{0.1421(S^4 + 101S^3 + 150S^2 + 5000S)}.$$

The new numerator and denominator polynomials, for k=1, are given to MATLAB as

```
num2=50*[-.02019 1.877 13.215 100];
den2=.01421*[1 101 150 5000 0];
```

with the result labeled as "compensated $G(j\omega)$, $k = 1$" in the Nichols chart. This contour now can be translated upward to give a reasonable maximum value for the closed-loop frequency response. A 13-db shift, or $k = 10^{13/20} = 4.47$, gives the final contour. The polynomial `num2` was multiplied by 4.47 to get this curve.

   The `step` command of MATLAB was used to verify the desirability of the controller, with the result below. The overshoot is larger than might be desired, but it might be acceptable considering the challenge in finding a reasonable solution. The steady-state error is zero, as specified. Note that the response starts out in the negative direction, which is a result of the zero in the right-half plane of the plant.

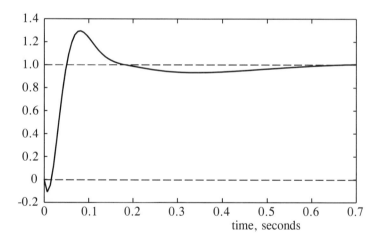

8.3.6 Approximate Correction for Digital Sampling

Digital control systems, as noted briefly in Section 8.2.7 (p. 552), normally employ samplers with holds. A sinusoidal signal of relatively low frequency thus treated becomes a series of steps that resembles a sine wave. The principal effect is a phase lag in the output of the sampler relative to its input. The phase lag equals 180° times the ratio of the sampling time to the period of the sine wave. If ten samples are taken per cycle, for example, the phase lag is 18°. If the open-loop frequency contour of a system closely passes the −1 point at this frequency, the sampler would degrade the relative or absolute stability of the closed-loop system. The amplitude of the signal emerging from the sampler is reduced from that of the input, but only slightly if there are ten or more samples per cycle, so for most purposes this effect may be neglected. If you have fewer than ten samples per cycle at a frequency critical to stability, you need to look more deeply into the subject, including the role of digital filtering used to address problems of signal noise.

EXAMPLE 8.19

The controller in the system of Example 8.16 (pp. 566-567) gives a sample-and-hold output, with a rate of two samples per second. Modify the Nichols contour approximately to account for the effect of the sampler, and estimate the revised value of the gain k that gives the same maximum peaking ratio $M_m = 1.5$ (3db).

Solution: The phase lag introduced by the sampler would be 180° at a frequency of $\omega = 2\pi/T = 2\pi/0.5 = 4\pi$ rad/s, or $\phi_{lag} = 180\omega/4\pi$ degrees. The phase angles of the various frequencies along the Nichols contour can be determined with the command `[m,p,w]=nichols(num,den)`, as done for example in Examples 8.17 and 8.18. Adding the phase lag from the sampler for the critical region near $-180°$ gives the contour indicated by the leftward shifted curve in the plot below. This curve has to be translated downward by about 7.5 db to give the desired peaking, so $k = 10^{-7.5/20} = 0.42$.

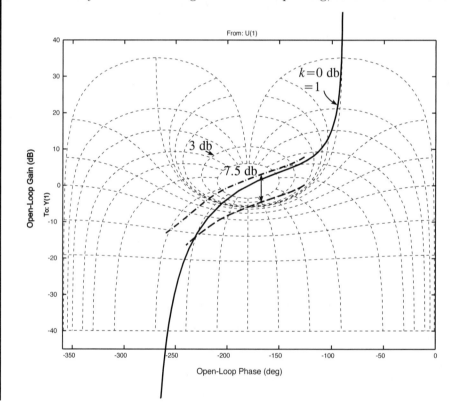

8.3.7 Special Roles for Bond Graphs in Control System Design

Relatively recently bond graphs, long used for general system design and analysis, have been used to advantage specifically in the design of control systems. A common tool for much of this work is the **bicausal bond graph** as developed by Gawthrop[4]. Gawthrop with others

---

[4]P.J. Gawthrop, "Bicausal Bond Graphs," *Proc. Int. Conf. on Bond Graph Modeling and Simulation (ICBGM'95)*, Society for Computer Simulation Simulation Series v 37 n 1, ed. F.E. Cellier and J.J.Granda, 1995, p. 83-88.

have applied this to their bond-graph-based **virtual actuator control**[5,6]. Other uses of this tool include the critical determinations of the **controllability** and **observability** of physical plants in terms of specific potential control inputs and measurement outputs. These determinations are associated intimately with the **invertibility** of a bond graph model of the plant, namely whether the roles of the chosen causal inputs and the causal outputs of the graph can be inverted[78]. This work also determines the minimum size of actuators and power needed for accomplishing a pre-determined control task. Yet another related development regards **input-output decouplability**[9].

8.3.8   Summary

This section has added frequency response methods to your arsenal of tools available for the design of feedback controllers. Which method or combination of methods works best for you depends on the particular situation. Frequency response methods are especially attractive when all you have to describe your plant is frequency-response data, for example found using an oscillator or a shaker table. It is possible to use the identification methods of Section 6.2 to get an analytical model from such data, making the analytical pole-placement methods such as root locus available, but that is not necessary. Design methods based on frequency response, particularly using the Nichols chart but also using the more basic Bode diagrams, provide a powerful predictive insight that warrants their use even when analytical models are known.

Polar or Nyquist plots of frequency response combine magnitude and phase information into a single curve, an advantage over the Bode plot. Although these plots can be used to predict quantitatively the behavior of closed-loop behavior, emphasis has been placed merely on the prediction of absolute stability via the Nyquist criterion. For more refined design and prediction, the alternative mapping called the Nichols chart is usually more convenient and has been emphasized. All these forms are readily available in the control toolbox of MATLAB and in the student version of MATLAB. They all can be used to determine the gain and phase margins of a system from its open-loop transfer function. These meassures, plus the maximum peaking of the closed-loop frequency response, often are cited by control engineers in describing the behavior and the robustness of their designs.

PI, PD, PID, phase lead, phase lag and phase lead-lag controllers are characterized in Figs. 8.23 and 8.25 (pp. 568, 571) in the form of Nichols contours. Their application is given in several examples, including the Guided Problem below. Small digital sampling delays introduce an effective phase delay equal to the ratio of the frequency to the sampling frequency, times $180°$.

Guided Problem 8.3

This problem involves the Nyquist stability criterion and the design use of the Nichols chart in an interesting case for which a phase lead controller outperforms a PD controller.

The control of an inverted pendulum in Guided Problem 8.2 (p. 553) has two parts. The first is the quite successful use of a PD controller to keep the pendulum upright. The

---

[5]P.J. Gawthrop, D. J. Ballance and D. Fink, "Bond graph based control with virtual actuators," *Proc. 13th European Simulation Symposium: Simulation in Industry*, SCS, ed. N. Giambiasi and C. Frydman, Marseille, France, Oct. 2001, pp. 813-817.

[6]P.J. Gawthrop and D. J. Ballance, "Virtual Actuator Control of Mechanical Systems," *Proc. ICBGM'05*, SCS, ed. F.E. Cellier and J.J. Granda, Jan.2005, pp. 233-238.

[7]Ngwompo, Scarvada, Thomasset, "Physical Model-Based Inversion in Control Systems Using Bond Graph Representation, Part 1 Theory," *Proc. IMechE Part I J. Systems and Control Engineering*, v.215, 2001, pp. 95-103.

[8]*Ibid, Part 2 Applications,"* pp. 105-112.

[9]Dauphin-Tanguy, Rahmani and Suer, *SIMPRA,* 1999.

second is the use of a feedback loop with another PD controller to allow the position, x, of the pendulum to be controlled as well. The second control involved a trade-off between the speed of response and the relative stability, with a result that begs improvement, if possible. Further, you were guided rather strongly in the key suggested step 7; now that you have further tools in your arsenal you can assume more initiative.

Leave the result of the first part of the contol unchanged. You also may retain the block diagram of step 5 (p. 557), except replace the proportional controller (which doesn't work, as step 6 revealed) with a lead controller. Choose the two parameters of this controller, find the resulting roots of the closed-loop performance, and compare to the results of the PD controller.

Suggested Steps:

1. The plant of this system has the force $-F$ as its input, the position x as its output, and a transfer function equal to the product of the transfer functions given in the two right-most blocks in the block diagram of the original step 5. Verify the instability of proportional control on this plant through the use of a Nyquist plot. Pay particular attention to the infinite segments of the Nyquist contour, which are not given by MATLAB.

2. Examine the Nichols chart contour for the same transfer function, and determine qualitatively how the controller must change this contour to achieve stability. Does the phase lead controller potentially satisfy the requirement? (Note that sometimes the MATLAB program `nichols` gives phase angles that are 360° larger than desired, with the result that the M and N contours are not available. You may therefore want to refer to these contours as given in Fig. 8.22 (p. 566). This difficulty will vanish when the contoller is inserted.)

3. Choose the size of the controller (the parameter α) and an approximate value for the "nose" frequency of the controller, through examination of the plant contour and the controller contour as given in Fig. 8.23 (p. 568). You may wish to magnify the plot through the use of the `axis` command.

4. Deduce the value of the frequency parameter z that follows from your choices, and then write the transfer function for the resulting controller. At this point the value of the gain k is not yet determined; set it at 1.

5. Find the complete open loop transfer function, using your controller, and plot its contour using `nichols`. Repeat this step and the last if you are not satisfied with the result.

6. Examine the contour to see how many decibels it should be raised or lowered in order to give the smallest maximum magnitude to the closed-loop frequency response. Adjust the gain, k, correspondingly, and repeat the contour to verify your choice.

7. Find the characteristic equation for your closed-loop system, and use the command `roots` to determine its dynamic behavior. Compare with the results for the PD controller as given in step 7 on page 558.

8. Deduce the closed-loop transfer function, and plot the step response. Is the result qualitatively consistent with your earlier findings?

PROBLEMS

8.17 The following Bode plots represent the plants in unity-feedback systems with a proportional controller. A gain margin of 2.5 is specified. Estimate the associated phase margins, the gains of the controller and the steady-state errors. Work from the given Bode plots directly, as though they represented experimental data.

(a) Bode plot of Example 7.6 (p. 433).

(b) Bode plot of Example 7.9 (p. 437).

(c) Bode plot for $G_1(S)$ in Example 7.11 (p. 439).

(d) Bode plot of part(c) in Fig. 7.13 (p. 447).

8.18 The following Bode plots represent the plants in unity-feedback systems with a proportional controller. A phase margin of $30°$ is specified. Estimate the gain of the controller and the steady-state error. Work directly from the plots, as though they represent experimental data.

(a) Bode plot of Example 7.6 (p. 433).

(b) Bode plot of Example 7.9 (p. 437).

(c) Bode plot for $G_1(S)$ in Example 7.11 (p. 439).

(d) Bode plot of Example 7.12 (p. 440).

8.19 Hand sketch Nyquist plots for the systems represented by the following Bode plots. Indicate whether they would be stable if placed as the forward path in unity feedback systems.

(a) Bode plot of Example 7.7 (p. 434).

(b) Bode plot of Example 7.8 (p. 435).

(c) Bode plots (both) of Example 7.10 (p. 438).

(d) Bode plot of Problem 7.21 (p. 451).

(e) Bode plot of Problem 7.22 (p. 452).

(f) Bode plot of part (c) of Fig. 7.13 (p. 447).

8.20 Consider the system represented by the block diagram

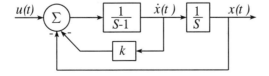

in which $\dot{x}(t)$ is a velocity and $x(t)$ is its integral, a position.

(a) For the case with $k = 0$ (zero velocity feedback), plot the Nyquist contour, including any infinite radius segments, and deduce from there the number of any destabilizing poles in the right-half plane.

(b) State the net number of clockwise or counterclockwise encirclements of the -1 point that the contour modified by velocity (derivative) feedback must exhibit in order to represent a stable system.

(c) Plot the Nichols contour for the system with $k = 0$.

(d) Plot the Nichols contour for other values of k; determine what value gives stability with an approximate peaking of 3 db.

(e) Confirm the stability with a Nyquist plot.

8.21 Use MATLAB to plot the Nichols contour for the system

$$G(S) = \frac{k}{(1 + S/10)(1 + S/50)(1 + S/200)}$$

with $k = 1$. If $G(S)$ is the forward path in a unity-feedback system, estimate the gain k for which the maximum peaking is 6 db. (Note: This is the $G(S)$ of Example 7.6, pp. 432-433.)

8.22 Use MATLAB to plot the Nichols contour for the plant

$$G(S) = \frac{1}{(1 + S/10)(1 + 0.0013S + S^2/10,000)}$$

(which is the system of Example 7.7, p. 434). Specify a lag controller with $\alpha = 10$ that would allow the output to better track the input than would a proportional controller. Report the resulting steady-state error.

8.23 Carry out the above problem, substituting a PI controller.

8.24 Carry out the above problem, substituting a PID controller.

8.25 A PID controller is applied to a particular plant in Example 8.18 (p. 572). Lag-lead control is not as effective, but substitute such a controller and at least achieve a fairly small steady-state error and plausible dynamics. The following steps are suggested:

(a) Secure a Nichols contour of the plant.

(b) Examine the contour with an eye toward finding an acceptable lead controller. The nose of a lead controller works best where the contour is most steeply sloped, so pay most attention in your selection of the nose point to the region below and to the left of the -1 point. You obviously will need the maximum value of α available, which is assumed to be 10. Note that the maximum phase shift of the controller is about 55°, so you can't go too far down the contour.

(c) Determine the frequency of your selected nose point, using MATLAB.

(d) Deduce the value of z, and write the transfer function for the lead compensator, leaving $k = 1$. Multiply this by the transfer function of the plant to get the transfer function of the combination. Secure a new Nichols contour for the combination, and see if it looks promising. (The lag portion to come will tend to push the contour back to the left somewhat, so it is best to leave some margin. Also, issue the "hold on" command so you do not have to repeat the original plant contour.)

(e) The phase lag compensator comes next. Choose a nose frequency that is at least ten times (and perhaps more) smaller than that for the phase lead part of the controller, so as not to undo what already has been accomplished. The largest value of β available (10) should be used.

(f) Determine p in essentially the same way you found z, write the transfer function of the lag controller, and multiply it by the result of step (d) to get the overall open-loop transfer function. Plot its Nichols contour.

(g) The final design step is to increase the gain as high as relative stability permits, in order to reduce the steady-state error. This also makes the system as fast as practicable. Report the maximum value of the closed-loop frequency response from examination of the final open-loop Nichols contour.

(h) Write the closed-loop transfer function, and insert it into the `step` command to get a better comparison with the result of the PID controller in Example 8.18.

8.26 In Fig. 8.24 (p. 568) (and Example 8.10, p. 549), a mass-spring system is controlled with unity feedback and a phase lead controller. If the control implementation is changed from continuous analog to digital sampling with a zero-order hold signal sent to the plant, estimate from the plot the minimum sampling rate necessary to keep the Nichols contour from causing a peaking greater than 3 db. Ignore any digital filtering that might in fact require a higher sampling rate.

SOLUTION TO GUIDED PROBLEM

Guided Problem 8.3

1. The following commands define the plant and give the Nyquist plot at the top of the next page (except for the infinite-radius circles, the text and the dashed lines):

```
num=32.2*[-1 -15 50 750]; den=[1 50 650 0 0];
nyquist(num,den)
axis([-5 5 -5 5])
```

The -1 point is encircled twice, clockwise. Since there are no poles of the plant transfer function in the right-half plane, this means that a unity-feedback system with this plant has two right-half-plane poles, and is unstable. If the segment of the locus that is moving rightward toward the origin were pushed downward, passing below instead of above the -1 point, there would be no encirclements and the closed-loop system would be stable. This idea is suggested by the dashed lines. Specifically, the contour would have to have its phase angle reduced below $-180°$ in the vicinity of the -1 point.

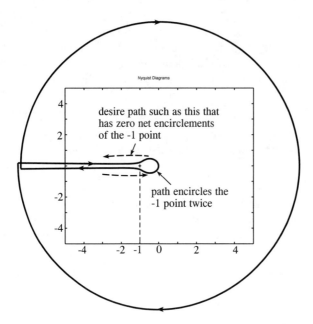

2. The Nichols plot below (apart from the text and the dashed line) follows from the subsequent commands

```
nichols(num,den)
axis([90 270 -20 20])
```

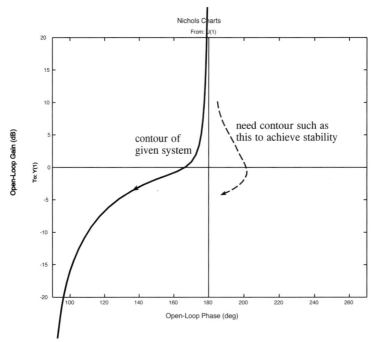

When a Nichols contour resides exclusively in the first and/or second quadrants, MATLAB (version 5.3) assigns positive values to the phase angles, which unfortunately makes the M and N contours unavailable. You should interpret the phase angles as being 360° less than the reported values, so the −1 point is at the center of the diagram. The type of contour needed to achieve stability is indicated by the dashed line; the phase lead needed is small enough to be provided by a phase-lead controller, as long as the nose of the controller is placed appropriately.

3. The large size lead controller with $\alpha = 10$ is chosen. With this controller, the nose point on the contour gets pushed downward by 10 db as well to the right by about $55°$. This suggests choosing a point at about 5 db and $-185° = 175°$. The commands

   ```
   [mag,phase,w]=nichols(num,den);
   [log10(w),20*log10(mag),phase]
   ```

   produce a table of the plotted points, from which the value $\log_{10} \omega_{nose} = 0.8$ or $\omega_{nose} = 10^{0.8} = 6.3$ rad/s can be inferred.

4. From Fig. 8.23, $a = \sqrt{10}$ and $\Omega_{nose} \equiv \omega_{nose}/z = a$. Therefore, $z = 6.3/\sqrt{10} = 2.0$ rad/s. The transfer function of the controller becomes

   $$G_c(S) = k\frac{S+2}{S+20}.$$

5. The open-loop transfer function including the controller becomes

   $$G(S) = \frac{32.2k(-S^4 - 17S^3 + 20S^2 + 850S + 1500)}{S^5 + 70S^4 + 1650S^3 + 13,000S^2 + 0S + 0}.$$

   With this $G(S)$, MATLAB gives the contour on the right below:

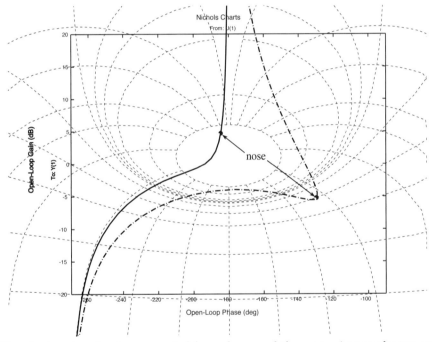

   The plant contour has been repeated from above, and the nose points on the two contours are indicated.

6. Raising the contour would worsen the high-fequency peak where the contour passes directly below the -1 point. Lowering the contour would worsen the low-frequency peaking where the contour also passes somewhat close to the 6 db closed-loop contour. The best compromise might be to lower the contour by about 1 db, setting $k = 0.9$. For present purposes, however, $k = 1$ is retained, so no new contour is found.

7. The characteristic equation of the closed-loop system becomes

   $$S^5 + 37.8S^4 + 1102.6S^3 + 13,644S^2 + 27,370S + 48,300 = 0.$$

   The `roots` command of MATLAB gives the characteristic values

   $-9.84 \pm 24.75i$
   -16.1057
   $-1.01 \pm 1.79i$

The PD controller, as reported on page 557, produced a somewhat slower and significantly less damped low-frequency oscillation. The advantage of the lead controller results from the limit it places on the magnitude of the controller gain for high frequencies. The higher phase lead that the PD controller has at these frequencies is wasted.

8. The denominator of the closed-loop system comprises the characteristic polynomial above, and the numerator is the same as for the open-loop system. The use of **step** gives the plot (solid line)

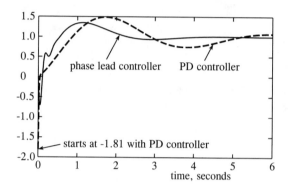

which represents the position of the bottom of the pendulum given a commanded step displacement. Notice that the bottom must start by moving in the opposite direction from the commanded final change, which should be intuitive (try it with a ruler or other stick). The slower oscillation can be seen to be likely more critical than the faster oscillation, even though it is somewhat more damped, because of its effect on the settling time.

The dashed line represents the inferior control provided by the PD controller (from Guided Problem 8.2). The instantaneous jump to -1.81 at $t = 0$ would require an infinite force, and is not realistic.

Chapter 9

Extended Modeling

This chapter greatly extends the domain of physical and engineering systems amenable to modeling and analysis by the unified methods presented in this book. It does so without applying analytical methods beyond the reach of undergraduate engineering programs. Certain more sophisticated topics in modeling are deferred to later chapters.

Section 9.1 allows the moduli of transformers and gyrators to vary, becoming functions of the displacement at either end of the elements or functions of remote displacements. Such parameter modulations occur quite often. Modelers also often wish to employ two-port elements that, like one-port resistances, do not conserve energy. The **activated bond** is introduced in Section 9.2 as an elemental non-conservative coupler.

Compliances, inertances and resistances are allowed, starting in Section 9.3, to have two or more bonds. Such **multiport fields**, which may be linear or nonlinear, are especially adept at treating components that involve two different energy domains. Examples include electromechanical and mechanical/thermal devices. They also can be helpful in analyzing classes of two-port devices with a single energy domain, such as real electrical transformers and mechanical members with more than one degree of freedom.

Systems involving magnetic fields have been viewed previously from the relatively remote electric domain of volts and amperes, and sometimes force and velocity. Entering the domain of magnetic forces and fluxes directly, as shown in Section 9.5, permits a more detailed and explicit consideration of the physics involved, including magnetic hysteresis. The tools of this section and the multiport fields of the previous two sections aid in the analysis of selected electromechanical motors, which is the subject of Section 9.6.

The static couplers considered thus far, including transformers, gyrators and resistance fields, are *reversible* in the thermodyanic sense, generating no entropy. An **irreversible coupler**, introduced in Section 9.7, conserves energy like the transformer and gyrator but, unlike them, generates entropy. Irreversible couplers expedite the modeling of systems with heat conduction, including the heat generated by friction and electrical resistance. The proper flow or generalized velocity for heat conduction is identified, and thermal machines and thermal compliances are introduced. Models with mass transfer or convection, on the other hand, are deferred to Chapter 11.

9.1 Modulated Transformers

The definition of the ideal machine in Chapter 2 permits its modulus to be non-constant, a function of one or more variables. The two most common cases are a transformer modulated by a *remote* displacement, and a transformer modulated by a *local* displacement. Local

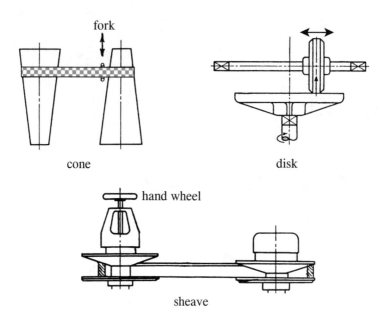

Figure 9.1: Adjustable mechanical drives

displacements are the displacement variables on the two sides of the transformer; any other displacement is called remote. The distinction is significant.

9.1.1   Remotely Modulated Transformers

The mechanical friction drives shown in Fig. 9.1 can be modeled as ideal transformers with moduli that can be changed by adjusting the fork, the handwheel or the axial postion of the wheel. (The disk-and-wheel drive may be familiar because of its common usage on self-propelled lawn mowers.) The transmissions in vehicles are in the same category, although the modulations usually are discrete rather than continuous. The modulus of electrical transformers (or "variacs") can be changed by adjusting a slide-wire that varies the turns ratio. Simple purely fluid transformers do not exist, but the moduli of many mechanical-to-fluid pumps and motors are adjustable; one example is the variable swash-plate pump/motor shown in Fig. 9.2. In all these cases the modulating variable is a remote displacement, as contrasted to the rotational angle of one of the drive shafts, the electrical charge on one side of the transformer, or the angle of the pump/motor shaft or the integral of the volume flow. These later displacements are *local*.

Some devices mimic transformers but are not; they behave like transformers only as long as the modulus is held constant, at any value, but *require significant power or energy to effect a change in modulus*. An example is given in Guided Problem 9.9. For the present, you should simply recognize that a true transformer is strictly a two-port device in terms of power and energy. Any modulation must be achieved without any *essential* generalized force, power or energy. A third port would be needed to represent a modulating force and its associated power and energy, which the transformer does not have. Consider the examples of Figs. 9.1 and 9.2. Is there any *essential* force required to effect the modulation? A real device always presents some frictional force to overcome, but this is theoretically infinitely reducible and can be modeled separately (by an added junction with appended resistance element) if it is deemed important, leaving a transformer as the core element.

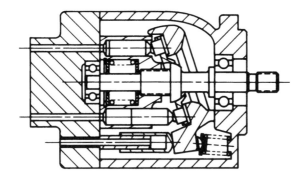

Figure 9.2: Pump/motor with adjustable displacement, swash-plate type
Reprinted with permission and courtesy of Eaton Corporation

The control of large power flows from motors, IC engines, turbines and electrical and fluid power supplies depends upon the existence of physical devices that very closely approximate the ideal remotely modulated transformer. Such devices are modeled, therefore, by transformers, with resistances added as appropriate to account for the minor losses.

9.1.2 Locally Modulated Transformers

The cam drive system shown in Fig. 9.3, when idealized to neglect friction and elasticity, is an example of a *locally modulated* transformer. The transformer modulus is the ratio of $\dot{x}$ to $\dot{\phi}$, and this ratio is a function of the position (angle) of the cam, or $T = T(\phi)$; ϕ is one of the two local displacements. Alternatively, you could specify $T = T(x)$; this is an unwise choice for most applications, however, since there are two values of ϕ for most values of x.

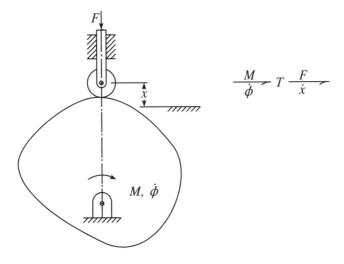

Figure 9.3: Drive with cam and roller follower

The following example is similar.

EXAMPLE 9.1

Show that the slider-crank mechanism pictured below can be modeled as a locally-modulated transformer, and express its modulus as a function of the crank angle, ϕ.

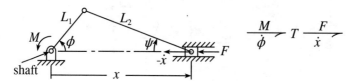

Solution: The output position x is a function of the input angle, ϕ, and therefore the output velocity $\dot{x}$ can be expressed as the product of a function of ϕ and $\dot{\phi}$, qualifying the device as a locally modulated transformer. The calculation of $T(\phi)$ is as follows. The geometry requires

$$L_1 \sin \phi = L_2 \sin \psi,$$
$$x = L_1 \cos \phi + L_2 \cos \psi.$$

From the first equation,

$$\cos^2 \psi \equiv 1 - \sin^2 \psi = 1 - (L_1/L_2)^2 \sin^2 \phi,$$

so that the second equation can be written in terms of ϕ as

$$x = x(\phi) = L_1 \cos \phi + L_2 \sqrt{1 - (L_1/L_2)^2 \sin^2 \phi}.$$

The modulus of the transformer is the ratio of $\dot{x}$ to $\dot{\phi}$:

$$\dot{x} = T(\phi)\dot{\phi}.$$

This modulus can be found by differentiating the expression for x above, with the result

$$T(\phi) = - \left[1 + \frac{L_1 \cos \phi}{L_2 \sqrt{1 - (L_1/L_2)^2 \sin^2 \phi}} \right] L_1 \sin \phi.$$

The drive with the circular cam and roller follower pictured in Fig. 9.4 acts just like the slider-crank mechanism. As shown by the dashed lines, the effective length of the crank, L_1, is the eccentricity of the cam, and the effective length of the connecting rod, L_2, is the sum of the radii of the cam and the follower. This assumes the cam follower maintains contact with the cam.

9.1.3   Increase in the Order of a Model Due to Modulation

The order of a model with no modulated transformers or gyrators equals the number of independent energy-storage elements. When a transformer modulus is modulated locally, however, the order often increases by one. This means that an additional first-order differential equation is needed. One case of this type is examined below, followed by a counter case with no such increase, and a third case in which there is a choice. All employ the slider-crank mechanism.

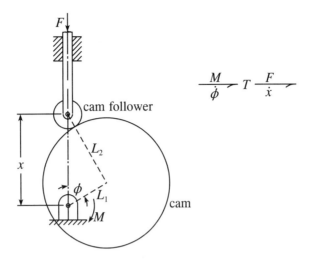

Figure 9.4: Drive with circular cam and roller follower

EXAMPLE 9.2

A slider-crank mechanism is driven from a flywheel with significant inertia, and drives a load comprising a series connection of a spring and a dashpot. Other inertias are to be neglected.

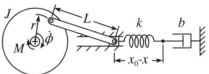

Model the system with a bond graph, and find the state differential equations. Identify the order of the system (the number of independent differential equations).

Solution: The bond graph, annotated with the usual integral causality, is

The effort of the 0-junction is labeled F, for convenience. The energy storage state variables p and x produce in standard fashion the differential equations

$$\frac{dp}{dt} = M - TF = M - \frac{T}{C}x,$$

$$\frac{dx}{dt} = T\dot{\phi} - \frac{1}{R}F = \frac{T}{I}p - \frac{1}{RC}x.$$

The modulus T in the right sides of these equations is given by the result of Example 9.1 as a function of the angle ϕ. To make the equations solvable, therefore, ϕ must be added to the list of state variables, which becomes p, x and ϕ. The associated state differential equation can be seen from the annotations on the bond graph to be

$$\frac{d\phi}{dt} = \frac{1}{I}p.$$

The model, then, is third order: one for each of the energy storage elements, and one for the modulated transformer.

In general, a modulated transformer will add one order, unless the argument on which it depends is already a state variable or a function of state variables. This exception is exemplified in the following:

EXAMPLE 9.3

The slider-crank drive from the previous two examples is retained, but the crank is attached to ground through a spring (and therefore cannot be rotated indefinitely) and a substantial mass is placed at the right end:

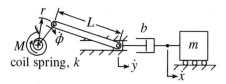

Again, model the system with a bond graph, and find the state differential equations. Note the order of the system.

Solution: The bond graph with the usual annotations is

$$
\begin{array}{c}
M \\
\xrightarrow{\quad} 1 \xrightarrow{\quad} T \xrightarrow[\dot y]{\quad} 0 \xrightarrow{\quad} R \\
\phi/C \;\Big|\;(\phi) \qquad\qquad (\dot p)\Big| p/I = \dot x \qquad I = m \quad C = 1/k \quad R = b \\
C \qquad\qquad\qquad\quad I
\end{array}
$$

The compliance (spring) renders the angle ϕ a state variable. Therefore, the argument of the function $T(\phi)$ is calculable, and the second-order model

$$
\frac{d\phi}{dt} = \frac{\dot y}{T(\phi)} = \frac{1}{TI}p + \frac{1}{RT^2(\phi)}\left(M - \frac{\phi}{C}\right),
$$

$$
\frac{dp}{dt} = \frac{1}{T(\phi)}\left(M - \frac{\phi}{C}\right).
$$

is complete.

Sometimes there is a *choice* of the order of a model for a given system. The **minimum order** of a model is the smallest order that can be achieved by choosing among alternative state variables. Consider the following variation on the application of the slider-crank mechanism:

EXAMPLE 9.4

Model the slider-crank system below considering $\dot\phi$ (rather than M) as the causal input. Also, negelect inertia. Write the state differential equation(s) and identify the order of your model.

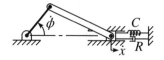

Solution: The bond graph reveals a single energy-based state variable:

$$S_f \vdash \frac{M = T(F_C + F_R)}{\dot{\phi}} \; T \vdash \frac{F_C + F_R}{} \; \underset{\dot{x} = T\dot{\phi}}{1}$$

with $F_C = x/C$, C, $\dot{x}$, $F_R = R\dot{x}$, R.

The differential equation for this x is

$$\frac{dx}{dt} = T\dot{\phi}, \qquad T = T(\phi).$$

The modulation $T = T(\phi)$ introduces ϕ as the second state variable, with the differential equation

$$\frac{d\phi}{dt} = \dot{\phi}(t).$$

This equation appears to be an identity, but since $\dot{\phi}(t)$ is given, it must be integrated to get ϕ. The simulation model therefore is second order.

The *mimimum order* of the model actually is *one* rather than two. This is because it is possible to express T as a function of x rather than ϕ, eliminating the need to compute ϕ:

$$T = T(x).$$

In order to carry this out, the expression for $x = x(\phi)$ (found in Example 9.1, p. 586) must be inverted to give an equation of the form $\phi = \phi(x)$. The inversion is so awkward algebraically, however, that almost surely it is wiser to stick with the second-order model.

9.1.4 Dependent Inertance with a Locally Modulated Transformer

An over-causal model with its dependent energy-storage element and its differential causality can be treated either of two ways, as described in Sections 6.1 and 6.2. First, you may implement the differential causality directly. Second, you may eliminate the dependency through the use of an equivalent bond graph with one fewer inertance. These approaches can be applied to models with locally modulated transformers, as the following pair of examples illustrates.

EXAMPLE 9.5

Consider a slider-crank mechanism with inertias at both ends:

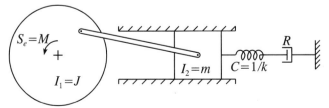

Model the system with a bond graph, apply causal strokes in the usual fashion, and write the state differential equations.

Solution: The bond graph reveals that one of the two inertances must be assigned differential causality. The solution below is based on choosing integral causality for I_1, forcing differential causality for I_2:

Note particularly the annotations for $\dot{x} = T\dot{\phi}$ and $I_2\ddot{x}$, which are mandated by the causal strokes. ($\dot{x}$ and $\ddot{x}$ are temporary variables used for algebraic convenience.) The differential equations for the two energy storage elements with integral causality become directly

$$\frac{dq}{dt} = \dot{x} - \frac{1}{RC}q = \frac{T}{I_1}p - \frac{1}{RC}q,$$

$$\frac{dp}{dt} = M - T\left(\frac{1}{C}q + I_2\ddot{x}\right).$$

You need to find $\ddot{x}$ as a function of the state variables p and/or q in order to place the right side of the second equation in solvable form. Since $\dot{x} = (T/I_1)p$ and $\dot{\phi} = p/I_1$, you can write

$$\ddot{x} = \frac{T}{I_1}\frac{dp}{dt} + \frac{dT}{d\phi}\frac{d\phi}{dt}\frac{1}{I_1}p = \frac{T}{I_1}\frac{dp}{dt} + \frac{dT}{d\phi}\frac{1}{I_1^2}p^2.$$

Substituting this result, and collecting the two terms in dp/dt, gives

$$\frac{dp}{dt} = \frac{1}{1 + (I_2/I_1)T^2}\left(M - \frac{T}{C}q - \frac{TI_2}{I_1^2}\frac{dT}{d\phi}p^2\right).$$

Adding the following third state differential equation to the first and fourth equations above gives a solvable set with the state variables q, p and ϕ:

$$\frac{d\phi}{dt} = \frac{1}{I_1}p.$$

EXAMPLE 9.6

Find a set of state differential equations for the example above, except this time reduce the number of inertances so as to preclude any differential causality.

Solution: You have a choice of relating the inertance energies to either $\dot{\phi}$ or $\dot{x}$. Electing $\dot{\phi}$, the total kinetic energy becomes

$$T = \frac{1}{2}I_1\dot{\phi}^2 + \frac{1}{2}I_2\dot{x}^2 = \frac{1}{2}(I_1 + T^2I_2)\dot{\phi}^2, \qquad T = T(\phi).$$

Were T a constant, you could simply define an equivalent inertance equal to $I_1 + T^2I_2$.

If you try this now, however, you get an inertance which is a function of a displacement. The scheme for writing differential equations does not permit this; all inertances must either be constants or (if nonlinear) functions of the local generalized *velocity*. To accommodate this restriction, you insert a new transformer indicated as T_1:

You could choose to use *any* constant inertance, but assume you keep the original I_1 as shown. The energy stored is $\frac{1}{2}T_1^2 I_1 \dot{\phi}^2$. Setting this equal to the proper value as given by the equation above for T, there results

$$T_1 = \sqrt{1 + T^2 I_2 / I_1}, \qquad T = T(\phi).$$

Examination of the annotated bond graph gives the complete set of state differential equations:

$$\frac{dq}{dt} = \frac{1}{I_1}\frac{T}{T_1}p - \frac{1}{RC}q,$$

$$\frac{dp}{dt} = \frac{1}{T_1}\left(M - \frac{1}{C}Tq\right),$$

$$\frac{d\phi}{dt} = \frac{1}{I_1 T_1}p.$$

The result of this method is somewhat simpler than that produced by the method used in Example 9.5, particularly due to the absence of the derivative $dT/d\phi$. It also is easier to find and provides a more concise perspective of the essential dynamics of the system. The first method has one advantage, however: it is less sophisticated.

The general method, applied above to generalized kinetic energies or inertances, also applies to generalized potential energies or compliances with differential causality, usually with less difficulty. There are the same two choices: using differential causality on the original bond graph where necessary, or restructuring the graph through an energy analysis to eliminate dependent compliances.

9.1.5 Inertance Dependent on Local Displacement; Case Study

The technique employed in Example 9.6 above can be applied also to the broader class of models in which an inertance depends on a local displacement. Specifically, the introduction of a modulated transformer into a bond for an inertance allows its generalized kinetic energy to be a function of the local displacement, while retaining the requirement that the inertance itself not be a function of any displacement. The procedure is best understood through a case study.

The rigid rocker shown in Fig. 9.5 rolls without slipping on a cylindrical surface of radius R. Since the instantaneous center of the rocker changes with the attitude ϕ, the inertance of the rocker is not constant but depends on ϕ. The integral $p = \int M\, dt$, where M is the moment acting on the rocker, is not a function of the state of the system, so any attempt to *define* the inertance I so that the velocity $\dot{\phi}$ equals p/I will fail. Instead, the insertion of a locally modulated transformer allows the inertance to be kept constant.

The approach starts by writing the kinetic energy T as a function of the generalized displacement (in the present case, the angle ϕ) and the generalized velocity (the angular velocity $\dot{\phi}$):

$$T = \tfrac{1}{2}\left[J + m\left(b^2 + R^2\phi^2\right)\right]\dot{\phi}^2. \tag{9.1}$$

J is the rotational inertia of the rocker about its center of mass; the first term on the right represents the rotational kinetic energy. The mass of the rocker is m; the other term represents the translational kinetic energy. Note that the displacement ϕ appears in this term. Upon examination of equation (9.1) it might seem reasonable to *define* the inertance as $I = J + m(b^2 + R^2\dot{\phi}^2)$, but if this is done the moment is *not* $I\ddot{\phi}$ and the angular velocity $\dot{\phi}$ is *not* p/I with $\dot{p} = M$. Therefore, you should not make this definition. Rather, you define

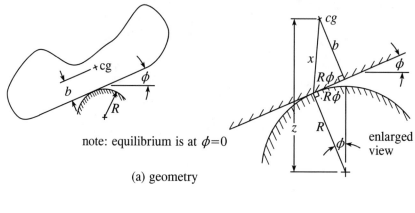

note: equilibrium is at $\phi = 0$

(a) geometry

$$C \xrightarrow[\displaystyle \dot{\phi}]{\phi/C} 1 \xrightarrow[\displaystyle p/IT]{-\phi/C} T \xrightarrow[\displaystyle \dot{q}=p/I]{\boxed{\dot{p}}} I$$

(b) bond graph

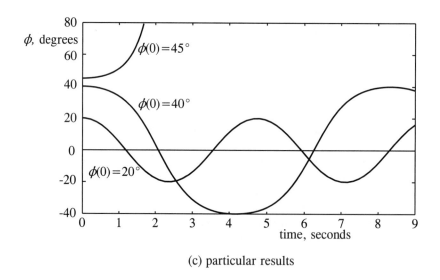

(c) particular results

Figure 9.5: Rigid member with planar surface rocking on a cylinder

a new flow $\dot{q}_c$ and transformer modulus T such that the inertance is defined as a *constant*:

$$\dot{q}_c = T\dot{\phi} \qquad T = \sqrt{1 + \frac{m}{J}(b^2 + R^2\phi^2)}. \tag{9.2}$$

This gives the bond graph of part (b) of the figure with the simple constant inertance

$$I = J. \tag{9.3}$$

Since this inertance is a constant, the flow $\dot{q}_c$ equals p/I where p is the generalized momentum associated with $\dot{q}_c$, and p is the generalized force acting thereon.

Suppose equilibrium occurs when the planar surface of the rocker is horizontal, that is $\phi = 0$. The gravitational potential energy equals the mass times the elevation of the center of mass, or

$$\mathcal{V} = mg[(R+b)(\cos\phi - 1) + R\phi\sin\phi], \tag{9.4}$$

so that the moment is

$$M = \frac{\phi}{C} = \frac{d\mathcal{V}}{d\phi} = mg(R\phi\cos\phi - b\sin\phi). \tag{9.5}$$

Causal strokes now may be applied to this graph in standard fashion, giving as the state differential equations

$$\frac{dp}{dt} = -\frac{1}{TC}\phi, \tag{9.6a}$$

$$\frac{d\phi}{dt} = \frac{1}{TI}p, \tag{9.6b}$$

where T is as given in equation (9.2). If you prefer to find a single differential equation in terms of the angle ϕ, the second derivative of ϕ can be found from equation (9.6b) by substituting equation (9.6a) into the right side:

$$\frac{d^2\phi}{dt^2} = \frac{1}{TI}\frac{dp}{dt} - \frac{p}{IT^2}\frac{dT}{d\phi}\frac{d\phi}{dt} = \frac{1}{T^2IC}\phi - \frac{1}{T}\frac{dT}{d\phi}\left(\frac{d\phi}{dt}\right)^2$$

$$= \frac{-m}{J + m(b^2 + R^2\phi^2)}\left[g(R\phi\cos\phi - b\sin\phi) + R^2\phi\left(\frac{d\phi}{dt}\right)^2\right]. \tag{9.7}$$

Solutions for various initial conditions are plotted in part (c) of the figure. The rocker is unstable if it is given too large an initial displacement or velocity.

9.1.6 Summary

The power in machinery, vehicles, etc. is largely controlled through the use of remotely modulated transformers. By definition, these elements have only two power ports and two bonds; the modulation is effected without any energy transfer through the modulating variable. Locally modulated transformers model the way many engineering devices, particularly mechanisms, transmit generalized force and motion in a variable way depending on the generalized position of the device. Also, the kinetic energy of many engineering devices, again particularly mechanisms, depends on their positional state. The standard scheme for writing state variable differential equations requires that inertances have constant moduli, or at most a dependence on a local velocity. Use of the modulated transformer therefore greatly extends the domain of devices you can model. Sometimes you start with knowledge of the constraint, and sometimes you start with knowledge of the stored energy.

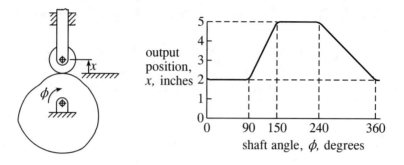

Figure 9.6: The cam system of Guided Problem 9.1

The modulation of a transformer usually adds one order, that is one first-order differential equation, to the equaivalent model with no modulation. There is no increase, however, if the variable that would be added already is a state variable, or could be expressed as a function of these variables. Sometimes you have a choice of which set of variables to use; a set with the minimum order is not necessarily the most practical.

Guided Problem 9.1

This first problem is intended to emphasize the primitive idea that locally modulated transformers display the same type of reciprocal behavior as transformers with constant moduli.

The drive shaft of a cam system pictured in Fig. 9.6 rotates at a constant 180 rpm, while the linear ouput displacement moves as plotted. The load force is a constant 200 lb, and the forces of inertia and friction are to be neglected. Sketch a plot of the torque on the drive shaft as a function of the angle of the shaft, labeling the values $\phi = 45°$, $120°$, $180°$ and $270°$.

Suggested Steps:

1. Draw a bond graph, labeling the moment, angular velocity, force and linear velocity.

2. Sketch a plot of $\dot{x}$ vs. ϕ. Note that this plot comprises well-defined straight-line segments connected by short smooth curved segments that are relatively poorly defined.

3. Write an expression for the conservation of energy, and solve for the moment. Alternatively, use your knowledge about ideal transformers.

4. Combine the results of steps 2 and 3 to find the desired moment.

Guided Problem 9.2

Experiencing the determination of the modulus of the transformer in this geometric problem is highly valuable.

An aircraft landing on the deck of a carrier extends a hook that catches a cable stretched across the deck between two pulleys, as shown in Fig. 9.7. The cables extend below deck, where they are attached to an "arresting engine" that for present purpuses can be modeled as a giant adjustable dashpot in parallel with a giant (air) spring. (More refined details of such an engine are given in Problem 5.60, p. 321, but are neglected here.) The motion of the cable forces oil through an orifice, the area of which is continuously varied to shape

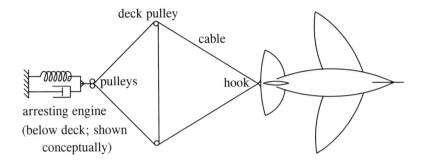

Figure 9.7: Aircraft arresting system for Guided Problem 9.2

the deceleration profile of the aircraft appropriately. This gives the effective dashpot. The displacement of the oil also compresses the air in a tank, providing the effective spring. The aircraft lands under full throttle, just in case the hook doesn't catch. Since the deceleration takes only a couple of seconds, the aircraft engine isn't shut down until afterward. You are asked to model the system, and write a set of state differential equations.

The problem is continued in Problems 9.13 and 9.14 (p. 602) with simulations and design of a position-dependent damping profile to give desirable performance.

Suggested Steps:

1. The geometry of the pulleys makes the velocity of the cable below deck less than the velocity of the aircraft. Represent the relationship by a locally modulated transformer. To find its modulus, define the two displacements at each end, express the geometric interrelationship algebraically, and compute the derivative of this relation.

2. Draw a bond graph model for the system, including the inertia of the aircraft, its thrust and drag, the cable transformer, the resistance of the variable "dashpot" and the compliance of the "spring." Define whatever parameters are needed.

3. Apply causal strokes, define state variables and apply the usual annotations to the graph.

4. Write the differential equation(s), making sure that the right side(s) of your equation(s) are in terms of the state variable(s).

Guided Problem 9.3

This problem demonstrates that dependent compliances can be treated in essentially the same way as dependent inertances.

The familiar slider-crank mechanism with series spring-dashpot load, this time with negligible inertias, is driven by a significantly flexible shaft characterizable by a compliance C_s. This system therefore can be represented by the bond graph

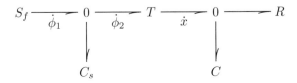

Use of one of the known relationships $x = x(\phi_2)$ or $T = T(\phi_2)$ is preferred over use of the more awkward inverse relationships. Define a set of state variables and write the state differential equations. Do this two ways: employ the differential causality that results from the given bond graph, and combine the two compliances into a single compliance.

Suggested Steps:

1. Observe that both compliances cannot be given integral causality; one of them must be considered a dependent energy storage. (This problem would vanish, ironically, if the inertia of the load was not neglected.)

2. First, carry out the instructions of the causal strokes including the differential causality. Either compliance can be given integral causality; presume you choose C_s. Therefore, differential causality is applied to C in the process of completing the causual assignments. The state variables become the q_s for the compliance C_s and ϕ_2 for the modulated transformer; their time derivatives can be written and circled on the relevant bonds. Apply the usual notation for the effort on the bond for C_s. Label all the remaining efforts and flows on the graph consistent with the dictates of the causal strokes. Be especially careful with the flow on the bond for C, which should equal C times the time derivative of the effort. The terms T and $dT/d\phi_2$ may be left as they are; in an actual simulation it would be necessary to substitute the detailed functions of ϕ_2.

3. Write the differential equations for the two state variables, using the causal strokes and the effort and flow notations in the usual way.

4. The differential equation for q_s is integrable, simplifying the procedure. Carry out the integration, and substitute the resulting expression for q_s into the differential equation for ϕ_2. Collect terms with the common factor $d\phi_2/dt$. The result is a first-order differential equation with dependent variable ϕ_2 and independent variables ϕ_1 and $\dot{\phi}_1$.

5. Proceed to the second method. Define the efforts of the two 0-junctions as e_s and e, respectively, and write an expression for the total stored energy as a function of these efforts.

6. The total stored energy can be represented by a single compliance bonded through a transformer to either of the two 0-junctions. The following steps assume the left-hand junction is chosen. Draw this bond graph, label the new compliance as C' and the new transformer as T'.

7. Equate the energy stored in C' to the sums of the energies stored in the original C_s and C. (This is most easily accomplished through energy expressions using efforts.) Any value of C' can be chosen; $C' = C$ is acceptable. Evaluate the necessary consequent value of T'. Also, relate $dT'/d\phi_2$ to $dT/d\phi_2$.

8. Redraw the new bond graph, apply causal strokes and annotate the efforts and flows using integral causality, in the usual way. Write the differential equation in terms of the state variable q on the bond for C.

9. A differential equation for ϕ_2 is needed, since the moduli of both transformers are functions thereof. $\dot{\phi}_2$ equals none of the flows on the graph, complicating the matter. The effort on the left side of the transformer T equals the actual effort on the actual

compliance C_s, however, and the difference $\dot{\phi}_1 - \dot{\phi}_2$ is proportional to the time derivative of this effort. Write the corresponding differential equation, and integrate it to get q as a function of ϕ_1, ϕ_2 and T'.

10. Substitute the result of step 9 into the differential equation of step 8. The result should be a first-order differential equation in ϕ_2.

11. The differential equation should contain derivative coefficients $dT/d\phi_2$ and $dT'/d\phi_2$. Use the results of step 7 to express the latter in terms of the former.

12. Rearrange the differential equation for comparison to the result of the first method, as found in step 4. They should agree.

PROBLEMS

9.1 Sketch-plot the transformer modulus (*i.e.,* speed ratio) for the adjustable cone friction drive below, as a function of the position of the idler shaft or wheel. (Get the endpoints and the shape about right.)

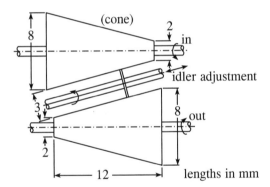

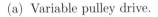

9.2 Identify which of the three devices below could be approximated as remotely modulated transformers, which have two power ports only. Justify your conclusions.

(a) Variable pulley drive.

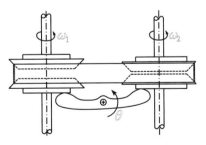

(b) Electrical transformer
with variable coil size
and fixed magnetic circuit.

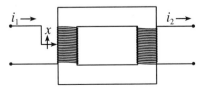

(c) Electrical transformer with fixed coils and variable magnetic circuit.

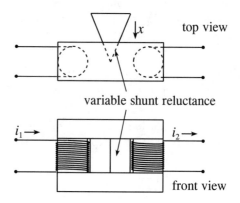

9.3 The hydraulic power supply on the left below provides a nearly constant pressure source of hydraulic fluid to drive several loads. One of these, as shown, comprises a "modulated hydrostatic transformer" (MHT) through which a load is driven. The MHT comprises two back-to-back hydraulic pump/motors with displacements D_1 and D_2 which are adustable by means of the control lever labeled with the angle θ. In particular, $D_1 = D_0 \sin\theta$ and $D_2 = D_0 \cos\theta$.

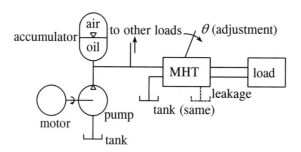

Define key variables and parameters and give a model structure. The rotating member within the MHT (not shown) has significant rotational inertia, and there is some external leakage (to the reservoir) at both ends of the MHT. Internal leakage may be neglected.

9.4 The Scotch-yoke mechanism below has a crank arm 0.05 m long and drives a load compression force of 500 N for the rightward stroke and a load tension force of 100 N for the leftward or return stroke. Find the modulus of the transformer which can model the system, and plot the moment on the crank as a function of its angle. Neglect friction.

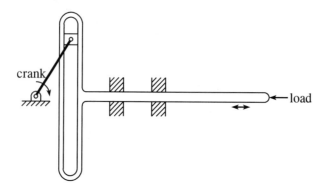

9.5 A motor with a torque-speed characteristic as plotted below drives a cam with inertia J_c, which in turn drives a follower system with inertia J_f, spring constant K and damping coefficient B. The shape of the cam can be described by a known function $\phi_f = f_f(\phi_c)$. Other effects may be neglected.

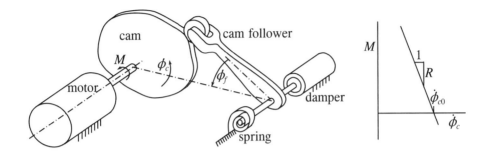

(a) Form a bond graph model of the system, and relate the moduli of its elements to the given information.

(b) Define state variables, and give the order of the system.

(c) Write a set of state differential equations in a form suitable for simulation.

9.6 Consider the slider-crank mechanism with the series RC load as shown in Example 9.2 (p. 587), but with all inertances neglected. Compute and plot the driving moment and the displacements of both ends of the spring as functions of time, assuming the effective crank starts in the fully extended position, runs with $\dot\phi = $ constant $= 10$ rad/s for $90°$, and then stops dead. The parameters are as follows: $r = 6$ in, $L = 10$ in, $k = 20$ lb/in and $b = 1.0$ lb s/in. All masses are negligible.

9.7 The effort e_1 of the system represented by the bond graph below is specified, the modulus of the transformer is a function of the displacement q_1, and all other parameters are invariant. Identify the order of the system, define state variables, and write a complete set of first-order differential equations.

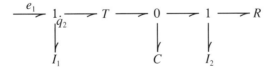

9.8 The bond graph of the preceding problem models a hydraulic positive-displacement pump with shaft angular velocity $\dot q_1$ and torque e_1, volumetric displacement $T = T_0 + T_1 \sin(6q_1)$ that includes sinusoidal pulsations due to the individual pistons (T_0 and $T_1 \ll T_0$ are constants), rotor inertance I_1, fluid inertance I_2, a fluid compliance C due to an accumulator and a load resistance R. Repeat the problem for the case in which the accumulator and its compliance are eliminated.

9.9 A Scotch yoke is driven from a flywheel and in turn drives a mass, linear spring and linear dashpot as shown.

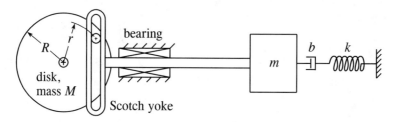

(a) Model the system with a bond graph, and express all moduli in terms of the given parameters.

(b) Compare your model to that given for the system shown in Example 9.5 (pp. 589-590). Comment on the applicability of the solutions deduced for that system.

9.10 A cylindrical disk of mass m_d and radius r is mounted on a shaft with eccentricity e to form a cam drive with a flat-faced follower, as shown below. The follower system has total mass m_f, and a linear spring with rate k and a linear dashpot with coefficient b are attached.

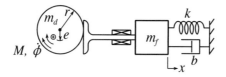

(a) Model the system with a bond graph, and state the moduli of the elements. Rotation of the drive shaft is resisted by a linear resistance, R, but friction between the cam and the follower may be neglected.

(b) Apply causal strokes. Can integral causality be employed on all of the energy storage elements?

(c) Write a set of state differential equations. Warning: Special care must be taken to account for differential causality in the face of a non-constant transformer modulus. The results should be nonlinear. Also, make sure the number of first-order differential equations equals the number of state variables.

(d) Report the order of your model, and the *minimum* order of the system.

9.11 Continue the preceding problem using the parameters $m_d = 1$ kg, $R = 0.1$ N·m·s, $r = 0.1$ m, $e = 0.03$ m, $m_f = 3$ kg, $k = 100$ N/m and $b = 5$ N·s/m. Carry out a simulation assuming that the applied moment is a constant $M = 2.0$ N·m. Find the minimum value of the spring precompression necessary for the cam to remain in continuous contact with the follower. (You are not asked to analyze the behavior for higher torques and speeds.) Plot the angular velocity $\dot{\phi}$ and the normal contact force between the cam and the follower as functions of time. Hint: Use of the MATLAB® `diff` function can expedite the plotting of the contact force.

9.12 A plastic disk of radius b, thickness w and density ρ_p rolls back and forth without slipping on a flat horizontal surface. A steel cylinder of radius a, length L and density ρ_s is pressed into a hole in the disk centered a distance c from its center. The equilibrium position is shown.

To save you work, the mass of a solid plastic disk (without the hole) is defined as m_p, and the excess mass for the steel cylinder is defined as m_s:

$$m_p = \pi w b^2 \rho_p; \qquad m_s = \pi a^2 (L\rho_s - w\rho_p).$$

The center of mass of the system is defined to be at a distance r from the center of the disk,

$$r = cm_s/(m_p + m_s),$$

and the mass moment of inertia of the system about the center of mass is, from the parallel axis theorem,

$$\bar{I} = m_p \left(\frac{1}{2}b^2 + r^2 \right) + m_s \left[\frac{1}{2}a^2 + (c - r)^2 \right].$$

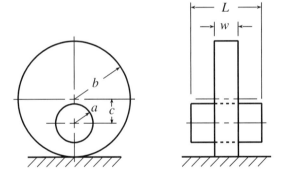

(a) Express the gravity potential energy of the system as a function of the angle of rotation from the equilibrium, ϕ.

(b) Express the kinetic energy of the system as a function of $\dot{\phi}$ and ϕ.

(c) Draw a bond graph for the system that employs a constant inertance, and define this inertance and other moduli therein.

(d) Write state differential equations.

(e) Simulate the motions following release from rest at the angles of $10°$, $60°$ and $120°$ (relative to equilibrium) for the parameters $b = 3$ in, $w = a = 1$ in, $L = 2$ in, $c = 1.5$ in, $\rho_p g = 0.0430$ lb/in^3 and $\rho_s g = 0.283$ lb/in^3.

(f) Qualitatively identify any limitations of your model for very large angles or large velocities.

(g) *If you have access to a physical model of this system,* verify the predicted cycle times experimentally.

9.13 Carry out a simulation for the aircraft landing on the deck of a carrier, as described in Guided Problem 9.2. The aircraft weighs 35,000 lbs, the thrust also is 35,000 lbs. The initial velocity is 180 ft/s, at which the drag is 5000 lbs. The drag is proportional to the square of the speed. The two deck pulleys shown in Fig. 9.7 are 100 ft apart. The arresting engine essentially comprises a set of pulleys to reduce the velocity of a large piston to which the cables are attached, an orifice through which oil is forced due to the motion of the piston, and an air tank that gets compressed by the displacement of the oil. Therefore, the force of the dashpot on the cable can be approximated as $a\dot{s}^2$, where s is the displacement of the cable where it is attached to the dashpot.

(a) Determine the coefficient a which brings the aircraft to a halt in a distance of 300 ft. Let the spring constant be fixed at 300 lb/ft. Plot the velocity as a function of time. Does the result represent satisfactory performance?

(b) *Design problem.* It is possible for the coefficient a to be varied as a function of the cable position s (by continuously varying the area of the orifice). Through trial-and-error or other means propose and demonstrate a desirable function $a = a(s)$. Hint: A solution with a good terminal "flare" will require reducing the spring constant, k.

9.14 Repeat the above problem using the more refined model of the arresting engine given in Problem 5.60 (pp. 321-322). The diameter of the hydraulic piston is 1 ft and the initial volume of the air in the tank can be set at 13 ft^3. You will need to choose the charging (initial) pressure, P_0, in the tank. The cross-head assembly weighs 2000 lbs not including the pulleys, each pulley weighs 75 lbs, and the hydraulic fluid weighs 50 lb/ft^3. Neglect the mass of the cable. For part (a), find an acceptable combination of P_0 and the fixed effective fixed orifice area for the fluid restriction, A_0. For part (b), choose a reasonable value for P_0 (of necessity smaller than in part (a)) and plot the orifice area as a function of the cable displacement. A solution to Problem 5.60 should be provided to you.

SOLUTIONS TO GUIDED PROBLEMS

Guided Problem 9.1

1. $\dfrac{M}{\dot{\phi}}\longrightarrow$ CAM DRIVE $\dfrac{F}{\dot{x}}\longrightarrow$

2.
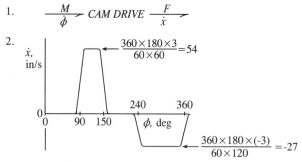

3.   Given: $\dot{\phi}=360\times180/60 = 1080$ deg/s; $F =200$ lbs; $\dot{x}$ as plotted above.

4. $M=F\dot{x}/\dot{\phi}$

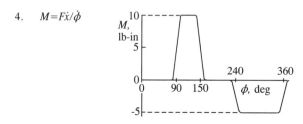

Guided Problem 9.2

1.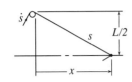

$$s = \sqrt{x^2 + L^2/4}$$

$$\dot{s} = \frac{x\dot{x}}{s} = T\dot{x}$$

Therefore, $T = T(x) = \dfrac{x}{\sqrt{x^2 + L^2/4}}$

2-3.

F_t is the thrust of the engines
F_d is the air drag on the aircraft
I is the mass of the aircraft $= m$
R is the (variable) resistance of the dashpot
C is the compliance of the spring

4. $\dfrac{dp}{dt} = F_t - F_d - T(x)\left(\dfrac{RT(x)}{I}p + \dfrac{1}{C}s\right)$

$\dfrac{dx}{dt} = \dfrac{1}{I}p$

$\dfrac{ds}{dt} = \dfrac{T(x)}{I}p$

Guided Problem 9.3

1-2.

3. $\dfrac{dq_s}{dt} = \dot{\phi}_1 - \dot{\phi}_2$

$\dfrac{d\phi_2}{dt} = \dfrac{1}{T}\left[\dfrac{q_s}{TRC_s} + \dfrac{C}{C_s}\left(\dfrac{1}{T}\dfrac{dq_s}{dt} - \dfrac{1}{T^2}\dfrac{dT}{d\phi_2}\dfrac{d\phi_2}{dt}q_s\right)\right]$

$= \left(\dfrac{1}{T^2RC_s} - \dfrac{C}{C_sT^3}\dfrac{dT}{d\phi_2}\dfrac{d\phi_2}{dt}\right)q_s + \dfrac{C}{T^2C_s}\dfrac{dq_s}{dt}$

4. $q_s = \phi_1 - \phi_2$. Substituting this into the second equation,

$\left[1 + \dfrac{C}{C_sT^3}\dfrac{dT}{d\phi_2}(\phi_1 - \phi_2) + \dfrac{C}{T^2C_s}\right]\dfrac{d\phi_2}{dt} + \dfrac{1}{T^2RC_s}\phi_2 = \dfrac{1}{T^2RC_s}phi_1 + \dfrac{C}{T^2C_s}\dot{\phi}_1$

5. $\mathcal{V} = \frac{1}{2}C_se_s^2 + \frac{1}{2}Ce^2$

6-7.

$\mathcal{V} = \dfrac{1}{2}C'\left(\dfrac{e_s}{T'}\right)^2 = \dfrac{1}{2}C_se_s^2 + \dfrac{1}{2}C\left(\dfrac{e_s}{T}\right)^2$

Therefore, $\dfrac{C'}{T'^2} = C_s + \dfrac{C}{T^2}$ or, letting $C' = C$,

$\dfrac{1}{T'^2} = \dfrac{C_s}{C} + \dfrac{1}{T^2}$. Computing derivatives, this also

gives $\dfrac{-2}{T'^3}\dfrac{dT'}{d\phi_2} = \dfrac{-2}{T^3}\dfrac{dT}{d\phi_2}$ or $\dfrac{dT'}{d\phi_2} = \dfrac{T'^3}{T^3}\dfrac{dT}{d\phi_2}$

8.

$$\frac{dq'}{dt} = T'\dot{\phi}_1 - \frac{T'^2}{T^2}\frac{q'}{RC}$$

9. $\dot{\phi}_1 - \dot{\phi}_2 = C_s \dfrac{de_s}{dt} = C_s \dfrac{d}{dt}\left(\dfrac{T'q'}{C}\right)$

Integrating, $\phi_1 - \phi_2 = \dfrac{C_s T'}{C}q'$, giving $q' = \dfrac{C}{C_s T'}(\phi_1 - \phi_2)$.

10. $\dfrac{dq'}{dt} = \dfrac{C}{C_s T'}(\dot{\phi}_1 - \dot{\phi}_2) - \dfrac{C}{C_s T'^2}\dfrac{dT'}{dt}(\phi_1 - \phi_2) = T'\dot{\phi}_1 - \dfrac{T'}{T^2 RC_s}(\phi_1 - \phi_2)$

11. Substitution of the relations between T' and T and between $dT'/d\phi_2$ and $dT/d\phi_2$ found in step 7 into the above equation gives

$$\left[\frac{1}{T^2} + \frac{C_s}{C} + \frac{1}{T^3}\frac{dT}{D\phi2}(\phi_1 - \phi_2)\right]\frac{d\phi_2}{dt} + \frac{1}{T^2 RC}\phi_2 = \frac{1}{T^2}\dot{\phi}_1 + \frac{1}{T^2 RC}\phi_1$$

which, when multiplied by C/C_s, is identical to the result from the first method (step 4).

9.2   Activated Bonds

The bond graph models employed thus far have demanded the conservation of energy, except for the one-port sources and sinks (resistances) where energy is explicitly created or dissipated. The junction structures, including simple bonds, junctions, transformers and gyrators, have been strictly conservative. Occasionally one wishes to model a coupler that violates the conservation of energy, however. This need may arise because the details of the actual system are both too complex and too unimportant to justify faithful representation; an overall simplification suits the purpose better. The **activated bond** is the simplest non-conservative coupler.

The use of such a coupler is illustrated in Fig. 9.8. An instrument measures a pressure or a velocity, etc. and thereby stimulates an electromechanical or hydraulic control system. The power drained from the system by the instrument likely is trivial, but including it in the model would introduce unnecessary complexity in the form of a much higher order model, etc. Nevertheless, the measured *signal* plays a major role in the behavior of the system; it cannot be ignored.[1] You want the system to power the instrument, but you do not want the instrument to drain power from the system model. Activating the bond between the system and the instrument accomplishes this objective. The activation and its directivity is indicated by the arrow in the middle of the bond.

9.2.1   Definition and Application

The role of activated bonds can be grasped through their application to electronic amplifiers, which often are modeled as non-energy-conservative devices. The power supply is ignored,

---

[1]The term "signal" is used by systems and control engineers to focus on a single variable, excluding its conjugate variable and the associated power from immediate and possibly ultimate consideration. Arrows are employed in block diagrams (see for example any of the several examples in Chapter 8) to indicate the directivity of signals. The arrows in block diagrams and signal flow graphs are closely related to the arrows on activated bonds. These relationships are examined in Section 6.3.

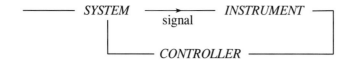

Figure 9.8: Instrument with activated coupling

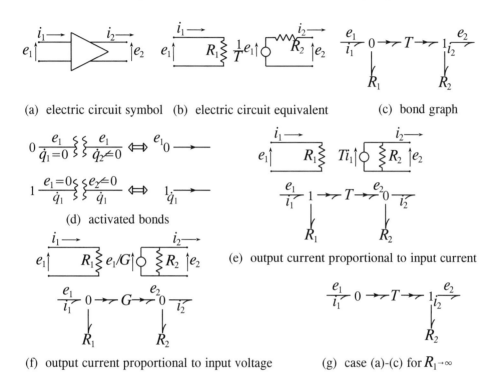

(a) electric circuit symbol   (b) electric circuit equivalent   (c) bond graph

(d) activated bonds

(e) output current proportional to input current

(f) output current proportional to input voltage   (g) case (a)-(c) for $R_1 \to \infty$

Figure 9.9: Activated bonds and electronic amplifiers

so what is actually a three-port device is treated as though it were a two-port device. The circuit symbol for an electrical amplifier is a triangle, as shown in part (a) of Fig. 9.9. A model in the form of an equivalent circuit is given in part (b), showing a voltage gain, T, and input and output resistive impedances R_1 and R_2, respectively. In the model, the power is produced by a dependent voltage source as if out of nothing. (The power supply is not directly described.) Further, any effect of the output current on the input is neglected. The bond graph for this voltage amplifier is shown in part (c) of the figure. The bonds on both sides of the transformer (either side alone will do) are **activated**, as indicated by the full arrows drawn near their centers.

The meaning of the activation is suggested in part (d) of the figure. An activated bond is a partially broken power bond. On the upstream (left) side of a bond emanating from a 0-junction the flow (current) is zero, and thus the power is zero. On the downstream (right) side of the bond the flow (current) is not zero, but equals whatever the downstream side of the system demands. The efforts, on the other hand, are identical, as for a regular bond. Thus the bond itself creates or annihilates power; it does not conserve energy.

For an activated bond emanating from a 1-junction, the flow is identical at both ends; the effort is zero at the upstream end but not at the downstream end. One application is to a current amplifier, as shown in part (e) of the figure. An amplifier in which the output current is proportional to the input voltage, using an activated bond and a gyrator, is shown in part (f) of the figure.

If it is desired to neglect the effect of the input impedance, that is to assume that the amplifier draws no power at all from its input circuit, the resistance R_1 can be simply removed. Examples are shown in part (g) of the figure. It is important to retain the upstream 0 or 1-junction in the bond graph; otherwise there is ambiguity regarding whether the efforts or the flows at the two ends of the activated bonds are equal. Sometimes one needs to insert a 0- or 1-junction into a bond graph purely to avoid this kind of ambiguity. If the effect of the output impedance of the amplifier is to be ignored, the resistance R_2 and its bond also may be removed, and the junction there also may be removed if nothing but the load is attached.

Fluid flow with an irreversible jet can be modeled with the help of activated bonds, as the following example demonstrates.

EXAMPLE 9.7

The water tank shown below is filled with water from a pipe that discharges below the waterline. The system shown on the right differs by the removal of the section of the pipe downstream of the restriction, which becomes a nozzle. Model the two systems with bond graphs.

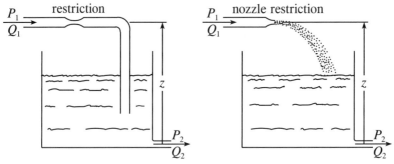

Solution: The first system can be modeled by the bond graph on the left below. The resistance to flow of the restriction is represented by an R element, the gravity head $\rho g h$ is represented by an S_e element, the gravity compliance of the tank is represented by a C element and the interconnection between the pipe and the tank is represented by a 0-junction and its attached bonds. Note that the pressure P_1 is affected by both the height z and the pressure P_2, which is proportional to the depth of the water in the tank.

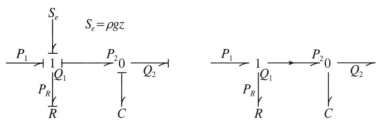

For the second system, which is modeled above right, the pressure P_1 is no longer affected by either the height z or the pressure P_2. Rather, P_1 is affected only by the resistance, R. On the other hand, the tank is still receiving the flow Q_1. Energy is being dissipated in the free fall of the water between the level of the nozzle and the level of the water surface in the tank, and the tank is receiving power of magnitude $P_2 Q_1$. This power is referenced to the level of the bottom of the tank.

The new situation is represented in the bond graph by removal of the effort source for the gravity head, and by activation of the bond interconnecting the two junctions, as indicated by the full arrow. This activation cuts off all recognition by P_1 of downstream effects; the effort on the upstream end of an activated bond driven by a flow (*i.e.* emanating from a 1-junction) is zero. It maintains the flow interconnection of a regular bond, however, and allows the effort on the downstream side to be whatever it needs to be to satisfy conditions there. The power $P_2 Q_1$ is generated by the bond itself.

9.2.2 Causality

The causal stroke applied to an activated bond unambiguously shows whether the equality of effort or of flow is being enforced:

$$\text{flow equality:} \quad 1 \longrightarrow \qquad \Longleftrightarrow \qquad \vdash\!\!\longrightarrow \quad \text{(zero effort on left side)}$$
$$\text{effort equality:} \quad 0 \longrightarrow \qquad \Longleftrightarrow \qquad \longrightarrow\!\!\dashv \quad \text{(zero flow on left side)}$$

The application of causal strokes and the writing of state differential equations follows in a normal manner, as illustrated by the following examples. In general, the differential equations for a system with activated bonds are the same as they would be without the bond activation, except for the deletion of the effort or the flow at the upstream end of each activated bond.

EXAMPLE 9.8

Causal strokes have been applied to the bond graph of the first system in Example 9.7, assuming the pressure P_1 is the input variable and the pressure P_2 is the output variable. Apply causal strokes to the bond graph for the second system, assuming the same input and output variables, and write and compare the state differential equations for the two systems.

Solution: Note that integral causality is used for the compliance element in the first system. If the same is done for the second system, the causality of the activated bond becomes as shown at the right.

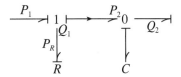

This is compatible with flow equality at both ends of the activated bond, as given above, and zero effort on the left side of the bond. This is precisely what is intended. Both systems are first order and have a single differential equation with state variable V. In both cases this equation can be written initially as

$$\frac{dV}{dt} = Q_1 - Q_2 = \frac{P_R}{R} - Q_2, \quad R = R(P_R).$$

The difference lies in the pressure P_R. In the case with the submerged pipe, the bond graph reveals that P_R is the weighted sum of the efforts on three bonds, as follows:

$$P_R = P_1 + \rho g h - P_2 = P_1 + \rho g h - V/C.$$

In the case with the cut-off pipe, however, only one bond determines P_R:

$$P_R = P_1$$

The resulting differential equation is changed only by the deletion of the additional terms in the equation for P_R resulting from the forced setting to zero of the effort at the left end of the activated bond.

The representation of gear friction is simplified through the use of activated bonds, as the next example demonstrates.

EXAMPLE 9.9

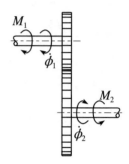

A gear pair often is assumed to have a constant efficiency, η.

Assuming that the driving power is from the left side,

$$M_2\dot{\phi}_2 = \eta M_1 \dot{\phi}_1.$$

Thus, the relations between the two speeds and the two torques can be expressed in terms of η and a transformer modulus, T_ϕ:

$$\dot{\phi}_2 = T_\phi \dot{\phi}_1,$$
$$M_1 = \frac{T_\phi}{\eta} M_2 = T_M M_2; \qquad T_M = T_\phi/\eta.$$

Represent this model by a bond graph.

Solution: There are in effect two constant transformer moduli: T_ϕ is the ratio of the speeds, and T_M is the ratio of the moments. This leads to either of the two alternative models below, depending on whether the input variables are $\dot{\phi}_1$ and M_2 or M_1 and $\dot{\phi}_2$:

Note that should the direction of the power be reversed, the equations must be inverted and the moduli of the transformers changed. Certain other mechanical drives such as pulley and chain drives can be represented similarly. When $\eta = 1$, the two transformer moduli become identical, so the activated pair can be reduced to a single ordinary or bilateral transformer.

9.2.3 Summary

An activated bond is like a simple bond except that the effort or the flow at one end is defined as zero. The power at this end therefore also is zero; an arrow is placed in the middle of the bond directed away from this end. Activated bonds permit simple representation of models that purposefully violate the conservation of energy. They are particularly useful for modeling amplifiers, the attachment of low-power transducers to high-power systems, free-falling liquids and the friction in drive systems such as gear pairs.

The activation of a bond changes the differential equations simply by the deletion of terms associated with the bond variable that is forced to equal zero.

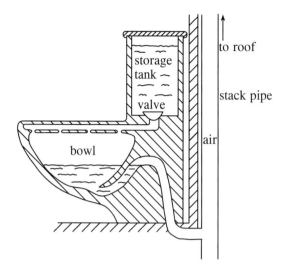

Figure 9.10: Toilet for Guided Problem 9.4

Guided Problem 9.4

This is an excellent problem to cut your teeth on. It uses all of the basic elements including an activated bond, and deserves some nonlinearity; yet the result should be fairly simple. It is continued in Problem 9.21 (p. 613).

Consider the traditional toilet system shown in Fig. 9.10. Model the basic system, neglecting any refill of the tank or the bowl. Define variables, and identify any elements which ought to be nonlinear, showing the functional dependence. Write a set of state differential equations, defining parameters as needed. The valve can be considered to open widely, but there are flow restrictions around the rim of the bowl. Swirl may be neglected. The stack pipe is so large that water never forms a plug in it. For simplicity it may be assumed that whenever the channel between the bowl and the stack pipe is flowing, it is full of water.

Suggested Steps:

1. Note that the state of the bowl has no effect on the state of the tank. Model the tank accordingly. Define needed parameters, and identify a key nonlinearity.

2. Note that the flow out of the bowl is impeded by the restricted area of the channel and possibly by the inertia of the flow. On the other hand, there is a distinct siphoning action. Complete the model of the rest of the system, defining parameters as needed, and note the key nonlinearities.

3. Apply causal strokes to your bond graph and write the differential equations. All that should be missing from your result are numbers for the fixed parameters and specific functions for the nonlinear parameters, although you ought to be able to give reasonable forms for the resistances.

Guided Problem 9.5

The electric circuit of Fig. 9.11 illustrates the use of a bond graph model of some complexity, including meshes and activated bonds. The modeling and the reduction is given; you are asked to write a set of state-variable differential equations. Assume that node a is grounded.

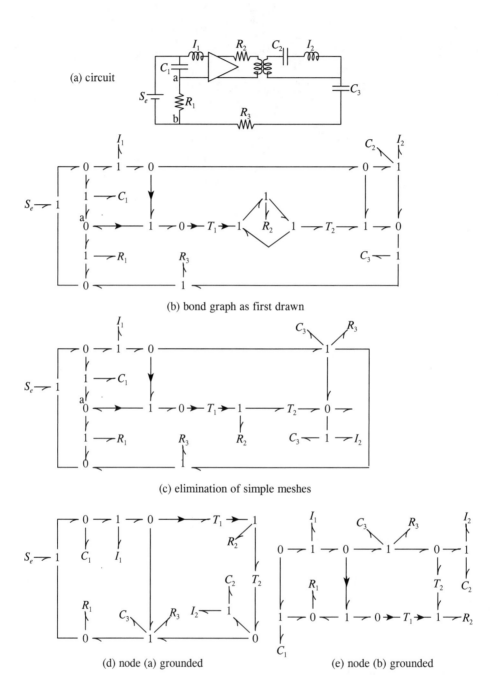

Figure 9.11: Guided Problem 9.5

Suggested Steps:

1. Copy the bond graph of part (d) of the figure. Apply integral causality. Temporarily, overlook the presence of the bond activation. Note that the graph is undercausal. Apply the usual annotations on the bonds with integral notation, which includes the definitions of state variables. Annotate the input effort as e_i.

2. Add a bond with a virtual compliance to the 0-junction on the left side of the activated transformer, as befits an undercausal bond graph. Label its flow with a circled zero and its effort as e, in the usual manner. Applying integral causality to this compliance then dictates the causalities of all remaining bonds. Note that the resulting causalities on the bonds for the activated transformer are consistent with the bond activation and, in fact, are required by it. The use of the virtual compliance therefore is not mandatory, but it resolves in routine fashion the usual difficulty associated with undercausal bond graphs, which the graph retains despite the activation.

3. Annotate all of the bonds with the proper symbols as dictated in the usual manner by the causal strokes. The flow on the bond to the left of the activated transformer should be designated as 0, as the consequence of the activation. Some of the annotations may be too long to fit in the confines of the graph, so you may wish to define one or more temporary symbols. In particular, it is suggested that you define the effort on the 0-junction toward the lower-right part of the graph as e_1.

4. Write the five first-order state differential equations in terms of variables labeled in steps 2 and 3. The causal strokes identify the proper terms, and the power convention arrows determine the proper signs, as always. Note that these equations are not solvable until the effort e on the virtual compliance is defined in terms of the five state variables and the input voltage.

5. Write the algebraic equation associated with the virtual compliance. This equation should contain the unknown e in two places. Collect these two terms, and solve for e as a function of the state and state variables. Substitution of this result where indicated in the state differential equations renders these equations solvable.

PROBLEMS

9.15 Draw a bond graph for the electrical amplifier shown below, which has voltage gain g and resistive input and output impedances R_1 and R_2, respectively.

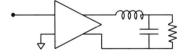

9.16 A DC motor with constant field, torque-to-current ratio G, constant applied voltage e_0 and armature resistance R drives its load through a gear train with speed reduction ratio T and efficiency η. Draw a bond graph of the system, and find a general relation between the output torque and speed of the gear reducer in terms of the given parameters.

9.17 Show that a bond graph model for the system of the previous problem can avoid the use of activated bonds merely by changing the effective value of R, presuming that the i of the graph need not represent the actual electric current.

9.18 A stream with given flow Q_i feeds a dammed reservoir. Some of the water flows through a power-generating turbine, and some overflows a sluice gate at the top of the dam. Both of these flows exit to a body of water so large that its level is invariant. Draw a bond graph of the system, employing elements with as simple characteristics as possible. (It is suggested that your resistance element for the sluice gate be defined so that its characteristic goes through the origin.) The case with no flow over the sluice gate may be omitted from consideration.

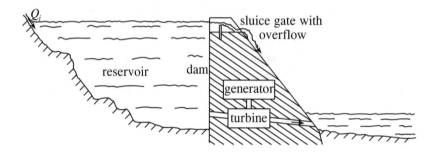

9.19 Many types of transistors used in amplifiers can be modeled by the bond graph below. Find the associated transmission matrix.

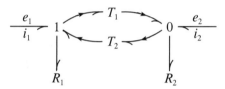

9.20 A decorative water display comprises an upper pond fed by a water fountain, a waterfall overflow into a lower pond, and a hidden recirculating pump, as pictured below. You are asked to prepare for an analysis to determine how quickly the system approaches steady-state when started from rest.

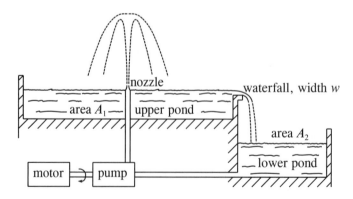

(a) Define appropriate variables and parameters, and model the system with a bond graph. Make appropriate simplifications, but do not ignore key nonlinearities. Relate the parameters of the bond graph to the direct physical parameters which are given or which you have defined.

(b) Write a set of state differential equations that describes the dynamics.

9.21 *Design project.* Carry out the simulation of the toilet given in Guided Problem 9.4 (pp. 609, 614-615). Choose parameters by trial-and-error or any other means until you are satisfied with the response; report the physical dimensions implied by these parameters. Then add a refill feature to your design and simulation. (Note that conventional toilets have a separate refill flow.) Other features may be added, if desired. Considerations should include but not necessarily be limited to the time required, the vigor of the flush and the water usage. Governmental regulations have reduced water consumption in stages from 5 gallons per flush to 3.5 to 1.5 gallons per flush, requiring very careful design.

9.22 *Design project.* A small passenger vehicle is to be powered by an IC engine connected through a 2:1 reduction spur gear pair to a variable displacement pump. Hydraulic flow from a reservoir through the pump goes to hydraulic motor/pumps geared by other 2:1 reductions to the axles of the four drive wheels. A hydraulic accumulator also is attached to the pressure line from the pump in order to store excess energy from the pump and energy from braking. This allows the engine to be smaller than it would have to be otherwise. The throttle is adjusted automatically to keep the torque of the engine virtually constant, which improves fuel efficiency. The volumetric displacement of the pump is varied automatically, within limits, to attempt to keep the pressure in the accumulator constant. The driver controls the volumetric displacement of the hydraulic motors, effectively determining the thrust force on the vehicle. When this displacement is made negative, the motors become pumps, and the resulting braking energy is stored in the accumulator rather than being dissipated. A relief valve prevents the pressure in the accumulator from exceeding 3500 psi for safety, although such a pressure should not be reached in normal operation except during braking. The speed of the engine is limited automatically to 500 rad/s, and the torque of the engine is allowed to decrease in order to prevent its speed from falling below 100 rad/s.

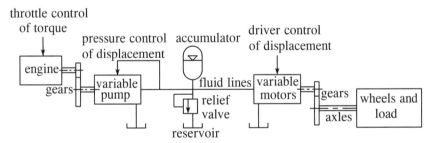

Your ultimate task is the specification of the size of the accumulator and the maximum displacements of the pump and the motors. Other parameters are given. The performance specifications are for you to choose, within the range of possibilities. You will defend your design based on its predicted performance.

The vehicle complete with engine (but not the accumulator, the driver and the payload) weighs 2500 lbs. The engine characteristics and fuel consumption are as plotted in the statement for Problem 5.65 (p. 335). The air drag, tire properties and accessory power are as given below that figure. (Other parameters or features stated there, including the weight of the vehicle, the axle gear ratio, drive train losses and the possibility of changing the size of the engine do not apply here.) The accumulator will weigh 20 pounds for each gallon (231 in^3) of total internal volume.

The gear pairs have an efficiency of 99%. The hydraulic pump and motors can be approximated as having a friction torque equal to a constant plus an added torque proportional to speed; the constant equals aD and the linear resistance equals bD, where D is the *maximum* volumetric displacement and $a = 20$ lb/in^2, $b = 0.50$ lb s/in^2. They also suffer leakage corresponding to a constant leakage resistance of c/D, where $c = 400$ lb s/in^2. Note that since one side of each machine is at the pressure of the reservoir, or zero psig, there is no operational difference between internal and external leakage.

The work can proceed in the following phases. You may be asked to work in a small group.

(a) Model the system with a bond graph, neglecting the gear losses and the auxiliary power consumption. Consider the accumulator to be a simple nonlinear compliance, and neglect minor inertances and compliances. Evaluate all parameters or functions in your model except for the key accumulator volume and machine volumetric displacements.

(b) Choose an operating torque for the engine and nominal operating pressure and charging pressure for the accumulator. Use this information, plus other information given above, to choose tentatively the control of the volumetric displacement of the pump and its maximum value.

(c) The engine speed $\dot{\phi}_e$ is controlled by the fuel throttle. When the accumulator pressure is low, the engine should be driven fast, and when it is high, the engine should idle. Choose an automatic control function that relates the engine speed to the pressure.

(d) Choose a maximum desired thrust on the vehicle, and with information developed earlier choose a tentative maximum volumetric displacement of the motors (or the sum of their displacements).

(e) Develop a computer simulation of the vehicle, and demonstrate its behavior starting from rest on a level road with maximum thrust and a full accumulator. You will have to assume some volume for the accumulator, and may prefer to keep it large enough so it doesn't empty before the vehicle reaches highway speeds. (This avoids your having to change the model.)

(f) Simulate, as you think appropriate, various schedules of acceleration, braking and hill climbing, in order to refine your tentative choices for the volume of the accumulator and the maximum displacements of the pump and motors.

(g) Repeat the steps above, including the consideration of the power loss to the accessories and the friction in the gears.

(h) Report your results according to whatever instructions are given to you. Include a discussion on the merits and demerits of the basic scheme.

SOLUTIONS TO GUIDED PROBLEMS

Guided Problem 9.4

1.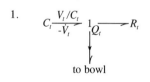

V_t is the volume of the water in the tank.
Q_t is the volume flow rate *out from* the tank; $Q_t = -\dot{V}_t$.
$C_t = A_t/\rho g$, where A_t is the tank area.
R is the nonlinear resistance to the flow, which occurs largely in the small orifices around the rim of the bowl.

2.

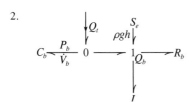

The flow Q_t enters the bowl, which has a compliance such that $P_b = P_b(V_b)$ is not linear. This pressure equals the pressure due to the resistance of the outlet channel, R_b, plus the pressure due to the inertance of the outlet channel, I, minus the constant siphon head, $\rho g h$. R_b is likely close to a nonlinear resistance of the Bernoulli type.

3.

$$\frac{dV_t}{dt} = -Q_t(P_{Rt}); \quad P_{Rt} = V_t/C_t$$

$$\frac{dV_b}{dt} = Q_t(P_{Rt}) - Q_b; \quad Q_t = \frac{1}{I}p$$

$$\frac{dp}{dt} = P_b(V_b) + \rho g h - P_{rb}(Q_b)$$

Suggested: $Q_t(P_{Rt}) \simeq c_1\sqrt{P_{Rt}}; \qquad P_{Rb}(Q_b) \simeq c_2 Q_b^2$

Guided Problem 9.5

1-2.

3.

4. $$\frac{dq_1}{dt} = -\frac{1}{R_1 C_1}q_1 + \frac{1}{R_1}e_i \qquad \frac{dp_1}{dt} = \frac{1}{C_1}q_1 - e$$

$$\frac{dq_2}{dt} = \frac{1}{I_2}p_2 \qquad\qquad\qquad \frac{dp_2}{dt} = e_1 - \frac{1}{C_2}q_2$$

$$\frac{dq_3}{dt} = \frac{1}{I_1}p_1$$

5. Setting the flow on the virtual compliance equal to zero,

$$0 = \frac{p_1}{I_1} + \frac{p_2}{I_2} - \frac{T_2}{R_2}\left[T_1 e - T_2\left(\frac{q_1}{C_1} - e_i + \frac{q_3}{C_3} + \frac{R_3 p_1}{I_1} - e\right)\right]$$

Solving for e,

$$e = \frac{T_2}{T_1 + T_2} \left[\frac{q_1}{C_1} - e_i + \frac{q_3}{C_3} + \left(R_3 + \frac{R_2}{T_2^2} \right) \frac{p_1}{I_1} + \frac{R_2}{T_2^2} \frac{p_2}{I_2} \right]$$

This gives $e_1 = \dfrac{T_1}{T_1 + T_2} \left[-e_i + \dfrac{q_1}{C_1} + \dfrac{q_3}{C_3} + \left(R_3 - \dfrac{R_2}{T_1 T_2} \right) \dfrac{p_1}{I_1} - \dfrac{R_2}{T_1 T_2} \dfrac{p_2}{I_2} \right]$

These terms are substituted into the differential equations above.

9.3   Linear Multiport Fields

Starting in this section, resistances, inertances and compliances will be allowed to have more than one bond. An **implicit multiport element** is largely a shorthand for a field of one-port elements of a single type (I, R or C) interconnected by bonds, junctions and perhaps transformers. The **explicit multiport field**, shown in Fig. 9.12, is the new class of elements. The subscript f stands for *field.*. These elements can model an energy conversion device that stores both electric and mechanical energy, for example, or an essentially multiport device such as the real electrical transformer or the analogous rigid body with two degrees of freedom, as discussed in Section 9.3.1 below.

Multiport inertance, compliance and resistance fields often are assumed to be linear. Linearity allows matrix representations to be useful, including the concepts of *mutual inertance*, *mutual compliance*, *mutual resistance*, and permits certain standard bond graph equivalences to be employed. These fields are introduced first. Nonlinear fields, introduced in the subsequent section, are particularly powerful in representing components with energy transduction.

9.3.1   Linear Two-Port Inertance: The Electric Transformer

The case of the real electric transformer with two coils and the surprisingly analogous inertive rigid body with two degrees of freedom (and small displacements only) will serve to introduce the linear inertance and resistance fields.

A conventional real electrical transformer acts very differently from the idealized transformer, for example not operating at all for steady-state or DC conditions, and operating poorly at very high frequencies. It stores energy in the magnetic fluxes that pass through magnetically permeable material (most often iron) and are generated by the currents i_1 and i_2 in the two coils. The electric circuit symbol for the transformer, given in Fig. 9.13 part (a), includes two parallel lines that suggest the magnetically permeable core. The model also is indicated as a two-port inductive field, I_f. Neglecting any resistance, the constitutive equations can be written in the form

$$\begin{bmatrix} e_1 \\ e_2 \end{bmatrix} = \begin{bmatrix} I_{11} & I_{12} \\ I_{12} & I_{22} \end{bmatrix} \frac{d}{dt} \begin{bmatrix} i_1 \\ i_2 \end{bmatrix}. \tag{9.8}$$

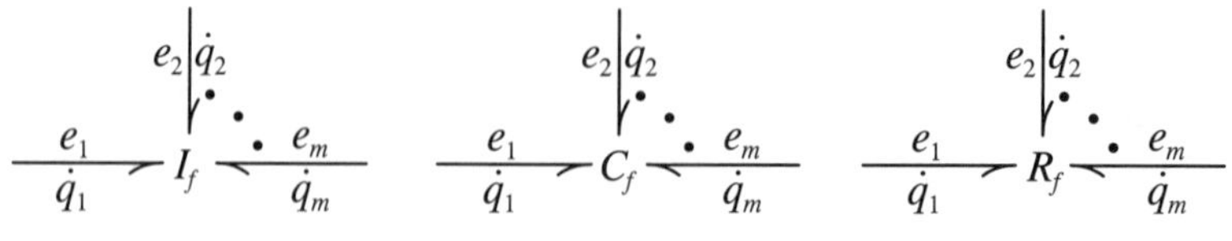

Figure 9.12: Bond Graph Notation for Multiport Fields

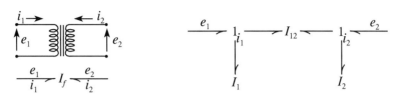

(a) circuit diagram and bond graph symbols   (b) bond graph with mutual inertance

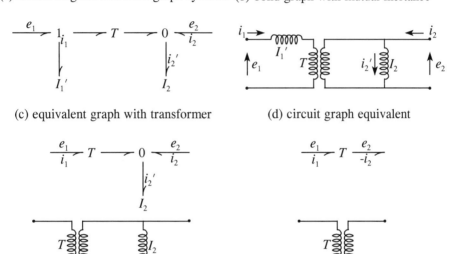

(c) equivalent graph with transformer   (d) circuit graph equivalent

(e) reduction with perfect flux linkage   (f) added reduction for $I_2 \to \infty$ or $\omega \to \infty$

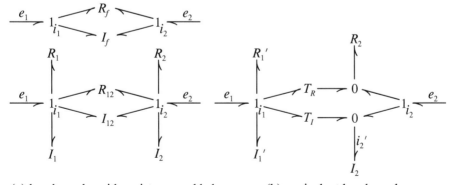

(g) bond graphs with resistances added   (h) equivalent bond graph

Figure 9.13: Linear model for the electric transformer

I_{11} and I_{22} are known as **self-inductances**, and I_{12} as the **mutual inductance**. It is this mutual inductance, due to the magnetic flux linkage through the common path for the two coils, that makes the transformer work. Note that both currents are defined as positive in the symmetrical inward directions, and that the matrix must be symmetric. The energy stored in the magnetic field is

$$T = \int (e_1 i_1 + e_2 i_2)dt = \frac{1}{2}I_1 i_1^2 + I_{12}i_1 i_2 + \frac{1}{2}I_2 i_2^2. \tag{9.9}$$

The currents i_1 and i_2 are appropriate flows or generalized velocities, as with the ideal transformer. The inductances I_1 and I_2 then become common one-port inertances. It is convenient to represent I_{12} as a special two-ported **mutual inertance**, as shown in part (b) of the figure. Its respective efforts are $e_1 = I_{12}di_2/dt$ and $e_2 = I_{12}di_1/dt$. (This special definition, distinguished from a general field by the absense of the subscript *f*, is not commonly recognized in the literature.)

A reticulated bond graph with a mutual inertance is awkward to implement causally in order to find the corresponding state differential equations. It serves very well, however, as a bridge between the energy expression and a corresponding bond graph with only one-port inertances that is indeed in a convenient form for application of the causal strokes and the writing of the differential equations. This latter bond graph is given in part (c) of the figure. It is a universal equivalent to the graph of part (b), assuming the parameters I_1' and T are properly determined.

For the equivalence to apply, the only requirement is that the two bond graphs must represent the same energy in terms of their boundary variables. This new graph represents the energy

$$T = \frac{1}{2}I_1' i_1^2 + \frac{1}{2}I_2 i_2'^2, \tag{9.10}$$

which equals the energy of equation (9.9) if and only if

$$i_2' = i_2 + Ti_1, \tag{9.11a}$$

$$T = \frac{I_{12}}{I_2}, \tag{9.11b}$$

$$I_1' = I_1 - T^2 I_2 = I_1 - \frac{I_{12}^2}{I_2}. \tag{9.11c}$$

You may verify this claim by direct substitution. Equation (9.11a) gives the junction structure of Fig. 9.13 part (c), which includes an ideal transformer. Equation (9.10) shows that two inertances I_1' and I_2 should be attached to their respective 1-junctions. This bond graph model can be converted back into the circuit diagram of part (d) of the figure, which includes an ideal transformer, a series inductor and a parallel inductor. This diagram in fact appears commonly in textbooks for electrical engineers. Electrical engineers are taught how to write differential equations for circuits like this upon examination of the circuit diagram; you may accomplish the same result by working with the bond graph directly.

The best possible performance of an electrical transformer results when the mutual inductance I_{12} is maximized. The theoretical limit of I_{12} corresponds to perfect flux linkage between the two coils. For this value, I_1' vanishes, and T has its maximum value consistent with the physical fact that I_1' cannot be negative:

$$I_{12} = \sqrt{I_1 I_2} \qquad T = \sqrt{I_1/I_2}. \tag{9.12}$$

The corresponding bond graph and circuit diagram equivalent are shown in part (e) of the figure. The series inductance has been eliminated, which becomes important if the device is to operate well at high frequency or high current.

The current passing through the inductor I_2 in part (e) decreases as the frequency of excitation is increased or as the value of I_2 is increased. If this current is sufficiently decreased, the energy $\frac{1}{2}I_2 i_2^2$ becomes negligible, the inertance I_2 can be erased from the graph, and the device virtually reduces to an ideal transformer as shown in part (f) of the figure, as is normally desired. The designer of a transformer that is to operate at 60 Hz would make I_2 large enough so that this condition is more or less satisfied at that frequency. He does this by adding iron to the core. This explains why power transformers typically are so heavy. Some secondary effects are addressed in Section 9.5.3 (p. 656-659).

Energy dissipation has been ignored thus far. Linear dissipation may be approximated by the quadratic expression

$$\mathcal{P} = R_1 i_1^2 + 2R_{12}i_1 i_2 + R_2 i_2^2, \tag{9.13}$$

where R_1 and R_2 refer primarily to the electrical resistances of the two coils and R_{12} is a **mutual resistance** associated largely with the eddy-current losses in the core. This description leads directly to the general bond graph model of part (g) of the figure. The mutual resistance can be replaced by the same type of bond graph structure employed for the mutual inertance, however, giving the equivalent bond graph of part (h). The analogous equations to (9.9) and (9.10) are

$$\mathcal{P} = R_1 i_1^2 + R_2 i_2^2, \tag{9.14a}$$

$$i_2 = i_2 + T_R i_1, \tag{9.14b}$$

$$T_R = \frac{R_{12}}{R_2}, \tag{9.14c}$$

$$R_1 = R_1 - T_R^2 R_2 = R_1 - \frac{R_{12}^2}{R_2}. \tag{9.14d}$$

Much of the complexity of this model evaporates if R_{12} is neglected.

9.3.2 Linear Two-Port Inertance: The Rigid Inertive Floating Link

An instructive direct analogy exists between the transformer without dissipation and a "floating link" of rigid material, pictured in Fig. 9.14, in which the two pivot points are moved virtually in a fixed direction parallel to one another, and only within a small maximum displacement, so the axis drawn between them remains nearly perpendicular to the direction of the motions.

The two velocities $\dot{x}_1$ and $\dot{x}_2$ are analogous to the two electric currents, and the inductance energies become analogous to the kinetic energies of the mass m in vertical translation and of the rotational inertia mr^2 (where r is the radius of gyration) in rotation about the center of mass (cm):

$$\mathcal{T} = \frac{1}{2}m\dot{y}_{cm}^2 + \frac{1}{2}mr^2\dot{\phi}^2 \tag{9.15}$$

Substitution of the kinematical constraints

$$\dot{y}_{cm} = a + b\dot{y}_1 + \frac{a}{a+b}\dot{y}_2; \quad \dot{\phi} = \frac{1}{a+b}(\dot{y}_2 - \dot{y}_1), \tag{9.16}$$

into the equation for energy of the mutual inertance gives

$$I_1 = \frac{m(b^2 + r^2)}{(a+b)^2}; \qquad I_2 = \frac{m(a^2 + r^2)}{(a+b)^2}; \qquad I_{12} = \frac{m(ab - r^2)}{(a+b)^2}. \tag{9.17}$$

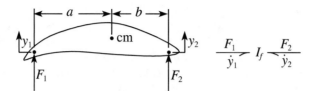

(a) link with pivot points

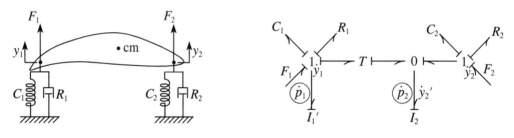

(b) equivalent bond graphs (compare to Fig. 9.13)

(c) springs and dashpots added (equivalent to   (d) associated bond graph
vehicle model of Fig. 6.15 (p. 381), but no mesh)

Figure 9.14: Rigid inertive floating link

The bond graphs and equivalent circuits for the electric transformer (as given in Fig. 9.13) *also apply to the floating link.* The equivalent bond graph with the transformer includes the parameters

$$I_1' = \frac{mr^2}{a^2 + r^2}; \qquad T = \frac{ab - r^2}{a^2 + r^2}. \qquad (9.18)$$

The analogy to maximizing the flux linkage occurs when the mass is concentrated at a point (the mass center, of course), reducing the rotational inertia (and the radius of gyration, r) to zero. The subsequent analogy to neglecting the current i_2 by making I_2 or the frequency large corresponds to making the mass so large or the frequency so high that the mass center does not move appreciably. By not moving, it acts as a pivot point; the floating link becomes an ideal lever, which indeed is properly modeled by an ideal transformer. Is your feeling for the behavior of a real electrical transformer enhanced?

In parts (c) and (d) of the figure, springs C_1 and C_2 and dashpots R_1 and R_2 are added to the pivot points of the floating link. This model could represent a vehicle with springs and shock absorbers at its front and rear axles. The resulting motion involves heave and pitch modes. The absense of a bond graph mesh, unlike the bond graph developed for the same system (called a vehicle body) in Section 6.2.7 (p. 381), simplifies subsequent analysis.

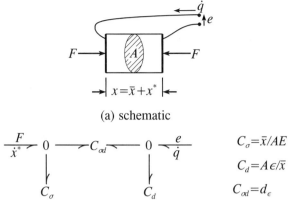

(a) schematic

$$\frac{F}{\dot{x}^*} \multimap 0 \longrightarrow C_{\sigma d} \longleftarrow 0 \multimapdot \frac{e}{\dot{q}} \qquad C_\sigma = \bar{x}/AE$$

$$C_d = A\epsilon/\bar{x}$$

$$C_{\sigma d} = d_\epsilon$$

(b) bond graph with mutual compliance

$$\vdash \frac{F}{\dot{x}^*} \multimap 0 \longrightarrow \dashv T \longrightarrow \dashv 1 \multimapdot \frac{e}{\dot{q}} \dashv \qquad T = C_d/C_{\sigma d}$$

$$C_\sigma{}' = C_\sigma - C_{\sigma d}{}^2/C_d$$

(c) equivalent bond graph for computation

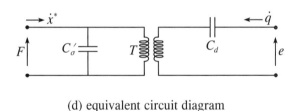

(d) equivalent circuit diagram

Figure 9.15: Piezoelectric transducer

9.3.3 Linear Two-Port Compliance: The Piezoelectric Transducer

A piezoelectric crystal stores energy by virtue of both its mechanical stress or strain and its electric field or electric displacement (charge). The effects are coupled. It is presumed here that the stress or strain and the electric field or charge is imposed in only one direction, as shown in Fig. 9.15. The constitutive relation is further assumed to be linear:

$$\begin{bmatrix} S \\ D \end{bmatrix} = \begin{bmatrix} 1/E & d_\varepsilon \\ d_\varepsilon & \varepsilon \end{bmatrix} \begin{bmatrix} \sigma \\ E \end{bmatrix}. \tag{9.19}$$

The strain is S and the relative displacement of the two conductive surfaces is $\Delta x = x_0 S$, where x_0 is the thickness of the crystal; the tensile stress is σ and the associated force is $F = A\sigma$, where A is the cross-sectional area; the electric field strength is E and the applied voltage is $e = x_0 E$; the electric displacement is D and the electric charge is $q = AD$. The mechanical compliance of the material is represented by the reciprocal of Young's modulus, $1/E$, the electrical compliance (dielectric constant) by ε, and the piezoelectric coupling by d_ε.

The device is used as a transducer. This means that either a mechanical force or displacement is imposed and an electrical voltage or current is produced, or that an electrical voltage or current is imposed and a mechanical motion or force is produced. Such devices are used, for example, in crystal microphones and pressure transducers (to convert pressure or force to voltage) and in some vibration and valve control systems (to convert voltage to force and small motion).

A bond graph model aids understanding. The graph of part (b) of the figure shows three separate energy storage elements: C_σ for strain energy, C_d for electric energy, and the **mutual compliance** $C_{\sigma d}$ for the piezoelectric or coupled energy. The stored energy can be written

$$\mathcal{V} = \frac{1}{2}C_\sigma F^2 + C_{\sigma d}Fe + \frac{1}{2}C_d e^2. \tag{9.20}$$

It is awkward to write equations directly from this graph, since there are three compliances but only two independent energy storages, so integral causality cannot be applied to all three compliances. This problem vanishes upon conversion to an equivalent bond graph with only two compliances, each with a single port. One of two possible versions is shown in part (c) of Fig. 9.15; the other has a mirror image structure, with the 1-junction on the left and the 0-junction on the right. This bond graph equivalence adapts the same idea as the equivalence employed in Figs. 9.13 and 9.14 for the inertances. The relationships for the modulus of the new transformer and the modulus of the modified mechanical compliance C_σ also are given in the figure. These are universal, and can be used whenever a graph of the form of part (b) of the figure is converted to a graph of the form of part (c). The equivalent circuit diagram is shown in part (d).

The bond graph or circuit diagram reveals potential limitations to piezoelectric transducers. At low frequency the compliance C_d can be seen to reduce the mechanical response to a voltage e, but not to current $i = \dot{q}$. At high frequency the compliance C_σ drains off flow, producing a break frequency above which the transducer largely fails to operate. This break frequency is raised as the mutual compliance $C_{\sigma d}$ is increased, analogous to the effect of increasing the flux linkage in an electrical transformer or the concentration of mass at a single point in the floating link. In the theoretical limit of maximum mutual compliance (not achieved with real crystals), C_σ would vanish altogether, so that at sufficiently high frequencies the transducer would act just like an ideal transformer with modulus T. This objective, only partially realized in practice, is the ideal goal for the design of a piezoelectric transducer.

9.3.4 Linear Two-Port Compliance: The Piezomagnetic (Magnetostrictive) Transducer

A piezomagnetic or magnetostrictive material is like a piezoelectric material except energy is stored in a magnetic field instead of an electric field. Consider the system shown in Fig. 9.16 part (a), in which the magnetic field is generated by a coil, and the core is made of a magnetostrictive material which is fixed at one end. The constitutive relation is

$$\begin{bmatrix} S \\ B \end{bmatrix} = \begin{bmatrix} 1/E & d_\mu \\ d_\mu & \mu \end{bmatrix} \begin{bmatrix} \sigma \\ H \end{bmatrix}. \tag{9.21}$$

where B is the magnetic flux density, H is the magnetizing force, μ is the magnetic permeability coefficient and d_μ is the piezomagnetic coefficient. As with the piezelectric transducer, $\Delta x = x_0 S$ and $F = A\sigma$. The magnetomotive force $F_m = x_0 H$ (which neglects the reluctance of the magnetic circuit other than the plunger) becomes analogous to the voltage e, and the magnetic flux $\phi = AB$ becomes analogous to the current $\dot{q}$. F_m is directly proportional to the actual applied electric current, however; specifically, $F_m = 4\rho N\dot{q}$. Thus, the bond graph in part (b) of the figure is analogous to that in part (b) of the preceding figure for the piezoelectric transducer, except that the coil is represented by the added gyrational

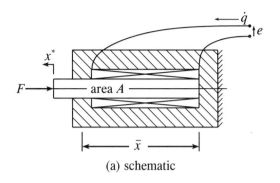

(a) schematic

$$\frac{F}{\dot{x}^*} \; 0 \longrightarrow C_{12} \longrightarrow 0 \; \frac{\bar{x}H}{AB} \; G \; \frac{e}{\dot{q}}$$

$$C_1 \quad\quad\quad\quad C_2$$

$C_1 = \bar{x}/EA$
$C_2 = A\mu/\bar{x}$
$C_{12} = d_\mu$
$G = 4\pi N$

(b) bond graph with mutual compliance

$$\frac{F}{\dot{x}^*} \; 0 \longrightarrow T \longrightarrow 1 \; \frac{\bar{x}H}{AB} \; G \; \frac{e}{\dot{q}}$$

$$C_1' \quad\quad\quad\quad C_2$$

$T = C_2/C_{12}$
$C_1' = C_1 - C_{12}/C_2$

(c) equivalent bond graph with transformer

$$\frac{F}{\dot{x}} \; 0 \longrightarrow G' \longrightarrow 0 \; \frac{AB}{\bar{x}H} \; T' \; \frac{e}{\dot{q}}$$

$$C_1' \quad\quad\quad\quad I_2$$

$T' = G$
$G' = T$
$I_2 = C_2$

(d) bond graph with dualized region

$$\frac{F}{\dot{x}^*} \; 0 \longrightarrow G'' \longrightarrow 0 \; \frac{e}{\dot{q}}$$

$$C_1' \quad\quad\quad\quad I_2'$$

$G'' = G'T' = GT$
$I_2' = C_2/T^2$

(e) final reduced bond graph

Figure 9.16: Piezomagnetic (magnetostrictive) transducer

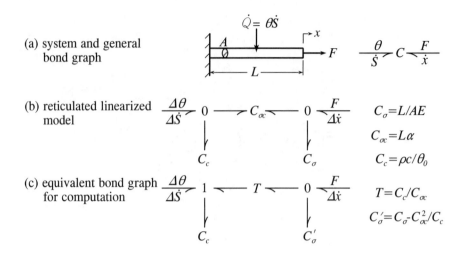

(a) system and general bond graph

(b) reticulated linearized model

$C_\sigma = L/AE$

$C_{\sigma c} = L\alpha$

$C_c = \rho c/\theta_0$

(c) equivalent bond graph for computation

$T = C_c/C_{\sigma c}$

$C'_\sigma = C_\sigma - C_{\sigma c}^2/C_c$

Figure 9.17: Thermoelastic rod

coupling with modulus $G = 4\rho N$. Similarly, the equivalent bond graph of part (c) contains only two energy storage elements, which are causally independent. To simplify this result, the 1-junction with its compliance can be *dualized* to leave a 0-junction with an inertance; the transformer thus becomes a gyrator and vice-versa, as shown in part (d). Finally, the values of the gyrator and inertance moduli can be modified to allow the modulus of the transformer to become unity, so the transformer can be replaced by a simple bond, as shown in part (e). This is the simplest representation possible, unless one or more of the remaining effects are neglected. Note that, unlike the piezoelectric transducer and the thermoelastic rod, the piezomagnetic transducer is essentially *gyrational*.

9.3.5   Linear Two-Port Compliance: The Thermoelastic Rod

The thermoelastic rod stores energy by virtue of both its mechanical stress or strain and its temperature or entropy changes. Presume that the stress or strain is imposed in only one direction, as shown in Fig. 9.17. The constitutive relation is further assumed to be linear:

$$\begin{bmatrix} S \\ \Delta s \end{bmatrix} = \begin{bmatrix} 1/E & \alpha \\ \alpha & \rho c/\theta_0 \end{bmatrix} \begin{bmatrix} \sigma \\ \Delta\theta \end{bmatrix}. \tag{9.22}$$

The strain, S, stress, σ, and Young's modulus, E, are the same as with the piezoelectric transducer: $\Delta x = x_0 S$ and $F = A\sigma$. The change of temperature from some reference value is $\Delta\theta$. The change of entropy per unit mass is Δs, so the total change in entropy is $\Delta S = \rho A x_0 \Delta s$. The **thermal expansion coefficient**, α, represents the thermoelastic effect, whereby not only does heating cause expansion, but stretching causes cooling.

As with the piezoelectric transducer, the behavior can be understood better in terms of bond graphs. The graph of part (b) of the figure shows the analogous three separate energy storage elements: C_σ for strain energy as before, C_c for thermal energy, and the mutual compliance $C_{\sigma c}$ for the thermoelastic or coupled energy. The stored energy can be written in the same form as before,

$$\mathcal{V} = \frac{1}{2}C_\sigma F^2 + C_{\sigma c}F\Delta\theta + \frac{1}{2}C_c\Delta\theta^2. \tag{9.23}$$

This graph is converted, as before, to the equivalent form shown in part (c) of the figure, in order to permit integral causality. This graph reveals directly that

$$F = \frac{1}{C'_\sigma}(\Delta x - \frac{1}{T}\Delta S), \qquad (9.24a)$$

$$\Delta\theta = \frac{1}{C_c}\Delta S - \frac{1}{T}F, \qquad (9.24b)$$

demonstrating that inward heat flow reduces the tensile force (or increases compressive force) and that tensile force decreases temperature.

Equation (9.23) expresses the energy in terms of efforts. Alternatively, the energy could be expressed in terms of generalized displacements,

$$\mathcal{V} = \mathcal{V}(\Delta x, \Delta S), \qquad (9.25)$$

and the results of equation (9.23) found in the traditional general manner given in the following section (9.4):

$$\Delta\theta = \frac{\partial \mathcal{V}}{\partial \Delta S}; \qquad F = \frac{\partial \mathcal{V}}{\partial x}. \qquad (9.26$$

Implementation of this approach employs the inverse of equation (9.22). The initial bond graph with a mutual compliance has 1-junctions in place of 0-junctions. This approach is often useful, as Guided Problem 9.6 demonstrates.

9.3.6 Linear Multiport Compliances: Generalized Linear Media*

The piezoelectric, thermoelastic and piezomagnetic media considered above are special cases of more general media. First, they were restricted to unidirectional strain and stress only. In general, there are 3 normal and 3 shear stresses and strains; the elastic behavior can be represented by a symmetric 6×6 array of elastic coefficients, leading to 6 one-port compliances and 15 mutual compliances for the 21 independent coefficients. Second, adding directionality to the dielectric and magnetic permeability coefficients gives two 3×3 arrays with 3 two-port and 6 mutual compliances each. (More information on magnetic variables and parameters is given in Section 9.5.) Finally, thermal expansion also can be added, to give an additional one-port compliance. The couplings between these four different types of fields can be described by 6 dielectric coefficients, 18 piezoelectric coefficients, 18 piezomagnetic coefficients, 6 thermal expansion coefficients, 3 pyro-electric coefficients and 3 pyro-magnetic coefficients, all representable by *mutual* compliances.

The complete result is a symmetric 13×13 array of coefficients, as shown in Table 9.1 using matrix tensor notation. Stress is represented by the symbol T, temperature increments by $\Delta\theta$, and entropy increments by $\Delta\sigma$. The 13 diagonal elements are representable by one-port compliances, and the 78 independent off-diagonal elements are representable by mutual compliances. This gives the bond graph of Fig. 9.18, in which the 78 bonds interconnecting the 13 0-junctions implicitly represent the mutual elements.

Fortunately in engineering one always vastly simplifies this extremely complicated general case. Probably no real material exhibits significant effects for all the coefficients or even a majority of them, and if it did the material itself or at least most of its effects would be ignored.

Table 9.1   Sub-matrix coefficients for energy storage in a medium with
elastic, electric, magnetic and thermal energies

$$
\begin{bmatrix} S_{ij} \\ D_m \\ B_m \\ \Delta s \end{bmatrix} = \begin{bmatrix} s_{ijkl}^{E,H,\theta} & d_{ijk}^{H,\theta} & d_{ijk}^{E,\theta} & \alpha_{ij}^{E,H} \\ & \varepsilon_{mk}^{T,H,\theta} & m_{mk}^{T,\theta} & p_m^{T,H} \\ (\text{symmetric}) & & \mu_{mk}^{T,E,\theta} & p_m^{T,E} \\ & & & \rho c^{E,H,T}/\theta \end{bmatrix} \begin{bmatrix} T_{kl} \\ E_k \\ H_k \\ \Delta\theta \end{bmatrix}
$$

The subscripts i, j, k, l, m refer to Cartesian directions.

sub-matrix and name of coeff.		symmetric?	dimensions	independ. coeff.
$s_{ijkl}^{E,H,\theta}$	elastic	yes	6×6	21
$\varepsilon_{mk}^{T,H,\theta}$	dielectric	yes	3×3	6
$\mu_{mk}^{T,E,\theta}$	magnetic permeability	yes	3×3	6
$\rho c^{E,H,T}/\theta$	specific heat	not applic.	1×1	1
$d_{ijk}^{H,\theta}$	piezoelectric	no	6×3	18
$d_{ijk}^{E,\theta}$	piezomagnetic	no	6×3	18
$\alpha_{ij}^{E,H}$	thermal expansion	no	6×1	6
$m_{mk}^{T,\theta}$	magneto-dielectric	no	3×3	9
$p_m^{T,H}$	pyro-electric	no	3×1	3
$p_m^{T,E}$	pyro-magnetic	no	3×1	3
c_f	energy storage	yes	13×13	91

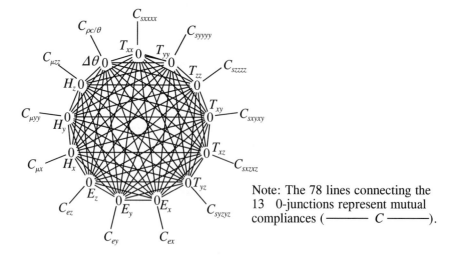

Note: The 78 lines connecting the
13 0-junctions represent mutual
compliances (———— C ————).

Figure 9.18: Compliance field for energy storage in a volume of a substance

9.3.7 Reticulation of General Linear Multiport Fields*

Breedveld[2] has developed a **congruence canonical form** for representing a linear n-port C, I or R field with n one-port C, I or R elements interconnected with a "weighted" junction structure comprising 0- and 1-junctions and $(n^2 - n)/2$ transformers. Although not unique, his structure appears to be as simple as any. The two-port case is identical to that employed repeatedly above. The three-port case is shown in Fig. 9.19 part (c) for the compliance field and in Fig. 9.20 part (b) for the inertance field.

The dual junction structures also can be used; these are shown in Fig. 9.19 part (b) and Fig. 9.20 part (c). (The options shown in part (b) of each figure are somewhat easier to manipulate. The two cases are perfect duals of one another, as are the two cases shown in part (c) of each figure.) The graphs are not symmetric with respect to the three ports; there are therefore six options for each field. The two-port fields also are not symmetric, as you have seen, and possesses two options. Integral causality is shown in each case; differential causality also may be used.

These field reticulations are helpful because they expose the essential transformational couplings between the parts. If, for example, the effort e_1 in Fig. 9.20 part (b) is made an independent input (changing the causality of its bond and the bond for C_1), its contribution to e_1 is simply $T_{31}e_1$ and its contribution to e_2 is simply $T_{21}e_1$. Also, state variables may be associated with each energy storage element and the corresponding state differential equations can be found by the standard methods.

Some of the bond graphs with mutual elements above employ 1-junctions, while others employ 0-junctions. Not infrequently, the most convenient energy expression is in terms of displacement instead of effort, or momentum instead of flow. Similarly, the most convenient energy dissipation expression might be in terms of effort rather than flow. As a result, the type of junction to which the expression corresponds is switched. Determination of these equivalences is straightforward. An example is developed step-by-step in Guided Problem 9.6 below.

9.3.8 Summary

The energy stored or dissipated in linear models is represented by a sum of terms, some of which are proportional to the square of an effort, momentum, displacement or flow variable and others of which are proportional to the product of a pair of variables. Those terms of the energy associated with each separate variable are directly representable, as before, by standard single-ported compliances, inertances and resistances. Those associated with a pair of variables are directly representable by special double-ported or *mutual* compliances, inertances or resistances.

It is possible to write state differential equations directly from bond graphs that contain both self and mutual energy storage elements, but the number of independent differential equations is less than the number of elements, requiring some use of differential causality. The recommended alternative is to construct equivalent bond graphs which contain only single-ported compliances and inertances, plus ideal transformers. These equivalent graphs not only permit simple interpretation in the form of differential equations, but also reveal in familiar terms the inherent transformational structure of the coupled storage mechanism. The process of working from the stored energy expression through a bond graph with mutual elements and then to an equivalent graph without these kinds of elements and finally a set of state differential equations can be simpler than the apparently more direct methods that have been employed previously.

---

[2]P.C. Breedveld, "Decomposition of multiport elements in a revised multibond graph notation," *J. Franklin Institute*, v. 318 n. 4 pp. 253-273, 1984.

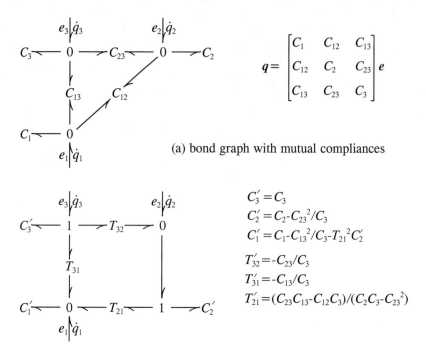

(a) bond graph with mutual compliances

$$q = \begin{bmatrix} C_1 & C_{12} & C_{13} \\ C_{12} & C_2 & C_{23} \\ C_{13} & C_{23} & C_3 \end{bmatrix} e$$

$C_3' = C_3$
$C_2' = C_2 - C_{23}{}^2/C_3$
$C_1' = C_1 - C_{13}{}^2/C_3 - T_{21}{}^2 C_2'$

$T_{32}' = -C_{23}/C_3$
$T_{31}' = -C_{13}/C_3$
$T_{21}' = (C_{23}C_{13} - C_{12}C_3)/(C_2C_3 - C_{23}{}^2)$

(b) equivalent bond graph (dual of Breedveld structure)

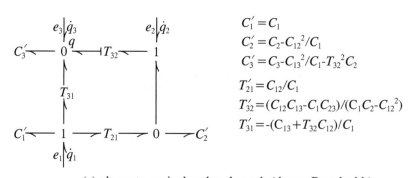

$C_1' = C_1$
$C_2' = C_2 - C_{12}{}^2/C_1$
$C_3' = C_3 - C_{13}{}^2/C_1 - T_{32}{}^2 C_2$

$T_{21}' = C_{12}/C_1$
$T_{32}' = (C_{12}C_{13} - C_1 C_{23})/(C_1 C_2 - C_{12}{}^2)$
$T_{31}' = -(C_{13} + T_{32}C_{12})/C_1$

(c) alternate equivalent bond graph (due to Breedveld)

Figure 9.19: Reticulation of a general linear three-port C field

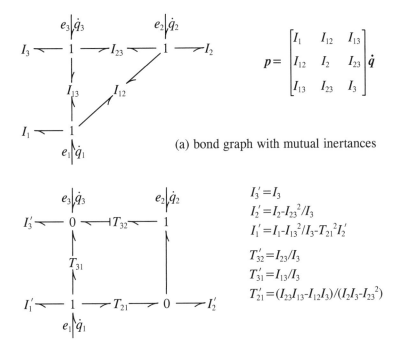

$$p = \begin{bmatrix} I_1 & I_{12} & I_{13} \\ I_{12} & I_2 & I_{23} \\ I_{13} & I_{23} & I_3 \end{bmatrix} \dot{q}$$

(a) bond graph with mutual inertances

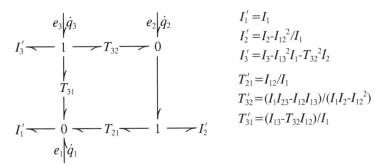

$I_3' = I_3$
$I_2' = I_2 - I_{23}^2/I_3$
$I_1' = I_1 - I_{13}^2/I_3 - T_{21}'^2 I_2'$

$T_{32}' = I_{23}/I_3$
$T_{31}' = I_{13}/I_3$
$T_{21}' = (I_{23}I_{13} - I_{12}I_3)/(I_2I_3 - I_{23}^2)$

(b) equivalent bond graph (due to Breedveld)

$I_1' = I_1$
$I_2' = I_2 - I_{12}^2/I_1$
$I_3' = I_3 - I_{13}^2 I_1 - T_{32}'^2 I_2$

$T_{21}' = I_{12}/I_1$
$T_{32}' = (I_1I_{23} - I_{12}I_{13})/(I_1I_2 - I_{12}^2)$
$T_{31}' = (I_{13} - T_{32}'I_{12})/I_1$

(c) alternate equivalent bond graph (dual of Breedveld structure)

Figure 9.20: Reticulation of a general linear three-port I field

Guided Problem 9.6

Find the general bond graph that is equivalent to the graph shown below, but that contains no mutual element and can be given integral causality. The parameters C_1, C_2 and C_{12} are constants. (This graph is used in Guided Problem 9.7 in the following section. Note its difference from the graphs of Figs. 9.15, 9.16 and 9.17.)

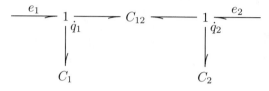

Suggested Steps:

1. Write the total energy stored in the graph elements in terms of q_1 and q_2.

2. Propose a graph with either of the forms

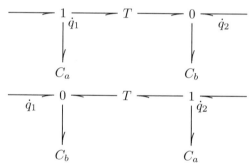

   and also find its energy in terms of q_1 and q_2.

3. Equate the two energy expressions for arbitrary values of q_1 and q_2, so as to find the constant parameters C_a, C_b and C_{12}.

4. Apply integral causality to the new equivalent graph.

PROBLEMS

9.23 A linear spring has two independent external attachments, as shown. Model this system including a linear two-port compliance, find the stored spring energy as functions of the displacements and the associated constitutive matrix, and find an equivalent reticulated bond-graph model with energy storage elements that have only one port.

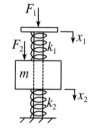

9.24 A yoke shown opposite is rigidly attached to a rotating shaft at a right angle. A small mass m slides frictionlessly on a small rod attached to the yoke as shown; the mass is loosely constrained by springs with a net spring rate k. The yoke has moment of inertia I_ϕ about the shaft.

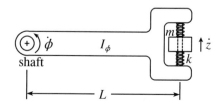

(a) Write an expression that approximates the kinetic energy of the system, using $\dot{\phi}$ and $\dot{z}$ as the velocity variables.

(b) Find a linearized model for small displacements of z, identifying self and mutual inertances as well as a compliance. Draw a corresponding bond graph. (Note: When $\dot{\phi}$ is large, the nonlinear behavior of this system can be significant for fairly modest displacements. Problems of this type are addressed in Chapter 12.)

(c) Find an equivalent bond graph with no mutual inertance.

(d) Find the natural frequency of the model when the moment in the shaft is zero.

9.25 Oil-cooled transformers tend to be noisy. (You may have heard them buzzing on telephone poles.) Eddy current and hysteresis losses are proportional to the square of the rate of change of the magnetic flux. Magnetostrictive action of the nickel/iron core causes a dilational fluid flow Q. There is an effective compressibility of the oil, and the acoustic pressure P acts against an acoustic impedance with resistance R_2 and inertance I_3. It is proposed that these effects can be represented in a summary fashion in the bond graph model below.

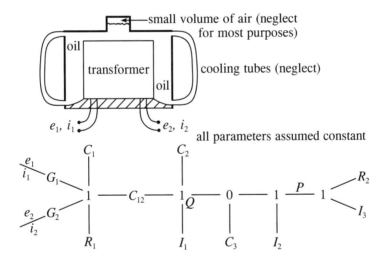

(a) Identify qualitatively and briefly the meanings of the various parameters (other than R_2 and I_3 which are identified already).

(b) Identify what you consider to be the most glaring omission in this model.

(c) Simplify the given graph in preparation for part (d), defining any new elements in terms of what is replaced. Apply causal strokes, identify input variables, and relate the output pressure P to the state variables. (This part can be done independently of parts (a) and (b).)

(d) Write in solvable form the set of state differential equations corresponding to your graph of part (c).

9.26 An electrical transformer has three coils wrapped around a core. Model this system with a multiport inductance, neglecting losses. Find an equivalent model without multiport elements (hint: see Fig. 9.20). Then, reduce this model for the special case of perfect flux linkage, which implies that the values of the mutual inductances are maximized. Finally, introduce resistive losses for the three coils.

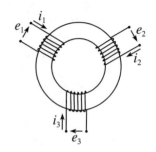

9.27 You are asked to evaluate the feasibility of basing a **solar engine** on the thermal expansion of a *metal*, avoiding some of the problems associated with fluids. Since the strain and percent change in absolute temperature are small, linearized characteristics can be used to model the behavior. Consider a unit volume of metal, for example in the shape of a rod, with uniform state, which undergoes a thermodynamic cycle by being alternately exposed to and shaded from sunlight.

(a) Represent the strain, thermoelastic and thermal energies of the metal in the form associated with a linear model. Draw the corresponding bond graph, and convert this graph into an equivalent graph with only two energy storages, one associated with the strain, S, and the other with the changes in entropy, $\delta\sigma$.

(b) Make the metal undergo a Carnot cycle, with the processes 1 to 4 below. Find the corresponding stress (T) - strain diagram, and the thermal efficiency of the cycle.

(c) Define the ratio of the actual thermal expansion coefficient to the maximum theoretically possible as ϵ. Evaluate ϵ for two or three metals for which you can find data (such as aluminum, iron, copper or magnesium), and compare to an ideal gas (such as air or a monatomic gas).

(d) Find the ratio of the energy converted per cycle to the maximum energy stored during the cycle (at the end of process 1) in terms of ϵ.

(e) Assume that the frictional loss per cycle is a fixed fraction, f, of the maximum strain energy. Find an expression for the *mechanical efficiency* in terms of ϵ and f.

(f) Draw practical conclusions about the feasibility of solar engines based on the expansion of metals from the results of the steps above.

SOLUTIONS TO GUIDED PROBLEMS

Guided Problem 9.6

1. $\mathcal{V} = \dfrac{1}{2C_1}q_1^2 + \dfrac{1}{C_{12}q_1q_2} + \dfrac{1}{2C_2}q_2^2$

2. The two graphs are mirror images of one another. For the left-hand graph,

$$\mathcal{V} = \frac{1}{2C_a}q_1^2 + \frac{1}{2C_b}(Tq_1 + q_2)^2 = \frac{1}{2}\left(\frac{1}{C_a} + \frac{T^2}{C_b}\right)q_1^2 + \frac{T}{C_b}q_1q_2 + \frac{1}{2C_b}q_2^2$$

3. Equating the three respective coefficients gives, directly,

$$\frac{1}{C_1} = \frac{1}{C_a} + \frac{T^2}{C_b}; \qquad \frac{1}{C_{12}} = \frac{T}{C_b}; \qquad \frac{1}{C_2} = \frac{1}{C_b}$$

from which $T = \dfrac{C_2}{C_{12}}; \qquad C_a = \dfrac{C_1}{1 - C_1 C_2 / C_{12}^2}; \qquad C_b = C_2$

4.

9.4 Nonlinear Multiport Fields

Nonlinear multiport compliances and inertances are particularly helpful in modeling a rich assortment of power and signal transducers.

9.4.1 Nonlinear Compliance Fields: Generic Relationships

The compliance field with stored potential energy

$$\mathcal{V} = \mathcal{V}(q_1, q_2, \ldots, q_m) \tag{9.27}$$

has as its net input power or energy flux

$$\mathcal{P} = \sum_{i=1}^{m} e_i \dot{q}_i = \frac{d\mathcal{V}}{dt} = \sum_{i=1}^{m} \frac{\partial \mathcal{V}}{\partial q_i} \frac{dq_i}{dt}. \tag{9.28}$$

Since this result applies regardless of the magnitude of the individual $\dot{q}_i$, the key result is the **theorem of virtual work**:

$$e_i = \frac{\partial \mathcal{V}}{\partial q_i}. \tag{9.29}$$

This equation computes e_i given the values of q_i, $i = 1, 2, \ldots, m$ and thus is consistent with integral causality:

It should be noted that the theorem of virtual work also applies directly to a one-port compliance; the partial derivative in equation (9.29) becomes an ordinary derivative.

9.4.2 Examples with Geometrically Varied Capacitance

Any capacitor that has its capacitance varied mechanically serves as an electromechanical transducer, that is, a device that transforms electrical energy to mechanical form or vice-versa. As a result, the force that the electric field places on the movable member can do work. Similarly, mechanical work done externally on the movable member affects the electrical field. The device becomes a transducer that converts electrical energy to mechanical energy, and vice-versa. It may be useful either as a sensor (instrument transducer) or as an actuator (power transducer).

Two examples are given below. The mechanical motion in the second case is actually the motion of a fluid. The further example of a capacitace microphone is given as Guided Problem 9.7 at the end of the section.

EXAMPLE 9.10

The upper plate of a parallel-plate capacitor is movable in the direction normal to
the planes of the plates:

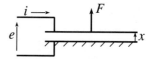

Model the system with a bond graph, and relate the voltage e across the capacitor and
the force F on the plate as functions of the gap width, x, and the electric charge, q. You
may assume that the capacitance is given by

$$C = \frac{\epsilon A}{x},$$

in which ϵ is the dielectric constant, A is the area of the plates, and fringing effects of
the electrostatic field around its edges are neglected.

Solution: The bond graph is $\vdash\!\!\frac{e_1}{i}\!\!\rightarrow C \frac{F}{\dot{x}}\!\!\dashv$.

The electrostatic energy $\mathcal{V}$ can be represented in the standard quadratic form

$$\mathcal{V} = \frac{1}{2C}q^2 = \frac{x}{2\epsilon A}q^2.$$

Thus, from equation (9.29), the voltage e and the force F are

$$e = \frac{\partial \mathcal{V}}{\partial q} = \frac{q}{C} = \frac{xq}{A\epsilon},$$

$$F = \frac{\partial \mathcal{V}}{\partial x} = \frac{q^2}{2\epsilon A} = \frac{\epsilon A e^2}{2x^2}.$$

Note that the applied force F and velocity $\dot{x}$ are in the same direction, so the power
product $F\dot{x}$ is positive when mechanical power flows *into* the compliance.

Different substances have different dielectic constants, and thus the capacitance of a
fixed-plate capacitor changes when different materials are moved into the gap. This is the
basis of the device analyzed in the following example, which could be used to sense the
level of a liquid such as water. It also could be used as the heart of a pump which has no
solid moving parts; the author has suggested this concept for use in microelectromechanical
systems.

EXAMPLE 9.11

Two immiscible fluids with different dielectric constants, either or both of which are
liquids, are separated by a meniscus that lies in the gap between two fixed plates of a
capacitor:

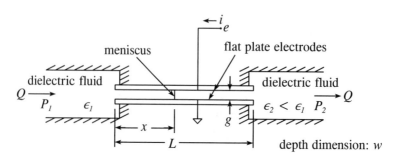

Model the system with a bond graph. Then, relate the voltage e across the capacitor to the electric charge q and the position x of the meniscus, assuming the parameters L, g, w, ϵ_1 and ϵ_2 and neglecting the effects of electrostatic fringing. Also, relate the horizontal force F that acts on the meniscus, producing the pressure drop $P_1 - P_2$, to the same variables and parameters.

Solution: The bond graph below employs the force F and velocity $\dot{x}$ to describe the fluid within the gap. They are related to the more conventional pressures P_1 and P_2 and flow Q by the transformer, the modulus of which is the reciprocal of the area of the channel.

$$P_2 \uparrow \quad 1 \underset{Q}{\overset{}{\rightharpoonup}} T \xrightarrow{\dfrac{F}{\dot{x}}} C \leftharpoondown \dfrac{e}{i} \qquad T = 1/wg$$
$$P_1 \downarrow$$

The meniscus will be assumed to be planar; this simplifies the model and tends to be approximately correct because both the surface tension and electrostatic forces encourage it. The capacitor then can be viewed as two capacitors in parallel, one containing the first dielectric fluid and the other containing the second dielectric fluid. The total capacitance is the sum of the two. (Both have a common potential, and currents or charges that sum, and therefore they can be bonded to a common 0-junction.) Thus,

$$C = \frac{\epsilon_1 wx}{g} + \frac{\epsilon_2 w(L - x)}{g}$$

where x and $L - x$ are the respective lengths of the two slugs of fluid within the gap. Note that the channel has a rectangular cross-section.

The potential energy becomes

$$\mathcal{V} = \mathcal{V}(q, x) = \frac{1}{2C(x)} q^2 = \frac{gq^2/2}{w[\epsilon_2 L + (\epsilon_1 - \epsilon_2)x]}.$$

Therefore,

$$e = \frac{\partial \mathcal{V}}{\partial q} = \frac{gq}{w[\epsilon_2 L + (\epsilon_1 - \epsilon_2)x]},$$

$$F = \frac{\partial \mathcal{V}}{\partial x} = \frac{-g(\epsilon_1 - \epsilon_2)q^2}{2w[\epsilon_2 L + (\epsilon_1 - \epsilon_2)x]^2} = -\frac{w}{2g}(\epsilon_1 - \epsilon_2)e^2.$$

Note that the equations correspond to the standard use of integral causality.

9.4.3 Nonlinear Multiport Inertances

The analogous situation to a multiport compliance that depends on more than one generalized displacement is a multiport inertance that depends on more than one generalized momentum. Thus for an inertance with m ports,

$$\mathcal{T} = \mathcal{T}(p_1, p_2, \ldots, p_n). \tag{9.30}$$

The time rate of change of the energy is the net power input:

$$\frac{d\mathcal{T}}{dt} = \sum_{i=1}^{m} e_i \dot{q}_i = \frac{\partial \mathcal{T}}{\partial p_1} e_1 + \frac{\partial \mathcal{T}}{\partial p_2} e_2 + \ldots + \frac{\partial \mathcal{T}}{\partial p_m} e_m, \tag{9.31}$$

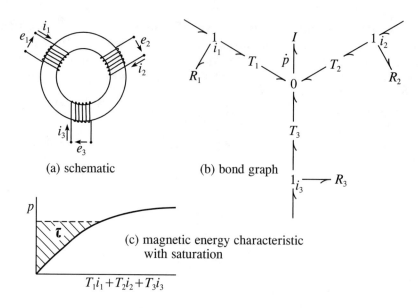

(a) schematic   (b) bond graph

(c) magnetic energy characteristic
with saturation

Figure 9.21: Three-coil transformer with perfect flux linkage and saturation

which gives

$$\dot{q}_i = \frac{\partial T}{\partial p_i},$$ (9.32

consistent with admittance causality:

An electrical transformer with magnetic saturation is an example. Only the very special case of a three-coil transformer with perfect flux linkage is detailed here. (Problem 9.26, p. 632, deals with the linear version of this case.) What in general would be a very complicated model collapses to require only a single one-port inertance, representing the existence of only one independent magnetic flux. This inertance is nonlinear, representing the magnetic saturation, as illustrated in Fig. 9.21. Perfect flux linkage means that the magnetomotive forces of the three coils sum together. The individual magnetomotive forces are proportional to the currents and the number of turns of the respective coils, as with the linear model. The nonlinearity imposes a virtual limit to the magnetic flux and to the energy that can be stored, as indicated in the plot given in part (c) of the figure. Note that there is no such phenomenon in classical mechanics, by its very definition.

Models with nonlinear multiport inertances satisfying equation (9.30) tend to require more sophisticated techniques than are presented in the present chapter and may not be encountered often. More is offered in Chapter 12.

9.4.4 Inertances Dependent on Holonomic Constraints

A different type of nonlinear inertance occurs often, however. This type has kinetic energy that is modulated not only by a momentum, but also by one or more displacements.

For example, the kinetic energy of a fluid flow may depend on a generalized momentum and on a remote displacement, namely the width of the channel. The generalized kinetic energy for the major class of systems being considered is represented in the form

$$\mathcal{T} = \mathcal{T}(p_1, q_2, q_3, \ldots, q_m). \tag{9.33}$$

The variables in parentheses are treated as state variables. To be consistent with integral causality, they are the time integrals of the causal inputs to the energy-storage element. Accordingly, the mixed causality employed is indicated by the causal strokes below:

$$e_1 = \dot{p}_1 \quad \overset{e_2 \mid \dot{q}_2}{\underset{\dot{q}_1}{\longrightarrow}} \; I \; \overset{e_3}{\underset{\dot{q}_3}{\longrightarrow}} \; \underset{e_m \mid \dot{q}_m}{}$$

The dependence on one or more displacements has led some to call this an "*IC* element."

Only the first bond has generalized momentum as its input. The inertance can be said to refer to p_1 or $\dot{q}_1$ but at the same time be modulated by $q_2, \ldots, q_m$. Taking the time derivative of equation (9.33) in the standard fashion,

$$\frac{d\mathcal{T}}{dt} = \sum_{i=1}^{m} e_i \dot{q}_i = \frac{\partial \mathcal{T}}{\partial p_1} e_1 + \frac{\partial \mathcal{T}}{\partial q_2} \dot{q}_2 + \ldots + \frac{\partial \mathcal{T}}{\partial q_m} \dot{q}_m \tag{9.34}$$

Therefore, comparing terms,

$$\dot{q}_1 = \frac{\partial \mathcal{T}}{\partial p_1} \tag{9.35a}$$

$$e_i = \frac{\partial \mathcal{T}}{\partial q_i} \qquad i = 2, \ldots, m \tag{9.35b}$$

Note that if a displacement affects the generalized kinetic energy, and that displacment is changed, power must flow into or out from the inertance through the bond associated with that displacement or generalized velocity. The inertance is said to interrelate the variables of the various bonds by a **holonomic constraint**. The contrast to a **nonholonomic constraint** is made below.

There are holonomic constraints that do not fit equation (9.33) and are beyond the scope of this chapter.[3]

The example below models the **solenoid**, which is the most common type of electro-mechanical actuator. It also can be used in the inverse way to sense electrically the position of a mechanical member. Another example of a holonomic modulation of an inertance is given in Guided Problem 9.9 at the end of this section.

---

[3]The approach here is a special case of the Hamiltonian method presented in a Chapter 12. Other cases also are treatable by the Lagrangian method also presented in Chapter 12.

EXAMPLE 9.12

A solenoid, pictured in cross-section below, can be described as a variable-geometry inductor. It comprises a coil with inductance dependent on the mechanical position of a movable ferromagnetic prismatic core.

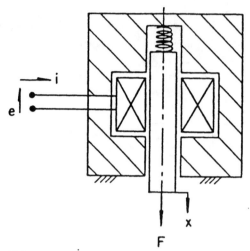

Model the solenoid by a bond graph with the preferred causalities, and relate the causal output electric current i and mechanical force F to the input displacement x and voltage e. The dependence of the inductance on x, that is the function $I(x)$, can be assumed as known.

Solution: The *local* generalized displacement, q, is the time integral of the electric current, which is the local flow. The local generalized momentum, p, is the integral of the voltage. The generalized kinetic energy can be written in the form

$$T = \frac{1}{2I(x)}p^2,$$

in which x is a *non-local* displacement. The causal bond graph is

$$\underset{i=\dot{q}}{\overset{e=\dot{p}}{\rule{3cm}{0pt}}}\!\!\!\!\dashv I \xleftarrow{} \underset{\dot{x}}{\overset{F}{\rule{2cm}{0pt}}}\vdash$$

Substitution of the equation for T above into equation (9.29) gives the results

$$i = \dot{q} = \frac{p}{I(x)}$$

$$F = -\frac{p^2}{2[I(x)]^2}\frac{dI}{dx} = -\frac{i^2}{2}\frac{dI}{dx}$$

In a particular design, the detailed task now is to evaluate $I(x)$. More information is given in the following section (9.5), and in Section 12.3.4. Or with luck you could assign this job to a specialist.

9.4.5   Inertances Dependent on Nonholonomic Constraints

Consider a constant inertance I_c driven through a transformer with modulus $T(x)$ that depends on a *remote* displacement, x:

$$\frac{e = T\dot{p}}{\dot{q} = p/TI_c} \mid T \overbrace{\frac{\int e\,dt = \dot{p}}{\dot{q}_c = T\dot{q} = p/I_c}} \mid I_c \qquad\qquad T = T(x)$$

Although a variation in x changes the stored energy, the power flow associated with the change equals the product of the generalized velocity $\dot{q}$ and effort e exclusively; no generalized force must act on the displacement x itself. This kind of relationship is an example of a **nonholonomic constraint**. It is the same type considered early in Section 9.1.

An expression for the kinetic energy in a particular holonomic system may be identical to an expression for the kinetic energy in a radically different nonholonomic system. Obviously, some information is needed in addition to the expressions for kinetic energy. Although the distinction may be tricky to understand, it is critical. From a formal perspective, a holonomic system has the same number of generalized coordinates as degrees of freedom; a nonholonomic system has fewer degrees of freedom than generalized coordinates. The number of generalized coordinates in a model is the minimum number of displacement variables needed to specify its displacement state. The number of degrees of freedom can be less: it is the number of *independent* small changes in the displacement variables that are geometically possible.

In the case at hand, the displacements $\dot{q}$, $\dot{q}_c$ and x are generalized coordinates, but the transformer introduces a constraint that reduces the number of degrees of freedom. Specifically, Δq_c is made dependent on Δq, while still requiring values of q, q_c and x to describe the displacement state.

To highlight the distinction between homonomic and nonholonomic systems, observe that the generalized kinetic energy of the bond graph above can be written as

$$\mathcal{T} = \frac{1}{2}I_c\dot{q}_c^2 = \frac{1}{2}T^2(x)I_c\dot{q}^2. \tag{9.36}$$

This is the same as the following expression for the generalized kinetic energy of the holonomic variable-geometry inductor or solenoid of Example 9.12 above,

$$\mathcal{T} = \frac{1}{2}I(x)i^2, \tag{9.37}$$

if the associations $\dot{q} = i$ and $T^2(x)I_c = I(x)$ are made. Nevertheless, the bond graph above is incompatible with the bond graph of the solenoid, and the general holonomic relation of equation (9.32) cannot represent a nonholonomic system. The difference is fundamental, however; an attempt to represent the variable-geometry inductor by the nonholonomic bond graph above, for example, would result in a solenoid that produced no mechanical force whatsoever. On the other hand, without nonholomonic systems the control of power would be extremely difficult, for such things as adjustable mechanical transmissions would not exist. The student reading an advanced textbook on dynamics might conclude incorrectly that holonomic systems are the norm, and nonholonomic systems are rare and exotic. The reverse is actually closer to the truth.

Looking deeper at these contrasting examples, Equation (9.33) expresses the energy in terms of the generalized *momentum*. It is not possible, in fact, to express the energy of the nonholonomic system in terms of its generalized momentum, which is $\int e\,dt = \int T(x)\dot{p}\,dt$ (except in the trivial case in which x is a constant, which voids all coupling between the inertances and x.) Therefore, the two bond graphs themselves, or energy relations expressed in terms of generalized *momenta* as opposed to generalized velocities, are sufficient to distinguish the two types of constraints.

The following example represents conceptually the nonholonomic coupling between a flywheel used to store energy for propelling a vehicle, and the vehicle drive wheel. It bears a superficial resemblance to a variable speed flywheel with a holonomic constraint given in Guided Problem 9.9, but their behaviors are dramatically different.

EXAMPLE 9.13

It is proposed to store kinetic energy for a bicycle or other vehicle in a flywheel. The flywheel is driven from a drive wheel by a transmission with a continuously adjustable speed ratio that is a known function of the axial displacement of the flywheel, x:

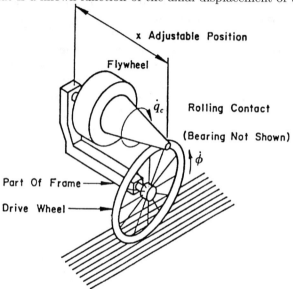

Identify whether the system is holonomic, and model it with a bond graph. Then, find the associated equations, assuming that the moment M applied to the drive wheel and the position x of the flywheel are specified.

Solution: The bond graph below represents the fixed inertia of the flyweel by I_c. The speed ratio between the angular velocities of the drive wheel ($\dot{\phi}$) and the flywheel ($\dot{q}_c$) is represented by the modulus of the transformer, $T(x)$. This transformer constrains infinitesimal displacements of q_c without reducing the number of generalized coordinates (three, for q, q_c and x) necessary to describe the positional state of the system.

$$\frac{M}{\dot{\phi}} \rightharpoonup T \; \frac{\overset{\textstyle \overset{.}{p}}{\textstyle \bigcirc}}{\dot{q}_c = p/I_c} \rightharpoonup I_c$$

The associated differential equation and output variable, as the annotations indicate, are

$$\frac{dp}{dt} = \frac{1}{T(x)} M,$$

$$\dot{\phi} = \frac{1}{I_c T(x)} p,$$

Note that adjustment of the friction drive requires zero essential force and energy, consistent with a nonholonomic modulation and inconsistent with a holonomic modulation. Minor forces due to the accelerations or friction in the adjustment mechanism could, if desired, be modeled separately.

Heat transfer and thermal expansion can change the modulus of an inertance. This change can be represented as a nonholonomic contraint that modulates a transformer bonded to a fixed inertance. An example is offered in Guided Problem 9.10. Heat transfer is possibly the most ubiquitous (and as such under-recognized) of all nonholonomic constraints.

9.4.6 Summary

Potential energy storages which are irreducibly a function of two or more generalized displacements can be represented by multiport compliances, regardless of whether the model is linear or nonlinear. If the stored energy is known as an expressed function of these displacements, as in the examples given for variable capacitance, the efforts or generalized forces are readily computed as the gradient of the function in the direction of the respective displacements. This computation is consistent with integral causality and thus can be incorporated into the writing of differential equations for a larger system into which the compliance field is imbedded.

The bond graph formalism employed in this book (except in Chapter 12) requires that one-port inertances be either constants or at most functions of the local generalized velocity. Multiport inertances can be functions of the momenta on each of its bonds. In a different case, a kinetic energy that depends partly on a local displacement can be represented by a displacement-modulated transformer in series with such an inertance, as introduced in Section 9.1. A kinetic energy that depends partly on a non-local displacement can be accommodated the same way if and only if the constraint is nonholonomic. Otherwise, it can be represented as a multiport inertance field, in much the same way as compliance fields are treated. The energy should be expressed as a function of the primary momentum and the remote displacement or displacements.[4]

Holonomic constraints produce generalized forces on all bonds of the multiport inertance. Significant power is apt to flow through any of the bonds. Nonholonomic constraints have no essential efforts associated with their influence on the kinetic energy through their modulation of the transformer. No power is required to effect this modulation, apart from any parasitic friction or other minor nonessential forces that you might wish to model separately. The contrast is illustrated in the comparison of the systems in Example 9.13 above and Guided Problem 9.9 below.

Guided Problem 9.7

A capacitance microphone is based on a parallel plate capacitor; as shown in Fig. 9.22, one electrode is a diaphragm of effective area A and mass m that moves with a uniform displacement, x, while the other electrode is rigid and perforated to allow free passage of air. The mean separation of the plates is d, to give a capacitance of $C = \varepsilon A/(d + x)$. The effective stiffness of the diaphragm is k, which includes any effect of compression of the air in the volume behind the diaphragm. The voltage applied is $\bar{e} + e^*$, where $\bar{e}$ is constant; the associated electric charge is $\bar{q} + q^*$ where $\bar{q}$ is constant. Write differential equations that model the dynamics for small changes $x \ll d$, $e^* \ll \bar{e}$ and $q^* \ll \bar{q}$.

Suggested Steps:

1. Write an expression for the energy stored ($\mathcal{V}$), including both the electrostatic energy and the elastic energy. Employ the generalized displacement pair V (equal to Ax)

---

[4]In Chapter 12, the bond graph for the holonomic systems considered in the present section becomes a special case of what is called a Hamiltonian bond graph.

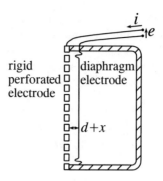

Figure 9.22: Capacitance microphone of Guided Problem 9.7

and q as the variables. Note that when $x = 0$ the diaphragm has a strain deflection, which you could call $-\bar{x}$, due to the charge $\bar{q}$.

2. Calculate the pressure $P^* = \partial \mathcal{V}/\partial V$ and voltage $e^* + \bar{e} = \partial \mathcal{V}/\partial q$. Note that the constant terms in the first derivative must cancel, and the constant terms in the second derivative equal $\bar{e}$.

3. Complete the linearization started in step 2 by retaining only first order terms in V and q; the higher order terms are relatively small and may be neglected. Represent the results by the bond graph below. Evaluate the compliances as functions of the constant parameters.

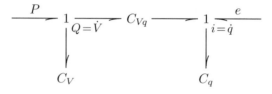

4. Add the inertia of the diaphram to this graph. Find its value through use of its kinetic energy.

5. Apply the bond graph equivalences found in Guided Problem 9.6 to eliminate the mutual compliance in favor of a transformer. Apply integral causality to the graph.

6. Write state variable differential equations for the system, and relate the causal outputs to the state variables.

Guided Problem 9.8

The movable plate of a variable-plate capacitor, shown in Fig. 9.23, is supported mechanically so that it can move freely in the direction *parallel* to the plane of the plates. (This concept has been used in some microelectromechanical systems.) Represent the resulting coupling between the electric and mechanical variables by an annotated bond graph. Find approximate relations for this coupling, neglecting the fringing effects of the electrostatic field.

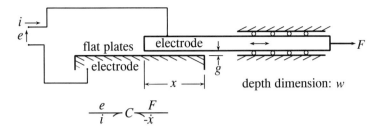

Figure 9.23: Guided Problem 9.8

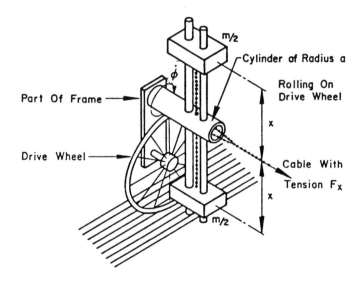

Figure 9.24: Guided Problem 9.9

Suggested Steps:

1. Write an approximate expression for the electrostatic energy of the field, as functions of the displacement x and the electric charge q. The examples above should indicate how this can be done.

2. Find the force F and voltage e by computing the appropriate derivatives, as indicated by equation (9.29) (p. 633).

Guided Problem 9.9

A flywheel with adjustable inertia comprises two blocks, each with mass $m/2$, which move radially on rods welded perpendicularly to the shaft, as shown in Fig. 9.24. The blocks are held by cords that pass over pulleys; the centers of mass can be drawn inward or released outward with this cord. Model the system by a bond graph, define state variables, and find the state differential equations. The blocks may be idealized as point masses, all other inertias may be neglected, and all friction may be neglected.

Suggested Steps:

1. Define the adjustable radius, x, and the angle of the shaft, ϕ, as the generalized displacements. Write the kinetic energy of the idealized system in terms of these displacements and associated generalized momenta. Note that the kinetic energy for radial velocity of the blocks can be expressed in a term separate from the kinetic energy for rotation.

2. Draw a bond graph, apply causality, and identify the effort and flow variables. The torque on the shaft and the force on the cord may be considered as independent variables.

3. Write the differential equations using the standard techniques, except that equation (9.35) (p. 637) is used for the two-port inertance.

4. Compare the practicality of this system with that of Example 9.13 (p. 640) as a means of storing and extracting energy from a flywheel. What is the essential difference in their behaviors?

Guided Problem 9.10

The temperature-regulated valve shown in Fig. 9.25 is used with a highly unsteady flow, so the flow inertance I as well as the flow resistance R are of interest. Consider that these parameters depend on the temperature of the expandable member in a known way. The thermal compliance of this member is C, which can be considered to be a constant. Its temperature is θ, the temperature of the environment is θ_e, and there is a thermal conductance H between the two. Find differential equations which if solved would give the volume flow in response to the pressure drop ΔP and the environmental temperature. Neglect any heating or cooling of the expandable member caused by the flow.

Suggested Steps:

1. Identify the state variables.

2. Formulate the key constraint and identify whether it is holonomic or nonholonomic.

3. Draw a bond graph and label the variables.

4. Write the differential equations. (Since you are not given the forms $I(\theta)$ and $R(\theta)$ and are not asked to model these, the equations are not yet in a form ready for solution. This could be done as an extension to the problem, however.)

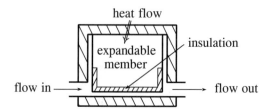

Figure 9.25: Guided Problem 9.10

PROBLEMS

9.28 The slider-crank mechanism, shown opposite, lifts a weight, mg, vertically. The weights of the other members may be neglected by comparison. Thus the representation

$$\frac{M}{\dot{\phi}} \rightharpoonup T \xleftarrow{mg} \frac{mg}{\dot{y}} \rightharpoonup S_e$$

can be used, in which $T = T(\phi)$. Alternatively, however, from the point of view of the shaft the device looks like a rotary spring: $\dfrac{M}{\dot{\phi}} \rightharpoonup C$.

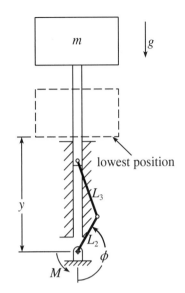

(a) Write the potential energy of the spring as a function of ϕ, and determine the compliance relation $M = M(\phi)$.

(b) Find a linearized model valid for small displacements from a nominal angle $\phi = \overline{\phi}$, including evaluation of the linearized compliance C^* and any other parameter(s).

9.29 A movable-plate capacitor has its moving member attached to a spring, as shown opposite. The spring exerts zero force when $x = x_0$. For the voltage e being constant, find any equilibrium positions x and determine their stability. You may express the answers on a sketch-plot of the electric and spring forces, rather than algebraically. Also, find the minimum value of k necessary for stability at a particular value of the equilibrium x, and the associated ratio x_0/x.

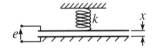

9.30 An *electrostatic* motor, shown in cross-section opposite, has a fixed cathode of radius r, a rotor of thickness t comprising alternate 90° segments of high dielectric constant $k_H \epsilon_0$ and low dielectric constant $k_L \epsilon_0$, and an outer anode with four equal segments. A high constant voltage v_0 is attached to terminals A-A and a zero voltage to terminals B-B; every quarter revolution of the rotor the voltages are switched. The unit has thickness w. The clearances may be assumed to be very small.

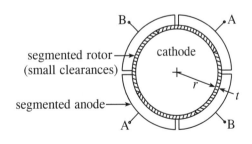

segmented rotor → (small clearances)

segmented anode →

cathode

(a) Estimate the potential energy stored in the electric fields of the motor, valid for up to 90° of rotation.

(b) Find the torque, M, generated by the motor in terms of the information given.

(c) Find the electric current required by the motor.

9.31 The electromechanical balance below has a position-dependent capacitor and inductor with values as given (which neglect fringing effects). The motion is very small; $x << d$.

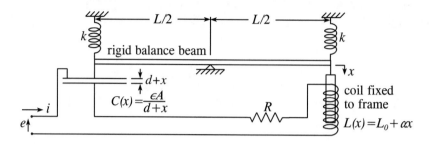

(a) Model the system with a bond graph, and evaluate all moduli in terms of the given information.

(b) Find the describing state differential equations.

(c) For $i = i_0 \sin \omega t$, determine the frequency ω for which the time-average of position x equals zero when no external force is applied to the beam.

9.32 Identify whether the constraints in the five systems given in Problems 9.1, 9.2 and 9.3 (pp. 597-598) are holonomic or nonholonomic.

9.33 A liquid-filled tank of uniform area A_T is drained through a tube of uniform area A_L and length L. Initially, a valve located a distance L_0 downstream of the tank is closed, and the section of the tube downstream of this valve is empty. The valve is opened abruptly, giving a flow equal to $a\sqrt{\Delta P}$, where ΔP is the pressure drop across the valve and a is a constant. No other frictional pressure drop need be considered, and the liquid-air interface remains virtually normal to the axis of the drain line as it moves to the right.

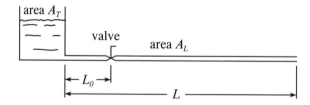

(a) Model the system with a bond graph.

(b) Find a solvable set of differential equations, which if solved would give the behavior of the system valid until the liquid-air interface reaches the downstream end of the tube.

9.34 An incompressible fluid of density ρ flows through a long narrow channel with width, x, that can be varied by the application of a force, F. Model the system. Find differential equations that interrelate the flow Q, the position x, the velocity $\dot{x}$, the force F and the pressures P_1 and P_2. Viscosity and Bernoulli losses may be neglected. The velocity of each fluid particle is virtually horizontal (*i.e.* $w\dot{x} << Q/wx$).

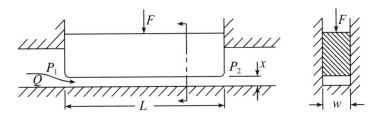

9.35 A dielectric fluid with density ρ and dielectric constant ϵ is pumped from a lower reservoir to a reservoir h higher. The environment is air, which has a dielectric constant $\epsilon_0 < \epsilon$. This is accomplished by placing a voltage v_0 across a pair of plate electrodes of width w and separation d. As soon as the fluid reaches the top of the right-hand plate the voltage is discontinued, but inertia keeps the fluid moving upward for a while. The fluid that rises above the top of the right-hand plate falls back into an upper reservoir; the fluid remaining between the plates falls back to the lower reservoir. The cycle is then repeated.

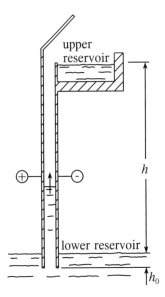

Notes: Parts (a)-(c) can be done independently of one another. Viscous wall-shear effects may be neglected, but not all other resistive or inertial effects may be.

(a) Model the system when the surface of the liquid is rising between the plates. Evaluate the parameters of your model, or alternatively give expressions for their energy in terms of defined state variables.

(b) Model the system after the surface has broached the upper edge of the right-hand plate but before it has stopped rising. Give parameters as in (a).

(c) Model the system when the surface is settling down between the plates. Again, give the parameters.

(d) If you have completed part (a), find a set of state differential equations to describe that part of the cycle. Otherwise, find the equations for part (c) for part credit or part (b) for the least credit.

9.36 A cylindrical piston with dimensions as shown is (i) supported submerged on a spring with stiffness k in an incompressible liquid of density ρ, or (ii) floats on the liquid with the equilibrium position as shown. In both cases the cylinder is only slightly larger in diameter than the piston. Viscous effects may be neglected.

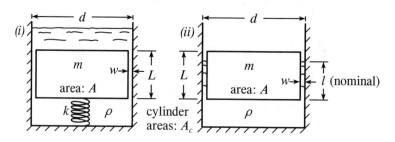

(a) Find the effective mass, stiffness and natural frequency in case (i), including the effect of the kinetic energy of the liquid in the annular region.

(b) Define state variables and find differential equations for case (ii) in a form suitable for solution.

(c) Linearize the equations of part (b), combine them to leave only a position variable, and determine the response to an initial condition which is not far from equilibrium.

SOLUTIONS TO GUIDED PROBLEMS

Guided Problem 9.7

1. $\mathcal{V} = \dfrac{d+x}{2\epsilon A}(\bar{q}_q^*)^2 + \dfrac{1}{2}k(x-\bar{x})^2 = \dfrac{d+V/A}{2\epsilon A}(\bar{q}+q^*)^2 + \dfrac{1}{2}k\left(\dfrac{V}{A}-\bar{x}\right)^2$

2. $P = \dfrac{\partial \mathcal{V}}{\partial V} = \dfrac{1}{2\epsilon A^2}(\bar{q}^2 + 2\bar{q}q^* + q^{*2}) + \dfrac{1}{2}k\left(\dfrac{V}{A}-\bar{x}\right)$

    $\bar{e}+e^* = \dfrac{\partial \mathcal{V}}{\partial q} = \dfrac{d+V/A}{\epsilon A}(\bar{q}+q^*)$

    Canceling the constant terms,

    $P = \dfrac{\bar{q}}{\epsilon A^2}q^* + \dfrac{1}{2\epsilon A^2}q^{*2} + \dfrac{k}{A}V$

    $e^* = \dfrac{d}{kA}q^* + \dfrac{\bar{q}}{\epsilon A^2}V + \dfrac{1}{\epsilon A^2}Vq$

3. Neglecting the higher-order terms in $q*$ and V, and comparing to the bond graph,

    $P = \dfrac{\bar{q}}{\epsilon A^2}q^* + \dfrac{k}{A}V = \dfrac{1}{C_{Vq}}q^* + \dfrac{1}{C_v}V$

    $e^* = \dfrac{d}{\epsilon A}q^* + \dfrac{\bar{q}}{\epsilon A^2}V = \dfrac{1}{C_q}q^* + \dfrac{1}{C_{Vq}}V]$

    Therefore, $C_V = \dfrac{k}{A}$; $C_q = \dfrac{d}{\epsilon A}$; $C_{Vq} = \dfrac{\bar{q}}{\epsilon A^2}$

4.

$$\frac{1}{2}IQ^2 = \frac{1}{2}m\dot{x}^2; \text{ therefore, } I = m\left(\frac{\dot{x}}{Q}\right)^2 = \frac{m}{A^2}$$

5.

where $T = \dfrac{C_q}{C_{Vq}} = \dfrac{Ad}{\overline{q}};$ $C_V' = \dfrac{C_V}{1 - C_V C_q C_{Vq}^2} = \dfrac{k/A}{1 - kd\epsilon A^2/\overline{q}^2}$

6. The state variables are noted in the graph above, from which the differential equations are

$$\frac{dp}{dt} = P - \frac{1}{C_v}V - \frac{T}{C_q}q^*$$

$$\frac{dV}{dt} = \frac{1}{I}p$$

$$\frac{dq}{dt} = i + TQ = I + \frac{t}{I}p$$

and the causal outputs are $Q = \dot{V} = \dfrac{1}{I}p;$ $e^* = \dfrac{1}{C_q}q^*.$

Guided Problem 9.8

1. $\mathcal{V} = \dfrac{gq^2}{2\epsilon wx}$

2. $e = \dfrac{\partial \mathcal{V}}{\partial q} = \dfrac{gq}{\epsilon wx}$: $F = -\dfrac{\partial \mathcal{V}}{\partial x} = \dfrac{gq^2}{2\epsilon wx^2} = \dfrac{\epsilon we^2}{2g}$

Guided Problem 9.9

1. $\mathcal{T} = \frac{1}{2}m[(x\dot{\phi})^2 + \dot{x}^2]$. However, the momentum for the radial velocity is $p_x = m\dot{x}$, and the momentum for the rotational velocity is $p_\phi = mx^2\dot{\phi}$. Therefore,

$$\mathcal{T} = \frac{1}{2mx^2}p_\phi^2 + \frac{1}{2m}p_x^2$$

2.

3. $\dot{\phi} = \dfrac{\partial \mathcal{T}}{\partial p_\phi} = \dfrac{p_\phi}{mx^2}$

$e = \dfrac{\partial \mathcal{T}}{\partial x} = -\dfrac{p_\phi^2}{mx^3}$

The differential equations are

$$\frac{dp_\phi}{dt} = M,$$

$$\frac{dp_x}{dt} = -F_x - e = -F_x + \frac{p_\phi}{mx^3}$$

$$\frac{dx}{dt} = \frac{1}{I_x} p_x$$

Note that $\dot{x} = p_x/m$.

4. This holonomic device is impractical compared with the nonholonomic system of Example 9.13, in that a large energy is required to change x. The adjustment of x in that example requires no essential force or energy, which is the idea behind the design, and therefore it is represented by a modulated transformer. All practical variable transmissions can be modeled essentially as nonholonomic modulated transformers.

Guided Problem 9.10

1. The entropy of the expandable member, S, and a momentum for the flow, p, are proper state variables. The need for a constant inertance (to satisfy the requirements of our scheme for writing differential equations) affects the choice of p, however, as noted below.

2. Both the resistance to the flow, $R = R(\theta, Q)$, and the inertance, $I = I(\theta)$, are assumed to depend only on the temperature, θ, which is a direct function of S. Our scheme for writing differential equations requires that we employ a *constant* inertance, however. This could be $I_0 = I(\theta_0)$, which is connected to the flow Q by a transformer with modulus $T = T(\theta) = \sqrt{I(\theta)/I(\theta_0)}$. But does this transformer represent a nonholonomic constraint, which would deny any further coupling between the fluid flow and the entropy (or temperature)?

The literal answer to this question is *no*. This follows from the observation that the generalized velocity TQ is not *integrable*, that is $\int TQ\,dt$ is a quasi-coordinate not representable by a function other than its being a time integral of a state function. Further, an accurate accounting of the kinetic energy of the fluid in the gap includes the effect of the *velocity* at which the gap opens or closes. This gives a direct force on the moving member dependent on both this velocity and the fluid flow. The problem is not included in the restricted class represented by equation (9.33) (p. 637).

Nevertheless, the velocity at which the gap opens or closes is likely to be so small that its effects on the kinetic energy of the fluid are negligible. The change in the gap due to the strain of the expandable member that results from the fluid-induced force on it also is likely to be negligible. Therefore, the modeler can be quite justified to treat the coupling as though it were nonholonomic. No further coupling between the thermal and fluid parts of the system is recognized by this approximation. This interpretation is consistent with the problem statement.

The bond graph has two pieces:

4. $\dfrac{dS}{dt} = H\left(\dfrac{\theta_e}{\theta} - 1\right),$

$\dfrac{dp}{dt} = \dfrac{1}{T}(\Delta P - RQ) = \dfrac{1}{T}\left(\Delta P - \dfrac{R}{I_0 T}p\right),$

$R = R(\theta, Q); \quad T = T(\theta); \quad \theta = \dfrac{S}{C}; \quad Q = \dfrac{1}{I_0 T}p; \quad I = I(\theta_0).$

9.5   Magnetic Circuits

Magnetic fields and fluxes are central to the operation of many devices considered earlier in this chapter and in earlier chapters. In these considerations the role of the fields and fluxes are viewed, through the keyhole as it were, by their effect on voltages and currents and sometimes also forces and velocities. Thus, an electrical transformer is viewed as a two-port inductor, and a solenoid is viewed as a position-dependent inductor. This simple view has advantages, and its use will continue when appropriate. There are times, however, when one wishes to get a little deeper into the magnetics, in order to understand the behavior better or to recognize important secondary phenomena. Magnetic and electric fields, as represented by the Maxwell's equations, are inseparable and three-dimensional, and both produce complex force fields. Most of the detail is far too complex to present here, but it is nevertheless posssible to give a concise view sufficient for approximate analysis of widely occuring geometries of engineering significance. This section focuses on systems with fixed coils, different from the translating coil device and the DC motor considered earlier.

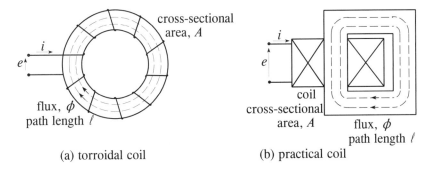

Figure 9.26: Electric coils with magnetic core

9.5.1   Magnetic Effort, Flow and Compliance

Most engineering devices with significant magnetic effects are connected to electric circuits with coils. Consider a coil wrapped around a magnetic core, as shown in Fig. 9.26. Ideally, the shape of the core would be a perfect torroid and the coil would be tightly and uniformly wrapped around the entire periphery, as shown on the left, but the more practical and commonly encountered geometry pictured on the right gives almost the same result. The electrical current produces what is called a magnetomotive force (mmf) M, proportional to the current and the number of coils. This relation is consistent with the gyrational coupling

$$\frac{e}{i} \quad G \quad \frac{M}{\dot{\phi}} \qquad\qquad G = n \qquad\qquad (9.38)$$

where n is the number of turns in the coil. The units of the mmf M clearly are amp-turns. The conjugate variable to M must have the dimensions of volts (or volts/turn or weber/s, Wb/s) and is called $\dot{\phi}$, the rate of change of **magnetic flux**, ϕ, with units Wb. This flux is proportional to the **magnetic field strength**, H, which as you would expect is proportional to M and inversely proportional to the length of the magnetic path, ℓ:

$$H = M/\ell \text{ amp-turn/meter.} \qquad\qquad (9.39)$$

It is here that the torroidal shape would be preferred, since its area would be uniform and the length of the flux path more nearly uniform, but for practical purposes the errors of the squarish geometry tend to average out and are not of major significance. The flux-per-unit-area is defined as B, or $B = \phi/A$, and for a linear core material B is proportional to H with the coefficient μ, known as the **magnetic permeability**:

$$B = \mu H; \qquad \phi = \mu A H = \frac{\mu A}{\ell} M. \tag{9.40}$$

The coefficient $\mu A/\ell$ is known as the **permeance** of the core, and its better-known reciprocal as its **reluctance**. Reluctance is to magnetic flux and magnetomotive force what resistance is to electric current and voltage, but one should not carry this analogy too far since resistance dissipates energy and reluctance stores it. The stored energy of the entire core of volume $A\ell$ is

$$\mathcal{V} = A\ell \int H \, dB = \int M \, d\phi = \int M \dot{\phi} \, dt, \tag{9.41}$$

which can be represented by a bond-graph compliance with modulus equal to the permeance:

$$\frac{e}{i} \longrightarrow G \frac{M}{\dot{\phi}} \longrightarrow C \qquad C = \mu A/\ell \tag{9.42}$$

Relevant dimensions and SI units are summarized in Table 9.2, using the format of Table 1.1 in Chapter 1 (p. 11) that separates primary and derived quantities. No new primary dimensions or units are needed; all are derived. In fact, it can be seen that the coulomb, or unit electrical charge, is not really a primary dimension after all but can be expressed in terms of newtons and seconds. The unit T stands for *tesla*.

Table 9.2 Dimensions and Units Related to Magnetic Circuits

name	symbol	dimensions	SI units
primary set of consistent dimensions and units:			
force	f	F	newton, N
length	L	L	meter, m
time	t	t	seconds, s
electric charge	q	Q	coulombs, $C \equiv \sqrt{N \cdot s}/\sqrt{4\pi \times 10^{-7}}$
derived dimensions and units:			
electric current	$i = \dot{q}$	Q/t	$C/s \equiv A$ (amperes)
electric potential	e	$F \cdot L/Q$	$N \cdot m/C \equiv V$ (volts)
energy	$\mathcal{V}$ or $\mathcal{T}$	$F \cdot L$	$N \cdot m \equiv J$ (joules)
magnetic flux	ϕ	$F \cdot L \cdot t/Q$	$N \cdot m \cdot s/C \equiv V \cdot s \equiv Wb$ (weber)
flux density	B	$F \cdot t/Q \cdot L$	$N \cdot s/C \cdot m \equiv V \cdot s/m^2 \equiv Wb/m^2 \equiv T$
magnetomotive force	M	Q/t	$C/s \equiv A$ (ampere-turns)
magnetic field strength	H	$Q/t \cdot L$	$C/s \cdot m \equiv A/m$ (ampere-turns/meter)
magnetic permeability	μ	$F \cdot t^2/Q^2$	$N \cdot s^2/C^2 \equiv T \cdot m/A \equiv H/m$ (henrys/m)
permeability, free space	μ_0	same	same; value $\equiv 4\pi \times 10^{-7}$
magnetic compliance	C	$F \cdot L \cdot t^2/Q^2$	$N \cdot m \cdot s^2/C^2 \equiv V \cdot s/A \equiv H$ (henrys)

In bond graph terms, the magnetomotive force M becomes the effort, the flux rate $\dot{\phi}$ becomes the flow, and magnetic energy becomes a compliance element relating ϕ to M. To relate this to previous representations of magnetic circuits, note that the combination $\longrightarrow G \longrightarrow C$ can be telescoped into an equivalent *inertance*, $\frac{e}{i} \longrightarrow I$, showing that the entire system acts as an inductance or ideal inductor of magnitude $I = G^2 C = n^2 \mu A/\ell$ henries, assuming linearity. The analysis produces the magnitude of the inductance in terms of its geometric and property descriptions.

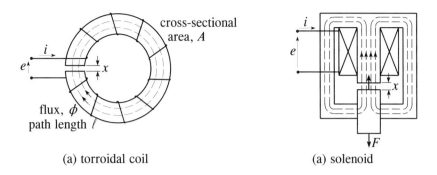

Figure 9.27: Magnetic circuits with air gaps

9.5.2 Generation of Mechanical Forces

Electromechanics requires that mechanical forces be placed on moving members. Forces in systems with fixed coils are associated with changes in reluctance due to the displacement of a member. Consider the effect of the air gap in the magnetic circuit shown on the left in Fig. 9.27. The mmf for the closed loop equals the mmf for the part with a metal core with its geometric properties A, ℓ_m and μ_m, plus the mmf for the air gap, with its properties A, gap-width x, and μ_0 (the common symbol for the permeability of free space, which air approximates). Since the magnetic flux ϕ is common to the two segments of the loop,

$$M = \left(\frac{\ell}{\mu_m A} + \frac{x}{\mu_0 A} \right) \phi. \tag{9.43}$$

The value of μ_m is vastly larger than μ_0, as discussed below, assuming the metal is ferromagnetic. Therefore, the insertion of the air gap vastly increases M and the magnetic energy for a given magnetic flux. The inverse of the magnetic compliance becomes, from equation (9.42),

$$\frac{1}{C} = \frac{\ell}{\mu_m A} + \frac{x}{\mu_0 A}. \tag{9.44}$$

A magnetically equivalent circuit is shown on the right in Fig. 9.27. The difference is that the magnetic path involves two separate rigid bodies so that the air gap can be varied mechanically. An important detail is the use of symmetry to minimize friction forces on the sliding member. Now, the energy in the compliance becomes a function of *two* displacements, the magnetic ϕ and the mechanical x, which defines a two-port compliance:

$$\frac{e = \dot{p}}{i} \dashv G \vdash \frac{M}{\dot{\phi}} \quad C \quad \frac{F}{\dot{x}} \vdash$$

The device has become a solenoid. The energy in the magnetic field is

$$\mathcal{V} = \frac{1}{2C(x)} \phi^2 = \frac{1}{2} \left(\frac{\ell}{\mu_m A} + \frac{x}{\mu_0 A} \right) \phi^2. \tag{9.45}$$

The magnitude of the force in the x-direction, in Newtons, can be computed from equation (9.29) (p. 633):

$$\boxed{F_x = \frac{\partial \mathcal{V}}{\partial x} = \frac{\phi^2}{2\mu_0 A}.} \tag{9.46}$$

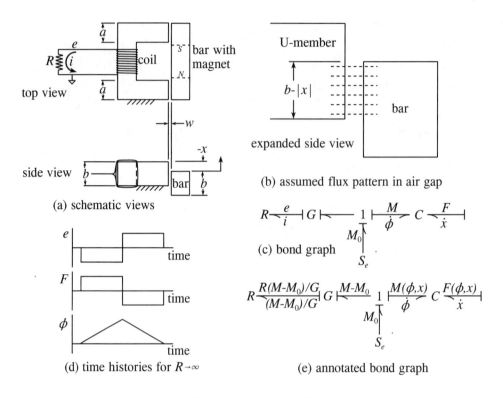

Figure 9.28: Magnetic pick-off transducer

This is the attractive force between parallel surfaces with a normal magnetic field, which exists in many practical devices. The relation $F_x = \partial V / \partial x$ also applies in more complex situations where x is not in a direction normal to the solid surfaces.

A solenoid was viewed as an inertance dependent on a displacement in Example 9.12 (p. 638), but no means was given to evaluate that inertance. The present model above also can be converted to an inertance, using the standard equivalence $I = G^2C$, which for the present geometry gives $I = n^2/(\ell/\mu_m A + x/\mu_0 A)$. The results agree. This particular inertance or compliance does not however apply to the geometry pictured in Example 9.12, which does not involve an air gap between parallel surfaces with normal flux lines. The treatment of magnetic energy as a position-dependent compliance readily can be adapted to this and many other geometries, fortunately.

9.5.3   First-Order Treatment of Permanent Magnets: A Case Study

Accurate representation of permanent magnets is perhaps the most complex issue in electro-mechanics. Nevertheless, a permanent magnet can be approximated to a first order simply as a constant mmf, or effort source, which can be inserted into a magnetic curcuit through use of a 1-junction. This procedure is illustrated in the following case study.

The magnetic pick-off transducer shown in Fig. 9.28 comprises a U-shaped magnetic path of a magnetically permeable material containing a coil with n turns and an electrical resistor, and a bar containing a permanent magnet that passes by at a constant distance w and velocity $\dot{x}$ (constrained and driven mechanically by means not pictured). When the bar passes by, magnetic flux surges around the loop, generating a pulse in the voltage e. The objective is to model the system in order to find both e and the component of the magnetic force acting on the bar in the x-direction. Only a first-order analysis is expected.

The first step is to relate the flux ϕ in the magnetic loop to the position x and to the net mmf produced by the permanent magnet and modified by the coil. For the first-order analysis, the drop in mmf across the two air gaps will be assumed to be so large that other mmf drops through the vastly more permeable ferromagnetic material is negligible by comparison. Further, the flux will be approximated as being perpendicular to the parallel facing surfaces of the two members, as pictured in part (b) of the figure. (This neglects what is known as magnetic fringing, as discussed later in this section.) The gradient of the mmf drop across each gap is $dM/ds = M/2w$, where M is the total mmf drop of the circuit produced by the permanent magnet and modified by the permanent magnet, s is the position along the path, and w is the gap width. The factor 2 results from the presence of two air gaps in series around the loop. The area through which the flux passes can be seen to be $A = a(b-|x|)$ when $|x| < b$, and zero when $|x| > b$. The flux therefore is approximated as

$$\phi = \mu \frac{dM}{ds} A \simeq \frac{1}{2w}\mu_0 M a(b - |x|), \qquad |x| < b. \tag{9.47}$$

The magnetic energy in the field, which is limited to the air gaps in this approximation, becomes

$$\mathcal{V} = \left(\int_0^\phi F \, d\phi \right)_{x=const.} = \frac{2w}{a\mu_0(b - |x|)} \frac{\phi^2}{2}. \tag{9.48}$$

The bond graph in part (c) of the figure represents this energy with the two-port compliance element. In standard fashion for such compliances, the mmf M can be teased back by computing $M = \partial \mathcal{V}/\partial \phi$, reproducing an application of equation (9.43). More importantly, the mechanical force acting on the bar in the x-direction can be found as

$$F = \frac{\partial \mathcal{V}}{\partial x} = \frac{w\,\text{sign}(x)}{a\mu_0(b - |x|)^2}\,\phi^2. \tag{9.49}$$

The total mmf across the air gaps equals the mmf of the permanent magnet, M_0, minus the mmf that drives the electrical coil. The entire system, then, can be modeled as shown in part (c) of the figure, which represents the permanent magnet by an effort source, the summation of the mmf's by a 1-junction, the coil by a gyrator with $G = n$, and the electrical resistor by a one-port resistance.

If the system is used as signal transducer, the resistance R would be large. In the limit $R \to \infty$, $i \to 0$ and $M \to M_0$. The flux ϕ is then given by equation (9.47) with $M = M_0$, and the voltage and mechanical force become

$$e = G\dot{\phi} = -\frac{n}{2w}\mu_0 a M_0(b - |x|)\,\text{sign}(x)\dot{x}, \tag{9.50}$$

$$F = \frac{a\mu_0 M_0^2}{4w}\,\text{sign}(x). \tag{9.51}$$

Plots of these are given in part (d) of Fig. 9.28 for the case of the velocity $\dot{x}$ being constant.

When $R \neq \infty$, the annotated bond graph of part (e) of the figure leaves equation (9.49) for the force and the relation $e = n\dot{\phi}$ unchanged but gives a differential equation for ϕ:

$$\frac{d\phi}{dt} = \frac{R(M_0 - M)}{G^2} = \frac{R}{n^2}\left(M_0 - \frac{2w}{\mu_0 a(b - |x|)}\phi \right). \tag{9.52}$$

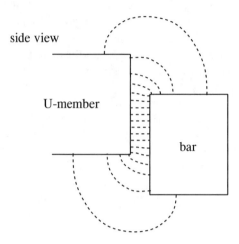

Figure 9.29: Flux fringing for the magnetic pick-off transducer

9.5.4   Flux Fringing and Leakage

Secondary magnetic effects, which can be very important, include flux fringing, flux leakage, eddy-current lossess, magnetic saturation and magnetic hysteresis. The first two are addressed in this sub-section. The topic of flux leakage includes treatment of multiple magnetic loops, which also may occur for reasons other than leakage.

Magnetic flux distributes itself in space so as to minimize energy for a given overall mmf, often presenting a complex vector field. The case study example above illustrates this. The flux that would actually occur in and around the air gaps is shown approximately in Fig. 9.29 (the actual distribution would be three-dimensional). The deviation from the straight, parallel field lines assumed in the first-order analysis is fairly significant in this case, since the width of the gap is not very small compared to the other dimensions. It is greatest near the edges of the field, leading to the name "fringing." The effective cross-sectional area for the flux is increased. This increase may not be as large as you would imagine from the drawing, but it does reduce the magnetic-mechanical coupling somewhat, and smoothes the changes in variables particularly where $x \simeq b$. The discussion of vector fields in Section 12.1 defines fields like this, and can help you approximate them and their energy through hand sketching, which can be quite adequate for many purposes of design and analysis. Fancy finite-element and other computing packages are available for accurate analysis. The details are beyond the scope of the present discussion.

A magnetic path can bifurcate into two or more paths. The mmf at the junction is common to both paths, while the flux divides, the effort and flow variables bonded to a 0-junction describe. The construction of an electrical transformer is shown in part (a) of Fig. 9.30, and the equivalent magnetic circuit with a path representing flux leakage is shown in part (b). An equivalent circuit diagram and its bond graph is repeated from Fig. 9.13 (p. 617) in part (c) of the figure. The nominal transformer modulus is T. The series inductance I_1' is caused by the flux leakage, and produces a voltage drop for large currents or frequencies. The shunt inductance I_2 prevents operation at low frequencies. The magnetic circuit for the transformer is given in part (d) of the figure, including an equivalent simplification resulting from defining the mmf on the lower 0-junction as zero. Note that the permeance C refers to one-half of the entire loop, and C_m is the relatively small permeance of the leakage path.

A design process could proceed by choosing acceptable values of these parameters, de-

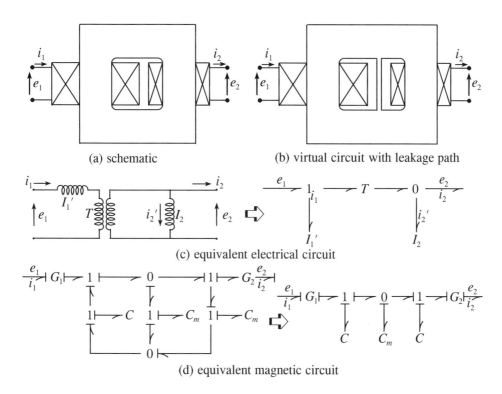

(a) schematic   (b) virtual circuit with leakage path

(c) equivalent electrical circuit

(d) equivalent magnetic circuit

Figure 9.30: Electric Transformer with Coupled Magnetic Flux Loops

termining the equivalent values of C, C_m, $n_1 = G_1$ and $n_2 = G_2$, identifying the magnetic properties including the permeability of the core material and the maximum allowable flux density due to saturation, and finally translating this information into the geometry of the core. The resistance of the coil and the space that it would occupy also can be critical, but this relatively straightforward issue is assumed to be handled separately.

The parameters in the equivalent electrical circuit can be related to the parameters in the magnetic circuit many ways. One strategy evaluates and compares the two transmission matrices. With the current i_1 defined as positive inward and the current i_2 as positive outward, the transmission matix of the equivalent electrical circuit, defined in the conventional way so that $[e_1\ i_1]^T = \mathbf{T}[e_2\ i_2]^T$, is

$$\mathbf{T} = \begin{bmatrix} 1 & I_1's \\ 0 & 1 \end{bmatrix} \begin{bmatrix} T & 0 \\ 0 & 1/T \end{bmatrix} \begin{bmatrix} 1 & 0 \\ 1/I_2s & 1 \end{bmatrix} = \begin{bmatrix} T + I_1'/TI_2 & I_1's/T \\ 1/TI_2s & 1/T \end{bmatrix}. \tag{9.53}$$

The corresponding transmission matrix for the magnetic circuit is

$$\mathbf{T} = \begin{bmatrix} 0 & G_1 \\ 1/G_1 & 0 \end{bmatrix} \begin{bmatrix} 1 & 1/Cs \\ 0 & 1 \end{bmatrix} \begin{bmatrix} 1 & 0 \\ C_ms & 1 \end{bmatrix} \begin{bmatrix} 1 & 1/Cs \\ 0 & 1 \end{bmatrix} \begin{bmatrix} 0 & G_2 \\ 1/G_2 & 0 \end{bmatrix}$$
$$= \begin{bmatrix} (G_1/G_2)(1 + C_m/C) & G_1G_2C_ms \\ (2/G_1G_2Cs)(1 + C_m/2C) & (G_2/G_1)(1 + C_m/C) \end{bmatrix}. \tag{9.54}$$

Equating the respective elements of these two matrices,

$$T = \frac{G_1/G_2}{1 + C_m/C}; \qquad I_2 = \frac{G_2^2(C + C_m)}{2(1 + C_m/C)} \qquad I_1' = \frac{G_1^2 C_m}{1 + C_m/C}. \tag{9.55}$$

EXAMPLE 9.14

Electrical transformers tend to weigh more than desired. They also can distort signals, and produce an undesirable power factor (lag in the input current relative to the voltage). To investigate the critical design factors, use the model above to find the weight of the transformer in terms of the permeability (μ), maximum flux density allowed due to saturation (B_m) and mass density (ρg) of the core material, the design frequency of operation (ω), the power output assuming a resistive load ($\mathcal{P}$) and the ratio of the magnitude of the shunt current allowed through the inductance I_2 compared to the current to the load (r). Also find the effect on the power factor, in both cases neglecting the small effect of flux leakage. Finally, determine the distortion in the ratio of the voltages (e_2/e_1), not neglecting the flux leakage. Draw a few conclusions regarding design.

Solution: Let voltages e and currents i refer to their amplitudes, so that the output power is $\mathcal{P} = e_2 i_2 / 2$. When the current peaks, the flux density should peak at its maximum allowed value, or $B_m = \mu M_{max} = \mu n_2 i_2 / \ell$, where ℓ is one-half the complete path length for the loop. The magntude of the maximum shunt current is $i_{sh} = e_2 / I_2 \omega$, and $I_2 \simeq n_2^2 \mu A / 2\ell$ (neglecting the leakage ratio C_m / C). Rearranging the first equation gives

$$n_2 = \frac{B_m \ell}{\mu i_2},$$

and combining the second and third equations gives

$$n_2^2 = \frac{2 e_2 \ell}{\mu A \omega i_{sh}}.$$

Multiplying these two results, and setting $i_{sh} = r I_2$,

$$n_2^3 = \frac{\ell^2}{A} \frac{2 B_m e_2}{\mu^2 r i_2^2 \omega}$$

It is a good idea to check that this result is unitless, using Table 9.2 (p. 652).

The length becomes $\ell = n_2 \mu i_2 / B_m$, and the weight is

$$W = 2\ell A \rho g = 2 \frac{A}{\ell^2} \ell^3 \rho g = \frac{n_2^3 \mu i_2}{Bm}^3 = \frac{4 e_2 i_2 \mu \rho g}{r \omega B_m^2} = \frac{8 \mathcal{P} \mu \rho g}{r \omega B_m^2}$$

This reveals that allowing the ratio r to become large may be necessary to achieve a reasonable weight, particularly if the design frequency is low. Also, a large maximum flux density is desirable, but a large permeativity is not. Small permeativity requires a large number of coils, however, increasing the wire length and decreasing wire diameter to fit the space available, which could give unacceptable electrical resistance. The net result is a compromise.

If the load resistance is defined as $e_2 / i_2 = R$, the transmission matrix gives

$$e_1 = \left[\left(\frac{T + I_1'}{T I_2} \right) R + \frac{I_1' s}{T} \right] i_2$$

$$i_1 = \left[\frac{R}{T I_2 s} + \frac{1}{T} \right] i_2.$$

Neglecting the leakage-produced I_1', these can be combined to give the input admittance

$$\frac{i_1}{e_1} = \frac{1}{RT^2}\left(1 + \frac{R}{I_2 j\omega}\right) = \frac{1}{RT^2}\left(1 + \frac{r}{j}\right).$$

The right-most result, which follows from substituting some of the above equations, shows that at the design frequency the phase advance is always the same. It seems unlikely that a value of r greater than 1 would be acceptable. The tranmission matrix equations also can be combined, without neglecting the flux leakage terms, to give

$$\frac{e_2}{e_1} = \frac{1/T}{1 + (I_1'/T^2 I_2)(1 + I_2 s/R)} \simeq \frac{1/T}{1 + 2(C_m/C)(1 + C_m/C)(1 + j/r)},$$

where again some algebra has been omitted. With no flux leakage, therefore, the ratio of voltages is ideal. Otherwise, the resulting phase shift at the design frequency actually suffers from a *small* value of r, suggesting that its value also should be a compromise. Much depends on the application, the requirements for a power transformer differing from those of a signal transformer.

9.5.5 Eddy-Current Losses

Eddy-current losses and **hysteresis losses** together produce the **core losses** in a magnetic circuit. Eddy currents are induced in closed paths in a conducting material subject to a changing magnetic field. They dissipate energy as heat. The corresponding mmf drop per unit volume is proportional to the rate of change of the local flux density, dB/dt, and the power dissipated is proportional to its square. Since the flow in a bond graph is the total flux rate $\dot\phi = A$, the dissipation can be modeled by placing a 1-junction and a resistance in the path of the flux, or using 1-junctions that already are employed to afix magnetic permeance, as shown in part (a) of Fig. 9.31 for a segment of length ℓ and area A.

A locus of the flux (a displacement) versus the mmf (an effort) is plotted in part (b) of the figure for sinusoidal behavior. At virtually zero frequency the locus follows the straight line, which represents the permeance (compliance). The resistance is playing no role compared to the compliance. Note that the usual convention in this book of plotting effort on the ordinate and displacement on the abcissas is reversed, in deference to the opposite convention in electromagnetics. At a significant frequency the locus becomes an ellipse, which becomes fatter as the frequency is increased. The power in is $M\,d\phi/dt$, and the net energy is its time integral, $\int M\,d\phi$. The energy lost in one cycle, therefore, equals the area inscribed by the ellipse.

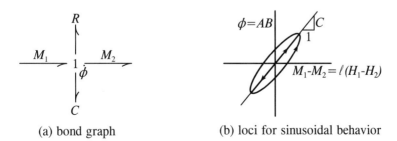

(a) bond graph (b) loci for sinusoidal behavior

Figure 9.31: Model and behavior of a linear magnetic core with eddy-current losses

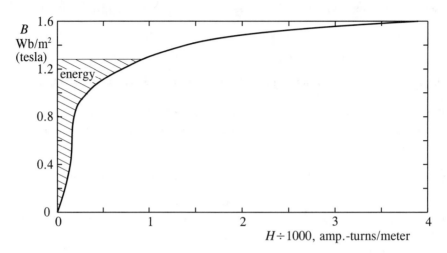

Figure 9.32: Normal magnetization curve for a typical core laminate

Eddy current losses normally are sharply reduced by laminating conductive cores into thin sheets aligned in the direction of the magnetic flux. The sheets are coated with insulating layers. The effective area for the magnetic flux is reduced by what is called a **stacking factor**. Bundles of insulated wires and sometimes even powdered cores are used for protection at very high frequencies.

9.5.6   Saturation and Hysteresis

The energy that can be stored in a magnetic field of a ferromagnetic material is effectively limited by saturation. A typical $B - H$ characteristic of a magnetically "soft" material used in laminated cores is shown in Fig. 9.32. Recall that H is the effort, and B is the displacement, so that energy is $\mathcal{V} = \int H\,dB$, indicated by the cross-sectioned area in the plot. The behavior assuming uniform flux density can be represented in a bond graph by a nonlinear compliance (permeance) instead of the constant compliance with slope equal to a constant permeability, known as the normal or initial magnetization characteristic.

EXAMPLE 9.15

Determine the mmf necessary for the electrical relay opposite to produce a force of 80 newtons on the armature. The material is characterized in Fig. 9.32, a stacking factor of 0.9 can be assumed, and fringing increases the effective area in the air gaps by 8%.

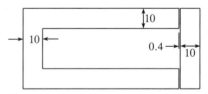

dimensions in millimeters; thickness also 10mm

Solution: The force in each air gap is 40 newtons, and the effective area of each gap is $0.01 \times 0.01 \times 1.08 = 1.08 \times 10^{-4}$ m^2. The flux needed to produce this force is, from equation (9.45), $\phi = \sqrt{2\mu_0 AF} = \sqrt{2 \times 4\pi \times 10^{-7} \times 1.08 \times 10^{-4} \times 40} = 1.042 \times 10{-4}$ Wb. The effective cross-sectional area in the metal is $0.9 \times 0.01 \times 0.01 = 0.9 \times 10^{-4}$ m^{-4}, so the flux density there is $1.042 \times 10{-4}/0.9 \times 10{-4} = 1.16 Wb/m^2$. The mmf across each air gap is $1.16 \times 0.4 \times 10^{-3}/4\pi \times 10^{-7} = 369$ ampere-turns. The plot gives approximately 550 ampere-turns per meter in the metal, and the path length is 0.18 m, so the its mmf is $550 \times 0.18 = 99$ ampere-turns. The mmf for the entire loop becomes $2 \times 369 + 99 = 837$ ampere-turns.

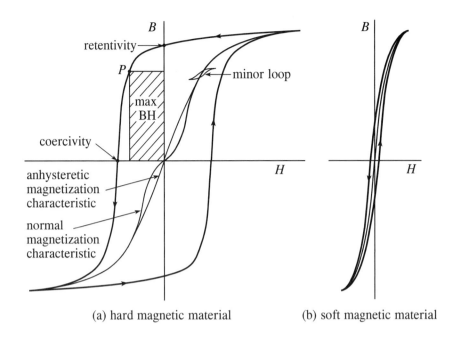

(a) hard magnetic material   (b) soft magnetic material

Figure 9.33: Hysteresis loops for hard and soft ferromagnetic materials

The normal magnetization curve applies only to DC conditions. A more realistic curve for most purposes, called the **anhysteretic characteristic**, results when an AC dither is superimposed on the DC excitation. If the mmf of a magnetic core is cycled *slowly*, the actual flux will respond with a loop, as represented for hard and soft ferromagnetic materials in Fig. 9.33. The anhysteretic charcteristic applies to both hard and soft materials, although for clarity it is shown only in the former. The cusps of hysteretic locus usually are assumed to lie on the normal magnetization curve, though in fact a small deviation therefrom must occur. The energy dissipated each cycle equals the inscribed area of the loop. For a given material, this loss often is approximated empirically as being proportional to a power of the maximum value of B in the loop, assuming the mean value is zero. The power, called the **Steinmetz exponent**, can vary from 1.5 to over 2.0 and often is taken as 1.6. If a cycle of small amplitude has a non-zero mean value, a minor hystersesis loop results that, as shown in the figure, has a slope known as the **incremental permeability** that is significantly less than the slope of the normalized magnetization curve. Clearly, this is tough to model. If a cycle is carried out at a high enough frequency to excite significant eddy-current losses, the loci balloon-out sideways; the increased area of the loop represents the increased dissipation, but that has been modeled above simply.

Hard magnetic materials are not normally cycled at large amplitude but rather are used as permanent magnets. The **retentivity** and **coercivity** are defined as the value of B that remains when H is removed, and the value of negative H needed to reduce B to zero. Permanent magnets retain their oomph best for long storage periods if H is reduced to zero, usually through use of a "keeper" bar. Soft magnetic materials have much smaller values of retentivity and coercivity.

The point P is located where the negative product BH is maximized. This point has special significance, since if the permanent magnet produces a flux in an air gap, with no other mmf drops, the associated mechanical force is maximized when the magnet operates at this point. This is because the force equals $(\mu_0/g)|M|\phi$, where g is length of the gap, and the $|M|$ and ϕ of the gap equal the $|M|$ and ϕ of the magnet. Operation at this point may be achieved by proper selection of the ratio of the area of the gap to its length in view of the values of B and H in the magnet and the ratio of the area of the magnet to its length.

Displacement-effort hysteretic behavior has been represented in bond graphs by the RC element.[5] The symbol for this new element recognizes the simultaneous storage and dissipation of energy. For the magnetic domain, the combination of hysteresis and eddy-current behavior is represented by the causal bond graph

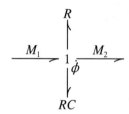

where R represents the eddy-current dissipation as described above. The causal strokes treat the RC element with the same integral causality applied to C elements; it degenerates to the C element when a function $M = M(\phi)$ is defined. Hysteresis, however, means that the slope $dM/d\phi$ satisfies a more general function of M, ϕ and the sign of $\dot{\phi}$, which implies a kind of history dependence. Other hysteretic pheneomena includes elastic-plastic stress-strain.

Bond graph approaches to magnetic hysteresis have been given by Karnopp[6] and Masson et al.[7] The author prefers to adapt the basic model of Jiles and Atherton[8], because it appears to be related better to the underlying physics and is more widely accepted. As interpreted and modified by the author it is also relatively simple. The following discussion is intended to give a practical procedure and coding, without the physical justification. It assumes that you have at least one experimental hysteresis plot as a basis for parameterization.

The anhysteretic magnetization characteristic is represented by the Langevin function:

$$M_{an} = M_s \mathcal{L}(x); \quad x \equiv (H + \alpha M_g)/a; \quad \mathcal{L}(x) \equiv \coth(x) - 1/x. \tag{9.56}$$

M_{an} equals M_g, the **magnetization** of the material, in the absence of hysteresis. (The subscript g has been added by the author to avoid confusion with other uses of the symbol M, particularly the magnetomotive force.) M_s, a and α are constant material parameters. The first derivative of the Langevin equation, which will be needed, is

$$\mathcal{L}^1(x) = \frac{-1}{\sinh^2 x} + \frac{1}{x^2}; \quad \mathcal{L}^1(0) = \frac{1}{3}, \tag{9.57}$$

The relations between the magnetic field, B, the **effective field**, B_e, the mmf H and the magnetization M_g are

$$B = \mu_0(H + M_g); \quad B_e = \mu_0(H + \alpha M_g) = \mu_0 a x. \tag{9.58}$$

[5]See for example D. Gaude, H. Morel, B. Allard, H. Yahoui and J.-P. Masson, "Bond Graph Model of the Induction Motor", 1999 International Conference on Bond Graph Moleling and Simulation, *SCS Simulation Series* v 31 n 1, p./17-322.

[6]D. Karnopp, "Computer Models of Hysteresis in Mechanical and Magnetic Components," *J. Franklin Institute*, v 316 n 5, 1983, pp. 405-415.

[7]J.P. Masson, H Fraisse, H. Morel and B. Allard, "Contribution to the Modeling of Electrical Engineering Systems with Bond Graph Representation," 1997 International Conference on Bond Graph Modeling and Simulation, *Proc. 1997 Western Multiconference*, SCS, pp. 331-336.

[8]D.C. Jiles and D.L. Atherton, "Theory of Ferromagnetic Hysteresis," *J. Applied Physics*, v 55 n 6, 1984, pp. 2115-2120.

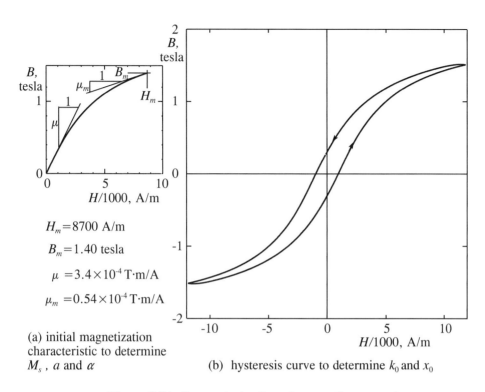

H_m=8700 A/m

B_m=1.40 tesla

μ =3.4×10⁻⁴ T·m/A

μ_m =0.54×10⁻⁴ T·m/A

(a) initial magnetization
characteristic to determine
M_s , a and α

(b) hysteresis curve to determine k_0 and x_0

Figure 9.34: Parameterization of magnetic properties

The values of M_s, a and α can be deduced from experimental data in the form of the anhysteretic magnetization characteristic. See part (a) of Fig. 9.34. The data used are the values of the maximum mmf for the given characteristic, H_m, the associated magnetic flux, B_m, the permeability at the origin (slope of the curve), $\mu \equiv (dB/dH)_{B=0}$, and the incremental permeability at the maximum condition, $\mu_m \equiv (dB/dH)_{B=B_m}$.

The permeability at the origin is related to the derivative $\mathcal{L}^1(0)$ as given in equation (9.56). After a little algebra, the result can be written as

$$\frac{3a}{M_s} = \frac{1}{1 - \mu_0/\mu} - (1 - \alpha). \tag{9.59}$$

For the state at (B_m, H_m),

$$\left(\frac{B_m}{\mu_0} - H_m\right)\frac{1}{M_s} = \mathcal{L}(x_m); \qquad x_m = \frac{H_m(1 - \alpha) + \alpha B_m/\mu_0}{a}. \tag{9.60}$$

$$\frac{a}{M_s\mathcal{L}^1(x_m)} = \frac{1}{(\mu_m/\mu_0) - 1} + \alpha. \tag{9.61}$$

To aid in solution, equations (9.59) and (9.60) can be combined to eliminate the ratio a/M_s, and equations (9.59 and 9.61) also can be combined to eliminate the same ratio, giving the respective results

$$\alpha = \frac{1}{x_m - 3\mathcal{L}(x_m)}\left[\frac{3\mathcal{L}(x_m)}{B_m/\mu_0 H_m - 1} - \frac{x_m}{\mu/\mu_0 - 1}\right], \tag{9.62a}$$

$$\alpha = \frac{1}{1 - 3\mathcal{L}^1(x_m)}\left[\frac{3\mathcal{L}^1(x_m)}{\mu_m/\mu_0 - 1} - \frac{1}{\mu/\mu_0 - 1}\right]. \tag{9.62b}$$

These equations can be solved iteratively by selecting various values of x_m until the two resulting values of α agree. Convergence is rapid. Once α and x_m are known, equations (9.60) and (9.59) give M_s and a.

9.5.7   Simulation with Hysteresis*

Simulation with hysteresis loops can be based on Jiles' and Atherton's differential equation, the third and presumed final version of which is given by Jiles[9] as

$$\frac{dM_g}{dH} = \frac{M_{an} - M_g}{\delta k M_s - [\alpha/(1-c)](M_{an} - M_g)} + c \frac{dM_{an}}{dH}, \tag{9.63}$$

in which k redefined to make it dimensionless. This can be rearranged, using the relations above between B, B_e, H, M_g, M_s, x, $h \equiv H/a$ and $b \equiv B/a\mu_0$ to give

$$\frac{dx}{dt} = \frac{1 - c + cQ/\delta}{c_0(1 - c + Q/\delta)} \frac{dh}{dt}; \quad c_0 \equiv 1 - \frac{\alpha c M_s \mathcal{L}^1(x)}{a}; \quad Q \equiv \frac{\alpha}{k}\left[\frac{a(x-h)}{\alpha M_s} - \mathcal{L}(x)\right]. \tag{9.64}$$

Hysteresis results from setting $\delta = 1$ for an increasing field and $\delta = -1$ for a decreasing field.

In the first paper, cited in footnote 8, the term with coefficient c is missing. In its place, k usually decreases with increasing $|x|$, although the authors give no formula for this. In this book the first model is used in the various simulations, so $c = 0$ is substituted into equations (9.63), (9.64) and (9.67). The following variation in k is employed:

$$k = k_0/(1 + |x/x_0|), \tag{9.65}$$

When $c = 0$, the width of the hysteresis loop at $x = 0$ is proportional to, and determines, k_0. The values of x_0 and c affect the shapes of the cusps, and increasing c narrows the hystersis loop. Whether the first or the third version produces better matching to observed hysteresis loops in general is unclear; the advantage of the third may be largely theoretical.

EXAMPLE 9.16

Determine the parameters M_s, a, α, k_0 and x_0 from the anhysteretic magnetization characteristic plotted in part (a) of Fig. 9.34. (This data corresponds approximately to 1% manganese steel.)

Solution: From the plot in part (a), approximate values of the slopes are $\mu = 3.4 \times 10^{-4}$ T·m/A and $\mu_m = 0.54 \times 10^{-4}$ T·m/A, $H_m \simeq 8700$ A/m and $B_m \simeq 1.40$ T (tesla). With these values, equations (9.62a) and (9.62b) give, respectively,

$$\alpha = \frac{0.0236\mathcal{L}(x_m) - 0.00371 x_m}{x_m - 3\mathcal{L}(x_m)},$$

$$\alpha = \frac{0.0715\mathcal{L}^1(x_m) - 0.00371}{1 - 3\mathcal{L}^1(x_m)}.$$

After four or five iterations with these equations set equal, the values $x_m = 3.15$, $\alpha = 0.0041$, $M_s = 1.61 \times 10^6$ A/m, and $a = 4200$ A/m result. (The individual numbers are very sensitive to the measured values, but the net effect is not.)

Simulation results for $k_0 = 6 \times 10^{-4}$ and $x_0 = 10$ are given in part (b) of Fig. 9.34. This compares well with known data; other values were tried first and then adjusted. The slender hysteresis loop is affected little by x_0. The code, listed below, can be downloaded from the internet; see page 1018.

---

[9] Jiles, *Magnetism and Magnetic Materials,* Chapman and Hall, 1991. Another version is given in Jiles, D.C, and Atherton, D.L., "Theory of Magnetic Hysteresis," *J. Mag. & Mag. Mat.,* v. 61, pp 48 -60 (1986).

Simulation is especially simple when $H(t)$ is known. The following MATLAB® coding produces two loops:

```
% script file 'hysteresism.m' to run function file 'hysteresis.m'
clear
global c1 c2 x0 h0
alpha=.0041; a=4200; mu0=4*pi*1e-7; Ms=1.61e6; k0=6e-4;
c1=alpha/k0; c2=a/alpha/Ms; b=a*mu0/alpha; x0=10; xm=4;
X=xm; Langevin; h0=xm-L/c2;
OPTIONS=ODESET('RelTol',1e-9);
[t,x]=ode45('hysteresis',[-pi/2 11*pi/2],-xm,OPTIONS);
for i=1:size(t); h(i)=h0*sin(t(i)); end
H=a*h; B=b*(x-(1-alpha)*h');
plot(H,B)

function f=hysteresis(t,x) % function file called above
global c1 c2 x0 h0
h=h0*sin(t); dhdt=h0*cos(t);
X=x;
Langevin % call script file below
Q=c1*(1+abs(x/x0))*(c2*(x-h)-L);
f1=dhdt/(1+Q);
if dhdt>0; f=f1; else; f=dhdt/(1-Q); end

% script file 'Langevin' which has multiple uses:
if X==0; L=0; L1=1/3; % L1 is the derivative dL/dx
else; L=cosh(X)/sinh(X)-1/X; L1=-1/(sinh(X))^2+1/X/X; end
```

The amplitude of $h(t)$ chosen in the coding above corresponds to the initial condition on the magnetization curve for which $x = -4$. Actually, the precise value of the initial condition is not critical, since the locus rapidly approaches the limit-cycle hysteresis path; the initial portion that is not on this path is removed from the plot, but if included would be barely visible in its lower left-hand corner.

For general simulation, one wishes to use integral causality for the RC element. Thus, $\dot{\phi}$ or $\dot{B}$ is the causal input, and the mmf $M = H\ell$ or H is the causal output. Using the dimensionless variables x, b and h, this means $\dot{b}$ is the causal input and h is the causal output. The relations of equation (9.58) require that

$$x = (1 - \alpha)h + \alpha b, \tag{9.66}$$

which when substituted into equation (9.64) give

$$\dot{h} = (1/r)\dot{b}; \qquad r = 1 - \frac{1}{\alpha} + \frac{1}{\alpha c_0}\left[1 - \frac{1}{\delta/Q + 1/(1-c)}\right]. \tag{9.67}$$

Since an integration scheme such as a MATLAB ode routine is to be used, $\dot{h}$ is then integrated to give the desired h.

Sounds simple, but there are two catches that are far from obvious. The foremost difficulty is that the fundamental equation (9.63) does not apply everywhere, although it applies almost everywhere. (In fact, it applies everywhere in the case study above in which $h(t)$ and $\dot{h}(t)$ are specified as sine waves.) It cannot produce a remotely reasonable solution for those usually short segments of a simulation for which $\dot{b}$ and Q have the same sign.

This circumstance happens quite often right after a reversal of the field, and a satisfactory solution, found by the author, fortunately is simple: r is multiplied by $-\frac{1}{2}$ during these segments.[10] Further discussion is given below.

The second potential catch is that a simulation can get hung up if r becomes zero or very close to it. The simulation can be attempting to evaluate $0/0$ or something close to it. The fix is simply to put a floor under how small the magnitude of r can be permitted to get. Done with a little care, the resulting error is trivial.

As a simple case study, consider the coil with a resistance R excited by a known time-dependent voltage $e_0(t)$ and wrapped around a core of the same material assumed above, as shown along with a causal bond graph in part (a) of Fig. 9.35. The part of the system external to the RC element gives

$$\dot{B} = \frac{1}{GA}e_0 - \frac{R\ell}{G^2A}H. \tag{9.68}$$

When nondimensionalized and combined with equation (9.67) for the RC element, the following differential equation is produced:

$$\frac{dh}{dt} = \frac{R\ell}{G^2A\mu_0}\frac{1}{r}\left[\frac{G}{Ra\ell}e_0(t) - h\right]. \tag{9.69}$$

This is soluble if the sign of r is reversed at the intervals indicated above. The factor of $\frac{1}{2}$ is chosen to make the small segments involved better resemble the data; this part of the model might be improved. It is helpful to add b as a second state variable (with the differential equation $\dot{b} = r\dot{h}$), if only to expedite the plotting of b as well as h.

MATLAB coding is given below for an example in which $e_0(t) = 64\sin(1000\pi t)$ volts, $t < .001$ seconds, and $e_0 = 0$ subsequent to this half-cycle, as plotted in part (c) of the figure. The material is the same manganese steel considered above; other parameters can be read from the top part of the program. Plots of the results are given in parts (b) and (c) of Fig. 9.35. Note that in this case, unlike the earlier example, both b and h (and δ) reverse simultaneously. The reversal of b is dictated by the part of the sytem external to RC element, and must happen before the state reaches the anhysteretic magnetization characteristic, which has the locus Q=0. Before this, $Q < 0$, and afterward, $Q > 0$. This requires r to drop abruptly in magnitude at the moment of reversal, but remain positive. Its value must continue to decline until the state Q=0 is reached, at which moment $r = 0$ and the slope db/dh is unity (seen as virtually horizontal in the b, h plot with its greatly different scales for the two axes). Any other scenario is incompatible with the governing differential equation before the reversal and after the initial magnetization curve is broached, and which joins these two regions with a reasonable segment. This segment itself, however, is incompatible with the differential equation, which would have r become negative. It agrees very nicely, however, with the equations if the sign of r produced by the equation is reversed, and divided by some coefficient close to 2 (any number between 1 and about 4 produces plausible results).

---

[10] Jiles and Atherton (footnote 8) emphasize approximate analytical solutions for equation (9.63) (with c=0) for the two values of δ. They then stitch these solutions together, assuming that the cusps lie on the anhysteretic magnetization curve, to produce "major" limit-cycle hysteresis loops. This appears to the present author to violate the differential equation in the immediate vicinity of the cusps, and to produce hysteresis loops for given values of k_0 that are too narrow. To produce "minor" hysteresis loops they acheive closure only by tweaking some coefficients. This tweaking violates the differential equation *everywhere*, if only by a little bit. More important, the result cannot be interpreted for dynamic simulation without inferring knowledge of the future. The difference between "major" and "minor" hysteresis loops is arbitrary; general loci are not necessarily even "loops." The present author prefers to retain the differential equation without modification where it makes physical sense, and to modify it in the small but critical domains in which it demonstrably fails altogether. The new model requires no foreknowledge and is vastly simpler.

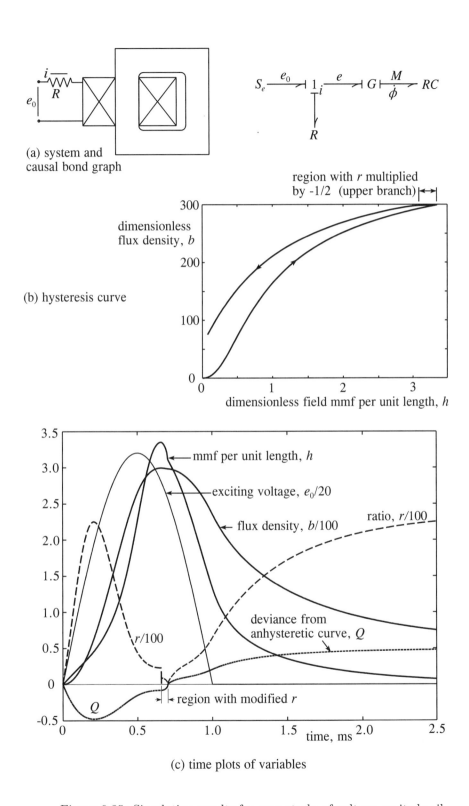

(a) system and causal bond graph

(b) hysteresis curve

(c) time plots of variables

Figure 9.35: Simulation results for case study of voltage-excited coil

In the earlier case study in which H was forced to vary sinusoidally in time, b does not reverse sign when h does but it continues to rise and reaches the state for which $Q = 0$ virtually at its peak value (actually slightly later). This gives the cusp a different shape (not represented by the approximate solutions of Jiles and Atherton). Between the two reversals, then, $r < 0$, and the differential equation applies without change. Considering both case studies, therefore, the sign of r should be changed only when $\dot{b}$ and Q have the same sign, or when their product is positive. This explains the reasoning behind the proposed application of the theory.

In the case study with the voltage input, the test for the reversal is delayed a few microseconds in order not to be triggered by the arbitrarily assumed initial conditions. The floor in r is invoked for only 0.15 μs at $t = 0.658$ ms and for 1.2 μs at $t = 0.701$ ms. The segment for which r is multiplied by $-\frac{1}{2}$ is between these two times. The code follows, except for the subprogram "Langevin" which is listed above. The code can be downloaded from the Internet (see page 1018).

```
% program coilm.m for simulation of the response of a simple coil with
% manganese iron core to an applied voltage
global c1 c2 c3 c4 x0 alpha delta
R=4; ell=.1; G=100; A=.0001; alpha=.0041; mu0=4*pi*1e-7; a=4200;
Ms=1.61e6; ell=.1; x0=10; k0=6e-4;
c1=alpha/k0; c2=a/alpha/Ms; c3=G/R/a/ell; c4=R*ell/G/G/A/mu0; delta=1;
OPTIONS=ODESET('Reltol',2.5e-14,'InitialStep',5e-6);
[t,x]=ode45('coil',[0 .003],[0 0],OPTIONS);
plot(a*x(:,1),mu0*a*x(:,2)) % gives B,H plot. Could request time plots.
ylabel('B, tesla'); xlabel('H, A/m');


%function f=coil(t,x) % function file for master program above
global c1 c2 c3 c4 x0 alpha delta
% x(1) is h; x(2) is b;
if t<.001; e0=64*sin(1000*pi*t); else; e0=0; end
f(2)=c4*(c3*e0-x(1)); % from equations (9.67) and (9.69)
X=(1-alpha)*x(1)+alpha*x(2); % equation (9.66); X is the x in the text
Langevin % short subprogram given above to compute L
Q=c1*(1+abs(X/x0))*(c2*(X-x(1))-L); % second equation (9.64)
if t>.00001 & f(2)>0; delta=-1; end % reversal of field
r=1-1/alpha/(1+delta/Q); % equation (9.67)
if f(2)*Q>0; r=-r/2; end % special reversal of r; note eqn. (9.69)
if abs(r)<1; r=1; end % floor under |r| (r>0 always in this problem)
f(1)=f(2)/r; % First of equations (9.67)
f=f';
```

9.5.8 RC Element with Mechanical Port*

The RC element, like the C element for conservative magnetic materials, can accommodate a second port for mechanical force F and velocity $\dot{y}$:

$$\vdash\!\!\frac{M}{\dot{\phi}}\!\!\longrightarrow RC \frac{F}{\dot{y}}\!\!\dashv$$

The partial derivative term in Equation (9.46) (p. 653) gives the force in the direction of motion, although the term at the far right side applies only if the two surfaces are parallel to one another, normal to the magnetic flux and the motion. An example is developed in

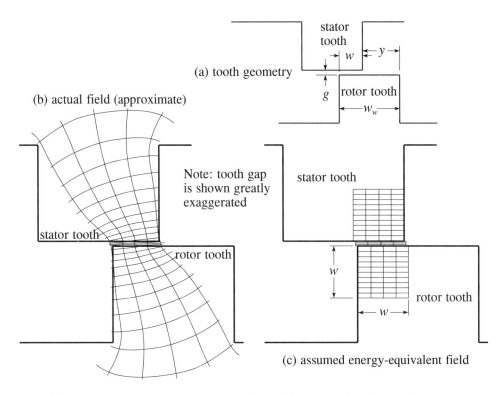

Figure 9.36: Roughly energy-equivalent field for teeth of a stepping motor

Guided Problem 9.12. Fringing can be considered by restricting the potential energy in the equation to the gap itself, and the force equated to its gradient as the middle term specifies.

The force is considerably more difficult to find when the motion is not parallel to the lines of flux in the presence of hysteresis. Such a displacement changes the pattern of the flux distribution, usually in a complicated way; the field is both nonlinear and two or three dimensional. The methods of this chapter are nominally restricted to uniform fields. Nevertheless, an approximate model might be constructed from such fields.

The example of two rectangular teeth of a stepping motor is shown in Fig. 9.36. The opposing teeth are assumed to have the same size. The magnetic field typically is highly saturated within the teeth, particularly near the air gap when the overlap between the two teeth is small. This pattern is approximated in part (b) of the figure, in which each rectangle contains the same magnetic energy. Saturation is employed purposely to keep the motor reasonably compact. The field within the gap can be represented by a linear two-port C element. (In the case study of the magnetic pick-off in Section 9.5.3, this is the only element assumed to have an mmf, greatly simplifying the analysis.) Neglecting hysteresis, the field within a tooth also can be approximated by a two-port C element, but nonlinear, and by a two-port RC element otherwise. The actual field would be extremely complex and difficult to analyze, being highly nonlinear (and hysteretic) as well as two or three dimensional and dependent on the relative position of the two teeth. An accurate analysis would require a complicated finite-element model which is well beyond the scope of this book and most or all available finite-element software. Nevertheless, a crude model that includes saturation and hysteresis is possible and is given now.

This very simplified model for the magnetic behavior of a tooth comprises a single uniform rectangular region with length and width dependent on the tooth offset y, defined

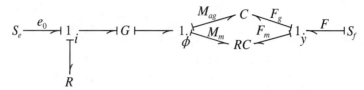

Figure 9.37: Bond graph for interaction between two teeth of a stepping motor

in part (a) of the figure. This approximately equivalent magnetic field, shown in part (c), need not resemble the actual field as long as it has the same total flux and approximates its energy. The width and length of the rectangle, w, are assumed to equal the overlap distance $ww - |y|$ plus a fringing correction that ranges as high as $4.5g$ for each end of the gap:

$$w = w_w - 0.9525|y| + 5.2g - 1.7ge^{-|y|/1.623g}; \qquad \frac{dw}{dy} = (-0.9525 + 1.7e^{-|y|/1.623g})\operatorname{sign}(y).$$
(9.70)

Fringing effects can be quite significant for two-dimensional fields and depend partly on non-local geometry (the presence of other teeth). The estimate of equation (9.70) is developed as part of a discussion on linear fields in Section 12.1.8.

Variations in field saturation within the teeth adds credence to the assumption of square equivalent fields, depending on the degree of saturation employed. When the overlap is small, w also is small, and the field density is highly saturated, giving a very high mmf per unit length. As the overlap increases, the total field increases, but the field density, degree of saturation and mmf decrease. When the opposing teeth are fully aligned, the region of the assumed element overfills the entire tooth, which is assumed to have a height less than its width. One cannot expect particularly accurate results, but semi-quantitative changes in the flux and force patterns indicated for changes in the assumed excitation, for example, should be useful for purposes of design and analysis.

The field density in the metal base to which the teeth are attached is half or less than that within the tooth element (and the gap element), which because of the degree of saturation means that its mmf is a small fraction of that in the tooth itself. It is neglected here. The element for the gap will be assumed to have the same width and therefore the same flux density, B, as the teeth elements. The height of the gap is set at the fixed separation distance, g.

The lesser force produced when the teeth are not overlapping is more complicated but is not addressed. In usual practice with stepping motors the excitation normally is shut off in this case anyway, avoiding the problem.

The dependence of the width of the two tooth elements and the gap element on the tooth position y produces three forces that sum to give the total force. The equal elements for the two teeth can be combined into the single two-port compliance element. With the neglect of hysteresis, these forces are given by energy gradients of the form $F_{yi} = \partial \mathcal{V}_i/\partial y$. Hysteresis is introduced by substituting an RC element for the tooth compliance element, which combines the compliances of the two teeth by multiplying by two. The result is the bond graph of Fig. 9.37, in which the position y is assumed to be an input variable.

Apart from the added complexity of simulating the hysteresis, there is the problem of computing the energy in an RC element, which has not yet been defined. The energy per unit volume is assumed to equal the area to the left of the anhysteretic magnetization curve $B = B(H_{im})$, which still exists for the RC element, just as for the C element:

$$\mathcal{V}_m = \int_0^{H_a} H_a\, dB = \mu_0 a^2 \int_0^{h_a} dh_a\, db. \qquad (9.71)$$

Here $h \equiv H/a$ and $b \equiv B/a\mu_0$ as before, and the subscript a stands for "anhysteretic." To carry out the integration it is necessary to express h_a and b as functions of x_a. From equations (9.56) and (9.66) (p. 662, p. 665),

$$h_a = x_a - c_1 \mathcal{L}(x_a), \tag{9.72a}$$

$$b = x_a + c_0 \mathcal{L}(x_a). \tag{9.72b}$$

$$c_0 \equiv (1 - \alpha)M_s/a; \qquad c_1 \equiv \alpha M_s/a. \tag{9.72c}$$

Therefore, the energy per unit volume for the element becomes

$$\mathcal{V}_m = \mu_0 a^2 \int_0^{x_a} (x - c_1 \mathcal{L})(dx + c_0 \, d\mathcal{L})$$

$$= \mu_0 a^2 \left[\frac{x_a^2}{2} - c_0 c_1 \int_0^{\mathcal{L}_a} \mathcal{L} \, d\mathcal{L} - c_1 \int_0^{x_a} \mathcal{L}(x) \, dx + c_0 \int_0^{x_a} x \mathcal{L}^1(x) dx \right]$$

$$= \mu_0 a^2 \left[\frac{x_a^2}{2} - c_0 c_1 \int_0^{\mathcal{L}_a} \mathcal{L} \, d\mathcal{L} - c_1 \int_0^{x_a} \mathcal{L}(x) \, dx + c_0 x_a \mathcal{L}(x_a) - c_0 \int_0^{x_a} \mathcal{L}(x) \, dx \right]$$

$$= \mu_0 a^2 \left\{ \frac{x_a^2}{2} - \frac{c_0 c_1}{2} [\mathcal{L}(x_a)]^2 + c_0 x_a \mathcal{L}(x_a) + (c_0 + c_1) \log_e \left[\frac{\sinh(x_a)}{x_a} \right] \right\}, \tag{9.73}$$

where x_a must be found as a function of b by iterative solution of equation (9.72b), and $\mathcal{L}(x)$ is given in equation (9.56).

The energy density for the air-gap element is of course much simpler:

$$\mathcal{V}_g = \frac{1}{2} \mu_0 a^2 b^2. \tag{9.74}$$

Note that the nondimensionalizations $h = H/a$ and $b = B/a\mu_0$ are also applied to this air gap element, for convenience, even though the value of a is associated with the metal. The total energy stored in the air gap is the product of $\mathcal{V}_g$ and the volume of the element, which is to be expressed as a function of the state variables ϕ and y:

$$\mathcal{V}_{tg} = g w w_d \mathcal{V}_g = \frac{g \phi^2}{2\mu_0 w_d w}. \tag{9.75}$$

The force produced by this energy is

$$F_g = \frac{\partial \mathcal{V}_{tg}}{\partial y} = \frac{\phi^2}{2\mu_0 w_d w} \left(\frac{dg}{dy} - \frac{g}{w} \frac{dw}{dy} \right), \tag{9.76}$$

where w and g are appropriate functions of y.

The total energy stored in the two metal elements and its resulting force become

$$\mathcal{V}_{tm} = 2w^2 w_d \mathcal{V}_m, \tag{9.77}$$

$$F_m = 4w w_d \mathcal{V}_m \frac{dw}{dy} + 2w^2 w_d \frac{\partial \mathcal{V}_m}{\partial y}. \tag{9.78}$$

Here w is the same function of y as above, and $\mathcal{V}_m$ is a function of both ϕ and y. The $\mathcal{V}_m$ of equation (9.73) is a function of x_a, which in turn can be found as a function of b by solving equation (9.72b) iteratively. Then,

$$b = \frac{\phi}{w w_d a \mu_0}; \qquad \frac{\partial b}{\partial y} = \frac{-\phi}{a \mu_0 w_d w^2} \frac{dw}{dy}. \tag{9.79}$$

Thus, the derivative $\partial \mathcal{V}_m / \partial y$ in equation (9.78) is found as follows:

$$\frac{\partial \mathcal{V}_m}{\partial y} = \frac{d\mathcal{V}_m}{dx_a}\frac{dx_a}{db}\frac{\partial b}{\partial y} = \frac{d\mathcal{V}_m}{dx_a}\frac{-\phi}{[1 + c_0 \mathcal{L}^1(x)]a\mu_0 w_d w^2}\frac{dw}{dy}, \tag{9.80}$$

in which $\mathcal{L}^1(x)$ is given in equation (9.57). Finally, from equation (9.73),

$$\frac{d\mathcal{V}_m}{dx_a} = \mu_0 a^2 \left\{ x_a - c_1 \mathcal{L}(x_a)[1 + c_0 \mathcal{L}^1(x_a)] + c_0 x_{im} \mathcal{L}^1(x_a) \right\}. \tag{9.81}$$

To summarize, the total force is the sum of the F_g of equation (9.76) and the F_m of equation (9.78), with substitutions from equations (9.80) and (9.81).

EXAMPLE 9.17

Approximate and plot the steady-state force acting between the teeth shown in Fig. 9.36 as a function of the position, assuming some overlap. The magnetic circuit is completed by a core of virtually infinite permeability excited by a coil with 100 turns and a resistance of 7 ohms that is excited with a constant 1.5 volts. Assume the magnetic properties $\alpha = 1.7585 \times 10^{-4}$, $M_s = 1.0024 \times 10^6$ A/m, $a = 74.31$ A/m, $k_0 = 5.0 \times 10^{-5}$ and $x_0 = 7$ (approximate values for silicon steel; see Problem 9.45). The teeth dimensions are $w_w = 2$ mm and depth $w_d = 20$ mm.

Solution: A bond graph for the system is given in Fig. 9.37 (p. 670). Defining the state variables as the total flux, ϕ, and the dimensionless mmf in the metal, h, the state differential equations become

$$\frac{d\phi}{dt} = \frac{1}{G}\left[e_o(t) - \frac{Ra}{G}(gb + 2wh)\right],$$

$$\frac{dh}{dt} = \frac{1}{ra\mu_0 w_d w}\left[\frac{d\phi}{dt} - \frac{b}{w}\frac{dw}{dt}\right].$$

The first equation follows from the bond graph and the definitions given earlier; the second follows from the definition $b = \phi / a\mu_0 w_d w$ and equation (9.67) (p. 665). One strategy would be to place the teeth in a particular position, which sets the width and heighth w at a fixed value, specify zero initial conditions, and run until equilibrium is reached. In this way the equilibrium field could be established for the particular position, and the process would be repeated for as many positions as desired. The method actually employed below is to sweep the relative tooth position and resulting value of w slowly enough that virtual equilibrium is maintained. Specifically, the position is fixed with two opposite corners of the opposing teeth facing each other, and the coil is activated. The transient response of the field is observed in order to estimate how fast a subsequent sweep of the postion can be justified. The plot of the mmf H below reveals that equilibrium is reached in about two milliseconds:

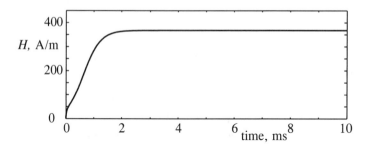

The entire sweep is then set at 100 times longer, or 0.2 seconds, which should give results quite close to the steady-state. (In the coding below, the velocity itself is ramped up, to minimize errors at the start.) A plot of the field B vs. the mmf H (with the initial ramp-up removed) shows significant saturation and some hysteresis:

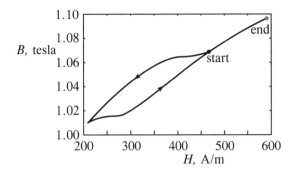

The force is computed by running a subsequent program based on the equations above, also given below. The results are plotted as a function of position, rather than time:

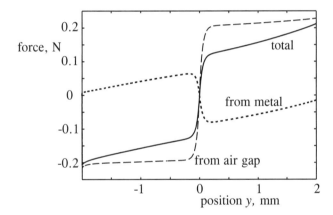

The permeability of silicon iron is so high that without magnetic saturation there would be little mmf associated with it; most of the force would be due to field changes in gap. This force would be approximately uniform as the offset from the centered position is increased. For the given excitation the force due to field changes in the gap remains roughly constant with changes in position, but saturation produces a counter force from changes in the field in the metal. The net effect is a force that is greatest for the least overlap of the teeth. This feature is consistent with published results[a]. The hysteresis also introduces some asymmetry with respect to position.

The field cannot exceed 1 tesla by much without a large increase in the mmf. The mmf is very sensitive to small changes in the electrical resistance about the value of 7 ohms used. The use of 8 ohms, for example, gives distinctly less force, illustrating why saturation is employed conventionally. Reducing the resistance to 6.5 ohms increases the mmf greatly, but the resulting field strength and total force increase very little, although the force becomes even less uniform.

The code below can be downloaded from the Internet (See page 1018). It is adapted in Section 9.6.5 to represent some dynamics of a DC stepping motor.

---

[a]See Fig. 3.9 of Kenjo and Sugawara, *Stepping Motors and Their Microprocessor Controls*, Clarendon Press, Oxford, 1994.

```
%Master file ex9_17m.m for Example 9.17
global alpha c1 a c2 G R mu0 g ww wd x0 t0 V0 e0
alpha=1.7585e-4; k=5e-5; c1=alpha/k; a=74.31; Ms=1.0024e6; c2=a/alpha/Ms;
G=100; R=7; mu0=4*pi*1e-7; g=.000025; ww=.002; wd=.02; c=1; x0=7;
t0=.01; V0=.02; e0=1.5; x00=[0 0]; % zero initial conditions
options=odeset('RelTol',1e-6,'InitialStep',1e-10); % gives flexibility
[t,x]=ode45('ex9_17de',[0 .21],x00,options); % x(1) is flux, x(2) is h
plot(t(1:342),x(1:342,2)*a) % gives plot page 672 (first 10 milliseconds)


function f=ex9_17de(t,x) % as called by the ode routine above
global alpha c1 a c2 G R mu0 g ww wd x0 t0 V0 e0
if t<t0; Y=ww; w=.0475*Y+5.2*g-1.7*g*exp(-Y/1.785/g); dwdt=0;
else;
    y=V0*(t-t0)*(1-exp(-(t-t0)/t0))-ww; Y=abs(y);
    w=ww-0.9525*Y+5.2*g-1.7*g*exp(-Y/1.623/g); % equation (9.70)
    V=V0*(1+((t-t0)/t0-1)*exp(-(t-t0)/t0));
    dwdt=V*(1.7/1.623*exp(-Y/1.623/g)-.9525)*sign(y); % from eq'n (9.70)
end
b=x(1)/a/mu0/wd/w; % definition as given above
X=(1-alpha)*x(2)+alpha*b; % equation (9.66)
if X<600; Langevin; else; L=1; end % call program on p. 665
Q=c1*(c2*(X-x(2))-L)*(1+abs(X/x0)); % equation (9.64)
f(1)=(e0-R/G*a*(g*b+2*w*x(2)))/G; % first state differential eq'n above
dbdt=1/a/mu0/wd/w*(f(1)-x(1)*dwdt/w); % derivative of b above
if dbdt>=0; delta=1; else; delta=-1; end
r=1-1/alpha/(1+delta/Q); % equation (9.67)
if dbdt*Q>0; r=-r/2; end
if abs(r)<1; r=1; end % note that r>0 everywhere
f(2)=dbdt/r; % second state differential equation above
f=f';


% script file 'force'; execute after simulation above
c0=(1-alpha)*Ms/a; c3=1/c2; s=size(t); mu0=4*pi*1e-7;
for i=1:s(1,1)
    if t(i)<t0; y(i)=-ww;
    else; y(i)=V0*(t(i)-t0)*(1-exp(1-t(i)/t0))-ww; end
    Y=abs(y(i)); phi=x(i,1); w=ww-.9525*Y+5.2*g-1.7*g*exp(-Y/g/1.785);
    w=ww-.9525*Y+5.2*g-1.7*g*exp(-Y/g/1.785);
    dwdy=.9525*(exp(-Y/1.785/g)-1)*sign(y(i));
    Fg(i)=-phi^2/2/mu0/wd*g/w/w*dwdy; % force due to field in gap
    b(i)=x(i,1)/a/mu0/wd/w; % definition of dimensionless field density
    X=b(i)/(1+c0); Langevin; ft=X+c0*L; fp=1+c0*L1;
    while abs(b(i)-ft)>.01; % Newton-Raphson inversion of eq'n (9.71b)
        X=X+(b(i)-ft)/fp; Langevin; ft=X+c0*L; fp=1+c0*L1;
    end % the following line is equation (9.72)
    Vm=mu0*a*a*(X*X/2-c0*c3/2*L*L+c0*X*L+Ms/a*log(sinh(X)/X));
    dVmdx=mu0*a*a*(X-c3*L*(1+c0*L1)+c0*X*L1); dVdmb=dVmdx/(1+c0*L1);
    Fm(i)=(4*w*Vm*wd-2*phi/a/mu0*dVdmb)*dwdy; % eq's (9.78), (9.80)
    F(i)= Fg(i)+Fm(i); % the total force is gap plus metal forces
end
plot(x(:,2)*a,b*mu0*a) % gives hysteresis plot (including ramp up)
plot(y,Fg,y,Fm,y,F) % plots the two forces and their sum
```

9.5.9 Summary

The effort variable in a magnetic curcuit is the magnetomotive force (mmf) M, which can be produced by a current i in a fixed coil of n turns with the relation $M = ni$. The conjugate flow variable is the rate of change of the magnetic flux, $\dot{\phi}$, which when multiplied by n gives the voltage in the coil. Thus, the coil can be modeled as a gyrator with modulus $G = n$.

The displacement ϕ, or magnetic flux, also can be related to M by the properties of the magnetic circuit. The flux-per-unit-area of a particular magnetic material, $B = \phi/A$, is a function of the magnetic field strength, $H = M/\ell$, where ℓ is the length of the magnetic path that resides within the material. For air or a vacuum, $B = \mu_0 H$, where μ_0 is the permeability of free space. For "soft" ferromagnetic materials the relation $B = \mu H$ is also most commonly used, but the permeability μ is much greater than μ_0. As a result, a portion or the whole of a magnetic circuit often is approximated by a compliance, $C = \mu A/\ell$.

Looking more closely, the relation between B and H for a ferromagnetic material is hysteretic in nature, dissipating as well as storing energy. This is represented by the RC element, as described in some detail to allow you to carry out dynamic simulations. Further, eddy currents generated inside magnetic materials produce rate-dependent dissipation; these can be modeled by a magnetic resistance, R. Electrically insulated laminations of ferromagnetic material often are employed to minimize this effect.

An attractive mechanical force is produced in the members facing each other across an air gap, giving a second bond to the magnetic C and RC elements if this gap is variable. This results from the fact that, for a given flux, the magnetic energy in the circuit increases when the gap increases. By holding the flux conceptually constant, the force can be computed using equation (9.46) (p. 653), since the air gap is linear magnetically and the potential energy in the equation can be restricted to that of the gap itself, regardless of whether saturation occurs in the metal. The phenonenon of flux fringing can spread out the flux and weaken the energy and the force.

When it is not practical to compute the force from the right-most term in equation (9.46), the more general alternative $F_x = \partial \mathcal{V}/\partial x$ can be used. This is the usual occurence when the displacement x is not normal to the solid surfaces. More information about magnetic fields is given in Section 12.1. When hysteresis exists, this equation still applies, but it is necessary to evaluate the energy $\mathcal{V}$ as a function of the field ϕ using the anhysteretic magnetization characteristic. Detailed equations for doing this have been given, and their use illustrated.

Guided Problem 9.11

A given permanent magnet with a cross-sectional area of 0.0006 m^2 and length $\ell = 0.2 m$, pictured in part (a) of Fig. 9.38, is used to pick up magnetically soft iron bars. The point P of its B, H characteristic, defined in Fig. 9.33 and the associated text (p. 661), has $H_P = -36,000$ A/m and $B_P = 1.0$ tesla, representing one of the strongest classical metallic magnet materials, Alnico 5. Soft iron flared sections can be added to the poles of the magnet so as to present a different flux area for the air gaps. Determine the area of each air gap that maximizes the force on a bar when the gap is 0.25 cm, and find that force. Neglect fringing, flux leakage and any mmf drops in the bar and in the flared sections.

Suggested Steps:

1. The force is maximized when the permanent magnet operates at the point P, as explained on pp. 661-662. Determine the desired values of the total flux ϕ and the mmf drop M_{ag} for each air gap.

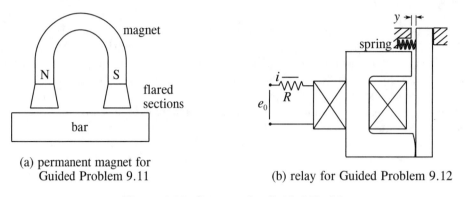

(a) permanent magnet for
Guided Problem 9.11

(b) relay for Guided Problem 9.12

Figure 9.38: Systems for Guided Problems

2. Relate ϕ to the flux density B_{ag} in the air gap and the physical dimensions of the gap.

3. Relate M to the value of H_{ag} in the air gap and the physical dimensions of the gap.

4. Relate B_{ag} to H_{ag} in the air gap.

5. Combine the results of steps 2-4 to solve for the desired area.

6. The force for a single air gap, which is doubled to get the total force on the bar, is given by equation (9.46) (p. 653).

Guided Problem 9.12

Simulate the closing of the magnetic relay pictured in part (b) of Fig. 9.38, which has an air gap of approximately constant area $A = 0.0001$ m^2 and initial length $y = 0.001$ m. The effective mass of the moving part is 8 grams (referred to motion at the air gap), and the electric and magnetic circuit components otherwise are the same as in the case study with the simulation code given in Section 9.5.6 with results plotted in Fig. 9.35. Motion is resisted by a spring with an essentially constant force over the allowed range of displacement. The applied voltage is to be sinusoidal with an amplitude of 64 volts, with the negative parts removed. Choose the frequency and the spring force such that the relay closes and opens reliably in reponse to the cyclical voltage, but not much more slowly than necessary.

Suggested Steps:

1. Model the system with a bond graph, and apply causality. Neglect flux leakage and eddy currents. Include the air gap in the RC element to preclude the causal conflict that would arise from adding a separate two-port C element.

2. Write appropriate equations to describe the RC element, noting that the magnetic flux is common for the ferromagnetic and air gap portions of the magnetic circuit, while the mmfs sum. Reduction to dimensionless variables will aid subsequent coding.

3. Define state variables and write a set of state differential equations.

4. Mechanical stops must be provided to keep the armature within bounds. This can be tricky; a large spring-like reaction force is wanted only when the velocity is increasing the out-of-boundedness, in order to prevent bounce by presenting zero coefficients of restitution.

5. Write code for MATLAB ode45. You may adapt the coding in the case study above (p.74).

6. The spring force should be about half of the maximum force that the magnetic field provides. Run the code and estimate by trial-and-error the maximum spring force that the magnetic field produces, and set the spring force accordingly.

7. Now run the code for various frequencies to discover how fast the relay can oscillate reliably.

8. Make any final adjustments, and report your result with a B, H or b, h plot and a plot of the position as a function of time.

PROBLEMS

9.37 In designing circuits with permanent magnets it is often convenient to set the area of air gaps equal to the area of the magnet. Determine the ratio of the length of the magnet to the length of the gap, ℓ/g, which equalizes these areas and maximizes the force for (i) Alnico 5 ($B_P = 1.0$ T, $H_P = -36,000$ A/m) and (ii) a modern rare-earth magnet ($B_P = 0.5$ T, $H_p = -330,000$ A/m). Assume no other significant drops in mmf.

9.38 The plunger of the solenoid shown opposite slides in plastic sleeves of thickness g for purposes of centering and reducing friction. The total mmf drop across these sleeves is so much larger than the mmf drops through the ferromagnetic parts that the latter may be neglected for purposes of an approximate analysis. (a) Estimate the force exerted on the plunger by the magnetic field. (b) Draw a bond graph model for the system, and deduce a set of state differential equations.

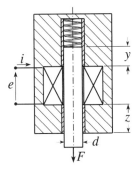

9.39 "E-I" torque motors such as shown on the left below, named because of their shape, are employed as the electromechanical component of classical electrohydraulic servo valves. A permanent magnet provides the bulk of the magnetic flux through the two paths, while coils, operated normally in push-pull fashion, increase the force in one of the outer air gaps and decrease the force in the other, placing a torque on the upper member, rapidly causing it to rotate through a small angle. (This rotation affects the hydraulic portion of the system, which is not shown here; its effect is to be neglected.) The rotation is resisted by a spring-like support in the center of this member.

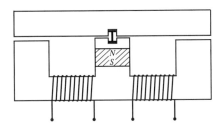

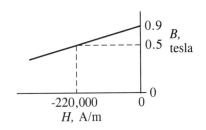

You are asked to draw a bond graph model for the system, neglecting mmf drops through the ferromagnetic solid parts in deference to the much larger mmf drops across the air gaps. Also, give approximate expressions for magnetic compliances in terms of dimensions that you define. The permanent magnet shown is of the rare-earth type, and has the characteristic plotted to the right above.

9.40 Simulate and plot the hysteretic B, H locus for the case study on pages 666-668 for the limit cycle in which the given sinusoidal voltage is continued indefinitely rather than stopped after one-half a cycle. (Two or three complete cycles is enough.) Hint: Little change needs to be made to the given coding, but a statement that resets δ to 1 at the appropriate times must be added.

9.41 Repeat the above problem adding eddy-current losses with magnetic resistances of (i) 1000 ohms^{-1} and (ii) 10,000 ohms^{-1}. Extend the run to approach the limit cycle.

9.42 Repeat Problem 9.40 for the limit-cycle that results when the one millisecond semi-sinusoidal pulse is repeated every two milliseconds, that is, setting all voltages to zero that otherwise would be less than zero.

9.43 Simulate and compare minor hysteresis loops of the RC element of Fig. 9.35 (p. 667) as follows:

(a) Adapt the code on page 665 (which gives the hysteresis plot in Fig. 9.34, p. 663) for the specified sinusoidal field $h = 2.85 + 0.37 \sin(1000\pi t)$, and plot the resulting hysteresis loop.

(b) Adapt the code on page 668 for the voltage-excited system of Fig. 9.35 for the roughly comparable case of $e_0 = 48 + 6.4 \sin(1000\pi t)$ V, and plot the resulting hysteresis loop.

(c) The two cases above demonstrate characteristically different shaped hysteresis plots; the difference persists even if the resistance R in part (b) is made very small. To understand the difference, plot and compare the two voltages e on the left sides of the gyrators. (Hint: Use the MATLAB command `diff`.) Also, plot what the voltage of the voltage-driven system would have to be in order to produce the behavior of the field-driven system.

9.44 The solution to Example 9.17 assumes infinite permeance for everything except the air gap and the metal within the tooth itself.

(a) Augment the bond graph and give the augmented differential equations for the addition of a rectangular magnetic element.

(b) Repeat the simulation and force analysis of the example, assuming the added element is also silicon steel with a cross-sectional area of 0.0001 m^2 and length 0.05 m.

9.45 The magnetic parameters used in Example 9.17 for a typical hot-rolled silicon-iron alloy were deduced from the following plot, adapted from A. E. Berkowitz and E. Kneller, *Magnetism and Metalurgy*, Academic Press, New York and London, 1969, v 1 p. 47. (The anhysteretic characteristic is assumed.) (Note: The author also tried using the third version of the Jiles and Atherton model for this hysteresis loop, but was not able to produce quite as good a match. Neither model produces quite the right tilt at the cusps.)

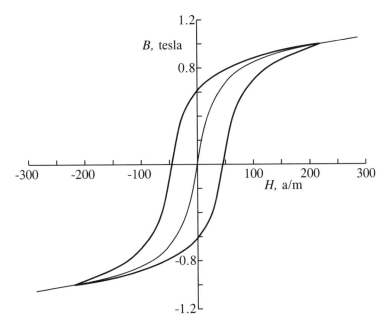

Deduce these parameter values independently from the plot. Your values likely will differ somewhat. The value x_0 is somewhat arbitrary; assume $x_0 = 7$ (although a smaller value could do better).

SOLUTIONS TO GUIDED PROBLEMS

Problem 9.11

1. $\phi = A_m B_P$; $M_{ag} = \frac{1}{2} H_m \ell$.

2. $\phi = A_{ag} B_{ag}$.

3. $M_{ag} = H_{ag} g$.

4. $B_{ag} = \mu H_{ag}$.

5. $A_m B_P = A_{ag} B_{\lceil ag} = A_{ag} \mu_0 H_{ag} = A_{ag} \mu_0 M_{ag}/g = A_{ag} \mu_0 H_m \ell/2g$.

   Therefore, $\dfrac{A_{ag}}{A_m} = \dfrac{2g B_P}{\mu_0 \ell |H_m|} = \dfrac{2 \times 0.0025 \times 1}{4\pi \times 10^{-7} \times 0.2 \times 36000} = 1.105.$

   $A_{ag} = 1.105 \times 0.0006 = 0.000663 \text{ m}^2.$

6. $F = \dfrac{2\phi^2}{2\mu_0 A_g} = 432 \text{ N}.$

Problem 9.12

1.

2. The total mmf M comprises the sum of the mmf for the ferromagnetic path, as in the case study, plus the mmf for the air gap, which is $\dfrac{y\phi}{\mu_0 A} = \dfrac{yB}{mu_0} = ayb.$

   The mechanical force equals (from equation (9.46)), $F = \dfrac{\phi^2}{2\mu_0 A} = \dfrac{AB^2}{2\mu_0} = \dfrac{Aa^2\mu_0 b^2}{2}.$

3. The bond graph gives $\dfrac{d\phi}{dt} = \dfrac{1}{G}\left[e_0(t) - \dfrac{RM}{G}\right]$, or $\dfrac{db}{dt} = \dfrac{1}{a\mu_0 G}\left[e_0(t) - \dfrac{R}{G}(lah - yab)\right]$

   and $\dfrac{dp}{dt} = F - F_0 - \dfrac{R_y}{I}p - \dfrac{1}{C}y.$

4. The mechanical stops are represented by the R_y and C elements. The forces of these elements are invoked only when they are resisting further deflection of the stops.

5. The coefficients `c5=A*A*A*mu0/2; C=1e-6; Ry=400` are added to the global statements. The code differences from the case study in Fig. 9.35 and the associated text (omitting unchanged statements) are

```
e0=64*sin(500*pi*t); % Sine wave with period of 4 ms.
if e0<0; e0=0; end % This leaves the upper half of the sine wave only.
if x(3)>0; x3=x(3); else x3=0; end % prevents air gap from becoming negative
f2=c4*(c3*e0-x(1)-x3*x(2)/ell); % Includes the mmf of the air gap
if t>.00001 & f2<0; delta=-1; end
if t>.00001 & f2>0; delta=1; end
f(2)=f2;
f(1)=f(2)/r;
f(3)=x(4)/I;
f(4)=-c5*x(2)^2+F0; % Force when stops are not in play
if f(4)<0 & f(3)<0 & x(3)<0; f(4)=f(4)-x(3)/C-Ry*f(3); end % inner stop
if f(4)>0 & f(3)>0 & x(3)>.001; f(4)=f(4)-(x(3)-.001)/C-Ry*f(3); end
```

   The complete files, `GP9_12m.m` and `GP9_12.m`, can be downloaded from the Internet (see page 1018).

6. The spring initially pushes the relay outward, and the magnetic field barely pulls it back to zero for $F_0 \equiv 50$ N. Therefore, the spring force is set to $F_0 = 25$ N in the subsequent runs.

7. The speed is limited by the available force, the gap length and the effective mass. When the period is set at 2 ms there is more than enough time for the switching to take place, so this time was used.

8.

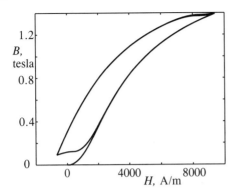
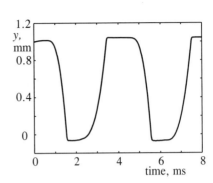

9.6 Electric Motors

Models of electric motors that include the effects of all the pole pieces and magnetic hysteresis would be very complex, and beyond most needs for dynamic analysis. The simple models of DC motors and induction motors given in earlier chapters should normally suffice. The following more elaborate models of selected motor types nevertheless suggest what can be done when necessary and can be useful in their own right, particularly when there is dynamic control. These are a synchronous motor, a brushless DC motor, a three-phase induction motor, and stepping motors. Only in the last case does the modeling delve into the magnetic circuit in the manner of the preceding section. There is also a brief discussion regarding single-phase induction motors.

The discussion starts by introducing vector bonds and transformers. These bond graph elements are used widely in the literature but have been avoided thus far in this book in favor of more highly reticulated models. They are a convenience here, however, and also are used in Chapters 11 and 12.

9.6.1 Vector Bonds and Transformers

A bundle of bonds of similar type is represented widely in the literature by a **vector bond**, a shorthand notation drawn with two parallel lines:

$$\left.\begin{array}{c} \dfrac{e_1}{\dot{q}_1} \\[4pt] \vdots \\[4pt] \dfrac{e_n}{\dot{q}_n} \end{array}\right\} \quad \Longleftrightarrow \quad \frac{\mathbf{e}}{\dot{\mathbf{q}}}$$

Two vector bonds can be joined by a **vector transformer**:

$$\frac{\mathbf{e_1}}{\dot{\mathbf{q}_1}} \quad \mathbf{T} \quad \frac{\mathbf{e_2}}{\dot{\mathbf{q}_2}}$$

This element is defined partly by a transformer matrix $\mathbf{T}$ so that

$$\dot{\mathbf{q}}_2 = \mathbf{T}\dot{\mathbf{q}}_1; \qquad \dot{\mathbf{q}}_1 = \mathbf{T}^{-1}\dot{\mathbf{q}}_2, \tag{9.82}$$

analogous to a scalar T. The rest of the definition, like that of scalar transformers, requires the conservation of power:

$$\mathcal{P} = \mathbf{e_1}^T \dot{\mathbf{q}}_1 = \mathbf{e_2}^T \dot{\mathbf{q}}_2. \tag{9.83}$$

This requires

$$\mathbf{e_1} = \mathbf{T}^T \mathbf{e_2}; \quad \mathbf{e_2} = (\mathbf{T}^T)^{-1} \mathbf{e_1}, \tag{9.84}$$

which can be verified by substituting this equation and equation (9.82) into equation (9.83). In certain, $\mathbf{e_2} = \mathbf{T}\mathbf{e_1}$; this condition (which requires T=1 for the scalar transformer) occurs when $\mathbf{T}^T = \mathbf{T}^{-1}$.

The elements of the transformer matrix $\mathbf{T}$ can be either constants or functions of a remote variable. The following example employs both a modulated vector transformer and two modulated scalar gyrators.

9.6.2   Synchronous Motor/Brushless D.C. Motor

These motors differ not in their basic structure but in the way they are driven. The simple version considered has three stator coils distributed evenly around the periphery and one permanent magnet fixed to the rotor, described as two pole-pairs. Some fancier versions have rotor coils driven by DC currents instead of permanent magnets, or four or more pole-pairs. The synchronous machine is driven by three-phase power: three voltages of equal frequency and amplitude but shifted in phase from one another by 120°. The magnitude or phase of the voltage may be controlled to prevent slippage and minimize power. The speed also can be controlled by changing the excitation frequency. The brushless DC motor is driven in a more complex manner to allow for position or speed control. The three voltages no longer need to be equal in amplitude, but the requirement that they continuously sum to zero is maintained, leaving two independent inputs. The absense of brushes in these motors virtually eliminates the Coloumb friction that plagues ordinary DC motors, and makes them superb actuators for dynamic control of mechanical systems.

An idealized bond-graph model for the motors is given in Figure 9.39. This model, consistent with that originally proposed by Sahm[11], assumes ideal magnetic circuits, not accounting properly for the secondary eddy current and hysteresis losses. A first step in accounting for these losses in a more physical way is given by Morel et al.[12] The vector transformer $\mathbf{T}$ converts the (a,b,c) voltages and currents of the stator windings into a set of imaginary voltages and currents (d,q,0). The d-coordinate is associated with a fictitious winding aligned along the magnetic axis of a permanent magnet. The fictitious q-coordinate winding is oriented 90° ahead. The voltage and power of the 0-coordinate vanish identically, consistent with the case of a star-connected, symmetrical winding operated such that $e_a + e_b + e_c = 0$ at every moment. This property can be seen from the matrix for the transformer that links the real and fictitious windings, defined so that $\mathbf{i}_{d,q,0} = \mathbf{T}\mathbf{i}_{a,b,c}$ and $\mathbf{e}_{a,b,c} = \mathbf{T}^T \mathbf{e}_{d,q,0}$:

$$\mathbf{T} = \begin{bmatrix} \cos(n_p\theta) & \sin(n_p\theta) & 1/\sqrt{2} \\ \cos(n_p\theta - 2\pi/3) & \sin(n_p\theta - 2\pi/3) & 1/\sqrt{2} \\ \cos(n_p\theta + 2\pi/3) & \sin(n_p\theta + 2\pi/3) & 1/\sqrt{2} \end{bmatrix}. \tag{9.85}$$

This matrix satisfies the special condition $\mathbf{T}^T = \mathbf{T}^{-1}$ noted above, so that $\mathbf{e}_{d,q,0} = \mathbf{T}^T \mathbf{e}_{a,b,c}$ and $\mathbf{i}_{a,b,c} = \mathbf{T}^T \mathbf{i}_{d,q,0}$. The angle of the rotor is $\theta = \int \omega \, dt$, and the number of pole-pairs is

---

[11] D. Sahm, "A Two-Axis Bond Graph Model of the Dynamics of Synchronous Electrical Machines," *J. Franklin Institute,* v 308 n3, 1979, pp. 205-218.

[12] H. Morel, Ph. Lautier, B. Allard, J.P. Masson and H Fraisse, "A Bond Graph Model of the Synchronous Motor," *Proc. of 1997 Western Multiconference, ICBGM'97,* v 29 n 1, ed. J. Granda and G. Dauphin-Tanguy, Society for Computer Simulation International.

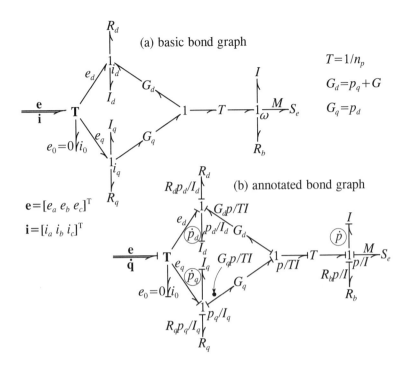

Figure 9.39: Bond graph model of a synchronous motor/brushless DC motor

$n_p = 2$. (Two magnets at right angles to one another, and six rather than three equally spaced stator coils, would give $n_p = 4$, etc.)

The electromechanical coupling is represented by the scalar transformer, for which $T = 1/n_p$, and the two gyrators, which are highly modulated as noted in the figure. The G that is one of two parts of the modulus of the gyrator G_q represents the constant flux of the permanent magnet. It acts much like the magnet in the DC motor introduced in Section 5.2, including placing a drag on the rotor (in conjuction with the electrical resistance) that dissipates more energy than may be desired.

The annotated bond graph shows the state variables p, which is the angular momentum of the rotor, θ, and the electrical momenta p_d and p_q. The input variables are, for the brushless DC motor, any two of the three voltages e_a, e_b and e_c. The output variables of interest in general are $\omega = p/I$, θ, $i_d = p_d/I_d$, $i_q = p_q/I_q$ and the three actual currents i_a, i_b and i_c. The following state-variable differential equations result:

$$\frac{dp}{dt} = n_p \left[\frac{p_d p_q}{I_q} - \frac{(G + p_q)p_d}{I_q} \right] \tag{9.86a}$$

$$\frac{d\theta}{dt} = \frac{1}{I}p \tag{9.86b}$$

$$\frac{dp_d}{dt} = e_d - \frac{R_d}{I_d}p_d + \frac{n_p(G + p_q)}{I}p \tag{9.86c}$$

$$\frac{dp_q}{dt} = e_q - \frac{R_q}{I_q}p_q - \frac{n_p}{I}p_d p \tag{9.86d}$$

When operated as a synchronous machine, the three-phase excitation is of the form

$$e_a = e_0 \sin(\omega t + \alpha), \quad e_b = e_0 \sin(\omega t - 2\pi/3 + \alpha), \quad e_c = e_0 \sin(\omega t + 2\pi/3 + \alpha), \quad (9.87)$$

which allows for the possibility of a control variable α, and satisfies the more general requirement

$$e_a(t) + e_b(t) + e_c(t) = 0. \tag{9.88}$$

The synchronous machine produces a steady or nearly steady torque on the rotor only when the rotor is turning at the speed ω/n_p. Startup therefore poses a major problem. Large synchronous machines classically solve this problem by adding windings to the motor that allows it to operate partly like an induction motor. Modern control of power circuits allows, instead, the frequency to accelerate, or the phase shift α to be varied.

EXAMPLE 9.18

Design a control scheme to accelerate a synchronous motor from rest, and then convert it to a generator. The machine is connected to a 220 v three-phase 60-cycle line, has two pole-pairs, and has the parameters $G = 0.75$ N·m/amp, $I_d = I_q = 0.013$ H, $R_d = R_q = 0.6$ ohms, $I = 0.10$ kg·m^2 and $b = 0.005$ N·m·s. It is connected to a three-phase 220 volt line. When acting as a generator it is driven by a torque of 60 N·m. Simulate and plot the speed of the rotor and the electrical energy required for acceleration and produced by generation.

Solution: One way to start the motor is to synchronize the frequency and phase of the three-phase electrical signal to the rotation of the shaft, that is to make the rotation of the electrical field remain approximately fixed relative to the rotor. There are two electrical cycles per revolution of the shaft (associated with the two pole-pairs), so this means substituting twice the rotation angle of the shaft, or 2θ, for what otherwise would be ωt, where $\omega = 60 \times 2\pi$ rad/s. It should be noted that 220 V three-phase means that the peak voltage on any one phase is $220\sqrt{2/3}$ V. In addition, it is helpful to shift the phase by an additional angle; -45° is used here. This causes the rotor to accelerate to the desired speed in 0.223 s, during which time it rotates through only 4.35 revolutions, and "slips" another 22.4 revolutions. (See Problem 9.46 for relevant simulation files.)

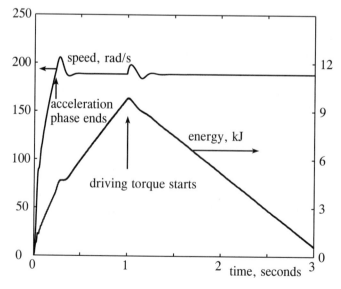

When the rotor is up to speed, the electrical signal is fixed to the desired 60 cycles. Care must be taken not to introduce a discontinuity in phase, for the machine has a limited ability to tolerate shocks and can crash easily. The plot shows a brief transient overspeed. Some damping is needed to achieve the desired steady-state. This was achieved by adding the phase angle $-0.06(\Omega - \omega/2)$, where Ω is the rotational velocity of the shaft, that is a phase shift proportional to the rate of change of the slippage. See Problem 9.46 (p. 693).

It was decided to convert to a generator starting one second after startup. The 60 N·m torque is applied starting at the point, with no other change. Another brief but satisfactory transient perturbation in the speed can be seen. The plot of total energy reverses in direction, demonstrating generation. (Energy is plotted instead of power, which has a large ripple.) The energy efficiency is only about 41%; losses occur mostly in the resistances and partly in the mechanical damping. Better design certainly is possible, and would be aided by replacing the permanent magnet with a controllable DC coil and by using a model that better reflects the reality of the magnetic circuit.

Designing a control for a brushless DC motor is relatively tricky. Junko and Donaire[13] use advanced bond graph methods to achieve this objective using the same motor model as above.

9.6.3   Three-Phase Induction Motors

Most motors sold today are three-phase induction motors, which are preferred because of their economy and efficiency in sizes from fractional horsepower to thousands of horsepower particularly for commercial application wherever three-phase power is available. They have a stator winding of the same type as the synchronous and brushless DC motors above. The rotor is most commonly of the **squirrel-cage** type, consisting of an array of conducting bars embedded around the magnetically soft rotor, and short-circuited at each end by conducting end rings. There are also **wound-rotor** types that have the same number of poles as the stator, and slip rings with brushes that permit additional resistance to be inserted into the rotor circuit. In both cases the field in the rotor is induced by the field in the stator (eliminating the need for a commutator) and the slip between the rotation of the stator field and the rotation of the rotor. Therefore, the machine produces no torque without slip. Nevertheless, normal operating speed is roughly constant, at roughly 90% or 95% of the synchronous speed. As with the motors above, synchronous speed equals the frequency of the excitation divided by the number of pole pairs on the stator. The starting torque may be low or high, but usually a peak torque occurs at an intermediate speed.

The bond graph model in part (a) of Fig. 9.40 has been given by Karnopp[14] and is consistent with much older models. This is incorporated into the model of part (b) of the figure by expanding the two-port inductance fields so as to employ only one-port inertances, as in Section 9.3.1, recognizing the number of poles with a distinct transformer as in the preceding subsection, and adding a vector transformer somewhat similar to that in equation (9.85) to represent the necessary transformation from the three-phase stator variables to the two-phase rotor variables. Modeling is being done at the level of electric circuits rather than the more basic magnetic circuits, skirting the issues of magnetic saturation and hysteresis in the interest of simplicity. Energy losses are represented by the resistances. The induction acts like a transformer, the modulus of which should be approximately 1;

---

[13]S. Junco and A. Donaire, "BG-Supported Synthesis of Speed- and Position-Tracking Controllers for Brushless DC-Motor Drives," Proc. *2005 International Conference on Bond Graph Modeling and Simulation,* Society for Modeling and Simulation International Simulation Series, v 37 n 1, pp. 245-251.

[14]Dean Karnopp, "Understanding Induction Motor State Equations Using Bond Graphs," *2003 Int. Conf. on Bond Graph Modeling and Simulation,* ed. F. E. Cellier and J. J. Granada, Simulation Series, v 35 n 2, The Society for Modeling and Simulation International, pp. 269-273.

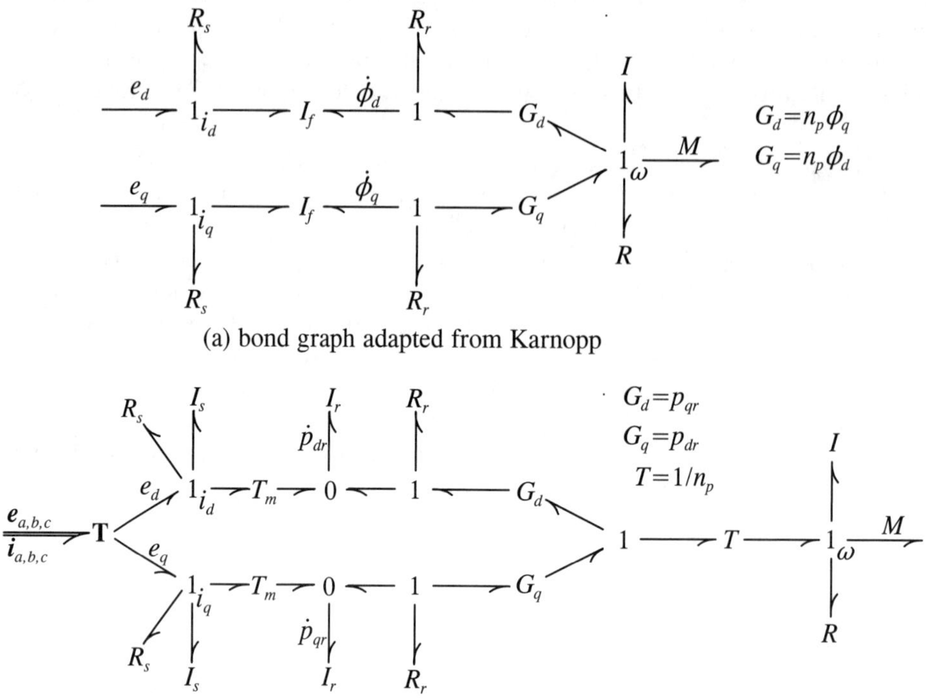

(a) bond graph adapted from Karnopp

(b) completed and reticulated bond graph

Figure 9.40: Bond graph model of three-phase induction motor

its leakage flux should be small, so the values of I_L are small compared with those of the primary inductances, I_I, which are proportional to the size of the machine.

The vector transformer is different from that used for the synchronous machine. Induction occurs only when there is slip between the rotor speed and the synchoronous or field speed. There is no salient rotor pole; the induced field in the rotor rotates at a different rate from the rotor itself. It rotates at the same absolute rate as the stator field, so that the torque on the rotor is continuous. Two vector transformers normally are represented to act in tandem. The first converts the stator field from the variables of the 120°-displaced winding to an equivalent orthogonal three-phase form. The second converts this to the (d, q, o) form. See for example S. Junco [15]. Cascading these two transformers,

$$\mathbf{T}^T = \begin{bmatrix} \cos\theta & \sin\theta & 0 \\ -\sin\theta & \cos\theta & 0 \\ 0 & 0 & 1 \end{bmatrix} \sqrt{\frac{2}{3}} \begin{bmatrix} 1 & -1/2 & -1/2 \\ 0 & -\sqrt{3}/2 & \sqrt{3}/2 \\ 1/\sqrt{2} & 1/\sqrt{2} & 1/\sqrt{2} \end{bmatrix}. \tag{9.89}$$

The variables with the subscript 'o' carry no power and as with the analysis of the synchronous machine may be dropped.

The angle θ, which varies slowly and little, is difficult to compute during a simulation. Fortunately, however, for time scales greater than a cycle the result is virtually the same regardless of its value. As a result, the author has used a fixed angle, such as 0, $\pi/4$ or $\pi/2$.

[15]Sergio Junco, "Real – and Complex – Power Bond Graph Modeling of the Induction Motor," *1999 International Conference on Bond Graph Modeling and Simulation,* Simulation Series, v 31 n 1, The Society for Computer Simulation International, pp. 323-328.

EXAMPLE 9.19

A three-phase induction motor with two pole-pairs is driven by a three-phase 220 V line. It may be modeled as in Fig. 9.40, with the primary inductances $I_I = 0.1$ H, the "leakage" inductances $I_L = 0.05 I_I$, transformer modulus 1, and all four resistances $R = 2$ ohms. The load includes an inertia $I_r = 0.1$ kg·m². Simulate the start-up of the motor under no additional load. Then, using other simulations as appropriate, find and plot the steady-state torque-speed characteristic of the motor.

Solution: Causal strokes and appropriate annotations are first placed on the bond graph:

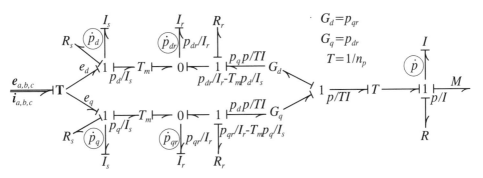

This gives five state differential equations

$$\frac{dp}{dt} = \frac{1}{T}\left[p_{dr}\left(\frac{p_{qr}}{I_r} - \frac{T_m p_q}{I_s}\right) - p_{qr}\left(\frac{p_{dr}}{I_r} - \frac{T_m p_d}{I_s}\right)\right] - \frac{R}{I}p - M,$$

$$\frac{dp_{dr}}{dt} = \frac{p_{qr}p}{TI} - R_r\left(\frac{p_{dr}}{I_r} - \frac{T_m p_d}{I_s}\right),$$

$$\frac{dp_d}{dt} = e_d - \frac{R_s p_d}{I_s} - T_m \frac{dp_{dr}}{dt},$$

$$\frac{dp_{qr}}{dt} = -\frac{p_{dr}p}{TI} - R_r\left(\frac{p_{qr}}{I_r} - \frac{T_m p_q}{I_s}\right),$$

$$\frac{dp_q}{dt} = e_q - \frac{R_s p_q}{I_s} - T_m \frac{dp_{qr}}{dt}.$$

The value of θ is assumed to be an arbitray constant, as suggested above. Note that $\mathbf{e}_{d,q,0} = \mathbf{T}^T \mathbf{e}_{a,b,c}$, using the $\mathbf{T}^T$ given in equation (9.89), and $e_0 = 220\sqrt{2/3}$.

Simulations for the startup with $M = 0$ and three values of the load resistance R are given below. (The relevant code can be downloaded from the Internet; see page 1018.)

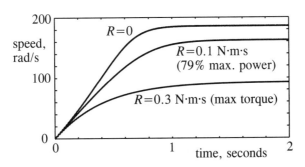

The terminal speeds of these runs can be multiplied by the respective values of R to get the steady-state torques. From these, plus a few others, the torque-speed curve below can be found:

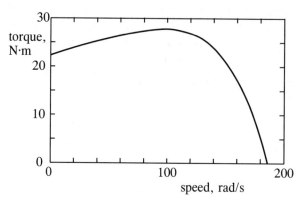

Note that a simulation would fail if the value of the torque, M, was assumed to be greater than the initially unknown start-up torque. There is a single intersection between the load characteristic $M = R\dot{\phi}$ and the motor characteristic, however, so the entire characteristic can be found by assuming various values of R rather than M.

In practice, the electromechanical dynamics may be so fast as to be negligible relative to the effect of the mechanical inertia. In this case, the model above is useful principally in deducing the steady-state torque-speed characteristic, which would replace the entire model except for the mechanical inertia and load.

9.6.4   Single-Phase Induction Motors

Most small induction motors are single-phase. A motor constructed with a single peripherally symmetric stator winding and a squirrel-cage rotor would operate very nicely near its synchronous speed, rotating in either direction. There would be zero starting torque, however; such a characteristic is given in Fig. 2.10 (p. 23) with the label "main winding." In practice, something must be done to produce a starting torque. The torque-speed characteristic first given in Fig. 2.4 (p. 17) and repeated many other places in this book is typical of a **shaded-pole motor**, which is used only in very small motors used for such applications as small fans, timing devices and relays. One-half of each of the salient poles in this motor is surrounded by a heavy, short-circuited winding called a shading pole. Induced currents in these coils cause the flux in half of the pole to lag the flux in the other half. The resulting periodic shift in flux from the unshaded to the shaded half of each pole produces a low starting torque.

Most single-phase induction motors use an auxiliary stator winding to produce the starting torque. In fractional-horsepower **split-phase** motors the current in the auxiliary winding, and thus the field it produces, is made out-of-phase with that in the main winding by virtue of a higher resistance-to-reactance ratio than the main winding. **Capacitor motors** achieve the phase lag by inserting a series capacitor into the auxiliary circuit. The cheapest and most reliable versions use compromise values of capacitance and inductance in the auxiliary to buy a modest amount of starting torque without greatly reducing torque sensitivity at operating speed. The resulting torque-speed characteristic usually resembles that of the shaded-pole motor as mentioned above and reproduced frequently in this book. Fancier versions increase the capacitance and inertance of the secondary to increase the starting torque, but then cut out this circuit with a centrifugal switch when the speed reaches

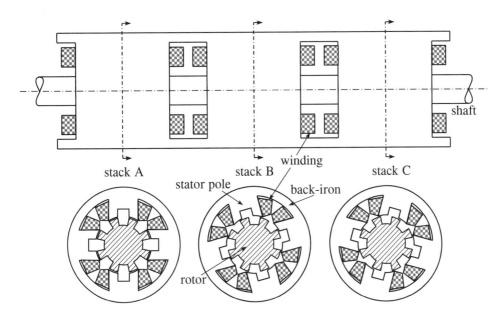

Figure 9.41: A three-stack variable reluctance stepping motor

perhaps 75% of operating speed. The torque-speed characteristic of such a **capacitor-start**, induction run machine is given in Fig. 2.10 (p. 23). Sometimes there are two capacitors in parallel in the auxiliary circuit, one of which is switched off by a centrifugal switch.

Gaude et al.[16] give a progress report on an investigation into bond-graph modeling and simulation of one-phase induction motors, delving into the magnetic domain including hysteresis. The model includes the effects, individually and collectively, of 20 bars in a squirrel-cage rotor. Thus far they had not introduced an auxiliary winding, so the simulation presented starts at about 75% of full speed. They do give a bond graph that includes this winding, however. For most purposes such an extremely detailed high-frequency model is not needed; as with other motors it may suffice simply to ignore the electromechanical dynamics and represent the motor simply by its steady-state torque-speed characteristic.

9.6.5 Stepping Motors

Modern control techniques with DC motors have diverted some of the market away from stepping motors, since the motion is typically smoother. Nevertheless, stepping motors retain a large share of the market for position-control actuators, because they give essentially zero error with only open-loop control. A stepping motor also usually has a detent effect that prevents motion in the absence of a command.

There are three basic types of stepping motors, and many variations[17]. These are called the variable-reluctance, permanent magnet and hybrid types.

A three-stack variable-reluctance motor is shown in Fig. 9.41. The rotor and stator have the same number of teeth; in stack B they are offset relative to stack A by one-third of the

---

[16]D. Gaude, H. Morel, B. Allard, H. Yahoui and J.-P. Masson, "Bond Graph Model of the Induction Motor," *1999 International Conference on Bond Graph Modeling and Simulation,* ed. J.J. Granda and F.E. Cellier, Simulation Series v 31 n1, The Society for Computer Simulation, pp. 317-322.

[17]See Takashi Kenjo, *Stepping Motors and Their Microprocessor Controls,* second ed., Clarendon Press, Oxford, 1994, and/or Paul Acarnley *Stepping Motors, A Guide to Theory and Practice,* fourth ed., The Institution of Electrical Engineers, London, 2002.

tooth pitch, and in stack C by minus one-third of the tooth pitch. Adjacent windings have opposite polarity relative to the axis of the rotor; the two resulting field loops in an activated stack are completed through the surrounding "back iron". The stacks are activated one at a time. Sequential activation in the order of stacks A,B,C,A,B,C,··· produces clockwise stepping; A,C,B,A,C,··· produces counterclockwise stepping. For the eight-toothed design shown, therefore, one revolution results from 24 sequential steps, or 15° per step.

EXAMPLE 9.20

The tooth widths of the motor of Fig. 9.41 equal four fifths the tooth spaces. Adapt the linear single-tooth model and coding of Example 9.17 (p. 674) to simulate the response to a single step excitation, starting from rest, assuming the parameters there apply per tooth. The effective mass at the radius of the air gaps (5.73 mm) is 8.5 grams per tooth.

Solution: The bond graph of Fig. 9.37 (p. 670) applies, with an inertance substituted for the flow source at its right end. The linear output position is y. State variables must be added to the given function file to represent this position and its velocity derivative. These depend on the forces ($F = ma$), so it is necessary to incorporate the sub-program *force.m* into the function file. Most of the vectors used can be reduced to scalars. The

initial conditions are all zero except for the position, which is minus three-quarters of the width of a tooth, w_w. The resulting motion of the tooth is plotted opposite; $y/w_w = 1$ corresponds to $360/16 = 20°$.

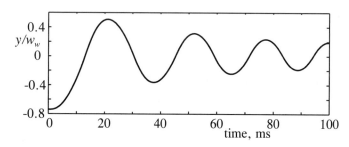

The mmf and field undergo two cycles per motion cycle. (See page 1018 for downloadable MATLAB code.)

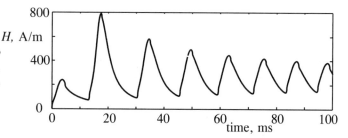

The damping is less than what would be desired and is rather typical of the behavior of stepping motors without added dampers. This damping is due partly to the hysteresis, which can be seen in the $B - H$ plot (with the large ramp-up omitted for clarity):

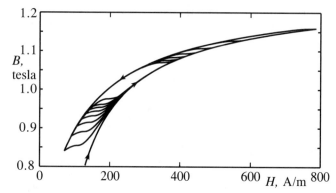

Eddy currents would add some additional damping, as would mechanical friction, but an added damper would be highly desirable.

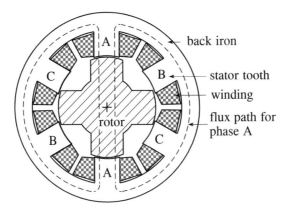

Figure 9.42: Cross-section of a single-stack variable-reluctance stepping motor

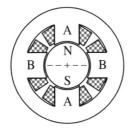

Figure 9.43: Elementary permanent-magnet stepping motor

Single-stack variable reluctance motors can be made by placing a different number of teeth on the rotor than on the stator. The three-phase example shown in Fig. 9.42 has six stator teeth and four rotor teeth, producing a step size of 30°. Windings on opposing stator teeth have opposite polarity relative to the rotor axis, so the field passes in one rotor tooth, out the opposing tooth, and is completed through the surrounding stator "back" iron. Flux leakage produces some minor mutual coupling between the phase windings. The clockwise sequence A,B,C,A,B,C,··· produces counterclockwise rotation; clockwise rotation results from A,C,B,A,C,B,···. It is also possible to operate with two phases on and one phase off, rather than only one phase on, which produces desirably more damping than that found in Example 9.20.

Multiple teeth can be placed on each pole. Doubling the numbers of teeth on the example motor cuts the step size in half, and tripling results in one third. It is also possible to have more rotor teeth than stator teeth. A three-phase (six-pole) design with three teeth on each pole (18 total stator teeth) and twenty rotor teeth produces a step angle of 6°. With six teeth on each pole and 44 rotor teeth there are 132 steps per revolution. Four phases (eight poles) also are used; five stator teeth per pole (40 teeth total) and fifty rotor teeth produce a step angle of 1.8°.

Permanent-magnet stepping motors have smooth rotors with permanent magnet poles. The elementary version shown in Fig. 9.43 has four wound stator poles, with two phases, which must be able to be excited with either polarity. For example, exciting phase B with its two poles with one polarity causes a 90° rotation clockwise, while exciting with the opposite polarity causes a 90° rotation counterclockwise. Many variants of this design concept are emerging because of the availability of small rare-earth magnets.

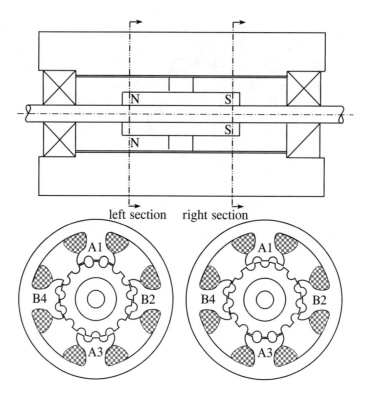

Figure 9.44: Hybrid stepping motor with two phases and four poles

Hybrid stepping motors have two toothed rotors and stators on the same shaft, with a permanent magnet polarizing one rotor with respect to the other, as pictured in Fig. 9.44. The teeth on one rotor are offset by half a tooth relative to the other rotor. There are usually eight wound poles on the stator, each connected to two to six teeth; the illustration has four wound poles, each connected to three teeth. One phase excites opposite poles with opposite polarity (relative to the machine axis) with enough current to balance the mmf produced by the permanent magnet on one of the poles, eliminating flux there, and to increase the mmf and the resulting flux on the other pole. Thus, when phase A is excited, pole A1 on the left rotor and pole A3 on the right rotor pull the rotor to the position shown, while poles A3 on the left and A1 on the right have negligible flux. Torques are produced by the permanent-magnet-induced fluxes in poles B2 and B4, but these cancell each other. When phase A is turned off and phase B is excited, the rotor turns one quarter of a tooth pitch clockwise or counterclockwise, depending on the polarity of the excitation. Then, to keep the motion proceeding in the same direction, phase B could be turned off and phase A turned on again, but with the opposite polarity from before, moving the rotor another one quarter of a tooth, or $6°$.

A bond graph for a hybrid stepping motor is given in Fig. 9.45. This describes a single phase but is useful in predicting the behavior of the entire machine, presuming only one phase is active at a time. The subscript "PM" refers to the permanent magnet; the other subscripts refer to the poles as designated on Fig. 9.44. Note the over-causality, which is ameliorated by neglecting hysteresis in the permanent magnet. See Problem 9.52.

Eight poles are usually used in order to achieve a force-balance rotor. The most common version has five teeth per stator pole (40 teeth total) and 50 teeth on the rotor, for 200 steps per revolution or $1.8°$ per step. Five-phase motors also are popular; a version with

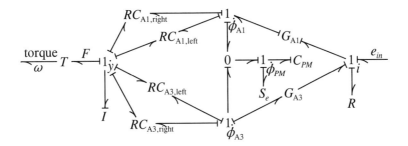

Figure 9.45: Bond graph for one phase of a hybrid stepping motor

forty teeth on the stator and fifty teeth on the rotor produces relatively smooth motion comprising 500 steps per revolution. Stacked designs also are made to increase torque.

Hybrid motors are preferred over variable-reluctance motors when small step size and a somewhat higher torque per unit volume is desired, or when a detent action is wanted when the power fails or is shut off. Variable-reluctance motors are preferred when rapid and large rotations and therefore large step sizes are wanted, and the existence of a somewhat smaller mechanical inertia is an advantage.

9.6.6   Summary

Two or more parallel bonds are often designated in the literature with a single double-line vector bond, using vector notation for the effort and flow variables. One use for such a bond is the representation of a three-phase electrical line, with its three voltages and three currents.

Bond graph models at the level of electrical voltages and currents are given for a permanent-magnet synchronous motor or brushless D. C. motor and for a three-phase induction motor. In the latter case the phase shift between the stator and rotor fields has been assumed arbitrarily, since it appears to have no effect on transient responses for time scales longer than individual cycles, which normally is the only interest for even rather small load inertias. The steady-state load characteristics developed by this model are unaffected by the assumption, and suffice for most loads to represent the motor for simulation purposes, as assumed in earlier chapters.

Stepping motors have the advantage over continuous-angle machines of simple open-loop control, and often a detent action even when the power fails. There are three broad categories: variable-reluctance motors with doubly salient teeth and exciting coils, permanent-magnet motors with magnets with poles transverse to the axis of rotation attached to the rotor, and hybrid machines that have doubly salient teeth and exciting coils like the variable-reluctance machines but also a permanent magnet that polarizes one rotor with respect to a second rotor. Bond graph models for the variable-reluctance and hybrid machines have been given, and the former simulated to show the behavior for a single step, including the resulting oscillations and magnetic hysteresis.

PROBLEMS

9.46 The function file ex9_18.m for the simulation of the startup of the synchronous motor/generator of Example 9.18 (pp. 684-685) is downloadable, but the script file needed to supply the data, set the initial conditions, request the ode simulation and specify the plot is not. Provide this file and carry out the simulation. Note the use of a flag, F.

9.47 The steady-state torque-speed characteristic of the three-phase induction motor found in Example 9.19 (p. 687-688) is approximated by the polynomial expression $M = -9.08 \times 10^{-8}\dot{\phi}^4 + 1.756 \times 10^{-5}\dot{\phi}^3 - 0.00135\dot{\phi}^2 + 0.1065\dot{\phi} + 22.20$. Simulate the start-up of the unloaded motor with the specified rotational inertance, and compare the resulting speed-vs-time plot with that given. Conclude whether the simpler model would suffice for inertias of this size or larger.

9.48 Write coding that produces the result of Example 9.19 (pp. 687-688). Run for the load resistance $R = 0.1$ N·m·s.

9.49 The function file ex9_20.m for the simulation of transient tooth motion for a stepping motor (Example 9.20), which produced the plots shown on page 690, is downloadable, but the script file needed to supply the data, set the initial conditions, request the ode simulation and specify the plot is not. Provide this file and carry out the simulation.

9.50 Apply the extended model of Problem 9.44 to the system of Example 9.20 (p. 690).

  (a) Draw a bond graph.

  (b) Write and execute the code to produce modified plots of the same variables.

9.51 The programs hysteresism.m and hysteresis.m (p. 665) assume that the field h is specified, and compute all results of interest with a single state variable. Other simulations require at least two state variables and more extensive coding. Specifying the electric current in the coil would specify the field. Does this mean that general simulations with specified current as a function of time are simpler to execute? Determine any key limitations.

9.52 Define state variables and write state differential equations for the hybrid stepping motor, based on the bond graph of Fig. 9.45 (p. 693). Note the differential causality.

9.7   Irreversible Couplers and Thermal Systems

An energy bond is a *reversible* coupler; energy can flow in either direction without degradation as well as loss. The same is true of transformers and gyrators, although this requirement has not been emphasized previously. The only irreversible elements we have considered are resistances. Energy "dissipated" in a resistance does not vanish, of course; it is transformed into heat, and thermal energy has yet to be considered. This omission is rectified in this section.

Thermal energy is qualitatively different from the gravity, elastic, kinetic, electrostatic and electromagnetic energies considered thus far. Rather than being based on macroscopic coherent generalized displacements or velocities, it is based on the microscopic incoherent (random) motions of atoms and molecules. As a result, engineering devices can convert thermal energy to the other forms of energy only partially, as described by the second law of thermodynamics and the concept of irreversibility. On the other hand, the other forms are readily convertable to thermal energy. Since energy is *conserved*, as expressed by the first law of thermodynamics, thermal energy has been excluded from the energy bookkeeping. That exclusion sometimes is unacceptable, directly so if you are interested in temperatures, or indirectly so if temperature affects other key parameters.

Irreversibilities are introduced here with the examples of heat conduction and friction.

The energy fluxes in these cases are analogous to the other types of power flows in that they can be represented by simple bonds with a single effort and a single flow. The introduction of mass transfer complicates the representation, however; this subject is deferred to Chapter 12, as is the accurate evaluation of the thermodynamic properties of gasses and liquids.

9.7.1 Effort and Flow Variables

A quantity of heat is labeled as Q in most thermodynamics texts; here the script $\mathcal{Q}$ is used to distinguish it from the volume flow rate of a fluid. The rate of heat transfer, which is an energy flux[18] with the dimensions of power, is given as $\dot{\mathcal{Q}}$.

The first need is to identify the effort or generalized force associated with heat conduction. Stop a moment before reading the next paragraph and try to determine the proper physical variable. If you succeed your learning will be enhanced.

The correct answer is *absolute temperature*, which is indicated herein with the symbol θ, since T is already dedicated to the moduli of transformers. (θ is also used frequently in the literature to indicate temperature.) Temperature has precisely the characteristics of an effort: it is the traditional "force" driving the energy flux, and it is a symmetric scalar with respect to the control surface.

The second needed identification is tougher. What variable is the flow or generalized velocity? This time you might like to get a pencil and some paper and recall the definition of entropy and the relationship between effort, flow and power before reading on. The reward can be well worth the trouble.

The flow is the ratio of the energy flux $\dot{\mathcal{Q}}$ to the absolute temperature θ, or $(d\mathcal{Q}/dt)/\theta$. However, recall that the change of *entropy* of a system due to a reversible transfer of heat $\delta\mathcal{Q}$ is

$$dS = \frac{\delta\mathcal{Q}}{\theta}. \tag{9.90}$$

The flow is therefore the *rate* of entropy transfer, dS/dt, due to the heat conduction. This is called the **entropy flux** due to heat conduction and is labeled $\dot{S}$, since the entropy gained by one system because of its contact with another system is lost by that other system. Thus the heat flux $\dot{\mathcal{Q}}$ is

$$\dot{\mathcal{Q}} = \theta\dot{S}. \tag{9.91}$$

The symbols θ and $\dot{S}$ are entered in the updated table of efforts and flows given as Table 9.3. (A refinement appears in the use of the term α in the first two entries of the table. This is a coefficient equal to or greater than 1 which can recognize the effect of non-uniform velocity profiles.)

9.7.2 Heat Conduction

Heat conduction associated with a finite temperature drop is irreversible and generates entropy. To see in simple terms how this happens, and to introduce a new entropy-producing bond graph element, consider the uniform slab of material shown in Fig. 9.46 through which a steady flux of heat passes. The heat flux entering one face equals the heat flux leaving the other, so that

$$\theta_1\dot{S}_1 = \theta_2\dot{S}_2, \tag{9.92}$$

which gives an entropy generation rate of

$$\dot{S}_2 - \dot{S}_1 = \frac{\theta_1 - \theta_2}{\theta_2}\dot{S}_1 > 0. \tag{9.93}$$

---

[18]The term *flux* implies *per unit time* and also usually means *per unit area*. In this text the latter is dropped; the energy flux refers to the rate of energy transfer for an entire macro bond.

Table 9.3   Effort and Flow Analogies (Extended)

	generalized force or effort, e	generalized velocity or flow, $\dot{q}$
Fluid, incompressible		
general	$\rho u + P + \alpha \rho v^2/2 + \rho g z$	Q
less thermal	$P + \alpha \rho v^2/2 + \rho g z$	Q
approximate	P	Q
micro-bond	any of above	v
Mechanical, longitudinal		
general	$F_c + \rho A v^2/2$	$v = \dot{x}$
approximate	F_c	$v = \dot{x}$
micro-bond	$\sigma + \rho v^2/2$	$v = \dot{x}$
Mechanical, transverse		
rotation	M	$\dot{\phi}$
translation (shear)	F_s	v_s
micro-bond	τ	v_s
Electric circuit	e	i
Thermal, conduction	θ	$\dot{S}$

This is always non-negative, since the sign of $\dot{S}_1$ is the same as the sign of $\theta_1 - \theta_2$. In the linear special case, for example,

$$\theta_1 \dot{S}_1 = \theta_2 \dot{S}_2 = H(\theta_1 - \theta_2), \tag{9.94}$$

where the constant H is known as a coefficient of heat conductance. This gives

$$\dot{S}_2 - \dot{S}_1 = H \frac{(\theta_1 - \theta_2)^2}{\theta_1 \theta_2}, \tag{9.95}$$

which shows an irreversibility that grows with the square of the temperature difference.

Entropy is generated throughout the slab. Energy storage and energy dissipation, when they occur, also are distributed spatially. Nevertheless, *lumped models* are most commonly used for thermal as well as other systems, for reasons of computational simplicity. Only *spatial integrals* of the extensive properties are represented in lumped models. The resistance element, $\longrightarrow R$, is an example. **Distributed parameter** models are discussed in Chapter 11, and the problem of approximating distributed phenomena by lumped models is discussed in some detail in Section 12.1.

9.7.3   The Irreversibility Coupler

The energy-conservative but entropy-generating **lumped general irreversible coupler** is now introduced. Traditionally represented as

$$\frac{e_1}{\dot{q}_1} \longrightarrow RS \frac{e_2}{\dot{q}_2} \longrightarrow$$

this element often is called simply "the RS coupler." When applied to heat transfer, as indicated in Fig. 9.46, e_1 and e_2 are the temperatures θ_1 and θ_2, and $\dot{q}_1$ and $\dot{q}_2$ are the entropy fluxes $\dot{S}_1$ and $\dot{S}_2$, respectively.

The RS coupler is defined partly as a static coupler that conserves but does not store energy; the energy fluxes $e_1 \dot{q}_1$ and $e_2 \dot{q}_2$ are equal. In this sense it is the same as the

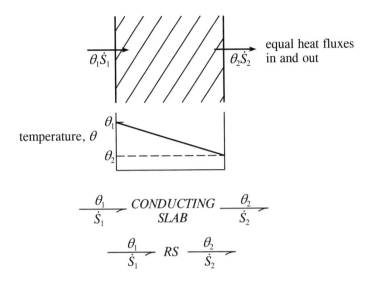

Figure 9.46: Steady heat conduction and the RS element

transformer element. It is very different in another sense, however; the transformer is used exclusively to represent reversible behavior while the RS coupler never represents reversibility except in the limit of zero energy flux. The transformer is reversible simply because it is never used when a variable entropy flux is involved. The same is true of gyrators. Therefore, *the RS coupler is always used in place of a transformer or a gyrator whenever either or both ports have an essential entropy flux.* In the specification of a particular RS coupler it is necessary to insure that the second law of thermodynamics is satisfied, which implies that entropy never vanishes. In the case of heat conduction, the constitutive relation in equation (9.94) satisfies this requirement, as revealed by the non-negative net entropy generation of equation (9.95).

9.7.4 Application to Friction

You can replace an R element with an RS coupler whenever you wish your model to retain the energy converted to thermal form. A transformer or a gyrator cannot be so used, because the flow on the output bond is an entropy flux. The modeling of energy conversion from mechanical to thermal form through friction, as the following example illustrates, is a major application.

Example 9.21

A shaft is supported by a journal bearing having some friction. Heat conduction along the shaft is neglected, compared with the heat conduction through the bearing support to the frame. The bearing support offers significant resistance to heat flow, and the frame is at a known environmental temperature θ_e.

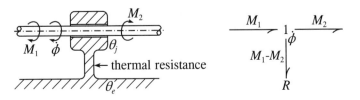

Model the system with a bond graph, including the heat fluxes, and write the corresponding equations. The resistance of the traditional bond graph model with dissipation, shown above right, can be approximated as a function of the form $R(\dot{\phi})$, or more accurately of the form $R(\dot{\phi}, \theta_j)$.

Solution: The bond graph opposite employs the element RS_1 to represent the irreversible conversion of mechanical power to the thermal form $\theta_j \dot{S}_j$, where θ_j is the temperature of the journal. The element RS_2 represents the subsequent heat conduction through the bearing support to the frame, which is at temperature θ_e. This cascade of elements details the reduction in the availability of the energy and its effect on the environment.

The constitutive relation

$$M_1 - M_2 = R\dot{\phi}$$
$$R = R(\dot{\phi})$$

implies that the frictional torque is independent of the thermal variables. In this case,

$$\theta_e \dot{S}_e = \theta_j \dot{S}_j = R\dot{\phi}^2,$$

so that if θ_e and $\dot{\phi}$ are known, $\dot{S}_e$ can be calculated. Note that $\dot{S}_e$ never can be negative; the second law is satisfied. If a thermal conductance H between the journal and the environment (as in equation (9.94), p. 696) is known, then θ_j and $\dot{S}_j$ can be calculated also.

If the friction in the journal depends significantly on the local temperature (as it well might since fluid viscosity depends on temperature), expressed as $R = R(\dot{\phi}, \theta_j)$, the equations above still allow for the computation of the temperature θ_j, the frictional torque, the heat transfer and the entropy fluxes.

The pressure drop of fluid passing through an orifice also can be treated as friction.

9.7.5   Use of 0- and 1-Junctions

A body can interact thermally with any number of other bodies, as suggested in Fig. 9.47. This is recognized by use of a 0-junction, or constant-temperature junction. On the other hand, a 1-junction rarely can be used for heat-conduction bonds, since this would imply a principle of conservation of entropy, in contradiction to the second law of thermodynamics. The only exception to this would be in the reticulation of a machine that is idealized as being reversible anyway.

You might in fact wish to create a three-port element to represent an ideal heat engine (IHE), a favorite of thermodynamicists. Such an element is reticulated in Fig. 9.48. The shaft port with its M and $\dot{\phi}$ is mechanical and has no entropy flux. The other two ports are thermal with entropy fluxes as flows. The requirement of reversibility forces these two entropy fluxes to sum to zero. This constraint is recognized explicitly by the 1-junction. The thermodynamic efficiency of the IHE is

$$\eta = \frac{M\dot{\phi}}{\theta_H \dot{S}} = \frac{(\theta_H - \theta_L)\dot{S}}{\theta_H \dot{S}} = 1 - \frac{\theta_L}{\theta_H} \tag{9.96}$$

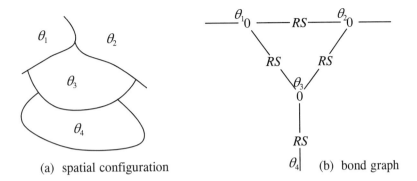

Figure 9.47: Heat interaction between regions separated by partially insulating interfaces

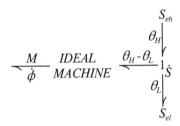

Figure 9.48: Ideal heat engine (IHE)

which agrees with the classical Carnot result.

An IHE with three or more thermal ports also can be defined; the sum of the entropy flows is zero.

9.7.6 Thermal Compliance

The simplest model for thermodynamic compliance refers to a fixed mass; there is no convection. Also, work interactions with the environment are neglected, which implies that either the volume of the body is constant or its surrounding pressure is negligible. This leaves only a heat transfer interaction, and is implied by use of the word "thermal" as opposed to the more general "thermodynamic." This thermal compliance is represented by a standard one-port compliance:

$$\frac{\theta}{\dot{S}} \quad C$$

A constitutive relation exists between S and θ, as it does between the effort and generalized displacement of any one-port compliance. The most common assumption is a constant specific heat, c, so that

$$\theta\,ds = du + P\,d(1/\rho) = du = c\,d\theta \tag{9.97}$$

where u is the specific internal thermodynamic energy and s is the specific entropy. Solving for θ,

$$\theta = \theta_0 e^{(S-S_0)/mc} \tag{9.98}$$

where m is the mass of the compliance substance and the subscript 0 refers to an arbitrary reference state. This nonlinear relation is plotted in Fig. 9.49.

The linearized model

$$\theta - \theta_0 = \frac{1}{C}(S - S_0) \quad \text{or} \quad \Delta\theta = \frac{1}{C}\Delta S, \qquad C = \frac{mc}{\theta_0}, \tag{9.99}$$

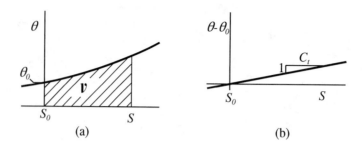

Figure 9.49: Thermal characteristic (equation (9.98))

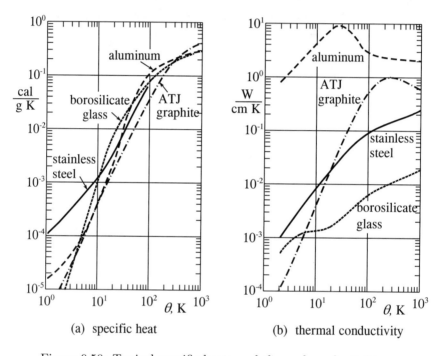

Figure 9.50: Typical specific heats and thermal conductivities

as shown in part (b) of the figure, can be used alternatively. This model corresponds to a specific heat that is proportional to absolute temperature, that is

$$c = C\theta/m, \tag{9.100}$$

Noble gases have constant specific heats, and the specific heats of liquids tend to be nearly independent of temperature. Also, the specific heats of most solids at moderate or high temperatures and of most other gases do not vary rapidly with temperature. These substances therefore agree or come close to agreeing with equation (9.98) for most practical purposes. On the other hand, the specific heats of solids at cryogenic temperatures decrease radically with decreasing temperature, even faster than the linear model of equations (9.99) and (9.100). Some typical behavior is given in part (a) of Fig. 9.50.

The thermal conductivity of the same solids, showing radical sensitivity to temperatures in the cryogenic range, is given in part (b) of Fig. 9.50. Such data is apt to be very sensitive to impurities and the crystaline form.

There is no such thing *thermal inertance*. One consequence of this profound fact is that purely thermal systems do not exhibit resonances.

Example 9.22

The case of a shaft with journal bearing with friction, given in Example 9.21, is now elaborated by including a path for heat conduction along the shaft, represented crudely by a heat transfer coefficient and an environmental temperature θ_3. The model also is made unsteady by including a lumped thermal compliance for the bearing. The bearing support has a fixed temperature θ_2. Represent this model with a bond graph. Then, annotate the graph with causal strokes and designations of the efforts and flows, leading to the associated differential equation(s).

Solution: The corresponding bond graph is

With the added causal strokes and annotations,

The associated differential equation is

$$\frac{dS}{dt} = \frac{R\dot{\phi}^2}{\theta(S)} - H_2\left[1 - \frac{\theta_2}{\theta(S)}\right] - H_3\left[1 - \frac{\theta_3}{\theta(S)}\right]; \qquad \theta = \theta_0 e^{(S-S_0)/mc}.$$

In general, the thermal conductances H_2 and H_3 and the specific heat c are functions of the temperatures.

9.7.7 Pseudo Bond Graphs for Heat Conduction

The assumptions that both the specific heats and the thermal conductivities of a system are invariant with temperature allow a striking computational simplification. This advantage exists, however, only if one substututes for the state variable S the state variable Q. Karnopp and Rosenberg (citation on page 14) represent this idea in the form of what they call a **pseudo bond graph** for heat conduction.

A pseudo bond graph for heat passing through a thermal resistance is shown in part (a) of Fig. 9.51. The heat flow, $\dot{Q}$, equals the temperature difference, $\theta_1 - \theta_2$, divided by the resistance, R. A pseudo bond graph for thermal energy storage is shown in part (b). The rate of change of temperature, $\dot{\theta}$, equals the difference between the heat flow in and the

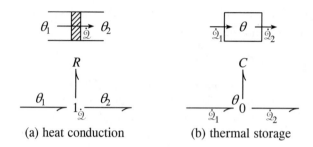

(a) heat conduction   (b) thermal storage

Figure 9.51: Pseudo bond graphs for heat conduction

heat flow out, $\dot{Q}_1 - \dot{Q}_2$, divided by the thermal compliance, C, which in turn equals mc. Causal strokes can be applied as with true bond graphs, and differential equations written with relative ease even for fairly complex combinations of resistances and compliances. The bond graphs and the associated equations still can be used if the thermal resistances or compliances are functions of temperature, but the advantage over the true bond graph is lost.

The designation *pseudo* recognizes that, unlike true bond graphs, the product of the effort, θ, and the pseudo-flow, $\dot{Q}$, is not the energy flux. The pseudo-flow itself is the energy flux. This poses a problem whenever a model for heat conduction, represented by a pseudo bond graph, is coupled to a model for a mechanical, fluid, electrical or hybrid system, represented by a true bond graph. Such couplings have been represented by *ad hoc* couplers defined to address the particular situation. One can write equations in terms of Qs rather than Ss regardless of which type of graph is drawn. The author therefore prefers to avoid pseudo bond graphs, except when there is an isolated or an extensive region with constant thermal resistances and specific heats. Pseudo bond graphs nevertheless are quite common in the literature.

Example 9.23

Convert the thermal conductance portion of the bond graph in the prior example to a quasi bond graph. Apply causal strokes, and annotate the graph variables assuming invariant thermal conductivities and specific heat. Finally, write the corresponding state differential equation(s).

Solution:

$$\frac{dQ}{dt} = R_1 \dot{\phi}^2 - \frac{1}{R_2}\frac{Q}{C'} - \theta_2 - \frac{1}{R_3}\frac{Q}{C'} - \theta_3; \quad C' = mc; \quad R_2 = \frac{1}{H_2}; \quad R_3 = \frac{1}{H_3}.$$

This is a simple first-order linear differential equation with constant coefficients. The associated time constant for transients is $\tau = C'/(H_1 + H_2)$. Should the specific heats or the thermal conductances not be assumed as invariant, the advantage of using Q or θ as the state variable vanishes.

9.7.8 Thermodynamic Coupling Between Mechanical and Thermal Energies

The coupling between mechanical and thermal effects is important to the modeling and analysis of many nominally mechanical systems. The thermal expansion of a solid rod has been modeled by a linear two-port compliance in Section 9.3.5, where it is seen that the temperature of a body is changed if it is significantly compressed or expanded in size. The modeling is now generalized to include nonlinear properties and heat transfer resulting from temperature change, with application to a gas. A hydraulic accumulator is used as a case study.

A hydraulic accumulator is based on the compression of a fixed mass of gas. In Section 3.5.4 (pp. 131-132) it is modeled crudely by a one-port compliance. It is noted there, however, that unless the compression and expansion takes place either very slowly (isothermal conditions) or very fast (adiabatic conditions), heat transfer effects are fairly significant. A better model recognizes both mechanical and thermal energy fluxes:

$$\vdash \frac{P}{-\dot{V}} \; C \; \frac{\theta}{\dot{S}} \dashv RS \vdash \frac{\theta_e}{} \; S_e$$

The volume flow rate of the hydraulic fluid is written as $(-\dot{V})$; the minus sign results from the definition of V as the volume of the *gas*, not the liquid which displaces it, in contrast to the earlier treatment. Assuming an ideal gas of mass m, the conventional thermodynamic assumptions are

$$PV = mR\theta, \tag{9.101a}$$

$$\mathcal{V} \equiv U = mc_v\theta, \tag{9.101b}$$

in which $\mathcal{V} \equiv U$ is the internal energy, c_v is the specific heat at constant volume and R is the gas constant. The crudest approximation being made is that the heat flux, $dQ/dt = \theta\dot{S}$, is proportional to the temperature difference between the temperature of the environment, θ_e, and the mean temperature of the gas:

$$\frac{dQ}{dt} = H(\theta_e - \theta). \tag{9.102}$$

The first law of thermodynamics requires

$$d\mathcal{V} \equiv dU = dQ - P\,dV. \tag{9.103}$$

Substituting equations (9.101) and (9.102) into equation (9.103),

$$mc_v\,d\theta = H(\theta_e - \theta)dt - (mR\theta/V)dV. \tag{9.104}$$

This can be rewritten in the final desired form as

$$\frac{d\theta}{dt} = \frac{H}{mc_v}(\theta_e - \theta) + \frac{R\theta}{c_vV}\frac{dV}{dt}. \tag{9.105}$$

In practice, dV/dt is treated as a causal input, that is a function of external variables, and is integrated to give V. Equation (9.105) becomes the state differential equation for the compliance, giving θ. Finally, the pressure P is calculated using equation (9.101a).

Problems in which a flow of gas enters a chamber are considered in Chapter 11, along with fancier equations of state including phase change.

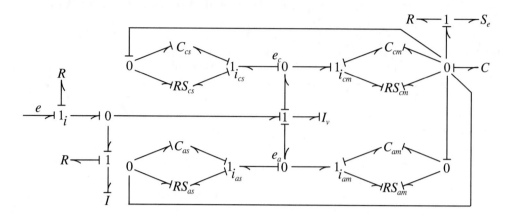

Figure 9.52: Bond graph for a lead-acid battery with thermal effects

9.7.9 Lead-Acid Battery; Use of Legendre Transformation

A bond-graph model for one cell of a lead-acid battery is given in Fig. 5.1 (p. 261), which assumes constant temperature. Rapid charging or discharging, such as in modern hybrid-powered vehicles, generates considerable heat, making the temperature significantly non-constant, however. Esperilla *et al* [19] accomodate this phenomenon by developing the bond graph shown in Fig. 9.52. The four one-port compliances of Fig. 5.1, which represent the primary and secondary electrochemical reactions of the annode and the cathode, have thermal ports added to become two-port compliances. RS elements also are introduced to represent heat transfer to the surroundings, with its entropy generation. The elements are nonlinear; partial details are given in the paper.

Notice that the four thermal bonds on the 2-port compliances are forced to have differential causality. This is the first time in this book that differential causality has been used on a multiport compliance. For nonlinear one-port compliances, however, differential causality is accomodated in Fig. 3.37 and the accompanying text (pp. 134-136) by employing a Legendre transformation, or using co-energy (complementary energy) rather than real energy. This concept is applied to the present case by Esperilla *et al* by taking Legendre transformations of the energies in these elements with respect to the thermal ports only[20]:

$$G \equiv \mathcal{L}(\mathcal{V}) = \mathcal{V} - \theta S, \tag{9.106}$$

or, since $d\mathcal{V} = e\,dq + \theta\,dS$,

$$dG = e\,dq - S\,d\theta. \tag{9.107}$$

G is treated as a state function, giving the differential causality result

$$\dot{S} \equiv \frac{d}{dt}(S); \qquad S = -\frac{\partial G}{\partial \theta}. \tag{9.107}$$

The inertance I_v is a virtual inertance added because the bond graph is undercausal, using the procedure detailed in Section 6.2. Thus, the use of $I_v = 0$ gives an algebraic equation in place of a differential equation. Esperilla *et al* preferred the option of giving a

---

[19]Esperilla, J.J., Felez, J. and Vera, C., "Thermal Model for Lead-Acid Batteries Using the Bond Graph Technique," *2005 International Conference of Bond Graph Modeling and Simulation,* Simulation Series, Society for Modeling and Simulation International (SCS), v 37 n 1

[20]The pressure is assumed to be constant. Otherwise, a third port for pressure-volume interactions can be added, and also given differential causality by expanding the definition of G to $G = \mathcal{V} - \theta S + Pv$.

small non-zero value to I_v, converting the equation into an approximate differential equation. In fact, they also add two further quasi-virtual inertances, but these are not necessary to eliminate algebraic equations.

9.7.10   Summary

Previous to this section thermal energy is treated as "dissipated," or useless, and accordingly it is omitted from the energy bookkeeping. Instead, it simply vanished in resistance elements. This approach is conventional in the elementary analysis of dynamic systems, but fails whenever temperature affects those properties of a system that are of interest or is of interest *per se*. The subject of this section is how to bookkeep thermal energy alone, or in conjuction with the other energies, in both static and dynamic systems. Cases with coupled mass flow and thermal effects are deferred until Chapter 11; heat conduction is the only mode of heat transfer considered here.

The use of absolute temperature as the effort or generalized force for the transfer of thermal energy requires that the flow or generalized velocity be the associated entropy flux. Thermal energy thereby becomes a generalized potential energy. There is no such thing as a generalized thermal kinetic energy. Thermal energy is different from the other energies in that its convertibility to these other forms is restricted by the second law of thermodynamics. Heat conduction can be represented by the RS static coupler element. The coupler is like a transformer or gyrator in that it conserves energy, but unlike them in that it allows irreversibilities and the associated generation of entropy. The RS coupler also can represent the irreversible flow of energy from the other forms into the thermal form and is particularly useful in representing mechanical or fluid friction. Its constitutive relations depend on the particular application.

Should you wish to model a thermal system by assuming that its specific heats are constant and that its heat conductions are proportional to its temperature differences, you may substitute the heats as the state variables. (You can alternatively use the associated temperatures, since in this case they are proportional to the heats.) This substitution can be represented in the form of a pseudo bond graph for heat conduction. This is not a true bond graph, however, because its energy fluxes are not the products of its efforts and flows, and does not mate with true bond graphs using conventional couplers.

Guided Problem 9.13

This guided problem illustrates the organization of steady-state problems involving heat engines and heat conduction provided by the methods presented in this section.

The classical Carnot efficiency of the ideal heat engine operating between the temperatures θ_H and θ_L is confirmed in Fig. 9.48 and equation (9.96) (pp. 698-699). In order to make the model conform more closely to a real system, interpose a thermal resistance with conductance H between the warmer heat reservoir and the ideal heat engine. Find how the resulting thermal efficiency is affected by the output power, $\mathcal{P}_{out}$. Specialize the result for $\theta_L/\theta_H = 0.5$. Also, find the *maximum* $\mathcal{P}_{out}$ which is possible, and the associated thermal efficiency.

Suggested Steps:

1. Add an RS coupler to the bond graph model to represent the thermal resistance. Define symbols for all the efforts and flows, and place them on the graph.

2. Write a complete set of algebraic equations to represent the constraints of each of the elements in the model. Also, express the thermal efficiency as a function of appropriate

parameters and variables. Treat $\mathcal{P}_{out}$ as known. Make a list of the unknowns, and make sure their number equals the number of equations.

3. Eliminate all variables, solving for the efficiency. This is tricky; it may help to number the equations, and make a table in which the variables in each equation are indicated with check marks.

4. Verify that the thermal efficiency equals the Carnot efficiency in the limiting case $\mathcal{P}_{out} = 0$. If not, a mistake has been made.

5. Make the indicated specialization $\theta_L/\theta_H = 0.5$, and sketch-plot the efficiency vs. $\mathcal{P}_{out}$. The maximum power and its associated thermal efficiency should become apparent.

Guided Problem 9.14

This problem provides experience in the writing of differential equations when irreversibilities are present. It does not require much effort if the simulation part is omitted.

The fading of brakes for vehicles on mountainous roads is a serious problem; the phenomenon of "runaway" trucks on long downhill runs is the most common result. Consider a truck weighing 100,000 lbs traveling down a long 6 percent slope. The coefficient of friction of the brake materials is

$$\mu = \mu_0 - \mu_1(\Delta\theta)^2 \qquad \mu_0 = 0.25 \quad \mu_1 = 0.2 \times 10^{-6} \, °F^{-2}$$

where $\Delta\theta$ is the mean temperature of the brake material above the ambient. (Should the temperature become so high that this equation gives $\mu < 0$, the proper relation is $\mu = 0$.) The trucker attemps to keep the speed steady at 50 mph, and is successful for some time. The normal force squeezing the brake materials together cannot exceed eight times its initial value (when $\Delta\theta = 0$), however, and eventually the truck begins to accelerate. The effective mass of the heated material is 125 lbm, its specific heat is 0.30 Btu/lbm·°F (or 233.5 ft·lbf/lbm·°F), and the effective coefficient for heat transfer to the surroundings is 0.38 Btu/°F.

Write state differential equations for a model of the system. Determine how long the speed remains constant, and the maximum lower speed for which the truck would never become a "runaway" on this slope. Optional: simulate the subsequent acceleration.

Suggested Steps:

1. Draw a bond graph for a simple model. Define symbols for the variables and the parameters.

2. Write the state differential equations for the first phase, in which the speed is constant.

3. Find the critical temperature above which the brakes will not prevent acceleration. Solve the equations of step 2 analytically (this should be simple) to find the time at which this temperature is exceeded, and the acceleration phase begins. At this point one also can determine the critical speed below which the brakes would never have reached the critical temperature.

4. Write the differential equations in solvable form for the subsequent acceleration phase.

5. The equations of step 4 are not easy to solve analytically. Solve them numerically using the MATLAB simulator or any other simulation program. Make sure the effective coefficient of friction never becomes negative. Plot the resulting velocity as a function of time.

PROBLEMS

9.53 A thin-walled spherical vessel, 0.5 m in diameter is full of hot water at 100°C when it is placed in an environment with an air flow at 0°C such that the mean local overall heat transfer coefficient is 150 kJ/hr·m²°C. Determine the temperature of the water as it cools over time. Note: For water, $c = 4.18$ kJ/kg·K and $\rho = 997$ kg/m³.

9.54 An ideal heat engine (IHE) operates between a hot reservoir of temperature θ_H and a cold reservoir of temperature θ_L. A thermal resistance with conductance H is placed between the IHE and the cold reservoir, with $\theta_L/\theta_H = 0.5$. Find how the resulting thermal efficiency is affected by the output power, P_{out}. Also, find the maximum P_{out} possible, and the associated thermal efficiency. Compare your results with those of Guided Problem 9.8, which is similar except for the location of the thermal resistance.

9.55 Consider the slab of material below, which exchanges heat with environments to the left and the right through thin partially insulating layers. The thermal conductivities can be assumed to be constant at H_1 and H_2, respectively. The specific heat of the slab can be assumed to be constant, also.

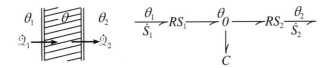

(a) Write a differential equation for the state of the model in terms of the conventional bond-graph state variable S. The temperatures θ_1 and θ_2 are specified independently.

(b) Repeat (a) substituting Q as the state variable. You may wish to represent your model with a pseudo bond graph. Which approach do you prefer?

9.56 A long piece of stainless steel of cross-sectional area A is used at cryogenic temperatures in the range 2 K to 50 K; typically, one end is much colder than the other. A simulation model consists of a series of alternating lumped thermal resistances and lumped thermal compliances for elements of length L. Choose a state variable and write a state differential equation for the ith such compliance using approximations for the thermal properties based on the data given in Fig. 9.50 (p. 700).

9.57 An assumed incompressible fluid at 120 °F flows through a valve, dropping in pressure from 3000 psi to 1000 psi. The heat generated initially goes into the fluid, but ultimately goes to the environment at 70 °F. The density of the fluid is 0.8×10^{-4} lb·s²/in⁴ and the specific heat is 0.50 BTU/lbm·°F. Find the heat generated, the temperature of the fluid immediately downstream, the entropy generation before the fluid cools, and the additional entropy generation due to the subsequent cooling.

(a) Draw a bond graph, showing the two fluid ports and one thermal port, and two irreversible elements. Label all efforts and flows.

(b) Find the heat generated. Note that 1 BTU=778.16 ft-lb and that one pound mass weighs one pound force in one standard gravity.

(c) Find the temperature rise and the temperature of the downstream fluid.

(d) Find the entropy fluxes at each of the two temperatures.

9.58 The bladder of a hydraulic accumulator is filled with an open-cell solid foam of mass m_f and specific heat c_f intended to inhibit circulation of the compressed gas, and therefore reduce heat transfer to the walls and increase the reversibility of a compression-decompression cycle. The gas satisfies the perfect gas law $PV_g = m_g R\theta_g$; the gas has mass m_g, specific heat c_{vg}, volume V_g and temperature θ_g. Heat transfer between the gas and the foam can be assumed to satisfy $Q = H(\theta_g - \theta_f)$, where H is a constant and θ_f is the temperature of the foam. The volume of the solid material in the foam does not change, although the voids expand in size so that the foam fills the entire space available.

Draw a bond graph model for the situation of a changing volume, but with no heat transfer to the walls. Define variables and parameters as appropriate, and write a (set of) state differential equation(s).

SOLUTIONS TO GUIDED PROBLEMS

Guided Problem 9.8

1-2.

$$\eta = \frac{\mathcal{P}_{out}}{\theta_H \dot{S}_H} \qquad (1)$$

$$\theta_H \dot{S}_H = \theta'_H \dot{S} \qquad (2)$$

$$\mathcal{P}_{out} = (\theta'_H - \theta_L)\dot{S} \qquad (3)$$

$$\theta_H \dot{S}_H = H(\theta_H - \theta'_H) \qquad (4)$$

3.

eq. no.	η	$\dot{s}_H$	$\dot{S}$	θ'_H
1	x	x		
2		x	x	x
3			x	x
4		x		x

Equation (1) shows that η depends only on the unknown $\dot{S}_H$, and equation (4) allows $\dot{S}_H$ to be expressed in terms of unknown θ'_H. You start therefore by combining equations (2) and (3) to eliminate $\dot{S}$ and get a relation between $\dot{S}_H$ and θ'_H. From equation (3), $\dot{S} = \mathcal{P}/\theta_H \dot{S}_H$.

Substituted into equation (2), this gives $\dfrac{\theta_L}{\theta'_H} = 1 - \dfrac{\mathcal{P}_{out}}{\theta_H \dot{S}_H}$.

Substituting equation (1) into this result gives an equation which can be solved for the desired result, η:

$$\frac{\mathcal{P}_{out}}{\eta} = H\left(\theta_H - \frac{\theta_L}{1-\eta}\right). \text{ Therefore, if we define } p \equiv \frac{\mathcal{P}_{out}}{H\theta_H},$$

$$\eta^2 - \left(p + 1 - \frac{\theta_L}{\theta_H}\right)\eta + p = 0$$

$$\eta = \frac{1}{2}\left(p + 1 - \frac{\theta_L}{\theta_H}\right) \pm \sqrt{\frac{p^2}{4} + \frac{1}{4}\left(1 - \frac{\theta_L}{\theta_H}\right)^2 - \frac{p}{2}\left(1 + \frac{\theta_L}{\theta_H}\right)}$$

4. When $\mathcal{P}_{out} = 0$, this gives $\eta = 1 - \dfrac{\theta_L}{\theta_H}$, which you know is correct.

5. When $\theta_L/\theta_h = 0.5$,

$$\eta = \frac{1}{2}\left(p + \frac{1}{2}\right) \pm \frac{1}{4}\sqrt{4p^2 - 12p + 1}$$

The plot of this result reveals that the efficiency declines with increasing output power, at first slowly, and then more rapidly. The maximum possible output power is $\mathcal{P}_{out} = 0.0858 H\theta_H$, which occurs when the efficiency is decreased to 29.3%.

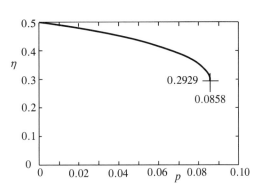

Guided Problem 9.9

1.

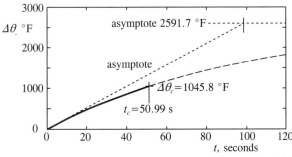

2. $\dfrac{d\mathcal{Q}_c^*}{dt} = \dot{\mathcal{Q}}_{in} - \dot{\mathcal{Q}}_{out}$, but use θ^* as the state variable: $\mathcal{Q}_c^* = mc\,\Delta\theta$.

$\dot{\mathcal{Q}}_{in} = W \sin\alpha\,\dot{x} = 100,000 \times \sin 6° \times (50 \times 88/60) = 776,542$ ft-lb/s,

$\dot{\mathcal{Q}}_{out} = H\,\theta^*$.

Therefore, $\dfrac{d\theta^*}{dt} = \dfrac{1}{mc}(W\sin\alpha\,\dot{x} - H\,\theta^*) = \dfrac{76,542}{125 \times 233.5} - \dfrac{0.38}{125 \times 0.30}\theta^*$,

or $(\tau s + 1)\theta^* = (98.68s + 1)\theta^* = 2591.7$.

3. The brake overheats when $\mu = \mu_c = \dfrac{1}{8}\mu_0$ so that $\dfrac{7}{8}\mu_0 = \mu_1(\theta_c^*)^2$

from which the critical temperature is $\theta_c^* = \sqrt{7\mu_0/8\mu_1} = 1045.8$ °F.

$\theta^* = 2591.7(1 - e^{-t/\tau})$; specifically, $2591.7(1 - e^{-t_c/\tau}) = 1045.8$ °F.

Therefore, $\dfrac{t_c}{\tau} = 0.5167$; $t_c = 50.99$ seconds.

The critical temperature never would be reached if $\dot{x} \leq \dfrac{1045.8}{2591.7} \times 50 = 20.18$ mph.

4. Above the critical temperature, the drag force produced by the brakes decreases with increasing temperature by the ratio $\dfrac{\mu_1 d}{\mu_2} = 8 - 6.4 \times 10^{-6}(\theta^*)^2$. Therefore, the differential equations become

$$\frac{d\theta^*}{dt} = \frac{1}{mc}\left\{W \sin\alpha[8 - 6.4 \times 10^{-6}(\theta^*)^2]\dot{x} - H\,\theta^*\right\}$$
$$= 0.35813[8 - 6.4 \times 10^{-6}(\theta^*)^2]\dot{x} - 0.010130\theta^*,$$

$$\frac{d\dot{x}}{dt} = \frac{W \sin\alpha}{W/g}[1 - 8 + 6.4 \times 10^{-6}(\Delta\theta)^2] = 3.3631[-7 + 6.4 \times 10^{-6}(\Delta\theta)^2].$$

If $\theta^* > \sqrt{\mu_2/\mu_1} = 1118\,°\mathrm{F}$, braking would cease altogether, so that

$\dfrac{d\dot{x}}{dt} = g \sin\alpha = 3.363\ \mathrm{ft/s^2}$. This state is never reached, however; it is approached asymptotically.

5. A simulation starting at $t = t_c$ and $\theta^* = \theta_c^*$ is given below. Most of the run-away acceleration develops within a mere 10 seconds after the critical temperature is reached.

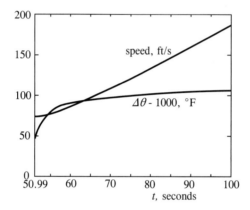

Chapter 10

Distributed-Parameter Models

All physical systems are distributed in space, so it is natural to seek models that also are distributed in space. A **distributed-parameter model** has at least one spatial independent variable, and possibly two or three, for example the Cartesian coordinates x, y and z. Dynamic models also have time as an additional independent variable. The lumped dynamic models considered exclusively thus far have only time as their one independent variable, leading to *ordinary* differential equations. The presence of two or more independent variables requires that distributed parameter models for dynamic systems be represented mathematically by *partial* differential equations.

A distributed-parameter model of a physical system is not necessarily more accurate than a lumped model of the same system. It also is not necessarily more difficult to handle than the lumped model. Distributed-parameter models should be thought of as alternatives which not infrequently produce superior results for a given degree of complexity.

Nonlinear distributed-parameter models are commonly addressed with numerical simulation, like lumped models. Rather than delve into the complexities of simulation with two independent variables, however, this book assumes linearity and taps directly into the powerful and revealing analytical consequences of the property of superposition. Certain nonlinear boundary conditions will be treated, nevertheless.

The models considered employ only one spatial variable. One-dimensional models nevertheless often can include two and three-dimensional effects. Further, complex systems often can be viewed as networks of one-dimensional distributed-parameter models joined at simple junctions. Most *phenomena* that occur in two and three dimensions also occur in one dimension, however, and are introduced here. Assemblages of one-dimensional models represent many two-and three-dimensional systems.

Section 10.1 introduces the contrast between lumped and distributed models and makes critical defintions regarding pure wave-like behavior. Graphical solutions are emphasized. Section 10.2 presents general linear representations for the class of models described as having only one power. Section 10.3 delves into the special nature of traveling waves, introducing such concepts as wave dispersion and phase, group and energy-propagation velocities. Section 10.4 provides bread-and-butter analytical procedures for dealing with one-power distributed-parameter models with known boundary conditions. The methods are extended to multi-power models in Section 10.5. Common dissipative processes are presented in Section 10.6, followed by a more general discussion of wave- scattering variables in Section 10.7, including practical simulation approaches. Section 10.8 considers internal, non-boundary excitations, and Section 10.9 introduces the concept of modal decomposition. Finally, a case study is presented in Section 10.10 that dramatizes how an exceedingly complex type of system can be addressed through the frequency domain.

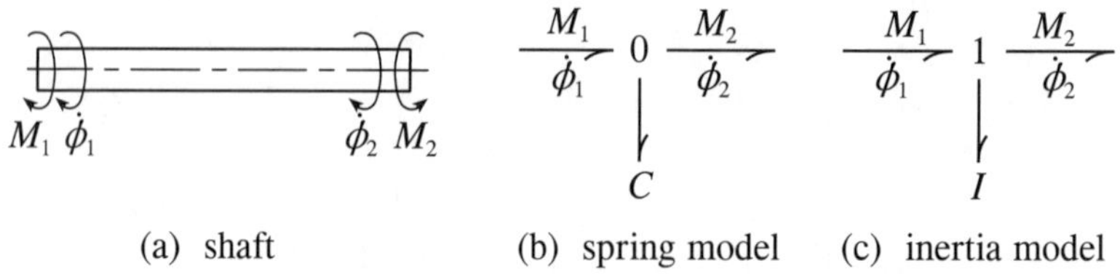

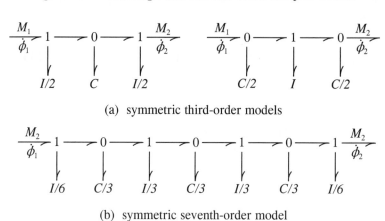

(a) shaft   (b) spring model   (c) inertia model

Figure 10.1: Rotating shaft and low-order lumped models

(a) symmetric third-order models

(b) symmetric seventh-order model

Figure 10.2: Higher-order lumped models for rotating shaft

10.1   Wave Models with Simple Boundary Conditions

The simplest and most commonly assumed distributed-parameter model is the pure bilateral-wave-delay model. Physical structures often so modeled include the rotating shaft, the rod with longitudinal motion, the acoustic tube, the stretched string with lateral vibrations, and the electic coaxial cable. The discussion below focuses on the simple wave propagation associated with this model. Attention is restricted to steady-state boundary conditions, apart from an initial excitation. Methods to treat dynamic boundary conditions are presented in Section 10.4.

10.1.1   Comparison of Lumped and Distributed Models

Consider the cylindrical shaft of Fig. 10.1 part (a), which is constrained to rotate about its axis. Perhaps one end is attached to a motor and the other end to a load. If the torque is high but the angular accelerations are low, you might use the simple lumped model given in part (b) of the figure, in which C represents the flexibility (compliance) of the entire shaft. On the other hand, if the torque is low and the angular accelerations high, the simple alternative lumped model of part (c) might be appropriate, in which I represents the inertia (inertance) of the entire shaft. For either the inertia or the compliance of the shaft to be negligible, as they are assumed to be in the two models, respectively, the frequencies of excitation must be low. For high excitation frequencies, *both* the torque and angular accelerations become large, and therefore both compliance and inertance are significant.

If both the compliance and inertance of the shaft are important, one of the alternative lumped models shown in part (a) of Fig. 10.2 might suffice. For quite high frequencies, however, waves can be seen to propagate back and forth along the shaft, and the model of part (b) might be proposed. This model has seven lumps, and is not particularly simple to handle analytically or computationally. Further, it does a rather poor job of representing the wave propagation. More lumps would improve the representation, but at the expense of additional complication.

As shown below, it is possible to characterize the shaft with a distributed-parameter model that has only two parameters, the **wave travel time** from one end of the shaft to the other, and the ratio of the angular velocity changes in these waves to the associated torque changes, called the **characteristic** or **surge admittance**. This model could be represented by a new kind of bond graph element, the **pure bilateral-wave-delay** element:

$$\frac{M_1}{\dot\phi_1} \rightharpoonup D \frac{M_2}{\dot\phi_2} \rightharpoonup$$

Analysis and simulation with this delay model can be considerably simpler than with a multi-lumped model, as may be seen intuitively and as will be demonstrated below. Further, the model is more accurate. Or is it, always?

Now consider that the shaft is extremely small and comprises a perfect single crystal with lattice planes perpendicular to the axis. Further, presume that the frequencies of excitation are exceedingly high, near the natural frequencies of the individual atoms within the lattice. The lattice vibrations in this periodic structure[1] modify the gross wave behavior markedly, changing the propagation velocity and introducing at least one absorption band. To model this behavior, you must return to a lumped model, using point masses for the atoms and lumped springs for the interatomic forces.

You have seen an example in which a lumped model is simpler than a distributed model for the same system, and a counter example in which a distributed model is simpler than a lumped model. You also have seen an example in which a distributed model is more accurate than a lumped model, and a counter example in which a lumped model is more accurate than a distributed model. The most appropriate type of model to use tends to change as the scale of the system is increased or decreased dramatically; there could be several reversals if the change is over a truly enormous range. The same observation applies to changes in the frequency of interest. There are no simple rules that dictate the choice of a lumped or a distributed model. Modeling, again, becomes partly an art, based on knowledge, experience and intuition.

Beware of the term *distributed-parameter systems*, since all physical systems are distributed over space. The term *distributed-parameter model* is clear, however. To a mathematician, nevertheless, a *system* is merely a set of equations. For example, a matrix partial differential equation often is called a distributed system.

10.1.2   The Pure Bilateral-Wave-Delay Model

The rotating shaft is represented well by a simple wave-like model, as found in Example 10.1 below. Other examples that can be so approximated are given in Examples 10.2-10.5. The common model for these cases can be represented by the bond graph of Fig. 10.3, which describes an infinitesimal segment of the structure having length dx; the independent position variable is x. If I' and C' are the inertance per unit length and the compliance per unit length, respectively, the inertance I_{dx} and the compliance C_{dx} are

$$I_{dx} = I' \, dx; \qquad C_{dx} = C' \, dx. \tag{10.1}$$

---

[1]See Leon Brillouin, *Wave Propagation in Periodic Structures*, first edition, McGraw-Hill, 1946; second edition, Dover Publications, 1953.

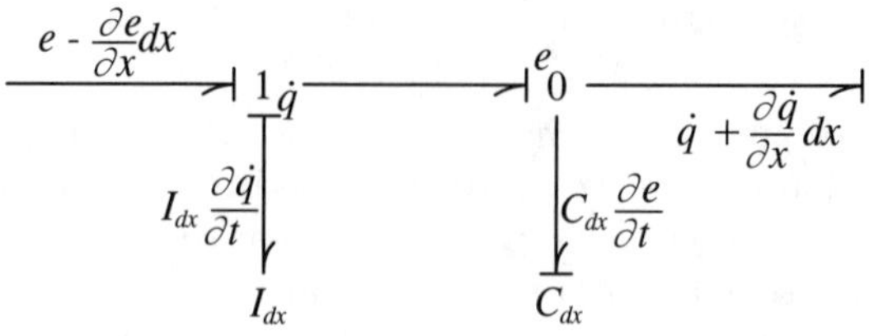

Figure 10.3: Bond-graph model for an infinitesimal segment of a simple wave-like element

Example 10.1

Identify the generic variables e and $\dot{q}$ and the parameters I' and C' for a uniform elastic rotating shaft.

Solution: For the shaft, e represents the moment, M, and $\dot{q}$ represents the angular velocity, $\dot{\phi}$, as before:

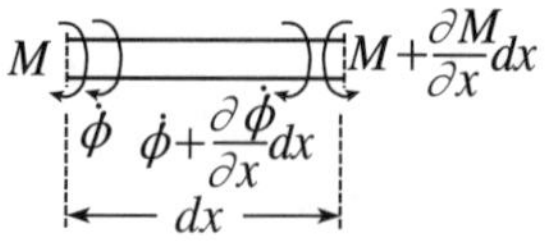

The moment due to inertia is the integral over the area of the radius times the acceleration times the density, and the moment of inertia I' is the ratio of this moment to the angular acceleration, or

$$I' = \frac{\int_0^a r(r\partial\dot{\phi}/\partial t)\rho(2\pi r\, dr)}{\partial\dot{\phi}/\partial t} = \frac{\rho\pi a^4}{2}.$$

The shear stress on a transverse element at a radius r is

$$\tau = \mu\frac{\partial u}{\partial x} = \mu r\frac{\partial\phi}{\partial x},$$

in which μ is the shear modulus. The moment that this stress produces on the shaft can be found by integrating the product of the stress and the moment arm (*i.e.*, the radius, r) over the area of the shaft, which has radius a:

$$M = \int_0^a r\tau(2\pi r\, dr) = 2\pi\mu\frac{\partial\phi}{\partial x}\int_0^a r^3 dr = \frac{\pi\mu a^4}{2}\frac{\partial\phi}{\partial x}.$$

The compliance is the ratio of the twist in the shaft to the moment, or

$$C' = \frac{\partial\phi/\partial x}{M} = \frac{2}{\pi\mu a^4}.$$

Example 10.2

Represent the approximate behavior of the longitudinal vibrations of a rod (or any slender prism) by the pure bilateral-delay model, defining the variables e and $\dot{q}$ and the parameters I and C. Comment on the accuracy of this model as compared with the torsional shaft of Example 10.1.

Solution: Longitudinal vibrations in a rod or any slender prism can be approximated in terms of the axial force, F, as the effort, and the longitudinal velocity, $v = \dot{q}$, as the flow:

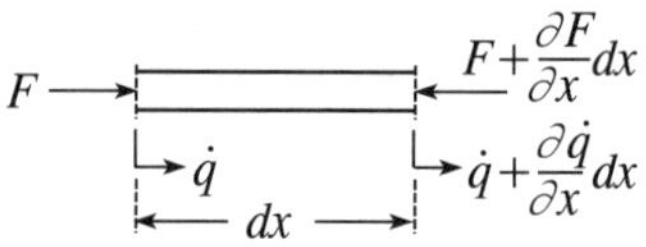

The inertance is simply the cross-sectional area, A, times the density; the compliance is the reciprocal of Young's modulus, E, times the area, or:

$$I' = \rho A; \qquad C' = 1/EA.$$

The phenomenon represented by Poisson's ratio, however, produces small transverse expansions and contractions. This implies the existence of small transverse velocities and associated kinetic energy. The gradient of the expansions also produces small stresses in the transverse direction, and consequent strains. As a result of these neglected secondary effects, the pure bilateral wave delay model is not nearly as accurate as it is for the twist of a shaft, which has no comparable effects. Note that the slenderer the rod or prism, the more accurate the model.

Example 10.3

Represent the approximate behavior of acoustic waves in a tube by a pure bilateral-delay model, identifying the efforts and flows and the inertance and compliance. Comment on any limitations in the model.

Solution: Acoustic waves in a rigid tube are almost the same as the longitudinal waves in the rod or prism of Example 10.2, except that one usually uses the pressure perturbations as the effort and the volume flow rate as the flow:

As a result, the inertance and compliance are

$$I' = \rho/A; \qquad C' = A/\beta.$$

In place of the Young's modulus there is the bulk modulus, β. The difference is that fluid does not expand laterally, unlike the solid, and as a result the secondary effects of the solid are missing in the fluid. On the other hand, however, a real fluid has viscous effects which introduce dissipation that is missing in the pure bilateral wave model.

Example 10.4

Represent the approximate behavior of the lateral vibrations of a stretched string by a pure bilateral-delay model, identifying the effort and flow and the inertance I' and the complinace C'.

Solution: The string is presumed to be flexible enough and stretched tight enough for its tension, T, to be considered uniform and constant. The effort is the lateral force, which equals the tension times the slope $-\partial y/\partial x$, where y is the lateral deflection. This slope is presumed to be small. The flow is the lateral velocity $\partial y/\partial t$.

The product of the effort and the flow is the propagated power, as always. The inertance becomes the mass per unit length, ρA, and the compliance becomes $1/T$.

Table 10.1   Examples Modeled by the Pure Bilateral Wave Delay

	e	$\dot{q}$	I'	C'	$c = 1/\sqrt{I'C'}$	$Y_c = \sqrt{C'/I'}$
rotating shaft	M	$\dot{\phi}$	$\rho\pi a^4/2$	$2/\pi\mu a^4$	$\sqrt{\mu/\rho}$	$2/\pi a^4\sqrt{\rho\mu}$
rod, longitudinal	F	v	ρA	$1/EA$	$\sqrt{E/\rho}$	$1/A\sqrt{\rho E}$
tube, acoustic	P	Q	ρ/A	A/β	$\sqrt{\beta/\rho}$	$A/\sqrt{\rho\beta}$
string, lateral	$-T\dfrac{\partial y}{\partial x}$	$\dfrac{\partial y}{\partial t}$	ρA	$1/T$	$\sqrt{T/\rho A}$	$1/\sqrt{\rho AT}$
coaxial cable	e	i	$\dfrac{\mu}{2\pi}\ln(\dfrac{r_o}{r_i})$	$\dfrac{2\pi\epsilon}{\ln(r_o/r_i)}$	$\sqrt{1/\epsilon\mu}$	$\dfrac{2\pi}{\ln(r_o/r_i)}\sqrt{\dfrac{\epsilon}{\mu}}$

Example 10.5

Represent the approximate behavior for the propagation of electromagnetic waves along coaxial cable, which is the simplest case of a pair of conductors. Identify the effort and the flow. Either compute or find in a reference the inertance I' and the compliance C'.

Solution: The variables for planar electromagnetic waves traveling along a pair of conductors are the voltage difference and the current. The compliance is the capacitance per unit length, and the inertance is the inductance per unit length.

The capacitance and inductance are (derived in Section 12.1.9, p. 947)

$$C' = \frac{2\pi\epsilon}{\ln(r_o/r_i)}; \qquad I' = \frac{\mu}{2\pi}\ln(r_o/r_i).$$

Here ϵ and μ are the dielectric constant and magnetic permeability, respectively, for the annular material, and r_i and r_o are the radii of the inner conductor and the inner radii of the outer conductor, respectively.

The results of the five examples above are assembled in Table 10.1, plus further consequences of the models as discussed in the following subsection.

10.1.3   Analysis of the Pure Bilateral-Wave-Delay Model

It is convenient to employ differential causality, as shown in Fig. 10.3 (p. 714). Solving for e and $\dot{q}$,

$$e = e - \frac{\partial e}{\partial x}dx - I_{dx}\frac{\partial \dot{q}}{\partial t}, \tag{10.2a}$$

$$\dot{q} = \dot{q} + \frac{\partial \dot{q}}{\partial x}dx + C_{dx}\frac{\partial e}{\partial t} \tag{10.2b}$$

With the help of equation (10.1) (p. 713) these give the differential equations

$$\boxed{\begin{aligned}\frac{\partial e}{\partial x} &= -I'\frac{\partial \dot{q}}{\partial t}, \\ \frac{\partial \dot{q}}{\partial x} &= -C'\frac{\partial e}{\partial t}.\end{aligned}} \tag{10.3}$$

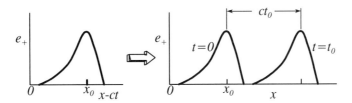

Figure 10.4: Arbitrary traveling wave

These equations can be combined to eliminate either e or $\dot{q}$. The latter case gives

$$\frac{\partial^2 e}{\partial x^2} = -I' \frac{\partial^2 \dot{q}}{\partial x \partial t} = I'C' \frac{\partial^2 e}{\partial t^2}. \tag{10.4}$$

There are two general solutions to this equation, which can be written as

$$e_+ = e_+(x - ct); \qquad e_- = e_-(x + ct). \tag{10.5}$$

The terms in parentheses represent the arguments of the functions. Imagine a pulse-like function e_+, such as illustrated in Fig. 10.4. The peak occurs when the argument $x - ct$ equals x_0. Therefore, if the function is plotted as a function of x, instead, the peak occurs when $x = x_0 + ct$. As time advances this peak moves to the right at the constant velocity c, which is known as the wave speed or **celerity**. The shape of the wave remains unchanged. The function e_- differs in that it represents a wave traveling to the left with the same wave celerity c. The complete solution is the sum of these two waves, which are transparent to one another.

To prove that these functions are indeed the solution, and to evaluate the wave celerity, you can substitute the proposed solution into the differential equation (10.4). The first and second derivatives of each function are indicated *with respect to their argument* $(x - ct)$ or $(x + ct)$ as e_+', e_+'' and e_-', e_-'', respectively. Therefore,

$$\frac{\partial^2 e}{\partial x^2} = e_+'' + e_-''; \qquad \frac{\partial^2 e}{\partial t^2} = c^2(e_+'' + e_-''). \tag{10.6}$$

With this notation the substitution gives

$$e_+'' + e_-'' = I'C' c^2 (e_+'' + e_-''), \tag{10.7}$$

which establishes the correctness of the solution and gives the wave celerity as

$$\boxed{c = 1/\sqrt{I'C'}.} \tag{10.8}$$

The consequent wave celerities of the various physical systems considered thus far are included in Table 10.1.

The bilateral waves of effort e_+ and e_- imply waves of flow, $\dot{q}$, also:

$$\dot{q} = \dot{q}_+(x - ct) + \dot{q}_-(x + ct) \tag{10.9}$$

It is necessary to understand these flows, because the end conditions usually are expressed in terms of flow as well as effort. Substitution of only the right-traveling waves into equation (10.3b) gives

$$\dot{q}_+' = C'c\,e_+'. \tag{10.10}$$

Thus $\dot{q}_+$ and e_+ are proportional to one another, with a ratio called the **characteristic admittance**, Y_c:

$$Y_c \equiv \frac{\dot{q}_+}{e_+} = \frac{\dot{q}_+'}{e_+'} = C'c = \sqrt{\frac{C'}{I'}}. \tag{10.11}$$

Substitution of the left-traveling waves into equation (10.3b) gives

$$\dot{q}_-' = -C'c\,e_-', \tag{10.12}$$

so that

$$\frac{\dot{q}_-}{e_-} = -Y_c. \tag{10.13}$$

The difference between the left and right-traveling waves, therefore, is the minus sign. This follows directly from the fact that $\dot{q}$ is defined as positive from left to right.

The characteristic admittances for the specific cases considered in Fig. 10.3 are entered into Table 10.1.

The reflections of waves at boundaries can be determined from the following relations, which result directly from the equations above:

$$e = e_+ + e_-,$$
$$\dot{q} = Y_c(e_+ - e_-). \tag{10.14}$$

The inverse of these relations also is useful:

$$e_+ = \frac{1}{2}(e + \dot{q}/Y_c),$$
$$e_- = \frac{1}{2}(e - \dot{q}/Y_c). \tag{10.15}$$

10.1.4   Fixed and Free Boundary Conditions

A procedure for treating fixed and free boundary conditions is now developed, starting with the case study of a musical instrument with a stretched string that is clamped (fixed) at each end. As shown in Fig. 10.5, the string is plucked at a location one-quarter of the distance from one end to the other. At the instant of release, the lateral velocity, $\dot{q}$, is zero for all x. Equations (10.15) require $e_+ = e_- = e/2$ everywhere at this instant. Recall that e is the lateral force, which equals the tension T times the slope of the string, $\partial q/\partial x$. Therefore, $\partial q_+/\partial x = \partial q_-/\partial x = (\partial q/\partial x)/2$. The displacements q, q_+ and q_- equal the integrals of these slopes over x, so you can conclude that, at the first instant at every location, $q_+ = q_- = q/2$. The bilateral waves are shown by the dashed line in part (a) of the figure.

As the waves q_+ and q_- propagate, reflected waves are continuously generated at the ends of the string in such a manner that the end conditions are satisfied. At each end $\dot{q} = 0$ and $q = 0$, which from equations (10.5) require $e_+ = e_-$, which in turn requires $q_+ = q_-$. (Note that if q_+ is the wave incident on the right end, q_- is the reflected wave there; if q_- is the wave incident on the left end, q_+ is the reflected wave there.) The consequent wave pattern and total string position is shown in parts (b) - (i) of the figure for the times $t = T/4$, $T/2$, $3T/4$, T, $5T/4$, $3T/2$, $7T/4$ and $2T$. It is helpful to include the virtual wave shapes beyond the ends of the string in order to anticipate the pattern of the reflected waves which satisfy the boundary condition $q = 0$. The position of the string at $t = 2T$ is identical to its position at $t = 0$, indicating a cycle time of $2T$.

Free boundary conditions can be treated in the same manner as fixed boundary conditions, except it is e, rather than q or $\dot{q}$, that is zero. Thus, for example, at each end $e_+ = -e_- = \dot{q}/2Y_c$. This means that the reflected wave still has the same amplitude as

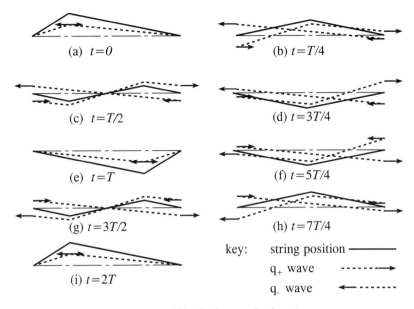

Figure 10.5: Plucked stretched string

the incident wave, but its sign is inverted. The difference is illustrated by a compression wave in a tube approaching either a blocked or an open end. The blocked end produces a reflected compression wave, whereas the open end produces a reflected rarefaction wave.

10.1.5 Fourier Analysis with Fixed or Free Boundary Conditions

Functions of *time*, t, are represented by a Fourier sum of sine and cosine waves in Section 7.4.1. With distributed parameter models there is a second independent variable, *position*, x; often it is useful also to represent a function of x by a Fourier series.

Continuing with the case study of the stretched string, with its fixed ends, Fig. 10.6 shows that the components of displacement $q(x)$ can be represented by the sine series

$$q(x) = \sum_{n=1}^{\infty} q_n \sin(n\pi x/L). \tag{10.16}$$

This is similar to the sine series for an odd periodic function of time, except that the interval is $0 < x < L$ rather than $-T/2 < t < T/2$. As a result, the lowest or fundamental harmonic has *one-half* a wavelength over the interval, rather than a whole wavelength of the periodic function of time. This is acceptable because, unlike before, the function does not have to repeat for intervals outside of its length, L.

Values of q_n can be found to match any given continuous function $q(x)$ that satisfies $q(0) = q(L) = 0$. Multiplying both sides of equation (10.16) by $\sin(m\pi x/L)$ and integrating over the interval $0 < x < L$, in standard Fourier fashion,

$$\int_0^L q(x) \sin\left(\frac{m\pi x}{L}\right) = \sum_{n=1}^{\infty} q_n \int_0^L \sin\left(\frac{n\pi x}{L}\right) \sin\left(\frac{m\pi x}{L}\right) dx \quad m = 1, 2, \ldots. \tag{10.17}$$

As before, the integral on the right side is non-zero only when $m \neq n$, and then it equals $Lq_n/2$. Therefore,

$$q_n = \frac{2}{L} \int_0^L q(x) \sin(n\pi x/L) \, dx. \tag{10.18}$$

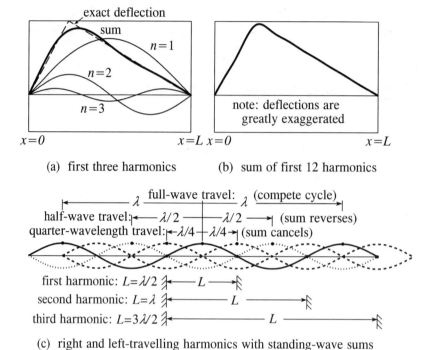

(a) first three harmonics   (b) sum of first 12 harmonics

(c) right and left-travelling harmonics with standing-wave sums

Figure 10.6: Fourier sine series decomposition of initial string displacement

For the initial position of the string as shown in Fig. 10.6,

$$q(x) = 4q_0 x/L; \qquad 0 < x < L/4,$$
$$= 4q_0(L - x)/3L; \quad L/4 < x < L. \qquad (10.19)$$

Substitution of this function into equation (10.18) gives, after cancelation of several terms,

$$\frac{q_n}{q_0} = \frac{32}{3(n\pi)^2} \sin\left(\frac{n\pi}{4}\right). \qquad (10.20)$$

The respective values of the first 12 harmonics are 0.764, 0.270, 0.085, 0, -0.031, -0.030, -0.016, 0, 0.009, 0.011, 0.006, 0. Part (a) of the figure shows the first three harmonics and their sum. Part (b) shows the sum of the first twelve harmonics, which is a reasonable approximation to the complete straight-line V shape.

Each harmonic produces a pair of bilateral propagating sinusoidal waves of one half the amplitude of the harmonic itself. These waves all propagate at the constant wave celerity, and the sum of each pair satisfies the end conditions. Part (c) of the figure is intended to show how the propagation of the component pairs of traveling waves sum to produce standing waves with the periods $T_n = 2L/cn$. The sinusoid depicted with a solid line represents the left and right-traveling waves of some harmonic at a moment when they are superimposed. The dashed lines represent the waves after they have propagated a distance of one quarter of a wavelength. They sum to zero, so the standing wave has advanced by one quarter of a cycle. The dotted lines represent the traveling waves after they have propagated a distance of one half a wavelength. They sum to exactly the negative of the original standing wave, and therefore the standing wave has advanced by one half cycle. The cycle is completed when the left- and right-traveling waves have moved a whole wavelength.

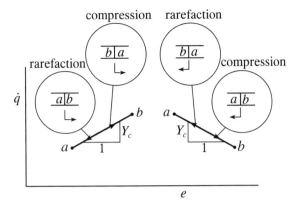

Figure 10.7: Waves in the hodograph plane

The wavelength of the nth harmonic of the stretched string is $2L/n$, and the wave speed is c; therefore, its frequency is $1/T_n = nc/2L$.

Musicians call the second harmonic[2] the first **overtone**; a frequency doubling is precisely one octave. The frequency of the third harmonic, by the same reasoning, is triple that of the first, and so forth. For example, if the fundamental frequency is 256 Hz (middle C), the first overtone is at 512 Hz (the next C higher), and the second overtone is at 1024 Hz (G above that). (If you raise the dampers for this higher C and G on a piano, by holding down their keys, and then strike middle C, these higher strings will resonate. Their reverberation is audible if the middle C is quickly damped. This phenomenon does not occur for non-harmonic notes.)

The quality of a musical tone is affected by its pattern of overtones, plus any randomness that may occur due to phenomena such as turbulence-induced vibrations of an air column. Different instruments produce different patterns. The plucked string itself is affected by the location of the plucking; were it to be plucked in the center, the second harmonic would be noticeably absent, which may be considered undesirable. This affects the way guitars and harps are played.

10.1.6   The Hodograph Plane

Sometimes the terminations at either end of a pure bilateral delay line can be expressed as static sources or resistances, that is by effort-flow characteristics. These cases are, in general, nonlinear and more complicated than the fixed and free boundary conditions considered above. They are addressed here for the simplest kind of wave: a discontinuity between two otherwise steady states such as shown in Fig. 10.7. Such longitudinal waves in rods and fluid tubes have four types: compression to the right, compression to the left, rarefaction to the right, and rarefaction to the left. "Compression" and "rarefaction" can be generalized for other media to mean an increase or decrease, respectively, in the effort, e.

It is convenient to represent the two conditions on either side of such a wave graphically, as points on a map with the generalized flow as the ordinate and the generalized effort as the abcissas. *Straight line segments drawn between the two states separated by a simple wave discontinuity must have slopes of $\pm Y_c$, in order to satisfy equations (10.10) and (10.12).* The vertical and horizontal components of the line segment represent the rightward-traveling

---

[2]Some authors identify as the first harmonic what is called here the second harmonic, and the second harmonic what is called here the third harmonic, etc. Their use of the word "harmonic" becomes essentially the same as the present use of the word "overtone."

waves $\dot{q}_+$ and e_+, respectively, if the slope is positive. If the slope is negative, they represent the leftward-travelling waves $\dot{q}_-$ and e_-, respectively. A map using these coordinates is known as a **hodograph plane**. The arrow drawn near the middle of the line segments points toward the new state, which is sweeping away the old state as the wave propagates.

Example 10.6

An acoustic tube is closed at its left end and open on its right end:

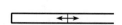

At any one time, presume there are only two states in existence, separated by a single wave that is traveling to the left or to the right, as suggested in the drawing. Show on a hodograph plane the sequence of waves that comprise a cycle.

Solution: The abscissa of the hodograph plane is the characteristic of the left end, and the ordinate is the characteristic of the right end. Start with, say, a rarefaction wave traveling to the left. This is represented in the hodograph plane below by the line segment in the first quadrant. The arrow is pointing toward the new state that is sweeping out the old state, at the left (and upper) end of the line segment.

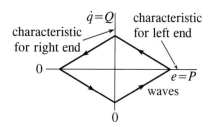

When the wave reaches the left end of the tube, it reflects; what had been the new state becomes the old, and the new state is found where the new surge characteristic intersects with the proper boundary characteristic. Since the new wave is traveling to the right, its line segment has a positive slope, and appears in the second quadrant. The new state is at the left (and lower) end of this line, toward which the arrow points. After this wave reaches the open right end of the tube, it reflects again, producing a compression wave with a new line segment (third quadrant). The wave that completes the cycle (fourth quadrant) is produced after this second leftward-traveling wave reaches the closed end of the tube. The states repeat themselves after every two full round trips of the wave. The oscillation never dies out and has a period of $4T$, where T is the time for a wave to propagate from one end to the other.

Example 10.7

An acoustic tube is blocked at both ends. It cannot contain a single wave discontinuity, since the flow is zero at both ends. It can, however, contain two symmetrical waves that move symmetrically. The pressure at the center of the tube remains constant. As suggested in the drawing below, the tube can be analyzed as two half-tubes, each of which has a zero flow at one end and a constant pressure at the other end. Show the hodograph analysis, assuming some compatible initial state, and find the period of a cycle.

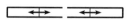

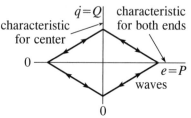

Solution: This is the same as Example 10.6, except that since the time for a wave to travel from one end of the half-tube to the other is $T/2$, the period is $2T$. The hodograph analysis is as shown:

Example 10.8

In a more interesting case, a hydraulic tube of length $L = 125$ in. has constant pressure of 2000 psi maintained at its right end by a large accumulator (tank with compressed gas inside). The left end also has an accumulator, with a pressure of 3000 psi. An adjustable valve is located immediately to the right of this accumulator, however, which when open gives a pressure drop proportional to the square of the flow and equal to 1000 psi when the flow rate is 20 in^3/s. The hydraulic fluid has a density of $\rho = 0.8 \times 10^{-4}$ lb s/in^4 and a bulk modulus of $\beta = 200,000$ lb/in^2. Two different cross-sectional areas of the tube are to be considered: $A = 0.04$ in^2 and $A = 0.10$ in^2.

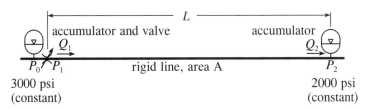

Find the pressures P_1 and the P_2 and flows Q_1 and Q_2 at the two ends of the tube as functions of time, following an initial quiescent state with the valve closed and a subsequent instantaneous opening.

Solution: The first step is to draw the pressure-flow characteristics for the two ends of the tube in the hodograph plane (Q vs. P). The right-end characteristic is a vertical line at 2000 psi. Initially, the valve at the left end is closed, so that there is no flow and the pressure everywhere in the tube is 2000 psi. After the valve is opened, the pressure-flow characteristic at the left end of the tube (just to the right side of the valve) is as plotted

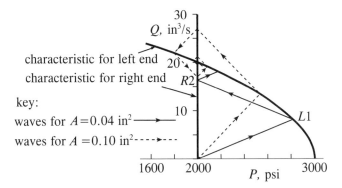

The pressure drop across the valve equals $3000 - P_1$, and the flow is proportional to the square root of this pressure drop, producing the parabolic-shaped characteristic. The surge admittances for the smaller and the larger tube areas are, respectively, $Y_s = A/\sqrt{\beta\rho} = 0.0100$ and 0.0250 $\text{in}^5/\text{lb·s}$.

A wave starts propagating from the left end of the tube toward the right the moment the valve opens. The line segment in the hodograph plane that connects the two states on either side of the wave therefore must have the slope Y_s. Further, the state in the process of being swept out has zero flow and 2000 psi, and the new state must be on the characteristic for the left boundary. This defines the new state, which is labeled $L1$. When the wave reaches the right side and reflects back, the original state is gone and the new state on the right side, labeled $R2$, lies at the intersection of the characteristic for the right side and the surge characteristic with slope $-Y_s$ that passes through state $L1$. The process continues until equilibrium is reached. The flows at the two ends are plotted below:

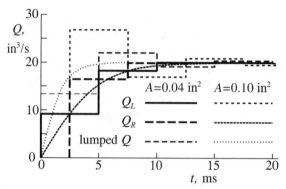

Note that the case with the smaller area approaches the equilibrium flow quickly, and without overshooting. The larger area produces an overshoot and an oscillation, although the oscillation decays fairly rapidly. Sometimes one wishes to design such a system to minimize oscillations.

It is instructive to compare this solution to the one that results if the compressibility of the fluid is neglected, leaving a lumped model with the same boundary characteristics at each end but just a fluid inertia in between. For the line area of 0.04 in^2,

$$P_1 = P_2 + \frac{\rho L}{A}\frac{dQ}{dt} = 2000 + 0.25\frac{dQ}{dt},$$

$$Q_1 = 20\sqrt{\frac{3000 - P_1}{1000}}.$$

Combining these equations leads to

$$\frac{dQ}{dt} = 4000 - 10Q^2,$$

which has the solution

$$Q = 20\tanh(200\,t).$$

The line area of 0.10 in^2 changes the coefficient 200 s^{-1} to 500 s^{-1}. Both solutions are also plotted above. It can be seen that the more precise wave solutions zig-zag about the lumped solutions, a typical occurence.

10.1.7   Summary

Spatially distributed phenomena are the stuff of elasticity, plasticity, fluid mechanics, heat and mass transfer and electromagnetics. Most conventional methods and associated computer packages are restricted to static or steady-state problems. They often employ finite element or finite difference techniques. In the analysis of complex engineering systems these tools can be used to aid in field lumping, that is the approximation of a phenomenon distributed over a spatial field by discrete or lumped models. This text employs lumped models exclusively, except for the present chapter. The more difficult or less obvious field lumping situations have been avoided (but see Section 12.1). Sometimes you may find it more practical to engage a specialist to carry out needed field lumping, rather than to become a universal expert yourself.

The dynamic interaction of inertance, compliance and resistance fields can be particularly troublesome. The behavior at low frequencies usually can be deduced by lumping these fields separately, even when they are coextensive. For high frequencies, however, partial differential equation models may be more practical. This subject has been introduced via a special class of systems: one-dimensional models exhibiting bilateral waves of constant wave speed. Examples include models of the rotating shaft, the translating rod, the acoustic behavior of fluid in tubes, the vibrating string and the electrical transmission line. These are characterized by two constant parameters: the wave speed or celerity and the surge impedance, which is the ratio of the effort of a wave to its flow.

One approach is to plot the effort or the displacement or flow versus position, and to decompose it into bilateral waves. The reflections of these waves are particularly easy to represent for fixed and free terminations. The waves also can be decomposed into Fourier components, resulting in an infinite set of natural frequencies. Purely sinusoidal behavior can be analyzed for unforced conservative systems and forced systems with static or dynamic boundary conditions by assuming a particular effort or flow and tracing it for one complete spatial cycle, noting that the pure wave-like medium merely advances or delays the phase by the angle $\omega L/c$.

Transient behavior has been addressed for the case in which the terminations of the wave-like system can be represented by steady-state characteristics. A hodograph plane, or plot of flow versus effort, is used. Waves are represented by line segments having a slope equal to plus or minus the characteristic admittance.

Guided Problem 10.1

Two simple bilateral wave media with the same wave celerity but different surge or characteristic admittances, Y_{si} and Y_{st}, are joined. They might be, for example, acoustic tubes of different diameter, or slender torsion shafts of different diameter. A wave approaches the interface from the side with surge admittance Y_{si}. Find the ratios of the amplitudes of the reflected and transmitted waves to the incident wave.

Suggested Steps:

1. Define the incident wave as $(e_+)_i$, the reflected wave as $(e_-)_i$ and the transmitted wave as $(e_+)_t$. Note that there is no wave $(e_-)_t$.

2. Find the effort e_i and flow $\dot{q}_i$ in the incident member as functions of the incident and reflected waves.

3. Find the effort e_t and flow $\dot{q}_t$ in the transmittance member as functions of the transmitted wave.

4. The efforts and flows of the two members must be equal at the interface. Therefore, equate the respective results of steps 2 and 3.

5. Solve the two equations from step 4 simultaneously for the reflection ratio $(e_-)_i/(e_+)_i$ and the transmission ratio $(e_+)_t/(e_+)_i$.

6. As a check, make sure that when $Y_{si} = Y_{st}$, which implies no discontinuity at all, the reflection ratio is zero and the transmission ratio is unity.

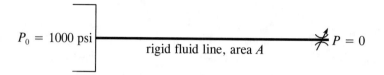

Figure 10.8: System for Guided Problem 10.2

Guided Problem 10.2

A rigid hydraulic tube connects a hydraulic chamber with a constant pressure of $P_0 = 1000$ psi to a valve with Bernoulli-type characteristics which is connected to a virtually zero-pressure region, as shown in Fig. 10.8. The valve is opened abruptly. Choose the area of the tube such that the duration of the transients of the pressure and the flow is minimized. The density of the fluid is 0.8×10^{-4} lb·s²/in⁴, its bulk modulus is 200,000 psi, and the equilibrium desired flow is $Q_0 = 30$ in³/s.

Suggested Steps:

1. Sketch the pressure-flow characteristics of the terminations of the fluid line in the hodograph plane. Numbers aren't needed at this point; it is sufficient to scale the plot using the symbols P_0 and Q_0. Note the final equilibrium state.

2. Sketch the initial wave on the hodograph plane, choosing an arbitrary surge impedance (inverse slope of the line segment) and taking care to get its proper direction of propagation. Then continue with the second or reflected wave.

3. It is possible for the surge impedance to be selected such that the second wave leaves the entire tube at the desired final flow, Q_0. Show graphically how this is done.

4. Evaluate the pressure and flow behind the initial wave in terms of P_0 and Q_0. Find the surge impedance in terms of these parameters.

5. Relate the surge impedance to the density, bulk modulus and tube area, and solve for the area as desired.

PROBLEMS

10.1 The strings for the lowest register on a piano are wrapped with heavy wire in order to increase their effective mass per unit length without increasing the tension force. Determine the effect on the wave speed and resonant frequency of quadrupling the effective mass in this manner. (What musical interval does this represent?)

10.2 One of the most common standard steel hydraulic tubing has an outer diameter of 0.5 inch and a wall thickness of 0.049 inch. Compute the wave speed and the characteristic admittance if this tubing is used to propagate power (a) with a hydraulic oil having a bulk modulus of 200,000 psi and a specific weight density of 50 lb/ft³, (b) with air at 100 psig and 70°F, (c) as a push/pull rod and (d) as a rotating shaft. The Young's and shear moduli for steel are 30×10^6 psi and 11×10^6 psi, respectively, and its weight density is 0.283 lb/in³. The effect of expansion of the tube caused by internal pressure may be neglected.

10.3 A virtually incompressible fluid of density ρ passes slowly through a tube with thin, radially compliant walls with mean radius r, thickness t, Young's modulus E and a Poisson's ratio of essentially zero. Determine the compliance and inertance per unit length, the wave speed and the characteristic admittance.

10.4 The properties of coaxial cables most commonly specified are the capacitance per unit length and the "impedance," which means surge or characteristic impedance, the reciprocal of the characteristic admittance. A particular typical cable is listed as having 67.2 pf/m (or $\mu\mu$ f/m) and 75 Ω. Determine the inductance per unit length and the wave speed; represent the latter as a fraction of the speed of light (3.00×10^8 m/s).

10.5 Placing thumb tacks on the hammers of a piano produces a "honky-tonk" sound. Explain the phenomenon qualitatively in terms of the harmonics or overtones.

10.6 A string in a musical instrument is plucked at its midpoint.

(a) Sketch the positions of the string for every one quarter of a wave-travel time, similar to the example of Fig. 10.5 (p. 719).

(b) Find the harmonic content of the resulting motion. Do you expect this is more or less desirable than the result given in the text for plucking the string at the location one quarter of the distance from one support to the other? Why?

10.7 A rigid tube through which fluid waves propagate bifurcates at a Y junction into two tubes with the same properties. Determine the reflection and transmission of a simple wave discontinuity which is incident on this junction. Hint: Note the similarity to Guided Problem 10.1.

10.8 A rod 100 in. long and 0.375 in. in diameter is suspended horizontally from above by two wires so that it is free to move in every direction except the vertical. The rod, which is observed to weigh 3.13 lbs, is then struck on one end by a hammer. Subsequent vibrations, monitored by a small accelerometer mounted on the other end, reveal harmonics at 1010 Hz and multiples thereof. Determine the Young's modulus of the material.

10.9 After the system described and analyzed in Example 10.8 (pp. 723-724) has reached its equilibrium state at 20 in^3/s, the valve opening is changed suddenly. Find the resulting waves, and plot the flows as a function of time.

(a) Assume that the valve is shut completely.

(b) Assume that the valve is closed partly to give an equilibrium flow of 5 in^3/s and then is shut completely.

(c) Generalize from the results the qualitative effect of slow as opposed to rapid closure of the valve.

SOLUTIONS TO GUIDED PROBLEMS

Guided Problem 10.1

1.

   waves:
   $$\xrightleftharpoons[(e_-)_i]{(e_+)_i} \Bigg| \xrightarrow{(e_+)_t}$$

   interface

2. From equation (10.14), $e_i = (e_+)_i + (e_-)_i$
   $$\dot{q}_i = Y_{ci}[(e_+)_i - (e_-)_i]$$

3. Similarly, $e_t = (e_+)_t$
   $$\dot{q}_t = Y_{ct}(e_+)_t$$

4-5. $(e_+)_t = (e_+)_i + (e_-)_i$
   $$Y_{ct}(e_+)_t = Y_{ct}[(e_+)_i + (e_-)_i] = Y_{ci}[(e_+)_i - (e_-)_i]$$

   Therefore, $(Y_{ci} + Y_{ct})(e_-)_i = (Y_{ci} - Y_{ct})(e_+)_i$

   which gives the reflection ratio $\dfrac{(e_-)_i}{(e_+)_i} = \dfrac{Y_{ci} - Y_{ct}}{Y_{ci} + Y_{ct}}$

   and the transmission ratio $\dfrac{(e_+)_t}{(e_+)_i} = 1 + \dfrac{Y_{ci} - Y_{ct}}{Y_{ci} + Y_{ct}} = \dfrac{2Y_{ci}}{Y_{ci} + Y_{ct}}$

6. If $Y_{ct} = Y_{ci}$, the above results give $\dfrac{(e_-)_i}{(e_+)_i} = 0$ and $\dfrac{(e_+)_t}{(e_+)_i} = 1$, as expected.

Guided Problem 10.2

1-2. neglecting wall shear:

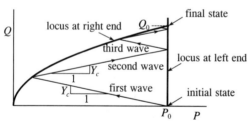

3.

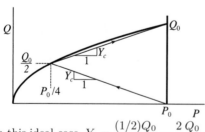

In this ideal case, $Y_c = \dfrac{(1/2)Q_0}{(3/4)P_0} = \dfrac{2}{3}\dfrac{Q_0}{P_0} = \dfrac{2}{3}\dfrac{30}{1000} = 0.020$ in^5/lb·s

but $Y_C = A/\sqrt{\rho\beta}$, so $A = Y_c\sqrt{\rho\beta} = \dfrac{50}{\sqrt{0.8 \times 10^{-4} \times 200,000}} = 12.5$ in^2

$$\frac{\text{lb·s}}{\text{in}^5}\sqrt{\frac{\text{in}^4}{\text{lb·s}^2}\cdot\frac{\text{in}^2}{\text{lb}}}$$

10.2   One-Dimensional Models

The pure bilateral-wave-delay model is a very special one-dimensional model. General linear one-dimensional models are represented mathematically in this section. Subcategories are described by the power number and whether local or global symmetry or anti-symmetry applies. Analysis of these models is deferred to later sections.

10.2.1   General Formulation

The operator $S \equiv \partial/\partial t$ is employed to represent differentiation with respect to time, t, as before. The model is represented in general with the following matrix equation which is first-order in the spatial variable x:

$$\boxed{\frac{\partial \mathbf{y}}{\partial x} = -\mathbf{A}(x, S)\mathbf{y} + \mathbf{B}(x)\mathbf{u}(x, t).} \tag{10.22}$$

The state variables are represented by the vector $\mathbf{y} = \mathbf{y}(x,t)$. When the square matrices $\mathbf{A}$ and $\mathbf{B}$ are not functions of x, the model is said to be **uniform** in x.

   The example of a uniform rod in torsion illustrates this form. If the state variables are chosen to be the torque or moment, M, and the angular velocity, $\dot{\phi}$, and no excitation exists for the region of x under consideration (this does not include the ends of the rod), equation (10.22) becomes either

$$\frac{\partial}{\partial x}\begin{bmatrix} M \\ \dot{\phi} \end{bmatrix} = -\begin{bmatrix} 0 & IS \\ CS & 0 \end{bmatrix}\begin{bmatrix} M \\ \dot{\phi} \end{bmatrix}, \tag{10.23a}$$

or equivalently

$$\frac{\partial}{\partial x}\begin{bmatrix} M \\ \phi \end{bmatrix} = -\begin{bmatrix} 0 & IS^2 \\ C & 0 \end{bmatrix}\begin{bmatrix} M \\ \phi \end{bmatrix}. \tag{10.23b}$$

The rotational mass moment of inertia per unit length is I, and the rotational compliance per unit length is C. Without the operator notation the equations become

$$\frac{\partial M}{\partial x} = -I\frac{\partial \dot{\phi}}{\partial t}, \tag{10.24a}$$

$$\frac{\partial \dot{\phi}}{\partial x} = -C\frac{\partial M}{\partial t}, \tag{10.24b}$$

or equivalently

$$\frac{\partial M}{\partial x} = -I\frac{\partial^2 \phi}{\partial t^2}, \tag{10.25a}$$

$$\frac{\partial \phi}{\partial x} = -CM. \tag{10.25b}$$

If the Laplace transform is taken of these equations *with respect to the independent variable* t, there results the respective forms

$$\frac{d}{dx}\begin{bmatrix} M \\ \dot{\phi} \end{bmatrix} = -\begin{bmatrix} 0 & Is \\ Cs & 0 \end{bmatrix}\begin{bmatrix} M \\ \dot{\phi} \end{bmatrix} + \begin{bmatrix} dM(x,0)/dx \\ d\dot{\phi}(x,0)/dx \end{bmatrix}, \tag{10.26a}$$

$$\frac{d}{dx}\begin{bmatrix} M \\ \phi \end{bmatrix} = -\begin{bmatrix} 0 & Is^2 \\ C & 0 \end{bmatrix}\begin{bmatrix} M \\ \phi \end{bmatrix} + \begin{bmatrix} dM(x,0)/dx \\ d\phi(x,0)/dx \end{bmatrix}. \tag{10.26b}$$

The derivative operator S can be seen to correspond to the Laplace notation and variable s except for the initial-condition terms, which usually are not of interest. Both notations serve the useful purpose of converting partial differential equations into ordinary differential equations.

An alternative approach to equation (10.22) employs a Heaviside operator $\mathcal{D} = \partial/\partial x$, or the corresponding Laplace operator:

$$\frac{\partial \mathbf{y}}{\partial t} = -\mathbf{A}(x, \mathcal{D})\mathbf{y} + \mathbf{B}(x)\mathbf{u}(x, t). \tag{10.27}$$

The particular example of the torsion rod becomes, for example,

$$\frac{\partial}{\partial t} \begin{bmatrix} M \\ \dot{\phi} \end{bmatrix} = - \begin{bmatrix} 0 & \mathcal{D}/C \\ \mathcal{D}/I & 0 \end{bmatrix} \begin{bmatrix} M \\ \dot{\phi} \end{bmatrix}. \tag{10.28}$$

There is no clear preference between the alternative forms in this particular case. Both are used in the literature. The first form is used exclusively in the remainder of this book, however, because it allows functions $\mathbf{A}(S)$ which are *irrational* (employing square roots, for example), very usefully expanding the domain of the models and permitting important three-dimensional effects to be included.

The pair of equations (10.24a) and (10.24b) or (10.25a) and (10.25b) can be combined to give

$$\frac{\partial^2 M}{\partial x^2} = ICS^2 M = IC\frac{\partial^2 M}{\partial t^2}. \tag{10.29}$$

Since this form of the model describes only M and not ϕ or $\dot{\phi}$ it can be called **incomplete**. Completeness, of course, has nothing to do with *exactness*, which never applies to a model. Adding the equation

$$\frac{\partial^2 \phi}{\partial x^2} = ICS^2\phi = IC\frac{\partial^2 \phi}{\partial t^2} \tag{10.30}$$

does not confer completeness, either, since the relation between M and ϕ is left undescribed. For general purposes, then, the matrix models are preferred.

10.2.2   One-Power Models

The torsion rod is an example of a **one-power model**; a transverse cut at any location defines a single effort and its conjugate flow, the product of which equals the instantaneous power. The connection between the two sides of the cut can be represented by a single bond. Whenever practical, the effort is defined as *symmetric* with respect to the cut, while the flow and the product of the two are *anti-symmetric*, oriented in either the plus-x or the minus-x direction. This is precisely as described in Chapter 2. The state vector comprises the effort on top and the flow below, or in general the symmetric variable y_s on top and the anti-symmetric variable y_a below:

$$\mathbf{y} = \begin{bmatrix} y_s \\ y_a \end{bmatrix}. \tag{10.31}$$

The general 1-power matrix $\mathbf{A}$ is partitioned with the following notation:

$$\mathbf{A} = \begin{bmatrix} W & Z \\ Y & X \end{bmatrix}. \tag{10.32}$$

In the usual case of effort and flow variables, the dimensions of Z are those of the gradient of effort divided by flow; Z is called the **series impedance per unit length**. The dimensions of Y are those of the gradient of flow divided by effort; Y is similarly called the **shunt admittance per unit length**. The symbols W and X, which experience much less interest and use for a reason you will see shortly, are chosen here simply because of their juxtaposition in the alphabet to Y and Z.

Occasionally one assumes that the effort or the flow of a one-power model is itself a constant, leaving only a single state variable. This is called a **degenerate one-power model**. Only W or X then exists.

10.2.3 Symmetric One-Power Models

A **locally symmetric** model has the same differential equation if written in the minus-x direction as it does if written in the plus-x direction. This is a consequence of the symmetry of a slice of the system of infinitesimal length dx. The symmetric variable appears the same regardless of the $+x$ or $-x$ perspective, but the anti-symmetric variable has its sign changed. Therefore, the differential equations in the two perspectives become, respectively,

$$\frac{\partial}{\partial x} \begin{bmatrix} y_s \\ y_a \end{bmatrix} = - \begin{bmatrix} W & Z \\ Y & X \end{bmatrix} \begin{bmatrix} y_s \\ y_a \end{bmatrix}, \tag{10.33a}$$

$$\frac{\partial}{\partial (-x)} \begin{bmatrix} y_s \\ (-y_a) \end{bmatrix} = - \begin{bmatrix} -W & Z \\ Y & -X \end{bmatrix} \begin{bmatrix} y_s \\ (-y_a) \end{bmatrix}. \tag{10.33b}$$

Local symmetry requires the two square matrices to be equal. Thus, the model exhibits local symmetry if and only if

$$W = X = 0. \tag{10.34}$$

A locally symmetric system is **globally symmetric** if it is locally symmetric and is uniform in x. Note that the uniform rod in torsion satisfies this condition, but a non-uniform rod usually does not, despite its local symmetry.

Since all degenerate one-power models include only W or X, they are necessarily anti-symmetric.

10.2.4 Multiple-Power Models

A transverse cut through a non-degenerate multiple-power model reveals two or more pairs of effort and flow variables. Each pair can be represented by its own bond; there is a bundle of parallel bonds. Each bond represents one of the powers. If one of the powers is not coupled to the others, it can be separated therefrom, reducing the complexity of the remaining model. It will be assumed that such separations have been made.

The state variables need not be effort and flow variables, as noted above; each conjugate pair of variables nevertheless includes one that is symmetric with respect to the transverse cut and one that is anti-symmetric. All the symmetric variables are collected together into a vector, $\mathbf{y}_s$, and all the anti-symmetric variables are collected into a second vector, $\mathbf{y}_a$. To be as orderly as possible, the elements in the two vectors are placed in the same order. Thus, for example, the second asymmetric element is the conjugate state variable paired to the second symmetric element. Finally, the two vectors are assembled into the overall state vector:

$$\mathbf{y} = \begin{bmatrix} \mathbf{y}_s \\ \mathbf{y}_a \end{bmatrix}. \tag{10.35}$$

The state differential equation can be partitioned, like the one-power model above, to give square matrix elements $\mathbf{W}$, $\mathbf{X}$, $\mathbf{Y}$ and $\mathbf{Z}$:

$$\frac{\partial}{\partial x} \begin{bmatrix} \mathbf{y}_s \\ \mathbf{y}_a \end{bmatrix} = - \begin{bmatrix} \mathbf{W} & \mathbf{Z} \\ \mathbf{Y} & \mathbf{X} \end{bmatrix} \begin{bmatrix} \mathbf{y}_s \\ \mathbf{y}_a \end{bmatrix} + \mathbf{B}\mathbf{u}(x,t). \tag{10.36}$$

The identification of locally symmetric and anti-symmetric cases given above for one-power models can be generalized directly to multiple-power models, with the result that symmetry exists if and only if $\mathbf{W}$ and $\mathbf{X}$ vanish. In other words,

$$\mathbf{A} = \begin{bmatrix} \mathbf{0} & \mathbf{Z} \\ \mathbf{Y} & \mathbf{0} \end{bmatrix} \qquad (10.37)$$

represents the general symmetric model.

Solutions to the state differential equations are given in the following sections, starting with the simpler cases. The basic approach and several of the results have been published by the author.[3]

10.2.5   Summary

The distributed-parameter models addressed in this chapter are linear and employ only one spatial variable as an independent variable. Nevertheless, certain two and three-dimensional phenomena can be included if the first-order canonical form of equation (10.22) is used. This form defines a square matrix $\mathbf{A}$ in which the dynamics is represented by use of the derivative operator S or the Laplace derivative operator s.

The models can be categorized according to the number of bonds necessary to represent the power which flows across any transverse cut. The state of a non-degenerate n-power model therefore is represented by a state vector with $2n$ elements. The symmetric variables are collected in the first subvector component of the state vector, and the asymmetric variables, ordered to match, form the balance of the state vector. This partitioning defines four submatrices of $\mathbf{A}$. One advantage of this scheme is that local or global symmetry is immediately apparent, corresponding to the vanishing of the upper-left and lower-right submatrices.

Guided Problem 10.3

Find a state differential equation modeling gravity waves with wavelength considerably greater than the mean depth. Assume a horizontal bottom and wave heights considerably smaller than the mean depth.

Suggested Steps:

1. Choose state variables, one symmetric and the other asymmetric. These could be efforts and flows, but need not be. Presume that the velocity of the water is largely horizontal and uniform over the depth, which follows from the long wavelength assumption.

2. Find the continuity relation between the state variables. Drawing two vertical control surfaces dx apart can be helpful.

3. Find the momentum relation between the state variables. Drawing a small horizontal contol volume can help.

4. Assemble your results in the canonical form.

---

[3]F.T. Brown, "A Unified Approach to the Analysis of Uniform One-Dimensional Distributed Systems," *ASME Transactions J. of Basic Engineering*, v 89 n 2, pp 423- 432, June 1967.

Guided Problem 10.4

The static bending of a uniform slender beam is presented in elementary textbooks by the model $dM/dx = F$, $d^2\eta/dx^2 = (1/EI)M$ and $dF/dx = -W(x)$, where M is the bending moment, F is the shear force, η is the lateral deflection, $W(x)$ is the applied loading per unit length, E is Young's modulus and I is the area moment of inertia of the cross-section. Introduce dynamics into this model with the simplest possible assumptions, and present it in the canonical form given above. The resulting model is known as the **Bernoulli-Euler beam**. Check for the symmetry of your result. Solutions are sought in Section 10.5.

Suggested Steps:

1. Draw an element of the beam. Label the variables appropriately.

2. Power is propagated along the beam by virtue of transverse force and motion and rotational moment and motion. Identify two pairs of symmetric amd anti-symmetric state variables, and collect them into a state vector **y**.

3. Place the given differential equations into the canonical form given above.

4. The simplest possible representation of the dynamics includes the inertia for lateral velocity but overlooks the inertia for rotational velocity. Write the corresponding differential equation, place it in the canonical form and complete the matrix **A** accordingly.

5. Identify the component submatrices **W**, **X**, **Y** and **Z**. The model is known to be symmetric, so your **W** and **X** should be zero.

PROBLEMS

10.10 A coaxial cable (with inductance and capacitance) has a series resistance per unit length R'. Find the elements of the associated matrix **A**.

10.11 Repeat the above problem, adding a shunt conductance G' per unit length.

10.12 The fluid in an acoustic tube has a viscosity μ. Find the elements of the associated matrix **A** assuming a resistance consistent with laminar steady flow, as can be deduced from equation (2.14) in Section 2.3 (p. 37).

10.13 A fluid-filled tube with series inertance I' per unit length and shunt compliance C' per unit length leaks through pores in its walls; the shunt conductance is G' per unit length. Evaluate the matrix **A**.

10.14 An acoustic muffler comprises a long central tube with numerous radial holes leading to otherwise sealed chambers, as pictured below.

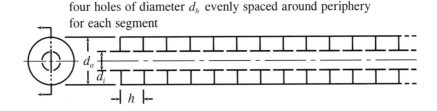

four holes of diameter d_h evenly spaced around periphery for each segment

(a) Construct a bond-graph model of one segment of this discrete system, using the pressure and the axial flow as the state variables. The orifices may be modeled as inertances, and the chambers as compliances; energy dissipation may be neglected.

(b) Evaluate the moduli of the bond graph elements of part (a). The orifices may be assumed to be through a sheet of zero thickness; see Section 12.1.8 (pp 946-947).

(c) Convert the discrete model into an approximately equivalent distributed-parameter model; give the elements of the matrix $\mathbf{A}$.

Continued in Problem 10.23.

10.15 Identify whether the models for the five preceding problems are locally or globally symmetric. (You need not solve the problems to answer this question.)

SOLUTIONS TO GUIDED PROBLEMS

Guided Problem 10.3

1. The choice here is the depth of the water, y (let Y_0 be the *mean* depth), and the horizontal velocity, v.

2.

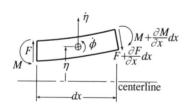

$$\frac{dy}{dt}dx = y(x)v(x) - y(x+dx)v(x+dx)$$

$$= -\frac{\partial}{\partial x}(yv)dx \simeq -y_0\frac{\partial v}{\partial x}dx - V_0\frac{\partial y}{\partial x}dx$$

The mean velocity will be assumed to be zero, so that $\dfrac{\partial y}{\partial t} = -y_0\dfrac{\partial v}{\partial x}$.

3. At the bottom:

$$P(x) \longrightarrow \qquad \longleftarrow P(x+dx)$$

If the dynamic heads are neglected (due to the zero mean velocity),

$$\frac{\partial P}{\partial x}dx = \rho g\frac{\partial y}{\partial x}dx = -(\rho\,dx)\frac{\partial v}{\partial t}.$$

4. $\dfrac{\partial}{\partial x}\begin{bmatrix} y \\ v \end{bmatrix} = -\begin{bmatrix} 0 & S/g \\ S/y_0 & 0 \end{bmatrix}\begin{bmatrix} y \\ v \end{bmatrix}$

Guided Problem 10.4

1.

2. The transverse displacement, η, is symmetric, whereas the transverse force, F, is anti-symmetric; these can form the first conjugate pair of variables since the associated power is $f\,\partial\eta/\partial x$. The moment, M, is symmetric, and the angle $\partial\eta\partial x$ is asymmetric; these can form the second pair since the associated power is $M\,\partial^2\eta/\partial x\partial t$. The resulting state vector is $[\eta,\ M,\ F,\ \partial\eta/\partial x]^T$.

3. $\dfrac{\partial}{\partial x}\begin{bmatrix} \eta \\ M \\ F \\ \partial\eta/\partial x \end{bmatrix} = -\begin{bmatrix} 0 & 0 & 0 & -1 \\ 0 & 0 & -1 & 0 \\ 0 & 0 & 0 & 0 \\ 0 & -1/EI_m & 0 & 0 \end{bmatrix}\begin{bmatrix} \eta \\ M \\ F \\ \partial\eta/\partial x \end{bmatrix} + \begin{bmatrix} 0 \\ 0 \\ -W(x) \\ 0 \end{bmatrix}$

This is valid only for the static case.

4. The net lateral force per unit length equals the mass per unit length times the lateral acceleration, or

$$\frac{\partial F}{\partial x}dx = (\rho A\,dx)\frac{\partial^2\eta}{\partial t^2} \quad \text{(omitting the effect of the external loading, } W(x)\text{)}.$$

Entering this into the equation above gives

$\mathbf{A} = \begin{bmatrix} 0 & 0 & 0 & -1 \\ 0 & 0 & -1 & 0 \\ \rho A S^2 & 0 & 0 & 0 \\ 0 & -1/EI_m & 0 & 0 \end{bmatrix}$ (This is used in Section 10.5, equation (10.119).)

5. $\mathbf{Z} = \begin{bmatrix} 0 & -1 \\ -1 & 0 \end{bmatrix}; \quad \mathbf{Y} = \begin{bmatrix} \rho A S^2 & 0 \\ 0 & -1/EI_m \end{bmatrix}; \quad \mathbf{W} = \mathbf{X} = \begin{bmatrix} 0 & 0 \\ 0 & 0 \end{bmatrix}.$

10.3 Wave Propagation

The concepts of phase and group velocities, wave number, attenuation factor and dispersion are introduced. This is done in the context of degenerate one-power models, in particular the case study of a heated incompressible fluid flowing at a constant velocity through a tube of uniform area. The waves in this case are propagated unilaterally in the downstream direction, simplifying the analysis. Frequency and transient reponses are calculated for three assumed wall conditions.

10.3.1 Simplest Model: Pure Transport

The simplest model for heated fluid flowing in a tube presumes a slug-flow velocity profile, as shown in Fig. 10.9, with negligible mixing of the fluid. Axial heat transfer is neglected, and the walls are assumed to be adiabatic. The behavior of the system is clear before any equations are written: temperature disturbances propagate, without distortion, at the velocity v of the fluid. This fact will be expressed mathematically.

The differential equation that describes the system can be written directly. This approach is inadequate to address more difficult problems, however, so a more powerful method will be used. When mass transport is involved, often it is best to employ the **substantial** or **material derivative** $\mathbf{d}/\mathbf{d}t$ which describes the behavior of a fixed quantity of mass:

$$\frac{\mathbf{d}}{\mathbf{d}t} = \frac{\partial}{\partial t} + \mathbf{v}\cdot\boldsymbol{\Delta} = \frac{\partial}{\partial t} + v\frac{\partial}{\partial x}. \tag{10.38}$$

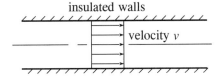

Figure 10.9: Pure transport of heated fluid in tube

The form on the far right is the scalar version for one spatial dimension. Since the temperature θ of any particle of fluid in pure transport is constant,

$$\frac{\mathrm{d}\theta}{\mathrm{d}t} = 0, \tag{10.39}$$

or, in the canonical operator form of equation (10.22) (p. 729),

$$\frac{\partial \theta}{\partial x} = -\frac{S}{v}\theta. \tag{10.40}$$

The solution of this equation can be found readily by retaining the time derivative in operational form:

$$\theta(x, t) = e^{-TS}\theta(0, t) = \theta(0, t - x/v); \tag{10.41a}$$

$$T \equiv x/v. \tag{10.41b}$$

The operator e^{-TS} is the pure delay operator, and e^{-Ts} is its Laplace counterpart. T is the delay time associated with the transport velocity. Equation (10.41) also can be generalized slightly to give

$$\theta(x_2, t) = \theta(x_1, t - T_{21}), \tag{10.42}$$

where T_{21} is the time delay required for a fluid particle to traverse the distance $x_2 - x_1$. This solution corresponds to what you should anticipate.

The response to the sinusoidal input disturbance

$$\theta(0, t) = \mathrm{Re}\,\theta_0(0)e^{j\omega t} \tag{10.43}$$

can be found by assuming a solution of the form

$$\theta(x, t) = \mathrm{Re}\,\theta_0(x)e^{j\omega t}. \tag{10.44}$$

Upon substitution into the differential equation, this frequency response becomes

$$\frac{\theta_0(x)}{\theta_0(0)} = e^{-jkx}, \tag{10.45}$$

$$k = \omega/v = 2\pi/\lambda. \tag{10.46}$$

The same result follows most directly from substituting $j\omega$ for S in equation (10.41). The symbol k is employed here in its traditional role as **wave number**, which is defined as the ratio of the frequency of a propagating sinusoid to the **phase velocity**, v_p. This also equals the ratio of 2π divided by the **wavelength**, λ. The phase velocity is defined as the velocity at which a sinusoidal wave propagates. In the present case, the phase velocity equals the transport velocity: $v_p = v$.

10.3.2   First Modification: Thermal Leakage to a Constant-Temperature Environment

Attenuation can be introduced into the above problem by allowing heat to flow through the walls to an environment that has a constant temperature over time, though it may vary as a function of x. Specifically, it will be assumed that the thermal conductivity-per-unit-length, G, is constant, and that all thermal storage in the walls or in the thin boundary layer associated with the conductivity is negligible. Thus, equation (10.39) is modified to become

$$A_f c_f \frac{\mathrm{d}\theta}{\mathrm{d}t}(x, t) = -G[\theta(x, t) - \theta_e(x)], \tag{10.47}$$

in which A_f is the area of the tube, c_f is the specific heat of the fluid and θ_e is the temperature of the environment. This equation represents the actual thermal energy transferred per unit length. Using equation (10.38), it gives

$$\frac{\partial \theta}{\partial x} = -\frac{1}{v}\frac{\partial \theta}{\partial t} - \frac{1}{\tau_f}(\theta - \theta_e); \tag{10.48a}$$

$$\tau_f \equiv A_f c_f / G, \tag{10.48b}$$

where τ_f is a time constant for heat transfer to or from the fluid.

If the temperature $\theta(0,t)$ is held constant for a while, a steady-state solution is approached that will be rather complicated if θ_e is a complicated function of x. To simplify matters it is useful to distinguish the steady-state solution $\theta_0(x)$ from an incremental unsteady solution $\theta_1(x,t)$:

$$\theta(x,t) = \theta_0(x) + \theta_1(x,t). \tag{10.49}$$

Two equations result, one that can be solved for $\theta_0(x)$,

$$\frac{\partial \theta_0}{\partial x} = -\frac{1}{\tau_f v}(\theta_0 - \theta_e), \tag{10.50}$$

and one which can be solved for $\theta_1(x,t)$:

$$\frac{\partial \theta_1}{\partial x} = -\left(\frac{S + 1/\tau_f}{v}\right)\theta_1. \tag{10.51}$$

The solution of the latter is always simple because it is independent of $\theta_e(x)$:

$$\theta_1(x,t) = e^{-TS}e^{-T/\tau_w}\theta(0,t) = e^{-T/\tau_w}\theta(0,\ t - x/v). \tag{10.52}$$

The two solutions are essentially identical, of course, if θ_e is uniform in position.

This result reveals that the sole effect of the conductance is an exponential attenuation of the temperature for increasing x. As an illustration, the response to an upward followed by a smaller downward change in the upstream temperature is plotted in Fig. 10.10.

10.3.3 Second Modification: Thermal Compliance in the Walls

Consider, finally, that the previously time-invariant environmental temperature becomes the time-varying temperature of the wall of the tube, $\theta_w(x,t)$. The conductances G and G' refer to the film coefficient associated with the thin boundary layer, which is presumed to be independent of position. The exterior of the wall is assumed to be adiabatic, or perfectly insulated. The wall itself is assigned a uniform thermal compliance $A_w c_w$ per unit length. Axial heat transfer in the wall is neglected, as emphasized in Fig. 10.11 by the probably hypothetical placement of insulating spacers. (Inclusion of axial heat transfer would require a two-power model.) These assumptions apply approximately to the hot water pipes in your residence, particularly when you turn on the faucet in the morning and wait for the water to become hot.

Equation (10.47) is modified only by the substitution of the variable wall temperature for the constant environmental temperature:

$$A_f c_f \frac{d\theta(x,t)}{dt} = -G\left[\theta(x,t) - \theta_w(x,t)\right]. \tag{10.53}$$

Equating the heat efflux from the fluid to the heat influx to the wall,

$$A_w c_w \frac{\partial \theta_w}{\partial t} = G(\theta - \theta_w), \tag{10.54}$$

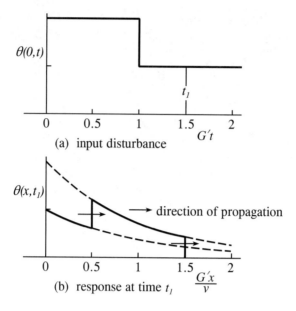

$\theta(0,t)$

0   0.5   1   1.5   2
$G't$
(a) input disturbance

t_1

$\theta(x,t_1)$

→ direction of propagation

0   0.5   1   1.5   2
$\dfrac{G'x}{v}$
(b) response at time t_1

Figure 10.10: Step responses of heated flow in tube, first modification

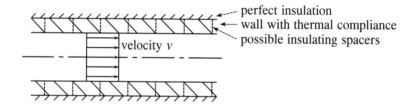

perfect insulation
wall with thermal compliance
possible insulating spacers

velocity v

Figure 10.11: Heated flow in tube, second modification

or

$$\theta_w = \frac{1}{1 + \tau_w S}\theta; \tag{10.55a}$$

$$\tau_w \equiv A_w c_w / G, \tag{10.55b}$$

where τ_w is a time constant for heat transfer to or from the wall. The combination of equations (10.53) and (10.55) with equation (10.38) can be written

$$\frac{\partial \theta}{\partial x} = -W\,\theta; \tag{10.56a}$$

$$W \equiv \frac{1}{v}\left[S + \frac{1}{\tau_f}\left(1 - \frac{1}{1 + \tau_w S}\right)\right]. \tag{10.56b}$$

Note that the use of the symbol W is consistent with its use in Section 10.2. The model is of necessity anti-symmetric.

The steady-state solution is trivial: a constant uniform temperature for everything. The solution of equation (10.56) for the incremental unsteady temperature is

$$\begin{aligned}
\theta_1(x,t) &= e^{-xW}\theta_1(0,t)\\
&= e^{-T/\tau_w}e^{-TS}e^{T/\tau_w(1+\tau_w S)}\theta_1(0,t).
\end{aligned} \tag{10.57}$$

This is rather complicated. The simplest case is the frequency response, which can be found by employing the substitution[4] $S \to j\omega$:

$$\theta_1 = \operatorname{Re} \theta_{10} e^{j\omega t}, \tag{10.58a}$$

$$\frac{\theta_{10}(x)}{\theta_{10}(0)} = e^{-W(j\omega)x} = e^{-(jk+\alpha)x}, \tag{10.58b}$$

$$k = \frac{\omega}{v}\left(1 + \frac{\tau_w/\tau_f}{1+\tau_w^2\omega^2}\right), \tag{10.58c}$$

$$\alpha = \frac{1}{\tau_f}\left(\frac{\tau_w^2\omega^2}{1+\tau_w^2\omega^2}\right). \tag{10.58d}$$

Here, k is the wave number, as before, and α is known as the **attenuation factor**. All phase information is contained in the wave number, which is traditionally plotted against frequency as shown in Fig. 10.12. For reasons that will become apparent later, equation (10.58c) is known as a **dispersion relation**. The phase velocity is precisely the slope of the chord drawn from the origin to the point of interest, or

$$v_p = \frac{\omega}{k} = \frac{\omega\lambda}{2\pi} = \frac{v}{1 + \tau_w/\tau_f(1+\tau_w^2\omega^2)}. \tag{10.59}$$

At very high frequencies this phase velocity approaches the fluid transport velocity. This should be anticipated, for at high frequencies the temperature of the wall would not fluctuate appreciably, due to its thermal compliance and the brevity of each half cycle. Thus the problem becomes virtually the same as the earlier problem with constant environmental temperature.

At very low frequencies, on the other hand, the temperatures of the wall and the adjacent fluid track each other nearly perfectly, so that the thermal resistance of the boundary layer becomes inconsequential. The phase velocity is less than the transport velocity because the wall as well as the fluid is alternately heated and cooled:

$$\underset{\omega\to 0}{L}\ \frac{v_p}{v} = \frac{1}{1+\tau_w/\tau_f}. \tag{10.60}$$

At intermediate frequencies the phase velocity changes continuously from its lower to its upper asymptotic values.

The attenuation of sinusoidal waves is represented exclusively by the attenuation factor, which is plotted in part (b) of the figure. At very high frequencies the attenuation, like the phase velocity, approaches that of the earlier problem with a constant environmental temperature. At lower frequencies the attenuation is reduced, approaching zero at zero frequency and infinite wavelength.

Transient responses are harder to deduce from equation (10.57). The first factor on the right clearly represents an attenuation that is exponential in position, and the second factor represents a pure delay corresponding to the transport velocity. The meaning of the third and final factor, called a **percolation operator**, is not immediately apparent, however. It can be evaluated by employing the one-sided time Laplace transform of the equation, presuming zero initial conditions. Using a bar over a symbol to represent its Laplace transform,

$$\overline{\theta}_1(x,s) = e^{-T/\tau_w} e^{-Ts} e^{T/\tau_w(1+\tau s)} \overline{\theta}_1(0,s). \tag{10.61}$$

---

[4]The rigor of this approach can be demonstrated, but its application can overlook the possiblilty that the response to an old transient disturbance still lingers. This problem occurs mostly in conservative systems, where transient responses do not decay.

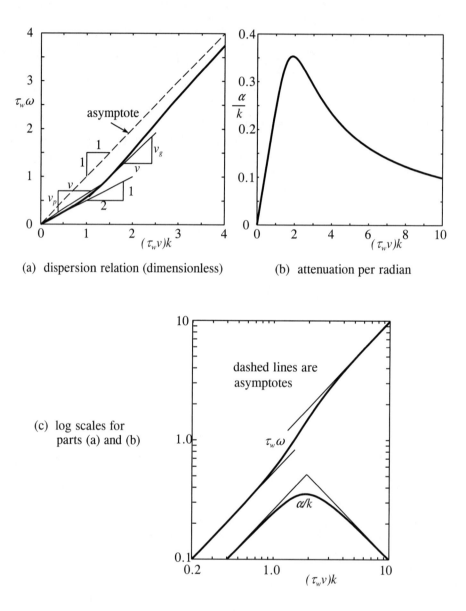

(a) dispersion relation (dimensionless)   (b) attenuation per radian

(c) log scales for
     parts (a) and (b)

Figure 10.12: Frequency behavior for heated flow in tube, second modification

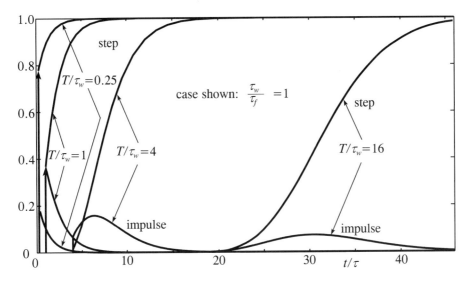

Figure 10.13: Impulse and step responses for heated flow in tube, second modification

The percolation term can be simplified by noting the following property of Laplace transforms, taken from Table 4.4 (p. 199):

$$\mathcal{L}^{-1}[f(a+s)] = e^{-at}\mathcal{L}^{-1}[f(s)]. \tag{10.62}$$

Here f is any function and a is any constant. Thus, after both sides of equation (10.61) are multiplied by e^{Ts}, the anti-transform becomes

$$\theta_1(x,\, t+x/v) = e^{-T/\tau_w} e^{-t/\tau_w} \mathcal{L}^{-1}\left[e^{T/\tau_w{}^2 s}\overline{\theta}_1(0,\, s-1/\tau_w) \right]. \tag{10.63}$$

When θ_1 is a unit impulse, $\overline{\theta}_1(0,s) = 1$, which is not a function of s at all, so $\overline{\theta}_1(0,\, s-1/\tau_w) = 1$ also. The inverse transform within equation (11.62) becomes, from transform pair #44 of Appendix C with $b = 1$,

$$\mathcal{L}^{-1}\left[e^{T/\tau_w{}^2 s} \right] = \frac{d}{dt}\left[I_0(2\sqrt{\frac{Tt}{\tau_w^2}}) \right] = \delta(t) + \sqrt{\frac{T}{\tau_w^2 t}} I_1\left(2\sqrt{\frac{Tt}{\tau_w^2}} \right), \tag{10.64}$$

in which I_0 and I_1 are modified Bessel functions. The impulse response here is stated as the time derivative of the step response. Both responses, with the other terms in equation (10.64) also accounted for, are plotted in Fig. 10.13. Note that the impulse and step terms propagate at the velocity of the fluid, but decay rapidly. The remainder of the response, which is its bulk except close to the source, propagates more slowly. This observation agrees with the earlier conclusions for the propagation of sinusoidal disturbances.

10.3.4 Dispersion and Absorption

A non-sinusoidal propagating wave is distorted by the dependencies of both the phase velocity and the attenuation factor on frequency. These two effects can be separated, at least conceptually, by first considering the hypothetical situation of the actual varying phase velocity but zero attenuation, and then later the hypothetical situation of constant phase velocity but the acutal attenuation. The first case is called **dispersive** but **non-absorptive**; the energy is propagated without loss but at a range of speeds depending on the frequency.

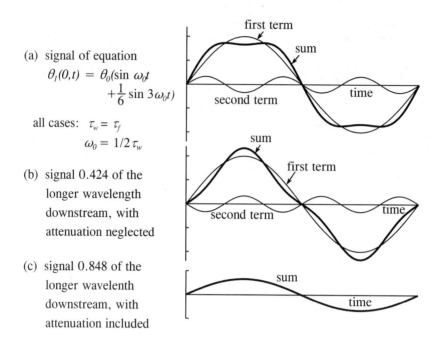

(a) signal of equation
$$\theta_1(0,t) = \theta_0(\sin \omega_0 t + \frac{1}{6}\sin 3\omega_0 t)$$

all cases: $\tau_w = \tau_f$
$\omega_0 = 1/2\,\tau_w$

(b) signal 0.424 of the longer wavelength downstream, with attenuation neglected

(c) signal 0.848 of the longer wavelenth downstream, with attenuation included

Figure 10.14: Attenuation and dispersion of heat flow in tube

To see how dispersion produces distortion, consider the following particular disturbance, which contains two sinusoidal components:

$$\theta_1(0,t) = \theta_0(\sin \omega_0 t + \frac{1}{6}\sin 3\omega_0 t). \qquad (10.65)$$

This is plotted in Fig. 10.14 part(a). The two components propagate at different speeds. At regular intervals in position the wave pattern at $x = 0$ reappears, but elsewhere the pattern is different. The pattern midway between these intervals, corresponding to a relative phase shift of 180° of the higher frequency components, is shown in part (b) of the figure. If three or more irrationally related frequency components were present, the wave profile at the origin never would be repeated.

To see how non-constant absorption produces distortion, presume that both frequency components of equation (10.64) are propagated with different attenuation factors. As shown in part (c) of the figure, the ratio of the two components changes as the wave propagates downstream; ultimately one virtually vanishes compared with the other, leaving a nearly pure sine wave. The particular case shown corresponds to the separated dispersion and attenuation of the heat transport problem above, the property τ_w/τ_f being set at unity and ω_0 being set at $1/2\tau_w$.

Any periodic signal can be represented by a Fourier series, and most aperiodic signals by a Fourier integral. Knowledge of the wave number and attenuation factor as functions of frequency then can be used to construct the response to these signals.

10.3.5 Group Velocity*

Engineers often are interested in narrow-band signals. Examples include AM and FM modulation in communication, and bursts of non-sinusoidal disturbances in various media. The velocity at which a *modulation* or a *burst* or a *group* propagates is *not* generally equal to the phase velocity for any frequency within its frequency band. Indeed, the **group**

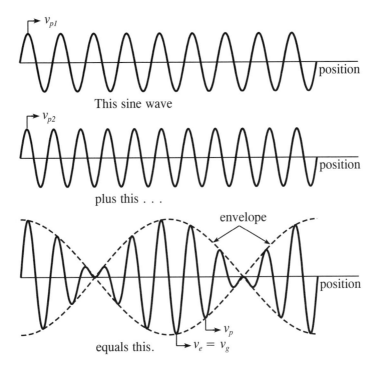

Figure 10.15: Concept of group velocity

velocity, defined as the velocity at which a disturbance propagates in the asymptotically limiting case of zero bandwidth, can be radically different from the phase velocity.

To see this, consider a wave comprising two frequency components $\omega + \Delta\omega/2$ and $\omega - \Delta\omega/2$ in which $\Delta\omega$ is very small, and the corresponding wave numbers $k + \Delta k/2$ and $k - \Delta k/2$:

$$\theta = \theta_0 \mathrm{Re} \left\{ e^{j[(\omega+\Delta\omega/2)t-(k+\Delta k/2)x]} + e^{j[(\omega-\Delta\omega/2)t-(k-\Delta k/2)x]} \right\}$$

$$= \mathrm{Re} \left[e^{j(\Delta\omega\, t/2 - \Delta k\, x/2)} + e^{-j(\Delta\omega\, t/2 - \Delta k\, x/2)} \right] e^{j(\omega t - kx)}$$

$$= 2\theta_0 \cos(\Delta\omega\, t/2 - \Delta k\, x/2) \cos(\omega t - kx). \qquad (10.66)$$

This result is plotted in Fig. 10.15. The first cosine term in equation (10.66) represents an *envelope* of the response with a long wavelength; the second cosine oscillates at the mean frequency of the two components. You may have experienced an analogous phenomenon in the "beating" of two musical tones that are at slightly different frequencies; musicians often tune their instruments to a known standard by eliminating the beating.

The two propagation velocities represented by equation (10.66) can be found by holding the respective phase angles in parentheses constant. The high- frequency wave within the envelope propagates at the phase velocity

$$\boxed{v_p = \frac{\omega}{k},} \qquad (10.67)$$

and the envelope propagates at the velocity

$$v_{env} = \frac{\Delta\omega}{\Delta k}. \qquad (10.68)$$

The group velocity v_g is defined as the limit of this velocity as $\Delta\omega \to 0$, or as the wavelength approaches infinity:

$$\boxed{v_g = \lim_{\Delta\omega \to 0} \frac{\Delta\omega}{\Delta k} = \frac{d\omega}{dk}.}\tag{10.69}$$

Returning to the example given in Fig. 10.12 (p. 740), you can see that this velocity is the local *slope* of the dispersion plot of ω as a function of k, whereas the phase velocity is the slope of the chord drawn from the origin.

Dispersion implies a difference between v_p and v_g. If $v_g > v_p$, as with the example of the heated fluid traveling down the tube, the dispersion is called **anomolous**, whereas if $v_g < v_p$ it is called **normal**. These terms originated in the field of optics where normal dispersion is the usual case; their implication is misleading in many other media and structures, including the particular example here.

Had a wave-train been chosen with a more complicated composition within the same narrow band, the envelope of the wave would have been more complicated, but the short-wavelength oscillation bounded by the envelope would remain. In the limit of vanishingly small bandwidth, this complicated envelope would travel without distortion at the same group velocity as the simpler envelope. For a somewhat wider frequency band the envelope distorts slowly as it propagates at virtually the same speed. Thus the group velocity is an important property of a distributed medium or structure. Its importance is enhanced further when, in the following section, the velocity at which *energy* propagates through conservative distributed-parameter systems is seen to equal the group velocity.

10.3.6 Summary

The use of the derivative (Heaviside) and Laplace operator notations have been illustrated for a particular degenerate one-power model. The simplest version demonstrates unilateral wave propagation without energy absorption or dispersion; the more general versions demonstrate both absorption and dispersion in their unilateral waves. Dispersion means that the velocity at which a sinusoid propagates, called the phase velocity, depends on its frequency, and the shape of non-sinusoids therefore is continuously being distorted. The wave number $k(\omega)$ of a sinusoid equals $2\pi/\lambda$, where $\lambda(\omega)$ is the wavelength; the phase velocity equals ω/k. The group velocity equals $d\omega/dk$. The group velocity is the speed at which the envelope of a narrow-band wave propagates. The propagation of complicated wave forms can be calculated either by Fourier decomposition or through the use of Laplace transforms, which are typically of non-polynomial type.

PROBLEM

10.16 Surface waves in water of density ρ and surface tension T in a narrow tank of uniform depth h can be shown to obey the dispersion relation

$$\omega^2 = \left(gk + \frac{Tk^3}{\rho} \right) \tanh kh,$$

where g is the acceleration of gravity.

(a) "Deep-water waves" are said to exist if increasing the depth has very little effect on the phase or group velocities. Determine an approximate depth necessary for this condition, as a function of the wavelength, and find the phase velocity as a function of the wavelength and the parameters. Also, for water at room temperature ($\rho = 1000$) kg/m^3, $T = 0.075$ N/m), find the wavelength and frequency below which surface tension is more important, and above which gravity is more important.

(b) The motion of "shallow water waves" penetrates to the bottom without any decrease in amplitude, unlike deep-water waves which have negligible motion at the bottom. (See Guided Problem 10.3, pp. 732, 734.) Repeat (a) for this case.

(c) Find the ratio of the group velocity to the phase velocity for four cases: shallow-water gravity waves, shallow-water capillary waves, deep-water gravity waves and deep-water capillary waves.

(d) Describe qualitatively what you would would expect if you throw a stone into a deep pond and observe the ripples. Note that the disturbance is an impulse in both space and time and therefore contains all wavelengths equally. Suggestion: It is helpful to find and plot the phase and group velocities as functions of the wavelength. Phenomena to consider also include the attenuation due to spreading and the nonlinear effects of the shorter wavelengths.

10.4 One-Power Symmetric Models

The fundamental concepts of propagation operator, characteristic impedance and causal and transmission matrices are introduced or extended in this section to apply to general one-power symmetric models. Emphasis is placed on the propagation of waves and on boundary-value problems. Applications are given to heat conduction, and to pure wave-like behavior as represented by the uniform rod in torsion. Finally, exponentially tapered systems are analyzed.

10.4.1 Wave Behavior

The general symmetric uniform one-power model with no internal excitation is, from Section 10.2.3 (p. 731),

$$\frac{\partial}{\partial x}\begin{bmatrix} y_s \\ y_a \end{bmatrix} = -\begin{bmatrix} 0 & Z(S) \\ Y(S) & 0 \end{bmatrix}\begin{bmatrix} y_s \\ y_a \end{bmatrix}. \tag{10.70}$$

Substitution of the second component equation into the first gives

$$\frac{\partial^2 y_s}{\partial x^2} = YZ\, y_s \quad \text{or} \quad \frac{\partial^2 y_s}{\partial(\sqrt{YZ}x)^2} = y_s. \tag{10.71}$$

The general solution becomes

$$y_s(x,t) = e^{\gamma x}y_{s1}(t) + e^{-\gamma x}y_{s2}(t),$$
$$y_a(x,t) = Y_c[-e^{\gamma x}y_{s1}(t) + e^{-\gamma x}y_{s2}(t)], \tag{10.72}$$

in which the **propagation operator** $\gamma(S)$ and **characteristic admittance** $Y_c(S)$, also known as the **surge admittance**, are defined as

$$\gamma(S) = \sqrt{YZ},$$
$$Y_c(S) = \sqrt{Y/Z}, \tag{10.73}$$

and y_{s1} and y_{s2} are any functions of time. Equations (10.70), (10.72) and (10.73) are employed below repeatedly. The special case of Y_c for the pure bilateral wave delay has been given in Section 10.1 (Equation (10.11), p 718). The reciprocal of the characteristic admittance, known as the **characteristic impedance** or the **surge impedance**, often is used instead.

For frequency response calculations, the substitution $S \to j\omega$ gives, as before, a wave number k and attenuation factor α:

$$\boxed{\gamma(j\omega) = jk + \alpha.}$$
(10.74)

Thus the phase velocity becomes

$$\boxed{v_p = \frac{\omega}{k} = \frac{\omega}{\operatorname{Im}\gamma(j\omega)}.}$$
(10.75)

These results also are of considerable general utility.

Example 10.9

Find the propagation operator, characteristic admittance and phase velocity for the torsion rod of Example 10.1 (p. 714). You may use the information for C' and I' given in that example. Also, observe the nature of the propagating waves from the results.

Solution: The symmetric and asymmetric variables are, respectively, $y_s = M$ and $y_a = \dot{\phi}$. From Example 10.1 or Table 10.1 (p. 716),

$$Z(S) = I'S = (\pi r^4 \rho/2)S,$$

$$Y(S) = C'S = (2/\pi r^4 \mu)S,$$

in which r is the radius of the rod, ρ is its mass density and μ is its shear modulus. With these substitutions, equations (11.73) give the propagation operator and characteristic admittance as follows:

$$\gamma = \sqrt{C'I'}S = \sqrt{\frac{\rho}{\mu}}S,$$

$$Y_c = \sqrt{\frac{C'}{I'}} = \frac{2}{\pi r^4 \sqrt{\rho\mu}}.$$

The Y_c agrees with equation (10.11) (p. 718). Equation (10.75) then shows that the phase velocity is a *constant* (consistent with equation (10.8), p. 717):

$$v_p = \frac{\omega}{\operatorname{Im}\gamma(j\omega)} = \frac{1}{\sqrt{I'C'}} = \sqrt{\frac{\mu}{\rho}}.$$

As a result, the operator $e^{-\gamma x}$ represents a *pure time delay*, and the operator $e^{\gamma x}$ represents a *pure time advance*. Waves travel at the same speed in both directions; equation (11.72) reveals that the entire behavior can be represented as the sum of such right- and left-traveling waves. The model is non-dispersive, and the group velocity equals the phase velocity. Further, Y_c is a constant, so waves of $\dot{\phi}$ propagate in phase with the corresponding waves of M.

Example 10.10

Find the propagation operator, characteristic admittance, phase velocity and attenuation factor for heat conduction. Assume one spatial dimension with no internal heat source or sink, for which the differential equations can be written

$$\frac{\partial}{\partial x}\begin{bmatrix} \theta \\ q \end{bmatrix} = -\begin{bmatrix} 0 & 1/k_\theta \\ CS & 0 \end{bmatrix}\begin{bmatrix} \theta \\ q \end{bmatrix}.$$

Note that this uses the heat flux q as the asymmetric variable y_a, rather than the flow variable which is the entropy flux, largely because it is more traditional. The thermal conductivity is k_θ, and C is the specific heat; both are assumed to be invariant.

Also, observe the nature of the propagating waves from the results, including a comparison of the group velocity with the phase velocity.

Solution: Only one energy storage mechanism is present, as indicated by the single time derivative operator $S \equiv \partial/\partial t$ in the given differential equation. Equations (10.73) give the propagation operator and characteristic admittance

$$\gamma(S) = \sqrt{CS/k_\theta},$$

$$Y_c(S) = \sqrt{k_\theta CS}.$$

The phase velocity v_p and attenuation factor α are defined for sine waves. Letting $S \to j\omega$,

$$\gamma(j\omega) = \sqrt{Cj\omega/k_\theta} = jk + \alpha; \qquad k = \alpha = \sqrt{\frac{C\omega}{2k_\theta}}.$$

The traditional dispersion plot of frequency as a function of wave number (with the constant coefficient C/k_0 for the ordinate) is given below:

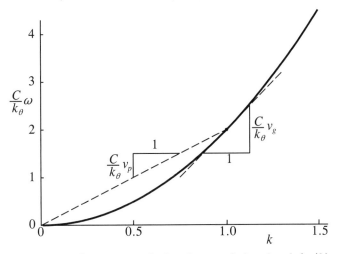

The phase and group velocities equal the slopes of the chord (ω/k) and of the curve ($d\omega/dk$), respectively:

$$v_p = \frac{\omega}{k} = \sqrt{\frac{2k_\theta\omega}{C}},$$

$$v_g = \frac{d\omega}{dk} = 2v_p.$$

Heat conduction behaves very differently from the pure bilateral-wave-delay model. Since energy cannot slosh back and forth between different storages, the system is non-oscillatory in the usual sense, and possesses monatonic (continuously increasing or decreasing) reponses to step disturbances. It is highly dispersive and absorptive. The equality of k and α implies that the envelope of the decaying sinusoidal response for unilateral waves (that is either θ_1 or θ_2 alone), as shown below, falls off by the factor $e^{-2\pi}$ per wavelength.

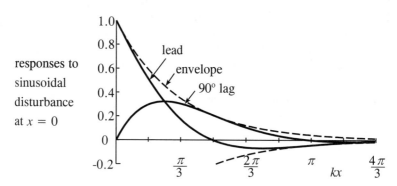

responses to sinusoidal disturbance at $x = 0$

The wavelength is inversely proportional to the square root of frequency. The heat flux is not in phase with the temperature variations, but leads by a constant 45°, which is the phase angle of Y_c. The ratio of the heat flux variations to the temperature variations increases with the wave number and therefore with the frequency, as can be seen by the plot of the magnitude of Y_c given below:

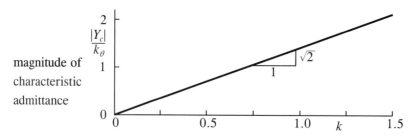

magnitude of characteristic admittance

Example 10.11

Continue the example of the infinite rod above by finding the temperature at a location x to the right of the point where an impulse disturbance in temperature is imposed. Also, find the reponse when the disturbance is a step in time. The Laplace transform table given in Appendix C may be used.

Solution: You can examine the rightward propagation of transient waves by setting the amplitude of the leftward waves in equation (10.72) (p. 745) to zero. The Laplace transform of the response is

$$\theta(x, s) = e^{-\sqrt{Cs/k_\theta}\,x}\theta_2(s).$$

The responses to the unit impulse ($\theta_2(s) = \theta_0$) and step ($\theta_2(s) = \theta_0/s$) are given in Appendix C by transform pairs #38 and #39, respectively:

$$\text{impulse response}: \quad \theta(x, t) = \theta_0\frac{1}{2}\sqrt{\frac{C}{\pi k_\theta}}xt^{-3/2}e^{-Cx^2/4k_\theta t},$$

$$\text{step response}: \quad \theta(x, t) = \theta_0\text{erfc}\sqrt{\frac{Cx^2}{4k_\theta t}}.$$

Here "erfc" stands for the complementary error function, which is widely tabulated. These results are plotted below. Note that a single dimensionless curve characterizes each case completely; it can be scaled for any distance or time. Notice also that the tails of the responses are extremely long. This feature of many distributed-parameter models cannot be modeled with great accuracy using only a few lumped elements.

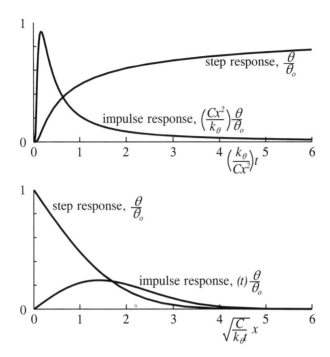

The differential equation given for heat conduction in Example 10.10 above, and also used in Example 10.11, also is known as the **diffusion equation**. This classical equation describes the diffusion of the concentration of a component through a mixture. Specifically, θ becomes the concentration per unit length and q becomes the flow of the component of interest. With this interpretation, k_θ becomes the permittivity and C becomes unity.

10.4.2   Energy Velocity in Conservative Media*

The time-averaged power propagated through a nonabsorptive medium by a sinusoidal wave can be respresented as the product of the time-averaged energy density and the **velocity of energy propagation**, v_e:

$$\overline{\mathcal{P}} = (\overline{\mathcal{T}} + \overline{\mathcal{V}})v_e. \tag{10.80}$$

The overbars imply the time averaging. For non-dispersive media this velocity equals the phase velocity, v_p, which also equals the group velocity, v_g. For dispersive media, however, the energy velocity equals the group velocity, which can differ markedly from the phase velocity:

$$\boxed{v_e = v_g.} \tag{10.81}$$

This relation does not apply in the presence of dissipation, for which the concept of an energy velocity has nebulous meaning. For example, the group velocity for light traveling

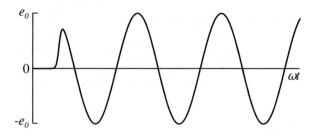

Figure 10.16: Function used to find the mean potential energy in the compliance $C = C(\omega)$

through a material medium[5] in a dissipation band is known to exceed the speed of light in a vacuum, c. Energy, of course, never travels faster than c.

A proof of equation (10.81) now will be given. Since the medium is conservative, the series impedance $Z(j\omega)$ and shunt admittance $Y(j\omega)$ must be pure imaginary functions. Therefore, $Z(S)$ and $Y(S)$ can be represented as

$$Z(S) = IS = I_0 S + I_1 S^3 + I_2 S^5 + I_3 S^7 + \cdots, \qquad (10.82a)$$

$$Y(S) = CS = C_0 S + C_1 S^3 + C_2 S^5 + C_3 S^7 + \cdots. \qquad (10.82b)$$

Consider the traveling wave

$$e = e_0 \sin(\omega t - kx), \qquad (10.83a)$$

$$\dot{q} = \dot{q}_0 \sin(\omega t - kx). \qquad (10.83b)$$

The series expansions require

$$\dot{q}_0 = Y_c e_0; \quad Y_c = \sqrt{\frac{C_0 - C_1\omega^2 + C_2\omega^4 - C_3\omega^6 + \cdots}{I_0 - I_1\omega^2 + I_2\omega^4 - I_3\omega^6 + \cdots}}, \qquad (10.84)$$

and the mean power becomes

$$\mathcal{P} = \overline{e\dot{q}} = e_0\dot{q}_0\overline{\sin^2(\omega t - kx)} = \frac{1}{2}e_0\dot{q}_0 = \frac{1}{2}Y_c e_0^2. \qquad (10.85)$$

The plan is to solve equation (10.80) for v_e, so the time-averaged energy density, $\overline{\mathcal{V}} + \overline{\mathcal{T}}$ is needed next. To find $\overline{\mathcal{V}}$, the compliance C is excited with an effort $e(t)$ that, at the initial time, equals zero and has zero derivatives $d^m e/dt^m$ for all $m \leq n$, where n is twice the largest subscript of interest in the expansion for C given in equation (10.82b). For larger time it is required that

$$e \to e_0 \sin \omega t; \qquad t \to \infty. \qquad (10.86)$$

Such a function is illustrated in Fig. 10.16. Its details for small values of time greater than zero are immaterial; only the conditions at $t = 0$ and as $t \to \infty$ matter.

The flow $\dot{q}$ becomes

$$\dot{q} = \sum_{i=0}^{n} C_i \frac{d^{i+1} e}{dt^{i+1}}. \qquad (10.87)$$

---

[5]Leon Brillouin, *Wave Propagation and Group Velocity*, Academic Press, New York, 1960. Brillouin shows that equation (11.81) applies when $C = C(\omega)$ is a slowly varying function of ω and strongly implies that it does not hold otherwise. The proof herein shows that it applies for any real functions $C = C(\omega)$ and $I = I(\omega)$.

This allows the potential energy to be written as

$$\mathcal{V}(t) = \int_0^t e\dot{q}\,dt = \sum_{i=0}^{n} C_i \int_0^t e\frac{d^{i+1}e}{dt^{i+1}}\,dt. \tag{10.88}$$

The term for $i = 0$ can be integrated directly. The term for $i = 1$ can be integrated by parts. The term for $i = 2$ can be integrated by parts twice, and so on for larger values of i. The result is

$$\mathcal{V}(t) = \left\{ \frac{1}{2}C_0 e^2 + C_1\left[e\frac{d^2e}{dt^2} - \frac{1}{2}\left(\frac{de}{dt}\right)^2\right] + C_2\left[e\frac{d^4e}{dt^4} - \frac{de}{dt}\frac{d^3e}{dt^3} + \frac{1}{2}\left(\frac{d^2e}{dt^2}\right)^2\right]\right.$$

$$\left. +C_3\left[e\frac{d^6e}{dt^6} - \frac{de}{dt}\frac{d^5e}{dt^5} + \frac{d^2e}{dt^2}\frac{d^4e}{dt^4} - \frac{1}{2}\left(\frac{d^3e}{dt^3}\right)^2\right] + \cdots\right\}\Bigg|_{t=0}^{t}. \tag{10.89}$$

Substituting the values of $e(t)$ for $t = 0$ and $t \to \infty$, and noting the time averages for the square of sine waves, there results

$$\overline{\mathcal{V}} = \frac{1}{4}C^*e_0^2; \qquad C^* \equiv C_0 - 3C_1\omega^2 + 5C_2\omega^4 - 7C_3\omega^6 + \cdots. \tag{10.90}$$

A similar process gives the mean kinetic energy

$$\overline{\mathcal{T}} = \frac{1}{4}I^*\dot{q}_0^2; \qquad I^* \equiv I_0 - 3I_1\omega^2 + 5I_2\omega^4 - 7I_3\omega_3^6 + \cdots. \tag{10.91}$$

The substutution of equations (10.85), (10.90) and (10.91) into (10.80), noting that the phase velocity is $v_p = 1/\sqrt{IC}$, gives

$$\frac{v_e}{v_p} = \frac{2/v_p}{C^*/Y_c + I^*Y_c} = \frac{2}{(C^*/C) + (I^*/I)}. \tag{10.92}$$

In the nondispersive case $C = C_0$ and $I = I_0$, $C^* = C$ and $I^* = I$ confirming that $v_e = v_p$.

The general dispersion relation

$$k = \omega\sqrt{IC} = \omega/v_p \tag{10.93}$$

gives the group velocity

$$v_g = \frac{d\omega}{dk} = \frac{1}{\dfrac{1}{v_p} - \dfrac{\omega}{v_p^2}\dfrac{dv_p}{d\omega}}. \tag{10.94}$$

However,

$$\frac{dv_p}{d\omega} = -\frac{1}{2}v_p\left[\frac{1}{C}\frac{dC}{d\omega} + \frac{1}{I}\frac{dI}{d\omega}\right], \tag{10.95}$$

so that equation (11.94) yields

$$\frac{v_g}{v_p} = \frac{1}{1 + \dfrac{\omega}{2C}\dfrac{dC}{d\omega} + \dfrac{\omega}{2I}\dfrac{dI}{d\omega}} = \frac{2}{\left(1 + \dfrac{\omega}{C}\dfrac{dC}{d\omega}\right) + \left(1 + \dfrac{\omega}{I}\dfrac{dI}{d\omega}\right)}. \tag{10.96}$$

The terms in parentheses can be evaluated through the use of the expansion for C and I; comparison with the definitions of C^* and I^* show that they are identical, respectively, to C^*/C and I^*/I. Therefore the right sides of equations (10.92) and (10.96) are equal, proving that $v_e = v_g$ as was to be shown.

10.4.3   Boundary-Value Problems; Transmission Matrices

The response of distributed-parameter systems has been computed thus far in terms of *waves*. More commonly, however, you are confronted with certain *end conditions* and wish to find the remaining end conditions, with little interest in the internal behavior. For the problem with hot water transport in Section 10.3, only one possible input existed — the inlet temperature — and the remaining end condition was in fact a single wave. For the torsion rod and the heat conduction or diffusion problem, however, or in fact any non-degenerate one-power model or indeed any general linear two-port system, four different combinations of independent and dependent end conditions (at $x = 0$ and $x = L$) have direct causal meaning. Each of these can be represented by a causal matrix, as with lumped models. Placing the independent or input variables on the right and the dependent or output variables on the left, these are

$$
\begin{aligned}
\text{admittance matrix}: \quad & \begin{bmatrix} y_a(0) \\ y_a(L) \end{bmatrix} = \mathbf{Y} \begin{bmatrix} y_s(0) \\ y_s(L) \end{bmatrix}, \\[2mm]
\text{impedance matrix}: (\mathbf{Z} = \mathbf{Y}^{-1}) \quad & \begin{bmatrix} y_s(0 \\ y_s(L) \end{bmatrix} = \mathbf{Z} \begin{bmatrix} y_a(0) \\ y_a(L) \end{bmatrix}, \\[2mm]
\text{adpedance matrix}: \quad & \begin{bmatrix} y_a(0) \\ y_s(L) \end{bmatrix} = \mathbf{G} \begin{bmatrix} y_s(0) \\ y_a(L) \end{bmatrix}, \\[2mm]
\text{immitance matrix}: \quad & \begin{bmatrix} y_s(0) \\ y_a(L) \end{bmatrix} = \mathbf{H} \begin{bmatrix} y_a(0) \\ y_s(L) \end{bmatrix}.
\end{aligned}
\tag{10.97}
$$

Only two more combinations of the variables exist:

$$
\begin{aligned}
\text{transmission matrix}: \quad & \begin{bmatrix} y_s(0) \\ p_a(0) \end{bmatrix} = \mathbf{T} \begin{bmatrix} y_s(L) \\ y_a(L) \end{bmatrix}, \\[2mm]
\text{inverse of } \mathbf{T}: \quad & \begin{bmatrix} y_s(L) \\ y_a(L) \end{bmatrix} = \mathbf{T}^{-1} \begin{bmatrix} y_s(0) \\ y_a(0) \end{bmatrix}.
\end{aligned}
\tag{10.98}
$$

These last two relations are acausal, in the sense that the two variables on the right sides cannot be varied independently in a real experiment; as you have seen almost from the beginning, it is necessary for one variable to be specified from the right and another to be specified from the left. These causalities also have meaning in the assembly of equations for numerical simulation, as you have seen.

Transmission matrices nevertheless have one very useful property: if two or more two-port devices are connected end-to-end, the transmission matrix of the overall cascade equals the product of the individual transmission matrices. Recall that if the cascade is oriented left-to-right, the order of the matrix factors corresponds to the order of the corresponding elements, as shown in Fig. 10.17. The reverse order applies for the inverse transmission matrices.

In practice it is convenient to compute the transmission matrix representation of a distributed-parameter model first, and then convert it to one of the causal forms if necessary. The conversion formulas, which can be derived readily, are

$$
\mathbf{Y} = \frac{1}{T_{12}} \begin{bmatrix} T_{22} & -\Delta \\ 1 & -T_{11} \end{bmatrix}; \quad \mathbf{Z} = \frac{1}{T_{21}} \begin{bmatrix} T_{11} & -\Delta \\ 1 & -T_{22} \end{bmatrix};
$$

$$
\mathbf{G} = \frac{1}{T_{11}} \begin{bmatrix} T_{21} & \Delta \\ 1 & -T_{12} \end{bmatrix}; \quad \mathbf{H} = \frac{1}{T_{22}} \begin{bmatrix} T_{12} & \Delta \\ 1 & -T_{21} \end{bmatrix};
\tag{10.99a}
$$

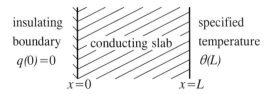

$$\frac{p_1}{\dot{q}_1} \ \text{2-port A} \ \frac{p_a}{\dot{q}_a} \ \text{2-port B} \ \frac{p_b}{\dot{q}_b} \ \text{2-port C} \ \frac{p_c}{\dot{q}_c} \ \cdots \ \frac{p_2}{\dot{q}_2}$$

$$\frac{p_1}{\dot{q}_1} \ \text{2-port N} \ \frac{p_2}{\dot{q}_2}$$

$$\begin{bmatrix} p_1 \\ \dot{q}_1 \end{bmatrix} = \mathbf{T}_A \begin{bmatrix} p_a \\ \dot{q}_a \end{bmatrix}; \qquad \begin{bmatrix} p_a \\ \dot{q}_a \end{bmatrix} = \mathbf{T}_B \begin{bmatrix} p_b \\ \dot{q}_b \end{bmatrix}; \qquad \begin{bmatrix} p_b \\ \dot{q}_b \end{bmatrix} = \mathbf{T}_C \begin{bmatrix} p_c \\ \dot{q}_c \end{bmatrix}; \quad \text{etc.}$$

$$\begin{bmatrix} p_1 \\ \dot{q}_1 \end{bmatrix} = \mathbf{T}_N \begin{bmatrix} p_2 \\ \dot{q}_2 \end{bmatrix}; \qquad \mathbf{T}_N = \mathbf{T}_A \cdot \mathbf{T}_B \cdot \mathbf{T}_C \cdots$$

Note: same form of result for 4-ports, 6-ports, etc.: 4-port

Figure 10.17: Cascade of two-port elements

$$\mathbf{T} = \begin{bmatrix} T_{11} & T_{12} \\ T_{21} & T_{22} \end{bmatrix}; \quad \Delta \equiv \det \mathbf{T}. \tag{10.99b}$$

The transmission matrix for the general model of equation (10.70) (p. 745) can be found by evaluating y_s and y_a as given in equation (10.72) at $x = 0$ and $x = L$ in terms of $y_{s1}(t)$ and $y_{s2}(t)$, and inverting one of these relations to eliminate y_{s1} and y_{s2}. The result, which you should verify, is of considerable utility:

$$\begin{bmatrix} y_s(0) \\ y_a(0) \end{bmatrix} = \underbrace{\begin{bmatrix} \cosh\Gamma & (1/Y_c)\sinh\Gamma \\ Y_c\sinh\Gamma & \cosh\Gamma \end{bmatrix}}_{\text{transmission matrix, } \mathbf{T}} \begin{bmatrix} y_s(L) \\ y_a(L) \end{bmatrix}; \quad \Gamma = \gamma L. \tag{10.100}$$

Applications now will be illustrated.

Example 10.12

A slab of conducting material of thickness L is insulated on its left surface and excited on its right surface by a sinusoidally varying temperature:

insulating
boundary — conducting slab — specified
$q(0)=0$ — temperature
$\theta(L)$
$x=0$ — $x=L$

Find the temperature at the left side of the slab as a function of frequency, and plot its real (in-phase) part imaginary (90° out-of phase) parts. (Use of a complex plane for a frequency response is known, in the controls literature including Chapter 8, as a **Nyquist plot**.)

Solution: The given information is in the form suitable for using the immittance matrix **H**:

$$\begin{bmatrix} \theta(0) \\ q(L) \end{bmatrix} = \mathbf{H} \begin{bmatrix} q(0) \\ \theta(L) \end{bmatrix} = \mathbf{H} \begin{bmatrix} 0 \\ \theta(L) \end{bmatrix}.$$

Only H_{12} and H_{22} are needed, which from equation (10.99) are

$$H_{12} = \Delta/T_{22} = \text{sech}\,\Gamma; \qquad H_{22} = -T_{21}/T_{22} = -Y_c \tanh\Gamma.$$

The following identities are particularly useful for frequency response calculations:

$$\sinh(\alpha + j\beta) \equiv \sinh\alpha\cos\beta + j\cosh\alpha\sin\beta,$$
$$\cosh(\alpha + j\beta) \equiv \cosh\alpha\cos\beta + j\sinh\alpha\sin\beta.$$

From the solution to Problem 10.10 (p. 747),

$$\Gamma = j\Omega + \Omega; \qquad \Omega = kL = L\sqrt{\frac{C\omega}{2k_\theta}}.$$

Therefore, the ratio of the temperature at the left to the excitation temperature on the right is

$$\frac{\bar\theta(0, j\omega)}{\bar\theta(L, j\omega)} = \text{sech}\,\Gamma = 1/\cosh\Gamma = \frac{1}{\cosh\Omega\cos\Omega + j\sinh\Omega\sin\Omega}$$

$$= \frac{\text{sech}\,\Omega\,\sec\Omega}{\sqrt{1 + (\tanh\Omega\tan\Omega)^2}}e^{-j\tan^{-1}(\tanh\Omega\tan\Omega)}.$$

This result is plotted below, using the complex plane or Nyquist coordinates. Note that the phase lag in the response doesn't reach $180°$ until a frequency at which the amplitude has decayed to a small value.

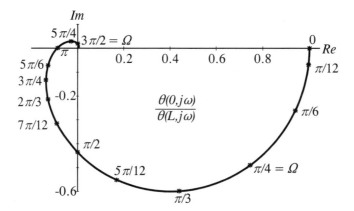

Example 10.13

Consider a torsion rod of length L, for which

$$\Gamma \equiv \gamma L = (L/v_p)S = TS; \quad T = L/v_p = L\sqrt{CI} = L\sqrt{\rho/\mu}.$$

Here, T is the time a wave takes to propagate from one end of the rod to the other. The rod is terminated in a linear ideal rotary dashpot with a constant resistance R, so that $M(L) = R\dot\phi(L)$:

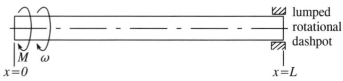

A constant torque $M(0)$ is suddenly applied at the left end. Plot the consequent angular velocity $\dot{\phi}(0)$ at this driving point as a function of time for the cases $RY_c = 0$, $1/3$, 1 and 2.

Solution: The transmission matrix relation is

$$\begin{bmatrix} M(0) \\ \dot{\phi}(0) \end{bmatrix} = \begin{bmatrix} \cosh TS & (1/Y_c)\sinh TS \\ Y_c\sinh TS & \cosh TS \end{bmatrix} \begin{bmatrix} R\dot{\phi}(L) \\ \dot{\phi}(L) \end{bmatrix}.$$

Solving for $\dot{\phi}(0)$,

$$\dot{\phi}(0) = Y_c\frac{RY_c\sinh TS + \cosh TS}{RY_c\cosh TS + \sinh TS}M(0) = Y_c\frac{1 + a\,e^{-2TS}}{1 - ae^{-2TS}}M(0)$$

$$= Y_c(1 + 2a\,e^{-2TS} + 2a^2e^{-4TS} + 2a^3e^{-6TS} + \cdots)M(0),$$

$$a \equiv \frac{1 - RY_c}{1 + RY_c}.$$

The successive operators on the right side of the equation represent pure delays of, respectively, zero, $2T$ or one round-trip delay, $4T$ or two round-trip delays, $6T$ or three round-trip delays, etc. The amplitude of each successive term or wave is attenuated more than its predecessor if $R > 0$, since then $a < 1$. The requested responses are plotted below. Note that for the special case of $R = 1/Y_c$, that is when the resistance of the termination equals the surge impedance, no waves at all are reflected from the termination; $a = 0$. It is as if the rod were infinitely long.

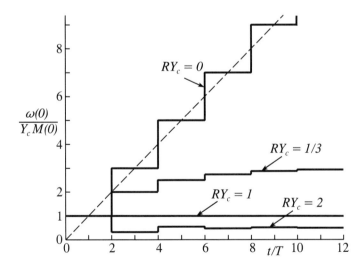

Back in the heyday of analog electronics it was very difficult to achieve pure time delays of electrical signals, and a mechanical torsion rod with the reflectionless termination sometimes was employed, with appropriate input and output electromechanical transducers, to do the job. Today, important analogs to this ideal termination remain in the design of hydraulic and acoustic systems.

Example 10.14

A torsion rod is free
at its left end and
is terminated by a
(stiff) rotary spring
at its right end:

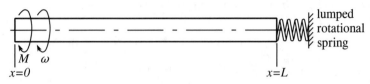

This system has no damping and possesses an infinity of resonant frequencies. Determine these frequencies as functions of the spring compliance C_t, the characteristic admittance Y_c for torsion of the rod and the wave delay time, T. Also, compare the result when the spring is very compliant to that of a lumped model that neglects the flexibility of the rod.

Solution: For the lumped termination comprising the spring,

$$M(L) = \frac{1}{C_t S}\dot{\phi}(L) = \frac{1}{C_t}\phi(L).$$

Since $M(0) = 0$, the transmission relation becomes

$$\begin{bmatrix} 0 \\ \dot{\phi}(0) \end{bmatrix} = \begin{bmatrix} \cosh TS & (1/Y_c)\sinh TS \\ Y_c \sinh TS & \cosh TS \end{bmatrix} \begin{bmatrix} \dot{\phi}(L)/C_t S \\ \dot{\phi}(L) \end{bmatrix}.$$

For a non-trivial solution to exist, the first scalar equation requires

$$\coth TS = -C_t S/Y_c.$$

This can happen only at the resonant or natural frequencies ω_i,

$$\coth Tj\omega_i = -j\cot T\omega_i = -C_t j\omega_i/Y_c,$$

$$\text{or} \qquad \cot T\omega_i = C_t\omega_i/Y_c.$$

The two sides of this transcendental equation are plotted below for three values of the parameter C_t/TY_c, which represents the ratio of the compliance of the spring termination to the total rotational compliance of the rod. Solutions are represented by the intersections. When this parameter is very small (*i.e.*, very stiff termination), the natural frequencies are seen to approach $(n-1/2)\pi$, $n = 1,\ 2\ 3\ldots$, which is the limit for the shaft clamped at the right end. When the parameter is very large (*i.e.*, very flexible termination), the first natural frequency becomes so low that the higher natural frequencies likely become unimportant. Since $\cot T\omega_1 \to 1/T\omega_1$, the first natural frequency approaches $1/\sqrt{ILC_t}$, which is the natural frequency of the *lumped* mass-spring model in which the rod is the rotational inertia, (IL), and the spring is the termination spring (C_t).

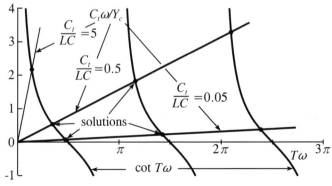

Virtually identical analyses to the last two examples above can be carried out for analogous systems represented by pure wave delay models, such as those highlighted in Section 10.1.

10.4.4 Exponentially Tapered Systems*

The simplest behavior for a non-uniform locally symmetric system occurs when there is an exponential taper:

$$Z = Z_0 e^{ax}; \qquad Y = Y_0 e^{-ax}. \tag{10.101}$$

Note that the product YZ is independent of x. Such a model can represent, for example, a torsion rod or a push rod or an acoustic tube for which the diameter is exponentially tapered; the nominal wave speed of the medium is independent of x. Systems with other tapers might well be approximated by a sequence of exponential segments.

Elimination of y_a in favor of y_s gives

$$\frac{d^2 y_s}{dx^2} = Z_0 Y_0 \, y_s + a \frac{dy_s}{dx}. \tag{10.102}$$

Assumption of the particular solution

$$y_s = e^{\mu x} y_{s0} \tag{10.103}$$

gives the following eigenvalues and general solution:

$$\mu_{1,2} = \frac{a}{2} \pm b; \qquad b = \sqrt{\frac{a^2}{4} + Z_0 Y_0}; \tag{10.104a}$$

$$\begin{bmatrix} y_s(x,t) \\ Z_0 y_a(x,t) \end{bmatrix} = \begin{bmatrix} e^{ax/2} & 0 \\ 0 & e^{-ax/2} \end{bmatrix} \begin{bmatrix} -\left(\frac{a}{2} - b\right) e^{bx} & -\left(\frac{a}{2} - b\right) e^{-bx} \end{bmatrix} \begin{bmatrix} y_{s1}(t) \\ y_{s2}(t) \end{bmatrix} \tag{10.104b}$$

Equation (10.104b) can be inverted to solve for $y_{s1}(t)$ and $y_{s2}(t)$ as functions of $y_s(x,t)$ and $Z_0 y_a(x,t)$. The result then can be substituted back into the special case of equation (10.104b) for which $x = 0$. The final result is the transmission matrix relation

$$\begin{bmatrix} y_s(0,t) \\ Z_0 y_a(0,t) \end{bmatrix} = \mathbf{T} \begin{bmatrix} y_s(x,t) \\ Z_0 y_a(x,t) \end{bmatrix}, \tag{10.105a}$$

$$\mathbf{T} = \begin{bmatrix} \cosh bx + \frac{a}{2b} \sinh bx & \frac{1}{b} \sinh bx \\ \frac{Z_0 Y_0}{b} \sinh bx & \cosh bx - \frac{a}{2b} \sinh bx \end{bmatrix} \begin{bmatrix} e^{-ax/2} & 0 \\ 0 & e^{ax/2} \end{bmatrix}. \tag{10.105b}$$

The right-most square matrix in equation (10.105b) represents a transformer action with transformer modulus $e^{-ax/2}$. The first square matrix represents both the delay that would occur in the absence of the taper and the distortion produced by the taper. The distortion is large for low frequencies and long wavelengths, and small for high frequencies and short wavelengths; note that the distortion coefficient $a/2b \to 0$ as $\omega \to \infty$. It is the transformer action that makes water waves grow as they approach the beach, that concentrates the sound in the historical hearing tubes, and makes the tool mounted at the small end of an exponentially tapered rod used in ultrasonic machining vibrate with a much larger amplitude than the large driven end.

These examples can be modeled by the pure bilateral wave model with nominal wave celerity $c_0 = 1/\sqrt{IC} = $ constant and nominal wave number $k_0 = \omega/c_0 = \lambda_0/2\pi$. For frequency response,

$$b = j\frac{\omega}{c_0} \sqrt{1 - \left(\frac{ac_0}{2\omega}\right)^2}. \tag{10.106}$$

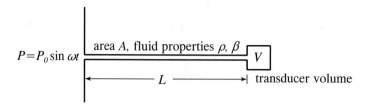

Figure 10.18: System for Guided Problem 10.5

For frequencies below $ac_0/2$ this is a real number; waves do not propagate, and the distortion is considerable. For frequencies above this critical frequency, on the other hand, b is an imaginary number; waves propagate and grow or decay according to the transformer action. The distortion coefficient in equation (10.105b) becomes

$$\frac{a}{2b} = \frac{1}{\sqrt{1-(2\omega/ac_0)^2}} = \frac{1}{\sqrt{1-(2k_0/a)^2}} = \frac{1}{\sqrt{1-(4\pi/\lambda_0 a)^2}}. \tag{10.107}$$

This reveals that little distortion exists for $a\lambda_0 \ll 4\pi$. For example, if the taper is a factor of two in one wavelength, $a\lambda_0 = \ln 2$ and $a/2b = j0.0552$, which is fairly small; shorter wavelengths produce yet less distortion.

10.4.5   Summary

The bilateral waves in symmetric one-power models propagate in a way similar to the unilateral waves in degenerate one-power models. The information on the speed and attenuation of waves is contained in the propagation operator, γ, which can be employed as a temporal or Laplace operator or as a complex function of frequency. A wave of the anti-symmetric variable is related to the corresponding wave of the symmetric variable by the characteristic or surge impedance, Y_c, which if not constant also can be treated as a temporal or Laplace operator or as a complex function of frequency. Energy propagates at the group velocity when there is no dissipation.

Boundary value problems are most commonly addressed starting with the transmission matrix that interrelates the states at the two ends. Matrices for elements that are cascaded can be multiplied directly to give an overall transmission matrix. This acausal matrix can be converted, if desired, into any of four causal matrices, depending on which boundary variables are specified and which are to be found. Both operator and frequency approaches and both uniform and exponentially tapered media have been illustrated.

The pure bilateral wave model is based on an inertance per unit length and a compliance per unit length. A variety of systems can be approximated by this model, as detailed in Section 10.1; the example of a torsion rod has been given here. The model for one-dimensional diffusion or heat transfer is based on a compliance per unit length and a resistance per unit length.

Waves amplify or attenuate in tapered systems, but only if their wavelength is short relative to the taper.

Guided Problem 10.5

Capacitive-type pressure transducers typically have a significant internal volume, V, which introduces fluid compliance. Often they are placed at the end of an instrument line, as shown in Fig. 10.18. Presume that the pressure of interest is $P_0 \sin \omega t$ and exists at the opposite end of the line from the transducer. Also presume that the line is rigid and has a length L and an internal area A, the density and bulk modulus for the fluid are ρ and β, and the volume has no gas bubbles and is virtually rigid mechanically. Find the ratio of the measured pressure to the pressure P.

Suggested Steps:

1. Represent the system by a bilateral pure delay line with a terminal compliance. Relate the compliance to the volume and the fluid properties.

2. Find the transmission matrix as a function of $S \to j\omega$.

3. Substitute the relation for the flow into the transducer volume, expressed as a function of the pressure there, into the transmission matrix relation. Then, use this relation to find the frequency transfer function between the two pressures. Report the result as a magnitude ratio and a phase angle.

Guided Problem 10.6

A flexible shaft of radius $r = 1.0$ in., length $L = 40$ in., shear modulus $\mu = 11.5 \times 10^6$ psi (steel) and weight density $\rho g = 0.283$ lb/in^3 is driven at its left end by an applied moment $M = M_0 \sin \omega t$, $M_0 = 200$ in-lb. and is terminated at the right end by a flywheel of the same material having a radius of 5.0 in. and a thickness of 0.10 in.; a viscous damping moment with resistance $R = 2000$ in-lb-s is applied to the flywheel. Find the driving-point admittance $\dot{\phi}_0(0)/M_0$ and the transfer admittance $\dot{\phi}_0(L)/M_0$ as a function of frequency. Plot their magnitudes and phase angles for frequencies up to 6000 Hz.

Suggested Steps:

1. Write the relation between the moment applied to the flywheel to its angular velocity.

2. Write the transmission matrix relationship between the two ends. Substitute the known applied moment at the left end and the result from step 1 at the right end.

3. Solve for the desired admittances. At this point your work ought to be in terms of symbols.

4. Evaluate the parameters numerically, substitute into the results of step 3, and plot. Linear axes are suggested rather than the usual Bode axes in problems dominated by traveling waves.

PROBLEMS

10.17 A particular linear uniform one-dimensional one-power distributed-parameter system has known constant values for its phase velocity, attenuation factor and characteristic admittance.

(a) Determine what you can about its series impedance per unit length and shunt admittance per unit length.

(b) A known compliance with an initial effort e_0 discharges into one end of an infinitely long length of this system, which initially has a uniformly zero effort. Determine the response of the effort at a distance L from that end.

10.18 A uniform linearly elastic tube containing water has an effective inertance to fluid motion I', compliance C' and resistance R', all per unit length and constant. In addition, the tube is pierced with many small holes to produce a constant leakage conductance, G', per unit length. (The resulting model is the same as in Problem 10.13, p. 733.)

(a) Define appropriate variables and find linear describing differential equations in standard form, neglecting the time-mean velocity and other second-order or nonlinear effects.

(b) Show that for given values of I', C' and R', a particular value of G' causes the characteristic impedance to be constant. Find this value, and investigate its effect on wave propagation (phase velocity, group velocity, dispersion, distortion, attenuation).

(c) With the condition of part (b), show that a particular value of the resistive termination, R_t, eliminates wave reflections. Also, find this value.

10.19 A steel torsion rod 500 in. long is driven at the left end and terminated at the right end by a linear viscous damper with coefficient 0.07 in.·lb·s. (The properties of steel are noted in Guided Problem 10.6 above.) The left end starts at rest and is given a step change in angular velocity to 1800 rpm. Find and plot the moment in the rod at the driving end and the angular velocity at the right end as functions of time for rod diameters of (a) 0.250 in. (b) 0.375 in.

10.20 A 0.250 in. torsion rod of the preceding problem is unconstrained at the left end and terminated at the right end by a linear rotary spring with rate 7.74 in lb/rad. Find the eigenvalue equation for the system and determine the first four natural frequencies.

10.21 The torsion rod of the preceding problem is driven sinusoidally from the left end. Determine the ratio of the moment to the angular velocity at the driving end.

10.22 A long torsion rod of length L, wavespeed c and characteristic impedance Z_c is clamped at its left end and terminates in a flywheel of rotational inertance I_f at its right end.

(a) Find the eigenvalue equation for the resonant frequencies.

(b) Approximate the first three natural frequencies in the special case for which the rotational inertance of the entire rod equals I_f.

(c) Evaluate the lowest eigenvalue when the rotational inertance of the rod approaches zero. Then compare this result with that for a lumped model that ignores the inertance of the rod in favor of its compliance.

10.23 The muffler of Problem 10.14 (p. 733) has length $L = 2$ ft and joins together two infinitely long tubes with the same diameter d_i as the perforated inner tube within the muffler. The resonant (Helmholtz) frequency of all the side chambers is $\omega_n = 2\pi \times 100$ rad/s, and the ratio of the outer to the inner diameter is $(d_o/d_i) = 10$. A sinusoidal wave with frequency ω, wave speed $c = 1100$ ft/s and pressure amplitude P_+ is incident to the muffler at one end, and a consequent wave P_L emerges at the other end. Approximating the muffler as a uniform continuous structure, find and plot the magnitude ratio $|P_L/P_+|$ as a function of $\Omega \equiv \omega/\omega_n$. Comment on the performance of the muffler.

10.24 A coaxial cable is terminated in a parallel combination of a resistance and a capacitance. The value of the resistance equals the surge impedance, so that if the capacitance were absent there would be no wave reflections and no standing waves. The termination can be corrected at any given frequency for the presence of the capacitance by placing an open-ended *stub* of the same cable in parallel with the load, as shown opposite.

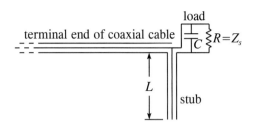

The proper length of the stub, L, depends on the load capacitance, C. This example of "stub matching" is practical, since most coaxial cables are used only for a very narrow band of frequencies. (Note: Stub matching has been used to prevent reflections and resonances due to compliance and inertance in fluid transmission lines and acoustic ducts, also.)

(a) Consider the stub by itself, with zero current at its open end. Find the ratio of the voltage to the current at the driven end in terms of the length of the stub and its surge impedance and wave speed.

(b) Place the impedance from part (a) in parallel with the actual load. Set the resulting impedance equal to the surge impedance to find the proper length of the stub.

10.25 A magnetostrictive driver oscillates the large end of a steel exponentially tapered ultrasonic machine tool at the frequency 30kHz. A motion amplification of 40 is desired, and the large end of the taper has a diameter of one inch. Make reasonable choices for the length and diameter at the small end of the taper. The taper should not be unnecessarily long.

SOLUTIONS TO GUIDED PROBLEMS

Guided Problem 10.5

1. $\dfrac{P}{Q} \longrightarrow D \dfrac{P_t}{Q_t} \longrightarrow C \qquad C = V/\beta$

2. $\mathbf{T} = \begin{bmatrix} \cos T\omega & Z_c \sinh T\omega \\ (1/Z_c)\sinh T\omega & \cos T\omega \end{bmatrix}; \qquad T = L/c = L\sqrt{\rho/\beta}; \quad Z_c = \sqrt{\beta\rho}$

3. $Q_{t0} = Cj\omega P_{t0}$

   The first equation in the transmission relation becomes

   $P_0 = [\cos T\omega + (Cj\omega)jZ_c \sin T\omega]P_{t0}$

   so that $P_{t0} = P_0/(\cos T\omega - C\omega Z_c \sin T\omega); \quad P_t = P_{t0}\sin\omega t.$

   The magnitude ratio therefore is $\left|\dfrac{P_{t0}}{P_0}\right| = \left|\dfrac{1}{(\cos T\omega - C\omega Z_c \sin T\omega)}\right|$

   and the phase angle is $0°$ when $P_{t0}/P_0 > 0$ and $\pm 180°$ when $P_{t0}/P_0 < 0$.

Guided Problem 10.6

1. $M(L) = (I_f s + R)\dot\phi(L)$, or $M_0(L) = (jI_f\omega + R)\dot\phi_0(L)$, where I_f is the moment of inertia of the flywheel.

2. $$\begin{bmatrix} M_0 \\ \dot\phi_0(0) \end{bmatrix} = \begin{bmatrix} \cos T\omega & jZ_c \sin T\omega \\ (j/Z_c)\sin T\omega & \cos T\omega \end{bmatrix} \begin{bmatrix} I_f j\omega + R \\ 1 \end{bmatrix} \dot\phi_0(L)$$

3. $$\frac{\dot\phi_0(L)}{M_0} = \frac{1}{(I_f j\omega + R)\cos T\omega + jZ_c \sin T\omega}$$

   $$\frac{\dot\phi_0(0)}{M_0} = \frac{(I_f j\omega + R)(j/Z_c)\sin T\omega + \cos T\omega}{(I_f j\omega + R)\cos T\omega + jZ_c \sin T\omega}$$

   With MATLAB it is not necessary to write expressions for the real and imaginary parts separately, or otherwise simplify the equations above.

4. $I_f = \frac{1}{2}mr_f^2 = \frac{1}{2}(\rho\pi r_f^2 w)r_f^2 = \frac{1}{2}\frac{0.283}{386}\pi(0.1)(5)^4 = 0.0720$ lb·in.·s^2.

   $R = 2000$ lb·in.·s.

   $$T = \frac{L}{c} = L\sqrt{\frac{\rho}{\mu}} = 40\sqrt{\frac{0.283}{386 \times 11.5 \times 10^6}} = 3.194 \times 10^{-4} \text{ seconds}$$

   $$Z_c = \frac{\pi a^4}{2}\sqrt{\rho\mu} = \frac{\pi(1.0)^4}{2}\sqrt{\frac{0.283}{386}11.5 \times 10^6} = 144.2 \text{ lb·in.·s.}$$

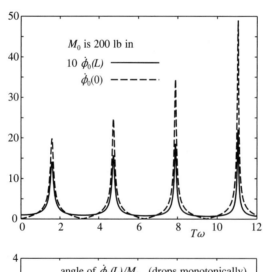

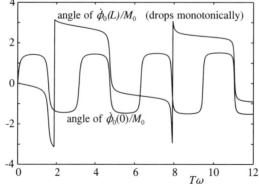

10.5 Multiple-Power Models

Systems such as flexible fluid-carrying tubes, parallel and counter-flow heat exchangers and beams with lateral vibrations can be represented by one-dimensional distributed-parameter models with more than a single power. The structure of such models has been given in Section 10.2.4 (pp. 731-732). The general boundary-value problem is addressed here, followed by some examples. (Wave behavior and internal excitations are discussed in the reference cited in footnote 1 on page 713.)

10.5.1 Symmetric and Anti-Symmetric Variables and Models

A conjugate pair of symmetric and anti-symmetric variables have been employed in modeling one-power one-dimensional models. Multiple-power models, except in degenerate cases, employ a conjugate pair of symmetric and anti-symmetric variables for each power. When these are chosen as true power-factoring variables (efforts and flows) the sum of the products of the respective symmetric and anti-symmetric variables is the total power. This choice is not mandatory, however, as before: the variables can be merely proportional to the efforts and flows or their time derivatives or integrals, as long as they satisfy the criteria of symmetry and anti-symmetry. The symmetric variables are collected together in a vector, $\mathbf{y}_s$, and the anti-symmetric variables are collected together in a second vector, $\mathbf{y}_a$, employing the same order with respect to the powers conveyed. Thus, repeating equation (10.35),

$$\mathbf{y} = \begin{bmatrix} \mathbf{y}_s \\ \mathbf{y}_a \end{bmatrix}. \tag{10.108}$$

The example of a Bernoulli-Euler beam has been given in Guided Problem 10.4 (pp. 733, 734-735). The general differential equation is, repeating equation (10.36),

$$\boxed{\frac{\partial}{\partial x} \begin{bmatrix} \mathbf{y}_s \\ \mathbf{y}_a \end{bmatrix} = - \begin{bmatrix} \mathbf{W} & \mathbf{Z} \\ \mathbf{Y} & \mathbf{X} \end{bmatrix} \begin{bmatrix} \mathbf{y}_s \\ \mathbf{Y}_a \end{bmatrix} + \mathbf{Bu}(x,t).} \tag{10.109}$$

When the model is locally symmetric, $\mathbf{W} = \mathbf{X} = \mathbf{0}$. In this case, the transmission matrix can be shown to be[6]

$$e^{\mathbf{A}x} = \begin{bmatrix} \cosh\sqrt{\mathbf{ZY}}\,x & \mathbf{Z}(\sqrt{\mathbf{YZ}})^{-1}\sinh\sqrt{\mathbf{YZ}}x \\ \mathbf{Y}(\sqrt{\mathbf{ZY}})^{-1}\sinh\sqrt{\mathbf{ZY}}\,x & \cosh\sqrt{\mathbf{YZ}}x \end{bmatrix}. \tag{10.110}$$

The matrix elements of this partitioned matrix can be found in terms of the matrix products $\mathbf{ZY}$ and $\mathbf{YZ}$ and their identical eigenvalues. This can be accomplished using Sylvester's theorem or the Cayley-Hamilton theorem[7]; the example of the lateral vibration of the slender beam is carried out below. Note that for the one-power case, this equation reduces to the familiar form of equation (10.100) (p. 753).

Another advantage of the ordering of the variables for the symmetric case results if you wish to examine the propagation of waves, or use wave-scattering variables. The characteristic values of $\mathbf{A}$ are desired. The coefficients of the odd-powered terms in the characteristic polynomial vanish, as can be seen from Bocher's formulas.[8] Further, since

$$\mathrm{trace}(\mathbf{ZY})^m = \mathrm{trace}(\mathbf{YZ})^m \qquad m = 1, 2, 3, \ldots, \tag{10.111}$$

---

[6]See footnote 3 (p. 732) for a reference to a derivation.

[7]See for example L.A. Pipes, *Matrix Methods for Engineering*, Prentice-Hall, Inc, Englewood Cliffs, NJ, 1963.

[8]ibid.

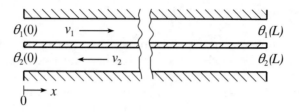

Figure 10.19: Counterflow heat exchanger

the $2n$ (or less) nonzero eigenvalues of $\mathbf{A}$ can be shown to be equal to the *square roots* of the n (or less) nonzero eigenvalues of $\mathbf{ZY}$ or $\mathbf{YZ}$. This relationship exists regardless of whether $\mathbf{Z}$ and $\mathbf{Y}$ are square (partial degeneracies). As a result, matrix and numerical manipulations are kept to a minimum.

10.5.2   Case Study of a Degenerate System: A Counterflow Heat Exchanger

A totally degenerate model, for which all symmetric or all asymmetric variables have been purged, has $\mathbf{A} = \mathbf{X}$ or $\mathbf{A} = \mathbf{W}$, respectively, and cannot be symmetrical. The simple counterflow heat exchanger shown in Fig. 10.19 can be represented by such a degenerate two-power uniform one-dimensional model. The degeneracy results from converting the otherwise variable fluid velocities into assumed constant parameters, v_1 and v_2, leaving the temperatures θ_1 and θ_2 as the only state variables. To simplify matters, it is assumed that the thermal compliance of the fluids per unit length, C_1 and C_2, are so much larger than the compliances of the walls that the latter can be neglected. The thermal conductivity between the two tubes, per unit length, is G. The outer walls are assumed to be perfect insulators. The model becomes

$$\frac{d}{dx}\begin{bmatrix} \theta_1 \\ \theta_2 \end{bmatrix} = -\begin{bmatrix} -\dfrac{G/C_1 + S}{v_1} & \dfrac{G}{C_1 v_1} \\ -\dfrac{GF}{C_2 v_2} & \dfrac{G/C_2 + S}{v_2} \end{bmatrix}\begin{bmatrix} \theta_1 \\ \theta_2 \end{bmatrix}. \tag{10.112}$$

The Cayley-Hamilton method of evaluating a function of a square matrix will be illustrated for the transmission matrix $\exp(-\mathbf{W}L)$. The eigenvalues of $\mathbf{W}$ are defined as μ_1 and μ_2. A function of an $n \times n$ matrix can be expressed as a polynomial in the matrix of order $n-1$. Since $\mathbf{W}$ is second order,

$$e^{-\mathbf{W}L} = a_0\mathbf{I} + a_1\mathbf{W}. \tag{10.113}$$

The problem is to find the coefficients a_0 and a_1. According to the Cayley-Hamilton theorem, scalar equations corresponding to equation (10.113) apply, in which an eigenvalue is substituted for $\mathbf{W}$. Therefore,

$$e^{-\mu_1 L} = a_0 + a_1\mu_1, \tag{10.114a}$$

$$e^{-\mu_2 L} = a_0 + a_1\mu_2. \tag{10.114b}$$

Inverting this result gives the coefficients:

$$\begin{bmatrix} a_0 \\ a_1 \end{bmatrix} = \frac{1}{\mu_2 - \mu_1}\begin{bmatrix} \mu_2 & -\mu_1 \\ -1 & 1 \end{bmatrix}\begin{bmatrix} e^{-\mu_1 L} \\ e^{-\mu_2 L} \end{bmatrix}. \tag{10.115}$$

The causal inputs are $\theta_1(0)$ and $\theta_2(L)$. The transfer function between the latter and the output $\theta_2(0)$, for example, is the lower-right element of the transmission matrix $\exp(-\mathbf{W}L)$, or T_{22}, which is

$$T_{22} = a_0 + a_1 W_{22} = \frac{1}{\mu_2 - \mu_1} \left[(\mu_2 - W_{22})e^{-\mu_1 L} - (\mu_1 - W_{22})e^{-\mu_2 L} \right]. \tag{10.116}$$

The eigenvalues are

$$\mu_1 = l_1 - l_2; \qquad \mu_2 = l_1 + l_2, \tag{10.117a}$$

$$l_1 = \frac{1}{2} \left[S \left(\frac{1}{v_2} - \frac{1}{v_1} \right) + \frac{G}{C_2 v_2} - \frac{G}{C_1 v_1} \right], \tag{10.117b}$$

$$l_2 = \sqrt{l_3^2 - \frac{G^2}{C_1 C_2 v_1 v_2}}, \tag{10.117c}$$

$$l_3 = \frac{1}{2} \left[S \left(\frac{1}{v_2} + \frac{1}{v_1} \right) + \frac{G}{C_2 v_2} + \frac{G}{C_1 v_1} \right]. \tag{10.117d}$$

With these substitutions the transfer function reduces to

$$\frac{\theta_2(0)}{\theta_2(L)} \equiv T_{22} = \frac{e^{-l_1 L}}{l_2} [l_2 \cosh(l_2 L) - l_3 \sinh(l_2 L)]. \tag{10.118}$$

Should the frequency response be desired, the substitution $S \to j\omega$ gives the result directly. Workable transfer functions for the direct evaluation of transient responses are harder to achieve, although approximations combining the ratios of polynomials and pure delays can be made to apply at high frequencies, and ratios of polynomials without pure delays at low frequencies. Fourier decomposition might be the best route in complicated cases.

10.5.3 Case Study of a Symmetric Model: The Bernoulli-Euler Beam

The **Bernoulli-Euler** model of the slender beam with lateral vibration has been introduced in Guided Problem 10.4 (p. 733, 734-735). It assumes the classical relation between the bending moment and the curvature of the centerline of a slender beam, as presented in elementary strength-of-materials courses, assuming a shear-free rotation of each differential element. It relates the gradient of the moment to the shear force. Finally, it restricts inertial forces to the product of the mass per unit length and the lateral acceleration. With the mass density, Young's modulus, cross-sectional area, and area moment of inertia of the cross section being defined respectively as ρ, E, A and I_m, the state differential equation becomes [9], repeating from the solution to Guided Problem 10.4,

$$\frac{d}{dx} \begin{bmatrix} \eta \\ M \\ F \\ \partial \eta / \partial x \end{bmatrix} = - \begin{bmatrix} 0 & 0 & 0 & -1 \\ 0 & 0 & -1 & 0 \\ \rho A S^2 & 0 & 0 & 0 \\ 0 & -1/EI_m & 0 & 0 \end{bmatrix} \begin{bmatrix} \eta \\ M \\ F \\ \partial \eta / \partial x \end{bmatrix}. \tag{10.119}$$

---

[9] A somewhat more accurate model, known as the **Timoshenko beam**, results from including the effects of rotational inertia and shear-induced deflections. This model is described by the following, in which G is the shear modulus and $k^2 = G\kappa^2/[E/(1-\nu^2)]$ where ν is Poisson's ratio and κ^2 is the Timoshenko shear coefficient:

$$\frac{d}{dx} \begin{bmatrix} \eta \\ \frac{\partial \psi}{\partial x} \\ \frac{\partial \eta}{\partial x} \\ \psi \end{bmatrix} = \begin{bmatrix} 0 & 0 & \frac{1}{EI_m} & 0 \\ 0 & 0 & -\frac{k^2 GA}{EI_m} & \frac{\rho I s^2 + k^2 GA}{EI_m} \\ \frac{\rho s^2}{k^2 G} & 1 & 0 & 0 \\ 0 & 1 & 0 & 0 \end{bmatrix} \begin{bmatrix} \eta \\ \frac{\partial \psi}{\partial x} \\ \frac{\partial \eta}{\partial x} \\ \psi \end{bmatrix}.$$

Note that $\mathbf{W}$ and $\mathbf{X}$ vanish, verifying the local symmetry of the beam or, assuming the symmetry is known, serving as a partial check of the model.

The four eigenvalues of $\mathbf{A}$, which equal the square roots of the eigenvalues of $\mathbf{ZY}$ or $\mathbf{YZ}$, are

$$\mu = \left(-\frac{\rho A S^2}{E I_m} \right)^{1/4}. \tag{10.120}$$

It is convenient to label the principal eigenvalue, which is positive real if $S \to j\omega$, as μ_1. This allows the following nondimensionalization of the state differential equations:

$$\frac{d}{dx} \begin{bmatrix} \eta \\ \frac{M}{\mu_1^2 E I_m} \\ \frac{F}{\mu_1^3 E I_m} \\ \frac{1}{\mu_1} \frac{\partial \eta}{\partial x} \end{bmatrix} = -\mu_1 \begin{bmatrix} 0 & 0 & 0 & -1 \\ 0 & 0 & -1 & 0 \\ -1 & 0 & 0 & 0 \\ 0 & -1 & 0 & 0 \end{bmatrix} \begin{bmatrix} \eta \\ \frac{M}{\mu_1^2 E I_m} \\ \frac{F}{\mu_1^3 E I_m} \\ \frac{1}{\mu_1} \frac{\partial \eta}{\partial x} \end{bmatrix}. \tag{10.121}$$

The transmission matrix can be found using equation (10.110), with each quadrant evaluated using the Cayley-Hamilton method as illustrated above. (The reader might well check one of the quadrants.) The result is

$$\begin{bmatrix} \eta \\ \frac{M}{\mu_1^2 E I_m} \\ \frac{F}{\mu_1^3 E I_m} \\ \frac{1}{\mu_1} \frac{\partial \eta}{\partial x} \end{bmatrix}_{x_0} = \frac{1}{2} \begin{bmatrix} k_1 & k_2 & k_4 & k_3 \\ k_2 & k_1 & k_3 & k_4 \\ k_3 & k_4 & k_1 & k_2 \\ k_4 & k_3 & k_2 & k_1 \end{bmatrix} \begin{bmatrix} \eta \\ \frac{M}{\mu_1^2 E I_m} \\ \frac{F}{\mu_1^3 E I_m} \\ \frac{1}{\mu_1} \frac{\partial \eta}{\partial x} \end{bmatrix}_{x}, \tag{10.122a}$$

$$k_1 \equiv \cosh \mu_1 (x - x_0) + \cos \mu_1 (x - x_0);$$

$$k_2 \equiv \cosh \mu_1 (x - x_0) - \cos \mu_1 (x - x_0);$$

$$k_3 \equiv \sinh \mu_1 (x - x_0) + \sin \mu_1 (x - x_0);$$

$$k_4 \equiv \sinh \mu_1 (x - x_0) - \sin \mu_1 (x - x_0). \tag{10.122b}$$

You now may substitute whatever boundary conditions are imposed, and solve for the remaining boundary variables. For an example, see Guided Problem 10.7 below.

10.5.4   Summary

Boundary-value problems for one-dimensional distributed-parameter models have been addressed generally. Advantages accrue from segregating the symmetric and asymmetric elements in the vector of state variables. If the model is locally symmetric, two of the four quadrants of $\mathbf{A}$ vanish, the eigenvalues can be found from treating a matrix of half the dimension, and the transmission matrix is simplified. The evaluation of functions of matrices using the Cayley-Hamilton theorem has been illustrated.

The example of a simple counterflow heat exchanger typifies a totally degenerate two-power model, and the example of the Bernoulli-Euler beam typifies a multi-powered locally symmetric model.[10]

---

[10] A fuller description of the Bernoulli-Euler example is given in Section 10.7.6, where a new kind of wave is shown to decay spatially while propagating no power.

Guided Problem 10.7

Estimate the bending natural frequencies of a slender beam of length L that is cantilevered at one end and free at the other end.

Suggested Steps:

1. Since the beam is slender, the Bernoulli-Euler model likely is acceptable except for the higher modes that have short wavelengths.

2. Identify which of the boundary variables in equation (10.122a) are zero. With this information, find a 2×2 matrix function of $\mu_1 L$ which has a zero determinant.

3. Write the characteristic equation, and use trigonometric identities to simplify.

4. Represent the resulting equation as one simple trigonometric function of $\mu_1 L$ equal to another. Then, sketch-plot the two functions and estimate their intersections, which represent the respective modes.

5. Solve for the modal frequencies in terms of the parameters of the beam.

PROBLEMS

10.26 A slender beam is free at both ends. Find an equation for its natural frequencies, and find the first four values. (This result is used in Section 10.10.)

10.27 A slender uniform beam is cantilevered at one end and is free at the other end, apart from the attachment of a compact mass equal to the total mass of the beam itself. (a) Find the first three resonant frequencies of the system in terms ofthe dimensions and material properties of the beam. (b) Compare the first resonant frequency to the frequency found from a simple analysis that ignores the dynamics of the beam altogether, instead considering it as a pure lateral spring. Show that the results agree in the limit in which the mass of the beam becomes insignificant compared to the mass of the end. (c) Estimate the effective lumped mass and compliance of the beam using a field lumping technique valid when this mass is small compared to the mass at the end. Determine the corresponding natural frequency. Compare this result with the distributed-parameter answer. Is this an improvement to the estimate of part (b)?

10.28 A layer of fresh water over denser salt water frequently slows boats in Norweigian fiords and elsewhere. To investigate this phenomenon, consider the simpler problem of the linearized one-dimensional wave motion of two equally thick stratified layers, with small-amplitue perturbations having wavlengths considerably longer that the total depth, h. (a) Select appropriate variables, and find the corresponding matrix differential equation. (b) Find the phase velocities of both the "fast" and "slow" wave solutions, presuming $\rho_1 - \rho_2)/\rho_1 << 1$.

10.29 A long slender beam is supported by a very short cantilevered flexure. A given moment on the flexure produces the same slope at its end as would the entire beam, were the beam similarly cantilevered. Closely estimate the first three natural frequencies. You may assume the linear Bernoulli-Euler model for the beam, and employ any relevant results in the text, which also can supply the nomenclature.

$x{=}0$ $x{=}L$

10.30 Hydroelectric turbines are fed with water through long tunnels, called "penstocks," bored through the rock and lined with reinforced concrete so as to be quite rigid. A sudden reduction in the flow admitted to the turbine causes a large waterhammer wave to propagate upstream and subsequently to reverberate. In order to limit the pressures of these waves, the installation of a relatively small flexible air hose throughout the length of the penstock is proposed. The compliant hose is kept under enough pressure to prevent its collapse. It contains frequent porous plugs to restrict the axial flow of air.

(a) Construct an equivalent bond graph model or lumped circuit for an element Δx of the system. Do not include nonlinear or relatively unimportant phenomena. Relate the parameters of your model to the density and bulk modulus of the water, the mean absolute pressure and the ratio of specific heats of the air, the cross-sectional areas of the tunnel and the tube, and the compliance and resistance of the tube. The latter two may be left unrelated to physical parameters for purposes of this problem.

(b) Write a differential equation model for the system. Hints: Identify the symmetric and anti-symmetric variables, and write the partial differential equation in the recommended form. Note which elements of the $\mathbf{A}$-matrix represent the coupling between the tunnel and the tube. Are the elements of $\mathbf{W}$ and $\mathbf{X}$ indeed zero?

(c) Find the eigenvalues for wave propagation in terms of frequency and the parameters of your model. Plot the resulting phase velocity and attenuation factors for the special case in which the compliances of the water, the air and the wall of the tube are equal. Hints: Form the matrix product $\mathbf{ZY}$ or $\mathbf{YZ}$, and find its eigenvalues. The eigenvalues of interest are the square-roots of these. When the three compliances are equal, the eigenvalues can be characterized in terms of two time constants (an I/R and an RC time constant), allowing the definition of dimensionless eigenvalues and frequency. Thus a single plot for each attenuation factor and phase velocity can be made.

(d) Discuss the effectiveness of this solution. Hint: Note the effect of the air tube on the surge impedance of the system, particularly for infinite frequency or abrupt surges, as well as the propagation characteristics. Ought the air tube be made larger, or more flexible?

SOLUTION TO GUIDED PROBLEM

Guided Problem 10.7

1.-2. Cantilevered at left end ($x = 0$): $\eta(0) = (\partial\eta/\partial x)(0) = 0$.

Free at right end: $M(L) = F(L) = 0$.

$$\text{Therefore,} \quad \begin{bmatrix} 0 \\ M/\mu_1 EI_m \\ F/\mu_1^3 EI_m \\ 0 \end{bmatrix}_{x=0} = \frac{1}{2} \begin{bmatrix} k_1 & k_2 & k_4 & k_3 \\ k_2 & k_a & k_3 & k_4 \\ k_3 & k_4 & k_1 & k_2 \\ k_4 & k_3 & k_2 & k_1 \end{bmatrix} \begin{bmatrix} \eta \\ 0 \\ 0 \\ (1/\mu)(\partial \eta/\partial x) \end{bmatrix}_{x=L},$$

from which
$$\begin{bmatrix} 0 \\ 0 \end{bmatrix} = \frac{1}{2} \begin{bmatrix} k_1 & k_3 \\ k_4 & k_1 \end{bmatrix} \begin{bmatrix} \eta \\ (1/\mu_1)(\partial \mu/\partial x) \end{bmatrix}_{x+L},$$

$$k_1 \equiv \cosh \mu_1 L + \cos \mu_1 L,$$
$$k_3 \equiv \sinh \mu_1 L + \sin \mu_1 L,$$
$$k_4 \equiv \sinh \mu_1 L - \sin \mu_1 L.$$

3. The determinant of this square matrix must equal zero, or $k_1^2 - k_3 k_4 = 0$.
   Expanding: $\cosh^2 \mu_1 L + 2 \cosh \mu_1 L \cos \mu_1 L + \cos^2 \mu_1 L - \sinh^2 \mu_1 L + \sin^2 \mu_1 L = 0$.
   Use of the identities $\cosh^2 x - \sin^2 x \equiv 1$; $\cos^2 x + \sin^2 x \equiv 1$
   reduces this result to $1 + \cosh \mu_1 L \cos \mu_1 L = 0$.

4.-5. The result above can be cast as $\cos \mu_1 L = -\dfrac{1}{\cosh \mu_1 L}$.

These functions can be plotted (a quick sketch will do) to get approximate answers:

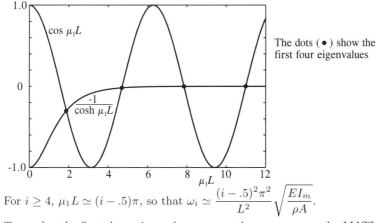

The dots (•) show the first four eigenvalues

For $i \geq 4$, $\mu_1 L \simeq (i - .5)\pi$, so that $\omega_i \simeq \dfrac{(i-.5)^2 \pi^2}{L^2} \sqrt{\dfrac{EI_m}{\rho A}}$.

To resolve the first three eigenvalues accurately, you can use the MATLAB function `fzero`. Start with the m-file

```
function y=beam(x)
y=1+cosh(x).*cos(x);
```

which you can save as `beam.m`. Then, the command `xzero=fzero('beam',1.6)` gives the zero which is closest to 1.6, namely 1.8751. The guess 5 for the second zero produces 4.7124, and the guess 8 for the third zero produces 7.8548. The respective frequencies are proportional to the squares of these numbers: $\omega_i = (\mu_{1i}L)^2 \sqrt{EI_m/\rho A}$.

10.6 Models of Dissipative Processes

The viscosity in fluid transmission lines and the resistivity in electric transmission lines dissipate energy, causing waves to attenuate and disperse. The simplest model, valid only at low frequencies, assumes a constant resistance and a constant inertance. Models valid at higher frequencies account for boundary-layer effects, also known as the skin effect in electric lines. Heat transfer produces similar effects in a gas through its coupling to the compliance. Also, various classical phenomenological models of liquid and solid viscoelastic substances are given and analyzed.

The simulation of coupled storage and dissipation fields also is addressed below, without the coupling between two different energy storage types that produces wave propagation. The amount of computation necessary is reduced drastically through the use of lumped models. Application to traveling waves is deferred until Section 10.7.

10.6.1   The Uniform Constant IRC Model

This model gives an impedance per unit length of $Is + R$ and an admittance per unit length of Cs, where the values of I, R and C are assumed to be constants. Usually, the static or zero-frequency values are used.

The pressure gradient in laminar steady flow in a circular tube is $8\mu Q/\pi a^4$, where Q is the volume flow, a is the radius and μ is the viscosity. This gives a static resistance per unit length of $R = 8\mu/\pi a^4$, assuming use of the pressure as the effort and symmetric state variable and the volume flow as the flow and asymmetric state variable. The corresponding inertance per unit length is $I = 4\rho/3\pi a^2$ assuming the parabolic steady-flow velocity profile, and the compliance is $C = \sqrt{\beta/\rho}$, as given before. For turbulent flow, the simplest assumption is a constant friction factor, f, and linearization about some mean flow rate, Q_0. This gives a resistance $R = f\rho Q_0/\pi^2 a^5$. The inertance is somewhat larger than $\rho/\pi a^2$, depending on the nonuniformity of the velocity profile, but the difference is of little significance in most cases.[11]

For an electric transmission line, R is the resistance per unit length, I is the inductance per unit length and C is the capacitance per unit length; the details depend on whether the configuration is a coaxial cable, a twisted pair or a lecher line (two parallel wires), etc.

The constant IRC model gives the following results:

$$\text{propagation operator}: \quad \gamma = \sqrt{ZY} = \sqrt{(IS + R)CS}, \qquad (10.123a)$$

$$\text{characteristic admittance} \quad Y_c = \sqrt{Y/Z} = \sqrt{\frac{CS}{IS + R}}. \qquad (10.123b)$$

For sinusoidal signals, the substitution $S = j\omega$ gives

$$\gamma(\omega) = \sqrt{jRC\omega - IC\omega^2} = jk + \alpha, \qquad (10.124a)$$

$$k = \frac{\omega}{\sqrt{2}v_{p0}} \left[\sqrt{1 + \left(\frac{1}{\tau\omega}\right)^2} + 1 \right]^{1/2}, \qquad (10.124b)$$

$$\alpha = \frac{\omega}{\sqrt{2}v_{p0}} \left[\sqrt{1 + \left(\frac{1}{\tau\omega}\right)^2} - 1 \right]^{1/2}, \qquad (10.124c)$$

$$Y_c = \sqrt{\frac{Cj\omega}{Ij\omega + R}} = \frac{Y_{c0}}{\sqrt{1 - j/\tau\omega}}$$

$$= \frac{Y_{c0}}{\sqrt{2[1 + (1/\tau\omega)^2]}} \left[\sqrt{\sqrt{1 + (1/\tau\omega)^2} + 1} + j\sqrt{\sqrt{1 + (1/\tau\omega)^2} - 1} \right], \qquad (10.124d)$$

$$v_{p0} = 1/\sqrt{IC}; \qquad \tau = I/R; \qquad Y_{c0} = \sqrt{C/I}. \qquad (10.124e)$$

Normalized values of the phase velocity $v_p = \omega/k$ and characteristic admittance are plotted in Fig. 10.20 as a function of the dimensionless frequency $\tau\omega$. (The abscissa $\sqrt{8}\,\tau\omega$ is used to

---

[11]The factors for the Reynolds numbers of 2500, 10^4, 10^5, 10^6 and 10^7 are, respectively, 1.113, 1.049, 1.020, 1.012 and 1.008.

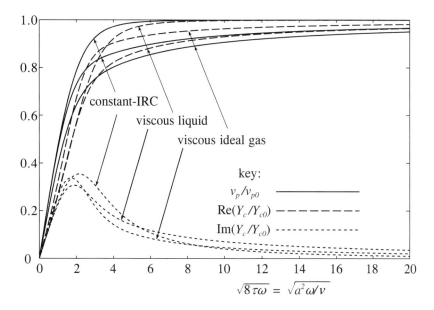

Figure 10.20: Properties of dissipative models of transmission lines

allow comparison to other models discussed below.) Boundary-value problems can be solved by substituting values of $\Gamma = \gamma L$ and Y_c into the transmission matrix relation (equation (10.100)). Note that as $\omega \to \infty$, waves travel at the speed v_{p0} and suffer an attenuation factor of $\alpha(\infty) = RY_{c0}/2$.

For other than sinusoids the operator notation can be used, and Laplace transforms employed. Exact analytical solutions do not exist, however, and accurate sum or product series solutions likely are too complicated to justify in view of the crudeness of the model itself and the availability of better models such as those given below. It is instructive to see the result of the simplest high-frequency approximate solution, however, despite the fact that the model itself is poor at high frequencies. For this case,

$$\lim_{S \to \infty} \gamma(S) = \lim_{S \to \infty} \sqrt{(IS + R)CS} = \sqrt{IC}S \left(1 + \frac{R}{2IS}\right), \qquad (10.125a)$$

$$\lim_{S \to \infty} Y_c(S) = \lim_{S \to \infty} \sqrt{\frac{CS}{IS + R}} = \sqrt{\frac{C}{I}} \left(1 - \frac{R}{2IS}\right), \qquad (10.125b)$$

from which it can be seen that

$$v_p = v_{p0}; \qquad \alpha = RY_{c0}/2; \qquad Y_c = Y_{c0}\left(1 - \frac{\alpha v_{p0}}{LS}\right). \qquad (10.126)$$

which means that waves travel at the same speed as though there were no resistance and decay at a fixed exponential rate.

Consider the simple boundary-value problem in which the flow at the right end is blocked, the effort at the left end is specified and the flow at the left end is to be found. The transmission matrix relation is

$$\begin{bmatrix} e(0) \\ \dot{q}(0) \end{bmatrix} = \begin{bmatrix} \cosh\Gamma & (1/Y_c)\sinh\Gamma \\ Y_c\sinh\Gamma & \cosh\Gamma \end{bmatrix} \begin{bmatrix} e(L) \\ 0 \end{bmatrix}. \qquad (10.127)$$

With T defined as the wave delay time L/v_{p0}, the result becomes

$$\frac{\dot{q}(0)}{e(0)} = Y_c \tanh\Gamma = Y_c \frac{e^{TS}e^{\alpha L} - e^{-TS}e^{-\alpha L}}{e^{TS}e^{\alpha L} + e^{-TS}e^{-\alpha L}} = Y_c \frac{1 - e^{-2TS}e^{-2\alpha L}}{1 + e^{-2TS}e^{-2\alpha L}}$$

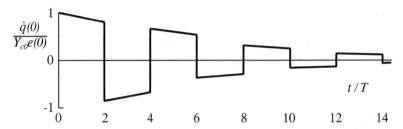

Figure 10.21: Driving point flow for applied step of effort, constant IRC line with blocked end

$$= Y_{c0} \left(1 - \frac{\alpha L}{TS} \right) \left[1 - 2e^{-2TS} e^{-2\alpha L} + 2e^{-4TS} e^{-4\alpha L} - 2e^{-6TS} e^{-6\alpha L} + \cdots \right]. \qquad (10.128)$$

Thus, as plotted in Fig. 10.21, the flow at the left end is admitted and ejected every round-trip wave travel time in steps that decrease exponentially, with fixed slopes between the steps. A more exact model, however, would smooth out the front ends of the steps and place long tails on them, as you will see.

10.6.2   Fluid Line Dynamics

A rigid cylindrical tube is assumed to contain a compressible viscous fluid with kinematic viscosity $\nu \equiv \mu\rho$ and zero *bulk* viscosity (the Stokes condition) and no radiation effects. The equation of state for a liquid is taken to be $dP = \beta\, d\rho/\rho$, and for a gas to be $dP/P = d\rho/\rho + d\theta/\theta$. Small perturbations are assumed for the gas. Swirl modes are ignored. Torroidal modes are neglected, which limits the frequencies to $\nu\omega/\rho c_0^2 \ll 1$, which for most systems is a very high limit. The Navier-Stokes equations are simplified under these assumptions to give

$$\nu \frac{1}{r} \frac{\partial}{\partial r} \left(r \frac{\partial v_x}{\partial r} \right) - \frac{\partial v_x}{\partial t} = \frac{1}{\rho} \frac{\partial P}{\partial x}. \qquad (10.129)$$

The same inequality plus $\sigma(\gamma-1)\omega a/c_0 \ll 1$ also allows axial heat transfer to be neglected. Here, $\sigma = \mu c_p/k$ is the Prandtl number (c_p is specific heat and k is thermal conductivity) and γ is the ratio of specific heats, c_p/c_v. For a gas, this allows the energy equation to be approximated as

$$\frac{1}{r} \frac{\partial}{\partial r} \left(r \frac{\partial \theta}{\partial r} \right) - \frac{\sigma}{\nu} \frac{\partial \theta}{\partial t} = -\frac{(\gamma-1)\theta_0}{\gamma\nu P_0} \frac{\partial P}{\partial t}. \qquad (10.130)$$

Small amplitude laminar disturbances are assumed, and the line must be long enough so radial end effects are negligible. Isothermal walls are assumed for the gas.

The time Laplace transform of equation (10.129) is

$$\frac{1}{r} \frac{\partial}{\partial r} \left(r \frac{\partial \overline{v}_x}{\partial r} \right) - \frac{s}{\nu} \overline{v}_x = \frac{1}{\mu} \frac{\partial \overline{P}}{\partial x}, \qquad (10.131)$$

where the overbar indicates a transformed variable. The boundary condition is that $\overline{v}_x = 0$ at $r = a$. Any forcing function will do. Perhaps the simplest solution of this equation is

$$\overline{v}_x = \overline{v}_{x0} \left[J_0 \left(jr\sqrt{\frac{s}{\nu}} \right) - J_0 \left(ja\sqrt{\frac{s}{\nu}} \right) \right], \qquad (10.132)$$

where $\overline{v}_{x0}$ is any constant with proper units, and J_0 is the Bessel function for the first kind and zeroth order. This solution corresponds to the particular forcing function

$$\frac{1}{\mu} \frac{\partial \overline{P}}{\partial x} = \overline{v}_{x0} s J_0 \left(ja\sqrt{\frac{s}{\nu}} \right). \qquad (10.133)$$

It is necessary to find the transform of the total mass through a cross-section of the tube:

$$\overline{Q} = 2\pi\rho \int_0^a r\overline{v}_x\, dr$$

$$= -\overline{v}_{x0}\pi a^2 \left[J_0\left(ja\sqrt{\frac{s}{\nu}}\right) - \frac{2}{j\sqrt{s/\nu}} J_1\left(ja\sqrt{\frac{s}{\nu}}\right) \right]. \tag{10.134}$$

The series impedance is the ratio of the pressure gradient from equation (10.133) to the flow from equation (10.134):

$$Z(s) = -\frac{\partial\overline{P}/\partial x}{\overline{Q}} = \frac{(\rho/\pi a^2)s}{1 - \dfrac{2J_1(ja\sqrt{s/\nu})}{ja\sqrt{s/\nu}J_0(ja\sqrt{s/\nu})}}. \tag{10.135}$$

The shunt admittance of the line for a liquid is simply Cs, where $C = \pi a^2/\beta$. For a gas, however, it is necessary to apply equation (10.130) to a short section of a line over which the pressure can be assumed to be uniform. This and some other details are given elsewhere[12], with the result

$$Y(s) = C_g s \left[1 + 2(\gamma-1)\frac{J_1(ja\sqrt{\sigma s/\nu}}{ja\sqrt{\sigma s/\nu}J_0(ja\sqrt{\sigma s/\nu})} \right], \tag{10.136a}$$

$$C_g = \pi a^2/\gamma P_0, \tag{10.136b}$$

in which P_0 is the absolute pressure.

The series impedance as given in equation (10.135) can be interpreted in terms of a frequency-dependent resistance and a frequency-dependent inertance, and the shunt admittance of equation (10.136) can be interpreted in terms of a frequency-dependent conductance $G(\omega)$ and a frequency-dependent compliance $C(\omega)$:

$$Z(j\omega) = R(\omega) + j\omega I(\omega), \tag{10.137a}$$

$$Y(j\omega) = G(\omega) + j\omega C(\omega). \tag{10.137b}$$

These are plotted in Fig. 10.22. The resistance and the conductance increase indefinitely with frequency; the dynamic boundary layers, within which the shear action and the temperature gradient are concentrated, shrink ever thinner. As the velocity profile becomes more slug-like, the inertance approaches the simple limit of $\rho/\pi a^2$. As the temperature profile becomes more uniform, the compliance approaches the simple limit of C_g.

The wave speed and characteristic admittance can be found in the usual way for sinusoidal signals. The results are compared with those of the constant-IRC model in Fig. 10.20. The constant-IRC model is represented using the value of $\tau = I/R$ consistent with the infinite-frequency inertance $I = \rho/\pi a^2$ and the zero-frequency resistance $R = 8\mu/\pi a^4$, which has been assumed widely.

Transient responses can be calculated at least three basically different ways. First, Fourier decomposition can utilize the frequency response equations directly. Second, inverse Laplace transforms can be taken of analytical approximations, including certain truncated series and product expansions, to the appropriate transfer functions. Superposition may be invoked in parallel with this method. Third, numerical methods can be applied in the time domain. The literature is extensive and yet fragmentary. The focus below and in Section 10.7 is on a particularly powerful numerical approach involving field lumping.

---

[12]F. T. Brown,"The Transient Response of Fluid Lines," *ASME Transactions, J. of Basic Engineering,* v 84 n 4, December 1962, pp 547-553.

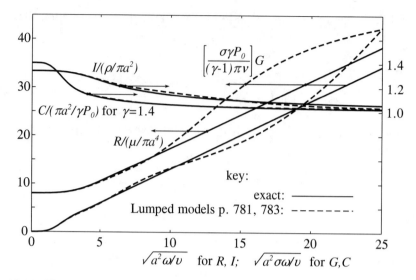

Figure 10.22: Frequency-dependent resistance, inertance, conductance and compliance of laminar flow in a tube

10.6.3   The Skin Effect in Electrical Conductors

Unsteady fields in a conductor push the current towards its surface. The concentration of current, which at high frequencies is limited to a thin boundary layer, increases the effective resistance. It also affects the inertance. This phenomenon is known as the **skin effect**.

The design of high-voltage transmission lines, for example, is affected by this phenomenon. Large conductors must be used to avoid excessive fields and arcing that otherwise would result in the air next to the surface. The skin effect renders the central part of the conductor as superfluous. Hollow tubes are used in consequence, saving weight and cost.

The internal impedance per unit length of a circular wire of radius a, conductivity σ and magnetic permeability μ is[13], in Laplace transform form,

$$Z(s) = \frac{1}{\pi a^2 \sigma} \frac{(ja\sqrt{\sigma\mu s})\, J_0(ja\sqrt{\sigma\mu s})}{2 J_1(ja\sqrt{\sigma\mu s})}. \tag{10.138}$$

The resulting resistance and inductance, as defined in equation (10.137), are plotted in Fig. 10.23 as a function of dimensionless frequency. Nearly all of the inductance in a conducting *pair* represents energy that lies *outside* of the conductors and is not included here.

10.6.4   The Maxwell Model

This simple viscoelastic model applies to shear motions but is shown in part (a) of Fig. 10.24 by analogy as though it applies to longitudinal motion. The material deflects elastically and flows like a viscous liquid. Given a step of strain, the stress decays exponentially with the time constant or **relaxation time** of $\tau = \eta/\mu$. The Maxwell model well describes simple liquids, where $\tau << 10^{-12}$ seconds, which is too small for direct measurement. More significantly, it applies to short chain polymer liquids; an example is polymerized castor oil, for which $\mu = 1.2 \times 10^7 \mathrm{dynes/cm}^2$ and $\eta = 18 \mathrm{dynes/cm}^2$, giving $\tau = 1.5 \times 10^{-6}$ seconds.

[13]S. Ramo and J. R. Whinnery, *Fields and Waves in Modern Radio*, 2d ed., John Wiley & Sons Inc., New York, 1953, pp 230-253

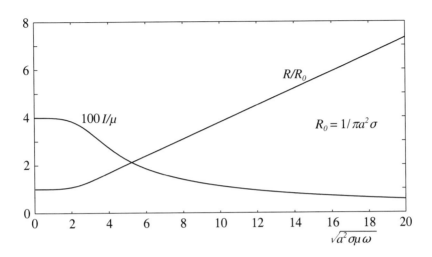

Figure 10.23: Electrical resistance and inductance of circular wire

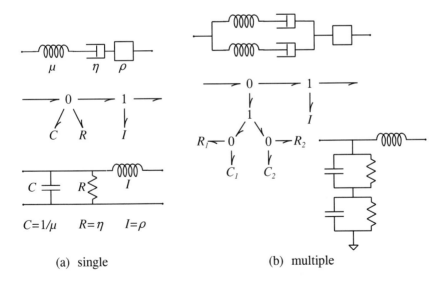

(a) single

(b) multiple

Figure 10.24: The Maxwell model

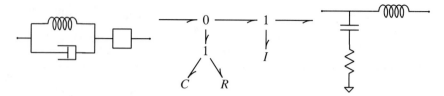

Figure 10.25: The Voigt model

The impedance and admittance per unit length are, respectively,

$$Z(S) = IS; \quad Y(S) = CS + \frac{1}{R}; \quad C = \frac{1}{\mu}; \quad R = \nu; \quad I = \rho. \tag{10.139}$$

The frequency propagation operator becomes

$$\gamma(j\omega) = \sqrt{Y(j\omega)Z(j\omega)} = \sqrt{\left(j - \frac{\omega}{\omega_0}\right)\frac{\rho\omega}{\nu}}; \qquad \omega_0 \equiv \frac{\mu}{\eta}. \tag{10.140}$$

The most useful range is $(\omega/\omega_0) \ll 1$, for which

$$\gamma(j\omega) \simeq (1 + j)\sqrt{\frac{\rho\omega}{2\eta}}, \tag{10.141}$$

which is like diffusion or heat conduction.

A multiple Maxwell model, shown in part (b) of the figure, is more flexible in its description. The material remains a liquid.

10.6.5   The Voigt Model

This model, shown in Fig. 10.25, is the simplest that could describe viscoelastic behavior, either shear or longitudinal, of a solid. A step of stress produces a strain that grows gradually toward an asymptotic limit. The impedance per unit length comprises that of an inertance equal to the density, as with the Maxwell model; the admittance per unit length is

$$Y(S) = \frac{1}{R + 1/CS}. \tag{10.142}$$

Multiple Voigt models can be employed, also.

10.6.6   The Linear Elastic Solid

This model applies either in shear or longitudinally; in the latter case the material sometimes is called a **firmoviscous liquid**. As can be seen from Fig. 10.26, a step of stress produces an immediate elastic deformation followed by an additional deformation that approaches an asymptotic limit with a relaxation time constant. The series impedance per unit length is the same as with the other models above. The shunt admittance per unit length is

$$Y(S) = C_1 S + \frac{1}{R + 1/C_2 S}. \tag{10.143}$$

The resulting frequency propagation operator is

$$\gamma(j\omega) = \sqrt{\frac{-(C_1 + C_2)I\omega^2 - jRC_1C_2I\omega^3}{1 + jRC_2\omega}}. \tag{10.144}$$

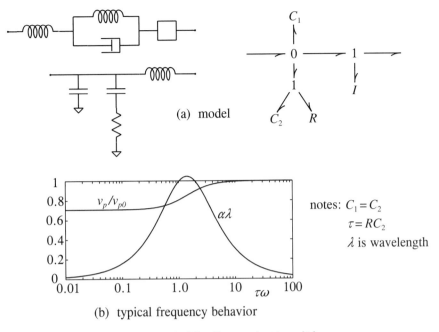

(a) model

notes: $C_1 = C_2$

$\tau = RC_2$

λ is wavelength

(b) typical frequency behavior

Figure 10.26: The linear elastic solid

The phase velocity becomes, for the limits of zero frequency and infinite frequency, respectively,

$$\underset{\omega \to 0}{L}\, v_p = \underset{\omega \to 0}{L}\, \frac{\omega}{\mathrm{Im}\gamma(j\omega)} = \frac{1}{\sqrt{(C_1 + C_2)I}}, \tag{10.145a}$$

$$\underset{\omega \to \infty}{L}\, v_p = \underset{\omega \to \infty}{L}\, \frac{\omega}{\mathrm{Im}\gamma(j\omega)} = \frac{1}{\sqrt{C_1 I}}. \tag{10.145b}$$

An example of the phase velocity and the log of the attenuation ratio per wavelength is given in part (b) of the figure. The material is characterized by a transition band within which the phase velocity changes from a lower to a higher value, and the damping exhibits a large peak.

Related models often contain such **absorption bands** within which the attenuation and a transition of the phase velocity are concentrated.

10.6.7 Simulation of Coupled Storage and Dissipation Fields

The first simulation considered is that of one-port implicit storage and dissipation fields, that is the **impedance field** or RI field and the **admittance field** or RC field, in the absence of the complementary storage field. These fields are not wave-like. Later, the complete coupled IRC field, which combines wave-like and non-wave-like features, is considered. Impedance fields are typified by equation (10.135) for fluid flow in a channel with viscous effects and by equation (10.138) for electric current with resistance effects. Admittance fields are typified by equation (10.136) for the effects of heat transfer on a fluid compliance and by the equations for Y(s) given in equations (10.139), (10.142) and (10.143) for the Maxwell, Voigt and linear elastic solid models, respectively. It should be remembered that the very concept of dissipation represents a discounting of thermal energy; to be more complete one should refer to irreversible rather than dissipative processes and fields.

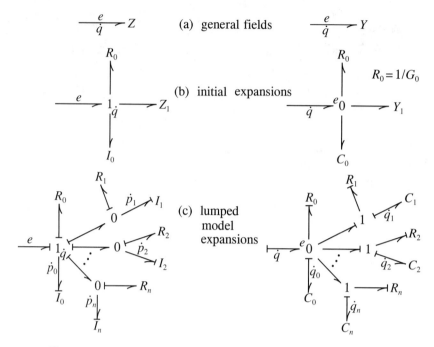

Figure 10.27: Reticulation of impedance and admittance fields

As a first step, an impedance or admittance field is conveniently expanded into a sum of three terms:

$$Z(S) = R_0 + I_0 S + Z_1(S); \quad R_0 = \underset{S \to 0}{L} \; Z(S); \quad I_0 = \underset{S \to \infty}{L} \; \frac{Z(S)}{S}, \tag{10.146a}$$

$$Y(S) = G_0 + C_0 S + Y_1(S); \quad G_0 = \underset{S \to 0}{L} \; Y(S); \quad C_0 = \underset{S \to \infty}{L} \; \frac{Y(S)}{S}. \tag{10.146b}$$

These three terms are indicated as three elements in the bond graphs of part (b) of Fig. 10.27. The third terms, $Z_1(S)$ and $Y_1(S)$, represent the coupling of the fields.

Assume the flow on the impedance and the effort on the admittance experience a unit step change in time. The resulting force and flow are, respectively,

$$h_e(t) = \mathcal{L}^{-1} \left[\frac{R_0}{S} + I_0 + \frac{Z_1(S)}{S} \right], \tag{10.147a}$$

$$h_{\dot{q}}(t) = \mathcal{L}^{-1} \left[\frac{G_0}{S} + C_0 + \frac{Y_1(S)}{S} \right]. \tag{10.147b}$$

The terms which result from Z_1 and Y_1 are called **weighting functions** and are labeled below as $h_{e1}(t)$ and $h_{\dot{q}1}(t)$, respectively. Due to the definitions of R_0 and G_0 they are asymptotic to zero as $t \to \infty$. Due to the definitions of I and C they contain no impulse at $t = 0$, although they may jump to infinite magnitude. They decay monatonically with time, and their time integrals are finite. The force and flow, respectively, can be represented in the time domain by the convolutions

$$e(t) = h_e(t) * \frac{d\dot{q}}{dt} = R_0 \dot{q} + I_0 \frac{d\dot{q}}{dt} + h_{1e}(t) * \frac{d\dot{q}}{dt}, \tag{10.148a}$$

$$\dot{q}(t) = h_{\dot{q}}(t) * \frac{de}{dt} = G_0 e + C_0 \frac{de}{dt} + h_{1\dot{q}}(t) * \frac{de}{dt}. \tag{10.148b}$$

The weighting functions can be found once for all time. Convolution can be carried out, during time simulation of the model, to give the desired effort or flow.

Although this method is potentially very accurate, it is inefficient computationally. A more efficient technique often results from approximating the weighting functions by a finite series of decaying exponentials:

$$h_{1e}(t) \simeq \sum_{i=1}^{n} R_i e^{-t/\tau_i}; \qquad h_{\dot{q}1} \simeq \sum_{i=1}^{n} (1/R_i) e^{-t/\tau_i}. \tag{10.149}$$

This is tantamount to the lumped model shown in part (c) of Fig. 10.27, in which n junctions are each bonded to a resistance and an inertance or compliance. The time constants are

$$\tau_i = I_i/R_i \quad \text{and} \quad \tau_i = R_i C_i. \tag{10.150}$$

This model reduces the inertance or the admittance to a system with $n+1$ simple first-order differential equations. For example, the inertance gives

$$\frac{dp_0}{dt} = e - \sum_{0}^{n} \frac{R_i}{I_i} p_i; \qquad \frac{1}{I_0} p_0 = \dot{q}, \tag{10.151a}$$

$$\frac{dp_i}{dt} = R_i \left(\dot{q} - \frac{1}{I_i} p_i \right), \qquad i = 1, \cdots, n. \tag{10.151b}$$

The time integral of the weighting function equals $\sum_{i=1}^{n} I_i$, which should be set equal to $I(0) - I(\infty)$. For the admittance, $\sum_{i=1}^{n} C_i$ is set equal to $C(0) - C(\infty)$. This leaves $2n - 1$ additional constants to be determined so that the approximations (10.149) and (10.150) are acceptable.

An alternative expansion to that given in equation (10.149) and the corresponding bond graph occasionally gives simpler results. The Voigt and linear elastic solid models, for example, are already in simple calculable forms as represented by their respective bond graphs; the use of equation (10.147) and what follows is unnecessary and complicating. The Maxwell model is a degenerate case; use of equation (10.147) reveals no coupling between storage and dissipation fields.

10.6.8 Example of Viscous Effects in Laminar Flow in a Tube

The ratio of the pressure gradient to the flow for the laminar flow in a tube of length L, as given by the Laplace transform equation (10.135), can be recast as

$$Z(s) = \frac{\rho L s}{\pi a^2} \left[1 + \frac{2}{\mathcal{I}_1(z) - 2} \right], \tag{10.152a}$$

$$\mathcal{I}_1(z) \equiv z J_0(z)/J_1(z); \qquad z \equiv j\sqrt{a^2 s/\nu}. \tag{10.152b}$$

The zero-frequency resistance and the infinite-frequency inertance are

$$R_0 = \lim_{s \to 0} Z(s) = \frac{8\nu\rho L}{\pi a^4}, \tag{10.153a}$$

$$I_0 = \lim_{s \to \infty} \left[\frac{Z(s)}{s} \right] = \frac{\rho L}{\pi a^2}. \tag{10.153b}$$

Introduction of these results into equation (10.152a) gives

$$\frac{Z_1(s)}{s} = \frac{2\rho}{\pi a^2}\left[\frac{1}{\mathcal{I}_1(z)-2} + \frac{4}{z^2}\right]. \tag{10.154}$$

The weighting function is

$$h_{1e}(t) = \mathcal{L}^{-1}\left[\frac{Z_1}{s}\right] = \frac{1}{2\pi j}\oint \frac{Z_1(s)}{s}e^{st}ds, \tag{10.155}$$

where the closed path of integration in the complex plane encompasses all the poles of $Z_1(s)/s$. This has been carried out by Zielke[14], using residue theory, to give

$$h_{1e}(t) = \frac{R_0}{2}\sum_{i=1}^{\infty}e^{-\eta_i^2\tau}; \qquad \tau \equiv \frac{\nu t}{a^2}. \tag{10.156}$$

The process also is illustrated in the subsection that follows.

Equation (10.156) is in the form of equation (10.149) with $R_i = R_0/2$ and $\tau_i = a^2/\nu\eta_i^2$, but unfortunately the series does not converge rapidly for small values of time t or dimensionless time τ. Zielke gives the first 20 values of η_i, but only the first six have any significant effect for $\tau > 0.01$:

$$h_{1e}(\tau) = (R_0/2)[e^{-26.37\tau} + e^{-70.85\tau} + e^{-135.02\tau} + e^{-218.92\tau}$$

$$+e^{-322.55\tau} + e^{-445.93\tau}]; \quad \tau > 0.01. \tag{10.157}$$

For $\tau < 0.01$ the series

$$\mathcal{I}_1(jy) = y + \frac{1}{2} + \frac{3}{8y} + \frac{3}{8y^2} + \frac{63}{128y^3} + \dots, \tag{10.158}$$

gives

$$h_{1e}(\tau) = \frac{R_0}{2}\left[\frac{1}{2\sqrt{\pi}}\frac{1}{\sqrt{\tau}} - \frac{5}{4} + \frac{15}{8\sqrt{\pi}}\sqrt{\tau} + \frac{15}{16}\tau + \frac{45}{64\sqrt{\pi}}\tau^{3/2}\right]; \quad \tau < 0.01. \tag{10.159}$$

The complete weighting function is plotted in Fig. 10.28.

At the first instant, the step change in flow produces an infinitely thin shear layer and therefore an infinite shear stress and infinite weighting function. This corresponds to the infinite resistance at infinite frequency, which also produces an infinitely thin dynamic boundary layer. As time progresses the boundary layer quickly grows, and the wall shear and weighting function decays. Steady-state is approached asymptotically; while the resistance goes to R_0, the weighting function approaches zero, since the steady-state resistance is not included in it.

An approximate model that corresponds to the lumped type given in Fig. 10.27 and equations (10.49)–(10.51) with $n = 3$ is given by Trikha.[15] The integral condition given above requires $I_1 + I_2 + I_3 = I_0/3$. The first term on the right side of equation (10.157) is used as is, using up two of the remaining $2n - 1 = 5$ conditions and insuring that the approximation is within 2% of the proper value for all $\tau > 0.1$. For the remaining three

---

[14]W. Zielke, "Frequency-Dependent Friction in Transient Pipe Flow," *ASME Transactions, J. Basic Engineering*, v 90 n 1, March 1968, pp 109-115

[15]A.K. Trikha, "An Efficient Method for Simulating Frequency-Dependent Friction in Transient Liquid Flow," *ASME Transactions, J. Fluids Engineering*, v 97 n 1, March 1975, pp 97-105

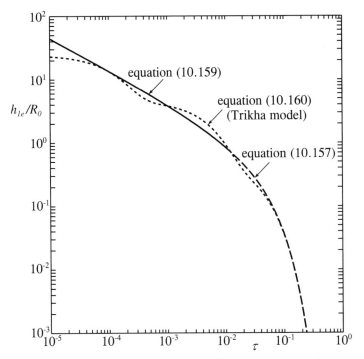

Figure 10.28: Weighting function for impedance of laminar flow in a tube

conditions, the approximation is set equal to the function for three values of τ somewhat close to 0.0001, 0.001 and 0.01. The resulting weighting function is

$$h_{1e}(\tau) \simeq (R_0/2)[e^{-26.4\tau} + 8.1\,e^{-200\tau} + 40.0\,e^{-8000\tau}], \qquad (10.160)$$

which corresponds to

$$R_1 = 0.5\,R_0, \quad R_2 = 4.05\,R_0, \quad R_3 = 20\,R_0;$$
$$I_1 = 0.1515\,I_0, \quad I_2 = 0.162\,I_0, \quad I_3 = 0.02\,I_0. \qquad (10.161)$$

The approximation is compared to the original function in Fig. 10.28. Frequency-domain comparisons of the resulting resistance and inertance are plotted in Fig. 10.22.

More recently, Schohl[16] has given a five-term approximation to the weighting function:

$$h_{1e}(\tau) = (R_0/2)[1.051e^{-26.65\tau} + 2.358e^{-100\tau} + 9.021e^{-669.6\tau} + 29.47e^{-6497\tau}$$
$$+ 79.55e^{-57990\tau}]. \qquad (10.162)$$

This corresponds to

$$R_1 = .5255\,R_0; \quad R_2 = 1.179\,R_0; \quad R_3 = 4.5105\,R_0; \quad R_4 = 14.735\,R_0;$$
$$R_5 = 39.775\,R_0; \qquad (10.163a)$$
$$I_1 = 0.15775\,I_0; \quad I_2 = 0.09432\,I_0; \quad I_3 = 0.05389\,I_0; \quad I_4 = 0.018144\,I_0;$$
$$I_5 = 0.005487\,I_0. \qquad (10.163b)$$

This is the model of choice when high accuracy is desired; the match to the exact weighting function is so good over the range $\tau \geq 10^{-5}$ that it is not plotted separately in Fig. 10.28.

---

[16]G.A. Schohl, "Improved Approximate Method for Simulating Frequency-Dependent Friction in Transient Laminar Flow," *ASME Transactions, J. of Fluids Engineering*, v 115 n 3, 1993, pp. 420-424.

Nevertheless, it is not exact in the limit of $\tau \to \infty$, and the sum of the five Is is less than $I_0/3$ by over one percent (presumably due to error for $\tau < 10^{-5}$).

The models are applied to traveling waves in Section 10.7.5.

10.6.9 Example of Heat Transfer Effects in a Cylindrical Accumulator

The compressibility of a gas is significantly affected by heat transfer. As first noted in Section 3.1.5, for very rapid changes in volume or mass there is no time for significant heat to transfer despite temperature changes, and the effective incremental bulk modulus equals the ratio of the specific heats times the mean absolute pressure: γP_0. For very slow changes in the volume or mass, on the other hand, heat transfer to or from the walls keeps the temperature of the gas virtually constant, and the effective incremental bulk modulus equals the mean pressure, P_0. The intermediate case where heat transfer is important but not complete is modeled for a gas accumulator in Section 9.7.8. The following is more accurate for the special case addressed.

Consider a long cylindrical container with heat conduction, but negligible convection, to isothermal walls. The transfer function between a change in pressure and the resulting change in specific density is given by the admittance of equation (10.136). The corresponding weighting function $h_{1\dot{q}}$ is found below, along with an approximate lumped model which can be used for dynamic simulation.

Equation (10.136) applied to a tube of length L can be written

$$Y(s) = C_g L s \left[1 + \frac{2(\gamma - 1)}{\mathcal{I}_1(z)}\right]; \qquad z \equiv ja\sqrt{\frac{\sigma s}{\nu}}; \qquad C_g = \frac{\pi a^2 L}{\gamma P_0}. \tag{10.164}$$

The low-frequency conductance and the high-frequency compliance are

$$G_0 = \lim_{s \to 0} Y(s) = \lim_{s \to 0} C_g s[1 + \gamma - 1] = 0, \tag{10.165a}$$

$$C_0 = \lim_{s \to \infty} \left[\frac{Y(s)}{s}\right] = C_g. \tag{10.165b}$$

Introduction of these results into equation (10.146b) gives

$$\frac{Y_1(s)}{s} = \frac{2(\gamma - 1)C_g}{\mathcal{I}_1(z)}. \tag{10.166}$$

The weighting function is

$$h_{1\dot{q}}(t) = \mathcal{L}^{-1}\left[\frac{Y_1(s)}{s}\right] = \frac{1}{2\pi j}\oint \frac{Y_1(s)}{s} e^{st} ds = \sum_{k=1}^{\infty} \text{Res}\left\{\left[\frac{Y_1(s)}{s}\right] e^{st}, s_k\right\}. \tag{10.167}$$

The closed path of integration in the complex plane must encompass all the poles of $Y_1(s)/s$ which are the zeros of $\mathcal{I}_1(z)$ with $s_j = -\nu z_k^2/a^2\sigma$. The first ten roots are -2.4048, -5.5201, -8.6537, -10.1735, -13.3237, -16.4706, -19.6159, -22.7601, -25.9037, -29.0468. The weighting function becomes

$$h_{1\dot{q}}(\tau) = \frac{4\nu C_0(\gamma - 1)}{a^2 \sigma} \sum_{k=1}^{\infty} e^{-z_j^2 \tau}; \qquad \tau \equiv \frac{\nu t}{a^2 \sigma}. \tag{10.168}$$

Although this series is absolutely convergent for all $\tau > 0$, the convergence is not fast enough to make this a practical result for small values of τ. For small values, the series given by equation (10.158) can be used, with the result

$$h_{1\dot{q}} = \frac{(\gamma - 1)\nu C_0}{a^2 \sigma}\left[\frac{2\tau^{-1/2}}{\sqrt{\pi}} - 1 - \frac{\tau^{1/2}}{2\sqrt{\pi}} - \frac{\tau}{4} - \frac{25\tau^{3/2}}{48\sqrt{\pi}} - \cdots\right]. \tag{10.169}$$

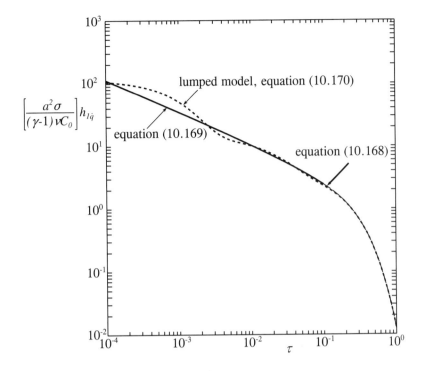

Figure 10.29: Weighting function for admittance of an ideal gas in a tube

This result, used for $\tau < 0.1$, covers 61% of the dynamic portion of Y_1, leaving 39% for equation (10.168) used for $\tau > 0.1$.

An approximate lumped model of the type given in Fig. 10.27 and equations (10.149)-(10.151) with $n = 3$ now will be given. The first term of equation (10.166) is retained, like in the Trikha model above but for a stronger reason; it represents 69% of the total dynamic effect. This gives, with $R_i = 1/G_i$,

$$G_1 = 4.000\frac{(\gamma - 1)\nu C_0}{a^2\sigma}; \quad G_2 = 10.39\frac{(\gamma - 1)\nu C_0}{a^2\sigma}; \quad G_3 = 100.3\frac{(\gamma - 1)\nu C_0}{a^2\sigma}; \qquad (10.170a)$$

$$C_1 = 0.6917(\gamma - 1)C_0; \quad C_2 = 0.208(\gamma - 1)C_0; \quad C_3 = 0.1003(\gamma - 1)C_0. \qquad (10.170b)$$

This model is compared with the continuous model in Fig. 10.29. It matches the complete model very well for $\tau \geq 0.009$, fairly well for $0.002 < \tau < 0.009$ and roughly for $0.00008 < \tau < 0.002$. It does not match for smaller values of τ but does give the exact integral from 0 to ∞. The region $\tau < 0.002$ represents 9.9% of the integral, and the region $\tau < 0.00008$ represents 2.0% of the integral.

Frequency-domain comparisons of the lumped and continuous conductance and compliance are included in Fig. 10.22.

10.6.10   Summary

The impedance of a coupled general inertance-resistance field comprises both frequency-dependent resistance and inertance. The examples of the viscous boundary layer in a tube and the skin effect in a wire have been examined. The admittance of a coupled general compliance-resistance field comprises both frequency-dependent conductance and compliance. The example of a gas in a tube with isothermal walls has been examined. Frequency responses can be carried out fairly directly through the substitution $s \to j\omega$. In most cases, however, it is not practical to carry out the corresponding inverse Laplace transformations to find temporal responses. Usually it is desirable to separate the impedance or admittance into three parts: a simple inertance or compliance, a simple resistance or conductance, and the inherently coupled portion of the two fields. The coupled portion can be characterized by a weighting function which represents its response to a step disturbance. Its effect can be computed by convolution with the time derivative of the imposed rate of change of flow (acceleration) or the rate of change of effort. Much less computation is involved, however, if this coupled portion is approximated by a small number of lumped resistances and inertances or compliances; conventional Runge-Kutta simulation then applies.

The wave-like behavior that results when a coupled storage and dissipation field is itself coupled with a storage field of the complementary type, or with another coupled storage and dissipation field of the complementary type, also can be addressed fairly directly when only pure frequency disturbances or a sum of pure frequency disturbances are being considered. This has been done for the degenerate case of the constant IRC model, and tools necessary for the fluid line, the electrical conductor, and the viscoelastic models known by the names of Maxwell, Voigt and the linear elastic solid have been developed. The problem of the temporal simulation of such a combination is addressed in the following section under the heading of the quasi-method of characteristics.

Guided Problem 10.8

A pressure difference across the ends of a circular tube accelerates an incompressible fluid from rest. The tube is 1.0 meter long and has a diameter of 2.0 cm; the fluid has a density of 900 kg/m^3 and a kinematic viscosity of 1.0×10^{-4} m^2/s (SAE 10W oil at 23°C).

(a) Find and plot the pressure difference as a function of time which produces a constant acceleration of 5 m/s^2. Neglect end effects.

(b) Apply a constant pressure difference of the magnitude which, in the absence of wall shear, would give an acceleration of 5 m/s^2. Model the system with a set of ordinary differential equations.

(c) Simulate and plot the response to part (b).

Suggested Steps:

1. Evaluate R_0, I_0 and $e_{1e}(t)$ for the tube.

2. Address part (a) through the use of equation (10.148a). It is instructive to plot each term separately, as well as the sum, for a duration of 0.2 seconds.

3. Write the differential equations for the lumped model approximation of Fig. 10.27, using the Trikha parameters given in equation (10.161). The applied pressure equals the ratio of the acceleration to the inertance I_0.

4. Carry out the simulation for 0.5 seconds using MATLAB®. A straight line representing the motion of part (a) can be added for comparison, as can the constant inertance-resistance model.

PROBLEM

10.31 The viscoelastic model below applies roughly to materials like acrylic plastic. Characterize the behavior of the materials by showing how the wave speed and attenuation factor vary with frequency.

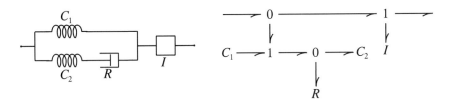

(a) Find the series impedance and shunt admittance per unit length as a function of s.

(b) Find the propagation operator, and substitute $j\omega$ for the s.

(c) Solve for $v_p = \omega/\mathrm{Im}\,\gamma$ and $\alpha = \mathrm{Re}\,\gamma$.

(d) Choose some ratio C_2/C_1 (unity will do), and sketch normalized versions of v_p and α as functions of a dimensionless ω.

SOLUTION TO GUIDED PROBLEM

Guided Problem 10.8

1.
$$R_0 = \frac{8\nu\rho L}{\pi a^4 g_c} = \frac{8 \times 10^{-4} \times 900 \times 1}{\pi(.01)^4 \times 1} = 22.92 \times 10^6 \ \frac{\mathrm{N \cdot s}}{\mathrm{m}^5}$$

$$\frac{\mathrm{m}^2}{\mathrm{s}} \frac{\mathrm{kg}}{\mathrm{m}^3} \frac{\mathrm{m}}{\mathrm{m}^4} \frac{\mathrm{N\,s}^2}{\mathrm{kg\,m}}$$

$$I_0 = \frac{\rho L}{\pi a^2 g_c} = \frac{900 \times 1}{\pi(.01)^2 \times 4} = 2.865 \times 10^6 \ \frac{\mathrm{N \cdot s}^2}{\mathrm{m}^5}$$

$$\frac{\mathrm{kg}}{\mathrm{m}^3} \frac{\mathrm{m}}{\mathrm{m}^2} \frac{\mathrm{N\,s}^2}{\mathrm{kg\,m}}$$

$$\tau = \frac{\mu t}{a^2} = \frac{1 \times 10^{-4}t}{(.01)^2} = 1 \times t, \quad t \text{ in seconds}$$

2. Weighting function: $h_{1e}(t) = \dfrac{R_0}{2}\left[\dfrac{1}{2\sqrt{\pi}}\dfrac{1}{\sqrt{t}} - \dfrac{5}{4} + \dfrac{15}{8\sqrt{\pi}}\sqrt{t} + \dfrac{15}{16}t + \dfrac{45}{64\sqrt{\pi}}t^{3/2}\right]; \quad t \le 0.01 \text{ s}$

$$= \frac{R_0}{2}\left[e^{-a_1 t} + e^{-a_2 t} + e^{-a_3 t} + e^{-a_4 t} + e^{-a_5 t}\right]; \quad t \ge 0.01 \text{ s}$$

$a_1 = 26.37 \ \mathrm{s}^{-1}; \ a_2 = 70.85 \ \mathrm{s}^{-1}; \ a_3 = 135.02 \ \mathrm{s}^{-1}; \ a_4 = ?$

Resistance R_0 term: $R_0 Q = R_0 \dot{Q} t = 47.0 \times 10^3 t$ Pa

Inertance I_0 term: $I_0 \dot{Q} = 4.50 \times 10^3$ Pa

h_{ie} term: $H_{1e}(t) * \dot{Q} = \dot{Q}\int_0^t h_{1e}(\tau)\,d\tau$

$$\dot{Q}\int_0^t h_{1e}(\tau)\,d\tau = \frac{R_0\dot{Q}}{2}\left[\frac{2}{2\sqrt{\pi}}\sqrt{t} - \frac{5}{4}t + \frac{2 \times 15}{3 \times 8\sqrt{\pi}}t^{3/2} + \frac{15}{16 \times 2}t^2 + \frac{45 \times 2}{64\sqrt{\pi} \times 5}t^{5/2}\right], \quad t \le 0.01 \text{ s}$$

$$\dot{Q}\int_0^t h_{1e}(\tau)\,d\tau = \text{integral above for } t = 0.01 \text{ s} + \frac{r_0\dot{Q}}{2}\sum_{i=1}^5\frac{1}{a_i}\left(e^{-0.01a_i} - e^{-a_it}\right),\quad t > 0.01 \text{ s}$$

```
% script file for Guided Problem 10-8
global I0 I1 I2 I3 R0 R1 R2 R3 Qdot
R0=22.918e6; I0=2.865e6; Qdot=1.5708e-3;
a1=26.37; a2=70.85; a3=135.02; a4=218.92; a5=322.55; a6=445.93;
I1=.1515*I0; I2=.162*I0; I3=.02*I0; R1=.5*R0; R2=4.05*R0; R3=20*R0;
% weighting function part of pressure for part (b):
t1=[0:.001:.01]; t2=[.011:.001:.2]; t3=[.001:.001:.19]; t=[t1 t2];
Pw1=Qdot*R0/2*(sqrt(t1/pi)-1.25*t1+30/24/sqrt(pi)*t1.^(1.5)+15/32*t1.^2);
Pw1=Pw1+Qdot*R0/2*90/320/sqrt(pi)*t1.^(2.5);
Pw10=Pw1(11);
ex=exp(-.01*a1)*(1-exp(-a1*t3))/a1+exp(-.01*a2)*(1-exp(-a2*t3))/a2;
ex=ex+exp(-.01*a3)*(1-exp(-a3*t3))/a3+exp(-.01*a4)*(1-exp(-a4*t3))/a4;
ex=ex+exp(-.01*a5)*(1-exp(-a5*t3))/a5+exp(-.01*a6)*(1-exp(-a6*t3))/a6;
Pw2=Pw10*ones(size(t3))+Qdot*R0/2*ex;
PI=Qdot*I0*ones(size(t)); PR=Qdot*R0*t;
plot(t,PI,t,PR,t,[Pw1 Pw2],t,PI+PR+[Pw1 Pw2])
title('Pressure required for constant acceleration of liquid in tube')
xlabel('time, seconds'); ylabel('Pressure, Pa')
```

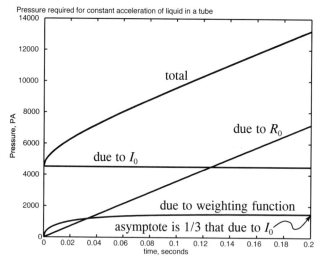

3.

$$\frac{dp_0}{dt} = P - \frac{R_0 + R_1 + R_2 + R_3}{I_0}p_0 + \frac{R_1}{I_1}p_1 + \frac{R_2}{I_2}p_2 + \frac{R_3}{I_3}p_3$$

$$\frac{dp_1}{dt} = \frac{R_1}{I_0}p_0 - \frac{R_1}{I_1}p_1$$

$$\frac{dp_2}{dt} = \frac{R_2}{I_0}p_0 - \frac{R_2}{I_2}p_2$$

$$\frac{dp_3}{dt} = \frac{R_3}{I_0}p_0 - \frac{R_3}{I_3}p_3$$

$P = I_0\dot{Q}(0)$ (also=$R_0 Q_{final}$); note that $Q_{final} = (I_0/R_0)\dot{Q}(0) = \tau\dot{Q}(0)$, where τ is a time constant. For the constant $I - R$ model, $Q = (1 - e^{t/\tau})Q_{final}$.

4.

```
% simulation for flow, part (c), solving the differential equations above
[t,x]=ode45('GP108',[0 .5],[0 0 0 0]);
tau=I0/R0; Qf=tau*Qdot;
t4=[0:.005:.5];
Q=Qf*(1-exp(-t4/tau));
plot(t,x(:,1)/I0,t,Qdot*t,t4,Q)
title('Response of fluid line to step change in pressure')
xlabel('time, seconds'); ylabel('flow rate, m^3/s')

function f=GP108(t,x) % function file called above
global I0 I1 I2 I3 r) R0 R1 R2 R3 Qdot
P=I0*Qdot;
f(1)=P-(R0+R1'+R2+R3)/I0+R1/I1*x(1)+R1/I1*x(2)+R2/I2*x(3)+R3/I3*x(4);
f(2)=R1/I0*x(1)-R1/I1*x(2);
f(3)=R2/I0*x(1)-R2/I2*x(3);
f(4)=R3/I0*x(1)-R3/I3*x(4);
f=f';
```

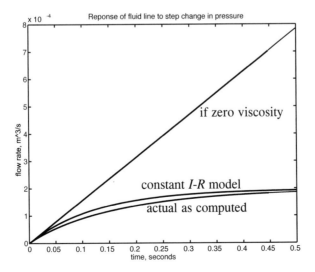

The constant $I - R$ model is extremely simple and gives the proper steady-state flow. The deviation of the actual velocity profile across the tube from the steady-state parabolic shape reduces the acceleration of the flow noticeably, however.

10.7   Wave-Scattering Variables

Bilateral waves in linear media pass through each other without interaction, as has been seen in Section 10.1 for the special case of pure wave-like models. This makes them particularly useful for analyzing distributed-parameter models. One application you have seen employs the hodograph plane (Section 10.1.6). Waves also can be seen in some of the more general equations in Section 10.4, but the emphasis there was quickly placed on the power-factor variables as seen at boundaries. The presentation below generalizes the earlier results. **Wave-scattering variables** actually are a fundamental alternative to power-factor variables and can be adapted to non-wave-like models, although that is not done here.

Since the various wave-scattering variables of a particular uniform linear model, $\mathbf{w} = [w_1, w_2, \cdots, w_n]^T$, act independently of one another, they satisfy the differential equation

$$\frac{dw_i}{dx} = -\mu_i w_i; \quad \mu_i = \mu_i(S) \qquad i = 1, \cdots, n, \tag{10.171}$$

assuming that there are no internal excitations. The solution of this equation is

$$w_i(x, t) = e^{-\mu_i x} w_i(0, t) \qquad i = 1, \cdots, n. \tag{10.172}$$

The simplicity of this result commends its use as long as the transformation itself is not too awkward, which sometimes it is. Internal excitations can be introduced, also.

The special case of pure wave-like behavior, such as discussed earlier, has the special form

$$\mu_i = S/c_i \quad \text{or} \quad \mu_i = -S/c_i \qquad i = 1, \cdots, n, \tag{10.173}$$

in which the c_i are constants and S is the time derivative operator. In these cases, equation (10.172) represents waves traveling one way or the other at constant velocities, c_i. The use of wave-scattering variables in this case often is called **the method of characteristics**.

Models that exhibit underlying wave-like behavior with superimposed non-wave-like features, such as are introduced by resistance in the fluid and electrical lines and gradual nonuniformities in the lines, can be analyzed approximately by an adaptation that has been called the **quasi-method of characteristics**.

The basic method of characteristics applies to semi-linear, quasi-linear and nonlinear models as well as linear models, but the focus here is on linear models only.

10.7.1   The Transformation

The linear partial differential equation

$$\frac{\partial \mathbf{y}}{\partial x} = -\mathbf{A}\mathbf{y} + \mathbf{B}u(x, t) \tag{10.174}$$

will be assumed, in which $\mathbf{A}$ is a function of S, x and t. In the restrictive case of greatest interest, $\mathbf{A}$ is a function of s only. In this case it will be presumed that the eigenvalues μ_i of the equation

$$(\mathbf{A} - \mu_i \mathbf{I})\mathbf{v}_i = \mathbf{0} \quad i = 1, \cdots, n \tag{10.175}$$

are distinct. The eigenvectors $\mathbf{v}_i$ can be collected in any order to form a square modal matrix, $\mathbf{M}$. The wave-scattering variables, $\mathbf{w}$, are defined by

$$\mathbf{y} = \mathbf{M}\mathbf{w}; \qquad \mathbf{w} = \mathbf{M}^{-1}\mathbf{y}. \tag{10.176}$$

Substitution of this transformation into equation (10.174) gives

$$\frac{\partial \mathbf{w}}{\partial x} = -\mathbf{C}\mathbf{w} + \mathbf{M}^{-1}\mathbf{B}u(x, t), \tag{10.177}$$

in which $\mathbf{C}$ is the diagonal matrix

$$\mathbf{C} = \mathbf{M}^{-1}\mathbf{A}\mathbf{M} = \begin{bmatrix} \mu_1 & 0 & \cdot & \cdot \\ 0 & \mu_2 & \cdot & \cdot \\ \cdot & \cdot & \cdot & \cdot \\ \cdot & \cdot & \cdot & \mu_n \end{bmatrix}. \tag{10.178}$$

In the unexcited case ($\mathbf{u} = \mathbf{0}$) this corresponds to equation (10.171), and has the solution of equation (10.172). More specialized cases are considered next.

10.7.2   Single-Power Uniform Symmetric Models

As shown earlier, this common case corresponds to

$$A = \begin{bmatrix} 0 & Z \\ Y & 0 \end{bmatrix}. \tag{10.179}$$

The eigenvalues are

$$\mu_{1,2} = \pm\sqrt{YZ} = \pm\gamma, \tag{10.180}$$

in which $\gamma = \gamma(s)$ is the familiar propagation operator. Therefore,

$$\mathbf{C} = \begin{bmatrix} \gamma & 0 \\ 0 & -\gamma \end{bmatrix}, \tag{10.181}$$

and the solution in the unexcited case is

$$w^+(x) \equiv w_1(x) = e^{-\gamma(x-x_0)} w_1(x_0), \tag{10.182a}$$

$$w^-(x) \equiv w_2(x) = e^{\gamma(x-x_0)} w_2(x_0). \tag{10.182b}$$

The eigenvectors can be multipied by any constant; we will choose their upper elements to equal $1/\sqrt{2Y_c}$, where, as before,

$$Y_c = \sqrt{Y/Z}. \tag{10.183}$$

This gives

$$\mathbf{M} = \frac{1}{\sqrt{2}} \begin{bmatrix} 1/\sqrt{Y_c} & 1/\sqrt{Y_c} \\ \sqrt{Y_c} & -\sqrt{Y_c} \end{bmatrix}; \quad \mathbf{M}^{-1} = \frac{1}{\sqrt{2}} \begin{bmatrix} \sqrt{Y_c} & 1/\sqrt{Y_c} \\ \sqrt{Y_c} & -1/\sqrt{Y_c} \end{bmatrix}. \tag{10.184}$$

If the effort-and-flow variables are used in $\mathbf{y}$, the power transmitted is

$$\mathcal{P} = \frac{w^{+2}}{2} - \frac{w^{-2}}{2}. \tag{10.185}$$

Thus, w^+ represents the amplitude of waves carrying energy in the positive-x direction, and w^- represents the amplitude of waves carrying energy in the minus-x direction. For this symmetrical single-power type, wave-scattering variables could be called *difference-factoring* variables to contrast them with *product-factoring* variables.

10.7.3   The Method of Characteristics for Pure Bilateral Waves

The simplest case is the familiar pure bilateral wave model

$$Z = Is; \quad Y = Cs, \tag{10.186}$$

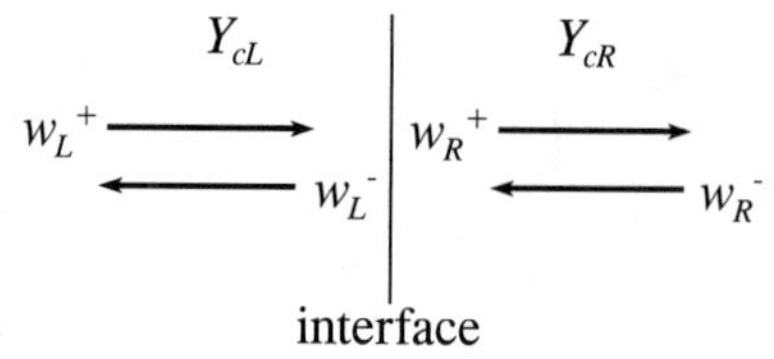

Figure 10.30: Transmitted and reflected scattering variables at a discontinuity

with I and C being constants, for which

$$\gamma = S/c; \qquad c = 1/\sqrt{IC} \qquad Y_c = \sqrt{C/I}. \tag{10.187}$$

This corresponds to equation (10.173) with $c_1 = c_2$, that is

$$\begin{bmatrix} w^+(x + x_0) \\ w^-(x + x_0) \end{bmatrix} = \begin{bmatrix} e^{-(x-x_0)s/c} & 0 \\ 0 & e^{(x-x_0)s/c} \end{bmatrix} \begin{bmatrix} w^+(x_0) \\ w^-(x_0) \end{bmatrix}. \tag{10.188}$$

The scattering variables emanate from either or both ends because of changes in the boundary conditions or because an incident wave is reflected. If x runs from left to right, one can readily show that the right-to-left scattering variable emanating from the left end is

$$w^+ = \sqrt{\frac{2}{Y_c}}\, \dot{q} + w^- = \sqrt{2Y_c}\, e - w^-. \tag{10.189}$$

Thus, for example, if $\dot{q} = 0$ at this end (blocked end), an incident scattering variable is reflected by a scattering variable of equal amplitude; the **reflection coefficient** is 1. If on the other hand the boundary condition is $e = 0$, the reflection coefficient is -1. Similarly, the left-to-right scattering variable emanating from the right end is

$$w^- = -\sqrt{\frac{2}{Y_c}}\, \dot{q} + w^+ = \sqrt{2Y_c}\, e - w^+. \tag{10.190}$$

The junction of two line segments with different characteristic admittances, Y_{cL} on the left and Y_{cR} on the right, as shown in Fig. 10.30, is another circumstance of frequent interest. Using the subscripts L and R to represent the left and right sides, respectively,

$$\begin{bmatrix} w_L^+ \\ w_L^- \end{bmatrix} = \mathbf{M}_L^{-1}\mathbf{M}_R \begin{bmatrix} w_R^+ \\ w_R^- \end{bmatrix} = \begin{bmatrix} a & b \\ b & a \end{bmatrix} \begin{bmatrix} w_R^+ \\ w_R^- \end{bmatrix}, \tag{10.191a}$$

$$a = \frac{1}{2}\left(\sqrt{\frac{Y_{cL}}{Y_{cR}}} + \sqrt{\frac{Y_{cR}}{Y_{cL}}} \right), \tag{10.191b}$$

$$b = \frac{1}{2}\left(\sqrt{\frac{Y_{cL}}{Y_{cR}}} - \sqrt{\frac{Y_{cR}}{Y_{cL}}} \right). \tag{10.191c}$$

Solving for the causal output waves,

$$\begin{bmatrix} w_R^+ \\ w_L^- \end{bmatrix} = \begin{bmatrix} c_t & c_{rR} \\ c_{rL} & c_t \end{bmatrix} \begin{bmatrix} w_L^+ \\ w_R^- \end{bmatrix}. \tag{10.192a}$$

$$\textbf{reflection coefficient}: c_{rL} = -c_{rR} = \frac{b}{a} = \frac{Y_{cL} - Y_{cR}}{Y_{cL} + Y_{cR}} \tag{10.192b}$$

$$\text{transmission coefficient}: c_t = \frac{2}{\sqrt{Y_{cL}/Y_{cR}} + \sqrt{Y_{cR}/Y_{cL}}} \qquad (10.192c)$$

When the two characteristic admittances are equal, that is when there is no discontinuity in the characteristic admittance, $T = 1$ and $R = 0$, as should be expected. When the right side is blocked so that $Y_{cR} = 0$, the reflection coefficient for the scattering variables on the left becomes $+1$, and when $Y_{cR} \to \infty$, it becomes -1, again as expected.

Waves are usually thought of as *changes* or discontinuites in the scattering variables, which propagate with them. Waves therefore emanate from a particular location because of reflections or transmissions of incident waves and/or because of *changes* imposed in the effort or flow. The reflection and transmission coefficients of waves are the same as with the scattering variables themselves.

In the classical method of characteristics, waves are often represented graphically in a position-time plot. In the linear case under consideration, the waves become straight lines of slope equal to the inverse of the wave speed. In acoustic applications, compression waves are represented by solid lines and rarefaction waves by dashed lines; this can be generalized to apply to waves in which the effort increases and decreases, respectively. An acoustic application is given in Fig. 10.31. A piston at the left end of a tube suddenly starts moving at a velocity v_0, generating a compression wave. This wave is partially reflected and transmitted at a discontinuity in the area of the tube and is reflected completely at the open end. Waves that return to the left end are reflected completely there, also, although with a reflection coefficient of opposite sign. The complete diagram can be constructed and magnitudes determined from knowledge of the reflection and transmission coefficients and the amplitude of the original wave. Pressure and flow histories at a particular location are extracted and plotted in part (b) of the figure.

A segment of uniform bilateral delay line can be represented with the delay bond graph symbol:

$$\frac{e_1}{\dot{q}_1} \longrightarrow D \frac{e_2}{\dot{q}_2} \longrightarrow$$

Acoustic tubing tees and other common-effort junctions can be represented by 0-junctions. Compliances can be modeled by short but fat delay lines. Inertances can be approximated by short but narrow delay lines. Whole systems, particularly networks of quasi-delay lines such as hydraulic tubing systems, can be represented in this manner. Resistances can be lumped at the interfaces between delay lines with reasonable accuracy if the lines are short, as they are apt to be. Such a model is attractive if waves in all the delay lines are synchronized, that is if all the delay lines have the same time delay. Usually this would require rather short but numerous delay lines. Updating of the waves is done once per wave delay time, simply using the reflection and transmission coefficients at each interface.

The reflection and transmission coefficients at a junction between two delay lines connected by a series resistance R can be found from the following generalization of equation (10.191a):

$$\begin{bmatrix} w_L^+ \\ w_L^- \end{bmatrix} = \mathbf{M}^{-1} \begin{bmatrix} 1 & R \\ 0 & 1 \end{bmatrix} \mathbf{M} \begin{bmatrix} w_R^+ \\ w_R^- \end{bmatrix}. \qquad (10.193)$$

The resulting reflection and transmission coefficients are

$$c_r = \frac{Y_{ci} - Y_{ct} + RY_{ci}Y_{ct}}{Y_{ci} + Y_{ct} + RY_{ci}Y_{ct}}, \qquad (10.194a)$$

$$c_t = \frac{2}{\sqrt{Y_{ci}/Y_{ct}} + \sqrt{Y_{ct}/Y_{ci}} + R\sqrt{Y_{ci}Y_{ct}}}, \qquad (10.194b)$$

where the subscript i refers to the delay line with the incident wave and the subscript t refers to the delay line with the transmitted wave. Reflection and transmission coefficients

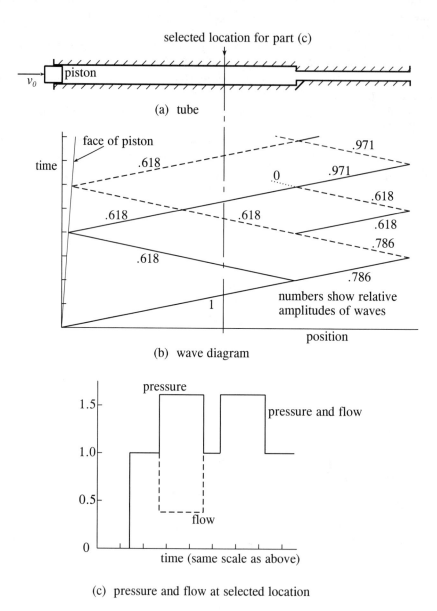

Figure 10.31: Acoustics solution by the graphical method of characteristics

for interfaces with shunt resistances and for junctions of three or more delay lines can be found using similar techniques.

The plots of waves in the hodograph plane given in Section 10.1.6 represent a related graphical technique. This approach is particularly well suited to problems in which the end conditions can be represented by static effort-flow characteristics, either linear or nonlinear.

10.7.4   Generalization to Non-Symmetric Cases

The method of characteristics is useful for hyperbolic models, which implies that the eigenvalues μ_i are of the form

$$\mu_i = \tau_i S, \tag{10.195}$$

in which the τ_i are real and distinct. With the new derivative operator

$$D_i = \partial/\partial x + \tau_i \, \partial/\partial t, \tag{10.196}$$

equation (10.177) becomes n uncoupled ordinary first-order differential equations describing how the wave-scattering variables change along the characteristics:

$$D_i w_i = [\mathbf{M}^{-1}\mathbf{B}u(x,t)]_i \qquad i = 1, \cdots, n. \tag{10.197}$$

In the absence of internal excitation this reduces to $w_i = $ constant along its trajectory or characteristic in space-time, which describes a velocity of $-1/\tau_i$.

For the one-power system with effort and flow variables, the generally non-symmetric case

$$\mathbf{A} = \begin{bmatrix} W'S & IS \\ CS & X'S \end{bmatrix} \tag{10.198}$$

gives

$$\tau_{1,2} = \frac{W' + X'}{2} \pm \sqrt{\left(\frac{W' - X'}{2}\right)^2 + IC}, \tag{10.199}$$

which describes waves traveling at different speeds in the two directions. As a special example, the equations for planar waves in a fluid, including the possibility of a mean velocity are

$$\text{conservation of momentum}: \quad \frac{\partial P}{\partial x} + V\rho\frac{\partial V}{\partial x} + \rho\frac{\partial V}{\partial t} = 0, \tag{10.200a}$$

$$\text{conservation of mass}: \quad \frac{\partial}{\partial x}(\rho A V) + A\frac{\partial \rho}{\partial t} = 0, \tag{10.200b}$$

$$\text{equation of state}: \quad dP = \beta\,d\rho/\rho. \tag{10.200c}$$

For the effort $e = P$ and flow $\dot{q} = V$, these equations give, for the acoustic linearization about the velocity V_0,

$$\mathbf{A} = \frac{1}{c^2 - V_0^2} \begin{bmatrix} -V_0 & \beta \\ 1/\rho & -V_0 \end{bmatrix} S, \qquad c \equiv \sqrt{\beta/\rho}, \tag{10.201a}$$

$$\tau_{1,2} = \frac{-1}{c + V_0}, \frac{1}{c - V_0}. \tag{10.201b}$$

The modal matrix is the same as given by equation (10.184), with $Y_c = 1/\sqrt{\beta\rho}$.

10.7.5   The Quasi Method of Characteristics

Dissipation introduced into an otherwise pure wave-like system nearly always renders the differential equation as non-hyperbolic[17], greatly complicating temporal simulation. If the

---

[17]One special exception is the constant-I-R-C-G line with $RC = GI$, in which waves propagate with attenuation but no dispersion.

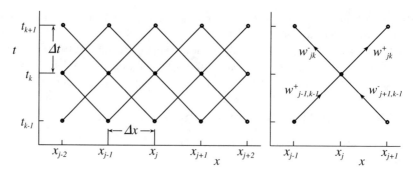

Figure 10.32: Space-time grid for the quasi method of characteristics

dissipation does not dominate the behavior, however, there may be good reason to isolate the terms that destroy hyperbolicity, and to treat them as disturbances on an otherwise hyperbolic system analyzed using the method of characteristics. The resulting hybrid procedure has been called the **quasi method of characteristics** [18]. If implemented using the lumped-model approximation of the previous section, it becomes reasonably efficient computationally.

Acoustic waves propagated in a rigid circular tube of constant diameter are considered below. Either a viscous hydraulic fluid having constant compressibility or an ideal viscous gas which also exhibits the effects of heat transfer to an isothermal wall are assumed. The series impedance $Z(s)$ and shunt admittance $Y(s)$ are expanded as in equations (10.146). The lumped approximations given in part (c) of Fig. 10.27 will be used for both the viscous and the heat transfer effects.

Segregating that part of the pressure gradient due to viscosity as P', and that part of the gradient in flow due to heat transfer as Q', the equations of motion become

$$\frac{\partial P}{\partial x} = -I_0 \frac{\partial Q}{\partial t} + P', \tag{10.202a}$$

$$\frac{\partial Q}{\partial x} = -C_0 \frac{\partial P}{\partial t} + Q'. \tag{10.202b}$$

The propagation of the waves w^+ and w^- then can be written as

$$w^+ \text{ wave}: \quad \left(\frac{\partial}{\partial x} + \frac{1}{c} \frac{\partial}{\partial t} \right) (Y_c P + Q) = Y_c P' + Q', \tag{10.203a}$$

$$w^- \text{ wave}: \quad \left(\frac{\partial}{\partial x} - \frac{1}{c} \frac{\partial}{\partial t} \right) (Y_c P - Q) = Y_c P' - Q', \tag{10.203b}$$

$$Y_c = \sqrt{C_0/I_0}. \tag{10.203c}$$

These equations are applied to a space-time grid as shown in Fig. 10.33. Grid points are identified with numbered subscripts, e.g., x_j, x_{j+1}. The angled lines represent the characteristics in the absence of heat transfer or viscous effects. They have the slope $\pm 1/c$, so the time increments used in the solution are $\Delta t = \Delta x/c$. The solution times also are identified with numbered subscripts, e.g., t_k, t_{k+1}. The pressures, velocities and waves associated with grid points will be distinguished with two subscripts, the first for position

---

[18]F.T. Brown, "A Quasi Method of Characteristics With Application to Fluid Lines With Frequency-Dependent Wall Shear and Heat Transfer," *ASME Transactions, J. of Basic Engineering*, v 91 n 2, June 1969, pp 217-227

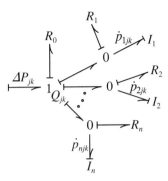

Figure 10.33: Lumped model for the viscous effects

and the second for time. For example, $w^+{}_{jk}$ represents the wave traveling in the positive-x direction which emanates from the grid point at x_j and t_k, and becomes incident on the grid point at x_{j+1} and t_{k+1}. The best choice of length of the segments is a compromise betweeen accuracy and computational efficiency. It should be shorter than the smallest wavelength of serious interest. An exception exists when there are no viscous and heat-transfer effects; the grid then can be as large as the boundary conditions permit.

The viscosity-induced pressure drop over a segment Δx in the x-direction is defined as ΔP, and the heat-transfer-induced flow drop is defined as ΔQ. This gives

$$w^+ \text{ wave}: \quad P' = -\Delta P/\Delta x; \quad Q' = -\Delta Q/\Delta x, \quad (10.204a)$$

$$w^- \text{ wave}: \quad P' = -\Delta P/(-\Delta x); \quad Q' = -\Delta Q/(-\Delta x), \quad (10.204b)$$

Following the characteristics from x_{j-1}, t_{k-1} and x_{j+1}, t_{k-1} to the common x_j, t_j, equations (10.203) now give the approximations

$$w^+ \text{ wave}: \quad (Y_c P_{jk} + Q_{jk}) - (Y_c P_{j-1,k-1} + Q_{j-1,k-1})$$
$$= -\frac{Y_c}{2}(\Delta P_{j-1,k-1} + \Delta P_{jk}) - \frac{1}{2}(\Delta Q_{j-1,k-1} + \Delta Q_{jk}), \quad (10.205a)$$

$$w^- \text{ wave}: \quad (Y_c P_{jk} - Q_{jk}) - (Y_c P_{j+1,k-1} - Q_{j+1,k-1})$$
$$= \frac{Y_c}{2}(\Delta P_{j+1,k-1} + \Delta P_{jk}) - \frac{1}{2}(\Delta Q_{j+1,k-1} + \Delta Q_{jk}). \quad (10.205b)$$

The model for the ΔPs is taken from part (c) of Fig. 10.27 (p. 778), with the inertance I_0 removed, and is shown in Fig. 10.33. The pressure drop ΔP_{jk} across the segment centered at x_j and t_k due to the viscous effects is

$$\Delta P_{jk} = \sum_{i=0}^{n} R_i Q_{jk} - \sum_{i=1}^{n} \frac{R_i}{I_i} p_{ijk} \quad (10.206)$$

Either the Trikha values of R_i and I_i with $n = 3$ or the Schohl values with $n = 5$ can be used; the former is given by equations (10.153) and (10.161) and the latter is given by equations (10.153) and (10.163), in both cases with the length L replaced by Δx. The need now is to express the momenta p_{ijk} within equation (10.206) by functions of the state variables at the preceding time. The differential equations for the momenta are given by

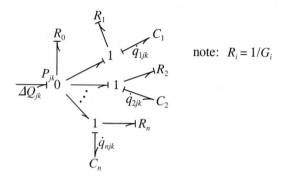

Figure 10.34: Lumped model for the effects of heat transfer in an ideal gas

equations (10.151), from which

$$\frac{dp_{ijk}}{dt} = R_i\left(Q_{jk} - \frac{1}{I_i}p_{ijk}\right) \qquad i = 1,\cdots,n. \tag{10.207}$$

This equation can be integrated approximately over the time interval from $t = (k-1)\Delta t$ to $k\Delta t$ by assuming that Q_j varies linearly; i.e., $Q_j(t) = Q_{j,k-1} + (t/\Delta t)(Q_{jk} - Q_{j,k-1})$. The result is

$$p_{ijk} = a_i p_{ij,k-1} + b_i Q_{jk} + c_i Q_{j,k-1}, \tag{10.208a}$$

$$a_i = e^{-\Delta t/\tau_i}; \quad b_i = I_i\left[1 - \frac{\tau_i}{\Delta t}(1 - a_i)\right]; \quad c_i = I_i\left[\left(1 + \frac{\tau_i}{\Delta t}\right)(1 - a_i) - 1\right], \tag{10.208b}$$

$$\tau_i \equiv I_i/R_i. \tag{10.208c}$$

When substituted into equation (10.206) this gives

$$\Delta P_{jk} = A_1 Q_{jk} - A_2 Q_{j,k-1} - p_{0j,k-1}, \tag{10.209a}$$

$$p_{0j,k-1} = \sum_{i=1}^{n}(a_i/\tau_i)p_{ij,k-1}, \tag{10.209b}$$

$$A_1 = R_0 + \sum_{i=1}^{n} R_i(1 - b_i/I_i); \qquad A_2 = \sum_{i=1}^{n}(c_i/\tau_i). \tag{10.209c}$$

The case with heat transfer effects, also as adapted from part (c) of Fig. 10.27, is modeled as shown in Fig. 10.34. The flow difference across the segment centered at x_j and t_k due to these effects, analogous to equation (10.206), is

$$\Delta Q_{jk} = \sum_{i=0}^{n} G_i P_{jk} - \sum_{i=1}^{n} \frac{G_i}{C_i} q_{ijk}. \tag{10.210}$$

For most applications, the values of $G_i = 1/R_i$ and C_i given by equations (10.164), (10.165) and (10.170), with the length L replaced by Δx, likely are adequate, although an improved version with $n = 4$ or $n = 5$ might be preferred. The displacements within this equation satisfy

$$\frac{dq_{ijk}}{dt} = G_i\left(P_{jk} - \frac{1}{C_i} q_{ijk}\right), \qquad i = 1,\cdots,n, \tag{10.211}$$

which when discretized becomes

$$q_{ijk} = e_i q_{ij,k-1} + f_i P_{jk} + g_i P_{j,k-1}, \tag{10.212a}$$

$$e_i = e^{-\Delta t/\tau_{hi}}; \quad f_i = C_i \left[1 - \frac{\tau_{hi}}{\Delta t}(1 - e_i)\right]; \quad g_i = C_i \left[\left(1 + \frac{\tau_{hi}}{\Delta t}\right)(1 - e_i) - 1\right]; \tag{10.212b}$$

$$\tau_{hi} \equiv C_i/G_i. \tag{10.212c}$$

When substituted into equation (10.210) this gives

$$\Delta Q_{jk} = B_1 P_{jk} - B_2 P_{j,k-1} - q_{0j,k-1}, \tag{10.213a}$$

$$q_{0j,k-1} = \sum_{i=1}^{n} (e_i/\tau_{hi}) q_{ij,k-1}, \tag{10.213b}$$

$$B_1 = G_0 + \sum_{i=1}^{n} G_i(1 - f_i/C_i); \qquad B_2 = \sum_{i=1}^{n} g_i/\tau_{hi}. \tag{10.213c}$$

When the index j represents a right-hand boundary condition (so that $j+1$ does not exist), equation (10.205b) is replaced by a boundary condition. When j represents a left-hand boundary (so that $j-1$ does not exist), equation (10.205a) is replaced by a boundary condition. For non-boundary nodes, equations (10.205a) can be added to and subtracted from (10.205b), to give, respectively,

$$P_{jk} = P_{0jk} - \frac{1}{4}(\Delta P_{j-1,k-1} - \Delta P_{j+1,k-1})$$
$$- \frac{1}{4Y_c}(\Delta Q_{j-1,k-1} + 2\Delta Q_{jk} + \Delta Q_{j+1,k-1}), \tag{10.214a}$$

$$Q_{jk} = Q_{0jk} - \frac{Y_c}{4}(\Delta P_{j-1,k-1} + 2\Delta P_{jk} + \Delta P_{j+1,k-1})$$
$$- \frac{1}{4}(\Delta Q_{j-1,k-1} - \Delta Q_{j+1,k-1}), \tag{10.214b}$$

$$P_{0jk} \equiv \frac{1}{2}(P_{j-1,k-1} + P_{j+1,k-1}) + \frac{1}{2Y_c}(Q_{j-1,k-1} - Q_{j+1,k-1}), \tag{10.214c}$$

$$Q_{0jk} \equiv \frac{Y_c}{2}(P_{j-1,k-1} - P_{j+1,k-1}) + \frac{1}{2}(Q_{j-1,k-1} + Q_{j+1,k-1}). \tag{10.214d}$$

The substitution of equations (10.209) and (10.213) give

$$\left(1 + \frac{B_1}{2Y_c}\right) P_{jk} = P_{0jk} - \frac{1}{4}(\Delta P_{j-1,k-1} - \Delta P_{j+1,k-1})$$
$$- \frac{1}{4Y_c}(\Delta Q_{j-1,k-1} + \Delta Q_{j+1,k-1} - 2B_2 P_{j,k-1} - 2q_{0j,k-1}), \tag{10.215a}$$

$$\left(1 + \frac{Y_c A_1}{2}\right) Q_{jk} = Q_{0jk} - \frac{1}{4}(\Delta Q_{j-1,k-1} - \Delta Q_{j+1,k-1})$$
$$- \frac{Y_c}{4}(\Delta P_{j-1,k-1} + \Delta P_{j+1,k-1} - 2A_2 Q_{j,k-1} - 2p_{0j,k-1}). \tag{10.215b}$$

An example simulation of a hydraulic transient is shown in Fig. 10.35. The valve is opened suddenly to its final state, starting the transient. The problem and its parameter

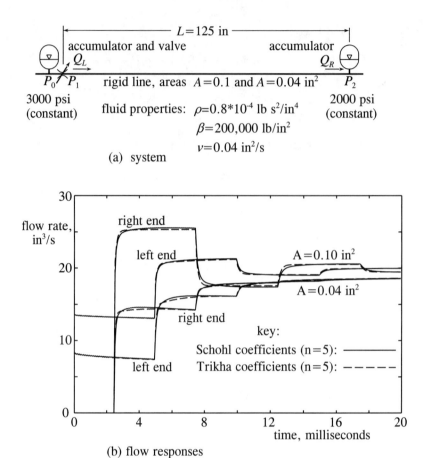

Figure 10.35: Example of a hydraulic transient

values are the same as those in Example 10.8 (pp. 723-724) except for the introduction of viscosity. The fluid is SAE 10W at 118°F. Note the smoothing and long tails, which result from the viscous effects. The results with the Schohl values ($n = 5$) are nearly exact; those with the Trikha values ($n = 3$) are inferior but adequate for most purposes.

The simulations employed $m = 41$ nodes to divide the tube into 40 segments. Using a larger number makes negligible difference. The boundary condition at the right end is $P_{m,k} = P_R = 2000$ psi, which when combined with equation (10.205b) gives

$$Q_{mk} = \frac{1}{1 + Y_c A_1/2}\left[Q_{m-1,k-1} + Y_c(P_{m-1,k-1} - P_R)\right.$$
$$\left. + \frac{Y_c}{2}(A_2 Q_{m,k-1} + p_{0m,k-1} - \Delta P_{m-1,k-1}).\right. \tag{10.216}$$

The boundary condition at the left end is

$$P_{1k} = P_L - \frac{(P_L - P_R)Q_{1k}^2}{Q_{eq}^2}, \tag{10.217}$$

where $P_L = 3000$ psi and the flow at equilibrium (in the case of zero viscosity) is $Q_{eq} = 20$

in^3/s. When combined with equations (10.205b) and (10.207) this gives

$$\frac{Y_c(P_L - P_R)}{Q_{eq}^2}Q_{1k}^2 + \left(1 + \frac{Y_c A_1}{2}\right)Q_{1k} - Q_{2,k-1}$$
$$+ Y_c\left[P_{2,k-1} - P_L - \frac{1}{2}(A_2 Q_{1,k-1} + p_{01,k-1} - \Delta P_{2,k-1})\right] = 0.$$

(10.218)

This equation is quadratic in Q_{1k}.

The Reynolds numbers are well within the range for laminar flow. The determination of good weighting functions for turbulent flow is a major unsolved problem. Rapidly accelerating a flow delays its transition to turbulence, so the laminar assumption can be valid for rather large Reynolds numbers.

The following MATLAB® code can be downloaded from the Internet; see page 1018. It employs three-dimensional arrays for p_{ijk}.

```
% script file viscoustubem. for quasi method of characteristics
% The variables are the pressure, P(j,k), and the volume flow rate, Q(j,k).
% The first index, j, is the position, which goes from 1 to m;
% the second index, k, times the time increment dt is the time
clear
m=41; A=0.1; L=125; % tube and sectioning parameters
rho=8e-5; beta=200000; nu=.04; % fluid parameters
PL=3000; PR=2000; Qeq=20; % upstream and downstream pressures, ss flow
C=A/beta; I0=rho/A; vp=1/sqrt(I0*C); Yc=sqrt(C/I0); %wave parameters
dt=L/vp/(m-1); % time increment
R0=8*nu*rho*pi*L/A^2/(m-1); % steady state resistance per segment
n=5; % n=3; % order of Schohl model; Trikha model alternative
R=R0*[.5255,1.179,4.5105,14.735,39.775]; % equation (10.163a) (Schohl)
I=I0*L/(m-1)*[.15775,.09432,.05389,.018144,.005487]; % equation (10.163b)
%R=R0*[.5,4.05,20]; % equation (10.161) (simpler Trikha alternative)
%I=I0*L/(m-1)*[.1515,.162,.02]; % equation (10.161) (Trikha continued)
tau=[I./R]; % equation (208c); equations (208b) follows:
a=exp(-dt./tau); b=I.*(1-tau/dt.*(1-a)); c=I.*((1+tau/dt).*(1-a)-1);
a1=a./tau; % for use in equation (209b)
A1=R0+R*(1-b./I)'; A2=c*(1./tau)'; bb=1/(1+A1*Yc/2); % equation (10.209c)
% initial conditions:
d1=Qeq^2/Yc^2/(PL-PR)-2*PR; d2=PR^2-PL*Qeq^2/(PL-PR)/Yc^2;
P(1,1)=-d1/2+sqrt(d1^2/4-d2); Q(1,1)=Yc*(P(1,1)-PR);
for j=2:m; P(j,1)=PR; Q(j,1)=0; dP(j,1)=0; p(:,j,1)=zeros(n,1); end
t(1)=0; p(:,1,1)=b'*Q(1,1);
for k=2:8*(m-1) % to get eight wave delay times
    % boundary conditions:
    P(m,k)=PR; % condition at right end
    t(k)=(k-1)*dt; % time in seconds
    c2=1+A1*Yc/2; c1=Yc*(PL-PR)/Qeq^2; c4=c2/2/c1;
    c3=Yc*(P(2,k-1)-PL+(dP(2,k-1)-A2*Q(1,k-1)-a1*p(:,1,k-1))/2)-Q(2,k-1);
    Q(1,k)=-c4+sqrt(c4^2-c3/c1);
    P(1,k)=PL-(PL-PR)*(Q(1,k)/Qeq)^2;
    p(:,1,k)=a'.*p(:,1,k-1)+b'*Q(1,k)+c'*Q(1,k-1);
    Q(m,k)=bb*(Q(m-1,k-1)+Yc*P(m-1,k-1)-Yc*P(m,k));
```

```
Q(m,k)=Q(m,k)-Yc/2*bb*(dP(m-1,k-1)-A2*Q(m,k-1)-a1*p(:,m,k-1));
p(:,m,k)=a'.*p(:,m,k-1)+b'*Q(m,k)+c'*Q(m,k-1);
% equations for all nodes:
for j=1:m; if k==2,dP(j,k-1)=A1*Q(j,k-1); end; end
%equations for internal nodes:
for j=2:m-1 % the first two equations below are (10.214c), (10.214d)
    P0(j,k)=(P(j-1,k-1)+P(j+1,k-1)+(Q(j-1,k-1)-Q(j+1,k-1))/Yc)/2;
    Q0(j,k)=(Yc*(P(j-1,k-1)-P(j+1,k-1))+Q(j-1,k-1)+Q(j+1,k-1))/2;
    P(j,k)=P0(j,k)-(dP(j-1,k-1)-dP(j+1,k-1))/4; % eqn (10.214a)
    Q(j,k)=bb*(Q0(j,k)-Yc/4*(dP(j-1,k-1)+dP(j+1,k-1)-2*(A2*Q(j,k-1)
        +a1*p(:,j,k-1)))); % (10.214b)
    dP(j,k)=A1*Q(j,k)-A2*Q(j,k-1)-a1*p(:,j,k-1); % eqn (10.209a
    p(:,j,k)=(a'.*p(:,j,k-1)+b'*Q(j,k)+c'*Q(j,k-1)); % eqn (10.208a)
end
end
plot(t,Q(1,:)) % plots the left-end volume flow rate
%plot(t,Q(m,:)) % plots the right-end volume flow rate
xlabel('time, seconds'); ylabel('flow rate, in^3/s')
```

10.7.6   The Bernoulli-Euler Beam

Wave-scattering variables can be applied also to multiple-power models. The example of the Bernoulli-Euler beam is examined to illustrate the methods and certain results peculiar to such models. The basic model, repeated from equation (10.121) (Section 10.5.3), is

$$
\frac{d}{dx}
\begin{bmatrix}
\eta \\
M \\
\frac{F}{\mu_1^2 EI_m} \\
\frac{F}{\mu_1^3 EI_m} \\
\frac{1}{\mu_1}\frac{\partial\eta}{\partial x}
\end{bmatrix}
= -\mu_1
\begin{bmatrix}
0 & 0 & 0 & -1 \\
0 & 0 & -1 & 0 \\
-1 & 0 & 0 & 0 \\
0 & -1 & 0 & 0
\end{bmatrix}
\begin{bmatrix}
\eta \\
M \\
\frac{F}{\mu_1^2 EI_m} \\
\frac{F}{\mu_1^3 EI_m} \\
\frac{1}{\mu_1}\frac{\partial\eta}{\partial x}
\end{bmatrix}.
\tag{10.219}
$$

Here, μ_1 is the principal of the four eigenvalue solutions of

$$
\mu = \left(-\frac{\rho A S^2}{EI_m}\right)^{1/4},
\tag{10.220}
$$

which is positive real when $S \to j\omega$. The other three eigenvalues are

$$
\mu_2 = -\mu_1; \qquad \mu_3 = j\mu_1; \qquad \mu_4 = -j\mu_1.
\tag{10.221}
$$

Waves in the Bernoulli-Euler beam are of two types, neither of which *dissipate* energy: highly dispersive waves which propagate energy, and **evanescent waves** which *decay but propagate no energy*. These two types of waves exist in a variety of systems, motivating this examination. The secondary purpose is to illustrate the transformation to wave-scattering variables for a multi-powered model.

Arrangement of the resulting eigenvectors of $\mathbf{A}$ as defined by equation (10.219) into four columns give the modal matrix and its inverse:

$$
\mathbf{M} = \frac{1}{2}
\begin{bmatrix}
1 & 1 & 1 & 1 \\
1 & 1 & -1 & -1 \\
1 & -1 & j & -j \\
1 & -1 & -j & j
\end{bmatrix},
\tag{10.222a}
$$

$$\mathbf{M}^{-1} = \frac{1}{2}\begin{bmatrix} 1 & 1 & 1 & 1 \\ 1 & 1 & -1 & -1 \\ 1 & -1 & j & -j \\ 1 & -1 & j & -j \end{bmatrix}. \tag{10.222b}$$

The corresponding matrix $\mathbf{C}$ is diagonal, as can be verified through the use of equation (10.178) (p. 789):

$$\frac{d}{dx}\begin{bmatrix} w_1 \\ w_2 \\ w_3 \\ w_4 \end{bmatrix} = -\mu_1 \begin{bmatrix} 1 & 0 & 0 & 0 \\ 0 & 1 & 0 & 0 \\ 0 & 0 & -j & 0 \\ 0 & 0 & 0 & j \end{bmatrix}\begin{bmatrix} w_1 \\ w_2 \\ w_3 \\ w_4 \end{bmatrix}. \tag{10.223}$$

The inverse of the transmission matrix relation is simply

$$\begin{bmatrix} w_1 \\ w_2 \\ w_3 \\ w_4 \end{bmatrix}_x = \begin{bmatrix} e^{-\mu_1(x-x_0)} & 0 & 0 & 0 \\ 0 & e^{\mu_1(x-x_0)} & 0 & 0 \\ 0 & 0 & e^{j\mu_1(x-x_0)} & 0 \\ 0 & 0 & 0 & e^{-j\mu_1(x-x_0)} \end{bmatrix}\begin{bmatrix} w_1 \\ w_2 \\ w_3 \\ w_4 \end{bmatrix}_{x_0}. \tag{10.224}$$

Consider the behavior of the first "wave," w_1. Equation (10.224) shows that this "wave" decays exponentially with increasing x. This does not mean that any dissipation of energy is taking place: there is no dissipation mechanism in the model. You can find out what is happening by setting $w_2 = w_3 = w_4 = 0$ and evaluating the power propagated, which for sine wave bevavior is

$$\mathcal{P} = F\dot{\eta} + m\frac{\partial\dot{\eta}}{\partial x} = \frac{F}{\mu_1^3 EI_m}j\omega\eta[\mu_1^3 EI_m] + \frac{M}{\mu_1^2 EI_M}j\omega\mu_1\left(\frac{1}{\mu_1}\frac{\partial\eta}{\partial x}\right)[\mu_1^3 EI_m]$$

$$= 2\left(\frac{j\omega w_1}{2}\frac{w_1}{2}[\mu_1^3 EI_m]\right) = \frac{1}{2}j[\omega\mu_1^3 EI_m]w_1^2. \tag{10.225}$$

The factor j reveals that this power is *pure imaginary*; the force and the velocity and the moment and the angular velocity are each 90° out of phase with each other, and no power is propagated. This is what is known as an **evanescent mode**. The "wave" w_2 behaves the same way, except its amplitude grows exponentially with increasing x.

Consider the third "wave," w_3, in the same way. Here, the eigenvalue is pure imaginary, which implies a real propagating wave. The phase velocity is

$$v_{p3} = \frac{\text{Im}\,\mu}{\omega} = \left(\frac{\rho A}{\omega^2 EI_m}\right)^{1/4}. \tag{10.226}$$

This is highly dispersive; the velocity is inversely proportional to the square root of frequency. Setting $w_1 = w_2 = w_4 = 0$ and evaluating the propagated power,

$$\mathcal{P} = 2\left(\frac{1}{2}\frac{j\omega w_3}{2}\frac{jw_3}{2}[\mu_1^3 EI_m]\right) = -\frac{1}{2}[\omega\mu_1^3 EI_m]w_3^2. \tag{10.227}$$

The power is being propagated to the left. Note that it is **equipartitioned** between the force-times-velocity and moment-times-angular-velocity types of power, a common feature of models of this type. The wave w_4 is similar to w_3, except it is propagated in the positive-x direction.

10.7.7   Summary

The state differential equations are considerably simplified along space-time trajectories that follow the propagation of waves, particularly in the linear case when waves traveling at different speeds or in opposite directions are transparent to one another. The corresponding use of wave-scattering variables isolates each wave and its corresponding eigenvalue.

Single-power uniform symmetric models have been addressed in general. For transient calulations in the special case of pure bilateral waves without dispersion, the method of characteristics implements these variables. Calculations involve the repeated application of reflection and transmission coefficients at boundary and internal junctions of line segments. Advantages accrue from wave synchronization if these segments are chosen so that all delay times are multiples of a common delay time.

The method of characteristics demands that the defining differential equations be hyperbolic, which implies the absence of wave dispersion. In cases where dissipation introduces significant but not overriding dispersion, efficient transient calculations result by separating out the hyperbolic or pure-wave-like terms, and applying the method of characteristics to these. The effects of the remaining terms are then treated as superimposed disturbances. The procedure, called the quasi method of characteristics, has been developed for the cases of viscous losses in circular tubes and heat-transfer effects in an ideal gas in circular tubes, which can occur separately or together.

Wave-scattering variables also can describe the behavior of multiple-power models. The example of the Bernoulli-Euler beam illustrates the appearance of evanescent modes, which decay exponentially with increasing distance from a boundary and propagate no steady-state power; energy oscillates back and forth spatially over each cycle. In the next section it is seen that the examples of a stretched string and a Bernoulli-Euler beam on elastic foundations also produce evanescent waves relative to a moving load.

Guided Problem 10.9

Develop the quasi method of characteristics for the linear elastic solid as given in Section 10.6.6 (pp. 776-777), for which $C_1 = C_2$.

(a) Adapt the model of the gaseous line with viscosity and heat transfer. (The result is considerably simpler.)

(b) Carry out a simulation for the response at the free right end of a rod made from such a material to a torsional step of 1 in.· lb at the left end. The shear modulus at infinite frequency is 4×10^5 psi, the weight density is 0.05 lb/in^3, the diameter is 1 in. and the length is 80 in. Consider relaxation times of 0.02 s, 0.002 s and 0.0002 s to span the behaviors of interest. Include three or four round-trips of the front end of the leading wave.

Suggested Steps:

1. Identify appropriate effort and flow variables, and establish the values of the parameters I and C_1 (define per-unit-length), C_2 and R (define per segment) relating to the bond graph model given in Fig. 10.26 (p. 777), letting the number of nodes be n (that is, $n-1$ segments). Leave τ unspecified at this point. Also, find the nominal phase velocity and characteristic admittance (neglecting C_2 and R).

2. There is only one R–C pair and time constant. Evaluate the various coefficients defined in equations (10.212b) and (10.213c).

3. Write the equations for the non-boundary nodes by adapting equation (10.215) with equations (10.213a), (10.213b), (10.214c) and (10.214d).

4. Write the given initial and boundary conditions for the moments.

5. Write the equations for the boundary conditions for the angular velocities.

6. To carry out the simulation it is necessary to choose the spatial and temporal grid. Solve for the nominal wave speed and delay time from one end of the rod to the other, and compare the latter to the decay time. It is reasonable to employ as few as four spatial grid points (although more are suggested) as long as a wave does not suffer more than a few percent distortion when traveling between adjacent points. The proper time interval Δt follows automatically.

7. Write the code, carry out the simulation and plot the resulting angular velocity at the right end of the rod. Comment on differences between the three cases.

PROBLEMS

10.32 Find the reflection and transmission coefficients for a junction of three pure bilateral-wave transmission lines. The characteristic admittances of the line with the incident wave is Y_{ci} and the characteristic admittances of the other two lines are Y_{ct1} and Y_{ct2}. The result can be applied directly to a variety of problems.

(a) Equate the efforts at the junction as seen by the three transmission lines in terms of the incident, reflected and two transmitted waves. This gives two equations with three unknowns.

(b) Sum the flows at the junction appropriately in terms of the waves. This gives a third independent equation with the same variables.

(c) Solve for the reflection coefficient and the two transmission coefficients in terms of the three characteristic admittances.

(d) Check your results against equations (10.192) for the special case in which one of the downstream characteristic admittances is zero.

10.33 Simulate a constant$-I$–R–C model of a one-dimensional distributed-parameter system, using the quasi method of characteristics. Consider in particular the torsion rod, as in Guided Problem 10-9 above; use the same values of the shear modulus, weight density, radius and length (but of course not the relaxation time). For the principal case, consider that the resistance per unit length times the length equals the surge impedance ($\sqrt{I/C}$). Apply a step change in moment at the left end of 1 in.·lb, and assume the right end is fixed. Plot the angular velocity at the driving end. This is the same situation as for the plot of Fig. 10.21 (p. 772). The model assumed there applies only to large values of $I\omega/R$, however, and a smaller value of R was used. Also consider a second case with a smaller resistance that more closely resembles the plot of Fig. 10.21. How do the results for your relatively exact solution differ, qualitatively, from those of the more approximate method?

SOLUTION TO GUIDED PROBLEM

GUIDED PROBLEM 10.9

1. Letting the radius $r = 0.5$ in., the length $L = 80$ in., the shear modulus $G = 4 \times 10^5$ lb/in^2, and the weight density $\gamma = 0.05$ lb/in^3, and defining the time constant $\tau = RC_2$, the parameters are $I = \pi\gamma r^4/2g$, where g is the acceleration of gravity in in/s^2, $C_1 = 2/\pi Gr^4$, $C_2 = C_1 L/(n-1)$ and $R = \tau/C_2$. Also, the nominal phase velocity is $v_p = 1/\sqrt{IC_1}$ and the nominal characteristic admittance is $Y_c = \sqrt{C_1/I}$.

2. $e_1 = e^{-\Delta t/\tau}$; $f = C_2[1 - (\tau/\Delta t)(1-e)]$ where $e \equiv e^{-\Delta t/\tau}$; g1=$C_2(1-e) - f$; $B_1 = (1 - f/C_2)/R$; and $B_2 = g1/\tau$.

3. $M_{j,k} = \left[M_{0jk} - \dfrac{Y_c}{4}\left(\Delta\omega_{j-1,k-1} + \Delta\omega_{j+1,k-1} - 2B_2 M_{j,k-1} - \dfrac{2e}{\tau}q_{j,k-1} \right) \right] \dfrac{1}{1 + B1/2Y_c}$,

    $\omega_{jk} = \omega_{0jk} - \dfrac{1}{4}(d\omega_{j-1,k-1} - d\omega_{j+1,k-1})$,

    $q_{j,k} = eq_{j,k-1} + fM_{j,k} + g1 M_{j,k-1}$,

    $M_{0jk} = \dfrac{1}{2}\left[M_{j-1,k-1} + M_{j+1,k-1} + \dfrac{1}{Y_c}(\omega_{j-1,k-1} - \omega_{j+1,k-1}) \right]$,

    $\omega_{0jk} = \dfrac{1}{2}\left[Y_c(M_{j-1,k-1} - M_{j+1,k-1}) + \omega_{j-1,k-1} + \omega_{j+1,k-1} \right]$.

    For all nodes and $k > 2$, $d\omega_{j,k-1} = B1 * M_{j,k-1} - B_2 M_{j,k-2} - e_1 q_{j,k-2}$.

    For all nodes and $k = 2$, $d\omega(j,1) = B1 M_{j,1}$.

4. The initial conditions are all zero except at the left end, for which $M_{1,1} = 1$ in.·lb and $\omega_{1,1} = Y_c * M_{1,1}$. The boundary condition at the left is $M_{1,k} = M_{1,1}$ and at the right is $M_{n,k} = 0$ for all $k \geq 2$.

5. Left end: $\omega_{1,k} = \omega_{2,k-1} + (Y_c + \dfrac{B1}{2})M_{1,k} - Y_c M_{2,k-1} + \dfrac{1}{2}(d\omega_{2,k-1} - B_2 M_{1,k-1} + e_1 q_{1,k-1})$.

    Right end: $\omega_{n,k} = \omega_{n-1,k-1} + Y_c M_{n-1,k-1} - \left(Y_c + \dfrac{B1}{2} \right) M_{n,k}$

    $+ \dfrac{1}{2}(B_2 M_{nk-1} - d\omega_{n-1,k-1} - e_1 q_{n,k-1})$.

6. The ratio $\Delta t/\tau$ should not exceed about 0.1 for good accuracy. Note that $\Delta t = L/(n-1)v_p$. The smaller the value of τ, the larger the number of segments $n-1$ are needed, and the longer the simulation will take to run. Some trial-and-error should clarify the compromise.

7. The symbol W is used for ω, and dt for Δt. Others should be obvious.

```
% downloadable script file GP10_9.m for Guided Problem 10.9
% The variables are the moment, M, and the angular velocity, W.
% The first index, j, is the position, which goes from 1 to n;
% the second index, k, is the time, which goes from 1 to m.
tau=.002; n=41; m=8*n; % chosen for the first simulation
r=.5; L=80; G=4e5; gamma=.05; g=32.174; C1=2/pi/G/r^4; C2=C1*L/(n-1);
I=gamma/2/g*pi*r^4; vp=1/sqrt(I*C1); Yc=sqrt(C1/I); R=tau/C2; T=L/vp;
dt=T/(n-1); e=exp(-dt/tau); e1=e/tau; f=C2*(1-tau/dt*(1-e));
g1=C2*(1-e)-f; B1=1/R*(1-f/C2); B2=g1/tau; b1=1+B1/Yc/2;
ratio=dt/tau % prints out critical accuracy parameter
% initial and boundary conditions:
M(1,1)=1; W(1,1)=Yc*M(1,1);
for j=2:n; M(j,1)=0; W(j,1)=0; end
t(1)=0; q(1,1)=0; q(n,1)=0;
for k=2:m; M(1,k)=M(1,1); M(n,k)=0; end
for k=2:m
    t(k)=(k-1)*dt;
```

```
%equations for left and right ends:
q(1,k)=e*q(1,k-1)+f*M(1,k)+g1*M(1,k-1);
q(n,k)=e*q(n,k-1)+f*M(n,k)+g1*M(n,k-1);
for j=1:n % except these apply to all nodes:
    if k>2; dW(j,k-1)=B1*M(j,k-1)-B2*M(j,k-2)-e1*q(j,k-2);
    else; dW(j,k-1)=B1*M(j,k-1);
    end
end
W(1,k)=W(2,k-1)+(Yc+B1/2)*M(1,k)-Yc*M(2,k-1)+dW(2,k-1)/2;
W(1,k)=W(1,k)-B2*M(1,k-1)/2+e1*q(1,k-1)/2;
W(n,k)=W(n-1,k-1)+Yc*M(n-1,k-1)-(Yc+B1/2)*M(n,k)+B2*M(n,k-1)/2;
W(n,k)=W(n,k)-dW(n-1,k-1)/2-e1*q(n,k-1)/2;
%equations for internal nodes:
for j=2:n-1
    MO(j,k)=(M(j-1,k-1)+M(j+1,k-1)+(W(j-1,k-1)-W(j+1,k-1))/Yc)/2;
    WO(j,k)=(Yc*(M(j-1,k-1)-M(j+1,k-1))+W(j-1,k-1)+W(j+1,k-1))/2;
    dW1=dW(j-1,k-1)+dW(j+1,k-1)-2*B2*M(j,k-1)-2*e1*q(j,k-1);
    M(j,k)=(MO(j,k)-dW1/4/Yc)/b1;
    W(j,k)=WO(j,k)-(dW(j-1,k-1)-dW(j+1,k-1))/4;
    q(j,k)=e*q(j,k-1)+f*M(j,k)+g1*M(j,k-1);
end
end
plot(t,W(n,:)) % plots the right-end angular velocity
```

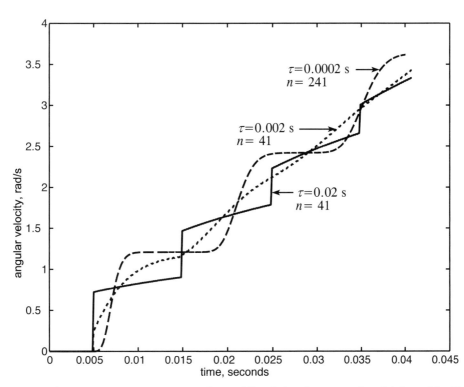

The case of $\tau = 0.002$ seconds is near the middle of the absorption band (plotted in Fig. 10.26 on p. 777); after three or four cycles the oscillation is nearly fully decayed. The case of $\tau = 0.02$ seconds has waves propagate at $v_p = v_{p0}$ with small attenuation but no spreading of the wavefront. The case of $\tau = 0.0002$ seconds has waves propagate at $v_p = v_{p0}/\sqrt{2}$, also with little attenuation but with significant spreading of the wavefront. A comparison of this solution with that of a purely lumped model, given in F. T. Brown, "Hybrid Lumped-and

Delay-Bond Modeling and Simulation," *2005 Int. Conf. on Bond Graph Modeling and Simulation*, Society for Modeling and Simulation International, Simulation Series v 37 n 1, pp. 67-72, shows that the latter exhibits significant spurious oscillations except when $\tau\omega$ is very small. The elimination of these oscillations is the principal benefit of the present technique.

10.8 Internal Excitation*

A general formulation and solutions for multiple-power models with internal excitations are presented. The treatment also applies to single-power models. Special considerations for symmetric systems are given. The methods and results are illustrated for the examples of heat conduction, a stretched string on an elastic foundation, a Bernoulli-Euler beam on an elastic foundation, and, in a Guided Problem, a counterflow heat exchanger.

10.8.1 Model Formulation and General Solution

The general linear uniform model, as noted before, can be represented as

$$\frac{\partial \mathbf{y}}{\partial x} = -\mathbf{A}(S)\mathbf{y} + \mathbf{B}\mathbf{u}(x,t). \tag{10.228}$$

To find the general solution, let

$$\mathbf{y}(S,x) = e^{-\mathbf{A}(S)x}\mathbf{f}(S,x). \tag{10.229}$$

This satisfies equation (10.228) if

$$\mathbf{B}\mathbf{u}(S,x) = e^{-\mathbf{A}(S)x}\frac{\partial \mathbf{f}(S,x)}{\partial x}, \tag{10.230}$$

or

$$\mathbf{f}(S,x) = \mathbf{f}(S,x_0) + \int_{x_0}^{x} e^{\mathbf{A}(S)\xi}\mathbf{B}\mathbf{u}(S,\xi)d\xi. \tag{10.231}$$

Note that

$$\mathbf{f}(S,x_0) = e^{\mathbf{A}x_0}\mathbf{y}(S,x_0). \tag{10.232}$$

Substitution of equation (10.232) into (10.231) and the result into (10.230) gives

$$\mathbf{y}(S,x) = e^{-\mathbf{A}(S)(x-x_0)}\mathbf{y}(S,x_0) + \int_{x_0}^{x} e^{\mathbf{A}(S)(\xi-x)}\mathbf{B}\mathbf{u}(S,\xi)d\xi. \tag{10.233}$$

Equations (10.231) and (10.233) may seem somewhat puzzling, since the excitation $\mathbf{B}\mathbf{u}(s,x)$ for $\xi > x$, namely the entire excitation for values of x to the right of the point of observation, is missing in the integral form. These excitations nevertheless produce an effect. The explanation is that only leftward propagating consequences of these excitations can be felt at the point of observation, unlike excitations for $\xi < x$, and these waves also must be registered at the left-hand boundary at $x = x_0$. Thus, the non-integral boundary terms on the right side of equation (10.231) contain all the appropriate information about the excitation $\mathbf{B}\mathbf{u}(S,\xi)$ for $\xi > x$. The relations are acausal both in the operator or generalized frequency, S, and in the position, x. The integral terms are necessary because rightward-traveling phenomena might never be felt at x_0.

The number of independent boundary conditions that exist equals the order of the matrices. For the general linear case with the left end set at $x = 0$ and the right end at $x = L$, the boundary conditions can be collected into the form

$$\mathbf{M}(S)\mathbf{y}(S,0) + \mathbf{N}(S)\mathbf{y}(S,L) = \mathbf{z}(S), \tag{10.234}$$

in which $\mathbf{z}(s)$ includes both time-varying boundary conditions or inputs and initial conditions. This statement permits feedback or coupling effects between the boundary conditions at the two ends; if none exists, it separates into two expressions, one for each end:

$$\mathbf{M}'(S)\mathbf{y}'(S,0) = \mathbf{z}_0(S), \qquad (10.235a)$$

$$\mathbf{N}'(S)\mathbf{y}''(S,L) = \mathbf{z}_L(S). \qquad (10.235b)$$

Here $\mathbf{y}'(S,0)$ and $\mathbf{y}''(S,L)$ have order which sum to the order of $\mathbf{y}$.

The result of evaluating equation (10.233) at $x = 0$ and $x = L$, and substituting the results into equation (10.234), is

$$\mathbf{y}(S,x) = e^{-\mathbf{A}x}\left[\mathbf{M}(S) + \mathbf{N}(S)e^{-\mathbf{A}L}\right]^{-1}$$

$$\times \left[\mathbf{z}(S) - \mathbf{N}(S)\int_0^L e^{\mathbf{A}(\xi-L)}\mathbf{B}\mathbf{u}(S,\xi)d\xi\right] + \int_0^L e^{\mathbf{A}(\xi-x)}\mathbf{B}\mathbf{u}(S,\xi)d\xi. \qquad (10.236)$$

This is for $\xi < x$ only; the integral term gets dropped for $\xi > x$. When the excitation $\mathbf{u}(S,x)$ is zero for all $0 < x < L$, this equation reduces to $\mathbf{y}(x) = e^{\mathbf{A}x}\mathbf{y}(0)$. It is convenient to recast this result in the form

$$\mathbf{y}(S,x) = e^{-\mathbf{A}x}\left[\mathbf{M}(S) + \mathbf{N}(S)e^{-\mathbf{A}L}\right]^{-1}\mathbf{z}(S) + \int_0^L \mathbf{G}(S,x|\xi)\mathbf{B}\mathbf{u}(S,\xi)d\xi. \qquad (10.237)$$

The first term on the right is the response at x to the disturbances at the boundaries represented by $\mathbf{z}(S)$. The second term is the response at x to the disturbances within the boundaries. Here the **Green's function** $\mathbf{G}(S,x|\xi)$ appears in its traditional role as the response at x to a *unit spatial impulse* disturbance at ξ. The summation or integration of all the impulse responses for the same x but different ξ give the complete response[19]. For the present case,

$$\mathbf{G}(S,x|\xi) = e^{-\mathbf{A}(S)x}[\mathbf{I} - \mathbf{F}(S)]e^{\mathbf{A}(S)\xi}; \qquad \xi < x,$$

$$= -e^{-\mathbf{A}(S)x}\mathbf{F}(S)e^{\mathbf{A}(s)\xi}; \qquad \xi > x, \qquad (10.238a)$$

$$\mathbf{F}(S) = \left[\mathbf{M}(S) + \mathbf{N}(S)e^{-\mathbf{A}(S)L}\right]^{-1}\mathbf{N}(S)e^{-bfAx(S)L}. \qquad (10.238b)$$

Equation (10.236) or (10.237) is a formal statement of considerable generality, which in practice proves unnecessarily cumbersome when its components can be assembled on an *ad hoc* basis. Nevertheless, it illustrates one of the three basic classes of analytical approaches advanced in this chapter for the solution of distributed-parameter models. The second class of approaches is the modal or eigenvalue expansion, and the third is wave methods. Equations (10.236) or (10.237) also give a most important result regarding the singularities of transfer functions from any excitation to any response.

If $\mathbf{A}(S)$ is an **entire function**[20] of S, $e^{\mathbf{A}(S)x}$ also is an entire function as can be seen from its series expansion. If, further, $\mathbf{M}(S)$ and $\mathbf{N}(S)$ are entire functions of S or can become entire functions by manipulation of equation (10.234), then $[\mathbf{M}(S) + \mathbf{N}(S)\exp(-\mathbf{A}L)]^{-1}$ (which appears in equations (10.236) and (10.237)) is an entire function, since the sum of products of entire functions are entire functions, and the determinant and all elements of

---

[19] Green's functions generally refer to unit impulses in space or impulses in both space and time. They are commonly used for solving two- and three-dimensional problems in distributed media, such as in acoustics and electromagnetics. For a detailed discussion see, for example, Morse and Feshback, *Methods of Theoretical Physics*, McGraw-Hill, 1953, v1, chapter 7.

[20] An entire function of a complex variable is one that is analytic for all finite values of S, such as a polynomial in S.

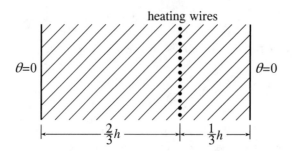

Figure 10.36: Problem with internal heating of a slab

the adjoint of entire functions are entire functions. Thus, if $\mathbf{z}(S,x)$ and $\mathbf{Bu}(S,x)$ also are ratios of entire functions, as they usually are, any transfer function for the system can be expressed as a ratio of entire functions. Such **meromorphic transfer functions** have poles as their only singularites in the finite s-plane. Knowledge of this fact is important in view of the fact that individual terms in practical expressions for transfer functions often are non-meromorphic.

10.8.2   Example of Green's Functions

A one-dimensional thermal slab has heating wires imbedded at the location $x = x_0$, as shown in Fig. 10.36. The objectives are to find the distributions of temperature and of the thermal energy fluxes. The boundary conditions are zero temperatures at $x = 0$ and $x = L$.

Using the state vector $[\theta \quad q]'$, the boundary conditions can be represented by

$$\mathbf{M} = \begin{bmatrix} 1 & 0 \\ 0 & 0 \end{bmatrix}; \qquad \mathbf{N} = \begin{bmatrix} 0 & 0 \\ 1 & 0 \end{bmatrix}; \qquad \mathbf{z} = \mathbf{0}. \tag{10.239}$$

The internal heat generation q_0 can be represented by use of the dirac delta function as

$$\mathbf{Bu}(\xi) = \begin{bmatrix} 0 \\ \delta(\xi - x_0)q_0 \end{bmatrix}. \tag{10.240}$$

Thus equation (10.115) gives

$$\begin{bmatrix} \theta(S,x) \\ q(S,x) \end{bmatrix} = \int_0^L \mathbf{G}(S,x|\xi)\mathbf{Bu}(S,\xi)\,d\xi = \begin{bmatrix} G_{12}(S,x|x_0) \\ G_{22}(S,x|x_0) \end{bmatrix} q_0. \tag{10.241}$$

The ubiquitous formula for symmetric one-power systems

$$e^{\mathbf{A}x} = \begin{bmatrix} \cosh\gamma x & \frac{1}{Y_c}\sinh\gamma x \\ Y_c\sinh\gamma x & \cosh\gamma x \end{bmatrix}, \tag{10.242}$$

along with equation (10.239), permits equation (10.238) to be evaluated for the Green's function. After some algebra, including use of well-known identities regarding hyperbolic trigonometric functions, there results

$$G_{12}(x|x_0) = \frac{\sinh\gamma x}{Y_c}\frac{\sinh\gamma(L-x_0)}{\sinh\gamma L},$$

$$G_{22}(x|x_0) = -\cosh\gamma x\frac{\sinh\gamma(L-x_0)}{\sinh\gamma L}, \qquad x < x_0; \tag{10.243a}$$

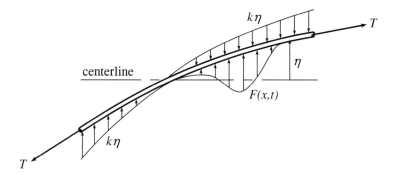

Figure 10.37: Stretched string on an elastic foundation

$$G_{12}\left(x|x_0\right) = \frac{\sinh\gamma(L-x)}{Y_c}\frac{\sinh\gamma x_0}{\sinh\gamma L},$$

$$G_{22}\left(x|x_0\right) = \cosh\gamma(L-x)\frac{\sinh\gamma x_0}{\sinh\gamma L}, \qquad x > x_0. \qquad (10.243b)$$

Use of the Green's function also is illustrated in Guided Problem 10.10.

10.8.3 Stretched Strings and Beams on Elastic Foundations with Moving Loads

Vehicles on guideways motivate the construction of models involving elastic foundations and moving loads. The simplest case is the stretched string; a more realistic case is the Bernoulli-Euler beam. In both cases the foundation will be considered as linear and elastic, with spring rate k per unit length.

A segment of the stretched string is shown with an exaggerated deflection and an arbitrary loading in Fig. 10.37. The beam, not shown, is similar. The differential equations in terms of the deflections η are

$$\text{string}: \quad -T\frac{\partial^2\eta}{\partial x^2} + \rho\frac{\partial^2\eta}{\partial t^2} + k\eta = F(x,t); \qquad (10.244a)$$

$$\text{beam}: \quad EI\frac{\partial^4\eta}{\partial x^4} + \rho\frac{\partial^2\eta}{\partial t^2} + k\eta = F(x,t). \qquad (10.244b)$$

The tension in the string, T, is assumed to be constant. The objective is to find the steady-state solutions that are approached when the force moves at a steady velocity. This velocity therefore becomes the phase velocity $v_p = \omega/k'$ for the homogeneous segments ahead of and behind the moving load, where k' is the wave number. For these segments, therefore, you can write

$$\eta = \eta_0 e^{-(jk'+\alpha)(x-v_pt)} = \eta_0 e^{-(jk'+\alpha)(x-\omega t/k')}. \qquad (10.245)$$

Substitution of this solution into equations (10.244) with $F = 0$ gives

$$\text{string}: \quad (-T + \rho v_p^2)(\alpha^2 - k'^2 + 2jk'\alpha) + k = 0; \qquad (10.246a)$$

$$\text{beam}: \quad EI(jk'+\alpha)^4 + \rho v_p^2(jk'+\alpha)^2 + k = 0. \qquad (10.246b)$$

For the stretched string, it is apparent that either k' or α must vanish. Defining the critical phase velocity v_{pc} and critical wave number k'_c as

$$v_{pc} \equiv \sqrt{T/\rho}, \qquad (10.247a)$$

$$k'_c \equiv \sqrt{k/T}, \tag{10.247b}$$

the solutions become

$$\text{subcritical velocity}: \quad k' = 0; \quad \alpha = \frac{\pm k'_c}{\sqrt{1 - (v_p/v_{pc})^2}}; \quad v_p < v_{pc},$$
$$\tag{10.248a}$$

$$\text{supercritical velocity}: \quad \alpha = 0; \quad k' = \frac{\pm k'_c}{\sqrt{(v_p/v_{pc})^2 - 1}}; \quad v_p > v_{pc}.$$
$$\tag{10.248b}$$

The beam also has a critical phase velocity above which $\alpha = 0$:

$$v_{pc} \equiv (4EIk/\rho^2)^{1/4}, \tag{10.249a}$$

$$k'_c \equiv \left(\frac{k}{EI}\right)^{1/4}. \tag{10.249b}$$

Below this velocity, however, the wave number does not vanish. The results are

$$\text{subcritical velocity}: \quad \alpha = \pm k'_c \sqrt{\frac{1}{2}\left[1 - \left(\frac{v_p}{v_{pc}}\right)^2\right]}, \tag{10.250a}$$

$$k' = k'_c \sqrt{\frac{1}{2}\left[1 + \left(\frac{v_p}{v_{pc}}\right)^2\right]}, \tag{10.250b}$$

$$\text{supercritical velocity}: \quad \alpha = 0, \tag{10.250c}$$

$$k' = k'_c \frac{v_p}{v_{pc}} \sqrt{1 \pm \sqrt{1 - \left(\frac{v_{pc}}{v_p}\right)^4}}. \tag{10.250d}$$

For the supercritical case, note the existence of *two* wave numbers for any given phase velocity, one larger and the other smaller than the critical wave number.

Some particular solutions with a point force are shown in Fig. 10.38. For the subcritical cases the waves are evanescent with respect to the traveling load; there is no steady-state propagation of energy. Therefore, the force is vertical and does no work. The solutions are symmetric with respect to the force.

The supercritical cases are more subtle to understand, because waves propagate energy away from the load, which therefore does work and must be angled. As an aid, the dispersion diagrams, shown in Fig. 10.39, can be constructed by substituting $v_p = \omega/k'$ in equations (10.248b), (10.250b) and (10.250d). The phase velocity is the slope of the chord from the origin to the condition of interest, revealing the single solution with its asymptotic limit in the case of the stretched string and the double solution in the case of the beam. The critical condition for the beam is at the point where the chord is tangent to the characteristic.

The velocity at which energy propagates in a dissipationless medium equals the group velocity, as is shown in Section 10.4.2, which in turn equals the slope of the dispersion characteristic, $d\omega/dk'$. In the string this velocity is always less than the phase velocity, revealing that energy is only propagated *backward* from the point of loading. This explains why the associated example in Fig. 10.38 has no motion upstream of the load. On the other hand, the solution with the shorter wavelength in the beam propagates faster than the phase velocity, carrying energy ahead of the moving load; the long wavelength solution propagates more slowly than the phase velocity, carrying energy backward.

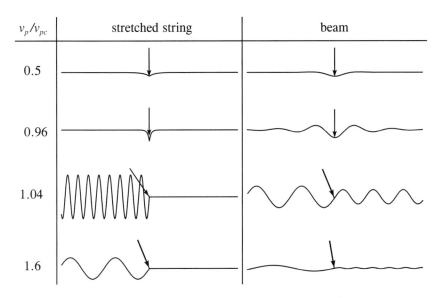

Figure 10.38: Steady-state deflections of a stretched string and a beam on elastic foundations for a force traveling from left to right

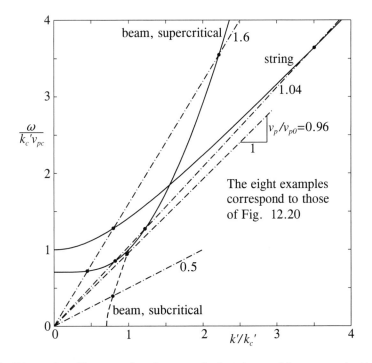

Figure 10.39: Dispersion diagrams for the stretched string and beam on elastic foundations

The final problem is the determination of the amplitude of the motions for a given point load. To accomplish this, the differential equations (10.244) can be integrated an infinitessimal distance across the load, from $\beta = 0^-$ to $\beta = 0^+$ with the definition $\beta \equiv x - v_p t$:

$$\text{string}: \int_{0^-}^{0^+} \left[\left(-T + \rho v_p^2\right) \frac{d^2\eta}{d\beta^2} + k\eta \right] d\beta = - \int_{0^-}^{0^+} F_0 \delta(\beta) d\beta; \qquad (10.251a)$$

$$\text{beam}: \int_{0^-}^{0^+} \left(EI \frac{d^4\eta}{d\beta^4} + \rho v_p^2 \frac{d^2\eta}{d\beta^2} + k \right) d\beta = - \int_{0^-}^{0^+} F_0 \delta(\beta) d\beta. \qquad (10.251b)$$

In both cases, η is continuous across the load; for the beam, $d\eta/d\beta$ also is continuous. Therefore the integrations give, respectively,

$$\left(-T + \rho v_p^2\right) \left[\left(\frac{d\eta}{d\beta}\right)_{0^+} - \left(\frac{d\eta}{d\beta}\right)_{0^-} \right] = -F_0, \qquad (10.252a)$$

$$EI \left[\left(\frac{d^3\eta}{d\beta^3}\right)_{0^-} - \left(\frac{d^3\eta}{d\beta^3}\right)_{0^+} \right] = -F_0. \qquad (10.252b)$$

For the subcritical case of the stretched string, the result is

$$\left(\frac{d\eta}{d\beta}\right)_{0^-} = -\alpha\eta_0; \qquad \left(\frac{d\eta}{d\beta}\right)_{0^+} = \alpha\eta_0; \qquad (10.253a)$$

$$\eta_0 = \frac{F_0}{2\alpha(-T + \rho v_p^2)} = \frac{-F_0}{2\sqrt{kT[1 - (v_p/v_{pc})^2]}}, \qquad (10.253b)$$

where η_0 is the maximum deflection that occurs directly under the load. The corresponding result for the beam is

$$\eta_0 = \frac{-F_0 k_c'}{2k\sqrt{2[1 - (v_p/v_{pc})^2]}}. \qquad (10.254)$$

In the supercritical cases the sine wave solutions give the maximum deflections

$$\text{string}: \quad \eta_0 = \frac{-F_0}{\sqrt{kT[(v_p/v_{pc})^2 - 1]}}; \qquad (10.255a)$$

$$\text{beam}: \quad \eta_0 = \frac{-F_0 k_c'}{2k\sqrt{(v_p/v_{pc})^4 - 1}\sqrt{(v_p/v_{pc})^2 - \sqrt{(v_p/v_{pc})^4 - 1}}}. \qquad (10.255b)$$

In all cases the deflections become infinite in the limiting case of the critical velocity. This velocity could be broached by accelerating through it at a rapid enough rate for the motion to remain within acceptable limits. It is much like an undamped resonator broaching a critical frequency, or an aircraft broaching the velocity of sound. Studies of this sort would be more complex but could be carried out through use of the Green's function.

10.8.4   Summary

The response to internal excitations can be represented generally by a spatial convolution of the distributed disturbance with the response to a unit spatial impulse; the latter is known as a Green's function. If the disturbance is sharply localized, however, an alternative procedure is to divide the model into pieces and treat the excitation in terms of boundary conditions. This has been illustrated for both fixed and moving point excitations. The examples of the stretched string and Bernoulli-Euler beam on elastic foundations show evanescent waves relative to a moving load for velocities below a critical velocity, and energy-bearing waves for higher velocities. Only for the beam do some of these propagate ahead of the load.

Guided Problem 10.10

A uniform rod of length L is insulated over all surfaces except for its right end, which is kept at a constant temperature. The thermal conductivity is k_θ and the specific heat is C.

(a) Find the Green's function for the temperature $\theta(x)$ resulting from a localized heat source q_0 at $x = x_0$.

(b) Use the result of part (a) to find the transfer function between a heat input that is uniform between the locations $x_0 = L/3$ and $x_0 = 2L/3$ and the temperature at x. Interpret this for the heat source varying sinusoidally in time at frequency ω. Plot a Nyquist (polar) plot of the heat flux and the temperature for locations $x = L/6$, $x = L/2$ and $x = 5L/6$.

(c) Note the difficulty for finding the temperature distribution if the heat input is a step function of time.

Suggested Steps:

1. This problem differs from that analyzed earlier in this section by the substitution of an adiabatic left end for an isothermal left end. Change equation (10.239) accordingly and find the results corresponding to equations (10.243).

2. To find the transfer function for the spatially distributed heat source, carry out an integration as directed by equation (10.237). Note that it is necessary to distinguish the regions $x < x_0$ and $x > x_0$.

3. For the frequency response, one uses the conventional substitution $s \to j\omega$. MATLAB can readily make the plot.

4. One can write the Laplace transform of the step function easily, but not its inverse. Further, the Fourier transform is not defined. The step response is found much more easily using the method of the following section; see Guided Problem 10.11 (p. 826).

SOLUTION TO GUIDED PROBLEM

Guided Problem 10.10

1. Boundary conditions: $q(0) = 0$:
$$\underbrace{\begin{bmatrix} 0 & 1 \\ 0 & 0 \end{bmatrix}}_{M} \begin{bmatrix} \theta \\ q \end{bmatrix}_{x=0} = \begin{bmatrix} 0 \\ 0 \end{bmatrix}$$

$$\theta(L) = 0 : \quad \underbrace{\begin{bmatrix} 0 & 0 \\ 1 & 0 \end{bmatrix}}_{N} \begin{bmatrix} \theta \\ q \end{bmatrix}_{x=L} = \begin{bmatrix} 0 \\ 0 \end{bmatrix}$$

Define $\Gamma \equiv \gamma L$
$$F(s) = [M + Ne^{-AL}]^{-1}Ne^{-AL}$$

$$= \begin{bmatrix} 0 & 1 \\ \cosh\Gamma & -(\sinh\Gamma)/Y_c \end{bmatrix} \begin{bmatrix} 0 & 0 \\ 1 & 0 \end{bmatrix} \begin{bmatrix} \cosh\Gamma & -(\sinh\Gamma)/Y_c \\ -Y_c\sinh\Gamma & \cosh\Gamma \end{bmatrix}$$

$$= \frac{1}{-\cosh\Gamma} \begin{bmatrix} -(\sinh\Gamma)/Y_c & -1 \\ -\cosh\Gamma & 0 \end{bmatrix} \begin{bmatrix} 0 & 0 \\ \cosh\Gamma & -(\sinh\Gamma)/Y_c \end{bmatrix} = \begin{bmatrix} 1 & -(\tanh\Gamma)/Y_c \\ 0 & 0 \end{bmatrix}$$

For $\xi < x$, $G(s, x|\xi) = e^{-Ax}[I - F(s)]e^{A\xi}$

$$= \begin{bmatrix} \cosh\gamma x & -(\sinh\gamma x)/Y_c \\ -Y_c\sinh\gamma x & \cosh\gamma x \end{bmatrix} \begin{bmatrix} 0 & \tanh\Gamma \\ 0 & 1 \end{bmatrix} \underbrace{\begin{bmatrix} \cosh\gamma\xi & \sinh\gamma\xi)/Y_c \\ Y_c\sinh\gamma\xi & \cosh\gamma\xi \end{bmatrix}}$$

$$= \begin{bmatrix} '' & '' \\ '' & '' \end{bmatrix} \begin{bmatrix} \tanh\Gamma\sinh\gamma\xi & (\tanh\Gamma\cosh\gamma\xi)/Y_c \\ Y_c\sinh\gamma\xi & \cosh\gamma\xi \end{bmatrix}$$

$$= \begin{bmatrix} \text{don't need} & \cosh\gamma\tanh\Gamma\cosh\xi)/Y_c - (\sinh\gamma x\cosh\gamma\xi)/Y_c \\ \text{don't need} & -\sinh\gamma x\tanh\Gamma + \cosh\gamma x\cosh\gamma\xi \end{bmatrix}$$

Noting the following identities: $\cosh(\alpha - \beta) \equiv \cosh\alpha\cosh\beta - \sinh\alpha\sinh\beta$,

$$\sinh(\alpha - \beta) \equiv \sinh\alpha\cosh\beta - \cosh\alpha\sinh\beta,$$

$$g(s, x|\xi) = \begin{bmatrix} \text{don't need} & [(\cosh\gamma\xi)/Y_c\cosh\gamma L]\sinh\gamma(L_\xi) \\ \text{don't need} & (\sinh\gamma\xi/\cosh\gamma L)\sinh\gamma(L - \xi) \end{bmatrix}$$

2. $Bu(\xi) = \begin{bmatrix} 0 \\ 0 \end{bmatrix}$ for $\xi < \dfrac{l}{3}$ and $\xi > \dfrac{2L}{3}$; $Bu(\xi) = \begin{bmatrix} 0 \\ 3q_0/L \end{bmatrix}$ for $\dfrac{L}{3} < \xi < \dfrac{2L}{3}$.

For $x \geq \dfrac{2}{3}L$, $\begin{bmatrix} \theta \\ q \end{bmatrix} = \dfrac{3q_0}{L}\displaystyle\int_{L/3}^{2L/3} \begin{bmatrix} G_{12} \\ G_{22} \end{bmatrix} d\xi = \dfrac{3q_0}{L} \begin{bmatrix} \frac{\sinh\gamma(L-x)}{\gamma Y_c\cosh\gamma L}\sinh\gamma\xi \\ \frac{\cosh\gamma(L-x)}{\gamma\cosh\gamma L}\sinh\gamma\xi \end{bmatrix}_{L/3}^{2L/3}$

$$= \dfrac{3q_0}{L} \begin{bmatrix} \frac{\sinh\gamma(L-x)}{\gamma Y_c\cosh\gamma L}\left(\sinh\frac{2\gamma L}{3} - \sinh\frac{\gamma L}{3}\right) \\ \frac{\cosh\gamma(L-x)}{\gamma\cosh\gamma L}\left(\sinh\frac{2\gamma L}{3} - \sinh\frac{\gamma L}{3}\right) \end{bmatrix}$$

but $\sinh\dfrac{2\gamma L}{3} - \sinh\dfrac{\gamma L}{3} = \sinh\left(\dfrac{\gamma L}{2} + \dfrac{\gamma L}{6}\right) - \sinh\left(\dfrac{\gamma L}{2} - \dfrac{\gamma L}{6}\right) = 2\cosh\dfrac{\gamma L}{2}\sinh\dfrac{\gamma L}{6}$

Finally, $\begin{bmatrix} \theta \\ q \end{bmatrix} = \dfrac{6q_0}{\gamma L}\dfrac{\cosh(\gamma L/2)\sinh(\gamma L/6)}{\cosh\gamma L} \begin{bmatrix} [\sinh\gamma(L - x)]/Y_c \\ \cosh\gamma(L - x) \end{bmatrix}$; $x > \dfrac{2}{3}L$

For $\xi > x$, $G(s, x|\xi) = -e^{-Ax}Fe^{A\xi}$

$$= \begin{bmatrix} \cosh & -(\sinh\gamma x)/Y_c \\ -Y_c\gamma x & \cosh\gamma x \end{bmatrix} \underbrace{\begin{bmatrix} -1 & (\tanh\Gamma)/Y_c \\ 0 & 0 \end{bmatrix} \begin{bmatrix} \cosh\gamma\xi & (\sinh\gamma L)/Y_c \\ Y_c\sinh\gamma\xi & \cosh\gamma\xi \end{bmatrix}}$$

$$= \begin{bmatrix} '' & '' \\ '' & '' \end{bmatrix} \begin{bmatrix} -\cosh\gamma\xi + \tanh\Gamma\sinh\gamma\xi & (\sinh\gamma\xi - \tanh\Gamma\cosh\gamma\xi)/Y_c \\ 0 & 0 \end{bmatrix},$$

$$= \begin{bmatrix} \text{don't need} & -\cosh\gamma x(\sinh\gamma\xi - \tanh\Gamma\cosh\gamma\xi)/Y_c \\ \text{don't need} & \sinh\gamma x(\sinh\gamma\xi - \tanh\Gamma\cosh\gamma\xi) \end{bmatrix},$$

$$= \begin{bmatrix} \text{don't need} & (\cosh\gamma x/Y_c\cosh\gamma L)\sinh\gamma(L - \xi) \\ \text{don't need} & -(\sinh\gamma x/\cosh\gamma L)\sinh\gamma(L - \xi) \end{bmatrix},$$

For $x \leq \dfrac{1}{3}L$, $\begin{bmatrix} \theta \\ q \end{bmatrix} = \displaystyle\int_{l/3}^{2L/3} 3q_0\left[(\cosh\gamma x/Y_c psh\sinh\gamma(l - xi) - (\sinh/\cosh)\sinh\gamma(L - \xi)\right] d\xi,$

$$= \dfrac{3q_0}{\cosh} \begin{bmatrix} -(\cosh\gamma x)Y_c \\ \sinh \end{bmatrix} \cosh(L - \xi)|_{L/3}^{2L},$$

$$= \dfrac{3q_0}{\gamma L\cosh\gamma L} \begin{bmatrix} -(\cosh\gamma x)/Y_c \\ \sinh\gamma x \end{bmatrix} \left(\cosh\dfrac{\gamma L}{3} - \cosh\dfrac{2\gamma L}{3}\right)$$

$$= \dfrac{6q_0\sinh(\gamma L/2)\sinh(\gamma L/6)}{\gamma L\cosh\gamma L} \begin{bmatrix} (\cosh\gamma x)/Y_c \\ -\sinh\gamma x \end{bmatrix}.$$

For $\dfrac{L}{3} \le x \le \dfrac{2L}{3}$, $\begin{bmatrix} \theta \\ q \end{bmatrix} = \dfrac{3q_0}{L} \left\{ \displaystyle\int_{L/3}^{x} \begin{bmatrix} G_{12,\xi<x} \\ G_{22,\xi<x} \end{bmatrix} d\xi + \int_{x}^{2L/3} \begin{bmatrix} G_{12,\xi>x} \\ G_{22,\xi>x} \end{bmatrix} d\xi \right\}$

$= \dfrac{3q_0}{\gamma L \cosh \gamma L} \left[\begin{matrix} \frac{\sinh \gamma(L-x)}{Y_c}(\sinh \gamma x - \sinh \gamma L/3) & -\frac{\cosh \gamma x}{Y_c}(\cosh \gamma L/3 - \cosh \gamma(L-x)) \\ \cosh \gamma(L-x)(\sinh \gamma x - \sinh \gamma L/3) & \sinh \gamma x(\cosh \gamma L/3 - \cosh \gamma(L-x)) \end{matrix} \right]$

3. For $s = j\omega$; $\gamma = jK + \alpha$; $k = \alpha = \sqrt{\dfrac{C\omega}{2k_{theta}}}$; $Y_c = \sqrt{k_\theta C j\omega}$,

$\gamma L = (j+1)\sqrt{\dfrac{C\omega}{2k_\theta}} L = (j+1)\sqrt{\Omega}$, where $\Omega \equiv \dfrac{c^2}{2k_{theta}}$; $Y_c = k_\theta \gamma = \dfrac{k_\theta}{L}\gamma L$.

Define $Th \equiv \dfrac{\theta k_\theta}{L q_0}$ and $Q \equiv \dfrac{q}{q_0}$

MATLAB plot of frequency responses, Nyquist coordinates:

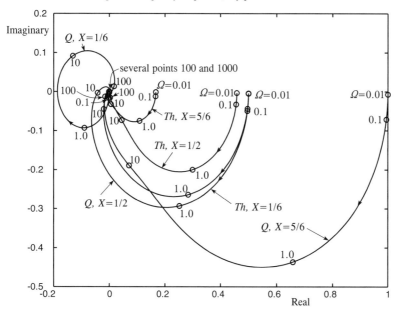

See Guided Problem 10.11 (p. 826) for the step response problem, which is found readily using the modal method.

10.9   Modal Decomposition

Modal decompositions of linear distributed-parameter models, when they can be found, effect descretization in a particularly intelligent way: frequency content below some cut-off is accurately represented, whereas the content above the cut-off is disregarded. Arbitrary initial conditions and distributed forcing functions are readily accommodated. The basic idea behind this classic technique posits that the solution to the unforced system can be represented by a sum of products of functions of position and functions of time, called a **separation of variables**, for some state function y:

$$y = \sum_\ell y_\ell(t)\psi_\ell(x). \tag{10.256}$$

A set of functions $\psi_\ell(x)$ must be found that both satisfies the end or boundary conditions and is orthogonal over the length, that is

$$\int_0^L \psi_\ell(x)\psi_m(x)\,dx = 0, \qquad m \neq \ell. \tag{10.257}$$

Each member of this modal set is substituted into the partial differential equation that describes the model as viewed by the variable y. The result is an ordinary differential equation, with its own forcing function. Truncation of the infinite number of modes is tantamount to approximation by a lumped model.

Only special partial differential equations are separable in the form of equation (10.256). Even if solutions of this form exist, furthermore, the solution that satisfies the given boundary and initial conditions generally cannot be constructed from a superposition of these solutions unless the orthogonality condition of equation (10.257) happens to be satisfied. Fortunately, however, many of the classical equations for distributed-parameter models satisfy these requirements for boundary conditions of common interest. The pure-bilateral-delay model (or the wave equation), the linear diffusion or heat-conduction model and the Bernoulli-Euler beam model are examples, depending on the end conditions.

10.9.1   General Procedure

The first step in carrying out a modal analysis of a distributed-parameter model is to find a function $\psi_\ell(x)$ that satisfies both the governing differential equation and the boundary conditions and satisfies the orthogonality condition. This is illustrated in the subsections that follow; for the present it is assumed that this critical step has been accomplished. The next step depends on whether an initial-value problem or a forced-reponse problem is being addressed.

For initial-value problems, one solves for $y_\ell(t)$ in terms of the initial conditions $y_\ell(0)$, $dy_\ell/dt)(0)$, $\ldots, (d^n y_\ell/dt^n)(0)$ which are as yet unknown. The number of independent initial conditions equals the spatial order of the partial differential equation, n. Therefore,

$$y(x, 0) = \sum_\ell y_\ell(0)\psi_\ell(x),$$

$$\cdot$$
$$\cdot \tag{10.258}$$
$$\cdot$$

$$\left(\frac{\partial^n y}{\partial t^n}\right)(x, 0) = \sum_\ell \frac{dy_\ell}{dt^n}\psi_\ell(x).$$

Multiplying both sides of these equations by $\psi_m(x)$ and integrating over the length of the system gives

$$\int_0^L y(x, 0)\psi_m(x)\,dx = \sum_\ell y_\ell(0) \int_0^L \psi_\ell(x)\psi_m(x)\,dx,$$

$$\cdot$$
$$\cdot \tag{10.259}$$
$$\cdot$$

$$\int_0^L \left(\frac{\partial^n y}{\partial t^n}\right)(x, 0)\psi_m\,dx = \sum_\ell \frac{d^n y_\ell}{dt^n}(0) \int_0^L \psi_\ell(x)\psi_m(x)\,dx.$$

The orthogonality condition of equation (10.257) eliminates all terms on the right side except for the one with $\ell = m$. With the definition of the constant b_m,

$$b_m \equiv \int_0^L [\psi_m(x)]^2 \, dx, \tag{10.260}$$

and switching the indices ℓ and m, equations (10.259) therefore require

$$y_\ell(0) = \frac{1}{b_\ell} \int_0^L y(x,0)\psi_\ell(x) \, dx,$$

$$\cdot$$
$$\cdot$$
$$\cdot$$

$$\frac{d^n y_\ell}{dt^n}(0) = \frac{1}{b_\ell} \int_0^L \frac{\partial^n y}{\partial t^n}(x,0)\psi_\ell(x) \, dx. \tag{10.261}$$

Substitution of these results into the expression for $y_\ell(t)$ in terms of the initial conditions, and the resulting $y_\ell(t)$ into equation (10.256), produces the desired solution.

When the system is forced, the differential equation becomes of the form

$$\sum_\ell \phi_\ell \left[S, y_\ell(t) \right] \psi_\ell(x) = F(x,t), \tag{10.262}$$

where $F(x,t)$ is the forcing. As with the initial-value problem, both sides of equation (10.262) are multiplied by $\psi_m(x)$, and the product is integrated over the length of the system:

$$\int_0^L \sum_\ell \phi_\ell \left[S, y_\ell(t) \right] \psi_\ell(x)\psi_m(x) \, dx = \int_0^L F(x,t)\psi_m(x) \, dx. \tag{10.263}$$

The orthogonality condition then gives a set of ordinary differential equations:

$$\phi_\ell \left[S, y_\ell(t) \right] = \frac{1}{b_\ell} \int_0^L F(x,t)\psi_\ell(x) \, dx \equiv f_\ell(t). \tag{10.264}$$

These equations are solved for the $y_\ell(t)$, and the results are substituted into equation (10.256) to give the complete solution.

Various categories of forcing can be distinguished. For a simple frequency response,

$$F_\ell(t) = F_{l0} \sin \omega t. \tag{10.265}$$

For a periodic excitation, one can use a Fourier series:

$$F_\ell(t) = \sum_i F_{\ell i} e^{j\omega_i t}. \tag{10.266}$$

For an impulse or step excitation, respectively,

$$F_\ell(t) = F_0 \delta(t) \qquad \text{or} \qquad F_\ell(t) = F_0 \mu_\ell(t), \tag{10.267}$$

either Laplace transforms can be used, or the problem can be converted into initial-value form. For a more general aperiodic excitation, either a Fourier transform (if it is defined) or a Laplace transform can be used.

10.9.2   Sinusoidal Modes: Example of a Plucked String

The simplest orthogonal set of functions comprise sine or cosine waves

$$\boxed{\psi_l(x) = \sin(k_l x + \alpha_l) \qquad l = 1, 2, \ldots,} \tag{10.268}$$

which applies if only even-ordered partial derivatives of y with respect to x appear in each of the non-forcing terms of the linear partial differential equation of the model, and if the boundary conditions permit.

The example of a plucked string will be started by considering a general conservative 1-power uniform symmetric model, which is then specialized, as the various steps require, for a stretched string clamped at both ends and plucked at a point along its length. This problem also is addressed from more specialized perspectives in Sections 10.1.4 and 10.1.5, allowing comparisons of methods and results.

The general 1-power uniform symmetric model has the form

$$\frac{\partial}{\partial x}\begin{bmatrix} y_s \\ y_a \end{bmatrix} = -\begin{bmatrix} 0 & Z(S) \\ Y(S) & 0 \end{bmatrix}\begin{bmatrix} y_s \\ y_a \end{bmatrix} + \begin{bmatrix} u_s(x,t) \\ u_a(x,t) \end{bmatrix}. \tag{10.269}$$

This can be viewed from the perspective of either the symmetric or the asymmetric variable. The symmetric deflection of the string, η, is chosen, since it is involved directly in the initial state and the end conditions [21]. The general result is

$$\frac{\partial^2 y_s}{\partial x^2} - Z(S)Y(S)y_s = -Z(S)u_a(x,t) + \frac{\partial u_s}{\partial x}(x,t) \equiv F(x,t). \tag{10.270}$$

Equations (10.256) and (10.268) give the presumption

$$y_s = \sum_\ell y_{s\ell}(t)\sin(k_\ell x + \alpha_\ell). \tag{10.271}$$

The boundary conditions determine the values of k_ℓ and α_ℓ. The stretched string has

$$y_s(0) = y_s(L) = 0, \tag{10.272}$$

which requires

$$k_\ell = \ell\pi/L; \qquad \alpha_l = 0. \tag{10.273}$$

Next, equation (10.271) is substituted with equations (10.272) into equation (10.270) to get

$$\sum_\ell (-)\left[\left(\frac{\ell\pi}{L}\right)^2 + Z(S)Y(S)\right]y_{s\ell}(t)\sin\left(\frac{\ell\pi x}{L}\right) = F(x,t). \tag{10.274}$$

The possibility of analytic solution depends on $Z(S)Y(S)$.

The initial-value problem type is chosen, for which equation (10.274) becomes

$$\left[\left(\frac{\ell\pi}{L}\right)^2 + Z(S)Y(S)\right]y_{s\ell}(t) = 0. \tag{10.275}$$

The stretched string obeys the pure bilateral delay model, for which

$$Z(S)Y(S) = I'C'S^2 = (\rho A/T)S^2, \tag{10.276}$$

---

[21] For the string, η is th time integral of the *flow* in Example 10.4 (p. 715), which makes $Z(S) = C' = 1/T$ and $Y(S) = I'S^2 = \rho A S^2$, referring to Table 10.1 (p. 716). The forcing is $F(x,t) = -Z(S)u_a(x,t) = -f(x,t)/T$, where $f(x,t)$ is the applied lateral force distribution.

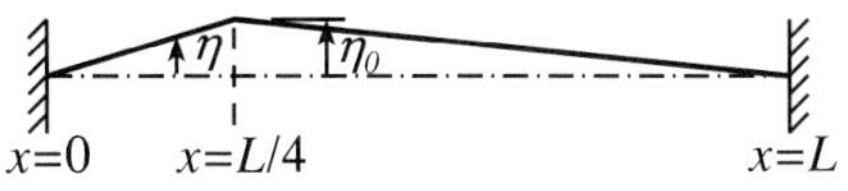

Figure 10.40: Initial condition for the plucked string

where ρA is the mass per unit length of the string, and T is the tension, which is assumed to be constant. Substitution into equation (10.275) gives the solution

$$y_{s\ell}(t) = \frac{\sin(\omega_\ell t + \beta_\ell)}{\sin \beta_\ell} y_{s\ell}(0); \qquad \omega_\ell^2 = \frac{1}{IC}\left(\frac{\ell\pi}{L}\right)^2, \qquad (10.277)$$

and equation (10.261) gives

$$b_\ell = L/2. \qquad (10.278)$$

The modal decomposition is given by equation (10.261):

$$y_{s\ell}(0) = \frac{2}{L}\int_0^L y_s(x,0)\sin\left(\frac{\ell\pi x}{L}\right)dx, \qquad (10.279a)$$

$$\frac{dy_{s\ell}}{dt}(0) = \frac{2}{L}\int_0^L \frac{\partial y_s}{\partial t}(x,0)\sin\left(\frac{\ell\pi x}{L}\right)dx. \qquad (10.279b)$$

The time derivative of equation (10.277), evaluated at $t = 0$, is

$$\frac{dy_{s\ell}}{dt}(0) = \omega_\ell \cot \beta_\ell\, y_{s\ell}(0), \qquad (10.280)$$

which gives

$$\beta_\ell = \tan^{-1}\left[\frac{\omega_\ell f_\ell(0)}{(dy_{s\ell}/dt)(0)}\right]. \qquad (10.281)$$

The final solution results from substituting equation (10.281) into (10.277), and the result plus equations (10.273) into equation (10.271).

The stretched string is plucked at a location one-quarter along its length, as shown in Fig. 10.40. The initial condition is thus

$$\eta(x,0) = \eta_0\left(\frac{4x}{L}\right); \qquad 0 < x < \frac{L}{4}, \qquad (10.282a)$$

$$= \frac{4}{3}\eta_0\left(1 - \frac{x}{L}\right); \qquad \frac{L}{4} < x < L. \qquad (10.282b)$$

from which equations (10.279) and (10.280) give, after some cancellation of terms (and in agreement with equation (10.20)),

$$\eta_\ell(0) = \frac{32}{3\ell^2\pi^2}\eta_0\sin\left(\frac{\ell\pi}{4}\right), \qquad (10.283a)$$

$$\frac{d\eta_\ell}{dt}(0) = 0, \qquad (10.283b)$$

$$\beta_\ell = \pi/2. \qquad (10.283c)$$

Selected harmonics are plotted in Fig. 10.6 (p. 720). Equation (10.277) now gives

$$\eta_\ell(t) = \eta_\ell(0)\cos\omega_\ell t, \qquad (10.284)$$

which when substituted into equation (10.271) yields the complete solution

$$\eta = \eta_0\sum_\ell \frac{32}{3\ell^2\pi^2}\sin\left(\frac{\ell\pi}{4}\right)\sin\left(\frac{\ell\pi x}{L}\right)\cos\left(\sqrt{\frac{T}{\rho A}}\frac{\ell\pi t}{L}\right). \qquad (10.285)$$

10.9.3   Sinusoidal Modes: Example of a Struck Simply-Supported Beam

As a second example with sinusoidal modes, consider a Bernoulli-Euler model of a beam which is simply pinned at its ends and is struck impulsively at its center. Later, in Guided Problem 10.12, you will consider how this beam responds to a point load which moves at constant velocity from one end to the other; this could represent the response of a bridge to traffic.

The differential equation in terms of the deflection of the beam, η, from Section 10.5.3, is

$$\frac{\partial^4 \eta}{\partial x^4} - \mu_1^4 \eta = \frac{F(x,t)}{EI}. \tag{10.286}$$

Here μ_1 is the principal root of

$$\mu = \left(-\frac{\rho A S^2}{EI} \right)^{1/4}, \tag{10.287}$$

where ρ is the density of the beam, E is Young's modulus, A is its cross-sectional area and I is the area moment of inertia for bending.

The boundary conditions require $\eta = 0$ and zero moment, which implies $\partial^2 \eta / \partial x^2 = 0$. These are satisfied by the orthogonal set

$$\psi_\ell = \sin(\ell \pi x / L). \tag{10.288}$$

Note that many end conditions, even if conservative, do not produce a simple orthogonal set. The presence of the evanescent or exponential non-propagating solution, in particular, voids a sinusoidal set of orthogonal functions. Another acceptable pair of end conditions is $\partial \eta / \partial x = 0$ and zero force, which implies $\partial^3 \eta / \partial x^3 = 0$; these satisfy $\psi_\ell = \cos(\ell \pi x / L)$.

Substitution of equation (10.288) into (10.286) gives

$$\left[\left(\frac{\ell \pi}{L} \right)^4 - \mu_1^4 \right] \sin \left(\frac{\ell \pi x}{L} \right) = \frac{F(x,t)}{EI}. \tag{10.289}$$

Multiplying both sides of this equation by $\sin(m \pi x / L) dx$ and integrating over the interval $0 < x < L$, as generalized in equation (10.264), gives the orthogonality result

$$\left[\left(\frac{\ell \pi}{L} \right)^4 + \frac{\rho A}{EI} S^2 \right] \eta_\ell(t) = F_\ell(t). \tag{10.290}$$

With the impulsive force at the center of the beam,

$$F(x,t) = F_0 \delta(t) \delta(x - L/2), \tag{10.291}$$

the modal forcing becomes

$$F_\ell(t) = \frac{2 F_0}{LEI} \sin \frac{\ell \pi}{2} \delta(t). \tag{10.292}$$

Note that the even-numbered modes produce no respose, as should be expected since these modes have a node at the center of the beam where the force is imposed. For $\ell = 1, 5, 9 \ldots$ there results $F_l(t) = (2F_0/LEI)\delta(t)$, and for $\ell = 3, 7, 11, \ldots$ there results $F_l(t) = -(2F_0/LEI)\delta(t)$. The response now can be found by taking the Laplace transforms

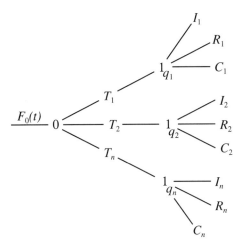

Figure 10.41: Bond graph for modal decomposition of a distributed-parameter model

of equations (10.290) and (10.292), solving for $\mathcal{L}[F_\ell(t)]$, and finding the inverse transform, with the result

$$\eta_\ell(t) = (-1)^{(\ell-1)/2} \frac{2F_0 L}{\pi^2 \sqrt{\rho A E I}} \frac{1}{\ell^2} \sin\left(\sqrt{\frac{EI}{\rho A} \frac{\ell^2 \pi^2}{L^2}}\, t\right) \quad \ell = 1, 3, 5, \ldots. \tag{10.293}$$

Therefore, the second harmonic present has nine times the frequency of the first and one-ninth its amplitude, the third harmonic present has twenty-five times the frequency of the first and one-twenty-fifth its amplitude, etc. A stretched string, by comparison, would have a much richer harmonic content.

It is possible to convert this problem into an initial-value problem. The impulse has as its immediate consequence a spatial pulse of velocity, $\partial \eta / \partial t$, located at the center of the beam. There is no subsequent forcing. The same solution results.

10.9.4 Bond-Graph Modeling

The examples given above and many others produce modes that are second-order in time and therefore representable by an inertance bonded to a compliance. Further, the total value of the variable representing the behavior equals a weighted sum of the corresponding variables for the individual modes. This suggests the construction of bond graph models. The details of these models depend on what variables are of particular interest, and what variables represent the forcing.

The case of driving-point variables is shown in Fig. 10.41. The forcing is presumed to be concentrated at $x = x_0$, $i.e.$

$$F(x, t) = \delta(x - x_0) F_0(t), \tag{10.294}$$

and the variable of interest is the generalized velocity $\dot{q}$ (or its time integral, the generalized displacement q) at the same location. The transformer T_ℓ has as the inverse of its modulus $\psi_\ell(x_0)$, which when multiplied by $F_0(t)$ gives the modal forcing, $b_\ell \phi_\ell$. (See equation (10.264).) The modal forcing in turn produces the modal velocity, $\dot{q}_\ell(t)$, which when multiplied by the inverse of T_ℓ gives the contribution of the ℓth mode to the overall velocity $\dot{q}(x_0, t)$. Note that in general,

$$b_\ell \phi_\ell = I_\ell \ddot{q}_\ell + \frac{1}{C_\ell} q_\ell. \tag{10.295}$$

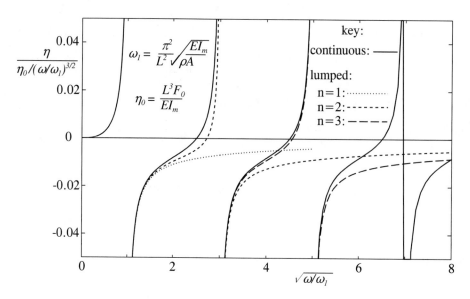

Figure 10.42: Frequency response of Bernoulli-Euler beam and truncated modal models

For the stretched string with $q = \eta$,

$$b_\ell \phi_\ell = \left[\left(\frac{\ell \pi}{L} \right)^2 + \frac{\rho A}{T} S^2 \right] \frac{L}{2}, \tag{10.296}$$

so that

$$I_\ell = \frac{\rho A L}{2T}; \qquad \frac{1}{C_l} = \frac{\ell^2 \pi^2}{2L}. \tag{10.297}$$

For the simply supported Bernoulli-Euler beam,

$$I_\ell = \frac{\rho A L}{2}; \qquad \frac{1}{C_\ell} = \frac{\ell^4 \pi^4 EI}{2L^3}. \tag{10.298}$$

The bond graph leads to a set of ordinary differential equations for an arbitrary forcing. The set would be truncated in practice. The added resistances, as shown in the figure, could be used to match known damping near resonances. Small damping should produce only a minor error in the otherwise perfect modal decomposition.

The frequency response of the Bernoulli-Euler beam is compared to single, double and triple-mode models in Fig. 10.42. The truncated models can be seen to represent the behavior quite reasonably up to the highest resonant frequency included.

Differential equations which are first-order in time produce first-order or exponential modes representable by a resistance bonded to an inertance or a compliance. The most common example is the diffusion or heat-conduction model, which is the subject of Guided Problem 10.11.

10.9.5 Modes with Effort-Free Boundaries: Example of a Bernoulli-Euler Beam

A distributed-parameter element often is but one component in a larger system. One would like to have a one-size-fits-all modal model of the element. This model must admit motion everywhere, which means that it must not have any constrained nodes, including at its ends.

The cleanest way to do this is to set all efforts at its boundaries equal to zero. The sum of the modes still can present a fixed boundary, should that be desired. The method also does not preclude efforts from being applied virtually at the ends, if desired, as is illustrated below. An added benefit of zero boundary forces and moments for a mechanical element is its inclusion of the rigid-body motion according to $F = ma$ and $M = I\alpha$. The whole scheme is nicely represented by bond graphs.

Application of this plan to a Bernoulli-Euler model of a slender beam, as described by the differential equation (10.286), proscribes zero moment and zero shear force at the two ends of each of its modal components:

$$\frac{\partial^2 \psi_\ell}{\partial x^2}(0) = \frac{\partial^3 \psi_\ell}{\partial x^3}(0) = 0, \tag{10.299a}$$

$$\frac{\partial^2 \psi_\ell}{\partial x^2}(L) = \frac{\partial^3 \psi_\ell}{\partial x^3}(L) = 0. \tag{10.299b}$$

Note that applied lateral forces are proportional to the *fourth* derivative of ψ_ℓ, and are not precluded at the ends. The homogeneous differential equation gives the general solution

$$\psi_\ell = A \cosh \mu_{1\ell} + B \cos \mu_{1\ell} + C \sinh \mu_{1\ell} + D \sin \mu_{1\ell}. \tag{10.300}$$

Substitution of this into equations (10.299a) gives $B = A$ and $D = C$. Subsequent substitution into equations (10.299b) gives $C/A = -(\cosh \mu_{1\ell}L - \cos \mu_{1\ell}L)/(\sinh \mu_{1\ell}L - \sin \mu_{1\ell}L)$ and the eigenvalue equation

$$\cosh \mu_{1\ell}L \cos \mu_{1\ell}L = 1. \tag{10.301}$$

An alternative (and simpler) way to deduce equation (10.301) is to substitute the zero boundary conditions into equation (10.122) (p. 766) and set the resulting 2×2 determinant equal to zero. There is an eigenvalue at 0, which represents rigid-body motion. The first three non-zero eigenvalues are 4.73004, 7.85320 and 10.99561. The first seven digits or more of subsequent eigenvalues satisfy $\mu_{1\ell}L = (2\ell + 1)\pi/2$. The associated natural frequencies are, from equation (10.287),

$$\omega_\ell^2 = \frac{(\mu_{1\ell}L)^4 EI}{\rho A L^4}. \tag{10.302}$$

The mode shape functions become

$$\psi_\ell = a_\ell(\cosh \mu_{1\ell}x + \cos \mu_{1\ell}x) - b_\ell(\sinh \mu_{1\ell}x + \sin \mu_{1\ell}x);$$

$$a_\ell = \frac{\sinh \mu_{1\ell}L - \sin \mu_{1\ell}L}{\sqrt{\cosh \mu_{1\ell}L \sinh \mu_{1\ell}L}}; \quad b_\ell = \frac{\cosh \mu_{1\ell}L - \cos \mu_{1\ell}L}{\sqrt{\cosh \mu_{1,\ell}L \sinh \mu_{1\ell}L}}, \tag{10.303}$$

which are scaled to give simple results below. The first four modal shapes are plotted in Fig. 10.43. These functions happily are orthogonal over the interval $0 \le x \le L$, which is necessary for the method to work.

The zero eigenvalue gives $d^4\psi_0/dx^4 = 0$, from which

$$\psi_0 = c_1 + c_2 x + c_3 x^2 + c_4 x^3. \tag{10.304}$$

The boundary conditions of equations (10.299) simplify this to $\psi_0 = c_1 + c_2 x$, which represents a combination of rigid-body translation and rotation. It is helpful to separate these motions into two modes: translation, for which $\psi_{00} = 1$, and rotation about the center of the beam, for which $\psi_0 = x - L/2$. Adding these modes to those above does not alter the orthogonality.

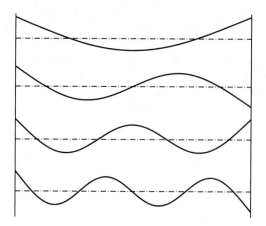

Figure 10.43: The first four bending mode shapes

For the forced response, equation (10.262) becomes

$$\sum_{\ell}(\mu_{1,\ell}^4 - \mu_1^4)\eta_\ell\psi_\ell = \frac{F(x,t)}{EI}. \tag{10.305}$$

This equation is multiplied by ψ_m and integrated from $x = 0$ to $x = L$, in the standard fashion of equation (10.263). The force will be assumed to comprise concentrated forces F_j and/or moments M_k at discrete locations $x = x_j$ and $x = x_k$, respectively, that is

$$F(x,t) = \sum_j F_j(t)\delta(x - x_j) + \sum_k \lim_{\Delta x \to 0}\{\frac{-M_j}{\Delta x}[\delta(x - x_j) - \delta(x - x_j - \Delta x)]\}. \tag{10.306}$$

The orthogonality condition eliminates all terms for $m \neq l$, with the result

$$(\mu_{1\ell}^4 - \mu_1^4)\eta_\ell \int_0^L \psi_l^2\,dx = \frac{1}{EI}\sum_j F_j(t)\psi(x_j) + \frac{1}{EI}\sum_k M_k(t)\frac{d\psi}{dx}(x_k). \tag{10.307}$$

For the rigid-body modes in translation and rotation, respectively, this gives the expected results

$$\eta_0 = \frac{1}{mS^2}\sum_j F(j)(t); \quad m = \rho AL, \tag{10.308a}$$

$$\omega_0 = \frac{1}{JS^2}\left[\sum_j F_j(t)\left(x_j - \frac{L}{2}\right) + \sum_k M_k(t)\right]; \quad J = \frac{1}{12}mL^2. \tag{10.308b}$$

For the other modes, the result is

$$\eta_\ell = \frac{1}{k_\ell + m_\ell S^2}\left[\sum_j F_j(t)\psi_\ell(x_j) + \sum_k M_k(t)\frac{d\psi_\ell}{dx}(x_k)\right]; \quad m_\ell = \rho AL\frac{b_\ell}{L}; \quad k_\ell = m_\ell\omega_\ell^2, \tag{10.309}$$

in which m_ℓ is known as the **modal mass** and k_ℓ is the **modal stiffness**, and the ratio of modal mass to the total mass is b_ℓ/L, where b_ℓ is given by equation (10.260):

$$\frac{b_\ell}{L} = \int_0^1 \psi_\ell^2(X)\,dX; \quad X \equiv \frac{x}{L}. \tag{10.310}$$

The ψ_ℓ for this integration is given by equation (10.303). Numerical integrations have been carried out by the author with the results given below:

ℓ	1	2	3	4 and greater
$\dfrac{m_\ell}{m} = \dfrac{b_\ell}{L}$	1.03545	0.99845	1.00007	1.00000

The entire model is very simply represented by the bond graph of Fig. 10.44, in which the inertances are the respective modal masses and the compliances are the reciprocals of the respective modal stiffnesses. The transformer moduli for the bending modes are the respective values of ψ_ℓ and $d\psi_\ell/dx$ as given by equation (10.303) and evaluated at the respective locations x_j and x_k where the forces and/or moments are applied. Only one applied force and one moment are shown, but any number are possible. The beauty of this method is underscored when these forces are associated with models of other components attached to the beam and themselves represented by bond graphs.

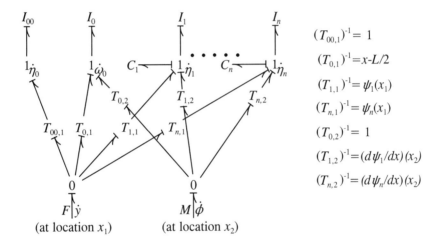

$$(T_{00,1})^{-1} = 1$$
$$(T_{0,1})^{-1} = x\text{-}L/2$$
$$(T_{1,1})^{-1} = \psi_1(x_1)$$
$$(T_{n,1})^{-1} = \psi_n(x_1)$$
$$(T_{0,2})^{-1} = 1$$
$$(T_{1,2})^{-1} = (d\psi_1/dx)(x_2)$$
$$(T_{n,2})^{-1} = (d\psi_n/dx)(x_2)$$

Figure 10.44: Bond graph representation of excited beam using force and moment-free boundary modes

10.9.6 Avoiding Differential Causality

Sometimes you may wish to impose a velocity at a particular location, rather than a force. Or, more likely, you may wish to append an inertia directly to a particular point along a distributed member. Either of these choices produces differential causality, assuming your modes are resonant, which can be extremely awkward to handle. An example, which employs the beam treated earlier, is illustrated in part (a) of Fig. 10.45.

A better solution, suggested by Margolis[22], is to add only the compliance of the first higher mode that otherwise is being dropped. As illustrated in part (b) of the figure, this eliminates the need for differential causality. The addition of an extra mode minus its

[22]D.L. Margolis, "Bond Graphs for Distributed System Models Admitting Mixed Causal Inputs," *ASME J. Dynamic Systems, Measurement and Control*, v 102 n 2, June 1980, pp. 94-100.

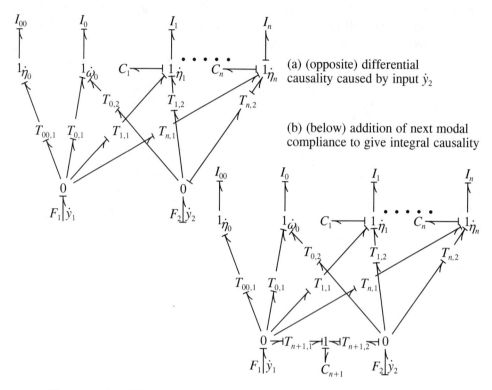

Figure 10.45: Velocity inputs and a way to preclude differential causality

inertance introduces no error more significant than eliminating the mode altogether. Should two or more velocities be specified, or inertances added, an equal number of additional compliances can be added; the model becomes under-causal, but this condition is easier to treat than differential causality.

10.9.7   Summary

Modal decomposition replaces a model which is distributed in space with an infinite assemblage of lumped models, ordered in their frequency content. The truncation of the number of modes results in a fairly predictable break frequency below which the behavior is well represented. Retaining all modes with frequencies less than two to four times the maximum frequency of interest likely is sufficient. Complicated initial conditions and complicated internal forcing functions are accommodated quite simply, compared for example with the use of Green's functions. A modal model based on zero boundary forces is most easily imbedded within a larger system.

The method is sharply restricted to linear differential equations amenable to a product separation of variables and which, together with the boundary conditions, give an orthogonal set of spatial functions. Modal decomposition strictly admits zero dissipation, with the important exception of **proportional damping** as discussed in Appendix B.2.4 (pp. 1026-1027). In practice, however, little loss of fidelity is suffered if a small amount of arbitrary damping is added subsequently in order to match known behavior. There is no restriction on the damping in attached components.

The use of bond graphs directs causal choices that are particularly helpful in assembling complex systems, including structuring the model so as to preclude differential causality.

Guided Problem 10.11

Consider again the problem of Guided Problem 10.10 part (c) in which a uniform thermal rod of length L which is insulated over all surfaces except for its right end, which is kept at a constant temperature equal to the uniform initial temperature. The rod is subjected to uniform heating $Q(x,t) = q_0$ per unit volume between the locations $x = L/3$ and $x = 2L/3$. The thermal conductivity is k_θ and the specific heat is C. Find the temperature history of the rod. Plot the temperature distribution for $t = 0.05t_0$, $0.10t_0$, $0.20t_0$, $0.50t_0$ and $\infty \cdot t_0$, where $t_0 = CL^2/k_\theta$ is a characteristic time.

Suggested Steps:

1. Add the heat input to the differential equations, such as given in the problem statement for Example 10.10 (pp. 747-748). Combine the two equations to leave the temperature as the only dependent variable. Time differentiation may be indicated by the operator s.

2. Find the end conditions in terms of $\theta(0)$, $(\partial\theta/\partial x)(0)$, $\theta(L)$ and $(\partial\theta/\partial x)(L)$.

3. The differential equation is of second order in space, so try $\psi_l = \sin(k_l x + \alpha)$ or $\psi_l = \cos(k_l x + \alpha)$. Evaluate k_l and α using the results of step 2.

4. Substitute an assumed product solution $\theta = \sum_l \psi_l \theta_l(t)$ into the differential equation, using the result of step 3.

5. Multiply both sides of the result of step 4 by $\psi_m\,dx$ and integrate over the length of the rod. Note which terms cancel. Define and evaluate b_l.

6. Solve the remaining ordinary differential equation for $\theta_l(t)$.

7. Use MATLAB or other software to make the plots. Note that the steady-state plots comprise two straight lines and one parabola.

8. You may wish to construct a bond graph to represent the modal decomposition and the differential equations. The heat flow q could be used as the asymmetric variable, even though the result is a pseudo bond graph (see Section 9.7.7, pp. 701-702).

Guided Problem 10.12

A story from the European theater of World War II has a heavy tank cross a light bridge that collapses the moment the tank successfully reaches the other side. Probe the plausibility of this story by considering a single-span bridge of length L traversed at a constant speed v by a point force of magnitude F_0. Represent the bridge by a uniform Bernoulli-Euler beam, simply pinned at its ends, with properties ρ, A E and I.

Suggested Steps:

1. Write a differential equation for the beam in terms of the deflection $\eta(x,t)$ and the loading $F_0\delta(x - Vt)$.

2. Find the modal decomposition and the modal loading.

3. Solve for the deflection of the bridge while the tank is still on it.

4. Specialize your answer to step 3 for the case of the first critical velocity, which is $V_c = (\pi/L)\sqrt{EI/\rho A}$.

5. Find the time at which the largest deflection occurs. If the bridge fails at this time, does the tank also fall?

6. Compare the deflection of step 5 with the largest deflection that would occur if the tank moved very slowly, which is $F_0 L^3/48EI$. Might the bridge survive if the tank moved slowly?

PROBLEMS

10.34 Find a modal model of a uniform bar with longitudinal (tension and compression) motion only, assmuing free ends. The bar has density ρ, cross-sectional area A, Young's modulus E and length L and is slender enough to disregard the effects of Poisson's ratio. Represent the model with a bond graph that includes an input force at one-third the distance from one end to the other. Determine the moduli of its elements.

10.35 The model of the previous problem is excited by a given velocity at the one-third location, instead of a force. The output motion at the far end is of interest. Draw a bond graph, apply causality, and indicate how derivative causality can be avoided.

10.36 A stretched uniform vibrating string of length L, tension T and total mass m has an added mass attached to its center equal to $m/2$.

(a) Model the system with a modal bond graph. Include the first four modes, and establish its parameters.

(b) Use the bond graph to establish the first four natural frequencies.

(c) Determine these frequencies from an exact analysis. Suggestion: apply equation (10.100) (p. 753) to one half of the the string.

10.37 A uniform Bernoulli-Euler beam is supported by springs with dashpots at its left end and in the center, and supports a mass-spring-dashpot system at its right end, as shown. Give a bond-graph modal model, and establish a consistent causality.

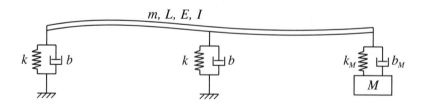

10.38 A uniform Bernoulli-Euler beam of mass m, length L and stiffness EI is supported by springs so soft they do not affect its bending dynamics. A controllable light motor mounted at an arbitrary location on the beam rotates a balanced rotor with mass M and mass moment of inertia J.

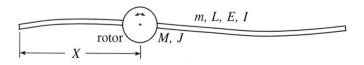

(a) Give a modal model in bond graph form for the beam with the added controllable rotor. Determine the parameters of the model.

(b) The rotor is oscillated clockwise and counterclockwise so that its angle is $\theta_0 \sin \omega_m t$. Without carrying out the details, show how the linear and angular velocities of the center of mass could be closely estimated.

(c) The motor now is given an arbitrary time history. Give a bond graph with integral causality appropriate for digital simulation. Treat any problems of differential causality that arise.

SOLUTIONS TO GUIDED PROBLEMS

Guided Problem 10.11

1. $\dfrac{\partial q}{\partial x} = -CS\theta + q_{in}(x,t); \qquad \dfrac{\partial \theta}{\partial x} = -\dfrac{1}{k_\theta} q.$

   Combining these two equations, $\dfrac{\partial^2 \theta}{\partial x^2} = -\dfrac{1}{k_\theta}\dfrac{\partial q}{\partial x} = \dfrac{CS}{k_\theta}\theta - \dfrac{1}{k_\theta}q_{in}(x,t).$

   The excitation is $q_{in}(x,t) = 0$ for $x < L/3$ and $x > 2L/3$, and is $3q_0/L$ for $L/3 < x < 2L/3$.

2. $q(0) = 0$, therefore $\dfrac{\partial \theta}{\partial x}(0) = 0$; $\theta(L) = 0$.

3. Use $\psi_\ell = \cos k_\ell x$, since this gives $\dfrac{\partial \psi_\ell}{\partial x}(0) = 0$ (*i.e.* $\alpha = 0$)

   $\psi_\ell(L) = 0 = \cos(k_\ell L)$, therefore $k_\ell = \dfrac{\pi}{2L}(1,\ 3,\ 5,\ 7,\cdots)$, or $k_\ell L = \dfrac{\pi}{2}k,$, where $k = 1, 3, 5, 7, \cdots$

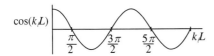

4. $\theta = \displaystyle\sum_\ell \cos k_\ell x\,\theta_\ell(t)$

   $\displaystyle\sum_\ell -k_\ell^2\,x\theta_\ell(t) = \sum_\ell \dfrac{C}{k_\theta\,\ell}\,x\,\dfrac{d\theta_\ell(t)}{dt} - \dfrac{1}{k_\theta}q_m(x,t)$

5. $-\displaystyle\int_0^L \sum_\ell k_\ell^2 \cos k_\ell x \cos k_m x\,\theta_\ell(t)\,dx$

   $= \displaystyle\int_0^L \sum_\ell \dfrac{C}{k_\theta}\cos k_\ell x \cos k_m x \dfrac{d\theta_e ll}{dt}\,dx - \dfrac{1}{k_\theta}\int_0^L q_{in}(x,t)\cos k_m x\,dx$

   This gives $-\dfrac{L}{2}k_m^2\theta_m(t) = \dfrac{CL}{2k_\theta}\dfrac{d\theta_\ell}{dt} - \dfrac{3q_0}{Lk_\theta k_m}\underbrace{\left(\sin\dfrac{2k_m L}{3} - \sin\dfrac{k_m L}{3}\right)}_{2\cos k_m L/2\,\sin(k_m L/6)\ \text{(optional)}}$

6. $\left(C_\theta k_m^2\right)\dfrac{d\theta_m}{dt} + \theta_m = \dfrac{2}{Lk_m^2}\dfrac{6q_0}{Lk_\theta k_m}\cos\dfrac{k_m L}{2}\sin\dfrac{k_m L}{6}$

   Therefore, $\theta_m = \dfrac{12q_0}{k_\theta L^2 k_m^3}\cos\dfrac{k_m L}{2}\sin\dfrac{k_m L}{6}\left[1 - e^{-tk_\theta k_m^2/C}\right]\cos k_m x.$

   Let $Th \equiv \dfrac{\theta k_\theta}{Lq_0};\quad t_0 = \dfrac{Cl^2}{k_\theta};\quad k_m L = \dfrac{\pi}{2}k,\ k = 1\,3,\ 5,\ 7,\cdots$

   $$\boxed{\,Th = \dfrac{96}{\pi^3}\sum_{k\ \text{odd}}\cos\left(\dfrac{\pi k}{4}\right)\sin\left(\dfrac{\pi k}{12}\right)\left[1 - e^{-(/2)^2 t/t_0}\right]\cos\left(\dfrac{\pi k}{2}\dfrac{x}{L}\right)\,}$$

7. Coding the boxed equation above gives the following plots:

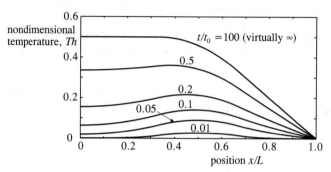

8. The pseudo bond graph opposite treats thermal resistance as though it were energy dissipation, which of course it is not. This is a result of using heat flux rather than entropy flux as the flow variable. The modes also are different from the others illustrated in this section in that they are not resonant; there are no inertances. The single input is distributed over a region rather than being located at a discrete point. The output transformers vary with the location of interest.

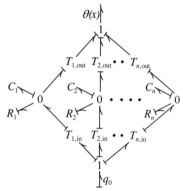

Guided Problem 10.12

1. The differential equation is $\dfrac{\partial \eta}{\partial x^4} + \rho A S^2 \eta = \dfrac{F_0}{EI} \delta(x - vt)$.

2. Separation of variables: $\eta = \sum_{\ell} \psi_\ell(x) \eta_\ell(t) = \sum_{\ell} \sin\left(\dfrac{\ell \pi x}{L}\right) \eta_\ell(t)$

$$\sum_{\ell} \left[\left(\dfrac{\ell \pi}{L}\right)^4 + \dfrac{\rho A S^2}{EI}\right] \sin\left(\dfrac{\ell \pi x}{L}\right) \eta_\ell(t) = \dfrac{F_0}{EI} \delta(x - vt)$$

$$\int_0^L \sum_{\ell} \left[\left(\dfrac{\ell \pi}{L}\right)^4 + \dfrac{\rho A S^2}{EI}\right] \eta_\ell(t) \sin\left(\dfrac{\ell \pi x}{L}\right) \sin\left(\dfrac{m \pi x}{L}\right) dx = \int_0^L \dfrac{F_0}{EI} \delta(x - vt) \sin\left(\dfrac{m \pi x}{L}\right) dx$$

$$\left[\left(\dfrac{m \pi}{L}\right)^4 + \dfrac{\rho A S^2}{EI}\right] \eta_m(t) \dfrac{L}{2} = \dfrac{F_0}{EI} \sin\left(\dfrac{m \pi v t}{L}\right)$$

$$\eta_m(s) = \dfrac{F_0}{EI} \dfrac{m \pi v/L}{s^2 + (m \pi v/L)^2} \dfrac{2}{L} \dfrac{1}{\rho A s^2/EI + (m \pi/L)^4}$$

$$= \dfrac{2 F_0 \pi m v}{\rho A L^2} \dfrac{1}{S^2 + b_\ell^2} \dfrac{1}{S^2 + \omega_\ell^2}; \quad b_m = \dfrac{m \pi v}{L}; \quad \omega_m = \dfrac{m^2 \pi^2}{l^2}\sqrt{\dfrac{EI}{\rho A}}$$

$$\eta_m(t) = \dfrac{2 F_0 \pi m v}{\rho A L^2} \left[-\dfrac{\sin \omega_m t}{\omega_m (\omega_m^2 - b_m^2)} + \dfrac{\sin b_m t}{b_m (\omega_m^2 - b_m^2)}\right]$$

3.

$$\boxed{\eta = \sum_{\ell} \sin\left(\dfrac{\ell \pi x}{L}\right) \dfrac{2 F_0 \pi m v}{\rho A L^2} \left[-\dfrac{\sin \omega_\ell t}{\omega_\ell (\omega_\ell^2 - b_\ell^2)} + \dfrac{\sin b_\ell t}{b_\ell (\omega_\ell^2 - b_\ell^2)}\right]}$$

4. When $v = V_c = \dfrac{\pi}{L}\sqrt{\dfrac{EI}{\rho A}}$, $b_1 = \dfrac{\pi^2}{L^2}\sqrt{\dfrac{EI}{\rho A}} = \omega_1$, so the second term above becomes infinite for the first mode ($\ell = 1$). Backing up,

for $\ell = 1$, $\quad \eta_1(S) = \dfrac{2F_0\pi^2}{L^2\rho AL}\sqrt{\dfrac{EI}{\rho A}}\dfrac{1}{(S^2 + \omega_1^2)^2}\sin\left(\dfrac{\pi x}{L}\right)$

$$\eta_1(t) = \dfrac{2F_0\pi^2}{\rho AL^3}\sqrt{\dfrac{EI}{\rho A}}\dfrac{1}{2\omega_2^3[\sin\omega_1 t - \omega_1 t\cos\omega_1 t]}\sin\left(\dfrac{\pi x}{L}\right)$$

Define $\tau \equiv \dfrac{t}{T}$, where $T = \dfrac{L}{V_c}$; $\quad \omega_1 t = \omega_1 T\tau = \dfrac{\omega_1 L}{V_c}\tau = \pi\tau$

Therefore, $\eta_1(t) = \dfrac{F_0\pi^2}{\rho AL^3}\sqrt{\dfrac{EI}{\rho A}}\left(\dfrac{\rho A}{EI}\right)^{3/2}\dfrac{L^6}{\pi^6}[\sin\pi\tau - \pi\tau\cos\pi\tau]\sin\left(\dfrac{\pi x}{L}\right)$

The static deflection for the load at the center is $\eta_0 = \dfrac{F_0 L^3}{48EI}$

$$\boxed{\dfrac{\eta_1}{\eta_0} = \dfrac{48}{\pi^4}[\sin\pi\tau - \pi\tau\cos\pi\tau]\sin\left(\dfrac{\pi x}{L}\right)}$$

For $\ell \geq 2$, $\dfrac{\eta_\ell}{\eta_0} = \dfrac{48EI}{F_0 L^3}\dfrac{2F_0\pi\ell}{\rho AL^2}\dfrac{\pi}{L}\sqrt{\dfrac{IE}{\rho A}}\left[-\dfrac{\sin\omega_\ell t}{\omega_\ell(\omega_\ell^2 - b_\ell^2)} + \dfrac{\sin b_\ell t}{b_\ell(\omega_\ell^2 - b_\ell^2)}\right]\sin\left(\dfrac{\ell\pi x}{L}\right)$

where $\omega_\ell = \dfrac{\ell^2\pi^2}{L^2}\sqrt{\dfrac{EI}{\rho A}}$; $\quad b_\ell = \dfrac{\ell\pi}{L}\dfrac{\pi}{L}\sqrt{\dfrac{EI}{\rho A}} = \dfrac{\omega_\ell}{\ell}$; $\quad b_\ell t = \ell\pi\tau$

$\dfrac{\eta_\ell}{\eta_0} = \dfrac{96EI\pi^2\ell}{L^6\rho A}\sqrt{\dfrac{EI}{\rho A}}\left(\dfrac{L^2}{\pi^2\ell}\right)^3\left(\dfrac{\rho A}{EI}\right)^{3/2}\left[-\dfrac{\sin(\ell^2\pi\tau)}{\ell(\ell^2 - 1)} + \dfrac{\sin(\ell\pi\tau)}{\ell^2 - 1}\right]\sin\left(\dfrac{\ell\pi x}{L}\right)$

$$\boxed{\dfrac{\eta_\ell}{\eta_0} = \dfrac{96}{\pi^4\ell^2(\ell^2 - 1)}\left[-\dfrac{\sin(\ell^2\pi\tau)}{\ell} + \sin(\ell\pi\tau)\right]\sin\left(\dfrac{\ell\pi x}{L}\right)}$$

The total normalized deflection is $\dfrac{\eta}{\eta_0} = \dfrac{1}{\eta_0}\displaystyle\sum_\ell \eta_\ell$.

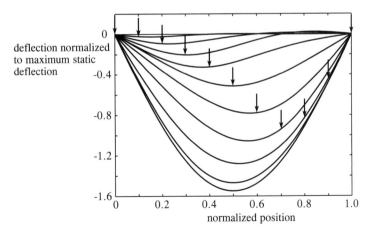

The curves above represent the bridge deflection normalized to the maximum static deflection (when the tank is in the center of the bridge) for each tenth of the total traverse. The arrows show the positions of the tank. Since the maximum deflection occurs just as the tank reaches the far side, the scenario of a collapse with the tank surviving is plausible but very lucky.

6. The maximum deflection is about 1.55 times the maximum static deflection. Therefore, no failure would have occurred had the tank crawled over the bridge.

10.10   Complex Compound Systems: A Case Study*

The assembly of complex compound systems in the form of a computer model can quickly lead to fatal round-off errors. How to avoid this problem while efficiently assembling extremely complicated systems comprising complex distributed elements, lumped elements and junctions is illustrated for a particular class of problems. The resulting code successfully treated elaborate models on a modest early vintage microcomputer. The presentation is summary in nature; further details appear in references, especially the dissertation of Tentarelli.[23]

The class of problems addressed is the propagation of noise and vibration in complex hydraulic tubing or piping systems. A single section of tubing is treated as a seven-power problem: one power for the fluid, one for the axial motion of the walls, four for lateral flexure in two planes similar to the Bernoulli-Euler beam and one for twisting about the axis. Curved as well as straight sections are addressed. Four different types of coupling mechanisms between the different powers are considered: bends, tees and other piping discontinuities cause unbalanced presssure forces on the tubing and the fluid, the Poisson effect couples extensional tubing motion with fluid pressure, the Bourdon effect couples bending with fluid pressure in curved sections with non-circular cross-sections, and fluid viscosity generates friction forces with couple fluid motion with axial wall motion. A tall order!

10.10.1   A Model for Uniform Straight Tubing

A Cartesian coordinate system is used with z aligned along the axis of the tube. The straight-pipe equations uncouple nicely into four groups: axial motion, flexure in the y-z plane, flexure in the x-z plane, and twisting about the z axis. The notation of Tentarelli is used. The time derivative operator s is employed, which in practice becomes $j\omega$, since only frequency responses are practical to find for such a complex problem. The high-frequency limitations of the model are discussed by Tentarelli in his dissertation.

The symmetric and asymmetric variables for the fluid power are the pressure, p, and the fluid displacement, v, respectively. The corresponding variables for the axial wall motion are the force, f_z, and the wall displacement, u_z. The parameters are the fluid and wall densities, ρ_f and ρ_p, the cross-sectional areas of the fluid and the wall, A_f and A_p, the inner and outer radii, r_i and r_0, the bulk modulus of the fluid, β, Young's modulus E and Poisson's ratio ν. The state differential equation is, in standard Lapace transform notation,

$$\frac{\partial}{\partial z}\begin{bmatrix} p \\ f_z \\ v \\ u_z \end{bmatrix} = (-)\begin{bmatrix} 0 & 0 & Z_1 & Z_2 \\ 0 & 0 & Z_3 & Z_4 \\ Y_1 & Y_2 & 0 & 0 \\ Y_3 & Y_4 & 0 & 0 \end{bmatrix}\begin{bmatrix} p \\ f_z \\ v \\ u_z \end{bmatrix}; \qquad (10.311a)$$

$$Z_1 = -\frac{\rho_f s^2}{C(s)}; \quad Z_2 = \rho_f\left(1 + \frac{1}{C(s)}\right)s^2;$$

$$Z_3 = -A_f Z_2; \quad Z_4 = -(\rho_p A_p s^2 + Z_3); \qquad (10.311b)$$

---

[23]Forbes T. Brown and Stephen C. Tentarelli, "Dynamic Behavior of Complex Fluid-Filled Tubing Systems - Part 1:Tubing Analysis," *J. of Dynamic Systems, Measurement and Control,* v 123 n 1 (March 2001) pp 71-77; Stephen C. Tentarelli and Forbes T. Brown, " · · · · - Part 2: System Analysis," *ibid,* pp 78-84; F.T. Brown and S.C. Tentarelli, "Analysis of Noise and Vibration in Complex Tubing Systems with Fluid-Wall Interaction," *Proceedings of the 43rd National Conference on Fluid Power,* October, 1988; and Stephen C. Tentarelli, *Propagation of Noise and Vibration in Complex Hydraulic Tubing Systems,* Ph.D. dissertation, Lehigh University, Bethlehem, PA, 1989.

$$Y_1 = \frac{1}{\beta^*} : \quad Y_2 = -\frac{2\nu}{EA_p}; \quad Y_3 = -\frac{2\nu}{E}\frac{r_i^2}{(r_0^2 - r_i^2)}; \quad Y_4 = -\frac{1}{EA_p}; \tag{10.311c}$$

$$C(s) = \frac{2}{jr_i\sqrt{s/\nu}}\frac{J_1(jr_i\sqrt{s/\nu})}{J_0(jr_i\sqrt{s/\nu})} - 1; \tag{10.311d}$$

$$\beta^* = \frac{\beta\beta_c}{\beta + \beta_c}; \quad \beta_c = \frac{0.5E(r_0^2 - r_i^2)}{(1+\nu)r_0^2 + (1-\nu)r_i^2}. \tag{10.311e}$$

Notice the confirmation of local symmetry, according to equation (10.37) (p. 732). The function $C(s)$ will be recognized as virtually an inversion of the $Z(s)$ of equation (10.135) (p. 773) that describes viscous behavior; the symbol ν therein is kinematic viscosity and is not to be confused with Poisson's ratio. The compliance of the wall reduces the effective bulk modulus of the fluid from β to β^*. Notice that wall dilation shortens the tube via the Poisson's ratio effect.

Flexure in the y-z plane is treated by a model equivalent to the Timoshenko beam given in footnote 9 on page 765. The symmetric variables are the displacement u_y and the moment m_x, and the conjugate asymmetric variables are the force f_y and angular displacement ψ_x; the subscripts indicate the axis. The axis of the observer is in the y coordinate, so flexures in the x-z plane are seen differently: f_x and ψ_y are symmetric, and u_x and m_y are asymmetric. New parameters are the shear modulus, G, the Timoshenko shear coefficient, k, and the area moment of inertia for bending, I_p. The equations are

$$\frac{\partial}{\partial z}\begin{bmatrix} u_y \\ m_x \\ f_y \\ \psi_x \end{bmatrix} = (-)\begin{bmatrix} 0 & 0 & Z_5 & 1 \\ 0 & 0 & -1 & Z_6 \\ Y_5 & 0 & 0 & 0 \\ 0 & Y_6 & 0 & 0 \end{bmatrix}\begin{bmatrix} u_y \\ m_x \\ f_y \\ \psi_x \end{bmatrix}; \tag{10.312a}$$

$$\frac{\partial}{\partial z}\begin{bmatrix} f_x \\ \psi_y \\ u_x \\ m_y \end{bmatrix} = (-)\begin{bmatrix} 0 & 0 & Z_7 & 0 \\ 0 & 0 & 0 & Z_8 \\ Y_7 & -1 & 0 & 0 \\ 1 & Y_8 & 0 & 0 \end{bmatrix}\begin{bmatrix} f_x \\ \psi_y \\ u_x \\ m_y \end{bmatrix}; \tag{10.312b}$$

$$Z_5 = -\frac{1}{kGA_p}; \quad Z_6 = -\rho_p I_p s^2; \quad Z_7 = -(\rho_p A_p + \rho_f A_f)s^2; \quad Z_8 = -\frac{1}{EI_p}; \tag{10.312c}$$

$$Y_5 = Z_7; \quad Y_6 = Z_8; \quad Y_7 = Z_5; \quad Y_8 = Z_6; \tag{10.312d}$$

$$k = \frac{3}{4}\frac{(r_0^4 - r_i^4)}{(r_0^3 - r_i^3)}\frac{(r_0 - r_i)}{(r_0^2 - r_i^2)} \simeq 0.5. \tag{10.312e}$$

Twisting about the z-axis is modeled by

$$\frac{\partial}{\partial z}\begin{bmatrix} m_z \\ \psi_z \end{bmatrix} = (-)\begin{bmatrix} 0 & Z_9 \\ Y_9 & 0 \end{bmatrix}\begin{bmatrix} m_z \\ \psi_z \end{bmatrix}, \tag{10.313a}$$

$$Z_9 = -\rho_p J s^2; \quad Y_9 = -\frac{1}{GJ} \tag{10.313b}$$

where J is the polar area moment of inertia.

The transmission matrix of equation (10.110) (p. 763) with $z - z_0$ substituted for x is evaluated numerically for each frequency of interest. For example, the upper right quadrant of $e^{\mathbf{A}(z-z_0)}$ is

$$\mathbf{Z}(\sqrt{\mathbf{YZ}})^{-1}\sinh(\sqrt{\mathbf{YZ}}(z_0 - z)) = \mathbf{Z}\left[a_0\mathbf{I} + a_1\mathbf{YZ} + \cdots + a_{n-1}(\mathbf{YZ})^{n-1}\right], \tag{10.314a}$$

$$\frac{1}{\sqrt{\mu_i}}\sinh[\sqrt{\mu_i}(z_0 - z)] = a_0 + a_1\mu_1 + \cdots + a_{n-1}\mu_i^{n-1}, \quad i = 1,\ldots,n. \tag{10.314b}$$

10.10.2 A Model for Uniformly Curved Tubing

The axes are oriented so that the bend, which has the radius R, is in the y-z plane. Coupling between axial motion and flexure in this plane is produced by pressure waves and by the Bourdon effect. This latter effect is produced by the ovalization of a tube which occurs in the bending process. Flexure causes the cross-sectional area to change and thereby displace fluid; conversely, fluid pressure attempts to unbend the tube. The result is a four-power model with eight state variables. Flexure in the x-z plane is coupled with twisting about the z-axis and results in a separate three-power model with six state variables. A coordinate q is introduced that follows the bent axis of the tube. The equations are, respectively,

$$
\frac{\partial}{\partial s}
\begin{bmatrix} p \\ f_z \\ u_y \\ m_x \\ v \\ u_z \\ f_y \\ \psi_x \end{bmatrix}
= (-)
\begin{bmatrix}
0 & 0 & 0 & 0 & Z_1 & Z_2 & 0 & 0 \\
0 & 0 & 0 & 0 & Z_3 & Z_4 & 1/R & 0 \\
0 & 0 & 0 & 0 & 0 & -1/R & Z_5 & 1 \\
0 & 0 & 0 & 0 & 0 & 0 & -1 & Z_6 \\
Y_{14} & Y_2 & 1/R & -Y_{11} & 0 & 0 & 0 & 0 \\
Y_3 & Y_4 & 1/R & 0 & 0 & 0 & 0 & 0 \\
A_f/R & -1/R & Y_5 & 0 & 0 & 0 & 0 & 0 \\
-Y_{12} & 0 & 0 & -Y_{13} & 0 & 0 & 0 & 0
\end{bmatrix}
\begin{bmatrix} p \\ f_z \\ u_y \\ m_x \\ v \\ u_z \\ f_y \\ \psi_x \end{bmatrix}
$$
(10.315a)

$$
\frac{\partial}{\partial s}
\begin{bmatrix} f_x \\ \psi_y \\ m_z \\ u_x \\ m_y \\ \psi_z \end{bmatrix}
= (-)
\begin{bmatrix}
0 & 0 & 0 & Z_7 & 0 & 0 \\
0 & 0 & 0 & 0 & -Y_{13} & -1/R \\
0 & 0 & 0 & 0 & 1/R & Z_9 \\
Y_7 & -1 & 0 & 0 & 0 & 0 \\
1 & Y_8 & -1/R & 0 & 0 & 0 \\
0 & 1/R & Y_9 & 0 & 0 & 0
\end{bmatrix}
\begin{bmatrix} f_x \\ \psi_y \\ m_z \\ u_x \\ m_y \\ \psi_z \end{bmatrix},
$$
(10.315b)

$$
Y_{11} = \frac{-1}{\pi Rbc} Y_{12}; \quad Y_{12} = \frac{b^2 - c^2}{Ebh^2}\left[\sqrt{3(1-\nu^2)} - \frac{Rh}{bc};\right]
$$

$$
Y_{13} = \frac{c\sqrt{12(1-\nu^2)}}{2\pi Eb^2h^2R}; \quad Y_{14} = Y_1 - \frac{1}{R}\left(1 - \frac{b^2}{c^2}\right)\left(1 - \frac{2Rh}{bc\sqrt{12(1-\nu^2)}}\right)Y_{12}.
$$
(10.315c)

The coefficients $Z_1, Z_2, \cdots, Z_9$ and $Y_1, Y_2, \cdots, Y_9$ are the same as for a straight tube. The coefficients Y_{11} and Y_{12} reflect the fluid-wall coupling due to the Bourdon effect. Y_{13} is the effective bending stiffness for a curved and ovalized tube and is applicable both in the plane of the bend and out of the plane of the bend. The coefficients $Y_{11} \cdots Y_{14}$ are as evaluated by Reinwald[24], who used an algebraic approximation to a Fourier series solution for an assumed elliptical cross-section given by Clark et al.[25] The thickness or the wall is $h = r_0 - r_i$, the semi-minor axis for an assumed elliptical cross-section (measured to the center of the tube wall) is b, and the corresponding semi-major axis is c. When the tube is circular, $(c/b) = 1$ and $Y_{11} = Y_{12} = 0$ and $Y_{14} = Y_1$. When, also $R \to \infty$, the equations reduce to those for straight tubes (equations (10.312) and (10.313)).

Note that in all the differential equations above, $\mathbf{W} = \mathbf{X} = \mathbf{0}$, which implies local symmetry and considerably simplifies the analysis.

[24] Reinwald, C., 1989, "The Bourdon Coupling between Internal Pressure and Bending of Curved Noncircular Hydraulic Tubes," Master's thesis, Lehigh University, Bethlehem, PA.

[25] Clark, R. A., Gilroy, R. I. and Reissner, E., 1952, "Stresses and Deformations of Toroidal Shells of Elliptical Cross Section with Applications to the Problems of Bending of Curved Tubes and the Bourdon Gage," ASME J. Applied Mech., v 74, pp. 37-48.

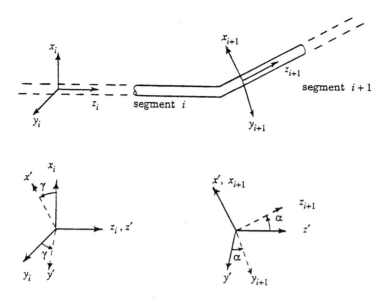

Figure 10.46: Coordinate systems for cascaded piping segments

10.10.3 The Rotation Matrix

Two tubing or pipe segments joined together at an abrupt angle are shown in Fig. 10.46. Each segment is given its own coordinate system, with z always pointing along the longitudinal direction and an arbitrary orientation assigned for the x and y axes, unless the segment is curved. The transformation of axes from one segment to the next results from two consecutive rotations, γ and α, as depicted in the figure. The state vector at the end of segment I is transformed to the state vector at the beginning of segment $i+1$ by the rotation matrix $\mathbf{T}_R$:

$$
\begin{bmatrix} p \\ F_z \\ u_y \\ m_x \\ v \\ u_z \\ f_y \\ \psi_x \\ f_x \\ \psi_y \\ m_z \\ u_x \\ m_y \\ \psi_z \end{bmatrix}_{z_i}
=
\begin{bmatrix}
1 & 0 & 0 & 0 & 0 & 0 & 0 & 0 & 0 & 0 & 0 & 0 & 0 & 0 \\
a_2 & c_\alpha & 0 & 0 & 0 & 0 & s_\alpha & 0 & 0 & 0 & 0 & 0 & 0 & 0 \\
0 & 0 & a_5 & 0 & 0 & a_6 & 0 & 0 & 0 & 0 & 0 & s_\alpha & 0 & 0 \\
0 & 0 & 0 & c_\gamma & 0 & 0 & 0 & 0 & 0 & 0 & a_8 & 0 & a_7 & 0 \\
0 & 0 & s_\alpha & 0 & 1 & a_1 & 0 & 0 & 0 & 0 & 0 & 0 & 0 & 0 \\
0 & 0 & s_\alpha & 0 & 0 & c_\alpha & 0 & 0 & 0 & 0 & 0 & 0 & 0 & 0 \\
a_3 & a_6 & 0 & 0 & 0 & 0 & a_5 & 0 & s_\gamma & 0 & 0 & 0 & 0 & 0 \\
0 & 0 & 0 & 0 & 0 & 0 & 0 & c_\gamma & 0 & a_7 & 0 & 0 & 0 & a_8 \\
a_4 & a_8 & 0 & 0 & 0 & 0 & a_7 & 0 & c_\gamma & 0 & 0 & 0 & 0 & 0 \\
0 & 0 & 0 & 0 & 0 & 0 & 0 & s_\gamma & 0 & a_5 & 0 & 0 & 0 & a_6 \\
0 & 0 & 0 & 0 & 0 & 0 & 0 & 0 & 0 & 0 & c_\alpha & 0 & s_\alpha & 0 \\
0 & 0 & a_7 & 0 & 0 & a_8 & 0 & 0 & 0 & 0 & 0 & c_\gamma & 0 & 0 \\
0 & 0 & s_\gamma & 0 & 0 & 0 & 0 & 0 & 0 & 0 & a_6 & 0 & a_5 & 0 \\
0 & 0 & 0 & 0 & 0 & 0 & 0 & 0 & 0 & s_\alpha & 0 & 0 & 0 & c_\alpha
\end{bmatrix}
\begin{bmatrix} p \\ f_z \\ u_y \\ m_x \\ v \\ u_z \\ f_y \\ \psi_x \\ f_x \\ \psi_y \\ m_z \\ u_x \\ m_y \\ \psi_z \end{bmatrix}_{z_{i+1}}
$$

$$(10.316a)$$

$$a_1 \equiv c_\alpha - 1; \quad a_2 \equiv -A_f a_1; \quad a_3 \equiv -A_f a_5; \quad a_4 \equiv -A_f a_8; \tag{10.316b}$$

$$a_5 \equiv c_\gamma c_\alpha; \quad a_6 \equiv -c_\gamma s_\alpha; \quad a_7 \equiv -s_\gamma c_\alpha; \quad a_8 \equiv s_\gamma s_\alpha; \tag{10.316c}$$

$$c_\alpha \equiv \cos\alpha; \quad c_\gamma \equiv \cos\gamma; \quad s_\alpha \equiv \sin\alpha; \quad s_\gamma \equiv \sin\gamma. \tag{10.316d}$$

Notice that $\mathbf{T}_R$ reduces to a unit diagonal matrix when z_i and z_{i+1} are colinear.

The "node" defined at each segment interface is labeled and its location is specified with respect to a set of Cartesian reference axes chosen arbitrarily by the user. The program automatically assigns a coordinate system to each segment and calculates the rotation angles α and γ. By convention, x_{1+1} points in the direction of the vector cross-product $\mathbf{z}_i \times \mathbf{z}_{i+1}$. Uniformly curved segments supercede this convention, since the model requires the x-axis to be normal to the plane of the bend pointing in the positive-θ direction.

10.10.4 The Transmission Matrix for a Cascade of Tubes

A series connection of pipes and fittings can be modeled by multiplying their respective transmission matrices with appropriate interposed rotation matrices. A system comprised of two simple piping segments joined by an elbow fitting, for example, gives

$$\mathbf{u}(z_1) = \mathbf{T}_1\mathbf{T}_R\mathbf{T}_L\mathbf{T}_2\,\mathbf{u}(z_3), \tag{10.317}$$

where $\mathbf{u}(z_1)$ and $\mathbf{u}(z_3)$ denote the state vectors at the two ends of the system. The matrix product $\mathbf{T}_1\mathbf{T}_R\mathbf{T}_L\mathbf{T}_2$ represents the transmission matrix of the first piping segment followed by a rotation matrix for a bend, a lumped inpedance matrix for the additional mass of an elbow fitting, and finally the transmission matrix for the second piping segment. The positions of $\mathbf{T}_R$ and $\mathbf{T}_L$ may be reversed without effect.

The traditional numerical approach evaluates each matrix for a specific frequency, and then multiplies the cascade of transmission matrices to give an overall numerical transmission matrix. The boundary conditions then would be introduced and the resulting 14 linear algebraic equations solved for the unknown variables. This approach quickly develops a serious problem with round-off error, however. Some of the matrix elements tend to be very large numbers, since they involve hyperbolic sines and cosines of arguments that grow with both frequency and the length of the tubing segments. The routine that solves the system of equations (Gauss-Jordan with partial pivoting was used) tends to subtract large numbers of nearly equal value, losing information. Fatal round-off errors were experienced whenever the sum of the exponents of the diagonal elements equaled the number of significant digits on the computer. The net effect is a scheme that works only for very low frequencies, for which the matrix elements are not large. Double-precision helps, but not dramatically, since the problem escalates rapidly with frequency. The use of fewer or more elements does not alter the basic pattern.

The solution to this problem, which can be generalized to other classes of models, was inspired by Kalnins.[26] Each branch (*i.e.*, series connection without junctions such as tees) is divided into n segments. Each segment has its own transmission matrix, $\mathbf{T}_i$, which itself may be the product of subcomponent transmission matrices. The magnitudes of the matrix elements for each segment are much smaller than the magnitudes of the matrix elements for a single cascaded transmission matrix that represents the entire branch, because the effective length is reduced. Adjusting the number of segments, n, effects a compromise between computational time and the propensity for round-off errors.

The state vectors at the ends of the branch are related by

$$\mathbf{u}(z_1) = \mathbf{T}_1\mathbf{T}_2\ldots\mathbf{T}_n\,\mathbf{u}(z_{n+1}). \tag{10.318}$$

Since the boundary conditions always specify half the variables at each end, it is convenient to separate the state vectors, $\mathbf{u}(z_1)$ and $\mathbf{u}(z_{z+1})$, into the knowns, $\mathbf{u}_1(z_1)$ and $\mathbf{u}_1(z_{n+1})$, and

[26]Kalnins, A., 1964, "Analysis of Shells of Rotation Subjected to Symmetrical and Nonsymmetrical Loads," ASME *J. of Applied Mechanics*, v 31, Sept., pp 467-476.

the unknowns, $\mathbf{u}_2(z_1)$ and $\mathbf{u}_2(z_{n+1})$, as follows:

$$\mathbf{u}^*(z_1) \equiv \begin{bmatrix} \mathbf{u}_1(z_1) \\ \mathbf{u}_2(z_1) \end{bmatrix} = \mathbf{F}_1 \mathbf{u}(z_1), \qquad (10.319a)$$

$$\mathbf{u}^*(z_{n+1}) \equiv \begin{bmatrix} \mathbf{u}_1(z_{n+1}) \\ \mathbf{u}_2(z_{n+1}) \end{bmatrix} = \mathbf{F}_{n+1} \mathbf{u}(z_{n+1}). \qquad (10.319b)$$

The asterisks represent reordered state vectors, and $\mathbf{F}_1$ and $\mathbf{F}_{n+1}$ are non-singular reordering matrices. Substituting equations (10.319) into (10.318), and then expanding, gives

$$\mathbf{u}^*(z_1) = \mathbf{T}_1^* \mathbf{u}(z_1), \qquad T_1^* = \mathbf{F}_1 \mathbf{T}_1,$$
$$\mathbf{u}(z_2) = \mathbf{T}_2 \mathbf{u}(z_3),$$

$$\cdot$$
$$\cdot$$
$$\cdot$$

$$\mathbf{u}(z_{n-1}) = \mathbf{T}_{n-1} \mathbf{u}(z_n),$$
$$\mathbf{u}(z_n) = \mathbf{T}_n^* \mathbf{u}^*(z_{n+1}), \qquad \mathbf{T}_n^* = \mathbf{T}_n \mathbf{F}_{n+1}^{-1}. \qquad (10.320)$$

If $\mathbf{T}_1^*$ and $\mathbf{T}_n^*$ are written as partitioned matrices,

$$\mathbf{T}_1^* = \begin{bmatrix} \mathbf{T}_{11} & \mathbf{T}_{12} \\ \mathbf{T}_{13} & \mathbf{T}_{14} \end{bmatrix}; \qquad \mathbf{T}_n^* = \begin{bmatrix} \mathbf{T}_{n1} & \mathbf{T}_{n2} \\ \mathbf{T}_{n3} & \mathbf{T}_{n4} \end{bmatrix}, \qquad (10.321)$$

the equations (10.320) can be written as a global transmission matrix for the entire branch as follows:

$$(10.322)$$

This is a set of $m \times m$ linear algebraic equations, where m equals $14 \times n$. The state vector on the left is known and the state vector on the right is unknown. A set of routines in the IMSL software package[27] were used for solution. Any suitable numerical procedure can substitute. The bandedness of the equation suggests more efficient schemes, but when junctions are included this potential advantage disappears. Many zero elements remain, nevertheless.

With a global transmission approach the segments always can be made sufficiently short to avoid round-off problems, although at the cost of increased computational cost.

---

[27] J.J. Dongarra et al., *LINPACK User's Guide*, SIAM, Philadelphia, 1979.

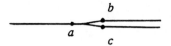

Figure 10.47: Three-port junction with parallel legs

10.10.5   Three-Port Junctions

The general case of a three-port junction with parallel legs, shown in Fig. 10.47, can be represented by 21 simple relations between the variables. The points a, b and c are shown as distinct, but they occupy the same position and share the same coordinate axes. The fluid pressure and the wall displacements are common, while the forces, moments and fluid displacements sum to zero. In matrix form,

$$\begin{bmatrix} \mathbf{u}(z_a) \\ \mathbf{0} \end{bmatrix} = \mathbf{B} \begin{bmatrix} \mathbf{u}(z_b) \\ \mathbf{u}(z_c) \end{bmatrix}, \qquad (10.323a)$$

$$\mathbf{B} =$$

$$\begin{bmatrix}
1 & 0 & 0 & 0 & 0 & 0 & 0 & 0 & 0 & 0 & 0 & 0 & 0 & 0 & 0 & 0 & 0 & 0 & 0 & 0 & 0 & 0 & 0 & 0 & 0 & 0 & 0 \\
a & 1 & 0 & 0 & 0 & 0 & 0 & 0 & 0 & 0 & 0 & 0 & 0 & 0 & 0 & 1 & 0 & 0 & 0 & 0 & 0 & 0 & 0 & 0 & 0 & 0 & 0 \\
0 & 0 & 1 & 0 & 0 & 0 & 0 & 0 & 0 & 0 & 0 & 0 & 0 & 0 & 0 & 0 & 0 & 0 & 0 & 0 & 0 & 0 & 0 & 0 & 0 & 0 & 0 \\
0 & 0 & 0 & 1 & 0 & 0 & 0 & 0 & 0 & 0 & 0 & 0 & 0 & 0 & 0 & 0 & 0 & 1 & 0 & 0 & 0 & 0 & 0 & 0 & 0 & 0 & 0 \\
0 & 0 & 0 & 0 & 1 & b & 0 & 0 & 0 & 0 & 0 & 0 & 0 & 0 & 0 & 0 & 0 & 0 & 1 & 0 & 0 & 0 & 0 & 0 & 0 & 0 & 0 \\
0 & 0 & 0 & 0 & 0 & 1 & 0 & 0 & 0 & 0 & 0 & 0 & 0 & 0 & 0 & 0 & 0 & 0 & 0 & 0 & 0 & 0 & 0 & 0 & 0 & 0 & 0 \\
0 & 0 & 0 & 0 & 0 & 0 & 1 & 0 & 0 & 0 & 0 & 0 & 0 & 0 & 0 & 0 & 0 & 0 & 0 & 1 & 0 & 0 & 0 & 0 & 0 & 0 & 0 \\
0 & 0 & 0 & 0 & 0 & 0 & 0 & 1 & 0 & 0 & 0 & 0 & 0 & 0 & 0 & 0 & 0 & 0 & 0 & 0 & 0 & 0 & 0 & 0 & 0 & 0 & 0 \\
0 & 0 & 0 & 0 & 0 & 0 & 0 & 0 & 1 & 0 & 0 & 0 & 0 & 0 & 0 & 0 & 0 & 0 & 0 & 0 & 0 & 1 & 0 & 0 & 0 & 0 & 0 \\
0 & 0 & 0 & 0 & 0 & 0 & 0 & 0 & 0 & 1 & 0 & 0 & 0 & 0 & 0 & 0 & 0 & 0 & 0 & 0 & 0 & 0 & 0 & 0 & 0 & 0 & 0 \\
0 & 0 & 0 & 0 & 0 & 0 & 0 & 0 & 0 & 0 & 1 & 0 & 0 & 0 & 0 & 0 & 0 & 0 & 0 & 0 & 0 & 0 & 1 & 0 & 0 & 0 & 0 \\
0 & 0 & 0 & 0 & 0 & 0 & 0 & 0 & 0 & 0 & 0 & 1 & 0 & 0 & 0 & 0 & 0 & 0 & 0 & 0 & 0 & 0 & 0 & 0 & 0 & 0 & 0 \\
0 & 0 & 0 & 0 & 0 & 0 & 0 & 0 & 0 & 0 & 0 & 0 & 1 & 0 & 0 & 0 & 0 & 0 & 0 & 0 & 0 & 0 & 0 & 0 & 0 & 1 & 0 \\
0 & 0 & 0 & 0 & 0 & 0 & 0 & 0 & 0 & 0 & 0 & 0 & 0 & 1 & 0 & 0 & 0 & 0 & 0 & 0 & 0 & 0 & 0 & 0 & 0 & 0 & 0 \\
1 & 0 & 0 & 0 & 0 & 0 & 0 & 0 & 0 & 0 & 0 & 0 & 0 & 0 & 0 & b & 0 & 0 & 0 & 0 & 0 & 0 & 0 & 0 & 0 & 0 & 0 \\
0 & 0 & 1 & 0 & 0 & 0 & 0 & 0 & 0 & 0 & 0 & 0 & 0 & 0 & 0 & 0 & b & 0 & 0 & 0 & 0 & 0 & 0 & 0 & 0 & 0 & 0 \\
0 & 0 & 0 & 0 & 0 & 1 & 0 & 0 & 0 & 0 & 0 & 0 & 0 & 0 & 0 & 0 & 0 & b & 0 & 0 & 0 & 0 & 0 & 0 & 0 & 0 & 0 \\
0 & 0 & 0 & 0 & 0 & 0 & 0 & 1 & 0 & 0 & 0 & 0 & 0 & 0 & 0 & 0 & 0 & 0 & b & 0 & 0 & 0 & 0 & 0 & 0 & 0 & 0 \\
0 & 0 & 0 & 0 & 0 & 0 & 0 & 0 & 0 & 1 & 0 & 0 & 0 & 0 & 0 & 0 & 0 & 0 & 0 & b & 0 & 0 & 0 & 0 & 0 & 0 & 0 \\
0 & 0 & 0 & 0 & 0 & 0 & 0 & 0 & 0 & 0 & 0 & 1 & 0 & 0 & 0 & 0 & 0 & 0 & 0 & 0 & 0 & b & 0 & 0 & 0 & 0 & 0 \\
0 & 0 & 0 & 0 & 0 & 0 & 0 & 0 & 0 & 0 & 0 & 0 & 0 & 1 & 0 & 0 & 0 & 0 & 0 & 0 & 0 & 0 & 0 & b & 0 & 0 & 0
\end{bmatrix}$$

$$(10.323b)$$

$$a \equiv -A_f; \qquad b \equiv -1. \qquad (10.323c)$$

The ordering of variables in the state vectors is the same as in equation (10.316). Note that a tee can be created by adding a right-angle bend immediately after the junction.

For the example shown in Fig. 10.48, $a = 2$, $b = 3$ and $c = 4$, and segments are attached to each leg. The procedure used to give equation (10.322) now gives the global transmission matrix equation

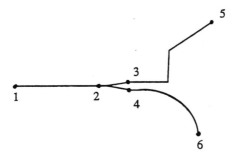

Figure 10.48: Junction with pipe reaches added

$$
\begin{bmatrix}
\mathbf{u}_1(z_1) \\
\mathbf{0} \\
\hline
\mathbf{0} \\
\hline
\mathbf{0} \\
-\mathbf{T}_{31}\,\mathbf{u}_1(z_5) \\
-\mathbf{T}_{33}\,\mathbf{u}_1(z_5) \\
-\mathbf{T}_{41}\,\mathbf{u}_1(z_6) \\
-\mathbf{T}_{43}\,\mathbf{u}_1(z_6)
\end{bmatrix}
=
\begin{bmatrix}
\begin{array}{c|c|c|c|c}
\begin{matrix}\mathbf{0}\\ \hline -\mathbf{I}\end{matrix} & \mathbf{T}_1 & \mathbf{0} & \mathbf{0} & \mathbf{0} \\
\hline
\begin{matrix}\mathbf{0}\\ \hline \mathbf{0}\end{matrix} & -\mathbf{I} & & \mathbf{0} & \\
\hline
\begin{matrix}\mathbf{0}\\ \hline \mathbf{0}\end{matrix} & \mathbf{0} & & \mathbf{0}\;\;\mathbf{0} & \\
 & & -\mathbf{I} & \mathbf{T}_{32}\;\;\mathbf{0} & \mathbf{0} \\
\begin{matrix}\mathbf{0}\\ \hline \mathbf{0}\end{matrix} & & & \mathbf{T}_{34}\;\;\mathbf{0} & \\
\hline
\mathbf{0}\;\;\mathbf{0}\;\;\mathbf{0} & & -\mathbf{I} & \mathbf{0}\;\;\mathbf{T}_{42} \\
 & & & \mathbf{0}\;\;\mathbf{T}_{44}
\end{array}
\end{bmatrix}
\begin{bmatrix}
\mathbf{u}_2(z_1) \\
\hline
\mathbf{u}(z_2) \\
\hline
\mathbf{u}(z_3) \\
\hline
\mathbf{u}(z_4) \\
\mathbf{u}_2(z_5) \\
\mathbf{u}_2(z_6)
\end{bmatrix}
\qquad (10.324)
$$

The example shown in Fig. 10.49 has a closed-loop structure, with $a = 3$, $b = 5$ and $c = 2$. Its global transmission matrix is

$$
\begin{bmatrix}
\mathbf{u}_1(z_1) \\
\mathbf{0} \\
\hline
\mathbf{0} \\
\hline
\mathbf{0} \\
\hline
\mathbf{0} \\
\hline
\mathbf{0}
\end{bmatrix}
=
\begin{bmatrix}
\begin{array}{c|c|c|c|c}
\begin{matrix}\mathbf{0}\\ \hline -\mathbf{I}\end{matrix} & \mathbf{0} & \mathbf{T}_1 & \mathbf{0} & \mathbf{0} \\
\hline
\begin{matrix}\mathbf{0}\\ \hline \mathbf{0}\end{matrix} & -\mathbf{I} & & \mathbf{0} & \\
\hline
\begin{matrix}\mathbf{0}\\ \hline \mathbf{0}\end{matrix} & \mathbf{0}\;\;\mathbf{0} & \mathbf{B} & \mathbf{0}\;\;\mathbf{0} \\
\hline
\begin{matrix}\mathbf{0}\\ \hline \mathbf{0}\end{matrix} & -\mathbf{I} & \mathbf{0} & \mathbf{0} & \mathbf{T}_3 \\
\hline
\begin{matrix}\mathbf{0}\\ \hline \mathbf{0}\end{matrix} & \mathbf{0} & \mathbf{0} & \mathbf{T}_4 & -\mathbf{I}
\end{array}
\end{bmatrix}
\begin{bmatrix}
\mathbf{u}_2(z_1) \\
\hline
\mathbf{u}(z_3) \\
\hline
\mathbf{u}(z_2) \\
\hline
\mathbf{u}(z_5) \\
\hline
\mathbf{u}(z_4)
\end{bmatrix}
\qquad (10.325)
$$

The basic format can be expanded to handle any number of branches and loops.

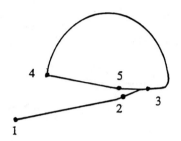

Figure 10.49: Closed loop pipe system

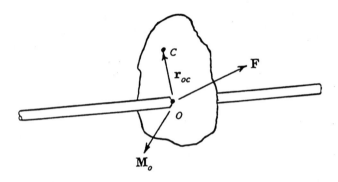

Figure 10.50: Rigid mass attached to a segment of pipe

10.10.6   Added Lumped Impedances

Fitting masses, various valves, accumulators, clamps and boundary conditions usually can be well represented by lumped models in the form of transmission matrices.

A rigid mass attached to a segment of pipe, as shown in Fig. 10.50, relates the forces in the tubing wall that are immediately upstream and downstream of the mass by

$$\mathbf{F}_{up} = -m(\mathbf{a}_0 - \mathbf{r}_{0c} \times \alpha) + \mathbf{F}_{down}, \tag{10.326}$$

where m is the mass, $\mathbf{r}_{0c}$ is the vector from point 0 to the center of mass with Cartesian co-ordinates a, b, c, $\mathbf{a}_0$ is the acceleration of point 0 and α is the angular acceleration. Similarly, the moments are related by

$$\mathbf{M}_{up} = -\mathbf{I}_c \alpha + m\mathbf{r}_{0c} \times (\mathbf{r}_{0c} \times \alpha) - m\mathbf{r}_{0c} \times \mathbf{a}_0 + \mathbf{M}_{down}, \tag{10.327}$$

where $\mathbf{I}_c$ is the inertia matrix about the center of mass, defined in terms of the coordinate axes which are fixed by the pipe system at the location where the mass is attached. In practice it is convenient to specify a diagonal matrix $\mathbf{I}_{cp}$ containing the principal moments of inertia and a matrix $\mathbf{C}$ comprising direction cosines, and to generate $\mathbf{I}_c$ in the appropriate coordinate axes by using the rotational transformation $\mathbf{I}_c = \mathbf{C}\mathbf{I}_{cp}\mathbf{C}^T$. If the axes for $\mathbf{I}_c$ and $\mathbf{I}_{cp}$ are denoted as $x\,y\,z$ and $x_p\,y_p\,z_p$, respectively, then

$$\mathbf{C} = \begin{bmatrix} \mathbf{x} \cdot \mathbf{x}_p & \mathbf{x} \cdot \mathbf{y}_p & \mathbf{x} \cdot \mathbf{z}_p \\ \mathbf{y} \cdot \mathbf{x}_p & \mathbf{y} \cdot \mathbf{y}_p & \mathbf{y} \cdot \mathbf{z}_p \\ \mathbf{z} \cdot \mathbf{x}_p & \mathbf{z} \cdot \mathbf{y}_p & \mathbf{z} \cdot \mathbf{z}_p \end{bmatrix}, \tag{10.328}$$

where $\mathbf{x}\,\mathbf{y}\,\mathbf{z}$ and $\mathbf{x}_p\,\mathbf{y}_p\,\mathbf{z}_p$ are unit vectors in the directions of $x\,y\,z$ and $x_p\,y_p\,z_p$, respectively.

Equations (10.326) and (10.327) are implemented by a transmission matrix $\mathbf{Z}_m$, which relates the state vectors upstream and downstream of the lumped mass as

$$\mathbf{u}_{up} = \mathbf{Z}_m \mathbf{u}_{down}, \tag{10.329a}$$

$$\mathbf{Z}_m = \begin{bmatrix}
1 & 0 & 0 & 0 & 0 & 0 & 0 & 0 & 0 & 0 & 0 & 0 & 0 & 0 \\
0 & 1 & 0 & 0 & 0 & a_1 & 0 & -ba_1 & 0 & aa_1 & 0 & 0 & 0 & 0 \\
0 & 0 & 1 & 0 & 0 & 0 & 0 & 0 & 0 & 0 & 0 & 0 & 0 & 0 \\
0 & 0 & ca_1 & 1 & 0 & -ba_1 & 0 & a_2 & 0 & a_3 & 0 & 0 & 0 & a_4 \\
0 & 0 & 0 & 0 & 1 & 0 & 0 & 0 & 0 & 0 & 0 & 0 & 0 & 0 \\
0 & 0 & 0 & 0 & 0 & 1 & 0 & 0 & 0 & 0 & 0 & 0 & 0 & 0 \\
0 & 0 & a_1 & 0 & 0 & 0 & 1 & ca_1 & 0 & 0 & 0 & 0 & 0 & -aa_1 \\
0 & 0 & 0 & 0 & 0 & 0 & 0 & 1 & 0 & 0 & 0 & 0 & 0 & 0 \\
0 & 0 & 0 & 0 & 0 & 0 & 0 & 0 & 1 & -ca_1 & 0 & a_1 & 0 & ba_1 \\
0 & 0 & 0 & 0 & 0 & 0 & 0 & 0 & 0 & 1 & 0 & 0 & 0 & 0 \\
0 & 0 & -aa_1 & 0 & 0 & 0 & 0 & a_4 & 0 & a_5 & 1 & ba_1 & 0 & a_2 \\
0 & 0 & 0 & 0 & 0 & 0 & 0 & 0 & 0 & 0 & 0 & 1 & 0 & 0 \\
0 & 0 & 0 & 0 & 0 & aa_1 & 0 & a_3 & 0 & a_6 & 0 & -ca_1 & 1 & a_5 \\
0 & 0 & 0 & 0 & 0 & 0 & 0 & 0 & 0 & 0 & 0 & 0 & 0 & 1
\end{bmatrix} \tag{10.329b}$$

$$a_1 \equiv -ms^2; \quad a_2 \equiv -I_x s^2; \quad a_3 \equiv -I_{xy} s^2;$$

$$a_4 \equiv -I_{xz} D^s; \quad a_5 \equiv -I_{yz} s^2; \quad a_6 \equiv -I_y s^2. \tag{10.329c}$$

The force $\mathbf{F} = [f_x f_y f_z]^T$ applied to a pipe along a line of action $\mathbf{d} = [d_x d_y d_z]^T$, due to a linear mechanical impedance Z_{ui} such as a spring, mass or dashpot and a pipe displacement $\mathbf{U} = [u_x u_y u_z]^T$, becomes

$$\mathbf{F} = Z_u \mathbf{d}\,\mathbf{d}^T \mathbf{U}. \tag{10.330}$$

For a simple spring, $Z_u = -k$, whereas for a simple dashpot, $Z_u = -rs$, and so forth for other directed impedances. A connection to a rigid mass m instead of ground gives

$$Z_u = \frac{-(k+rs)ms^2}{k+rs+ms^2}. \tag{10.331}$$

More general cases allow any linear impedance acting in the direction of $\mathbf{d}_i$, and any rotational impedance with a specific line of action, to give

$$\mathbf{F} = \sum_i Z_{\mathbf{u}i} \mathbf{d}_i \mathbf{d}_i^T \mathbf{U}, \tag{10.332a}$$

$$\mathbf{M} = \sum_i A_{\psi i} \mathbf{d}_i \mathbf{d}_i^T \psi. \tag{10.332b}$$

A simple model of a gas-filled accumulator is shown by electrical analogy in Fig. 10.51. The inertance I and resistance R_1 represent the constriction of the neck of the accumulator. The compliance C_1 represents the adiabatic compliance (compressibility) of the gas. The parallel combination $C_1 + C_2$ represents the isothermal compressibility, which requires heat transfer to or from the walls. The resistance R_2 represents the thermal resistance to this

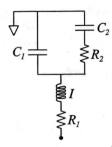

Figure 10.51: Model of a gas accumulator

conduction. The transmission matrix of such an accumulator that is attached to a pipe segment is

$$
\begin{bmatrix} p \\ f_z \\ u_y \\ v \\ u_z \\ f_y \\ f_x \\ u_z \end{bmatrix}_{upstr}
=
\begin{bmatrix}
1 & 0 & 0 & 0 & 0 & 0 & 0 & 0 \\
d_z z_5 & 1 & d_z d_y z_6 & 0 & d_z^2 z_6 & 0 & 0 & d_z d_x z_6 \\
0 & 0 & 1 & 0 & 0 & 0 & 0 & 0 \\
z_3 & 0 & d_y z_4 & 1 & d_z z_4 & 0 & 0 & d_x z_4 \\
0 & 0 & 0 & 0 & 1 & 0 & 0 & 0 \\
d_y z_5 & 0 & d_y^2 z_6 & 0 & d_y d_z z_6 & 1 & 0 & d_y d_x z_6 \\
d_x z_5 & 0 & d_x d_y z_6 & 0 & d_x d_z z_6 & 0 & 1 & d_x^2 z_6 \\
0 & 0 & 0 & 0 & 0 & 0 & 0 & 1
\end{bmatrix}
\begin{bmatrix} p \\ f_x \\ u_y \\ v \\ u_z \\ f_y \\ f_x \\ u_x \end{bmatrix}_{downstr}
\tag{10.333a}
$$

$$
z_3 = \frac{1}{BA_f s},
$$

$$
z_4 = -\frac{IA_c s}{BA_f},
$$

$$
z_5 = A_c \left[\frac{R_1}{B} + \frac{1}{BC_1 s} \left(1 - \frac{C_2}{C_2 + C_1 + C_2^2 R_2 s} \right) - 1 \right],
$$

$$
z_6 = -IA_c^2 s^2 (z_5 + 1),
$$

$$
B = R_1 + \frac{1}{C_1 s} + Is - \frac{C_2}{(C_1 C_2 + C_1^2)s + C_1 C_2^2 R_2 s^2},
\tag{10.333b}
$$

where d_x, d_y, d_z are direction cosines defining a unit vector that points from the attachment point to the accumulator in the direction of the neck, A_c is the cross-sectional area of the neck, and A_f is the internal cross-sectional area of the pipe. This model does not include the mechanical inertia of the accumulator, which can be added separately.

A valve or orifice representable by a linear flow resistance, R, in series with a linear flow inertance, I, can be accommodated simply. The pressure drop across the restriction produces a mechanical force on the walls which also must not be overlooked. Assuming the force acts in the axial direction, the result is

$$
\begin{bmatrix} p \\ f_z \\ v \\ u_z \end{bmatrix}
=
\begin{bmatrix}
1 & 0 & A_f(Rs + Is^2) & -A_f(Rs + Is^2) \\
0 & 1 & A_f^2 Rs & -a_f^2 Rs \\
0 & 0 & 1 & 0 \\
0 & 0 & 0 & 1
\end{bmatrix}
\begin{bmatrix} p \\ f_z \\ v \\ u_z \end{bmatrix}.
\tag{10.334}
$$

The series fluid resistance and inertance also can represent the mechanical inertia and resistance of a fluid motor with a rigidly connected load. The inertance and resistance must be expressed in terms of equavalent fluid impedances.

10.10.7 Boundary Conditions

Boundary conditions at the ends of pipe segments are treated by separating the specified variables from the unspecified variables, and employing the reordering matrices of equation (10.319). The most common mechanical constraints are free end, pinned end (allowing rotations), cantilevered end, end with applied force and end with applied displacement. The most common fluid terminations are constant pressure, zero flow (closed), applied pressure and applied fluid displacement. These end conditions can be combined with impedances such as lumped masses or accumulators by introducing these elements just before the termination. The variables associated with each of these end conditions are noted below, and two examples of the rearranging relation are given. Tentarelli gives further examples.

A total of seven variables must be specified to define the end conditions completely. A free end has zero for the values of the five variables F_x, F_y, m_x, m_y and m_z, and the sixth condition $f_z = 0$ if the end is open or $f_z = A_f p$ if it is closed. A pinned end has zero for the values of the six variables u_x, u_y, u_z, m_x, m_y and m_z. A cantilevered end has zero for the six variables u_x, u_y, u_z, ψ_x, ψ_y and ψ_z. The seventh condition for all three cases is: if the end is open, $p = 0$; if the end is closed, $v = u_z$; for the pressure excitation $P_0 \sin(\omega t + \phi)$, $p = P_0(\cos\phi + j\sin\phi)$ and for the fluid displacement excitation $V_0 \sin(\omega t + \phi)$, $v = V_0(\cos\phi + j\sin\phi)$. Note that the complex magnitude is used for applied excitations, since the model deals specifically with sinusoidal motion. Linear displacement excitation would give $u_x = U_{x0}(\cos\phi_1 + j\sin\phi_1)$, $u_y = U_{y0}(\cos\phi_2 + j\sin\phi_2)$ and $u_z = U_{z0}(\cos\phi_3 + j\sin\phi_3)$. An axial force excitation would give $f_z = F_{z0}(\cos\phi + j\sin\phi)$.

The first example is a pinned and closed end, which allows rotational motion. If one elects to solve for external force applied to the tube by the pin, f_{ext}, rather than f_z, the reordering relation would be

$$
\begin{bmatrix} 0 \\ 0 \\ 0 \\ 0 \\ 0 \\ 0 \\ 0 \\ p \\ f_{ext} \\ f_y \\ \psi_x \\ f_x \\ \psi_y \\ \psi_z \end{bmatrix} =
\begin{bmatrix}
0 & 0 & 1 & 0 & 0 & 0 & 0 & 0 & 0 & 0 & 0 & 0 & 0 & 0 \\
0 & 0 & 0 & 1 & 0 & 0 & 0 & 0 & 0 & 0 & 0 & 0 & 0 & 0 \\
0 & 0 & 0 & 0 & 1 & -1 & 0 & 0 & 0 & 0 & 0 & 0 & 0 & 0 \\
0 & 0 & 0 & 0 & 0 & 1 & 0 & 0 & 0 & 0 & 0 & 0 & 0 & 0 \\
0 & 0 & 0 & 0 & 0 & 0 & 0 & 0 & 0 & 0 & 1 & 0 & 0 & 0 \\
0 & 0 & 0 & 0 & 0 & 0 & 0 & 0 & 0 & 0 & 0 & 1 & 0 & 0 \\
0 & 0 & 0 & 0 & 0 & 0 & 0 & 0 & 0 & 0 & 0 & 0 & 1 & 0 \\
1 & 0 & 0 & 0 & 0 & 0 & 0 & 0 & 0 & 0 & 0 & 0 & 0 & 0 \\
-A_f & 1 & 0 & 0 & 0 & 0 & 0 & 0 & 0 & 0 & 0 & 0 & 0 & 0 \\
0 & 0 & 0 & 0 & 0 & 0 & 1 & 0 & 0 & 0 & 0 & 0 & 0 & 0 \\
0 & 0 & 0 & 0 & 0 & 0 & 0 & 1 & 0 & 0 & 0 & 0 & 0 & 0 \\
0 & 0 & 0 & 0 & 0 & 0 & 0 & 0 & 1 & 0 & 0 & 0 & 0 & 0 \\
0 & 0 & 0 & 0 & 0 & 0 & 0 & 0 & 0 & 1 & 0 & 0 & 0 & 0 \\
0 & 0 & 0 & 0 & 0 & 0 & 0 & 0 & 0 & 0 & 0 & 0 & 0 & 1
\end{bmatrix}
\begin{bmatrix} p \\ f_z \\ 0 \\ 0 \\ v \\ 0 \\ 0 \\ f_y \\ \psi_x \\ f_x \\ \psi_y \\ 0 \\ 0 \\ \psi_z \end{bmatrix}.
\tag{10.335}
$$

The second example is a cantilevered end with pressure excitation:

$$
\begin{bmatrix} p \\ 0 \\ 0 \\ 0 \\ 0 \\ 0 \\ 0 \\ f_z \\ m_x \\ v \\ f_y \\ f_x \\ m_z \\ m_y \end{bmatrix}
=
\begin{bmatrix}
1 & 0 & 0 & 0 & 0 & 0 & 0 & 0 & 0 & 0 & 0 & 0 & 0 & 0 \\
0 & 0 & 1 & 0 & 0 & 0 & 0 & 0 & 0 & 0 & 0 & 0 & 0 & 0 \\
0 & 0 & 0 & 0 & 0 & 1 & 0 & 0 & 0 & 0 & 0 & 0 & 0 & 0 \\
0 & 0 & 0 & 0 & 0 & 0 & 0 & 1 & 0 & 0 & 0 & 0 & 0 & 0 \\
0 & 0 & 0 & 0 & 0 & 0 & 0 & 0 & 0 & 1 & 0 & 0 & 0 & 0 \\
0 & 0 & 0 & 0 & 0 & 0 & 0 & 0 & 0 & 0 & 0 & 1 & 0 & 0 \\
0 & 0 & 0 & 0 & 0 & 0 & 0 & 0 & 0 & 0 & 0 & 0 & 0 & 1 \\
0 & 1 & 0 & 0 & 0 & 0 & 0 & 0 & 0 & 0 & 0 & 0 & 0 & 0 \\
0 & 0 & 0 & 1 & 0 & 0 & 0 & 0 & 0 & 0 & 0 & 0 & 0 & 0 \\
0 & 0 & 0 & 0 & 1 & 0 & 0 & 0 & 0 & 0 & 0 & 0 & 0 & 0 \\
0 & 0 & 0 & 0 & 0 & 0 & 1 & 0 & 0 & 0 & 0 & 0 & 0 & 0 \\
0 & 0 & 0 & 0 & 0 & 0 & 0 & 0 & 1 & 0 & 0 & 0 & 0 & 0 \\
0 & 0 & 0 & 0 & 0 & 0 & 0 & 0 & 0 & 0 & 1 & 0 & 0 & 0 \\
0 & 0 & 0 & 0 & 0 & 0 & 0 & 0 & 0 & 0 & 0 & 0 & 1 & 0
\end{bmatrix}
\begin{bmatrix} p \\ f_z \\ 0 \\ m_x \\ v \\ 0 \\ f_y \\ 0 \\ f_x \\ 0 \\ m_z \\ 0 \\ m_y \\ 0 \end{bmatrix}.
\qquad (10.336a)
$$

$$
p = P_0(\cos \phi + j \sin \phi). \qquad (10.336b)
$$

Points internal to the system that are clamped rigidly can be treated by adding extremely large masses or stiff springs, but this can lead to computational trouble. Alternatively, the constrained displacement variables may be set equal to zero. If the forces of a clamp at z_i need not be found, one can simply reduce the global transmission matrix equation by eliminating the known displacements at z_i, the unknown clamp forces at z_i, and the associated rows and columns of the matrix, reducing its order and decreasing computational cost.

Extending the procedure to the clamped tee pictured in Fig. 10.47 requires elimination of those rows in the global matrix pertaining to $\mathbf{F}_{aR}$, $\mathbf{F}_b$, $\mathbf{F}_c$ in the left-hand vector and the rows pertaining to the relations $\mathbf{U}_2 = \mathbf{U}_3$, $\mathbf{0} = \mathbf{U}_3 - \mathbf{U}_4$ as given in the $\mathbf{B}$ matrix. It also requires elimination of those columns that multiply $\mathbf{U}_a$, $\mathbf{U}_b$ $\mathbf{U}_c$, $\mathbf{F}_3$ and $\mathbf{F}_4$ in the right-hand vector.

10.10.8　Power Flow and Energy Density

The present models incorporate dissipative mechanisms, including viscous dissipation due to wall shear and flow restrictions, material damping (implemented by introducing an imaginary component of Young's modulus equal to one to two tenths of a percent of the real component) and sound radiation from clamps and terminations. Therefore, it is possible to create a map of time-average power flow of a system, and to distinguish the one fluid and six mechanical modes of propagation. Such a map may aid a designer who wishes to minimize audible sound. The designer interested in identifying regions with high potential for noise and vibration also can secure values of the time-average energy density at the various nodes of the system. Details and illustrations are given by Tentarelli.

10.10.9　Computational and Experimental Example

Several simple and complicated tubing systems are addressed computationally and experimentally by Tentarelli. Three of the systems are included in the ASME papers. One of these, a complex system which used standard 0.5-in. hydraulic tubing, is pictured in Fig. 10.52 and is addressed here. It contained two tubing bends with Bourdon coupling, a 90° elbow fitting and a tee. All were joined by several straight segments that were oriented at random angles. The system was supported near two of its terminations by a 150 kg vice.

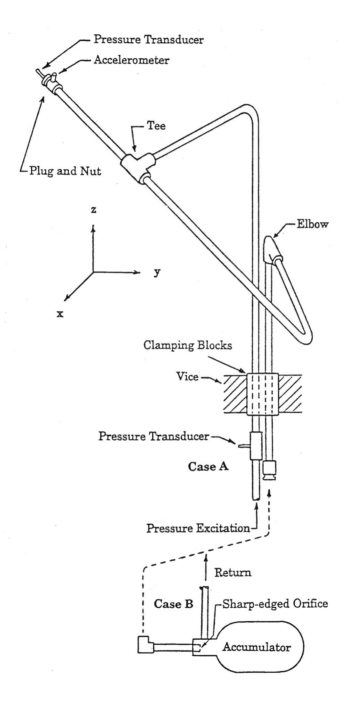

Figure 10.52: Complex experimental system

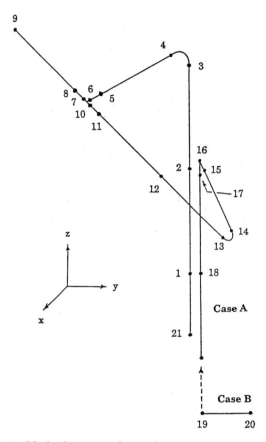

Figure 10.53: Node diagram of complex system used by software

One end of the system was supplied with a pressure excitation. Another end was capped and unrestrained. The third termination had two different conditions, designated as cases A and B. In case A the termination was capped. In case B, an orifice and a one-gallon accumulator, which approximated a constant-pressure boundary condition, were present. A node diagram is given in Fig. 10.53.

The pressure at node 9 and the displacement normal to the plane of the tee at node 9 are the variables of interest. All data are presented as transfer functions relating the sinusoidal magnitude of the output variables and the pressure excitation signal. The computational and experimental results for the pressure in case A are given in Fig. 10.54. Part (a) of the figure shows very good agreement for up through 1000 Hz, and part (b) shows general success up to 5000 Hz. The broader peaks at the lower frequencies are predominantly fluid resonances; the narrow spikes are likely due to fluid/structure coupling. The peaks in the experiment would be sharper and closer to the computation if noise in the denominator signal had been eliminated. This noise appears to be the principal source of experimental error. At the higher frequencies, the behavior represents a very intimate coupling between fluid and mechanical phenomena.

A net fluid flow exists in case B; the resistance of the orifice was estimated from steady-flow data. The pressure at node 9 is plotted in Fig. 10.55. The peaks are dramatically smaller than in case A as a result. The relatively broad peaks again are probably predominantly fluid resonances, although in general the fluid/structure coupling appears even stronger than in case A. The displacement at node 9 normal to the plane of the tee is plot-

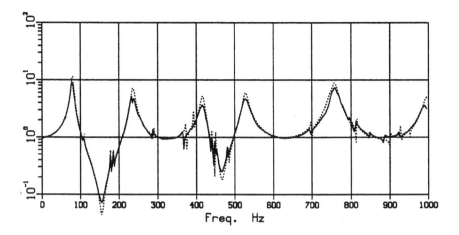

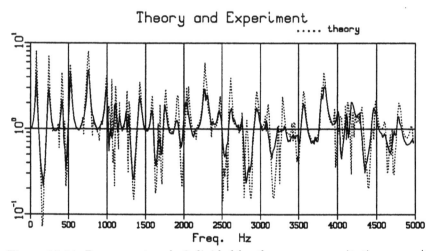

Figure 10.54: Pressure at node 9 divided by the pressure excitation, case A

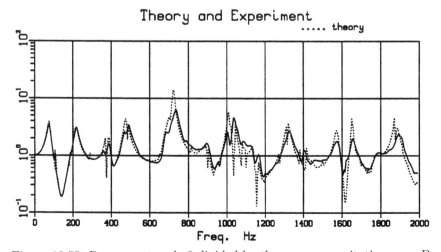

Figure 10.55: Pressure at node 9 divided by the pressure excitation, case B

ted in Fig. 10.56. The model successfully identified twelve resonant frequencies below 100 Hz. and almost 30 or more from 100-1000 Hz. Transverse sensitivity in the accelerometer appeared to be a significant source of experimental error. All model parameters were as determined from direct measurement.

Any model ultimately fails when frequencies are raised without limit. Even when the details are off, however, a model may still accurately reflect the pattern of modal shapes, energies and densities.

10.10.10   Summary

The use of transmission matrices allows series combinations of distributed and lumped elements to be reduced. For complex linear dynamic models the individual transmission matrices are best evaluated numerically as a function of frequency. Direct multiplication of the transmission matrices often causes fatal round-off errors, however, particularly for high frequencies. This problem can be avoided through the use of what has been called a global transmission matrix, which increases the dimensionality of the representation to include

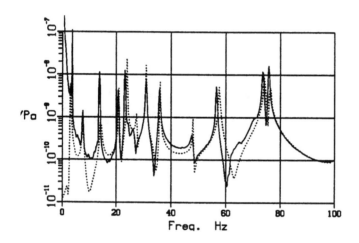

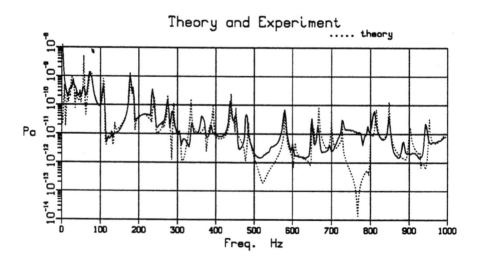

Fig. 10.56 Displacement at node 9 divided by the pressure excitation, case B

state variables at selected internal locations and segregates the known from the unknown variables. Global transmission matrices also have been found to represent models with junctions.

The propagation of noise and vibration in complex hydraulic tubing systems has been used to illustrate these ideas. A software package is referred to that allows the user to construct a wide variety of three-dimensional system models, including the effects of coupling between the fluid and the solid behavior and the effect of added masses, lumped fluid impedances and various soft and hard geometric constraints and boundary conditions. Individual tubing segments may be either straight or curved, and the use of rotation matrices allows each segment to be described using a simple local coordinate system. The couplings between the fluid and solid motions include the forces due to viscosity, the Poisson effect and the Bourdon effect, as well as the common accelerations that the two media experience.

Chapter 11

Thermodynamic Systems

Irreversibilities were introduced in Section 9.7 with the examples of heat conduction and friction. The energy fluxes in these cases are analogous to the other power flows considered, in the sense that they can be represented by simple bonds with a single effort and a single flow. Mass transfer complicates the energy flux, however; a new **convection bond** is introduced in Section 11.1 to address this. The bond has two efforts and a single flow. It represents the entropy fluxes as well as the mass flows and thereby accounts for the thermodynamic availability as well as the conservation of energy. Implementation for heat interactions and junctions for convection bonds is given in Section 11.2. Steady-state analysis can be carried out at this point, as illustrated in Section 11.3, analogous to the steady-state analysis of simple systems in Chapter 2 and parts of later chapters.

Thermodynamic energy storage is a potential energy representable by a compliance field, and coherent kinetic energy by an inertance field. Special elements that represent these fields are presented in Section 11.4 This is followed in Section 11.5 by a detailed consideration of practical ways to represent thermodynamic state functions for real substances with computational models, foregoing table look-up alternatives. Examples given include simulations of the unsteady behavior of an air compressor, a refrigeration cycle and a distributed condenser tube. The chapter ends in Section 11.6 with an introduction to systems with unsteady chemical reaction.

11.1   The Convection Bond and Compressible Flow

The mechanical energy in the incompressible flows considered thus far is decoupled from the thermal energy, and the latter has been discounted as "dissipation." No such decoupling exists for compressible flow, however, since a change in temperature changes the density of the fluid and therefore its velocity, etc. It is necessary to bookkeep both the mechanical and thermal forms of energy. Sometimes you might opt to keep track of thermal energy and power even when you presume incompressible flow.

A new kind of compound bond, called the **convection bond**, is introduced below. This leads to a generalization of the two-port irreversible or RS element, and later to a special heat interaction element and generalizations of the 1- and 0-junctions. Ultimately, thermodynamic energy is represented by a multiport C element and is extended to accommodate chemical reaction.

11.1.1   Flow Through a Port

Consider a micro-port of area dA through which a compressible or incompressible fluid is flowing. The volume and mass flow rates, respectively, are

$$dQ = v\,dA, \tag{11.1a}$$
$$dm = \rho v\,dA, \tag{11.1b}$$

where v is the fluid velocity and ρ is its mass density. The total energy flux which crosses a fixed control surface that cuts this micro-port is the sum of a "flow-work" power and a transport of energy. The flow-work power is

$$d\mathcal{P}_f = Pv\,dA = \frac{P}{\rho}d\dot{m}, \tag{11.2}$$

in which P is the local absolute pressure. This is the same power that a piston with force $P\,dA$ and velocity v would transmit. The energy transport comprises the transport of internal energy (per unit mass: u), kinetic energy (per unit mass: $v^2/2$) and gravity energy (per unit mass: gz), assuming the absence of capillarity and electric and magnetic field effects:

$$d\mathcal{P}_t = (u + v^2/2 + gz)d\dot{m}. \tag{11.3}$$

The total energy flux is therefore

$$d\mathcal{P} = d\mathcal{P}_f + d\mathcal{P}_t = \left(\frac{P}{\rho} + u + \frac{v^2}{2} + gz\right)\frac{d\dot{m}}{dA}dA. \tag{11.4}$$

The total power for a complete flow channel or port of cross-sectional area A is the central interest. Equation (11.4) is integrated across the channel to evaluate this power. The simplest assumptions are that P, ρ, u, v and z are constants over the cross-section. In this case, the integral becomes

$$\mathcal{P} = \left(\frac{P}{\rho} + u + \frac{v^2}{2} + gz\right)\dot{m}. \tag{11.5}$$

With one exception these assumptions are usually acceptable, particularly if the area of the cross-section is not large as in pipe flow, etc. The exception is the velocity, v. If the mean velocity across the channel is defined as $\bar{v}$, the integral can be written as

$$\mathcal{P} = \left(\frac{P}{\rho} + u + \alpha\frac{\bar{v}^2}{2} + gz\right)\dot{m}$$
$$= \left(P + \rho u + \alpha\rho\frac{\bar{v}^2}{2} + \rho gz\right)Q, \tag{11.6}$$

where α depends on the shape of the velocity profile. The lower right-hand term in parentheses appears as the effort in the expanded list of effort and flow analogies given in Table 11.1, while Q appears as the flow; this is the usual choice for incompressible flow if thermal energy is to be included in the bookkeeping. The transport of kinetic energy is $\int \frac{1}{2}\rho v^2 v\,dA = \frac{1}{2}\alpha\rho\bar{v}^2 Q$. Since $Q = \bar{v}A$, this gives

$$\alpha = \frac{\int v^3 dA}{\bar{v}^3 A}. \tag{11.7}$$

A flat or slug-like profile, with $v = \bar{v}$, gives $\alpha = 1$. A parabolic profile in a round tube, such as occurs in fully developed laminar flow, gives $\alpha = 2$. Nearly all other cases are

Table 11.1   Effort and Flow Analogies (Extended)

	generalized force or effort, e	generalized velocity or flow, $\dot{q}$
Fluid, incompressible:		
general	$\rho u + P + \alpha\rho\bar{v}^2/2 + \rho g z$	Q
less thermal	$P + \alpha\rho\bar{v}^2/2 + \rho g z$	Q
approximate	P	Q
micro-bond	any of above	v
Fluid, compressible:		
general	$h_0 + g z$	$\dot{m}$
approximate	h	$\dot{m}$
micro-bond	either above	ρv
Mechanical, longitudinal:		
general	$F_c + \rho A v^2/2$	$v = \dot{x}$
approximate	F_c	$v = \dot{x}$
micro-bond	$\sigma + \rho v^2/2$	$v = \dot{x}$
Mechanical, transverse:		
rotation	M	$\dot{\phi}$
translation (shear)	F_s	v_s
micro-bond	τ	v_s
Electric circuit:	e	i
Thermal, conduction:	θ	$\dot{S}$

intermediate between these values; turbulent flow in most channels is apt to give a value closer to 1.

The enthalpy, h, is defined as $P/\rho + u$. The **stagnation enthalpy**, h_0, is $h + \frac{1}{2}\alpha\bar{v}^2$. This is the enthalpy that the stream would have if its velocity were slowed reversibly until the kinetic energy term ($\frac{1}{2}\alpha\bar{v}^2$) became negligible. Therefore, the power can be written in the condensed form

$$\mathcal{P} = (h_0 + g z)\dot{m}. \tag{11.8}$$

The gravity term is omitted below, since in most instances it has little or no significance. It can be reintroduced easily.

The density of a compressible flow may change from point to point along a streamline, particularly if it is in a channel of varying area, even if the flow is steady. In this instance the volume flow rate $Q = \dot{m}/\rho$ varies from point to point also. This discourages the choice of Q as the flow or generalized velocity, unlike with assumed incompressible flow. The mass flow rate, $\dot{m}$, on the other hand, is constant along a channel if the flow is steady. Thus, $\dot{m}$ is chosen as the flow variable when compressibility is assumed. It also can be used when the flow is considered incompressible, if you so prefer.

Equation (11.7) indicates that, neglecting gravity effects, the proper effort or generalized force to be paired with $\dot{m}$ is the stagnation enthalpy, h_0, giving

$$\frac{h_0}{\dot{m}}$$

These effort and flow variables are added to Table 11.1. Several authors have used this representation, but it contains a very significant ambiguity that demands resolution. An explanation follows.

11.1.2   The Convection Bond

The conjugate pairs of effort and flow M and $\dot{\phi}$, F and $\dot{x}$, P and Q, and e and i describe work interactions. The conjugate pair θ and $\dot{S}$ describes a heat interaction. The problem with the pair h_0 and $\dot{m}$ is that it describes a *sum* of a work interaction and a heat interaction, with unspecified proportions. Heat and work are not universally convertable. The meaning thus is non-unique.[1]

Complete description of the interaction requires two variables in place of the single h_0, assuming that no electrical, magnetic, capillary, chemical or other special effects exist. Recall the **state postulate** in thermodynamics: the state of a pure substance in the absence of these special effects is specified uniquely by *two* independent intensive variables. The most commonly used pair is temperature and pressure, although these are not independent in the two-phase liquid-vapor region. Other commonly used variables include density (or specific volume), quality (in the liquid-vapor two-phase region), enthalpy, entropy and the Gibbs function. Discussion of the Gibbs function is deferred to Sections 11.4 and 11.6, where it will be used in connection with thermal and chemical energies. The **available enthalpy**, h_a, and the **unavailable enthalpy**, h_u, are a conceptually satisfying pair of efforts or generalized forces, but they are not so computationally convenient and are not emphasized.[2]

The choice of which pair of intensive variables to use in describing the state of a pure fluid substance is ultimately a practical matter that depends on computational requirements and the availability of state data. The most convenient pair for most purposes is P_0, h_0. The fact that a **convection bond** has two independent variables to describe the effort while a single variable describes the flow is indicated by adding a dashed line on the effort side of the bond:

$$\underset{\dot{m}}{\underline{\underline{P_0, h_0}}}\longrightarrow$$

Note that P_0 is not directly an effort, because the product $P_0\dot{m}$ is not an energy flux.[3] Rather, h_0 is the proper effort, and the value of P_0 qualifies the character of this effort. Knowledge of P_0 and h_0 is sufficient to define other stagnation state properties, including density. When combined with the mass flow rate ($\dot{m}$) they are sufficient also to define the velocity and all the actual (non-stagnation) state properties.

The choice of P_0 and h_0 to describe the effort is based partly on considerations of causality, which become critical when unsteady behavior (Sections 11.4 and 11.5) is addressed. One of these, together with $\dot{m}$, must form a bilateral causal pair. The enthalpy h_0 is physically causal only in the actual direction of the flow, unlike the usual effort variables on simple bonds. It could be said that the causality of enthalpy moves with the flow; it reverses whenever the flow reverses. The pressure P_0, fortunately, suffers no such limitation. Therefore, P_0 and the mass flow $\dot{m}$ form the needed bilateral causal pair, like the P_0 and

---

[1] The same is true when the internal energy u is allowed to appear in the effort for incompressible flow, as it does in one row of Table 11.1. Recall, however, that in past chapters u was dropped with the explanation that it is decoupled from the work, and conversion from work to heat was labeled as "dissipation."

[2] **Availability** is defined as the *useful work potential* of the fluid, which is the work delivered when the fluid undergoes a reversible process from its current state to the state (temperature and pressure, for example) of the environment. The **available enthalpy**, more conventionally known as the **stream availability**, is shown in elementary thermodynamics textbooks to be

$$h_a = (h_0 - h_0^*) - \theta^*(s - s^*),$$

where the asterisk ($*$) indicates the state of the environment. The product $h_a\dot{m}$ is the theoretical maximum rate of work the flow could produce. The balance of the enthalpy, $h_0 - h_a$, is the **unavailable enthalpy**, h_u. The energy flux $h_u\dot{m}$ represents only part of a heat interaction.

[3] When it is desired to break h_0 into the sum of two actual efforts, the Gibbs function, g, and the product θs appear to be the best choice, as discussed in Section 11.6.

volume flow Q of a bond for incompressible flow; the traditional causal stroke is used to indicate the causalities of P_0 and $\dot{m}$ only. For example, the causal stroke in the example

$$\vdash\!\!\frac{P_0, h_0}{\dot{m}}\!\!\rightharpoonup$$

indicates that $\dot{m}$ is specified from the left and P_0 is specified from the right. No indication of the causality of h_0 is indicated or needed.

The convection bond is compatible with the previously existing stable of bond graph elements, although certain logical generalizations are needed. Certain new primitive elements are necessary. Certain additional combinations of primitive elements also will be defined as new macro-elements, for purposes of computational convenience. The result will be a comprehensive although simple scheme for representing lumped models of thermofluid and hybrid systems.

11.1.3 The RS Element for Fluid Flow

The RS element is used in Section 9.7 to represent a thermal resistance and a mechanical resistance with thermal power included. Its application is now extended to the flow of a fluid through a valve, or any other device, that drops its pressure without any heat interaction or storage of energy:

$$\frac{P_{01}, h_0}{\dot{m}}\!\!\rightharpoonup RS \frac{P_{02}, h_0}{\dot{m}}\!\!\rightharpoonup$$

The stagnation enthalpy and the total power remain constant, but the total pressure decreases; *i.e.,* if $\dot{m} > 0$, then $P_{01} > P_{02}$. The unavailable enthalpy increases at the expense of the available enthalpy. Further, the rate of entropy generation, which is the difference between the entropy efflux and the entropy influx, must be positive:

$$\dot{S}_g = (s_2 - s_1)\dot{m} > 0. \tag{11.9}$$

The fundamental property relation

$$\theta\,ds = dh - dP/\rho \tag{11.10}$$

often is helpful in computing an entropy change.

It is awkward to carry the subscript o wherever it applies, so this subscript is dropped below, and stagnation properties are assumed.

The actual calculations are made the same way you treated flow through a valve in your study of thermodynamics. Usually, the enthalpy and the pressures are known, so all other properties can be computed. The flow $\dot{m}$ would have to be computed from knowledge of the mechanics of the flow, expressible functionally as

$$\dot{m} = \dot{m}(P_1, P_2, h_1). \tag{11.11}$$

Finding such functions in general is beyond the scope of this text, but two cases are used repeatedly. The first is incompressible inviscid flow through an idealized orifice, for which Bernoulli's equation model gives

$$\dot{m} = A_m \sqrt{2\rho(P_1 - P_2)}, \tag{11.12}$$

where A_m is the effective minimum area of the orifice or the *vena contracta*. The second is compressible flow of an ideal and inviscid gas with constant specific heats through a converging orifice, which is given in Guided Problem 11.1 below.

Sometimes the mass flow rate, the upstream enthalpy and one of the two pressures are specified and the other pressure is to be found. This also is readily accomplished in most cases once the details of equation (11.11) are established.

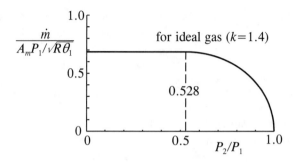

Figure 11.1: Inviscid compressible flow through a nozzle

11.1.4 Summary

The state of a pure substance is specified by any two independent intensive variables. Specification of the state of a flowing fluid requires a third variable to describe the rate of flow. The mass flow rate is chosen for this third variable, since in steady flow it is invariant along a channel of non-uniform area. The two intensive variables describe the effort associated with the flow. For most purposes the stagnation enthalpy and the stagnation pressure are chosen. The stagnation enthalpy is the true effort; its product with the mass flow rate gives the overall energy flux (neglecting any gravity effects). The pressure serves not only to properly qualify this effort, but it also acts as the causal effort variable. A convection bond is distinguished from a simple bond by the addition of a dashed line to its effort side.

Flow through a restriction or throttle drops the stagnation pressure and generates entropy but leaves the stagnation enthalpy unchanged. This irreversible process is represented by a convection-bond adaptation of the RS element.

Guided Problem 11.1

In this problem you will compare the flows of an inviscid incompressible fluid (as given by equation (11.12)) and an inviscid compressible ideal gas through the same simple, smoothly convergent nozzle. The latter is given in textbooks on compressible flow as

$$\dot{m} = \begin{cases} \dfrac{A_m P_1}{\sqrt{R\theta_1}} \sqrt{\dfrac{2k}{k-1} \left[\left(\dfrac{P_2}{P_1}\right)^{2/k} - \left(\dfrac{P_2}{P_1}\right)^{(k+1)/k} \right]}, & \text{if } \dfrac{P_2}{P_1} \geq r, \\[2em] \dfrac{A_m P_1}{\sqrt{R\theta_1}} \sqrt{k \left[\dfrac{2}{k+1}\right]^{(k+1)/(k-1)}}, & \text{if } \dfrac{P_2}{P_1} \leq r, \end{cases}$$

$$r \equiv \left(\dfrac{2}{k+1}\right)^{k/(k-1)},$$

where P_1 and θ_1 refer to the upstream stagnation state, R is the gas constant and k is the ratio of specific heats, c_p/c_v, which is assumed constant. P_2 refers to the absolute static pressure at the throat, which equals the downstream stagnation pressure since the walls expand abruptly (without "pressure recovery") downstream. The dimensionless mass flow rate, as determined by the above equation, is plotted as a function of P_2/P_1 in Fig. 11.1 for air ($k = 1.4$). For pressure ratios below the indicated critical ratio, the flow is independent of the downstream pressure and is said to be **choked**.

(a) Place equation (11.12) for incompressible flow into the same form, and plot the result for air directly onto the graph in Fig. 11.1 for comparison. Observe the nature of the error.

(b) Expand the compressible-flow function for small (first-order) differences in the pressure drop, $\Delta P = P_2 - P_1$, and show that this approximation agrees with equation (11.12).

(c) Find the rate of entropy generation for the ideal gas.

(d) (Optional.) Find the reduction in the available enthalpy across the nozzle, $h_{a1} - h_{a2}$. Assume the temperature of the environment is the same as θ_1, and the pressure of the environment is the same as P_2.

Suggested Steps:

1. Factor out from equation (12.12) the term $A_m P_1 / \sqrt{R\theta_1}$, noting that $P_1 = \rho R \theta_1$. The result is a simple function of P_2/P_1. Roughly plot this into Fig. 11.1 to compare this relation for incompresible flow to that for compressible flow.

2. Replace P_2 in the equation given for compressible flow by $P_1 - \Delta P$. Get each of the pressure-ratio terms under the square root into the form $(1 - a\Delta P)^x$, which for small ΔP can be approximated as $1 - xa\Delta P$. Simplify and compare, again using $P = \rho R\theta$ if necessary.

3. To find the rate of entropy generation, start by finding θ_2, knowing that $h = h(\theta)$ for an ideal gas, and $h_1 = h_2$ for a steady-flow nozzle. Then employ equation (11.10), noting also the ideal gas law.

4. Use footnote 2 and the results above to find the reduction in the available enthalpy.

PROBLEMS

11.1 Consider incompressible flow in a long uniform tube of area A in which the velocity profile can be approximated as parabolic.

(a) Verify that the factor α in Table 11.1, as computed using equation (11.7), equals 2.

(b) The inertance of the fluid per unit length, relative to the volume flow, can be written as $\gamma\rho/A$. The factor γ is not in general equal to α, even though both factors are based on the same profile. Develop a formula for γ in terms of the velocity profile $v(r)$, where r is radius, and show that in the parabolic case $\gamma = 4/3$.

11.2 Air flows at 100 psig through a channel with a 10-foot change in elevation at velocities (i) 10 ft/s and (ii) 1000 ft/s and at temperatures of (a) 70°F and (b) 470°F. Compare the magnitudes of the four terms that contribute to the effort in equation (11.6). Comment on the significance of uncertainty regarding the factor α. (You will need some data on the thermodynamic state. It is important that you evaluate the efforts relative to some reasonable reference environmental state, which can be taken as 0 psig and 70°F.)

SOLUTION TO GUIDED PROBLEM

Guided Problem 11.1

1. Incompressible approximation:

$$\dot{m} = A_m \sqrt{2\rho\,\Delta P} = \frac{A_m P_1}{\sqrt{R\theta_1}}\sqrt{\frac{R\theta_1}{P_1}2\rho\frac{P_1 - P_2}{P_1}} = \frac{A_m P_1}{\sqrt{R\theta_1}}\sqrt{2\left(1 - \frac{P_2}{P_1}\right)}$$

As shown in the plot below, this approximation is fairly accurate for $(P_2/P_1) > 0.9$, but increasingly overestimates the flow for smaller pressure ratios.

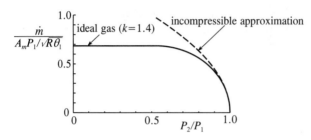

2. For small $\Delta P/P_1$,

$$\left(\frac{P_2}{P_1}\right)^{2/k} = \left(1 - \frac{\Delta P}{P_1}\right)^{2/k} \simeq 1 - \frac{2}{k}\frac{\Delta P}{P_1}$$

$$\left(\frac{P_2}{P_1}\right)^{(k-1)/k} = \left(1 - \frac{\Delta P}{P_1}\right)^{(k-1)/k} \simeq 1 - \frac{k+1}{k}\frac{\Delta P}{P_1}$$

Therefore, $\dfrac{\dot{m}}{A_m P_1/\sqrt{R\theta_1}} \simeq \sqrt{\dfrac{2k}{k+1}\left[\left(1 - \dfrac{2}{k}\dfrac{\Delta P}{P_1}\right) - \left(1 - \dfrac{k+1}{k}\dfrac{\Delta P}{P_1}\right)\right]}$

$$\simeq \sqrt{\frac{2k}{k-1}\left[(1-1) + \left(-\frac{2}{k} + 1 + \frac{1}{k}\right)\right]\frac{\Delta P}{P_1}} = \sqrt{\frac{2\Delta P}{P_1}} = \sqrt{2\left(1 - \frac{P_1}{P_2}\right)}$$

This approximation is identical to the incompressible result of step 1.

3. Since $h_1 = h_2$ and $h = h(\theta)$, $\theta_2 = \theta_1$ (constant temperature).

$$ds = \frac{dh}{\theta} - \frac{dP}{\rho\theta} = 0 - R\frac{dP}{P};\ \text{integrated, this gives } s_2 = s_1 + R\ln\left(\frac{P_1}{P_2}\right)$$

Finally, $\dot{S}_g = \dot{m}(s_2 - s_1) = \dot{m}R\ln\left(\dfrac{P_2}{P_1}\right)$

4. $h_{s1} = h_1 - h* -\theta_1(s_1 - s*) = -\theta_1(s_1 - s*)$

   $h_{s2} = h_2 - h* -\theta_2(s_2 - s*) = -\theta_1(s_2 - s*)$

   Therefore, $h_s - h_{s2} = \theta_1(s_2 - s_1) = R\theta_1\ln\left(\dfrac{P_1}{P_2}\right)$

11.2   Heat Interaction and Junctions

The RS element applied to the flow of a fluid has a pressure drop but no heat transfer. Another element with two fluid ports is now introduced that has just the inverse, heat transfer (through a third port) but no pressure drop. It can be used to represent an ideal **heat exchanger**. There are two versions: a special case that is reversible, and a general case that is not. In the second case the irreversibility is kept to a theoretical minimum consistent with the boundary conditions.

   The 0- and 1-junctions are used with convection bonds in almost the same way as they are used with simple bonds. The presence of two effort variables for a convection bond introduces complications, however. The merging of two fluid streams, for example, is in general an irreversible process and therefore cannot be represented by a 0-junction alone.

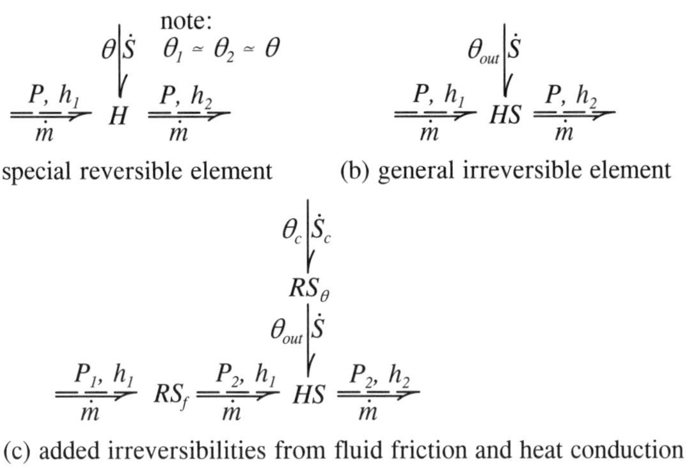

(a) special reversible element   (b) general irreversible element

(c) added irreversibilities from fluid friction and heat conduction

Figure 11.2: Heat interaction elements

11.2.1 The Reversible Heat Interaction Element

This element, abbreviated as the **H element**, is shown in Fig. 11.2 part (a). There are convection bonds for the inlet and outlet flows and a simple bond for the heat conducted to or from the fluid. Reversibility requires, in addition to the absence of friction, that heat is transferred across no more than an infinitesimal temperature difference. Thus this element is very special: the temperatures for the three bonds must be virtually the same. There are two ways this can occur. In the first case, only an infinitesimal amount of heat is transferred. This describes a micro-element: an infinite number of such elements would be needed to represent a significant heat interaction. In the second case, the entering and leaving fluids are liquid-vapor mixtures, and since the pressure doesn't change the two fluid temperatures would be the same even if a large quantity of heat is transferred. It is still necessary, however, for the temperature at the heat conduction port or bond to be no more than infinitesimally different from the fluid temperature. The element therefore implies infinite thermal conductivity for the heat conduction path.

For the first case in which only an infinitesimal amount of heat is transferred, the entropy flux becomes infinitesimal; *i.e.*, $\dot{S} \to d\dot{S}$, $\dot{S}_2 - \dot{S}_1 \to d\dot{S}$, $h_2 - h_1 \to dh$, etc. The conservation of energy requires

$$\dot{m}\,dh = \theta\,d\dot{S}. \tag{11.13}$$

Reversibility requires that the change of entropy flux of the fluid, $\dot{m}\,ds$, equals the entropy flux of the heat transfer:

$$\dot{m}\,ds = d\dot{S}. \tag{11.14}$$

Combining these relations gives

$$dh = \theta\,ds. \tag{11.15}$$

The thermodynamic identity

$$dh = dP/\rho + \theta\,ds, \tag{11.16}$$

when compared to equation (11.15), gives the important result

$$dP = 0, \tag{11.17}$$

proving that the input and output pressures are identical.

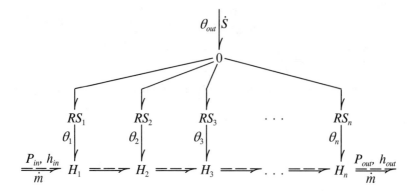

Figure 11.3: Decomposition of an *HS* element into infinitesimal *H* and *RS* elements

11.2.2   The General Heat Interaction Element

An ideal heat exchanger usually has a varying temperature along its length and can be represented by this more general element, abbreviated as the ***HS* element** and shown in Fig. 11.2 part (b). This element is irreversible, since the two fluid boundary temperatures are unequal. The conduction-bond temperature, labeled θ_{out}, is defined to equal whichever of θ_1 or θ_2 represents the temperature of the downstream or outlet flow. (The flow could move in either direction.)   Additional irreversibilities can be added in the form of *RS* elements. In Fig. 11.2 part (c), the element RS_f on one of the fluid ports represents fluid friction with its pressure drop, and the element RS_θ on the heat conduction port represents resistance to conduction with its temperature drop.

The *HS* element can be understood better by decomposing it into an infinite cascade of infinitesimal *H* elements with coupled *RS* elements, as shown in Fig. 11.3. The first conclusion is that the stagnation pressure P_{in} equals the stagnation pressure P_{out}, since for each *H* element, $dP = 0.$[4] The second conclusion is that the irreversibility and its associated entropy generation results from the assumption of a single bond with a single temperature for the heat interaction with the environment. Since this temperature is at one end of the interval between θ_{in} and θ_{out}, the heat fluxes through the various *RS* elements and through any added external *RS* element are all in the same direction, in or out.

---

[4]The sophisticated reader familiar with the classical Rayleigh line solution may initially wince at this result. The Rayleigh line assumes zero wall shear, for which heat transfer indeed forces both the static and stagnation pressures to change along the flow path whenever the Mach number is significantly different from zero. The same result would follow from the present analysis, moreover, if the static temperature were used in place of the stagnation temperature in the equations for energy balance and reversibility. Proper use of an entropy balance and proper use of momentum balance are redundant (see Guided Problem 11.2). Thus, momentum balances are not needed in energy-based modeling, although sometimes they are convenient, particularly for finding the forces of constraint.

The question becomes whether the static or the stagnation temperature should be used for the model of a reversible heat exchanger. The author considers the stagnation temperature to be far more useful, since the walls of a real adiabatic channel are at that temperature. The Rayleigh model corresponds, as it were, to walls that move at the same velocity as the bulk fluid. With stationary walls, irreversibility demands that wall shear result from heat transfer. In cooling, for example, the molecules that give up part of their kinetic energy to the walls also impart an impulse force to them.

A real heat exchanger has additional wall shear associated with the friction irreversibility. This effect can be represented by placing an *RS* element in series with the *H* or *HS* element, as shown above. Conventional analysis of a channel with heating or cooling and wall shear uses a friction factor for the latter, resulting in a wall shear that is essentially independent of the heat transfer but a pressure drop that is not independent. The model recommended here is rather the reverse: in the simplest case, the pressure drop is independent of the heat transfer but the wall shear is not. The pressure rise in cooling indicated by the Rayleigh line never even becomes a theoretical possibility, and the analyst is not misled.

Assuming the fluid is not in the two-phase region, only the specific heat of the fluid, c_p, is needed to represent the H and HS elements computationally. Since the total pressure is constant,

$$dh = c_p \, d\theta. \tag{11.18}$$

Thus if c_p is assumed to be constant over the temperature range θ_1 to θ_2, where the subscripts 1 and 2 refer to the inlet and outlet, respectively,

$$h_2 - h_1 = c_p(\theta_2 - \theta_1). \tag{11.19}$$

The conservation of energy requires

$$(h_2 - h_1)\dot{m} = \dot{S}\theta_2. \tag{11.20}$$

In practice one usually knows h_1, P and $\dot{m}$ and either $\dot{S}$ or $\dot{S}\theta_2$. (Other cases are discussed later.) If $\dot{S}\theta_2$ is known, equation (11.20) is the only computation necessary. If $\dot{S}$ is known, you can combine equations (11.19) and (11.20) and solve for θ_2:

$$\theta_2 = \frac{\theta_1}{1 - \dot{S}/c_p\dot{m}}. \tag{11.21}$$

Therefore,

$$h_2 = h_1 + \frac{(\dot{S}/\dot{m})\theta_1}{1 - \dot{S}/c_p\dot{m}}. \tag{11.22}$$

It is necessary in this case to find θ_1, which is a state function of h_1 and P.

The rate of entropy generation, defined as $\dot{S}_g$, may be of interest, particularly if you wish to understand the irreversibility or loss of availability. The balance of entropy fluxes is

$$s_2\dot{m} = s_1\dot{m} + \dot{S} + \dot{S}_g. \tag{11.23}$$

Solving the above relations for $\dot{S}_g$ gives

$$\dot{S}_g = c_p\dot{m}\left[\frac{\theta_1}{\theta_2} - 1 - \ln\left(\frac{\theta_1}{\theta_2}\right)\right] \simeq \frac{c_p\dot{m}}{2}\left(\frac{\theta_1}{\theta_2} - 1\right)^2. \tag{11.24}$$

11.2.3 The *HRS* Macro Element

The RS element in a model of a real heat exchanger likely generates more entropy than the HS element with its $\dot{S}_g$ from equation (11.24). The combination of the HS element with a conduction RS element is so nearly universal that it is practical to represent it with a *macro element*, called here the **HRS element**. As shown in Fig. 11.4, the HRS element allows the boundary temperature for the heat conduction, θ_c, to be much different from either θ_1 or θ_2. Further, θ_c can be taken as an assumed or independent variable, in place of $\dot{S}$ or the product $\theta_2\dot{S}$ as assumed above. The parameters now are c_p and the thermal conductance of the RS element, which is defined as H. The conservation of energy gives the results

$$h_2 = h_1 + \theta_c\dot{S}_c/\dot{m}, \tag{11.25}$$

$$\dot{S}_c = \frac{1 - \theta_1/\theta_c}{1/H + 1/c_p\dot{m}}, \tag{11.26}$$

$$\dot{S}_g = c_p\dot{m}\left[\frac{\theta_1}{\theta_2} - 1 - \ln\left(\frac{\theta_1}{\theta_2}\right)\right] + H\frac{(\theta_c - \theta_2)^2}{\theta_c\theta_2}. \tag{11.27}$$

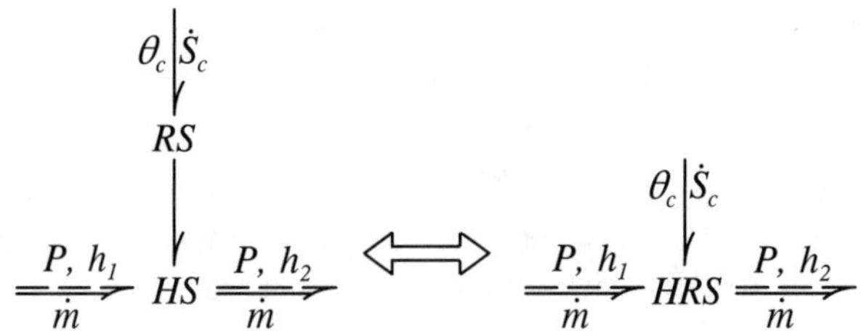

$$\theta_c \Big|\dot{S}_c$$

RS

$$\theta_c \Big|\dot{S}_c$$

$$\underset{\dot{m}}{\overset{P,\,h_1}{=\!\!=\!\!\Rightarrow}} HS \underset{\dot{m}}{\overset{P,\,h_2}{=\!\!=\!\!\Rightarrow}} \Longleftrightarrow \underset{\dot{m}}{\overset{P,\,h_1}{=\!\!=\!\!\Rightarrow}} HRS \underset{\dot{m}}{\overset{P,\,h_2}{=\!\!=\!\!\Rightarrow}}$$

Figure 11.4: The *HRS* macro element

The first term on the right comes directly from equation (11.24) for the *HS* component within the *HRS* element, and the second term, for the *RS* component, comes from equation (9.95) (p. 696). As before, it is necessary to use property data to find θ_1 and θ_2, from knowledge of the enthalpies and the pressure, before equation (11.27) can be evaluated. Also as before, the total pressures at the inlet and outlet are equal, *i.e.*, $P_2 = P_1$. A pressure difference caused by friction can be introduced by adding a convection *RS* element.

EXAMPLE 11.1

A flow of 280 lbm/sec of exhaust gas from a gas turbine enters a water boiler at 800°F. Saturated liquid water (already preheated) enters at a pressure of 500 psia and temperature of 467°F. Subsequent pressure drops can be neglected. If the effective integrated heat transfer coefficient is 300 Btu/s·°F (independent of the flow rate), at what rate should the water be pumped in so it leaves as saturated vapor? The specific heat of the gas is 0.280 Btu/lbm R, and the enthalpy of vaporization of the water is 756 Btu/lbm. Organize your work with a bond graph. Also, find the rate of entropy generation.

Solution: You can start by drawing a bond graph comprising an *HS* element, an *RS* element, an *H* element and bonds, and labelling the input and output variables. Then, as shown on the right below, combine the *HS* and *RS* elements to form an *HRS* element:

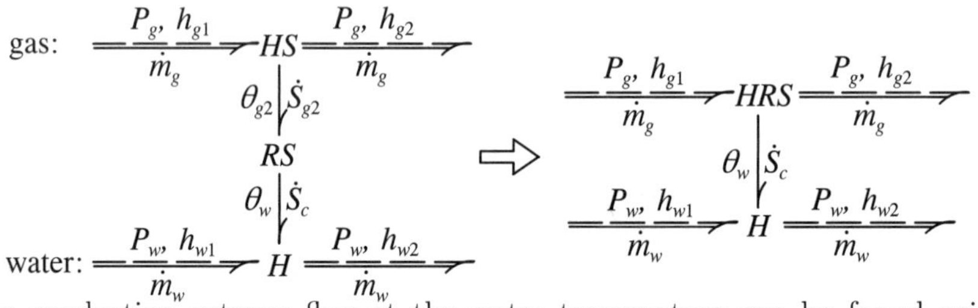

gas: $\quad \underset{\dot{m}_g}{\overset{P_g,\,h_{g1}}{=\!\!=\!\!\Rightarrow}} HS \underset{\dot{m}_g}{\overset{P_g,\,h_{g2}}{=\!\!=\!\!\Rightarrow}}$

$$\theta_{g2} \Big|\dot{S}_{g2}$$

RS

$$\theta_w \Big|\dot{S}_c$$

water: $\quad \underset{\dot{m}_w}{\overset{P_w,\,h_{w1}}{=\!\!=\!\!\Rightarrow}} H \underset{\dot{m}_w}{\overset{P_w,\,h_{w2}}{=\!\!=\!\!\Rightarrow}}$

$\Rightarrow$

$$\underset{\dot{m}_g}{\overset{P_g,\,h_{g1}}{=\!\!=\!\!\Rightarrow}} HRS \underset{\dot{m}_g}{\overset{P_g,\,h_{g2}}{=\!\!=\!\!\Rightarrow}}$$

$$\theta_w \Big|\dot{S}_c$$

$$\underset{\dot{m}_w}{\overset{P_w,\,h_{w1}}{=\!\!=\!\!\Rightarrow}} H \underset{\dot{m}_w}{\overset{P_w,\,h_{w2}}{=\!\!=\!\!\Rightarrow}}$$

The conduction entropy flux at the water temperature can be found using equation (12.26):

$$\dot{S}_c = \frac{1 - \theta_{g1}/\theta_w}{(1/H) + (1/c_{pg}m_g)} = -\frac{1 - (800 + 460)/(467 + 460)}{(1/300) + (1/0.260 \times 280)} = 21.04 \text{ Btu/R s}$$

The heat transfer now can be found, followed by the flow of the water and the temperature of the efflux gas:

$$\dot{Q} = \theta_w \dot{S}_c = (467 + 460)21.04 = 19{,}510 \text{ Btu/s}$$

$$\dot{m}_w = \frac{\dot{Q}}{h_{fg}} = \frac{19{,}510}{756} = 25.8 \text{ lbm/sec}$$

$$\theta_{g2} = \theta_{g1} - \dot{Q}/c_{pg}m_g = 800 - 19150/0.260(280) = 532°F$$

Finally, you can find $\dot{S}_g$:

$$\dot{S}_g = c_{pg}\dot{m}_g\left[\frac{\theta_{g2}}{\theta_{g1}} - 1 - \ln\left(\frac{\theta_{g2}}{\theta_{g1}}\right)\right] + \frac{(\theta_{g2} - \theta_w)^2}{\theta_{g2}\theta_w}$$

$$= 0.260(280)\left[\frac{532 + 460}{800 + 460} - 1 - \ln\left(\frac{532 + 460}{800 + 460}\right)\right]$$

$$+\ 300\frac{(532 - 467)^2}{(532 + 460)(467 + 460)}$$

$$= 3.30\ \text{Btu/R s}$$

11.2.4   The 0-Junction for Convection Bonds

This junction is *defined* such that all bonds joined to it represent a common thermodynamic state; *both* the enthalpy and the pressure are common. As with simple-bonded 0-junctions, energy is neither dissipated nor stored; the powers on the respective bonds, with signs indicated by the power-convention half-arrows, sum to zero. A mixture of simple and convection bonds is not allowed for any 0-junction.

The junction represents the merging and/or dividing of separate convection channels. The dividing of a flow from one into two or more channels is the simplest case, since all the flows necessarily have a common pressure and enthalpy. As shown in Fig. 11.5, this is the same structure as a junction for an incompressible flow represented by simple bonds and a 0-junction. (In both cases the dynamic pressures for the separate channels can be different, requiring special treatment; these differences are neglected here.)

11.2.5   Merging Streams: The $0S$ Junction

The merging of two or more streams is inherently more complicated. In general, the merging streams will have the same pressure (again neglecting any differences in the dynamic pressure or head), but they will not have the same enthalpy or temperature. Consequently, a 0-junction alone cannot represent this merging except for the very special case in which the streams happen to have a common state. The merging of two streams at different temperatures is irreversible. Mixing with heat transfer takes place, and entropy is generated.

A bond graph can represent the merger of two streams by separating the heat transfer operation from the mixing operation. As shown in Fig. 11.6 part (a), an HS element is used to equalize the two temperatures which, since the two pressures are constrained to be equal, equalizes the two states. (When two streams at different pressures are to merge, at least one of them must first pass through a resistance to equalize their pressures; for incompressible flow this resistance is a 1-junction with a shunt R element, and for compressible flow it is an RS element.) The 0-junction then can be used to represent the physical mixing. If you don't know in which directions the three flows may be heading, three HS elements can insure proper accounting of the irreversiblility, as shown in part (b). The HS element(s) attached between outflow bonds simply have no effect, since the associated temperature

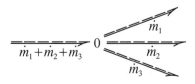

Figure 11.5: Use of a 0-junction to represent bifurcating flows

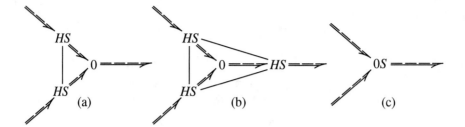

Figure 11.6: Merging flows with irreversibility

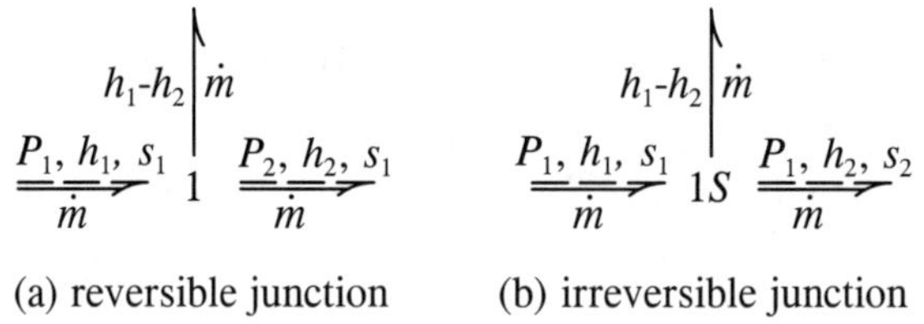

$$h_1\text{-}h_2 \downarrow \dot{m} \qquad\qquad h_1\text{-}h_2 \downarrow \dot{m}$$

(a) reversible junction   (b) irreversible junction

Figure 11.7: Reversible and irreversible 1-junctions

differences are zero.[5]

    The combination shown in Fig. 11.6 part (b) occurs so frequently that the macro $0S$ **junction** is created to represent it. The $0S$ junction, shown in part (c), applies for any number of joined bonds and channels. This, the second of the macro elements, has the added advantage of simplifying computation. The conservation of energy gives

$$h_{\text{out}} = \frac{\sum \dot{m}_{\text{in}} h_{\text{in}}}{\sum \dot{m}_{\text{in}}}. \tag{11.28}$$

The entropy generation can be deduced from the HS elements of the more primitive graph. Equation (11.26) applies for each of these elements, if the subscripts are adjusted accordingly.

11.2.6   The 1-Junction and 1S-Junction with Convection

1-junctions are defined to force a common flow on all attached bonds. Since the convection bond has only a single flow variable, unlike its effort variables, the problem experienced with the 0-junction does not exist. Two types of 1-junctions are useful, nevertheless. The basic 1-junction is reversible; the entropies of the convection bonds are equal. The 1S-junction is irreversible; the *pressures* on the convection bonds are defined as being equal. In both cases, shown in Fig. 11.7, simple bonds as well as convection bonds can be joined to the junction. This is the essential way parts of systems represented by simple bonds are connected mechanically to other parts represented by convection bonds. The vertical bonds in the examples of Fig. 11.7 are *simple* bonds; they have no enthalpy or pressure or thermodynamic state associated with it, but rather *scalar* potentials equal in *magnitude* to the enthalpy difference[6] $h_1 - h_2$.

---

[5]This is one reason for defining the effort on the thermal conduction bond of HS elements as the temperature of the *outlet* fluid stream.

[6]The scalar potential on the simple bond can also be said to equal the difference of the *available* enthalpies, $h_{a1} - h_{a2}$. This follows from the inherent reversibility of the junction, which requires the unavailable enthalpies, h_{u1} and h_{u2}, to be equal.

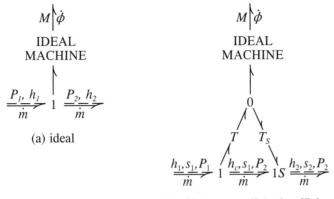

(a) ideal

(b) with known adiabatic efficiency

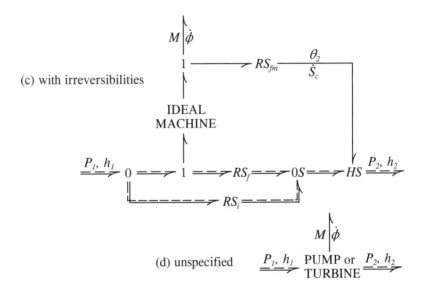

(c) with irreversibilities

(d) unspecified

Figure 11.8: Pump or turbine for compressible flow

An ideal (lossless, reversible) motor or turbine operating with a compressible flow can be represented by a combination of the 1-junction and a simple-bonded ideal machine, as shown in Fig. 11.8 part (a). Losses are added in the bond graphs of parts (b) and (c). In part (b), the losses are represented in a summary fashion; in part (c), losses are detailed according to their physical origin. The power convention arrows shown give positive moduli for the various elements when the machine is acting as a turbine. The arrows can be left as is when the machine acts as a compressor, although you probably would rather retain positive moduli by inverting the arrows on certain bonds. In both cases, the IDEAL MACHINE element could be replaced by a simple transformer if overall behavior resembles that of a positive-displacement machine; the effects of mechanical and fluid friction and fluid leakage are represented elsewhere.

Were the model of part (b) 100% efficient, the transformer modulus T would be unity and the elements $1S$ and T_S would be removed. The state of the fluid leaving the compressor, particularly its enthalpy $h_c = h_2$, would be determined by the requirement that its entropy equal s_1 and its pressure equal P_2. The product of the enthalpy drop (for a turbine) and the mass flow would equal the output mechanical power. If the IDEAL MACHINE is represented by a transformer of modulus T_m, the torque on the shaft would be the product of the enthalpy drop and T_m, and the speed of the shaft would equal $\dot{m}/T_m$. Thus, the modulus T_m would be the ratio D/v_1, where D is the volumetric displacement per radian of shaft rotation and v_1 is the specific volume of the fluid entering the machine.

An actual machine will suffer mechanical losses due to mechanical friction, fluid losses due to fluid friction (*i.e.* pressure drops through orifices, etc.) and fluid losses due to internal leakage. Heat interactions with the surroundings also can produce irreversibilities, but the present discussion assumes that these are being treated separately, which is relatively easy to do. Instead, the overall machine is treated as being adiabatic. In the summary model of part (b), the considered overall losses are described conventionally by the **adiabatic efficiency** of the machine, written here as η_{ad}. This efficiency is the ratio of the output power to the input power; for the turbine, this means the ratio of the actual turbine work to the isentropic work, or

$$\eta_{ad} = \frac{h_1 - h_2}{h_1 - h_c}. \tag{11.29}$$

The irreversibility occurs exclusively within the $1S$ element. Following the efforts from the 1-junction through the two transformers to the $1S$-junction, you can see that

$$\frac{1}{TT_S}(h_1 - h_c) = h_2 - h_c = -(h_1 - h_2) + (h_1 - h_c). \tag{11.30}$$

Substitution of equation (12.29) gives the result

$$\boxed{\frac{1}{TT_S} = 1 - \eta_{ad}.} \tag{11.31}$$

An additional condition must be specified in order to determine the individual values of T and T_S. This condition depends on the nature of the machine. A positive-displacement-like motor can be described partly by its **volumetric efficiency**, η_v, defined as the ratio of the actual shaft speed to the speed it would have if there were no fluid leakage or slippage. In this case the effort and flow on the bond between the 0-junction and the transformer T_m equal $(h_1 - h_2)/\eta_v$ and $\eta_v \dot{m}$, respectively. Thus, $(T - 1/T_S) = \eta_v$, and

$$T = \frac{\eta_v}{\eta_{ad}}; \qquad T_S = \frac{1}{\eta_v(1/\eta_{ad} - 1)}. \tag{11.32}$$

The model shown in part (c) of Fig. 11.8 details the causes of losses in a turbine, allowing the adiabatic efficiency to vary depending on the state. Leakage passes through the element RS_i. Were the machine a pump or compressor, the $0S$ element would be placed on the inlet side rather than the outlet; you might also prefer to reverse some of the power convention half-arrows. For a positive-displacement machine, the simple-bonded ideal machine would be a transformer with constant modulus. The element RS_{fm} represents mechanical friction, and the element RS_f represents frictional losses in the fluid. Note that the heat generated by friction is carried off in the fluid. For most dynamic machines the simple-bonded ideal machine does not resemble a fixed-modulus transformer,[7] but nevertheless can be defined by a transformer heavily modulated by two independent variables: either M or $\dot{\phi}$ and either $h_1 - h_2$ or $\dot{m}$.

---

[7]The simple-bonded ideal dynamic machine may be represented better by a modulated gyrator. The element RS_i is not needed. Details for Eulerian turbomachines with incompressible flow have been given by H.M. Paynter, "The Dynamics and Control of Eulerian Turbomachines," *J. Dynamic Systems, Measurement and Control, ASME Trans.*, v. 94 n. 3, pp. 198-205, 1972.

In some cases it may be better to employ overall empirical relations regarding the behavior of a fluid machine, and leave the bond graph model in the general form of part (d) of the figure. Most of these matters are beyond the scope of this text. Nevertheless, the model of part (b) is used in the example below, and the model of part (c) is given within a case study in Section 11.3. Guided Problem 11.2 below serves in part as preparation.

EXAMPLE 11.2

A piston compressor has a volumetric displacement D per radian, and is known to give an adiabatic efficiency of 80% and a volumetric efficiency of 95%. Draw a bond graph of the system, give the moduli of its transformers, and find an expression for the output enthalpy in terms of variables that could be found from knowledge of the state of the inlet fluid and the outlet pressure.

Solution: The bond graph of Fig. 11.8 applies, except that the power flows in the opposite direction through four of the bonds, because that graph deals with a turbine instead of a compressor. The power direction arrows on these bonds have been reversed in order to keep their moduli positive:

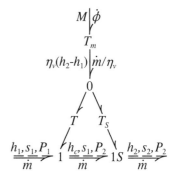

The flow on the bond below the ideal positive displacement transformer T_m would be $\dot{m}$ if the volumetric efficiency were 100%. The flow actually is larger, specifically $\dot{m}/\eta_v$, since the shaft has to rotate faster to supply the leakage as well as the through-flow. Since the conservation of energy requires the total mechanical power to equal the total change in fluid energy increase per unit time, the effort on this bond is $\eta_v(h_2 - h_1)$. The modulus of T_m is the ideal ratio of the mass flow pumped to the shaft speed; this equals the volumetric displacement per radian, as in an incompressible-flow pump, times the density of the flow that enters the compressor, which is $1/v_1$.

The definition of the adiabatic efficiency (the inverse of equation (11.29)),

$$\eta_{ad} = \frac{h_c - h_1}{h_2 - h_1},$$

gives the working formula

$$h_2 = \frac{h_c}{\eta_{ad}} - \left(\frac{1}{\eta_{ad}} - 1\right) h_1,$$

where h_1 is given, and h_c is found from its known entropy $s_c = s_1$ and known pressure $P_c = P_2$. Further, the effort side of the transformers T and T_S give

$$T = \frac{\eta_v(h_2 - h_1)}{h_c - h_1} = \frac{\eta_v}{\eta_{ad}},$$

$$T_S = \frac{\eta_v(h_2 - h_1)}{h_2 - h_c} = \frac{\eta_v(h_2 - h_1)}{(h_2 - h_1) - (h_c - h_1)} = \frac{\eta_v}{1 - \eta_{ad}}.$$

The relations for the flow sides of the transformers confirm these results. As a partial check, note that when both efficiencies equal unity, $T \to 1$ and $T_S \to \infty$. This would correctly eliminate the path for T_S.

11.2.7   Summary

Heat conduction has been represented previously by the simple-bonded RS element. A means also is needed to transfer heat between a flowing fluid and a solid. This transfer is represented by the H, HS and HRS elements. All three have two convection bonds, representing the inlet and outlet flows, and one conduction bond. The H element is reversible; all three bonds have the same temperature, which implies either that the heat is transferred at an infinitesimal rate or that the fluid enters and leaves in the two-phase region. The irreversible HS element allows the fluid to enter and exit at different temperatures, but like the H element assigns the temperature of the fluid efflux port to the conduction port also. Additional thermal resistance to heat conduction can be represented by bonding a conduction RS element to the conduction port. The HRS macro element represents this combination.

Causal strokes can be added to the bonds, consistent with the constraints, and the causal output variables computed as functions of the causal input variables and the parameters of the element. The rate of entropy generation, which never is negative, also can be computed as a measure of the irreversibility.

1-junctions, in which all bonds have a common flow, can be used to join systems described using simple bonds with systems described using convection bonds. This allows the 1-junction to represent the reversible heart of a fluid machine. The net power flow on a single simple mechanical bond attached to the 1-junction equals the difference between the power flows of its two convection bonds. A transformer on this simple bond represents the ratio of the mass flow and the mechanical generalized velocity. Friction and leakage losses can be added with proper use of R, RS, HS and additional junction elements. Irreversibilities also can be represented by use of the $1S$-junction, which has a pair of convection bonds with equal pressures rather than equal entropies.

0-junctions require identical efforts on all their bonds, and therefore cannot be used with mixed simple and convection bonds, since convection bonds have two effort variables rather than one. A convection 0-junction can directly represent a flow which is being divided between two or more output channels. The merging of two streams, however, is an irreversible process. It can be modeled by adding HS bonds to a convection 0-junction, or by employing the macro equivalent $0S$ junction.

Guided Problem 11.2

A known mass flow $\dot{m}_1$ of air at a known temperature θ_1 is augmented in a mixing chamber by a mass flow of air $\dot{m}_2$, as shown in Fig. 11.9. This second mass flow first passes through a simple nozzle of throat area A with known upstream temperature θ_2 and a known pressure P_2, which is high enough relative to P_1 to produce choked flow. The flow leaving the mixing chamber passes through a positive displacement motor with known volumetric displacement per radian, D, relative to its inlet. The torque on the shaft, M, is known. The motor can be assumed to behave isentropically, and has a known exhaust pressure, P_e (which should be atmospheric pressure). The air may be approximated as an ideal gas with invariant specific heats.

You are asked to find relations for the pressure P_1, mass flow $\dot{m}_2$, shaft speed $\dot{\phi}$ and outlet temperature, θ_e.

Suggested Steps:

1. Represent the system model with a bond graph and label all the pertinent variables on the bonds, recognizing when different bonds have the same generalized efforts or flows.

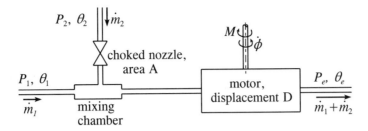

Figure 11.9: The configuration of Guided Problem 11.3

Place causal strokes on all bonds on the graph, consistent with the given information. This graph should guide you through the following steps.

2. Find the transformer modulus for the motor in terms of known quantities except for an unknown temperature and pressure.

3. Use the result of step 2 to relate the ratio of the inlet and outlet temperatures of the motor to the known shaft moment, M.

4. The ratio of the inlet to outlet temperatures of the motor also is related to the known ratio of inlet to outlet pressures, since the machine is assumed to be isentropic. Determine this relationship.

5. Combine the results of steps 3 and 4 to find the inlet pressure, P_1.

6. Find the mass flow rate through the nozzle, $\dot{m}_2$, using the choked flow relationship given in Guided Problem 11.1 (p. 856).

7. Find the temperature of the flow leaving the mixing chamber.

8. Find the outlet temperature, θ_e.

9. Find the shaft speed, $\dot{\phi}$.

PROBLEMS

11.3 A complicated distributed-parameter model is required to represent the dynamics of a counterflow heat exchanger. For most modeling purposes a much simpler steady-state model suffices, however. You are asked to find the basic equations for such a model. The mass flows in the two directions are the same; so are the specific heats. There is a thermal conductivity G per unit length between the two streams. Longitudinal heat conduction may be neglected, as may pressure drops.

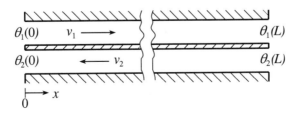

(a) Draw a bond graph representing an element of the heat exchanger dx long. The complete heat exchanger represents an infinite cascade of such micro elements. Define variables on your graph.

(b) Write equations which give the output temperatures in terms of the input temperatures, the mass flow and fixed parameters. This is your basic model.

(c) Write an expression for the rate of entropy generation. (This need not be in a simple form, but could contain integrals.)

(d) Show how frictional pressure drops affect your model.

11.4 An open feedwater heater for a steam power plant mixes a small amount of steam tapped from an interstage of the turbine with cold water pressurized by a pump, to give hot water. The pressures in the mixing process may be approximated as constant, but as a practical matter the flow of steam must be regulated by a valve, introducing some additional irreversibility to what you likely considered in an introductory thermodynamics course. In practice, also, the state of the output hot water will not be saturated liquid, unlike what you may have assumed previously, but will lie in the compressed liquid region. It is desired to develop a computational model that can readily adapt to small changes in the operating states.

(a) Draw a bond graph that models the valve and the mixer, using RS and $0S$ elements. Reticulate the $0S$ element into a combination of HS and 0 elements.

(b) Write expressions for the changes in the enthalpies of the water and the vapor in terms of the temperature downstream of the valve θ_v, the temperatures of the cold and hot water θ_c and θ_h, the specific heats of the vapor and the liquid (c_{pv} and c_{pl}) and the enthalpy of vaporization, h_{fg}.

SOLUTION TO GUIDED PROBLEM

Guided Problem 11.2

1.

2. $T = \dfrac{\dot\phi}{\dot m_1 - \dot m_2} = \dfrac{\dot\phi}{\rho_3 Q_3} = \dfrac{1}{\rho_3 D} = \dfrac{R\theta_3}{P_3 D} = \dfrac{R\theta_3}{P_1 D}$

3. $M = \dfrac{1}{T}(h_3 - h_e) = \dfrac{c_p(\theta_3 - \theta_4)}{T} = \dfrac{R}{(1 - 1/k)}\dfrac{P_1 D}{R\theta_3}(\theta_3 - \theta_e)$

$\quad = \dfrac{P_1 D}{(1 - 1/k)}\left(1 - \dfrac{\theta_e}{\theta_3}\right)$

4. Isentropic flow: $\dfrac{\theta_e}{\theta_3} = \left(\dfrac{P_e}{P_3}\right)^{(1-1/k)} = \left(\dfrac{P_e}{P_1}\right)^{(1-1/k)}$

5. Equate ratios θ_e/θ_3 : $\quad \left(\dfrac{P_e}{P_3}\right)^{(1-1/k)} = 1 - \dfrac{1-1/k}{DP_1}M$

   Therefore, $\quad M = \dfrac{DP_1[1 - (P_e/P_1)^{(1-1/k)}]}{1-1/k}$ which can be solved for P_1.

6. $\dot{m}_2 = \dfrac{AP_2}{\sqrt{R\theta_2}}\sqrt{k\left(\dfrac{2}{k+1}\right)^{(k+1)/(k-1)}}$

7. From equation (11.28), $\theta_3 = \dfrac{\dot{m}_1\theta_1 + \dot{m}_2\theta_2}{\dot{m}_1 + \dot{m}_2}$

   (Note that the air leaving the nozzle has the same stagnation enthalpy as the air entering, and thus also has the same temperature, θ_2.)

8. Substitute the P_1 found from a numerical evaluation in step 5 into the equation from step 4 to find θ_e.

9. $\dot{\phi} = T(\dot{m}_1 - \dot{m}_2) = \dfrac{R\theta_3}{P_1D}(\dot{m}_1 + \dot{m}_2)$ (All desired results now can be evaluated.)

11.3 Case Study with Quasi-Steady Flow*

The double-rotor two-lobed compressor pictured in Fig. 11.10, often called a "**Roots blower**," is a very simple positive-displacement machine that is used to increase gas pressures up to a factor of about two.[8] It typically runs dry with no oil mist. Metal-to-metal friction is avoided without excessive leakage by using a small clearance between the two rotors, and between the rotors and the housing. External gears provide the necessary synchronization. In position (A), the chamber on the right side of rotor A is still in communication with the inlet port. By position (B) the chamber has been isolated from both the inlet and the outlet, but remains at constant volume and thus at constant pressure. In position (C) a surge of high-pressure fluid flows back into the chamber, charging it to the outlet pressure. In position (D) the fluid that was in the chamber is being pushed out, or discharged through the outlet.

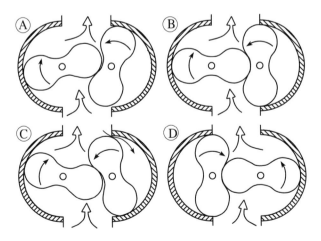

Figure 11.10: Operating cycle of a two-impeller straight-lobe rotary compressor

---

[8]Staged Roots compressors sometimes produce air pressures in excess of 100 psi.

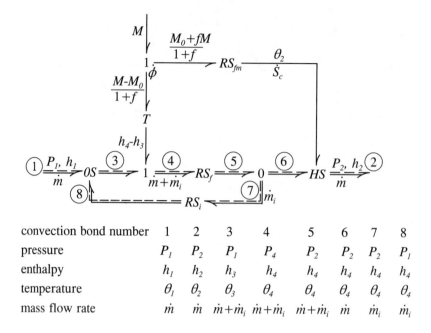

convection bond number	1	2	3	4	5	6	7	8
pressure	P_1	P_2	P_1	P_4	P_2	P_2	P_2	P_1
enthalpy	h_1	h_2	h_3	h_4	h_4	h_4	h_4	h_4
temperature	θ_1	θ_2	θ_3	θ_4	θ_4	θ_4	θ_4	θ_4
mass flow rate	$\dot{m}$	$\dot{m}$	$\dot{m}+\dot{m}_i$	$\dot{m}+\dot{m}_i$	$\dot{m}+\dot{m}_i$	$\dot{m}$	$\dot{m}_i$	$\dot{m}_i$

Figure 11.11: Bond graph for the rotary compressor

11.3.1   Bond Graph Model

The bond graph of Fig. 11.11 can be used to model the time-average behavior of the compressor. It is adapted from Fig. 11.8(c), which is primarily for a motor rather than a pump, by reversing the power-convention arrows through the transformer to avoid confusion, and by interchanging the $0S$ and 0-junctions, since the leakage flow through the RS_f element is from the right (outlet) to the left (inlet). The inlet bond is labeled with the subscript 1 and the outlet bond with the subscript 2. Other bonds are labeled with circled numbers which also are used as subscripts. The table below the graph explicitly identifies constraints between bond variables that are implicit in the definitions of the various elements; the number of variables is reduced from the outset by using these identifications. The information which follows will be interpreted to quantify the elements T, RS_{fm}, RS_i and RS_f, and to predict the mass flow, outlet temperature and shaft torque as a function of the shaft speed, inlet conditions and outlet pressure.

The machine displaces $D = 25.0/2\pi$ ft³/rad (its overall size is little over a cubic yard), draws in atmospheric air ($P_1 = 14.7$ psia, $\theta_1 = 70°$F, $R = 0.3704$ psia·ft³/lbm·R, $k = 1.4$) which can be considered an ideal gas, and normally operates in the range 200-600 rpm. The torque loss due to gear and bearing friction is $M_0 = 173$ ft·lb plus the fraction $f = 0.05$ of the balance of the moment. The fluid leakage around the rotor from the outlet to the inlet is controlled mostly by viscosity, since the leakage passages are relatively slender and long. As a result, the leakage is almost proportional to the pressure drop; it is determined to be 51 scfm for each psi. ("scfm" means *standard* cubic feet per minute, *i.e.* volume flow at atmospheric pressure and temperature.) This is all the quantitative information that is needed.

The transformer modulus equals the volumetric displacement per radian times a density. Since the given displacement D corresponds to the density in the inlet chamber, the density ρ_3 should be used, giving

$$T = \rho_3 D = \frac{P_1 D}{R\theta_3}, \tag{11.29}$$

from which

$$\dot{m} + \dot{m}_i = T\dot{\phi} = \frac{P_1 D\dot{\phi}}{R\theta_3}, \tag{11.30}$$

$$M = M_0 + (1+f)T(h_4 - h_3) = M_0 + (1+f)\frac{P_1 D(h_4 - h_3)}{R\theta_3}$$
$$= M_0 + (1+f)\frac{P_1 D(\theta_4/\theta_3 - 1)}{(1 - 1/k)}. \tag{11.31}$$

The fluid 1-junction implies the isentropic relation

$$\frac{\theta_4}{\theta_3} = \frac{h_4}{h_3} = \left(\frac{P_4}{P_1}\right)^{(k-1)/k}. \tag{11.32}$$

11.3.2 Irreversibilities

The mechanical friction energy is converted to heat in the element RS_{fm} at the rate

$$\dot{Q} = \theta_2 \dot{S}_c = \frac{(M_0 + fM)\dot{\phi}}{1 + f}, \tag{11.33}$$

causing the air to increase its enthalpy and temperature in the HS element according to

$$\frac{\theta_2}{\theta_4} = \frac{h_2}{h_4} = 1 + \frac{(M_0 + fM)\dot{\phi}}{(1+f)h_4\dot{m}} = 1 + \frac{(1 - 1/k)(M_0 + fM)\dot{\phi}}{(1+f)R\theta_4\dot{m}}. \tag{11.34}$$

The leakage flow through the element RS_i is given in the form

$$\dot{m}_i = \rho_1 a(P_2 - P_1); \qquad a = \frac{51}{60} \text{ ft}^3/\text{sec}; \qquad \rho_1 = \frac{P_1}{R\theta_1}. \tag{11.35}$$

The mixing of this flow with the air from the intake increases the temperature and enthalpy according to equation (11.28):

$$\frac{\theta_3}{\theta_1} = \frac{h_3}{h_1} = \frac{\dot{m} + (h_4/h_1)\dot{m}_i}{\dot{m} + \dot{m}_i} = \frac{\dot{m} + (\theta_4/\theta_1)\dot{m}_i}{\dot{m} + \dot{m}_i}. \tag{11.36}$$

As a pumping chamber begins to communicate with the discharge port, which is at a lower pressure, flow surges from the chamber to the port. This presents an inherent irreversibility representable as a frictional or throttling loss. The steady-flow equivalency for this irreversibility occurs in the bond-graph element RS_f. It can be approximated by the classic problem in which a constant-volume chamber, with initial conditions indicated by the subscript 0, is charged through a restriction by a mass m_c with conditions indicated by the subscript c, until equilibrium is reached at mass $m_0 + m_c$ and conditions indicated by the subscript f. The conservation of energy for the chamber as a control volume requires $\Delta E = h_c m_c$, or

$$(m_0 + m_c)u_f - m_0 u_0 = m_c h_c. \tag{11.37}$$

Since $du = c_v d\theta$ and $dh = c_p d\theta$, this gives

$$m_0 c_v(\theta_f - \theta_0) = m_c(c_p \theta_c - c_v \theta_f). \tag{11.38}$$

The enthalpies h_c and h_f are equal, so that $\theta_c = \theta_f$. Solving for the ratio of the temperatures and noting the definition of k, this gives

$$\frac{\theta_f}{\theta_0} = \frac{1}{1 - (k-1)m_c/m_0}. \tag{11.39}$$

The ideal gas relation requires

$$\frac{m_c + m_0}{m_0} = \frac{\rho_f}{\rho_0} = \frac{P_f \theta_0}{P_0 \theta_f}. \tag{11.40}$$

The final state, indicated by the subscript f, is the state on bonds 5, 6 and 7, so that $\theta_f = \theta_4$ and $P_f = P_2$. The initial state, indicated by subscript i, is the state on bond 3, so that $\theta_0 = \theta_3$ and $P_0 = P_1$. Combining equations (11.39) and (11.40) now gives

$$\frac{\theta_4}{\theta_3} = \frac{1 + (k-1)P_2/P_1}{k}, \tag{11.41}$$

where P_f and P_i are the corresponding pressures.

11.3.3   Computation of Results

Equations (11.30), (11.31), (11.32), (11.34), (11.35), (11.36) and (11.41) comprise seven equations in the seven unknowns $\dot{m}$, $\dot{m}_i$, θ_2, θ_3, θ_4, P_4 and M. The variables P_1, P_2, θ_1 and $\dot{\phi}$ and the parameters D, a, M_0, f, R and k are assumed to be given. Equation (11.35) gives $\dot{m}_i$ directly. Substitution of equation (11.41) into (11.31) gives the torque on the shaft:

$$M = M_0 + (1 + f)D(P_2 - P_1). \tag{11.42}$$

The mass flow $\dot{m}$ can be found by solving equations (11.30) and (11.41) for θ_3 and θ_4, respectively, and substituting the results into equation (11.36) to give the quadratic

$$\left(\frac{\dot{m}}{\dot{m}_i}\right)^2 - \left[\frac{D\dot{\phi}}{a(P_2 - P_1)} - 1\right]\left(\frac{\dot{m}}{\dot{m}_i}\right) + \frac{D\dot{\phi}}{a(P_2 - P_1)k}\left[1 + (k-1)\frac{P_2}{P_1} - k\right] = 0. \tag{11.43}$$

Equation (11.34) now can be solved for θ_2, with the help of equation (11.41):

$$\theta_2 = \frac{P_1 D\dot{\phi}}{kR(\dot{m} + \dot{m}_i)}\left[1 + (k-1)\frac{P_2}{P_1}\right] + \frac{(fM + M_0)(1 - 1/k)\dot{\phi}}{(1 + f)R\dot{m}}. \tag{11.44}$$

The efficiency, η, of the machine is of special interest. Efficiency is defined as the ratio of the power that would be required to achieve the same mass flow rate and output pressure if the machine were reversible, to the actual power required. In the reversible machine the balance of mechanical and fluid powers requires

$$\text{reversible power} = (h_2 - h_1)\dot{m} = \frac{Rm\theta_1}{1 - 1/k}\left(\frac{\theta_2}{\theta_1} - 1\right). \tag{11.45}$$

The isentropic relation of equation (11.32) applies, with $\theta_2 = \theta_4$ and $\theta_1 = \theta_3$. Thus,

$$\eta = \frac{Rm\theta_1}{(1 - 1/k)M\dot{\phi}}\left[\left(\frac{P_2}{P_1}\right)^{(1 - 1/k)} - 1\right]. \tag{11.46}$$

The mass flow, shaft torque, inlet temperature and efficiency for the given parameters are plotted in Fig. 11.12 for two shaft speeds. You must pay careful attention to the units for the various variables in carrying out this type of computation. The results closely resemble data from an actual Roots compressor.

The entropy generation in the elements RS_{fm}, HS, RS_f, RS_i and $0S$ can be computed and compared. This information could aid a designer interested in potential improvements.

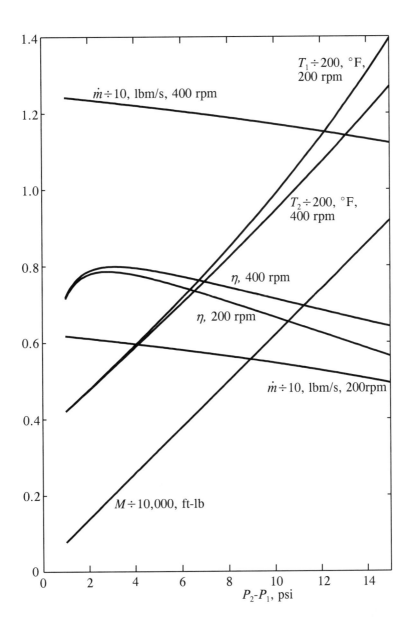

Figure 11.12: Model results for the rotary compressor

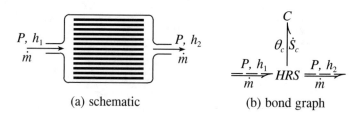

(a) schematic   (b) bond graph

Figure 11.13: Thermal regenerator

The irreversibility in RS_f is inherent in the design and could not be removed, for example, whereas the others could be mitigated.

The tools now at your disposal should allow you to represent and reduce most lumped engineering models for the steady-state behavior of thermodynamic engineering systems. Consideration of models of unsteady or dynamic systems is next. This will have a side benefit of allowing equilibria to be calculated without solving numerous simultaneous algebraic equations, as to some extent had to be done above.

11.4   Thermodynamic Compliance and Inertance

The *coherent* kinetic energy of a substance is represented by inertance elements, with direct implications for macroscopic momentum. The *incoherent* kinetic energy of the molecules plus the energy due to intermolecular forces is the thermodynamic energy. From the classical macroscopic viewpoint, taken here, this energy can be represented as a function of generalized displacements. Thus it becomes a potential energy described in bond graphs by compliance elements. In particular, the thermodynamic energy of a pure substance in a control volume, neglecting the effects of capillarity and electric and magnetic fields, can be changed by heat transfer, mass transport and the work of volume change.

11.4.1   Application of a Simple Thermal Compliance

The simplest model for thermodynamic compliance refers to a fixed mass; there is no convection. Also, work interactions with the environment are neglected, which implies either that the volume of the body is constant or its surrounding pressure is negligible. This leaves only a heat transfer interaction. This degenerate thermodynamic compliance is called a *thermal compliance* in Section 9.7.6 (pp. 699-701), and it is represented by a standard one-port compliance:

$$\vdash \frac{\theta}{\dot{S}} \rightharpoonup C$$

The thermal **regenerator** pictured in Fig. 11.13 illustrates a use of the simple thermal compliance in the presence of fluid flow. This device comprises a porous mass m_r of metal or ceramic which is heated or cooled by fluid passing through the pores. In a typical application to a thermal engine, the material is heated by hot exhaust gases during one portion of the cycle, and in turn heats cool inlet gases during another portion of the cycle. The mass of the gas within the pores is probably small enough to allow neglect of its thermal capacity.

A bond graph model for the regenerator is given in part (b) of the figure. This model assumes that the entire mass of metal or ceramic is at a uniform temperature, θ. Considering the HRS element and the causal stroke on the compliance, the differential equation becomes (from equation (11.26), p. 861)

$$\frac{dS_c}{dt} = \frac{\theta_1/\theta_c - 1}{1/H + 1/c_p\dot{m}}, \tag{11.47}$$

in which θ_1 is the known temperature of the inlet gas,

$$\theta_1 = \theta_1(h_1), \tag{11.48}$$

H is the heat transfer coefficient, c_p is the specific heat of the gas, and $\dot{m}$ is the mass flow. From equation (9.98) (p. 699), θ_c is

$$\theta_c = \theta_0 e^{(S-S_0)/m_r c}, \tag{11.49}$$

in which c is the specific heat of the mass m_r. The enthalpy of the outlet gas is, from equation (11.25) (p. 861),

$$h_2 = h_1 - \frac{\theta \dot{S}}{\dot{m}} = h_1 - \frac{\theta_1 - \theta}{\dot{m}/H + 1/c_p}. \tag{11.50}$$

The flow through the regenerator is driven by a small pressure drop. This likely could be represented by a single RS element placed on either side of the HRS element; one on each side would improve the accuracy of the model slightly by making it symmetric. These elements do not change the enthalpies. For the whole regenerator, then, the independent variables are P_1, P_3 and h_1. The system is first order.

11.4.2 Causality

The introduction of compliance leads to differential equation models; as always, causal strokes are very helpful in constructing such models.

Convection bonds have two effort variables; the choice used here, as noted in Section 11.1.2 (pp 854-855), is P, h (the *stagnation* pressure and enthalpy) and $\dot{m}$. Recall that the enthalpy h is causal physically in the actual direction of the flow. (There is a special exception with theoretical rather than practical significance.[9]) Since at any point in a simulation this direction is known, no further specification of its causality is needed. This leaves the other effort variable, P, which together with the flow, $\dot{m}$, forms a bilateral causal pair just like the effort and flow of a simple bond. The direction of the causality is indicated on a bond graph by the usual causal stroke. Thus, for example,

$$\vdash\!\!\frac{h, P}{\dot{m}}\!\!\!\!-$$

indicates that $\dot{m}$ is specified from the left and P is specified from the right.

The elements introduced earlier in this chapter impose various constraints on causality. The convection RS element imposes the same mass flow on both bonds, therefore allowing three possible causal combinations:

$$---\!|RS|\!--\!\!\!- \qquad ---\!|RS---| \qquad \vdash\!--- RS|\!--\!\!\!-$$

---

[9]For a **C**-element with a convection bond, the enthalpy h is causally determined by the element itself regardless of the direction of $\dot{m}$. In practice, however, the **C**-element is replaced by the **CS**-element, described in Section 11.4.4 below, which does not suffer this anomaly.

The thermal conduction bond of the H and HS elements must have $\dot{S}$ as its causal input, since the temperature is constrained by the specification of the enthalpy on the convection bond with the incoming flow. This leaves only two possible causalities for these elements:

$$\theta \bigg|\dot{S}_c \qquad\qquad\qquad \theta \bigg|\dot{S}_c$$
$$\text{------} | HS \text{ -----} | \qquad\qquad | \text{----} HS | \text{----}$$

On the other hand, the HRS macro element allows the conduction temperature to be set independently; either θ_c or $\dot{S}_c$ could be the causal input:

$$\theta_c \bigg|\dot{S}_c \qquad\qquad\qquad \theta_c \bigg|\dot{S}_c$$
$$\text{------} | HRS \text{ -----} | \qquad\qquad \text{-----} | HRS \text{ -----} |$$

Note that the causality of the convection bonds on these graphs can be reversed to give two more possibilities.

The pressure of the 0 and $0S$ junctions can be specified by only one bond, just like the simple 0-junction.

11.4.3 General Thermodynamic Compliance

The more general thermodynamic energy of a system or control volume in which the volume and/or the mass can change is now addressed. The state of the substance will be assumed to be homogeneous throughout the control volume. If its volume, V, changes by an amount dV, the system does work $P\,dV$ on its surroundings. If mass dm enters, the system energy increases by $h\,dm$, where h is the stagnation enthalpy. If heat is conducted across the control surface into the control volume, the energy increases by $\theta\,ds_c$ where ds_c is the inward entropy displacement due to the heat. Denoting the total energy by $\mathcal{V}$, the first law of thermodynamics requires

$$\frac{d\mathcal{V}}{dt} = h\dot{m} + \theta\dot{S}_c - P\dot{V}. \tag{11.51}$$

This form suggests a three-port compliance element with one convection bond for the mass transport and two simple bonds for the heat and work, respectively:

$$\theta \bigg|\dot{S}_c$$
$$| \frac{h, P}{\dot{m}} \text{---} \mathbf{C} \text{---} \frac{-P}{\dot{V}} |$$

If the practice with simple compliances is followed, the generalized displacements are used as the state variables. The time integral of the mass flow $\dot{m}$ is the total mass, m; the integral of $\dot{V}$ is the volume, V. The integral of the conduction $\dot{S}_c$ is only part of the entropy of the system, however, since additional entropy is transported along with the mass flow $\dot{m}$. Thus the implied third state variable is the *total* entropy of the system, S:

$$S = \int (\dot{S}_c + s\dot{m})dt. \tag{11.52}$$

The total compliance energy therefore becomes a function of S, V and m:

$$\mathcal{V} = \mathcal{V}(S, V, m), \tag{11.53}$$

from which

$$\frac{d\mathcal{V}}{dt} = \frac{\partial \mathcal{V}}{\partial S}\frac{dS}{dt} + \frac{\partial \mathcal{V}}{\partial V}\frac{dV}{dt} + \frac{\partial \mathcal{V}}{\partial m}\frac{dm}{dt}$$

$$= \left(\frac{\partial \mathcal{V}}{\partial S}\right)_{V,m} \dot{S}_c + \left(\frac{\partial \mathcal{V}}{\partial V}\right)_{S,m} \dot{V} + \left[\left(\frac{\partial \mathcal{V}}{\partial S}\right)_{V,m} s + \left(\frac{\partial \mathcal{V}}{\partial m}\right)_{V,S}\right]\dot{m} \qquad (11.54)$$

A comparison of equations (11.51) and (11.54) reveals that[10]

$$\theta = \left(\frac{\partial \mathcal{V}}{\partial S}\right)_{V,m}, \qquad (11.55a)$$

$$-P = \left(\frac{\partial \mathcal{V}}{\partial V}\right)_{S,m}, \qquad (11.55b)$$

$$h = \left(\frac{\partial \mathcal{V}}{\partial S}\right)_{V,m} s + \left(\frac{\partial \mathcal{V}}{\partial m}\right)_{V,S}. \qquad (11.55c)$$

In practice it is simpler to use the specific quantities

$$u \equiv \frac{\mathcal{V}}{m} \qquad \frac{1}{\rho} \equiv \frac{V}{m} \qquad s \equiv \frac{S}{m}, \qquad (11.56)$$

so that

$$u = u(s, \rho), \qquad (11.57)$$

and equations (11.55) are replaced by

$$\theta = \left(\frac{\partial u}{\partial s}\right)_{\rho}$$

$$P = \rho^2 \left(\frac{\partial u}{\partial \rho}\right)_{s}$$

$$h = u + \frac{P}{\rho}. \qquad (11.58)$$

Although this computational scheme is ideal theoretically, its direct implementation assumes the availability of analytic state functions of the form of equation (11.57). They are not generally available, however. As a result, it is necessary to guess the entropy and iterate until the error is sufficiently small. This iterative method is not presented here in favor of an alternative procedure based on a different set of state variables. This new procedure is aided by the definition of a particular macro element, described next.

11.4.4   The CS Macro Element

The general thermodynamic **C** element virtually always occurs in combination with other primitive elements. As a practical matter, a particular combination of elements called the **CS** (macro) element can be employed as the basic building block for dynamic models, replacing also the H, HS and HRS elements. This approach eliminates the iteration

---

[10]The definition of Gibb's function

$$g \equiv h - \theta s$$

allows equation (11.55c) to be recast as

$$g = \left(\frac{\partial \mathcal{V}}{\partial m}\right)_{V,S}$$

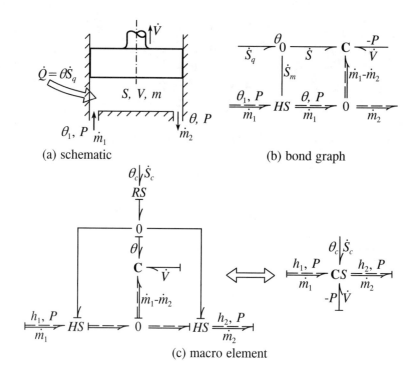

(a) schematic   (b) bond graph

(c) macro element

Figure 11.14: Fluid chamber with interactions

associated with the primitive **C** element, by using temperature, θ, as a state variable in place of S. The other two state variables, m and V, are unchanged.

Consider a fluid-filled chamber with an input port and an output port, such as the cylinder of an engine shown in Fig. 11.14. A bond graph model is shown in part (b) of the figure. The input flow $\dot{m}_1$, which enters at the lower left, is likely at a different temperature from the fluid in the cylinder, so an HS element is needed to equalize the temperature. The resulting entropy flux $\dot{S}_m$ is combined with whatever external entropy flux is associated with heat conduction ($\dot{S}_q$ at the upper left) to give $\dot{S}_\theta$. The output flow $\dot{m}_2$ is at the lower right, and the mechanical bond for volume change is at the upper right.

This combination of elements occurs repeatedly in the modeling of thermo-fluid systems. Usually, a thermal conductance RS element is added to the upper 0-junction to recognize conduction resistance and allow temperature causality for the heat conduction. Provision often is wanted to allow the material flows to reverse direction, also. The result is an expanded combination, shown in part (c) of the figure. This is the combination designated as the **CS macro element**. The causal strokes on both the detailed and the macro representations imply that V and $\dot{m} = \dot{m}_1 - \dot{m}_2$ are integrated to give the state variables V and m, respectively. The causal strokes on the primitive **C** element do not dictate S as the third state variable, although they suggest its use. In the associated differential equation, which is consistent with equation (11.52), $\dot{S} = \dot{S}_\theta + s\dot{m}$, or

$$\dot{S} = \frac{\theta_c \dot{S}_c}{\theta} + \frac{(h_{in} - h)\dot{m}_{in}}{\theta} + (\dot{m}_{in} - \dot{m}_{out})s. \tag{11.59}$$

The subscripts "in" and "out" refer to either bonds 1 or 2, depending on the direction of the flow. The causal thermal input at the top could be either θ_c, $\dot{Q} = \theta_c \dot{S}_c$, or $\dot{S}_c$, so no causal stroke is shown.

The un-elaborated $\mathbf{CS}$ element, on the other hand, does not by itself suggest that S should be a state variable. It allows the temperature, θ, to substitute, which is precisely what is needed to allow the practical computation of thermodynamic state properties from available analytical state functions (deduced from empirical data). The other two state variables (m, and when the volume varies, V) remain.

There are two commonly available forms for empirically-based state formulas of a single-phase pure substance. The most widely available are "P-v-T" formulas of the form

$$\boxed{P = P(v, \theta),} \tag{11.60a}$$

which are used in conjunction with specific heat relations at zero density ($\rho \equiv 1/v = 0$) that have the form

$$\boxed{c_v^0 = c_v^0(\theta).} \tag{11.60b}$$

These formulas are closely associated with the fundamental experiments. Some common substances, including water, ammonium and certain refrigerants, are available in the alternative form of the Helmholtz free energy, $\Psi \equiv U - \theta S$, as a function of density and temperature:

$$\boxed{\psi \equiv \Psi/m = \psi(v, \theta).} \tag{11.61}$$

Note that both equations (11.60a) with (11.60b) and (11.61) employ the same two independent variables (v and θ), and therefore the two forms can be treated similarly.

Expressions for the properties of pressure, entropy, internal energy, and enthalpy can be deduced from either of these forms, as will be shown in Section 11.5. Thus, for example, equations of the form

$$u = u(v, \theta) \tag{11.62}$$

can be derived, where $v = 1/\rho m$. Since $v = V/m$, this equation requires

$$\dot{u} = \frac{\partial u}{\partial v}\left(\frac{\dot{V}}{m} - \frac{v}{m}\dot{m}\right) + \frac{\partial u}{\partial \theta}\dot{\theta}. \tag{11.63}$$

The conservation of total energy, U, can be expressed as

$$\dot{U} = m\dot{u} + u\dot{m} = \dot{Q} - P\dot{V} + \sum_i h_i \dot{m}_i. \tag{11.64}$$

The substitution of equation (11.63) into equation (11.64) gives, after some manipulation including use of the definition $h \equiv u + Pv$, the major result[11]

$$\boxed{\frac{d\theta}{dt} = \frac{1}{m\partial u/\partial \theta}\left[\dot{Q} + \sum_i (h_i - h)\dot{m}_i + \left(P + \frac{\partial u}{\partial v}\right)(v\dot{m} - \dot{V})\right].} \tag{11.65}$$

Equation (11.65) applies also to an incompressible substance (approximation of a liquid or solid) if the term $\partial u/\partial \theta$ (in the denominator) is set equal to the specific heat, c, and the term $\partial u/\partial v$ is set equal to zero.

Applying equation (11.65) to the saturated mixture region requires some elaboration. Formulas of the form

---

[11]The results of equations (11.65) and (11.72) and derivations were first submitted by the author for publication in the *Proceedings of the Institution of Mechanical Engineers Part I* and appeared as "Non-Iterative Evaluation of Multiphase Thermal Compliances in Bond Graphs," in v 216 n —1 (2002), pp. 13-20.

$$\boxed{P \equiv P_{sat} = P_{sat}(\theta)} \tag{11.66}$$

have been deduced for many substances from the basic relations of equations (11.60a) and (11.60b) or (11.61). The thermodynamic identities

$$h = h_g - (s_g - s)\theta, \tag{11.67a}$$

$$s_g - s = (v_g - v)\left(\frac{dP_{sat}}{d\theta}\right), \tag{11.67b}$$

$$h = u + Pv \tag{11.67c}$$

(the second being the Clapeyron relation), where the subscript g refers to saturated vapor, give

$$u = u_g + \left(P_{sat} - \theta\frac{dP_{sat}}{d\theta}\right)(v_g - v). \tag{11.68}$$

To evaluate the term v_g, state formulas of the form

$$\boxed{v_g = v_g(\theta)} \tag{11.69}$$

also have been deduced for many substances, based also on the fundamental relations. The term u_g can be deduced in the form

$$u_g = u_g(v_g, \theta) \tag{11.70}$$

by combining equation (11.69) with equation (11.62).

Equation (11.68) gives

$$\frac{\partial u}{\partial v} = \theta\frac{dP_sat}{d\theta} - P_{sat}, \tag{11.71a}$$

$$\frac{\partial u}{\partial \theta} = \left(\frac{\partial u_g}{\partial v_g} + P_{sat} - \theta\frac{dP_{sat}}{d\theta}\right)\frac{dv_g}{d\theta} - \theta\frac{d^2P_{sat}}{d\theta^2}(v_g - v). \tag{11.71b}$$

Substitution of these results into equation (10.65) gives the desired differential equation for the time derivative of the state variable, θ, in the saturated mixture region:

$$\boxed{\begin{aligned}\frac{d\theta}{dt} &= \frac{1}{denom}\left[\dot{Q} + \theta\frac{dP_{sat}}{d\theta}(v\dot{m} - \dot{V}) + \sum_i(h_i - h)\dot{m}_i\right], \\ denom &= m\left[\left(\frac{\partial u_g}{\partial v_g} + P_{sat} - \theta\frac{dP_{sat}}{d\theta}\right)\frac{dv_g}{d\theta} + \frac{\partial u_g}{\partial \theta} - \theta\frac{d^2P_{sat}}{d\theta^2}(v_g - v)\right].\end{aligned}} \tag{11.72}$$

The temperature (θ) and pressure (P) are nominally assumed to be uniform over the volume, consistent with lumped modeling; for mixed saturation states, u, h and s need not be uniform but, for example, could be affected by gravity separation. More general approximations are possible, but can lead to computational instabilities, and are beyond the scope of the presentation here.

11.4.5   Computations for the Ideal Gas

A simple very special case is the ideal gas with assumed constant specific heats, for which

$$s - s_0 = c_v \ln\left(\frac{\theta}{\theta_0}\right) - R\ln\left(\frac{v_0}{v}\right). \tag{11.73}$$

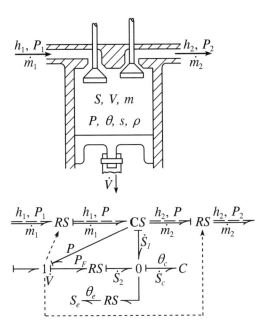

Figure 11.15: Piston-cylinder compressor and model

The subscript 0 refers to any convenient reference state. Thus, s is known in terms of v and θ, and no iteration is needed to carry out the scheme in which S is a state variable. In place of equation (11.58) (p. 879) there results

$$\theta = \theta_0 \exp\left[\frac{s - s_0 + R\ln(v_0/v)}{c_v}\right], \tag{11.74a}$$

$$P = R\theta/v, \tag{11.74b}$$

$$h = c_v(\theta - \theta_0) + Pv. \tag{11.74c}$$

These equations can be used to address the following case study.

11.4.6 Case Study: A Piston-Cylinder Compressor

A one-cylinder air compressor, shown in Fig. 11.15, has an assumed sinusoidal piston motion. For simplicity of presentation, the inlet and outlet valves are assumed to open to their full extent whenever the pressure on their left side exceeds that on their right side, and to be closed otherwise, thus passing flow only in the rightward direction, like ideal check valves. When open, the pressure drops across them are assumed to be represented by the equations given in Guided Problem 11.1 (p. 856). In the bond-graph model shown in part (b) of the figure, these valves are represented by the two convection RS elements. The magnitude of the friction force between the piston and the cylinder is modeled as a constant, giving an equivalent pressure, P_F, that heats the piston and cylinder walls, which are assumed to be uniform in temperature. Heat flows between the gas in the cylinder and the metal of the piston and cylinder, with an assumed constant heat conductance associated with a thin thermal boundary layer. A separate larger constant heat conductance is applied between the walls and the surroundings. The inlet and outlet pressures are assumed to be given and constant.

The internal energy of the fluid in the cylinder and the thermal conductance of the boundary layer is represented by the **CS** element. The thermal compliance of the metal is represented by a C element. The frictional conversion of mechanical power to thermal energy is represented by a simple-bonded RS element; the fact that thermal fluxes to the C element can come from both the fluid and the friction, at the same temperature, is represented by a 0-junction. The synchronization of the motions of the valves and the piston is indicated by dashed lines.

The causal strokes to the **CS** element produce the first-order state differential equations

$$\frac{dV}{dt} = V_1 \sin(\dot{\phi}t); \qquad\qquad V(0) = V_0, \tag{11.75a}$$

$$\frac{dm}{dt} = \dot{m}_1 - \dot{m}_2. \tag{11.75b}$$

In this example, the third state variable for the **CS** element is chosen to be S, rather than θ (either would be a good choice), so the third differential equation becomes, from equation (11.59) (p. 880),

$$\frac{dS}{dt} = H\left(\frac{\theta_c}{\theta} - 1\right) + \frac{h_1 - h}{\theta}\dot{m}_1 + \dot{m}s. \tag{11.75c}$$

The final state differential equation is associated with the energy storage in the walls of the cylinder, namely the C element:

$$\frac{dS_c}{dt} = H\left(\frac{\theta}{\theta_c} - 1\right) - H_e\left(1 - \frac{\theta_e}{\theta_c}\right) + \frac{P_F|\dot{V}|}{\theta_c}. \tag{11.75d}$$

Before these can be evaluated, the variables $\dot{m}_1$, $\dot{m}_2$, θ_1, θ, θ_c, $\dot{S}_1$ and s need to be related to the state variables. The mass fluxes are associated with the RS elements. The flow $\dot{m}_1$ is set at zero when $P > P_1$, and $\dot{m}_2$ is set at zero when $P < P_2$. Otherwise, assuming unchoked flow (this condition can be checked after simulation and corrected if necessary), from Guided Problem 11.1:

$$\dot{m}_1 = \frac{A_1 P_1}{\sqrt{R\theta_1}}\sqrt{\frac{2k}{k-1}\left[\left(\frac{P}{P_1}\right)^{2/k} - \left(\frac{P}{P_1}\right)^{1+1/k}\right]}, \tag{11.76a}$$

$$\dot{m}_2 = \frac{A_1 P}{\sqrt{R\theta}}\sqrt{\frac{2k}{k-1}\left[\left(\frac{P_2}{P}\right)^{2/k} - \left(\frac{P_2}{P}\right)^{1+1/k}\right]}, \tag{11.76b}$$

$$P_2 = \rho R\theta_3. \tag{11.76c}$$

Temperatures are unchanged across convection RS elements, since enthalpy is a function of temperature. The other two temperatures are given by the standard relations for constant specific heats:

$$\theta = \theta_0 \exp\left[\frac{s - s_0 + R\,\ln(\rho/\rho_0)}{c_v}\right], \tag{11.77a}$$

$$\theta_c = \theta_0 \exp\left[\frac{S_c - S_{c0}}{m_c c}\right], \tag{11.77b}$$

The entropies s_0 and S_{c0} are defined at the arbitrary reference temperature θ_0, and may be set equal to zero. The density ρ_0 also is defined at that temperature and the pressure P_1; the density ρ is

$$\rho = \frac{m}{V}, \tag{11.78a}$$

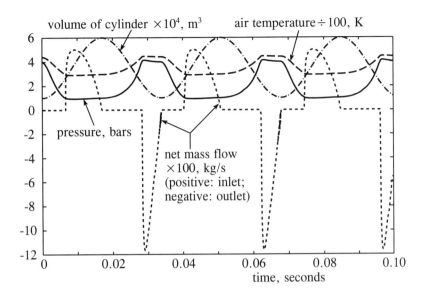

Figure 11.16: Simulation results for piston-cylinder compressor

Similarly,

$$s = \frac{S}{m}.\qquad(11.78b)$$

Results of a particular simulation are plotted in Fig. 11.16. Such a simulation can give good estimates of the amount of flow compressed and the energy efficiency of the process. It can show how the valve areas affect this performance, and what timing of the valves ought to be used. It can predict how hot the parts will become due to friction, and how both friction and heat transfer between the gas and the metal and the surroundings affect the performance. More refined questions can be addressed by elaborating the model to include profiles of the valve opening as a function of the position of the piston, substituting a more realistic slider-crank geometrical constraint, improving the friction model, adding an output tank, applying a torque rather than a velocity and adding a modulated inertia (including, presumably, the effect of a flywheel), etc. None of these modifications are beyond the scope of what you have done before, although as a systems engineer you might wish to consult with specialists to establish some of the parameters of the model, such as friction and heat transfer coefficients.

The inlet and output pressures are 1.0 bar and 4.0 bar, respectively. The maximum and minimum volumes of the cylinder are 0.1 and 0.6 liters, respectively. The shaft turns at 185 rad/s, the inlet air is at 293 K, the effective areas of both valves are 4.0 cm^2, the heat transfer coefficient H and H_e are 0.012 and 0.120 Joules/m^2/K, respectively, the mass of the walls is 1.0 kg and their specific heat is 0.45 Joules/kg K. The initial temperature of the walls is 744 K, close to equilibrium under the given circumstances of an assumed friction force of 10 N. (More external cooling would be desirable.) The initial condition of the cylinder is minimum volume with the air at 440 K and density 3.15 kg/m^3, which is close to the steady-state values for that part of the cycle.

The plotted results actually modeled the air as a mixture of nitrogen, oxygen and argon with very accurately modeled properties, as described in Section 11.5.3 below. This fancier simulation is detailed in Guided Problem 11.4 at the end of the next section. (It adapts equation (11.65) with temperature as the state variable rather than equation (11.75c) which

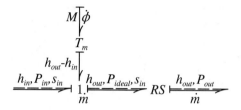

Figure 11.17: Simple model for the joining of the convection and mechanical bonds of a quasi-steady compressor

uses entropy. Using entropy with the more accurate model of the air would have required repeated iteration in the evaluation of state properties.) The assumption of an ideal gas with constant specific heats, as in the equations above, is considerably simpler to implement. The results are not plotted because they are virtually indistinguishable from those in the figure.

11.4.7   Treatment of a Quasi-Steady-State Fluid Machine

The piston-cylinder system above treats a fluid machine as a device that goes through a cycle for each revolution of its shaft. Often the cycle time for one revolution is a tiny fraction of the time required for a major change to occur in the thermodynamic properties of a system outside of the fluid machine. The refrigeration system to be detailed in Section 11.5.6 is an example. The shaft makes thousands of revolutions while the refrigerant masses and temperatures gradually change during start-up. If these relatively slow changes are the center of interest, rather than the rapid changes over each rotation of the shaft, it makes sense to employ a quasi-steady model for the compressor. Quasi-steady models for turbines and compressors are discussed in Section 11.2.6 (pp. 864-866), and the example of a Roots blower is developed in Section 11.3 (pp. 871-876).

Consider the simple model of a compressor shown in Fig. 11.17, in which the irreversibilites are restricted to those of an equivalent virtual series flow restriction that is represented by the element RS. Internal and external leakage is neglected, and any mechanical friction losses are assumed to occur above the ideal compressor element T_m, and not to affect the flow. The transformer modulus is the volumetric displacement of the machine, per radian of rotation, times the specific volume of the upstream fluid. The mass flow is computed as the product of the shaft speed and this modulus, and the torque is computed as the the product of the difference $h_{out} - h_{in}$ and this modulus. The enthalpy h_{in} is causal from the upstream part of the system, but how is h_{out} determined?

The 1-junction is a reversible as well as conservative element. Therefore, the entropies s of the upstream and downstream flows are the same. Further, the pressure P_{ideal} can be found from knowledge of $\dot{m}$ and P_{out}, the latter being a result of the state of the compliance element. Therefore, the state of the fluid for the convection bond to the right of the 1-junction is to be found from knowledge of its entropy and pressure. This is awkward, since the available state functions assume knowledge of the temperature and the specific volume (or density) instead.

Example 11.2 in Section 11.2.6 (p. 867) gives an alternative model, in which the volumetric and adiabatic efficiencies of the compressor are assumed to be known. In this case, it was found that

$$h_{out} = h_c/\eta_{ad} - (1/\eta_{ad} - 1)h_{in}, \tag{11.79}$$

in which h_{in} is known, and h_c corresponds to a fluid state having the known entropy s_{in}

and known pressure P_{out}. P_{out} is different from (less than) the P_{ideal} of Fig. 11.17, but the task of finding h_c is essentially the same as the task of finding h_{out}; only the simple step of equation (11.79) is added to find h_{out}.

This situation generally requires iteration: one starts with a guess of the proper temperature and specific volume (probably using the most recent known values), and computes the associated entropy and pressure. A comparison of these values with the actual values produces a second iteration.

The iterative procedure may be expedited by making use of the gradients $\partial P/\partial v$, $\partial P/\partial \theta$, $\partial s/\partial P$ and $\partial s/\partial \theta$, all of which can be computed as functions of v and θ. A Newton-Raphson iteration becomes

$$\begin{bmatrix} v \\ \theta \end{bmatrix}_{i+1} = \begin{bmatrix} v \\ \theta \end{bmatrix}_i - \begin{bmatrix} \partial P/\partial v & \partial P/\partial \theta \\ \partial s/\partial v & \partial s/\partial \theta \end{bmatrix}^{-1} \begin{bmatrix} P_i - P \\ s_i - s \end{bmatrix}, \qquad (11.80)$$

in which the thermodynamic identity $\partial s/\partial v \equiv \partial P/\partial \theta$ can be used to reduce computation. Since the partial derivatives are usually nearly constant over the narrow range of consideration, only one such iteration is usually necessary to reduce the errors to within acceptable limits. Equation (11.80) does not work under all circumstances, however. The entropy of a saturated vapor depends largely on temperature, and the second or later iterations may work better if the terms $\partial s/\partial v \equiv \partial P/\partial \theta$ in equation (11.80) are set equal to zero.

The process, including the iteration, is illustrated in Example 11.5 (on p. 903) and in the case study given in Section 11.5.7 (on p. 912).

EXAMPLE 11.3

The compressor of Example 11.2 (p. 867) operates on an assumed ideal gas with inlet temperature θ_{in} and a pressure P_{in} and outlet pressure P_{out}. Find the outlet enthalpy, h_{out}, as is necessary in simulating a system that includes such a compressor.

Solution: The plan is to find the temperature θ_c, use this to find h_c, and then employ equation (11.79) to find h_{out}. The state c has the same entropy as the inlet, and an isentropic process satisfies $Pv^k = $ constant, where $k = c_p/c_v$, and $Pv = R\theta$. Therefore,

$$\frac{\theta_c}{\theta_{in}} = \frac{P_c v_c}{P_{in} v_{in}} = \frac{P_c}{P_{in}} \left(\frac{P_{in}}{P_c} \right)^{1/k} = \left(\frac{P_c}{P_{in}} \right)^{1-1/k}.$$

From equations (11.74b) and (11.74c) (p. 883),

$$h_c = c_v(\theta_c - \theta_0) + P_c v_c = (c_v + R)\theta_c - c_v\theta_0 = c_p\theta_c - c_v\theta_0,$$

where θ_0 is the reference temperature for which $s = 0$. Finally, equation (11.79) gives

$$h_{out} = c_p\theta_{in} \left\{ \frac{1}{\eta_{ad}} \left[\left(\frac{P_{out}}{P_{in}} \right)^{1-1/k} - 1 \right] + 1 \right\} - \frac{c_p}{k}\theta_0.$$

EXAMPLE 11.4

The compressor of Fig. 11.17 operates on ideal gas, and the RS element is based on a virtual orifice of effective area A. The flow through this orifice can be assumed not to be choked. Develop a procedure for finding h_{out} given P_{out}, $\dot{\phi}$, D and the inlet conditions. (This procedure is more complex than that found in the previous example; you are not asked to carry it out, which in practice would be done numerically.)

Solution: The mass flow is given by $\dot{m} = (D/V_{in})\dot{\phi}$, and the pressure P_{ideal} must be found from the orifice relation

$$\dot{m} = \frac{A P_{ideal}}{\sqrt{R\theta_{ideal}}} \sqrt{\left(\frac{2k}{k-1}\right) \left[\left(\frac{P_{out}}{P_{idea;}}\right)^{2/k} - \left(\frac{P_{out}}{P_{ideal}}\right)^{(k+1)/k}\right]}.$$

The isentropic process across the 1-junction requires (as in equation (11.32), p. 873, for the case study of a rotary compressor)

$$P_{ideal} = P_{in} \left(\frac{\theta_{ideal}}{\theta_{in}}\right)^{k/(k-1)}.$$

Substituting this into the previous equation gives a result in the form

$$\dot{m} = \dot{m}(\theta_{ideal}, \theta_{in}, P_{out}, \text{constants}).$$

Since everything in this equation except θ_{ideal} is known, it can in principle be solved for θ_{ideal}, after which the answer is

$$h_{out} = h_{ideal} = c_p \theta_{ideal} - c_v \theta_0.$$

In practice, θ_{ideal} must be found iteratively, presumably through the use of a Newton-Raphson procedure.

11.4.8   Pseudo Bond Graphs for Compressible Thermofluid Systems

Pseudo bond graphs for non-compressible media have an advantage over true bond graphs when dealing with the special case of constant specific heats and constant thermal conductivities, as pointed out in Section 9.7. The concept was proposed by Karnopp.[12] The convection bond graph is a relatively new development; the major bond graph representation for compressible thermofluid systems in the literature is an extension of the pseudo bond graph. A pseudo bond graph for a variable volume with heat and mass fluxes followed by a fluid restriction is shown in part (b) of Fig. 11.18; a convection bond graph for the same model is shown in part (c). The term *pseudo* recognizes that the products of the indicated efforts and flows are not the actual energy fluxes, unlike *true* bond graphs. The energy flux into the compliance is $\dot{E}$, which includes the mechanical power $-\dot{E}_m = -P\dot{V}$ that is communicated through the modulated flow source S_f and the diagonal bond, not the horizontal bond with flow $\dot{V}$. Karnopp writes, "This signal interaction is the price we pay for connecting a pseudo bond graph with a true bond graph."

The purpose of the graph is to aid in the writing of state differential equations, not particularly to explicate the energy fluxes. The causal strokes indicate that the state variables

---

[12]Karnopp, Dean C., "State Variables and Pseudo Bond Graphs for Compressible Thermofluid Systems," *Trans. ASME, J. Dynamic Systems, Measurement and Control*, v 101, n3, pp 201-204, Sept 1979.

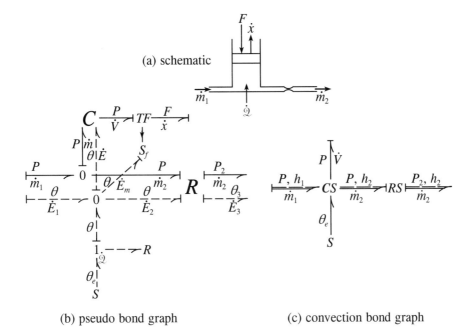

(a) schematic

(b) pseudo bond graph   (c) convection bond graph

Figure 11.18: Comparison of pseudo and convection bond graphs

for the compliance are the mass m of the fluid therein, the total energy E of this mass, and the volume V. Dividing m and E by V gives the density, ρ and the specific energy, u, which are sufficient to define the thermodynamic state. The implication of the graph is that the pressure, P, and the temperature, θ, are deduced from the knowledge of ρ, u. The available state formulas, as noted in Section 11.4.4 above (p. 881), unfortunately, use θ in place of u as an independent variable. Thus, the same kind of iteration needed for dealing with ρ, s as independent variables must be employed. It is largely for this reason that pseudo bond graphs are not emphasized in this book. Their relative awkwardness in joining with other parts of a model represented by true bond graphs also contributes to this decision.

11.4.9 Fluid Inertance and Area Change With Compressible Flow

The *incoherent* kinetic energy of fluid molecules has been included in the fluid compliance. The *coherent* kinetic energy of a fluid is represented in an *inertance*. The use of proper inertance elements permits direct dealing with actual as opposed to stagnation properties.

The use of stagnation properties for the pressure, enthalpy and temperature suppresses most problems associated with fluid inertia in *steady* flow. In unsteady flow, the kinetic energy of a compressible flow is much less apt to be significant than the kinetic energy of an incompressible flow, since its density typically is less and its thermodynamic or potential (compliance) energy dominates. This is fortunate, since when the fluid in a channel has both significant kinetic and potential energies, wave propagation results. Lumped modeling is then apt to become either inaccurate or complex, as has been suggested in Chapter 10. Nevertheless, there are problems where inertia is indeed important. In these cases wall shear resistance usually is also significant.

Rather than represent the inertial effects and the resistance effects with separate elements, the author recommends combining them into a single element, called the *IRS* element[13], since the resistance effects are irreversible. This combination effectively repre-

---

[13]F.T. Brown, "Kinetic Energy in Convection Bond Graphs," *2003 International Conference on Bond*

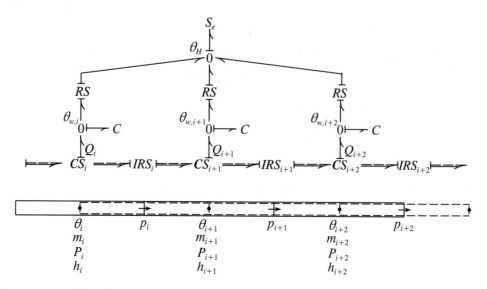

Figure 11.19: Bond graph for a segment of a channel with thermodynamic and kinetic energies and resistance

sents the momentum balance of the flow to which it applies. A long channel such as the condenser of a refrigeration system, with both fluid compliance and inertance energies, is thus represented by a cascade of alternating CS and IRS elements, as indicated in Fig. 11.19. The control volumes for the CS elements represent the entire region, as do the control volumes for the IRS elements; the two types are both overlapping and staggered. They lead automatically to what in the non-bond graph literature is called the use of **staggered grids**, which has evolved into the standard method. Heat interaction to a uniform temperature surroundings also is shown in the bond graph. The variables used in computations for the CS elements are the temperature, θ, mass $m = AL/v$ where v is specific volume, static pressure, P, and enthalpy, h. The greater the number of these lumped elements, the more accurate the model. Extensions to wave-scattering methods is anticipated but has not yet been developed. (The methods of Chapter 10 generally require linearization, which is not intended here.) Finite-element analysis is also a significant prospect, particularly in its potential to treat two and three dimensions.[14] The discussion below is restricted to lumped modeling, however.

Control volumes for the IRS element are shown in Fig. 11.20, allowing for changes in the area of the channel. The total momentum of the flow within the control volume is defined as p_i, which is treated as a state variable. The momenta of equal lengths of channel, L, but centered at the left and right faces of the control volume are labeled as p_i^+ and p_i^-, respectively. In practice, the interpolations

$$p_i^- = (p_{i-1} + p_i)/2; \qquad p_i^+ = (p_i + p_{i+1})/2, \qquad (11.81)$$

Graph Modeling and Simulation, ed. F.E. Cellier and J.J.Granda, The Society for Modeling and Simulation International (SCS), Simulation Series, v 35, n 2, pp. 191-197.

[14]See E.P. Farenthold and M. Venkatamaran, "Eulerian Bond Graphs for Fluid Continuum Dynamics Modeling," *J. Dynamic Systems, Measurement and Control, Trans. ASME*, v 118 (1996) pp. 48-57, and subsequent non-bond-graph developments: E.P.Farenthold and J.C. Koo, "Discrete Hamilton's Equations for Viscous Compressible Fluid Dynamics," *Computer Methods in Applied Mechanics and Engineering*, v178 (1999) pp. 1-22, and J.C. Koo and E.P. Farenthold, " Discrete Hamilton's Equations for Arbitrary Lagrangian-Eulerian Dynamics of Viscous Compressible Flows," *ibid*, v189 iss 3 (2000) pp. 875-900.

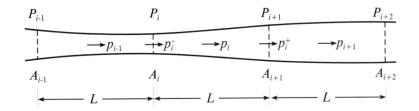

Figure 11.20: Control volumes for the IRS element

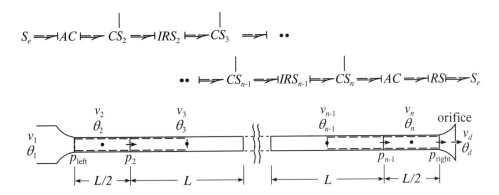

Figure 11.21: Model for ends of a tube, also introducing the AC element

are used. The respective cross-sectional areas of the left and right faces are A_i and A_{i+1}, the respective pressures are P_1 and P_{i+1}, and the respective densites are $\rho_i = 1/v_i$ and $\rho_{i+1} = 1.v_{i+1})$. With this notation, the momentum equation becomes

$$\frac{dp_i}{dt} = \left(\frac{A_i + A_{i+1}}{2}\right)(P_i - P_{i+1}) - F_{w,i} + \frac{(P_i^+)^2}{\rho_i A_i L^2} - \frac{(p_i^+)^2}{\rho_{i+1} A_{i+1} L^2}, \tag{11.82}$$

where $F_{w,i}$ represents the leftward-directed wall shear force.

The thermodynamic state of the fluid leaving the element CS_i for the element CS_{i+1} can be taken as the average of the mean states for the two elements. In another extreme, called **upwinding**, the fluid in the element CS_i is assumed to be well mixed, so its efflux has its mean properties. Upwinding, or at least partial upwinding, usually is necessary for sufficient relative or even absolute computational stability. An upwinding coefficient, F, can be defined such that the values of h_{in} and h_{out} used in equations (11.66) and (11.72) are given by

$$h_{in} = Fh_{i-1} + (1 - F)h_i; \qquad h_{out} = Fh_i + (1 - F)h_{i+1}. \tag{11.83}$$

For rightward-traveling flow, complete upwinding or mixing corresponds to $F = 1$, while zero upwinding corresponds to $F = 0.5$. For reverse flow, the respective values are $F = 0$ and $F = 0.5$. Some trial-and-error may be in order.

The staggered and overlapping nature of the control volumes produces special considerations at the ends of a channel. A uniform-area tube is shown in Fig. 11.21 that is fed by a constant-property source at its left end and is terminated in an orifice at its right end. The upstream end of the tube in the figure is located at what would be the center of the control volume for the element designated as CS_2, and at the left end of the control volume designated as IRS_2. This full-length CS_2 element therefore is replaced by a half-length element, so both the CS_2 and IRS_2 elements have the same location for their left ends. The downstream end of the tube is located at the downstream side of the element IRS_n, giving $n - 1$ equal-sized control volumes. Another half-length element CS_{n+1} is needed, giving a total of n CS elements.

A new **area change element**, AC, is introduced at both ends. This simple element is defined as reversible, with equal input and output flow rates and stagnation enthalpies. Using the causal strokes, assuming flow from left to right, assuming negligible dynamic head on the upstream side, and designating the interface between the area change element and the tube with the subscript "left," the upstream area change element is described by

$$s_{left} = s_1, \tag{11.84a}$$
$$P_{left} = (4P_2 - P_3)/3, \tag{11.84b}$$
$$V_{left}^2/2 = h_1 - h_{left}, \tag{11.84c}$$

where V_{left} is the velocity of the fluid at the interface. Equations (11.84a) and (11.84b) establish the thermodynamic state at the interface, from which h_{left} can be determined. Equation (11.84b) applies an approximate linear extrapolation. Equation (11.84c) then establishes the fluid velocity at the interface, which along with the known fluid density establishes the momentum flux there. These relations can be used either for steady-state analysis or dynamic simulation, and permit direct engagement with actual as opposed to stagnation properties.

The state variables v_n and θ_n near the right end of the tube are defined at a distance $L/4$ from the actual end. The state at the end can be estimated from the linear extrapolations

$$\theta_{right} = (4\theta_n - \theta_{n-1})/3, \tag{11.85a}$$
$$v_{right} = (4v_n - v_{n-1})/3. \tag{11.85bx}$$

The orifice is represented by an RS element. Its flow is to be calculated from the equation on page 856, which requires the stagnation upstream state. Therefore, another area change element is interposed. The entropy of the stagnation state equals that of the right-end state, and its enthalpy equals the stagnation enthalpy of the right-end state, a mirror image of the first area-change element. Implementation of the area-change elements requires iteration.

Examples with nonlinear simulations employing a refrigerant are given in Section 11.5.10, following the development of procedures to evaluate thermodynamic state variables.

11.4.10 Summary

The thermodynamic energy in a relatively uniform region of space is represented by the compliance element **C**. For the special case of a virtually incompressible solid or liquid body of fixed identity, its entropy suffices as the only independent variable needed. Thus the compliance has only one port, entropy is the generalized displacement, entropy flux is the generalized velocity, and temperature is the generalized effort. The generalized energy of heat flux, represented by a simple bond, is the product of the temperature and the entropy flux. This topic was addressed in Section 9.7.

A body of a pure substance with fixed identity but changing density also has its volume as its second generalized displacement, and rate of change of volume as its corresponding generalized velocity. The associated effort is minus its pressure. Thus a second simple bond is added to the compliance. When the mass of the pure substance is no longer fixed, that is when there is a control volume with mass entering or leaving, a third generalized velocity is added: the rate of change of mass. This requires a convection bond with enthalpy as its power-factor effort, so the general **C** element has three ports.

Irreversibilites due to heat transfer and fluid mixing can be represented in a bond graph by adding HS, 0 and conduction RS elements to the **C** element. This combination is so commonly useful it is represented by the **C**S macro element. Irreversibilites due to throttling can be appended by adding convection RS elements. Causal strokes can be applied to the bonds; on the effort side of a convection bond, the causality refers to the pressure. Causally, enthalpy and entropy are causal in the direction of the flow, so no special causal mark for them is needed on the bond graph.

The thermodynamic energy of a pure substance can be represented, on a continuum basis per unit mass, as a function of any two independent variables. The most natural pair is specific volume (or its reciprocal, mass density) and entropy. All other continuum intensive variables, such as temperature, pressure and enthalpy, can be deduced therefrom. This scheme works well for ideal gases, but for more accurate models of most substances it requires an awkward computational iteration. It is better to substitute temperature for the entropy as the state variable (making entropy a derived variable). The associated differential equations for a **C**S element have been found, but their evaluation depends on the details of the representation of the properties of the substance. This is the subject of the next section.

The joining of a fluid flow to a mechanical machine such as a compressor or turbine involves a 1-junction across which the entropy is common. Knowledge of entropy rather than enthalpy requires an iterative procedure for which a strategy is given. In some cases a $1S$ junction is helpful to represent the irreversible aspects of the coupling. The pair of convection bonds for this junction have the same pressure.

The kinetic energy of a compressible flow in a channel can be modeled by the use of IRS elements, which also include the resistance effects of wall shear and gradual changes in cross-sectional area. Abrupt changes in area can be treated through use of the AC element. Kinetic energy can be neglected in reasonable models for many systems. When it cannot, and particularly when long channels are involved as in heat exchangers, wave propagation may become an important consideration, and require the use of an extensive cascade of alternating CS and IRS elements.

Guided Problem 11.3

Piston-cylinder compressors (see Fig. 11.22) often have two stages with an interstage cooler to improve the efficiency of the process. You are asked to develop a model including a set of state differential equations for this system. Make assumptions similar to those for the single-stage compressor modeled above, and add an externally insulated load tank. Replace the air with helium for simplicity. Apply a moment to a crank shaft rather than a particular motion. You are not asked to work out the details of the geometry, however; simply note general transformer relations and a general modulated inertia.

Suggested Steps:

1. Draw a bond graph for your model. It is suggested that you ignore the volume of fluid within the intercooler. Apply causality.

2. Label the graph with the appropriate variables, noting that enthalpies remain constant across convection RS elements and pressures remain constant across **C**S and HRS elements.

3. Write three differential equations for each **C**S element (two for fixed volume), one differential equation of each one-port C element and two differential equations for the variable inertia element, in terms of convenient variables.

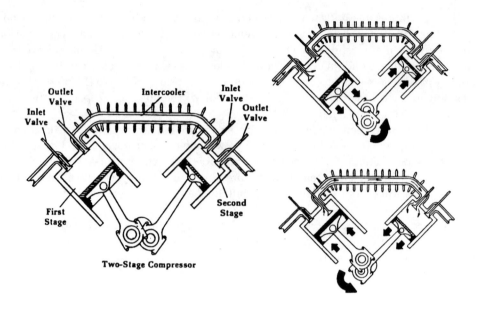

Figure 11.22: Piston-cylinder compressors of Guided Problem 11.4
Reprinted courtesy of Parker-Hannifin Corporation

4. Relate the "convenient variables" of step 3 to the state variables. This requires appropriate constitutive relations for the non-storage elements, consistent with the indicated causality, plus state relations for an ideal gas.

PROBLEMS

11.5 Show that, when equation (11.65) (p. 881) is applied to an ideal gas ($Pv = R\theta$ and constant specific heats) for an isentropic expansion, the classical formula $Pv^k = $ constant results.

11.6 A fixed-volume chamber containing an ideal gas is connected to a cylinder with piston of area A_c to form a catapult that accelerates a mass m_c. Draw a bond graph, define state variables, and write state differential equations. Neglect all effects of friction and heat transfer.

11.7 Repeat the above problem, placing a flow restriction of effective area A_r between the chamber and the cylinder.

11.8 Repeat Problem 11.6 when the chamber is charged with a saturated steam-water mixture. Assume the needed properties of the fluid are known; do not evaluate the associated partial derivatives in equation (11.72) (p. 882).

11.9 Repeat Problem 11.7 when the chamber is charged with a saturated steam-water mixture. Gravity separation allows only vapor to enter the orifice. Assume the needed properties of the fluid are known; do not evaluate the associated partial derivatives in equations (11.65) and (11.72) (pp. 881, 882).

11.10 Repeat Example 11.3 (p. 887), considering the machine to be a turbine rather than a compressor.

11.11 Solve Example 11.4 (p. 888) in the case where the virtual orifice is known to be choked (implying rather large losses). (See Guided Problem 11.1, p. 856, for the orifice flow equation.)

11.12 Equations (11.84) for the area-change element of the left end of the tube assume flow from left to right. Determine what relations could be used in the case of reverse flow. Linear extrapolations may be employed when appropriate. Note that P_1 would be established by external conditions.

SOLUTION TO GUIDED PROBLEM

Guided Problem 11.3

1-2.

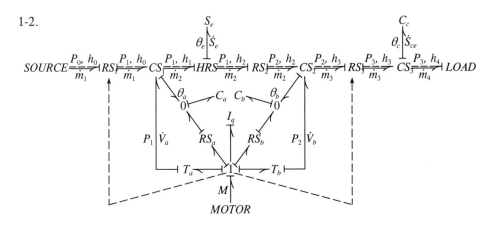

CS_1	thermodynamic compliance, mixing and heat conductance, gas in cylinder a
CS_b	thermodynamic compliance, mixing and heat conductance, gas in cylinder b
CS_c	thermodynamic compliance and mixing, gas in storage tank
RS_1	flow resistance, inlet valve
RS_2	flow resistnace, interstage valve
RS_3	flow resistance, outlet valve
HRS	interstage heat exchanger with conductance
S_e	environmental thermal reservoir

C_a	thermal compliance, walls of cylinder a
C_b	thermal compliance, walls of cylinder b
C_c	thermal compliance, walls of tank
I_q	inertance of moving parts (modulated)
RS_a	frictional "dissipation," cylinder a
RS_b	frictional "dissipation," cylinder b
T_a	slider-crank constraint, piston a
T_b	slider-crank constraint, piston b

3. gas, cylinder a: $\dfrac{dV_a}{dt} = T_1(\phi)\dot{\phi}$

$$\frac{dm_m}{dt} = \dot{m}_1 = \dot{m}_2$$

$$\frac{dS_a}{dt} = -\dot{S}_1 + c_p(1 - \theta_0/\theta_1)\dot{m}_1 + \dot{m}_a s_1$$

cylinder b:

$$\frac{DV_b}{dt} = T_2(\phi)\dot{\phi}$$

$$\frac{dm_b}{dt} = \dot{m}_2 - \dot{m}_3$$

$$\frac{dS_b}{dt} = -\dot{S}_2 + c_p(1 - \theta_0/\theta_3)\dot{m}_2 + \dot{m}_b s_3$$

gas, tank:

$$\frac{dm_c}{dt} = \dot{m}_3 - \dot{m}_4$$

$$\frac{dS_c}{dt} = -\dot{S}_{cc} + c_p(1 - \theta_3/\theta_4)\dot{m}_3 + \dot{m}_c s_4$$

walls, cylinder a: $$\frac{dS_{ca}}{dt} = \dot{S}_1 + \frac{P_{fa}|T_a\dot{\phi}|}{\theta_{ca}}$$

walls, cylinder b: $$\frac{dS_{cb}}{dt} = \dot{S}_2 + \frac{P_{fb}|T_b\dot{\phi}|}{\theta_{cb}}$$

walls, tank: $$\frac{dS_{cc}}{dt} = \dot{S}_{cc}$$

inertance: $$\frac{d}{dt}\left(\frac{1}{2}I_q\dot{\phi}\right) = M\dot{\phi} - P_{fa}|T_a\dot{\phi}| - P_{fb}|T_b\dot{\phi}| - T_a P_1 - T_b P_2$$

therefore: $$\frac{d\dot{\phi}}{dt} = \frac{1}{I_q}\left[M - \frac{1}{2}\frac{dI_q}{dt}\dot{\phi}^2 - \frac{P_{fa}|T_a\dot{\phi}| + P_{fb}|T_b\dot{\phi}| + T_a P_1 + T_b P_2}{\dot{\phi}}\right]$$

also: $$\frac{d\phi}{dt} = \dot{\phi}$$

There are therefore 13 state variables (V_a, V_b, m_a, m_b, m_c, S_a, S_b, S_c, S_{ca}, S_{cb}, S_{cc}, $\dot{\phi}$, ϕ). The parameters $T_a(\phi)$, $T_b(\phi)$, $\dot{m}_4$, c_p, P_{fa} and P_{fb} (pressures required to balance friction), $I_q(\phi)$ and M are presumed known. This leaves 17 other variables for which algebraic relations are needed: $\dot{m}_1$, $\dot{m}_2$, $\dot{m}_3$, $\dot{S}_1, \dot{S}_2, \dot{S}_{cc}$, s_1, s_3, s_4, P_1, P_2, θ_1, θ_2, θ_3, θ_4, θ_{ca} and θ_{cb}.

4. 10 state relations (R, ρ_0, θ_0, m_{ca}, m_{cb}, c_v, c, s_0, S_{ca0}, S_{cb0} and S_{cc0} are presumed known; the last four are probably set at zero):

$$P_1 = \rho_1 R\theta_1 = m_a R\theta_1/V_a$$

$$P_2 = \rho_3 R\theta_3 = m_b R\theta_3/B_b$$

$$s_1 = S_a/m_a; \quad s_3 = S_b/m_b; \quad s_4 = S_c/m_c$$

$$\theta_1 = \theta_0 \exp\left[\frac{s_1 - s_0 + R(\rho_1/\rho_0)}{c_v}\right] = \theta_0 \exp\left[\frac{s_1 - s_0 + Rm_a/V_a\rho_0}{c_v}\right]$$

$$\theta_3 = \theta_0 \exp\left[\frac{s_1 - s_0 + R(\rho_3/\rho_0)}{c_v}\right] = \theta_0 \exp\left[\frac{s_3 - s_0 + Rm_b/V_b\rho_0}{c_v}\right]$$

$$\theta_1 = \theta_0 \exp\left[\frac{s_4 - s_0 + R(\rho_4/\rho_0)}{c_v}\right] = \theta_0 \exp\left[\frac{s_1 - s_0 + Rm_c/V_c\rho_0}{c_v}\right]$$

$$\theta_{ca} = \theta_0 \exp\left[\frac{S_{ca} - S_{ca0}}{m_{ca}c}\right]$$

$$\theta_{cb} = \theta_0 \exp\left[\frac{S_{cb} - S_{cb0}}{m_{cb}c}\right]$$

The outlet valve of cylinder a and the inlet valve of cylinder b are redundant in this model, since the energy of the small mass of gas in the intercooler is neglected. Thus there are three valve relations (k, $A_1(\phi)$, $A_2(\phi)$, $A_3(\phi)$ and θ_0 are assumed known):

$$\dot{m}_1 = \frac{A_1(\phi)P_0}{\sqrt{R\theta_0}}\sqrt{\frac{2k}{k-1}\left[\left(\frac{P_1}{P_0}\right)^{2/k} - \left(\frac{P_1}{P_0}\right)^{1+1/k}\right]}$$

$$\dot{m}_2 = \frac{A_2(\phi)P_1}{\sqrt{R\theta_2}} \sqrt{\frac{2k}{k-1}\left[\left(\frac{P_2}{P_1}\right)^{2/k} - \left(\frac{P_2}{P_1}\right)^{1+1/k}\right]}$$

$$\dot{m}_3 = \frac{A_3(\phi)P_2}{\sqrt{r\theta_3}} \sqrt{\frac{2k}{k-1}\left[\left(\frac{P_3}{P_2}\right)^{2/k} - \left(\frac{P_3}{P_2}\right)^{1+1/k}\right]}$$

Intercooler relation (H, c_p and θ_e are assumed known; the absence of a thermal compliance can be criticized):

$$\theta_2 = \theta_1 + \frac{\theta_e \dot{S}_e}{\dot{m}_2 c_p}; \qquad \dot{S}_e = \frac{1 - \theta_1/\theta_e}{(1/H) + (1/c_p \dot{m}_2)}$$

Three conduction relations (H_a, H_b and H_c are assumed known):

$$S_1 = H_a(\theta_{ca}/\theta_1 - 1)$$
$$S_2 = H_b(\theta_{cb}/\theta_3 - 1)$$
$$S_{cc} = H_c(\theta_{cc}/\theta_4 - 1)$$

This completes 17 equations. New variables P_3 and θ_{cc} have been introduced, which have the following state relations (m_{cc} and S_{cc0} are assumed to be known; the latter probably is set at zero):

$$P_3 = \rho R\theta_4 = m_c R\theta_4 / V_c$$
$$\theta_{cc} = \theta_0 \exp\left[\frac{S_{cc} - S_{cc0}}{m_{cc}c}\right]$$

11.5   Evaluation of Thermodynamic Properties

The thermodynamic properties of a wide variety of substances are available in analytical forms that represent fits to experimental data. The most widely available form is that of the "P – v – T" relation and the specific heat relation, as given in equations (11.60) and (11.61), repeated below as equations (11.86) and (11.87). The next most widely available form is that of the Helmholtz relation, equation (11.62) (equation (11.99) below). This section presents practical means by which either of these forms can be used to develop analytical formulas for other thermodynamic properties such as internal energy, entropy, enthalpy, etc. The state is assumed implicitly to be in thermodynamic equilibrium. Illustrations and details are given for oxygen, nitrogen, air, water and two common refrigerants. Case studies and examples give simulation code for several systems.

Non-iterative evaluation of explicit formulas stands in sharp contrast to the iteration conventionally required. The author has avoided using the most commonly used data source for refrigerants[15] for this reason. Simulation is thereby greatly expedited, with some cost of preparation for each substance. The NIST database represents a larger number of substances, however.

11.5.1   The Most Commonly Available Analytical Form for State Properties

Thermodynamic state functions were originally developed and are still available largely as "P-v-T" relations of the form

$$\boxed{P = P(\rho, \theta) \ \text{or} \ P = P(v, \theta), \quad v = 1/\rho,}$$ (11.86)

plus specific heat relations at $\rho = 0$ of the form

---

[15]NIST Thermodynamic Properties of Refrigerants and Refrigerant Mixtures Database (REF_PROP), Version 4.0, National Institute of Standards and Technology, Gaithersburg, MD, 1994.

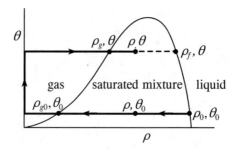

Figure 11.23: Path of integration in the evaluation of entropy or internal energy

$$c_v^0 = c_v^0(\theta). \tag{11.87}$$

The first consideration below is the calculation of specific entropy, s, from these functions. It uses the density ρ rather than its reciprocal, the specific volume, v (but could be transformed to use v).

The fundamental "T ds" relation

$$ds = \frac{1}{\theta}du + \frac{P}{\theta}d\left(\frac{1}{\rho}\right), \tag{11.88}$$

with substitutions for the definition of c_v and the identity

$$\left(\frac{\partial u}{\partial (1/\rho)}\right)_\theta = \theta\left(\frac{\partial P}{\partial \theta}\right)_\rho - P, \tag{11.89}$$

gives

$$ds = \frac{c_v}{\theta}d\theta - \frac{1}{\rho^2}\left(\frac{\partial P}{\partial \theta}\right)_\rho d\rho. \tag{11.90}$$

The entropy is defined to be zero at some arbitrary state ρ_0, θ_0, which often is taken at the triple point with 100% liquid (e.g. 0°C for liquid water). Equation (11.85) must be integrated along any path from this state to the desired state ρ, θ in order to get the desired s. Since c_v is known only for zero density, integration is chosen along the path shown in Fig. 11.23 for which temperature change occurs only at zero density. Adding and subtracting a term $-R\ln\rho$ from equation (11.90), this integral becomes

$$s = \int_{\theta_0}^{\theta}\frac{c_v(\theta)}{\theta}d\theta - R\int_{\rho_0}^{\rho}\frac{d\rho}{\rho} + \int_0^\rho \frac{1}{\rho^2}\left[\rho R - \left(\frac{\partial P}{\partial \theta}\right)_\rho\right]_\theta d\rho$$

$$- \int_0^{\rho_0}\frac{1}{\rho^2}\left[\rho R - \left(\frac{\partial P}{\partial \theta}\right)_\rho\right]_{\theta_0} d\rho. \tag{11.91}$$

Note that the second integrand also is integrated from ρ to 0 on the lower leg and from 0 to ρ on the upper leg, but these two (infinite) terms precisely cancel each other and thus are not represented. The remaining term can be integrated outright. The right-most integral minus the term $R\ln\rho_0$ is a constant, and is represented here as $-s_0$. Thus,

$$s = \int_{\theta_0}^{\theta}\frac{c_v^0(\theta)}{\theta}d\theta - R\ln\rho + \int_0^\rho \frac{1}{\rho^2}\left[\rho R - \left(\frac{\partial P}{\partial \theta}\right)_\rho\right]_\theta d\rho + s_0. \tag{11.92}$$

This is the desired result, the implementation of which is discussed below. The integrand of the second integral vanishes for the special case of an ideal gas, leaving a formula which may be familiar.

The corresponding equation for the specific internal energy, u, can be found by starting with the expression $u = u(\theta, 1/\rho)$, taking the derivative, substituting equation (11.84), and integrating as was done for s, to give

$$u = \int_{\theta_0}^{\theta} c_v^0(\theta)\, d\theta + \int_0^{\rho} \frac{1}{\rho^2} \left[P - \theta \left(\frac{\partial P}{\partial \theta} \right)_\rho \right] d\rho + u_0. \tag{11.93}$$

The P-v-T relations apply only in the single phase region. This includes, on its boundary, both the saturated vapor and saturated liquid states. These states can be identified from a relation in the form of equation (11.86a) because, at any particular temperature, they are the only two states having the same pressure. It is nevertheless awkward to have to calculate the saturation states this way every time they are needed. Fortunately, equations of the forms

$$\rho_f = \rho_f(\theta),$$

$$\rho_g = \rho_g(\theta), \tag{11.94}$$

are widely available.

The pressure, P, is given directly as a function of the density or specific volume. The specific enthalpy, h, can be computed from its definition

$$h \equiv u + Pv \equiv u + P/\rho. \tag{11.95}$$

The plan for evaluating s in the two-phase or saturation mixture region starts with finding s_g, namely the value of s when $\rho = \rho_g$ at the desired temperature, by substituting the density ρ_g from equation (11.94b) as the limit of integration in equation (11.92). The balance of the answer, $s - s_g$ (which is negative), is then added. This final term is based on the Clapeyron relation

$$\left(\frac{s - s_g}{1/\rho - 1/\rho_g} \right)_\theta = \left(\frac{\partial P}{\partial \theta} \right)_{\text{sat}}, \tag{11.96}$$

which is derived in most thermodynamics textbooks. Equations of the form

$$P_{\text{sat}} = P_{\text{sat}}(\theta) \tag{11.97}$$

have been deduced from the P-v-T relations and are widely available to expedite the evaluation of the right side of equation (11.96).

Changes in enthalpy in the two-phase region associated with changes in density at constant temperature simply equal the changes in entropy times the temperature. Thus, in the two-phase region,

$$h = h_g + (s - s_g)\theta. \tag{11.98}$$

W.C. Reynolds[16] has collected and interpreted data on many substances in the form of equations (11.86), (11.87), (11.94a) and (11.97). Stewart et al.[17] have done the same for many regrigerants and several other substances. These equations are generally very complex, involving many coefficients with many significant digits. Relations of the form of

---

[16]William C. Reynolds, *Thermodynamic Properties in SI*, Department of Mechanical Engineering, Stanford University, Stanford CA, 1979.

[17]R.B. Stewart et al., *ASHRAE Thermodyanmic Properties of Refrigerants*, American Society of Heating, Refrigeration and Air-Conditioning Engineers, 1988.

equation (11.94b) are conspicuously missing from Reynolds, but are included in Stewart, who also carries out the integrations to give the entropy (equation (11.92)) and enthalpy (equation (11.95 with equations (11.86 and (11.93) substituted) for specific substances. The author has taken some liberties in order to make a relatively unified presentation.

11.5.2   Helmholtz Analytical Form for State Properties

The properties of a few substances, including water, ammonium and certain common refrigerants, are available in the form of the Helmholtz free energy $\Psi \equiv U - \theta S$ as a function of the density and the temperature:

$$\boxed{\psi \equiv \Psi/m = \psi(\rho, \theta).}$$
(11.99)

The arguments of this formulation are the same ρ, θ use in the 'P–v–T' and c_v^0 formulation. The Helmholtz form is more convenient, however, since the entropy, internal energy and pressure are given by simple derivatives (without the confusion of constants of integration):

$$\boxed{\begin{aligned} s &= -\left(\frac{\partial \psi}{\partial \theta}\right)_\rho, \\ u &= \left[\frac{\partial(\psi\tau)}{\partial \tau}\right]_\rho, \\ P &= \rho^2 \left(\frac{\partial \psi}{\partial \rho}\right)_\tau. \\ \tau &\equiv 1/\theta \end{aligned}}$$
(11.100)

These replace equations (11.92), (11.93) and (11.86), respectively.

Application to water is given in Section 11.5.6 below. Application to R12, R22 and R123 is included in a special journal volume.[18]

11.5.3   Application to Gases

The specific heats c_v^0 of noble gases such as argon, helium and neon are constants. For nitrogen and oxygen, Reynolds gives

$$c_v^0 = \sum_{i=1}^{7} G_i \theta^{i-4} + G_8 e^{\beta/\theta} \left(\frac{\beta/\theta}{e^{\beta/\theta} - 1}\right)^2.$$
(11.101)

in which the coefficients are constants included in the MATLAB® file `Gasdata.m` that appears in Appendix D. For the pressure-volume-temperature relation, Reynolds gives

$$\begin{aligned} P = {}& \rho R\theta + \rho^2 \left[A_1\theta + A_2\theta^{1/2} + \sum_{i=3}^{5} A_i\theta^{3-i} \right] + \rho^3 \sum_{i=6}^{9} A_i\theta^{7-i} \\ & + \rho^4 \sum_{i=10}^{12} A_i\theta^{11-i} + \rho^5 A_{13} + \rho^6(A_{14}/\theta + A_{15}/\theta^2) + \rho^7 A_{16}/\theta \\ & \rho^8(A_{17}/\theta + A_{18}/\theta^2) + \rho^9 A_{19}/\theta^2 + [\rho^3(A_{20}/\theta^2 + A_{21}/\theta^3) \\ & \rho^5(A_{22}/\theta^2 + A_{23}/\theta^4) + \rho^7(A_{24}/\theta^2 + A_{25}/\theta^3) + \rho^9(A_{26}/\theta^2 + A_{27}/\theta^4) \\ & + \rho^{11}(A_{28}/\theta^2 + A_{29}/\theta^3) + \rho^{13}(A_{30}/\theta^2 + A_{31}/\theta^3 + A_{32}/\theta^4)]e^{-\gamma\rho^2} \end{aligned}$$
(11.102)

---

[18]*Fluid Phase Equilibria*, v 80 (1992), ed. J. F. Fly, Elsevier Science Publishers B.V., Amsterdam.

This relation also applies to hydrogen, methane, oxygen, nitrogen and air; the coefficients for air are given in `gasdata.m`.

For temperatures well above the saturation state, equation (11.102) can be approximated as an ideal gas, which means neglecting all terms beyond the first on its right side. The second integral in equation (11.92) also vanishes, leaving for the entropy

$$
\begin{aligned}
s_h = & -R\ln\left(\frac{\rho}{\rho_0}\right) - \frac{G_1}{3}\left(\frac{1}{\theta^3} - \frac{1}{\theta_0^3}\right) - \frac{G_2}{2}\left(\frac{1}{\theta^2} - \frac{1}{\theta_0^2}\right) - G_3\left(\frac{1}{\theta} - \frac{1}{\theta_0}\right) \\
& + G_4\ln\left(\frac{\theta}{\theta_0}\right) + G_5(\theta - \theta_0) + \frac{G_6}{2}(\theta^2 - \theta_0^2) + \frac{G_7}{3}(\theta^3 - \theta_0^3) \\
& + G_8\left[\frac{\beta/\theta}{e^{\beta/\theta} - 1} - \frac{\beta/\theta_0}{e^{\beta/\theta_0} - 1} + \frac{\beta}{\theta} - \frac{\beta}{\theta_0} - \ln\left(\frac{e^{\beta/\theta} - 1}{e^{\beta/\theta_0} - 1}\right)\right] + s_0.
\end{aligned} \qquad (11.103)
$$

The internal energy becomes, from equation (11.93),

$$
\begin{aligned}
u_h = \int_{\theta_0}^{\theta} c_v^0\, d\theta = & -\frac{G_1}{2}\left(\frac{1}{\theta^2} - \frac{1}{\theta_0^2}\right) - G_2\left(\frac{1}{\theta} - \frac{1}{\theta_0}\right) + G_3\ln\left(\frac{\theta}{\theta_0}\right) \\
& + G_4(\theta - \theta_0) + \frac{G_5}{2}(\theta^2 - \theta_0^2) + \frac{G_6}{3}(\theta^3 - \theta_0^3) \\
& + \frac{G_7}{4}(\theta^4 - \theta_0^4) - G_8\beta\left[\frac{1}{e^{\beta/\theta} - 1} - \frac{1}{e^{\beta/\theta_0} - 1}\right] + u_0.
\end{aligned} \qquad (11.104)
$$

The inclusion of more than the first term in equation (11.102) augments the relations for entropy and internal energy as follows:

$$
\begin{aligned}
s = & s_h - \rho\left(A_1 + \frac{A_2}{2\sqrt{\theta}} - \frac{A_4}{\theta^2} - \frac{2A_5}{\theta^3}\right) - \frac{\rho^2}{2}\left(A_6 - \frac{A_8}{\theta^2} - \frac{2A_9}{\theta^3}\right) \\
& - \frac{\rho^3}{3}\left(A_{10} - \frac{A_{12}}{\theta^2}\right) + \frac{\rho^5}{5}\left(\frac{A_{14}}{\theta^2} + \frac{2A_{15}}{\theta^3}\right) + \frac{\rho^6}{6}\frac{A_{16}}{\theta^2} \\
& + \frac{\rho^7}{7}\left(\frac{A_{17}}{\theta^2} + \frac{2A_{18}}{\theta^3}\right) + \frac{\rho^8}{8}\frac{2A_{19}}{\theta^3} + W_1\left(\frac{2A_{20}}{\theta^3} + \frac{3A_{21}}{\theta^4}\right) \\
& + W_2\left(\frac{2A_{22}}{\theta^3} + \frac{4A_{23}}{\theta^5}\right) + W_3\left(\frac{2A_{24}}{\theta^3} + \frac{3A_{25}}{\theta^4}\right) + W_4\left(\frac{2A_{26}}{\theta^3} + \frac{4A_{27}}{\theta^5}\right) \\
& + W_5\left(\frac{2A_{28}}{\theta^3} + \frac{3A_{29}}{\theta^4}\right) + W_6\left(\frac{2A_{30}}{\theta^3} + \frac{3A_{31}}{\theta^4} + \frac{4A_{32}}{\theta^5}\right) + s_m - s_0,
\end{aligned} \qquad (11.105)
$$

$$
\begin{aligned}
u = & u_h + \rho*\left(\frac{A_2\sqrt{\theta}}{2} + A_3 + \frac{2A_4}{\theta} + \frac{3A_5}{\theta^2}\right) + \frac{\rho^2}{2}\left(A_7 + \frac{2A_8}{\theta} + \frac{3A_9}{\theta^2}\right) \\
& + \frac{\rho^3}{3}\left(A_{11} + \frac{2A_{12}}{\theta}\right) + \frac{\rho^4}{4}A_{13} + \frac{\rho^5}{5}\left(\frac{2A_{14}}{\theta} + \frac{3A_{15}}{\theta^2}\right) + \frac{\rho^6}{6}\frac{2A_{16}}{\theta} \\
& + \frac{\rho^7}{7}\left(\frac{2A_{17}}{\theta} + \frac{3A_{18}}{\theta^2}\right) + \frac{\rho^8}{8}\frac{3A_{19}}{\theta^2} + W_1\left(\frac{3A_{20}}{\theta^2} + \frac{4A_{21}}{\theta^3}\right) \\
& + W_2\left(\frac{A_{22}}{\theta^2} + \frac{5A_{23}}{\theta^4}\right) + W_3\left(\frac{3A_{24}}{\theta^2} + \frac{4A_{25}}{\theta^3}\right) + W_4\left(\frac{3A_{26}}{\theta^2} + \frac{5A_{27}}{\theta^4}\right) \\
& + W_5\left(\frac{3A_{28}}{\theta^2} + \frac{4A_{29}}{\theta^3}\right) + W_6\left(\frac{3A_{30}}{\theta^2} + \frac{4A_{31}}{\theta^3} + \frac{5A_{32}}{\theta^4}\right),
\end{aligned} \qquad (11.106a)
$$

$$
W_1 = \frac{1}{2\gamma}\left(1 - e^{-\gamma\rho^2}\right) \qquad (11.106b)
$$

$$W_{i+1} = \frac{-\rho^{2i}}{2}e^{-\gamma\rho^2} + \frac{iW_i}{\gamma} \qquad 1 < i < 5 \tag{11.106c}$$

The term s_m is the entropy of mixture, related to the reference state.

Equations (11.103)–(11.104) for an ideal gas at low (literally zero) pressure are evalutated in the MATLAB function file gas.m listed in Appendix D. (This file, and several others cited in this Section, can be downloaded from the Internet; see page 1018.) The derivatives $\partial s/\partial T$ and $\partial u/\partial T$ also are evaluated, to aid in the evaluation of the terms in equation (11.80) and similar endeavors.

Gaseous air at higher pressures usually can be treated as a mixture of nitrogen, oxygen and argon. Coefficients for the defining relations are given in the downloadable file gasdata.m. The entropy and internal energy can be calculated by summing the contributions of each component, using equations (11.103) and (11.104) as computed using gas.m. Since equation (11.102) applies to the gas mixture of air, the corrections of equations (11.105)–(11.106) can be applied directly to the mixture rather than to the individual components. Proper constants s_m and u_m for the mixture should be added, replacing the individual s_0 and u_0. A downloadable function m-file air.m uses the equations above and their derivatives to compute P, u, $h = u + P/\rho$, s, $\partial u/\partial T$, $\partial u/\partial v$, $\partial s/\partial T$, $\partial s/\partial v \equiv \partial P/\partial T$ and $\partial P/\partial v$, as functions of θ and ρ. Note that the first two partial derivative terms are needed to evaluate the differential equations for temperature, as given by equation (11.65) (p. 881), and the additional derivatives can be useful in evaluating a state when a pair of variables other than θ and ρ (or $v = 1/\rho$) are known, such as P and s, for they allow the terms of a Newton-Raphson iteration to be evaluated analytically, giving rapid convergence. This procedure is discussed in Section 11.4.7 (p. 887) above; the derivatives are used in equation (11.80). Note that air.m can be used for pure nitrogen or oxygen, also, by setting the mass weighting coefficients Fo, Fn and the gas constant R appropriately.

A separate script file spech.m computes the specific heat of the air mixture, given the temperature, needed for estimating the flow through a restriction.

EXAMPLE 11.5

A piston-cylinder compressor draws in atmospheric air (temperature 293 K, density 1.205 kg/m^3) and charges an insulated tank from its initial atmospheric state. The tank has volume 0.1 m^3. The compressor is driven at 1750 rpm, has a volumetric displacement of 1×10^{-5} m^3/rev, an adiabatic efficiency of 80%, and a volumetric efficiency of 95%. Simulate and plot the resulting pressure and temperature in the tank for 60 seconds.

Solution: The system can be represented by the bond graph below:

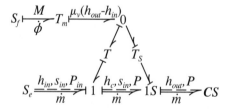

Two state variables are needed, one for the mass of air in the tank and the other for its temperature. Calling the shaft speed $\dot{\phi}$, the volumetric displacement per radian D and the specific volume of the atmospheric air v_{in}, the first differential equation becomes

$$\frac{dm}{dt} = \frac{D\dot{\phi}}{0.95 v_{in}}.$$

The second differential equation is adapted from equation (11.65) (p. 881):

$$\frac{d\theta}{dt} = \frac{1}{m\partial u/\partial \theta}\left[h_{out} - h + \left(P + \frac{du}{dv}v\right)\right]\frac{dm}{dt}.$$

The enthalpy of the air leaving the compressor, h_{out}, is found from equation (11.67). The other terms in the differential equation are fluid properties in the tank that depend on the state variables m and θ. They can be determined by running the routine air.m with the appropriate coefficients given by gasdata.m. A downloadable function file that computes the differential equations and the master program follow.

```
function f=ex11_5(t,x) % diff. equations for compressor/tank
% charging from atmospheric air
global V DPH0 D vin sin hin vc Tc
gasdata;
v=V/x(1); % specific volume of air in tank
[P]=air(x(2),1/v); % call 'air' just to get the pressure, P.
% Then call 'air' to get the properties of state c:
[Pc,uc,hc,sc,dudTc,dudvc,dsdTc,dsdvc,dPdTc,dPdvc]=air(Tc,1/vc);
E=(P/Pc-1)^2+(sin/sc-1)^2; % self-scaled error index
n=0; % start a count of the number of iterations
while E>1e-8 % iterate until error is acceptably small
    if n==0 % that is, for the first iteration only
        M=[dPdvc dPdTc;dPdTc dsdTc]; % matrix for Newt.-Raphson
        q=M\[Pc-P;sc-sin];
        vc=vc-q(1); % improved estimate, spec. volume, state c
        Tc=Tc-q(2); % improved estimate, temperature, state c
    else % following the suggestion on page 892
        vc=vc+(P-Pc)/dPdvc;
        Tc=Tc+(sin-sc)/dsdTc;
    end  % Now the revised estimate of the state c is found:
    [Pc,uc,hc,sc,dudTc,dudvc,dsdTc,dsdvc,dPdTc,dPdvc]=air(Tc,1/vc);
    E=(P/Pc-1)^2+(sin/sc-1)^2; % new (reduced) error
    n=n+1
end
hout=hc/.8-(1/.8-1)*hin; % from equation (11.67)
f(1)=DPH0*D/vin*.95; % first differential equation
f(2)=((hout-h+(P+dudv)*v)*f(1))/x(1)/dudT; % second diff. eqn.
f(3)=P; % to enable plotting of P
f=f';
t % to see the progress of the simulation on the screen

% ex11_5m.m, master program for simulation of an air compressor charging a
% tank from atmospheric air
% x(1) is the mass of air in the tank; x(2) is its temperature
global V DPH0 D vin hin sin vc Tc
V=.1; DPH0=1750*2*pi/60; D=1e-4; vin=1/1.205; T0=293;
vc=1; Tc=300; % non-critical trial values to get started
gasdata % imports the many coefficients that describe air
[P,u,hin,sin]=air(T0,1/vin); % finds the needed hin and sin
[t,x]=ode45('ex11_5',[0 60],[V/vin T0 0]);
k=size (t);
i=k(1); % for use below to find the pressure, P
```

```
    for j=1:i-1
        dt(j)=t(j+1)-t(j); % values of time increments
        P(j)=(x(j+1,3)-x(j,3))/dt(j) % P is time derivative of x(3)
        t1(j)=(t(j+1)+t(j))/2; % center times of time intervals
    end
    plot(t,x(:,2),t1,P/1000) % plots the temperature and pressure
```

The plot that results from the final instructions above is given below (annotations added). The high temperature reached shows the importance of the omitted heat transfer.

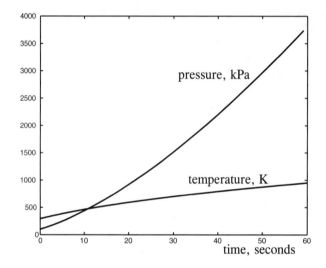

The model or air used above is dry. Atmospheric air normally is considered a mixture of dry air and water vapor. The ratio of the mass of water vapor to the mass of the dry air is called the **absolute** or **specific humidity**, ω:

$$\omega \equiv \frac{m_v}{m_a}. \tag{11.107}$$

Assuming a perfect gas, the specific humidity also can be expressed in terms of the total pressure, P, and the partial pressures of its components, P_a and P_v:

$$\omega = \frac{P_v/R_v}{P_a/R_a} = 0.622\frac{P_v}{P_a} = \frac{0.622P_v}{P - P_v}. \tag{11.108}$$

When the partial pressure of the water, P_v, equals the saturation pressure of water, P_g, the air is said to be **saturated**; it can hold no more water. The **relative humidity**, ϕ, is the ratio of the actual mass of water in the air to this maximum, or

$$\phi = \frac{m_v}{m_g} = \frac{P_v}{P_g} = \frac{\omega P}{(0.622 + \omega)P_g}. \tag{11.109}$$

Should the temperature of a given mixture cool beyond the temperature at which the relative humidity reaches 1.00, or 100%, known as the **dew point**, the excess water condenses out.

Extensive properties such as the total entropy and enthalpy equal the sums of the contributions from the various components of the mixture, including water vapor and, when it exists, liquid water. A treatment for water is given in Section 11.5.5.

11.5.4 Application to Refrigerants

Refrigerants change phase in their useful applications. Many refrigerants, including R-12 (the "freon" workhorse that is now outlawed for new equipment due to its alleged effect on ozone in the upper atmosphere), R-22 (a somewhat less offensive halogenated hydrocarbon that is scheduled to be phased out over the next 25 years) and R-134a (an emerging popular refrigerant[19] that does not attack ozone) have been characterized by the Martin-Hou equation of state, in which the specific volume v is the reciprocal of the density, ρ:

$$P = \frac{R\theta}{v-b} + \sum_{i=2}^{5} \frac{1}{(v-b)^i}\left(A_i + B_i\theta + C_i e^{-K\theta/\theta_c}\right) + \frac{A_6 + B_6\theta + C_6 e^{-K\theta/\theta_c}}{e^{\alpha v}(1 + ce^{\alpha v})}. \tag{11.110}$$

The saturation pressure for most refrigerants is given by

$$\ln P_{\text{sat}} = F_1 + \frac{F_2}{\theta} + \frac{F_3}{\theta^2} + F_4\ln\theta + F_5\theta + F_6\theta^2 + F_7\theta^3$$
$$+ \frac{F_8(F_9-\theta)}{\theta}\ln[(F_9-\theta)F_{10}], \tag{11.111}$$

in which $\log_{10}$ sometimes is substituted for ln. The density of the saturated liquid for most refrigerants is given by

$$\rho_f = \sum_{i=1}^{5} D_i X^{(i-1)/3} + D_6 X^{1/2} + D_7 X^2; \quad X = 1 - \theta/\theta_c, \tag{11.112}$$

and the density of the saturated vapor for most refrigerants is given by

$$\rho_g = \rho_c(\theta/\theta_c)^{E_{25}}\exp\left[\sum_{i=-10}^{13} E_{i+11}X^{i/3}\right]. \tag{11.113}$$

In Section 11.5.6 below a procedure is given for generating the coefficients E_i when they are not available, or when greater accuracy is desired than is available. Available relations for the specific heat for most refrigerants can be put into the form

$$c_v^0 = \sum_{i=1}^{7} G_i\theta^{i-3}. \tag{11.114}$$

To compute the entropy, a procedure similar to that used to give equation (11.92) (p. 898) can be used. Instead of adding and subtracting the term $-R\ln\rho$, however, $R\ln(v-b)$ is added and subtracted; the term ρR in the integrand is replaced by $R/(v-b)$, and $d\rho/\rho^2$ becomes $-dv$. Note also that the partial derivative of P with respect to θ for constant v must be computed from equation (11.110).

The entropy for the vapor states of substances described by equations (11.110) and (10.114) is

$$s = R\ln(v-b) + \sum_{i=2}^{5} \frac{B_i}{1-i}(v-b)^{1-i} - \frac{B_6}{\alpha e^{\alpha v}}$$
$$+ \frac{K}{\theta_c}e^{-K\theta/\theta_c}\sum_{i=2}^{5}\frac{C_i(v-b)^{1-i}}{(i-1)} - G_1/(2\theta^2) - G_2/\theta + G_3\ln(\theta) + G_4\theta$$
$$+ G_5\theta^2/2 + G_6\theta^3/3 + G_7\theta^4/4 + Y, \tag{11.115}$$

---

[19]Properties of R-134a are given by D.P. Wilson and R.S. Basun in "Thermodynamic Properties of a New Stratospherically Safe Working Fluid – Refrigerant 134a," *ASHRAE Transactions*, v. 94 Pt. 2 (1988), pp. 2095-2118.

where Y is a constant of integration that includes the datum state entropy, s_0. Entropy states for the saturated vapor, s_g, are calculated using the specific volume, v_g. Entropy states for a saturated mixture then can be computed from the Clapeyron relation,

$$s = s_g - (v_g - v)\frac{dP_{sat}}{d\theta}, \tag{11.116}$$

in which $dP_{sat}/d\theta$ is found by differentiating equation (11.111). The enthalpy for the vapor states of substances described by equations (11.110) and (11.114) is

$$h = \sum_{i=2}^{5}[A_i(v-b)^{(1-i)}/(i-1)] + JA_6/(ue^{uv})$$

$$+ J(1+K\theta/\theta_c)e^{-K\theta/\theta_c}\sum_{i=2}^{5}\frac{C_i(v-b)^{(1-i)}}{i-1} + JPv - G_1/\theta$$

$$.... + G_2\ln(\theta) + G_3\theta + G_4\theta^2/2 + G_5\theta^3/3 + G_6\theta^4/4 + G_7\theta^5/5 + X, \tag{11.117}$$

where X is the constant of integration that includes the datum state enthalpy.

In the two-phase saturated mixture region, the enthalpy is computed from

$$h = h_g - \theta(s_g - s). \tag{11.118}$$

Reynolds gives the coefficients for evaluating equation (11.112) for the density of the saturated liquid. From the Clapeyron relation,

$$s_f = s_g - \frac{dP}{d\theta}\left(v_g - \frac{1}{\rho_f}\right), \tag{11.119}$$

so that

$$h_f = h_g - \theta(s_g - s_f), \tag{11.120}$$

and the quality becomes

$$x = 1 - \frac{s_g - s}{s_g - s_f}. \tag{11.121}$$

These calculations are included in the downloadable master function file `propref.m`.

Data for R-12 and R-134a are given in the script files `R12data` and `R134a` in Appendix D (p. 1018). These are called by the program `refprop`, listed subsequently; one chooses to blank out the call to whichever file is not to be used. The subroutine `vapphase`, which also is called whenever the substance is in its vapor phase, is listed next. The programs evaluate all the critical properties and their several needed derivatives.

11.5.5 Application to Water

The Helmholtz free energy for water has been given by Keenan et al.[20] in the form

$$\psi = \psi_0(\theta) + R\theta[\ln\rho + \rho Q(\rho,\tau)], \tag{11.122a}$$

$$\psi_0 = \sum_{i=1}^{6} C_i/\tau^{i-1} + C_7\ln\theta + C_8\ln\theta/\tau \tag{11.122b}$$

$$Q = (\tau - \tau_c)\sum_{j=1}^{7}(\tau - \tau_{aj})^{j-2}\left[\sum_{i=1}^{8}A_{ij}(\rho - \rho_{aj})^{i-1} + e^{-E\rho}\sum_{i=9}^{10}A_{ij}\rho^{i-9}\right] \tag{11.122c}$$

$$\tau \equiv 1000/\theta. \tag{11.120d}$$

---

[20]J.H. Keenan et al., *Steam Tables*, Int. ed., 2d ed, Wiley, New York, 1978.

Although the International Association for the Properties of Water and Steam (IAPWS-IF1997) has replaced this model as the international standard for calculations in the power industry with a vastly more complex set of formulations,[21] the Keenan model should suffice for most purposes. Substitution into equations (11.95) and (11.92) gives

$$P = \rho R\theta \left[1 + \rho Q + \rho^2 \left(\frac{\partial Q}{\partial \rho} \right)_\tau \right], \tag{11.123a}$$

$$s = -R \left[\ln \rho + \rho Q - \rho \tau \left(\frac{\partial Q}{\partial \rho} \right)_\rho \right] - \frac{d\psi_0}{d\theta}, \tag{11.123b}$$

$$h = R\theta \left[\rho \tau \left(\frac{\partial Q}{\partial \tau} \right)_\rho + 1 + \rho Q + \rho^2 \left(\frac{\partial Q}{\partial \rho} \right)_\tau \right] + \frac{d(\psi\tau)}{d\tau}. \tag{11.123c}$$

The partial derivatives above are best carried out analytically.

The saturated mixture region can be addressed by using equations (11.118)–(11.121), which apply to any substance. The needed relations for the saturation pressure, density of the saturated liquid and approximate density of the saturated vapor are

$$P_{\text{sat}} = P_c \exp \left\{ \left(\frac{\theta_c}{\theta} - 1 \right) \sum_{i=1}^8 F_i \left[a(\theta - \theta_c) \right]^{i-1} \right\}, \tag{11.124a}$$

$$\rho_f = \rho_c \left[1 + \sum_{i=1}^8 D_i (1 - \theta/\theta_c)^{i/3} \right], \tag{11.124b}$$

$$\rho_g = \left(\frac{\theta}{\theta_c} \right)^{E_1} \exp \left[\sum_{i=2}^{15} E_i \left(1 - \frac{\theta}{\theta_c} \right)^{(i-1)/3} \right]. \tag{11.124c}$$

Coefficients for equations (11.122) and (11.124) are given in the downloadable program `waterdat.m` at the end of Appendix D. The equations are evaluated in the program `propwat`, which calls the program `steam.m`. Equation (11.124b) is due to Reynolds (footnote 16 p. 899), and equation (11.124) is due to the author. This last equation is accurate within 0.06% compared to the core model, producing modest errors in the derived properties. It could be improved.

11.5.6 Saturated Vapor Density for Other Substances

Reynolds does not give the needed relations for the density of the saturated vapor for the many substances he describes, unlike Stewart.[22] Wilson and Basun[23] also do not give the saturation vapor density for refrigerant R134a. A procedure is now given for finding such a relation in the form of equation (11.113) with $E_{25} = 0$. This form is essentially the same as that for water given in equation (11.124c). Then, in Example 12.6, coding is given with application to refrigerant R134a. This coding is readily adapted to other substances. The author also chose to replace the four coefficients given by Stewart for R12 with a more accurate set of nine coefficients found by this method.

---

[21] Wolfgang Wagner, *Properties of Water and Steam,* Springer-Verlag, 1998.

[22] R.B. Stewart et al., *ASHRAE Thermodyanmic Properties of Refrigerants,* American Society of Heating, Refrigeration and Air-Conditioning Engineers, 1988.

[23] D.P. Wilson and R.S. Basun, "Thermodynamic Properties of a New Stratospherically Safe Working Fluid – Refrigerant 134a," *ASHRAE Transactions,* v. 94 Pt. 2 (1988), pp. 2095-2118.

The procedure is as follows:

1. Choose a set of temperature ratios θ/θ_c over the range of interest, such as from 200K to slightly below the critical temperature.

2. Define the vector $y = (1 - \theta/\theta_c)^{1/3}$ corresponding to the values from step 1.

3. Use whatever relation is available to establish the vector of the saturation pressures corresponding to the values from step 1.

4. Use whatever equation of state that is available to establish a vector of the densities of the saturated vapor corresponding to the values from step 1. This requires the use of a Newton-Raphson iteration, which can be done without use of an analytical derivative.

5. Choose the values of i to be used in the model of equation (11.113). Define the corresponding vector $z = \ln(\rho_g/\rho_c) = \sum_{i=-10}^{13} E_{i+11} y_i$.

6. Set up the matrix A such that $AE = z$, where E is the vector of non-zero values of E_{i+11}. To be solvable, the number of rows in A must equal or exceed the number of columns; it is best to let the number exceed in order to get a least-squares best fit. Thus, the number of temperatures in step 1 best exceeds the number of values of E.

7. The desired answer can be found by the least-squares solution for the vector E using the MATLAB command E=A\b. It is wise to check your answer.

EXAMPLE 11.6
Determine coefficients E_{i+11} for the density of the saturated vapor of refrigerant R134a. Use the data given by Wilson and Basun as reproduced in the downloadable file R134adata.m listed in Appendix D (p. 1041), not including the answer of course which also is given on that page. Use the first nine values of the *subscript* used by Stewart for refrigerant R22, namely E_7, E_{11}, E_{14}, E_{20}, E_{21}, E_{22}, E_{23} and E_{24}.

Solution: Note that the matrix A is of the form

$$A = \begin{bmatrix} y_1^{-4} & 1 & y_1, & y_1^3 & y_1^9 & y_1^{10} & y_1^{11} & y_1^{12} & y_1^{13} \\ y_2^{-4} & 1 & y_2 & y_2^3 & y_2^9 & y_2^{10} & y_2^{11} & y_2^{12} & y_2^{13} \\ & & & & \cdot & & & & \\ & & & & \cdot & & & & \\ & & & & \cdot & & & & \\ y_n^{-4} & 1 & y_n & y_n^3 & y_n^9 & y_n^{10} & y_n^{11} & y_n^{12} & y_n^{13} \end{bmatrix}$$

The number of temperatures chosen (step 1) is three larger, at 12. The coding (which can be downloaded from the Internet; see page 1018) follows:

```
% file ex11_6.m gets coefficients for eq'n (11.113) (sat. vapor density)
global TC Pc Rc R B A2 A3 A4 A5 B2 B3 B5 C2 C3 C5 F1 F2 F5 F6 F8 F9
R134adata % gets data for description of refrigerant R134a
f0=10; v1=.004; % f0 is allowed error
th=[370,365,360,350,330,310,290,270,250,230,210,200]; % chosen temp's
y=(1-th/TC).^(1/3); % step 2; step 3 below based on eq'n (11.113)
Ps=exp(F1+F2./th+F5*th+F6*th.*th+F8*(F9-th)./th.*log(F9-th)); % step 3
for i=1:12 % run Newton-Raphson iteration, step 4.
    if i>2; f0=.04; end % tighten up on allowed error
    if i>1; v1=1.2*v(i-1); end % initial guesses for v(i)
    v0=v1; ex10_12P; f1=P-Ps(i); % call subroutine below for eq. of state
```

```
    v(i)=1.01*v1; v0=v(i); ex11_6P; f=P-Ps(i);
    while abs(f)>f0 % start actual iteration after set-up above
        r=f/f1; % ratio of new error to previous error
        v(i)=v1/(1-1/r)+v(i)/(1-r); % linear interpolation/extrapolation
        v0=v(i); ex10_12P; f1=f; % compute improved pressure
        f=P-Ps(i); % new error
    end
end
z=log(1/Rc./v); % step 5
A=[(y.^(-4))',ones(12,1),y',(y.^3)',(y.^9)',(y.^(10))',(y.^(11))'];
A=[A,(y.^(12))',(y.^(13))']; % step 6
E=A\z'; % step 7

% script file ex11_6P.m called above. Based on equation(11.110).
P=R*th(i)/(v0-B)+1/(v0-B)^2*(A2+B2*th(i)+C2*exp(-Ck*th(i)/TC));
P=P+1/(v0-B)^3*(A3+B3*th(i)+C3*exp(-Ck*th(i)/TC));
P=P+A4/(v0-B)^4+1/(v0-B)^5*(A5+B5*th(i)+C5*exp(-Ck*th(i)/TC));
```

11.5.7   Case Study: A Refrigeration Cycle

A crude model for the dynamics of a refrigeration cycle is given in Fig. 11.24. The evaporator is represented by the compliance element CS_1, the heater (which in practice assures that the compressor does not pump liquid) by element CS_2, the normally superheated inlet region of the condenser by elements CS_3 and CS_4 and the normally saturated-mixture region of the condenser by CS_5.   Thermal compliances for the metal shells of these components are

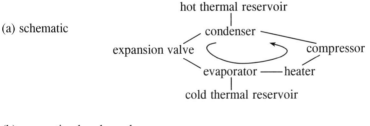

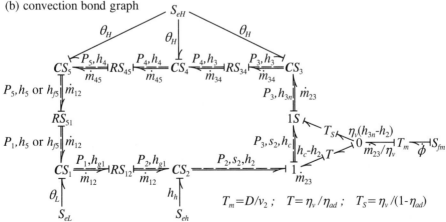

Figure 11.24: Simple model of a refrigeration cycle

omitted, for simplicity, but in practice could be added easily. Known temperature reservoirs are assumed for the hot and cold sides, and the heater power is given. An orifice, represented by the element RS_1, accomplishes the expansion process. Elements RS_2, RS_4 and RS_5 also are treated as virtual orifices (with much larger throat areas) to represent small frictional flow losses. The compressor is driven by a constant-speed shaft, and the moduli T and T_S are set to give a constant adiabatic efficiency, as described in Example 11.2 (p. 867) and Section 11.4.7 (pp. 886-889). The refrigerant is R-12, the properties of which are given in the downloadable computer file R12data.m in Appendix D. The evaporator is assumed to discharge only saturated vapor, because of gravity separation. The condenser discharges only saturated liquid whenever there is any, also assuming a gravity separation, but when the chamber is in the superheated region it emits vapor at its mean state. The assumed parameters of the system are listed in Table 11.1 below.

The state of the system before start-up has a uniform pressure throughout. Since the evaporator is at the temperature of the cold reservoir, this means that all liquid is located there; it was blown through the expansion valve when the compressor was last shut down. After start-up, however, considerable refrigerant slowly migrates through the compressor to the condenser, which finally enters the saturated mixture region and approaches the steady-state equilibrium. The temperatures are seen to approach their equilibrium values more quickly.

Table 11.1 Parameter values, refrigeration cycle

parameter	symbol	value
mass of refrigerant	MT	6 kg
specific heat ratio	K	1.3
volumetric displacement, compressor	D	2.305×10^{-4} m^3/rad
shaft speed, compressor	DPH0	28 rad/s
adiabatic efficiency, compressor	Eta	0.75
volume, chamber 1	V1	0.0060 m^3
volume, chamber 2	V2	0.0003 m^3
volume, chamber 3	V3	0.0006 m^3
volume, chamber 4	V4	0.0006 m^3
volume, chamber 5	V5	0.0060 m^3
heat cond. coefficient, chamber 1	CH1	622 W/K
heat cond. coefficient, chamber 3	CH3	4.07 W/K
heat cond. coefficient, chamber 4	CH4	13.895 W/K
heat cond. coefficient, chamber 5	CH5	849.3 W/K
temperature of cold reservoir	TEL	260 K
temperature of hot reservoir	TEH	295 K
heater power	HH	588.4 W
area, expansion valve	A51	1.246×10^{-6} m^2
area restriction, chambers 1 to 2	A12	2.385×10^{-4} m^2
area restriction, chambers 3 to 4	A34	3.762×10^{-4} m^2
area restriction, chambers 4 to 5	A45	3.790×10^{-4} m^2

The main MATLAB program for the simulation, called `refgmst`, is given below. (It can be downloaded from the Internet; see page 1018.) It defines the parameters, including running `R12data`, establishes initial conditions and calls the basic MATLAB integrator ode23. The state variables comprise the refrigerant masses in each of the first four chambers (the mass in the fifth equals the known total mass minus the first four masses) and the temperatures in each of the five chambers.

```
% refgmst.m, master program for simulating refrigerator start-up
% Five lumped thermal compliances assumed: one for the evaporator,
% one for the heater, and three for the condenser.
% The state variables are:
% x(1) is the mass in the evaporator
% x(2) is the mass in the heater
% x(3) is the mass in the condenser, superheated inlet section
% x(4) is the mass in the condenser, superheated exit section
% The mass in the condenser, nominal saturated mixture section,
% is m5 which equals the total mass minus the sum of the above.
% x(5) is the temperature in the section with mass x(1)
% x(6) is the temperature in the section with mass x(2)
% x(7) is the temperature in the section with mass x(3)
% x(8) is the temperature in the section with mass x(4)
% x(9) is the temperature in the section with mass m5
% The compressor runs at a constant speed DPH0, and has a fixed
% adiabatic efficiency Eta.
% The expansion  comprises flow through an orifice, either liquid
% (Bernoulli's equation) or unchoked compressible flow.
global MT K C D DPH0 E00 TEL TEH Vc Tc Eta
global HH V1 V2 V3 V4 V5 A12 A34 A45 A51 CH1 CH3 CH4 CH5
A12=2.385e-4; A34=3.762e-4; A45=3.79e-4; A51=1.246e-6; C=.35506;
V1=6e-3; V2=.3e-3; V3=.6e-3; V4=.6e-3; V5=6e-3; Eta=.75; HH=588.4;
vh=.099513; vg=.08539; x2=V2/vg; x3=V3/vh; x4=V4/vh; x5=V5/vh;
x1=(MT-x2-x3-x4-x5); K=1.3; C=.355; D=2.305e-4; DPH0=28; E00=1e-12;
CH1=622; CH3=4.07; CH4=13.895; CH5=849.3; TEL=260; TEH=295; MT=6;
Tc=260; Vc=.08; % values to start the simulation
R12data % gets the coefficients that describe the refrigerant
x0=[x1 x2 x3 x4 TEL TEL TEH TEH TEH]; % initial conditions
[t,x]=ode23s('refrgde',[0 14],x0); % integrator for stiff systems
```

The bulk of the programming for the refrigerator resides in the function m-file `refrgde`, which gives the differential equations. This program starts by computing the five specific volumes from the mass state variables. It then communicates these values and the corresponding five temperatures to the routine `propref` and receives back all the needed (and some unneeded) derivative state variables for each of the five chambers. The mass flow rate through the compressor (`dm23`) then can be computed:

```
function f=refrgde(t,x) % differential equations for refgmst.m
global MT K C D DPH0 E00 TEL TEH Vc Tc Eta
global HH V1 V2 V3 V4 V5 A12 A23 A34 A45 A51 CH1 CH2 CH3 CH4 CH5
m5=MT-x(1)-x(2)-x(3)-x(4); % mass in chamber 5;
v1=V1/x(1); v2=V2/x(2); v3=V3/x(3); v4=V4/x(4); v5=V5/m5;
[P1,h1,hg1,s1,sg1,vf1,vg1,hf1,dsdv1,dsdT1,dudv1,dudT1,dPs1,d2Ps1,
```

```
    dvdg1,dsg1,dugdv1,dugdT1,dPdv1,dPdT1]=refprop(v1,x(5));
[P2,h2,hg2,s2,sg2,vf2,vg2,hf2,dsdv2,dsdT2,dudv2,dudT2,dPs2,d2Ps2,
    dvdg2,dsg2,dugdv2,dugdT2,dPdv2,dPdt2]=refprop(v2,x(6));
[P3,h3,hg3,s3,sg3,vf3,vg3,hf3,dsdv3,dsdT3,dudv3,dudT3,dPs3,d2Ps3,
    dvdg3,dsg3,dugdv3,dugdT3,dPdv3,dPdT3]=refprop(v3,x(7));
[P4,h4,hg4,s4,sg4,vf4,vg4,hf4,dsdv4,dsdT4,dudv4,dudT4,dPs4,d2Ps4,
    dvdg4,dsg4,dugdv4,dugdT4,dPdv4,dPdt4]=refprop(v4,x(8));
[P5,h5,hg5,s5,sg5,vf5,vg5,hf5,dsdv5,dsdT5,dudv5,dudT5,dPs5,d2Ps5,
    dvdg5,dsg5,dugdv5,dugdT5,dPdv5,dPdT5]=refprop(v5,x(9));
dm23=D*DPH0/v2; % the flow pumped by the compressor;
```

The program continues with the tricky job of finding the enthalpy $h_{ideal} \equiv$ hc that would exist downstream of the compressor were that machine operating reversibly. As described in Section 11.4.7 (p. 887), the entropy and the pressure of this state equal s_2 and P_3, respectively; a Newton-Raphson iteration with full evaluation of derivatives in the appropriate state regime (superheated or saturated mixture) is carried out. The actual enthalpy h3n of the outflow from the compressor, resulting from an adiabatic efficiency of 75%, is then computed:

```
% Start by using values of Vc and Tc from the previously
% determined state as the first trial:
[Pc,hc,hgc,sc,sgc,vfc,vgc,hfc,dsdvc,dsdTc,dudvc,dudTc,dPsc,d2Psc,
    dvdgc,dsgc,dugdvc,dugdTc,dPdvc,dPdTc]=refprop(Vc,Tc);
E=(Pc-P3)^2*1e-12+(sc-s2)^2*1e-4; % weighted error value
vc=Vc;
while E>E00 % iterate to get vc and Tc for Pc=P3 and sc=s2
    M=[dPdvc dPdTc;dsdvc dsdTc];
    q=M\[Pc-P3;sc-s2];
    vc=vc-q(1);
    Tc=Tc-q(2);
    [Pc,hc,hgc,sc,sgc,vfc,vgc,hfc,dsdvc,dsdTc,dudvc,dudTc,dPsc,
    d2Psc,dvdgc,dsgc,dugdvc,dugdTc,dPdvc,dPdTc]=refprop(vc,Tc);
    E=(Pc-P3)^2*1e-12+(sc-s2)^2*1e-4;
end
h3n=hc/Eta+(1-1/Eta)*h2; % compressor enthalpy (eqn. (11.79))
```

The gravity separation that sometimes takes place at the outlet of the condenser is then recognized. Next, the program orifice is called repeatedly to compute the flows through the four actual or virtual fixed orifices, including the expansion valve. (This downloadable program, listed in Appendix D, is more general than is needed for the present simulation; it permits the inflow to be a saturated mixture or a compressed liquid as well as a saturated liquid or a superheated vapor.) This leads directly to the differential equations for the rates of change of mass in chambers 1 – 4, that is the first four state variables (note that the rate of change in the fifth chamber equals minus the sum of the other four):

```
if v5<vg5
    h5e=hf5; % sets the enthalpy of condenser outflow to that for
    % saturated liquid
else
    h5e=h5;
end
```

```
[dm12]=orifice(A12,P1,P2,v1,v2,vf1,vf2,vg1,vg2,x(5),x(6));
[dm34]=orifice(A34,P3,P4,v3,v4,vf3,vf4,vg3,vg4,x(7),x(8));
[dm45]=orifice(A45,P4,P5,v4,v5,vf4,vf5,vg4,vg5,x(8),x(9));
[dm51]=orifice(A51,P5,P1,v5,v1,vf5,vf1,vg5,vg1,x(9),x(5));
f(1)=dm51-dm12;
f(2)=dm12-dm23;
f(3)=dm23-dm34;
f(4)=dm34-dm45; % Completes first four differential equations.
f5=-f(1)-f(2)-f(3)-f(4); % rate of change of mass in chamber 5
% (The mass in chamber 5 is not a state variable)
```

Finally, the five differential equations for the chamber temperatures are given. Recall, from Section 11.4.4, that different equations are used for the superheated vapor and saturated mixture regions. Which to use is determined by comparing the specific volume v with the specific volume for saturated vapor vg at the same temperature:

```
den=x(1)*((dugdv1+P1-x(5)*dPs1)*dvdg1+dugdT1-x(5)*d2Ps1*(vg1-v1));
f(5)=(CH1*(TEL-x(5))+(h5e-h1)*dm51-(hg1-h1)*dm12
    +x(5)*dPs1*v1*f(1))/den;
if v2>vg2
    f(6)=(HH+(hg1-h2)*dm12+(P2+dudv2)*v2*f(2))/dudT2/x(2);
else
    den=x(2)*((dugdv2+P2-x(6)*dPs2)*dvdg2+dugdT2
        -x(6)*d2Ps2*(vg2-v2));
    f(6)=(CH1*(TEL-x(6))+(hg1-h2)*dm12+x(6)*dPs2*v2*f(2))/den;
end
f(7)=(CH3*(TEH-x(7))+(h3n-h3)*dm23+(P3+dudv3)*v3*f(3))/dudT3/x(3);
f(8)=(CH4*(TEH-x(8))+(h3-h4)*dm34+(P4+dudv4)*v4*f(4))/dudT4/x(4);
if v5>=vg5
    f(9)=(CH5*(TEH-x(9))+(h4-h5)*dm45-(h5e-h5)*dm51
        +(P5+dudv5)*v5*f5)/dudT5/m5;
else
    den=m5*((dugdv5+P5-x(9)*dPs5)*dvdg5+dugdT5-x(9)*d2Ps5*(vg5-v5));
    f(9)=(CH5*(TEH-x(9))+(h4-h5)*dm45-(h5e-h5)*dm51
        +x(9)*dPs5*v5*f5)/den;
end
f=f';
Vc=vc;
```

Selected results are plotted in Fig. 11.25. Temperature equilibrium is approached long before mass equilibrium; only the former occurs (virtually) in the 14 seconds shown. The condenser reaches saturation state at about three seconds, and holds there for awhile, rapidly alternating between discharging liquid and vapor. After eight seconds the discharge is pure liquid. Were less refrigerant used, mass equilibrium would occur sooner, and the blow-down process following shut-down also would be faster.

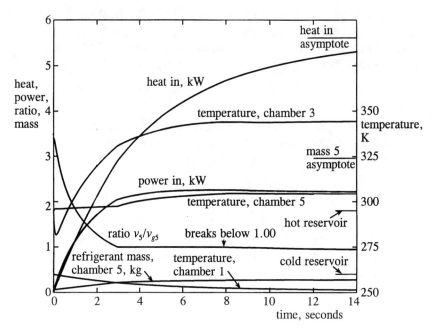

Figure 11.25: Start-up of the refrigeration system

11.5.8   Application to the Liquid Region

Two different approaches can evaluate properties in the liquid region. The first uses either the 'P–v–T' or the Helmholtz formulation, which may or may not apply in this region for a particular substance, depending upon the data used in its generation. The second approach uses a simple model including a specific heat and a bulk modulus. The first approach is most apt for states near the critical point, where the specific heat and bulk modulus are far from constant. The simpler second approach is usually satisfactory for lower temperatures.

Direct use of a P–v–T equation can give an inaccurate pressure, since in the liquid region pressure is extremely sensitive to small changes in density or specific volume. For a discussion and procedure to increase accuracy, see Reynolds.[24]

The simpler model gives the pressure as the sum of the saturation pressure at the same temperature plus a pressure difference due to compression with a constant bulk modulus, β:

$$P = P_{sat} + \beta \ln\left(\frac{v_{sat}}{v}\right). \tag{11.125}$$

The internal energy, u, equals the internal energy of the saturated liquid at the same temperature, defined as $u_{f\theta}$, plus the work done in compressing the liquid:

$$u = u_{f\theta} + \int_{v_f}^{v} P\,dv = u_{f\theta} + \beta\left[v - v_f + v\ln\left(\frac{v_f}{v}\right)\right]. \tag{11.126}$$

The entropy is unchanged in an assumed reversible process, and the enthalpy can be found from its definition $h = u + Pv$. The term $\partial u/\partial\theta$, which is used in the differential equation for temperature (equation (11.65), p. 881), is assumed to be a constant specific heat, c, values of which usually are available. The values of c_p and c_v are assumed to be the same, which is exactly true only for incompressible substances.

---

[24]See footnote 16 (p. 899).

The value of the bulk modulus may not be readily available but can be estimated from data for saturated liquid. Defining the internal energy at the saturated state that has the same specific volume, v (rather than $u_{f\theta}$, which is at the same temperature) as u_{fv}, the internal energy is approximately

$$u = u_{fv} + c(\theta - \theta_{fv}),\tag{11.127}$$

where θ_{fv} is the temperature of that saturated state. Equating the right sides of equations (11.126) and (11.127) gives, after a little algebra,

$$\beta = \frac{2[u_{fv} - u_{f\theta} + c(\theta - \theta_{fv})]}{v(1 - v/v_{f\theta})^2}.\tag{11.128}$$

A sample calculation uses tabular data for refrigerant R12 at the temperatures of $-20°C$ and $20°C$:

$$\beta = \frac{2[17,720 - 54,440 + 977(20 - -20)]}{0.0006855(1 - 0.0006855/0.0007525)} = 0.689 \times 10^9 \text{ Pa } = 126,000 \text{ psi.}\tag{11.129}$$

This value is almost half that of water, showing that the liquid is nearly incompressible for most purposes. As a practical matter, a simulation involving occasional compressed liquid states with a bulk modulus this high could be impeded by the "stiffness" of the differential equation set. Little error would be introduced by the expedient of decreasing the assumed bulk modulus. Also, recognize that the effective bulk modulus is apt to be significantly lower than its theoretical value, anyway, because of the unaccounted effects of the expansion of the container due to pressure-induced stresses and the microscopic vapor bubbles that usually reside in surface cracks.

11.5.9 Considerations of Reversing Flows

The differential equations for temperature include terms for the enthalpy fluxes entering and leaving the control volume. When a flow reverses, the differential equation itself needs to be changed, since the source of one or more fluxes changes. The switching can be implemented through the use of *if* statements.

Sometimes these *if* statements result in a false simulation, however. This occurs when the different evaluations of a flow rate within a single time step, as organized by the Runge-Kutta integration algorithm, produce different signs. This would not be meaningful physically. You are urged to keep a vigil for this phenomenon, and to take corrective action should it occur. The first telltale sign may be a distinct sensitivity of a result to the step sizes (or error index) used. You also can display the running values of a flow on the screen, and directly witness a rapidly alternating sign.

A simple remedy for this phenomenon is to put the the flow rates or the enthalpy fluxes themselves through a simple smoothing filter. The model is reasonable as long as the assumed time constant for such a filter is shorter than the residence time of the chamber. For the time constant to be effective, on the other hand, it should be larger than the time steps. This procedure does not guarantee elimination of the phenomenon, but it radically curtails any errors than might otherwise accrue.

The use of upwinding in channels with alternating CS and IRS elements also requires changes in the energy equation upon flow reversal, as noted in Section 11.4.9.

$$S_e \Rightarrow \dashv AC \mapsto\!\!\!\succ CS_2 \Rightarrow\!\!\dashv IRS_2 \mapsto\!\!\!\succ CS_3 \Rightarrow\!\!\dashv \quad \bullet\bullet$$

$$\bullet\bullet \mapsto\!\!\!\succ CS_{n-1} \Rightarrow\!\!\dashv IRS_{n-1} \mapsto\!\!\!\succ CS_n \Rightarrow\!\!\dashv AC \Rightarrow\!\!\dashv RS \mapsto\!\!\!\succ S_e$$

Control volumes for **CS** elements are shown by solid lines;
control volumes for **IRS** elements are shown by dashed lines.

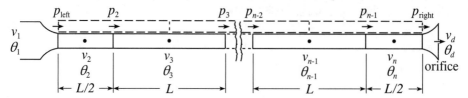

Figure 11.26: Fluid line with two-phase fluid for dynamic simulation

11.5.10   Simulation with Fluid Kinetic Energy

Simulation including the effects of kinetic energy is illustrated below with the example of a long fluid line containing refrigerant R12, as introduced in Section 11.4.9 (pp. 889-892). Figure 11.21 is reproduced in Fig. 11.26, with the overlapping control volumes for the CS and IRS elements shown in parallel rather than superimposed, for clarity. The 10-meter-long tube is divided into 40 equal control volumes for IRS elements and 41 control elements for CS elements; the first and last CS elements have half the length of the others. The radius of the tube is 5.0 mm and the area of the downstream orifice is 5.0 mm^2. The refrigerant downstream of the orifice is a saturated vapor at room temperature (293 K). The refrigerant in the tube initally is at the same pressure, but its temperature is 10 K higher. The upstream refrigerant, which is at a temperature of 345 K and the considerably higher pressure of 1.4 MPa, is admitted to the upstream end of the tube starting at the time $t = 0$. Heat transfer is neglected, for simplicity, although the coding includes the modeling of effective thermal conductivities in the sometimes turbulent flow. (This is a critical computation when heat transfer is included, which it must be, for example, in modeling the dynamic behavior of a condenser in a refrigeration system.) Wall shear is included in the analysis, however, because it plays a critical role regardless of heat transfer.

Selected results of the simulation are plotted in Fig. 11.27. The initial pressure wave reaches the right end of the tube in only 0.062 seconds, but the equilibrium transport time for fluid to travel the length of the tube is 1.93 seconds, and it takes roughly 3 seconds for equilibrium to be reached. A smaller orifice would increase the latter two times.

The MATLAB files listed below can be downloaded from the Internet. The first, the master file that sets up the MATLAB ODE45 simulation, is largely self-explanatory. Note that the first 42 state variables are the masses in the CS elements, the second 42 are the respective temperatures, and the final 41 are the momenta of the IRS elements. If heat transfer to the walls and then to a constant-temperature environment is added, an additional 42 state variables can be added for the wall temperatures. Note also that the amount of output data, which could be huge, is sharply limited by retaining results only at the ends of each time interval of one millisecond.

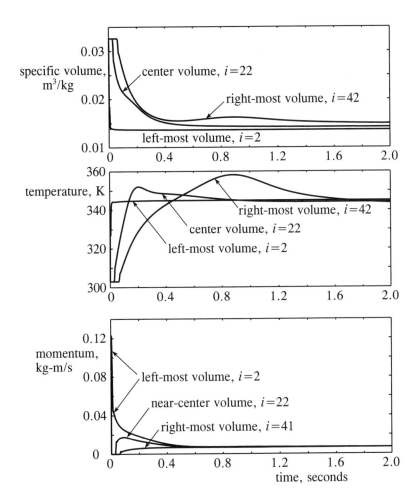

Figure 11.27: Selected simulation results for adiabatic flow in the refrigerant-filled tube

```
% shocktubem.m, master program for simulation of shock tube with refrigerant
% Use with function file shocktube.m and files for refrigerant
% x(1) to x(n) are the massess in the sections 1 to n
% x(n+1) to x(2*n) are the temperatures in the sections 1 to n
% x(2*n+1) to x(3*n-1) are the momenta in the sections 1 to n-1
global F V n A0 P0D v0D vf0D vg0D T0D
n=42; % first section is upstream; there are 40 full-length sections in tube
for i=1:n; V(i)=pi*.005^2*10/(n-2); end % volumes of most sections
V(2)=V(1)/2; V(n)=V(2); % except sections 2 and n are half-length
A0=10*.5e-6; % area of downstream orifice; properties there follow:
P0D=5.64630e5; v0D=.03092; vf0D=7.5212e-4; vg0D=0.03092; T0D=273.0;
vh=.032605; TEH=303; % initial specific volume and temperature in tube
x0=[V(1)/.01360 V(2)/vh V(1)/vh*ones(1,n-3) V(n)/vh]; % initial masses
Tp=[345 TEH*ones(1,n-1)]; % initial temperatures including upstream
x0=[x0 Tp 0*ones(1,n-1)]; % completed initial state vector
R12data % for refrigerant R12; alternatively could call R134adata
F=3/4; % upwinding factor: 1 is complete upwinding, 0.5 is none
```

```
options=odeset('RelTol',1e-8,'AbsTol',.0001,'InitialStep',.000001);
dataZ(1,:)=x0;
for i=1:5000 % sets large number of sequential short runs to compress output
    i % run-time indicator of progress
    [t,x]=ode45('shocktube',.001,x0,options);
    s=size(t);
    x0=x(s(1,1),:);
    dataZ(i+1,:)=x0; % output data at millisecond time increments
end
```

The function file called above starts by calling the refrigerant property files in order to establish all the relevant properties in all of the CS elements. The program handles vapor, saturated mixture and liquid phases, although only the vapor phase exists in the example.

```
function f=shocktube(t,x) % differential equations for shocktubem.m
global F V n A0 POD v0D vf0D vg0D T0D
FF=1-F; % for convenience
beta=6e6; % effective bulk modulus for the liquid phase
r=.005; L=V(1)/pi/r/r; R=pi*r^4/8/L; A=pi*r*r; % dimensions of tube
for i=1:n
    v(i)=V(i)/x(i); % specific volumes
end
v(n+1)=(4*v(n)-v(n-1))/3; % at the right end
for i=1:n
    temp(i)=x(n+i); % temperatures, to aid one's memory
end
temp(n+1)=(4*temp(n)-temp(n-1))/3; % at the right end
for i=1:n+1
    [P(i),h(i),hg(i),s(i),sg(i),vf(i),vg(i),hf(i),dsdv(i),dsdT(i),dudv(i),
        dudT(i),dPs(i),d2Ps(i),dvg(i),dsg(i),dugdv(i),dugdT(i),dPdv(i),
        dPdT(i)]=refprop(v(i),temp(i));
    if v(i)<vf(i)
        P(i)=P(i)+beta*(1-v(i)/vf(i));
    end
end % establishes all thermodynamic states except at left end of tube
```

Next, the states at the left end of the tube, which are not state variables, are found by invoking the conditions for the area change element (AC) given by equation (11.84) (p. 892). This requires a two-variable Newton-Raphson iteration, which employs analytical derivatives that are included in the return to the call for the program refprop:

```
P0=(4*P(2)-P(3))/3; % pressure at left end of tube
vleft=(4*v(2)-v(3))/3; % first guess at specific volume at left end of tube
Tleft=(4*temp(2)-temp(3))/3; % ditto for temperature
[Pleft,hleft,hgl,sleft,sgl,vfl,vgl,hfl,dsdvl,dsdTl,dudvl,dudTl,dPsl, d2Psl,
    dvgl,dsgl,dugdvl,dugdTl,dPdvl,dPdTl]=propref2(vleft,Tleft);
Perror=Pleft-P0; % initial error
serror=sleft-s(1); % initial error
while abs(Perror)>1 | abs(serror)>.001 % iteration to find state at left end
    den=dPdvl*dsdTl-dPdTl*dsdvl;
    deltav=1*(dPdTl*serror-dsdTl*Perror)/den;
    deltaT=1*(dsdvl*Perror-dPdvl*serror)/den;
```

```
    vleft=vleft+deltav;
    Tleft=Tleft+deltaT;
    [Pleft,hleft,hgl,sleft,sgl,vfl,vgl,hfl,dsdvl,dsdTl,dudvl,dudTl,dPsl,
            d2Psl,dvgl,dsgl,dugdvl,dugdTl,dPdvl,dPdTl]=propref2(vleft,Tleft);
    Perror=Pleft-P0;
    serror=sleft-s(1);
end % completes Newton-Raphson iteration
Vleft=sqrt(2*(h(1)-hleft)); % fluid velocity at left end of tube
mfleft=A*Vleft/vleft; % mass flow rate at left end of tube
```

The biggest challenge, perhaps, is to properly treat the area-change element upstream of the orifice, which is necessitated by an orifice-flow equation based on stagnation upstream properties. The difficulty is that the difference between the enthalpy at the downstream end of the tube and the stagnation enthalpy depends on the fluid velocity, which in turn depends on the flow rate, which depends on the stagnation properties. Rather than carry out a nested set of two Newton-Raphson iterations, the author chose to estimate the velocity from a simple linear extrapolation of the fluid momenta in the last two IRS elements. This should be reasonably accurate, and the integration scheme uses the resulting computation of the orifice flow to correct whatever error results.

It is still tricky to determine a first estimate of the specific volume and the temperature of the stagnation state from the assumed enthalpy and entropy, so that the Newton-Raphson iteration will converge reliably. The difference between the entropies at the tube end and the stagnation state is zero, so that

$$\Delta s = 0 \simeq \frac{\partial s}{\partial \theta}\Delta\theta + \frac{\partial s}{\partial v}\Delta v, \tag{11.130}$$

which reduces the number of unknowns to one:

$$\Delta\theta \simeq -\frac{\partial s/\partial v}{\partial s/\partial \theta}\Delta v. \tag{11.131}$$

The difference in pressure similarly can be related to Δv:

$$\Delta P = \frac{\partial P}{\partial \theta}\Delta\theta + \frac{\partial P}{\partial v}\Delta v \simeq \left(\frac{\partial P}{\partial v} - \frac{\partial P}{\partial \theta}\frac{\partial s/\partial v}{\partial s/\partial \theta}\right)\Delta v \tag{11.132}$$

It is now necessary to relate the known change in enthalpy, Δh, to Δv. Derivatives of h are not returned by refprop (although they could be, and are in fact computed in the code vapphase), so a little maneuvering is needed, based on $h \equiv u + Pv$:

$$\Delta h \simeq \Delta u + P\Delta v + v\Delta P \simeq \frac{\partial u}{\partial \theta}\Delta\theta + \left(\frac{\partial u}{\partial v} + P\right)\Delta v + v\Delta P. \tag{11.133}$$

Introducing equations (11.131) and (11.132) gives, after a little rearrangement,

$$\Delta h \simeq \left[\frac{\partial u}{\partial v} + P + v\frac{\partial P}{\partial v} - \left(\frac{\partial u}{\partial \theta} + v\frac{\partial P}{\partial \theta}\right)\frac{\partial s/\partial v}{\partial s/\partial \theta}\right]\Delta v. \tag{11.134}$$

This equation is inverted to solve for Δv given Δh. All the terms within the square brackets are returned by refprop.

The coding, including a single Newton-Raphson iteration, follows. The final line of code calls the program orifice (also given in Appendix D) to compute the mass flow through the orifice.

```
hdyn=1/2*((3*x(3*n-1)-x(3*n-2))/2/(x(n)+x(n-1)/2))^2; % change in h
dvstag=hdyn/(dudv(n+1)+P(n+1)+v(n+1)*dPdv(n+1)-(dudT(n+1)+v(n+1)*dPdT(n+1))
     *dsdv(n+1)/dsdT(n+1)); % first estimate of change in v
vstag=v(n+1)+dvstag; % first estimate of stagnation v
Tstag=temp(n+1)-dsdv(n+1)/dsdT(n+1)*dvstag; % associated stagnation temp
[Pstag,hstag,hgs,sstag,sgs,vfs,vgs,hfs,dsdvs,dsdTs,dudvs,dudTs,dPss,d2Pss,
     dvgs,dsgs,dugdvs,dugdTs,dPdvs,dPdTs]=propref2(vstag,Tstag);
herror=hstag-h(n+1)-hdyn; % error in first estimate of h
serror=sstag-s(n+1); % error in first estimate of s (actually 0)
while abs(serror)>.01 | abs(herror)>.1 % set up Newton-Raphson
     dhdvs=dudvs+Pstag+vstag*dPdvs; dhdTs=dudTs+vstag*dPdTs;
     den=dhdvs*dsdTs-dhdTs*dsdvs;
     deltav=-(dsdTs*herror-dhdTs*serror)/den;
     deltaT=-(dhdvs*serror-dsdvs*herror)/den;
     vstag=vstag+deltav;
     Tstag=Tstag+deltaT;
     [Pstag,hstag,hgs,sstag,sgs,vfs,vgs,hfs,dsdvs,dsdTs,dudvs,dudTs,dPss,
          d2Pss,dvgs,dugs,dugdvs,dugdTs,dPdvs,dPdTs]=propref2(vstag,Tstag);
     herror=hstag-h(n+1)-hdyn;
     serror=sstag-s(n+1);
end
mfright=orifice(A0,Pstag,P0D,vstag,v0D,vfs,vf0D,vgs,vg0D,Tstag,T0D);
```

The differential equations for the masses in the CS elements now can be evaluated:

```
f(1)=0; % mass 1 is constant
f(2)=mfleft-x(2*n+2)/L; % differential equation for mass in first volume
for i=3:n-1
    f(i)=(x(2*n+i-1)-x(2*n+i))/L; % d.e's for masses in other volumes
end
f(n)=x(3*n-1)/L-mfright; % d.e. for mass in final volume
```

The momentum equations are more involved, since they require knowledge of the wall shear forces. The first step in modeling these forces is to estimate an equivalent single-phase viscosity of the fluid, should the fluid become a saturated mixture. Linear interpolations for the viscosity and thermal conductivity of saturated mixtures are assumed,

$$\mu = \mu_f(1-\alpha) + \mu_g, \tag{11.135a}$$
$$k = k_f(1-\alpha) + k_g\alpha, \tag{11.135b}$$

where μ_f, k_f and μ_g, k_g are the saturated liquid and saturated vapor values, respectively, at the same temperature, and α is the volume fraction of vapor:

$$\alpha = (1 - v_f/v)/(1 - v_f/v_g). \tag{11.136}$$

Data given by Stoeker and Jones[25] for refrigerant R12 lead to the empirical relations

$$\mu_f = -1.331 \times 10^{-10}T^2 + 2.1984 \times 10^{-8}T^2 - 2.4666 \times 10^{-6}T + 2.6644 \times 10^{-4}, \tag{11.137a}$$
$$\mu_g = 6.25 \times 10^{-12}T^3 - 2.5 \times 10^{-10}T^2 + 4.25 \times 10^{-8} + 1.18 \times 10^{-5}. \tag{11.137b}$$
$$k_f = 0.0784 - 3.67 \times 10^{-4}T \tag{11.137c}$$
$$k_g = 1.1410^{-7}T^2 + 5.063 \times 10^{-5}T + 0.0083. \tag{11.137d}$$

The Reynold's numbers, designated NRe, also are established. The coding is

[25]W.F. Stoeker and F.W. Jones, *Refrigeration and Air Conditioning*, second ed., McGraw-Hill, 1982.

```
for i=1:n % set flags that identify the fluid phase in each CS element
    if vg(i)==0; alpha(i)=1; X(i)=1;
    elseif v(i)>vf(i);  alpha(i)=(1-vf(i)/v(i))/(1-vf(i)/vg(i));
    else; alpha(i)=0;
    end
    if alpha(i)>1; alpha(i)=1; end
    if vg~=0
        X(i)=alpha(i)*v(i)/vg(i);
        if X(i)>1; X(i)=1; end
    else; X(i)=1;
    end
    T1(i)=x(n+i)-273.15; T2(i)=T1(i)^2; T3(i)=T1(i)*T2(i);
    % for liquid phase,
    mul(i)=-1.331e-10*T3(i)+2.1984e-8*T2(i)-2.4666e-6*T1(i)+2.6644e-4;
    muv(i)=6.25e-12*T3(i)-2.5e-10*T2(i)+4.25e-8*T1(i) +1.18e-5; % vapor phase
    kl(i)=-3.67e-4*T1(i)+.0784; % liquid phase
    kv(i)=1.14e-7*T2(i)+5.063e-5*T1(i)+.0083; % vapor phase
    mu(i)=mul(i)*(1-alpha(i))+muv(i)*alpha(i); % viscosity
    k(i)=kl(i)*(1-alpha(i))+kv(i)*alpha(i); % thermal conductivity
end
mup(2)=(2*mu(2)+mu(3))/3; % viscosity at center of IRS element 2
mup(n-1)=(2*mu(n)+mu(n-1))/3; % viscosity at center of IRS element n-1
NRe(n-1)=abs(2*x(3*n-1))/L/pi/r/mup(n-1); % for use in momentum eq'n
for i=3:n-2
    mup(i)=(mu(i)+mu(i+1))/2; % viscosities at centers of other IRS elements
    xp(i)=(x(i)+x(i+1))/2; % masses at centers of IRS elements
    NRe(i)=abs(2*x(2*n+i))/L/pi/r/mup(i); % for use in dp/dt only (not Q)
end
```

The wall shear for laminar flow is assumed to agree with the classical assumption for fully developed flow in a round tube. A model due to Levy[26] is used for the turbulent regime. The respective equations are

$$F_w = 8\mu vp/r^2, \qquad |Re| < 2000, \tag{11.138a}$$
$$F_w = 0.395 \mathrm{sign}(p)(L\mu)^{0.2}|p|^{1.8}/(r^{0.8}m), \qquad |Re| > 2000. \tag{11.138b}$$

The momentum equations are coded as

```
pm(2)=L*mfleft; % momentum at left side of IRS element 2 (left end of tube)
for i=3:n-1
    pm(i)=(x(2*n+i-1)+x(2*n+i))/2; % momenta at left sides of other elements
end
if NRe(2)>2000 % criterion for turbulent flow
    Fw=sign(x(2*n+2))*.395*(L*(2*mu(2)+mu(3))/3)^.2*abs(x(2*n+2))^1.8/(x(2)
        +x(3)/2)/r^.8;
else
    Fw=A*(2*mu(2)+mu(3))*(2*v(2)+v(3))/9/R*x(2*n+2)/L; % for laminar flow
end
f(2*n+2)=A*(Pleft-P(3))-Fw+mfleft*Vleft-(x(2*n+2)+x(2*n+3))^2/4/x(3)/L;
for i=3:n-2
```

---

[26] Levy, Salomon, *Two-Phase Flow in Complex Systems*, John Wiley and Sons, 1999.

```
        vp(i)=(v(i)+v(i+1))/2;
        if NRe(i)>2000
            Fw=sign(x(2*n+i))*.395*(L*mup(i))^0.2*abs(x(2*n+i))^1.8/xp(i)/r^.8;
        else
            Fw=A*mup(i)*vp(i)/R*x(2*n+i)/L;
        end
        vp(i)=(v(i)+v(i+1))/2;
        f(2*n+i)=A*(P(i)-P(i+1))-Fw+pm(i)^2/x(i)/L-pm(i+1)^2/x(i+1)/L;
end
if NRe(n-1)>2000
    Fw=sign(x(3*n-1))*.395*(L*(mu(n-1)+2*mu(n))/3)^.2*abs(x(3*n-1))^1.8
        /(x(n-1)/2+x(n))/r^.8;
else
    Fw=A*(mu(n-1)+2*mu(n))*(v(n+1)+2*v(n))/9/R*x(3*n-1)/L;
end
f(3*n-1)=A*(P(n-1)-P(n+1))-Fw+(x(3*n-2)+x(3*n-1))^2/4/x(n-1)/L;
f(3*n-1)=f(3*n-1)-mfright^2*v(n+1)/A;
```

Finally, the energy equations, posed as differential equations for the temperatures, must be evaluated. This completes the program:

```
f(n+2)=((hleft-h(2))*mfleft-2*FF/3*(h(3)-h(2))*x(2*n+2)/L+(P(2)
    +dudv(2))*v(2)*f(2))/dudT(2)/x(2); % d.e. for T(2);
f(n+3)=(((1+2*F)/3*(h(2)-h(3))*x(2*n+2)-FF*(h(4)-h(3))*x(2*n+3))/L+(P(3)
    +dudv(3))*v(3)*f(3))/dudT(3)/x(3); % d.e. for T(3);
for i=4:n-2
    f(n+i)=((F*(h(i-1)-h(i))*x(2*n+i-1)-FF*(h(i+1)-h(i))*x(2*n+i))/L+(P(i)
        +dudv(i))*v(i)*f(i))/dudT(i)/x(i); % d.e.'s for other T's
end
f(2*n-1)=((F*(h(n-2)-h(n-1))*x(3*n-2)-4/3*FF*(h(n)-h(n-1))*x(3*n-1))/L
    +(P(n-1)+dudv(n-1))*v(n-1)*f(n-1))/x(n-1)/dudT(n-1);
f(2*n)=(((h(n-1)-h(n))*(4*F-1)/3*x(3*n-1)+(h(n+1)-h(n))*2*(1-F)*mfright*L)/L
    +(P(n)+dudv(n))*v(n)*f(n))/x(n)/dudT(n); % d.e. for T(n)
f=f'; % req'd inversion of vector of time derivatives of the state vector
```

11.5.11   Heat Transfer Equations for Three-Phase CS Elements

There is no universally accepted model for heat transfer to or from two-phase flows. Models cited by Holman[27] for vapor, condensing and liquid flows in horizontal tubes are given below. They assume no slip between the velocities of the two components.

Application is made to the fluid transient problem above, rendering it essentially as a condenser in a refrigeration system. This illustrates complex behavior, with all three phases occurring at the same time. The bond graph model given previously in Fig. 11.19 (p. 890) applies. The fluid transfers heat to the walls of the tube, which then transfer heat to the environment, which has a fixed temperature of 303 K. Axial heat conduction in the walls is neglected. The coding is added to the files shocktubem.m and shocktube.m listed above and renamed as condenserm.m and condenser.m. These are included in the list of Internet downloadable files on page 1018 but are not reproduced here. The heat transfer coefficient for the outside of the tube is assumed to be 30 J/s·K·m. The value for each full-length section is designated as CH. The thermal compliance of the walls is taken as a representative

---
[27] J.P. Holman, *Heat Transfer*, fifth ed., McGraw-Hill, 1981.

173 J/K·m. The value for section i is designated as CC(i). The compressibility of the liquid refrigerant is exaggerated to keep the equations from becoming too stiff, with little effect on the results. The temperatures of the individual wall sections are added to the vector of state variables. These programs may be adapted to answer problems 11.19 and 11.20 below.

Selected results are plotted in Fig. 11.28. The temperatures of the fluid and the walls at equilibirum, and of the environment, are plotted in part (a) of the figure as a function of location. The specific volumes are plotted in part (b). Zones of vapor, saturated mixture and pure liquid are revealed. They befit a practical condenser, in that the final volume contains pure liquid. The flow rate is 0.0254 kg/s, and it takes a fluid particle 9.58 seconds to travel from one end to the other. The total heat transfer to the environment is 3.37 kJ/s. The plots also show effects of the transitions in the liquid and vapor Reynolds numbers between laminar and turbulent states (see below). The equilibrium state vector (needed for Problems 11.19 and 11.20) is given in the downloadable file condenserstate.m.

The fluid temperatures of every other section are plotted in part (c) of the figure for the first three-quarters of a second. The first condensate appears quickly, at only 0.04 seconds and about one meter downstream, as the hot vapor interacts with the cold walls. The interface between superheated vapor and saturated mixture then moves downstream, reaching the end at roughly 0.7 seconds. Pure liquid first appears in the third section from the end at about 2.7 seconds and quickly spreads to the last four sections. It then quickly recedes, as the tube walls get hotter, remaining only in the final volume. It may appear as though equilibrium has been reached by the end of ten seconds, but the equilibirum momentum and mass flow must be the same in all sections; the spread of the values for the 40 IRS sections doesn't fall below 0.5% until over 40 seconds have passed.

Three Reynolds numbers apply to the liquid, vapor and mixture for the i th element:

$$Re_{f,i} = \frac{2(1-x_i)|\dot{m}_i|}{\pi r \mu_{f,i}}; \quad Re_{g,i} = \frac{2x_i|\dot{m}_i|}{\pi r \mu_{g,i}}; \quad Re_{m,i} = Re_{f,i} + \frac{\mu_{g,i}}{mu_{f,i}}\sqrt{\frac{v_{g,i}}{v_{f,i}}}Re_{g,i}, \quad (11.139)$$

in which x_i is the quality of the mixture. The Prandtl numbers for the liquid and the vapor phases are

$$Pr_{f,i} = c_{pf,i}\mu_{f,i}/k_{f,i}; \quad Pr_{g,i}\mu_{pg,i}\mu_g, i/k_{g,i}. \quad (11.140)$$

The values used for the specific heats are $c_{pf} = 977$ J/kg·K and $c_{pg} = 867$ J/kg·K. The thermal conductivities k_f and k_g are coded above already. The heat flow in the supercritical and superheated vapor regions with $Re_{g,i} > 20,000$ is

$$Q_i = 0.023Pr_{g,i}^{0.3}Re_{g,i}^{0.8}k_{f,i}\pi L(\theta_i - \theta_{w,i}), \quad (11.141)$$

where $\theta_{w,i}$ is the wall temperature of the ith element. The heat flow in the fully turbulent two-phase regions $Re_{g,i} > 20,000$ and $Re_{f,i} > 5000$ is

$$Q_i = 0.026Pr_{f,i}^{1/3}Re_{m,i}^{0.8}k_{m,i}\pi L(\theta_i - \theta_{Pw,i}). \quad (11.142)$$

In the partially turbulent two-phase region with $Re_{g,i} > 20,000$ and $Re_{f,i} < 5000$,

$$Q_i = [0.023Pr_{g,i}^{0.3}Re_{g,i}^{0.8}(1 - Re_{f,i}^{0.8}/5000)k_{g,i} + 0.026Pr_{f,i}^{1/3}Re_{m,i}^{0.8}Re_{f,i}^{0.8}k_{f,i}/5000]\pi L(\theta_i - \theta_{w,i}). \quad (11.143)$$

In the dual case of $Re_{g,i)<20,000}$ and $Re_{f,i} > 5000$,

$$Q_i = [0.023Pr_{f,i}^{0.3}Re_{f,i}^{0.8}(1 - Re_{g,i}^{0.8}/20,000)k_{g,i} + 0.026Pr_{f,i}^{1/3}(Re_{m,i}^{0.8}/20,000)k_{f,i}]\pi L(\theta_i - \theta_{w,i}). \quad (11.144)$$

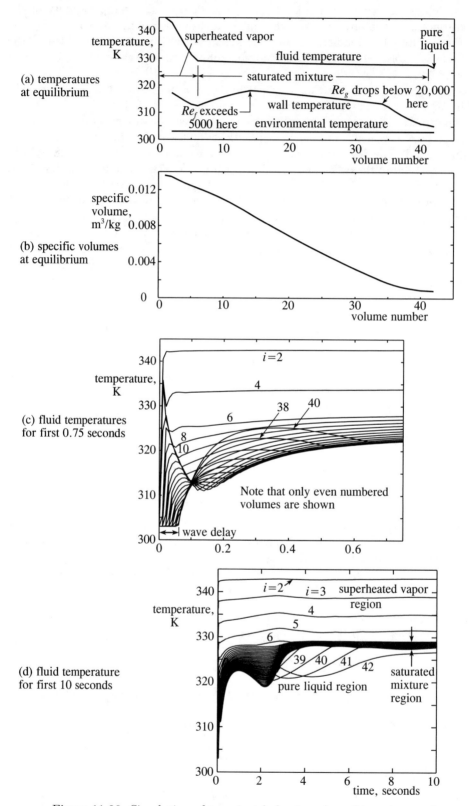

Figure 11.28: Simulation of transient behavior of a refrigerator condenser

In the laminar two-phase region with $Re_{g,i} < 20,000$ and $Re_{f,i} < 5000$,

$$Q_i = \text{sign}(\theta_i - \theta_{w,i})0.055\pi r L \left[\left(1 - \frac{v_i}{V_{g,i}}\right) 9.81 \frac{k_{f,i}^3 h_{fgp}}{2\mu_{f,i} r} \right] (\theta_i - \theta_{w,i})^{0.75}, \qquad (11.145)$$

in which the modified latent heat is

$$h_{fgp} = h_{g,i} - h_{f,i} + 0.375 c_{p,f}(\theta_i - \theta_{w,i}). \qquad (11.146)$$

If none of the conditions above apply, the fluid is single-phase and laminar; the formula used is

$$Q_i = 4\pi k_i L(\theta_i - \theta_w, i). \qquad (11.147)$$

11.5.12 Summary

The thermodynamic properties of substances are measured in terms of pressure-volume-temperature relations and specific heat relations, the latter at asymptotically small pressures. The independent variables are temperature and specific volume (or density). Pressure, enthalpy, entropy and internal energy are treated as derivative properties. The data for a wide variety of pure substances are available in the form of complex algebraic equations, with many coefficients but far less data than required to represent a a table of properties. This permits direct computation to replace table look-up with its interpolation. The data for several common substances have been converted to a Helmholtz formulation, from which the derivative properties can be deduced in a similar but somewhat simpler manner. This section has presented the necessary basic analytical equations. Applications to ideal gases, with air as a specific example, multiphase pure substances with refrigerants as a specific example, and compressed liquids have been discussed. MATLAB coding is given in Appendix D for the cases of air (and its components oxygen and nitrogen), water and refrigerants R-12 and R134a.

The assumption of thermodynamic equilibrium overlooks certain highly complex dynamic processes that occur in very rapid phase change. The bubble nucleation and growth that limits the rate of flashing (rapid boiling) is an example.

Examples of an air compressor, a refrigeration system and a long tube with a nonlinear transient response have been used to illustrate the complete process of writing differential equations and evaluating the various terms, including MATLAB coding.

Guided Problem 11.4

The simulation equations for the air piston compressor presented in Section 11.4.6 (pp. 883-886) assumes an ideal gas with constant specific heats. Carry out a more precise simulation by representing the air as a mixture of nitrogen, oxygen and argon, as described in Section 11.5.3 (pp. 900-902). The downloadable script files gasdata.m and spech.m and the function file air.m, all of which are listed in Appendix D, may be used.

Suggested Steps:

1. Define a set of state variables, using the bond graph given in Fig. 11.15 (p. 883) as a guide. Use the temperatures of the gas and the walls, rather than their entropies that were used in Section 11.4.6. Using the gas temperature, in particular, precludes the need for iteration in dealing with the properties.

2. Write the corresponding set of state differential equations.

3. Code the differential equations in a function m-file that calls on the available property files as necessary.

4. Write a master program that calls the integrator ode23s, which in turn calls your function file for the differential equations. The master program can call the file gasdata.m once, and transfer the results via a "global" statement. System parameters and initial conditions are given on the same page as the results plot.

5. Carry out a simulation for the same 0.1 seconds shown in Fig. 11.16 (p. 885), and plot the results. The results should be the same, since this is the procedure used to produce that figure.

PROBLEMS

11.13 A tank charged with air discharges into the atmosphere through a nozzle of effective area 0.0001 m^3. The tank has a volume of 0.5 m^3 and is equipped with a check valve that prevents reverse flow. The initial pressure of the air is 7.0 MPa absolute (about 1000 psi gage), and its temperature is in equilibrium with the 300 K surroundings, to give a specific volume of 0.01219 m^3/kg. Heat flows into the tank with the conduction coefficient 1.0 W/K. Simulate the discharge for a period of 120 seconds. Plot the resulting mass and temperature as a function of time. Check to make sure the air remains in the gaseous phase. (The condition at the moment the check valve closes is the greatest threat to this assumption; you can merely check that the temperature of the air does not fall below the dew point at atmospheric pressure, 81.8K.)

11.14 The refrigerator simulated in Section 11.5.6 is running at equilibrium when it is shut off. Find the subsequent distribution of mass and the temperatures until equilibrium is reached. To simplify the model and simulation, ignore all volumes and masses except for the principal condenser and evaporator (which contain almost all of the refrigerant anyway). Since the compressor is not running, its flow is zero; the problem is essentially a blow-down from one tank to another through a fixed orifice. Assume that the two tanks have the same volume of 0.006 m^3, and start with equal masses of 3 kg. The condenser starts at 304.3K and the evaporator at 251.5K.

11.15 Model a steam catapult for use on an aircraft carrier as a single variable chamber that acts on a piston of area 0.024 m^2 to push an aircraft of mass 16000 kg. The aircraft engine is producing a thrust of 100,000 N. The chamber initially contains a saturated mixture at 600 K with specific volume 0.002 m^3/kg. Assume the take-off is so fast that no water enters or leaves the chamber, and there is negligible heat transfer. Assume an equilibrium flashing process, and neglect friction. Simulate the take-off for the length of the catapult, which is 100 m. Determine the necessary initial size of the chamber if the final velocity is to be 55 m/s. (Note: Problem 11.8, p. 894, is a start.)

11.16 The fluid in a chamber is normally a saturated mixture with a free surface, but sometimes enters the compressed-liquid region. Show what coding changes or additions would be needed for that chamber, using the coding for the refrigeration cycle in Section 11.5.6 as reference. Assume a pure substance.

11.17 The flow `dm12` normally entering a chamber 2 from chamber 1 and the flow `dm23` normally leaving chamber 2 to enter chamber 3 may reverse direction. Assume that chamber 2 is in the saturated mixture region, and that the port from chamber 1 is above the level of the liquid and the port to chamber 3 is below that level. Write code for the differential equation for the temperature of chamber 2.

11.18 Simulate the blowdown of the adiabatic tube filled with refrigerant R12 analyzed in Section 11.5.10 when the entering flow at its left end is suddenly discontinued. This requires only a single change in the code for the function file `shocktube.m`, and the substitution of appropriate initial conditions in the master file `shocktubem.m`. These initial conditions are given in the downloadable file `shocktubess.m`; see page 1018. Plot the temperature and specific volume for the middle control volume for a period of three seconds.

11.19 The transient simulated in Section 11.5.10 assumes adiabatic conditions. Although effectively insulating the ouside of the tube is practical, preventing heat transfer to its walls would be virtually impossible. Repeat the simulation using the downloadable file `condenserm.m` and `condenser.m`, setting the external heat flow to zero. You may stop the simulation at 0.75 seconds, since the computation is slow, although 10 seconds is better. Compare the results to those of Figs. 11.27 and/or 11.28, as appropriate.

11.20 Simulate the blowdown of the condenser tube with refrigerant R12 analyzed in Section 11.5.11 when the flow entering at its left end is discontinued suddenly. This requires only a single change in the code for the downloadable function file `condenser.m`, and mimimal changes in the master program `condenserm.m`. The initial equilibrium states are available in the downloadable file `condenserss.m`. Stop the simulation when the flow would otherwise go negative; the program does not properly treat negative or intermittently negative flow. Plot the temperatures of the various sections and specific volume of the the fluid in the central section during this blowdown, and identify any phase changes.

SOLUTION TO GUIDED PROBLEM

Guided Problem 11.4

1. The first two state variables can be the same as before: the volume of the cylinder, V, and the mass of the air therein, m. The next two state variables are chosen as the temperature of the air in the cylinder, θ, and the temperature of the walls of the cylinder, θ_c. (Alternatively, you could stick with the entropy of the walls of the cylinder, S_c, but for the air the use of equation (11.65) (p. 881), which precludes the need for iteration, demands the use of its temperature.)

2.
$$\frac{dV}{dt} = V_1 sin(\dot{\phi}t); \qquad V(0) = V_0,$$

$$\frac{dm}{dt} = \dot{m}_1 - \dot{m}_2,$$

$$\frac{d\theta}{dt} = \frac{1}{m\partial u/\partial\theta}\left[H(\theta_c - \theta) + (h_{in} - h)\dot{m}_1 + \left(P + \frac{\partial u}{\partial v}l\right)(v\dot{m} - \dot{V})\right],$$

$$\frac{d\theta_c}{dt} = F * |\dot{V}| + H(\theta - \theta_c).$$

The flows through the valves, $\dot{m}_1$ and $\dot{m}_2$, are given by equations (11.76) (p. 884, not reproduced here), which are approximations since they assume a constant value for the specific heat. This value, for the mixture of air, is computed in a script m-file `Specheat.m` which is given in Appendix D.

3. The files below can be downloaded from the internet:

```
function f=GP11_4(t,x) % derivatives for simulation of an
% air compressor
% x(1) is the volume of the cylinder
% x(2) is the mass of the air in the cylinder
% x(3) is the absolute temperature of that mass
% x(4) is the absolute temperature of the cylinder wall
% dm1 and dm2 are the mass flows into and out from the
% cylinder, respectively
global P1 T1 H1 P2 Cm C Hc Hco Dphi PF V1 a1 R E1 E2 A0 T
rho=x(2)/x(1); % density of the air
[P,u,h,s,dudT,dudv]=feval('air',x(3),rho); % finds air properties
T=x(3); specheat; % finds cvm (slightly varying specific heat)
k=R/cvm+1; E3=2/k; E4=1+1/k; A=a1*sqrt(2*k/(k-1)/R);
Pr1=P/P1; Pr2=P2/P; % pressure ratios across valves
if Pr1<1; dm1=A0*P1*sqrt((Pr1^E1-Pr1^E2)/T1);else;dm1=0;end;
if Pr2<1; dm2=A*P*sqrt((Pr2^E3-Pr2^E4)/T); else; dm2=0; end;
f(1)=V1*Dphi*sin(Dphi*t); % first differential eqn., for volume
f(2)=dm1-dm2; % second differential equation, for mass
f(3)=(Hc*(x(4)-T)+(H1-h)/T*dm1+(P+dudv)*(f(2)/rho-f(1)))/x(2)/dudT;
f(4)=(Hc*(T-x(4))-Hco*(x(4)-T1)+PF*abs(f(1)))/Cm/C; % for walls
f(5)=P; % to enable the printing or plotting of P (see below)
f=f';
t % to let you see the progess of an extended simulation on screen
```

The various constants are defined in the master program below; values are taken from those listed on page 885. The fifth differential equation is given to enable the plotting of the pressure, P. (Since P is the derivative, a simple numerical differentiation of x(5) is necessary. This is a way of plotting any variable other than a state variable without having to recompute it as a function of state variables.) The statements that follow the ode command produce the plots of Fig. 11.16 (p. 885).

4.
```
% file {\tt GP11_4m.m}, the master program for a simple piston air
% compressor that treats the air as a mixture of gases
global P1 T1 H1 P2 Cm C Hc Hco Dphi PF V1 a1 R E1 E2 A0 T
P1=1e5; T1=293; H1=452540; P2=4e5; Cm=1; C=.45; Hc=.012; Hco=.12;
Dphi=185; PF=2000; V1=2.5e-4; a1=4e-4; V0=1e-4; m0=3.15e-4;
T=T1; gasdata; specheat; k1=1+R/cvm; E1=2/k1; E2=1+1/k1;
A0=a1*sqrt(2*k1/(k1-1)/R);
[t,x]=ode23('GP11_4',[0 .1],[V0 m0 440 744 0]);
t1=t(1:size(t)-1); % time vector shortened by one
P=diff(x(:,5))./diff(t); % P is the derivative of x(5)
dm=diff(x(:,2))./diff(t); % mass flow is the derivative of x(2)
plot(t,x(:,1)*1e4,t,x(:,3)/100,t1,P/1e5,t1,dm*100) % plots volume,
% temperature, pressure and mass flow, respectively, as scaled
```

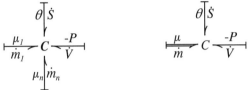

(a) previous version,   (b) reticulated version,
pure substance   pure substance

(c) C element, more than   (d) vector bond version
one species   of (c)

Figure 11.29: Thermodynamic energy storage

11.6 Systems with Chemical Reaction*

The dynamics of chemical reactions ("chemical kinetics") depends on transport phenomena such as mixing and diffusion mechanisms as well as the local properties of composition, temperature and density. This section introduces the subject, with emphasis on lumped models, which implies the ideal case of well-mixed reactants. Much of the bond graph notation used is that of the author.[28]

11.6.1 Energy of a Pure Substance

The thermodynamic energy of a pure substance has been represented by a three-ported compliance, as shown in Fig. 11.29 part (a), where $\dot{S}_q$ is the *conduction* entropy flux. Recalling equations (11.53) and (11.55) and footnote 10 (pp. 878-879),

$$\mathcal{V} = \mathcal{V}(S, V, m), \tag{11.148}$$

$$\theta = \left(\frac{\partial \mathcal{V}}{\partial S}\right)_{V,m}, \tag{11.149a}$$

$$-P = \left(\frac{\partial \mathcal{V}}{\partial V}\right)_{S,m}, \tag{11.149b}$$

$$h = \theta s + g, \tag{11.149c}$$

$$g = \left(\frac{\partial \mathcal{V}}{\partial m}\right)_{V,S}. \tag{11.149d}$$

Focusing on the Gibbs function g is especially helpful when chemical reactions are involved. This can be done in a bond graph by separating the terms θs and g with a 1-junction, as shown in part (b) of the figure. The bond with effort θs and flow $\dot{m}$ is then connected to a bond with effort θ and flow $s\dot{m}$ by an ideal transformer having modulus s. This modulus can change in time but may not change drastically.

---

[28]F.T. Brown, "Convection Bond Graphs," *Journal of the Franklin Institute*, v. 328 n. 5/6, 1991, pp. 871-886.

The flow $s\dot{m}$ is the *convection entropy flux* due to the mass transport into (or out of) the system. When this is added to the *conduction entropy flux* $\dot{S}_q$ through the use of a zero junction, the result is the *total entropy flux*, $\dot{S}$. The product $\theta s\dot{m}$ is the energy flux directly associated with the convection entropy flux. The product $g\dot{m}$ is the total "free energy" flux which conveys work but no entropy whatsoever. Note that its bond and the bond for the effort θs are simple bonds.

11.6.2   Energy of Multiple Species

When more than one substance is present, equation (11.148) is expanded to

$$\mathcal{V} = \mathcal{V}(S, V, m_1, m_2, m_3, \ldots), \tag{11.150}$$

so that

$$\theta = \left(\frac{\partial \mathcal{V}}{\partial S}\right)_{V, m_1, m_2, \ldots}, \tag{11.151a}$$

$$-P = \left(\frac{\partial \mathcal{V}}{\partial V}\right)_{S, m_1, m_2, \ldots}, \tag{11.151b}$$

$$\mu_i = \left(\frac{\partial \mathcal{V}}{\partial m_i}\right)_{V, S, m_{j \neq i}}. \tag{11.151c}$$

The **C** element of part (c) of the figure represents this relationship, in which each μ_i is a partial Gibbs function known as a **chemical potential**, although the constraints between S and $m_1, \ldots, m_n$ are not as yet shown.

A bundle of bonds of similar type is represented widely in the literature by a vector bond, drawn with two solid parallel lines, as introduced in Section 9.6.1 (pp. 681-682):

$$\left.\begin{array}{c} \dfrac{\mu_1}{\dot{m}_1} \\[1.5em] \vdots \\[1.5em] \dfrac{\mu_n}{\dot{m}_n} \end{array}\right\} \quad\Longleftrightarrow\quad \dfrac{\boldsymbol{\mu}}{\dot{\mathbf{m}}}$$

This notation permits the graph of part (c) of Fig. 11.29 to be represented as shown in part (d). The vector nature of $\boldsymbol{\mu}$ and $\dot{\mathbf{m}}$ is acknowledged through the use of bold-faced type.

11.6.3   Stoichoimetric Coefficients and Reaction Forces

A chemical reaction of arbitrary complexity can be described by

$$\sum_{i=1}^{n} \nu_i' \rho \rightleftharpoons \sum_{i=1}^{n} \nu_i'' \bar{\rho}_i, \tag{11.152}$$

where $\bar{\rho}_i$ is the molar density and ν_i' and ν_i'' are the **stoichoimetric coefficients** for the ith species. For example,

$$3\mathrm{H} \rightleftharpoons \mathrm{H}_2 + \mathrm{H}$$

gives $\nu_\mathrm{H}' = 3$, $\nu_\mathrm{H}'' = 1$, $\nu_{\mathrm{H}_2}' = 0$, $\nu_{\mathrm{H}_2}'' = 1$. Similarly,

$$2\mathrm{NO} + \mathrm{O}_2 \rightleftharpoons 2\mathrm{NO}_2$$

gives $\nu_\mathrm{NO}' = 2$, $\nu_\mathrm{NO}'' = 0$, $\nu_{\mathrm{O}_2}' = 1$, $\nu_{\mathrm{O}_2}'' = 0$, $\nu_{\mathrm{NO}_2}' = 0$, $\nu_{\mathrm{NO}_2}'' = 2$. The first consideration here is a reaction in a closed chamber (no mass flows across its boundaries) of volume V,

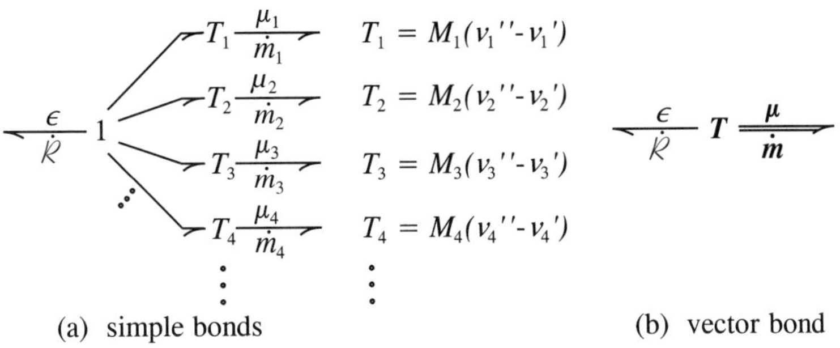

$$T_1 \xrightarrow{\frac{\mu_1}{\dot{m}_1}} \quad T_1 = M_1(\nu_1{''}\text{-}\nu_1{'})$$

$$T_2 \xrightarrow{\frac{\mu_2}{\dot{m}_2}} \quad T_2 = M_2(\nu_2{''}\text{-}\nu_2{'})$$

$$\xleftarrow[\dot{\mathcal{R}}]{\epsilon} 1 \qquad T_3 \xrightarrow{\frac{\mu_3}{\dot{m}_3}} \quad T_3 = M_3(\nu_3{''}\text{-}\nu_3{'}) \qquad\qquad \xleftarrow[\dot{\mathcal{R}}]{\epsilon} T \xrightarrow{\frac{\mu}{\dot{m}}}$$

$$T_4 \xrightarrow{\frac{\mu_4}{\dot{m}_4}} \quad T_4 = M_4(\nu_4{''}\text{-}\nu_4{'})$$

(a) simple bonds (b) vector bond

Figure 11.30: Stoichiometric coefficients and reaction force

which could vary in time. A cylinder of an engine with its valves closed is an example. The molar densities are thus

$$\bar{\rho}_i = \frac{m_i}{V M_i}, \tag{11.153}$$

where the M_i are the respective molar masses.

A reaction rate $\dot{\mathcal{R}}$ is defined such that for each species,

$$\dot{m}_i = M_i(\nu_i'' - \nu_i')\dot{\mathcal{R}}. \tag{11.154}$$

Note that the signs are chosen so that if $\dot{\mathcal{R}}$ is positive, the reaction proceeds from left to right. Note also that the stoichoimetric coefficients are constrained to satisfy the conservation of mass

$$\sum_{i=1}^{n} \dot{m}_i = 0. \tag{11.155}$$

The constraints of equation (11.154) are represented by the bond graph of Fig. 11.30, in which part (b) gives a vector-bond-graph shorthand for part (a). The effort ϵ, known as a **reaction force** or **chemical affinity**, results naturally:

$$\epsilon = \sum_{i=1}^{n} M_i(\nu_i' - \nu_i'')\bar{\mu}_i = -\sum_{i=1}^{n} M_i(\nu_i'' - \nu_i')\bar{\mu}_i. \tag{11.156}$$

The reaction force will be evaluated for the ideal gas and the ideal solution. For the ideal gas,

$$\bar{\mu}_i = \bar{h}_i - \theta \bar{s}_i, \tag{11.157}$$

in which the enthalpy $\bar{h}_i$ is a function of temperature only, while $\bar{s}_i$ is also a function of pressure. Further, the T ds relation

$$\theta \, ds = dh - v \, dP \tag{11.158}$$

mandates

$$\left(\frac{\partial \bar{s}_i}{\partial P_i}\right)_\theta = -\frac{\bar{v}_i}{\theta} = -\frac{R}{P_i}, \tag{11.159}$$

so that, assuming also the equation of state $P_i v_i = R\theta_i$,

$$\bar{\mu}_i = \bar{\mu}_i^0 + R\theta \int_{P_0}^{P_i} \frac{dP_i}{P_i} = \bar{\mu}_i^0 + R\theta \ln\left(\frac{P_i}{P_0}\right), \tag{11.160}$$

in which $\bar{\mu}_i^0$ is the value of $\bar{\mu}_i$ at temperature θ and reference pressure P_0, which usually is taken as one atmosphere. A similar result applies to the ideal solution:

$$\bar{\mu}_i = \bar{\mu}_i^0 + R\theta \, \ln\left(\frac{\bar{\rho}_i}{\bar{\rho}_0}\right). \tag{11.161}$$

Here $\bar{\rho}_i$ is the molar density of the ith species of solute or solvent, and $\bar{\mu}_i^0$ is the value of $\bar{\mu}_i$ at temperature θ and concentration $\bar{\rho}_0$.

The reaction force or affinity for the ideal gas now becomes

$$\varepsilon = -\sum_{i=1}^{n} (\nu_i'' - \nu_i')\left[\bar{\mu}_i^0 + R\theta \, \ln\left(\frac{P_i}{P_0}\right)\right]$$
$$= -\sum_{i=1}^{n} (\nu_i'' - \nu_i')\bar{\mu}_i^0 - R\theta \, \ln\left[\prod_{i=1}^{n}\left(\frac{P_i}{P_0}\right)^{(\nu_i'' - \nu_i')}\right]. \tag{11.162}$$

The product $\varepsilon\dot{\mathcal{R}}$ is an energy flux that generates entropy, and is a measure of irreversibility.

11.6.4   Chemical Equilibrium

For chemical equilibrium at temperature θ, no entropy is generated even in the presence of a virtual displacement $\delta\mathcal{R}$, so that $\varepsilon = 0$. Thus,

$$\sum_{i=1}^{n}(\nu_i'' - \nu_i')\bar{\mu}_i^0 = -R\theta \, \ln\left[\prod_{i=1}^{n}\left(\frac{P_{ie}}{P_0}\right)^{(\nu_i'' - \nu_i')}\right], \tag{11.163}$$

where the subscript e indicates equilibrium. With the definitions

$$K_g \equiv \prod_{i=1}^{n}\left(\frac{P_i}{P_0}\right)^{(\nu_i'' - \nu_i')}, \tag{11.164a}$$

$$K_{ge} \equiv \prod_{i=1}^{n}\left(\frac{P_{ie}}{P_0}\right)^{(\nu_i'' - \nu_i')}, \tag{11.164b}$$

substitution of equation (11.163) into (11.162) gives

$$\varepsilon = -R\theta(\ln K - \ln K_e) = -R\theta \, \ln\left(\frac{K}{K_e}\right) \tag{11.165}$$

or

$$\frac{K}{K_e} = \exp\left(-\frac{\varepsilon}{R\theta}\right). \tag{11.166}$$

The subscript g has been dropped from K_g and K_{ge} here, because the same result follows for the ideal solution if one substitutes the following K_s and K_{se}, respectively:

$$K_s \equiv \prod_{i=1}^{n}\left(\frac{\bar{\rho}_i}{\bar{\rho}_0}\right)^{(\nu_i'' - \nu_i')}, \tag{11.167a}$$

$$K_{se} \equiv \prod_{i=1}^{n}\left(\frac{\bar{\rho}_{ie}}{\bar{\rho}_0}\right)^{(\nu_i'' - \nu_i')}. \tag{11.167b}$$

K_e (and K_{se} and K_{ge}) is known as an **equilibrium constant**. It is a readily calculable function of temperature based on the thermodynamic properties of the components. For the ideal gas, for example, inverting equation (11.163) gives

$$K_{ge} = \exp\left[\frac{1}{R\theta}\sum_{i=1}^{n}(\nu_i' - \nu_i'')\bar{\mu}_i^0\right]. \tag{11.168}$$

Knowledge of the equilibrium constant permits calculation of the partial pressures or concentrations of the components at equilibrium. In general this is awkward, as the direct computation of the equilibrium of any type of system often is awkward because of the simultaneous nonlinear algebraic equations. The approach here, as with non-chemical systems, is to simulate the behavior in time. Equilibrium is achieved ultimately in the absence of disturbances; if the details of the dynamics are of no interest, convergence to the equilibrium state can be hastened by assuming artificially high reaction rates.

11.6.5 Reaction Rates

The actual reaction rates are given at least approximately by the **law of mass action**, which specifies that for a one-step reaction,

$$\dot{\mathcal{R}} = V\left[k_f\prod_{i=1}^{n}\left(\frac{\bar{\rho}_i}{\bar{\rho}_0}\right)^{\nu_i'} - k_b\prod_{i=1}^{n}\left(\frac{\bar{\rho}_i}{\bar{\rho}_0}\right)^{\nu_i''}\right], \tag{11.169}$$

in which k_f and k_b are the forward and backward **specific reaction-rate coefficients**, respectively. The reaction rate vanishes at equilibrium, so that the ratio of these two coefficients is

$$\frac{k_f}{k_b} = \prod_{i=1}^{n}\left(\frac{\bar{\rho}_{ie}}{\bar{\rho}_0}\right)^{(\nu_i'' - \nu_i')} \equiv K_{se}, \tag{11.170}$$

where, as before, the subscript e designates the equilibrium condition. As noted, this ratio is identical to the equilibrium constant for the solution, K_{se}. By also employing the definition of K_s from equation (11.167a), there results

$$\dot{\mathcal{R}} = Vk_f\prod_{i=1}^{n}\left(\frac{\bar{\rho}_i}{\bar{\rho}_0}\right)^{\nu_i'}\left(1 - \frac{K_s}{K_{se}}\right). \tag{11.171}$$

The same form results for the ideal gas,

$$\xi = Vk_f\prod_{i=1}^{n}\left(\frac{P_i}{P_0}\right)^{\nu_i'}\left(1 - \frac{K_g}{K_{ge}}\right), \tag{11.172}$$

and the subscripts s and g may be removed.

Like K_e, k_f is a function of temperature which depends on the particular reaction. Normally it is given the form

$$k_f = B\theta^\alpha e^{-E/R\theta}. \tag{11.173}$$

E is called the **activation energy**; the **Arrhenius** factor $e^{-E/R\theta}$ is the fraction of the total number of molecular collisions between the reacting chemical species for which the energy is sufficient to induce a reaction. The balance of the right side is proportional to the number of collisions. Along with the **frequency factor** B and the quantity α, E is a constant parameter determined by the nature of the reaction. They are commonly determined empirically; α is most frequently set equal to 0, $\pm\frac{1}{2}$ or 1.

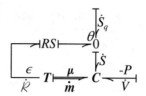

Figure 11.31: Reaction in a closed container

Substitution of equation (11.166) into (11.171) for the solution or (11.172) for the ideal gas reveals, respectively,

$$\dot{\mathcal{R}} = V k_f \prod_{i=1}^{n} \left(\frac{\bar{\rho}_i}{\bar{\rho}_0}\right)^{\nu'_i} \left[1 - e^{-\varepsilon/R\theta}\right], \qquad (11.174a)$$

$$\dot{\mathcal{R}} = V k_f \prod_{i=1}^{n} \left(\frac{P_i}{P_0}\right)^{\nu'_i} \left[1 - e^{-\varepsilon/R\theta}\right], \qquad (11.174b)$$

as the constitutive relation that determines the reaction rate $\dot{\mathcal{R}}$ as a function of the reaction force or affinity, the temperature and the densities or partial pressures of the component species. Note that the factor in square brackets vanishes for $\varepsilon = 0$, equals $\varepsilon/R\theta$ for values thereof very small compared to unity (near equilibrium), and reaches unity asymptotically for large values of $\varepsilon/R\theta$.

11.6.6 Models of Reactions without Mass Flows

The energy flux $\varepsilon \dot{\mathcal{R}}$ is converted irreversibly to heat, much like mechanical work is converted to heat by friction. Like the friction example, therefore, this conversion is represented in the bond graph by an RS element, as shown in Fig. 11.31. The generation of entropy is determined by the required conservation of energy: $\varepsilon \dot{\mathcal{R}} = \theta \dot{S}$. The boundary bond on the top of the graph permits heating or cooling of the system. The boundary bond on the right permits the volume to change, as in the cylinder of an engine.

It has been assumed tacitly that the chemical reaction is a one-step process. Most important reactions are not. For example, the reaction

$$2NO + 2H_2 \rightleftharpoons N_2 + 2H_2O$$

taken with $\nu'_{NO} = 2$, $\nu''_{NO} = 0$, $\nu'_{H_2} = 2$, $\nu''_{H_2} = 0$, $\nu'_{N_2} = 0$, $\nu''_{N_2} = 1$, $\nu'_{H_2O} = 0$, $\nu''_{H_2O} = 2$ gives an incorrect reaction rate if interpreted with the equations above. Two separate reactions in fact take place, for which the results apply separately:

$$2NO + H_2 \rightleftharpoons N_2 + H_2O_2$$

$$H_2O_2 + H_2 \rightleftharpoons 2H_2O$$

The second reaction takes place much faster than the first; often it is assumed to be in equilibrium. Any number of elemental reactions can be accommodated without such an approximation, however, by implementing the bond graph of Fig. 11.32. One reaction rate $\dot{\mathcal{R}}_j$ and corresponding element RS_j is used for each elemental reaction.

The actual reaction rate in a combuster, for example, can be much slower than the equations above would give, because the properties are not uniform over the volume due to diffusion and mixing phenomena. In such cases empirically determined rate coefficients

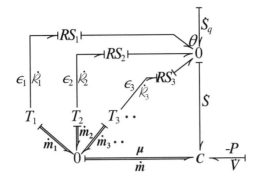

Figure 11.32: Multi-step reaction in a closed container

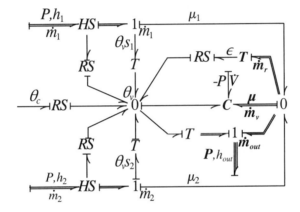

Figure 11.33: Single-step reaction with two in-flows and one out-flow

might be used. Frequently, also, reactions occur so quickly that it matters little what coefficients are used, as long as they are not too small. In these cases it is the rate of the mass flows into and out of the system that controls the dynamics.

11.6.7 Models of Reactions with Mass Flows

A system with two in-flows and one out-flow is shown in the bond graph of Fig. 11.33. The HS elements are employed as in a non-reacting $\mathbf{C}S$ element. The rates of change of masses in the control volume are $\dot{\mathbf{m}}_v$, and the mass fluxes undergoing reaction are $\dot{\mathbf{m}}_r$; in the absence of in-flows or out-flows these would be the same. The heart of the graph – the $\mathbf{C}$ and the loop with the two zero junctions, the two transformers and the RS element – is not affected by the mass influxes or effluxes. Causal strokes are applied in the usual manner to direct the routine writing of state equations. The addition of further mass fluxes can be accommodated directly.

The output bond with a solid line between two parallel dashed lines is a generalization of the convection bond for two or more component species. An absence of internal diffusion is assumed in the definition. The number of state variables equals two plus the number of species present, or five in the example shown, including the three mass flows, pressure and one convected property. The causal stroke applies to the mass fluxes and the pressure. Note that this bond can be reduced to a simple convection bond if the species are assumed to retain their proportions subsequently due to an absence of further chemical reaction or phase change.

The system as represented can be defined as a single macro-element for computational purposes. A complicated system could include several such macros. Distributed-parameter systems, such as a single spherical lump of coal, might be represented by several such macro-elements. The discussion above plus the examples given in Section 11.5 for non-reacting systems indicate how this can be done, but the detailed development of software is unfinished business.

11.6.8   Summary

The energy of a system comprising a well-mixed pure substance has been represented as a function of its total entropy, total volume and total mass. When a mixture of potentially reactive substances is involved, the only difference is that the masses of the individual species are needed. The effort on the convection bond remains the total enthalpy, but this is decomposed into the thermal term θs and the free energy term $\sum_i \mu_i$, where μ_i is the partial Gibbs function $(\partial \mathcal{V}/\partial m_i)_{V,S,m_{j \neq i}}$. The stoichoimetric coefficients that constrain the changes in the composition, assuming a known reaction, can be represented by a family of transformers that relate the individual μ_i and $\dot{m}_i$ to a common reaction rate $(\dot{\mathcal{R}})$ and reaction force (ε). The product $\varepsilon\dot{\mathcal{R}}$ represents the irreversible generation of heat and entropy. The assumption $\dot{\mathcal{R}} = 0$ allows the conditions for chemical equilibrium to be deduced.

A dynamic model for a single reaction in a closed container is completed by adding an RS element to regulate the rate of the reaction and provide a location for the entropy generation. The most commonly assumed model has been presented: the law of mass action. Multi-step reactions can be modeled by introducing separate families of transformers, each with its own RS element to control the rate of its particular reaction. Influx and efflux of reactants and product can be treated in the same way they have been for non-reacting systems.

The assumption of well-mixed species may not apply in practice except for small spatial domains. Individual trays in a distillation column often are designed to make the assumption a fair estimate. At the other extreme, a combustion process usually is controlled largely by complex turbulence and diffusion phenomena that are well beyond the scope of this introduction to chemical kinetics.

Chapter 12

Topics in Advanced Modeling

The topics in this chapter fill some important gaps and should be included in the repertoire of an accomplished analyst of physical dynamic systems. They are advanced in a relative sense only and should be accessible to most undergraduate seniors and engineering graduates.

Energy is distributed over space, or over a **field**. In most of this text energy is approximated by functions of discrete numbers of variables. The associated process of **field lumping** is examined both conceptually and practically in Section 12.1. The focus is on the lumping of the one-port resistance, compliance and inertance elements that have been assumed often with too little consideration of their meaning or values.

Section 12.2 discusses how to imbed models of complex devices or processes that are both non-dynamic and nonconservative into models of larger systems. This allows machinery, for example, to be modeled based on overall observed behavior, omitting details of little interest, rather than on a detailed analysis.

Models based on a detailed analysis have been structured thus far either from direct inspection of the system in question, or from simple operations on restricted statements of energy stored and dissipated. Occasionally, however, a particularly difficult problem arises in which it is unclear how to apply the direct method for a critical portion of a system, and the limitations of the basic energy methods prevent their use also. More powerful energy methods, associated with the names of Lagrange and Hamilton, are the subject of Sections 12.3–12.6. These methods also may be preferred for some simpler cases, since they demand little insight from the modeler.

Lagrangian and Hamiltonian methods were developed within classical mechanics. The development here starts in Section 12.3 with a derivation of Lagrange's equations that applies to energy-based systems in general rather than just classical dynamics systems. Then, in Section 12.4, these equations are applied to give a **Lagrangian bond graph**. The method forces one to distinguish two fundamentally different types of constraint, the holonomic and the nonholonomic. Their difference is the subject of Section 12.5, completing a discussion initiated in Section 9.4. Finally, the Hamiltonian method is presented in Section 12.6.

12.1 Field Lumping

This book emphasizes *lumped models* (with the exception of Chapter 11) described as compliances, inertances and resistances interconnected by a junction structure. Nevertheless, energy and the mechanisms that dissipate it actually are distributed over space. The process by which the energy storage or power dissipation over some region or field of space is approximated in terms of a small number of variables is called **field lumping**. The state of a one-port R, C or I field, for example, is described in terms of a single variable: a characteristic generalized displacement, velocity, force or momentum. In addition to their number of ports, fields can be categorized as scalar, vector or tensor, and also as nodic or non-nodic.

This book has engaged implicitly and almost routinely in field lumping, without identifying it as such. In a typical example, the energy in a coil spring results from the torsion and twisting strain of the wire, which in turn results from the shear stress and strain at each point. To evaluate the "spring constant" k or the lumped compliance $C = 1/k$ from first principles, the strain energy is integrated over space and set equal to $\frac{1}{2}kx^2$ or $\frac{1}{2C}x^2$. In another example, the resistance to a flow of a viscous fluid through a tube is associated with the integral of the power dissipation from the center of the tube to its periphery and from somewhat outside one end of the tube to somewhat outside the other end.

Some of the basic concepts and methods are organized in this section, and they are extended particularly with regard to nodic vector fields.

12.1.1   Scalar Fields

The **scalar field** is defined by the assumption of a common effort, flow, momentum or displacement over the entire spatial extent of the field. Translation of an incompressible solid or liquid, that is motion with a common displacement, velocity and momentum per unit mass, is a familiar example. The result is overall or lumped inertance (*e.g.*, mass), resistance, or, for motion in a gravity field, an effort (force) source as a degenerate or special type of compliance. Another familiar example is the compressibility field of a fluid in which the pressure is assumed to be uniform over the extent of the defined control volume. The result is a lumped compliance. A thermal field in which the temperature is assumed to be uniform over the extent of the defined control volume similarly results in a lumped compliance. A lumped resistance results from the assumption of uniform flow of a fluid through a homogeneous cylindrical porous plug with uniform pressures over its two ends. In all of these cases the lumped resistance, compliance or inertance equals a uniform property, on a per unit mass or volume or length, times the total mass or volume or length.

12.1.2   Rigid-Body Vector Fields

The gravity potential energy of a solid or fluid in rigid body in a uniform gravity field is expressed as the product of its mass and the elevation of its center of mass. This center is in fact *defined* so that the spatial integral of the potential energy, $\int gz\,dm$, equals mgz_{cm}. Similarly, the kinetic energy of the body in planar motion is represented by two lumps, one with linear inertia (mass) and linear velocity and the other with rotational inertia and rotational velocity. Here, the center of mass and the rotational inertia J are *defined* so as to make this lumping work; that is, so that the kinetic energy is

$$\mathcal{T} = \frac{1}{2}\int v^2 dm = \frac{1}{2}mv_{cm}^2 + \frac{1}{2}J\dot{\phi}^2. \tag{12.1}$$

12.1.3   The Role of Approximations

The approximation that one or at most a few common variables apply to all regions of a field is inherent in any definition of a lumped element. To the extent that a solid body is not rigid, for example, the very ideas of a center of mass fixed to the body and a constant rotational inertia become invalid. The spring compliance becomes an inaccurate description if the spring is deflected so fast that the axial force (or the torsion on the wire) is non-uniform along its length due to the mass and the acceleration. The resistance of a tube to flow becomes nebulous if the dissipation of energy is affected by the acceleration of the fluid; this phenomenon indeed sometimes occurs because the velocity profile across the tube, which strongly affects the dissipation, can be markedly altered by acceleration.

Compliance, inertance and resistance fields can be coextensive, that is occupy the same space and involve the same matter at the same time. The lumping of these fields nevertheless is accurate if the fields do not interact with each other. The kinetic energy and the gravity energy of a truly rigid body, for example, do not interact. The compliance energy of the spring does not interact much with its inertance energy as long as the latter is much smaller than the former (or vice-versa). The resistance field of the fluid-filled tube is not affected much by the coextensive inertance field if the latter has little energy, and the effect of the resistance field on the inertance field is limited in any case, permitting at least rough approximation. You must watch out for coupling between the inertance and compliance fields in the tube, however, which can produce wave-like behavior.

The couplings of fields usually can be reduced to an acceptable degree if the spatial region of interest is reticulated into multiple fields of small extent. The solid body that is not quite rigid, for example, might be represented as having a single deflection variable with a single stiffness or compliance; this implies two inertance fields and one compliance field. If that model still is not good enough, the body could be modeled with several inertances with several overlapping compliances. In certain cases it might be appropriate to allow the fields to interact continuously, as in the distributed-parameter models of Chapter 11. The whole business of modeling ultimately becomes an art which is strongly dependent on the objectives and skills of the modeler.

12.1.4 Nodic Fields

A field is called **nodic** if the flow or generalized displacement therein has zero divergence. Physically, this means that the flow or displacement is like that of an incompressible fluid with no sources or sinks other than at boundary ports. Mathemetically, this "continuity" relation is stated by

$$R \text{ or } I \text{ field:} \qquad \nabla \cdot \dot{\mathbf{q}} = \mathbf{0}, \qquad (12.2a)$$

$$C \text{ field:} \qquad \nabla \cdot \mathbf{q} = \mathbf{0}. \qquad (12.2b)$$

The rigid body motion discussed above corresponds to a nodic field. So does the incompressible flow of the liquid through a porous plug, and the motion of an elastic solid in pure shear, such as pure torsion. On the other hand, any kind of material *dilation or compression* violates the nodic condition, as does the thermal compliance. Nodic fields are generally incompatible with uniform effort, and exclude most fields described by tensors, such as complex stress-strain fields in solids.

Fields of most common types can be analyzed numerically with a wide variety of finite-element and finite-difference software packages. The objective here is not to replace these packages, except in the simplest cases when a quick approximation can be made directly. Rather, the aim is a deeper understanding of the role and meaning of field lumping in the context of the modeling of complex systems.

The focus here is on *vector* nodic fields. Vector fields are intermediate in complexity between scalar fields, which require little exposition, and tensor fields, which are illustrated in Section 9.3.6 (pp. 625-626). In particular, the focus is on linear fields in which the vector flow or displacement is in the direction of the gradient of the scalar effort or momentum and is uniformly proportional to it:

$$R \text{ field:} \qquad \dot{\mathbf{q}} = -k \, \nabla e, \qquad (12.3a)$$

$$C \text{ field:} \qquad \mathbf{q} = -k\,\nabla e, \tag{12.3b}$$

$$I \text{ field:} \qquad \dot{\mathbf{q}} = -k\,\nabla p. \tag{12.3c}$$

Bold-face notation implies a vector field.

For the R field, k is known as a **conductivity**, which is the reciprocal of a **resistivity**. For an electrostatic C field, k is the dielectric constant; for a magnetic C field it is the magnetic permeability. For an inertive I field, k is known as the **susceptivity**, which is the reciprocal of the density.

Substitution of equations (12.2) into (12.3) gives the well-known **Laplace's equation**:

$$R \text{ or } C \text{ field:} \qquad \nabla^2 e = 0 \quad \text{or} \quad \frac{\partial^2 e}{\partial x^2} + \frac{\partial^2 e}{\partial y^2} + \frac{\partial^2 e}{\partial z^2} = 0, \tag{12.4a}$$

$$I \text{ field:} \qquad \nabla^2 p = 0 \quad \text{or} \quad \frac{\partial^2 p}{\partial x^2} + \frac{\partial^2 p}{\partial y^2} + \frac{\partial^2 p}{\partial z^2} = 0. \tag{12.4b}$$

Together with equations (12.2) these require that a set of equipotential (e or p) lines or surfaces form an orthonormal network with a set of flow or displacement ($\dot{q}$ or q) lines or surfaces (across which no displacement or flow occurs). An example with ports at the two ends of a channel with square cross-section is pictured in Fig. 12.1 part (a). The spacing between the flow or displacement surfaces is chosen to equal the spacing between the potential surfaces, resulting in an array of cubes-in-the-small.

The element can be represented by the bond graphs of part (b). In the special case where either e_1 or e_2 is zero, the element becomes a simple one-port inertance, resistance or compliance. The ratio of the integrated flow or displacement through the channel to the difference between the efforts or momenta at the two ends is, in general, known as the **field transmittance**, τ. For a resistive field this transmittance becomes the **field conductance**, $G = 1/R$; for a compliance field it becomes the **field compliance**, C; for an inertance field it becomes the **field susceptance**, $1/I$.

12.1.5   Planar Vector Nodic Fields

A **planar field** is the special case with uniform depth, w. Consider first the special case of a square resistance planar field with flow from left to right, as shown in part (c) of Fig. 12.1. The conductance of this field is the conductivity, k, times the depth: $G = kw$. Divide this square into four equal squares, as shown. Each small square has one-half the flow and one-half the potential drop of the large square. Therefore, the conductance of each small square, which equals the ratio of its flow to its potential drop, is the same as for the large square. This process of subdivision can be repeated. The same result would apply to the other transmittances (compliance and susceptance). The conclusion is that *all squares in a planar vector nodic field have the same transmittance.*

A particular planar field is sketched in part (d) of the figure. The flow lines and the equipotential lines form an orthonormal net of "squares." Note that if each "square" were subdivided into four parts, these smaller regions would be closer to perfect squares. Therefore, the conductance of each subdivision is the same as every other, assuming they are drawn as nearly square as possible. There are ten "squares" in series for each stream tube, or sequence of squares from one port to the other; therefore, the transmittance of one stream tube is one-tenth that of one square, or $kw/10$. There are three such stream tubes, so the overall transmittance is $(3/10)kw$. In general, the overall transmittance of such a field is

$$\tau = \lambda k w, \tag{12.5}$$

in which λ, called the **form factor**, is the ratio of the number of squares in the direction transverse to the flow or displacement to the number of squares in the direction parallel to

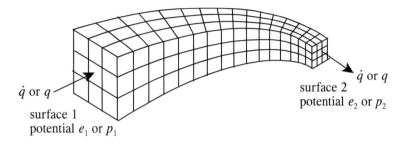

(a) three-dimensional two-port field

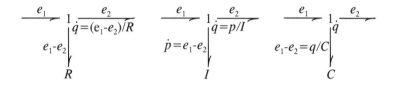

(b) bond-graph models

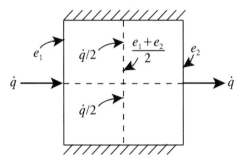

(c) conductance of planar squares

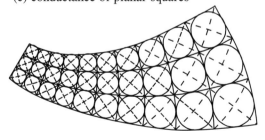

(d) sketching methods for planar fields

Figure 12.1: Two- and three-dimensional fields

the flow or displacement; in the present example, $\lambda = 3/10$. Fractions of squares may be accounted for greater accuracy.

A sophisticated analytical technique called **conformal mapping** is sometimes successful in mapping a curved region into a rectangular one, so that the form factor is determined analytically. Much more common today is the use of numerical computer routines for estimating the solution to Laplace's equation, to the same end. Most popular among these are finite element methods. For rapid and intuitive approximate analysis, however, it is hard to beat trial-and-error sketching of the field, particularly if it is planar, as suggested in the figure. Methods for sketching the orthogonal net of equipotential and flow or displacement lines are suggested in part (d) of the figure. If you use a pencil and a good eraser you usually can do a rather good job, and get a better grasp of the nature of the field than would result from using a computer solution. Note that lines drawn as diagonals of the "squares" also form an orthonormal set, and circles inscribed in the "squares" touch all four sides. You start by guessing either the flow/displacement lines or the equipotential lines; in the case shown, guessing the flow/displacement lines is probably easier. Further methods to aid in estimation are given below.

Estimates of the energy or dissipated power stored in a field also can be deduced from the sketched solution. For a single square, the power dissipated in the resistive case is $\dot{q}(e_1 - e_2) = kw(e_1 - e_2)^2 = (1/kw)\dot{q}^2$, the energy stored in the compliance case is $\frac{1}{2}kw(e_1 - e_2)^2 = (1/2kw)q^2$, and the energy stored in the inertance case is $\frac{1}{2}kw(p_1 - p_2)^2 = (1/kw)\dot{q}^2$. The potential differences, flows and displacements here refer to individual squares, of course, but are the same for all squares in a field. Therefore, these power dissipations or energy storages are the same for all squares, large and small. Thus, small squares have a higher power dissipation or energy storage *density* than large squares. Further, the power dissipated or energy stored for an entire field equals the power dissipated or energy stored per square times the total number of squares.

Using potential differences, flows and displacements that represent the entire field, the power dissipated or the energy stored becomes

$$\text{resistance:} \quad \mathcal{P} = \dot{q}(e_1 - e_2) = \tau(e_1 - e_2)^2 = \frac{1}{\tau}\dot{q}^2; \qquad R = \frac{1}{\tau}, \qquad (12.6a)$$

$$\text{compliance:} \quad \mathcal{V} = \frac{1}{2}q(e_1 - e_2) = \frac{1}{2}\tau(e_1 - e_2)^2 = \frac{1}{2\tau}q^2; \qquad C = \tau, \qquad (12.6b)$$

$$\text{inertance:} \quad \mathcal{T} = \frac{1}{2}\dot{q}(p_1 - p_2) = \frac{1}{2}\tau(p_1 - p_2)^2 = \frac{1}{2\tau}\dot{q}^2; \qquad I = \frac{1}{\tau}. \qquad (12.6c)$$

When τ is based on a transverse form factor, $1/\tau$ refers to the corresponding longitudinal form factor, and vice-versa. Thus, the inertance or the resistance of a field has the same form factor as the compliance of the same field, but in the transverse direction.

12.1.6   Estimating Upper and Lower Bounds for Field Transmittances*

Trial-and-error in the estimation of field transmittances may be reduced, and upper and lower bounds of the form factor estimated, through the use of methods given by Paynter.[1] He applies the **Thompson Principle** to give an estimate of τ which, if done with reasonable care, is on the low side of the actual value. He also applies the **Dirichlet Principle** similarly to give an estimate on the high side.

The Thompson Principle uses an assumed distribution of the displacement or the flow. The total energy storage or power dissipation associated with this distribution is estimated;

---

[1]H.M. Paynter, *Analysis and Design of Engineering Systems*, M.I.T. Press, 1961. This section may be skipped without loss of continuity.

errors in the assumed field produce an over-estimate in the energy or power. This happens because the actual flow or displacement distributes itself so as to minimize the energy or power in the field.

Using the subscript T to indicate the estimate with an assumed resistance field, equation (12.6a) gives

$$\tau_T \equiv \frac{(\Sigma \dot{q})^2}{\mathcal{P}_T} < \frac{(\Sigma \dot{q})^2}{\mathcal{P}} = \frac{\Sigma \dot{q}}{e_1 - e_2} = \tau, \tag{12.7}$$

showing that τ_T is indeed a lower bound for τ. The same result applies to the compliance and inertance fields.

The Dirichlet Principle uses an assumed distribution of the constant-potential lines (e or p). Again, the consequent field would have a greater energy or power dissipation than the actual field. Using the subscript D, the expression analogous to equation (12.7) is

$$\tau_D \equiv \frac{\mathcal{P}_D}{(e_1 - e_2)^2} > \frac{\mathcal{P}}{(e_1 - e_2)^2} = \frac{\Sigma \dot{q}}{(e_1 - e_2)} = \tau, \tag{12.8}$$

showing that τ_D is indeed an upper bound for τ. The Thompson-Dirichlet results can be summarized as follows:

$$\tau_T < \tau < \tau_D, \tag{12.9a}$$

$$\lambda_T < \lambda < \lambda_D. \tag{12.9b}$$

Example 12.1

Find a close overestimate of the inertance of a planar field of incompressible fluid of density ρ, between the left and right boundaries as plotted below. The pressures along each boundary are uniform, and viscous effects are to be neglected.

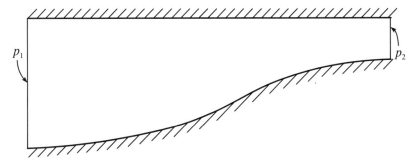

Solution: The Thompson principle, when applied approximately, gives the desired overestimate. First, streamlines are guessed; those shown below divide the field into four channels that each carry one-quarter of the total flow. This comprises the Thompson estimate of the flow distribution. Each channel now is carefully subdivided into squares (estimated below using circles), without reference to the other channels:

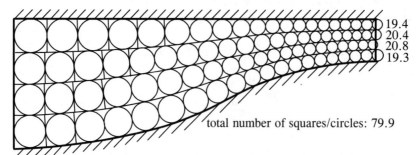

19.4
20.4
20.8
19.3

total number of squares/circles: 79.9

If the original streamlines were chosen without error, the resulting equipotential lines for the four channels would line up; clearly, the estimate was not perfect. At this point the work could be erased and a more informed estimate attempted. If instead the decision is to proceed, the kinetic energy of the assumed field is written

$$\mathcal{T}_T = \frac{1}{2}\frac{\rho}{w}\sum_{i=1}^{k}\dot{q}_i^2 = \frac{1}{2}\frac{\rho}{w}k\left(\frac{\dot{q}}{4}\right)^2.$$

This recognizes that each square is supposed to have the same flow, $\dot{q}/4$. The Thompson susceptance therefore becomes

$$\tau_T = \frac{\dot{q}^2/2}{\mathcal{T}_T} = \frac{16w}{\rho k}.$$

Here, k is the total number of squares, including fractions thereof; in the present case it is $k = 79.9$. It may be seen that errors in the original assumptions of the streamlines should always result, if care is taken, in an excess of squares, so therefore τ_T is an underestimate of the correct τ. The inertance estimate I_T is the reciprocal of $\mathcal{T}_T$, so that

$$I > I_T = 4.99\frac{\rho}{w}.$$

Example 12.2

Find a close underestimate of the inertance of the fluid field addressed in the above problem.

Solution: The Dirichlet principle, applied approximately, gives the desired result. This starts with a guess of the equipotential lines. The potential drop is arbitrarily subdivided into eighteen equal intervals. For each interval, squares or circles are drawn and counted; there are many fractions of squares to count also, giving a total of about $k = 64.2$:

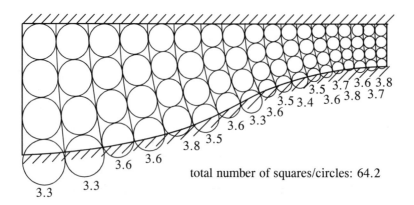

3.5 3.7 3.6 3.8
3.5 3.4 3.6 3.8 3.7
3.6
3.6 3.3 3.6
3.8 3.5
3.6 3.6
3.3 3.3
3.3

total number of squares/circles: 64.2

The kinetic energy of the assumed field becomes

$$\mathcal{T}_D = \frac{1}{2}\frac{\rho}{w}\sum_{i=1}^{k}(\Delta p_i)^2 = \frac{1}{2}\frac{\rho}{w}k\left(\frac{p_1 - p_2}{18}\right)^2,$$

with the result

$$\tau_D = \frac{\mathcal{T}_D}{(p_1 - p_2)^2/2} = \frac{k\rho}{324w}.$$

The inertance I is the reciprocal of the transmittance or susceptance, τ. Substituting the respective numbers of squares, k, gives

$$I < I_D = 5.05\frac{\rho}{w}.$$

Note that the difference between this overestimate and the underestimate found in the previous example is little over 1%.

12.1.7 Three-Dimensional Vector Nodic Fields

The three-dimensional equivalent of the square is the *cube*. Since a cube is the same as a square of thickness equal to its other two edge lengths, defined as L, its transmittance is

$$\tau_i = kL_i \tag{12.10}$$

The field comprises m stream tubes arrayed in a cluster instead of a sheet; fractions of such stream tubes can be included. The flow through each stream tube in a resistive field, and therefore through each cube, is the same. The product $\Delta e_i L_i$ or $\Delta p_i L_i$ for the ith cube therefore is the same for all cubes. Choosing any particular stream tube, the total potential drop between the ports of the field therefore is

$$e_1 - e_2 = \frac{\dot{q}}{km}\sum_{i=1}^{n}\frac{1}{L_i} \tag{12.11}$$

where n is the total number of cubes in the stream tube. The field conductance thus becomes

$$G = \frac{\dot{q}}{e_1 - e_2} = \frac{km}{\sum_{i=1}^{n}1/L_i} \tag{12.12}$$

Doubling all dimensions, while keeping the same shape, doubles the conductance. Thus, it is traditional to represent the conductance as a product of the conductivity, k, a **characteristic length**, L, which can be any dimension of the field, and a form factor, λ:

$$G = \lambda L k \tag{12.13}$$

Considering compliance and inertance fields also, this result can be generalized to give

$$\tau = \lambda L k \tag{12.14}$$

Notice that, unlike for the planar field, the form factor depends on the choice of characteristic length. The energy dissipated or stored in each cube is proportional to the linear size of the cube, like the transmittance. Equations (12.6) (p. 942) apply to the three-dimensional field as well as the planar field.

Rarely is it practical for the modeler actually to sketch cubes for a three-dimensional field; the analysis above can be interpreted in more practical terms. In most cases the

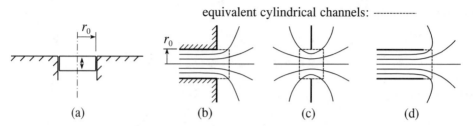

Figure 12.2: Inertances of cylindrical disk and holes

simplest and best approach is to estimate the flow lines so as to permit an estimate of the energy storage or power dissipation at each point in the field as a function of the total flow or displacement. This energy or power then is integrated over the field to give the total energy stored or power dissipated. The result is then equated to the proper power or energy of equation (12.6). Finally, the transmittance is deduced. This procedure is tantamount to application of the Thompson procedure, and consequently gives a resulting τ on the low side of the actual value. Starting with an estimate of the potential lines is an alternative: this application of the Dirichlet procedure gives an overestimate of τ.

12.1.8   Near-Spherical Fields

A major problem in acoustics is to establish the force required to vibrate a circular disk of radius r_0 in an infinite planar baffle facing a semi-infinite body of incompressible and inviscid fluid of density ρ, as pictured in part (a) of Fig. 12.2. The load impedance, or ratio of the force to the volume flow rate displaced by the disk, is inertive, with inertia $I = 0.85\rho/\pi r_0$, as reported by Beranek.[2] This inertia equals that of a cylinder of the same fluid with radius r_0 and length $0.85r_0$. It gives the impedance of a flat speaker disk in an infinite baffle when the frequency is low enough that the wavelength in the medium is much longer than the diameter of the disk. It also gives the end correction to the inertance of the flow through a circular tube that ends in an infinite baffle, as shown in part (b) of the figure. Thus, for example, when flow passes through a circular hole in a rigid sheet of virtually zero thickness, as shown in part (c) of the figure, the apparent inertia that the flow sees is $2 \times 0.85\rho/\pi r_0$. This is critical when the mean flow rate is zero, since the acoustic resistance is virtually zero in this case, leaving no other impedance to flow. The design of acoustic mufflers hinges on this fact. Similar corrections often are important in hydraulics. A companion problem removes the baffle, as pictured in part (d) of the figure. The result, also given by Beranek, is $I = 0.613\rho\pi r_0$, or an equivalent cylinder of fluid of length $0.613r_0$.

These results follow from integrating the kinetic energy of the fluid over the entire field, and setting it equal to $\frac{1}{2}IQ^2$, where $Q = \pi r_0^2 v_0$ in which v_0 is the velocity of the disk. This requires establishing the field response to a point source with flow dQ, and integrating over all point sources within the disk. The vector nature of the flow field makes this a formidable task. The objective of the following is to establish a relatively simple approximation, which also can be applied to similar problems.

Consider a hemispherical surface with the radius of the disk, placed directly on the disk, as pictured in Fig. 12.3. Assuming that the fluid touching the disk has the velocity v_0, the mean velocity of the fluid touching the hemispherical surface is $v_0/2$, since this surface has twice the area. The average velocity of the fluid between the two surfaces must be something close to $(3/4)v_0$ which, considering that the volume of this region is $V = (2/3)\pi r_0^3$, gives a kinetic energy of roughly $\frac{1}{2}\rho V v_0^2 = (3/16)\pi\rho r_0^3 v_0^2$. Outside the

---

[2]Leo L. Beranek, *Acoustics*, 2nd edition, American Institute of Physics, Inc., New York, NY, 1986.

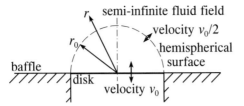

Figure 12.3: Approximate field for analysis of the apparent fluid inertia seen by a disk

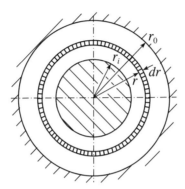

Figure 12.4: Cylindrical field

surface $(r/r_0 > 1)$, volumetric continuity requires that the velocity be $(r_0/r)^2(v_0/2)$, so the energy of the semi-infinite field is given by

$$\mathcal{T} = \frac{1}{2} \int_r^\infty \rho \left(\frac{r_0^2 v_0}{2r^2} \right)^2 2\pi r^2 \, dr = \frac{1}{4} \pi \rho r_0^4 v_0^2 \int_{r_0}^\infty \frac{1}{r^2} \, dr = \frac{1}{4} \pi \rho r_0^3 v_0^2. \tag{12.15}$$

The total kinetic energy becomes about $(7/16)\pi \rho r_0^3 v_0^2$, which is equivalent to a cylinder of the fluid with radius r_0 and length $(7/8)r_0 = 0.875r_0$, very close the actual $0.85r_0$. Carrying out the same procedure in the absence of the baffle gives the length $(5/8)r_0 = 0.625$, very close the the actual $0.613r_0$.

12.1.9 Near-Cylindrical Fields

The simplest two-dimensional linear field lies between two circles, as pictured in Fig. 12.4, and gives for example the capacitance and the inductance per unit length of a coaxial cable. The component differential ring of width dr shows field lines; its form factor in the radial direction is $d\lambda_r = (1/2\pi r)dr$. Integrating this between r_i and r_0 gives the overall radial form factor

$$\lambda_r = \frac{1}{2} \int_{r_i}^{r_0} \frac{1}{r} dr = \frac{1}{2\pi} \ln(r_0/r_i). \tag{12.16}$$

The inductance per unit length is therefore $I' = \mu/2\pi) \ln(r_0/r_i)$. The capacitance is most easily understood in terms of the transverse form factor, λ_p, which is the number of squares in the peripheral direction divided by the number in the radial direction. This is the inverse of λ_r, and gives the capacitance per unit length $C' = 2\pi\epsilon/\ln(r_0/r_i)$. These results are reported in Table 10.1 (p. 716).

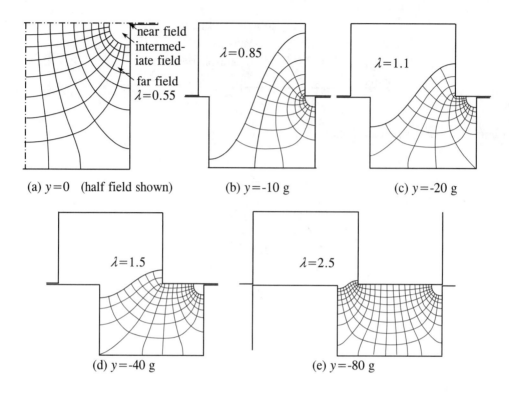

(a) $y=0$ (half field shown)   (b) $y=-10$ g   (c) $y=-20$ g

(d) $y=-40$ g   (e) $y=-80$ g

Figure 12.5: Approximate fringing electric fields for teeth in stepping motor problem

For a wedge of angle θ, the transverse form factor is proportional to θ. The form factor for a half-circle is used below: $\lambda_r = (1/\pi)\ln(r_0/r_i)$.

The radial form factor is unbounded as $r_0 \to \infty$; there is no limit to how large I' could get, or how small C' could get. This applies to two-dimensional fields in general, and is categorically different from three-dimensional fields. For example, the total inductance for a sphere in open space equals the limit $I = 2\mu r_0$, or the inertance for radial pulsations of a sphere in an infinite bath of an incompressible liquid is $I = 2\rho r_0$ (see Problem 12.5). This unboundedness complicates the estimation of form factors for fringing in essentially two-dimensional plate capacitors, as an example.

The simulation of the fields and forces of the teeth in a stepping motor, as carried out in Sections 9.5.8 (pp. 668-674) and 9.6.5 (p. 690), requires at least a rough estimate of the fringing of the field in the gap between two interacting teeth. The estimate used is based on the following analysis, which serves as an example of how one can proceed without invoking fancy software for analyzing fields.

The tooth spaces are assumed to be five-fourths of the tooth widths, which are 80 times the tooth gaps, and the tooth depths are five-sixths of the tooth widths. The position for perfect alignment of the teeth is shown in part (a) of Fig. 12.5. The form factor in the longitudinal direction is estimated using a mass flow or inertance analogy, giving an end correction or equivalent extension of the field within the gap into the region outside. This is easier to visualize than the transverse form factor, despite the fact that the latter applies most conventionally to the capacitance. The resulting end correction is the same in either case. The field outside of the gap itself is divided into three regions: near to the gap, intermediate distance, and far. The boundary of the field on its left side is midway to

	Positions $|y|/g$				
	0	10	20	40	80
near field	0.25	0.53	0.53	0.53	0.53
intermediate field	0.95	1.47	1.47	1.47	1.47
far field	0.55	0.85	1.1	1.5	2.5
total	1.75	2.85	3.1	3.5	4.5

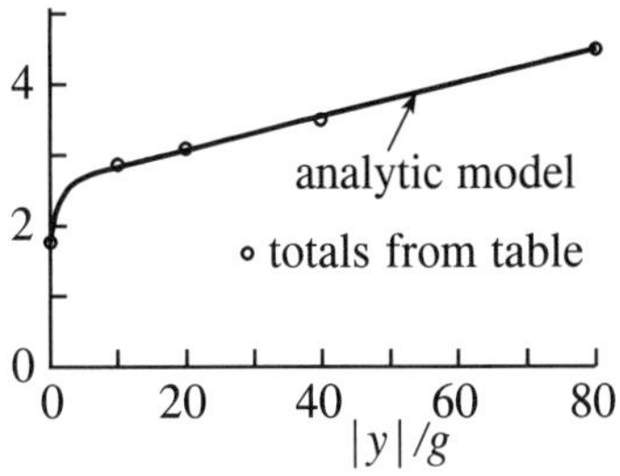

Figure 12.6: Values of the fringing form factors for the stepping motor tooth problem

the next tooth gap on that side; the region on the other side of this line is ascribed to the fringing of that tooth.

The half-circle comprising the near field (only half of which is shown, due to symmetry) has volume $\pi g^2 w_d/8$, where w_d is the depth or axial length of the teeth. If the velocity of the imaginary fluid in the gap is v_0, the average velocity on the quarter-circle is $(2/\pi)v_0$, and the average velocity within the near field is about $0.82v_0$. This gives a kinetic energy of $\frac{1}{2}\rho(\pi g^2 w_d/8)(.82v_0)^2 = \frac{1}{2}\rho g w_d \ell_1 v_0^2$, contributing to the fringing extension $\ell_1 = 0.26g$.

The intermediate field can be approximated as being cylindrical. Its inner radius is $g/2$, and its outer radius is taken, somewhat arbitrarily, as $10g$, or one-eighth of the tooth width. The longitudinal form factor for this field is therefore $(1/\pi)\ln(20) = 0.95$, which directly contributes the fringing extension $\ell_2 = 0.95g$. The far field is estimated by trial-and-error sketching; the field shown took about fifteen minutes to draw using a basic graphics program (pencil and eraser would do as well) and gives a longitudinal form factor of 0.55 (half that for the half-field shown), contributing $\ell_3 = 0.55g$ to the fringing extension. The total extension therefore is about $(0.25 + 0.95 + 0.55)g = 1.75g$, and double this when both ends of the air gap are considered.

The same treatment is applied to four other tooth positions in parts (c) – (f) of the figure. The various values are tabulated in Fig. 12.6, and the resulting totals are plotted. Based mostly on this, the total width of the equivalent gap with thickness g is approximated by

$$w = g(1.0475|y|/g + 5.2 - 1.7e^{|y|/1.623}. \tag{12.17}$$

The part of this that represents the fringing extension at one end of the gap is plotted for comparison to the tabulated values. The exponential term is used both to satisfy the condition for $y = 0$ and to make $(dw/dy)_{y=0} = 0$. Equation (12.17) is used in equation (9.70)(p. 670) and the codes for Example 9.17 (p. 674) and Example 9.20 (p. 690). It is probably more accurate than necessary.

12.1.10 Multiport Vector Nodic Fields

The linear vector fields considered thus far have two ports. Sometimes one wishes to model a linear field with three or more ports, which requires only a modest conceptual extension of what has been done already. An example of a three-port field is shown in Fig. 12.7 part (a). Assuming a resistive field (inertances could be readily substituted), a **tee model**, so called because of its shape, is shown in part (b), and a **delta model**, again named for its shape, is shown in part (c).

The first of two approaches starts by assuming that port 3 is blocked, and evaluates the resistance between the remaining ports 1 and 2. This resistance can be seen to equal $R_1 + R_2$. Next, port 2 is assumed blocked, and the resistance $R_2 + R_3$ is computed. Again,

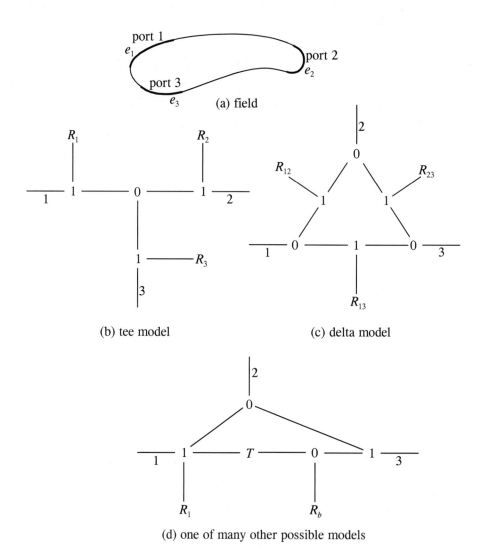

(a) field

(b) tee model   (c) delta model

(d) one of many other possible models

Figure 12.7: Three-port resistive field

the resistance $R_1 + R_3$ is computed in a similar manner. Finally, the individual resistances R_1, R_2 and R_3 are deduced from the three sums.

The second approach starts by assuming ports 1 and 2 are connected to the same effort, leaving a two-port field with conductance $(1/R_{13}) + (1/R_{23})$. Then ports 1 and 3 are similarly connected, leaving a two-port field with conductance $(1/R_{12}) + (1/R_{23})$. Connecting ports 2 and 3 together gives a two-port field with conductance $(1/R_{12}) + (1/R_{13})$. Knowledge of the three conductances permits the individual conductances or resistances to be computed. The tee model parameters can be converted to the delta model parameters, or vice-versa, if desired.

There are many other equivalent three-parameter models for a three-port field. Part (d) of the figure shows one of these. Equivalence may be established most generally by equating the power dissipation or energy storage of the respective models.

Access to an actual piece of hardware may permit three individual experiments to be carried out which determine the component parameters of a model, following the same logic as the analysis. Most experienced engineers employ a combination of analytical and experimental methods in modeling. The process of modeling from purposeful experiment has become known as **experimental identification**.

12.1.11   Summary

Field lumping is the essence of lumped modeling. The major source of error in a carefully executed lumped model results from neglect of couplings between different types of fields. The resulting error can be reduced, at the expense of added complexity, by choosing smaller spatial domains for the fields.

Scalar fields with a uniform effort are assumed frequently, and are relatively simple to understand and compute. Emphasis therefore has been placed on the next more difficult type of field, the linear nodic vector field. In the usual application of this field the vector generalized velocity or displacement is nodic, that is, acts like an incompressible medium without internal sources or sinks. Simply organized nodic vector fields such as rigid body motion and displacement of a rigid body or a liquid in a gravity field can be addressed analytically, as you have seen previously.

Special attention has been devoted to the case in which the effort or generalized momentum acts as a scalar potential; the flow is proportional to the gradient of this potential. The resulting field satisfies Laplace's equation. Examples include resistance fields in which the flow of a fluid or of an electric current is impeded by a resistivity, compliance fields in which the displacement of electric charge is impeded by the dielectric phenomenon or the magnetic displacement is impeded by the permeability phenomenon, and inertance fields in which the acceleration of a fluid is impeded by its inertance. Although precise analytic methods are available in some cases and computer methods are widely available, emphasis has been placed on approximate analytical and field-sketching methods, since these methods impart an understanding which aids the design process more directly, and give quick estimates. Approximations employing simple spherical and cylindrical fields have been highlighted.

Nodic vector fields with more than two ports can be treated as combinations of pairs of fields with two ports. The results can be organized in terms of bond graph equivalences. More complicated tensor fields such as are associated with the strain of an elastic solid are in general beyond the scope of this text, but simple approximate cases such as the classic bending of beams or torsion in rods and coil springs can be accommodated readily, either through the standard development of their constitutive relations or through evaluation of their potential energy storage. See also the tensor fields in Section 9.3.6.

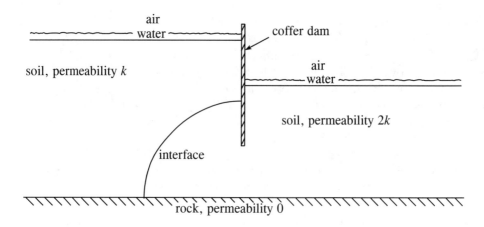

Figure 12.8: Guided Problem 12.1

Guided Problem 12.1

As sketched in Fig. 12.8, a coffer dam holds back earth and water at an excavation site. The **permeability** (conductivity) of the soil to water seepage in the region to the left of the given dividing line is k; in the region to the right it is $2k$. The area is undergirded by impermeable rock. Estimate the seepage rate under the coffer dam per running foot in terms of k.

Suggested Steps:

1. Photocopy the figure, perhaps doubling its size. Assume that the dividing line between the two soils is an equipotential surface. (The problem is in fact arranged so that this is a line of virtually uniform pressure.) Guess a streamline above which half the water flows. Take each of the two regions above and below this streamline and sketch further estimated streamlines which separate their respective regions into two equal-flow subregions. This process of subdivision can be repeated further, but as a practical matter you might do this only for the soil on the left, and there only once more.

2. Starting from the line dividing the two soils, sketch candidate constant-pressure lines, attempting to form squares with the flow lines.

3. Critique your results, paying special attention to whether the aspect ratios of your would-be "squares" are close to 1:1 and the flow and equipotential lines intersect perpendicularly to each other. Erase and re-sketch a few times until you are satisfied. (This is not getting as fancy as use of the Thompson or Dirichlet principles, which together are directed at estimating error.) Pay special attention to the region immediately below the coffer barrier, where a near-singularity exists that produces very small "squares."

4. Find the form factors λ for each of the two soil regions.

5. Find the conductances of the two regions in terms of k.

6. Relate the total flow to the two conductances. (Note that the pressure drops add, and the flows are common.)

7. State a compatible set of units for k, Q and ΔP.

PROBLEMS

12.1 Estimate the resistance of the *fuse link* of uniform thickness t. The constitutive relation (equation (12.3a)) is $\mathbf{J} = -\sigma\,\nabla\mathbf{e}$, where σ is the electrical conductivity.

12.2 Consider that the upper and lower surfaces of the fuse link above represent the electrodes of a capacitor, with depth w, and the material within has dielectric constant ϵ. Neglecting end corrections, how does the capacitance relate to the resistance of the fuse link?

12.3 Estimate the inertance referred to the velocity of the piston below. The system is axisymmetric and is filled with a fluid of density ρ.

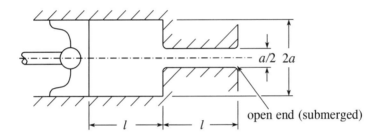

12.4 An impeller pump of the type used for cooling the steam in a power plant includes an inlet convergent section, a row of stator vanes, a row of rotating vanes, a second row of stator vanes and a diffuser section, with the rotating shaft contained within a fixed central column. (The axis drawn horizontally below actually is vertical.) The transient produced when the pump is started and the water rises from its initial level has the potential of producing an unacceptable waterhammer effect, so it is important to know the inertia of the water.

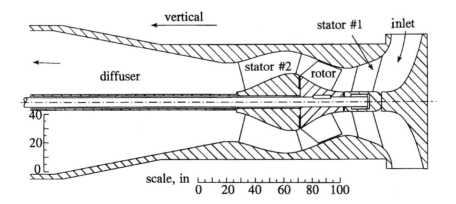

(a) Estimate this inertia for the system shown as a function of the depth of the water, starting at its initial level at the bottom of the diffuser and ending when the water has reached the top of the diffuser.

(b) Construct a calculable model that accommodates the bond-graph restriction to constant inertances. Define the moduli of all elements.

12.5 Determine by integration that the inertance seen by a sphere of radius r_0 pulsating in an infinite incompressible fluid bath of density ρ, defined in terms of the volumetric flow, is $I = 2\rho r_0$.

12.6 Two electrically conducting cylinders of radius r_0 face each other with a small gap $g \ll r_0$ to form a capacitor. Estimate the radius r_e of an equivalent disk of thickness g that includes the fringing effect, or for which the capacitance is approximately $C = \pi \epsilon r_e^2 / g$. Hint: Integration of a layer shaped like the surface of the wedge of an orange is suggested, since the field lines lie in planes that include the axis of the cylinders.

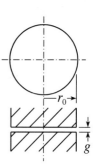

SOLUTION TO GUIDED PROBLEM

Guided Problem 12.1

1-3. The following sketch represents about ten minutes of draw-and-erase-and-draw-again:

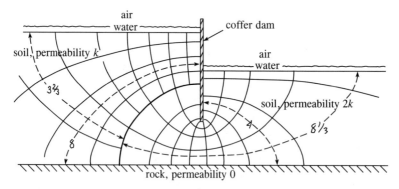

4. On the left side, the form factor is about $\lambda_1 = \dfrac{8}{3\frac{2}{3}}$; on the right side, $\lambda_2 = \dfrac{4}{8\frac{1}{3}}$.

5. The conductances are the transmittances $\tau_1 = k\lambda_1$ and $\tau_2 = 2k\lambda_2$.

6. $Q = \dfrac{1}{R}\Delta P = \dfrac{1}{(1/\tau_1) + (1/\tau_2)}\Delta P = \dfrac{2}{3}k\Delta P.$

7. $[Q] = \dfrac{\text{ft}^3/\text{hr}}{\text{ft}} = \dfrac{\text{ft}^2}{\text{hr}}; \quad [\Delta P] = \text{psi}; \quad [k] = \dfrac{\text{ft}^3}{\text{hr psi}}.$

12.2 Nonconservative Couplers

Engineers often wish to model highly complex two-port devices in an approximate summary fashion, without reticulating them into a junction structure with one-port elements attached. Examples include dynamic hydraulic machinery, such as the pumps and turbines pictured in Fig. 12.9. One port is the mechanical shaft; the other is the net fluid power with its common input and output flow and its pressure rise or drop. This section considers general static models of the form

$$\frac{e_1}{\dot{q}_1}\!-\text{STATIC 2-PORT}\,\frac{e_2}{\dot{q}_2}\!- \qquad \text{or} \qquad \frac{e_1}{\dot{q}_1}\!-\, R_f\,\frac{e_2}{\dot{q}_2}\!-$$

in which R_f is called an explicit two-port **resistance field**. The impeller pump will be used as a case study.

12.2.1 Causal Relations

Four possible causal patterms are possible:

$$\text{impedance}: \vdash\!\frac{e_1}{\dot{q}_1}\!-\, R_f\,\frac{e_2}{\dot{q}_2}\!-\!\dashv \quad e_1 = e_1(\dot{q}_1, \dot{q}_2); \quad e_2 = e_2(\dot{q}_1, \dot{q}_2) \qquad (12.18a)$$

$$\text{admittance}: \frac{e_1}{\dot{q}_1}\!-\!\dashv R_f\!\vdash\!\frac{e_2}{\dot{q}_2}\!- \quad \dot{q}_1 = \dot{q}_1(e_1, e_2); \quad \dot{q}_2 = \dot{q}_2(e_1, e_2) \qquad (12.18b)$$

$$\text{immittance}: \vdash\!\frac{e_1}{\dot{q}_1}\!-\, R_f\!\vdash\!\frac{e_2}{\dot{q}_2}\!- \quad e_1 = e_1(\dot{q}_1, e_2); \quad \dot{q}_2 = \dot{q}_2(\dot{q}_1, e_2) \qquad (12.18c)$$

$$\text{adpedance}: \frac{e_1}{\dot{q}_1}\!-\!\dashv R_f\,\frac{e_2}{\dot{q}_2}\!-\!\dashv \quad \dot{q}_1 = \dot{q}_1(e_1, \dot{q}_2); \quad e_2 = e_2(e_1, \dot{q}_2) \qquad (12.18d)$$

The environment of the coupler determines which causal pattern to use. As an example, consider an induction motor driving an impeller pump which in turn forces water through a load:

$$\text{IND. MOTOR}\,\frac{M}{\dot{\phi}}\!-\text{IMPELLER PUMP}\,\frac{P}{Q}\!-\text{HYD. LOAD}$$

If transient behavior is of interest, the inertias of the motor, the attached pump rotor, and the fluid being pumped are important. This gives the bond graph shown in Fig. 12.10. Application of integral causality shows that the pump should be represented by the impedance causality given in equations (12.18a). The manufacturer of the pump almost surely provides characteristics in largely graphical form.

12.2.2 Equilibrium

Equilibrium is determined, as a first step, using graphical characteristics. Each of the relationships within equations (12.18) can be generalized to the form $R(x_1, x_2, x_3) = 0$, where R can be read as "relation of." Represented graphically, this means a family of characteristics; one variable is plotted on the abscissa, another on the ordinate, while the third becomes a discrete parameter. Two such families of curves are needed to characterize the pump. There are many choices, any of which is acceptable for purposes of finding equilibrium. Two choices are particularly convenient, however, since they match with the ways the characteristics of the motor and the load are represented.

   You are accustomed to representing a hydraulic load with a pressure-flow $(P - Q)$ characteristic; presume the load curve is as plotted in part (a) of Fig. 12.11. Similarly, you are accustomed to representing the motor with a torque-speed $(M - \dot{\phi})$ characteristic;

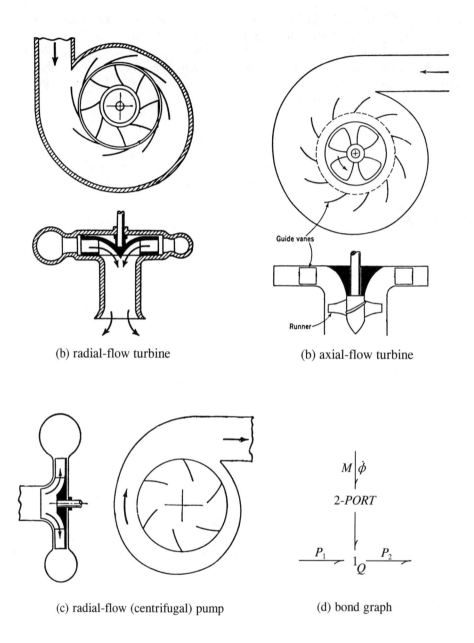

(b) radial-flow turbine

(b) axial-flow turbine

(c) radial-flow (centrifugal) pump

(d) bond graph

Figure 12.9: Examples of dynamic pumps and motors
Adapted from Engineering Applications of Fluid Mechanics
*by J.C. Hunsaker and B.G. Rightmire, copyright 1947,
with permission from The McGraw-Hill Companies*

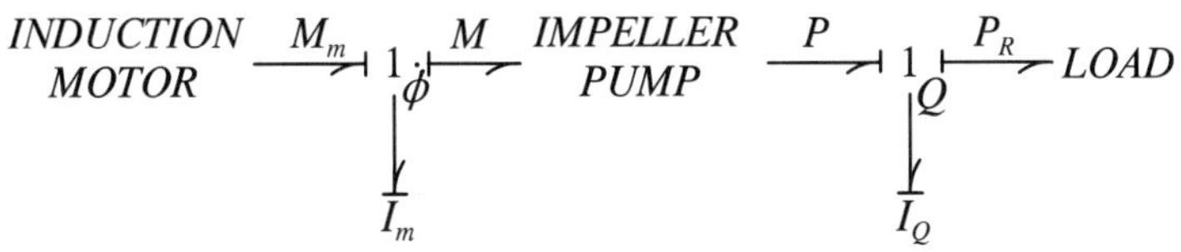

INDUCTION M_m 1 M IMPELLER P 1 P_R LOAD
MOTOR ϕ PUMP Q

I_m I_Q

Figure 12.10: Bond graph for impeller pump with dynamic source and load

presume the motor characteristic is as plotted in part (b) of the figure. The same coordinates can be used to represent the pump. The first of these therefore is either $R_1(P, Q, \dot{\phi}) = 0$ or $R_1(P, q, M) = 0$; the former is more conventional, and also is shown in part (a) of the figure. The second of these is either $R_2(M, \dot{\phi}, Q) = 0$ or $R_2(M, \dot{\phi}, P) = 0$; the former is closer to most commonly available information, and is also shown in part (b) of the figure.

The resulting equilibrium state for the system can be found in two steps. First, the intersections of the load characteristic with the various $P - Q$ curves are cross-plotted onto the $M - \dot{\phi}$ plot, as shown by the dashed line. Second, the equilibrium is noted as the intersection of this line and the motor characteristic, as noted by the cross. This result also is cross-plotted onto the $P - Q$ plot, as shown by the cross there. This is a general procedure, which applies to finding the equilibrium of any two-port device that connects a source to a resistive load.

Pump manufacturers rarely report characteristics in a form as complete as those given in Fig. 12.10, unfortunately. More likely, they give a single $P - Q$ curve for a particular speed $\dot{\phi}$, and a single $M - \dot{\phi}$ curve for a particular flow Q. They also may give results on efficiency, although this information is redundant, as will be shown. Although this situation is not particularly defensible, at least without further explanation, it results from the manufacturer's implicit assumption that the application engineer can very closely extrapolate the given information. How can you rise to this challenge without using a more specialized approach than necessary? As a systems engineer you cannot be expected to become expert on all types of engineering devices.

12.2.3 Dimensional Analysis

It is assumed that the inlet pressure ("suction head") is high enough to preclude cavitation, and that bearing friction is negligible. For a set of geometrically similar pumps of different sizes, the state is defined by the dimensional parameters of any characteristic length (typically the diameter of the impeller), the fluid density ρ and the viscosity μ, combined with the variables P, Q and $\dot{\phi}$. Thus the impedance relation of equation (12.18a) becomes

$$M = M(\dot{\phi}, Q, L, \rho, \mu), \tag{12.19a}$$

$$P = P(\dot{\phi}, Q, L, \rho, \mu). \tag{12.19b}$$

Various nondimensional groups can be formed from the dimensional variables and parameters. Perhaps the most useful is

$$\frac{M}{\rho L^5 \dot{\phi}^2} = f\left(\frac{Q}{L^3 \dot{\phi}}, \frac{\rho Q}{L\mu}\right), \tag{12.20a}$$

$$\frac{P}{\rho L^2 \dot{\phi}^2} = g\left(\frac{Q}{L^3 \dot{\phi}}, \frac{\rho Q}{L\mu}\right), \tag{12.20b}$$

where f and g represent as yet unknown functions of their two arguments. Results using this representation apply regardless of the size of the pump or the density or viscosity of the fluid.

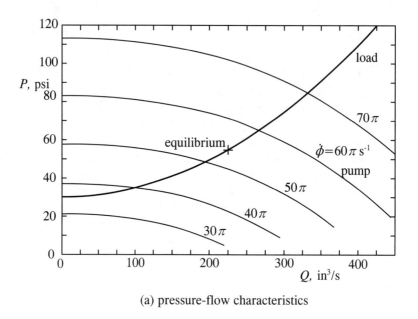

(a) pressure-flow characteristics

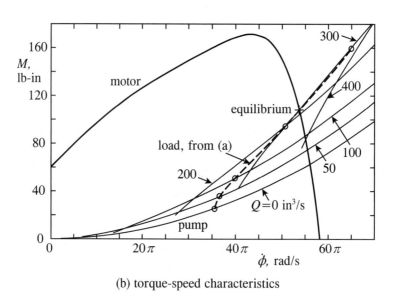

(b) torque-speed characteristics

Figure 12.11: Steady-state matching of motor, impeller pump and load

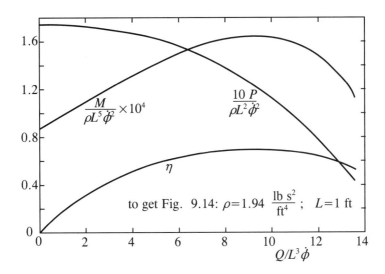

Figure 12.12: Dimensionless behavior of impeller pump assuming Eulerian similitude

The group $\rho Q/L\mu$ is a Reynolds number, which is proportional to the ratio of forces due to inertia to the forces due to viscosity. Most fluid machinery is operated at so large a Reynolds number, that is so large a flow or so small a viscosity, etc., that forces due to viscosity are virtually negligible. Assuming this applies to the pump, equations (12.20) specialize to the case of **Eulerian similitude**:

$$\frac{M}{\rho L^5 \dot{\phi}^2} = f'\left(\frac{Q}{L^3 \dot{\phi}}\right), \tag{12.21a}$$

$$\frac{P}{\rho L^2 \dot{\phi}^2} = g'\left(\frac{Q}{L^3 \dot{\phi}}\right). \tag{12.21b}$$

Here, f' and g' represent functions of a single variable or the group $Q/L^3\dot{\phi}$, which often is called the **flow coefficient** of the pump. The flow coefficient is proportional to the ratio of the velocity of the fluid at some location to the velocity of the tip (for example) of the impeller.

The efficiency of the pump, η, is the ratio of the output power to the input power, or

$$\eta \equiv \frac{PQ}{M\dot{\phi}} = \frac{(Q/L^3\dot{\phi})g'(Q/L^3\dot{\phi})}{f'(Q/L^3\dot{\phi})} = h\left(\frac{Q}{L^3\dot{\phi}}\right). \tag{12.22}$$

Therefore, if a manufacturer gives you just two curves or equations, each interpretable in terms of any two of the three equations (12.21a), (12.21b) or (12.22, you can deduce the third curve or equation directly, and go on to plot the two families of curves in Fig. 12.11 and the equations in any of the forms above. The relations used in the present case are given in Fig. 12.12. The characteristic length used in Fig. 12.11 is $L = 1$ ft, and the fluid density is $\rho = 1.94$ lb s^2/ft^4.

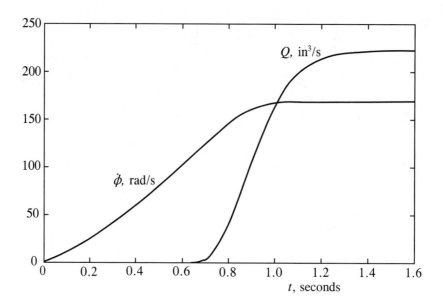

Figure 12.13: Simulation of start-up of impeller pump system

12.2.4   Dynamic Simulation

For the unsteady case of Fig. 12.10, the operating points on the motor characteristic and the pump characteristic (in part (b) of Fig. 12.11) no longer coincide; one is directly above the other, as required by the left-hand 1-junction, and the difference is the moment that accelerates or decelerates the motor. Similarly, the operating points on the load characteristic and the pump characteristic (in part (a)) no longer coincide; one is directly above the other (because of the right-hand 1-junction), and the pressure difference causes the fluid to accelerate.

The assumption of Eulerian similitude allows the state differential equations to be written with the only nonlinear functions for the pump being the f' and g' of equations (12.18). The causal strokes dictate

$$\frac{dp_m}{dt} = M_m - M = M_m(\dot{\phi}) - \rho L^5 \dot{\phi}^2 \, f'\left(\frac{Q}{L^3 \dot{\phi}}\right), \qquad (12.23a)$$

$$\frac{dp_Q}{dt} = P - P_R = \rho L^2 \dot{\phi}^2 \, g'\left(\frac{Q}{L^3 \dot{\phi}}\right) - P_R(Q). \qquad (12.23b)$$

The prominence of Q and $\dot{\phi}$ on the right sides of these equations suggests their use as the state variables in place of $p_m = I_m \dot{\phi}$ and $p_Q = I_Q Q$. Thus the equations become

$$\frac{d\dot{\phi}}{dt} = \frac{1}{I_m} \left[M_m(\dot{\phi}) - \rho L^5 \dot{\phi}^2 \, f'\left(\frac{Q}{L^3 \dot{\phi}}\right), \right] \qquad (12.24a)$$

$$\frac{dQ}{dt} = \frac{1}{I_Q} \left[\rho L^2 \dot{\phi}^2 \, g'\left(\frac{Q}{L^3 \dot{\phi}}\right) - P_R(Q) \right]. \qquad (12.24b)$$

This result is implemented for a starting transient in Fig. 12.13. The following equations are used to represent the static characteristics of the motor, pump and load, which conform closely to the plotted characteristics in Figs. 12.11 and 12.12:

$$M_m = M_0 + M_1\dot{\phi} - M_4\dot{\phi}^4 - M_5\dot{\phi}^5 - M_6\dot{\phi}^6, \tag{12.25a}$$

$$q \equiv (Q/L^3\dot{\phi}) \times 10^4 \tag{12.25b}$$

$$g' = 0.174 - 0.000328\,q^2 - 0.0000281\,q^3, \tag{12.25c}$$

$$\eta = 0.2 - 0.0236\,q + 0.00158\,q^2 - 0.000053\,q^4, \tag{12.25d}$$

$$f' = \frac{q\,g'}{\eta} \times 10^{-4} \tag{12.25e}$$

$$P_R = P_0 + P_2 Q^2, \quad P_0 = 30\ \text{lb/in}^2, \quad P_2 = 5 \times 10^{-4}\ \text{lb s}^2/\text{in}^8, \tag{12.25f}$$

$$M_0 = 60\ \text{lb in}, \quad M_1 = 1.2\ \text{lb in s}, \quad M_4 = 1.453 \times 10^{-6}\ \text{lb in s}^4,$$

$$M_5 = -1.82 \times 10^{-8}\ \text{lb in s}^5, \quad M_6 = 6.35 \times 10^{-11}\ \text{lb in s}^6. \tag{12.25g}$$

The other parameters are taken as $L = 1$ ft, $\rho = 1.94$ lb·s^2/ft^2, $I_m = 0.05$ lb·ft·s^2 and $I_Q = 8310$ lb·s^2/ft^5. (The last number corresponds to a uniform channel of cross-sectional area 0.007 ft^2 and length 30 ft, which by itself produces a load characteristic close to that given.)

The simulation assumes that fluid is not allowed to flow in the reverse direction, presumably through the use of a check valve. Therefore the flow remains at zero until the rotor has sufficient speed to overcome the static back pressure of 60 psi. Thereafter, the flow accelerates very rapidly, and the equilibrium found previously is reached in less than 1.5 seconds from the time the motor is started.

12.2.5   Linear Couplers

A *linear* static coupler, defined in terms of variables that are symmetric with respect to the coupler,

$$\frac{e_1}{\dot{q}_1} \multimap R_f \multimapdotbothA \frac{e_2}{\dot{q}_2},$$

can be represented by an **impedance matrix Z** or an **admittance matrix Y**:

$$\begin{bmatrix} e_1 \\ e_2 \end{bmatrix} = \mathbf{Z} \begin{bmatrix} \dot{q}_1 \\ \dot{q}_2 \end{bmatrix}; \qquad \begin{bmatrix} \dot{q}_1 \\ \dot{q}_2 \end{bmatrix} = \mathbf{Y} \begin{bmatrix} e_1 \\ e_2 \end{bmatrix}. \tag{12.26}$$

This element also is known as a two-port **resistance field**, as designated by the traditional subscript f.

In order to represent the linear static coupler by a more detailed bond graph, it is convenient to employ the special two-port mutual resistance, designated here as R_{12}, as introduced in Section 9.3.1 (p. 619):

$$\frac{e_1}{\dot{q}_1} \multimap R_{12} \multimapdotbothA \frac{e_2}{\dot{q}_2} \qquad e_1 = R_{12}\dot{q}_2, \qquad e_2 = R_{12}\dot{q}_1. \tag{12.27}$$

Its describing equations *appear* to be very similar to those of the gyrator:

$$\frac{e_1}{\dot{q}_1} \multimap G \multimap \frac{e_2}{\dot{q}_2} \qquad e_1 = G\dot{q}_2, \qquad e_2 = G\dot{q}_1. \tag{12.28}$$

The gyrator, however, has *anti-symmetric* power conventions on its two bonds, whereas the mutual resistance has *symmetric* power conventions. This is a radical difference, since it makes the gyrator conserve energy and the mutual resistance consume the power

$$\mathcal{P} = e_1\dot{q}_1 + e_2\dot{q}_2 = 2R_{12}\dot{q}_1\dot{q}_2. \tag{12.29}$$

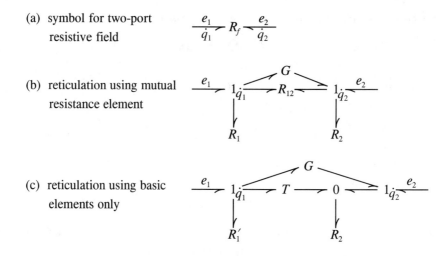

(a) symbol for two-port resistive field

(b) reticulation using mutual resistance element

(c) reticulation using basic elements only

Figure 12.14: Representations of the general linear static coupler

The mutual resistance is strictly symmetric with respect to its two ports; the gyrator is strictly anti-symmetric.

The linear static coupler can be represented as shown in part (b) of Fig. 12.13, from which you can see that

$$e_1 = R_1\dot{q}_1 + (R_{12} + G)\dot{q}_2, \tag{12.30a}$$

$$e_2 = R_2\dot{q}_2 + (R_{12} - G)\dot{q}_1, \tag{12.30b}$$

or

$$\mathbf{Z} = \begin{bmatrix} R_1 & R_{12} + G \\ R_{12} - G & R_2 \end{bmatrix}. \tag{12.31}$$

Three special categories of this coupler deserve attention. The **reciprocal coupler**, or coupler that satisfies **reciprocity**, has **transfer impedances** $\mathbf{Z}_{12}$ and $\mathbf{Z}_{21}$ that equal one another. This happens if and only if

$$G = 0, \tag{12.32}$$

so the gyrator is absent. The mutual resistance is inherently reciprocal, whereas the gyrator is not. Whole classes of systems, including ordinary electric circuits, are reciprocal (gyrators never are needed to model ordinary electric circuits).

Symmetric couplers are by definition unchanged if they are reversed end-for-end. Thus, not only are their transfer impedances equal to one another, the **self-impedances** Z_{11} and Z_{22} also equal one another, or

$$G = 0, \qquad R_1 = R_2. \tag{12.33}$$

Symmetry therefore imposes reciprocity, but not vice-versa.

The third category is the **passive coupler**, defined so that it never generates energy. The power of the general linear static coupler is

$$\mathcal{P} = e_1\dot{q}_1 + e_2\dot{q}_2 = [R_1\dot{q}_1 + (R_{12} + G)\dot{q}_2]\dot{q}_1 + [R_2\dot{q}_2 + (R_{12} - G)\dot{q}_1]\dot{q}_2$$
$$= R_1\dot{q}_1^2 + 2R_{12}\dot{q}_1\dot{q}_2 + R_2\dot{q}_2^2. \tag{12.34}$$

This is never negative, regardless of the values of $\dot{q}_1$ and $\dot{q}_2$, if and only if [3]

$$R_1 \geq 0, \qquad R_2 \geq 0, \qquad R_{12} \leq \sqrt{R_1 R_2} \qquad (12.35)$$

Violating the third condition gives $\mathcal{P} < 0$ for $\dot{q}_1 = -\dot{q}_2$. Note that the presence of the gyrator has no effect on passivity, which is to be expected since the gyrator conserves energy.

The bond graph of part (c) of Fig. 12.13 is equivalent to that of part (a) and avoids the use of the mutual resistance. The equivalence can be established, and the proper values of T and R_1' found, by equating the power dissipation in the two models:

$$\begin{aligned} R_1 \dot{q}_1^2 + 2R_{12}\dot{q}_1\dot{q}_2 + R_2\dot{q}_2^2 &= R_1'\dot{q}_1^2 + R_2(T\dot{q}_1 + \dot{q}_2)^2 \\ &= (R_1' + R_2 T^2)\dot{q}_1^2 + 2TR_2\dot{q}_1\dot{q}_2 + R_2\dot{q}_2^2. \end{aligned} \qquad (12.36)$$

The result is

$$T = R_{12}/R_2, \qquad (12.37a)$$
$$R_1' = R_1 - R_{12}^2/R_2. \qquad (12.37b)$$

Note that the conditions for passivity become simply

$$R_1' \geq 0, \qquad R_2 \geq 0, \qquad (12.38)$$

which should be apparent from the bond graph.

12.2.6 Summary

The determination of equilibrium and the unsteady behavior of models with general static couplers have been illustrated. Such a coupler can be characterized by two relations, each of which interrelates a different set of three of the four effort and flow variables. Each relation can be represented graphically as a family of curves. There are four causal interpretations, depending on which pair of variables is considered as independent, leaving the other pair as dependent.

The four causal interpretations for linear unbiased models can be represented by 2×2 matrices of constants. (Two acausal transmission matrices also are defined; they are developed in Guided Problem 12.3.) Many linear couplers are reciprocal, which means that their two transfer impedances or two transfer admittances are equal. Examples include all two-port electric circuits comprised of linear components. A linear coupler is symmetric, end-for-end, if it is reciprocal and its two self-impedances or two self-admittances are equal. A static coupler is passive if it cannot generate power. All these conditions can be represented as special cases of the bond-graph models given in Fig. 12.13.

Dimensional analysis is a powerful tool for deducing what format or formats can be used to characterize a complex static coupler most simply, depending on the knowledge of what parameters and variables are relevant.

Guided Problem 12.2

This problem investigates the characteristics of a fairly general nonlinear coupler, and its matching to a source and load.

A small boat is driven by a propeller attached directly to the drive shaft of an IC engine. The characteristics of a family of geometrically similar propellers are plotted in Fig. 12.15 part (a). The efficiency of the propeller is η, its diameter is d, the shaft has moment M

---

[3]Stated most succinctly, passivity applies if and only if the matrix $\mathbf{Z}$ is positive semi-definite.

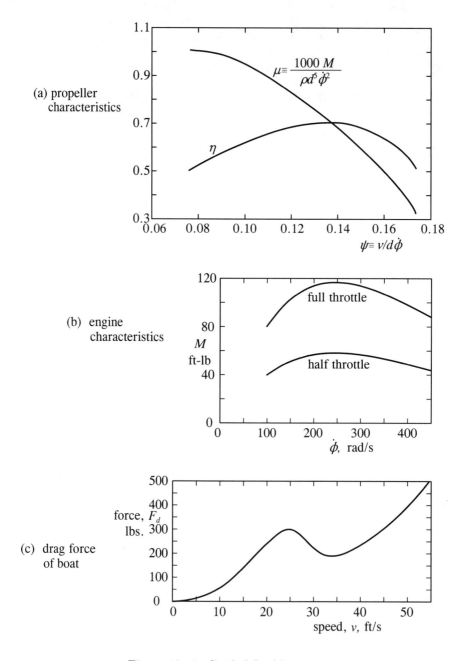

(a) propeller characteristics

$$\mu \equiv \frac{1000\,M}{\rho d^5 \dot{\phi}^2}$$

η

$\psi \equiv v/d\dot{\phi}$

(b) engine characteristics

full throttle

M
ft-lb

half throttle

$\dot{\phi}$, rad/s

(c) drag force of boat

force, F_d
lbs.

speed, v, ft/s

Figure 12.15: Guided Problem 12.2

and angular velocity $\dot{\phi}$, the boat travels at velocity v, and the water has density $\rho = 1.94$ lb s^2/ft^4. The characteristic of the engine at full throttle is given in part (b) of the figure; the torque may be assumed to be proportional to the percent throttle. To prevent excessive wear, the engine speed should not be allowed to exceed 450 rad/s. The drag of the boat, F_d, which includes a planing phenomenon, is given in part (c) of the figure.

(a) Estimate the diameter of the propeller, within the given family, that gives maximum steady speed of the boat. Determine this speed and the angular velocity of the shaft.

(b) Superimpose on part (c) of the figure sketch-plots of the thrust force F_t vs. velocity, for both full and one-half throttle, assuming the diameter found in part (a). (Identify three or four points for each curve. The purpose is to enable investigations of other loading conditions, such as occur during acceleration.) Also, estimate the steady speed that results for one-half throttle.

(c) Comment on whether the family of propellers described seems appropriate for the particular boat. (Alternative shapes are available with different ratios of the "pitch" to the diameter.) Also, might there be some merit in choosing a larger or smaller diameter propeller within the given family? Explain.

Suggested Steps:

1. If the family of propellers described is close to appropriate, the speed of the boat will be greatest when the engine produces maximum power or something very close to it. Determine the associated torque and angular velocity. Also, the efficiency of the propeller should not deviate radically from its maximum.

2. Get a rough idea of how fast the boat might go by assuming that the efficiency is maximized when the velocity is maximized. Do this by finding the corresponding value of ψ and μ. From μ you can get the tentative diameter, and then from ψ you can get the speed. Since you know the power and the efficiency you also can find the thrust force. Plot this point on the force-velocity diagram. Is the point near the actual drag-speed characteristic? If so, you are close to a solution; if the deviation is very great the proposed family of propellers likely is inappropriate.

3. Choose an additional two or three values of ψ, and tabulate the corresponding values of μ, d, v, η and thereby the thrust force F_t. Plot these forces on the force-speed diagram and find the desired solution at the intersection with the drag curve. Use interpolation to find the diameter.

4. Choose three or four points on each of the two engine characteristics. For the chosen diameter, tabulate the consequent values of μ, and thereby the corresponding values of ψ and η. This should give the points to plot on the force-speed diagram, and the steady-state speed for one-half throttle. Do the results suggest that a hysteresis phenomenon exists for some range of throttle positions?

5. To answer the second question in part (c), consider the fuel consumption and the possible need to travel frequently at speeds slower than the maximum. Does this suggest a different strategy for optimization?

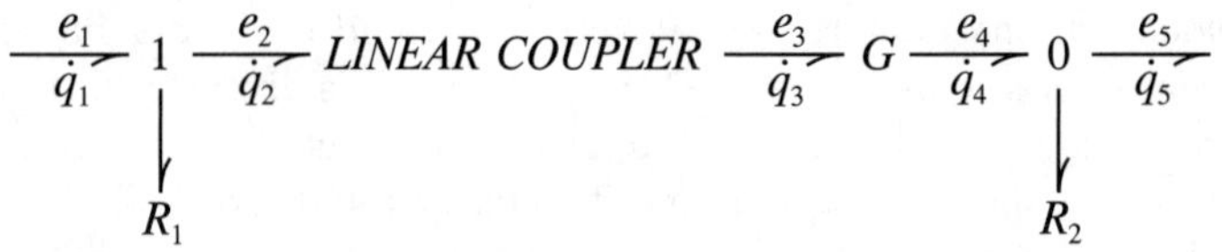

$$\xrightarrow[\dot{q}_1]{e_1} \; 1 \; \xrightarrow[\dot{q}_2]{e_2} \; LINEAR \; COUPLER \; \xrightarrow[\dot{q}_3]{e_3} \; G \; \xrightarrow[\dot{q}_4]{e_4} \; 0 \; \xrightarrow[\dot{q}_5]{e_5}$$

(with R_1 below the 1 junction and R_2 below the 0 junction)

Figure 12.16: Cascade of static elements for Guided Problem 12.3

Guided Problem 12.3

This problem applies the acausal transmission matrix, introduced in Chapter 2, to the linear static coupler. It is particularly useful when one wishes to reduce a cascade comprising a coupler and one or more other elements.

Recall that the transmission matrix is defined as follows; note in particular that unlike the causal matrices above, the power conventions of the two bonds are anti-symmetrical:

$$\xrightarrow[\dot{q}_1]{e_1} \; LINEAR \; COUPLER \; \xrightarrow[\dot{q}_2]{e_2} \qquad \begin{bmatrix} e_1 \\ \dot{q}_1 \end{bmatrix} = \begin{bmatrix} T_{11} & T_{12} \\ T_{21} & T_{22} \end{bmatrix} \begin{bmatrix} e_2 \\ \dot{q}_2 \end{bmatrix}$$

Find the four elements of the transmission matrix in terms of the four elements of the impedance matrix. Also, find the criteria for reciprocity and for symmetry in terms of the elements of the transmission matrix. Finally, find the overall transmission matrix for the cascade shown in Fig. 12.16.

Suggested Steps:

1. Expand the matrix relations for the impedance matrix and the transmission matrix into scalar algebraic equations. Change the sign of $\dot{q}_2$ on one of these, so the equations refer to the same variables despite the different directions of the power conventions on the right-hand bonds.

2. Compare the equations of step 1 to find the elements of the transmission matrix in terms of the elements of the impedance matrix. (You also might wish to do the inverse, and keep both results on file for future use.)

3. To find the criterion for reciprocity in terms of the elements of the transmission matrix, define the determinant of the transmission matrix as Δ, evaluate it in terms of the elements of the impedance matrix, and substitute the condition for reciprocity.

4. To find the criterion for symmetry, note that reciprocity must apply and that $T_{11} = T_{22}$. Alternatively, invert the transmission matrix, and change the signs of the two flows in the result. Since a symmetric system looks the same when inverted end-for-end, setting the new transmission matrix equal to the original transmission matrix gives the criteria for symmetry.

5. Find the transmission matrices of the combinations of elements on either side of the coupler element in Fig. 12.16.

6. Multiply the three transmission matrices, in the same left-to-right order in which they appear, to get the overall transmission matrix. (Note that one could, if desired, convert the result back to impedance matrix form.)

PROBLEMS

12.7 Assume that the load of the system shown and analyzed in Fig. 12.11 (p. 958) and the associated text is a nozzle located at some elevation above the pump, as in our water sprinkler system. Assume that the nozzle is lowered to the same distance below the pump that originally it was above the pump. The length of the piping, etc., is unchanged.

(a) Find the new equilibrium operation point, assuming no pulley drive, using the induction motor.

(b) Write linearized equations for the characteristics of the motor, pump and the load about the operating point.

12.8 A pulley drive is placed between the induction motor and the pump of the system characterized in Fig. 12.11 (p. 958). Determine the pulley ratio that maximizes the flow, and plot the corresponding equilibrium state. Losses in the pulley system may be neglected. Hint: Maximum flow means maximum power into the pump.

12.9 A propulsion system for a boat comprises an IC engine, ducts and an internal pump that takes water ingested in a forward-facing port and forces it out, at a higher velocity, from a backward-facing port. (No external propeller is needed, increasing safety and allowing operation in very shallow water.) The system has been analyzed to give the characteristics plotted below and on the next page. The drag of the boat is represented by its force-speed characteristic plotted below. The pump and duct system is described by two families of curves, one relating the propulsion force, craft speed and shaft speed, as plotted on the same coordinates. The other family of characteristics, plotted on the next page, interrelates the pump-shaft torque, pump-shaft speed and the speed of the craft. The engine is described, for the throttle position of interest, by the indicated engine torque-speed characteristic plotted on the same axes. The engine is connected to the pump by a gear pair, of undetermined ratio, that can be assumed to be frictionless.

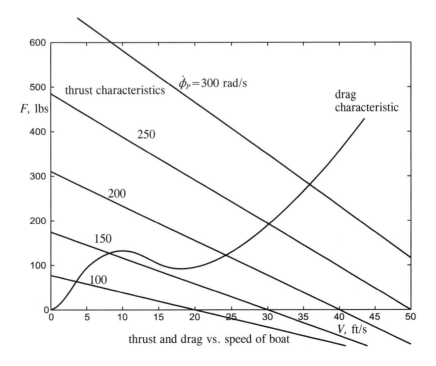

thrust and drag vs. speed of boat

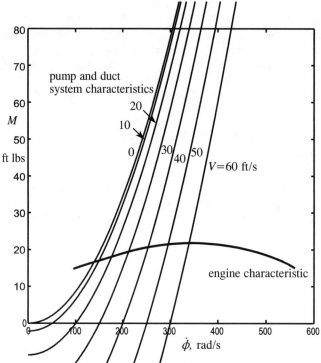

pump torque vs. pump speed, and engine torque vs. engine speed

Determine the maximum speed that the boat can go, and the corresponding gear ratio. Start by drawing a bond-graph model.

12.10 Given the model to the right and the characteristics of the individual components below, find the characteristics for the overall system. (It is suggested that you sketch with the aid of tracing paper.)

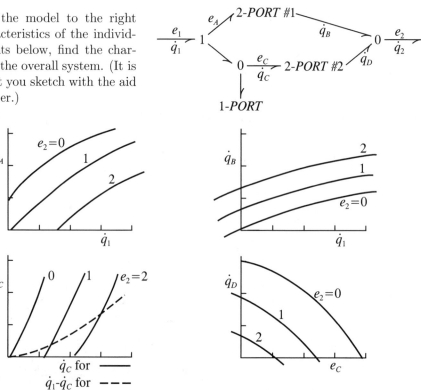

12.11 A standard open-center symmetric hydraulic spool valve is connected to a symmetric hydraulic ram as shown below. The pressure from the power supply is fixed at 1500 psi, and the annular area of the ram is 2 in^2. The differential pressure across the two components is P, the flow is Q and displacements of the spool and ram from their centered positions are y and x, respectively. The position of the spool depends on the balance of an applied force F_s and a spring-like flow-induced reaction force in the valve, which acts like a spring with a stiffness of 379 lb/in. The pressure-flow characteristics of the valve and the load are plotted below. (Note: The valve is shown at about two-thirds of full scale, whereas the ram is shown at only one-third of full scale.)

(a) Draw a bond graph for the system, using the word element SPOOL VALVE for the valve. Include the fact that F_s depends only on y (and vice-versa).

(b) Find the equilibrium velocity $\dot{x}$ and force F for the spool force $F_s = 0.68$ lb.

(c) Linearize the characteristics for small perturbations F_s^* about the equilibrium. Find the resulting motion of the load if its effective mass corresponds to a weight of 1000 lbs and $F_s^* = 0.1 \sin 20t$ lbs, where t is in seconds.

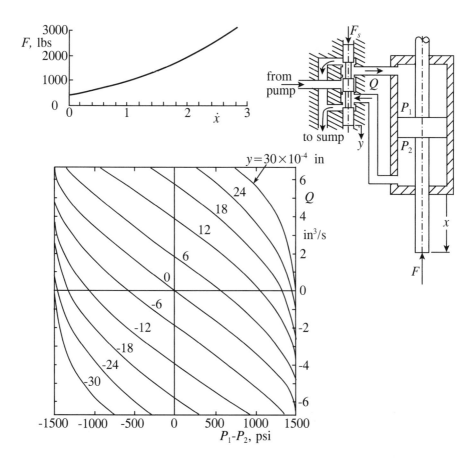

12.12 For the linear coupler represented by the three impedance matrices below,

$$\text{(i)} \ \mathbf{Z} = \begin{bmatrix} 2 & 1 \\ 1 & 2 \end{bmatrix} \quad \text{(ii)} \ \mathbf{Z} = \begin{bmatrix} 2 & 2 \\ 0 & 2 \end{bmatrix} \quad \text{(iii)} \ \mathbf{Z} = \begin{bmatrix} 2 & 3 \\ 3 & 2 \end{bmatrix}$$

(a) Determine whether the systems are reciprocal, passive and/or symmetric.

(b) If the systems are electric circuits, state whether they could comprise a network of resistors, or a network of resistors with a voltage amplifier or power supply.

(c) Find the parameters of the equivalent bond graph with a mutual resistance, as shown in part(b) of Fig. 12.14 (p. 962).

(d) Find the parameters of the equivalent bond graphs with basic elements, as shown in part (c) of Fig. 12.14.

12.13 Linear two-port systems have been described in terms of the transmission matrix $\mathbf{T}$, the impedance matrix $\mathbf{Z}$ and the admittance matrix $\mathbf{Y}$. Following the causal definitions given in equation (12.18) (p. 955), an immitance matrix $\mathbf{K}$ and adpedance matrix $\mathbf{L}$ also can be defined.

(a) Relate the variables of a linear two-port element by means of (i) the immitance matrix, and (ii) the adpedance matrix.

(b) For each of the four combinations of two-port elements below, one of the four types of causal matrix simply sum to give the overall causal matrix of the same type. Determine which matrix applies in each case.

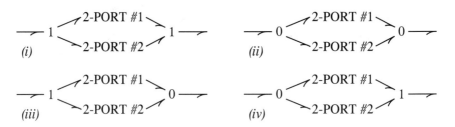

(c) Various transformations relate the transmission and causal matrices. As an example, determine the elements of the immitance matrix in terms of the elements of the transmission matrix. Assume all power conventions are directed from left to right.

SOLUTIONS TO GUIDED PROBLEMS

Guided Problem 12.2

1. The maximum engine power occurs at or very close to its maximum allowable speed, at which $\dot{\phi} = 450$ rad/s, $M = 88$ ft-lb. The maximum efficiency of the engine is about $\eta = 0.708$. (Numerical values in this solution were determined by carefully scaling a blown-up copy of the given plots.)

2. $\psi \equiv \dfrac{v}{d\dot{\phi}} = 0.14$; $\mu \equiv \dfrac{1000M}{\rho d^5 \dot{\phi}^2} = 0.688$.

   For the conditions of maximum engine power, therefore,

   $$d = \left(\frac{1000M}{0.688\rho\dot{\phi}^2}\right)^{0.2} = \left[\frac{1000 \times 88}{0.688 \times 1.94 \times (450)^2}\right]^{0.2} = 0.799 \text{ ft} = 9.59 \text{ in.}$$

   $v = 0.14 \times 0.799 \times 450 = 50.3$ ft/s.

   $$F_{thrust} = \frac{\eta P_{max}}{v} = \frac{0.708 \times 450 \times 88}{50.3} = 557 \text{ lb.}$$

This point is plotted below, and is seen to be close to the load curve but not to lie on it.

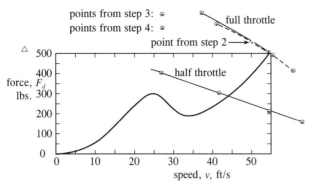

3. The case above is given as the first row in the table below:

ψ	$\mu = 0.224/d^5$	d, ft	v, ft/s	η	$F_t = \eta M\dot\phi/v$, lb
0.14	0.688	0.7990	50.34	0.708	557
0.12	0.835	0.7686	41.50	0.687	656
0.15	0.600	0.8211	55.42	0.685	490
0.16	0.505	0.8499	61.19	0.639	414

Connecting the four points on the plot shows a solution at about $v = 54.4$ ft/s, from which $F_t \simeq 505$ lb, $d = 0.8166$ ft, $\psi \simeq 0.148$, $\mu \simeq 0.617$.

4. Using $d = 0.8166$ ft and choosing three different engine speeds for the two throttle conditions,

$\dot\phi$, rad/s	M, ft-lb	μ	ψ	η	v, ft/s	F_t, lbs
450	88	0.6171	0.148	0.694	54.4	505.2
	44	0.3085	0.1735	0.512	63.76	159.0
400	98	0.8698	0.1145	0.675	37.40	707.5
	49	0.4349	0.1661	0.599	54.25	216.4
350	53.5	0.6202	0.1478	0.695	42.24	308.1
300	56.85	0.8970	0.1104	0.661	27.05	416.8

These points also are plotted on the force-speed diagram above to give thrust-speed characteristics for the two throttle conditions. The equilibrium speed for half-throttle is about 44.3 ft/s. It can be seen that a similar characteristic for somewhat less than half throttle would give three equilibrium speeds; the outer two are stable and the center is unstable. This indicates a distinct hysteresis; no speed in the range between about 25 to 35 feet per second could be maintained in steady-state.

5. The fuel efficiency of the engine is greatest at distinctly less than full throttle. When the consideration of miles per gallon of fuel is of considerable concern, therefore, one might prefer to emphasize performance at less than full speed. This implies shifting the speed at which the propeller has maximum efficiency to distinctly less than full speed, which can be done by using a smaller diameter. Unfortunately, however, this can require the engine to exceed its desirable maximum speed, causing excessive wear. A compromise could be sought, but it is hard to beat a propeller diameter for which maximum engine speed produces maximum boat speed. The only very practical way to increase efficiency at the lower speeds would be to change the style of the propeller, which largely means changing its pitch-to-diameter ratio. (The given family of propellers has a nominal ratio of 1:1.)

Guided Problem 12.3

1. The transmission matrix relation gives $e_1 = T_{11}e_2 + T_{12}\dot q_2$ \hfill (1)
$$\dot q_1 = T_{21}e_2 + T_{22}\dot q_2 \tag{2}$$

The impedance relation, using the same $\dot q_2$ as above, is

$$e_1 = Z_{11}\dot q_1 + Z_{12}(-\dot q_2) \tag{3}$$

$$e_2 = Z_{21}\dot{q}_1 + Z_{22}(-\dot{q}_2) \tag{4}$$

2. Equation (4) can be rewritten as $\dot{q}_1 = \dfrac{1}{Z_{21}}e_2 + \dfrac{Z_{22}}{Z_{21}}\dot{q}_2$ (5)

   Comparing equations (2) and (5): $T_{21} = 1/Z_{21}$; $T_{22} = Z_{22}/Z_{21}$ (6)
   Substitution of equation (5) into equation (3) gives

   $$e_1 = Z_{11}\left[\frac{1}{Z_{21}}e_2 + \frac{Z_{22}}{Z_{21}}\dot{q}_2\right] - Z_{12}\dot{q}_2,$$

   which when compared to equation (1) reveals $T_{11} = \dfrac{Z_{11}}{Z_{21}}$; $T_{12} = \dfrac{Z_{11}Z_{22}}{Z_{21}} - Z_{12}$.

3. Reciprocity applies when $1 = \dfrac{Z_{12}}{Z_{21}} = \dfrac{1}{Z_{21}}\left(\dfrac{Z_{11}Z_{22}}{Z_{21}} - T_{12}\right)$

   $$= \frac{Z_{11}}{Z_2}\frac{Z_{22}}{Z_{21}} - \frac{T_{12}}{Z_{21}} = T_{11}T_{22} - T_{12}T_{21} = \Delta.$$

4. For symmetry, $\Delta = 1$ *and* $T_{11} = T_{22}$.

   This can be derived by inverting the transmission matrix to give

   $$\begin{bmatrix} e_2 \\ \dot{q}_2 \end{bmatrix} = \frac{1}{\Delta}\begin{bmatrix} T_{22} & -T_{12} \\ -T_{21} & T_{11} \end{bmatrix}\begin{bmatrix} e_1 \\ \dot{q}_1 \end{bmatrix}$$

   and changing the signs of the two flows to give

   $$\begin{bmatrix} e_2 \\ -\dot{q}_2 \end{bmatrix} = \frac{1}{\Delta}\begin{bmatrix} T_{22} & T_{12} \\ T_{21} & T_{11} \end{bmatrix}\begin{bmatrix} e_1 \\ \dot{q}_1 \end{bmatrix}$$

   The system is symmetric if and only if this transmission matrix relation, which describes right-to-left behavior, equals the original transmission matrix relation, which describes left-to-right behavior. The criterion therefore is $\Delta = 1$ and $T_{22} = T_{11}$, as given above.

5. $\begin{bmatrix} e_1 \\ \dot{q}_1 \end{bmatrix} = \begin{bmatrix} 1 & R_1 \\ 0 & 1 \end{bmatrix}\begin{bmatrix} T_{11} & T_{12} \\ T_{21} & T_{22} \end{bmatrix}\begin{bmatrix} 0 & G \\ 1/G & 0 \end{bmatrix}\begin{bmatrix} 1 & 0 \\ 1/R_2 & 1 \end{bmatrix}\begin{bmatrix} e_5 \\ \dot{q}_5 \end{bmatrix}$

6. $\begin{bmatrix} e_1 \\ \dot{q}_1 \end{bmatrix} = \begin{bmatrix} T_{11} + R_1 T_{21} & T_12 + R_1 T_{22} \\ T_{21} & T_{22} \end{bmatrix}\begin{bmatrix} G/R_2 & G \\ 1/G & 0 \end{bmatrix}\begin{bmatrix} e_5 \\ \dot{q}_5 \end{bmatrix}$

   $$= \begin{bmatrix} (T_{11} + R_1 T_{21})G/R_2 + (T_{12} + R_1 T_{22})/G & (T_{11} + R_1 T_{21})G \\ T_{21}G/R_2 + T_{22}/G & T_{21}G \end{bmatrix}\begin{bmatrix} e_5 \\ \dot{q}_5 \end{bmatrix}$$

12.3   Lagrange's Equations for Holonomic Systems

Multiport energy storage with restricted non-constant moduli is addressed in Section 9.4. A generalization of this energy approach employs, as the state variables, the displacement and the momentum on each bond. The stored energy, known as the **Hamiltonian**, is written as

$$\mathcal{H} = \mathcal{H}(\mathbf{q}, \mathbf{p}) = \mathcal{T}(\mathbf{q}, \mathbf{p}) + \mathcal{V}(\mathbf{q}). \tag{12.39}$$

The associated state variable differential equations are known as **Hamilton's equations**. Unfortunately, however, it is far from obvious how to construct the Hamiltonian in this form for these more complex cases. The problem lies in the kinetic energy, which is much easier to express in terms of the generalized *velocities* and displacements, known as the Lagrangian generalized variables, rather than in terms of the Hamiltonian generalized *momenta* and displacements. State differential equations known as Lagrange's equations, on the other hand, employ the more attractive Lagrangian variables. These equations are deduced from the **Lagrangian**, an energy-like function defined as

$$\mathcal{L}(\dot{\mathbf{q}}, \mathbf{q}) = \mathcal{T}^*(\dot{\mathbf{q}}, \mathbf{q}) - \mathcal{V}(\mathbf{q}). \tag{12.40}$$

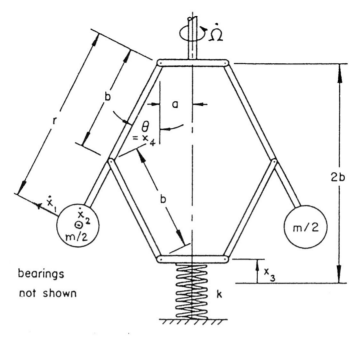

Figure 12.17: Flyball governor

The Lagrangian subsequently can be used to deduce the Hamiltonian, via what is known as the **Legendre transformation**. The consequent Hamilton's differential equations often have advantages, but these usually do not justify the effort necessary to complete the transformation. Because exceptions exist, however, they are presented in the final Section (12.6). The present section shows how the Lagrangian can be found for physical systems, and derives the consequent differential equations usable for simulation or other purposes.

12.3.1   Primitive Coordinates and Velocities

The key assumption in the energy-based modeling given herein is the *existence* of **primitive coordinates** x_i and **primitive velocities** $\dot{x}_i$ such that the stored energy can be given, at least theoretically, in the form

$$\mathcal{E} = \sum_{j=1}^{N} \mathcal{E}_j = \sum_{j=1}^{N} [\mathcal{T}_j(\dot{x}_j) + \mathcal{V}_j(x_j)]. \tag{12.41}$$

Thus each component potential energy $\mathcal{V}_j$ is a function of a single coordinate, and each component kinetic energy $\mathcal{T}_j$ is a function of a single velocity. The existence of $\mathcal{T}_j$ does not require that the corresponding $\mathcal{V}_j$ also exist, or vice-versa, however. In fact, $\mathcal{T}_j$ may exist and its associated coordinate x_j have no physical or geometric meaning; that is, $\dot{x}_j$ may not be *integrable* to give a coordinate x_j for which any $\mathcal{V}_j(x_j)$ would be physically meaningful. The simple flyball governor shown in Fig. 12.13 includes an example in which the x_j and $\dot{x}_j$ can be identified readily. The energy is modeled as

$$\mathcal{E} = \mathcal{T}_1(\dot{x}_1) + \mathcal{T}_2(\dot{x}_2) + \mathcal{V}_3(x_3) + \mathcal{V}_4(x_4), \tag{12.42a}$$

$$\mathcal{T}_1 = \frac{1}{2}m\dot{x}_1^2; \quad \mathcal{T}_2 = \frac{1}{2}m\dot{x}_2^2; \quad \mathcal{V}_3 = \frac{1}{2}kx_3^2; \quad \mathcal{V}_4 = mgr(1 - \cos x_4), \tag{12.42b}$$

where $\dot{x}_1$ is the component of the velocities of the masses $m/2$ resulting from $\dot{\theta}$, $\dot{x}_2$ is the component of these velocities that represents rotation about the vertical axis and is normal to x_1, x_3 is the vertical position of the collar and x_4 is the angle θ. The time integral of $\dot{x}_1$ can be seen to be proportional to the change in x_4. On the other hand, the time intergral of $\dot{x}_2$, which would be the *horizontal component* of the *path* traveled by one of the masses, bears no relationship to the instantaneous state of the system. It depends rather on the *history* of both states x_4 and $\dot{x}_2$; it is *path dependent*. The hypothetical placement of a spring with an energy $\mathcal{V}_2$, which would be a function of x_2, is utterly impossible. As a consequence, x_2 is called a **quasi coordinate**, as opposed to the other displacement variables, which are called **true coordinates**.

12.3.2   Holonomic Constraints

A set of **generalized coordinates** also is employed, which may be partly or entirely different from the set of primitive coordinates, and can be identified even if the primitive coordinates are not. The generalized coordinates $q_1, \ldots, q_n$ for a particular system are a *minimum* set of *true* coordinates or displacements which, together with the inherent geometric constraints, allow unique expression of the positional state of the system. The derivative $\dot{\mathbf{q}}$ is known as the **generalized velocities**. For example, the angles θ and Ω can serve as the generalized coordinates for the flyball governor; $\dot{\theta}$ and $\dot{\Omega}$ then become the generalized velocities.

"Geometric" constraints are expressed as position-dependent relationships between the generalized velocities and the usually more numerous primitive velocities. For the present, these constraints are assumed to be

$$\dot{x}_j = \sum_{i=1}^{n} \alpha_{ji}\dot{q}_i \quad j = 1, \ldots, N; \tag{12.43a}$$

$$\alpha_{ji} = \alpha_{ji}(\mathbf{q}). \tag{12.43b}$$

The flyball governor, for example, has the constraints

$$\dot{x}_1 = \alpha_{1\theta}\dot{\theta}; \qquad \alpha_{1\theta} = r; \tag{12.44a}$$

$$\dot{x}_2 = \alpha_{2\omega}\dot{\Omega}; \qquad \alpha_{2\theta} = a + r\sin\theta; \tag{12.44b}$$

$$\dot{x}_3 = \alpha_{3\theta}\dot{\theta}; \qquad \alpha_{3\theta} = 2b\sin\theta; \tag{12.44c}$$

$$\dot{x}_4 = \alpha_{4\theta}\dot{\theta}; \qquad \alpha_{4\theta} = 1. \tag{12.44d}$$

Constraints of the class known as nonholonomic force one or more of the generalized velocities to depend on the other generalized velocities, without reducing the number of generalized coordinates. The number of independent generalized velocities is known as the number of **degrees of freedom**. It also can be viewed as the maximum number of independent small or incremental changes in the positional state permitted by the constraints. If all constraints are holonomic, then the number of degrees of freedom equals the number of generalized coordinates. It is critical to determine whether any of the constraints in a system are nonholonomic, since the state differential equations are affected directly.

The constraints represented by equation (12.43) are holonomic, providing that all generalized coordinates are represented in $\mathbf{q}$ and none are masquerading alone in x. The practical problem is that sometimes one can establish a correct set of equations in the form of equations (12.41) and (12.43) but undercount one or two generalized coordinates by omitting them from $\mathbf{q}$. The result would be an incorrect model. This mistake cannot occur if equation (12.43) is *integrable*, that is when the constraint is of the form[4]

$$x_j = x_j(\mathbf{q}), \tag{12.45a}$$

$$\alpha_{ji} = \frac{\partial x_j}{\partial q_i}. \tag{12.45b}$$

In this case, x_j is not independent of $\mathbf{q}$, and therefore could not possibly represent a distinct generalized coordinate. Note that it is possible for a generalized coordinate to do double-duty as a primitive coordinate (*e.g.* $x_3 = q_2$). A constraint of the form of equation (12.43) is integrable if and only if

$$\sum_{i=1}^{n} \left(\frac{\partial \alpha_{jk}}{\partial q_i} - \frac{\partial \alpha_{ji}}{\partial q_k} \right) = 0. \tag{12.46}$$

This equation serves as a useful test when the integrability of a constraint is in doubt. For the flyball governor, x_1, x_3 and x_4 are relatable in this manner to θ. The constraint for the remaining primitive variable, x_2, is not integrable, however. Therefore it is not apparent, from the equations alone, whether x_2 should be recognized as a generalized coordinate. If it were a generalized coordinate it would be re-labeled as the third element in $\mathbf{q}$, and Lagrange's equations, derived below, would not apply.

In fact, however, x_2 cannot be a generalized coordinate because it is a quasi-coordinate rather than a true coordinate. You can discover this either of two ways: first, x_2 has no direct positional interpretation, as noted in the discussion above; second, the position of the system can be described completely by specifying only two displacements or coordinates, precluding the need for a third generalized coordinate.

A key property of holonomic constraints, as noted in Chapter 9, is that a change in energy of the system implies that the conjugate efforts associated with all the generalized velocities that participate in the constraint are non-zero; all these bonds convey power. This point is illustrated in Section 12.4. Detailed consideration of nonholonomic constraints is deferred to Section 12.5.

12.3.3   Lagrange's Equations

The equations of "motion" are now derived for the general holonomic case. The rate of change of energy $\mathcal{E}_j$ can be expressed as

$$\frac{d\mathcal{E}_j}{dt} = (Z_{aj} + Z_{bj})\dot{x}_j, \tag{12.47}$$

in which Z_{aj} represents external or nonconservative (dissipation) forces and Z_{bj} represents the internal forces of constraint. In the flyball governor the only external forces result from the moment M applied to the main shaft, and the force F applied to the collar. Dissipation is neglected, also, so most forces acting on the primitive velocities are internal. The total work done by the internal forces is zero, since the associated energies merely are conveyed between elements. Specifically, if you imagine the system undergoing an infinitesimal virtual displacement $\delta x_1, \ldots, \delta x_N$, known in the **calculus of variations** as a *variation* of $\mathbf{x}$,

$$\sum_{j=1}^{N} Z_{bj}\delta x_j = 0. \tag{12.48}$$

---

[4]More general constraints of the form $\mathbf{x}_j = \mathbf{x}_j(\mathbf{q}, t)$ are possible. The inclusion of time as an independent variable in holonomic constraint relations has been employed in classical mechanics without much added complexity, but is very awkard in more general cases. The author has not found it useful in practical modeling situations. Time is omitted as an independent variable for all homonomic constraints in this chapter.

Thus, if the virtual displacements from equilibrium were to occur at any time $t_0 \leq t \leq t_1$, this equation remains zero when integrated; substitution of the Z_{bj} found from equation (12.47) gives

$$\int_{t_0}^{t_1} \sum_{j=1}^{N} \left[\frac{1}{\dot{x}_j} \frac{d}{dt}(\mathcal{T}_j + \mathcal{V}_j) - Z_{aj} \right] \delta x_j dt = 0. \tag{12.49}$$

To make this equation useful, the complementary kinetic energy or **kinetic coenergy** $\mathcal{T}_j^*$ is *defined* as follows:

$$\mathcal{T}_j^* = \dot{x}_j \int_0^{\dot{x}_j} \frac{\mathcal{T}_j(y_j)}{y_j^2} dy_j. \tag{12.50}$$

Noting that $d\tau_j = \dot{x} dp$ and $d\tau_j^* = p \, d\dot{x}$, this gives

$$\frac{d^2 \mathcal{T}_j^*}{d^2 x_j^2} = \frac{dp}{d\dot{x}} = \frac{1}{\dot{x}_j} \frac{d\mathcal{T}_j}{d\dot{x}_j}. \tag{12.51}$$

In practice, one starts with $\mathcal{T}_j$ and computes $\mathcal{T}_j^*$ using equation (12.50). Throughout classical mechanics and many analogous domains the special case $\mathcal{T}_j = \frac{1}{2} I_j x_j^2$ applies, giving the simple result $\mathcal{T}_j^* = \mathcal{T}_j$. (As a result, texts in classical mechanics often fail to distinguish kinetic and cokinetic energies.) The term in equation (12.49) involving kinetic energies now can be recast as

$$\int_{t_0}^{t_1} \frac{1}{\dot{x}_j} \frac{d\mathcal{T}_j}{dt} \delta x_j dt = \int_{t_0}^{t_1} \frac{1}{\dot{x}_j} \frac{d\mathcal{T}_j}{d\dot{x}_j} \ddot{x}_j \delta x_j dt = \int_{t_0}^{t_1} \frac{d^2 \mathcal{T}_j^*}{d\dot{x}_j^2} \ddot{x}_j \delta x_j dt$$

$$= \int_{t_0}^{t_1} \frac{d}{dt}\left(\frac{d\mathcal{T}_j^*}{d\dot{x}_j} \right) \delta x_j dt = -\int_{t_0}^{t_1} \frac{d\mathcal{T}_j^*}{d\dot{x}_j} \delta \dot{x}_j dt + \frac{d\mathcal{T}_j^*}{d\dot{x}_j} \delta x_j \Big|_{t_0}^{t_1}. \tag{12.52}$$

With this substitution, equation (12.49) becomes

$$\int_{t_0}^{t_1} \sum_{j=1}^{N} \left[-\frac{d\mathcal{T}_j^*}{d\dot{x}_j} \delta \dot{x}_j + \left(\frac{d\mathcal{V}_j}{dx_j} - Z_{aj} \right) \delta x_j \right] dt + \sum_{j=1}^{N} \frac{d\mathcal{T}_j^*}{d\dot{x}_j} \delta x_j \Big|_{t_0}^{t_1} = 0 \tag{12.53}$$

Note that variations are treated mathematically like ordinary derivatives.

Equation (12.53) is not yet in a particularly useful form because the N components of δx_j are not independent of one another. The δx_j and $\delta \dot{x}_j$ above can be related to the virtual change of the generalized coordinates, $\delta \mathbf{q}$, through the use of equation (12.40):

$$\delta x_j = \sum_{i=1}^{n} \alpha_{ji} \delta q_i, \tag{12.54a}$$

$$\delta \dot{x}_j = \sum_{k=1}^{n} \left[\sum_{i=1}^{n} \frac{\partial \alpha_{ji}}{\partial q_k} \dot{q}_i \delta q_k + \alpha_{jk} \frac{d}{dt}(\delta q_k) \right]. \tag{12.54b}$$

With these substitutions, the integration by parts of equation (12.49) gives

$$\int_{t0}^{t1} \sum_{k=1}^{n} \left\{ \sum_{j=1}^{N} \left[\alpha_{jk} \frac{d}{dt}\left(\frac{d\mathcal{T}_j^*}{d\dot{x}_j} \right) + \frac{d\mathcal{T}_j^*}{d\dot{x}_j} \sum_{i=1}^{n} \left(\frac{\partial \alpha_{jk}}{\partial q_i} - \frac{\partial \alpha_{ji}}{\partial q_k} \right) \sum_{\ell=1}^{N} \beta_{i\ell} \dot{x}_\ell \right. \right.$$

$$\left. \left. + \alpha_{jk} \frac{d\mathcal{V}_j}{dx_j} - \alpha_{jk} Z_{aj} \right] \right\} \delta q_k \, dt = 0, \tag{12.55}$$

in which $[\beta_{i\ell}]$ is the inverse of $[\alpha_{\ell i}]$, i.e.,

$$\dot{q}_i = \sum_{\ell=1}^{N} \beta_{i\ell} \dot{x}_\ell. \tag{12.56}$$

The variations $\delta q_1, \ldots, \delta q_n$ of a holonomic system are independent, with no generalized geometric constraints between them, much like the modes of motion of a linear system. Thus, equation (12.55) produces n separate equations:

$$\sum_{j=1}^{N} \left\{ \frac{dT_j^*}{d\dot{x}_j} \left[\sum_{i=1}^{n} \left(\frac{\partial \alpha_{jk}}{\partial q_i} - \frac{\partial \alpha_{ji}}{\partial q_k} \right) \sum_{\ell=1}^{N} \beta_{i\ell} \dot{x}_\ell \right] + \alpha_{jk} \left[\frac{d}{dt} \left(\frac{dT_j^*}{d\dot{x}_j} \right) + \frac{\partial V_j}{\partial x_j} - Z_{aj} \right] \right\} = 0,$$

$$k = 1, \ldots, n. \tag{12.57}$$

Whenever the integrability condition of equation (12.46) is satisfied, equation (12.57) is simplified. In this case, the entire first half of equation (12.57) vanishes.

For the flyball governor, the equation becomes

$$(a + r \sin x_4)(m\ddot{x}_2 - Z_2) + mr(\cos x_4)\dot{x}_2 \dot{x}_4 = 0; \tag{12.58a}$$

$$mr\ddot{x}_1 - rZ_1 + 2b(\sin x_4)(kx_3 - Z_3) + mgr \sin x_4 - Z_4 = 0. \tag{12.58b}$$

The forces Z_j act directly on the primitive coordinates in the direction of their increase. They can be related to a corresponding set of generalized forces which act directly on the generalized coordinates. For the flyball governor, the force term in equation (12.58a) is the torque M_Ω on the shaft with displacement Ω, and the force terms in equation (12.58b) total to the moment M_θ acting on the member with displacement θ:

$$M_\Omega = (a + r \sin x_4)Z_2; \tag{12.59a}$$

$$M_\theta = rZ_1 + 2b(\sin x_4)Z_3 + Z_4. \tag{12.59b}$$

Even when primitive variables are employed in the modeling process, it is usually convenient to convert wholly to the generalized coordinates and velocities rather than use equations (12.55) or (12.57). In terms of the generalized coordinates and velocities, the energy of a system is

$$\mathcal{E} = T(\dot{\mathbf{q}}, \mathbf{q}, t) + V(\mathbf{q}, t), \tag{12.60}$$

where T and V are distinguished from T_j and V_j by the absence of subscripts. The relations for the kinetic coenergy T^* are

$$\frac{d}{dt} \left(\frac{\partial T^*}{\partial \dot{q}_k} \right) = \sum_{j=1}^{N} \left[\alpha_{jk} \frac{d}{dt} \left(\frac{dT_j^*}{d\dot{x}_j} \right) + \frac{dT_j^*}{d\dot{x}_j} \left(\sum_{j=1}^{n} \frac{\partial \alpha_{jk}}{\partial q_i} \dot{q}_i \right) \right]; \tag{12.61a}$$

$$\frac{\partial T^*}{\partial q_k} = \sum_{j=1}^{N} \frac{dT_j^*}{d\dot{x}_j} \sum_{i=1}^{n} \frac{\partial \alpha_{ji}}{\partial q_k} \dot{q}_i. \tag{12.61b}$$

With the customary definition of the **Lagrangian**,

$$\boxed{\mathcal{L}(\dot{\mathbf{q}}, \mathbf{q}) = T^*(\dot{\mathbf{q}}, \mathbf{q}) - V(\mathbf{q}),} \tag{12.62}$$

equation (12.57) can be recast in the relatively simple form

$$\int_{t_0}^{t_1} \sum_{k=1}^{n} \left[\frac{d}{dt} \left(\frac{\partial \mathcal{L}}{\partial \dot{q}_k} \right) - \frac{\partial \mathcal{L}}{\partial q_k} - e_k \right] \delta q_k \, dt = 0. \tag{12.63}$$

Since we are considering holonomic constraints only, as with equation (12.54) the various δ_k are independent, and this *unconstrained variational principle* or *extremum principle* is satisfied only when the term in brackets vanishes. This gives **Lagrange's equations**:

$$\frac{d}{dt}\left(\frac{\partial \mathcal{L}}{\partial \dot{q}_k}\right) - \frac{\partial \mathcal{L}}{\partial q_k} = e_k; \qquad k = 1, \ldots, n. \tag{12.64}$$

For the flyball governor, the Lagrangian is

$$\mathcal{L} = \frac{1}{2}m[r^2\dot{\theta}^2 + (a + r\sin\theta)^2\dot{\Omega}^2] - 2kb^2(1 - \cos\theta)^2 - mgr(1 - \cos\theta), \tag{12.65}$$

from which equation (12.64) gives

$$m(a + r\sin\theta)^2\ddot{\Omega} + 2mr(a + r\sin\theta(\cos\theta)\dot{\Omega}\dot{\theta} = e_\Omega = M_\Omega; \tag{12.66a}$$

$$mr^2\ddot{\theta} - mr(a + r\sin\theta)(\cos\theta)\dot{\Omega}^2 + 4kb^2(1 - \cos\theta)\sin\theta + mgr\sin\theta$$

$$= e_\theta = M_\theta = (2b\sin\theta)F. \tag{12.66b}$$

This deduction was simplified by the fact that $\mathcal{L}^* = \mathcal{L}$, as in all models in classical mechanics. The general relationship between $\mathcal{T}^*$ and $\mathcal{T}$ is the Legendre transformation

$$\sum_k \dot{q}_k p_k - \mathcal{T}^* = \mathcal{T}, \qquad p_k = \frac{\partial \mathcal{T}^*}{\partial \dot{q}_k}, \tag{12.67}$$

which follows from the relation between $\mathcal{T}_j^*$ and $\mathcal{T}_j$, as given by equation (12.50), and the identity

$$\sum_i \sum_k \beta_{ki}\alpha_{jk}\dot{x}_i \equiv \dot{x}_j. \tag{12.68}$$

In most cases of practical interest the following special forms apply, allowing the transformation to be made in terms of inertances rather than kinetic energies:

$$\mathcal{T} = \sum_{j=1}^{n} \sum_{j=1}^{n} \frac{1}{2}I_{ij}\dot{q}_i\dot{q}_j,$$

$$\mathcal{T}^* = \sum_{i=1}^{n} \sum_{j=1}^{n} \frac{1}{2}I_{ij}^* \, \dot{q}_i\dot{q}_j,$$

$$I_{ii} = I_{ii}(\dot{q}_i, \mathbf{q}),$$

$$I_{ii}^* = I_{ii}^* \, (\dot{q}_i, \mathbf{q}),$$

$$I_{ij}^* = I_{ij} = \text{constants}, \quad i \neq j. \tag{12.69}$$

For these cases, this gives

$$I_{ii}^* = \frac{1}{\dot{q}_i}\int_0^{\dot{q}_i} I_{ii}(f_i, \mathbf{q})df_i, \tag{12.70}$$

in which f_i is a dummy variable for q_i.

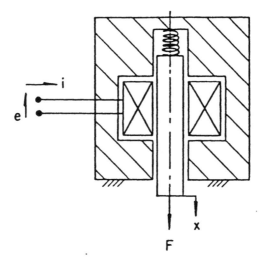

Figure 12.18: Variable-geometry inductor (solenoid)

12.3.4 Example of a Solenoid

Lagrange's equations may be used even if the primitive coordinates and velocitites are not known. It is enough to presume that they *exist*. The variable-geometry inductor or **solenoid** shown in Fig. 12.14 is an example. The energy can be represented as

$$\mathcal{E} = \frac{1}{2}I(i,x)i^2 + \frac{1}{2}m\dot{x}^2 + \frac{1}{2}k(x-x_0)^2, \qquad (12.71)$$

in which the first term represents the magnetic energy. This energy is a function of the position of the movable part, which makes the device work, and of the electric current, in a way that allows magnetic saturation to be recognized (unlike the model of a solenoid considered in Example 9.12 (p. 638)). This term is not readily decomposable to the form of equation (12.41) (p. 973), and need not be. It represents a field lumping, and may be identified experimentally. Theoretically, however, it could be broken down into a sum of terms of the form $\sum I_j(i)\alpha_j(x)$, each of which represents a small region in the field. The input power is of the form

$$\mathcal{P} = ei + F\dot{x}. \qquad (12.72)$$

The Lagrangian becomes

$$\mathcal{L} = \frac{1}{2}I^* i^2 + \frac{1}{2}m\dot{x}^2 - \frac{1}{2}k(x-x_0)^2; \qquad (12.73a)$$

$$I^* = \frac{1}{i}\int_0^i I(f,x)df. \qquad (12.73b)$$

Lagrange's equations (equation (12.64)) give, after a little calculation,

$$\left[I + \frac{1}{2}\frac{\partial I(i,x)}{\partial i}\right]\frac{di}{dt} + \frac{\dot{x}}{2}\left[i\frac{\partial I(i,x)}{\partial x} + \int_0^i \frac{\partial I(f,x)}{\partial x}df\right] = e; \qquad (12.74a)$$

$$m\ddot{x} - \frac{i}{2}\int_0^i \frac{\partial I(f,x)}{\partial x}\,df + k(x - x_0) = F. \qquad (12.74b)$$

The terms with $\partial I/\partial x$ represent the electromechanical coupling.

The special case of the solenoid for $I = I(x)$ (no saturation) was treated in Example 9.12 by a simpler method. In this case, equations (12.74) specialize to

$$I\frac{di}{dt} + \dot{x}i\frac{dI}{dx} = e, \qquad (12.75a)$$

$$m\ddot{x} - \frac{i^2}{2}\frac{dI}{dx} + k(x - x_0) = F, \qquad (12.75b)$$

which agree with the earlier result when it is noted that $e = dp/dt$. Both the more general solenoid problem and the flyball governor also can be handled by the simpler method; a comparison is offered for the latter in Section 12.6. Many other problems cannot be so treated, however, including the double pendulum given below as Problem 12.14, and the vibratory rate gyro discussed in Section 12.4.

12.3.5 Summary

Models or portions of models of physical systems which include an energy storage mechanism that depends on one or more displacements and on more than a single generalized momentum or velocity are difficult to deduce directly, and are not included in the special energy approaches of earlier chapters. Lagrangian analysis is the most direct energy-based apparoach available. Although this powerful method was developed originally in classical mechanics, the derivation given is general, and applies to any holomonic system in which the energy can be expressed as a function of a minimum set of generalized displacements or "coordinates" and their time derivatives, the generalized velocities. These become the Lagrangian state variables. The procedure has two steps: finding the energy-based Lagrangian as a function of the state variables, and routine application of the Lagrange's variables.

The number of degrees of freedom of a model is the maximum number of generalized velocities or infinitesimal displacements which the hard or "geometric" constraints allow to be set independently. The number of generalized coordinates of a model is the minimum number of displacements necessary to specify its "postition" or displacement state. When these two numbers are equal, the model is said to be holonomic.

The number of generalized coordinates of nonholonomic models is greater than the number of degrees of freedom. Nonholonomic models are discussed in more detail in Section 12.5.

PROBLEM

12.14 Two pendulums each comprise a point mass m and a massless link of length L. The first pendulum is pivoted from a fixed point, and the second is pivoted from the point mass of the first, as shown. Find a set of state differential equations for the system. The following steps are suggested:

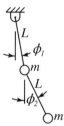

(a) Define as generalized coordinates the absolute angles of the two links. (You might prefer to use the *difference* between the two absolute angles as one of the coordinates.) Write the potential (gravity) energy of the system in terms of these angles and the fixed parameters. Do *not* make a small angle assumption.

(b) Write the kinetic energy of the system in terms of the generalized coordinates, their time derivatives and the fixed parameters.

(c) Apply Lagrange's equation to each of the generalized coordinates to get the two principal state differential equations. Note that the other two state differential equations are simply the identities $dq_1/dt = \dot{q}_1$ and $dq_2/dt = \dot{q}_2$.

12.4 Lagrangian Bond Graphs; Dissipation

12.4.1 Shorthand Notation

The region of a system treated by the Lagrangian method may be represented in a bond graph simply by the summary symbol $\mathcal{L}$, as shown in part (a) of Fig. 12.19.

This representation is particularly appropriate when the primitive coordinates are not identified specifically, as in the example of the solenoid given in the previous section. The generalized forces on the boundary bonds are indicated as causal inputs to the region, and the generalized velocities as causal outputs, consistent with Lagrange's equations that detail the behavior of the region. The solenoid is thus represented in part (b) of the figure. Two versions are shown; the one on the right excludes the mass and the spring from the Lagrange region. Note the imposition of differential causality on the inertance (mass).

12.4.2 Detailed Lagrangian Bond Graphs

When primitive coordinates are known, on the other hand, a detailed bond graph may be drawn. The rules for constructing and interpreting this graph substitute for the Lagrange's equations themselves. Two fundamental aspects of this optional approach are attractive. First, the meaning and behavior of the model is clearer to the modeler than the results of mere inspection of the Lagrange's equations. Second, the graph permits and even encourages the use of a different, non-Lagrangian set of state variables, which often produces a simpler set of state differential equations.

Construction of the graph starts by connecting 1-junctions for the generalized coordinates to 1-junctions for the primitive coordinates by transformers which have moduli identical to the α_{ji} of equation (12.43) (p. 974); 0-junctions are used to represent the summations. This scheme is shown in Fig. 12.19 part (c). The 1-junctions for the generalized coordinates are arrayed across the top, the 1-junctions for the primitive true coordinates are segregated on the left and the 1-junctions for the primitive quasi coordinates are displayed on the right. (You may prefer to array these elements differently.) The primitive inertances, compliances and resistances are bonded directly to these latter 1- junctions. Boundary bonds also can be connected either to the 1-junctions for the generalized velocities or the 1-junctions for the primitive velocities, as desired.

The graph is not yet complete, in general, despite its proper representation of the energy storage, dissipation, constraints and the boundary interconnections. Gyrators must be added. To deduce this, note that equation (12.57) represents the summation of the generalized forces around the kth 1-junction for the generalized coordinates. To simplify matters, attention is restricted to the special case in which the various α_{ji} are not functions of time; exceptions to this case are unnusual. With this restriction the equation can be written as

$$-\sum_{i=1}^{n} G_{ki}\dot{q}_i + \sum_{j=1}^{N} T_{jk}\left[\frac{d}{dt}\left(\frac{d\mathcal{T}_j^*}{d\dot{x}_j}\right) + \frac{\partial \mathcal{V}_j}{\partial x_j} - Z_{aj}\right] = 0, \quad k = 1, \cdots, n, \qquad (12.76a)$$

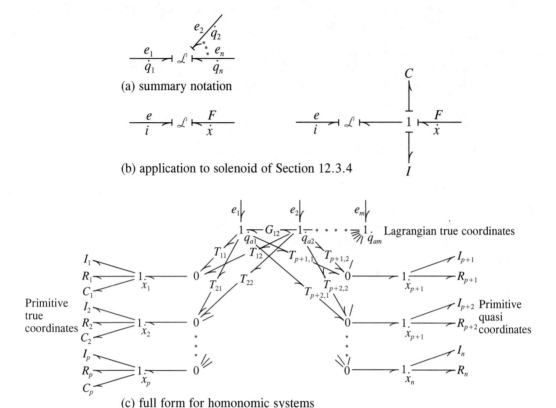

Figure 12.19: Bond graph representations for a holonomic Lagrangian domain

$$T_{jk} \equiv \alpha_{jk}, \tag{12.76b}$$

$$-G_{ki} = \sum_{j=1}^{N} \frac{dT^*_j}{d\dot{x}_j}\left(\frac{\partial \alpha_{jk}}{\partial q_i} - \frac{\partial \alpha_{ji}}{\partial q_k}\right) = \sum_{j=1}^{N}\left(I^*_j + \frac{\dot{x}_j}{2}\frac{dI^*_j}{d\dot{x}_j}\right)\dot{x}_j\left(\frac{\partial \alpha_{jk}}{\partial q_i} - \frac{\partial \alpha_{ji}}{\partial q_k}\right). \tag{12.76c}$$

The term in square brackets in equation (12.76a) represents the force on the transformer bond of the 1-junction for the jth primitive velocity. When multiplied by the factor T_{jk}, it gives the force on the bond that is on the other side of the transformer and is joined to the 1-junction for the kth generalized coordinate. The remaining term in the equation is supplied to the bond graph by adding the gyrators with moduli G_{ki} to connect the kth and ith 1-junctions for the generalized coordinates, as shown in the figure. The power convention half-arrow is directed toward the 1-junction indicated by the first subscript, or k in this case. The scheme works because G_{ik} is precisely the negative of G_{ki}:

$$G_{ki} = -G_{ik}. \tag{12.77}$$

The gyrators are ideal elements, introducing no energy or dissipation. They are necessary, however, to attach the generalized forces to the proper generalized coordinates. *They result exclusively from the quasi-coordinates*, and fail to exist if there are none.

Use of this graph precludes the need to write Lagrange's equations through direct implementation of equation (12.64) (p. 978). Instead, the moduli of the gyrators may be found

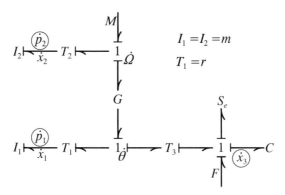

Figure 12.20: Bond graph for flyball governor

from equation (12.76c), and the sum of the generalized forces about each of the 1-junctions for the generalized coordinates set equal to zero. Each such sum is the respective Lagrange's equation. Alternatively, the application of causal strokes to the graph may suggest a different and often superior choice of state variables, as illustrated below. The corresponding state differential equations then can be found in the usual way.

12.4.3   Example of the Flyball Governor

This system, discussed in the previous section, can be represented by the Lagrangian bond graph given in Fig. 12.20. Recall that Ω and θ are the generalzied coordinates. Rather than draw the gravity energy as a compliance element connected to the $\dot{\theta}$ 1-junction by the transformer T_4, it is shown as a constant-force element connected to the $\dot{x}_3$ 1-junction; its energy equals $mgrx_3/2b$. (The primitive coordinate x_4 is unnecessary, in fact, although its use is natural.) The gyrator, as can be seen from the graph, represents the heart of the action of the governor. Neglecting the storage of potential and kinetic energies, the force F on the collar is seen to equal G/T_3 times the velocity of the collar, $\dot{x}_3$. The modulus of G is proportional to the angular velocity of the shaft, so the sensitivity as a speed measuring device is small when the speed is low.

The causal strokes reveal that the system is third order. The three corresponding first-order state differential equations could be the two Lagrange's equations (equations (12.66)) plus the identity $d\theta/dt = \theta$. Alternatively, use of the causal strokes give

$$\frac{dp_1}{dt} = \frac{1}{T_1}[G\dot{\Omega} - T_3(S_e + x_3/C - F)] = \frac{1}{r}\left[\frac{G}{T_2 m} - T_3(S_e + kx_3 - F)\right], \qquad (12.78a)$$

$$\frac{dp_2}{dt} = \frac{1}{T_2}(M - G\dot{\theta}) = \frac{1}{T_2}\left(M - \frac{G}{rm}p_1\right), \qquad (12.78b)$$

$$\frac{dx_3}{dt} = T_3\dot{\theta} = \frac{T_3}{rm}p_1, \qquad (12.78c)$$

$$T_2 = a + r\sin\theta = a + r\sqrt{1 - (1 - x_3/2b)^2}, \qquad (12.78d)$$

$$T_3 = 2b\sin\theta = 2b\sqrt{1 - (1 - x_3/2b)^2}, \qquad (12.78e)$$

$$G = mr\dot{\Omega}(a + r\sin\theta)\cos\theta = r(1 - x_3/2b)p_2, \qquad (12.78f)$$

$$S_e = \frac{mgr}{2b} + kx_0. \qquad (12.78g)$$

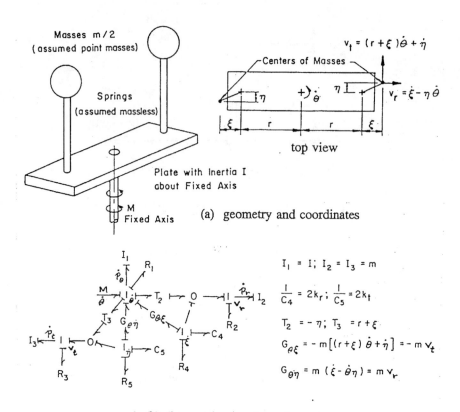

(a)　geometry and coordinates

(b)　Lagrangian bond graph

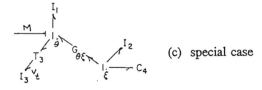

(c)　special case

Figure 12.21: Vibratory rate gyroscope

12.4.4　Example of a Vibratory Rate Gyro

This example, idealized in Fig. 12.21 illustrates the potential conceptual and computational advantages of employing a Lagrangian bond graph. Symmetrical geometry and initial conditions are assumed; only two symmetrical displacements are needed. The kinetic and potential energies are, respectively,

$$\mathcal{T} = \frac{1}{2}I\dot{\theta}^2 + \frac{1}{2}mv^2, \tag{12.79a}$$

$$v^2 = v_t^2 + v_r^2, \tag{12.79b}$$

$$v_t = (r + \xi)\dot{\theta} + \dot{\eta}, \tag{12.79c}$$

$$v_r = \dot{\xi} - \eta\dot{\theta}, \tag{12.79d}$$

$$\mathcal{V} = k_r\xi^2 + k_t\eta^2. \tag{12.79e}$$

The displacements η, ξ and θ are true coordinates (they represent directly measurable quantities) which also can be used as the Lagrangian or Hamiltonian generalized coordinates. The velocities v_r and v_t represent the time derivatives of quasi coordinates which are not directly measureable. The constraints between $\dot{\eta}$, $\dot{\xi}$ and $\dot{\theta}$ on one hand and v_r and v_t on the other hand, given above, are not integrable. The system is holonomic, however; a nonintegrable constraint is nonholonomic if and only if it relates two or more *true* coordinates.

The mathematical description above leads directly to the Lagrangian bond graph shown in part (b) of the figure. The two gyrator moduli, which can be computed using equation (12.76c) above, follow from the fact that T_2 and T_3 represent nonintegrable but holonomic constraints. (Nonholonomic constraints are discussed in the next section.)

By applying causal strokes and defining the momenta p_θ, p_r and p_t as shown, the equations of motion can be found directly, in the standard way, in terms of the state vector $[\eta, \xi, p_\theta, p_r d, p_t]^T$. The result is

$$\frac{d}{dt}\begin{bmatrix} \eta \\ \xi \\ p_\theta \\ p_r \\ p_t \end{bmatrix} = \begin{bmatrix} 0 & 0 & -(r+\xi)/I & 0 & 1/m \\ 0 & 0 & \eta/I & 1/m & 0 \\ k & 0 & 0 & 0 & 0 \\ 0 & -2k_r & 0 & 0 & p_\theta/I \\ -2k_t & 0 & 0 & -p_\theta/I & 0 \end{bmatrix}\begin{bmatrix} \eta \\ \xi \\ p_\theta \\ p_r \\ p_t \end{bmatrix} + \begin{bmatrix} 0 \\ 0 \\ M \\ 0 \\ 0 \end{bmatrix},$$

$$(12.80a)$$

$$k = 2k_t r + 2(k_t - k_r)\eta. \qquad (12.80b)$$

The matrix notation is suggested because the terms in the 5x5 matrix are, for most cases of interest, either nearly constant or nearly insignificantly small. With these variables, that is, the system is nicely nearly linear. The Lagrangian generalized velocities may be found, if desired, from the transformation

$$\dot{\theta} = p_\theta/I, \qquad (12.81a)$$

$$\dot{\xi} = p_r/m + \dot{\theta}\eta, \qquad (12.81b)$$

$$\dot{\eta} = p_t/m - \dot{\theta}(r + \eta), \qquad (12.81c)$$

which also is apparent from the bond graph.

Lagrange's variables and equations are an alternative to equation (12.80). Although they can be found without undue difficulty, they are considerably messier:

$$\{I + m[(r+\xi)^2 + \eta^2]\}\ddot{\theta} + m(r+\xi)\ddot{\eta} + m\eta\ddot{\xi} + 2m(r+\xi)\dot{\theta}\dot{\xi} + 2m\eta\dot{\theta}\dot{\eta} = M, \qquad (12.82a)$$

$$-m\eta\ddot{\theta} + m\ddot{\xi} - m(r+\eta)\dot{\theta}^2 - 2m\dot{\theta}\dot{\eta} + 2k_r\xi = 0, \qquad (12.82b)$$

$$m(r+\xi)\ddot{\theta} + m\ddot{\eta} - m\eta\dot{\theta}^2 + 2m\dot{\theta}\dot{\xi} + k_t\eta = 0. \qquad (12.82c)$$

The subsequent need to untangle the accelerations $\ddot{\theta}$, $\ddot{\xi}$ and $\ddot{\eta}$ as three separate functions of the velocities and displacements is awkward. Further, the resulting expressions (not shown) are considerably more nonlinear than equation (12.80) above, and consequently they develop less insight into the behavior, and cost more to implement computationally.

The device becomes usable as a rate gyro if the compliance C_5 is made so much smaller than C_4 that the displacement η can be neglected. In this case, C_5 and the transformer T_2 and the gyrator $G_{\theta\eta}$ can be erased from the bond graph, and the modulus $G_{\theta\xi}$ simplified, leaving the graph shown in part (c) of the figure. The displacement ξ is excited to vibrate at its natural frequency and with a controlled amplitude by an external driver (not shown). Then, if the rotational velocity $\dot{\theta}$ is steady, the torque M is seen to be virtually proportional to the product of $\dot{\theta}$ and the known ξ. Thus, measurement of M, perhaps by a strain gage, gives an indication of $\dot{\theta}$, creating a rate gyro. The heart of this behavior is the gyrational

986 CHAPTER 12. TOPICS IN ADVANCED MODELING

coupling $G_{\theta\xi}$. The use of a bond graph often permits such an insightful interpretation of the anticipated behavior of a system.

When the primitive coordinates are not known, as with the solenoid of the previous section, the procedure above does not apply. An "alternative Lagrangian bond graph" is available[5] but its conceptual and practical benefits probably do not justify the effort, so it is omitted from this text. Sometimes even the drawing of a detailed bond graph of the type described above produces relatively little benefit over direct application of Lagrange's equations and use of the summary symbol $\mathcal{L}$ in the bond graph of the larger overall system. An example comparison is given in Section 12.6. Generalizations on which method is superior are elusive.

12.4.5 Dissipation

Energy dissipative mechanisms may be added to bond graphs of any type through the use of resistances. Such elements must of course conform to the physics at least approximately; arbitrary modulation of a resistance can violate the intended passivity of a dissipation element, even converting it to an energy storage element. A chordal resistance of the rather general form

$$R = R(\dot{\mathbf{q}}, \mathbf{q}, \mathbf{t}), \tag{12.83}$$

which also is bounded as $\dot{\mathbf{q}} \to \infty$, is acceptable. This form can be shown to preclude a conservative force system.[6] No gyrators or other elements need to be added to a bond graph because of such a resistance.

12.4.6 Summary

A region of a system treated by Lagrangian methods can be represented in a bond graph of the entire system by a summary multiport element, $\mathcal{L}$, using admittance causality. If primitive variables can be identified for this region, use of an expanded Lagrangian bond graph may be an attractive alternative. This graph may increase the modeler's insight into the structure and behavior of the system. It also may result in the choice of primitive momenta and displacements as state variables and a simplification of the differential equations. As is shown in the next section, the graph also clarifies the distinction between holonomic and nonholonomic constraints.

The key steps in the construction of a Lagrangian bond graph are the identification of the primitive velocities and the associated structure of transformers and 0-junctions that represents their relation to the Lagrangian velocities, the attachment of the energy storage elements to the 1-junctions for the primitive velocities, and the evaluation of any non-zero moduli of gyrators which potentially interconnect the 1-junctions for the Lagrangian velocities. The last step uses equation (12.76c) (p. 982). Each Lagrange equation can be found by summing to zero the efforts on the bonds about the associated 1-junction. This is not necessary if the primitive variables are chosen as the state variables; rather, the standard procedure for writing differential equations for bond graphs is followed.

---

[5]F. T. Brown, "Lagrangian Bond Graphs," *J. Dynamic Systems, Measurement and Control*, v 94 n 3, pp 213-221, Sept. 1972.

[6]F. T. Brown, "Energy-Based Modeling and Quasi Coordinates," *ibid*, v103 n1, pp 5-13, March 1981.

PROBLEM

12.15 Establish a set of state differential equations for the general case of the **spinning top**. Key features of the dynamic structure and behavior should be made apparent. The rotational rate of the top about its axis of symmetry is $\dot\phi$, the angle of this axis from the vertical is θ, the angular rate at which the axis precesses about the vertical axis is $\dot\psi$, and the distance from the tip to the center of mass is r. The angles ϕ, θ and ψ are known as the **Euler angles**, which are commonly used to describe rotating bodies. The mass is m, the moment of inertia about the axis of symmetry is I_ϕ, and the moment of inertia about any centroidal axis normal to the axis of symmetry is I_θ.

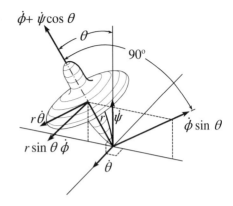

This usually is considered a difficult problem, so several manageable steps are suggested, both without and with use of a bond graph:

(a) The linear velocity of the center of mass can be expressed in terms of orthogonal components, one proportional to $\dot\theta$ and the other proportional to $\dot\psi$. Find these components.

(b) Write that part of the kinetic energy which results from the velocity of the center of mass as functions of the variables θ, $\dot\theta$, $\dot\phi$ and $\dot\psi$.

(c) To evaluate the part of the kinetic energy due to angular velocity, it is necessary to find the angular velocity of the top in three orthogonal directions. It is suggested that these directions be the axis of symmetry, the axis which defines $\dot\theta$ and a mutually perpendicular axis which must be tilted $90° - \theta$ from the vertical. Find these angular velocities, noting that $\dot\psi$ contributes (via direction cosines) to both the angular velocities in the direction of the axis of symmetry and the third axis described above.

(d) Find the kinetic energy due to angular velocity, and add it to the result of step 2 to get the total kinetic energy as a function of θ, $\dot\theta$, $\dot\phi$, $\dot\psi$ and the fixed parameters.

(e) Find the potential or gravity energy, which depends on θ.

(f) Write the Lagrangian in the standard form, noting that $\mathcal{T}^* = \mathcal{T}$.

(g) Apply Lagrange's equations to get three differential equations.

(h) To get first-order differential equations suitable for simulation, the identities $d\theta/dt = \dot\theta$, $d\psi/dt = \dot\psi$ and $d\phi/dt = \dot\phi$ are available, but not all are needed since the only displacement to appear in the Lagrangian is θ. Convert your differential equations to a complete set of first-order equations.

(i) If the Lagrangian bond graph is to be used, steps (f)–(h) are not necessary. Instead, use the results of steps (c) and (d) to define primitive coordinates. Do this by noting that the kinetic energy contributed by a primitive velocity is proportional to its square.

(j) Draw the Lagrangian bond graph. Start by placing and labeling 1-junctions for the Lagrangian velocities $\dot{\phi}$, $\dot{\theta}$ and $\dot{\psi}$ and the primitive velocities identified in step (i). Assume the *possible* existence of three gyrators interconnecting the the Lagrangian 1-junctions. Note that the potential energy can be represented simply as a nonlinear compliance bonded to the 1-junction for $\dot{\theta}$.

(k) Find the moduli for all transformers, which should be evident from the process used in step (i).

(l) Use equation (12.76c) to find the moduli of all gyrators. (Hint: only one gyrator is non-zero.)

(m) *Optional.* Lagrange's equations may be found by summing the efforts around each Lagrangian 1-junction. For this purpose, steps (c)–(l) are not as direct as the alternative steps (d)–(f), but serve as a check.

(n) Note that the 1-junction for $\dot{\phi}$ in the Lagrangian bond graph has only one bond, which therefore has zero effort. Thus one of the primitive velocities is a constant, and the system is seen to be simpler than might have been expected.

(o) Specialize the bond graph to the case of steady θ and $\dot{\psi}$ by deleting the inertances. Solve for the associated equilibrium rate of precession.

(p) For the general case, the bond graph indicates the use of the momenta p_1 and p_2 on the other two inertances as state variables, plus θ. Use the standard rules to write the corresponding state differential equations. Convert all variables to p_1, p_2 and θ. The equations are nonlinear; they describe a complex **nutating** motion.

(q) *Optional.* Specialize the results for small perturbations about $\theta = 90°$ so that the approximations $\sin\theta = 1$ and $\cos\theta = 0$ are justified. Describe the motion.

12.5   Nonholonomic Constraints

The Lagrangian methods presented thus far assume holonomic constraints. By definition, constraints for which the number of degrees of freedom equal the number of corresponding generalized coordinates are holonomic. Constraints for which the number of generalized coordinates exceed the number of degrees of freedom are nonholonomic. The distinction is not as subtle as it may appear.

Constraints represented by transformers with constant moduli are always holonomic, as are transformers modulated by a local generalized displacement. The special additional holonomic constraints introduced in Section 9.4 are more restrictive than those given in Section 12.3. In the introduction to the distinction given in Section 9.4 it is noted that an important class of nonholonomic constraint can be represented by a transformer with a modulus that is a function of a *remote* generalized displacement. It is necessary, however, that the modulation be effected without any power being associated with the rate of change of the generalized displacement. After all, a transformer is a two-port element with no provision for power entering from a third source due to a change in its modulus. In the previous section, however, transformers modulated by remote displacements have in fact been used to represent holonomic constraints. The sole difference between such a holonomic constraint and a corresponding nonholonomic constraint is that only the former must include gyrators to properly complete the system.

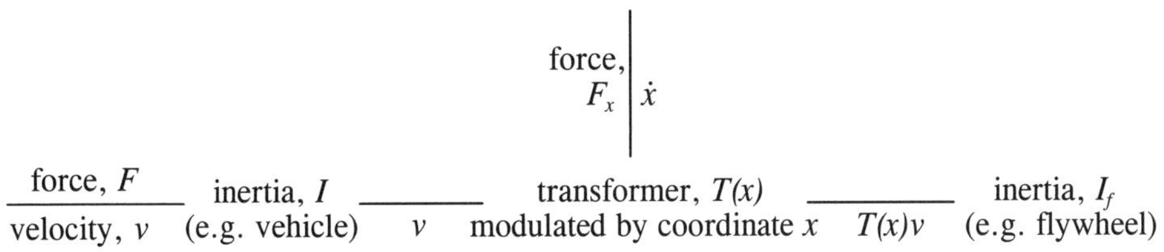

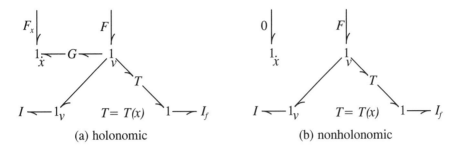

force,
F_x $\Big|\dot{x}$

$\dfrac{\text{force, } F}{\text{velocity, } v}$ ——— inertia, I ——— transformer, $T(x)$ ——— inertia, I_f

(e.g. vehicle) v modulated by coordinate x $T(x)v$ (e.g. flywheel)

Figure 12.22: Concept of variable-flywheel energy storage for vehicles

$F_x\Big|$ $F\Big|$ $0\Big|$ $F\Big|$

$1 \underset{x}{\leftarrow} G \leftarrow 1$ $1_{\dot{x}}$ 1_v

T

$I \leftarrow 1_v$ $T = T(x)$ $1 \longrightarrow I_f$ $I \leftarrow 1_v$ $T = T(x)$ $1 \longrightarrow I_f$

      (a) holonomic   (b) nonholonomic

Figure 12.23: Bond graphs for variable-flywheel systems

Another class of nonholonomic constraint employs RS elements such as for heat transfer; these can be represented also by "transformers" with moduli that depend on generalized velocites rather than just displacements. These statements deserve further explication.

12.5.1   Case Study: Variable-Flywheel Energy Storage for Vehicles

Two different systems may have identical energies in terms of equivalent coordinates and their derivatives (generalized velocities), and have identical transformational constraints between the various generalized velocities, and yet behave quite differently. Consider the class of systems described by the word bond graph of Fig. 12.22. This can represent a vehicle with kinetic energy stored in a flywheel. The transformational coupling between the speed of the vehicle, v, and the speed of the flywheel is modulated by the position of a lever with position x. It is especially important to recognize whether or not essential force and work is required to move the lever. (Force to overcome friction is not "essential" force.) The transformer has the modulus $T(x)$, so the kinetic energy of the flywheel due to its rotation is $(1/2)I_f[T(x)v]^2$, in which I_f is its rotational inertia. The kinetic energy of the vehicle, which has total mass I, is $(1/2)Iv^2$, which includes a contribution from the translation of the flywheel but not its rotation. Other details of a practical system, including dissipation, are neglected for clarity.

The standard Lagrangian method as presented earlier in this chapter produces the bond graph shown in Fig. 12.23 part (a). The gyrator results from the fact that the transformer modulus T represents a non-integrable constraint, rendering the integral of the velocity on the bond for I_f a quasi-coordinate as defined in Section 12.3. This gyrator implies that an essential (and large) force $F_x = Gv$ is necessary to change the transformer modulus. The power associated with this force, $F_x\dot{x} = Gv\dot{x}$, in fact equals the power $-\frac{d}{dt}\{\frac{1}{2}I_f[t(x)v]^2\}$ thereby induced to flow *out* of the inertia I_f. Therefore, assuming $F = 0$, *precisely one-half* of the power transferred to the inertia I, as a result of the displacement in x, comes from

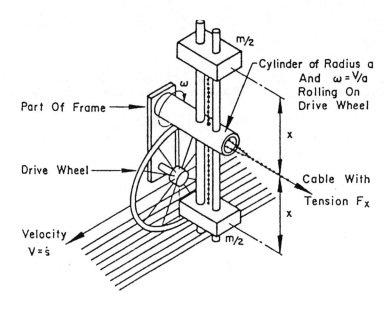

Figure 12.24: Example of holonomic variable flywheel

$F_x\dot{x}$, and the other half comes from the inertia I_f. This would be totally impractical.

An engineer would wish to eliminate the gyrator altogether, as shown in part (b) of the figure. In this case, no force or power would be required to change the modulus of the transformer; all the power delivered to I would come from I_f. The gyrator should be omitted, indeed, *if the integral of the velocity on the bond I_f is a true coordinate*, and therefore becomes a generalized coordinate. This means that the design is successful if and only if the constraint is nonholonomic, with one more generalized coordinate than there are degrees of freedom. How could this be done?

Consider the system shown in Fig. 12.24, which is also the subject of Guided Problem 9.9 (p. 643-644). The inertia of the flywheel is modulated by changing the radial location of the masses. The positional state of the mechanism is specified by x and the distance $s = \int v\,dt$, assuming no slipping of the wheels. The energy stored in the system[7] is specified uniquely in terms of x and s and their derivitives (velocities) $\dot{x}$ and v. The parameters are

$$I_f = m; \tag{12.84}$$

$$T(x) = x/a. \tag{12.85}$$

There are thus two generalized coordinates and two degrees of freedom. The system is holonomic, and is described properly by part (a) of Fig. 12.24. $F_x = Gv$ is simply the centrifugal force exerted by the masses on the cable.

Now consider the system of Fig. 12.25, which is first discussed in Example 9.13 (p. 640). A variable friction drive is substituted for the movable masses. Since the radius of the drum at the rolling contact point is chosen to be $r(x) = a^2/x$, which is the transformer modulus, the inertia and the stored energy are identical[8] to those of the holonomic system of Fig. 12.24. The number of degrees of freedom remains at two: the flywheel can rotate or translate. The number of generalized coordinates needed to describe the positional state

---

[7]The very small energy $\frac{1}{2}m\,\dot{x}^2$ and the corresponding inertial force are neglected for clarity.

[8]Again neglecting the energy $\frac{1}{2}m\dot{x}^2$.

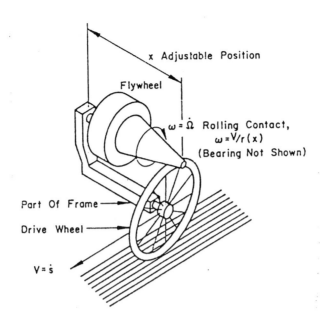

Figure 12.25: Example of nonholonomic variable flywheel

of the system, however, is increased to three, again assuming no slippage of the wheels: x, $s = \int v\, dt$ and $\Omega = \int \dot{\Omega}\, dt = \int T v\, dt$. The integral of the velocity on the bond I_f is Ω, but it is a *true* coordinate, unlike its counterpart in Fig. 12.24, with direct physical meaning (the angular position of the flywheel).

12.5.2   Analysis and Representation of Nonholonomic Systems

When the number of generalized coordinates, n, exceeds the number of degrees of freedom for small displacements, m, the n variations $\delta q_1, \ldots, \delta q_n$ are not in fact independent. They were assumed to be independent in the derivation of Lagrange's equations (equation 12.64) in Section 12.3, however; as a result these equations do not apply.

Primitive coordinates that are true have been distinguished from those that are quasi; the quasi coordinates are related to the true generalized coordinates by non-integrable constraints. This necessary condition is not sufficient; true generalized coordinates that are related to other true generalized coordinates by non-integrable constraints must be distinguished. These are the nonholonomic constraints. Fig. 12.26 shows the general scheme of the associated bond graph. The new true coordinates are arrayed across the bottom, with the identifying subscript b to contrast with the true coordinates at the top which are given the identifying subscript a. The transformers connecting the two domains represent the nonholonomic constraints. Additional domains of true coordinates connected by nonholonomic constraints could be added, with identifying subscripts c, d, etc.

Equation (12.43) (p. 974) represents holonomic constraints if all the generalized coordinates are included in $\mathbf{q}$. If, rather, one or more generalized coordinates are represented in $\mathbf{x}$ but not $\mathbf{q}$, this equation also represents a large subset of nonholonomic constraints. The domain of nonholonomic constraints can be expanded by employing the following more general form:

$$\sum_{k=1}^{n} B_{jk}(\dot{\mathbf{q}}, \mathbf{q}, t) dq_k + B_j(\dot{\mathbf{q}}, \mathbf{q}, t) dt = 0; \quad j = 1, \ldots, \ell. \tag{12.86}$$

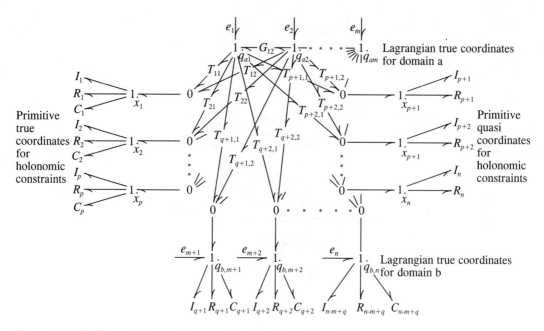

Figure 12.26: General form of Lagrangian bond graph including nonholonomic constraints

This relation connects the *differentials* of the qs. Time is held constant, for each point in the path, in the variational process. Hence the virtual displacements occurring in the variation must satisfy equations of constraint in the form

$$\sum_{k=1}^{n} B_{jk}\, \delta q_k = 0; \quad j = 1, \ldots, \ell. \tag{12.87}$$

This equation can be used to reduce the number of virtual displacements from the number of generalized coordinates to the number of degrees of freedom, *i.e.* $N - m = n - l$. The formal procedure followed is known as the method of **Lagrange undetermined multipliers**. Equation (12.87) also requires that

$$\lambda_j \sum_{k=1}^{n} B_{jk}\, \delta q_k = 0; \quad j = 1, \ldots, \ell, \tag{12.88}$$

where the λ_j are some undetermined constants or functions of time. This can be summed over j, and then integrated from point 0 to point 1:

$$\int_{t_0}^{t_1} \sum_{k=1}^{n} \sum_{j=1}^{\ell} \lambda_j\, B_{jk}\, \delta q_k dt = 0. \tag{12.89}$$

Adding this result to equation (12.63) (p. 977) gives

$$\int_{t_0}^{t_1} \sum_{k=1}^{n} \left[\frac{d}{dt}\left(\frac{\partial \mathcal{L}}{\partial \dot{q}_k} \right) - \frac{\partial \mathcal{L}}{\partial q_k} - e_k - \sum_{j=1}^{\ell} \lambda_j\, B_{jk} \right] \delta q_k\, dt = 0. \tag{12.90}$$

The $\delta q's$ are still not independent; they are related by the ℓ relations of equation (12.86). The first $n - \ell$ of these may be chosen independently, but the last ℓ are fixed by this equation.

The values of λ_j can be chosen arbitrarily, however. It is useful to choose the λ_js such that

$$\frac{d}{dt}\left(\frac{\partial \mathcal{L}}{\partial \dot{q}_k}\right) - \frac{\partial \mathcal{L}}{\partial q_k} - e_k - \sum_{j=1}^{\ell} \lambda_j B_{jk} = 0 \quad k = n-\ell,\, n-\ell+1,\ldots,n, \tag{12.91}$$

which become the equations of motion (the new Lagrange's equations) for the last ℓ of the q_k variables. With the λ_j so determined, equation (12.90) becomes

$$\int_{t_0}^{t_1} \sum_{k=1}^{n-\ell} \left[\frac{d}{dt}\left(\frac{\partial \mathcal{L}}{\partial \dot{q}_k}\right) - \frac{\partial \mathcal{L}}{\partial q_k} - e_k - \sum_{j=1}^{\ell} \lambda_j B_{jk}\right] \delta q_k\, dt = 0. \tag{12.92}$$

The only δqs involved here are the independent ones, so the function in square brackets vanishes for $k = 1,\ldots,n-\ell$. Combining this with equation (12.91) results in the complete set of Lagrange's equations for nonholonomic systems:

$$\boxed{\frac{d}{dt}\left(\frac{\partial \mathcal{L}}{\partial \dot{q}_k}\right) - \frac{\partial \mathcal{L}}{\partial q_k} = e_k + \sum_{j=1}^{\ell} \lambda_j B_{jk}; \qquad k = 1, 2, \ldots, n.} \tag{12.93}$$

There are n such equations, but we have added an additional ℓ unknowns in the form of the λs. The additional equations needed are precisely equations (12.86), which can be written as differential equations:

$$\boxed{\sum_{k=1}^{n} B_{jk}(\dot{\mathbf{q}}, \mathbf{q}, t)\dot{q}_k + B_j(\dot{\mathbf{q}}, \mathbf{q}, t) = 0; \qquad j = 1, \ldots, \ell.} \tag{12.94}$$

The partial derivatives in equation (12.93) for a particular value of k include only the effects of constraints within the holonomic domain that includes q_k. Therefore, gyrators are generated only because of constraints entirely *within* a domain. Constraints *between* domains are represented by the terms $\sum_{j=1}^{\ell} \lambda_j B_{jk}$ which are the generalized forces on the transformer bonds connecting the domains. *The bond graph of a system including nonholonomic constraints is identical to the bond graph of the same system with the constraints falsely considered as holonomic, except that the gyrators generated by the falsely identified constraints are missing.* Alternatively, one can restrict the use of the Lagrangian procedure to holonomic domains, and afterward simply interconnect these domains with transformers or equivalent elements to represent the nonholonomic constraints. Nonholonomic constraints are in a sense weaker than holonomic constraints, and typically produce simpler equations of motion.

Holonomic constraints do not always produce gyrators. If not, false identification of a constraint as holonomic nevertheless produces a correct bond graph, and the graph produces proper differential equations. On the other hand, the formal Lagrangian mathematical approach does not forgive such a mistake. This situation is illustrated in Problem 12.16 below.

12.5.3 Application to Case Study

The system of Fig. 12.25 has two degrees of freedom but three generalized coordinates:

$$\mathcal{L} = \frac{1}{2}I\dot{s}^2 + \frac{1}{2}I_f\dot{\Omega}^2, \tag{12.95}$$

$$q_a = \begin{bmatrix} x \\ s \end{bmatrix}, \tag{12.96a}$$

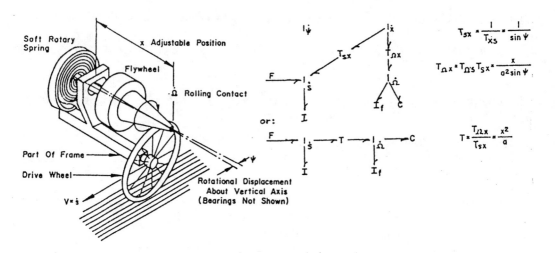

Figure 12.27: Elaboration of friction drive

$$q_b = \Omega, \tag{12.96b}$$

$$\dot{\Omega} = T\dot{s}; \qquad T = x/a^2. \tag{12.97}$$

The bond graph of Fig. 12.23 part (b) results directly; no gyrator exists because the only non-constant transformer interconnects the 1-junctions for two true coordinates. The equations of motion can be written upon inspection.

The system is expanded in Fig. 12.27 to include a fourth generalized coordinate, the angle ψ of the friction wheel, as a more practical control. In addition, energy is stored in a spring as a result of the rotation $\dot{\Omega}$. The true coordinates are

$$\mathbf{q}_a = \begin{bmatrix} x \\ \psi \end{bmatrix}; \tag{12.98a}$$

$$q_b = \Omega, \tag{12.98b}$$

$$q_c = s. \tag{12.98c}$$

The Lagrangian (excluding the spring) is

$$\mathcal{L} = \frac{1}{2}I\dot{s}^2 + \frac{1}{2}I_f\dot{\Omega}^2. \tag{12.99}$$

Both constraints are non-integrable and connect true coordinates (*i.e.* nonholonomic):

$$\dot{x} = T_{xs}\dot{s}; \qquad T_{xs} = \sin\psi, \tag{12.100a}$$

$$\dot{\Omega} = T_{\Omega s}\dot{s}; \qquad T_{\Omega s} = x/a^2. \tag{12.100b}$$

As a result there are no gyrators, and the equations of motion are decoupled.

Functionally similar nonholonomic systems result from substituting a pair of field-controlled DC motor-generators or a pair of variable hydraulic pump-motors for the variable friction drive.

12.5.4   Example of the Rolling Penny

The case of the large old British penny rolling on the barroom floor is given in Fig. 12.28. This is a favorite problem of dynamicists that illustrates an important class of nonholonomic systems. There are three degrees of freedom: roll, sidewise pitch and twist. The respective generalized velocities are chosen as the time derivatives of the traditional Euler angles ψ and ϕ and the complement of the third Euler angle, θ, namely $\alpha = \theta - 90°$. There are two additional generalized coordinates, however: the Cartesian coordinates ζ and η. Thus there are two more coordinates required to state the positional state of the penny than there are small allowable motions; the penny is assumed not to slide over the surface. It can move from one location to another by a variety of paths, nevertheless, each of which employs a different history of the three degrees of freedom.

The two groups of coordinates represent separate holonomic domains, so the transformational coupling between them produce no gyrators. The three degree-of-freedom coordinates produce one quasi-coordinate and one gyrator, as shown. Notice that the compliance is *negative*, producing the possibility of instability (the penny falling over). The penny is modeled as any axisymmetric planar solid with mass M, radius a and moment of inertia about a centroidal axis in its plane J. (The moment of inertia about a centroidal axis normal to its plane is $2J$.)

The differential equations for the system can be linearized for the special case of perturbations about straight-line rolling ($\alpha = 0$, $\dot{\alpha} = 0$, $\phi = 0$, $\dot{\phi} = 0$, $\dot{\psi} = $ constant), and the bond graph of part (c) of the figure drawn. The gyrator modulus comes only partly from the corresponding gyrator in the nonlinear bond graph. The inertance on the right side appears, when viewed from the left side of the gyrator, to be a compliance. Two compliances bonded to a common 1-junction can be coalesced into a single compliance by adding their stiffnesses, and the system is stable if and only if the sum is positive. As a result, the criterion of stability is

$$\frac{1}{C} + \frac{G^2}{I_b} > 0, \tag{12.101}$$

which gives

$$\dot{\psi} > \sqrt{\frac{g/a}{2 + 4J/Ma^2}}. \tag{12.102}$$

The example of a penny (a disk with $J = Ma^2/4$) gives $\dot{\psi} > \sqrt{g/3a}$, and the example of a hoop (with $J = Ma^2/2$) gives $\dot{\psi} > \sqrt{g/4a}$. For slower rotational speeds the disk or hoop wobbles and soon falls over.

Wheeled vehicles, boats in water and skates on ice have analogous nonholonomic constraints to the penny: they cannot nominally slide sideways (a restricted degree of freedom) but they can move to an equivalent position by, for example, traversing large arcs.

12.5.5   Irreversible Nonholonomic Constraints

The nonholonomic coefficients B_{jk} are allowed in equation (12.94) to be functions of $\dot{\mathbf{q}}$ as well as $\mathbf{q}$. These act like transformers computationally, but transformer moduli have not been allowed to depend on $\dot{\mathbf{q}}$, since such a modulation destroys reversibility. These "transformers" therefore are properly identified as RS elements.

Viewed in this way, heat transfer may be the most common class of nonholonomic constraint; the coefficients B_{jk} are functions of entropy fluxes. It is hard to misconstrue heat transfer as a holonomic constraint, since compliances are needed to represent the thermal energies, forcing the displacement entropies to be true and generalized coordinates. Further, the gyrators produced by holonomic constraints also require the presence of inertances, and

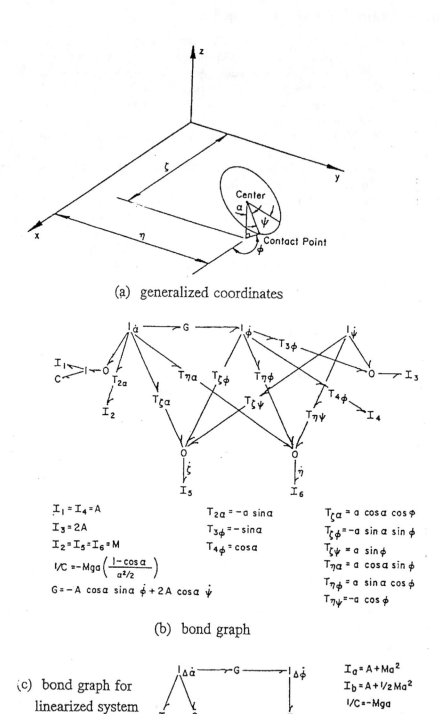

(a) generalized coordinates

(b) bond graph

(c) bond graph for
     linearized system

Figure 12.28: The rolling penny or disk

there is no such thing as a thermal inertance. The gyrators further require the modulation of the transformers to be remote, whereas in fact it is local. This helps explain the fact that heat transfer is outside the domain of the classical literature on nonholonomic constraints.

The nature of a nonholonomic constraint for heat transfer is illustrated in Problem 12.16, which by some stretch also illustrates a possible pitfall in using a Lagrangian approach while misconstruing an entropy as a primitive quasi-coordinate.

12.5.6 The Umbra-Lagrangian of Mukherjee

Amalendu Mukherjee and colleagues have developed a radial reformulation of Lagrangian-Hamiltonian mechanics aimed at greatly extending the range of nonholonomic and nonconservative system models that can be addressed conveniently. In addition to real time, t, Mukherjee introduces what he calls **umbra time**, η, which is allowed to vary even when t is fixed. The umbra-potential energy for a spring with time-varying stiffness becomes

$$V_c^*(t, x(\eta)) = \int_0^{x(\eta)} K(t)\, x(\xi)\, dx(\xi) = \frac{1}{2} K(t) x^2(\eta). \qquad (12.103)$$

The umbra-kinetic energy for a time-varying mass $m = m(t)$ is

$$\mathcal{T}_C^*(t, \dot{x}(\eta)) = \int_0^{\{} \dot{x}(\eta) m(t) \dot{x}(\xi)\, d\dot{x}(\xi) = \frac{1}{2} m(t) \dot{x}(\eta). \qquad (12.104)$$

Umbra potentials also are associated with generalized resistive fields:

$$\mathcal{V}_R^* = \int_0^{q(\eta)} e(t, \mathbf{q}(t), \dot{\mathbf{q}}(t)\, d\mathbf{q}(\eta). \qquad (12.105)$$

Unlike the Rayleigh potential, this includes both dissipative as well as gyroscopic or nonholonomic forces. The umbra-potential for an external force $F(t)$ is

$$\mathcal{V}_p^* = -F(t) x(\eta), \qquad (12.106)$$

and the total umbra-potential of a system comprises the sum of component umbra-potentials, $\mathcal{T}^*(t, \mathbf{q}(t), \dot{\mathbf{q}}(t), \{\mathbf{q}(\eta)\})$. This leads to an umbra-Lagrangian

$$\mathcal{L}^* = \mathcal{T}_C^*(t, \dot{\mathbf{q}}(\eta)) - \mathcal{V}^*(t, \mathbf{q}(t), \dot{\mathbf{q}}(t), \mathbf{q}(\eta)) \qquad (12.107)$$

and the umbra Lagrange's equation

$$\frac{d}{dt}\left[\lim_{\eta \to t} \frac{\partial \mathcal{L}^*}{\partial \dot{q}_i(\eta)} \right] - \lim_{\eta \to t} \frac{\partial \mathcal{L}^*}{\partial q_i(\eta)} = 0, \qquad i = 1, \cdots, n \qquad (12.108)$$

The work communicated to the author privately, which includes an umbra-Hamiltonian, goes far beyond this simple introduction and the scope of this chapter. As of this writing it has yet to appear in the archival mechanics literature, but some of it is presented in the bond graph literature[9], with an emphasis on bond graph interpretations.

---

[9] The following papers appear in the proceedings of the International Conference on Bond Graph Modeling and Simulation (ICBGM), Society for Modeling and Simulation International (SCS): Mukherjee, A. and Samantaray, A.K., "Umbra-Lagrange's Equations through Bond Graphs," 1997, pp.168-174; Mukharjee, A., "The Issue of Invariants for Motion for a General Class of Symmetric Systems through Bond Graphs and the Umbra-Lagrangian," 2001, pp. 295-304; Mukherjee, A., Rastogi, V. and Dasgupta, A., "A Procedure for Finding Invariants of Motions for a General Class of Unsymmetric Systems with a Gauge-Variant Umbra-Lagrangian Generated by Bond Graphs," 2005, pp. 11-16.

12.5.7   Summary

The generalized coordinates for a system are a minimum set of displacements necessary to define the positional state of the system. Holonomic constraints translate degrees of freedom, that is geometrically compatible independent infinitesimal displacements, into an equal number of generalized coordinates. Nonholonomic constraints are weaker: they produce or allow more generalized coordinates than degrees of freedom. As a consequence, the state differential equation for each generalized coordinate is augmented by terms including further dependent variables, equal in number to the nonholonomic constraints, known as Lagrange multipliers. Differential equations describing these constraints are needed to complete a set of solvable equations. Misidentification of the type of a constraint will result in an error.

The situation is clearer when described in terms of detailed Lagrangian bond graphs. The energy of the system is expressed in terms of a set of primitive coordinates or their derivatives, the primitive velocities. The holonomic constraints of the system are represented by a network of transformers and 0-junctions connecting the generalized coordinates to the primitive coordinates. The nonholonomic constraints are represented by transformers (in some cases RS elements) connecting the generalized coordinates to each other, which is a very different matter. Generalized coordinates are true coordinates, with direct physical or geometric meaning. Primitive coordinates are true coordinates if they have a compliance attached, or potentially could have a compliance attached. Otherwise, they are quasi-coordinates. Quasi coordinates produce gyrational couplings between the generalized coordinates.

If a generalized coordinate is not recognized as such but is treated merely as a primitive coordinate, the number of generalized coordinates is understated, and one or more erroneous gyrational couplings likely are produced between the remaining recognized true coordinates. The proper bond graph is found simply by deleting these gyrators. Proper identification of the generalized coordinate makes the transformer or transformers conecting it to other generalized coordinates represent a nonholonomic constraint. One should note that some coordinates do double-duty as both generalized and primitive coordinates.

Physical reasoning usually tells you whether a gyrator should be present or not; it has powerful consequences. Therefore, in practice this physical reasoning may be the most powerful evidence guiding you in distiguishing generalized from quasi coordinates, and holonomic from nonholonomic constraints.

Heat transfer produces a nonholonomic constraint but no gyrator. There is no confusion if a bond graph approach is used, but use of the formal Lagrangian mathematics requires careful attention to the Lagrange multipliers.

PROBLEM

12.16 A mass is supported in an adiabatic container of fluid by a slender reed, as pictured. The elasticity of the reed is great enough to allow vertical vibrations of the mass (with extension of the reed), and the thermal expansion coefficient of the reed and the thermal conductivity between the reed and the fluid combine to introduce damping. Write a set of state differential equations suitable for simulation, neglecting viscous forces. The reed may be characterized by the linear relation

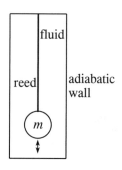

$$\begin{bmatrix} S \\ \Delta s_r \end{bmatrix} = \begin{bmatrix} 1/E & \alpha \\ \alpha & \rho_r c_r/\theta_0 \end{bmatrix} \begin{bmatrix} \sigma \\ \Delta \theta_r \end{bmatrix},$$

(equation (9.22)) in which S is the strain, Δs_r is the change in entropy about the nominal, σ is stress, $\Delta \theta_r$ is the change in temperature about the nominal absolute temperature of θ_0, E is Young's modulus, ρ_r is mass density and c_r is the specific heat at constant stress. The reed has length ℓ and cross sectional area A, and the mass has mass m. The conduction of heat between the reed and the fluid obeys a linear heat transfer coefficient H, and the mass and specific heat of the fluid are m_f and c_f, respectively. All coefficients may be assumed to be constants and the inertia of the reed may be neglected compared to that of the mass.

You are asked to employ the Lagrangian approach to this problem in order to understand its application to nonholonomic problems better and to see the role of heat transfer. Otherwise, it could be addressed more easily by the bond graph methods that more obviously allow a wider choice of variables.

Suggested Steps:

(a) Identify generalized coordinates and their derivatives suitable as Lagrangian variables, and write equations for the kinetic and potential energies in terms of these variables. Suggestion: include the entropies of the reed and the fluid as generalized coordinates.

(b) Write the Lagrangian of the model.

(c) Equation (12.94) for the nonholonomic heat transfer constraint can be written $B_r \dot{\sigma}_1 + B_f \Delta \dot{s} = 0$, and B_f can be set equal to 1 arbitrarily. Find B_r as a function of $\Delta \dot{s}$ and H.

(d) Continue the formal approach by deducing three Lagrange's equations from equation (12.93). Then, combine two of them to eliminate λ, which is not of real interest. Finally, add identities or whatever is necessary to write a set of first-order state differential equations.

(e) Next, start the bond graph approach by drawing the graph. Use the same variables $\Delta \dot{s}_r$, $\Delta \dot{s}_f$, $\Delta \theta_r$ and $\Delta \theta_f$, which should appear on the graph.

(f) Find the transformer modulus T for the RS element as a function of $\Delta \dot{s}_r$ and H (although leave the element as RS). Compare this with the B_r of step (c) above.

(g) Apply integral causality to all energy storage elements. Find the resulting state differential equations, and compare with those found from the Lagrangian approach (step (d)). Note: The mutual compliance with its impedance causality can be treated directly, without using a bond graph equivalence such as given in Section 9.3.5 or Guided Problem 9.6 (although these are alternatives that employ different state variables); the effort on each side equals the displacement on the other side divided by the mutual compliance.

(j) Note that no gyrator is generated in the bond graph regardless of whether the heat transfer is recognized as holonomic or nonholonomic. Therefore, the issue never even arises if the bond graph approach is used, which is an advantage. The RS element must represent a nonholomic constraint, however, because it is non-integrable and connects two true coordinates (Δs_r and Δs_f).

(k) Also note that the heat transfer produces damping, but it is nonlinear and vanishes in the case of vanishingly small disturbances. Therefore, no damping ratio is defined.

12.6   Hamilton's Equations and Bond Graphs*

The second most celebrated of the many energy methods developed for classical dynamics[10] and extended to other fields is the Hamilton equations of motion. These equations often are more elegant and insightful than Lagrange's equations. They also directly employ the momenta of inertances as state variables and thus represent a generalization of the methods employed throughout earlier chapters. Hamiltonian bond graphs are compatible with ordinary bond graphs. Unfortunately, however, in the cases of multiport inertances with non-constant moduli, which are the focus of present interest, the Hamilton equations usually require more effort to find than Lagrange's equations. As a result they are not so widely used in engineering. The major applications of the Hamiltonian method are in statistical mechanics, quantum mechanics and celestial mechanics.

12.6.1   Hamilton's Equations

The Hamiltonian is the total energy of the system expressed as a function of $\mathbf{p}$ and $\mathbf{q}$:

$$\mathcal{H} = \mathcal{H}(\mathbf{q}, \mathbf{p}) = \mathcal{T}(\mathbf{q}, \mathbf{p}) + \mathcal{V}(\mathbf{q}). \qquad (12.103)$$

To formulate the Hamiltonian in other than the simple cases addressed in earlier chapters one first finds the Lagrangian, which is easier to do. The Legendre transformation of equation (12.67), repeated here[11]

$$\mathcal{T} = \sum_{i=1}^{n} p_i \dot{q}_i - \mathcal{T}^*; \qquad p_i = \frac{\partial \mathcal{T}^*}{\partial \dot{q}_i} = \frac{\partial \mathcal{L}}{\partial \dot{q}_i}, \qquad (12.104)$$

is then employed. Substitution of this result into equation (12.103) gives

$$\mathcal{H} = \sum_{i=1}^{n} p_i \dot{q}_i - \mathcal{T}^* + \mathcal{V} = \sum_{i=1}^{n} p_i \dot{q}_i - \mathcal{L}. \qquad (12.105)$$

The derivative of $\mathcal{H}$ can be written as follows, noting that $\mathcal{L} = \mathcal{L}(\dot{\mathbf{q}}, \mathbf{q})$ and $\partial \mathcal{L}/\partial \dot{q}_i = p_i$:

$$d\mathcal{H} = \sum_{i=1}^{n} \left(p_i d\dot{q}_i + \dot{q}_i \, dp_i - p_i d\dot{q}_i - \frac{\partial \mathcal{L}}{\partial q_i} dq_i \right). \qquad (12.106)$$

The first and third terms cancel. The term $\partial \mathcal{L}/\partial q_i$ is, from Lagrange's equations (12.93) and the definition of p_i above,

$$\frac{\partial \mathcal{L}}{\partial q_i} = \frac{d}{dt}(p_i) - e_i - \sum_{i=1}^{n} \lambda_i B_{ji}. \qquad (12.107)$$

Therefore, equation (12.106) becomes

$$d\mathcal{H} = \sum_{i=1}^{n} \left[\left(e_i + \sum_{j=1}^{\ell} \lambda_i B_{ji} - p_i \right) dq_i + \dot{q}_i \, dp_i \right]. \qquad (12.108)$$

---

[10]The most cited references in the use of energy methods in classical dynamics are L. A. Pars, *A Treatise on Analytical Dynamics*, Heinemann, London, 1965; H. Goldstein, *Classical Dynamics*, Addison-Wesley, 1950; C. Lanczos, *the Variational Principles of Mechanics*, U. of Toronto Press, 1949; E. T. Whittaker, *A Treatise on Analytical Dynamics of Particles and Rigid Bodies*, Dover, 1944.

[11]The prior omission of any dependency on time in the constraints, and thus in $\mathcal{L}$ and $\mathcal{H}$, is continued.

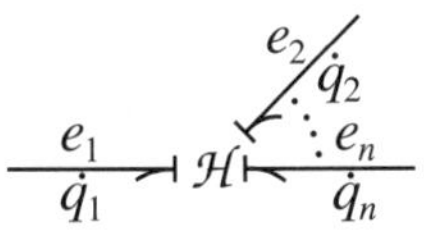

Figure 12.29: Summary representation of a Hamiltonian domain

This result can be compared with the derivative of equation (12.103), namely

$$dH = \sum_{i=1}^{n} \left[\frac{\partial H}{\partial q_i} dq_i + \frac{\partial H}{\partial p_i} dp_i \right], \tag{12.109}$$

to give the canonical Hamilton's equations

$$\boxed{\begin{aligned} \frac{dp_i}{dt} &= e_i + \sum_{j=1}^{\ell} \lambda_j B_{ji} - \frac{\partial H}{\partial q_i}; & i &= 1, \ldots, n, \\ \frac{dq_i}{dt} &= \frac{\partial H}{\partial p_i}; & i &= 1, \ldots, n. \end{aligned}} \tag{12.110}$$

These become the state differential equations. In the presence of nonholonomic constraints it is necessary to add equations (12.94), repeated here:

$$\boxed{\sum_{i=1}^{n} B_{ji}(\dot{\mathbf{q}}, \mathbf{q}) dq_i + B_i(\dot{\mathbf{q}}, \mathbf{q}) dt = 0; \qquad j = 1, \ldots, \ell.} \tag{12.111}$$

The results given in Section 12.5 are special cases of the above.

12.6.2   Hamiltonian Bond Graphs

The domain of a larger system treated by Hamilton's equations can be represented in a bond graph by a summary symbol H and admittance causality, just as a domain treated by Lagrange's equations can be represented by a summary symbol $\mathcal{L}$ and admittance causality. This is shown in Fig. 12.29.

Alternatively, one can draw a Hamiltonian bond graph[12], which reveals some of the structural detail of the domain, but usually less than a corresponding Lagrangian bond graph. There are two generic versions, one based on compliances and the other on inertances. These graphs are shown in Fig. 12.30 for the holonomic case with two independent generalized velocities. They represent the canonical equations directly, leaving unspecified only the actual function $H(\mathbf{p}, \mathbf{q})$. The gyrators are **symplectic**, which means having unity modulus.[13] A compliance seen through a symplectic gyrator is equivalent to an inertance with the same modulus, and vice-versa, explaining the duality of the two equivalent graph forms.

The graphs can be specialized to one generalized velocity or expanded to three or more; the central energy storage **C** or **I** remains common. In special cases (see below) the **C** or **I** can be subdivided into two or more energy storage elements, increasing the didactic value of the graph. Integral causality is used as shown, with $\mathbf{q}$ and $\mathbf{p}$ as state variables. Thus the standard procedure for writing the state differential equations from the graph is used.

---

[12]F. T. Brown, "Hamiltonian and Lagrangian Bond Graphs," *J. Franklin Institute*, v328 n5/6, 1991, pp 809-831.

[13]Symplectic gyrators were introduced by P.C. Breedveld in the *J. Franklin Institute*, v 314 n 1, pp 15-40, July 1982. He was motivated by the symplectic structure of the canonical transformations that result in Hamilton's equations. Although he shows only autonomous systems, his generic graphs (Fig. 7) are compatible with the Hamiltonian bond graphs herein.

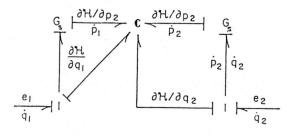

(a) compliance-based

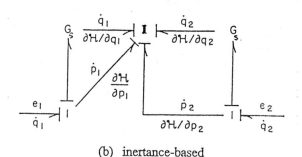

(b) inertance-based

Figure 12.30: General Hamiltonian bond graphs for two independent variables

12.6.3   Application to Flyball Governor

The flyball governor considered in Sections 12.3 and 12.4 by Lagrangian methods is now treated by the Hamiltonian method for comparison. Repeating equation (12.65),

$$\mathcal{L} = \frac{1}{2}m\left[r^2\dot{\theta}^2 + (a + r\sin\theta)^2\dot{\Omega}^2\right] - 2kb^2(1 - \cos\theta)^2 - mgr(1 - \cos\theta). \qquad (12.112)$$

From the definition of p_i given in equation (12.104),

$$p_\theta = \frac{\partial\mathcal{L}}{\partial\dot{\theta}} = mr^2\dot{\theta}; \qquad (12.113a)$$

$$p_\Omega = \frac{\partial\mathcal{L}}{\partial\dot{\Omega}} = m(a + r\sin\theta)^2\dot{\Omega}. \qquad (12.113b)$$

These equations are solved for $\dot{\theta}$ and $\dot{\Omega}$, which are substituted into equation (12.112) and, along with the result, substituted into equation (12.105) to give the Hamiltonian in terms of the proper variables. Alternatively, since $\mathcal{T} = \mathcal{T}^*$, one can use $\mathcal{H} = \frac{1}{2}\sum_k p_k\dot{q}_k + \mathcal{V}(q_k)$. In either case,

$$\mathcal{H} = \frac{p_\theta^2}{2\,mr^2} + \frac{p_\Omega^2}{2m(a + r\sin\theta)^2} + 2kb^2(1 - cos\theta)^2 + mgr(1 - \cos\theta). \qquad (12.114)$$

Finally, the canonical Hamilton's equations can be found using equations (12.110) and noting that $e_\theta = (2b\sin\theta)F$:

$$\frac{dp_\theta}{dt} = e_\theta - \frac{\partial\mathcal{H}}{\partial\theta} = (2b\sin\theta)F - 4kb^2(1 - \cos\theta)\sin\theta - mgr\sin\theta + \frac{p_\Omega^2 r\cos\theta}{m(a + r\sin\theta)^3},$$

$$(12.115a)$$

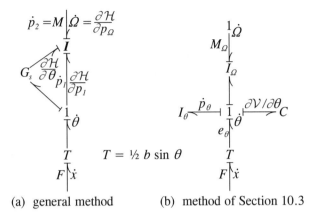

(a) general method   (b) method of Section 10.3

Figure 12.31: Hamiltonian bond graphs for the flyball governor

$$\frac{dp_\Omega}{dt} = e_\Omega - \frac{\partial \mathcal{H}}{\partial \Omega} = M_\Omega, \tag{12.115b}$$

$$\frac{d\theta}{dt} = \frac{\partial \mathcal{H}}{\partial p_\theta} = \frac{p_\theta}{mr^2}, \tag{12.115c}$$

$$\frac{\partial \Omega}{\partial t} = \frac{\partial \mathcal{H}}{\partial p_\Omega} = \frac{p_\Omega}{m(a + r \sin \theta)^2}. \tag{12.115d}$$

These equations are simpler, overall, than the corresponding Lagrange's equations of equations (12.66) (p. 978), to which the identities $d\theta/dt = \dot{\theta}$ and $d\Omega/dt = \dot{\Omega}$ must be added to give a complete set of first-order differential equations. In particular, the "cyclic" variable p_Ω is constant in the absence of an applied moment.

A Hamiltonian bond graph for the system is given in part (a) of Fig. 12.31. The symplectic gyrator associated with Ω is eliminated, since $\partial \mathcal{H}/\partial \Omega$ is identically zero. The information in the graph plus the Hamiltonian is sufficient to find the differential equations.

This problem happens to be barely simple enough to be treatable by the methods of Section 9.4, allowing a comparison of four different methods. As shown in the bond graph of part (b) of the figure, the kinetic energy can be separated into that of a simple inertance $I_\theta = mr^2$ for the $\dot{\theta}$ component of the velocities of the masses and a two-port inertance I_Ω for which the energy is a function of the momentum at one port and the displacement at the other:

$$\mathcal{T}_\Omega = \frac{1}{2} \frac{p_\Omega{}^2}{m(a + r \sin \theta)^2}. \tag{12.116}$$

The potential energy also is separable from the two-port energy storage, being only a function of θ:

$$\mathcal{V} = mgr(1 - \cos \theta) + 2kb^2(1 - \cos \theta)^2. \tag{12.117}$$

The causal strokes on the bond graph direct the writing of the state differential equations. Equation (12.115a) follows from the balance of efforts about the central 1-junction:

$$\frac{dp_\theta}{dt} = \frac{1}{T}F - \frac{\partial \mathcal{T}_\Omega}{\partial \theta} - \frac{\partial \mathcal{V}}{\partial \theta}. \tag{12.118}$$

Equations (12.115b) and (12.115c) are directly evident, and equation (12.115d) follows from

$$\frac{\partial \Omega}{\partial t} = \frac{\partial \mathcal{T}_\Omega}{\partial p_\Omega}. \tag{12.119}$$

Choosing between the four methods, when they apply, is a matter of taste. The method of Section 9.4 is the quickest and gives the nicest of the three different results, but requires more insight than either of the other methods. It is the method of choice when it is available to you. The general Hamiltonian procedure gives the same result but requires considerable work with more opportunities for making a mistake. The two Lagrangian methods (with and without the bond graph) are the most direct to carry out and therefore probably are second and third choice. Use of the Lagrangian bond graph leads to a nicer set of differential equations. The Lagrangian bond graph has only one-port inertances and therefore is probably more instructive than the bond graph of the method of Section 9.4, which has a two-port inertance, or the Hamiltonian bond graph, which has a three-port inertance.

12.6.4   Example of a Seating Valve Using Lagrange's Equations

The symmetrical two-dimensional seating valve of Fig. 12.32 virtually demands the use of energy methods. A Lagrangian analysis is presented first, assuming the heart of the system is conservative. A Hamiltonian analysis is presented next, followed finally by the introduction of dissipation into the heart of the system.

The fluid is incompressible, so the constraints between the flows Q_i, Q_0, Q_1 and the velocity $\dot{y}$ are represented by a structure of junctions and transformers as shown in the figure. The depth in the third dimension is w. The mass and linear spring are represented by the inertance I_m and compliance C_k. A "Bernoulli" resistor R is added to account for an assumed total head loss; P_i and P_o must be total pressures. Otherwise, the internal effects of viscosity are omitted at this stage. The graph applies only for $Q_0 > 0$. The pressure differences due to unsteady flow are neglected.

The effects of unsteady flow can be approximated by assuming that the kinetic energy of the flow in the two slots can be computed taking the velocities to be horizontal and independent of vertical location. The slots have length L and a variable height y. Therefore,

$$\mathcal{L} = \mathcal{T}^* = \mathcal{T} = 2w \left[\frac{1}{2}\rho y \int_0^L \left(\frac{Q_1/2 - wx\dot{y}}{wy} \right)^2 dx \right]$$

$$= \frac{\rho L}{4wy} \left[(Q_1 - Lw\dot{y})^2 + \frac{1}{3}(Lw\dot{y})^2 \right], \qquad (12.120)$$

which implicitly assumes that $y << L$, at least when the associated pressure drops are large.[14]

A Lagrangian bond graph can be inserted into the graph of part (b) of the figure to represent this energy. In this case there is little advantage in this strategy, however; the summary symbol $\mathcal{L}$ is inserted instead, as shown in part (c), and direct application of Lagrange's equations will be made. Causal strokes are added in the standard fashion, making e_1 and e_2 the causal inputs to the Lagrangian domain. For this to be possible, without causal conflict, the inertia I_m must be transferred to the domain, adding the term $\frac{1}{2}m\dot{y}^2$ to the right side of equation (12.120).

The two resulting Lagrange's equations each include terms linear in Q_1 and y. After these are separated, the result becomes

$$\ddot{y} = \frac{1}{1 + 6my/\rho L^3 w} \left[\frac{6y}{\rho L^3 w} \left(Lwe_Q + e_y \right) - \frac{1}{y}\dot{y}^2 + \frac{3}{Lwy}Q_i\dot{y} - \frac{3}{2L^2w^2y}Q_i^2 \right]; \qquad (12.121a)$$

---

[14]Inclusion of a vertical component of velocity, with linear variation from bottom to top, adds the small term $\rho w L\dot{y}^2/3$ to the right side of equation (12.120).

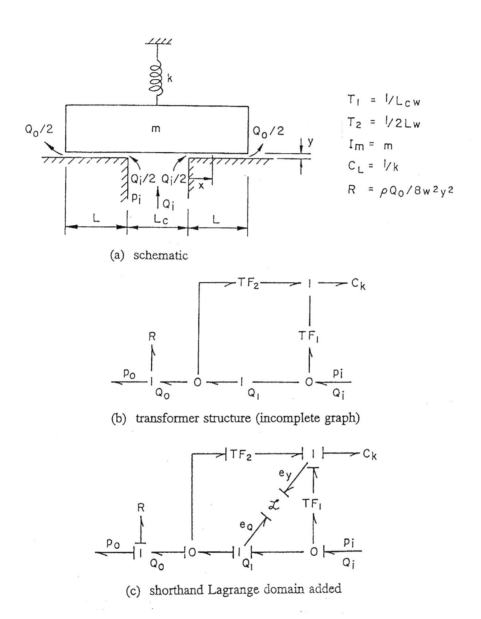

(a) schematic

$$T_I = 1/L_c w$$
$$T_2 = 1/2Lw$$
$$I_m = m$$
$$C_L = 1/k$$
$$R = \rho Q_0 / 8w^2 y^2$$

(b) transformer structure (incomplete graph)

(c) shorthand Lagrange domain added

Figure 12.32: Seating valve

$$\dot{Q}_1 = \frac{1}{1 + 6my/\rho L^3 w} \left[\frac{4wy}{\rho L} \left(2 + \frac{3my}{\rho L^3 w} \right) e_Q + \frac{6y}{\rho L^2} e_y - \left(\frac{2Lw}{y} + \frac{6m}{\rho L^2} \right) \dot{y} \right.$$

$$\left. + \left(\frac{4}{y} + \frac{6m}{\rho L^3 w} \right) Q_1 \dot{y} - \frac{3}{2Lwy} Q_1^2 \right]. \tag{12.121b}$$

These are converted into first-order state equations through the use of the identity

$$\frac{dy}{dt} = \dot{y}. \tag{12.122}$$

12.6.5   Hamilton's Equations for the Seating Valve

The Hamiltonian approach assumes the Lagrangian has been found as above, and proceeds to find the momenta

$$p_Q = \frac{\partial \mathcal{L}}{\partial Q_1} = \frac{\rho L}{2wy} (Q_1 - Lw\dot{y}); \tag{12.123a}$$

$$p_y = \frac{\partial \mathcal{L}}{\partial y} = \frac{\rho L^2}{2y} \left(-Q_1 + \frac{4}{3} Lw\dot{y} \right) + m\dot{y}. \tag{12.123b}$$

The inverse is next found:

$$\begin{bmatrix} Q_1 \\ Lw\dot{y} \end{bmatrix} = \frac{Lw/m}{1 + z} \begin{bmatrix} \frac{4}{3} + \frac{1}{3z} & 1 \\ 1 & 1 \end{bmatrix} \begin{bmatrix} Lwp_q \\ p_y \end{bmatrix}; \tag{12.124a}$$

$$z = \rho L^3 w / 6my. \tag{12.124b}$$

The Hamiltonian now can be expressed in terms of the momenta p_Q and p_y and the dimensionless displacement, $1/z$:

$$\mathcal{H} = p_Q Q_1 + p_y \dot{y} - \mathcal{T} + \mathcal{V} = \frac{1/m}{1 + z} \left[\left(\frac{2}{3} + \frac{1}{6z} \right) (Lwp_Q)^2 + Lwp_Q p_y + \frac{1}{2} p_y^2 \right]. \tag{12.125}$$

The Hamilton's equations become, finally,

$$Q_1 = \frac{\partial \mathcal{H}}{\partial p_Q} = \frac{Lw/m}{1 + z} \left[\left(\frac{4}{3} + \frac{1}{3z} \right) Lwp_Q + p_y \right], \tag{12.126a}$$

$$\dot{y} = \frac{\partial \mathcal{H}}{\partial p_y} = \frac{1/m}{1 + z} (Lwp_Q + p_y), \tag{12.126b}$$

$$\dot{q}_Q = e_Q - \frac{\partial \mathcal{H}}{\partial (\int Q_1 \, dt)} = e_Q, \tag{12.126c}$$

$$p_y = e_y - \frac{\partial \mathcal{H}}{\partial y} = e_y - \frac{1/\rho L^3 w}{(1 + z)^2} \left[(1 + 2z + 4z^2)(Lwp_Q)^2 + 6z^2(Lwq_Q p_y) + 3z^2 p_y^2 \right]. \tag{12.126d}$$

Two equivalent versions of the Hamiltonian bond graph are given in Fig. 12.33. If either of these optional graphs are used, the one given in part (b) likely would be preferred because the energy involved actually is kinetic and the graph contains one rather than two symplectic gyrators.

The Hamiltonian approach for this problem has no net advantage over the Lagrangian, which therefore is probably the method of choice, and without the highly reticulated bond graph. The Hamilton's equations are not simpler to code or faster to execute, and the analysis necessary to find them is considerably more laborious. This experience is believed to be most typical.

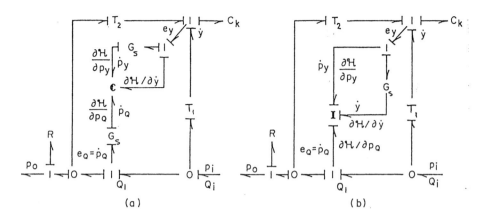

Figure 12.33: Hamiltonian bond graphs for the seating valve

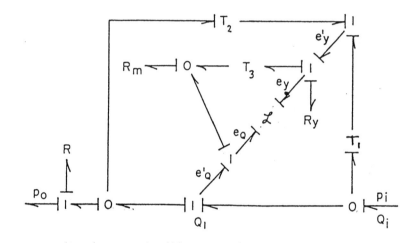

Figure 12.34: Dissipation added to the seating valve

12.6.6   Viscous Dissipation

Viscous dissipation can be added to the model by relating the power dissipated to the state variables. The approximation of a fully-developed parabolic velocity profile gives a lower bound for this power:

$$\mathcal{P} = \frac{24\mu}{y^3 w}\int_0^L \left(\frac{Q_1}{2} - wx\dot{y}\right)^2 dx = \frac{6\mu L}{y^3}\left[(Q_1 - Lw\dot{y})^2 + \frac{1}{3}(lW\dot{y})^2\right]$$

$$= R_m(Q_1 - Lw\dot{y})^2 + R_y\dot{y}^2. \tag{12.127}$$

This quadratic form directly suggests the bond graph resistances R_m and R_y shown in Fig. 12.34, from which

$$e_Q = e_Q' - \frac{6\mu L}{wy^3}(q_1 - Lw\dot{y}), \tag{12.128a}$$

$$e_y = e_y' - \frac{6\mu L^2 w}{wy^3}\left(-Q_1 + \frac{4}{3}Lw\dot{y}\right). \tag{12.128b}$$

Should one not use any bond graph, one can use one-half the quadratic function for the power dissipation, known as the **Rayleigh dissipation function**, to get the same results:

$$e_Q = E'_Q - \frac{\partial(\mathcal{P}/2)}{\partial Q_1}, \tag{12.124a}$$

$$e_y = e'_y - \frac{\partial(\mathcal{P}/2)}{\partial y}. \tag{12.124b}$$

12.6.7 Summary

Hamilton's equations and bond graphs represent a generalization of the equations and model types presented in earlier chapters. Often they are more insightful or require less computation than the Lagrange's equations. Usually these advantages are more than offset for engineering systems, however, by the inefficiency of their deduction. The awkwardness of inverting the relationship between the generalized momenta and the generalized velocities is the principal difficulty. The Hamiltonian bond graph substitutes for the general Hamilton's equations; its use is a matter of taste.

There are four steps in the Hamiltonian method, assuming a Lagrangian has been found already: find the generalized momenta from $p_i = \partial \mathcal{L}/\partial \dot{q}_i$, invert these relationships to find the generalized velocities as functions of the generalized momenta and the generalized coordinates, evaluate the Hamiltonian as a function of the generalized momenta and the generalized coordinates, and evaluate the Hamilton's equations.

PROBLEM

12.17 Find the Hamilton's equations for the spinning top of Problem 12.15, following the steps below:

(a) Use the Lagrangian found earlier to write the momenta p_ϕ, p_θ and p_ψ as functions of the Euler angles and their time derivatives.

(b) Invert the results of step (a) to find the time derivatives of the Euler angles as functions of the momenta and the Euler angles.

(c) Find the Hamiltonian as a function of the momenta and the Euler angles. This can be done using equation (12.105) or, more simply, since $T^* = T$, there results $\mathcal{H} = \mathcal{L} + 2\mathcal{V}$.

(d) Evaluate the Hamilton's equations using the Hamiltonian found in step (c).

(e) Compare the process and the results to those for the Lagrangian method, both with and without the Lagrangian bond graph.

Appendix A

Introduction to MATLAB®

MATLAB[1] is a commercial interactive software package that is widely used for numeric computation, data analysis and visualization. Its basic data element is an array that requires no dimensions, making it easy to use. It is especially well suited for the analysis of linear models. It performs symbolic math, although this feature is not used explicitly in this book. The recently released MATLAB & Simulink Student Version includes a full-featured version of MATLAB 7.1 and Simulink 6.3, as well as symbolic math functions. The most recent professional version at the time of writing is Version 7.1. Innumerable applications of MATLAB appear in a large number of "Toolboxes." Some of the commands in this book may require the Control Systems Toolbox or the Signal Processing Toolbox, depending on the version of MATLAB being used. The online help is extensive. A wide selection of texts present applications of MATLAB.[2]

Most of the use of MATLAB in this book regards simple-to-use commands described in a largely self-contained way in the following sections:

Topic	Sections
nonlinear simulation	3.6.2, 6.1.4, 9.5.7, 9.5.8, 11.5
simulation with Simulink	3.6.3, 4.1.8
transfer function transformations	4.1.6, 7.1.5, 7.2.2
linear simulation	4.1.7
discrete convolution	4.4.3
step, impulse responses; partial fraction expan.	4.5.6, 4.6.4, 7.1.6
Bode plots	7.1.2, 7.1.6
roots, factorization, eigenvalues, eigenvectors	7.1.5, 7.3.8, B.2.5
Fourier analysis	7.4.5
root locus, Nyquist and Nichols plots	8.1.6, 8.3.2-8.3.5
Quasi method of characteristics simulation	10.7.5
thermodynamic properties	11.5, Appendix D

A brief primer on the rudiments of MATLAB follows.

---

[1] For MATLAB and Simulink product information, contact The MathWorks, Inc., 3 Apple Hill Drive, Natick, MA, 01760-2098 USA. Tel: 508-647-7000; Fax: 508-647-7001; E-mail: info@mathworks.com; Web: www.mathworks.com.

[2] A list of hundreds of MATLAB-based books is maintained on the MathWorks Home Page, www.mathworks.com., including *Introduction to MATLAB 7 for Engineers* by Wm. J. Palm, McGraw-Hill, 2005; *MATLAB: An Introduction with Application, 2e* by Amos Gilat, John Wiley and Sons, Inc., and *Mastering MATLAB 7* by Hanselman and Littlefield, Prentice-Hall, 2005.

Scalar Calculations

MATLAB can be used like a simple calculator. For example, to evaluate $(5/\pi^2)\sin(3)$ you enter, in response to the MATLAB prompt (>>),

```
>> 5/pi^2*sin(3)
```

```
ans =
    0.0715
```

The answer is placed in a default variable "ans," and is placed on the screen as shown. Expressions are evaluated from left to right with power operation having the highest order of precedence, followed by multiplication and division, and trailing with addition and subtraction. Elementary math functions treated are listed in Table A.1

Table A.1	Elementary Math Functions
abs(x)	absolute value or magnitude of complex number
acos(x)	inverse cosine
acosh(x)	inverse hyperbolic cosine
angle(x)	angle of complex number
asin(x)	inverse sine
asinh(x)	inverse hyperbolic sine
atan(x)	inverse tangent
atan2(x,y)	four quadrant inverse tangent
atanh(x)	inverse hyperbolic tangent
ceil(x)	round towards plus infinity
conj(x)	complex conjugate
cos(x)	cosine
cosh(x)	hyperbolic cosine
exp(x)	exponential (e^x)
fix(x)	round towards zero
floor(x)	round towards minus infinity
imag(x)	complex imaginary part
log(x)	natural logarithm
log10(x)	common logarithm
real(x)	complex real part
rem(x,y)	remainder after division of x/y
round(x)	round towards nearest integer
sign(x)	signum (sign) function; returns 1, 0 or -1
sin(x)	sine
sinh	hyperbolic sign
sqrt(x)	square root
tan(x)	tangent
tanh(x)	hyperbolic tangent

Variables

Variables, such as the x,y used in the table above, can be defined by expressions. For example, the same answer as above results from

```
>> omega=3;
>> T=1;
>> x=5/pi^2*sin(omega*T)
```

except that it is stored in the variable x. The semicolons at the ends of the first two statements inhibit printing of their numerical values. The subsequent typing of

```
>> omega,T,x
```

gives the response

```
omega =
      3
T=
      1
x=
      0.0715
```

The variables that have been defined at any point in your session will be displayed if you enter

```
>>who
Your variables are
    omega   T   x
```

The instruction <whos> also gives details regarding the nature of the variables. Variables are case sensitive, must start with a letter, and are limited to 19 characters.

Complex Numbers

If you enter

```
>> a=2; b=3; c=4;
>> y=(-b+sqrt(b^2-4*a*c))/2/a;
```

the response to typing <y> is

```
y=
    -0.7500 + 1.1990i
```

where i is the unit imaginary number. Imaginary numbers can be entered by using the identifier <i> or <j>, as in standard mathematical notation:

```
p=5+2i; q=6-3j*sin(2);
```

Starting with Version 4, MATLAB does not require an asterisk between a number and the i or j.

Other predefined variables are <Inf> (for ∞) and <NaN> (for 0/0).

Arrays and Matrices

A row array can be entered as

```
r=[1 2 3];
```

and a column array as either of the forms

```
>> c=[1;2;3];
>> c=[1 2 3]';
```

The prime (') denotes transpose. Entry of a general array is illustrated by

```
A=[1 2 3;4 5 6]
A=
      1   2   3
      4   5   6
```

Elements in matrices may be imaginary or complex.

When arrays are added, subtracted, or multiplied with <+>, <->, <*>, they are treated as conventional matrices. On the other hand, **array multiplication, array division** and **array power**, indicated by the commands <.*>, <./>, <.^>, simply perform the indicated operation on the respective elements in the array, *i.e.,*

```
p=[4 5 6]; q=[3 2 1];
>>p.*q
ans=
    12  10  6
>>p./q
ans=
    1.3333   2.5000   6.0000
>>p.^q
ans=
    64  25  6
```

The dot and vector products are given by, respectively,

```
>> p*q'
ans=
    28
>>x=p'*q
x=
    12  8   4
    15  10  5
    18  12  6
```

The command `size(A)` returns a two-element vector stating the numbers of rows and columns in the matrix A, respectively. Thus, if A is any 2×3 matrix, `size(A)` returns the vector 2 3, `size(A,1)` returns the 2 and `size(A,2)` returns the 3. Further, the statement

```
y=x(size(A,1),:)
```

in which x is the 3×3 matrix defined above, returns the row vector

```
y=
    15  10  5
```

The `size` argument indicates the second row of x, and the wild card (:) indicates all columns.

To save time, the special matrices illustrated below are offered. Assuming A has been defined as a 3×3 matrix and B as a 2×4 matrix,

```
>>zeros(size(A))
ans=
    0  0  0
    0  0  0
    0  0  0
>>ones(size(B))
ans=
    1  1  1  1
    1  1  1  1
>>eye(size(A))
ans=
    1  0  0
    0  1  0
    0  0  1
```

The linear matrix equation

$$\mathbf{A}\mathbf{x} = \mathbf{b},$$

with $\mathbf{A}$ and $\mathbf{b}$ known, can be solved for $\mathbf{x}$ two ways, assuming $\mathbf{A}$ is square and nonsingular:

```
>>x=inv(A)*b;
>>x=A\b;
```

For example, if

$$\mathbf{A} = \begin{bmatrix} 1 & 2 & 3 \\ 4 & 5 & 6 \\ 7 & 8 & 0 \end{bmatrix}; \qquad \mathbf{b} = \begin{bmatrix} 14 \\ 32 \\ 23 \end{bmatrix},$$

either command will give

```
x=
    1.0000
    2.0000
    3.0000
```

The determinant of $\mathbf{A}$ is readily found, also:

```
>>det(A)
ans=
    27
```

The use of the backslash operator \ is preferred. It employs LU factorization, which is a modification of Gaussian elimination and produces greater accuracy with fewer internal multiplications.

When there are more equations than unknowns, so $\mathbf{A}$ has more rows than columns, the command A\b gives the *least squares solution*. This can be extremely useful. The command inv(A)*b, on the other hand, produces an error message.

Evaluating and Plotting Functions

The array operations on vectors are very useful in evaluating and plotting functions. As an example, the time response

$$x = (1/\cos\phi)e^{-\zeta\omega_n t}\cos(\sqrt{1-\zeta^2}\,\omega_n t + \phi); \qquad \phi = -\tan^{-1}(\zeta/\sqrt{1-\zeta^2}),$$

with $\zeta = 0.2$ and $\omega_n = 2\pi$ and 101 values of t evenly spaced between 0 and 4, can be evaluated as follows:

```
>>z=0.2; om=2*pi;
>>omd=sqrt(1-z^2)*om; phi=-atan(z/sqrt(1-z^2));
>>t=[0:.04:4];
>>x=exp(-z*om*t).*cos(omd*t+phi)/cos(phi);
```

Notice that only one array operation is needed; this operation is unnecessary for multiplying or dividing a scalar and an array.

The 101 resulting values of t and x are stored in the row arrays t and x, and will be displayed by entering these variables. You can secure a plot of x versus t as shown at the top of the next page:

```
>>plot(t,x)
>>xlabel('t'); ylabel('x');
>>title('homogeneous response with zero intial velocity')
```

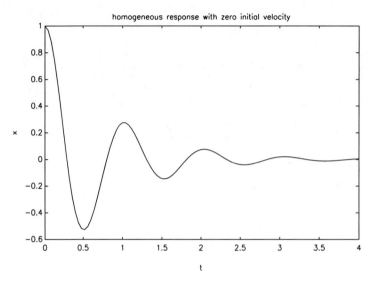

homogeneous response with zero initial velocity

Semilog and log-log plots to the base 10 also can be made; the commands are $<$semilogx(x,y)$>$, $<$semilogy(x,y)$>$, and $<$loglog(x,y) $>$. The command $<$grid$>$ draws grid lines, and the command $<$text(p1,p2,'text','sc')$>$ places the word "text" (or your substitute) at (p1,p2) in screen coordinates where (0.0,0.0) is the lower left and (1.0,1.0) is the upper right of the screen.

Fitting Curves to Data

The characteristic of a source, load, compliance, etc. often is observed experimentally, and you wish to represent it with a polynomial expansion for use in simulation, or perhaps just for curve plotting. This is an example of a task for which the MATLAB *help* feature is a sufficient guide. You find the information by clicking on "MATLAB Functions Listed by Category," then "Mathematics," then "Polynomial and interpolation functions," then "Polynomials," and finally "polyfit polynomial curve fitting." You learn that the command

        [p]=polyfit(x,y,n)

fits the coefficients of a polynomial $p(x)$ of degree n to the data, $p(x(i))$ to $y(i)$, in a least squares sense. The result is a row vector of length $n + 1$ containing the polynomial coefficients in descending powers:

$$p(x) = p_1x^n + p_2x^{n-1} + \cdots + P_nx + p_{n+1}.$$

A complete example is given. You also learn how to use the command

        f=polyval(p,z)

to compute the values of the polynomial for values of a vector z, which could be considerably longer than the vector x. The illustration includes use of the **plot** command for both the original data points and the resulting polynomial curve. You also are told that the algorithm used by **polyfit** makes use of the backslash operator, \, as described above.

The meaning and use of further optional output arguments of **polyfit** also are described.

Control Flow Commands

Statements can be executed conditionally with the **if** command. The simplest form is as follows:

```
    if condition statement;
    action statements;
    end;
```

A more complicated form is as follows:

```
    if condition statements;
    action statements;
    elseif condition statements;
    else action statements;
    end;
```

The `for` command allows statements to be repeated a specifed number of times. It is most commonly used relative to indices and often is used in conjuction with the `if` command. An example using `for`, `if`, `else` and `elseif` follows:

```
    n=5;
    for i = 1:n
        for j = 1:n
            if j > i
                A(i,j) = 2;
            elseif j == i
                A(i,j) = 1;
            else
                A(i,j) = 0;
            end;
        end;
    end;
```

The result is the matrix

$$A = \begin{bmatrix} 1 & 2 & 2 & 2 & 2 \\ 0 & 1 & 2 & 2 & 2 \\ 0 & 0 & 1 & 2 & 2 \\ 0 & 0 & 0 & 1 & 2 \\ 0 & 0 & 0 & 0 & 1 \end{bmatrix},$$

(which is small enough to have been entered directly with less fuss.)

The `while` command sets up conditional loops useful for such operations as Newton-Raphson iterations. For example, the value of x that satisfies $\cosh(x) = 2$ is returned from the following, which requires an initial guess (1.5 is used here) of the final answer (1.3170) (and presumes that the function `acosh(x)` was not available):

```
    xtrial=1.5; % initial guess
    error=cosh(xtrial)-2; % error produced by initial guess
    while error<1e-6 % sets up the iteration and allowed error
        derivative=sinh(xtrial); % derivative of the function
        xtrial=xtrial+error/derivative; % estimated new value of x
        error=cosh(xtrial)-2; % new error
    end
```

Script Files

When interactions with MATLAB are not particularly short or are repetitive, it is desirable to write programs off-line. Files executable by MATLAB are called **M-files**, since they must have the extension **.m**. The toolboxes such as the signals and systems toolbox and the symbolic math toolbox that come with the student version comprise a collection of M-files; these subprograms represent most of the ultimate power of MATLAB.

A **script** is one type of M-file. Scripts are ordinary ASCII text files and can be created off-line with nearly any text editor and stored on your disk. You will need to make sure that MATLAB has a path to the location of your file, which you might transfer to or write in the location `C:temp`. This can be accomplished within MATLAB by clicking on `file` and then `set path`. The procedure from here depends on the version of MATLAB. Back in the Command Window of MATLAB, you can open an existing m-file or write a new one after clicking `file`. The MATLAB Editor/Debugger should appear, and you can save the file to the location `C:temp` (or whatever location you have placed in the path).

Subsequent transfers between the Editor/Debugger and the Command Window are executed merely by clicking on their windows. Don't forget to save any changes you make in a file before attempting to run it. A script file can be run merely by typing its name, without extension or location, into the Command Window, and pressing `Return`.

Test the system by entering the following program:

```
x=[1:360];
z=exp(j*pi*x/180);
plot(imag(z),real(z))
```

If you name the program `myprog.m` and type "`myprog`" in response to the MATLAB prompt, you should see an ellipse; the imaginary part of the argument is plotted versus the real part. The plot can be annotated to your taste. The subsequent command

```
plot(x,imag(z))
```

plots one cycle of a sine wave.

Data Files

The command   save *filename* `x,y,A`

writes the content of the variables x, y A into a binary file with the name `filename.mat`. If no variables are specified, all workspace variables are saved. If no filename is specified, the default file name is `matlab.mat`.

The command   load *filename* `x,A`

retrieves the indicated data. The same defaults apply.

Function Files

There are two types of M-files: the script files described above, and **function files**. A function file is identified by the the word "function" at the beginning of its first line, followed by a statement in equation form that includes the name of the file. The example given in Section 3.7.3 starts with

```
function f = pendulum(t,x)
```

Unless special provisions are made, the arguments `t,x` are the only variables communicated from the main program to the function. The variable `f` is the only variable returned to the main program. All other variables defined in the function file are hidden from the main program. If more than one output variable is to be computed, the list should be placed in square brackets. As an example,

```
function [y,x,t] = step(num,den)
```

(see Section 4.6.4) is the first line of one of the hundreds of function M-files built into MATLAB. The variable names `y,x,t` can be changed when the function is called.

Functions can be evaluated through the use of the command

```
[y1,...,yn] = feval('function file name',x1,...,xn)
```

The variables x1, . . . , xn are communicated to the function file, and the variables y1, . . . , yn are communicated back to the calling file (or the command window). The left side can be omitted if no returns are intended.

Communication Between Files

An M-file can call another M-file. The statement in Section 3.6.2

```
[t,x] = ode23('pendulum',0,[0 3],[0 pi/18]);
```

calls the built-in function M-file ode23, which in turn calls the user-written function M-file pendulum. In this example, the parameter values L, W and I are redefined every time pendulum is called. This is not efficient.

Repetitive definitions or calculations can be eliminated through the use of **global variables**. Those functions that declare a particular name as a global variable, and the base workspace if it makes the same declaration, share a single copy of that variable. The simulation described in Section 3.6.2 becomes more efficient, therefore, if you write a script M-file named "pendulumsim" and a shortened function M-file as follows:

```
% Script M-file 'pendulumsim';
global L W I \% global variables
L = 1; W = 1; g = 32.2; I=L\^{ }2*W/(3*g);
options=odeset('RelTol', 1e-6);
[t,x] = ode23('pendulum',[0 3],[0 pi/18],options);
plot(t, x(:,2)*180/pi)
xlabel('time,seconds')
title('Angle of swinging pendulum')

function f = pendulum(t,x) \% Function M-file computes f
global L W I \% global variables
% The two differential equations follow:
f(1) = -W*(L/2)*sin(x(2)); % x(1) is the angular momentum
f(2) = x(1)/I; % x(2) is the angular velocity
f=f';
```

The entire computation and display follows from entering the name of the script M-file into the command window in response to the command prompt. The use of **global** also allows variables to be communicated from the function file to the command window or the script file, which can expedite data output.

Sometimes you will like to plot the *derivative* of a state vector, but the function f' is not communicated to the controlling script file. Further, the function M-file evaluates f several times per time interval; a plot of all these points would display undesired and misleading scatter. The best you can do is to plot the differences between successive values of the state variable, divided by the corresponding differences in time. The simplest way to accomplish this is by using the MATLAB function diff. For example, to plot the moment on the pendulum above, namely dx_1/dt, you could enter the command

```
plot(t(1:size(t)-1),diff(x(:,1))./diff(t))
```

Note that if the vector x(:,1) contains n elements, the vector diff(x(:,1) contains $n-1$ elements. The vector for the abscissa of the plot, which is the time t, must be shortened by one element for the plot routine to work. This explains the use of the size(t) command.

MATLAB Files Downloadable From the Internet

The following MATLAB files are listed or cited on the page numbers below in Chapters 9, 10 or 11. Thermodynamics files are also listed in Appendix D when noted below. All files below can be downloaded from the author's page at the website www.lehigh.edu/~inmem. Other files are listed in the Solutions Manual. All are chosen for their adaptability to a variety of related problems. They have been run and checked carefully, but the author plans to fix any errors that are discovered; please email him at ftb0@lehigh.edu or ftbmhb@aol.com. A general list of significant errata also will be maintained, should the need arise.

Location	page	file name	subject of file
Chap. 9	665	hysteresism.m	simulation of hysteresis given $h(t)$
		hysteresis.m	function file called by hysteresism.m
		Langevin	Langevin function (for nonlinear magnetics)
	668	coilm.m	for simulation of hysteresis given voltage
		coil.m	function file called by coilm.m
	674	ex9_17m.m	tooth field and force vs. position
		ex9_17de.m	function file called by ex9_17m.m
		force.m	script file to follow ex9_17m.m
	676	GP9_12m.m	simulation of relay closing
		GP9_12.m	function file called by GP9_12m.m
	684	ex9_18.m	function file for synchronous motor
	687	ex9_19.m	function file for induction motor
Chap. 10	799	viscoustube.m	quasi method of characteristics for viscous flow
	804	GP10_9.m	quasi method of charac. for linear elastic solid
Chap. 11	903	ex11_5m.m	air compressor charging tank
	903	ex11_5.m	function file called by ex11_5m.m
	908	ex11_6.m	coefficients for saturated vapor density
	909	ex11_6P.m	script file called by ex11_6.m
	911	refgmst.m	simulation of refrigeration cycle
	911	refrgde.m	function file called by refgmst.m
	917	shocktubem.m	simulation of adiabatic tube with refrigerant
	918	shocktube.m	function file called by shocktubem.m
	922	condenserm.m	simulation of refrigerant tube w. condensation
	922	conderser.m	function file called by condenserm.m
	923	shocktubess.m	equilibrium state vector for Problem 11.18
	923	condenserss.m	equilibrium state vector for Problem 11.20
	928	GP11_4m.m	simulation of air compressor cycles
	928	GP11_4.m	function file called by GP11_4m.m
Appen. D	902	gas.m	evaluates the properties of an ideal gas
	902	air.m	properties of air as mixture of gases
	902	gasdata.m	parameters for nitrogen, oxygen and argon
	902	specheat.m	specific heat of air as mixture of gases
	906	R12data.m	gives parameters describing refrigerant R12
	906	R134adata.m	gives parameters describing refrigerant R134a
	906	refprop.m	evaluates many properties of refrigerants
	906	vapphase.m	called by refprop.m
	907	waterdat.m	parameters defining a model of water
	907	propwat.m	evaluates the properties of multiphase water
	907	steam.m	evaluates the properties of water vapor
	912	orifice.m	flow of possibly multiphase fluid through orifice

PROBLEMS

A.1 Use MATLAB to find the magnitude and phase angles of the complex numbers (a) $3 - 4j$, (b) $0.5 + 2j$.

A.2 Find the real and imaginary parts of the following complex numbers: $e^{-.75j\pi}$, (b) $4e^{j(-5\pi/4)}$.

A.3 Give MATALB coding for the following (*but do not run*):

(a) If $d < 3$, set $r = 0$; otherwise, set $r = 2d$.

(b) If $x - y > 2$, print x and y.

(c) If the natural logarithm of x is between 2 and 10, increment the index m by one.

A.4 Enter the following matrices into MATLAB:

$$\mathbf{A} = \begin{bmatrix} 2 & 1 \\ 0 & -1 \\ 3 & 0 \end{bmatrix}; \quad \mathbf{B} = \begin{bmatrix} 1 & 3 \\ -1 & 5 \end{bmatrix}; \quad \mathbf{C} = \begin{bmatrix} 3 & 2 \\ -1 & -2 \\ 0 & 4 \end{bmatrix}; \quad \mathbf{b} = \begin{bmatrix} 3 \\ 2 \end{bmatrix}.$$

(a) Using MATLAB, compute the matrices $\mathbf{AB}$, $\mathbf{BC'}$, and $(\mathbf{A'C})^{-1}$.

(b) Using MATLAB, solve the equation $\mathbf{Bx} = \mathbf{b}$ for $\mathbf{x}$.

A.5 Plot the characteristic of the hydraulic source as given in Problem 2.15 (p. 55). On a second graph, plot the power produced by this source as a function of flow.

A.6 The IC engine in Problem 2.16 (p. 55) has the nonlinear torque-speed characteristic plotted in Figure 2.39 (p. 75).

(a) Read a few values of the shaft speed and the associated torque, and plot these points using the "o" argument of the plot command. Hold this plot.

(a) Make and plot three polynomial approximations to this characteristic: (i) a linear model (equation); (ii) a second-order (quadratic) model; (iii) a third-order (cubic) model. Plot these polynomials as essentially continuous curves on the same axes as the data points from part (a).

(b) Locate the points for which each of the models in part (b) gives maximum power. Hint: Type "help max" in the command window.

Appendix B

Classical Vibrations

Long before the state-space methods for modeling and analyzing dynamic systems given in this textbook were promulgated, engineers addressed the relatively narrow class of models comprising vibrating inertias interconnected by springs. Important applications were made to structures and machines. The methods that evolved are simple and insightful and still have relevance today. On the other hand they do not extend your capabilities and therefore are optional.

The approach in the tradition of dynamic systems, emphasized in this book, treats linear models with the first-order matrix differential equation

$$\dot{\mathbf{x}} = \mathbf{A}\mathbf{x} + \mathbf{B}u(t). \tag{B.1}$$

The approach in the tradition of classical vibrations, on the other hand, treats linear models with the second-order matrix differential equation

$$\mathbf{M}\ddot{\mathbf{x}} + \mathbf{C}\dot{\mathbf{x}} + \mathbf{K}\mathbf{x} = \mathbf{F}(t). \tag{B.2}$$

This latter approach has the advantage of dealing with matrices and vectors with half the dimension of those used in the corresponding first-order models. Also, $\mathbf{M}$ and $\mathbf{K}$ and often $\mathbf{C}$ are symmetric matrices, further simplifying the analysis, whereas $\mathbf{A}$ generally is asymmetric.

The purpose of this appendix is to explain and illustrate the application of equation (B.2). This is done for models with two or more degrees of freeedom (of fourth or higher order), since the differences for models with one degree of freedom are not significant. Usually it is assumed that the matrix M is invertible, which eliminates first, third and higher odd-ordered models from consideration, a serious restriction that motivates the tradition of system dynamics to use equation (B.1) instead.

B.1   Models with Two Degrees of Freedom

The tuned vibration absorber analyzed in Section 7.2.2 (pp. 458-460) is repeated in Fig. B.1 with the displacements of the centers of mass, x_1 and x_2, indicated as the principle variables. Direct application of Newton's law gives the equations of motion

$$\begin{bmatrix} m_1 + m_2 & 0 \\ 0 & m_3 \end{bmatrix} \begin{bmatrix} \ddot{x}_1 \\ \ddot{x}_2 \end{bmatrix} + \begin{bmatrix} k_1 + k_2 & -k_2 \\ -k_2 & k_2 \end{bmatrix} \begin{bmatrix} x_1 \\ x_2 \end{bmatrix} = \begin{bmatrix} F_e \\ 0 \end{bmatrix} \tag{B.3}$$

The first square matrix in this classical representation is known as the **mass matrix**; no **dynamical coupling** exists in this example since this matrix is diagonal. The second matrix

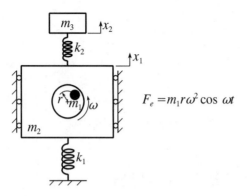

Figure B.1: System with vibration absorber

is known as the **stiffness matrix**; the non-diagonal terms here are said to impose **static coupling**. Were there no such coupling, the problem could be treated as two independent single-degree-of-freedom problems.

B.1.1 Normal Mode Vibration

In the unforced case ($F_e = 0$) the model can oscillate sinusoidally:

$$\begin{bmatrix} x_1 \\ x_2 \end{bmatrix} = \begin{bmatrix} x_{10} \\ x_{20} \end{bmatrix} \sin(\omega t + \phi). \tag{B.4}$$

Substitution of these assumed motions into equation (B.3) gives

$$\begin{bmatrix} [k_1 + k_2 - (m_1 + m_2)\omega^2] & -k_2 \\ -k_2 & (k_2 - m_3\omega^2) \end{bmatrix} \begin{bmatrix} x_{10} \\ x_{20} \end{bmatrix} = 0 \tag{B.5}$$

The determinant of the square matrix must vanish, giving the characteristic equation

$$\Delta \equiv (m_1 + m_2)m_3\omega^4 - [(k_1 + k_2)m_3 + k_2(m_1 + m_2)]\omega^2 + k_1k_2 = 0. \tag{B.6}$$

This equation is quadratic in terms of ω^2, giving two real roots. The two natural frequencies are the two square-roots of these roots, which are called ω_1 and ω_2 below, and are presumed to be positive.

The sinusoidal motions assumed above do not represent the general solutions, but rather are special cases called **normal modes**. If you substitute $\omega = \omega_1$ or $\omega = \omega_2$ into equation (B.5) you get the amplitude ratios

$$\left(\frac{x_{10}}{x_{20}}\right)_{\omega_1} = \frac{k_2}{k_1 + k_2 - (m_1 + m_2)\omega_1^2} = \frac{k_2 - m_3\omega_1^2}{k_2}, \tag{B.7a}$$

$$\left(\frac{x_{10}}{x_{20}}\right)_{\omega_2} = \frac{k_2}{k_1 + k_2 - (m_1 + m_2)\omega_2^2} = \frac{k_2 - m_3\omega_2^2}{k_2}. \tag{B.7b}$$

Thus the amplitudes of the two displacements have a fixed ratio that is different for the two modes, and along with the respective frequencies serves to define the modal motions.

The general unforced motion comprises a sum or superposition of the two modal motions:

$$x_1 = x_{11}\sin(\omega_1 t + \phi_1) + x_{12}\sin(\omega_2 t + \phi_2) \tag{B.8a}$$

$$x_2 = x_{21} \sin(\omega_1 t + \phi_1) + x_{22} \sin(\omega_2 t + \phi_2) \qquad (B.8b)$$

$$x_{11} = \left(\frac{x_{10}}{x_{20}}\right)_{\omega_1} x_{21}; \qquad x_{12} = \left(\frac{x_{10}}{x_{20}}\right)_{\omega_2} x_{22} \qquad (B.8c)$$

Thus there are two independent undetermined amplitudes, which can be taken to be x_{12} and x_{22}, and two independent phase angles, ϕ_1 and ϕ_2. These amplitudes and angles are determined by the initial conditions $x_1(0)$, $\dot{x}_1(0)$, $x_2(0)$ and $\dot{x}(0)$.

Unlike the general analysis above, the results of Sections 7.2.2 and 7.2.3 are specialized to the case of $m_3 = 0.2(m_1 + m_2)$ and $k_2 = 0.2k_1$. In this case, the natural frequencies are $\omega_1 = 0.80109\omega_0$ rad/s and $\omega_2 = 1.24830\omega_0$ rad/s, where $\omega_0 = \sqrt{k_1/(m_1 + m_2)}$. The modal shape ratios are $(x_{10}/x_{20})_{\omega_1} = -2.209$ and $(x_{10}/x_{20})_{\omega_2} = -6.791$. This means that if the system is set oscillating at $0.80109\omega_0$ rad/s, the larger mass exhibits simple sinusoidal motion $180°$ out-of-phase with the smaller mass, and with an amplitude 2.209 times larger than that of the smaller mass. At $1.24830\omega_0$, the motion is similar except the ratio of amplitudes is 6.791. In general, the unexcited motion represents a sum of these two modes and two frequencies, with proportions to suit the initial conditions.

B.1.2 Forced Harmonic Motion

The particular solution of the system described by equation (B.3) corresponding to the forcing function

$$F_e = m_1 r \omega^2 \cos \omega t \qquad (B.9)$$

has the same form as the special solutions of equation (B.4),

$$\begin{bmatrix} x_1 \\ x_2 \end{bmatrix} = \begin{bmatrix} x_{10} \\ x_{20} \end{bmatrix} \cos \omega t, \qquad (B.10)$$

although ω is now the forcing frequency rather than a natural frequency. Substitution of this solution into equation (B.3) gives

$$\begin{bmatrix} x_{10} \\ x_{20} \end{bmatrix} = \begin{bmatrix} k_1 + k_2 - (m_1 + m_2)\omega^2 & -k_2 \\ -k_2 & k_2 - m_3\omega^2 \end{bmatrix}^{-1} \begin{bmatrix} m_1 r \omega^2 \\ 0 \end{bmatrix}$$

$$= \frac{m_1 r \omega^2}{\Delta} \begin{bmatrix} k_2 - m_3\omega^2 \\ k_2 \end{bmatrix} \qquad (B.11)$$

where Δ is the determinant defined in equation (B.6). The result for $x_1(t)$ corresponds to the $G(S)$ used in Sections 7.2.2 and 7.2.3 but allows any set of physical parameters.

In the numerical example discussed above and plotted in Fig. 7.17, the motion of x_1 ceases when the excitation frequency is precisely $\sqrt{k_2/m_3} = \omega_0$. This is the purpose of the vibration absorber.

B.2 Higher-Order Models

B.2.1 Modal Motions

The general linear undamped and unexcited vibration problem with n degrees of freedom can be represented by the matrix equation

$$\mathbf{M}\ddot{\mathbf{x}} + \mathbf{K}\mathbf{x} = \mathbf{0}, \qquad (B.12)$$

in which $\mathbf{M}$ is the **inertia matrix** and $\mathbf{K}$ is the stiffness matrix. The matrices $\mathbf{M}$ and $\mathbf{K}$ are *symmetric*. Premultiplying equation (B.12) by $\mathbf{M}^{-1}$,

$$\ddot{\mathbf{x}} + \mathbf{A}\mathbf{x} = \mathbf{0}, \qquad \mathbf{A} = \mathbf{M}^{-1}\mathbf{K}. \qquad (B.13)$$

As with the two-degree-of freedom model, special modal solutions of this equation exhibit simple in-phase sinusoidal motions at the frequencies ω_i, where $i = 1, \cdots, n$. The assumption

$$\mathbf{x}(t) = \mathbf{x_i} \sin \omega_i t, \qquad (B.14)$$

where $\mathbf{x}_i$ is a vector of constants and θ_i is a constant, gives upon substitution into equation (B.12)

$$(\mathbf{A} - \omega_i^2 \mathbf{I})\mathbf{x}_i = \mathbf{0}. \qquad (B.15)$$

The characteristic equation

$$\Delta \equiv |\mathbf{A} - \omega_i^2 \mathbf{I}| = 0 \qquad (B.16)$$

gives n eigenvalues ω_i^2. Substitution of an eigenvalue into equation (B.13) gives the proportions of the corresponding **eigenvector**, $\mathbf{x}_i$. That is, the eigenvector can have any magnitude but the ratios of its members are fixed. The eigenvector describes the **modal shape** of the vibration at the special frequency ω_i.

It is convenient to fix the length of the eigenvector $\mathbf{x}_i$ at some value, and then represent smaller or larger versions of the same eigenvector as $c_i\mathbf{x}_i$, where c_i is the appropriate coefficient. This practice is employed below. Sometimes, the first element in $\mathbf{x}_i$ is set equal to 1, and the other elements scaled accordingly. At other times, particularly in standard software packages such as MATLAB that handle the extra computation easily, the *length* of the eigenvector is set equal to 1; this length is the square-root of the scalar product, $\mathbf{x}_i'\mathbf{x}_i$, where the prime ($'$) means transpose. This second option, unlike the first, handles the special case in which the first element in $\mathbf{x}_i$ is identically zero. Either scheme can be applied to the results below.

A model given an initial position corresponding to a particular mode shape, with zero velocity, or an initial velocity corresponding to the mode shape with zero position, will experience the motion of that mode exclusively. An arbitrary initial condition, on the other hand, likely excites all the modes, some more than others.

The eigenvectors possess a very important property that now will be established. For $\mathbf{x}$ equalling the ith eigenvector, equations (B.12) and (B.14) give

$$\omega_i^2 \mathbf{M}\mathbf{x}_i = \mathbf{K}\mathbf{x}_i. \qquad (B.17)$$

Premultiplying both sides by the transpose of $\mathbf{x}_j$,

$$\omega_i^2 \mathbf{x}_j'\mathbf{M}\mathbf{x}_i = \mathbf{x}_j'\mathbf{K}\mathbf{x}_i. \qquad (B.18)$$

Taking the transpose of this equation, noting the general theorem that the transpose of a product of three conformable matrices equals the products of the respective inverse matrices in reverse order, and noting that M and K are symmetric,

$$\omega_i^2 \mathbf{x}_i'\mathbf{M}\mathbf{x}_j = \mathbf{x}_i'\mathbf{K}\mathbf{x}_j \qquad (B.19)$$

Equation (B.18) also can be written with the roles of i and j reversed:

$$\omega_j^2 \mathbf{x}_i'\mathbf{M}\mathbf{x}_j = \mathbf{x}_i'\mathbf{K}\mathbf{x}_j. \qquad (B.20)$$

Equation (B.20) is now subtracted from equation (B.19) to yield

$$(\omega_i^2 - \omega_j^2)\mathbf{x}_i'\mathbf{M}\mathbf{x}_j = 0. \qquad (B.21)$$

As long as ω_i and ω_j are two *different* natural frequencies, it follows that

$$\mathbf{x}_i' \mathbf{M} \mathbf{x}_j = 0. \tag{B.22}$$

By the same approach it can be shown that

$$\mathbf{x}_i' \mathbf{K} \mathbf{x}_j = 0. \tag{B.23}$$

It is said that the eigenvalues are *orthogonal with respect to* $\mathbf{M}$ *and* $\mathbf{K}$.

On occasion, an eigenvalue is repeated. In this case there are two eigenvectors associated with a common eigenvalue; no unique mode exists corresponding to this frequency.

Finally, when $j = i$ the terms corresponding to the left sides of equations (B.22) and (B.23) do not vanish. Instead, we define the **generalized mass**, M_i, and the **generalized stiffness**, K_i, as follows:

$$M_i = \mathbf{x}_i' \mathbf{M} \mathbf{x}_i, \tag{B.24a}$$

$$K_i = \mathbf{x}_i' \mathbf{K} \mathbf{x}_i. \tag{B.24b}$$

B.2.2   The Initial Value Problem

The general solution to the initial value problem is given by the sum of all its modal components, each with its modal amplitude c_i and phase angle ϕ_i, or its components of amplitudes a_i and b_i:

$$\mathbf{x} = \sum_{i=1}^{n} \mathbf{x}_i c_i \sin(\omega_i t + \phi_i) = \sum_{i=1}^{n} \mathbf{x}_i (a_i \sin \omega_i t + b_i \cos \omega_i t). \tag{B.25a}$$

$$c_i = \sqrt{a_i^2 + b_i^2}; \qquad \phi_i = \tan^{-1}\left(\frac{b_i}{a_i}\right) \tag{B.25b}$$

The initial conditions are

$$\mathbf{x}(0) = \sum_{i=1}^{n} b_i \mathbf{x}_i, \tag{B.26a}$$

$$\dot{\mathbf{x}}(0) = \sum_{i=1}^{n} \omega_i a_i \mathbf{x}_i. \tag{B.26b}$$

The solution of equation (B.25) is completed by evaluating a_i and b_i, $i = 1, \cdots, n$, as functions of $\mathbf{x}(0)$ and $\dot{\mathbf{x}}(0)$. This is accomplished by multiplying both sides of equations (B.26) by $\mathbf{x}_j' \mathbf{M}$:

$$\mathbf{x}_j' \mathbf{M} \mathbf{x}(0) = \sum_{i=1}^{n} b_i \mathbf{x}_j' \mathbf{M} \mathbf{x}_i \tag{B.27a}$$

$$\mathbf{x}_j' \mathbf{M} \dot{\mathbf{x}}(0) = \sum_{i=1}^{n} \omega_i a_i \mathbf{x}_j' \mathbf{M} \mathbf{x}_i \tag{B.27b}$$

In view of the orthogonality relation of equation (B.22) and the definition of equation (B.24a), the right sides of equations (B.27a) and (B.27b) equal, respectively, $b_j M_j$ and $\omega_j a_j M_j$. Therefore, the desired result is

$$b_i = \frac{1}{M_i} \mathbf{x}_i' \mathbf{M} \mathbf{x}(0), \tag{B.28a}$$

$$a_i = \frac{1}{\omega_i M_i} \mathbf{x}_i' \mathbf{M} \dot{\mathbf{x}}(0). \tag{B.28b}$$

B.2.3   Forced Response

The coupled set of differential equations in the presence of a forcing function $\mathbf{F}(t)$ is written

$$\mathbf{M}\ddot{\mathbf{x}} + \mathbf{K}\mathbf{x} = \mathbf{F}(t). \tag{B.29}$$

The solution is expedited by tranforming the variables so as to achieve a set of *uncoupled* differential equations, one for each mode. The complete solution then becomes the sum of the individual modal solutions. The transformation is written

$$\mathbf{x} = \mathbf{P}\mathbf{y}. \tag{B.30}$$

To make this idea work, P has to be a very special matrix, called the **modal matrix**. The n columns of P are the n eigenvectors, $\mathbf{x}_i$:

$$\mathbf{P} = [\mathbf{x}_1\ \mathbf{x}_2\ \cdots\mathbf{x}_n]. \tag{B.31}$$

Note that the transposed matrix $\mathbf{P}'$ comprises the same eigenvectors arranged, instead, in rows.

Substitution of equation (B.30) into equation (B.29), and subsequent premultiplication by $\mathbf{P}'$, gives[1]

$$\mathbf{P}'\mathbf{M}\mathbf{P}\ddot{\mathbf{y}} + \mathbf{P}'\mathbf{K}\mathbf{P}\mathbf{y} = \mathbf{P}'\mathbf{F}. \tag{B.32}$$

The element on the ith diagonal element of $\mathbf{P}'\mathbf{M}\mathbf{P}$ is identically M_i, and all off-diagonal elements are identically zero. Similarly, the element on the ith diagonal element of $\mathbf{P}'\mathbf{K}\mathbf{P}$ is identically K_i, and off-diagonal elements are identically zero. These results follow directly from the orthogonal properties and definitions given above. Thus, equation (B.32) represents n uncoupled second-order differential equations

$$M_i\ddot{y}_i + K_i\ddot{y}_i = [\mathbf{P}'\mathbf{F}(t)]_i; \qquad i = 1, \cdots, n. \tag{B.33}$$

Once these equations are solved, the desired response $\mathbf{x}(t)$ is found by substitution into equation (B.30).

B.2.4   Modal Damping

The equation of motion with linear damping introduced becomes

$$\mathbf{M}\ddot{\mathbf{x}} + \mathbf{C}\dot{\mathbf{x}} + \mathbf{K}\mathbf{x} = \mathbf{F}(t). \tag{B.34}$$

Use of the modal matrix $\mathbf{P}$ and transformation $\mathbf{x} = \mathbf{P}\mathbf{y}$ *based on zero damping* gives

$$\mathbf{P}'\mathbf{M}\mathbf{P}\ddot{\mathbf{y}} + \mathbf{P}'\mathbf{C}\mathbf{P}\dot{\mathbf{y}} + \mathbf{P}'\mathbf{K}\mathbf{P}\mathbf{y} = \mathbf{P}'\mathbf{F}(t). \tag{B.35}$$

The matrices $\mathbf{P}'\mathbf{M}\mathbf{p}$ and $\mathbf{P}'\mathbf{K}\mathbf{P}$ are diagonal, as before, but in general $\mathbf{P}'\mathbf{C}\mathbf{P}$ is not; the result is modal coupling due to damping.

In some cases, nevertheless, $\mathbf{P}'\mathbf{C}\mathbf{P}$ is indeed diagonal, and the model is said to have **proportional damping**. This happens when

$$\mathbf{C} = \alpha\mathbf{M} + \beta\mathbf{K} \tag{B.36}$$

---

[1]The parallel use of the modal matrix in Section 7.3 for the first-order model $\dot{\mathbf{x}} = \mathbf{A}\mathbf{s}$ employs the inverse matrix $\mathbf{P}^{-1}$ in place of the present transpose matrix $\mathbf{P}'$. Only in the present case is the matrix $\mathbf{A}$ generally symmetric; as a consequence of this and the normalization of the component eigenvectors in $\mathbf{P}$, there results $\mathbf{P}^{-1} = \mathbf{P}'$. Herein, $\mathbf{P}'$ is used, since it is easier to find than $\mathbf{P}^{-1}$.

for any values of the constants α and β. The non-zero diagonal elements are

$$[\mathbf{P'CP}]_i = \alpha M_i + \beta K_i, \tag{B.37}$$

so that the separate modes are described by

$$M_i \ddot{y}_i + (\alpha M_i + \beta K_i)\dot{y}_i + K_i y_i = [\mathbf{P'F}(t)]_i, \tag{B.38}$$

or

$$\frac{1}{\omega_i^2}\ddot{y}_i + \frac{2\zeta_i}{\omega_i}\dot{y}_i + y_i = \frac{1}{K_i}[\mathbf{P'F}(t)]_i, \tag{B.39a}$$

$$\omega_i = \sqrt{K_i/M_i}; \qquad \zeta_i = \frac{1}{2}\left(\frac{\alpha}{\omega_i} + \beta\omega_i\right). \tag{B.39b}$$

Many systems are lightly damped. Light damping does not produce much modal coupling, even if equation (B.36) does not apply. Nevertheless, damping can be of critical importance in limiting the amplitude of resonances. It is often reasonable, therefore, to approximate damping by a proportional model. Nevertheless, the approach of the dynamic systems tradition, which is based on matrix first-order differential equations, has the advantage of giving perfect modal decoupling regardless of the form of the linear damping.

B.2.5 Example Using MATLAB®

A ten-story building is shaken by horizontal ground motion; the resulting motion is sought. The building, pictured in part (a) of Fig. B.2, is modeled by ten masses each equal to $m = 5 \times 10^5$ kg, ten springs for lateral displacements each equal to $k = 1 \times 10^9$ N/m, and ten corresponding damping coefficients each equal to $c = 1 \times 10^7$ N s/m. The ground is assumed to have infinite stiffnesses in translation and rotation. The equation of motion for the first floor is

$$m\ddot{x}_1 + 2c\dot{x}_1 + 2kx_1 = c\dot{x}_2 + kx_2 + c\dot{x}_0 + kx_0. \tag{B.40}$$

For the top floor the equation is

$$m\ddot{x}_{10} + c\dot{x}_{10} + kx_{10} = c\dot{x}_9 + kx_9, \tag{B.41}$$

and for floors $2 \le i \le 9$ it is

$$m\ddot{x}_i + 2c\dot{x}_i + 2kx_i = c(\dot{x}_{i+1} + x_{i-1}) + k(x_{i+1} + x_{i-1}). \tag{B.42}$$

The matrices M, C and K therefore become

$$\mathbf{M} = m\mathbf{I}, \tag{B.43a}$$

$$\mathbf{K} = k
\begin{bmatrix}
2 & -1 & 0 & 0 & 0 & 0 & 0 & 0 & 0 & 0 \\
-1 & 2 & -1 & 0 & 0 & 0 & 0 & 0 & 0 & 0 \\
0 & -1 & 2 & -1 & 0 & 0 & 0 & 0 & 0 & 0 \\
0 & 0 & -1 & 2 & -1 & 0 & 0 & 0 & 0 & 0 \\
0 & 0 & 0 & -1 & 2 & -1 & 0 & 0 & 0 & 0 \\
0 & 0 & 0 & 0 & -1 & 2 & -1 & 0 & 0 & 0 \\
0 & 0 & 0 & 0 & 0 & -1 & 2 & -1 & 0 & 0 \\
0 & 0 & 0 & 0 & 0 & 0 & -1 & 2 & -1 & 0 \\
0 & 0 & 0 & 0 & 0 & 0 & 0 & -1 & 2 & -1 \\
0 & 0 & 0 & 0 & 0 & 0 & 0 & 0 & -1 & 1
\end{bmatrix} \tag{B.43b}$$

$$\mathbf{C} = (c/k)\mathbf{K}. \tag{B.43c}$$

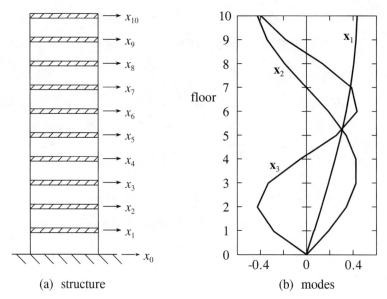

(a) structure   (b) modes

Figure B.2: Ten-story building and lowest-frequency mode shapes

This information can be entered into a MATLAB m-file `build.m` as follows:

```
m=5e5; k=1e9; c=1e7;
M=m*eye(10);
K=k*[2 -1 0 0 0 0 0 0 0 0;-1 2 -1 0 0 0 0 0 0 0;
    0 -1 2 -1 0 0 0 0 0 0;0 0 -1 2 -1 0 0 0 0 0;
    0 0 0 -1 2 -1 0 0 0 0;0 0 0 0 -1 2 -1 0 0 0;
    0 0 0 0 0 -1 2 -1 0 0;0 0 0 0 0 0 -1 2 -1 0;
    0 0 0 0 0 0 0 -1 2 -1;0 0 0 0 0 0 0 0 -1 1];
C=c/k*K;
A=inv(M)*K;
[P,W]=eig(A);
GM=P'*M*P; GK=P'*K*P; GC=P'*C*P; GF=P'(1,:)';
```

Notice the commands for the inverse of a matrix and for the modal matrix, P, and eigenvalue matrix, W. The matrix W is square and diagonal. If upon running MATLAB you respond to the prompt as follows,

```
>>build
>>diag(W)'
```

the response is

```
1.0e+003 *
Columns 1 through 7
   2.0000 1.0678 0.3961 0.0447 3.1099 4.2989 5.4614
Columns 8 through 10
   6.4940 7.3050 7.8223
```

The lowest-frequency mode is the fourth in the order chosen by MATLAB, the second lowest mode is the third, and the third lowest mode is the second.[2] The frequency of the lowest mode, in Hz, can be found by entering

```
EDU>> sqrt(W(4,4))/2/pi
ans =
1.0638
```

The eigenvectors corresponding to these three modes can be found by entering

```
>> [P(:,4) P(:,3) P(:,2)]
ans =
    0.0650    0.1894   -0.2969
    0.1286    0.3412   -0.4352
    0.1894    0.4255   -0.3412
    0.2459    0.4255   -0.0650
    0.2969    0.3412    0.2459
    0.3412    0.1894    0.4255
    0.3780    0.0000    0.3780
    0.4063   -0.1894    0.1286
    0.4255   -0.3412   -0.1894
    0.4352   -0.4255   -0.4063
```

These mode shapes are plotted in part (b) of Fig. B.2.

The elements M_i, K_i and C_i of the generalized mass, spring and damping vectors are the diagonal elements on the square diagonal matrices GM, GK and GC. The values for the lowest three modes can be displayed as follows:

```
>> [GM(4,4) GM(3,3) GM(2,2)]
ans =
   1.0e+005 *
       5.0000    5.0000    5.0000
>> [GC(4,4) GC(3,3) GC(2,2)]
ans =
   1.0e+006 *
       0.2234    1.9806    5.3390
>> [GK(4,4) GK(3,3) GK(2,2)]
ans =
   1.0e+008 *
       0.2234    1.9806    5.3390
>> [GF(4) GF(3) GF(2)]
ans =
       0.0650    0.1894   -0.2969
```

---

[2] The order of the modes varies depending on the version of MATLAB.

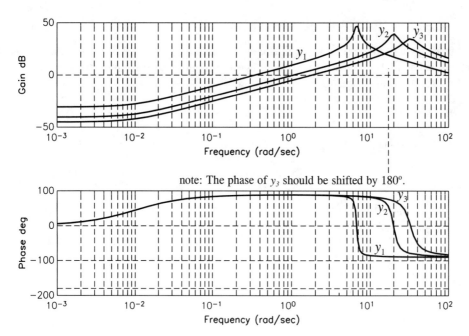

Figure B.3: Bode plots for the three lowest vibrational modes of the building

The information above can be assembled as follows to give the equations of motion for the first three modes:

$$5 \times 10^5 \ddot{y}_1 + 0.2231 \times 10^6 \dot{y}_1 + 0.2234 \times 10^8 y_1 = 0.0650(10^9 \dot{x}_0 + 10^7 x_0)$$
$$5 \times 10^5 \ddot{y}_2 + 1.9806 \times 10^6 \dot{y}_2 + 1.9806 \times 10^8 y_2 = 0.1894(10^9 \dot{x}_0 + 10^7 x_0)$$
$$5 \times 10^5 \ddot{y}_3 + 5.3390 \times 10^6 \dot{y}_3 + 5.3390 \times 10^8 y_3 = -0.2969(10^9 \dot{x}_0 + 10^7 x_0)$$

$$(B.44)$$

The actual displacements at the various floor levels are

$$x_i = x_0 + \sum_{j=1}^{10} \mathbf{P}_{ij} y_j \qquad (B.45)$$

The behavior of the modes as a function of frequency can be seen from Bode plots; the lowest three modes are given in Fig. B.3. The first of these plots will be displayed as a result of the following commands:

```
>> n1=[0.0650e9 0.0650e7]
>> d1=[5e5 0.2234e6 0.2234e8]
>> bode(n1,d1)
>> hold
```

The command hold allows the subsequent Bode plots to be superimposed on the first. Repeating this command releases the plots. Note that if the frequency content of an excitation does not exceed the resonant frequency of the third mode, there is little need to include more than the three modes in the solution.

B.2.6 Mode Reduction

The response of a high-order model is found routinely using digital computation, as the above example suggests. If the order is very high or there are a large number of cases to address, however, the computation can be costly. If the computation is done online in a controller, time and complexity become important. Moreover, often only a few modes are important, most often those with the lowest frequencies. These cases are candidates for mode reduction.

The essential idea is to employ the reduced modal matrix of dimensions $n \times m$ comprising the $m < n$ (usually $m << n$) eigenvectors of interest:

$$\mathbf{P} = \begin{bmatrix} x_{11} & x_{21} & \cdot & \cdot & \cdot & x_{m1} \\ & \cdot & \cdot & & & \cdot \\ \cdot & & \cdot & & & \cdot \\ \cdot & & \cdot & & & \cdot \\ x_{1n} & x_{2n} & \cdot & \cdot & \cdot & x_{mn} \end{bmatrix}. \qquad (B.46)$$

In the example above the three lowest-frequency modes as given could be used. The vectors $\mathbf{y}$ and $\mathbf{P'x}$ become of reduced length 3, and the matrices $\mathbf{P'MP}$, $\mathbf{P'CP}$ and $\mathbf{P'KP}$ become of reduced size 3×3. The analysis is carried out as before, and the transformation $\mathbf{x} = \mathbf{Py}$ restores the full displacement vector corresponding to the neglect of the missing modes.

In the example, the result of this reduced calculation would be identical to the truncation of equation (B.45) at $i = 3$ and the use of equations (B.44). MATLAB was not directed to use the reduced matrices $\mathbf{P'MP}$, $\mathbf{P'CP}$ and $\mathbf{P'KP}$ because the full computation required an insignificant amount of computer time. Were the order much higher and the matrices M, C and K fuller, however, or the calculation repeated a large number of times, the savings might be worthwhile.

PROBLEMS

B.1 A torsional system has three identically compliant shafts with stiffness K and three identical disks with inertance I.

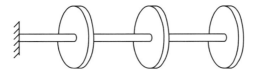

(a) Find the mass and stiffness matrices.

(b) Using MATLAB, find the natural frequencies and mode shapes.

(c) A torque $M(t)$ is applied at the free end. Find the modal equations of motion, and relate the displacement of the free end to these modes.

(d) Present a Bode plot for the response of the free end when $M(t)$ is a sinusoidal function.

B.2 The ten-story building considered above is buffeted by wind rather than an earthquake. The force on each floor is sinusoidal, due to the shedding of vortices, with an amplitude of 1 MN and frequency of one cycle in three seconds. Find the displacement amplitude of the top floor. Note: Only the lowest frequency mode is very significant, so consider only two modes.

Appendix C

Laplace Transform Pairs

The following is a highly selected list. For more extensive pairs, see for example Paul A. McCollum and Buck F. Brown, *Laplace Transform Tables and Theorems*, Holt, Rinehart and Winston, Inc., New York, 1965 and Ruel V. Churchill, *Operational Mathematics*, 2nd ed., McGraw-Hill, New York, 1958. Also note the last two entries in Table 4.4 (p. 199).

Note: n is a positive integer.

	$f(t)$	$F(s)$
1	$\delta(0)$	1
2	$u_s(t)$	$\dfrac{1}{s}$
3	t	$\dfrac{1}{s^2}$
4	$\dfrac{t^{n-1}}{(n-1)!}$	$\dfrac{1}{s^n}$
5	e^{-at}	$\dfrac{1}{s+a}$
6	te^{-at}	$\dfrac{1}{(s+a)^2}$
7	$\dfrac{1}{(n-1)!}t^{n-1}e^{-at}$	$\dfrac{1}{(s+a)^n}$
8	$\sin\omega t$	$\dfrac{\omega}{s^2+\omega^2}$
9	$\cos\omega t$	$\dfrac{s}{s^2+\omega^2}$
10	$\sinh\omega t$	$\dfrac{\omega}{s^2-\omega^2}$
11	$\cosh\omega t$	$\dfrac{s}{s^2-\omega^2}$
12	$\dfrac{1}{a}(1-e^{-at})$	$\dfrac{1}{s(s+a)}$

	$f(t)$	$F(s)$
13	$\dfrac{1}{b-a}(e^{-at} - e^{-bt})$	$\dfrac{1}{(s+a)(s+b)}$
14	$\dfrac{1}{b-a}(be^{-bt} - ae^{-at})$	$\dfrac{s}{(s+a)(s+b)}$
15	$\dfrac{1}{ab}\left[1 + \dfrac{1}{a-b}(be^{-at} - ae^{-bt})\right]$	$\dfrac{1}{s(s+a)(s+b)}$
16	$\dfrac{1}{a^2}(1 - e^{-at} - ate^{-at})$	$\dfrac{1}{s(s+a)^2}$
17	$\dfrac{1}{a^2}(at - 1 + e^{-at})$	$\dfrac{1}{s^2(s+a)}$
18	$e^{-at}\sin\omega t$	$\dfrac{\omega}{(s+a)^2 + \omega^2}$
19	$e^{-at}\cos\omega t$	$\dfrac{s+a}{(s+a)^2 + \omega^2}$
20	$\dfrac{\omega_n}{\sqrt{1-\zeta^2}}e^{-\zeta\omega_n t}\sin\omega_n\sqrt{1-\zeta^2}t$	$\dfrac{\omega_n{}^2}{s^2 + 2\zeta\omega_n s + \omega_n{}^2}$
21	$-\dfrac{1}{\sqrt{1-\zeta^2}}e^{-\zeta\omega_n t}\sin(\omega_n\sqrt{1-\zeta^2}t - \phi)$ $\phi = \tan^{-1}\dfrac{\sqrt{1-\zeta^2}}{\zeta}$	$\dfrac{s}{s^2 + 2\zeta\omega_n s + \omega_n{}^2}$
22	$1 - \dfrac{1}{\sqrt{1-\zeta^2}}e^{-\zeta\omega_n t}\sin(\omega_n\sqrt{1-\zeta^2}t + \phi)$ $\phi = \tan^{-1}\dfrac{\sqrt{1-\zeta^2}}{\zeta}$	$\dfrac{\omega_n{}^2}{s(s^2 + 2\zeta\omega_n s + \omega_n{}^2)}$
23	$1 - \cos\omega t$	$\dfrac{\omega^2}{s(s^2 + \omega^2)}$
24	$\omega t - \sin\omega t$	$\dfrac{\omega^3}{s^2(s^2 + \omega^2)}$
25	$\sin\omega t - \omega t\cos\omega t$	$\dfrac{2\omega^3}{(s^2 + \omega^2)^2}$
26	$\dfrac{1}{2\omega}t\sin\omega t$	$\dfrac{s}{(s^2 + \omega^2)^2}$
27	$t\cos\omega t$	$\dfrac{s^2 - \omega^2}{(s^2 + \omega^2)^2}$
28	$\dfrac{1}{\omega_2^2 - \omega_1^2}(\cos\omega_1 t - \cos\omega_2 t) \quad \omega_1{}^2 \neq \omega_2{}^2$	$\dfrac{s}{(s^2 + \omega_1{}^2)(s^2 + \omega_2{}^2)}$
29	$\dfrac{1}{2\omega}(\sin\omega t + \omega t\cos\omega t)$	$\dfrac{s^2}{(s^2 + \omega^2)^2}$

	$f(t)$	$F(s)$
30	$\dfrac{1}{(b-a)(c-a)}e^{-at} + \dfrac{1}{(a-b)(c-b)}e^{-bt}$ $+ \dfrac{1}{(a-c)(b-c)}e^{-ct}$	$\dfrac{1}{(s+a)(s+b)(s+c)}$
31	$\dfrac{-a}{(b-a)(c-a)}e^{-at} - \dfrac{b}{(a-b)(c-b)}e^{-bt}$ $- \dfrac{c}{(a-c)(b-c)}e^{-ct}$	$\dfrac{s}{(s+a)(s+b)(s+c)}$
32	$A\left[e^{-bt} + \dfrac{\sqrt{(b-a)^2+\omega^2}}{\omega}e^{-at}\sin(\omega t - \phi)\right]$ $A = \dfrac{1}{(b-a)^2+\omega^2} \qquad \phi = \tan^{-1}\left(\dfrac{\omega}{b-a}\right)$	$\dfrac{1}{(s+b)[(s+a)^2+\omega^2]}$
33	$Ae^{-ct} + Be^{-at}\sin(\omega t + \phi)$ $A = \dfrac{-c}{(a-c)^2+\omega^2} \quad B = \dfrac{1}{\omega}\sqrt{\dfrac{a^2+\omega^2}{(c-a)^2+\omega^2}}$ $\phi = \tan^{-1}\left(\dfrac{\omega}{-a}\right) - \tan^{-1}\left(\dfrac{\omega}{c-a}\right)$	$\dfrac{s}{(s+c)[(s+a)^2+\omega^2]}$
34	$\dfrac{1}{\omega}e^{-at}\sinh\omega t$	$\dfrac{1}{(s+a)^2-\omega^2}$
35	$\dfrac{1}{\sqrt{\pi t}}$	$\dfrac{1}{\sqrt{s}}$
36	$2\sqrt{\dfrac{t}{\pi}}$	$\dfrac{1}{s^{3/2}}$
37	$\dfrac{t^{a-1}}{\Gamma(a)} \quad a>0 \quad (\Gamma \text{ is Gamma function})$	$\dfrac{1}{s^a}$
38	$\dfrac{a}{2\sqrt{\pi t^3}}e^{-a^2/4t}$	$e^{-a\sqrt{s}} \quad (a>0)$
39	$\text{erfc}\left(\dfrac{a}{2\sqrt{t}}\right) \quad (\text{complementary error function})$	$\dfrac{1}{s}e^{-a\sqrt{s}} \quad a\geq 0$
40	$\dfrac{1}{\sqrt{\pi t}}e^{-a^2/4t}$	$\dfrac{1}{\sqrt{s}}e^{-a\sqrt{s}} \quad a\geq 0$
41	$J_0(2\sqrt{at}) \quad (\text{Bessel function})$	$\dfrac{1}{s}e^{-a/s}$
42	$\dfrac{1}{\sqrt{\pi t}}\cos 2\sqrt{at}$	$\dfrac{1}{\sqrt{s}}e^{-a/s}$
43	$\dfrac{1}{\sqrt{\pi t}}\cosh 2\sqrt{at}$	$\dfrac{1}{\sqrt{s}}e^{a/s}$
44	$\left(\dfrac{t}{a}\right)^{(b-1)/2}I_{b-1}(2\sqrt{at}) \quad (\text{modified Bessel fn.})$	$\dfrac{1}{s^b}e^{a/s} \quad b>0$

Appendix D

Thermodynamic Data and Computer Code

Complete programs that find the state variables corresponding to a specified density and entropy are listed below for dry air as a mixture of nitrogen, for oxygen and argon, for refrigerants R-12 and R134a, and for water. These files, along with many others in chapters 9-11, as listed on page 1018, may be downloaded from the author's page at the website www.lehigh.edu/~inmem. Comments or questions may be emailed to the author at ftb0@lehigh.edu or ftbmhb@aol.com. Data for refrigerant R-22 also are given below.

D.1 Programs and Data for Air and Components

The following MATLAB® files are discussed in Section 12.5.3:

```
function [s,dsdT,u,dudT]=gas(g1,g2,g3,g4,g5,g6,g7,g8,b,s0,T0g,u0,T)
% Evaluates the entropies s and dsdT and the internal energies u and dudT of
% an ideal (zero-pressure) gas, given the temperature. Based on equations
% (12.98) and (12.99).
y=b/T;
y0=b/T0g;
ey=exp(y)-1; ey0=exp(y0)-1;
T2=T*T; T3=T2*T; T4=T2*T2; T02=T0g*T0g; T03=T02*T0g; T04=T02*T02;
s=-g1/3*(1/T3-1/T03)-g2/2*(1/T2-1/T02)-g3*(1/T-1/T0g);
s=s+g4*log(T/T0g)+g5*(T-T0g)+g6/2*(T2-T02)+g7/3*(T3-T03)-g8*(y/ey-y0/ey0;
s=s+y-y0-log(ey/ey0))+s0;
dsdT=g1/T4+g2/T3+g3/T2+g4/T+g5+g6*T+g7*T2-g8/T*(y/ey)^2*(1+ey);
u=-g1/2*(1/T2-1/T02)-g2*(1/T-1/T0g)+g3*log(T/T0g)+g4*(T-T0g)+u0;
u=u+g5/2*(T2-T02)+g6/3*(T3-T03)+g7/4*(T4-T04)-g8*b*(1/ey-1/ey0);
dudT=g1/T3+g2/T2+g3/T+g4+g5*T+g6*T2+g7*T3+g8*exp(y)*(y/ey)^2;


function [P,u,h,s,dudT,dudv,dsdT,dsdv,dPdT,dPdv]=air(T,rho)
% Computes the indicated functions for air as a mixture of nitrogen, oxygen
% and argon, given the values of temperature and density. "n", "o" and "a"
% designate nitrogen, oxygen and argon, respectively. Based on equations
% (12.97), (12.100) and (12.101).
```

```
global G1n G2n G3n G4n G5n G6n G7n G8n Bn S0n T0n T2n T3n T4n U0n
global G1o G2o G3o G4o G5o G6o G7o G8o Bo S0o T0o T2o T3o T4o U0o
global Cva S0a T0a U0a Sm R Fn Fo ey0n ey0o
global Gamma A1 A2 A3 A4 A5 A6 A7 A8 A9 A10 A11 A12 A13 A14 A15 A16 A17
global A18 A19 A20 A21 A22 A23 A24 A25 A26 A27 A28 A29 A30 A31 A32
% Note that dsdv=dPdT (an identity), so one of these (presumably % dsdv) can
% be removed. They are left as a check; the coding is different.
r=rho; r2=r*r; r3=r*r2; r4=r2*r2; r5=r2*r3; r6=r3*r3; r7=r3*r4; r8=r4*r4;
r9=r4*r5; r10=r5*r5; r11=r5*r6; r12=r6*r6; r13=r6*r7; r14=r7*r7;
er=exp(-Gamma*r2);
W=ones(6); dW=ones(6);
W(1)=1/2/Gamma*(1-er);
dW(1)=r*er;
for i=1:5
   W(i+1)=(-r^(2*i)/2*er+i*W(i))/Gamma;
   dW(i+1)=(r^(2*i)*(Gamma*r-i/r)*er+i*dW(i))/Gamma;
end
T2=T*T; T3=T*T2; T4=T2*T2; T5=T2*T3; T6=T3*T3;
yn=Bn/T; yn0=Bn/T0n; yo=Bo/T; yo0=Bo/T0o; eyo=exp(yo)-1; eyn=exp(yn)-1;
P=r*R*T+r2*(A1*T+A2*sqrt(T)+A3+A4/T+A5/T2)+r3*(A6*T+A7+A8/T+A9/T2);
P=P+r4*(A10*T+A11+A12/T)+r5*A13+r6*(A14/T+A15/T2)+r7*A16/T+r8*(A17/T+A18/T2);
P=P+r9*A19/T2+(r3*(A20/T2+A21/T3)+r5*(A22/T2+A23/T4))*er+(r7*(A24/T2+A25/T3);
P=P+r9*(A26/T2+A27/T4)+r11*(A28/T2+A29/T3))*er+r13*(A30/T2+A31/T3+A32/T4)*er;
[sn,dsdTn,un,dudTn]
             =feval('gas',G1n,G2n,G3n,G4n,G5n,G6n,G7n,G8n,Bn,S0n,T0n,U0n,T);
[so,dsdTo,uo,dudTo]
             =feval('gas',G1o,G2o,G3o,G4o,G5o,G6o,G7o,G8o,Bo,S0o,T0o,U0o,T);
u=Fn*un+Fo*uo+(1-Fn-Fo)*(Cva*(T-T0a)+U0a);
u=u+r*(A2/2*sqrt(T)+A3+2*A4/T+3*A5/T2)+r2/2*(A7+2*A8/T+3*A9/T2);
u=u+r3/3*(A11+2*A12/T)+r4/4*A13+r5/5*(2*A14/T+3*A15/T2)+r6/6*2*A16/T;
u=u+r7/7*(2*A17/T+3*A18/T2)+r8/8*3*A19/T2+W(1)*(3*A20/T2+4*A21/T3);
u=u+W(2)*(3*A22/T2+5*A23/T4)+W(3)*(3*A24/T2+4*A25/T3);
u=u+W(4)*(3*A26/T2+5*A27/T4)+W(5)*(3*A28/T2+4*A29/T3);
u=u+W(6)*(3*A30/T2+4*A31/T3+5*A32/T4);
h=u+P/r;
s=-R*log(r)+Sm+Fn*sn+Fo*so+(1-Fn-Fo)*(Cva*log(T/T0a)+S0a);
s=s-r*(A1+A2/2/sqrt(T)-A4/T2-2*A5/T3)-r2/2*(A6-A8/T2-2*A9/T3);
s=s-r3/3*(A10-A12/T2)+r5/5*(A14/T2+2*A15/T3)+r6/6*A16/T2;
s=s+r7/7*(A17/T2+2*A18/T3)+r8/8*2*A19/T3+W(1)*(2*A20/T3+3*A21/T4);
s=s+W(2)*(2*A22/T3+4*A23/T5)+W(3)*(2*A24/T3+3*A25/T4);
s=s+W(4)*(2*A26/T3+4*A27/T5)+W(5)*(2*A28/T3+3*A29/T4);
s=s+W(6)*(2*A30/T3+3*A31/T4+4*A32/T5);
dudT=Fn*dudTn+Fo*dudTo+(1-Fn-Fo)*Cva;
dudT=dudT+r*(A2/4/sqrt(T)-2*A4/T2-6*A5/T3)-r2/2*(2*A8/T2+6*A9/T3);
dudT=dudT-r3/3*2*A12/T2-r5/5*(2*A14/T2+6*A15/T3)-r6/6*2*A16/T2;
dudT=dudT-r7/7*(2*A17/T2+6*A18/T3)-r8/8*(6*A19/T3)-W(1)*(6*A20/T3+12*A21/T4);
dudT=dudT-W(2)*(6*A22/T3+20*A23/T5)-W(3)*(6*A24/T3+12*A25/T4);
dudT=dudT-W(4)*(6*A26/T3+20*A27/T5)-W(5)*(6*A28/T3+12*A29/T4);
dudT=dudT-W(6)*(6*A30/T3+12*A31/T4*20*A32/T5);
dudr=A2/2*sqrt(T)+A3+2*A4/T+3*A5/T2+r*(A7+2*A8/T+3*A9/T2)+r2*(A11+2*A12/T);
dudr=dudr+r3*A13+r4*(2*A14/T+3*A15/T2)+r5*2*A16/T+r6*(2*A17/T+3*A18/T2);
```

```
dudr=dudr+r7*3*A19/T2+dW(1)*(3*A20/T2+4*A21/T3)+dW(2)*(3*A22/T2+5*A23/T4);
dudr=dudr+dW(3)*(3*A24/T2+4*A25/T3)+dW(4)*(3*A26/T2+5*A27/T4);
dudr=dudr+dW(5)*(3*A28/T2+4*A29/T3)+dW(6)*(3*A30/T2+4*A31/T3+5*A32/T4);
dudv=-r2*dudr;
dsdT=Fn*dsdTn+Fo*dsdTo+(1-Fn-Fo)*Cva/T;
dsdT=dsdT+r*(A2/4/T^(1.5)-2*A4/T3-6*A5/T4)-r2/2*(2*A8/T3+6*A9/T4);
dsdT=dsdT-2/3*r3*A12/T3-r5/5*(2*A14/T3+6*A15/T4)-r6/3*A16/T3;
dsdT=dsdT-r7/7*(2*A17/T3+6*A18/T4)-r8/8*6*A19/T4;
dsdT=dsdT-W(1)*(6*A20/T4+12*A21/T5)-W(2)*(6*A22/T4+20*A23/T6);
dsdT=dsdT-W(3)*(6*A24/T4+12*A25/T5)-W(4)*(6*A26/T4+20*A27/T6);
dsdT=dsdT-W(5)*(6*A28/T4+12*A29/T5)-W(6)*(6*A30/T4+12*A31/T5+20*A32/T6);
dsdr=-R/r-(A1+A2/2/sqrt(T)-A4/T2-2*A5/T3)-r*(A6-A8/T2-2*A9/T3);
dsdr=dsdr-r2*(A10-A12/T2)+r4*(A14/T2+2*A15/T3)+r5*A16/T2;
dsdr=dsdr+r6*(A17/T2+2*A18/T3)+r7*2*A19/T3;
dsdr=dsdr+dW(1)*(2*A20/T3+3*A21/T4)+dW(2)*(2*A22/T3+4*A23/T5);
dsdr=dsdr+dW(3)*(2*A24/T3+3*A25/T4)+dW(4)*(2*A26/T3+4*A27/T5);
dsdr=dsdr+dW(5)*(2*A28/T3+3*A29/T4)+dW(6)*(2*A30/T3+3*A31/T4+4*A32/T5);
dsdv=-r2*dsdr;
dPdT=r*R+r2*(A1+A2/2/sqrt(T)-A4/T2-2*A5/T3)+r3*(A6-A8/T2-2*A9/T3);
dPdT=dPdT+r4*(A10-A12/T2)-r6*(A14/T2+2*A15/T3)-r7*A16/T2;
dPdT=dPdT-r8*(A17/T2+2*A18/T3)-r9*2*A19/T3-r3*(2*A20/T3+3*A21/T4)*er;
dPdT=dPdT-(r5*(2*A22/T3+4*A23/T5)-r7*(2*A24/T3+3*A25/T4))*er;
dPdT=dPdT-(r9*(2*A26/T3+4*A27/T5)-r11*(2*A28/T3+3*A29/T4))*er;
dPdT=dPdT-r13*(2*A30/T3+3*A31/T4+4*A32/T5)*er;
dPdr=R*T+2*r*(A1*T+A2*sqrt(T)+A3+A4/T+A5/T2)+3*r2*(A6*T+A7+A8/T+A9/T2);
dPdr=dPdr+4*r3*(A10*T+A11+A12/T)+5*r4*A13+6*r5*(A14/T+A15/T2)+7*r6*A16/T;
dPdr=dPdr+8*r7*(A17/T+A18/T2)+9*r8*A19/T2;
dPdr=dPdr+(3*r2-2*Gamma*r4)*(A20/T2+A21/T3)*er;
dPdr=dPdr+(5*r4-2*Gamma*r6)*(A22/T2+A23/T4)*er;
dPdr=dPdr+(7*r6-2*Gamma*r8)*(A24/T2+A25/T3)*er;
dPdr=dPdr+(9*r8-2*Gamma*r10)*(A26/T2+A27/T4)*er;
dPdr=dPdr+(11*r10-2*Gamma*r12)*(A28/T2+A29/T3)*er;
dPdr=dPdr+(13*r12-2*Gamma*r14)*(A30/T2+A31/T3+A32/T4)*er;
dPdv=-r2*dPdr;
```

```
% script file gasdata.m: data for ideal gasses. "n", "o" and "a" % designate
% nitrogen, oxygen and argon, respectively.
global G1n G2n G3n G4n G5n G6n G7n G8n Bn S0n T0n T2n T3n T4n U0n
global G1o G2o G3o G4o G5o G6o G7o G8o Bo S0o T0o T2o T3o T4o U0o
global Cva S0a T0a U0a Sm Fn Fo ey0n ey0o g Ser
global Gamma A1 A2 A3 A4 A5 A6 A7 A8 A9 A10 A11 A12 A13 A14 A15 A16
global A17 A18 A19 A20 A21 A22 A23 A24 A25 A26 A27 A28 A29 A30 A31 A32 R
Gamma=5.97105475117183e-6; R=287.0686; A1=1.55623098409137E-1;
A2=1.25288666202326e1; A3=-2.92541568638838e2; A4=4.29432480725523e3;
A5=-5.58450959675108e5; A6=3.92054480883008e-4; A7=-4.40985641881347e-2;
A8=5.86387178724129e-4; A9=7.97411385439405e4; A10=9.88045320906742e-9;
A11=2.97999237261289e-4; A12=-6.81783040959070e-2; A13=2.02551630992042e-7;
A14=-1.62724281849497e-7; A15=-1.06340143152999e-4; A16=3.51428501875049e-10;
A17=-1.70388092279449e-13; A18=5.91103444646786e-11;
A19=-1.05363473794348e-14; A20=-7.32732651196979e4; A21=-5.42674649924748e5;
```

```
A22=-4.48935466142735e-1; A23=2.81453138446295e2; A24=-8.83132042791851e-7;
A25=-1.32229814838386e-5; A26=-2.16521865046609e-12;
A27=-1.47835008246593e-9; A28=-6.93219849301501e-19;
A29=6.06743598768355e-17; A30=-3.20538718135891e-24;
A31=-4.73178337355130e-23; A32=3.83950822306912e-22;
G1n=-218203.473713518; G2n=10157.3580096247; G3n=-165.50472165724;
G4n=743.17599919043; G5n=-5.14605623546025e-3; G6n=5.18347156760489e-6;
G7n=-1.05922170493616e-9; G8n=298.389393363817;
Bn=3353.4061; S0n=3288.2374; T0n=63.15; U0n=196622.81;
G1o=-1294427.11174062; G2o=59823.1747005341; G3o=-897.850772730944;
G4o=655.236176900400; G5o=-1.13131252131570e-2; G6o=3.49810702442228e-6;
G7o=4.21065222886885e-9; G8o=267.997030050139; Bo=2239.18105;
S0o=3271.4925; T0o=54.34; U0o=228267.4; Cva=312.192; S0a=2270.67; T0a=83.8;
U0a=149355.4; Sm=174.519; Fn=0.7553; Fo=.23146; Ser=0.001;
T2n=T0n*T0n; T3n=T0n*T2n; T4n=T2n*T2n; T2o=T0o*T0o; T3o=T0o*T2o; T4o=T2o*T2o;
yOo=Bo/T0o; yOn=Bn/T0n; eyOo=exp(yOo)-1; eyOn=exp(yOn)-1;
g=Fn*[G1n G2n G3n G4n G5n G6n G7n]+Fo*[G1o G2o G3o G4o G5o G6o G7o];
```

```
% file specheat.m; computes the specific heat of air, given its temperature.
% Oxygen and nitrogen components based on % equation (12.96).
global G1n G2n G3n G4n G5n G6n G7n G8n Bn Fn T
global G1o G2o G3o G4o G5o G6o G7o G8o Bo Fo Cva
cvn=G1n/T\^{ }3+G2n/T/T+G3n/T+G4n+G5n*T+G6n*T*T+G7n*T\^{ }3;
cvn=cvn+G8n*exp(Bn/T)*(Bn/T/(exp(Bn/T)-1))\^{ }2;
cvo=G1o/T\^{ }3+G2o/T/T+G3o/T+G4o+G5o*T+G6o*T*T+G7o*T\^{ }3;
cvo=cvo+G8o*exp(Bo/T)*(Bo/T/(exp(Bo/T)-1))\^{ }2;
cvm=Fn*cvn+Fo*cvo+(1-Fn-Fo)*Cva;
```

D.2   Programs and Data for Refrigerants R12 and R134a

The following files are discussed in Section 12.5.4:

```
% program R12data.m % data defining the model of refrigerant R-12
global A2 A3 A4 A5 B2 B3 B4 B5 C2 C3 C4 C5 G1 G2 G3 G4 G5 G6
global E6 E7 E11 E12 E13 E14 E20 E21 E22 E23 E24 F1 F2 F4 F5 F6 F8 F9
global D1 D2 D3 D4 D5 D6 D7 Rc VC S0 U0 T0 TC R B Ck
A2=-91.6210126; A3=.101049598; A4=-5.74640225e-5; A5=0;
B2=.0771136428; B3=-5.67539138e-5; B4=0; B5=4.08193371e-11;
C2=-1525.24293; C3=2.19982681; C4=0; C5=-1.66307226e-7;
D1=558.0845400; D2=854.4458040; D3=0; D4=299.40771103; D5=0;
D6=352.1500633; D7=-50.47419739; Rc=588.08;
E6=0; E7=-1.938269e-5; E11=.1099986; E12=-2.584661; E13=0; E14=-3.841820;
E20=-1112.466; E21=5373.373; E22=-10376.49; E23=9275.978; E24=-3225.634;
F1=93.3438056; F2=-4396.18785; F4=-12.4715223; F5=.0196060432;
F6=0; F8=0; F9=0; G1=0; G2=0; G3=33.89005260; G4=2.507020671;
G5=-.003274505926; G6=1.641736815e-6;
S0=894.48764+19.67+.9; U0=169701.87-185.54+190; T0=200; TC=385.17;
R=68.7480; B=.406366926e-3; Ck=5.475;
```

```
% program R134adata.m % data defining the model of refrigerant R-134a
global A2 A3 A4 A5 B2 B3 B4 B5 C2 C3 C4 C5 G1 G2 G3 G4 G5 G6
global E6 E7 E11 E12 E13 E14 E20 E21 E22 E23 E24 F1 F2 F4 F5 F6 F8 F9
global D1 D2 D3 D4 D5 D6 D7 Rc S0 U0 T0 TC R B Ck
A2=-119.5051; A3=.1447797; A4=-.1049005e-3; A5=-.6953905e-8;
B2=.1137590; B3=-.8942552e-4; B4=0; B5=.1269806e-9;
C2=-3531.592; C3=6.469248; C4=0; C5=-.2051369e-5; R=81.4881629;
D1=512.2; D2=819.6183; D3=1023.582; D4=-1156.757; D5=789.7191; D6=0; D7=0;
E6=0; E7=.1895892e-4; E11=-.9533431e-2; E12=-2.125941; E13=0; E14=-6.628752;
E20=473.2668; E21=-2457.424; E22=4533.927; E23=-3620.867; E24=1025.264;
F1=24.8033988+log(1000); F2=-3980.408; F4=0; F5=-.02405332; F6=.2245211e-4;
F8=.1995548; F9=374.8473; G1=0; G2=15821.70; G3=-5.257455-R;
G4=3.296570; G5=-.002017321; G6=0; VC=.0019524; Rc=512.2;
S0=1141.98; U0=222209; T0=200; TC=374.25; B=.3455467e-3; Ck=5.475;


function[P,h,hg,s,sg,vf,vg,hf,dsdv,dsdT,dudv,dudT,dPs,d2Ps,dvg,dsg dugdv,
                                  dugdT,dPdv,dPdT]=refprop(v,T)
% Program that finds properties of refrigerants R12 or R134a given v, T.
% Calls program vapphase.m below, and uses equations (12.106)-(12.108),
% (12.113) and (12.116)-(12.119).
global A2 A3 A4 A5 B2 B3 B4 B5 C2 C3 C4 C5 D1 D2 D3 D4 D5 D6 D7 T02 T03 T04
global G1 G2 G3 G4 G5 G6 E6 E7 E11 E12 E13 E14 E20 E21 E22 E23 E24
global F1 F2 F4 F5 F6 F8 F9 Rc R B U0 S0 T0 T2 T3 T4 ft ft0 ft2 ft3 Ck TC
T2=T*T; T3=T*T2; T4=T2*T2; T02=T0*T0; T03=T0*T02; T04=T02*T02;
ft=Ck/TC*exp(-Ck*T/TC); ft0=Ck/TC*exp(-Ck*T0/TC);
ft1=ft*TC/Ck; ft2=(1+Ck*T/TC)*exp(-Ck*T/TC); ft3=-ft*Ck*T/TC;
v01=1.369-B; v02=v01*v01; v03=v01*v02; v04=v02*v02;
x=1-T/TC;
if x<.01 % supercritical temperature (with some allowance)
   [P,h,s,dudv,dudT,dsdv,dsdT,dPdv,dPdT]=vapphase(v,T);
   temp=1 % possibly useful run-time flag for supercritical temperature
   hg=0; sg=0; vf=0; vg=0; hf=0; dPs=0; d2Ps=0; dvg=0; dsg=0; dudgv=0;
   dugdT=0; % These 11 property zeros are not to be used
else  % subcritical temperature
   y=E6*x^(-5/3)+E7*x^(-4/3)+E11+E12*x^(1/3)+E13*x^(2/3)+E14*x+E20*x^3;
   y=y+E21*x^(10/3)+E22*x^(11/3)+E23*x^4+E24*x^(13/3);
   vg=1/Rc*exp(-y);
      if vg<=v % superheated
      [P,h,s,dudv,dudT,dsdv,dsdT,dPdv,dPdT]=vapphase(v,T);
      hg=0; sg=0; vf=0; hf=0; dPs=0; d2Ps=0; dvg=0; dsg=0; dugdv=0;
      dugdT=0; % these zero values are not to be used
   else % saturated mixture or compressed liquid
      x2=x^(1/3);
             x3=x2*x2;
             x5=x2*x;
             rf=D1+D2*x2+D3*x3+D4*x+D5*x5+D6*sqrt(x)+D7*x*x;
             % density of liquid phase
      vf=1/rf;
      X=(v-1/rf)/(vg-1/rf); % quality of mixture
      v1g=vg-B; v2g=v1g*v1g; v3g=v1g*v2g; v4g=v2g^2; v5g=v2g*v3g;
      P=exp(F1+F2/T+F4*log(T)+F5*T+F6*T2+F8*(F9/T-1)*log(F9-T));
```

```
    hg=-G1*(1/T-1/T0)+G2*log(T/T0)+G3*(T-T0)+G4/2*(T2-T02)+G5/3*(T3-T03);
    hg=hg+G6/4*(T4-T04)+(A2+ft2*C2)/v1g+(A3+ft2*C3)/2/v2g;
    hg=hg+(A4+ft2*C4)/3/v3g+(A5+ft2*C5)/4/v4g+P*vg+U0;
    dPs=P*(-F2/T2+F4/T+F5+2*F6*T-F8*F9/T2*log(F9-T)-F8/T);
    d2Ps=P*(2*F2/T3+(F8-F4)/T2+2*F6+2*F8*F9/T3*log(F9-T));
    d2Ps=d2Ps+P*F8*F9/T2/(F9-T)+dPs^2/P;
    sg=-G1/2*(1/T2-1/T02)-G2*(1/T-1/T0)+G3*log(T/T0)+G4*(T-T0);
    sg=sg+G5/2*(T2-T02)+G6/3*(T3-T03)+R*log(v1g/v01)+B2*(1/v01-1/v1g);
    sg=sg+B3/2*(1/v02-1/v2g)+B4/3*(1/v03-1/v3g)+B5/4*(1/v04-1/v4g);
    sg=sg+C2*(ft/v1g-ft0/v01)+C3/2*(ft/v2g-ft0/v02)+C4/3*(ft/v3g-ft0/v03);
    sg=sg+C5/4*(ft/v4g-ft0/v04)+S0;
    ug=-G1*(1/T-1/T0)+G2*log(T/T0)+G3*(T-T0)+G4/2*(T2-T02)+G5/3*(T3-T03);
    ug=ug+G6/4*(T4-T04)+(A2+C2*ft2)/v1g+(A3+C3*ft2)/v2g/2;
    ug=ug+(A4+C4*ft2)/v3g/3+(A5+C5*ft2)/v4g/4+U0;
    hg=ug+P*vg;
    sm=(vg-v)*dPs;
    s=sg-sm;
    sf=sg-(vg-1/rf)*dPs;
    h=hg-T*sm; % enthalpy if saturated mixture
    sf=sg-dPs*(vg-1/rf); % entropy of saturated liquid
    hf=hg-T*(sg-sf); % enthalpy of saturated liquid
    u=h-P*v; % internal energy if saturated mixture
    dvg=vg/TC*(-5/3*E6*x^(-8/3)-4/3*E7*x^(-7/3)+1/3*E12*x^(-2/3));
    dvg=dvg+vg/TC*(2/3*E13*x^(-1/3)+E14+3*E20*x^2+10/3*E21*x^(7/3));
    dvg=dvg+vg/TC*(11/3*E22*x^(8/3)+4*E23*x^3+13/3*E24*x^(10/3));
    dsg=(R/v1g+B2/v2g+B3/v3g+B4/v4g+B5/v5g)*dvg;
    dsg=dsg-ft*(C2/v2g+C3/v3g+C4/v4g+C5/v5g)*dvg;
    dsg=dsg-Ck/TC*ft*(C2/v1g+C3/2/v2g+C4/3/v3g+C5/4/v4g);
    dsg=dsg+G1/T3+G2/T2+G3/T+G4+G5*T+G6*T2;
    dugdv=-(A2+C2*ft2)/v2g-(A3+C3*ft2)/v3g-(A4+C4*ft2)/v4g-(A5+C5*ft2)/v5g;
    dugdT=G1/T2+G2/T+G3+G4*T+G5*T2+G6*T3;
    dugdT=dugdT+ft3*(C2/v1g+C3/2/v2g+C4/3/v3g+C5/4/v4g);
    dsgdv=R/v1g+B2/v2g+B3/v3g+B4/v4g+B5/v5g;
    dsgdv=dsgdv-ft*(C2/v2g+C3/v3g+C4/v4g+C5/v5g);
    dsdT=dsg-dvg*dPs-(vg-v)*d2Ps;
    dPdv=0;
    dPdT=dPs;
    dsdv=dPs;
    dudv=T*dPs-P;
    dudT=dugdT+(P-T*dPs)*dvg-T*(vg-v)*d2Ps+dvg*dugdv;
  end
end


function [P,h,s,dudv,dudT,dsdv,dsdT,dPdv,dPdT]=vapphase(v,T)
% Program that finds the properties of vapor phase refrigerant R12 or R134a
% given its specific volume and temperature.
% Called by program propref.m above.
% Based on equations (11.110), (11.114), (11.115) and (11.117).
global B2 B3 B4 B5 T02 T03 T04 TC A2 A3 A4 A5 C2 C3 C4 C5
global G1 G2 G3 G4 G5 G6 R B U0 S0 T2 T3 T4 ft ft0 ft2 ft3 Ck
ft1=ft*TC/Ck;
```

```
v01=1.369-B; v02=v01*v01; v03=v01*v02; v04=v02*v02;
v1=v-B; v2=v1*v1; v3=v1*v2; v4=v2*v2; v5=v2*v3; v6=v3*v3;
s=-G1/2*(1/T2-1/T02)-G2*(1/T-1/T2)+G3*log(T/T0)+G4*(T-T0)+G5/2*(T2-T02);
s=s+G6/3*(T3-T03)+R*log(v1/v01)+B2*(1/v01-1/v1)+B3/2*(1/v02-1/v2);
s=s+B4/3*(1/v03-1/v3)+B5/4*(1/v04-1/v4)+C2*(ft/v1-ft0/v01);
s=s+C3/2*(ft/v2-ft0/v02)+C4/3*(ft/v3-ft0/v03)+C5/4*(ft/v4-ft0/v04)+S0
P=R*T/v1+(A2+B2*T+C2*ft1)/v2+(A3+B3*T+C3*ft1)/v3;
P=P+(A4+B4*T+C4*ft1)/v4+(A5+B5*T+C5*ft1)/v5;
h=-G1*(1/T-1/T0)+G2*log(T/T0)+G3*(T-T0)+G4/2*(T2-T02)+G5/3*(T3-T03);
h=h+G6/4*(T4-T04)+(A2+ft2*C2)/v1+(A3+ft2*C3)/2/v2;
h=h+(A4+ft2*C4)/3/v3+(A5+ft2*C5)/4/v4+P*v+U0
dPdT=R/v1+(B2-C2*ft)/v2+(B3-C3*ft)/v3+(B4-C4*ft)/v4+(B5-C5*ft)/v5;
dPdv=-R*T/v2-2*(A2+B2*T+C2*ft1)/v3-3*(A3+B3*T+C3*ft1)/v4;
dPdv=dPdv-4*(A4+B4*T+C4*ft1)/v5-5*(A5+B5*T+C5*ft1)/v6;
dudv=-(A2+ft2*C2)/v2-(A3+ft2*C3)/v3-(A4+ft2*C4)/v4-(A5+ft2*C5)/v5;
dhdv=dudv+P+v*dPdv;
dudT=G1/T2+G2/T+G3+G4*T+G5*T2+G6*T3;
dhdT=dudT+ft3*(C2/v1+C3/2/v2+C4/3/v3+C5/4/v4)+v*dPdT;
dsdv=R/v1+B2/v2+B3/v3+B4/v4+B5/v5-ft*(C2/v2+C3/v3+C4/v4+C5/v5);
dsdT=-Ck/TC*ft*(C2/v1+C3/2/v2+C4/3/v3+C5/4/v4)+G1/T3+G2/T2+G3/T+G4;
dsdT=dsdT+G5*T+G6*T2;
```

```
function [dm]=orifice(A,Pu,Pd,vu,vd,vfu,vfd,vgu,vgd,Tu,Td)
% Computes the flow through an orifice based on equation p. 856.
global C K
r=Pd/Pu;
if r<=1
   dmvap=A*C*Pu*sqrt((r^(2/K)-r^(1+1/K))/Tu);
   if Tu>381
      dm=dmvap;
   elseif vu>=vgu
      dm=dmvap;
   else
      xu=(vu-vfu)/(vgu-vfu);
      dmliq=A*sqrt(2/vfu*(Pu-Pd));
      dm=xu*dmvap+(1-xu)*dmliq;
   end
else
   r=1/r;
   dmvap=-A*C*Pu*sqrt((r^(2/K)-r^(1+1/K))/Td);
   if vd>=vgd
      dm=dmvap;
   else
      xd=(vd-vfd)/(vgd-vfd);
      dmliq=-A*sqrt(2/vfd*(Pd-Pu));
      dm=xd*dmvap+(1-xd)*dmliq;
   end
end
```

D.3 Data for Refrigerant R-22

Critical parameters:

Critical pressure: $P_c = 4977$ kPa; Critical volume: $v_c = 0.16478$ L/mol

Critical temperaure: $\theta_c = 369.16$ K; Molecular weight $= 86.469$

Entropy constant: $Y = 46.8883$ J/mol-K; Enthalpy constant: $X = 25990.02$ J/mol

All parameters[1] have units compatible with the critical parameters above; pressure is in kPa, density in mol/L and temperature in K. Coefficients not listed are zero. The value of J in equations (12.100) and (12.102) is 1000.

equation of state
equation (12.94)

$$
\begin{aligned}
R &= 0.83136998378 \times 10^{-2} \\
b &= 0.10796166646 \times 10^{-1} \\
A_2 &= -0.87466122549 \\
B_2 &= 0.87054394311 \times 10^{-3} \\
C_2 &= -0.88533742184 \times 10 \\
A_3 &= -0.18939998642 \times 10^{-1} \\
B_3 &= 0.14890632572 \times 10^{-3} \\
C_3 &= 0.16091656668 \times 10 \\
A_4 &= 0.13524290185 \times 10^{-1} \\
B_4 &= -0.37996244181 \times 10^{-4} \\
C_4 &= 0.0 \\
A_5 &= -0.11768746066 \times 10^{-2} \\
B_5 &= 0.30463870398 \times 10^{-5} \\
C_5 &= -0.58307411777 \times 10^{-2} \\
A_6 &= 0.94001896962 \times 10^{6} \\
B_6 &= -0.20757984460 \times 10^{4} \\
C_6 &= 0.0 \\
K &= 4.2 \\
u &= 0.10155456431 \times 10^{3} \\
c &= 0.0
\end{aligned}
$$

ideal gas heat capacity
equation (12.98)

$$
\begin{aligned}
G_1 &= 0.10183212 \times 10^{2} \\
G_2 &= 0.14697171 \\
G_3 &= -0.76354723 \times 10^{-4} \\
G_4 &= 0.0 \\
G_5 &= 0.28754121 \times 10^{5} \\
G_5 &= 0.0
\end{aligned}
$$

saturated liquid density
equation (12.96)

$$
\begin{aligned}
D_1 &= 0.6068821 \times 10 \\
D_2 &= 0.1012108 \times 10^{2} \\
D_3 &= 0.6807772 \times 10 \\
D_4 &= -0.4129719 \times 10 \\
D_5 &= 0.3792696 \times 10
\end{aligned}
$$

saturation pressure
equation (12.95) with $\log_{10}$

$$
\begin{aligned}
F_1 &= 0.25189356867 \times 10^{2} \\
F_2 &= -0.21362184178 \times 10^{4} \\
F_3 &= 0.0 \\
F_4 &= -0.78610312200 \times 10 \\
F_5 &= 0.39436902792 \times 10^{-2} \\
F_6 &= 0.0 \\
F_7 &= 0.0 \\
F_8 &= 0.44574670300 \\
F_9 &= 0.38116666667 \times 10^{-3} \\
F_{10} &= 1.8
\end{aligned}
$$

saturated vapor density
equation (12.97)

$$
\begin{aligned}
E_7 &= -0.1680024749 \times 10^{-4} \\
E_{11} &= 0.1344723347 \\
E_{12} &= -0.2845461873 \times 10 \\
E_{14} &= 0.1292966614 \\
E_{20} &= 0.3106835616 \times 10^{4} \\
E_{21} &= -.01189422522 \times 10^{5} \\
E_{22} &= 0.1914759823 \times 10^{5} \\
E_{23} &= -0.1472081631 \times 10^{5} \\
E_{24} &= 0.4527640563 \times 10^{4} \\
E_{25} &= 0.1306817379 \times 10^{3}
\end{aligned}
$$

---

[1] Taken from R.B. Stewart et al., *ASHRAE Thermodynamic Properites of Refrigerants,* American Society of Heating, Refrigeration and Air-Conditioning Engineers, 1988.

D.4 Programs and Data for Water

The coefficients C_i and A_{ij} for the Helmholtz free energy (equation (12.120), p. 930) are given in the script M-file below, and the properties are computed using this data in the two subsequent function M-files.

```
% file waterdat.m, data for the properties of water, from Keenan
% (footnote 17, p. 930); units in cgs
global C1 C2 C3 C4 C5 C6 C7 C8 R E tc
global A11 A12 A13 A14 A15 A16 A17 A21 A22 A23 A24 A25 A26 A27 A31 A32 A33
global A34 A35 A36 A37 A41 A42 A43 A44 A45 A46 A47 A51 A52 A61 A62 A67 A71
global A72 A81 A82 A91 A92 A93 A94 A95 A96 A97 A101 A102 A103 A104 A105 A106
global A107 D E1 E2 E3 E4 E7 E10 E13 E14 E15 F PC rc
C1=1857.065; C2=3229.12; C3=-419.465; C4=36.6649; C5=-20.5516; C6=4.85233;
C7=46; C8=-1011.249; R=.46151; E=4.8; tc=1.544912;
A11=29.492937; A12= -5.1985860; A13=6.8335354; A14=-0.1564104;
A15=-6.3972405; A16=-3.9661401; A17=-0.69048554; A21=-132.13917;
A22=7.7779182; A23=-26.149751; A24=-0.72546108; A25=26.409282;
A26=15.453061; A27=2.7407416; A31=274.64632; A32=-33.301902; A33=65.326396;
A34=-9.2734289; A35=-47.740374; A36=-29.142470; A37=-5.1028070;
A41=-360.93828; A42=-16.254622; A43=-26.181978; A44=4.3125840; A45=56.323130;
A46=29.568796; A47=3.9636085; A51=342.18431; A52=-177.31074; A61=-244.50042;
A62=127.48742; A71=155.18535; A72=137.46153; A81=5.9728487; A82=155.97836;
A91=-410.30848; A92=337.31180; A93=-137.46618; A94=6.7874983; A95=136.87317;
A96=79.847970; A97=13.041253; A101=-416.05860; A102=-209.88866;
A103=-733.96848; A104=10.401717; A105=645.81880; A106=399.17570;
A107=71.531353; PC=22.088; rc=0.3170;
D=[3.6711257 -28.512396 222.65240 -882.43852];
D=[D 2000.2765 -2612.2557 1829.7674 -533.50520];
E1=13; E2=-1.393188; E3=-6.418663; E4=18.46092; E7=-35.25809;
E10=219.5021; E13=-2199.413; E14=3765.631; E15=-1809.692;
F=[-741.9242 -29.72100 -11.55286 -0.8685635];
F=[F 0.1094098 0.439993 0.2520658 0.05218684];
```

```
function [P,h,vg,dvg,dudT,dudv,dugdT,dugdvg,dPs,d2Ps]=propwat(v,T)
% Computes the properties of water sufficient for equations (12.66) and
% (12.72), (pp. 904, 905).  Do not use outputs set to zero. The program
% requires that the script file waterdat.m be run first.  Also can output
% quality X, specific volume of liquid vf, entropies s and sg, enthalpy hg
% and energy ug. Easy to add some others, including u, uf, sf, hf. Note that
% units are different from those for the refrigerants; pressures are in MPa,
% energy densities are in kJ/kg, and specific volumes are in m^3/kg.
global C1 C2 C3 C4 C5 C6 C7 C8 R E tc
global A11 A12 A13 A14 A15 A16 A17 A21 A22 A23 A24 A25 A26 A27 A31 A32 A33
global A34 A35 A36 A37 A41 A42 A43 A44 A45 A46 A47 A51 A52 A61 A62 A67 A71
global A72 A81 A82 A91 A92 A93 A94 A95 A96 A97 A101 A102 A103 A104 A105 A106
global A107 D E1 E2 E3 E4 E7 E10 E13 E14 E15 F PC rc
t=1000/T; Tc=1000/tc;
```

```
x=1-tc/t;rho=1/v;
if x<.01 % supercritical temperature (with some allowance)
   [P,h,dudT,dudr]=steam(rho,T);
   dudv=-dudr/v/v;
   vg=0; dvg=0; dugdt=0; dugdvg=0; dPs=0; d2Ps=0; % not to be used
else % subcritical temperature
   y=E2*x^(1/3)+E3*x^(2/3)+E4*x+E7*x*x+E10*x^3+E13*x^4+E14*x^(13/3);
   y=y+E15*x^(14/3);
   vg=.003155*(t/tc)^(E1)*exp(-y);
       % specific volume of saturated vapor at T
   if v<=vg % saturated mixture or compressed liquid
      x1=x^(1/3); x2=x1*x1; x0=[x1 x2 x x*x1 x*x2 x*x x*x*x1 x*x*x2];
      rf=rc*(1+D*x0')*1000;
      vf=1/rf; % specific volume of saturated liquid at T
      if v>=vf % saturated mixture
         X=(v-vf)/(vg-vf); % quality of mixture
         t0=.65-.01*T+2.7316; t02=t0*t0; t03=t02*t0; t04=t02*t02;
         t05=t03*t02; t06=t03*t03;
         t1=[1 t0 t02 t03 t04 t05 t06 t04*t03];
         t2=-.01*[0 1 2*t0 3*t02 4*t03 5*t04 6*t05 7*t06];
         t3=.0001*[0 0 2 6*t0 12*t02 20*t03 30*t04 42*t05];
         P=PC*exp(.01*(Tc/T-1)*F*t1'); % saturation pressure
         dPs=P*.01*(-Tc/T/T*F*t1'+(Tc/T-1)*F*t2'); % derivative of Psat with
             % respect to T
         d2Ps=P*.01*(2*Tc/T^3*F*t1'-2*Tc/T/T*F*t2'+P*.01*(Tc/T-1)*F*t3');
         d2Ps=d2Ps+dPs^2/P; % second derivative of Psat with respect to T
         dvg=(-E1/T+1/Tc/3*(E2/x2+2*E3/x1+3*E4+6*E7*x))*vg;
         dvg=dvg+1/Tc/3*vg*(9*E10*x*x+12*E13*x^3);
         dvg=dvg+1/Tc/3*vg*(13*E14*x^(10/3)+14*E15*x^(11/3)); % derivative
             % of vg with respect to T
         [P,hg,sg,dudT,dudr]=steam(1/vg,T);
         ug=hg-P*1000*vg; % internal energy of vapor component
         sm=(vg-v)*sg;
         h=hg-T*sm; % enthalpy of mixture
         s=sg-sm; % entropy of mixure
         dugdT=dudT-dudr/1000/vg/vg*dvg; % deriv. of ug with respect to T
         dugdvg=dugdT/dvg; % derivative of ug with respect to vg
         dudT=0; dudv=0; % not intended to be used
      else % single phase (vapor or liquid), use steam.m
         [P,h,dudT,dudr]=steam(rho,T);
         dudv=-dudr/v/v;
         dvg=0; dudT=0; dugdT=0; dugdvg=0; dPs=0; d2Ps=0; % not to be used
      end
   end
end


function [P,h,s,dudT,dudr]=steam(rho,T) % computes properties of single-
% phase water.  Requires that script file waterdat.m be run first.  Intended
% as sub-program for propwat.m. If used directly, note some use of cgs units.
% Units of energy are J/g = kJ/kg.
```

```
global C1 C2 C3 C4 C5 C6 C7 C8 R E tc
global A11 A12 A13 A14 A15 A16 A17 A21 A22 A23 A24 A25 A26 A27 A31 A32 A33
global A34 A35 A36 A37 A41 A42 A43 A44 A45 A46 A47 A51 A52 A53 A54 A55 A56
global A57 A61 A62 A63 A64 A65 A66 A67 A71 A72 A73 A74 A75 A76 A77 A81 A82
global A83 A84 A85 A86 A87 A91 A92 A93 A94 A95 A96 A97 A101 A102 A103 A104
global A105 A106 A107
t=1000/T; r=rho/1000; % density in g/cm^3
t1=t-2.5; t2=t1*t1; t3=t1*t2; t4=t2*t2; t5=t3*t2;
r0=r-.634; r02=r0*r0; r03=r0*r02; r04=r02*r02; r05=r03*r02; r06=r03*r03;
r07=r04*r03; E0=exp(-E*r); tcc=t-tc;
r1=r-1; r2=r1*r1; r3=r1*r2; r4=r2*r2; r5=r3*r2; r6=r3*r3; r7=r4*r3;
a1=A11+A21*r0+A31*r02+A41*r03+A51*r04+A61*r05+A71*r06+A81*r07;
a1=a1+E0*(A91+A101*r);
a2=A12+A22*r1+A32*r2+A42*r3+A52*r4+A62*r5+A72*r6+A82*r7+E0*(A92+A102*r);
a3=A13+A23*r1+A33*r2+A43*r3+E0*(A93+A103*r);
a4=A14+A24*r1+A34*r2+A44*r3+E0*(A94+A104*r);
a5=A15+A25*r1+A35*r2+A45*r3+E0*(A95+A105*r);
a6=A16+A26*r1+A36*r2+A46*r3+E0*(A96+A106*r);
a7=A17+A27*r1+A37*r2+A47*r3+E0*(A97+A107*r);
Q=a1+tcc*(a2+t1*a3+t2*a4+t3*a5+t4*a6+t5*a7);
da1=A21+2*A31*r0+3*A41*r02+4*A51*r03+5*A61*r04+6*A71*r05+7*A81*r06;
da1=da1+E0*(A101-E*(A91+A101*r));
da2=A22+2*A32*r1+3*A42*r2+4*A52*r3+5*A62*r4+6*A72*r5+7*A82*r6;
da2=da2+E0*(A102-E*(A92+A102*r));
da3=A23+2*A33*r1+3*A43*r2+E0*(A103-E*(A93+A103*r));
da4=A24+2*A34*r1+3*A44*r2+E0*(A104-E*(A94+A104*r));
da5=A25+2*A35*r1+3*A45*r2+E0*(A105-E*(A95+A105*r));
da6=A26+2*A36*r1+3*A46*r2+E0*(A106-E*(A96+A106*r));
da7=A27+2*A37*r1+3*A47*r2+E0*(A107-E*(A97+A107*r));
dQdr=da1+tcc*(da2+t1*da3+t2*da4+t3*da5+t4*da6+t5*da7);
dQdt=a2+(t1+tcc)*a3+t1*(t1+2*tcc)*a4+t2*(t1+3*tcc)*a5+t3*(t1+4*tcc)*a6;
dQdt=dQdt+t4*(t1+5*tcc)*a7;
p0=C1+C2/t+C3/t/t+C4/t^3+C5/t^4+C6/t^5+C7*log(T)+C8*log(T)/t;
dp0dT=(C2/t+2*C3/t^2+3*C4/t^3+4*C5/t^4+5*C6/t^5+C7)/T+C8/t*(log(T)+1)/T;
dp0tdt=C1-C3/t/t-2*C4/t^3-3*C5/t^4-4*C6/t^5+C7*log(T)-C7-C8/t;
P=r*R*T*(1+r*Q+r*r*dQdr); % in N/cm^2 = MPa
u=R*T*r*t*dQdt+dp0tdt;
s=-R*(log(r)+r*Q-r*t*dQdt)-dp0dT;
h=u+P/r;
d2Qdt2=2*a3+(4*t1+2*tcc)*a4+(2*t1*(t1+3*tcc)+4*t2)*a5;
d2Qdt2=d2Qdt2+(3*t2*(t1+4*tcc)+5*t3)*a6;
d2Qdt2=d2Qdt2+(4*t3*(t1+5*tcc)+6*t4)*a7;
d2p0tdt2=2*C3/t^3+6*C4/t^4+12*C5/t^5+20*C6/t^6-C7/t+C8/t/t;
dudT=-R*r*t*t*d2Qdt2-t/T*d2p0tdt2;
d2Qdtdr=da2+(t1+tcc)*da3+t1*(t1+2*tcc)*da4+t2*(t1+3*tcc)*da5;
d2Qdtdr=d2Qdtdr+t3*(t1+4*tcc)*da6+t4*(t1+5*tcc)*da7;
dudr=1000*R*(dQdt+r*d2Qdtdr);
```

Index

RELATED TITLES

Dynamical Systems and Control
Firdaus E. Udwadia, H.I. Weber, and George Leitmann
ISBN: 0-415-30997-2